Operator Theory

A Comprehensive Course in Analysis, Part 4

Operator Theory

A Comprehensive Course in Analysis, Part 4

Barry Simon

AMS
American Mathematical Society
Providence, Rhode Island

2010 *Mathematics Subject Classification.* Primary 47-01, 34-01, 46-01;
Secondary 81Q10, 34L05, 35P05, 42C05, 46H05, 22B05.

For additional information and updates on this book, visit
www.ams.org/bookpages/simon

Library of Congress Cataloging-in-Publication Data

Simon, Barry, 1946–
Operator theory / Barry Simon.
pages cm. — (A comprehensive course in analysis ; part 4)
Includes bibliographical references and indexes.
ISBN 978-1-4704-1103-9 (alk. paper)
1. Mathematical analysis—Textbooks. 2. Operator theory. I. Title.

QA300.S5285 2015
515′.724—dc23

2015033104

♾ The paper used in this book is acid-free and falls within the guidelines
established to ensure permanence and durability.
Visit the AMS home page at `http://www.ams.org/`

10 9 8 7 6 5 4 3 2 1 20 19 18 17 16 15

To the memory of Cherie Galvez

extraordinary secretary, talented helper, caring person

and to the memory of my mentors,
Ed Nelson (1932-2014) and Arthur Wightman (1922-2013)

who not only taught me Mathematics
but taught me how to be a mathematician

Contents

Preface to the Series

Young men should prove theorems, old men should write books.

—*Freeman Dyson*, quoting G. H. Hardy[1]

Reed–Simon[2] starts with "Mathematics has its roots in numerology, geometry, and physics." This puts into context the division of mathematics into algebra, geometry/topology, and analysis. There are, of course, other areas of mathematics, and a division between parts of mathematics can be artificial. But almost universally, we require our graduate students to take courses in these three areas.

This five-volume series began and, to some extent, remains a set of texts for a basic graduate analysis course. In part it reflects Caltech's three-terms-per-year schedule and the actual courses I've taught in the past. Much of the contents of Parts 1 and 2 (Part 2 is in two volumes, Part 2A and Part 2B) are common to virtually all such courses: point set topology, measure spaces, Hilbert and Banach spaces, distribution theory, and the Fourier transform, complex analysis including the Riemann mapping and Hadamard product theorems. Parts 3 and 4 are made up of material that you'll find in some, but not all, courses—on the one hand, Part 3 on maximal functions and H^p spaces; on the other hand, Part 4 on the spectral theorem for bounded self-adjoint operators on a Hilbert space and det and trace, again for Hilbert space operators. Parts 3 and 4 reflect the two halves of the third term of Caltech's course.

[1]Interview with D. J. Albers, The College Mathematics Journal, **25**, no. 1, January 1994.

[2]M. Reed and B. Simon, *Methods of Modern Mathematical Physics, I: Functional Analysis*, Academic Press, New York, 1972.

While there is, of course, overlap between these books and other texts, there are some places where we differ, at least from many:

(a) By having a unified approach to both real and complex analysis, we are able to use notions like contour integrals as Stietljes integrals that cross the barrier.

(b) We include some topics that are not standard, although I am surprised they are not. For example, while discussing maximal functions, I present Garcia's proof of the maximal (and so, Birkhoff) ergodic theorem.

(c) These books are written to be keepers—the idea is that, for many students, this may be the last analysis course they take, so I've tried to write in a way that these books will be useful as a reference. For this reason, I've included "bonus" chapters and sections—material that I do not expect to be included in the course. This has several advantages. First, in a slightly longer course, the instructor has an option of extra topics to include. Second, there is some flexibility—for an instructor who can't imagine a complex analysis course without a proof of the prime number theorem, it is possible to replace all or part of the (non-bonus) chapter on elliptic functions with the last four sections of the bonus chapter on analytic number theory. Third, it is certainly possible to take all the material in, say, Part 2, to turn it into a two-term course. Most importantly, the bonus material is there for the reader to peruse long after the formal course is over.

(d) I have long collected "best" proofs and over the years learned a number of ones that are not the standard textbook proofs. In this regard, modern technology has been a boon. Thanks to Google books and the Caltech library, I've been able to discover some proofs that I hadn't learned before. Examples of things that I'm especially fond of are Bernstein polynomials to get the classical Weierstrass approximation theorem, von Neumann's proof of the Lebesgue decomposition and Radon–Nikodym theorems, the Hermite expansion treatment of Fourier transform, Landau's proof of the Hadamard factorization theorem, Wielandt's theorem on the functional equation for $\Gamma(z)$, and Newman's proof of the prime number theorem. Each of these appears in at least some monographs, but they are not nearly as widespread as they deserve to be.

(e) I've tried to distinguish between central results and interesting asides and to indicate when an interesting aside is going to come up again later. In particular, all chapters, except those on preliminaries, have a listing of "Big Notions and Theorems" at their start. I wish that this attempt to differentiate between the essential and the less essential didn't make this book different, but alas, too many texts are monotone listings of theorems and proofs.

(f) I've included copious "Notes and Historical Remarks" at the end of each section. These notes illuminate and extend, and they (and the Problems) allow us to cover more material than would otherwise be possible. The history is there to enliven the discussion and to emphasize to students that mathematicians are real people and that "may you live in interesting times" is truly a curse. Any discussion of the history of real analysis is depressing because of the number of lives ended by the Nazis. Any discussion of nineteenth-century mathematics makes one appreciate medical progress, contemplating Abel, Riemann, and Stieltjes. I feel knowing that Picard was Hermite's son-in-law spices up the study of his theorem.

On the subject of history, there are three cautions. First, I am not a professional historian and almost none of the history discussed here is based on original sources. I have relied at times—horrors!—on information on the Internet. I have tried for accuracy but I'm sure there are errors, some that would make a real historian wince.

A second caution concerns looking at the history assuming the mathematics we now know. Especially when concepts are new, they may be poorly understood or viewed from a perspective quite different from the one here. Looking at the wonderful history of nineteenth-century complex analysis by Bottazzini–Grey[3] will illustrate this more clearly than these brief notes can.

The third caution concerns naming theorems. Here, the reader needs to bear in mind Arnol'd's principle:[4] *If a notion bears a personal name, then that name is not the name of the discoverer* (and the related Berry principle: *The Arnol'd principle is applicable to itself*). To see the applicability of Berry's principle, I note that in the wider world, Arnol'd's principle is called "Stigler's law of eponymy." Stigler[5] named this in 1980, pointing out it was really discovered by Merton. In 1972, Kennedy[6] named Boyer's law *Mathematical formulas and theorems are usually not named after their original discoverers* after Boyer's book.[7] Already in 1956, Newman[8] quoted the early twentieth-century philosopher and logician A. N. Whitehead as saying: "Everything of importance has been said before by somebody who

---

[3]U. Bottazzini and J. Gray, *Hidden Harmony—Geometric Fantasies. The Rise of Complex Function Theory*, Springer, New York, 2013.

[4]V. I. Arnol'd, *On teaching mathematics*, available online at `http://pauli.uni-muenster.de/~munsteg/arnold.html`.

[5]S. M. Stigler, *Stigler's law of eponymy*, Trans. New York Acad. Sci. **39** (1980), 147–158.

[6]H. C. Kennedy, *Classroom notes: Who discovered Boyer's law?*, Amer. Math. Monthly **79** (1972), 66–67.

[7]C. B. Boyer, *A History of Mathematics*, Wiley, New York, 1968.

[8]J. R. Newman, *The World of Mathematics*, Simon & Schuster, New York, 1956.

did not discover it." The main reason to give a name to a theorem is to have a convenient way to refer to that theorem. I usually try to follow common usage (even when I know Arnol'd's principle applies).

I have resisted the temptation of some text writers to rename things to set the record straight. For example, there is a small group who have attempted to replace "WKB approximation" by "Liouville–Green approximation", with valid historical justification (see the Notes to Section 15.5 of Part 2B). But if I gave a talk and said I was about to use the Liouville–Green approximation, I'd get blank stares from many who would instantly know what I meant by the WKB approximation. And, of course, those who try to change the name also know what WKB is! Names are mainly for shorthand, not history.

These books have a wide variety of problems, in line with a multiplicity of uses. The serious reader should at least skim them since there is often interesting supplementary material covered there.

Similarly, these books have a much larger bibliography than is standard, partly because of the historical references (many of which are available online and a pleasure to read) and partly because the Notes introduce lots of peripheral topics and places for further reading. But the reader shouldn't consider for a moment that these are intended to be comprehensive—that would be impossible in a subject as broad as that considered in these volumes.

These books differ from many modern texts by focusing a little more on special functions than is standard. In much of the nineteenth century, the theory of special functions was considered a central pillar of analysis. They are now out of favor—too much so—although one can see some signs of the pendulum swinging back. They are still mainly peripheral but appear often in Part 2 and a few times in Parts 1, 3, and 4.

These books are intended for a second course in analysis, but in most places, it is really previous exposure being helpful rather than required. Beyond the basic calculus, the one topic that the reader is expected to have seen is metric space theory and the construction of the reals as completion of the rationals (or by some other means, such as Dedekind cuts).

Initially, I picked "A Course in Analysis" as the title for this series as an homage to Goursat's *Cours d'Analyse*,[9] a classic text (also translated into English) of the early twentieth century (a literal translation would be

[9]E. Goursat, *A Course in Mathematical Analysis: Vol. 1: Derivatives and Differentials, Definite Integrals, Expansion in Series, Applications to Geometry. Vol. 2, Part 1: Functions of a Complex Variable. Vol. 2, Part 2: Differential Equations. Vol. 3, Part 1: Variation of Solutions. Partial Differential Equations of the Second Order. Vol. 3, Part 2: Integral Equations. Calculus of Variations*, Dover Publications, New York, 1959 and 1964; French original, 1905.

"of Analysis" but "in" sounds better). As I studied the history, I learned that this was a standard French title, especially associated with École Polytechnique. There are nineteenth-century versions by Cauchy and Jordan and twentieth-century versions by de la Vallée Poussin and Choquet. So this is a well-used title. The publisher suggested adding "Comprehensive", which seems appropriate.

It is a pleasure to thank many people who helped improve these texts. About 80% was TEXed by my superb secretary of almost 25 years, Cherie Galvez. Cherie was an extraordinary person—the secret weapon to my productivity. Not only was she technically strong and able to keep my tasks organized but also her people skills made coping with bureaucracy of all kinds easier. She managed to wind up a confidant and counselor for many of Caltech's mathematics students. Unfortunately, in May 2012, she was diagnosed with lung cancer, which she and chemotherapy valiantly fought. In July 2013, she passed away. I am dedicating these books to her memory.

During the second half of the preparation of this series of books, we also lost Arthur Wightman and Ed Nelson. Arthur was my advisor and was responsible for the topic of my first major paper—perturbation theory for the anharmonic oscillator. Ed had an enormous influence on me, both via the techniques I use and in how I approach being a mathematician. In particular, he taught me all about closed quadratic forms, motivating the methodology of my thesis. I am also dedicating these works to their memory.

After Cherie entered hospice, Sergei Gel'fand, the AMS publisher, helped me find Alice Peters to complete the TEXing of the manuscript. Her experience in mathematical publishing (she is the "A" of A K Peters Publishing) meant she did much more, for which I am grateful.

This set of books has about 150 figures which I think considerably add to their usefulness. About half were produced by Mamikon Mnatsakanian, a talented astrophysicist and wizard with Adobe Illustrator. The other half, mainly function plots, were produced by my former Ph.D. student and teacher extraordinaire Mihai Stoiciu (used with permission) using Mathematica. There are a few additional figures from Wikipedia (mainly under WikiCommons license) and a hyperbolic tiling of Douglas Dunham, used with permission. I appreciate the help I got with these figures.

Over the five-year period that I wrote this book and, in particular, during its beta-testing as a text in over a half-dozen institutions, I received feedback and corrections from many people. In particular, I should like to thank (with apologies to those who were inadvertently left off): Tom Alberts, Michael Barany, Jacob Christiansen, Percy Deift, Tal Einav, German Enciso, Alexander Eremenko, Rupert Frank, Fritz Gesztesy, Jeremy Gray,

Leonard Gross, Chris Heil, Mourad Ismail, Svetlana Jitomirskaya, Bill Johnson, Rowan Killip, John Klauder, Seung Yeop Lee, Milivoje Lukic, Andre Martinez-Finkelshtein, Chris Marx, Alex Poltoratski, Eric Rains, Lorenzo Sadun, Ed Saff, Misha Sodin, Dan Stroock, Benji Weiss, Valentin Zagrebnov, and Maxim Zinchenko.

Much of these books was written at the tables of the Hebrew University Mathematics Library. I'd like to thank Yoram Last for his invitation and Naavah Levin for the hospitality of the library and for her invaluable help.

This series has a Facebook page. I welcome feedback, questions, and comments. The page is at `www.facebook.com/simon.analysis` .

Even if these books have later editions, I will try to keep theorem and equation numbers constant in case readers use them in their papers.

Finally, analysis is a wonderful and beautiful subject. I hope the reader has as much fun using these books as I had writing them.

Preface to Part 4

The subject of this part is "operator theory." Unlike Parts 1 and 2, where there is general agreement about what we should expect graduate students to know, that is not true of this part.

Putting aside for now Chapters 4 and 6, which go beyond "operator theory" in a narrow sense, one can easily imagine a book titled *Operator Theory* having little overlap with Chapters 2, 3, 5, and 7: almost all of that material studies Hilbert space operators. We do discuss in Chapter 2 the analytic functional calculus on general Banach spaces, and parts of our study of compact operators in Chapter 3 cover some basics and the Riesz–Schauder theory on general Banach spaces. We cover Fredholm operators and the Ringrose structure theory in normed spaces. But the thrust is definitely toward Hilbert space.

Moreover, a book like *Harmonic Analysis of Operators on Hilbert Space*[1] or any of several books with "non-self-adjoint" in their titles have little overlap with this volume. So from our point of view, a more accurate title for this part might be *Operator Theory—Mainly Self-Adjoint and/or Compact Operators on a Hilbert Space.*

That said, much of the material concerning those other topics, undoubtedly important, doesn't belong in "what every mathematician should at least be exposed to in analysis." But, I believe the spectral theorem, at least for bounded operators, the notions of trace and determinant on a Hilbert space, and the basics of the Gel'fand theory of commutative Banach spaces do belong on that list.

[1]See B. Sz.-Nagy, C. Foias, H. Bercovici, and L. Kérchy, *Harmonic Analysis of Operators on Hilbert Space*, second edition, revised and enlarged edition, Universitext, Springer, New York, 2010.

Before saying a little more about the detailed contents, I should mention that many books with a similar thrust to this book have the name *Functional Analysis*. I still find it remarkable and a little strange that the parts of a graduate analysis course that deal with operator theory are often given this name (since functions are more central to real and complex analysis), but they are, even by me[2].

One change from the other parts in this series of five books is that in them all the material called "Preliminaries" is either from other parts of the series or from prior courses that the student is assumed to have had (e.g., linear algebra or the theory of metric spaces). Here, Chapter 1 includes a section on perturbation theory for eigenvalues of finite matrices because it fits in with a review of linear algebra, not because we imagine many readers are familiar with it.

Chapters 4 and 6 are here as material that I believe all students should see while learning analysis (at least the initial sections), but they are connected to, though rather distinct from, "operator theory." Chapter 4 deals with a subject dear to my heart—orthogonal polynomials—it's officially here because the formal proof we give of the spectral theorem reduces it to the result for Jacobi matrices which we treat by approximation theory for orthogonal polynomials (it should be emphasized that this is only one of seven proofs we sketch). I arranged this, in part, because I felt any first-year graduate student should know the way to derive these from recurrence relations for orthogonal polynomials on the real line. We fill out the chapter with bonus sections on some fascinating aspects of the theory.

Chapter 6 involves another subject that should be on the required list of any mathematician, the Gel'fand theory of commutative Banach algebras. Again, there is a connection to the spectral theorem, justifying the chapter being placed here, but the in-depth look at applications of this theory, while undoubtedly a part of a comprehensive look at analysis, doesn't fit very well under the rubric of operator theory.

[2]See *Methods of Modern Mathematical Physics, I: Functional Analysis*, Academic Press, New York, 1972.

Chapter 1

Preliminaries

> Toto, we're not in Kansas anymore.
>
> —*Dorothy* (Judy Garland) in *The Wizard of Oz*[1]

Operator theory uses a number of ideas from earlier volumes. In these brief preliminaries, we recall a few of them which the reader might not be familiar with, and in preparation for our discussion of trace and determinant in infinite dimensions, we recall the theory of finite-dimensional determinants. Also, we discuss the theory of eigenvalue perturbation in finite-dimensional spaces.

1.1. Notation and Terminology

> A foolish consistency is the hobgoblin of little minds ... Is it so bad, then, to be misunderstood? Pythagoras was misunderstood, and Socrates, and Jesus, and Luther, and Copernicus, and Galileo, and Newton, and every pure and wise spirit that ever took flesh. To be great is to be misunderstood.
>
> —*Ralph Waldo Emerson* [**184**]

For a real number a, we will use the terms positive and strictly positive for $a \geq 0$ and $a > 0$, respectively. It is not so much that we find nonnegative bad, but the phrase "monotone nondecreasing" for $x > y \Rightarrow f(x) \geq f(y)$ is downright confusing so we use "monotone increasing" and "strictly monotone increasing" and then, for consistency, "positive" and "strictly positive." Similarly for matrices, we use "positive definite" and "strictly positive definite" where others might use "positive semi-definite" and "positive definite."

[1] The full quote from the 1939 movie is "Toto, I've got a feeling we're not in Kansas anymore." This is not from the original book. While the film was nominated for a best picture Academy Award, it lost to *Gone with the Wind*, a film with the same director!

Basic Rings and Fields.

$$\mathbb{R} = \textit{real numbers} \qquad \mathbb{Q} = \textit{rationals} \qquad \mathbb{Z} = \textit{integers}$$
$$\mathbb{C} = \textit{complex numbers} = \{x + iy \mid x, y \in \mathbb{R}\}$$

with their sums and products. For $z = x + iy \in \mathbb{C}$, we use $\operatorname{Re} z = x$, $\operatorname{Im} z = y$. $|z| = (x^2 + y^2)^{1/2}$.

Products. $X^n = n$-tuples of points in X with induced vector space and/or additive structure; in particular, $\mathbb{R}^n$, $\mathbb{Q}^n$, $\mathbb{Z}^n$, $\mathbb{C}^n$.

Subsets of $\mathbb{C}$.

- $\mathbb{C}_+ =$ upper half-plane $= \{z \mid \operatorname{Im} z > 0\}$;
  $\mathbb{H}_+ =$ right half-plane $= \{z \mid \operatorname{Re} z > 0\}$
- $\mathbb{Z}_+ = \{n \in \mathbb{Z} \mid n > 0\} = \{1, 2, 3, \dots\}$;
  $\mathbb{N} = \{0\} \cup \mathbb{Z}_+ = \{0, 1, 2, \dots\}$
- $\mathbb{D} = \{z \in \mathbb{C} \mid |z| < 1\}$; $\partial\mathbb{D} = \{z \in \mathbb{C} \mid |z| = 1\}$
- $\mathbb{D}_\delta(z_0) = \{z \in \mathbb{C} \mid |z - z_0| < \delta\}$ for $z_0 \in \mathbb{C}$, $\delta > 0$

Miscellaneous Terms.

- For $x \in \mathbb{R}$, $[x] =$ greatest integer less than x, that is, $[x] \in \mathbb{Z}$, $[x] \leq x < [x] + 1$
- $\{x\} = x - [x] =$ fractional parts of x
- $\sharp(A) =$ number of elements in a set A
- $\operatorname{Ran} f =$ range of a function f
- $\log(z) =$ natural logarithm, that is, logarithm to base e; if z is complex, put the cut along $(-\infty, 0]$, i.e., $\log(z) = \log(|z|) + i \arg(z)$ with $-\pi < \arg(z) \leq \pi$
- For sets A, B subsets of X, $A \cup B =$ union, $A \cap B =$ intersection, $A^c =$ complement of A in X, $A \setminus B = A \cap B^c$, $A \triangle B = (A \setminus B) \cup (B \setminus A)$
- For matrices, M_{12} means row one, column two
- $f \restriction K =$ *restriction* of a function to K, a subset of the domain of f

One-Point Compactification Notation.

- $X_\infty =$ one-point compactification of a locally compact space
- $C_\infty(X) =$ continuous functions vanishing at infinity
- $C(X_\infty) =$ continuous functions with a limit at infinity $(= C_\infty(X) + \{\alpha\mathbb{1}\})$

Given a set, S, a *relation*, R, is a subset of $S \times S$. For $x, y \in S$, we write xRy if $(x,y) \in R$ and $\sim xRy$ if $(x,y) \notin R$. A relation is called

$$\begin{aligned} \textit{reflexive} &\Leftrightarrow \forall\, x \in S, \ \text{ we have } xRx \\ \textit{symmetric} &\Leftrightarrow \forall\, x,y \in S, \ xRy \Rightarrow yRx \\ \textit{transitive} &\Leftrightarrow \forall\, x,y,z \in S, \ xRy \,\&\, yRz \Rightarrow xRz \end{aligned}$$

An *equivalence relation* is a relation that is reflexive, symmetric, and transitive.

If R is an equivalence relation on S, the *equivalence class*, $[x]$, of $x \in S$ is

$$[x] = \{y \in S \mid xRy\} \tag{1.1.1}$$

By the properties of equivalence relations,

$$\forall\, x,y \in S, \qquad \text{either } [x] = [y] \text{ or } [x] \cap [y] = \emptyset \tag{1.1.2}$$

The family of equivalence classes in S is denoted S/R.

We use "big oh, litttle oh" notation, e.g., $f(x) = O(|x|)$ or $f(x) = o(1)$. If the reader is not familiar with it, see Section 1.4 of Part 2A.

Banach Space Notation.

- X = Banach space, $\quad X^*$ = dual space
- $\sigma(X,Y)$ = Y-weak topology (Y are linear functionals acting on X, $\sigma(X,Y)$ is the weakest topology on X in which $x \mapsto \langle y, x\rangle$ is continuous for each y)
- $\mathcal{H}$ = Hilbert space (always complex; $\langle\,\cdot\,,\,\cdot\,\rangle$ is antilinear in the *first* factor)
- $\mathcal{L}(X)$ = bounded linear transformations from X to X
- $A \in \mathcal{L}(X)$, $\quad A^t \in \mathcal{L}(X^*)$ by $(A^t\ell)(x) = \ell(Ax)$
- $A \in \mathcal{L}(\mathcal{H})$, $\quad A^* \in \mathcal{L}(H)$ by $\langle A^*\varphi, \psi\rangle = \langle \varphi, A\psi\rangle$

The following are discussed in Chapter 2. We place their notation here.

- $\sigma(A)$ = spectrum of A
- $\sigma_d(A)$ = discrete spectrum of A
- $\operatorname{spr}(A) = \sup\{|\lambda| \mid A \in \sigma(A)\}$, spectral radius of A

1.2. Some Complex Analysis

Complex analysis is central to operator theory, so much so that we will almost always only consider operators on complex Banach spaces. The spectrum is natural as a subset of $\mathbb{C}$ and operator-valued entire analytic functions will be a major tool. We assume the reader is familiar with the basics of complex analysis—for example at the level of Chapters 2 and 3 of Part 2A, with some some knowledge of more advanced topics like Montel's and Vitali's theorems (Chapter 6 of Part 2A), fractional linear transformations (Chapter 7 of Part 2A), and infinite product expansions (Chapter 9 of Part 2A). Here we want to be explicit about a refined version of the Cauchy integral formula and about one result concerning algebroidal functions.

A *contour* is a closed rectifiable curve in $\mathbb{C}$. A *chain* is a finite formal sum and difference of contours. Given a chain, Γ, a finite collection of contours, $\{\gamma\}$ and $n(\gamma) \in \mathbb{Z}$ for each γ so formally,

$$\Gamma = \sum_{\gamma} n(\gamma)\gamma \tag{1.2.1}$$

we define the integral over Γ by

$$\oint_{\Gamma} f(z)\, dz = \sum_{\gamma} n(\gamma) \oint_{\gamma} f(z)\, dz \tag{1.2.2}$$

By $\text{Ran}(\Gamma)$, we mean $\bigcup_{\gamma} \text{Ran}(\gamma)$. If Γ is a chain and $z_0 \notin \text{Ran}(\Gamma)$, we define the *winding number* of γ about z_0 by

$$n(\Gamma, z_0) = \frac{1}{2\pi i} \oint \frac{dz}{z - z_0} \tag{1.2.3}$$

We say Γ is *homologous to zero* in Ω if and only if $\text{Ran}(\Gamma) \subset \Omega$ and $n(\Gamma, z_0) = 0$ for all $z_0 \notin \Omega$.

Theorem 1.2.1 (The Ultimate CIF). *Let Ω be a region, f analytic in Ω, and Γ a chain homologous to 0 in Ω. Then*

$$\frac{1}{2\pi i} \oint \frac{f(z)dz}{z - w} = n(\Gamma, w) f(w) \tag{1.2.4}$$

for all $w \in \Omega \setminus \text{Ran}(\Gamma)$.

Section 4.2 of Part 2A has an elegant proof of this result due to Dixon. We also need the existence of certain contours:

Theorem 1.2.2. *Let Ω be a region in $\mathbb{C}$ and K a compact subset of Ω. Then there exists a chain, Γ, homologous to 0 in Ω so that $K \cap \text{Ran}(\Gamma) = \emptyset$ and*

$$n(\Gamma, z_0) = \begin{cases} 1, & z_0 \in K \\ 0, & z_0 \notin \Omega \\ 0 \textit{ or } 1, & \textit{every } z_0 \notin \text{Ran}(\Gamma) \end{cases} \tag{1.2.5}$$

Γ is constructed in Section 4.4 of Part 2A. It has a finite number of line segments parallel to one coordinate axis or the other; see Figure 1.2.1. These theorems will be used heavily in Section 2.3.

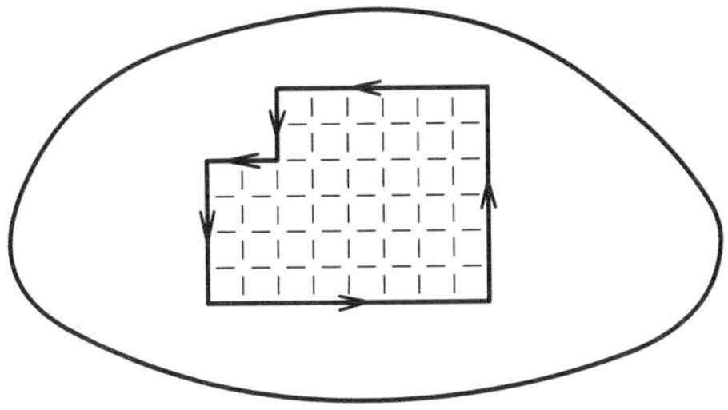

Figure 1.2.1. A mesh-defined contour.

The other complex variable result we'll need that the reader may not have seen concerns *algebroidal functions*, that is, $w = f(z)$ solving

$$w^n + a_{n-1}(z)w^{n-1} + \cdots + a_n(z) = 0 \tag{1.2.6}$$

where $\{a_j\}_{j=0}^{n-1}$ are analytic in some region Ω. Here is the result (it is proven in Section 3.5 of Part 2A where it is Theorem 3.5.2).

Theorem 1.2.3. *Let $a_0, a_1, \dots, a_{n-1}$ be analytic functions in a region Ω. Then there exists a discrete set, $D \subset \Omega$, so that all roots of (1.2.6) are analytic near any $z_0 \in \Omega \setminus D$. If $z_0 \in D$, the roots of (1.2.6) are given by the values of one or more Puiseux series about z_0.*

Remarks. 1. A *Puiseux series* is a convergent power series in a fractional power, that is, for $p \in \mathbb{Z}_+$,

$$\sum_{n=0}^{\infty} a_n(z - z_0)^{n/p} \tag{1.2.7}$$

2. The analyticity near $z_0 \in \Omega \setminus D$ is a local result. While germs of roots can be continued along any curve in $\Omega \setminus D$, in general, they define multivalued functions (hence, the Puiseux series).

3. There is an integer $q \leq n$ so that (1.2.6) has exactly q distinct solutions on $\Omega \setminus D$ and fewer distinct solutions at a point in D. Essentially, the Puiseux series coalesce at points in D.

We'll also need the following:

Theorem 1.2.4 (Implicit Function Theorem). *Let Ω be a region of $\mathbb{C}$ and $B \subset \mathbb{R}^\nu$. Let $F\colon \Omega \times B \to \mathbb{C}$ have $F(z,\beta)$ analytic in $z \in \Omega$ for each fixed β with $\beta \mapsto F(\cdot\,,\beta)$ continuous (respectively, C^k, real analytic) in β (locally uniform in z norm). Then if*

$$F(z_0, \beta_0) = 0, \qquad \frac{\partial F}{\partial z_0}(z_0, \beta) \neq 0 \tag{1.2.8}$$

for β near β_0, there is a function $z = g(\beta)$ with $z_0 = g(\beta_0)$ which is continuous (respectively, C^k, C^∞, real analytic) so that

$$F(g(\beta), \beta) = 0 \tag{1.2.9}$$

for all β near β_0 and $z = g(\beta)$ is the only solution near z_0.

Remarks. 1. This is Theorems 3.4.2 and 3.4.3 of Part 2A.

2. One uses

$$N(\beta) = \frac{1}{2\pi i} \oint_{|z-z_0|=\delta} \frac{\frac{\partial F}{\partial z}(z, \beta)}{F(z, \beta)}\, dz \tag{1.2.10}$$

to show that $F(\,\cdot\,, \beta)$ has a single zero near z_0 and then defines

$$g(\beta) = \frac{1}{2\pi i} \oint_{|z-z_0|=\delta} \frac{z\frac{\partial F}{\partial z}(z, \beta)}{F(z, \beta)}\, dz \tag{1.2.11}$$

1.3. Some Linear Algebra

Linear algebra is just the operator theory of finite-dimensional spaces. Thus, its results serve as motivation for the infinite-dimensional theory. In this section, we'll state the two main structural theorems for finite matrices—the Jordan normal form and the spectral theorem—and discuss finite-dimensional determinants. For the two big theorems, we'll state them for motivation—no need to prove them here since the infinite-dimensional results have proofs that do not rely on but do imply the finite-dimensional ones. Our treatment of infinite-dimensional determinants will rely on the finite-dimensional case, for which we therefore provide more details. We assume the reader is familiar with the definition of vector space, algebraic basis, finite dimension, invariance of dimension, matrix representations of operators, inner product space, unitary operator, finite-dimensional orthonormal basis, and signs of permutations (but see Section 1.7 of Part 1 and Problem 3).

Theorem 1.3.1 (Jordan Normal Form). *Let A be a linear transformation on a finite-dimensional vector space, V, of dimension n over $\mathbb{C}$. Then there exist k distinct complex numbers, $\{\lambda_j\}_{j=1}^k$ $(k \le n)$, operators $\{P_j\}_{j=1}^k, \{N_j\}_{j=1}^k$, so that*

(a) $$A = \sum_{j=1}^k \lambda_j P_j + N_j \tag{1.3.1}$$

(b) $$P_j P_k = P_k P_j = \delta_{jk} P_j \tag{1.3.2}$$

(c) $$P_j N_j P_j = N_j \tag{1.3.3}$$

(d) $$N_j^n = 0 \tag{1.3.4}$$

Remarks. 1. One can deduce this from Corollary 2.3.6 if one notes (by our determinant analysis below) that $\sigma(A)$ is finite and

$$\sum_{j=1}^{k} P_j = \mathbb{1} \tag{1.3.5}$$

where P_j are the spectral projections associated to the points of $\sigma(A)$. (1.3.5) comes from the fact that, by a simple argument, if R is large (so large that $(A-z)^{-1}$ is analytic on $\{z \mid |z| > R-1\}$), then $\frac{1}{2\pi i}\oint \frac{dz}{z-A} = \mathbb{1}$. Problems 9 and 10 have an alternate proof.

2. P_j's obeying (1.3.2) have

$$V = \text{Ran}(P_1) \dotplus \text{Ran}(P_2) \dotplus \cdots \dotplus \text{Ran}(P_k) \tag{1.3.6}$$

where $\dotplus$ means disjoint sum, that is, any $v \in V$ can be written uniquely as a sum of a v_j in $\text{Ran}(P_j)$. In particular, writing $X \dotplus Y$ means $X \cap Y = \{0\}$. The P_j's are projections called either *spectral projections* or *eigenprojections.*

3. (1.3.3) implies that N_j leaves $\text{Ran}(P_j)$ invariant and that $N_j \restriction \text{Ran}(P_i) = 0$ for all $i \neq j$.

4. We analyze nilpotents (operators with $N^\ell = 0$ for some $\ell \in \mathbb{Z}_+$) below. It is easy to see (Problem 4) there always exists $\varphi \in \text{Ran}(P_j)$ so $N_j\varphi = 0$; thus, $A\varphi = \lambda_j\varphi$ and the λ_j are precisely the eigenvalues of A. The N_j are called *eigennilpotents.*

5. If all N_j's are zero, then A can be diagonalized (and vice-versa). If nonzero N_j's occur for a given A, we say that A has a *Jordan anomaly.* (It is not hard to see the set of A's with a Jordan anomaly is a set of codimension 2, so "most" A's are diagonalizable.)

6. A single $\lambda_j P_j + N_j \restriction \text{Ran}(P_j)$ term is called a *Jordan block.*

7. $d_j \equiv \dim(\text{Ran}(P_j))$ is called the *algebraic multiplicity* of λ_j. $\dim\{u \mid Au = \lambda_j u\} \leq d_j$ is called the *geometric multiplicity* of λ_j.

Theorem 1.3.1 is supplemented by the following structure theorem for nilpotents whose proof is left to the reader (Problem 5):

Theorem 1.3.2 (Structure Theorem for Nilpotents). *Let N be an operator on a finite-dimensional vector space, V, of dimension n, so that*

$$N^\ell = 0 \tag{1.3.7}$$

for some $\ell \in \mathbb{Z}_+$. Then there exist $k_1, \dots, k_m \in \{1, 2, \dots\}$ so $\sum_{j=1}^m k_j = n$ and a basis $\{e_{jk}\}_{j=1,\dots,m; k=1,\dots k_j}$ of V so that

$$Ne_{jk} = \begin{cases} e_{j\,k-1} & \text{if } k \geq 2 \\ 0 & \text{if } k = 1 \end{cases} \tag{1.3.8}$$

Using this, one sees that V has a basis in which the matrix of A has λ_j's along the diagonal, λ_j repeated $\dim(P_j)$ times, and some number of 1's directly above the diagonal and otherwise all 0's.

If $\alpha \neq 0$ and $N^\ell = 0$, then

$$(\alpha\mathbb{1} + N)^{-1} = \sum_{j=0}^{\ell-1} (-1)^j \alpha^{-j-1} N^j \tag{1.3.9}$$

This can be checked either by direct multiplication or formally by expanding the geometric series, noting it terminates since $N^\ell = 0$. In particular, applying this to each Jordan block, we see that if $z \notin \{\lambda_1, \dots, \lambda_k\}$, then

$$(A - z)^{-1} = \sum_{j=1}^{k} (\lambda_j - z)^{-1} P_j + \sum_{m=1}^{n} (\lambda_j - z)^{-m-1} (-1)^m N^m \tag{1.3.10}$$

In particular, for $\varepsilon < \min_{j \neq k} |\lambda_j - \lambda_k|$,

$$P_j = \frac{1}{2\pi i} \oint_{|z-\lambda_j|=\varepsilon} \frac{dz}{z - A} \tag{1.3.11}$$

This will be useful in the next section and later in the book. Here we derived (1.3.11) from the Jordan normal form, but instead one can define P_j by (1.3.11) and use it as a tool in proving the Jordan normal form, as we do in Section 2.3.

* * * * *

Next, we turn to the finite-dimensional spectral theorem. We'll only discuss the complex case, although there is a real case (Problem 6).

If $\mathcal{H}$ is a finite-dimensional inner product space, A is called self-adjoint if

$$\langle \varphi, A\psi \rangle = \langle A\varphi, \psi \rangle \tag{1.3.12}$$

for all $\varphi, \psi \in \mathcal{H}$. If $\{e_j\}_{j=1}^n$ is an orthonormal basis, the matrix of any operator, T, is given by $t_{ij} = \langle e_i, Te_j \rangle$. Thus,

$$A \text{ self-adjoint} \Leftrightarrow a_{ij} = \bar{a}_{ji} \tag{1.3.13}$$

A matrix with $a_{ij} = \bar{a}_{ji}$ is called a self-adjoint matrix.

Theorem 1.3.3 (Spectral Theorem for Finite Dimensions). *Let A be a self-adjoint operator on a finite-dimensional inner product space. Then A has an orthonormal basis of eigenvectors, that is, there is an orthonormal basis, $\{x_j\}_{j=1}^n$, and $\{\lambda_j\}_{j=1}^n$ in $\mathbb{R}$, so that*

$$Ax_j = \lambda_j x_j \tag{1.3.14}$$

Remarks. 1. This is a special case of the analog for positive compact self-adjoint operators that we'll prove in Section 3.2. That proof shows the φ that maximizes $\langle \varphi, A\varphi \rangle$ among all φ with $\|\varphi\| = 1$ is an eigenvector. One then shows $[\varphi]^\perp$ is invariant and gets the full result by induction.

2. One can also deduce this from the Jordan normal form by proving λ_j is real, $P_j^* = P_j$, and $N_j = 0$.

3. An equivalent version is that for all self-adjoint matrices, A, there is a unitary matrix, U, so that UAU^{-1} is diagonal.

* * * * *

We next want to say something about rank-one perturbations of finite-dimensional self-adjoint operators. So assume that A is self-adjoint on a finite-dimensional inner product space, $\mathcal{H}$ and $\varphi \in \mathcal{H}, \varphi \neq 0$, and

$$B = A + \langle \varphi, \cdot \rangle \varphi \tag{1.3.15}$$

We want to suppose that for every nonzero eigenvector, ψ_j, of A with $A\psi_j = \alpha\psi_j$, we have that

$$\langle \varphi, \psi_j \rangle \neq 0 \tag{1.3.16}$$

This implies (Problem 1) that the eigenspaces all have multiplicity 1 so $A\psi_j = \alpha_j \psi_j$ with

$$\alpha_1 < \alpha_2 < \ldots < \alpha_d \tag{1.3.17}$$

The following is the fundamental theorem on rank-one perturbation of matrices:

Theorem 1.3.4. *Let A and B be self-adjoint operators on a finite-dimensional inner product space, $\mathcal{H}$, related by* (1.3.15) *where* (1.3.16) *holds. Let $\alpha_1 < \ldots < \alpha_d$ and $\beta_1 < \ldots < \beta_d$ be the eigenvalues of A and B, respectively. Then for $j = 1, 2, \ldots, d$ (with $\alpha_{d+1} = \infty$), we have that*

$$\alpha_j < \beta_j < \alpha_{j+1} \tag{1.3.18}$$

Remarks. 1. Included is that the β's are all simple eigenvalues.

2. Problem 1 has the general case where (1.3.17) or (1.3.16) might fail. All that happens is that (1.3.18) becomes $\alpha_j \leq \beta_j \leq \alpha_{j+1}$.

Proof. Here is an analyst's proof of this fact. Note first that

$$B\eta = \beta\,\eta; \quad \eta \in \mathcal{H}, \eta \neq 0 \Rightarrow \langle \varphi, \eta \rangle \neq 0 \tag{1.3.19}$$

for if $\langle \varphi, \eta \rangle = 0$, then $A\eta = \beta\eta$ violating (1.3.16). Suppose also that β is not an eigenvalue of A. Then

$$B\eta = \beta\eta \Rightarrow (A - \beta)\eta = -\langle \varphi, \eta \rangle \varphi \Rightarrow \eta = \langle -\varphi, \eta \rangle (A - \beta)^{-1} \varphi \tag{1.3.20}$$

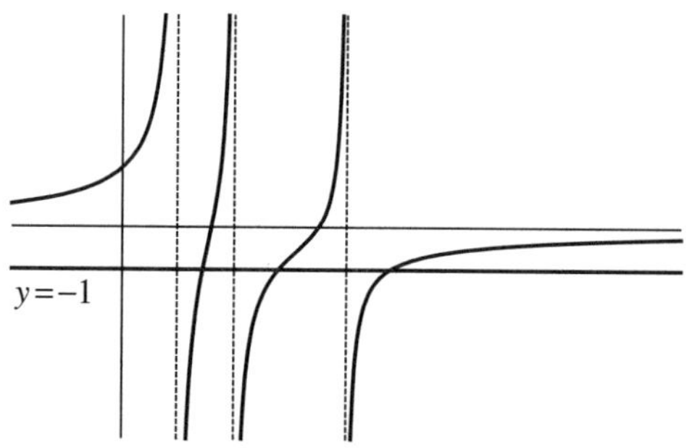

Figure 1.3.1. The graph of $F(z)$.

Taking inner products with φ and using $\langle\varphi,\eta\rangle\neq 0$, we see

$$B\eta=\beta\eta;\quad \beta\notin\{\alpha_1,\alpha_2,\dots\}\Rightarrow\langle\varphi,(A-\beta)^{-1}\varphi\rangle=-1 \tag{1.3.21}$$

Conversely, if $\langle\varphi,(A-\beta)^{-1}\varphi\rangle=-1$, then

$$\begin{aligned}(B-\beta)\big[(A-\beta)^{-1}\varphi\big]&=\varphi+\langle\varphi,(A-\beta)^{-1}\varphi\rangle\varphi\\&=0\end{aligned} \tag{1.3.22}$$

Thus, if

$$\beta\notin\{\alpha_1,\alpha_2,\dots\};\ \beta\text{ is an eigenvalue of }B\Leftrightarrow\langle\varphi,(A-\beta)^{-1}\varphi\rangle=-1 \tag{1.3.23}$$

Now let

$$F(z)=\langle\varphi,(A-z)^{-1}\varphi\rangle=\sum_{j=1}^{d}\frac{|\langle\varphi,\psi_j\rangle|^2}{\alpha_j-z} \tag{1.3.24}$$

Since $(\alpha_j-z)^{-1}$ is strictly monotone increasing and (by (1.3.16)) $F(z)\to\infty$ as $z\uparrow\alpha_j$, $F(z)\downarrow 0$ as $z\to-\infty$ with $F(z)\uparrow 0$ as $z\to\infty$ (see Figure 1.3.1), $F(z)=-1$ has exactly one solution in each (α_j,α_{j+1}), $j=1,2,\dots,d$. By (1.3.23), this implies (1.3.18). $\square$

Closely connected is the following whose proof we leave to Problem 2:

Theorem 1.3.5. *Let A be a self-adjoint operator on a finite-dimensional inner product space, $\mathcal{H}$, with eigenvalues $\alpha_1\leq\alpha_2\leq\dots\leq\alpha_d$. Let $\mathcal{H}_1\subset\mathcal{H}$ be a codimensional 1 subspace and let P be the projection onto $\mathcal{H}_1$. Let $B=PAP$ and let $\beta_1\leq\dots\leq\beta_{d-1}$ be its eigenvalues. Then for $j=1,\dots,d-1$*

$$\alpha_j\leq\beta_j\leq\alpha_{j+1} \tag{1.3.25}$$

Remarks. 1. The canonical example of this is to take $\mathcal{H}=\mathbb{C}^d$ so A is a $d\times d$ matrix and $\mathcal{H}_1=\{\varphi\in\mathbb{C}^d\mid\varphi_d=0\}$ in which case B is the $(d-1)\times(d-1)$ matrix obtained by removing the bottom row and the rightmost column.

2. This is related to the rank-one case because (Problem 2) if φ is a unit vector in $\mathcal{H}_1^\perp$, then $\lim_{\mu\to\infty}P(A+\mu\langle\varphi,\cdot\,\rangle\varphi-z)^{-1}P=(B-z)^{-1}$.

* * * * *

We next turn to finite-dimensional determinants. We begin by discussing exterior algebra (see also Section 3.8 of Part 1). If V is a finite-dimensional vector space of dimension n, its dual, V^*, is also of dimension n. Indeed, if $\{e_j\}_{j=1}^n$ is a basis for V, the *dual basis*, ℓ_j, defined by

$$\ell_j(e_i) = \delta_{ij} \tag{1.3.26}$$

is a basis for V^*. Given $A\colon V \to V$ linear, we define $A^{\mathrm{t}}\colon V^* \to V^*$, its *transpose*, by

$$(A^{\mathrm{t}}\ell)(V) = \ell(Av) \tag{1.3.27}$$

Given $V_1, V_2, \dots, V_k$ finite-dimensional vector spaces, their *tensor product*, $V_1 \otimes \cdots \otimes V_k$, is the set of maps

$$B\colon V_1^* \times \cdots \times V_k^* \to \mathbb{K}$$

($\mathbb{K}$ is shorthand for $\mathbb{R}$ or $\mathbb{C}$) which are multilinear, that is, linear in each argument when the others are fixed. Given $f_1 \in V_1, \dots, f_n \in V_n$, we define $f_1 \otimes \cdots \otimes f_n \in V_1 \otimes \cdots \otimes V_n$ by

$$(f_1 \otimes \cdots \otimes f_n)(\ell_1, \dots, \ell_n) = \prod_{j=1}^n \ell_j(f_j) \tag{1.3.28}$$

If $\{e_k^{(j)}\}_{k=1}^{d_j}$ is a basis for V_j, then $\{e_k^{(1)} \otimes e_{k^2}^{(2)} \otimes \cdots \otimes e_{k_n}^{(n)}\}_{1\le k_\ell \le d}$ is a basis for $V_1 \otimes \cdots \otimes V_n$ (proving $\dim(V_1 \otimes \cdots \otimes V_n) = d_1 d_2 \dots d_n$) for if $\{\ell_k^{(j)}\}_{k=1}^{d_j}$ are dual bases, then

$$(e_{k_1}^{(1)} \otimes \cdots \otimes e_{k_n}^{(n)})(\ell_{q_1}^{(1)}, \dots, \ell_{q_n}^{(n)}) = \delta_{k_1 q_1} \dots \delta_{k_n q_n} \tag{1.3.29}$$

proving the e's are independent, and for any $B \in V_1 \otimes \cdots \otimes V_n$, one has

$$B = \sum_{k_1 \dots k_n} B(\ell_{k_1}^{(1)}, \dots, \ell_{k_n}^{(n)}) e_{k_1}^{(1)} \otimes \cdots \otimes e_{k_n}^{(n)} \tag{1.3.30}$$

If A_j is a linear map of V_j to itself, we define $A_1 \otimes \cdots \otimes A_n\colon V_1 \otimes \cdots \otimes V_n \to V_1 \otimes \cdots \otimes V_n$ by

$$[(A_1 \otimes \cdots \otimes A_n)B](\ell_1, \dots, \ell_j) = B(A_1^t \ell, \dots, A_n^t \ell) \tag{1.3.31}$$

It is a unique linear map obeying

$$(A_1 \otimes \cdots \otimes A_n)(f_1 \otimes \cdots \otimes f_n) = A_1 f_1 \otimes \cdots \otimes A_n f_n \tag{1.3.32}$$

Matrix elements of $A_1 \otimes \cdots \otimes A_n$ in the tensor product basis are just products of matrix elements of A_j.

If $V_1 = V_2 = \cdots = V_n = V$, we use $\otimes^n V$ for $V \otimes \cdots \otimes V$ (n times). If $A\colon V \to V$, $\otimes^n A$ is $A \otimes \cdots \otimes A$ (n times). Note that

$$(\otimes^n A)(\otimes^n B) = \otimes^n (AB) \tag{1.3.33}$$

If $\pi_n \in \Sigma_n$ is a permutation of $\{1, \dots, n\}$, then we define $\sigma_\pi \colon \otimes^n V \to \otimes^n V$ by

$$\sigma_\pi(B)(\ell_1, \dots, \ell_n) = B(\ell_{\pi^{-1}(1)}, \dots, \ell_{\pi^{-1}(n)}) \tag{1.3.34}$$

Note (this is the reason for the π^{-1} rather than π)

$$\sigma_\pi \sigma_{\tilde{\pi}} = \sigma_{\pi\tilde{\pi}} \tag{1.3.35}$$

and

$$\sigma_\pi(f_1 \otimes \cdots \otimes f_n) = f_{\pi(1)} \otimes \cdots \otimes f_{\pi(n)} \tag{1.3.36}$$

Note that for any $A \in \mathcal{L}(V)$ and $\pi \in \Sigma_n$,

$$(\sigma_\pi)(\otimes^n A) = (\otimes^n A)\sigma_\pi \tag{1.3.37}$$

Recall each permutation $\pi \in \Sigma_n$ has a *sign*, $(-1)^\pi \in \{-1, 1\}$ defined in Problem 3. We define

$$\wedge^n(V) = \{B \in \otimes^n V \mid \forall \pi \in \Sigma_n,\ \sigma_\pi(B) = (-1)^\pi B\} \tag{1.3.38}$$

If we define on $\otimes^n V$,

$$\mathcal{A}_n = \frac{1}{n!} \sum_{\pi \in \Sigma_n} (-1)^\pi \sigma_\pi \tag{1.3.39}$$

it is exactly the projection onto $\wedge^n(V) \subset \otimes^n V$. By (1.3.37), $\otimes^n A$ maps $\wedge^n(V)$ to itself, and we use

$$(\otimes^n A) \restriction \wedge^n(V) \equiv \wedge^n(A) \tag{1.3.40}$$

More generally, we'll define $\wedge^n(A)$ on $\otimes^n \mathcal{H}$ by

$$\wedge^n(A) = (\otimes^n A)\mathcal{A} = \mathcal{A}(\otimes^n A) \tag{1.3.41}$$

By (1.3.33),

$$\wedge^n(A) \wedge^n (B) = \wedge^n(AB) \tag{1.3.42}$$

Given $f_1, \dots, f_n \in V$, define $f_1 \wedge \cdots \wedge f_n \in \wedge^n(V)$ by

$$f_1 \wedge \cdots \wedge f_n \equiv \sqrt{n!}\, \mathcal{A}_n(f_1 \otimes \cdots \otimes f_n) \tag{1.3.43}$$

$$= \frac{1}{\sqrt{n!}} \sum_{\pi \in \Sigma_n} (-1)^\pi f_{\pi(1)} \otimes \cdots \otimes f_{\pi(n)} \tag{1.3.44}$$

Where we use $\sqrt{n!}$ in (1.3.44), some authors use 1 (then (1.3.44) has $1/n!$) and others $n!$ (then (1.3.44) has no $n!$). It passes through all the calculations below. We used $\sqrt{n!}$ because of natural inner product considerations when V is an inner product space (see Section 3.8 of Part 1). Notice that

$$f_{\pi(1)} \wedge \cdots \wedge f_{\pi(n)} = (-1)^\pi f_1 \wedge \cdots \wedge f_n \tag{1.3.45}$$

In particular, if $\{e_j\}_{j=1}^d$ is a basis for V, $e_{j_1} \wedge \cdots \wedge e_{j_n}$ is zero if any two j_ℓ's are equal and is equal to $\pm$ the n-fold wedge product with the j's reordered.

It follows that $\{e_{j_1} \wedge \cdots \wedge e_{j_n} \mid j_1 < j_2 < \cdots < j_n\}$ is a basis for $\wedge^n(V)$. In particular, $\wedge^n(V) = \{0\}$ if $n > d$ and

$$\dim(\wedge^n(V)) = \binom{d}{n} \qquad \text{if } n \le d \tag{1.3.46}$$

Notice that $\wedge^d(V) = \mathbb{K}$ so $\wedge^d(A)$ is multiplication by a number.

Definition. Let V be a d-dimensional vector space over $\mathbb{K} = \mathbb{R}$ or $\mathbb{C}$ and $A \in \mathcal{L}(V)$. $\det(A) \in \mathbb{K}$ is the unique element in $\mathbb{K}$ with

$$\wedge^d(A)B = \det(A)B \tag{1.3.47}$$

for any $B \in \wedge^d(V)$.

This is, of course, a basis-independent definition.

Definition. Given a $d \times d$ matrix, $A = \{a_{ij}\}_{1 \le i,j \le d}$, of elements of $\mathbb{K}$, view it as a map of $\mathbb{K}^d$ to $\mathbb{K}^d$ and use $\det(A)$ for the determinant of that map. If $i, j \in \{1, \dots, d\}$, let $A^{(i,j)}$ be the $(d-1) \times (d-1)$ matrix with row i and column j removed. Its determinant is called the *minor*, $M_{ij}(A)$, of A. We define the *cofactor*, $C_{ij}(A)$, by

$$C_{ij}(A) = (-1)^{i+j} M_{ji}(A) \tag{1.3.48}$$

The matrix, $\{C_{ij}(A)\}_{1 \le i,j \le d}$, is called the *cofactor matrix* to A.

Theorem 1.3.6. *Let V be a finite-dimensional space of dimension d.* $\det(\,\cdot\,)$ *has the following properties:*

(a) $\det(AB) = \det(A)\det(B)$ $\qquad$ (1.3.49)

(b) $\det(\mathbb{1}) = 1$ $\qquad$ (1.3.50)

(c) *If A has matrix elements $\{a_{ij}\}_{1 \le i,j \le d}$ in some basis, then*

$$\det(A) = \sum_{\pi \in \Sigma_n} (-1)^\pi a_{1\pi(1)} a_{2\pi(2)} \cdots a_{d\pi(d)} \tag{1.3.51}$$

(d) *If A has matrix elements $\{a_{ij}\}_{1 \le i,j \le d}$ and $A^{(\pi)}$ has matrix elements $a_{\pi(i)j}$, then*

$$\det(A^\pi) = (-1)^\pi \det(A) \tag{1.3.52}$$

and similarly for permutations of the columns. In particular, if a matrix has two identical rows or columns, its determinant is 0.

(e) *If A is a $d \times d$ matrix and C its cofactor matrix, then*

$$CA = AC = \det(A)\mathbb{1} \tag{1.3.53}$$

(f) *A is invertible if and only if* $\det(A) \neq 0$.

Remark. $(-1)^\pi$ is defined in Problem 3.

Proof. (a) is just (1.3.42).

(b) Clearly, $\wedge^d(\mathbb{1}) = \mathbb{1}$ on $\wedge^d(V) = \mathbb{K}$, that is, multiplication by 1.

(c) By definition of the matrix of A in basis, $\{e_i\}_{i=1}^d$, $Ae_j = \sum_{i=1}^d a_{ij}e_i$. Thus,

$$\det(A)(e_1 \wedge \cdots \wedge e_d) = Ae_1 \wedge \cdots Ae_d \tag{1.3.54}$$

$$= \sum_{i_k=1,\dots,d\,(\text{for } k=1,\dots,d)} a_{i_1 1} \dots a_{i_d d} e_{i_1} \wedge \cdots \wedge e_{i_d} \tag{1.3.55}$$

Only terms with $i_1, \dots, i_d$ distinct, that is, with $i_\ell = \pi(\ell)$ for some permutation, π, have nonzero $e_{i_1} \wedge \cdots \wedge e_{i_d}$ and, by (1.3.45), those $e_{i_1} \wedge \cdots \wedge e_{i_d}$ are $(-1)^\pi e_1 \wedge \cdots \wedge e_d$. This leads to the analog of (1.3.51) but with $a_{1\pi(1)} \dots a_{d\pi(d)}$ replaced by $a_{\pi(1)1} \dots a_{\pi(d)d}$. Noting that

$$a_{i\pi(1)} \dots a_{d\pi(d)} = a_{\pi^{-1}(1)1} \dots a_{\pi^{-1}(d)d} \tag{1.3.56}$$

and $(-1)^\pi = (-1)^{\pi^{-1}}$, and $\sum_\pi$ and $\sum_{\pi^{-1}}$ is the same we get (1.3.51).

(d) is immediate from the group structure of Σ_d (i.e., summing over π or $\pi\pi_0$ for π_0 fixed is the same) and (1.3.51).

(e) If we fix j and collect all terms with $\pi(j) = k$, (1.3.51) becomes

$$\det(A) = \sum_{k=1}^d a_{jk} C(A)_{kj} \tag{1.3.57}$$

This is the diagonal part of $AC = \det(A)\mathbb{1}$. For the off-diagonal part, we note if $i \neq j$ and we replaced a_{jk} with a_{ik}, we get, by (1.3.57), the determinant of the matrix with row j equals row i. By interchanging them, (1.3.52) implies this determinant is 0. Thus, $AC = \det(A)\mathbb{1}$. Using the form of (1.3.51) with $a_{\pi(1)1} \dots a_{\pi(d)d}$, we get $CA = \det(A)\mathbb{1}$.

(f) If A is invertible, $\det(A)\det(A^{-1}) = 1$, so $\det(A) \neq 0$. If $\det(A) \neq 0$, then $C(A)\det(A)^{-1}$ is the inverse of A. □

Remarks. 1. That $C(A)\det(A)^{-1}$ is the inverse of A is *Cramer's rule*.

2. Section 3.10 has a basis-independent version of Cramer's rule using derivatives of determinants (note that $C(A)_{kj} = \frac{\partial \det(A)}{\partial a_{jk}}$).

Here is a connection between det and Jordan normal forms.

Theorem 1.3.7. *Let $A \in \mathcal{L}(V)$ with V a finite-dimensional complex vector space. Let*

$$A = \sum_{j=1}^k \lambda_j P_j + N_j \tag{1.3.58}$$

Let $d_j = \dim(\operatorname{Ran}(P_j))$, *the algebraic multiplicity of the eigenvalue* λ_j. *Then*

$$\text{(a)}\ \det(A) = \prod_{j=1}^{k} \lambda_j^{d_j} \tag{1.3.59}$$

$$\text{(b)}\ \det(z-A) = \prod_{j=1}^{k} (z-\lambda_j)^{d_j} \tag{1.3.60}$$

so the zeros of $z \mapsto \det(z-A) \equiv C_A(z)$ *are exactly the* λ_j*'s.*

(c) (Cayley–Hamilton Theorem)

$$C_A(A) = 0 \tag{1.3.61}$$

Remarks. 1. $C_A(z)$ is called the *characteristic polynomial* for A.

2. The constant term in $C_A(z)$ is $(-1)^d \det(A)$. Thus, if

$$C_A(z) = \sum_{j=0}^{k} d_j z^k \tag{1.3.62}$$

we see, by (1.3.59) that

$$A\biggl(\sum_{j=1}^{k} d_j A^{k-1}\biggr) = (-1)^{d+1} \det(A) \tag{1.3.63}$$

If $\det(A) \neq 0$, this realizes the inverse of A as a polynomial in A. This is the form in which Hamilton found his version of the theorem.

Proof. (a) Picking bases in which N_j is upper triangular, we see A has a matrix with λ_j along the diagonal d_j times and 0's below the diagonal. Only the identity permutation enters in (1.3.51), so (1.3.59) is immediate.

(b) $z - A = \sum_{j=1}^{k} (z-\lambda_j)P_j - N_j$ is the Jordan normal form for $z - A$, so (1.3.60) is just (1.3.59).

(c) N_j is a nilpotent operator on the d_j-dimensional space $\operatorname{Ran}(P_j)$, so by the structure theorem for nilpotents (Theorem 1.3.2), $N_j^{d_j} = 0$. Thus, if $\varphi \in \operatorname{Ran}(P_j)$,

$$(A-\lambda_j)^{d_j}\varphi = N_j^{d_j}\varphi = 0 \tag{1.3.64}$$

It follows that $C_A(A)\varphi = \prod_{j=1}^{k}(A-\lambda_j)\varphi = 0$. Since $V = \operatorname{Ran}(P_1) \dotplus \operatorname{Ran}(P_2) \dotplus \dots$, we see $C_A(A) = 0$. $\square$

Finally, we want to say something about the trace. An elegant approach is to define the *trace* by

$$\operatorname{Tr}(A) = \frac{d}{dz} \det(e^{zA})\bigg|_{z=0} \tag{1.3.65}$$

(Problem 7), but we'll use a more prosaic approach. Let $\{a_{ij}\}_{1\le i,j\le d}, \{b_{ij}\}_{1\le i,j\le d}$ be two matrices. If ab and ba represent the matrix products, we have

$$\sum_i (ab)_{ii} = \sum_{i,j} a_{ij}b_{ji} = \sum_{j,i} b_{ji}a_{ij} = \sum_j (ba)_{jj} \tag{1.3.66}$$

This is the first statement in

Theorem 1.3.8. *Fix a basis $\{e_j\}_{j=1}^d$ for V and define*

$$\mathrm{Tr}_{\{e\}}(A) = \sum_{i=1}^d a_{ij} \tag{1.3.67}$$

where a_{ij} is the matrix of A in basis e. Then

(a) $\mathrm{Tr}_{\{e\}}(AB) = \mathrm{Tr}_{\{e\}}(BA)$ (1.3.68)

(b) $\mathrm{Tr}_{\{e\}}(A)$ *is independent of choice of basis, call it* $\mathrm{Tr}(A)$.

(c) *If $\{\lambda_j\}_{j=1}^k$ are the eigenvalues of A with algebraic multiplicity d_j, then*

$$\mathrm{Tr}(A) = \sum_{j=1}^k d_j\lambda_j \tag{1.3.69}$$

Remark. (1.3.59) and (1.3.69), usually phrased as saying that $\det(A)$ and $\mathrm{Tr}(A)$ are the product and sum of all eigenvalues, counted up to algebraic multiplicity.

Proof. (a) was proven above.

(b) The matrices of A in a and $\tilde{a}$ in two bases, $\{e_i\}_{i=1}^d$ and $\{f_j\}_{j=1}^d$, are related by

$$\tilde{a} = t^{-1}at \tag{1.3.70}$$

where t is the change of basis matrix defined by

$$f_j = \sum_{i=1}^d t_{ij}e_i \tag{1.3.71}$$

(such a matrix exists since $\{e_i\}_{i=1}^d$ is a basis; moreover, $e_i = \sum_{j=1}^d s_{ji}f_j$ and t and s are inverses). (1.3.70) follows from

$$Af_j = \sum_{i=1}^d t_{ij}Ae_i = \sum_{i,k=1}^d t_{ij}a_{ki}e_k \tag{1.3.72}$$

but also

$$Af_j = \sum_{i=1}^d \tilde{a}_{ij}f_i = \sum_{i,k=1,\dots,d} \tilde{a}_{ij}t_{ki}e_k \tag{1.3.73}$$

so $at = t\tilde{a}$, which is (1.3.70).

Thus, if Tr_δ is the sum of the diagonal matrix elements in a matrix,

$$\mathrm{Tr}_\delta(\tilde{a}) = \mathrm{Tr}_\delta(t^{-1}at) \tag{1.3.74}$$

$$= \mathrm{Tr}_\delta(att^{-1}) \qquad \text{by (a)} \tag{1.3.75}$$

$$= \mathrm{Tr}_\delta(a) \tag{1.3.76}$$

(c) is immediate from passing to a basis where A has λ_j along the diagonal. $\square$

Notes and Historical Remarks. Muir [**484**] has a paper-by-paper historical discussion of the history of determinants. The Babylonians and Greeks knew how to solve simultaneous linear equations, but the modern era involving determinants dates to 1683 when the Japanese mathematician Seki (1642–1708) (whose name appears as Seki Tokakazu or Seki Kowa) published a book [**407**] that included them, and independently (indeed, Seki's work was unknown in Europe for many years), G. Leibniz wrote a letter to l'Hôpital (published with his correspondence much later [**435**]).

Leibniz wrote that $10+11x+12y=0$, $20+21x+22y=0$, $30+31x+32y$ had a solution because

$$10\cdot 21\cdot 32 + 11\cdot 22\cdot 30 + 12\cdot 20\cdot 31 = 10\cdot 22\cdot 31 + 11\cdot 20\cdot 32 + 12\cdot 21\cdot 30$$

so he knew something about what, in modern terms, would be expressed as $n+1$ equations in n unknowns have a solution only if a certain determinant vanishes (Problem 8). It is a stretch from this to say that Leibniz knew all about determinants, but still (1.3.51) is called *Leibniz's formula.*

G. Cramer (1704–52) was closer to a general theory and, in particular, had something essentially equivalent to Cramer's rule [**137**] for inverses (in the context of solving linear equations in two or three variables, not matrix multiplication). It is likely MacLaurin knew this rule as early as 1729 and, in any event, it appeared in his 1748 book [**463**] published two years after his death and two years before Cramer's work.

Important later developments are due to E. Bezout (1730–83) [**57**], A. Vandermonde (1735–96) [**711**], P. S. Laplace (1749–1827) [**427**], J.-L. Lagrange (1736–1827) [**419**], J. F. C. Gauss (1777–1855) [**224**] (who introduced the term "determinant" in a slightly different context), A. Cauchy (1789–1857) (who used it in its present context [**109**]), and a long series of papers spanning fifteen years by C. Jacobi (1804–51) [**339, 340, 341, 342, 343, 344**]. It was put into the context of matrix theory by A. Cayley [**111**].

The Cayley–Hamilton theorem is named after 1853 work of Hamilton [**287**] and 1858 work of Cayley [**111**]. Hamilton discussed the result without proof, and not in terms of matrices, for two special cases: three-dimensional rotations and multiplication of quaternions on $\mathbb{C}^2$ viewed as quaternions.

Cayley stated the general result, did a calculation verifying it for 2×2 matrices, and said he had done the calculation for 3×3 determinants. The first complete proof is by Frobenius in 1878 [**211**] who named the theorem after Cayley.

Leibniz's formula (1.3.51) is not used for computer calculations since it requires $n(n!)$ multiplications. Instead one decomposes a matrix into a product of triangular matrices whose determinants are just the product of their diagonal elements.

See the Notes to Section 1.7 of Part 1 for the history of the Jordan normal form and the finite-dimensional spectral theorem.

Problems

1. (a) Given a finite-dimensional self-adjoint operator, A with $A\psi_j = \alpha_j \psi_j$, $\alpha_1 \leq \alpha_2 \leq \ldots$, its eigenvalues and eigenvectors and $\varphi \in \mathcal{H}$, let $\mathcal{H}_1$ be the span of those ψ_j with $\langle \varphi, \psi_j \rangle \neq 0$ and $\mathcal{H}_2 = \mathcal{H}_1^\perp$ the span of those with $\langle \varphi, \psi \rangle = 0$ (if A has a degenerate eigenvalue, choose the ψ_j's so that, for that α, there is at most one ψ_j with $\langle \varphi, \psi_j \rangle \neq 0$). Prove that A leaves $\mathcal{H}_1$ and $\mathcal{H}_2$ invariant and that $A \restriction \mathcal{H}_2 = B \restriction \mathcal{H}_2$.

   (b) Apply Theorem 1.3.4 to $A \restriction \mathcal{H}_1$ and show that when combined with the unchanged eigenvalues for $A \restriction \mathcal{H}_2$, we get Theorem 1.3.4 in general with the only change being $\alpha_j < \beta_j < \alpha_{j+1}$ is replaced by $\alpha_j \leq \beta_j \leq \alpha_{j+1}$.

2. This problem will prove Theorem 1.3.5.

   (a) Let $\varphi \in \mathcal{H}$ obey $P\varphi = 0$, $\|\varphi\| = 1$ and $F(z) = \langle \varphi, (A - z)^{-1} \varphi \rangle$ as in (1.3.24). Prove that β is an eigenvalue of B if and only if $F(\beta) = 0$.

   (b) Conclude (1.3.25).

   (c) Let $\beta_j(\mu)$ be the eigenvalues of $A + \mu \langle \varphi, \cdot\, \varphi \rangle$. Prove that as $\mu \to \infty$, $\beta_j(\mu) \to \beta_j$ (the eigenvalues of B) for $j = 1, \ldots, d-1$ and $\beta_d(\mu) \to \infty$. Prove also that as $\mu \to -\infty$, $\beta_j(\mu) \to \beta_{j-1}$ for $j = 2, \ldots, d$ and $\beta_1(\mu) \to -\infty$. (*Hint*: Look at solutions of $F(z) = -\mu^{-1}$.)

3. (a) Let $P(x_1, \ldots, x_n) = \prod_{i<j}(x_i - x_j)$. For any permutation $\pi \in \Sigma_n$, prove $P(x_{\pi(1)}, \ldots, x_{\pi(n)}) = \pm P(x_1, \ldots, x_n)$, where $\pm$ is dependent on π but independent of x. Call it $(-1)^\pi$, the *sign of the permutation*, π. Thus,

$$P(x_{\pi(1)}, \ldots, x_{\pi(n)}) = (-1)^\pi P(x_1, \ldots, x_n) \tag{1.3.77}$$

   (b) Prove that $(-1)^{\pi\pi'} = (-1)^\pi (-1)^{\pi'}$.

(c) Let $i \neq j$. If (ij) is the permutation

$$(ij)(k) = \begin{cases} k, & k \neq i, j \\ j, & k = i \\ i, & k = j \end{cases} \tag{1.3.78}$$

Prove that

$$(-1)^{(ij)} = -1 \tag{1.3.79}$$

(d) If $(i_1, i_2, \dots, i_\ell)$ (for all unequal i's) is the permutation

$$(i_1, \dots, i_\ell) = \begin{cases} k, & k \notin \{i_1, \dots, i_\ell\} \\ i_{j+1}, & k = i_j,\ j = 1, 2, \dots, n-1 \\ i_1, & k = i_n \end{cases} \tag{1.3.80}$$

Prove that

$$(-1)^{(i_1,\dots,i_\ell)} = (-1)^{\ell-1} \tag{1.3.81}$$

(*Hint*: Prove that $(i_1, \dots, i_\ell)$ can be written as a product of $\ell - 1$ pairwise transpositions.)

4. Let $N \in \mathcal{L}(V)$ with $N^\ell = 0$. Prove that $\mathrm{Ker}(N) \neq \{0\}$. (*Hint*: By decreasing ℓ, suppose $N^{\ell-1} \neq 0$. What does N do to vectors in $\mathrm{Ran}(N^{\ell-1})$?)

5. Let N be a nilpotent operator on a finite-dimensional vector space, X. Suppose m is such that $N^m = 0$ but $N^{m-1} \neq 0$ (although $N^{m-1}x$ may be zero for some x, just not for all x). For $\ell = 1, 2, \dots, m$, define

$$Y_j = \{x \mid N^j x = 0,\ N^{j-1}x \neq 0\} \tag{1.3.82}$$

By picking a basis for $Y_1 \cup \{0\}$, then extending to a basis of $Y_1 \cup Y_2 \cup \{0\}, \cdots$, prove Theorem 1.3.2.

6. Let A be a real symmetric matrix but view it as acting on $\mathbb{C}^n$. Prove that one can choose the eigenvectors guaranteed by the complex spectral theorem as real and deduce a real spectral theorem.

7. Define trace by

$$\mathrm{Tr}(A) = \frac{d}{dz} \det(e^{zA}) \bigg|_{z=0} \tag{1.3.83}$$

where $e^A = \sum_{n=0}^\infty (n!)^{-1} A^n$.

(a) Prove $e^{z(A+B)} = e^{zA}e^{zB} + O(z^2)$ and conclude that

$$\mathrm{Tr}(A+B) = \mathrm{Tr}(A) + \mathrm{Tr}(B) \tag{1.3.84}$$

(b) If C is invertible, prove that $e^{z(CAC^{-1})} = Ce^{zA}C^{-1}$ and conclude that

$$\mathrm{Tr}(CAC^{-1}) = \mathrm{Tr}(A) \tag{1.3.85}$$

(c) Prove that any $C \in \mathcal{L}(V)$ is a sum of two invertible elements and conclude that (*cyclicity of the trace*)

$$\mathrm{Tr}(AC) = \mathrm{Tr}(CA) \tag{1.3.86}$$

for all C.

8. Consider n equations in $n-1$ unknowns

$$a_{jn} + \sum_{k=1}^{n-1} a_{jk} x_k = 0 \tag{1.3.87}$$

(a) Prove that, if there is a solution, then

$$\det\bigl((a_{jk})_{1\le j,k\le n}\bigr) = 0 \tag{1.3.88}$$

(b) Conversely, if $B^{(j)}$ is the $(n-1)\times(n-1)$, $(a_{\ell k})_{1\le \ell\le n; 1\le k\le n-1; \ell\ne j}$, and if (1.3.88) holds and $\det(B^{(j)}) \ne 0$ for all j, then (1.3.87) has a solution.

9. This problem will prove that any finite matrix has a basis in which it is upper-triangular providing a model for the results of Section 3.4.

(a) Prove that if A is an operator on a finite-dimensional complex vector space, there is a vector $v \ne 0$ and $\lambda \in \mathbb{C}$ so that $Av = \lambda v$. (*Hint*: Prove first that $\mathrm{Ker}(A-\lambda) \ne \{0\} \Leftrightarrow (A-\lambda)$ is not invertible and then look for zeros of $P(\lambda) = \det(A-\lambda)$.)

(b) If A is defined on a finite-dimensional space, X, a subspace Y is called *invariant* for A if $A[Y] \subset Y$. Prove if v, λ are given as in (a), then $\{\mu v \mid \mu \in \mathbb{C}\}$ is a one-dimensional invariant vector space for A.

(c) If Y is invariant for A, π the canonical projection from X to X/Y, prove that $\widetilde{A}\colon X/Y \to X/Y$ by $\widetilde{A}(\pi(x)) = \pi(Ax)$ is well-defined and linear.

(d) If $W \subset X/Y$ is invariant for $\widetilde{A}$, prove that $\pi^{-1}[W]$ is invariant for A and $\dim \pi^{-1}[W] = \dim W + \dim Y$.

(e) For any A, prove there is a *nest* of invariant subspaces. i.e., $0 = W_0 \subset W_1 \subset W_2 \subset \cdots \subset W_d = X$ (where $d = \dim(X)$) and $\dim(W_j) = j$. (*Hint*: Use (d) and induction.)

(f) Pick $e_j \in W_j \setminus W_{j-1}$ for $j = 1, \dots, d$. Prove that $\{e_j\}_{j=1}^d$ is a basis and that $Ae_j = \lambda e_j + \sum_{k=1}^{j-1} \alpha_{kj} e_k$.

(g) Conclude that A is upper-triangular in the basis.

10. This problem will use Problem 9 to get the Jordan normal form of A, an operator on a finite-dimensional complex vector space, X.

(a) In the two-dimensional case, if A is upper-triangular in $\{e_1, e_2\}$, so

$$Ae_1 = \lambda_1 e_1, \quad Ae_2 = \lambda_2 e_2 + \beta e_1 \tag{1.3.89}$$

prove that if $\lambda_1 \neq \lambda_2$, one can find α, so that A is diagonal in the basis $\{e_1, e_2 + \alpha e_1\}$.

(b) Generalize the argument in (a) and use Problem 9 to prove that for any operator A, there is a basis $\{e_j\}_{j=1}^d$ and $\{\lambda_j\}_{j=1}^d$ so that

$$Ae_j = \lambda_j e_j + \sum_{\{k \mid k<j,\, \lambda_k=\lambda_j\}} \alpha_{kj} e_k \tag{1.3.90}$$

(c) Prove that there are operators $\{P_j\}_{j=1}^M$, $\{N_j\}_{j=1}^M$ and $\{\mu_j\}_{j=1}^M$ so that

$$P_j P_k = \delta_{jk} P_j, \quad N_j P_k = P_k N_j = \delta_{jk} N_j, \quad N_j^{\dim(\operatorname{Ran} P_j)} = 0 \tag{1.3.91}$$

$$A = \sum_{j=1}^M \mu_j P_j + N_j \tag{1.3.92}$$

11. This problem will prove that if A is an operator on a finite-dimensional complex inner product space X, then

$$A = T + N \tag{1.3.93}$$

where T is normal (i.e., $T^* = T$), N is nilpotent, and the eigenvalues of A and T are the same with the same algebraic multiplicity. This serves as a model of results in Section 3.4.

(a) Prove there is an orthonormal basis in which A is upper-triangular. (*Hint*: Use Problem 9 and Gram–Schmidt.)

(b) Prove that the eigenvalues of A are the same (including algebraic multiplicity) as the diagonal elements of the matrix representation found in (a). (*Hint*: Consider $\det(A - \mu\mathbb{1})$.)

(c) Prove (1.3.93).

1.4. Finite-Dimensional Eigenvalue Perturbation Theory

Let V be a finite-dimensional complex vector space of dimension n. We want to consider operators $A(\beta) \in \mathcal{L}(V)$ depending on a parameter β. Most often, $\beta \in \Omega$, a region in $\mathbb{C}$ and A, will be analytic in β, but we will also consider the case $\beta \in (a, b) \in \mathbb{R}$ and $A(\beta)$ continuous or C^k or C^∞. We'll sometimes consider the "self-adjoint case," that is, V is a complex inner product space, $\Omega = \overline{\Omega}$ (complex conjugate, not closure), and

$$A(\beta)^* = A(\bar\beta) \tag{1.4.1}$$

We are concerned with how the eigenvalues and eigenprojections of $A(\beta)$ vary with β. We'll see in the general analytic case that the eigenvalues are algebroidal functions, so analytic, except at a discrete set where they are given by Puiseux series. The eigenprojections will, in general, have negative terms in a Laurent–Puiseux series.

One of our deepest results is a theorem of Rellich that in the self-adjoint case, eigenvalues and eigenprojections are analytic near any $\beta_0 \in \mathbb{R}\cap\Omega$. We'll also prove the Feynman–Hellmann theorem, a general formula for $\frac{dE(\beta)}{d\beta}$ in the self-adjoint, simple case that only requires that $A(\beta)$ be C^1.

It is reasonable to consider eigenprojections rather than eigenvectors. For if eigenprojections are analytic, picking $\varphi \in \operatorname{Ran}(P(\beta_0))$ and $\varphi(\beta) = P(\beta)\varphi$, we get analytic eigenvectors (at least if there aren't eigennilpotents). But if, say, $P(\beta)$ has a first-order pole at β_0, $(\beta - \beta_0)P(\beta)\varphi$ for suitable φ will be analytic at β_0. Put differently, eigenprojections are intrinsic to $A(\beta)$, while eigenvectors depend on choice of normalization.

It is also important to allow general analytic families rather than just the historically motivating case $A_0+\beta B$ since, as we'll see in Section 2.3 (see Theorem 2.3.11), the linear infinite-dimensional case for discrete eigenvalues can be reduced to a finite-dimensional problem but only with analytic rather than linear parameters.

We begin by noting that the eigenvalues $\lambda(\beta)$ are solutions of

$$\det\bigl(\lambda - A(\beta)\bigr) = 0 = \lambda^n + \sum_{j=0}^{n-1} a_j(\beta)\lambda^j \tag{1.4.2}$$

They are thus algebroidal functions, so by Theorem 1.2.3, we have the first part of

Theorem 1.4.1. *Let $A(\beta)$, $\beta \in \Omega$, be an analytic family of operators on an n-dimensional vector space, V, over $\mathbb{C}$. Then there exist $r_0 \le n$ and a discrete set $D \subset \Omega$ so that*

(a) *If $\beta \in \Omega \setminus D$, $A(\beta)$ has r_0 distinct eigenvalues, and if $\beta \in D$, it has fewer than r_0 distinct eigenvalues.*
(b) *For any $\beta_0 \in \Omega \setminus D$, there is a disk $\mathbb{D}_\delta(\beta_0)$ and r_0 analytic scalar functions, $\lambda_1(\beta), \dots, \lambda_{r_0}(\beta)$, and projection-valued functions, $P_1(\beta), \dots, P_{r_0}(\beta)$ so that $\lambda_j(\beta)$ are the eigenvalues of $A(\beta)$ for $\beta \in \mathbb{D}_\delta(\beta_0)$ and $P_j(\beta)$ the corresponding eigenprojections.*
(c) *Near any $\beta_0 \in D$, the eigenvalues of $A(\beta)$ are given by all the branches of one or more Puiseux series whose values at β_0 are the eigenvalues of $A(\beta_0)$.*
(d) *Near any $\beta_0 \in D$, the $P_j(\beta)$ are multivalued analytic functions with the same p-fold structure (i.e., return to themselves after continuation*

along p circuits about z_0) as λ_j, but, in general, the multivalued $P_j(\beta)$ may not be bounded as $\beta \to \beta_0$. Instead, one has for some $k > 0$,

$$\|P_j(\beta)\| \leq C(\beta - \beta_0)^{-k} \tag{1.4.3}$$

for all j and all β near to β_0. The $P_j(\beta)$ have Laurent–Puiseux series with at most finitely many negative terms.

Remark. Just as we'll use (1.3.11) to get analyticity of the $P_j(\beta)$ near any $\beta_0 \in \Omega \setminus D$, one can use

$$N_j = \frac{1}{2\pi i} \int_{|z-\lambda_j|=\varepsilon} \frac{(z-\lambda_j)\, dz}{z - A} \tag{1.4.4}$$

to get analyticity of the eigennilpotents.

Proof. (a), (b) For the eigenvalues, this is just Theorem 1.2.3 plus Remark 3 to that theorem. For any eigenvalue $\lambda_j(\beta_0)$ of $A(z_0)$, pick ε so $\varepsilon < \frac{1}{2}\min|\lambda_j(\beta_0) - \lambda_k(\beta_0)|$. Then $P_j(\beta_0)$ is given by (1.3.11). Since the invertible matrices are an open set (since B is invertible, $\det(B) \neq 0$ and $\det(\,\cdot\,)$ is continuous), for some δ and all $\beta \in \mathbb{D}_\delta(\beta_0)$, $z - A$ is invertible for all $z \in \partial\mathbb{D}_\varepsilon(\lambda_j(\beta_0))$. This plus (1.3.11) proves local analyticity of the $P_j(\beta)$.

(c) This is just the rest of Theorem 1.2.3.

(d) That $P_j(\beta)$ has the same p-fold structure as $\lambda_j(\beta)$ follows from the unique association of $P_j(\beta)$ to $\lambda_j(\beta)$. If we prove (1.4.3), by looking at $Q_j(\eta) \equiv P_j(\beta_0 + \eta P)$, we see $Q_j(\eta)$ has a Laurent series with a finite-order pole, and we get the claimed Laurent–Puiseux series. Thus, we are reduced to proving (1.4.3).

We claim first that for every compact $K \subset \Omega$ and $R > 0$, there is a C so that for $\beta \in K \setminus D$ and $z \notin \sigma(A(\beta))$ with $|z| \leq R$, we have that

$$\|(A(\beta) - z)^{-1}\| \leq C\text{dist}(z, \sigma(A(\beta)))^{-n} \tag{1.4.5}$$

For clearly,

$$|\det(z - A(\beta))| = \prod_{j=1}^{r_0} |z - \lambda_j(\beta)|^{m_j} \geq \text{dist}(z, \sigma(A(\beta)))^n$$

The cofactor matrix of $(z - A(\beta))$ is clearly bounded as β runs through K and z through $\{z \mid |z| \leq R\}$. Thus, by Cramer's rule, we have (1.4.5).

Next, we claim if

$$\varepsilon(\beta) = \min_{j \neq k, i \leq j, k \leq r_0} |\lambda_j(\beta) - \lambda_k(\beta)| \tag{1.4.6}$$

then for some $\Delta > 0$ and all β near β_0,

$$|\varepsilon(\beta)| \geq C|\beta - \beta_0|^{\Delta} \tag{1.4.7}$$

This is because the eigenvalues are given by different branches or values of different Puiseux series, so each difference is bounded from below by some $C_{jk}|\beta-\beta_0|^{\Delta_{jk}}$.

We can now prove (1.4.3). We write $P_j(\beta)$ by (1.3.11) with $\varepsilon=\frac{1}{2}\varepsilon(\beta)$. By (1.4.5) and (1.4.7), on the contour $\|(A(\beta)-z)^{-1}\|$ is bounded by $C|\beta-\beta_0|^{-n\Delta}$. Since contour sizes are bounded, we see (1.4.3) holds. □

Example 1.4.2 (Breaking a Jordan Anomaly). Let

$$A(\beta)=\begin{pmatrix} 0 & \beta \\ 1 & 0 \end{pmatrix} \tag{1.4.8}$$

Then the characteristic equation is $z^2-\beta=0$ or $z=\pm\beta^{1/2}$, that is,

$$\lambda_\pm(\beta)=\pm\beta^{1/2} \tag{1.4.9}$$

One easily finds if

$$\varphi_\pm(\beta)=\begin{pmatrix} \pm\beta^{1/2} \\ 1 \end{pmatrix} \tag{1.4.10}$$

then

$$A(\beta)\varphi_\pm(\beta)=\lambda_\pm(\beta)\varphi_\pm(\beta) \tag{1.4.11}$$

The hint that the eigenprojection will be singular is that $\varphi_+(0)=\varphi_-(0)$!

We claim that the eigenprojections are

$$P_\pm(\beta)=\begin{pmatrix} \frac{1}{2} & \pm\frac{1}{2}\beta^{1/2} \\ \pm\frac{1}{2}\beta^{-1/2} & \frac{1}{2} \end{pmatrix} \tag{1.4.12}$$

In Problem 1, the reader will derive (1.4.12). One can also verify by noting

$$\begin{gathered} P_\pm^2=P_\pm, \quad P_+P_-=0, \quad P_++P_-=\mathbb{1} \\ \lambda_+(\beta)P_+(\beta)+\lambda_-(\beta)P_-(\beta)=A(\beta) \end{gathered} \tag{1.4.13}$$

This example shows negative powers in the Laurent–Puiseux series for $P_j(\beta)$ can really occur. □

Example 1.4.3. Here are two reparametrizations of (1.4.8) that deflect potential guesses. If $A(\beta)=\left(\begin{smallmatrix} 0 & \beta^2 \\ 1 & 0 \end{smallmatrix}\right)$, then $\lambda_\pm(\beta)=\pm\beta$ and

$$P_\pm(\beta)=\begin{pmatrix} \frac{1}{2} & \pm\frac{1}{2}\beta \\ \pm\frac{1}{2}\beta^{-1} & \frac{1}{2} \end{pmatrix} \tag{1.4.14}$$

so polar singularities in $P_j(\beta)$ can occur even when eigenvalues are analytic (so long as eigenvalue degeneracy is involved).

If $A(\beta)=\left(\begin{smallmatrix} 0 & \beta^3 \\ \beta & 0 \end{smallmatrix}\right)$, then $\lambda_\pm(\beta)=\pm\beta^2$ and $P_\pm(\beta)$ is still given by (1.4.14). Thus, $P_j(\beta)$ can be singular even when $A(\beta_0)$ does not have a Jordan anomaly. □

Here is a deep and elegant result in finite-dimensional perturbation theory:

Theorem 1.4.4 (Rellich's Theorem). *Let $A(\beta)$, $\beta \in \Omega$, be a self-adjoint analytic family. Let $\beta_0 \in \mathbb{R} \cap \Omega$. Then the $\lambda_j(\beta)$ and $P_j(\beta)$ for β near β_0 have removable singularities at β_0 (i.e., are analytic at β_0 if given suitable limiting values).*

Remark. This result remains true if $A(\beta)$ is normal for all $\beta \in \mathbb{R} \cap \Omega$.

Proof. We first note (Problem 2) if $f(z)$ is a Puiseux zero about $z = 0$, all of whose branches are real on the real axis near 0, then f is a Taylor series and f is analytic at $z = 0$. Since $A(\beta)$ has only real eigenvalues for $\beta \in \Omega \cap \mathbb{R}$, the eigenvalue branches near β_0 are all real, so by the above fact, all eigenvalues λ_j are (single-valued) analytic functions.

This implies that the $P_j(\beta)$ are analytic in punctured disks $\mathbb{D}_\delta(\beta_0) \setminus \{\beta_0\}$ and, by (1.4.3), they have at worst polar singularities at $\beta = \beta_0$. If this potential singularity is not removable, we have

$$\lim_{\beta \to \beta_0} \|P_j(\beta)\| = \infty \tag{1.4.15}$$

But for β real, $P_j(\beta)$ is self-adjoint, so $\|P_j(\beta)\| = 1$, so (1.4.15) fails as $\beta \to \beta_0$ along the real axis. It follows that $P_j(\beta)$ has removable singularities at $\beta = \beta_0$. □

Corollary 1.4.5. *Let $A(\beta)$ be real analytic and self-adjoint on $(a, b) \subset \mathbb{R}$ acting on a space of dimension n. Then for any $[c, d] \subset (a, b)$, there are n functions, $f_1, \dots, f_n$, analytic in a neighborhood of $[c, d]$ so that $f_1(\beta), \dots, f_n(\beta)$ are the eigenvalues of $A(\beta)$ counting multiplicity.*

Proof. There are finitely many points $\{\beta_1, \dots, \beta_\ell\}$ in $[c, d]$, where there are fewer than r_0 eigenvalues. Pick $\beta_0 \notin \{\beta_1, \dots, \beta_\ell\}$, label the eigenvalues $f_1(\beta_0) \le f_2(\beta_0) \le \cdots \le f_n(\beta_0)$. By Rellich's theorem, one can analytically continue the function even through $\beta_1, \dots, \beta_\ell$. □

Finally, we turn to the C^k situations for simple eigenvalues.

Theorem 1.4.6. *Let $B \subset \mathbb{R}^\nu$ and let $A(\beta)$ be a continuous (respectively, C^k, C^∞) function with values in the operators on a finite-dimensional vector space, V. Let $\lambda(\beta_0)$ be a simple (i.e., algebraic multiplicity 1) eigenvalue $A(\beta_0)$. Then there is a neighborhood, $N \subset B$, of β_0 and a function $\lambda(\beta)$, $\beta \in N$, which is continuous (respectively, C^k, C^∞) so that $\lambda(\beta)$ is the only eigenvalue of $A(\beta)$ near $\lambda(\beta_0)$. If $P(\beta)$ is eigenprojection for $\lambda(\beta)$, the $P(\beta)$ is continuous (respectively, C^k, C^∞).*

Proof. Let

$$F(z,\beta) = \det\bigl(z - A(\beta)\bigr) \tag{1.4.16}$$

By hypothesis,

$$F\bigl(\lambda(\beta_0),\beta_0\bigr) = 0, \qquad \frac{\partial}{\partial z}\, F\bigl(\lambda(\beta_0),\beta_0\bigr) \neq 0 \tag{1.4.17}$$

(since the zero of $F(\,\cdot\,,\beta_0)$ is simple). Thus, Theorem 1.2.4 applies and $\lambda(\beta)$ is the only zero of $F(\,\cdot\,,\beta)$ near $\lambda(\beta_0)$.

Pick ε so $\mathrm{dist}(\lambda(\beta_0), \sigma(A(\beta_0))\setminus\{\lambda(\beta_0)\}) > \varepsilon$. $P(\beta_0)$ is given by (1.3.11). Since $A(\beta)$ is continuous for all β near β_0 and z with $|z-\lambda(\beta_0)| = \varepsilon$, $z - A(\beta)$ is invertible and the inverse is continuous (respectively, C^k, C^∞). Then (1.3.11) implies $P(\beta)$ is continuous (respectively, C^k, C^∞). □

Our final result is:

Theorem 1.4.7 (Feynman–Hellmann Theorem). *Let $A(\beta)$, a family of operators on a finite-dimensional complex vector space, be self-adjoint and C^1 for $\beta \in (a,b)$. Let $\lambda(\beta_0)$ be a simple eigenvalue for $A(\beta_0)$. For β near β_0, let $\lambda(\beta)$ be the unique eigenvalue of $A(\beta)$ near $\lambda(\beta_0)$. Suppose that*

$$A(\beta_0)\varphi_0 = \lambda(\beta_0)\varphi_0, \qquad \|\varphi_0\| = 1 \tag{1.4.18}$$

Then

$$\frac{d\lambda(\beta)}{d\beta}\bigg|_{\beta=\beta_0} = \left\langle \varphi_0, \frac{dA}{d\beta}\,(\beta_0)\varphi_0\right\rangle \tag{1.4.19}$$

Proof. Let $P(\beta)$ be the eigenprojection for $A(\beta)$ and eigenvalue $\lambda(\beta)$. Then $P(\beta)\varphi_0 \to \varphi_0$ so for β near β_0, $\|P(\beta)\varphi_0\| \neq 0$, and we can define for β near β_0,

$$\varphi(\beta) = \frac{P(\beta)\varphi_0}{\|P(\beta)\varphi_0\|} \tag{1.4.20}$$

Since $\|\varphi(\beta)\| = 1$, we have

$$\lambda(\beta) = \langle \varphi(\beta), A(\beta)\varphi(\beta)\rangle \tag{1.4.21}$$

$$0 = \frac{d}{d\beta}\,\|\varphi(\beta)\|^2 = \left\langle \varphi(\beta), \frac{d\varphi}{d\beta}\,(\beta)\right\rangle + \left\langle \frac{d\varphi}{d\beta}\,(\beta), \varphi(\beta)\right\rangle \tag{1.4.22}$$

Differentiating (1.4.21), we have

$$\frac{d\lambda}{d\beta}\bigg|_{\beta_0} = \left[\left\langle \frac{d\varphi}{d\beta}, A(\beta)\varphi(\beta)\right\rangle + \left\langle A(\beta)\varphi(\beta), \frac{d\varphi}{d\beta}\right\rangle + \left\langle \varphi(\beta), \frac{dA}{d\beta}\,\varphi(\beta)\right\rangle\right]\bigg|_{\beta_0} \tag{1.4.23}$$

Using (1.4.22) and $A(\beta_0)\varphi(\beta_0) = \lambda(\beta_0)\varphi(\beta_0)$, we see the first two terms on the right of (1.4.23) sum to zero, so we have (1.4.19). □

Notes and Historical Remarks. The standard textbook presentation of finite- and infinite-dimensional perturbation theory are the books of Friedrichs [**209**], Kato [**380**], Rellich [**558**], and Reed–Simon [**551**].

The series for the actual eigenvalues and eigenvectors are called Rayleigh–Schrödinger series after fundamental work of Rayleigh in 1894 (on sound waves) [**547**] and Schrödinger in 1926 (on quantum states) [**605**]. The first two terms, when $A(\beta) = A_0 + \beta V, A_0, V$ self-adjoint, E_j are the eigenvalues of A_0, V_{ij} are the matrix elements of V in a basis of eigenvectors of A_0, and E_1 is a simple eigenvalue, are

$$E_1(\beta) = E_1 + \beta V_{11} + \beta^2 \sum_{j\neq 1} \frac{|V_{j1}|^2}{E_1 - E_j} + O(\beta^3) \tag{1.4.24}$$

Rellich, in a series of five papers [**556**] in 1937–42, was the first to consider convergence questions. Sz.-Nagy [**689**] and Kato [**372**] clarified and enlarged this work. Kato did this work as a physics graduate student unaware of Rellich's work.

Rellich's theorem is from the first of his papers on perturbation theory. Its subtlety is seen by the fact that it is false for several complex parameters. For example, if

$$A(\beta, \gamma) = \beta \begin{pmatrix} 1 & 0 \\ 0 & -1 \end{pmatrix} + \gamma \begin{pmatrix} 0 & 1 \\ 1 & 0 \end{pmatrix} \tag{1.4.25}$$

the eigenvalues are

$$E_\pm(\beta, \gamma) = \pm\sqrt{\beta^2 + \gamma^2} \tag{1.4.26}$$

The Feynman–Hellmann theorem is named after papers of Feynman [**200**] and Hellmann [**312**] in the late 1930s. Feynman's paper was written when he was an undergraduate at MIT. The result appeared earlier in the 1930s in a paper by Güttinger [**277**] and a short text by Pauli [**512**] on quantum mechanics.

Problems.

1. (a) Let P be the eigenprojection associated to a simple eigenvalue, λ, for a matrix A. If $A\varphi = \lambda\varphi$ for a column vector and $\eta A = \lambda\eta$ for a vector η with $\eta\varphi = 1$, prove that $P = \varphi\eta$. (Note that $\eta\varphi$, as the product of a $1 \times n$ and $n \times 1$ matrices, is a number while $\varphi\eta$ is an $n \times n$ matrix.)

   (b) Use (a) to compute $P_\pm(\beta)$ in Example 1.4.2.

2. Let $f(z) = \sum_{n=0}^\infty a_n z^{n/p}$ converging for $|z|$ small. Suppose $f(x)$ is real for all small z in $\mathbb{R}$ and all branches of $f(z)$. Prove $a_n = 0$ if $n \neq mp$ and is real if $n = mp$. (*Hint*: Induction.)

1.5. Some Results from Real Analysis

Operator theory presupposes a fair amount of measure theory, theory of Banach spaces, and other results from Parts 1 and 3. Without attempting to be comprehensive, we want to focus on some of the more important aspects.

1.5.1. Basic Language and Techniques. We assume a familiarity with the basic notions of topology, including continuous and semicontinuous functions, weak topology defined by a family of functions, and the separation axioms—Hausdorff and normal spaces (see Sections 2.1 and 2.2 of Part 1). We will occasionally use (see Problem 4 of Section 2.2 of Part 1):

Theorem 1.5.1 (Tietze Extension Theorem). *Let E be a closed subset of a normal topological space, X. Let $g\colon E \to \mathbb{R}$ be a bounded continuous function. Then there exists a bounded continuous function $f\colon X \to \mathbb{R}$ so that $f \restriction E = g$.*

We assume familiarity with the notions of normed linear (NLS), Banach, and Hilbert spaces (Chapters 3 and 5 of Part 1), and that for linear maps $T\colon X \to Y$ between normed linear spaces, continuity is equivalent to a bound

$$\|Tx\|_Y \le C\|x\|_X \tag{1.5.1}$$

The smallest C defines a norm, $\|T\|$, on $\mathcal{L}(X,Y)$, the bounded linear maps of X to Y. $\mathcal{L}(X,Y)$ with this norm is a Banach space if and only if Y is a Banach space. $\mathcal{L}(X) \equiv \mathcal{L}(X,X)$ is the major object of study in this volume. We use $\mathbb{K}$ for $\mathbb{R}$ or $\mathbb{C}$, the field over which an NLS is a vector space. $\mathcal{L}(X,\mathbb{K}) \equiv X^*$ is the dual space for X.

We assume the reader knows about orthonormal (ON) bases and the Gram–Schmidt procedure (see Section 3.4 of Part 1). Occasionally, we will use ideas from the theory of locally convex spaces and tempered distributions (Chapter 6 of Part 1) and our examples often use the theory of Fourier transforms (also Chapter 6 of Part 1).

1.5.2. Hahn–Banach Weak and Weak-$*$ Topologies. The following is a special case of a stronger result proven in Theorems 5.5.1 and 5.5.5 of Section 5.5 of Part 1:

Theorem 1.5.2 (Hahn–Banach Theorem). *Let $Y \subset X$ be a subspace of a NLS. Let $\ell\colon Y \to \mathbb{K}$ obey*

$$|\ell(y)| \le C\|y\|_X \tag{1.5.2}$$

for all $y \in X$. Then there exists $L\colon X \to \mathbb{K}$ with $L \restriction Y = \ell$ so that for all $x \in X$,

$$|L(x)| \le C\|x\|_X \tag{1.5.3}$$

In particular, for all $x \in X$, there is $\ell \in X^*$ with $\|\ell\| = 1$ and $\ell(x) = \|x\|$. This guarantees X^* is large, so large that the $\sigma(X, X^*)$, the *weak topology* on X, is Hausdorff. Here, if Z is a set of linear functionals on X, the $\sigma(X, Z)$ topology is the weak topology defined by the functions $\{z(\cdot)\}_{z\in Z}$. The $\sigma(X^*, X)$ topology on X^* is called the *weak-$*$ topology*.

1.5.3. Compactness Results. We assume the reader is familiar with the basic facts about compact sets (see Section 2.3 of Part 1). We'll use two compactness criteria:

Theorem 1.5.3 (Ascoli–Arzelà Theorem). *Let X be a separable metric space and Y a complete metric space. Let $\mathcal{F}$ be a family of continuous functions from X to Y so that*

(i) *There is a compact set $C \subset Y$ so that for all $f \in \mathcal{F}$,* Ran$(f) \subset C$.
(ii) *$\mathcal{F}$ is equicontinuous, that is, $\forall\varepsilon > 0$, $\exists\delta$ so that $\forall f \in \mathcal{F}$,*

$$\rho_X(x, z) < \delta \Rightarrow \rho_Y(f(x), f(y)) < \varepsilon \tag{1.5.4}$$

Then the closure of $\mathcal{F}$ (in the uniform topology) is compact.

Remarks. 1. The uniform topology is defined by the metric $\rho_{\mathcal{F}(X,Y)}$

$$\rho_{\mathcal{F}(X,Y)}(f, g) = \sup_{x\in X} \rho_Y(f(x), g(x)) \tag{1.5.5}$$

2. The typical application has $Y = \mathbb{R}$ or $\mathbb{C}$. Then $\mathcal{F}(X, Y)$ is a vector space and (1.5.5) comes from a norm (distance from the 0 function). C is then a bounded set so condition (i) is uniform boundedness.

3. What we call equicontinuity is more properly uniform equicontinuity.

4. The proof is an application of the diagonalization trick (see the discussion in Section 1.5 of Part 1). Given a sequence in $\mathcal{F}$, one uses this trick to get convergence at a countable dense set in X. Equicontinuity then yields uniform convergence. One then uses a result of Weierstrass that uniform limits of continuous functions are continuous.

5. This is Theorem 2.3.14 of Part 1.

Theorem 1.5.4 (Banach–Alaoglu Theorem). *The unit ball in the dual, X^*, of a Banach space, X, is compact in the weak-$*$ (i.e., $\sigma(X^*, X)$) topology.*

The proof (see Section 5.8 of Part 1) is a simple consequence of Tychonoff's theorem on the compactness of arbitrary products of compact spaces in the product topology (see Section 2.7 of Part 1).

Finally, on the subject of compactness, we mention

Theorem 1.5.5 (Stone–Weierstrass Theorem). *Let X be a compact Hausdorff space and $C_{\mathbb{R}}(X)$ the real-valued continuous functions on X. Let $S \subset C_{\mathbb{R}}(X)$ obey*

(i) *S is an algebra, that is, $f, g \in S \Rightarrow f + g$, fg and $\lambda f \in S$ for all $\lambda \in \mathbb{R}$.*
(ii) *S strongly separates points, that is, for any $x, y \in X$ and $\alpha, \beta \in \mathbb{R}$, there exists $f \in S$ with $f(x) = \alpha$, $g(x) = \beta$.*

Then S is dense (in $\|\cdot\|_\infty$ norm) in $C_{\mathbb{R}}(X)$.

Remarks. 1. This is Theorem 2.5.2 of Part 1.

2. This is true for complex-valued functions, $C(X)$, if one adds the condition $f \in S \Rightarrow \overline{f} \in S$. The example $\mathfrak{A}(\overline{\mathbb{D}})$ of bounded continuous functions on $\overline{\mathbb{D}}$, analytic on $\mathbb{D}$ shows this complex conjugation condition is a necessary one.

1.5.4. Consequences of the Baire Category Theorem. The following results (proven in Section 5.4 of Part 1 as Theorems 5.4.9, 5.4.11, and 5.4.17) are useful:

Theorem 1.5.6 (Principle of Uniform Boundedness). *Let $\mathcal{F}$ be a subset of bounded linear maps from one Banach space, X, to another one, Y. Suppose that for each $x \in X$,*

$$\sup_{T \in \mathcal{F}} \{\|Tx\|_Y\} < \infty \tag{1.5.6}$$

Then

$$\sup_{T \in \mathcal{F}} \{\|T\|_{\mathcal{L}(X,Y)}\} < \infty \tag{1.5.7}$$

Theorem 1.5.7 (Open Mapping Theorem). *Let $T\colon X \to Y$ be a bounded linear map between Banach spaces. If $\operatorname{Ran}(T)$ is all of Y, then T is open, that is, if $A \subset X$ is open in X, then $T[A]$ is open in Y.*

Theorem 1.5.8 (Closed Graph Theorem). *Let T be a (not a priori bounded) linear map from a Banach space, X, to another Banach space, Y. If the graph of T is closed in $X \oplus Y$, then T is a bounded linear transformation.*

Remark. The graph of T is

$$\Gamma(T) = \{(x, Tx) \mid x \in X\} \tag{1.5.8}$$

1.5.5. Measure Theory. We assume familiarity with the basics of measure theory on compact and σ-compact spaces, as presented in the first ten sections of Chapter 4 of Part 1. This includes the basics of L^p-spaces including Hölder and Minkowski inequalities. We explicitly mention the following theorem, discussed in Section 4.7 of Part 1.

Theorem 1.5.9. *Let μ be a finite measure on $\mathbb{R}$. Then there is a function $w \in L^1(\mathbb{R}, dx)$ so that*

$$d\mu = w(x)\,dx + d\mu_{\text{sc}} + d\mu_{\text{pp}} \tag{1.5.9}$$

where $d\mu_{\text{pp}}$ is a pure point measure and $d\mu_{\text{sc}}$ singular continuous. This (Lebesgue) decomposition is unique.

Remarks. 1. μ_{pp} pure point means

$$\mu_{\text{pp}}(A) = \sum_{x\in A} \mu_{\text{pp}}(\{x\}) \tag{1.5.10}$$

2. μ_{sc} singular continuous means it is singular wrt Lebesgue measure (i.e., there A a Borel set in $\mathbb{R}$ so $|A| = 0$ and $\mu_{\text{sc}}(\mathbb{R}\setminus A) = 0$) and continuous (i.e., $\mu_{\text{sc}}(\{x\}) = 0$ for all x).

In Sections 6.8 and 6.9 on group representations, we'll assume results about Haar measure, as discussed in Section 4.19 of Part 1.

1.5.6. Sobolev Estimates. In Section 7.1, we'll need Sobolev estimates, as proven in Section 6.3 of Part 3; indeed, the following special case:

Theorem 1.5.10. *Let $\nu \in \mathbb{Z}_+$. Let*

$$\begin{aligned} q &\le \infty, && \nu = 1,2,3 \\ q &< \infty, && \nu = 4 \\ q &\le \frac{2\nu}{\nu-4}, && \nu > 5 \end{aligned}$$

Then there is a constant $c_{q,\nu}$ so for all $\varphi \in \mathcal{S}(\mathbb{R}^\nu)$, we have that

$$\|\varphi\|_q \le c_{q,\nu}[\|\Delta\varphi\|_2 + \|\varphi\|_2]$$

where $\|\cdot\|_r$ is the $L^p(\mathbb{R}^\nu, d^\nu x)$-norm.

1.5.7. Riesz Lemma. We'll use the following result especially in analyzing compact operators (in the context where Riesz introduced it):

Theorem 1.5.11 (The Riesz Lemma). *Let Y be a closed proper subspace of a normed linear space, X. Then for every $\varepsilon > 0$, there exists $x \in X$ with*

$$\|x\| = 1, \qquad \text{dist}(x, Y) \ge (1-\varepsilon)$$

Remarks. 1. This is Lemma 5.1.6 of Part 1.

2. $\text{dist}(x, Y)$ is defined by

$$\text{dist}(x, Y) = \inf\{\|x - z\| \mid z \in Y\} \tag{1.5.11}$$

Chapter 2

Operator Basics

> It is not so much whether a theorem is useful that matters, but how elegant it is.
>
> —*S. Ulam* (1909–1984) [**709**]

Big Notions and Theorems: Transpose, Adjoint, Strong Operator Topology, Weak Operator Topology, Projection, Compatible Projections, Self-adjoint Operators, Normal Operators, Unitary Operators, C^*-identity, Orthogonal Projection, Resolvent Set, Resolvent, Spectrum, Left and Right Shift, Banach Algebra, Nonempty Spectrum, Spectral Radius Formula, Quasinilpotent, Spectral Mapping Theorem, Dunford Functional Calculus, Spectral Projection, Spectral Localization Theorem, Discrete Spectrum, Algebraic Multiplicity, Geometric Multiplicity, Spectral Projections and Residues of the Resolvent, Positive Operator, Square Root Lemma, Absolute Value of an Operator, Partial Isometry, Initial and Final Subspaces, Polar Decomposition, Spectral Theorem

While our main topics will be special classes of operators on a Hilbert space, namely, compact operators in Chapter 3 and self-adjoint in Chapter 5, there is a language and basic tools that we need to present first. Much of the language works for operators on Banach spaces—indeed, for the elements of a Banach algebra—so we'll get a head start on Chapter 6 by putting much of Sections 2.2 and 2.3 in that even more general context.

Section 2.1 reprises some of the material from Part 1, especially Section 3.7 of Part 1, and provides more detail for some. An important object will be projections, that is, operators with $P^2 = P$. We'll see that even in the Hilbert space case, projections which might not be self-adjoint are important.

A key issue for extending results in operator theory from finite dimensions is the proper analog of eigenvalues. The right notion is spectrum whose study we begin in Section 2.2. Section 2.3 starts the discussion of functions of general operators. If f is analytic in a neighborhood of the spectrum, $\sigma(A)$, of an operator, we'll find a natural meaning for $f(A)$ so that $f(A)g(A) = fg(A)$. In particular, if $\sigma(A) = \sigma_1 \cup \sigma_2$ is a disjoint union of closed sets and $f(x) = \chi_{\sigma_1}(x)$, the characteristic function of σ_1, then f is analytic in a neighborhood of $\sigma(A)$, so $f(A) = P$ will obey $P^2 = P$ and $AP = PA$. We'll even prove that $\sigma(A \restriction \text{Ran}(P)) = \sigma_1$.

While much of Sections 2.1–2.3 applies to general operators on general Banach spaces, Section 2.4 concerns positive operators, A, on a Hilbert space. We'll show first there is a unique positive operator, B, with $B^2 = A$, and write $B = \sqrt{A}$. This will allow us to define $|A| = \sqrt{A^*A}$ for any bounded operator on a Hilbert space and write, for such operators, a decomposition $A = U|A|$ analogous to $z = e^{i\theta}|z|$ for $z \in \mathbb{C}$.

2.1. Topologies and Special Classes of Operators

Let X be a Banach space and $\mathcal{L}(X)$ the bounded linear operators from X to itself. If $A \in \mathcal{L}(X)$ and X^* is the dual of X, we define the *transpose* of A, denoted by $A^{\text{t}}\colon X^* \to X^*$ (we reserve "adjoint" and A^* for the slightly different Hilbert space object below), by

$$(A^{\text{t}}\ell)(x) = \ell(Ax) \tag{2.1.1}$$

$A \to A^{\text{t}}$ is linear and

$$(AB)^{\text{t}} = B^{\text{t}}A^{\text{t}} \tag{2.1.2}$$

It is easy to see (Problem 1) that

$$\|A^{\text{t}}\| = \|A\| \tag{2.1.3}$$

and that if $i\colon X \to X^{**}$ is the natural embedding, then

$$A^{\text{tt}}(i(x)) = i(Ax) \tag{2.1.4}$$

In case X is a Hilbert space, $\mathcal{H}$, there is map $j\colon \mathcal{H}^* \to \mathcal{H}$ by $\langle j(\ell), \varphi\rangle = \ell(\varphi)$. j is antilinear. For $A \in \mathcal{L}(\mathcal{H})$, one defines $A^* \in \mathcal{L}(\mathcal{H})$ by

$$A^* = jA^{\text{t}}j^{-1} \tag{2.1.5}$$

Equivalently, for all $\varphi, \psi \in \mathcal{H}$,

$$\langle \varphi, A\psi\rangle = \langle A^*\varphi, \psi\rangle$$

Thus, $A \mapsto A^*$ is antilinear and, by (2.1.2) and (2.1.3),

$$(AB)^* = B^*A^*, \qquad \|A^*\| = \|A\| \tag{2.1.6}$$

In addition to the norm topology on $\mathcal{L}(\mathcal{H})$, we have two other natural topologies: the *strong operator topology* where a net $A_\alpha \to A$ if and only if $A_\alpha x \to Ax$ in norm for all $x \in X$, and the *weak operator topology* where $A_\alpha \to A$ if and only if for all $x \in X$, $A_\alpha x \to Ax$ in the weak (i.e., $\sigma(X, X^*)$) topology; equivalently for all $x \in X$, $\ell \in X^*$, $\ell(A_\alpha x) \to \ell(Ax)$. We'll leave it to the reader to specify the actual topologies (Problem 2).

As with the weak topology on a Banach space (see Theorems 5.7.2 and 3.6.8 of Part 1), the strong operator and weak operator topologies are not metrizable, but if X and X^* are separable, their restriction to $\{A \in \mathcal{L}(X) \mid \|A\| \le 1\}$ is metrizable.

$A \mapsto A^{\rm t}$ is not, in general, continuous in the strong operator topology (Problem 3), but it is in the weak operator topology if X is reflexive (Problem 4). Restricted to $\{A \mid \|A\| \le 1\}$, $(A, B) \to AB$ is jointly continuous if the strong operator topology is used (Problem 5) but not for the full $\mathcal{L}(X)$. Even on the unit ball, products are not weakly continuous (Problem 6).

An operator, $A \in \mathcal{L}(X)$, is called *finite rank* if and only if $\dim(\mathrm{Ran}(A)) < \infty$. The dimension is called its *rank*, written $\mathrm{rank}(A)$

Proposition 2.1.1. (a) *If* $A \in \mathcal{L}(X)$ *is finite rank, with* $\mathrm{rank}(A) = n$, *then there exist linearly independent* $\{x_j\}_{j=1}^n$ *in* X *and linearly independent* $\{\ell_j\}_{j-1}^n$ *in* X^* *so that*

$$Ax = \sum_{j=1}^n \ell_j(x) x_j \tag{2.1.7}$$

(b) *If* $\mathrm{rank}(A) = n < \infty$, $A^{\rm t}$ *is also finite rank given by*

$$A^{\rm t}\ell = \sum_{j=1}^n \ell(x_j)\ell_j \tag{2.1.8}$$

so that

$$\mathrm{rank}(A^{\rm t}) = \mathrm{rank}(A) \tag{2.1.9}$$

Proof. (a) Let $\{x_j\}_{j=1}^n$ be a basis for $\mathrm{Ran}(X)$. Then their independence proves that Ax has the form (2.1.7), where the $\{\ell_j(\,\cdot\,)\}_{j=1}^n$ are linear. Since each $x_j \in \mathrm{Ran}(A)$, there exist y_j so $Ay_j = x_j$, so $\ell_j(y_k) = \delta_{jk}$ proving the independence of the ℓ_j's.

(b) If $\ell \in X^*$,

$$(A^{\rm t}\ell)(x) = \ell(Ax) = \sum_{j=1}^n \ell_j(x)\ell(x_j) \tag{2.1.10}$$

$$= \left[\sum_{j=1}^n \ell(x_j)\ell_j\right](x) \tag{2.1.11}$$

so (2.1.8) holds. Since the x_j are independent, the Hahn–Banach theorem implies there are $q_j \in X^*$, so $q_j(x_k) = \delta_{jk}$, that is, $A^t q_j = \ell_j$. It follows that

$$\operatorname{Ran}(A^t) = \operatorname{span}(\{\ell_j\}_{j=1}^n) \tag{2.1.12}$$

so (2.1.9) holds. □

The above also shows that

$$\operatorname{Ker}(A) = \{x \mid \ell_j(x) = 0;\ j = 1, \dots, n\} \tag{2.1.13}$$

so in terms of codimension (see Theorem 5.5.12 of Part 1),

$$\operatorname{codim}(\operatorname{Ker}(A)) = \operatorname{rank}(A) \tag{2.1.14}$$

which, in the finite-dimensional case, is the well-known

$$\dim(X) < \infty \Rightarrow \dim(\operatorname{Ker}(A)) + \dim(\operatorname{Ran}(A)) = \dim(X) \tag{2.1.15}$$

Projections are an important class of operators on a general Banach space.

Definition. A *projection* (aka *idempotent*) is an operator $P \in \mathcal{L}(X)$ with

$$P^2 = P \tag{2.1.16}$$

Proposition 2.1.2. *Let P be a projection. Then*

(a) *$Q = 1 - P$ is also a projection.*
(b) *$\operatorname{Ran}(P)$ and $\operatorname{Ran}(Q)$ are closed and overlap only in $\{0\}$.*
(c) *Any x can be uniquely written as a sum of $y \in \operatorname{Ran}(P)$ and $z \in \operatorname{Ran}(1-P)$.*
(d) *$\|Px\| + \|Qx\|$ is a norm equivalent to $\|\cdot\|$ on X.*
(e) *If $Y, Z \subset X$ are closed subspaces with $Y + Z = X$ and $Y \cap Z = \{0\}$, then there is a unique projection P with $\operatorname{Ran}(P) = Y$ and $\operatorname{Ran}(Q) = Z$.*

Remarks. 1. A closed subspace, $Y \subset X$, is said to have a (*closed*) *complement* if and only if there is a projection $P \in \mathcal{L}(X)$ with $\operatorname{Ran}(P) = Y$. That this is equivalent to the definition in Section 5.4 of Part 1 follows from (e) of the proposition. In a general Banach sense, not all subspaces have complements; see Problem 9 of Section 5.7 of Part 1.

2. If Y, Z are subspaces as in (e), then there is an indempotent, P, with $\operatorname{Ran}(P) = Y$, $\operatorname{Ran}(1 - P) = \operatorname{Ker}(P) = Z$, but P is not a bounded operator unless Y and Z are closed. For example, in ℓ^2, let Y be the subspace of sequences which are eventually zero which is dense but not closed. By a simple Zornification, one can find Z so $Y \dotplus Z = \ell^2$. Since neither Y nor Z is closed, P, the algebraic projection onto Y will be unbounded.

Proof. (a) $Q^2 = 1 - 2P + P^2 = 1 - P = Q$.

(b) $y = Px \Rightarrow (1-P)y = (P - P^2)x = 0$. On the other hand, $(1-P)y = 0 \Rightarrow y = Py \in \operatorname{Ran}(P)$. Thus,

$$\operatorname{Ran}(P) = \operatorname{Ker}(Q), \qquad \operatorname{Ran}(Q) = \operatorname{Ker}(P) \tag{2.1.17}$$

Since kernels of bounded operators are closed, $\operatorname{Ran}(P)$ and $\operatorname{Ran}(Q)$ are closed. If $y = Px = Qz$, then $y = Q^2 z = Qy = (1-P)Px = 0$.

(c) $x = Px + (1-P)x$ shows that any x is such a sum and uniqueness follows from $y_1 + z_1 = y_2 + z_2$ with $y_j \in \operatorname{Ran}(P)$, $z_j \in \operatorname{Ran}(Q) \Rightarrow y_1 - y_2 = z_2 - z_1 \in \operatorname{Ran}(P) \cap \operatorname{Ran}(Q) \Rightarrow y_1 - y_2 = z_2 - z_1 = 0$.

(d) Since P and Q are bounded

$$\|Px\| + \|Qx\| \leq (\|P\| + \|Q\|)\|x\| \tag{2.1.18}$$

with

$$\|x\| = \|Px + Qx\| \leq \|Px\| + \|Qx\|$$

so the norms are equivalent.

(e) Let $\widetilde{X} = Y \oplus Z$ with $\|(y,z)\|_{\widetilde{X}} = \|y\| + \|z\|$. By the standard construction of direct sum (see Section 5.1 of Part 1), $\widetilde{X}$ is a Banach space. Let S map $\widetilde{X}$ to X via

$$S(y,z) = y + z \tag{2.1.19}$$

Part (c) says that S is a bijection. Since

$$\|S(y,z)\|_X \leq \|(y,z)\|_{\widetilde{X}} \tag{2.1.20}$$

S is continuous. Thus, by the inverse mapping theorem (Theorem 5.4.11 of Part 1), there is an $\alpha > 0$ so that

$$\alpha\|(y,z)\|_{\widetilde{X}} \leq \|S(y,z)\|_X \tag{2.1.21}$$

Thus, if $x = y + z$, and $P(y+z) = y$, we have that $\|Px\| \leq \alpha^{-1}\|x\|$ so P is a BLT. Clearly, since $y = y + 0$ in the $Y + Z$ decomposition, $P^2 = P$ $\square$

Example 2.1.3. We say a closed subspace $Y \subset X$ is *complemented* if there is another closed subspace Z with $Y \cap Z = \{0\}$, $Y + Z = X$; equivalently, if Y is the range of a bounded projection. In a Hilbert space, every closed Y is complemented; take $Z = Y^\perp$. But for example, $c_0 = \{a \in l^\infty \mid \lim_{n\to 0} a_n = 0\}$ is not complemented in l^∞ (Problem 7), $H^1(\mathbb{D})$ is not complemented in $L^1(\partial\mathbb{D}, \frac{d\theta}{2\pi})$ (Problem 8), and $\mathfrak{A}(\mathbb{D})$, the functions analytic in $\mathbb{D}$ and continuous in $\overline{\mathbb{D}}$, restricted to $\partial\mathbb{D}$ is not complemented in $C(\partial\mathbb{D})$ (Problem 9). In fact, (see the Notes) it is a theorem that if X is a Banach space in which every closed subspace is complemented, then there is a Hilbert space inner product on X whose norm is equivalent to that of X.

Definition. If $A \in \mathcal{L}(X)$ and $Y \subset X$ is a subspace, we say Y is *invariant* for A if and only if $y \in Y \Rightarrow Ay \in Y$. We say P is a *reducing projection* for A if and only if $\mathrm{Ran}(P)$ and $\mathrm{Ran}(1-P)$ are both invariant for A.

The following is easy to see:

Theorem 2.1.4. $\mathrm{Ran}(P)$ *is invariant for* A *if and only if*

$$PAP = AP \tag{2.1.22}$$

Theorem 2.1.5. P *is reducing for* A *if and only if*

$$AP = PA \tag{2.1.23}$$

Remark. See Theorem 2.1.10 below for a further equivalence in the Hilbert space case.

Proof. (2.1.23) implies (2.1.22) and $A(1-P) = A - AP = A - PA = (1-P)A$, and so (2.1.22) for $(1-P)$. Thus, (2.1.23) implies P is reducing.

Conversely, if $PAP = AP$ and $(1-P)A(1-P) = A(1-P)$, then $PA(1-P) = P(1-P)A(1-P) = 0$, so $PA = PAP + PA(1-P) = PAP = AP$. □

If P, Q are two projections in $\mathcal{L}(X)$, interesting things happen if they commute:

Theorem 2.1.6. *If* P *and* Q *are projections in* $\mathcal{L}(X)$ *and* $PQ = QP$, *then* $R = PQ$ *is a projection and*

(a) $\mathrm{Ran}(R) = \mathrm{Ran}(P) \cap \mathrm{Ran}(Q)$ (2.1.24)

(b) $\mathrm{Ker}(R) = \mathrm{Ker}(P) + \mathrm{Ker}(Q)$ (2.1.25)

(c) $P + Q - PQ$ *is the projection onto* $\mathrm{Ran}(P) + \mathrm{Ran}(Q)$ (2.1.26)

In particular,

$$PQ = 0 \Leftrightarrow \mathrm{Ran}(P) \cap \mathrm{Ran}(Q) = \{0\} \tag{2.1.27}$$

Proof. $R^2 = PQPQ = P^2Q^2 = PQ = R$, so R is a projection. If $\varphi \in \mathrm{Ran}(P) \cap \mathrm{Ran}(Q)$, $P\varphi = \varphi = Q\varphi$, so $R\varphi = P(Q\varphi) = P\varphi = \varphi$. On the other hand, if $\varphi \in \mathrm{Ran}(R)$, $\varphi = PQ\varphi \in \mathrm{Ran}(P)$ and $\varphi = QP\varphi \in \mathrm{Ran}(Q)$.

For the second assertion, note $P\varphi = 0 \Rightarrow QP\varphi = 0$ so $\mathrm{Ker}(P) \subset \mathrm{Ker}(PQ)$ and thus, $\mathrm{Ker}(P) + \mathrm{Ker}(Q) \subset \mathrm{Ker}(PQ)$.

Given any φ, write

$$\varphi = PQ\varphi + (1-P)Q\varphi + (1-Q)P\varphi + (1-Q)(1-P)\varphi \tag{2.1.28}$$

and see $PQ\varphi = 0 \Rightarrow \varphi \in \mathrm{Ran}(1-P) + \mathrm{Ran}(1-Q) = \mathrm{Ker}(P) + \mathrm{Ker}(Q)$.

For the third assertion, let $S = P + Q - PQ$. All cross-terms are PQ, so

$$S^2 = P^2 + Q^2 + (PQ)^2 + 2PQ - 4PQ = S$$

So S is a projection.

Clearly, $\operatorname{Ran}(S) \subset \operatorname{Ran}(P) + \operatorname{Ran}(Q)$ and if $\varphi = P\varphi$, then $S\varphi = P\varphi + QP\varphi - PQ\varphi = P\varphi = \varphi$, so $\operatorname{Ran}(P), \operatorname{Ran}(Q) \subset \operatorname{Ran}(S)$, implying $\operatorname{Ran}(P) + \operatorname{Ran}(Q) \subset \operatorname{Ran}(S)$. □

Two commuting projections are called *compatible projections.*

Remark. Note that $P+Q-PQ = 1-(1-P)(1-Q)$. The reader should consider the geometric meaning of this formula.

The other classes of operators we want to single out next live on Hilbert spaces and rely on adjoints in their definition.

Definition. Let $\mathcal{H}$ be a Hilbert space. An operator $A \in \mathcal{L}(\mathcal{H})$ is called

(i) *self-adjoint* if $A = A^*$ (2.1.29)

(ii) *normal* if $AA^* = A^*A$ (2.1.30)

(iii) *unitary* if $U^*U = UU^* = \mathbb{1}$ (2.1.31)

Clearly, self-adjoint and unitary operators are normal. Any $A \in \mathcal{L}(\mathcal{H})$ can be written

$$A = B + iC, \qquad B^* = B, \qquad C^* = C \tag{2.1.32}$$

It is easy to see this is unique and B, C are given by

$$B = \tfrac{1}{2}(A + A^*), \qquad C = (2i)^{-1}(A - A^*) \tag{2.1.33}$$

We'll see later (Corollary 2.4.5) that any operator is a linear combination of four unitaries, although not in a unique way.

Proposition 2.1.7. *An operator obeys $U^*U = \mathbb{1}$ if and only if $\|U\varphi\| = \|\varphi\|$ for all φ. Such a U is unitary if and only if* $\operatorname{Ran}(U)$ *is all of $\mathcal{H}$.*

Proof. By polarization, $U^*U = \mathbb{1} \Leftrightarrow \forall\varphi,\ \langle \varphi, U^*U\varphi\rangle = \|\varphi\|^2$, but $\langle \varphi, U^*U\varphi\rangle = \langle U\varphi, U\varphi\rangle = \|U\varphi\|^2$.

Such a U is clearly one-one. If it is onto $\mathcal{H}$, it is a bijection, so it has a two-sided inverse, that is, V so $VU = UV = \mathbb{1}$. But then $U^* = U^*(UV) = (U^*U)V = V$, so $UU^* = \mathbb{1}$.

Conversely, if $UU^* = \mathbb{1}$, any φ is $U(U^*\varphi)$ so in $\operatorname{Ran}(U)$. □

Remark. The proof also shows U's obeying $U^*U = \mathbb{1}$ are inner product-preserving, that is,

$$U^*U = \mathbb{1} \Rightarrow \forall\varphi, \psi \in \mathcal{H},\ \langle U\varphi, U\psi\rangle = \langle \varphi, \psi\rangle$$

Theorem 2.1.8 (C^*-identity). *If $A \in \mathcal{L}(\mathcal{H})$, we have*

$$\|A^*A\| = \|A\|^2 \tag{2.1.34}$$

Proof. On the one hand, since $\|A\| = \|A^*\|$,

$$\|A^*A\| \le \|A^*\|\,\|A\| = \|A\|^2 \tag{2.1.35}$$

On the other hand,

$$\|A\|^2 = \Big(\sup_{\|\varphi\|=1} \|A\varphi\|\Big)^2 = \sup_{\|\varphi\|=1} \|A\varphi\|^2 = \sup_{\|\varphi\|=1} \langle \varphi, A^*A\varphi\rangle \le \|A^*A\| \tag{2.1.36}$$

□

A projection which is self-adjoint is called an *orthogonal projection* for the following reason:

Theorem 2.1.9. *Let $P \in \mathcal{L}(\mathcal{H})$ be a nonzero projection. Then the following are equivalent:*

(1) $P = P^*$
(2) $\|P\| = 1$
(3) $\operatorname{Ran}(P)$ *is orthogonal to* $\operatorname{Ran}(\mathbb{1} - P)$.

Proof. (1) ⇒ (2). If (1) holds,

$$\|P\|^2 = \|P^*P\| = \|P^2\| = \|P\|$$

so $\|P\| = 0$ or $\|P\| = 1$. Since we suppose $P \neq 0$, we must have $\|P\| = 1$.

(2) ⇒ (3). Let $\varphi \in \operatorname{Ran}(P)$, $\psi \in \operatorname{Ran}(\mathbb{1}-P)$ so for all $\lambda \in \mathbb{C}$, $P(\varphi+\lambda\psi) = \varphi$. Since $\|P\| = 1$, $\|P(\varphi + \lambda\psi)\| \le \|\varphi + \lambda\psi\|$. Thus, for all $\lambda \in \mathbb{C}$,

$$\|\varphi\|^2 \le \|\varphi + \lambda\psi\|^2 \tag{2.1.37}$$

If $\lambda = |\lambda| e^{i\eta}$, then (2.1.37) says

$$2\,|\lambda|\,\operatorname{Re}\big(e^{i\eta}\langle\psi,\varphi\rangle\big) + |\lambda|^2\,\|\psi\|^2 \ge 0$$

Divide by $|\lambda|$ and take $|\lambda| \downarrow 0$. Then for all η,

$$\operatorname{Re}\big(e^{i\eta}\langle\psi,\varphi\rangle\big) \ge 0 \Rightarrow \langle\psi,\varphi\rangle = 0$$

that is, $\operatorname{Ran}(P)$ is orthogonal to $\operatorname{Ran}(1 - P)$.

(3) ⇒ (1). For any $\varphi_1, \varphi_2 \in \mathcal{H}$, write $\eta_j = P\varphi_j$, $\psi_j = (\mathbb{1} - P)\varphi_j$. Then

$$\langle\varphi_1, P\varphi_2\rangle = \langle\eta_1 + \psi_1, \eta_2\rangle = \langle\eta_1, \eta_2\rangle = \langle\eta_1, \eta_2 + \psi_2\rangle = \langle P\varphi_1, \varphi_2\rangle$$

so $P = P^*$. □

Theorem 2.1.10. *Let $A \in \mathcal{L}(\mathcal{H})$ and P be an orthogonal projection. Then P is reducing for A if and only if* $\operatorname{Ran}(P)$ *is invariant for A and A^*.*

Proof. By Theorem 2.1.4, $\mathrm{Ran}(P)$ is reducing for A^* if and only if

$$PA^*P = A^*P \Leftrightarrow PAP = PA \tag{2.1.38}$$

by taking adjoints. Thus, by Theorem 2.1.5, this theorem is the same as

$$PAP = PA \ \&\ PAP = AP \Leftrightarrow PA = AP \tag{2.1.39}$$

which is immediate. □

Notes and Historical Remarks.

> Taking the Principle of the Excluded Middle from the mathematician ... is the same as ... prohibiting the boxer the use of his fists.
>
> —*David Hilbert* (1862–1943) as quoted in Reid [**552**][1]

See the Notes of Section 1.3 of this Part and Section 1.7 of Part 1 for background on the history of finite matrices as operators on $\mathbb{R}^n$ or $\mathbb{C}^n$. Of course, much of classical analysis concerning linear differential and integral equations was essentially dealing with operators, but without notions of spaces of functions, this viewpoint was hidden. Indeed, differential equations concern unbounded operators and the bounded operator theory needs to come first. Fortunately, integral equations define bounded operators and, as we'll see, were key to the development of operator theory. In the period from 1877–1901, George Hill (1838–1914) [**319**], Henri Poincaré (1854–1912) [**530**] (his capsule biography is in the Notes to Section 12.2 of Part 2B), Helge von Koch (1870–1914) [**718, 719**], and Alfred Dixon (1865–1936) [**164**] developed a theory of determinants of infinite matrices which, in the end, had limited influence.

Modern operator theory has its roots in a seminal paper of (Erik) Ivar Fredholm (1866–1927), a student of Mittag-Leffler: an announcement in 1900 [**205**] and a full paper in 1903 [**206**]. Fredholm was a Swedish mathematician and student of Mittag-Leffler. He published few papers beyond this pathbreaking work and none had its impact. Fredholm studied the integral equation

$$g(x) = f(x) + \lambda \int_a^b K(x,y)g(y)\,dy \tag{2.1.40}$$

where f and K are continuous functions, and a solution, g, is sought. He used an analogy to matrices and was able to define infinite determinants and solve (2.1.40) with essentially an infinite-dimensional analog of Cramer's rule. There was a discrete set of exceptional λ's for which the homogeneous

[1] Because of his use of indirect methods, beginning with his famous nonconstructive proof of his basis theorem in invariant theory, Hilbert was throughout his life battling the intuitionists—first Kronecker, then Brouwer and, for a period, Hilbert's pupil Weyl.

equation (i.e., with $f = 0$) had solutions, and except for these λ's, a unique g could always be found. In essence, if A was the operator,

$$(Ag)(x) = \int_a^b K(x,y)g(y)\,dy \tag{2.1.41}$$

Fredholm was looking at $(1 - \lambda A)^{-1}$. The special λ's are inverses of the eigenvalues of A, but in the early integral equation literature, they were called eigenvalues. Fredholm's work and influence will be seen throughout Chapter 3.

In 1901, Holmgren lectured on Fredholm's work and David Hilbert was so struck by the work that he totally shifted his attention to analysis, having made his reputation in algebra and geometry. According to Hellinger, at that time a student of Hilbert, when Hilbert announced his seminar would be devoted to the study of integral equations, he declared he expected to be able to use them to prove the Riemann hypothesis! Apparently, he hoped to realize the zeta function as a Fredholm determinant—something he did not succeed at!

While Hilbert was motivated by Fredholm's results, he did not follow his proofs. In a six-part masterwork [**317**], Hilbert set out extensions and, most notably, the spectral theorem. With his student Erhard Schmidt (1876–1959), whose capsule biography is in the Notes to Section 3.2, he analyzed compact self-adjoint operators.

David Hilbert (1862–1943) was a German mathematician, a student of Lindemann at Königsberg. Because of his wide-ranging interests and achievement, and because of his strong personality and wide array of students, Hilbert was one of the dominant figures at the turn of the twentieth century. While a student, he became close, long-lasting friends with Minkowski (a fellow student) and Hurwitz (a lecturer in Königsberg). He was initially an academic at Königsberg but Felix Klein brought him to Göttingen in 1895 where he spent the rest of his career.

His first major works were in algebra, particularly his work on invariant theory and codifying algebraic number theory. He was long interested in issues of axioms including his axioms for Euclidean geometry and in logic. His name is here because his work on integral equations gave birth to the notions of Hilbert space, compact operators, and the spectral theorem. He has rightfully been called the father of functional analysis.

His students include S. Bernstein, Courant, Dehn, Haar, Hamel, Hecke, Hellinger, Kellog, Schmidt, Steinhaus, Takagi, and Weyl, as well as the historian of mathematics, Otto Neugebauer, and the chess master, Emmanuel Lasker. He had a serious influence on others who spent time in Göttingen, most notably von Neumann. Reid [**552**] has a celebrated and readable biography of Hilbert.

Remarkably, Hilbert and his school mainly phrased their work in terms of quadratic forms, perhaps motivated by Hilbert's work in algebraic number theory. Of course, there is a one-one correspondence between bounded sesquilinear forms and bounded operators on a Hilbert space, but using forms obscures the structure of composition of operators.

Two books which culminated work of their authors, Riesz's 1913 work [**564**] and Banach's 1932 monograph [**44**], established firmly the operator theoretic point of view.

The history of transpose operators is intimately related to that of dual spaces (see especially Section 5.7 of Part 1). The abstract general definition dates to about 1930 in work of Banach [**43**], Schauder [**598**], and Hildebrandt [**318**]. But in special cases it appeared earlier, as early as Riesz' 1910 paper on L^p-duality [**563**]. He used the name *transpose* (or rather, its German equivalent) which is common but not universal. Conjugate operator, dual operator, or adjoint operator are also used.

von Neumann [**722, 726**] was the pioneer in the consideration of topologies on operators going beyond the norm topology. In the theory of operator algebras, additional topologies occur, such as the σ-strong ($\sum_{j=1}^\infty \|A_n - A)\varphi_j\| \to 0$ for all $\{\varphi_j\}_{j=1}^\infty$ with $\sum_{j=1}^\infty \|\varphi_j\| < \infty$), ultrastrong ($\|\cdot\|^2$ replaces $\|\cdot\|$ in σ-strong), and symmetrized strong ($\|(A_n - A)\varphi\| \to 0$ and $\|(A_n^* - A^*)\varphi\| \to 0$ for all φ).

We saw that in a Hilbert space, there are lots of projections of norm 1, namely, the orthogonal projections. In fact, this characterizes a Hilbert space; Kakutani [**365**] has proven if $\dim(X) \geq 3$ for an NLS and every two-dimensional subspace, Y, has a projection, P, with $Y = \operatorname{Ran}(P)$ and $\|P\| = 1$, then X is a Hilbert space. Lindenstrauss–Tzafriri [**450**] have proven if every closed $Y \subset X$ is the range of bounded projection, then X has an equivalent Hilbert space norm.

The study of which closed subspaces of a Banach space have complements has an extensive literature, so much so that it is regarded as a major theme in the theory of Banach spaces. In 1937, F. Murray [**486**] found the first example of an uncomplemented space (in $l^p, p > 1, p \neq 2$). Phillips [**521**] proved that c_0 does not have a complement in l^∞. That H^1 is not complemented in L^1 is a result of D. J. Newman in 1961 [**502**]. The proof in Problem 8 is due to Rudin [**582**]. As noted above, Lindenstrauss–Tzafriri [**450**] proved if X has the property that every closed subspace has a closed complement, then X has an equivalent norm which comes from an inner product. At the other extreme, Gowers–Murray [**267**] construct an infinite-dimensional Banach space in which the only pairs, Y, Z of closed complementary subspaces have Y or Z of finite dimension. For more, see the review article [**483**] or part of the books [**126, 308**].

Problems

1. (a) With X^* the dual of a Banach space, X, prove that

$$\|x\| = \sup_{\substack{\|\ell\|=1 \\ \ell \in X^*}} |\ell(x)|, \qquad \|\ell\| = \sup_{\substack{\|x\|=1 \\ x \in X}} |\ell(x)| \tag{2.1.42}$$

(b) For $A \in \mathcal{L}(X)$, prove that

$$\|A\| = \sup_{\substack{\|\ell\|=1,\, \ell \in X^* \\ \|x\|=1,\, x \in X}} |\ell(Ax)| \tag{2.1.43}$$

(c) For $B \in \mathcal{L}(X^*)$, prove that

$$\|B\| = \sup_{\substack{\|\ell\|=1,\, \ell \in X^* \\ \|x\|=1,\, x \in X}} |B\ell(x)| \tag{2.1.44}$$

(d) Prove that (2.1.3) holds.

2. (a) For a Banach space, X, consider the family of maps of $\mathcal{L}(X) \to X$ by $E_x(A) = Ax$. Let $\mathcal{T}$ be the weakest topology on $\mathcal{L}(X)$ in which all these maps are continuous when X is given the norm topology. Prove $A_\alpha \xrightarrow{s} A$ if and only if $\|A_\alpha x - Ax\| \to 0$ for all $x \in X$.

(b) Find a strong neighborhood base for $A = 0$.

(c) Do the same for the weak operator topology.

3. In an infinite-dimensional Hilbert space, $\mathcal{H}$, let $\{\varphi_n\}_{n=1}^\infty$ be an orthonormal basis. Let $A_n = (\varphi_n, \cdot\,)\varphi_1$. Prove that $A_n \to 0$ in the strong operator topology but that A_n^* has no limit in the strong operator topology.

4. If X is reflexive, prove that if $A_n \to A$ in the weak operator topology, then $A_n^t \to A_n^t$ in the weak operator topology.

5. If $A_\alpha \to A$, $B_\alpha \to B$ in the strong operator topology and $\sup_\alpha \|A_\alpha\| < \infty$, prove that $A_\alpha B_\alpha \to AB$ in the strong operator topology.

6. In a Hilbert space, $\mathcal{H}$, let $\{\varphi_n\}_{n=1}^\infty$ be an orthonormal basis. Let $A_n = (\varphi_n, \cdot\,)\varphi_1$. Prove that in the weak operator topology, $A_n \to 0$, $A_n^* \to 0$, $A_n^* A_n \to 0$, but that $A_n A_n^*$ has a nonzero limit.

7. This problem will show that $c_0 = \{a_n \in \ell^\infty \mid \lim_{n\to\infty} a_n = 0\} \subset \ell^\infty$ is a space with no closed complement.

(a) Let N be a countably infinite set. Prove there is an uncountable family of infinite subsets $\{A_\alpha\}$ so for all $\alpha \neq \beta$, $A_\alpha \cap A_\beta$ is finite. (*Hint*: Without loss, suppose $N = \mathbb{Q} \cap [0,1]$, the rationals in $[0,1]$ and that the index set is $\alpha \in [0,1]$. Pick A_α to be a sequence of distinct rationals converging to α.)

(b) Say that a Banach space, X has property W if there is a countable family $\{\ell_n\}_{n=1}^\infty \subset X^*$, so that for any $x \neq 0$ in X, there is an ℓ_n with $\ell_n(x) \neq 0$. Prove that any closed subspace of a Banach space with property W has property W and that ℓ^∞ has property W.

(c) If X is a Banach space, if Y, Z are complementary closed subspaces and X/Y is the quotient (see Section 5.1 of Part 1) and $\pi : X \to X/Y$ the canonical map, prove that $\pi \restriction Z$ is a continuous bijection with continuous inverse. Conclude that if X has property W, so does X/Y. (*Hint*: You'll need the open mapping theorem, see Theorem 5.4.10 of Part 1.) Then, conclude that if ℓ^∞/c_0 does not have property W, then c_0 has no closed complement.

(d) Prove that the norm on ℓ^∞/c_0 is given by

$$\|[\{a_n\}]\| = \lim_{n\to\infty} \sup|a_n|$$

(e) Let A_∞ be an uncountable family of subsets of $\mathbb{Z}_+$ with $A_\alpha \cap A_\beta$ finite if $\alpha \neq \beta$ (see part(a)). Let $\chi^\alpha \in \ell^\infty$ be the characteristic function of A_α and $[\chi^\alpha]$ its equivalence class in ℓ^∞/c_0. Prove that for all $\alpha \neq \beta$, $[\chi^\alpha] \neq [\chi^\beta]$ and that for any distinct $\alpha_1, \dots, \alpha_\ell$ and $c_1, \dots, c_l \in \mathbb{C}$, we have

$$\left\| \left[\sum_{j=1}^{\ell} c_j \chi^{\alpha_j} \right] \right\| = \max_{j=1,\dots,\ell} |c_j| \tag{2.1.45}$$

(f) Suppose $L \in [\ell^\infty/c_0]^*$ and $|L(\chi^{\alpha_j})| \geq \frac{1}{n}$ for $j = 1, \dots, \ell$ and $\alpha_1, \dots, \alpha_l$ distinct. Let

$$y = \left[\sum_{j=1}^{\ell} \frac{\overline{L(\chi^{\alpha_j})}}{L(\chi^{\alpha_j})} \chi^{\alpha_j} \right] \tag{2.1.46}$$

Prove that $\|y\| = 1$, $L(y) \geq \frac{\ell}{n}$ and conclude that $\ell \leq n\|L\|$ so $\{\alpha \mid L(\chi^\alpha) \geq \frac{1}{n}\}$ is finite.

(g) Prove that ℓ^∞/c_0 does not have property W and conclude that c_0 has no complement in ℓ^∞.

Remark. That c_0 has no complement is a result of Phillips [**521**] in 1940. The elegant proof above is from Whitley [**754**] although he notes that basic idea has appeared earlier. The fact in part (a) is well-known to set theorists. The clever proof is from Whitley's paper—he notes the proof is due to Artin–Krause.

8. This problem will show that $H^1(\mathbb{D})$ is not complemented in $L^1(\mathbb{D}, d\theta/2\pi)$. So suppose Q is a bounded projection on $L^1(\mathbb{D}, d\theta/2\pi)$ with range $H^1(\mathbb{D})$.
   (a) For $\varphi \in [0, 2\pi)$, let $(T_\varphi f)(z) = f(e^{-i\varphi}z)$. Prove that $\|T_\varphi^{-1} Q T_\varphi\| = \|Q\|$ and that $\varphi \mapsto T_\varphi^{-1} Q T_\varphi$ is continuous, so one can define
$$P = \int T_\varphi^{-1} Q T_\varphi \frac{d\varphi}{2\pi} \tag{2.1.47}$$
   as a Riemann integral with values in $\mathcal{L}(L^1)$.
   (b) Prove that if $e_n(e^{i\theta}) = e^{in\theta}$ then
$$\langle e_m, (T_\varphi^{-1} Q T_\varphi) e_n \rangle = e^{i(m-n)\varphi}(Qe_n) \qquad \left(\langle f, g \rangle \equiv \int \bar{f}(e^{i\theta}) g(e^{i\theta}) \frac{d\theta}{2\pi} \right)$$
   so that
$$Pe_n = \begin{cases} e_n, & n \geq 0 \\ 0, & n < 0 \end{cases} \tag{2.1.48}$$
   (c) Prove there is no continuous map on L^1 obeying (2.1.48). (*Hint*: Let $f_r(\theta) = (1-r)/(1+r^2-2r\cos\theta)$, the Poisson kernel. Compute by taking L^1-limits of polynomials and using $f_r(\theta) = \sum_{n=-\infty}^{\infty} r^{|n|} e^{in\theta}$ that $Pf_r = (1 - re^{i\theta})^{-1}$. Then compute $\|Pf_r\|_1$ and $\|f_r\|$ and conclude that $\lim_{r\uparrow 1} \|Pf_r\|/\|f_r\| = \infty$.)
   (d) Conclude that H^1 has no complement in L^1.

9. Let $\mathbb{A}(\overline{\mathbb{D}})$ be those functions $f \in C(\partial\mathbb{D})$ which are the values on $\partial\mathbb{D}$ of functions f analytic on $\mathbb{D}$ and continuous on $\partial\mathbb{D}$. Mimic the arguments in Problem 8 to prove that $\mathbb{A}(\overline{\mathbb{D}})$ does not have a complement in $C(\partial\mathbb{D})$. (*Hint*: In place of the Poisson kernels of (c) of Problem 8, look at Example 5.7.2 of Part 3.)

2.2. The Spectrum

For finite matrices, a special role is played by the eigenvalues. In this section, we'll begin the study of their replacement for general operators in the infinite-dimensional context. That eigenvalues are not sufficient can be seen by

Example 2.2.1. Let $\mathcal{H} = L^2([0,1], dx)$ and A the operator given by
$$(Af)(x) = xf(x) \tag{2.2.1}$$
A has no eigenvalues! For if $(A - \lambda)f = 0$, then for a.e. x, $(x - \lambda)f(x) = 0$. Thus, $f(x) = 0$ for a.e. x since $x - \lambda$ vanishes for a single value of λ. □

In the finite matrix case, if λ is not an eigenvalue, then $\operatorname{Ker}(A-\lambda) = \{0\}$. Thus, $A - \lambda$ is one–one. Moreover, $\dim \operatorname{Ran}(A - \lambda) = \dim(\mathcal{H})$, so since $\dim(\mathcal{H}) < \infty$, $\operatorname{Ran}(A - \lambda) = \mathcal{H}$, that is, $(A - \lambda)$ is a bijection. The inverse mapping theorem says that continuous bijections are invertible even in the infinite-dimensional case. This motivates the following definition.

Definition. Let $A \in \mathcal{L}(X)$ for a Banach space, X. The *resolvent set*, $\rho(A)$, is the set of those $\lambda \in \mathbb{C}$ so that there exists a bounded operator, $R_\lambda(A)$, with

$$(A - \lambda\mathbb{1})R_\lambda(A) = R_\lambda(A)(A - \lambda\mathbb{1}) = \mathbb{1} \tag{2.2.2}$$

R_λ is called the *resolvent* of A, written also as $(A - \lambda)^{-1}$. The *spectrum*, $\sigma(A)$, of A is defined by

$$\sigma(A) = \mathbb{C} \setminus \rho(A) \tag{2.2.3}$$

Example 2.2.1 (revisited). If $\lambda \notin [0, 1]$, we can define

$$(R_\lambda(A)f)(x) = (x - \lambda)^{-1}f(x) \tag{2.2.4}$$

It is bounded (by $1/\operatorname{dist}(\lambda, [0, 1])$) and easily seen to obey (2.2.2).

On the other hand, if there is a bounded R_λ obeying (2.2.2), then

$$\|(A - \lambda)R_\lambda f\| = \|f\| \tag{2.2.5}$$

so if $g = R_\lambda f$, then $\|g\| \le \|R_\lambda\|\|f\|$,

$$\|(A - \lambda)g\| = \|f\| \ge \|R_\lambda\|^{-1}\|g\| \tag{2.2.6}$$

so

$$\inf_{\|g\|=1} \|(A - \lambda)g\| > 0 \tag{2.2.7}$$

But, if $\lambda \in [0, 1]$ and $g = \chi_{[\lambda-\varepsilon,\lambda+\varepsilon]}$, we see

$$\|(A - \lambda)g\| \le \varepsilon\|g\|$$

so (2.2.7) fails, that is, $[0, 1] \notin \rho(A)$, so $\sigma(A) = [0, 1]$. □

Proposition 2.2.2. $\lambda \in \rho(A)$ *if and only if the following two conditions hold:*

(i) (2.2.7) *holds*

(ii) $\operatorname{Ker}(A^{\mathrm{t}} - \lambda) = \{0\}$ (2.2.8)

Remarks. 1. In the Hilbert space case, (2.2.8) can be written

$$\operatorname{Ker}(A^* - \bar{\lambda}) = \{0\} \tag{2.2.9}$$

2. (2.2.7) also implies

$$\operatorname{Ker}(A - \lambda) = \{0\} \tag{2.2.10}$$

3. Those λ for which $\operatorname{Ker}(A - \lambda) = \{0\} \neq \operatorname{Ker}(A^{\mathrm{t}} - \lambda)$, which lie in $\sigma(A)$ (since (2.2.8) fails), are called *residual spectrum*. These are precisely those λ for which $(A-\lambda)$ is injective but its range is not dense. If $(A-\lambda)$ is injective

and its range is dense but not all of X, we say λ lies in the *continuous spectrum*. Normal operators do not have any residual spectrum. Every point of $\mathbb{D}$ lies in the residual spectrum of the operator R of Example 2.2.5.

Proof. Suppose first $\lambda \in \rho(A)$. As in the example above, (2.2.5) implies (2.2.6), proving (i).

For $\ell \in X^*$, $\ell\big((A-\lambda)x\big) = 0$ for all x if and only if $(A^{\rm t} - \lambda)(\ell) = 0$. Thus,

$$\operatorname{Ker}(A^{\rm t} - \lambda) = \{\ell \mid \ell \restriction \operatorname{Ran}(A-\lambda) = 0\} \tag{2.2.11}$$

The Hahn–Banach theorem (see Theorem 5.5.12 of Part 1) implies

$$\operatorname{Ker}(A^{\rm t} - \lambda) = \{0\} \Leftrightarrow \operatorname{Ran}(A-\lambda) \text{ is dense in } X \tag{2.2.12}$$

proving (ii).

Conversely, if (i) and (ii) hold, as noted (i) implies (2.2.10), which implies that $A-\lambda$ is one–one. It is easy to see that (2.2.7) implies that $\operatorname{Ran}(A-\lambda)$ is closed (see Problem 1). Thus, by (ii) and (2.2.12), (i) and (ii) imply $\operatorname{Ran}(A-\lambda) = X$.

Therefore, (i) and (ii) imply $A-\lambda$ is a bijection and thus has a bounded inverse by the inverse mapping theorem (or directly from (2.2.7); see Problem 2). □

Remarks. 1. In the Hilbert space case, (2.2.11) becomes

$$\operatorname{Ker}(A^* - \bar\lambda) = \operatorname{Ran}(A-\lambda)^\perp \tag{2.2.13}$$

2. (2.2.7) is equivalent to $\operatorname{Ker}(A-\lambda) = \{0\}$ plus $\operatorname{Ran}(A-\lambda)$ is closed. Thus, the two conditions above can be thought of as the three conditions: $\operatorname{Ker}(A-\lambda) = \{0\}$, $\operatorname{Ran}(A-\lambda)$ dense, and $\operatorname{Ran}(A-\lambda)$ closed. These are equivalent to $A-\lambda$ being a bijection. The inverse mapping theorem (Theorem 5.4.14 of Part 1) then implies (2.2.7).

As a corollary, we have

Theorem 2.2.3. *Let $\mathcal{H}$ be a Hilbert space and $A \in \mathcal{L}(\mathcal{H})$ be self-adjoint. Then*

$$\sigma(A) \subset \mathbb{R} \tag{2.2.14}$$

Remark. We'll soon see that $\sigma(A) \subset \big[-\|A\|, \|A\|\big]$; see Theorem 2.2.9.

Proof. Let $\lambda = \alpha + i\beta$ with $\alpha \in \mathbb{R}$, $\beta \in \mathbb{R}\setminus\{0\}$. Then

$$\|(A-\lambda)\varphi\|^2 = \|(A-\alpha)\varphi\|^2 + |\beta|^2\|\varphi\|^2 \tag{2.2.15}$$

since $\langle (A-\alpha)\varphi, (i\beta)\varphi\rangle + \langle (i\beta)\varphi, (A-\alpha)\varphi\rangle = 0$ by the self-adjointness. Thus,

$$\|(A-\varphi)\varphi\| \geq |\beta|\|\varphi\| \tag{2.2.16}$$

proving (2.2.7).

Since $A = A^*$, (2.2.16) implies $\|(A^* - \bar{\lambda})\varphi\| \geq |\beta|\|\varphi\|$, so $\operatorname{Ker}(A^* - \bar{\lambda}) = \{0\}$. Thus, (i) and (ii) of the proposition hold. □

Theorem 2.2.4. *If X is reflexive, $\sigma(A^{\mathrm{t}}) = \sigma(A)$. For a Hilbert space*

$$\sigma(A^*) = \sigma(A)^{cc} = \{\bar{z} \mid z \in \sigma(A)\} \tag{2.2.17}$$

Proof. $R_\lambda(A^{\mathrm{t}}) = R_\lambda(A)^{\mathrm{t}}$ shows $\rho(A) \subset \rho(A^{\mathrm{t}})$. Since $A^{\mathrm{tt}} = A$ in the reflexive case, we get $\rho(A^{\mathrm{t}}) \subset \rho(A)$. Since the map of $\mathcal{H}^*$ to $\mathcal{H}$ is conjugate linear, $\sigma(A) = \sigma(A^{\mathrm{t}})$ implies (2.2.17). □

The following two examples illuminate the different ways a point can be in the spectrum:

Example 2.2.1 (revisited (again)). For A = multiplication by x in $L^2([0,1], dx)$ and $\lambda \in [0,1]$, $\operatorname{Ker}(A - \lambda) = \operatorname{Ker}(A^* - \lambda^*) = \{0\}$, but as we have seen, (2.2.7) fails. □

Example 2.2.5 (Right Shift and Left Shift)**.** Let $\mathcal{H} = \ell^2(\mathbb{Z}_+)$ the square integrable sequences $\{a_n\}_{n=1}^\infty$. Let

$$R(a_1, a_2, \dots) = (0, a_1, a_2, \dots), \qquad L(a_1, a_2, \dots) = (a_2, a_3, a_4, \dots) \tag{2.2.18}$$

called *right* and *left shift*, respectively. It is easy to see $\langle b, Ra\rangle = \langle Lb, a\rangle$, that is,

$$R^* = L \tag{2.2.19}$$

We'll prove shortly that for any operator, A, $\sigma(A) \subset \{\lambda \mid |\lambda| \leq \|A\|\}$, so $\sigma(R) = \sigma(L) \subset \overline{\mathbb{D}}$.

For $\lambda \in \mathbb{D}$, let

$$a^{(\lambda)} = (1, \lambda, \lambda^2, \dots) \tag{2.2.20}$$

then $La^{(\lambda)} = \lambda a^{(\lambda)}$, so λ is an eigenvalue of L. For $\lambda \in \partial\mathbb{D}$, $n = 1, 2, \dots$, let

$$a^{(\lambda,n)} = (1, \lambda, \dots, \lambda^n, 0, 0, \dots) \tag{2.2.21}$$

Then $\|(L - \lambda)a^{(\lambda,n)}\|_2 = |\lambda|^{n+1} = 1$ with $\|a^{(\lambda,n)}\| = (n+1)^{1/2}$, so

$$\inf_{\substack{b \in \ell^2 \\ \|b\|=1}} \|(L - \lambda)b\| \leq (n+1)^{-1/2} \Rightarrow \inf_{\substack{b \in \ell^2 \\ \|b\|=1}} \|(L - \lambda)b\| = 0 \tag{2.2.22}$$

so $\partial\mathbb{D} \in \sigma(L)$, that is, we have proven that

$$\sigma(R) = \sigma(L) = \overline{\mathbb{D}} \tag{2.2.23}$$

Notice $\|Ra\| = \|a\|$, so R has no eigenvalues in $\mathbb{D}$ (also in $\partial\mathbb{D}$) and $\inf_{b\in\ell^2,\,\|b\|=1}\|(R - \lambda)b\| \geq 1 - |\lambda|$. Thus, for $\lambda \in \mathbb{D}$, $(R - \lambda)$ is an injection onto a closed range, but the range is not dense, which is why $\lambda \in \sigma(R)$.

Shortly we'll prove spectra are closed, so in place of using (2.2.21), we can use this to go from $\mathbb{D} \subset \sigma(L) \subset \overline{\mathbb{D}}$ to $\sigma(L) = \overline{\mathbb{D}}$. □

We next turn to some general properties of $\sigma(A)$ and the resolvent. These hold in a wider context than $\mathcal{L}(X)$, which will become relevant in Chapter 6. We therefore define:

Definition. A *Banach algebra*, $\mathfrak{A}$, is a complex Banach space together with a product, that is, a map $(x,y) \mapsto xy$ obeying

(a) *distributivity*, that is, for each fixed x, $y \mapsto xy$ is a linear map, and for each fixed y, $x \mapsto xy$ is a linear map;
(b) *associativity*, that is, for each $x, y, z \in \mathfrak{A}$, $(xy)z = x(yz)$;
(c) *Banach algebra property*

$$\|xy\| \leq \|x\| \|y\| \tag{2.2.24}$$

Definition. A Banach algebra is said to have an *identity* if and only if there is $e \in \mathfrak{A}$ with $\|e\| = 1$ so that for all $x \in \mathfrak{A}$,

$$xe = ex = x \tag{2.2.25}$$

Section 6.1 will discuss how to add an identity to a Banach algebra if it doesn't already have one.

Definition. If $\mathfrak{A}$ is a Banach algebra with identity, we define

$$\rho(x) = \{\lambda \in \mathbb{C} \mid x - \lambda e \text{ has a two-sided inverse}\} \tag{2.2.26}$$

$$\sigma(x) = \mathbb{C} \setminus \rho(x) \tag{2.2.27}$$

known as the *resolvent set* and *spectrum* of x. We'll write $(x-\lambda)^{-1}$ for the inverse of $x - \lambda e$, the *resolvent* of x.

For any Banach space, $\mathcal{L}(X)$ is a Banach algebra and our previous notion of spectrum agrees with what we just defined. Thus, the general results we'll prove below for Banach algebras hold in any $\mathcal{L}(X)$. We'll define the notion of $\mathfrak{A}$-valued analytic function in terms of $\|\cdot\|$-convergent power series (but see Theorem 3.1.12 of Part 2A on the equivalence to weak analyticity).

Theorem 2.2.6. *Suppose $\mathfrak{A}$ is a Banach algebra with identity and $x \in \mathfrak{A}$ has a two-sided inverse. If $\|y\| < \|x^{-1}\|^{-1}$, then $x+y$ also has a two-sided inverse and*

$$(x+y)^{-1} = \sum_{n=0}^{\infty} x^{-1}(-yx^{-1})^n \tag{2.2.28}$$

Proof. Since $\|yx^{-1}\| < 1$, the series in (2.2.28) is uniformly convergent and one can interchange sums with products. If z is the infinite sum, we use

$$xz = \sum_{n=0}^{\infty} (-yx^{-1})^n = 1 - yz \tag{2.2.29}$$

so $(x+y)z = 1$. Similarly, since $z = \sum_{n=0}^{\infty}(-x^{-1}y)^n x^{-1}$, we see $z(x+y) = 1$. $\square$

Corollary 2.2.7. *If $\mathfrak{A}$ is a Banach algebra with identity, the set of invertible elements is open and the noninvertible elements are closed.*

Theorem 2.2.8. *If $\mathfrak{A}$ is a Banach algebra with identity, for any $x \in \mathfrak{A}$, $\sigma(x)$ is closed and $\lambda \mapsto (x-\lambda)^{-1}$ defined on $\rho(x)$ is analytic in λ.*

Proof. If $\lambda \in \rho(x)$ and $|\lambda - \mu| < \|(x-\lambda)^{-1}\|^{-1}$, by Theorem 2.2.6, $(x-\mu)$ is invertible, that is, $\mu \in \rho(x)$. Thus, $\rho(x)$ is open and $\sigma(x)$ is closed. By (2.2.28), for such μ,

$$(x-\mu)^{-1} = \sum_{n=0}^{\infty} (\mu-\lambda)^n \left[(x-\lambda)^{-1}\right]^{n+1} \tag{2.2.30}$$

is analytic in μ. □

Theorem 2.2.9. *If $\mathfrak{A}$ is a Banach algebra with identity, then*

(a) *$\sigma(x)$ is nonempty;*
(b) *$\sigma(x) \subset \{\lambda \mid |\lambda| \le \|x\|\}$.*

Proof. Let $f(\lambda) = (1 - \lambda x)$. Then

$$(x-\lambda) = (-\lambda)(1-\lambda^{-1}x) = (-\lambda) f(\lambda^{-1}) \tag{2.2.31}$$

so

$$\rho(x) \setminus \{0\} = \{\lambda^{-1} \mid 1 - \lambda x \text{ is invertible}\} \setminus \{0\}$$

If $|\lambda| < \|x\|^{-1}$, then $(1-\lambda x)^{-1} = \sum_{n=0}^{\infty} \lambda^n x^n$ is given by a convergent series (by Theorem 2.2.6), so $\{\lambda^{-1} \mid |\lambda| < \|x\|^{-1}\} \subset \rho(x)$ or $\{\lambda \mid \|x\| < |\lambda|\} \subset \rho(x)$, proving (b).

If $\mathbb{C} \setminus \{0\} \subset \rho(x)$, by (2.2.31), $f(\lambda^{-1})$ is invertible for all $\lambda \in \mathbb{C} \setminus \{0\}$. $f(0) = 1$ is also invertible, so $g(\lambda) = (1-\lambda x)^{-1}$ is an entire function.

By (2.2.31),

$$f(\lambda^{-1})^{-1} = (-\lambda)(x-\lambda)^{-1} \tag{2.2.32}$$

If $0 \notin \sigma(x)$, also, $(x-\lambda)^{-1}$ is bounded near $\lambda = 0$, so

$$\lim_{\mu\to\infty} \|f(\mu)^{-1}\| = \lim_{\mu\to\infty} |\mu|^{-1} \|(x-\mu^{-1})^{-1}\| = 0 \tag{2.2.33}$$

By Liouville's theorem, $f(\mu)^{-1} \equiv 0$, but this is impossible since $f(0)^{-1} = 1$. Thus, $\rho(x)$ cannot be all of $\mathbb{C}$, that is, $\sigma(x) \neq \emptyset$. □

Remarks. 1. To prove Liouville's theorem for $\mathfrak{A}$-valued functions, apply the scalar valued theorem to $\ell(f(\mu)^{-1})$ for $\ell \in \mathfrak{A}^*$ to see $\ell(f(\mu)^{-1})$ is constant for each ℓ, so, since $\mathfrak{A}^*$ separates points, $f(\mu)^{-1}$ is constant.

2. Problem 6 has a proof of the fundamental theorem of algebra from $\sigma(x) \neq \emptyset$ (not surprisingly, related to our Liouville theorem proof in Theorem 3.1.11 of Part 2A).

Theorem 2.2.10 (Spectral Radius Formula). *Let $x \in \mathfrak{A}$, a Banach algebra with identity. Then*

$$\operatorname{spr}(x) \equiv \lim_{n\to\infty} \|x^n\|^{1/n} \tag{2.2.34}$$

exists and

$$\sup\{|\lambda| \mid \lambda \in \sigma(x)\} = \operatorname{spr}(x) \tag{2.2.35}$$

Remarks. 1. $\operatorname{spr}(x)$ is called the *spectral radius* of x.

2. Once one knows the limit exists, (2.2.34) is essentially a vector-valued version of the Cauchy formula, that the radius of convergence of $(1-\lambda x)^{-1}$ is $(\limsup\|x^n\|^{1/n})^{-1}$ (see Theorem 2.3.1 of Part 2A).

Proof. Let $\alpha_n = \log\|x^n\|$. Since $\|x^{n+m}\| \leq \|x^n\|\|x^m\|$, we have, with $\alpha_0 = 0$,

$$\alpha_{n+m} \leq \alpha_n + \alpha_m \tag{2.2.36}$$

We now repeat the argument of Proposition 2.9.3 of Part 3. If k is fixed and $n = mk + r$, $0 \leq r \leq k-1$, then by (2.2.36),

$$\alpha_n \leq m\alpha_k + \sup_{0\leq r\leq k-1} \alpha_n \tag{2.2.37}$$

so

$$\limsup \frac{\alpha_n}{n} \leq \frac{\alpha_k}{k} \tag{2.2.38}$$

and thus,

$$\limsup \frac{\alpha_n}{n} \leq \inf \frac{\alpha_k}{k} \leq \liminf \frac{\alpha_k}{k} \tag{2.2.39}$$

proving that the limit exists; it can be $-\infty$ (i.e. $\operatorname{spr}(x)$ can be 0) but not ∞.

If $|\mu| < (\operatorname{spr}(x))^{-1}$, then $|\mu|^n\|x\|^n \leq C(1-\varepsilon)^n$ for $1-2\varepsilon = |\mu|\operatorname{spr}(x)$, so $1-\mu x$ is invertible so $\mu^{-1} \notin \sigma(x)$, that is,

$$\sup\{|\lambda| \mid \lambda \in \sigma(x)\} \leq \operatorname{spr}(x) \tag{2.2.40}$$

Conversely, if $r > \sup\{|\lambda| \mid \lambda \in \sigma(x)\}$, $(1-\mu x)^{-1}$ is analytic in a neighborhood of $\{\mu \mid |\mu| \leq r^{-1}\}$, so by a Cauchy estimate (see Theorem 3.1.8 of Part 2A),

$$\|x^n\| \leq C(r^{-n})^{-1} \tag{2.2.41}$$

so $\operatorname{spr}(x) \leq r$, that is, taking r to the sup,

$$\operatorname{spr}(x) \leq \sup\{|\lambda| \mid \lambda \in \sigma(x)\} \tag{2.2.42}$$

□

Remark. The proof assumed $\sigma(x) \neq \emptyset$ (since sup of the empty set is considered to be $-\infty$). We proved this in Theorem 2.2.9.

When we look at normal operators on a Hilbert space, because of the C^* identity, (2.1.34), one has $\operatorname{spr}(x) = \|x\|$.

Theorem 2.2.11 (Spectral Radius for Normal Operators). *Let A be a normal operator on a Hilbert space, $\mathcal{H}$. Then*

$$\sup\{|\lambda| \mid \lambda \in \sigma(A)\} = \|A\| \tag{2.2.43}$$

Proof. Suppose first $A = A^*$. Then by (2.1.34),

$$\|A^2\| = \|A^*A\| = \|A\|^2 \tag{2.2.44}$$

Thus, by induction,

$$\|A^{2^n}\|^{1/2^n} = \|A\| \tag{2.2.45}$$

Since the limit in (2.2.34) exists, we see $\operatorname{spr}(A) = \|A\|$.

If A is normal, by (2.1.34),

$$\begin{aligned} \|A^{2^n}\|^2 &= \|(A^*)^{2^n} A^{2^n}\| \\ &= \|(A^*A)^{2^n}\| \\ &= \|A^*A\|^{2^n} \qquad \text{(by (2.2.45))} \\ &= \|A\|^{2^{n+1}} \qquad \text{(by (2.1.34))} \end{aligned}$$

so (2.2.45) holds for A, which implies $\operatorname{spr}(A) = \|A\|$. □

If $\operatorname{spr}(x) = 0$, that is, $\lim\|x^n\|^{1/n} = 0$, we say x is *quasinilpotent*. Notice if x is quasinilpotent, for any $\lambda \in \mathbb{C}$, we have

$$(1 - \lambda x)^{-1} = \sum_{n=0}^{\infty} \lambda^n x^n \tag{2.2.46}$$

Here are two examples of such an object that is not nilpotent.

Example 2.2.12. On $\ell^2(\mathbb{Z}_+)$, let

$$A(x_1, x_2, \dots) = (0, \alpha_1 x_1, \alpha_2 x_2, \dots) \tag{2.2.47}$$

where $|\alpha_j| \to 0$. Then

$$\|A^n\| = \sup_j |\alpha_j \alpha_{j+1} \dots \alpha_{j+n-1}| \tag{2.2.48}$$

and it is easy to see (Problem 3) that

$$\lim\|A^n\|^{1/n} = 0 \tag{2.2.49}$$

So long as no α_j is 0, $\operatorname{Ker}(A^n) = \{0\}$ for all n. □

Example 2.2.13 (Volterra Integral Kernels). Let K be continuous on $[0,1] \times [0,1]$. The *Volterra integral operator*, V_K, is the map of $C([0,1])$ to itself given by

$$(V_K f)(x) = \int_0^x K(x,y) f(y)\, dy \tag{2.2.50}$$

(so the values $K(x,y)$ for $y > x$ are irrelevant!). Clearly,

$$\|V_K f\|_\infty \le \|K\|_\infty \|f\|_\infty \tag{2.2.51}$$

so $V_K \in \mathcal{L}(C[0,1])$. Notice that

$$(V_K^n f)(x) = \int_{0<y_1\le y_2\le\dots\le y_n\le x} K(x,y_n)K(y_n,y_{n-1})\dots K(y_2,y_1)f(y_1)\,dy_1\dots dy_n \tag{2.2.52}$$

so that

$$\|V_K^n f\|_\infty \le \|K\|_\infty^n \|f\|_\infty \sup_x[\mathrm{vol}_{[0,1]^n}\{(y_1,\dots,y_n) \mid 0\le y_1\le\dots\le y_n\le x\}]$$

and that volume by using permutations is $x^n/n!$, that is,

$$\|V_K^n\| \le \frac{\|K\|_\infty^n}{n!} \tag{2.2.53}$$

Since $(n!)^{1/n} \to \infty$, we see that $\|V_K^n\|^{1/n} \to 0$, that is, V_K is quasinilpotent. $\square$

Let us apply this to a problem in differential equations (see also Theorem 5.12.5 in Part 1). We are interested in solving

$$-u'' + Qu = 0, \qquad u(0) = 0, \qquad u'(0) = 1 \tag{2.2.54}$$

on $[0,1]$, where Q is a continuous function on $[0,1]$. By standard methods, we know such solutions are unique if they exist. Existence is more subtle, but we'll not only prove existence but get a convergent series solution:

Theorem 2.2.14. *Let $u_0(x) = x$ and let Q be continuous on $[0,1]$.*

(a) *If u solves*

$$u(x) = u_0(x) + \int_0^x (x-y)Q(y)u(y)\,dy \tag{2.2.55}$$

then u solves (2.2.54).

(b) *If $K(x,y) = (x-y)Q(y)$ and V_K is the associated Volterra integral operator, then* (2.2.55), *and so,* (2.2.54), *is solved by*

$$u = u_0 + \sum_{n=1}^\infty V_K^n u_0 \tag{2.2.56}$$

converging in $\|\cdot\|_\infty$. Moreover,

$$|u(x)| \le u_0(x)\exp\Big(\int_0^1 |Q(y)|\,dy\Big) \tag{2.2.57}$$

(c) *If $Q \ge 0$ on $[0,1]$, then the solution of* (2.2.54) *obeys $u(x) > 0$, $u'(x) > 1$ on all of $(0,1]$.*

Proof. (a) If u solves (2.2.55), then

$$u'(x) = 1 + \int_0^x Q(y)u(y)\,dy \tag{2.2.58}$$

since the boundary term is zero because $(x-y)|_{y=x} = 0$. Thus,

$$u''(x) = Q(x)u(x) \tag{2.2.59}$$

showing u solves $-u'' + Qu = 0$. By (2.2.55), $u(0) = 0$, and by (2.2.58), $u'(0) = 1$.

(b) Since V_K is quasinilpotent, this is immediate from (2.2.46). We get (2.2.57) from (2.2.53) if we note that $\|Qu\|_\infty \le \|u\|_\infty \int_0^1 |Q(y)|\,dy$ since $|x - y| \le 1$.

(c) Since $K(x,y) \ge 0$, each term in (2.2.58) is nonnegative, so $u(x) \ge x > 0$. By (2.2.58), $u'(x) \ge 1$ for all x. □

As a final topic, we are heading towards a simplest version of the spectral mapping theorem (see Theorem 2.3.2) that for any polynomial, P,

$$\sigma\big(P(x)\big) = \{P(\lambda) \mid \lambda \in \sigma(x)\} \tag{2.2.60}$$

As a preliminary, we note that

Lemma 2.2.15. *Let $x_1, \dots, x_n$ be mutually commuting elements of a Banach algebra with identity. Let $y = x_1 \dots x_n$. Then*

$$y \text{ is invertible} \Leftrightarrow x_1, \dots, x_n \text{ are each invertible} \tag{2.2.61}$$

Proof. If $x_1, \dots, x_n$ are invertible, $x_n^{-1} x_{n-1}^{-1} \dots x_1^{-1}$ is a two-sided inverse for y. For the converse, if y is invertible, then

$$x_1(x_2 x_3 \dots x_n) y^{-1} = y y^{-1} = 1 \tag{2.2.62}$$

$$y^{-1}(x_2 \dots x_n) x_1 = y^{-1} x_1 (x_2 \dots x_n) = y^{-1} y = 1 \tag{2.2.63}$$

But (Problem 4) if x_1 has a left and right inverse, they are equal, so x_1 has a two-sided inverse.

Since for any j, $y = x_j(x_1 x_{j-1} x_{j+1} \dots x_n)$, the same argument shows that x_j is invertible. □

Remark. $\Leftarrow$ in (2.2.61) did not depend on commuting x_j. Not only does our proof of $\Rightarrow$ depend on commuting, it is essential, as the next example shows.

Example 2.2.5 (revisited). $LR = 1$ is invertible, but as we saw, neither L nor R has a two-sided inverse. Essentially, L is a left inverse of R, but $(RL)(a_1 \dots) = (0, a_2, a_3, \dots)$ is not invertible, so R has no two-sided inverse.

This example shows $\sigma(xy)$ and $\sigma(yx)$ may not be equal for $x, y \in \mathfrak{A}$, a Banach algebra with identity, but (Problem 5) it is always true that

$$\sigma(xy) \setminus \{0\} = \sigma(yx) \setminus \{0\} \tag{2.2.64}$$

□

Theorem 2.2.16 (Spectral Mapping Theorem, First Version). *Let $\mathfrak{A}$ be a Banach algebra with identity. Then for any polynomial, P, and $x \in \mathfrak{A}$,* (2.2.60) *holds.*

Proof. Factor P into

$$P(X) = (X - \mu_1) \dots (X - \mu_n) \tag{2.2.65}$$

Then the same is true if the unknown X in (2.2.65) is replaced by $x \in \mathfrak{A}$. Clearly, $\{x - \mu_j\}_{j=1}^n$ commute, so by Lemma 2.2.15,

$$\begin{aligned} \lambda \in \sigma(P(X)) &\Leftrightarrow \text{for some } j,\ \mu_j \in \sigma(x) \\ &\Leftrightarrow \lambda \in P[\sigma(X)] \end{aligned}$$

□

Notes and Historical Remarks. In the first of his papers on integral equations, Hilbert [**317**] used the term "spectrum." It has been conjectured that he was motivated by an 1897 paper of Wirtinger [**764**] who, in discussing Hill's work on periodic differential operators, saw an analog of the bands there and the "band spectrum" found in the optical lines of molecules. It is serendipitous that when operator theory became important in quantum mechanics, Hilbert's spectrum was close to the actual physical spectrum. Hilbert also used "resolvent" (for $(1 - \lambda A)^{-1}$) in these papers.

The geometric series $(1 - \lambda A)^{-1} = \sum_{n=0}^{\infty} \lambda^n A^n$ is sometimes called the *Neumann series* after Carl Neumann's (1832–1925) use of it to solve the integral equation of potential theory in his 1877 book [**500**]. Dixon [**164**] used it in the context of infinite matrices.

The theorem that the spectrum of an operator is nonempty and its proof via Liouville's theorem is from Stone's spectral theory book [**670**].

Just as it is somewhat surprising how long it took for Banach spaces to be formally defined, it took even longer for Banach algebras (perhaps a sign that abstraction wasn't so natural in the early years of the twentieth century).

In the 1930s, von Neumann wrote a series of papers (some with Murray) on algebras of operators on a Hilbert space.

The first comprehensive paper defining the abstract theory was by Nagumo [**488**] in 1936. Only with Gel'fand's remarkable 1941 paper [**228**], the subject of Chapter 6, did the notion become firmly established in the mathematical canon. These early papers used the name "normed rings"—"Banach algebra" was first used by Ambrose [**15**] in 1945 and the name stuck.

The spectral radius formula (Theorem 2.2.10) is from Gel'fand's paper [**228**]. The Notes to the next section discuss the history of the spectral mapping theorem.

Problems

1. (a) If (2.2.7) holds, show that $\operatorname{Ran}(A-\lambda)$ is closed. (*Hint*: If $y_n = (A-\lambda)x_n$ and y_n is Cauchy, prove that x_n is Cauchy.)

   (b) If $\operatorname{Ran}(A-\lambda)$ is closed, prove that (2.2.7) holds. (*Hint*: Prove that $A-\lambda$ is invertible as a map of X to $\operatorname{Ran}(A-\lambda)$ which is a Banach space.)

2. Prove directly that (2.2.7) implies that $A-\lambda$ is a bijection of X to $\operatorname{Ran}(A-\lambda)$ whose inverse is bounded with norm equal to $(\inf_{\|g\|=1}\|(A-\lambda)g\|)^{-1}$.

3. (a) For the A of (2.2.47), prove that $\|A^n\| = \sup_j(\alpha_j\alpha_{j+1}\cdots\alpha_{j+n-1})$.

   (b) Prove (2.2.49).

4. In a Banach algebra with identity, suppose $\ell x = xr = 1$. Prove that $\ell = r$. (*Hint*: Look at $\ell x r$.)

5. (a) In a Banach algebra, let $\lambda \neq 0$ and suppose $xy-\lambda$ is invertible. Prove that $yx-\lambda$ is invertible with inverse given by

$$(yx-\lambda)^{-1} = -\lambda^{-1} + \lambda^{-1}y(xy-\lambda)^{-1}x \tag{2.2.66}$$

   This proves (2.2.64). (*Hint*: Show first that $(yx-\lambda)y(xy-\lambda)^{-1}x = yx$.)

   (b) In $\mathcal{L}(X)$, suppose λ is an eigenvalue of AB. If $\lambda \neq 0$, prove λ is also an eigenvalue of BA. (*Hint*: If $ABu = \lambda u$, prove that $Bu \neq 0$ and $(BA)Bu = \lambda Bu$.)

Remarks. 1. While (2.2.66) appears out of nowhere, if one notes $(yx-\lambda)^{-1} = -\sum_{n=0}^{\infty}\lambda^{-n-1}(yx)^n$ (for $|\lambda|$ large), one can deduce (2.2.66) directly for $|\lambda|$ large and then prove it in general.

2. (2.2.64) holds for many unbounded operators and is very useful. See Deift [**152**] for more about this.

6. Let $P(x) = x^n + a_{n-1}x^{n-1} + \cdots + a_0$. Define the companion matrix of P to be the $n \times n$ matrix

$$C = \begin{pmatrix} 0 & 0 & \cdots & 0 & -a_0 \\ 1 & 0 & \cdots & 0 & -a_1 \\ 0 & 1 & \cdots & 0 & -a_2 \\ \vdots & \vdots & \ddots & \vdots & \vdots \\ 0 & 0 & \cdots & 1 & -a_{n-1} \end{pmatrix} \tag{2.2.67}$$

(a) Prove that

$$\det\bigl(z\mathbb{1} - C\bigr) = P(z) \tag{2.2.68}$$

(b) Using $\sigma(C) \neq \emptyset$, prove that P has zeros.

2.3. The Analytic Functional Calculus

In the last section, we considered polynomials, $P(x)$, for $x \in \mathfrak{A}$ a Banach algebra with identity, and proved that $\sigma\bigl(P(x)\bigr) = P\bigl[\sigma(x)\bigr]$. In this section, we'll define $f(x)$ for some other functions and show $\sigma\bigl(f(x)\bigr) = f\bigl[\sigma(x)\bigr]$. Clearly, if f is analytic in a neighborhood of $\{\lambda \mid \lambda \le \text{spr}(x)\}$, the power series $\sum_{n=0}^{\infty} \frac{f^{(n)}(0)}{n!} x^n$ is convergent in $\|\cdot\|$ and can define $f(x)$. But we'll consider the much richer class of f analytic in a neighborhood of $\sigma(x)$ and define $f(x)$ so that

$$f(x)g(x) = fg(x) \tag{2.3.1}$$

We say this is much richer because if $\sigma = \sigma_1 \cup \sigma_2$, where $\text{dist}(\sigma_1, \sigma_2) > 0$, we can find U_1, U_2 open with $\sigma_j \subset U_j$ and $U_1 \cap U_2 = \emptyset$ and allow f's which are analytic on U_1 and U_2 with "no relation" between the pieces. In particular, if $f \equiv 1$ on U_1 and $f \equiv 0$ on U_2, $f(x)^2 = f(x)$ by (2.3.1), that is, we'll be able to define projections, P, commuting with x. We'll also prove that (2.3.1) plus a continuity property uniquely determine $f(x)$.

In the operator theory case (i.e, $\mathfrak{A} = \mathcal{L}(X)$), we'll prove

$$\sigma(A \restriction \text{Ran}(P)) = \sigma_1, \qquad \sigma(A \restriction \text{Ran}(1-P)) = \sigma_2 \tag{2.3.2}$$

We'll especially analyze the case where σ_1 is a single point, λ_0, that is, we'll analyze isolated points of $\sigma(A)$.

We'll need the language of chains and winding numbers as summarized in Section 1.2 of this volume and in Chapter 4 of Part 2A. For a compact set, $K \subset \Omega$, an open subset of $\mathbb{C}$, we'll be interested in chains, Γ, with $\text{Ran}(\Gamma) \subset \Omega \setminus K$, for which the winding numbers, $n(\Gamma, z)$, obey

$$\begin{aligned} n(\Gamma, z) &= 0 && \text{if } z \notin \Omega \\ n(\Gamma, z) &= 1 && \text{if } z \in K \\ n(\Gamma, z) &= 0 \text{ or } 1 && \text{if } z \in \Omega \setminus (K \cup \text{Ran}(\Gamma)) \end{aligned}$$

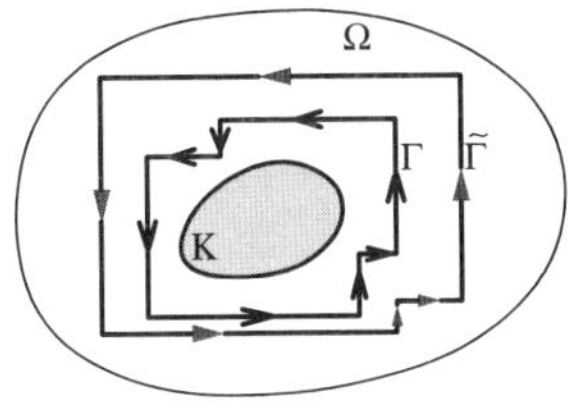

Figure 2.3.1. Graphical illustration of (2.3.3).

We'll call such a chain (Ω, K) *admissible.* In Section 4.4, especially Theorem 4.4.1, of Part 2A, we constructed admissible chains when Ω is connected. In the general case, using compactness, we can cover K by $\Omega_1, \dots, \Omega_\ell$ with each Ω_j open and connected, and $\Omega_j \cap \Omega_k = \emptyset$ if $j \neq k$. If Γ_j is $(\Omega_j, \Omega_j \cap K)$ admissible, $\Gamma_1 + \cdots + \Gamma_\ell$ is (Ω, K) admissible.

Given Γ which is (Ω, K) admissible, let $\widetilde{K} = \text{Ran}(\Gamma) \cup \{z \in \mathbb{C} \setminus \text{Ran}(\Gamma) \mid n(\Gamma, z) = 1\} \subset \Omega$. Let $\widetilde{\Gamma}$ be $(\Omega, \widetilde{K})$ admissible. Then $\widetilde{\Gamma}$ is (Ω, K) admissible, $\text{Ran}(\widetilde{\Gamma}) \cap \text{Ran}(\Gamma) = \emptyset$, and (see Figure 2.3.1)

$$z \in \text{Ran}(\Gamma) \Rightarrow n(\widetilde{\Gamma}, z) = 1, \qquad z \in \text{Ran}(\widetilde{\Gamma}) \Rightarrow n(\Gamma, z) = 0 \tag{2.3.3}$$

That is, Γ surrounds K and $\widetilde{\Gamma}$ surrounds Γ and K.

Theorem 2.3.1 (Analytic Functional Calculus). *Let $\mathfrak{A}$ be a Banach algebra with identity and $x \in \mathfrak{A}$. Let $\mathcal{F}(x)$ be the family of all functions, f, analytic in a neighborhood, N_f, of $\sigma(x)$.*

(a) *For $f \in \mathcal{F}(x)$, let Γ be $(N_f, \sigma(x))$ admissible. Then*

$$f(x) = \frac{1}{2\pi i} \oint_\Gamma f(z)(z - x)^{-1}\, dz \tag{2.3.4}$$

is independent of the choice of Γ.

(b) *If $f_1 = f_2$ on $N_{f_1} \cap N_{f_2}$, then $f_1(x) = f_2(x)$.*

(c) *If $f, g \in \mathcal{F}(x)$ and fg is defined on $N_f \cap N_g$ by $fg(z) = f(z)g(z)$, then $fg(x) = f(x)g(x)$.*

(d) *If $f_n, f_\infty \in \mathcal{F}(x)$ and $f_n \to f_\infty$ in the sense that there is a single open set U containing $\sigma(x)$ so that $f_n \to f_\infty$ uniformly on compact subsets of U, then $\|f_n(x) - f_\infty(x)\| \to 0$.*

(e) *If $f(z) = \sum_{n=0}^\infty a_n z^n$ has radius of convergence strictly bigger than* spr(x), *then*

$$f(x) = \sum_{n=0}^\infty a_n x^n \tag{2.3.5}$$

(f) *If $\lambda \in \rho(x)$ and $f(z) = (z - \lambda)^{-1}$, then $f(x) = (x - \lambda)^{-1}$.*

(g) *Moreover, if $G\colon \mathcal{F}(x) \to \mathfrak{A}$ is an algebra homomorphism with $G(z) = x$, where z is the function $f(z) = z$ and which obeys* (b) *and* (d), *then $G(f) = f(x)$.*

Remark. If x is $w_0\mathbb{1}$ with $w_0 \in \mathbb{C}$, then $\sigma(x) = \{w_0\}$ and (2.3.4) is $f(w_0)$ by the Cauchy integral formula.

Proof. (a) Let Γ_1, Γ_2 be two admissible contours. Then

$$n(\Gamma_1 - \Gamma_2, z) = n(\Gamma_1, z) - n(\Gamma_2, z)$$

is zero–zero if $z \in \mathbb{C} \setminus N_f$ and is one–one if $z \in \sigma(x)$. Thus, $\Gamma_1 - \Gamma_2$ is homologous to zero in $N_f \setminus \sigma(x)$. Since $f(z)(x-z)^{-1}$ is analytic in $N_f \setminus \sigma(x)$, the ultimate Cauchy theorem (Theorem 4.1.1 of Part 2A) implies that

$$(2\pi i)^{-1} \int_{\Gamma_1 - \Gamma_2} f(z)(z-x)^{-1}\, dz = 0 \tag{2.3.6}$$

implying the integral in (2.3.4) is independent of the admissible Γ chosen.

(b) Pick Γ which is $(N_{f_1} \cap N_{f_2}, \sigma(x))$ admissible. Then it is $(N_{f_1}, \sigma(x))$ and $(N_{f_2}, \sigma(x))$ admissible, so $f_1(x)$ and $f_2(x)$ can be defined using Γ. But then the integrals are identical.

(c) Let $U = N_f \cap N_g$. Find $\Gamma, \widetilde{\Gamma}$ admissible for $(U, \sigma(x))$ so (2.3.3) holds (as constructed before the statement of the theorem). By (a),

$$f(x) = \frac{1}{2\pi i} \oint_\Gamma f(z)(x-z)^{-1}\, dz, \qquad g(x) = \frac{1}{2\pi i} \oint_{\widetilde{\Gamma}} f(w)(x-w)^{-1}\, dw \tag{2.3.7}$$

so using

$$(z-x)^{-1}(w-x)^{-1} = \left[(z-x)^{-1}(w-z)^{-1} - (w-x)^{-1}(w-z)^{-1}\right] \tag{2.3.8}$$

we have

$$f(x)g(x) = \left(\frac{1}{2\pi i}\right)^2 \oint_{z\in\Gamma} \oint_{w\in\widetilde{\Gamma}} f(z)g(w)\, H(z,w,x)\, dz\, dw \tag{2.3.9}$$

where

$$H(z,w,x) = \left[(z-x)^{-1}(w-z)^{-1} - (w-x)^{-1}(w-z)^{-1}\right] \tag{2.3.10}$$

For the first term in (2.3.9), we use that, by (2.3.3), for $z \in \text{Ran}(\Gamma)$,

$$\frac{1}{2\pi i} \oint_{w\in\widetilde{\Gamma}} g(w)(w-z)^{-1}\, dw = g(z) \tag{2.3.11}$$

and for the second term that, by (2.3.3), for $w \in \text{Ran}(\widetilde{\Gamma})$,

$$\frac{1}{2\pi i} \oint_{z\in\Gamma} f(z)(w-z)^{-1}\, dz = 0 \tag{2.3.12}$$

Therefore,

$$f(x)g(x) = \frac{1}{2\pi i} \oint_{z\in\Gamma} f(z)g(z)(z-x)^{-1}\, dz = fg(x) \tag{2.3.13}$$

(d) Fix Γ which is $(U, \sigma(x))$ admissible. Since $\mathrm{Ran}(\Gamma)$ is compact and in U, $f_n(z) \to f(z)$ uniformly on Γ, so

$$\left\| \int f_n(z)(x-z)^{-1}\,dz - \int f_\infty(z)(z-x)^{-1}\,dz \right\| \leq \sup_{\mathrm{Ran}(\Gamma)} |f_n(z) - f_\infty(z)| \int \|(z-x)^{-1}\|\,dz$$

goes to zero since, by continuity, $\|(z-x)^{-1}\|$ is bounded on $\mathrm{Ran}(\Gamma)$.

(e) By (d), since the power series converge uniformly on compact subsets of its circle of convergence, it suffices to prove the result for $f(z) = z^n$. Let Γ be the circle of radius $1 + \mathrm{spr}(x)$. On Γ,

$$(z-x)^{-1} = \sum_{k=0}^{\infty} z^{-k-1} x^k \tag{2.3.14}$$

so if $f(z) = z^n$,

$$f(x) = \frac{1}{2\pi i} \sum_{k=0}^{\infty} \oint_{|z|=1+\mathrm{spr}(x)} z^{-k-1} z^n x^k\,dz = x^n \tag{2.3.15}$$

(f) Let $g(z) = (z - \lambda)$. Then $f(z)g(z) = g(z)f(z) \equiv 1$. Thus, by (b) and (c) (which says $g(x) = (x-\lambda)$, $(x-\lambda)f(x) = f(x)(x-\lambda) = 1$, so $f(x) = (x-\lambda)^{-1}$.

(g) To prove uniqueness, we note that, since for $z_0 \notin \sigma(x)$, $(z-z_0)(z-z_0)^{-1} = 1$, $G((\cdot - z_0)^{-1})$ is a two-sided inverse for $(x - z_0)$, that is, by (f), $G(x) = f(x)$ for f any simple pole outside $\sigma(x)$ and so, by the algebra homomorphism property and partial fraction expansion, it holds for any rational function whose poles are outside $\sigma(x)$.

If $f \in \mathcal{F}(x)$, pick Γ which is $(N_f, \sigma(x))$ admissible and let $U = \{z \mid z \notin \mathrm{Ran}(\Gamma), N(\Gamma, z) = 1\}$. By the ultimate Cauchy integral formula (Theorem 4.1.2 of Part 2A), for all $z \in U$,

$$f(z) = \frac{1}{2\pi i} \oint_\Gamma \frac{f(w)}{w-z}\,dw \tag{2.3.16}$$

showing that f is a limit of rational functions with poles in $\mathrm{Ran}(\Gamma)$ uniformly on compact subsets of U (approximate the integral by Riemann sums). Thus, $G(f) = f(x)$ for all $f \in \mathcal{F}(x)$. $\square$

Remark. See Problem 1 for a direct calculation of (f).

Theorem 2.3.2 (Spectral Mapping Theorem, Second Version). *Let $\mathfrak{A}$ be a Banach algebra with identity and $x \in \mathfrak{A}$. Let $f \in \mathcal{F}(x)$. Then*

$$\sigma(f(x)) = f\big[\sigma(x)\big] \equiv \{f(\lambda) \mid \lambda \in \sigma(x)\} \tag{2.3.17}$$

Proof. Suppose first $\lambda \notin f[\sigma(x)]$. Since $\sigma(x)$ is compact, $f[\sigma(x)]$ is compact, and so, closed. Thus, by continuity, there is U open with $U \subset N_f$ so that $\lambda \notin f[U]$. Thus,

$$g(z) = (f(z) - \lambda)^{-1} \tag{2.3.18}$$

is analytic on U, and so,

$$g(x)(f(x) - \lambda) = (f(x) - \lambda)g(x) = 1 \tag{2.3.19}$$

proving that $\lambda \notin \sigma(f(x))$.

Suppose next that $\lambda \in f[\sigma(x)]$, say $\lambda = f(\mu)$, where $\mu \in \sigma(x)$. Then $g(z) \equiv (f(z) - f(\mu))/(z - \mu)$ has a removable singularity at $z = \mu$ and so is analytic in N_f. We have

$$f(x) - f(\mu) = (x - \mu)g(x) = g(x)(x - \mu) \tag{2.3.20}$$

By Lemma 2.2.15, if $(f(x) - f(\mu))$ were invertible, so would $x - \mu$ be. Since $\mu \in \sigma(x)$, $f(x) - f(\mu)$ is not invertible, that is, $f(\mu) = \lambda \in \sigma(f(x))$. $\square$

Theorem 2.3.3. *Let $\sigma(x) = \sigma_1 \cup \sigma_2$, where σ_j are nonempty, disjoint closed sets. Then there is a neighborhood, U, of σ so that $U = U_1 \cup U_2$ with $\sigma_j \subset U_j$, U_1, U_2 open and disjoint. If*

$$f(z) = \begin{cases} 1, & z \in U_1 \\ 0, & z \in U_2 \end{cases} \tag{2.3.21}$$

then $f(x) \equiv p$ is a projection, with $p \neq 0$, $p \neq 1$. For any $\alpha, \beta \in \mathbb{C}$, we have

$$\sigma(xp + \alpha(1 - p)) = \sigma_1 \cup \{\alpha\} \tag{2.3.22}$$

$$\sigma(x(1 - p) + \beta p) = \sigma_2 \cup \{\beta\} \tag{2.3.23}$$

Proof. Since σ_1, σ_2 are compact and disjoint, $\delta \equiv \operatorname{dist}(\sigma_1, \sigma_2) > 0$. If $U_j = \{z \mid \operatorname{dist}(z, \sigma_j) < \frac{\delta}{2}\}$, the U_j are open and disjoint, with $\sigma_j \subset U_j$.

Since $f(z)^2 = f(z)$, we have $p^2 = p$ so p is a projection. By the spectral mapping theorem, $\sigma(p) = \{0, 1\}$, so $p \neq 0$, $p \neq 1$.

Let

$$g(z) = \begin{cases} z, & z \in \sigma_1 \\ \alpha, & z \in \sigma_2 \end{cases} \tag{2.3.24}$$

so $g(z) = zf(z) + \alpha(1 - f(z))$. Thus,

$$g(x) = xp + \alpha(1 - p) \tag{2.3.25}$$

By the spectral mapping theorem,

$$\sigma\big(xp + \alpha(1 - p)\big) = g(\sigma_1) \cup g(\sigma_2) = \sigma_1 \cup \{\alpha\}$$

The proof of (2.3.23) is identical. $\square$

Now, look at $\mathfrak{A} = \mathcal{L}(X)$ for a Banach space, X. Let $Y = \operatorname{Ran}(P)$, $Z = \operatorname{Ran}(1-P)$. By Proposition 2.1.2, the direct sum norm on $Y \oplus Z$ is equivalent to the norm of X. If $B \in \mathcal{L}(Y)$, $C \in \mathcal{L}(Z)$ and $B \oplus C \in \mathcal{L}(X)$ is defined by

$$(B \oplus C)(y+z) = By + Cz \tag{2.3.26}$$

It is easy to see (Problem 2) that

$$\sigma(B \oplus C) = \sigma(B) \cup \sigma(C) \tag{2.3.27}$$

Theorem 2.3.4 (Spectral Localization Theorem). *Let $A \in \mathcal{L}(X)$ for a Banach space, X. Suppose $\sigma(A) = \sigma_1 \cup \sigma_2$, where σ_1, σ_2 are closed and disjoint. Let P be defined as in Theorem 2.3.3. Then AP (respectively, $A(1-P)$) maps* Ran(P) *(respectively,* Ran$(1-P)$*) to itself and*

$$\sigma(AP \restriction \operatorname{Ran}(P)) = \sigma_1 \tag{2.3.28}$$

$$\sigma(A(1-P) \restriction \operatorname{Ran}(1-P)) = \sigma_2 \tag{2.3.29}$$

Proof. Pick $\alpha \in \sigma_2$. Then by (2.3.27),

$$\sigma\bigl(AP + \alpha(1-P)\bigr) = \sigma(AP \mid \operatorname{Ran}(P)) \cup \{\alpha\} \tag{2.3.30}$$

while on the other hand, by Theorem 2.3.3, it is equal to $\sigma_1 \cup \{\alpha\}$. Since $\alpha \notin \sigma_1$, we conclude

$$\sigma_1 \subset \sigma\bigl(AP \restriction \operatorname{Ran}(P)\bigr) \tag{2.3.31}$$

On the other hand, if $\alpha \in \sigma_1$, by Theorem 2.3.3,

$$\sigma(AP + \alpha(1-P)) = \sigma_1$$

But, by (2.3.30), which still holds by (2.3.27), we have

$$\sigma(AP \restriction \operatorname{Ran}(P)) \subset \sigma_1 \tag{2.3.32}$$

proving (2.3.28). The proof of (2.3.29) is the same. □

Here is what we can say about isolated points.

Theorem 2.3.5. *Let A be a Banach algebra with identity and $x \in \mathfrak{A}$. Let λ_0 be an isolated point of $\sigma(x)$, that is, $\lambda_0 \in \sigma(x)$ and* dist$(\lambda_0, \sigma(x)\setminus\{\lambda_0\}) \equiv \delta > 0$. *Let p be the projection of Theorem 2.3.3 with $\sigma_1 = \{\lambda_0\}$, and let*

$$d = g(x) \tag{2.3.33}$$

$$g(z) = \begin{cases} z - \lambda_0, & \operatorname{dist}(z, \lambda_0) < \frac{\delta}{2} \\ 0, & \operatorname{dist}(z, \sigma(x) \setminus \{\lambda_0\}) < \frac{\delta}{2} \end{cases} \tag{2.3.34}$$

Then,

(a) $dp = pd = d$. (2.3.35)

(b) *d is quasinilpotent, that is, $\|d^n\|^{1/n} \to 0$.*

(c) $xp = \lambda_0 p + d.$ (2.3.36)

(d) *The negative terms of the Laurent series of* $(z-x)^{-1}$ *near* $z = \lambda_0$ *are*

$$(z-x)^{-1} = (z-\lambda_0)^{-1}p + \sum_{n=1}^{\infty}(z-\lambda_0)^{-n-1}d^n \tag{2.3.37}$$

Remark. Thus, the *spectral projections*, p, are the residues of the poles of the resolvent.

Proof. (a) Let f be given by (2.3.21). Then $fg = g$, proving (2.3.35).

(b) Since g vanishes on $\sigma(x)$, $\sigma(d) = \{0\}$ by the spectral mapping theorem, so $\|d^n\|^{1/n} \to 0$ by the spectral radius formula.

(c) $g(z) + \lambda_0 f(z) = zf(z)$, proving (2.3.36).

(d) The negative Laurent coefficients of $(z-x)^{-1}$ are given by $\sum_{n=1}^{\infty} a_n(z-\lambda_0)^{-n}$, where

$$a_n = \frac{1}{2\pi i}\oint_{|z-\lambda_0|=\delta/2}(z-\lambda_0)^n(z-x)^{-1}\,dz = g_n(x)$$

where $g_n(z) = g(z)^n$, so $g_n = g(x)^n = d^n$. □

Definition. Let $A \in \mathcal{L}(X)$, the bounded operators on a Banach space, X. $\sigma_d(A)$, the *discrete spectrum* of A, is the set of $\lambda \in \sigma(A)$ which are isolated points of the spectrum for which the projection, $P = (2\pi i)^{-1}\oint_{|z-\lambda_0|=\delta/2}(z-x)^{-1}\,dz$, has a finite dimension, which is called the *algebraic multiplicity* of λ_0. The algebraic multiplicity is always at least as large as the geometric multiplicity (see below).

In (1.3.61), we showed that if a finite matrix, B, on a space of dimension ℓ has only 0 as an eigenvalue, then $B^\ell = 0$, that is, B is nilpotent. We thus have the following immediately from this and Theorem 2.3.5:

Corollary 2.3.6. *Let* $A \in \mathcal{L}(X)$ *and* $\lambda_0 \in \sigma_d(A)$. *Then*

$$\lambda_0 \notin \sigma(A(1-P) \restriction \operatorname{Ran}(1-P)) \tag{2.3.38}$$

$$AP = \lambda_0 P + N \tag{2.3.39}$$

$$N^{\dim(\lambda_0)} = 0 \tag{2.3.40}$$

In case X has total dimension $\ell < \infty$, it has only discrete spectrum, say $\{\lambda_j\}_{j=1}^J$. We have $P_jP_k = 0$ if $j \neq k$ and $\sum_{j=1}^J P_j = 1$, both by the functional calculus. Thus, $\sum_{j=1}^J \dim(\lambda_j) = \ell$. We have

$$A = \sum_{j=1}^{J}(\lambda_j P_j + N_j)$$

The analysis of nilpotent operators in Theorem 1.3.2 yields the Jordan normal form.

As noted, if z_0 is an isolated point of the spectrum of any operator, A, and P is given by (2.3.71) below, then $\dim(P)$ is called the algebraic multiplicity. If $\dim(P) = n < \infty$, it can be shown (Problem 3) that

$$\operatorname{Ran}(P) = \{\varphi \in X \mid (A - z_0)^n \varphi = 0\} \tag{2.3.41}$$

called *generalized eigenvectors*. The usual eigenspace, that is,

$$\{\varphi \in X \mid (A - z_0)\varphi = 0\} \tag{2.3.42}$$

has a dimension called the *geometric multiplicity* of z_0 as an eigenvalue. The P_j are called *spectral projections* or *eigenprojections* and the N_j are called *eigennilpotents*.

Since $A \restriction \operatorname{Ran}(P_j) = \lambda_j + N_j$, we see

$$\begin{aligned} \text{geo. mult. } (\lambda_j) &= \dim(\operatorname{Ker}(N_j) \cap \operatorname{Ran}(P_j)) \\ &= \operatorname{rank}(P_j) - \operatorname{rank}(N_j) && (2.3.43) \\ &\leq \operatorname{rank}(P_j) = \text{alg. mult. } (\lambda_j) && (2.3.44) \end{aligned}$$

by (2.1.15).

Here is a weaker version of (2.3.41) that doesn't require that z_0 be an isolated point of $\sigma(A)$ nor that the spectral projection be finite rank.

Proposition 2.3.7. *Let $A \in \mathcal{L}(X)$. Let U be the unbounded component of $\mathbb{C} \setminus \sigma(A)$. Let σ_1 be a closed subset of $\mathbb{C} \setminus U$ with $\partial\sigma_1 \subset \partial\sigma$. Let P be the projection*

$$P = (2\pi i)^{-1} \oint_\Gamma (z - A)^{-1}\, dx \tag{2.3.45}$$

where Γ surrounds σ_1 and $\Gamma \cap \sigma = \emptyset$. Let $z_0 \in \sigma_1$. Then

$$\overline{\{\varphi \mid \lim_{n\to\infty} \|(A - z_0)^n \varphi\|^{1/n} = 0\}} \subset \operatorname{Ran} P \tag{2.3.46}$$

Remark. Taking $z_0 = 0$ when A is compact (discussed in Chapter 3) will imply if $\lambda_0 \neq 0$, $\lambda_0 \in \sigma(A)$, then $P_{\lambda_0}\varphi = 0$ if φ is in the set on the left side of (2.3.46).

Proof. By replacing A by $A - z_0$, we can suppose $z_0 = 0$. For $|z| > \|A\|$, and any φ,

$$\begin{aligned} (A - z)^{-1}\varphi &= (-z)^{-1} (1 - z^{-1}A)^{-1} \varphi \\ &= -\sum_{n=0}^{\infty} z^{-1-n} A^n \varphi && (2.3.47) \end{aligned}$$

If also $\|A^n \varphi\|^{1/n} \to 0$, this function, call it $g(z)$, has an analytic continuation to $\mathbb{C} \setminus \{0\}$ and by analytic continuation for any $z_1 \neq 0$, $(A - z_1)g(z_1) = \varphi$. In particular, $g(z) = (A - z_1)^{-1} \varphi$ if $z_1 \in \Gamma$.

Since (2.3.47) converges uniformly on Γ, we can do the $(2\pi i)^{-1} \oint_\Gamma z^{-1-n}\, dz = \delta_{n0}$ integrals term by term and see that for such φ, $P\varphi = \varphi$. Since P is bounded, this extends to the closure. $\square$

If λ_0 is an isolated point of $\sigma(A)$, we can pick $f(z) = 0$ if $|z - \lambda_0| < \varepsilon$, $f(z) = (x-z)^{-1}$, $|z - \lambda_0| > 2\varepsilon$ where ε is picked with $\text{dist}(\lambda_0, \sigma(A) \setminus \{\lambda_0\}) > 2\varepsilon$. If $B = f(A)$, the functional calculus shows that

$$B(A - \lambda_0) = (A - \lambda_0)B = 1 - P_{\lambda_0} \tag{2.3.48}$$

where P_{λ_0} is the spectral projection for λ_0. B is sometimes called the *reduced resolvent*.

Theorem 2.3.8. *Let $A \in \mathcal{L}(X)$. If λ_0 is an isolated point of $\sigma(A)$, then*

$$\text{Ran}(1 - P_{\lambda_0}) \subset \text{Ran}(A - \lambda_0) \tag{2.3.49}$$

Proof. If $\varphi \in \text{Ran}(1 - P_{\lambda_0})$, by (2.3.48)

$$\varphi = (A - \lambda_0)B\varphi$$

lies in $\text{Ran}(A - \lambda_0)$. $\square$

An isolated point $\lambda_0 \in \sigma(A)$ is said to have *index one*, if $(A - \lambda_0)^2 x = 0 \Rightarrow (A - \lambda_0)x = 0$ (thus, if $\dim(\text{Ran}(P_{\lambda_0})) < \infty$, there is no eigennilpotent).

Corollary 2.3.9. *Suppose $A \in \mathcal{L}(X)$ and that $\lambda_0 \in \sigma_d(A)$ has index one. Then*

$$\text{Ker}(A - \lambda_0) \dotplus \text{Ran}(A - \lambda_0) = X \tag{2.3.50}$$

Proof. By the index-one hypothesis and (2.3.41)

$$\text{Ran}\, P_{\lambda_0} = \text{Ker}(A - \lambda_0) \tag{2.3.51}$$

Thus, by (2.3.49)

$$x = P_{\lambda_0} x + (1 - P_{\lambda_0})x \tag{2.3.52}$$

writes any x as a sum of $y \in \text{Ker}(A - \lambda_0)$ and $z \in \text{Ran}(A - \lambda_0)$. Therefore we need only prove that

$$\text{Ker}\big((A - \lambda_0)\big) \cap \text{Ran}\big((A - \lambda_0)\big) = \{0\} \tag{2.3.53}$$

Let x_0 be in the intersection, so for some y

$$(A - \lambda_0)x_0 = 0, \quad (A - \lambda_0)y = x_0 \tag{2.3.54}$$

Thus $(A - \lambda_0)^2 y = 0$ so $x_0 = (A - \lambda_0)y = 0$, by the index-one hypothesis. This proves (2.3.53). $\square$

One remarkable fact is that perturbation theory for points in the discrete spectrum can be reduced to a problem in finite-dimensional perturbation theory and thereby allow an easy extension of the results in Section 1.2. The key is the following:

Theorem 2.3.10. *Let P,Q be two bounded projections on a Banach space, X, with*

$$\|P-Q\|<1 \tag{2.3.55}$$

Then there exists an invertible $U\in\mathcal{L}(X)$ (with $U^{-1}\in\mathcal{L}(X)$) so that

$$Q=UPU^{-1},\qquad P=U^{-1}QU \tag{2.3.56}$$

Moreover,

(a) *If X is a Hilbert space and P,Q are orthogonal, the U can be chosen to be unitary.*

(b) *If $Q(\beta)$ is a $\|\cdot\|$-continuous (respectively, C^k, C^∞, analytic) function of β, for β so small that*

$$\|Q(\beta)-Q(0)\|<1 \tag{2.3.57}$$

one can choose $U(\beta)$ which is continuous (respectively, C^k, C^∞, analytic) so that

$$Q(\beta)=U(\beta)^{-1}Q(0)U(\beta) \tag{2.3.58}$$

Remarks. 1. The β parameter can lie in $\mathbb{R}^\nu$, a manifold, or for continuity, any topological space. "Analytic" can mean either real analytic or complex analytic.

2. Example 3.15.19 will discuss pairs of projections further, and again there, $A=P-Q$ will play a central role.

Proof. Define

$$A=P-Q,\quad R=A^2=P+Q-PQ-QP \tag{2.3.59}$$
$$=\mathbb{1}-[QPQ+(\mathbb{1}-Q)(\mathbb{1}-P)(\mathbb{1}-Q)]$$
$$=\mathbb{1}-[PQP+(\mathbb{1}-P)(\mathbb{1}-Q)(\mathbb{1}-P)] \tag{2.3.60}$$

by a direct calculation. Thus,

$$PR=P-PQP=RP \text{ and } RQ=QR \tag{2.3.61}$$

Define

$$\widetilde{U}=QP+(\mathbb{1}-Q)(\mathbb{1}-P),\qquad \widetilde{V}=PQ+(\mathbb{1}-P)(\mathbb{1}-Q) \tag{2.3.62}$$

By (2.3.60),

$$\widetilde{U}P=Q\widetilde{U},\qquad \widetilde{U}\widetilde{V}=\widetilde{V}\widetilde{U}=\mathbb{1}-R \tag{2.3.63}$$

Since $\|R\|<1$, $\mathbb{1}-R$ is invertible (indeed, $(\mathbb{1}-R)^{-1}=\mathbb{1}+R+R^2+\dots$). Since $\widetilde{U},\widetilde{V}$ commute with R, they commute with $(\mathbb{1}-R)^{-1}$, so $\widetilde{V}(\mathbb{1}-R)^{-1}=$

$(\mathbb{1}-R)^{-1}\widetilde{V}$ is a two-sided inverse for $\widetilde{U}$, and one could take $U=\widetilde{U}$ for the main result and for (b). But $\widetilde{U}$ is not unitary if P,Q are orthogonal projections on a Hilbert space, so we go a step further.

Since $\|R\|<1$, one can define $(\mathbb{1}-R)^{-1/2}$ by a convergent power series

$$(\mathbb{1}-R)^{-1/2}=\sum_{n=0}^{\infty}\frac{(-\frac{1}{2})(-\frac{1}{2})\dots(-n+\frac{1}{2})}{n!}(-R)^n \tag{2.3.64}$$

(where the $n=0$ term is $\mathbb{1}$). Now define

$$U=\widetilde{U}(\mathbb{1}-R)^{-1/2}=(\mathbb{1}-R)^{-1/2}\widetilde{U},\qquad V=\widetilde{V}(\mathbb{1}-R)^{-1/2}=(\mathbb{1}-R)^{-1/2}\widetilde{V} \tag{2.3.65}$$

Then

$$UV=VU=\mathbb{1},\qquad UP=QU \tag{2.3.66}$$

If $X=\mathcal{H}$, a Hilbert space, and $P=P^*$, $Q=Q^*$, then $R^*=R$ (by (2.3.59)), $(\widetilde{U})^*=\widetilde{V}$, so $U^*=V$ and U is unitary. □

Theorem 2.3.11. *Let $A(\beta)$ be a family of bounded operators on a Banach space, X, so that in $\|\cdot\|$, $A(\beta)$ is continuous (respectively, C^k, C^∞, analytic). Suppose λ_0 is a point of the discrete spectrum of $A(0)$. Then for β small, there exist a neighborhood, N, of λ_0 and invertible $U(\beta)$ with $U(0)=\mathbb{1}$ so that*

$$\widetilde{A}(\beta)=U(\beta)A(\beta)U(\beta)^{-1} \tag{2.3.67}$$

leaving $\operatorname{Ran}(P(0))$ *and* $\operatorname{Ran}(1-P(0))$ *setwise fixed, where $P(0)$ is the spectral projection associated to λ_0 and $A(0)$. All the spectrum of $\widetilde{A}(\beta)$ (and so of $A(\beta)$) in N is precisely all the spectrum of $\widetilde{A}(\beta)\restriction\operatorname{Ran}(P(0))$. Moreover, $U(\beta)$, $U(\beta)^{-1}$, and $\widetilde{A}(\beta)$ are continuous (respectively, C^k, C^∞, analytic) in β.*

Remarks. 1. Thus, the spectrum of $A(\beta)$ in N is given by the eigenvalues of a β-dependent family on a fixed finite-dimensional space (namely, $\operatorname{Ran}(P(0))$), so all the results of Section 1.2 apply. Notice that even if $A(\beta)$ is linear in β, $\widetilde{A}(\beta)$ will only be analytic, which is why one wants to do eigenvalue perturbation theory for analytic families.

2. If $A(\beta)$ is self-adjoint for $\beta\in[-\delta,\delta]$, then for small real β, $U(\beta)$ can be chosen unitary so $\widetilde{A}(\beta)$ is also self-adjoint and, for example, Rellich's theorem, Theorem 1.4.4, applies.

Proof. Since the invertible operators are open, if we pick ε so small that $\overline{\mathbb{D}}_\varepsilon(\lambda_0)\cap\sigma(A(0))=\{\lambda_0\}$, then for some δ small and $|\beta|<\delta$,

$$\partial\mathbb{D}_\varepsilon(\lambda_0)\cap\sigma(A(\beta))=\emptyset \tag{2.3.68}$$

Thus,

$$P(\beta) = \frac{1}{2\pi i} \oint_{|\lambda-\lambda_0|=\varepsilon} \frac{d\lambda}{\lambda - A(\beta)} \tag{2.3.69}$$

is continuous (respectively, C^k C^∞, analytic) in β. In particular, by shrinking δ, we can suppose that

$$|\beta| < \delta \Rightarrow \|P(\beta) - P(0)\| < 1$$

By Theorem 2.3.10, find $U(\beta)$ invertible (and unitary if $A(\beta)$ is self-adjoint) for $\beta \in [-\delta, \delta]$ since then $P(\beta)$ is self-adjoint, so that

$$P(\beta) = U(\beta)^{-1} P(0) U(\beta) \tag{2.3.70}$$

This implies that $A(\beta)$ given by (2.3.68) has $P(0)$ as its spectral projection for spectrum inside $\mathbb{D}_\varepsilon(\lambda_0)$.

Since $U(\beta)$ is continuous (respectively, C^k, C^∞, analytic), so is $\widetilde{A}(\beta)$.

By the spectral localization theorem (Theorem 2.3.4), all the spectrum of $\widetilde{A}(\beta)$ near λ_0 is spectrum of $\widetilde{A}(\beta) \restriction \text{Ran}(P(0))$. $\square$

Notes and Historical Remarks. That the formula

$$P = \frac{1}{2\pi i} \oint_{|z-z_0|=\varepsilon} \frac{dz}{z - x} \tag{2.3.71}$$

defines a projection when z_0 is an isolated point of $\sigma(x)$ goes back to Nagumo [**488**]. In an early 1913 book, Riesz [**564**] talked about analyticity of $A(1 - zA)^{-1}$ and mentioned that one could apply the calculus of residues, but didn't do so explicitly then! In 1930, Riesz [**567**] noted in the Hilbert space case that a decomposition of $\sigma(A)$ into disjoint closed sets induced a projection, i.e., Theorem 2.3.3.

The general functional calculus in a Banach algebra setting defined by (2.3.4) goes back to Gel'fand's great 1941 paper [**228**] that presented his theory of commutative Banach algebras, the subject of Chapter 6. These ideas were developed (in some cases rediscovered) in the United States in 1942–43 by Dunford [**173, 174**], Lorch [**457**], and Taylor [**698**]. In particular, Dunford stated and proved the general spectral mapping theorem and Lorch emphasized projections associated to splitting $\sigma(x) = \sigma_1 \cup \sigma_2$ into disjoint closed sets.

The functional calculus has been extended to analytic functions of several commuting operators, more generally, several elements of a commutative Banach algebra, using their joint spectrum (see Section 6.2). This is discussed by Arens–Calderón [**23**], Šilov [**625**], Taylor [**699**], Waelbroeck [**729**], and Zame [**771**].

The pairs of projections theorem and its use to turn infinite-dimensional perturbations of discrete spectrum to finite-dimensional perturbation theory is due to Kato, initially in a 1955 unpublished technical report [**374**]; the standard reference is now his book [**380**]; see also Baumgärtel [**47**]. A similar idea was earlier exploited by Sz.-Nagy [**689**].

Problems

1. Let $\lambda_0 \notin \sigma(x)$. Prove directly that if Γ is $(\mathbb{C}, \sigma(x))$ admissible and is chosen so $n(\Gamma, \lambda_0) = 0$, then

$$\frac{1}{2\pi i} \oint_\Gamma (z - \lambda_0)^{-1}(z - x)^{-1}\, dz = (\lambda_0 - x)^{-1}$$

(*Hint*: Write $(z-\lambda_0)^{-1}(z-x)^{-1}$ in terms of $(\lambda_0 - x)^{-1}(z-\lambda_0)^{-1}$ and $(\lambda_0 - x)^{-1}(z-x)^{-1}$ and use that $(2\pi i) \oint_\Gamma (z-x)^{-1}\, dx = \mathbb{1}$ (why?).)

2. Let Y, Z be two Banach spaces and $X = Y \oplus Z$. Given $B \in \mathcal{L}(Y)$, $C \in \mathcal{L}(Z)$, let $B \oplus C$ in $\mathcal{L}(X)$ be defined by (2.3.26).

   (a) Prove $(B_1 \oplus C_1)(B_2 \oplus C_2) = B_1B_2 \oplus C_1C_2$.

   (b) If $\lambda \in \rho(B) \cap \rho(C)$, prove that $\lambda \in \rho(B \oplus C)$.

   (c) Prove $(B \oplus C - \lambda) = [(B - \lambda) \oplus 1][1 \oplus (C - \lambda)]$ and conclude that if $\lambda \in \rho(B \oplus C)$, then $\lambda \in \rho(B) \cap \rho(C)$.

   (d) Conclude (2.3.27).

3. Let z_0 be an isolated point of $\sigma(A)$ for $A \in \mathcal{L}(X)$, with X a Banach space. Suppose $\mathrm{Ran}(P)$ has finite dimension. Prove (2.3.41) as follows:

   (a) If $\varphi \in \mathrm{Ran}(P)$, prove $(A - z_0)^{\dim(\mathrm{Ran}(P))}\varphi = 0$. (*Hint*: See Corollary 2.3.6.)

   (b) If $(A - z_0)^n\varphi = 0$, for some n, prove $\varphi \in \mathrm{Ran}(P)$. (*Hint*: Use the fact that there exists B with $B(A - z_0)(1 - P) = (1 - P)$ and $[B, A] = [B, P] = 0$ to prove $(1 - P)\varphi = 0$.)

4. Let X be a Banach space, A_n, $A \in \mathcal{L}(X)$, with $\|A_n - A\| \to 0$.

   (a) Let $\lambda \in \rho(A)$. For some $\varepsilon > 0$ and N, prove that $\mathbb{D}_\varepsilon(\lambda) \subset \rho(A_n)$ for all $n > N$.

   (b) Let $\lambda \in \sigma_d(A)$. Pick ε so small that $\sigma(A) \cap \overline{\mathbb{D}_\varepsilon(\lambda)} = \{\lambda\}$. Prove for some N and all $n > N$ that $\partial\mathbb{D}_\varepsilon(\lambda) \subset \rho(A_n)$ and the only spectrum of A_n is $\mathbb{D}_\varepsilon(\lambda)$ is discrete and the sum of the algebraic multiplicities of the points for A_n is the algebraic multiplicity for λ in $\sigma_d(A)$.

2.4. The Square Root Lemma and the Polar Decomposition

The bulk of the prior results of this chapter apply to operators on a general Banach space or even to elements of a general Banach algebra with identity. In distinction, this section applies *only* to the case of operators on a Hilbert space.

We'll look for an analog of $z = |z|\exp(i\arg z)$ for $z \in \mathbb{C}$. Our first step will be to define $|A|$ for $A \in \mathcal{L}(\mathcal{H})$. To do that, we'll first need to define what a positive operator is and then show that if $B \geq 0$, we can define $\sqrt{B}$ as the unique positive operator whose square is B. We can then define $|A| = \sqrt{A^*A}$. The polar decomposition will then say

$$A = U|A| \tag{2.4.1}$$

One might guess U unitary should be the analog of $\exp(i\arg z)$, and that is true if $\text{Ker}(A) = \{0\} = \text{Ker}(A^*)$. But if $\dim(\text{Ker}(A)) + \dim(\text{Ker}(A^*)) > 0$, examples will show there may not be a unitary U, so instead we'll define partial isometries and prove (2.4.1), called the *polar decomposition*, with U a partial isometry. With the additional condition that $\text{Ker}(U) = \text{Ker}(A)$, U will be unique. We'll end the section by analyzing the polar decomposition further when $A = A^*$ and use that analysis to prove a variant of the spectral theorem.

Definition. Let $\mathcal{H}$ be a Hilbert space and $A \in \mathcal{L}(\mathcal{H})$. We say A is *positive* (respectively, *strictly positive*), written $A \geq 0$ (respectively, $A > 0$), if and only if for all $\varphi \in \mathcal{H}$,

$$\langle \varphi, A\varphi \rangle \geq 0 \qquad \text{(respectively, } \langle \varphi, A\varphi \rangle > 0 \text{ if } \varphi \neq 0) \tag{2.4.2}$$

Remarks. 1. Some authors use "nonnegative" and "positive" where we use "positive" and "strictly positive."

2. We'll see below that $A \geq 0 \Rightarrow A = A^*$, but this argument uses the fact that $\mathcal{H}$ is a complex vector space. For real Hilbert spaces, one includes $A = A^*$ as part of the definition.

Theorem 2.4.1. *Let $A \in \mathcal{L}(\mathcal{H})$ be positive. Then*

(a) $A = A^*$

(b) *We have*

$$\|A\| = \sup_{\substack{\varphi \in \mathcal{H} \\ \|\varphi\| = 1}} \langle \varphi, A\varphi \rangle \tag{2.4.3}$$

Remarks. 1. (2.4.3) is closely related to $\|B\|^2 = \|B^*B\|$ in that, first, if one knew (we'll only prove that this is true using (2.4.3)) that there was a B with $B = B^*$ and $B^2 = A$, then $\|B^*B\|$ = right side of (2.4.3), while $\langle \varphi, A\varphi \rangle = \|B\varphi\|^2$, so $\|B\|^2$ = right side of (2.4.3). Second, the proofs are closely related.

2. We use the Schwarz inequality for semipositive inner products; see Problems 1 and 2.

Proof. (a) By sesquilinearity of $\varphi, \psi \to \langle \varphi, A\psi \rangle$, we have the analog of polarization, with $\mathcal{A}(\varphi) = \langle \varphi, A\varphi \rangle$,

$$\langle \varphi, A\psi \rangle = \tfrac{1}{4}\left(\mathcal{A}(\varphi+\psi) - \mathcal{A}(\varphi-\psi) + i\mathcal{A}(\psi+i\varphi) - i\mathcal{A}(\psi-i\varphi)\right) \tag{2.4.4}$$

Reality of $\mathcal{A}(\varphi)$ and $\mathcal{A}(e^{i\theta}\varphi) = \mathcal{A}(\varphi)$ imply that

$$\langle \varphi, A\psi \rangle = \overline{\langle \psi, A\varphi \rangle} = \langle A\varphi, \psi \rangle \tag{2.4.5}$$

proving that $A^* = A$.

(b) Since $\langle \varphi, A\varphi \rangle \geq 0$, we have a Schwarz inequality

$$|\langle \varphi, A\psi \rangle|^2 \leq \langle \varphi, A\varphi \rangle \langle \psi, A\psi \rangle \tag{2.4.6}$$

so

$$\sup_{\|\varphi\|=\|\psi\|=1} |\langle \varphi, A\psi \rangle| \leq \sup_{\|\varphi\|=1} \langle \varphi, A\varphi \rangle \tag{2.4.7}$$

Since $\|\eta\| = \sup_{\|\varphi\|=1} |\langle \varphi, \eta \rangle|$ and $\|A\| = \sup_{\|\psi\|=1} \|A\psi\|$, we see the left side of (2.4.7) is $\|A\|$, so (2.4.7) says

$$\|A\| \leq \sup_{\|\varphi\|=1} \langle \varphi, A\varphi \rangle \tag{2.4.8}$$

On the other hand, since $|\langle \varphi, A\varphi \rangle| \leq \|\varphi\| \|A\varphi\| \leq \|A\| \|\varphi\|^2$, we have

$$\sup_{\|\varphi\|=1} \langle \varphi, A\varphi \rangle \leq \|A\| \tag{2.4.9}$$

$\square$

Corollary 2.4.2. *If $A \in \mathcal{L}(\mathcal{H})$, $A \geq 0$, and $\|A\| \leq 1$, then $1 - A \geq 0$ and $\|1 - A\| \leq 1$.*

Proof. By (2.4.3), $\|A\| \leq 1$ implies that

$$\sup_{\|\varphi\|=1} \langle \varphi, A\varphi \rangle \leq 1 \tag{2.4.10}$$

so

$$\langle \varphi, (1-A)\varphi \rangle = \|\varphi\|^2 - \langle \varphi, A\varphi \rangle \tag{2.4.11}$$

$$= \|\varphi\|^2 \left(1 - \left\langle \frac{\varphi}{\|\varphi\|}, A\frac{\varphi}{\|\varphi\|} \right\rangle\right) \geq 0 \tag{2.4.12}$$

so $1 - A \geq 0$. Clearly, if $\|\varphi\| = 1$, (2.4.11) implies $\langle \varphi, (1-A)\varphi \rangle \leq 1$, so by (2.4.3), $\|1 - A\| \leq 1$. $\square$

Lemma 2.4.3. *Let $f(z)$ be defined on $\mathbb{D}$ by $f(z) = \sqrt{1-z}$, the analytic function with branch determined by $f(0) = 1$. Then*

$$f(z) = 1 - \sum_{n=1}^{\infty} c_n z^n \tag{2.4.13}$$

where

$$c_n \geq 0, \qquad \sum_{n=1}^{\infty} c_n = 1 \tag{2.4.14}$$

Remarks. 1. The key consequence is that the power series (2.4.14) converges on all of $\overline{\mathbb{D}}$ since $1 + \sum_{n=1}^{\infty} c_n = 2 < \infty$.

2. Since $f(z)^2 = 1 - z$ and the power series converge, we have

$$\left(\sum_{n=1}^{\infty} c_n z^n\right)^2 - 2\sum_{n=1}^{\infty} c_n z^n = -z \tag{2.4.15}$$

if $z \in \mathbb{D}$, or

$$c_1 = \tfrac{1}{2}, \qquad \sum_{\substack{k+\ell=m \\ k\geq 1, \ell \geq 1}} c_k c_\ell = 2c_m \text{ if } m \geq 2 \tag{2.4.16}$$

Proof. By taking derivatives of $(1-z)^{1/2}$, we see that

$$f^{(k)}(z) = \left(-\frac{1}{2}\right)\left(\frac{1}{2}\right)\left(\frac{3}{2}\right)\dots\left(\frac{2k-3}{2}\right)(1-z)^{\frac{1}{2}-k} \tag{2.4.17}$$

so $f^{(k)}(0) < 0$, that is, $c_n \geq 0$. Thus, by the monotone convergence theorem for sums,

$$\sum_{n=0}^{\infty} c_n = \lim_{t\uparrow 1} \sum_{n=1}^{\infty} c_n t^n = \lim_{t \uparrow 1}\bigl(1 - \sqrt{1-t}\bigr) = 1 \tag{2.4.18}$$

$\square$

Theorem 2.4.4 (Square Root Lemma). *Let $A \in \mathcal{L}(\mathcal{H})$ with $A \geq 0$. Then there exists a unique $B \geq 0$ so*

$$B^2 = A \tag{2.4.19}$$

Moreover, B is a norm limit of polynomials in A so, in particular,

$$[C, A] = 0 \Rightarrow [C, B] = 0 \tag{2.4.20}$$

Remarks. 1. $[C, D]$ is shorthand for $CD - DC$, called the *commutator* of C and D.

2. We'll write

$$B = \sqrt{A} \tag{2.4.21}$$

3. See Problem 11 for an alternate construction of $\sqrt{A}$ as a limit of polynomials in A.

Proof. By replacing A by $A/\|A\|$, we can suppose $\|A\| \leq 1$. In that case, by Corollary 2.4.2,

$$\|1 - A\| \leq 1 \tag{2.4.22}$$

So we can define

$$B = 1 - \sum_{n=1}^{\infty} c_n (1 - A)^n \tag{2.4.23}$$

where the sum converges in norm by (2.4.14). Thus, B is a limit of polynomials implying (2.4.20). Since this converges absolutely in norm, we can manipulate power series and use (2.4.16) to see that

$$B^2 = 1 - (1 - A) = A \tag{2.4.24}$$

That proves existence. To check uniqueness, suppose $C^2 = A$ also with $C \geq 0$. Since $[C, A] = 0$, we have $[C, B] = 0$, so $(C-B)(C+B) = C^2 - B^2 = 0$. Thus,

$$(C - B)C(C - B) + (C - B)B(C - B) = 0 \tag{2.4.25}$$

For any φ, we have with $\psi = (C - B)\varphi$,

$$\langle \psi, C\psi \rangle + \langle \psi, B\psi \rangle = 0 \tag{2.4.26}$$

so each is zero. Since $C \geq 0$, we have for all η that $|\langle \eta, C\psi \rangle|^2 \leq \langle \psi, C\psi \rangle \langle \eta, C\eta \rangle = 0$, so $C\psi = 0$, that is, we have shown

$$C(C - B) = B(C - B) = 0 \tag{2.4.27}$$

Thus, $(C - B)^2 = 0$, so for any φ,

$$\|(C - B)\varphi\|^2 = \langle \varphi, (C - B)^2 \varphi \rangle = 0$$

Thus, $B = C$. $\square$

Corollary 2.4.5. *Let $\mathcal{H}$ be a Hilbert space. Any $A \in \mathcal{L}(\mathcal{H})$ is a linear combination of four unitaries.*

Proof. By (2.1.6), A is a linear combination of two self-adjoint operators, so it suffices to prove that if $A = A^*$ and $\|A\| \leq 1$, then A is a linear combination of two unitaries.

If $\|A\| \leq 1$, $1 - A^2 \geq 0$, so

$$U_\pm = A \pm i\sqrt{1 - A^2} \tag{2.4.28}$$

can be defined. Clearly,

$$U_\pm^* = U_\mp, \qquad U_\pm U_\mp = \mathbb{1} \tag{2.4.29}$$

so $U_\pm$ are unitary and $A = \frac{1}{2}(U_+ + U_-)$ is a linear combination of two unitaries. $\square$

Definition. For any $A \in \mathcal{L}(\mathcal{H})$, we define $|A|$, the *absolute value* of A by

$$|A| = \sqrt{A^*A} \tag{2.4.30}$$

Remarks. 1. For L and R, the left and right shifts of Example 2.2.5, we see below that $\sqrt{L^*L} = \sqrt{RR^*}$ and $\sqrt{R^*R}$ are different. Thus, $|A|$ and $|A^*|$ in general may not be equal.

2. It is false in general (see Problem 3) that $|A+B| \le |A| + |B|$. It is even false (see Problem 4) that $\||A| - |B|\| \le \|A - B\|$.

We are heading towards a formula $A = U|A|$. The following example shows that U might not be unitary.

Example 2.4.6 (Example 2.2.5 (revisited)). Let L and R be the left and right shift of (2.2.18). Then $L = R^*$ and $R^*R = \mathbb{1}$, $L^*L = \mathbb{1} - \langle \delta_1, \cdot \rangle \delta_1$, where $\langle \delta_1, (a_1, a_2, \dots) \rangle = a_1$. Since $(R^*R)^2 = R^*R$ and $(L^*L)^2 = L^*L$, we see

$$|R| = \mathbb{1}, \qquad |L| = \mathbb{1} - \langle \delta_1, \cdot \rangle \delta_1 \tag{2.4.31}$$

If $R = U_R|R|$, then $U_R = R$, which is an isometry but not unitary since $\operatorname{Ran}(U_R) \neq \mathcal{H}$. It is not hard to show (Problem 5) that if $L = U_L|L|$, then

$$U_{L,\varphi} = L + \langle \delta_1, \cdot \rangle \varphi \tag{2.4.32}$$

for some φ since $\operatorname{Ran}|L|$ is orthogonal to δ_1. The choice we'll eventually make has $\varphi = 0$ and $\operatorname{Ker}(U_{L,\varphi}) = \{\beta \delta_1\}$. While $\varphi \neq 0$ has $\delta_1 \notin \operatorname{Ker}(U_{L,\varphi})$, it can be shown (Problem 6) that $\operatorname{Ker}(U_{L,\varphi}) \neq \{0\}$ for any $\varphi \neq 0$. □

Thus, we need a bigger class than unitaries.

Definition. A *partial isometry* is an operator $U \in \mathcal{L}(\mathcal{H})$ so that

$$U^*U = P_{\mathrm{I}}, \qquad UU^* = P_{\mathrm{F}} \tag{2.4.33}$$

are both (orthogonal) projections.

$$\operatorname{Ran}(P_{\mathrm{I}}) = \mathcal{H}_{\mathrm{I}}, \qquad \operatorname{Ran}(P_{\mathrm{F}}) = \mathcal{H}_{\mathrm{F}} \tag{2.4.34}$$

are called the *initial subspace* and *final subspace*, respectively.

Remark. The definition we give is equivalent to (Problem 10)

$$U = UU^*U \tag{2.4.35}$$

The names come from

Proposition 2.4.7. *If U is a partial isometry, we have*

(a) $\operatorname{Ker}(U) = \mathcal{H}_{\mathrm{I}}^{\perp}$
(b) $\operatorname{Ran}(U) = \mathcal{H}_{\mathrm{F}}$
(c) $U \restriction \mathcal{H}_{\mathrm{I}}$ *is a unitary map of* $\mathcal{H}_{\mathrm{I}}$ *onto* $\mathcal{H}_{\mathrm{F}}$.

Conversely, if U obeys (a)–(c), *then U is a partial isometry with initial subspace $\mathcal{H}_\text{I}$ and final subspace $\mathcal{H}_\text{F}$.*

Proof. Suppose first that U is a partial isometry. Then

$$\|P_\text{I}\varphi\| = \langle P_\text{I}\varphi, P_\text{I}\varphi\rangle = \langle \varphi, P_\text{I}\varphi\rangle = \langle \varphi, U^*U\varphi\rangle = \|U\varphi\|^2$$

Thus, $U\varphi = 0 \Leftrightarrow P_\text{I}\varphi = 0 \Leftrightarrow (1 - P_\text{I})\varphi = \varphi \Leftrightarrow \varphi \in \text{Ran}(1 - P_\text{I}) = \mathcal{H}_\text{I}^\perp$, since P_I is an orthogonal projection. This proves (a) and proves (c) once we prove (b).

To prove (b), we note, by (2.2.13), that $(\text{Ran}\, U)^\perp = \text{Ker}(U^*) = \text{Ker}(UU^*) = \text{Ker}(1 - P_\text{F})$, that is, $\mathcal{H}_\text{F} = \overline{\text{Ran}(U)}$. Since U is an isometry from $\mathcal{H}_\text{I}$ to $\text{Ran}(U)$, $\text{Ran}(U)$ is closed, so (b) is proven.

Conversely, suppose (a)–(c) hold. By (c), if $\varphi, \psi \in \mathcal{H}_\text{I}$ and $\eta \in \mathcal{H}_\text{I}^\perp$, then, since $U\eta = 0$,

$$\langle(\psi + \eta), U^*U\varphi\rangle = \langle U\psi, U\varphi\rangle \tag{2.4.36}$$
$$= \langle\psi, \varphi\rangle \tag{2.4.37}$$
$$= \langle\psi + \eta, \varphi\rangle \tag{2.4.38}$$

Here (2.4.37) follows from the fact that U is norm and so, by polarization, inner product converging on $\mathcal{H}_\text{I}$. Thus, if $\varphi \in \mathcal{H}_\text{I}$, $U^*U\varphi = \varphi = P_\text{I}\varphi$. If $\varphi \in \mathcal{H}_\text{I}^\perp$, $U^*U\varphi = 0 = P_\text{I}\varphi$, proving $U^*U = P_\text{I}$.

By (2.2.13), $\mathcal{H}_\text{F}^\perp = \text{Ran}(U)^\perp = \text{Ker}(U^*) = \text{Ker}(P_\text{F})$. On the other hand, if $\psi \in \mathcal{H}_\text{F}$, $\psi = U\varphi$ for some φ, so $UU^*\psi = UU^*U\varphi = UP_\text{I}\varphi = U\varphi = \psi = P_\text{F}\psi$. Thus, $UU^* = P_\text{F}$ and thus, U is a partial isometry. □

Theorem 2.4.8 (Polar Decomposition Theorem). *Let $A \in \mathcal{L}(\mathcal{H})$. Then there is a partial isometry, U, so that*

(a) $A = U|A|$ (2.4.39)

(b) $\mathcal{H}_\text{I}(U) = \overline{\text{Ran}(|A|)} = \text{Ker}(A)^\perp$ (2.4.40)

(c) $\mathcal{H}_\text{F}(U) = \overline{\text{Ran}(A)} = \text{Ker}(A^*)^\perp$ (2.4.41)

Moreover, (a), (b) *determine U uniquely.*

Remarks. 1. (2.4.39) is called the *polar decomposition* of A.

2. If we need to specify the A involved (e.g., we consider polar decompositions of several operators), we'll use U_A, as in

$$A = U_A|A| \tag{2.4.42}$$

3. We'll extend this to unbounded operators in Problem 21 of Section 7.5.

Proof. Define U on $\text{Ran}(|A|)$ by

$$U(|A|\varphi) = A\varphi \tag{2.4.43}$$

Since

$$\begin{aligned}\||A|\varphi\| &= \langle |A|\varphi, |A|\varphi\rangle = \langle \varphi, |A|^2\varphi\rangle = \langle \varphi, A^*A\varphi\rangle \\ &= \langle A\varphi, A\varphi\rangle = \|A\varphi\|^2\end{aligned} \tag{2.4.44}$$

we see $|A|\varphi = 0 \Rightarrow A\varphi = 0$, so U is well-defined on $\text{Ran}(|A|)$ and, by (2.4.44), for $\psi \in \text{Ran}(|A|)$,

$$\|U\psi\| = \|\psi\| \tag{2.4.45}$$

Extend U to $\overline{\text{Ran}(|A|)}$ by continuity so that (2.4.45) holds for all $\psi \in \overline{\text{Ran}(|A|)}$.

Set

$$U\psi = 0 \quad \text{if } \psi \in \text{Ran}(|A|)^\perp \tag{2.4.46}$$

Then U is a partial isometry with initial subspace $\overline{\text{Ran}(|A|)}$ and final subspace $\overline{\text{Ran}(A)}$.

By (2.2.13), $\text{Ker}(A^*)^\perp = \overline{\text{Ran}(A)}$, and by (2.4.44), $\text{Ker}(A)^\perp = \text{Ker}(|A|)^\perp = \overline{\text{Ran}(|A|)}$, completing the proof of (a)–(c).

To see that (a) and (b) uniquely determine ψ, note that (2.4.38) determines U on $\text{Ran}(|A|)$ and so on $\overline{\text{Ran}(|A|)}$, and (b) determines U on $\text{Ran}(|A|)^\perp$. □

Here are some additional properties of the polar decomposition for general A:

Theorem 2.4.9.

(a) $U_A^* A = |A|$ (2.4.47)

(b) $|A^*| = U_A|A|U_A^*$ (2.4.48)

(c) $U_{A^*} = U_A^*$

(d) $A = |A^*|U_A$

(e) $A^* = U^*AU^*, \quad A = UA^*U$ (2.4.49)

Remarks. 1. It is remarkable that the same U appears in both $A = U|A|$ and $A = |A^*|U$.

2. Thus $A = U\,|A| = |A^*|\,U$. One can show (Problem 16) that for any θ, $A = |A^*|^\theta\, U\, |A|^{1-\theta}$.

Proof. We'll use the notation $U = U_A$ in the proofs of (a) and (b).

(a) We have that $U^*A = U^*U|A| = |A|$ since $U^*U = 1$ on $\text{Ran}(|A|)$.

(b) We have that

$$\begin{aligned}(U|A|U^*)^2 &= U|A|U^*U|A|U^* \\ &= (U|A|)(|A|U^*) \qquad (2.4.50) \\ &= (U|A|)(U|A|)^* \\ &= AA^* \qquad (2.4.51)\end{aligned}$$

where (2.4.50) comes from $U^*U = 1$ on $\text{Ran}(|A|)$.

Since $\langle \varphi, U|A|U^*\varphi\rangle = \langle U^*\varphi, |A|U^*\varphi\rangle \geq 0$, uniqueness of positive square root implies (2.4.48).

(c) We have, by (2.4.48),

$$U_A^*|A^*| = U_A^*U_A|A|U_A^* = |A|U_A^* = A^* \tag{2.4.52}$$

since $U_A^*U_A = 1$ on $\text{Ran}(|A|)$. In addition, $U_A^* = 0$ on $\mathcal{H}_{\text{F}}(U_A) = \text{Ker}(A^*)$ (by (2.4.41)) $= \text{Ker}(|A^*|) = \overline{\text{Ran}(|A^*|)}$, so (2.4.52) is the polar decomposition for A^*.

(d) We have that $A = (A^*)^* = (U_{A^*}|A^*|)^* = |A^*|U_{A^*}^* = |A^*|U_A$.

(e) By (a), $U^*AU^* = |A|U^* = (U|A|)^* = A^*$, proving the first equation. Taking adjoints, we get the second. □

Finally, we want to turn to the polar decomposition when $A = A^*$. We need a lemma ((a)–(c) are preliminaries for (d) which we'll need below):

Lemma 2.4.10. *Let $C \geq 0$ and let P be the projection onto $\text{Ker}(C)^\perp$. Then for $\varepsilon > 0$,*

(a) $\|(C+\varepsilon)^{-1}\| \leq \varepsilon^{-1}$ (2.4.53)

(b) $\|C(C+\varepsilon)^{-1}\| \leq 1$ (2.4.54)

(c) $\lim_{\varepsilon\downarrow 0}\|C(C+\varepsilon)^{-1}C - C\| = 0$ (2.4.55)

For any $\varphi \in \mathcal{H}$

(d) $\lim_{\varepsilon\downarrow 0} C(C+\varepsilon)^{-1}\varphi = P\varphi$ (2.4.56)

Proof. It is easy to see (Problem 7) that $C \geq 0$ implies $C+\varepsilon$ is invertible and $(C+\varepsilon)^{-1} \geq 0$.

(a), (b) Let $D = \sqrt{C}$. Then

$$(D\varphi, (C+\varepsilon)^{-1}D\varphi) + \varepsilon(\varphi, (C+\varepsilon)^{-1}\varphi) = \|\varphi\|^2 \tag{2.4.57}$$

since $(C+\varepsilon)$ commutes with C and so with D. $D(C+\varepsilon)^{-1}D = C(C+\varepsilon)^{-1}$ and $(C+\varepsilon)(C+\varepsilon)^{-1} = 1$. This implies

$$\|C(C+\varepsilon)^{-1}\| = \|D(C+\varepsilon)^{-1}D\| = \sup_{\|\varphi\|=1} \langle D\varphi, (C+\varepsilon)^{-1}D\varphi\rangle \leq 1$$

and

$$\varepsilon\|(C+\varepsilon)^{-1}\| = \sup_{\|\varphi\|} \varepsilon\langle\varphi,(C+\varepsilon)^{-1}\varphi\rangle \le 1$$

proving (a) and (b).

(c) We have that $C(C+\varepsilon)^{-1} = \mathbb{1} - \varepsilon(C+\varepsilon)^{-1}$ so, by (2.4.54),

$$\|C(C+\varepsilon)^{-1}C - C\| = \varepsilon\|(C+\varepsilon)^{-1}C\| \le \varepsilon$$

proving (2.4.55).

(d) Since $\|C(C+\varepsilon)^{-1}\|$ is bounded uniformly in ε, it suffices to prove (2.4.56) for the dense set $\mathrm{Ran}(C) + \mathrm{Ker}(C)$ (since $\overline{\mathrm{Ran}(C)} = \mathrm{Ker}(C)^\perp$). If $\varphi \in \mathrm{Ker}(C)$, $C(C+\varepsilon)^{-1}\varphi = (C+\varepsilon)^{-1}C\varphi = 0$ and $P\varphi = 0$, so (2.4.56) is immediate.

If $\varphi = C\psi$, then by (2.4.55),

$$\lim C(C+\varepsilon)^{-1}\varphi = \lim C(C+\varepsilon)^{-1}C\psi = C\psi = \varphi = P\varphi \qquad \square$$

Theorem 2.4.11. *Let $A \in \mathcal{L}(\mathcal{H})$ with $A = A^*$. Let $U|A|$ be its polar decomposition and P the projection onto* $\mathrm{Ker}(A)^\perp = \overline{\mathrm{Ran}(A)}$. *Then*

(a) $U = U^*$ (2.4.58)

(b) *If $[B,A] = 0$, then $[B,U] = 0$ and, in particular,*

$$U|A| = |A|U \tag{2.4.59}$$

$$UA = AU \tag{2.4.60}$$

(c) $U^2 = P$ (2.4.61)

(d) $\mathcal{H} = \mathcal{H}_+ \oplus \mathcal{H}_0 \oplus \mathcal{H}_-$ (2.4.62)

where $\mathcal{H}_0 = \mathrm{Ker}(A) = \mathrm{Ker}(U)$ *and* $\mathcal{H}_\pm = \mathrm{Ker}(U \mp 1)$.

$$\varphi \in \mathcal{H}_\pm \Rightarrow U\varphi = \pm\varphi \tag{2.4.63}$$

(e) *If $P_\pm, P_0$ are the projections onto $\mathcal{H}_+, \mathcal{H}_-, \mathcal{H}_0$, we have*

$$(P_0 + P_\pm)\varphi = \varphi \Rightarrow \pm\langle\varphi, A\varphi\rangle \ge 0 \tag{2.4.64}$$

Remark. This is an analog of a Hahn decomposition for measures. If $A_\pm = AP_\pm$, then $A = A_+ - A_-$, $A_\pm \ge 0$, and $A_+A_- = A_-A_+ = 0$; see Problem 12.

Proof. (a) In general, $U_{A^*} = U_A^*$, so if $A = A^*$, (2.4.58) holds.

(b) We claim that for any φ,

$$U\varphi = \lim_{\varepsilon\downarrow 0} A(|A|+\varepsilon)^{-1}\varphi \tag{2.4.65}$$

for $UP = U$ so, by the lemma,

$$U\varphi = UP\varphi = \lim_{\varepsilon\downarrow 0} U|A|(|A|+\varepsilon)^{-1}\varphi = \text{RHS of (2.4.65)}$$

by the polar decomposition.

If B commutes with A, it commutes with $A^*A = A^2$ and so with $|A|$, and thus, with $A(|A|+\varepsilon)^{-1}$. By (2.4.65), B commutes with U.

(c) Since $U = U^*$, $U^2 = U^*U = P$.

(d) On $\text{Ker}(P)^\perp$, $U^2 = 1$, so on $\text{Ran}(P)$,

$$(U+1)(U-1) = U^2 - 1 = 0 \tag{2.4.66}$$

Thus, $\text{Ran}(U-1)\cap\text{Ran}(P) \subset \text{Ker}(U+1)$. It follows that if $\mathcal{H}_\pm = \text{Ker}(U\mp 1)$, then $\mathcal{H}_+ \oplus \mathcal{H}_- = \text{Ran}(P)$. On $\mathcal{H}_\pm$, $U\varphi = \pm\varphi$, so if $\varphi_\pm \in \mathcal{H}_\pm$, $\langle\varphi_+,\varphi_-\rangle = \langle U\varphi_+,\varphi_-\rangle = \langle\varphi_+,U\varphi_-\rangle = -\langle\varphi_+,\varphi_-\rangle$. Thus, $\mathcal{H}_+ \perp \mathcal{H}_-$, proving (d).

(e) If $\varphi = \varphi_+ + \varphi_- + \varphi_0$ with $\varphi_\pm \in \mathcal{H}_\pm$, $\varphi_0 \in \text{Ker}(U)$, then

$$\langle\varphi, A\varphi\rangle = \langle\varphi_+, A\varphi_+\rangle + \langle\varphi_-, A\varphi_-\rangle \tag{2.4.67}$$

since (2.4.62) holds and A commutes with P and with U. Moreover,

$$\langle\varphi_\pm, A\varphi_\pm\rangle = \langle\varphi_\pm, U|A|\varphi_\pm\rangle = \langle U\varphi_\pm, |A|\varphi_\pm\rangle = \pm\langle\varphi_\pm, |A|\varphi_\pm\rangle \tag{2.4.68}$$

proving (2.4.64). □

Notice, since (2.4.62) is a sum of eigenspaces for U, we have

$$P_\pm = \tfrac{1}{2}(1\pm U)U, \qquad P_0 = (1+U)(1-U) \tag{2.4.69}$$

In particular, if $[B,A]=0$, then $[B,U]=0$ (by (b) of the theorem) and so, by (2.4.69), $[B,P_\pm]=0$, $[B,P_0]=0$.

Remarkably, this last theorem provides an easy proof of a version of the spectral theorem. That this is a form of the spectral theorem will be discussed further in Chapter 5. First, we need a lemma.

Lemma 2.4.12. *Let A be a self-adjoint operator.*

(a) *Let Q be a projection so that $[Q,A]=0$ and $\varphi \in \text{Ran}(Q)$, $\varphi \neq 0 \Rightarrow \langle\varphi, A\varphi\rangle > 0$ (respectively, $\langle\varphi, A\varphi\rangle \geq 0$). Then, $\text{Ran}(Q) \subset \text{Ran}(P_+(A))$ (respectively, $\text{Ran}(Q) \subset \text{Ran}(P_+(A)) + \text{Ran}(P_0(A))$).*

(b) *If B is a second self-adjoint operator with $[B,A]=0$ and $B \geq A$, then $\text{Ran}(P_+(A)) \subset \text{Ran}(P_+(B))$ and $\text{Ran}(P_0(A)) + \text{Ran}(P_+(A)) \subset \text{Ran}(P_0(B)) + \text{Ran}(P_+(B))$.*

Proof. (a) We do the P_+ case. The $P_+ + P_0$ case is similar. By Theorem 2.4.11(b), $P_+(A)$ and $1-Q$ are compatible. If $\varphi \in \text{Ran}[1-P_+(A)] \cap \text{Ran}(Q)$, then $\langle\varphi, A\varphi\rangle \leq 0$ (since $\varphi \in \text{Ran}(P_0(A)) + \text{Ran}(P_-(A))$), and if $\varphi \neq 0$, $\langle\varphi, A\varphi\rangle > 0$ (by hypothesis). Thus, $\varphi = 0$. It follows from (2.1.27) that

$$Q(1-P_+(A)) = 0 \tag{2.4.70}$$

so $Q = P_+(A)Q \Rightarrow \text{Ran}(Q) \subset \text{Ran}(P_+(A))$.

(b) Again, we do the $P_+(A)$ case. If $\varphi \in \operatorname{Ran}(P_+(A))$, $\varphi \neq 0$, then $\langle \varphi, A\varphi \rangle > 0$. Since $B \geq A$, $\langle \varphi, B\varphi \rangle > 0$.

Since A commutes with B, $P_+(A)$ commutes with B by Theorem 2.4.11(b). Part (a) implies $\operatorname{Ran}(P_+(A)) \subset \operatorname{Ran}(P_+(B))$. □

Theorem 2.4.13 (Spectral Theorem, First Form). *Let $A \in \mathcal{L}(\mathcal{H})$ obey $A = A^*$. Then for any interval I of the form $[a,b)$, if $-\|A\| \leq a < b \leq \|A\|$, or $[a, \|A\|]$ if $-\|A\| < a < \|A\|$, there is a projection, P_I, commuting with A so that if $\{I_j^{(n)}\}_{j=-2^n}^{2^n-1}$ is $[\frac{\|A\|j}{2^n}, \frac{\|A\|(j+1)}{2^n})$, with $\|A\|$ included in the set if $j = 2^n - 1$ and if $x_j^{(n)}$ are arbitrary points in $\overline{I_j^{(n)}}$ then*

(a) *For all I, J, P_I and P_J are compatible and*

$$P_I P_J = P_J P_I = P_{I \cap J} \tag{2.4.71}$$

(b) *For each n and all $j, k \in \{-2^n, \dots, 2^n - 1\}$,*

$$P_j^{(n)} P_k^{(n)} = \delta_{jk} P_j^{(n)} \tag{2.4.72}$$

$$\sum_{j=-2^n}^{2^n-1} P_j^{(n)} = \mathbb{1} \tag{2.4.73}$$

(c) *If $I = [a,b)$ and $\varphi \in \operatorname{Ran}(P_I)$ with $\|\varphi\| = 1$, then*

$$a \leq \langle \varphi, A\varphi \rangle \leq b \tag{2.4.74}$$

(d) *For any $\ell \in 0, 1, 2, \dots,$*

$$\left\| A^\ell - \sum_{j=-2^n}^{2^n-1} (x_j^{(n)})^\ell P_j^{(n)} \right\| \leq \ell \|A\|^{\ell-1} 2^{-n} \tag{2.4.75}$$

so, in particular, the norm goes to zero as $n \to \infty$ for ℓ fixed.

Remark. (2.4.75) implies that (indeed, is equivalent to, as we'll see)

$$\varphi \in \operatorname{Ran}(P_j^{(n)}) \Rightarrow \|[A^\ell - (x_j^{(n)})^\ell]\varphi\| \leq \ell \|A\|^{\ell-1} 2^{-n} \|\varphi\| \tag{2.4.76}$$

Proof. For $b \in [-\|A\|, \|A\|]$, define

$$P_{<b} = P_+(b - A) \tag{2.4.77}$$

Two applications of Theorem 2.4.11(b), prove that the $\{P_{<b}\}$ are mutually commuting. If $b_1 > b_2$, then $b_1 - A > b_2 - A$, so $\operatorname{Ran}(P_{<b_2}) \subset \operatorname{Ran}(P_{<b_1})$ by the lemma. Thus, $\operatorname{Ran}(P_{<b})$ is increasing in b. Define

$$P_{\geq a} = 1 - P_{<a} \tag{2.4.78}$$

Then $\operatorname{Ran}(P_{\geq a})$ is decreasing in a. If $I = [a,b)$ with $a < b$, define

$$P_I = P_{\geq a} P_{<b} \tag{2.4.79}$$

(in case $b = |A|$ and we want $I = [a, b]$, we use $P_I = P_{\geq a}$). Since the $P_{<b}$ are mutually commuting, so are the P_I's.

(2.4.71) is easy to see (Problem 8), given the monotonicity properties, and it implies (2.4.72). (2.4.72) in turn implies that $I \cap J = \emptyset \Rightarrow P_{I \cap J} = P_I + P_J$ (see (2.1.32)) and this implies (2.4.73).

By construction, we have (2.4.74), that is,

$$\varphi \in \text{Ran}(P_{[a,b)}) \Rightarrow a\|\varphi\|^2 \leq (\varphi, A\varphi) \leq b\|\varphi\|^2 \tag{2.4.80}$$

Thus, on $\text{Ran}(P_{[a,b)})$, $A - a \geq 0$, and by (2.4.3),

$$\|(A - a)P_{[a,b)}\| \leq |b - a| \tag{2.4.81}$$

Similarly (looking at $b - A$),

$$\|(A - b)P_{[a,b)}\| \leq |b - a| \tag{2.4.82}$$

If $a \leq c \leq b$, we can write $c = \theta a + (1 - \theta)b$ for some $\theta \in [0, 1]$, and so, using $A - c = \theta(A - a) + (1 - \theta)(A - b)$, conclude that

$$\|(A - c)P_{[a,b)}\| \leq |b - a| \tag{2.4.83}$$

Writing

$$A^\ell - c^\ell = (A - c)\sum_{j=0}^{\ell-1} A^j c^{\ell-1-j} \tag{2.4.84}$$

we see if $|c| \leq \|A\|$, we have by (2.4.82) that

$$\|(A^\ell - c^\ell)P_{[a,b)}\| \leq \ell\|A\|^{\ell-1}|b - a| \tag{2.4.85}$$

which proves (2.4.76). This easily (Problem 9) yields (2.4.75). □

Notes and Historical Remarks. The polar decomposition for operators on a Hilbert space is due to von Neumann [**724**] in 1932. But it has an older pedigree for finite matrices, appearing first in a 1902 paper of Autonne [**38**] who proved a finite invertible matrix, A, can be written $A = UB$, where U is unitary and $B > 0$. This was rediscovered by Wintner and Murnaghan [**763**], apparently unaware of the earlier work. There is a close connection between the polar decomposition and the canonical expansion of Section 3.5, which in one form ("singular value decomposition") predates the polar decomposition; see the Notes to Section 3.5. The finite matrix work was restricted to A invertible, so U is unitary. For the case of compact A, where the polar decomposition will be an important tool, the case A is not invertible (and so U only a partial isometry) is relevant, so the form used by von Neumann will be critical.

While not stated in terms of a polar decomposition, Schmidt [**601**] in 1907, with followup by Picard [**522**] in 1910, considered integral kernels and numbers μ and pairs of functions φ, ψ with

$$\varphi(s) = \mu \int K(s,t)\psi(t)\,dt \tag{2.4.86}$$

$$\psi(s) = \mu \int K(t,s)\varphi(t)\,dt \tag{2.4.87}$$

If $A\eta(s) = \int K(s,t)\eta(s)\,ds$, then μ^{-1} is an eigenvalue of $|A|$.

von Neumann constructed $\sqrt{A}$ using the spectral theorem. The construction in Problem 11 is due to Visser [**717**] in 1937. He used it to prove the decomposition, $A = A_+ - A_-$, of Riesz, discussed below. The method in the text using the power series for $\sqrt{1-x}$ goes back at least to Ford [**203**], but I suspect it is much older.

The term partial isometry and definition via $U = UU^*U$ are due to Murray–von Neumann [**487**].

The decomposition in Theorem 2.4.11, $P_+ + P_- + P_0 = \mathbb{1}$, for self-adjoint operators, A, is due to Riesz [**567**] who thought of it as a kind of Hahn decomposition. Its use to prove the spectral theorem as we do in Theorem 2.4.13, is also from Riesz's paper.

It follows from this Riesz decomposition and (2.4.3) that for any self adjoint A,

$$\|A\| = \sup_{\substack{\varphi\in\mathcal{H} \\ \|\varphi\|=1}} |\langle\varphi, A\varphi\rangle| \tag{2.4.88}$$

Riesz uses this in the proof of his decomposition.

Problems

1. If $A > 0$, (2.4.8) is just the ordinary Schwarz inequality. Prove (2.4.8) by showing it for $A + \varepsilon\mathbb{1}$ and taking $\varepsilon \downarrow 0$.

2. Let $\mathcal{A}(\varphi, \psi)$ be a semidefinite sesquilinear form.

   (a) Prove $\mathcal{A}(\varphi,\varphi) = 0 \Rightarrow \mathcal{A}(\varphi,\psi) = 0$ for all ψ. (*Hint*: Consider $\mathcal{A}(\varphi + z\psi, \varphi + z\psi)$ for z small and complex.)

   (b) If $\mathcal{A}(\varphi,\varphi) \neq 0 \neq \mathcal{A}(\psi,\psi)$, prove that
   $$|\mathcal{A}(\varphi,\psi)|^2 \le \mathcal{A}(\varphi,\varphi)\mathcal{A}(\psi,\psi) \tag{2.4.89}$$
   (*Hint*: Look at the quadratic equation $t \mapsto \mathcal{A}(\varphi + t\psi, \varphi + t\psi)$.)

   (c) Prove (2.4.89) in all cases.

3. Let $A = \left(\begin{smallmatrix} 2 & 0 \\ 0 & 0 \end{smallmatrix}\right)$, $B = \left(\begin{smallmatrix} -1 & 1 \\ 1 & -1 \end{smallmatrix}\right)$. Is it true that $|A+B| \le |A| + |B|$?

4. Find an example of 2×2 matrices for which $\||A| - |B|\| \le \|A - B\|$ fails.

5. Prove that if L is left shift and if $L = V|L|$, then for some φ, $V = U_{L,\varphi}$ given by (2.4.32).

6. Let $U_{L,\varphi}$ be given by (2.4.32). Prove that $\text{Ker}(U_{L,\varphi})$ is always one-dimensional.

7. Let $C \geq 0$.

   (a) Prove $\|(C+\varepsilon)\varphi\| \geq \varepsilon\|\varphi\|$, $\text{Ker}(C+\varepsilon) = \{0\} = \text{Ker}(C^* + \varepsilon)$ and conclude that $(C+\varepsilon)$ is invertible so, in particular, $\text{Ran}(C+\varepsilon) = \mathcal{H}$.

   (b) Prove $(C+\varepsilon)^{-1} > 0$. (*Hint*: To look at $\langle\varphi, (C+\varepsilon)^{-1}\varphi\rangle$, suppose $\varphi = (C+\varepsilon)\psi$.)

8. If P_I is given by (2.4.79), confirm that $P_I P_J = P_{I\cap J}$.

9. Let B be an operator and $\{P_j\}_{j=1}^J$ be mutually commuting projections, each commuting with B. For $j \neq k$, suppose $P_j P_k = 0$ and that $\sum_{j=1}^J P_j = \mathbb{1}$. Prove that

$$\left\| B - \sum_{j=1}^J c_j P_j \right\| = \sup_{j=1,\dots,J} \|(B - c_j)P_j\|$$

10. (a) If U is a partial isometry, prove that

$$U = UU^*U \tag{2.4.90}$$

   (b) Prove that (2.4.90) implies that U^*U is a self-adjoint projection.

   (c) Prove that (2.4.90) implies UU^* is a self-adjoint projection.

11. This will provide an alternate proof that if $0 \leq A$ has $\|A\| \leq 1$, then there exist polynomials, $p_n(A)$, converging in norm to a positive operator B with $B^2 = A$. As in the preliminaries leading to Theorem 2.4.4, one has that $C = 1 - A$ has $0 \leq C \leq 1$ and $\|C\| \leq 1$.

   (a) Define D_n inductively by

$$D_0 = 0, \qquad D_{n+1} = \tfrac{1}{2}\,(C + D_n^2) \tag{2.4.91}$$

   Prove inductively that $\|D_n\| \leq 1$ and $[R, C] = 0 \Rightarrow [R, D_n] = 0$.

   (b) Prove that D_n is a polynomial in C with positive coefficients.

   (c) Prove that

$$\begin{aligned} D_{n+1} - D_n &= \tfrac{1}{2}\,(D_n^2 - D_{n-1}^2) \\ &= \tfrac{1}{2}\,(D_n - D_{n-1})(D_n + D_{n-1}) \end{aligned} \tag{2.4.92}$$

   (d) Prove inductively that $D_n - D_{n-1}$ is a polynomial in C with nonnegative coefficients, and conclude that $D_{n-1} \leq D_n$.

(e) For any φ, prove that $\lim_{n\to\infty}\langle\varphi, D_n\varphi\rangle$ exists.

(f) For any positive operator C, prove that

$$\|C\varphi\|^2 \le \langle\varphi, C\varphi\rangle\|C\| \tag{2.4.93}$$

(*Hint*: See (2.4.6) and the proof of (2.4.3).)

(g) Prove $D_n\varphi$ is $\|\cdot\|$-Cauchy, so $\text{s-lim}_{n\to\infty} D_n = D_\infty$ exists.

(h) Prove that D_∞ obeys

$$2D_\infty = C + D_\infty^2, \qquad 0 \le D_\infty \le 1 \tag{2.4.94}$$

(i) Let $B = 1 - D_\infty$. Prove that $B^2 = A$, $B \ge 0$, completing the proof of the existence of a positive square root which is a limit of polynomials in A.

12. Suppose A is self-adjoint, $A = A_+ - A_-$ with $A_\pm \ge 0$, $A_+A_- = A_-A_+ = 0$. This problem will prove that $A_\pm = AP_\pm(A)$, showing a uniqueness result that specifies $A_\pm$.

(a) Prove $P_\pm(A)$, $P_\pm(A_\pm)$, $P_0(A)$, $P_0(A_\pm)$ are compatible projections.

(b) Prove that $A^2 = A_+^2 + A_-^2$.

(c) Prove that

$$\text{Ker}(A_+) \cap \text{Ker}(A_-) = \text{Ker}(A) \tag{2.4.95}$$

(d) Prove that $\text{Ran}(A_\mp) \subset \text{Ker}(A_\pm)$ and then that

$$\text{Ker}(A_\pm)^\perp \subset \text{Ker}(A_\mp) \tag{2.4.96}$$

(e) Prove that

$$P_0(A) + P_+(A_+) + P_+(A_-) = \mathcal{H} \tag{2.4.97}$$

(*Hint*: See (2.4.95) and (2.4.96).)

(f) Prove $\text{Ran}(P_+(A)) \subset \text{Ran}(P_+(A_+))$ and $\text{Ran}(P_-(A)) \subset \text{Ran}(P_+(A_-))$. (*Hint*: Lemma 2.4.12.)

(g) Prove $P_+(A) = P_+(A_+)$, $P_-(A) = P_+(A_-)$.

(h) Prove $A_\pm = AP_\pm(A)$.

13. A main point of this problem is to prove

$$AA^* = A^*A \Rightarrow [U, |A|] = [U, A] = [U, U^*] = 0$$

so, in particular, U is a unitary operator from $\text{Ker}(A)$ to itself. But there are many more equivalences you will prove, including

$$AA^* = A^*A \overset{\boxed{1}}{\Longleftrightarrow} \|A\varphi\|^2 = \|A^*\varphi\|^2$$

$$\overset{\boxed{2}}{\Longleftrightarrow} |A| = |A^*|$$

$$\overset{\boxed{3}}{\Longrightarrow} [U, |A|] = 0$$

$$\overset{\boxed{4}}{\Longleftrightarrow} [U^*, |A|] = 0$$

$$\overset{\boxed{5}}{\Longleftrightarrow} [U, A] = 0$$

and

$$\|A\varphi\|^2 = \|A^*\varphi\|^2 \overset{\boxed{6}}{\Longrightarrow} \operatorname{Ker}(A) = \operatorname{Ker}(A^*)$$

and

$$\operatorname{Ker}(A) = \operatorname{Ker}(A^*) \overset{\boxed{7}}{\Longleftrightarrow} [U, U^*] = 0$$

(a) Prove $\boxed{1}$.

(b) Prove $\boxed{2}$ given $\boxed{1}$ by showing $AA^* = A^*A \Leftrightarrow |A| = |A^*|$.

(c) Prove $\boxed{3}$. (*Hint*: See (d) of Theorem 2.4.9.)

(d) Prove $\boxed{4}$.

(e) Prove $\boxed{5}$ (given $\boxed{4}$).

(f) Prove $\boxed{6}$.

(g) Prove $\boxed{7}$.

14. This problem will prove that $A \to \sqrt{A}$ is norm-continuous. Below, all operators (A, A_n, B) are nonnegative.

(a) For real numbers, $c, d \geq 0$, prove that

$$\left|\sqrt{c} - \sqrt{d}\,\right| \leq \sqrt{|c-d|} \tag{2.4.98}$$

(*Hint*: Prove $|\sqrt{c} - \sqrt{d}\,| = |c-d|/(\sqrt{c} + \sqrt{d}\,)$ and that $\sqrt{c} + \sqrt{d} \geq \sqrt{|c-d|}$.)

(b) If $0 \leq A \leq 1 - \varepsilon$, prove that

$$\|(1-A)^n - (1-(A+\varepsilon))^n\| \leq \|1-A\|^n - (\|1-A\| - \varepsilon)^n \tag{2.4.99}$$

(*Hint*: Prove first that $\|1-(A+\varepsilon)\| = \|1-A\| + \varepsilon$.)

(c) If $0 \leq A \leq 1 - \varepsilon$, prove that

$$\left\|\sqrt{A+\varepsilon} - \sqrt{A}\,\right\| \leq \sqrt{\varepsilon} \tag{2.4.100}$$

(*Hint*: Use $\sqrt{B} = 1 - \sum_{n=1}^\infty c_n (1-B)^n$ with c_n given by (2.4.13) and parts (a) and (b).)

(d) If $\varepsilon \leq B \leq 1$ and $\varepsilon \leq A \leq 1$, prove that (with c_j given by (2.4.13))

$$\begin{aligned} \left\|\sqrt{A} - \sqrt{B}\,\right\| &\leq \|A-B\| \sum_{j=1}^\infty j c_j (1-\varepsilon)^j \\ &= (2\varepsilon^{1/2})^{-1} \|A-B\| \end{aligned} \tag{2.4.101}$$

(e) If $0 \leq B \leq 1 - \varepsilon$ and $0 \leq A \leq 1 - \varepsilon$, prove that

$$\|\sqrt{A} - \sqrt{B}\,\| \leq 2\varepsilon^{1/2} + (2\varepsilon^{1/2})^{-1}\|A - B\| \tag{2.4.102}$$

(*Hint*: Use (c) and (d).)

(f) If $\|A\| \leq \frac{1}{2}$, $\|B\| \leq \frac{1}{2}$, prove that

$$\|\sqrt{A} - \sqrt{B}\,\| \leq 2\sqrt{\|A - B\|} \tag{2.4.103}$$

(*Hint*: Take $\varepsilon = \frac{1}{4}\|A - B\|$.)

(g) If $\|A_n - A\| \to 0$, prove that $\|\sqrt{A_n} - \sqrt{A}\,\| \to 0$.

15. Let $\mathcal{H}_1, \mathcal{H}_2$ be two Hilbert spaces and $A \in \mathcal{L}(\mathcal{H}_1, \mathcal{H}_2)$ a bounded linear map of $\mathcal{H}_1$ to $\mathcal{H}_2$. This problem will prove a polar decomposition

$$A = U|A| \tag{2.4.104}$$

with $|A| \in \mathcal{L}(\mathcal{H}_1)$ is self-adjoint and positive and U is a partial isometry from $\mathcal{H}_1$ to $\mathcal{H}_2$ (with the obvious extension of the notion on a single Hilbert space).

(a) Prove A^*A is a positive self-adjoint operator on $\mathcal{L}(\mathcal{H}_1)$ and let $|A| = \sqrt{A^*A}$.

(b) Verify a U exists, unique if $\overline{\text{Ran}|A|}$ is the initial space for U, so (2.4.104) holds.

16. Let A be a bounded operator. Let $\mathcal{H}_i = \text{Ker}(A)^\perp$, $\mathcal{H}_f = \text{Ker}(A^*)^\perp$ so $V \equiv U \restriction \mathcal{H}_i$ is a unitary map of $\mathcal{H}_i$ to $\mathcal{H}_f$.

(a) Prove that $\mathcal{H}_i$ (respectively, $\mathcal{H}_f$) is an invariant space for $|A|$ (respectively, $|A^*|$).

Let $B = |A| \restriction \mathcal{H}_i, \quad C = |A^*| \restriction \mathcal{H}_f$.

(b) Prove that $VBV^{-1} = C$.

(c) Prove for any bounded Borel function on $[0, \infty)$, $V\varphi(B)V^{-1} = \varphi(C)$.

Remark. This requires the functional calculus form of the spectral theorem; see Section 5.1.

(d) If $\varphi(x)$, $\psi(x) \equiv 1$ for two bounded Borel functions on $[0, \infty)$, prove that

$$\varphi(A^*)\, U\, \psi(A) = A \tag{2.4.105}$$

(e) Prove that for $0 < \theta < 1$, $\;|A^*|^\theta\, U\, |A|^{1-\theta} = A$.

Remarks. 1. This is a result of Gesztesy et al. [**241**].

2. This proof extends to the unbounded case once one has the polar decomposition (see Problem 21 of Section 7.5) and spectral theorem for such operators.

17. This problem illuminates the square root lemma. The result here is stronger in that it works on a general Banach space or for some non-self-adjoint operator on a Hilbert space, but it is much weaker in that it requires A be invertible.

    (a) Let A be a bounded operator on a Banach space, X. Suppose $\sigma(A) \cap (-\infty, 0] = \emptyset$. Prove there exists another operator B on X so that

    $$\sigma(B) \subset \mathbb{H}_+, \quad B^2 = A \tag{2.4.106}$$

    (b) Prove that B obeying (2.4.106) is unique.

Chapter 3

Compact Operators, Mainly on a Hilbert Space

Big Notions and Theorems: Compact Operators, Finite Approximable Operators, Completely Continuous Operators, Schauder's Theorem, $\mathrm{FA} \subset \mathrm{Com} \subset \mathrm{CC}$, $\mathrm{FA} = \mathrm{Com} = \mathrm{CC}$ for Hilbert Spaces, Concrete Hilbert–Schmidt Operators, Schauder Basis, Hilbert–Schmidt Theorem, Sturm–Liouville Operators, Integral Kernel, Sturm–Liouville Theorem, Riesz–Schauder Theorem, Fredholm Alternative, Ringrose–West Decompositions, Nest of Subspaces, Ringrose Structure Theorem, Volterra Nest, Singular Values, Canonical Expansion, Ky Fan Inequalities, Horn's Inequalities, Trace Class, Trace Norm, Invariance of Trace, Doubly Substochastic Matrix, Trace, $\mathrm{Tr}(AB) = \mathrm{Tr}(BA)$, Duals of $\mathcal{I}_\infty$ and $\mathcal{I}_1$, Trace Ideals, Hölder's Inequality for Matrices, Abstract Hilbert–Schmidt, Schur Basis, Schur–Lalesco–Weyl Inequality, Determinants of $1+\mathcal{I}_1$, Cramer's Rule, Plemelj–Smithies Formula, Fredholm Kernel, Fredholm Determinant, Fredholm Minor, Hadamard's Inequality, Mercer's Theorem, Lidskii's Theorem, Regularized Determinants, Renormalized Plemelj–Smithies Formulae, Hilbert–Fredholm Kernel, Weyl's Invariance Theorem, Analytic Fredholm Theorem, Min-Max Principle, Cokernel, Fredholm Operators, Index, Pseudo-inverse, Atkinson's Theorem, Dieudonné's Theorem, Homotopy Invariance of Index, Additivity of Index, Pairs of Projections, Toeplitz Operators, Aharonov–Casher Theorem, M. Riesz's Criterion, Rellich's Criterion

The central results in finite-dimensional operator theory are existence of determinant and trace, the spectral theorem, and the Jordan normal form (see Section 1.3). In this chapter, we'll discuss a class of infinite-dimensional operators that are so close to finite-dimensional operators that all these notions extend.

In Section 3.1, we'll present three possible notions of "close to finite-dimensional" for $A \in \mathcal{L}(X)$, where X is a Banach space:

(i) $\mathrm{Com}(X)$, the compact operators for which $\overline{A[X_1]}$ is norm compact (with $X_1 = \{x \in X \mid \|x\| \le 1\}$);

(ii) $\mathrm{CC}(X)$, the completely continuous operators that obey $x_n \xrightarrow{w} x \Rightarrow \|Ax - Ax_n\| \to 0$;

(iii) $\mathrm{FA}(X)$, the finite approximal operators that are norm limits of finite rank operators.

In general, $\mathrm{FA}(X) \subset \mathrm{Com}(X) \subset \mathrm{CC}(X)$, but in many cases, including the Hilbert space case, they are all equal, as we'll prove in Section 3.1. In all but Sections 3.1, 3.3, and 3.4, we'll consider the Hilbert space case only—not so much because every compact operator can be approximated by finite rank operators (although that will be useful), but because we have a complete analysis of compact self-adjoint operators and we have the polar decomposition.

Section 3.2 will discuss compact self-adjoint operators and prove that any such operator, A, has a complete orthonormal set of eigenvectors $\{\varphi_n\}_{n=1}^\infty$ so $A\varphi_n = \lambda_n\varphi_n$, where $|\lambda_n| \to 0$. This is an analog of the spectral theory for finite matrices. As an application, we present Sturm–Liouville theory.

Sections 3.3 and 3.4 will discuss an analog of Jordan normal forms (Riesz–Schauder theorem). We'll prove $\sigma(A)$ is countable with only 0 as a possible nonisolated point in $\sigma(A)$. If $\lambda_0 \neq 0$ is in $\sigma(A)$, $(A - \lambda)^{-1}$ will be meromorphic at λ_0 with finite rank residues. In particular, if $P_{\lambda_0} = (2\pi i)^{-1} \int_{|\lambda-\lambda_0|=\varepsilon} (\lambda - A)^{-1}\, d\lambda$ for small ε is the associated spectral projection, then $\dim(P_{\lambda_0}) < \infty$ and $AP_{\lambda_0} = \lambda_0 P_{\lambda_0} + D_{\lambda_0}$, where D_{λ_0} is nilpotent. Moreover, one can decompose $A = N + Q$ where N is normal with $r(N) = \sigma(N)$ and Q is quasinilpotent.

In Section 3.5 and the rest of the chapter (except for Section 3.15), we'll specialize to the Hilbert space case. The singular values of a compact operator, A, are the nonzero eigenvalues of $|A|$ but in decreasing order $\mu_1(A) \ge \mu_2(A) \ge \dots$. If $|A|$ is finite rank, we'll set $\mu_n(A) = 0$ for $n > \mathrm{rank}(|A|)$. By combining this definition and the polar decomposition, we'll prove there exist two orthonormal sets, $\{\varphi_n\}_{n=1}^\infty$ and $\{\psi_n\}_{n=1}^\infty$, so that

$$A = \sum_{n=1}^\infty \mu_n(A) \langle \varphi_n, \cdot \rangle \psi_n \tag{3.0.1}$$

called the canonical expansion.

Sections 3.6–3.8 push the analogy between trace and measure. The analogs of L^p space will be the set, $\mathcal{I}_p$, of compact operators with

$$\|A\|_p = \left(\sum_{n=1}^{\infty} \mu_n(A)^p\right)^{1/p} \tag{3.0.2}$$

finite. $\mathcal{I}_p$ is a two-sided ideal of $\mathcal{L}(\mathcal{H})$.

Section 3.7 discusses some aspects of the general theory, and Section 3.6 is on $\mathcal{I}_1$. For $A \in \mathcal{I}_1$, one can define

$$\operatorname{Tr}(A) = \sum_{n=1}^{\infty} \mu(A)\langle \varphi_n, \psi_n \rangle \tag{3.0.3}$$

We'll prove for any orthonormal basis $\{\eta_n\}_{n=1}^{\infty}$, $\sum_{n=1}^{\infty}\langle \eta_n, A\eta_n\rangle$ is absolutely convergent with the sum equal to $\operatorname{Tr}(A)$. We'll also see that $\operatorname{Tr}(AB) = \operatorname{Tr}(BA)$ and that $A \in \mathcal{I}_p \Leftrightarrow |A|^p \in \mathcal{I}_1$ and $\|A\|_p^p = \operatorname{Tr}(|A|^p)$. Section 3.8 deals with $\mathcal{I}_2$ and realizes $\mathcal{H}^* \otimes \mathcal{H}$ as an $\mathcal{I}_2$ space.

One purpose of Sections 3.9, 3.10, and 3.12 is to prove that if $\{\lambda_n(A)\}_{n=1}^N$ is a listing of the nonzero eigenvalues of a compact operator, counted up to algebraic multiplicity, then for $A \in \mathcal{I}_1$,

$$\operatorname{Tr}(A) = \sum_{n=1}^{N} \lambda_n(A) \tag{3.0.4}$$

Of course, for this to make sense, we need to know the sum is convergent. Section 3.9 will prove a general inequality

$$\sum_{n=1}^{N} |\lambda_n(A)|^p \le \sum_{n=1}^{\infty} \mu_n(A)^p \tag{3.0.5}$$

Section 3.10 will define $\det(1 + A)$ for $A \in \mathcal{I}_1$, which will lead to a proof of (3.0.4) and also an analog of Cramer's rule for $A \in \mathcal{I}_1$ and solutions of $f = g + Af$. Section 3.11 discusses Tr and det when A has an integral kernel.

The last four sections discuss specialized bonus topics including the theory of Fredholm operators (and their index) in Section 3.15.

3.1. Compact Operator Basics

In this section, we'll study three classes of operators, often the same, on a general Banach space, X.

Definition. $\mathrm{FA}(X)$, the *finite approximable operators* on X, is the closure in $\|\cdot\|$ in $\mathcal{L}(X)$ of the finite rank operators.

Definition. $\mathrm{Com}(X)$, the *compact operators*, is the set of $A \in \mathcal{L}(X)$ so that $\overline{A[X_1]}$ is compact in the norm topology. Here $\overline{\quad}$ is taken in norm and

$$X_1 = \{x \in X \mid \|x\| \leq 1\} \tag{3.1.1}$$

Definition. $\mathrm{CC}(X)$, the *completely continuous operators* on X, is the set of $A \in \mathcal{L}(X)$ with the property $\{x_n\}_{n=1}^\infty \subset X_1$ and $x_n \to x_\infty$ in $\sigma(X, X^*)$-topology implies $Ax_n \to Ax_\infty$ in norm.

We will first prove each of these is a two-sided, norm-closed ideal of $\mathcal{L}(X)$ and that $A \in \mathrm{FA}(X)$ (resp., $\mathrm{Com}(X)$) implies $A^t \in \mathrm{FA}(X^*)$ (resp., $\mathrm{Com}(X^*)$). Then we'll show $\mathrm{FA}(X) \subset \mathrm{Com}(X) \subset \mathrm{CC}(X)$, that $\mathrm{Com}(X) = \mathrm{CC}(X)$ if X is reflexive (with X^* separable), and finally, that if X has a Schauder basis (a notion we'll define and explore), then $\mathrm{FA}(X) = \mathrm{Com}(X)$.

Readers interested only in the Hilbert space case should look at Theorem 3.1.10 to see $\mathrm{Com}(\mathcal{H}) = \mathrm{CC}(\mathcal{H})$ and Proposition 3.1.12/Theorem 3.1.13 to see $\mathrm{FA}(\mathcal{H}) = \mathrm{Com}(\mathcal{H})$, and finally Theorem 3.1.1 to see they are all two-sided, norm-closed star ideals. Example 3.1.15 is also important for the Hilbert space case. We note that in the Hilbert space case, where we use $\mathcal{I}_\infty$ for $\mathrm{FA}(\mathcal{H}) = \mathrm{Com}(\mathcal{H}) = \mathrm{CC}(\mathcal{H})$, the only two-sided, norm-closed ideals in $\mathcal{L}(\mathcal{H})$ are $\{0\}$, $\mathcal{L}(\mathcal{H})$, and $\mathcal{I}_\infty$ (Problem 1).

Theorem 3.1.1. $\mathrm{FA}(X)$ *is a two-sided, norm-closed ideal in* $\mathcal{L}(X)$. *If* $A \in \mathrm{FA}(X)$, *then* $A^{\mathrm{t}} \in \mathrm{FA}(X)$.

Proof. If A is finite rank and $B \in \mathcal{L}(X)$, $\mathrm{Ran}(AB) \subset \mathrm{Ran}(A)$ and $\mathrm{Ran}(BA) \subset B[\mathrm{Ran}(A)]$ so AB and BA are finite rank. Thus, the finite rank operators are a two-sided ideal, so their closure is a norm-closed, two-sided ideal.

If A_n is finite rank and $\|A_n - A\| \to 0$, then A_n^{t} is finite rank (see (2.1.9)) and $\|A_n^{\mathrm{t}} - A^{\mathrm{t}}\| = \|A_n - A\|$ (by (2.1.3)), so $\|A_n^{\mathrm{t}} - A^{\mathrm{t}}\| \to 0$. □

A subset, S, of a metric space has $\overline{S}$ compact if and only if every sequence in S has a convergent subsequence. Thus, since X is complete, we have

Lemma 3.1.2. $A \in \mathcal{L}(X)$ *is compact if and only if for every sequence,* $\{x_n\}_{n=1}^\infty$ *in* X_1, *there is a subsequence* $\{x_{n(j)}\}_{j=1}^\infty$, *so* $Ax_{n(j)}$ *is Cauchy in norm.*

Theorem 3.1.3. (a) $\mathrm{Com}(X)$ *is a norm-closed subset of* $\mathcal{L}(X)$.

(b) $\mathrm{Com}(X)$ *is a two-sided ideal.*

(c) (Schauder's Theorem) *If* $A \in \mathrm{Com}(X)$, *then* $A^{\mathrm{t}} \in \mathrm{Com}(X^*)$.

Remarks. 1. It is easy to see (Problem 2) that given (c), if $A^{\mathrm{t}} \in \mathrm{Com}(X^*)$, then $A \in \mathrm{Com}(X)$.

2. Problem 3 has an alternate proof of (c).

Proof. (a) Let $\{x_n\}_{n=1}^\infty$ be a sequence in X_1, let $\{A_m\}_{m=1}^\infty$ be a sequence in $\mathrm{Com}(X)$ and $\|A_m - A_\infty\| \to 0$ for some $A_\infty \in \mathcal{L}(X)$.

By the lemma and $A_m \in \mathrm{Com}(X)$, we can find for any subsequence $\{x_{n(j)}\}_{j=1}^\infty$ of $\{x_n\}_{n=1}^\infty$ a subsubsequence, $\{x_{n(\tilde{n}(k))}\}_{k=1}^\infty$, so $A_m x_{n(\tilde{n}(k))}$ is Cauchy in k. By passing to successive subsequences and using the diagonalization trick (see the discussion in Section 1.5 of Part 1), we can find a fixed subsequence $\{x_{n(j)}\}_{j=1}^\infty$ so for each m, $A_m x_{n(j)}$ is Cauchy as $j \to \infty$.

Given ε, find m_0 so $\|A_{m_0} - A\| \le \varepsilon/3$. Since $x_{n(j)} \in X_1$, for all j, $\|A_{m_0}x_{n(j)} - Ax_{n(j)}\| \le \varepsilon/3$. Find J so $\|A_{m_0}x_{n(j)} - A_{m_0}x_{n(\ell)}\| \le \varepsilon/3$ if $j, \ell \ge J$. It follows, if $j, \ell \ge J$, then $\|Ax_{n(j)} - Ax_{n(\ell)}\| \le \varepsilon$, that is, $\{Ax_{n(j)}\}_{j=1}^\infty$ is Cauchy. Thus, by the lemma, $A \in \mathrm{Com}(X)$.

(b) Given the lemma, it is easy to see $A, B \in \mathrm{Com}(X)$, $C \in \mathcal{L}(X) \Rightarrow A + B, AC, CA \in \mathrm{Com}(X)$.

(c) Suppose $A \in \mathrm{Com}(X)$. Let $\ell \in (X^*)_1$. Then on $\overline{A[X_1]} = Q$, we have that $|\ell(y)| \le \|y\| \le \|A\|$ and $\|\ell(y) - \ell(\tilde{y})\| \le \|y - \tilde{y}\|$. It follows that $\{\ell \subset (X^*)_1\} \subset C(Q)$ is a uniformly equicontinuous family of continuous functions on a compact space.

By the Ascoli–Arzelà theorem (Theorem 2.3.14 of Part 1), given any sequence $\{\ell_n\}_{n=1}^\infty$ in $(X^*)_1$, there is $n(j)$, so $\ell_{n(j)}$ are Cauchy in the uniform norm on Q. Since $\ell(Ax) = (A^{\mathrm{t}}\ell)(x)$, $\{A^{\mathrm{t}}\ell_{n(j)}\}_{n=1}^\infty$ are uniformly Cauchy on X_1. Since $\sup_{x \in X_1} |\ell(x)| = \|\ell\|_{X^*}$, $\{A^{\mathrm{t}}\ell_{n(j)}\}_{j=1}^\infty$ are Cauchy in the $\|\cdot\|_{X^*}$-norm. By the lemma, A^{t} is compact. □

Example 3.1.4. Recall (see Lemma 5.1.6 of Part 1) the Riesz lemma says if Y is a proper closed subspace of an NLS, X, then there exists $x \in X$ so $\|x\| = 1$ and $\mathrm{dist}(x, Y) \ge \frac{1}{2}$. Given this, an inductive argument shows if $\dim(X) = \infty$, there exist $\{x_n\}_{n=1}^\infty \subset X_1$, so $\|x_n - x_m\| \ge \frac{1}{2}$ if $n \ne m$, and this implies X_1 is not compact. Thus, in an infinite-dimensional space, the identity map is never in $\mathrm{Com}(X)$, that is, $\mathrm{Com}(X)$ is a proper subset of $\mathcal{L}(X)$. □

As a consequence:

Theorem 3.1.5. *If X is an infinite-dimensional Banach space and $A \in \mathrm{Com}(X)$, then $0 \in \sigma(A)$.*

Proof. If $0 \in \rho(A)$, then there exists $B \in \mathcal{L}(\mathcal{H})$, so $BA = \mathbb{1}$. Since $\mathrm{Com}(X)$ is an ideal, $\mathbb{1}$ would be compact. Since it is not compact, $0 \notin \rho(A)$. □

Example 3.1.6. Let Ω be a compact metric space and $K\colon \Omega \times \Omega \to \mathbb{C}$ continuous. Let μ be a Baire probability measure on Ω. Let $X = C(\Omega)$ and

$A\colon X \to X$ by

$$(Af)(\omega) = \int K(\omega, \tilde{\omega}) f(\tilde{\omega})\, d\mu(\tilde{\omega}) \tag{3.1.2}$$

Since $\|Af\|_\infty \le \|K\|_\infty \|f\|_\infty$, $A \in \mathcal{L}(X)$. We claim $A \in \mathrm{Com}(X)$. Note first that K is uniformly continuous. Thus, for any ε, there is a δ with $\rho(\omega_1, \omega_2) < \delta \Rightarrow |K(\omega_1, \tilde{\omega}) - K(\omega_2, \tilde{\omega})| < \varepsilon$ for all $\tilde{\omega}$. In particular, if also $\|f\|_\infty \le 1$, then

$$|(Af)(\omega_1) - (Af)(\omega_2)| \le \varepsilon \tag{3.1.3}$$

Thus, $\{Af \mid \|f\|_\infty \le 1\}$ is uniformly bounded and uniformly equicontinuous, and so, by the Ascoli–Arzelà theorem (see Theorem 2.3.14 of Part 1), this set has compact closure, that is, $A \in \mathrm{Com}(X)$. □

Proposition 3.1.7. *Any $A \in \mathcal{L}(X)$ is continuous as a map of X to X when domain and range are given the $\sigma(X, X^*)$-topology.*

Proof. Immediate from $\ell \in X^* \Rightarrow A^{\mathrm{t}}\ell \in X^*$. □

Theorem 3.1.8. *$\mathrm{CC}(X)$ is a two-sided, norm-closed ideal in $\mathcal{L}(X)$.*

Proof. If X^* is separable, the $\sigma(X, X^*)$-topology on X_1 is given by a metric and $\mathrm{CC}(X)$ is just the continuous linear maps of X_1^* in the $\sigma(X, X^*)$-topology to X in the norm topology. The fact that a uniform limit of continuous functions is continuous implies $\mathrm{CC}(X)$ is norm-closed. Even if X^* is not separable, the same argument that proves a uniform limit of continuous functions is continuous proves $\mathrm{CC}(X)$ is norm-closed.

That $\mathrm{CC}(X)$ is a vector space is obvious. If $A \in \mathrm{CC}(X)$ and $B \in \mathcal{L}(X)$, $AB \in \mathrm{CC}(X)$ by Proposition 3.1.7 and $BA \in \mathrm{CC}(X)$ since B is continuous in norm. □

Theorem 3.1.9. *For any Banach space, X,*

$$\mathrm{FA}(X) \subset \mathrm{Com}(X) \subset \mathrm{CC}(X) \tag{3.1.4}$$

Proof. If $\mathrm{rank}(A) < \infty$, $A[X_1] \subset \{y \in \mathrm{Ran}(A) \mid \|y\| \le \|A\|\}$, a ball in a finite-dimensional space which is compact, so $A \in \mathrm{Com}(X)$. Since $\mathrm{Com}(X)$ is norm-closed, the closure of the finite rank operators lies in $\mathrm{Com}(X)$.

Let $A \in \mathrm{Com}(X)$ and $\{x_n\}_{n=1}^\infty \subset X_1$, so $x_n \to x_\infty$ in $\sigma(X, X^*)$-topology. By Proposition 3.1.7, $Ax_n \to Ax_\infty$ in $\sigma(X, X^*)$. Since any norm limit point is a weak limit point, Ax_∞ is the only norm limit point. Since a sequence in a compact metric space with a unique limit point converges, $Ax_n \to Ax_\infty$ in norm. Thus, $A \in \mathrm{CC}(X)$. □

Theorem 3.1.10. *If X is a reflexive Banach space with X^* separable (and, in particular, if X is a Hilbert space),*

$$\mathrm{Com}(X) = \mathrm{CC}(X)$$

Moreover, if $A \in \mathrm{Com}(X)$, $A[X_1]$ is norm-closed.

Proof. Since X is reflexive, the $\sigma(X, X^*)$-topology is the $\sigma(X^{**}, X^*)$- topology, so X_1 is $\sigma(X, X^*)$-compact by the Banach–Alaoglu theorem (see Theorem 5.8.1 of Part 1). Since X^* is separable, the $\sigma(X, X^*)$-topology on X_1 is metrizable (see Theorem 5.7.2 of Part 1). If $A \in \mathrm{CC}(X)$, A is continuous from $(X, \sigma(X, X^*))$ to $(X, \|\cdot\|)$, where (X, τ) has $\tau =$ topology. Thus, $A[X_1^*]$ is compact and so closed. Thus, $A \in \mathrm{Com}(X)$ and $A[X_1^*]$ is closed. □

Example 3.1.11. Let $X = \ell^1$. Then Example 5.7.3 of Part 1 showed $x_n \to 0$ in $\sigma(X, X^*) \Leftrightarrow \|x_n\| \to 0$ (for sequences only). Thus, $\mathbb{1} \in \mathrm{CC}(\ell^1)$, so $\mathrm{Com}(\ell^1) \neq \mathrm{CC}(\ell^1)$. □

Our final topic discusses when $\mathrm{FA}(X) = \mathrm{Com}(X)$. We begin with the Hilbert space case. Even though our general analysis below applies, the warmup is illuminating.

Proposition 3.1.12. *Let $A \in \mathrm{Com}(\mathcal{H})$ and $\{\varphi_n\}_{n=1}^\infty$ an orthonormal set. Then $\|A\varphi_n\| \to 0$ as $n \to \infty$.*

Proof. Since Bessel's inequality says $\sum_{n=1}^\infty |\langle\psi, \varphi_n\rangle|^2 < \infty$ for any ψ, we have $\varphi_n \xrightarrow{w} 0$ so $\|A\varphi_n\| \to 0$ since $A \in \mathrm{CC}(\mathcal{H})$. □

Theorem 3.1.13. *For a Hilbert space, $\mathcal{H}$, $\mathrm{FA}(\mathcal{H}) = \mathrm{Com}(\mathcal{H})$.*

Proof. We need only show that any $A \subset \mathrm{Com}(\mathcal{H})$ is a norm limit of finite rank operators. Pick $\{\varphi_n\}_{n=1}^\infty$ and projections, P_n, inductively as follows:

$$\|\varphi_1\| = 1, \qquad \|A\varphi_1\| \geq \tfrac{1}{2}\|A\| \tag{3.1.5}$$

$$P_n = \text{projection onto } \mathrm{span}(\varphi_1, \dots, \varphi_n), \quad n = 1, 2, \dots$$

$$\|\varphi_n\| = 1, \quad P_{n-1}\varphi_n = 0, \quad \|A\varphi_n\| \geq \tfrac{1}{2}\|A(1 - P_{n-1})\| \tag{3.1.6}$$

which is possible since $\|A(1 - P_{n-1})\| = \sup_{|\psi|=1,\, P_{n-1}\psi=0}\|A\psi\|$.

Since $P_{n-1}\varphi_n = 0$, $\{\varphi_n\}_{n=1}^\infty$ are orthonormal. By the proposition, $\|A\varphi_n\| \to 0$. By (3.1.6), $\|A - AP_n\| \to 0$, so A is the norm limit of the finite rank operators AP_n. □

It will be useful to abstract the argument of this theorem. We note that if Q is the projection onto $\mathrm{Ker}(A)^\perp$, $\|AQ\varphi\| = \|A\varphi\|$, while $\|Q\varphi\| \leq \|\varphi\|$. Thus, we can inductively pick $\{\varphi_n\}_{n=1}^\infty$ so $\varphi_n \perp \mathrm{Ker}(A)$. Thus,

Proposition 3.1.14. *Suppose $A \in \mathcal{L}(\mathcal{H})$ so that for any orthonormal set $\{\varphi_n\}_{n=1}^\infty$ in $\mathrm{Ker}(A)^\perp$, we have $\|A\varphi_n\| \to 0$. Then A is compact.*

Example 3.1.15 (Hilbert–Schmidt Integral Kernels)**.** In many examples of interest, X is some space of functions on a measure space (Ω, Σ, μ) and A is given in the form

$$(Af)(x) = \int K(x, y) f(y)\, d\mu(y) \tag{3.1.7}$$

where K is a function on $\Omega \times \Omega$ called the *integral kernel* for A (which we will sometimes write as A_K). This use of kernel has no relation at all to $\mathrm{Ker}(A)$! There is, of course, an issue of convergence of the integral.

Particularly simple is the case $X = L^2(\Omega, d\mu)$ when $K \in L^2(\Omega \times \Omega, d\mu \otimes d\mu)$. The A_K are called *concrete Hilbert–Schmidt operators* and K the *Hilbert–Schmidt kernel.* By the Schwarz inequality,

$$\int |K(x,y)f(y)|\,d\mu(y) \le \left(\int |K(x,y)|^2\,d\mu(y)\right)^{\frac{1}{2}} \left(\int |f(y)|^2\,d\mu(y)\right)^{\frac{1}{2}} \tag{3.1.8}$$

so the integral in (3.1.7) is absolutely convergent for a.e. x and all $f \in L^2(\Omega, d\mu)$, since Fubini's theorem implies $\int |K(x,y)|^2\,d\mu(y)$ is finite for a.e. x.

Squaring (3.1.8) and integrating $d\mu(x)$ shows

$$\|A_K f\|_{L^2(\Omega,d\mu)} \le \|K\|_{L^2(\Omega\times\Omega,d\mu\otimes d\mu)} \|f\|_{L^2(\Omega,d\mu)} \tag{3.1.9}$$

showing that $A_K \in \mathcal{L}(\mathcal{H})$ with $\|A_K\| \le \|K\|_{L^2}$. We want to show in fact that A_K is compact.

Pick any orthonormal basis $\{\varphi_n\}_{n=1}^\infty$ of $L^2(\Omega, d\mu)$. Then by Theorem 4.11.8 of Part 1, $\{\varphi_n(x)\varphi_m(y)\}_{n,m=1}^\infty$ is an orthonormal basis for $L^2(\Omega \times \Omega, d\mu \otimes d\mu)$, so K has an abstract Fourier series expansion

$$K(x,y) = \sum_{n,m=1}^{\infty} \alpha_{nm}\varphi_n(x)\varphi_m(y) \tag{3.1.10}$$

In particular, if

$$K_N(x,y) = \sum_{n,m=1}^{N} \alpha_{nm}\varphi_n(x)\varphi_m(y) \tag{3.1.11}$$

then by convergence of abstract Fourier expansions,

$$\|K - K_N\|_{L^2(\Omega\times\Omega,d\mu\otimes d\mu)} \to 0 \tag{3.1.12}$$

which implies that A_K is a norm limit in $\mathcal{L}(\mathcal{H})$ of the A_{K_N}. Since

$$A_{K_N} f = \sum_{n,m=1}^{N} \alpha_{nm}\langle \bar\varphi_m, f\rangle \varphi_n \tag{3.1.13}$$

each A_{K_N} is finite rank, proving A_K is compact. $\square$

In Theorem 3.8.6, we'll note that if $f, g \in L^2(\mathbb{R}^\nu)$ and $g(P)$ is defined by $g(P)\varphi = (g(k)\widehat{\varphi})^\vee$, then $f(X)g(P)$ has a Hilbert-Schmidt kernel.

We summarize in:

Theorem 3.1.16. *For any $K \in L^2(\Omega \times \Omega, d\mu \otimes d\mu)$, A_K given by* (3.1.7) *defines an operator in $\mathcal{L}(L^2(\Omega, d\mu))$ which obeys* (3.1.9) *and which is a compact operator.*

By integrating, it is easy to see

$$K^*(x,y) \equiv \overline{K(y,x)} \Rightarrow A_K^* = A_{K^*} \tag{3.1.14}$$

With that in hand, we get compactness of a larger class of integral operators; we need a definition: $L_\varepsilon^\infty(\mathbb{R}^\nu) = \{f \in L^\infty(\mathbb{R}^\nu) \mid \forall\varepsilon\, \exists R \text{ with } |f(x)| \le \varepsilon \text{ if } |x| > R\}$:

Theorem 3.1.17. *Let p and q be dual indices in $[1,\infty]$. Let $f \in L^p(\mathbb{R}^\nu, d^\nu x)$, $g \in L^q(\mathbb{R}^\nu, d^\nu x)$ with $g \in L_\varepsilon^\infty(\mathbb{R}^\nu)$ if $q=\infty$ and $f \in L_\varepsilon^\infty$ if $p=\infty$. Let*

$$(A_{f,g}h)(x) = \int f(x)g(x-y)h(y)\,d^\nu y \tag{3.1.15}$$

Then $A_{f,g}$ is a compact operator on $L^2(\mathbb{R}^\nu, d^\nu x)$.

Remarks. 1. The idea of the proof allows a general result. If Y is a Banach space of measurable functions on $\Omega \times \Omega$ so that for $K \in Y$, the integral operator given by (3.1.7) is a bounded operator with $\|A_K\| \le \|K\|_Y$ and if $Y \cap L^2(\Omega\times\Omega, d\mu\otimes d\mu)$ is dense in Y, then A_K is compact.

2. The same idea shows if Ω is compact in $\mathbb{R}^\nu$, $f \in L^1$, then $(A_f h)(x) = \int f(x-y)h(y)\,d^\nu y$ is compact as an operator from $L^2(\Omega, d^\nu x)$ to itself.

Proof. By Hölder and Young's inequality (see Sections 5.3 and 6.6 of Part 1),

$$\|A_{f,g}h\|_2 \le \|f\|_p\|g\|_q\|h\|_2 \tag{3.1.16}$$

Since $L^2 \cap L^p$ is L^p-dense in L^p and $L^2 \cap L^q$ is L^q-dense in L^q (if $q=\infty$, in L_ε^∞), we can find $f_n, g_n \in L^2$ so $\|A_{f,g} - A_{f_n,g_n}\| \to 0$. Since A_{f_n,g_n} is a Hilbert–Schmidt operator, we get the results by the fact that $\text{Com}(L^2)$ is norm-closed. $\square$

We complete this section with a mini-tutorial on Schauder bases and the proof that any Banach space with a Schauder basis has $\text{FA}(X) = \text{Com}(X)$ (a Banach space with $\text{FA}(X) = \text{Com}(X)$ is said to have the *approximation property*).

Definition. A *Schauder basis*, $\{x_n\}_{n=1}^\infty$, of a Banach space, X, is a family of vectors so that for any $x \in X$, there exists a unique sequence $\{\alpha_n\}_{n=1}^\infty \in \mathbb{K}^\infty$, so that

$$S_N(x) \equiv \sum_{n=1}^N \alpha_n x_n \to x \tag{3.1.17}$$

If $\|x_n\| = 1$ for all n, we say the basis is *normalized.* By replacing x_n by $x_n/\|x_n\|$, we can always suppose the basis is normalized, which we henceforth do.

We set $\ell_n(x) = \alpha_n$. The uniqueness implies the ℓ_n are linear. We'll see shortly that if the basis is normalized, each $\ell_n \in X^*$ and $\sup_n \|\ell_n\| < \infty$.

Example 3.1.18. In ℓ^p, $1 \le p < \infty$, it is easy to see that $\{\delta_n\}_{n=1}^\infty$, where $(\delta_n)_m = \delta_{nm}$, is a Schauder basis. By Corollary 5.8.5 of Part 3, $\{e^{2\pi i n \cdot}\}_{n=-\infty}^\infty$ is a Schauder basis for $L^p((0,1), dx)$, $1 < p < \infty$. The Haar basis of Example 4.6.1 of Part 3 is a Schauder basis of L^p, $1 \le p < \infty$. Problem 4 constructs a Schauder basis for $C([0,1])$. □

Given a Schauder basis, define

$$\|x\|_\sim = \sup_N \|S_N(x)\| \tag{3.1.18}$$

Since $S_N(x) \to x$, this sup is finite for each x and

$$\|x\| \le \|x\|_\sim \tag{3.1.19}$$

Lemma 3.1.19. *X is complete in $\|\cdot\|_\sim$.*

Proof. Let $x^{(m)}$ be a Cauchy sequence in $\|\cdot\|_\sim$. By (3.1.19), $x^{(m)}$ is Cauchy in $\|\cdot\|$, so there is $x^{(\infty)} \in X$ with $\|x - x^{(\infty)}\| \to 0$.

Since $|\ell_n(x)| = \|S_n(x) - S_{n-1}(x)\|$, $\ell_n(x^{(m)})$ is Cauchy in $\mathbb{K}$ for each n, so $\ell_n(x)$ has a limit $\beta_n^{(\infty)}$ and

$$S_N(x^{(m)}) \to \sum_{n=1}^N \beta_n^{(\infty)} x_n \tag{3.1.20}$$

Since $x^{(m)}$ is Cauchy in $\|\cdot\|_\sim$, the $\|\cdot\|$-convergence of $S_N(x^{(m)})$ to $x^{(m)}$ as $N \to \infty$ is uniform in m, so by (3.1.20), $\sum_{n=1}^N \beta_n^{(\infty)} x_n \to x^{(\infty)}$. Thus, by the uniqueness property of the Schauder basis expansion, $\beta_n^{(\infty)} = \ell_n(x^{(\infty)})$ and $S_N(x^{(\infty)}) = \sum_{n=1}^N \beta_n^{(\infty)} x_n$. By (3.1.20) and the Cauchy assumption in $\|\cdot\|_\sim$, $\|x^{(m)} - x^{(\infty)}\|_\sim \to 0$, that is, every Cauchy sequence converges. □

Theorem 3.1.20. *Let $\{x_n\}_{n=1}^\infty$ be a Schauder basis for a Banach space, X. Then*

(a) *For some $C > 0$,*

$$\sup_N S_N(x) \le C\|x\| \tag{3.1.21}$$

(b) *Each $\ell_n \in X^*$ and*

$$\sup_n \|\ell_n\| < \infty \tag{3.1.22}$$

(c) *The convergence of $S_N(x)$ to x is uniform on $\|\cdot\|$-compact subsets of X.*

(d) *If $A \in \mathrm{Com}(X)$, then $A_N x \equiv S_N(Ax)$ is finite rank and $\|A_N - A\| \to 0$. Thus, $\mathrm{Com}(X) = \mathrm{FA}(X)$.*

Proof. (a) By Corollary 5.4.15 of Part 1, (3.1.19), and Lemma 3.1.19, we get (3.1.21).

(b) Since $\ell_n = S_n - S_{n-1}$, we have

$$|\ell_n(x)| \le 2C\|x\| \tag{3.1.23}$$

(c) Given a compact set $S \subset X$, pick $x_1, \dots, x_\ell$, so that $\bigcup_{j=1}^{\ell} B_{\varepsilon/3C}(x_j)$ covers S. Then pick N_0 so $N > N_0 \Rightarrow \|S_N(x_j) - x_j\| < \varepsilon/3$ for $j = 1, \dots, \ell$. Then for any $x \in S$, pick j so $\|x - x_j\| < \varepsilon/3C$ and note that

$$\begin{aligned} \|S_N(x) - x\| &\le \|S_N(x_j) - x_j\| + \|S_N(x) - S_N(x_j)\| + \|x - x_j\| \\ &< \frac{\varepsilon}{3} + C\left(\frac{\varepsilon}{3C}\right) + \frac{\varepsilon}{3} \le \varepsilon \end{aligned}$$

proving uniform convergence.

(d) Let $S = \overline{A[X_1]}$. Then, by (c),

$$\|A_N - A\| \le \sup_{\|x\| \le 1} \|S_N(Ax) - Ax\| \to 0$$ □

Notes and Historical Remarks. The notions connected to compact operators came out of attempts to understand Fredholm's 1903 work [**206**] on integral equations of the form (2.1.40). We saw in Example 3.1.6 that $f \mapsto \int_a^b K(\,\cdot\,, y) f(y)\, dy$ defines a compact operator, but that was not noted or used by Fredholm. Indeed, it should be emphasized that Fredholm did his work before Fréchet had defined abstract metric spaces or the general notion of compact set!

Much of the early work done by Hilbert and his school was done on Hilbert space where $\text{FA}(\mathcal{H}) = \text{Com}(\mathcal{H}) = \text{CC}(\mathcal{H})$ and the differing definitions were used somewhat interchangeably. The German word "vollstetigkeit" was used. In German, "stetigkeit" means steadiness and is used for continuous, so "vollstetigkeit" means "completely continuous." Only in 1948 did Hille [**320**] suggest the English term "compact" which has become so standard that if you ask Google to translate "vollstetigkeit," it will say "compactness" even though "completely continuous" now has another meaning!

Among the high points in the early codification of these notions are:

(1) the appearance of infinite matrices (operators on ℓ^2) with $\sum_{j,k=1}^{\infty} |a_{jk}|^2 < \infty$, the discrete analog of Example 3.1.15 in 1906 in Part 4 of Hilbert's work [**317**];
(2) Schmidt's 1907 work [**601**] where he approximated continuous kernels with finite rank operators;

(3) Riesz's 1913 book [**564**] where he defined completely continuous by taking weakly convergent sequences in the unit ball to norm convergent sequences;
(4) Riesz's great 1918 paper [**566**] where he made a comprehensive spectral analysis of compact operators on $C([a,b])$ using ideas that will be discussed in Section 3.3.

Schauder's theorem that $A \in \text{Com}(X) \Rightarrow A^t \in \text{Com}(X^*)$ is from Schauder [**598**] in 1930.

Erhard Schmidt (1876–1959) was a student of Hilbert who, as we've noted, did fundamental work on compact operators. The terms Hilbert–Schmidt kernel, operator, and theorem are named in honor of their related fundamental work rather than any joint papers.

As we saw, there are spaces (like ℓ^1) with $\text{Com}(X) \neq \text{CC}(X)$. For a long time, it was an open question if $\text{FA}(X)$ always equals $\text{Com}(X)$. Finally, in 1973, Enflo [**185**] constructed a Banach space for which $\text{FA}(X) \neq \text{Com}(X)$. For other examples, which include $\mathcal{L}(\ell^2)$ (!), see Gowers–Maurey [**266, 267**], Pisier [**523, 524, 526**], Szankowski [**681**], and Tsirelson [**706**]. Recall the Banach–Mazur theorem (Theorem 5.0.1 of Part 1) says that any Banach space is isometrically isomorphic to a closed space in $C\big([0,1]\big)$ which has a Schauder basis (see Problem 4). Thus, by thinking of the range in a large space, we can get finite rank approximations of a sort for $A \in \text{Com}(X)$.

The notion of Schauder basis is from a 1927 paper of Schauder [**596**]. It included the construction in Problem 4. He considered Haar bases in [**597**]. In his original definition, Schauder required that the ℓ_n be continuous. That this isn't needed and the rest of Theorem 3.1.20 is due to Banach [**44**].

For much more on Schauder bases, see the books of Heil [**306**], Megginson [**472**], Singer[**656**], and Young [**770**].

These notions extend to operators between two Banach spaces (see Problems 5–8). In particular, if $\mathcal{H}$ is a Hilbert space, then $\text{Com}(\mathcal{H}, X) = \text{FA}(\mathcal{H}, X)$.

Problems

1. This problem will show that if $\mathcal{H}$ is a Hilbert space and $\mathcal{I}$ is a norm-closed, two-sided ideal in $\mathcal{L}(\mathcal{H})$, then $\mathcal{I}$ is $\{0\}$, $\mathcal{I}_\infty$, or all of $\mathcal{L}(\mathcal{H})$. It will need the spectral theorem as discussed in Chapter 5 (or as Theorem 2.4.13).

   (a) If $A \in \mathcal{I}$, $A \neq 0$, prove that for any $\varphi, \psi \in \mathcal{H}$, $\langle \varphi, \cdot \rangle \psi \in \mathcal{I}$. (*Hint*: If $\langle \gamma, A\eta \rangle \neq 0$, $B = \langle \gamma, \cdot \rangle \psi$, and $C = \langle \varphi, \cdot \rangle \eta$, compute BAC.)

   (b) If $\mathcal{I}$ is not $\{0\}$, prove $\mathcal{I}_\infty \subset \mathcal{I}$.

   (c) If there exists $A \in \mathcal{I}$, $A \notin \mathcal{I}_\infty$, prove $|A| \notin \mathcal{I}_\infty$ and $|A| \in \mathcal{I}$.

(d) If $A > 0$, $A \notin \mathcal{I}_\infty$, prove, for some $a > 0$, the spectral projection, $P_{(a,\|A\|]}$, is not compact.

(e) If P is a projection and not compact, prove that $UPV = \mathbb{1}$ for some $U, V \in \mathcal{L}(\mathcal{H})$.

(f) If $\mathcal{I} \neq \{0\}, \mathcal{I}_\infty$, conclude $\mathcal{I} = \mathcal{L}(\mathcal{H})$.

2. If $A^{\rm t} \in \text{Com}(X^*)$, prove that $A \in \text{Com}(X)$. (*Hint*: $A^{\rm tt} \restriction i[X] = i \circ A$ if i is the isometric embedding of X into X^{**}.)

3. This will provide an alternate proof that $A \in \text{Com}(X) \Rightarrow A^{\rm t} \in \text{Com}(X^*)$ (Schauder's theorem).

(a) If $\ell_n \to \ell$ in the $\sigma(X, X^*)$-topology with $\sup_n \|\ell_n\| \le 1$, prove that $\ell_n(x) \to \ell(x)$ uniformly for x in any compact subset of X.

(b) If $A \in \text{Com}(X)$ and $\ell_n \to \ell$ in $\sigma(X, X^*)$, prove that $\|A^{\rm t}\ell_n - A^{\rm t}\ell\| \to 0$.

(c) If $A \in \text{Com}(X)$, prove that $A^{\rm t}[X_1^*]$ is compact.

4. This problem will construct a Schauder basis of $C([0,1])$.

(a) If $x_1, \dots, x_n \in (0,1)$ are distinct and $y_0 \equiv 0 < y_1 < \dots < y_n < 1 \equiv y_{n+1}$ is their reordering in increasing order with $y_j = x_n$, then $f_{(x_1,\dots,x_n)}$ is the function with

$$f(y_k) = 0, \quad \text{all } k \neq j; \qquad f(y_j) = 1$$

and f is linear on each $[y_\ell, y_{\ell+1}]$, $\ell = 0, 1, \dots, n$. Prove that each $f_{(x_1,\dots,x_n)} \in C([0,1])$, with $\|f_{(x_1,\dots,x_n)}\|_\infty = 1$, and that the linear span of $\{f_{(x_1,\dots,x_\ell)}\}_{\ell=1}^n$ is the set of f in $C([0,1])$ which are 0 at $x = 0$ and $x = 1$ and piecewise linear on each $[y_\ell, y_{\ell+1}]$, $\ell = 0, 1, \dots, n$.

(b) Let $\{x_j\}_{j=1}^\infty$ be a dense set in $[0,1]$ of points distinct and not 0 or 1. If $g \in C([0,1])$, with $g(0) = g(1) = 0$, let $g_\ell(x)$ be the function which agrees with g at $0, 1$ and $\{x_j\}_{j=1}^\ell$ and is otherwise piecewise linear. Prove that $\|g - g_\ell\|_\infty \to 0$.

(c) Prove that $\{g_{(x_1,\dots,x_\ell)}\}_{\ell=1}^\infty$ is a Schauder basis of $\{g \in C([0,1]) \mid g(0) = g(1) = 0\}$.

(d) Let $h_0(x) \equiv 1$ and $h_1(x) = x$. Prove that $\{h_0, h_1, \{g_{(x_1,\dots,x_\ell)}\}_{\ell=1}^\infty\}$ is a Schauder basis of $C([0,1])$.

5. Let X, Y be two Banach spaces. Extend the notions of $\text{Com}(X)$, $\text{FA}(X)$, $\text{CC}(X)$ to maps $T\colon X \to Y$ so, for example,

$$\text{Com}(X, Y) = \{T \mid \overline{T[B_1^X]} \text{ is compact in } Y\}$$

Prove each of these spaces is a norm-closed vector space which is a two-sided ideal in the sense that for all $A \in \mathcal{L}(Y)$ and $B \in \mathcal{L}(X)$, T in one of three spaces implies ATB is in the same space.

6. If $A \in \mathrm{FA}(X,Y)$ (respectively, $\mathrm{Com}(X,Y)$), prove that $A^{\mathrm{t}} \in \mathrm{FA}(Y,X)$ (respectively, $\mathrm{Com}(Y,X)$).

7. For any pair of Banach spaces, prove that $\mathrm{FA}(X,Y) \subset \mathrm{Com}(X,Y) \subset \mathrm{CC}(X,Y)$.

8. Let $\mathcal{H}$ be a Hilbert space and Y a Banach space. Prove that $\mathrm{Com}(\mathcal{H},Y) = \mathrm{FA}(\mathcal{H},Y)$.

3.2. The Hilbert–Schmidt Theorem

In this section, we'll prove the following, which can be thought of as the spectral theorem for positive compact operators:

Theorem 3.2.1 (The Hilbert–Schmidt Theorem). *Let $A \in \mathcal{L}(\mathcal{H})$ be positive and compact. Then there is an orthonormal basis of eigenvectors whose eigenvalues accumulate only at 0; more specifically, there exist two orthonormal sets, $\{\varphi_n\}_{n=1}^N$ and $\{\psi_m\}_{m=1}^M$, whose union is an orthonormal basis so that*

$$A\varphi_n = \lambda_n\varphi_n, \quad \lambda_1 \geq \lambda_2 \geq \cdots > 0 \tag{3.2.1}$$

$$\lambda_n \downarrow 0 \quad \text{if } N = \infty \tag{3.2.2}$$

$$A\psi_m = 0 \tag{3.2.3}$$

Remarks. 1. We'll also have results if A is only normal (and compact); see Theorem 3.2.3.

2. Besides the proof here, there are at least two other proofs to mention: one from the general spectral theorem (see Problem 1) and one that uses the general spectral analysis of compact operators of the next section (see Problem 4 of Section 3.3). Our proof here is quite illuminating and yields min-max formulae which are widely useful.

After proving this theorem, we'll apply it to differential operators on an interval (Sturm–Liouville theory).

Lemma 3.2.2. (a) *If $\varphi_n \to \varphi_\infty$ weakly (with $\sup_n \|\varphi_n\| < \infty$) and $\psi_n \to \psi_\infty$ in norm, then $\langle \varphi_n, \psi_n \rangle \to \langle \varphi_\infty, \psi_\infty \rangle$.*

(b) *If A is compact and $\varphi_n \to \varphi_\infty$ weakly, then $\langle \varphi_n, A\varphi_n \rangle \to \langle \varphi_\infty, A\varphi_\infty \rangle$.*

Remarks. 1. If $\{\varphi_n\}_{n=1}^\infty$ is an orthonormal basis, $\varphi_n \to 0$ weakly, but $\langle \varphi_n, \varphi_n \rangle \equiv 1 \nrightarrow \langle 0, 0 \rangle = 0$, so norm convergence of one of the vectors in (a) is essential.

2. That $\sup_n \|\varphi_n\| < \infty$ follows from the uniform boundedness principle.

Proof. (a) We have that

$$\langle \varphi_\infty, \psi_\infty \rangle - \langle \varphi_n, \psi_n \rangle = \langle \varphi_\infty - \varphi_n, \psi_\infty \rangle + \langle \varphi_n, \psi_\infty - \psi_n \rangle \tag{3.2.4}$$

so

$$|\langle \varphi_\infty, \psi_\infty \rangle - \langle \varphi_n, \psi_n \rangle| \leq |\langle \varphi_\infty - \varphi_n, \psi_\infty \rangle| + \|\psi_\infty - \psi_n\| \to 0 \tag{3.2.5}$$

by the assumed weak and norm convergence.

(b) Since A is compact, $A\varphi_n \to A\varphi_\infty$ in norm, so this follows from (a). □

Proof of Theorem 3.2.1. On the closed unit ball, $\mathcal{H}_1$, in $\mathcal{H}$, define

$$a(\varphi) = \langle \varphi, A\varphi \rangle \tag{3.2.6}$$

By the lemma, $\varphi \mapsto a(\varphi)$ is continuous if $\mathcal{H}_1$ is given the weak topology. Since $\mathcal{H}_1$ is compact in this topology, a must take its maximum value on $\mathcal{H}_1$, so there exist $\varphi_1 \in \mathcal{H}_1$ so

$$\langle \varphi_1, A\varphi_1 \rangle = \sup_{\|\varphi\| \leq 1} \langle \varphi, A\varphi \rangle = \|A\| \tag{3.2.7}$$

by Theorem 2.4.1. If $\|\varphi_1\| < 1$, $\langle (\varphi_1/\|\varphi_1\|), A(\varphi_1/\|\varphi_1\|) \rangle \geq \langle \varphi_1, A\varphi_1 \rangle$, so without loss, take φ_1 to have norm 1.

Given $\psi \perp \varphi_1$, let $\eta_t = (\varphi_1 + t\psi)/\|\varphi_1 + t\psi\|$ for $t \in \mathbb{R}$. Since $\|\varphi_1 + t\psi\|^2 = 1 + t^2\|\psi\|^2$ (because $\langle \psi, \varphi_1 \rangle = 0$), we see that

$$\langle \eta_t, A\eta_t \rangle = \langle \varphi_1, A\varphi_1 \rangle + 2t \operatorname{Re}\langle \psi, A\varphi_1 \rangle + O(t^2) \tag{3.2.8}$$

Since $\langle \varphi_1, A\varphi_1 \rangle \geq \langle \eta_t, A\eta_t \rangle$ for all t, we must have $\operatorname{Re}(\langle \psi, A\varphi_1 \rangle) = 0$. Taking ψ to $i\psi$, we see $\langle \psi, A\varphi_1 \rangle = 0$, that is, $A\varphi_1 \in (\{\varphi_1\}^\perp)^\perp = \{\alpha\varphi_1 \mid \alpha \in \mathbb{C}\}$, that is, $A\varphi_1 = \alpha\varphi_1$. Since $\alpha = \langle \varphi_1, A\varphi_1 \rangle$, $\alpha = \|A\|$, that is, we have proven for any positive compact operator, A, there is φ_1 with $\|\varphi_1\| = 1$ and $A\varphi_1 = \|A\|\varphi_1$. Set $\lambda_1 = \|A\|$.

Define

$$\mathcal{H}^{(1)} = \{\varphi \mid \langle \varphi, \varphi_1 \rangle = 0\} \tag{3.2.9}$$

Then for any $\varphi \in \mathcal{H}^{(1)}$,

$$\langle \varphi_1, A\varphi \rangle = \langle A\varphi_1, \varphi \rangle = \lambda_1 \langle \varphi_1, \varphi \rangle = 0 \tag{3.2.10}$$

so $A\colon \mathcal{H}^{(1)} \to \mathcal{H}^{(1)}$. Define $A_1 = A \restriction \mathcal{H}^{(1)}$.

Clearly, A_1 is compact and positive, so as above, there is $\varphi_2 \in \mathcal{H}^{(1)}$ and $\lambda_2 \geq 0$, so $A\varphi_2 = A_1\varphi_2 = \lambda_2\varphi_2$. Proceeding inductively, we find $\varphi_1, \dots, \varphi_n$ so $A\varphi_j = \lambda_j\varphi_j$ with $\lambda_j = \|A_{j-1}\| \leq \|A_{j-2}\| = \lambda_{j-1}$, and then $\mathcal{H}^{(j)} = \{\varphi \mid \langle \varphi, \varphi_k \rangle = 0,\, k = 1, \dots, j\}$ and $\varphi_{j+1} \in \mathcal{H}^{(j)}$, so $A\varphi_{j+1} = \|A_j\|\varphi_{j+1}$.

As constructed, $\{\varphi_j\}_{j=1}^\infty$ is an orthonormal set, $\lambda_1 \geq \lambda_2 \geq \cdots \geq 0$ and $\|A \restriction \mathcal{H}^{(j)}\| = \lambda_{j+1}$. By Proposition 3.1.12, $\lambda_j \to 0$. Thus, if $\mathcal{H}^{(\infty)} = \bigcap_{j=1}^\infty \mathcal{H}^{(j)}$, then $\|A \restriction \mathcal{H}^{(\infty)}\| \leq \lambda_j$ for all j, and thus, $A \restriction \mathcal{H}^{(\infty)}$ is 0, that is,

$\mathcal{H}^{(\infty)} \subset \mathrm{Ker}(A)$. If $\{\psi_j\}_{j=1}^{M_1}$ is an orthonormal basis for $\mathcal{H}^{(\infty)}$, we almost have the prescribed families. If $\lambda_n > 0$ for all n, we have exactly as claimed. If $\lambda_N > 0$ but $\lambda_{N+1} = 0$, rename the $\{\varphi_j\}_{j=N+1}^{\infty}$ as ψ's (interlacing the original ψ's if they are infinite in number) and so get the claimed families. □

Theorem 3.2.1 has an extension to normal operators.

Theorem 3.2.3. *Let $A \in \mathcal{L}(\mathcal{H})$ be normal and compact. Then there is an orthonormal basis of eigenvectors whose eigenvalues accumulate only at 0; more specifically, there exist two orthonormal sets, $\{\varphi_n\}_{n=1}^{N}$ and $\{\psi_m\}_{n=1}^{M}$, whose union is an orthonormal basis so that*

$$A\varphi_n = \lambda_n \varphi_n, \qquad \lambda_n \neq 0 \tag{3.2.11}$$

$$|\lambda_n| \to 0 \qquad \text{if } N = \infty \tag{3.2.12}$$

$$A\psi_m = 0 \tag{3.2.13}$$

Problem 2 has a proof using the polar decomposition and Theorem 3.2.1. Problem 3 has a proof mimicking the proof of Theorem 3.2.1.

The proof of Theorem 3.2.1 provides a min-max characterization of the eigenvalues.

Theorem 3.2.4 (Max–Min Criterion). *Let $\lambda_1 \geq \lambda_2 \geq \dots$ be the eigenvalues of a positive compact operator, A, written in order counting multiplicity, with $\lambda_n = 0$ if A has fewer than n nonzero eigenvalues. Then*

$$\lambda_n = \inf_{\psi_1, \dots \psi_{n-1}} \Big(\sup_{\substack{\varphi \perp \psi_1, \dots, \psi_{n-1} \\ \|\varphi\| = 1}} \langle \varphi, A\varphi \rangle \Big) \tag{3.2.14}$$

Remarks. 1. By compactness, one can take min and max rather than inf and sup (see Problem 4).

2. We will extend this to general self-adjoint operators in Sections 3.14 and 7.8.

Proof. We have

$$\langle \varphi, A\varphi \rangle = \sum_{j=1}^{N} |\langle \varphi, \varphi_j \rangle|^2 \lambda_j \tag{3.2.15}$$

Since for any $\psi_1, \dots, \psi_{n-1}$, there exists φ in the span of $\{\varphi_1, \dots, \varphi_n\}$ with $\langle \psi_j, \varphi \rangle = 0$ and $\|\varphi\| = 1$,

$$\sup_{\substack{\varphi \perp \psi_1, \dots, \psi_{n-1} \\ \|\varphi\| = 1}} \langle \varphi, A\varphi \rangle \geq \lambda_n$$

for any $\psi_1, \dots, \psi_{n-1}$.

On the other hand, if $\psi_1 = \varphi_1, \dots, \psi_{n-1} = \varphi_{n-1}$, the sup is exactly taken if $\varphi = \varphi_n$ and $\langle \varphi_n, A\varphi_n \rangle = \lambda_n$. Thus,

$$\min\Big(\sup_{\substack{\varphi \perp \psi_1, \dots, \psi_{n-1} \\ \|\varphi\|=1}} \langle \varphi, A\varphi \rangle \Big) = \lambda_n$$

as required. □

Corollary 3.2.5. *If B is compact, there is $\varphi \in \mathcal{H}$, $\varphi \neq 0$, with $\|B\varphi\| = \|B\|\|\varphi\|$.*

Proof. Let $A = B^*B$. By the above, there is φ with $\|\varphi\| = 1$, so $\|B\varphi\|^2 = \langle \varphi, A\varphi \rangle = \|A\| = \|B\|^2$. □

The following is a main use for the min-max principle in that without it, it is not clear how to prove this fact for more than the lowest eigenvalue.

Corollary 3.2.6. *If $A \geq B \geq 0$ for compact operators A and B, we have that $\lambda_n(A) \geq \lambda_n(B)$ for each n.*

Proof. For any $\psi_1, \dots, \psi_{n-1}$,

$$\sup_{\substack{\varphi \perp \psi_1, \dots, \psi_{n-1} \\ \|\varphi\|=1}} \langle \varphi, B\varphi \rangle \leq \sup_{\substack{\varphi \perp \psi_1, \dots, \psi_{n-1} \\ \|\varphi\|=1}} \langle \varphi, A\varphi \rangle \tag{3.2.16}$$

Now take the inf and use (3.2.14). □

As a final topic, we want to apply this theorem to the theory of Sturm–Liouville operators. We'll consider solutions of $-u'' + Qu = \lambda u$. The more general $-(Pu')' + Qu = \lambda u$ case will be considered in Problem 5 and other boundary conditions in Problem 6 and Theorem 7.4.17. Here is what we'll prove:

Theorem 3.2.7 (Sturm–Liouville Theorem). *Let Q be a continuous function in $[0,1]$. Then there exists an orthonormal basis in $L^2([0,1], dx)$ of C^2-functions, $\{u_n\}_{n=1}^\infty$ obeying*

$$-u_n''(x) + Qu_n(x) = \lambda_n u_n(x), \qquad u_n(0) = u_n(1) = 0 \tag{3.2.17}$$

where $\lambda_1 \leq \lambda_2 \leq \cdots \lambda_n \leq \dots$ and $\lambda_n \to \infty$.

Remark. In Problem 7, the reader will add two additional facts: each λ_n is simple, that is, $\lambda_j < \lambda_{j+1}$, and any solution of (3.2.17) is one of the orthonormal basis.

In some sense, this theorem considers the operator, $B = -\frac{d^2}{dx^2} + Q(x)$ on L^2, which is not everywhere defined and not bounded. In a sense we'll study in Section 7.4, B is self-adjoint if one adds the boundary conditions $u(0) = u(1) = 0$ and carefully specifies the domain (or uses the theory of operator closure discussed in Section 7.1). In essence, the trick we'll use is to study $A = (B+c)^{-1}$ for c a large constant which is a bounded, compact, self-adjoint operator. While this is secretly what we are doing, we'll not need to specifically discuss B as a self-adjoint operator, although it will appear applied to C^2 functions.

By adding a constant to Q, we can suppose

$$Q(x) > 0, \qquad x \in [0,1] \tag{3.2.18}$$

which we henceforth do. By Theorem 2.2.14, there exist C^2-solutions $u_0(x), u_1(x)$ of

$$-u'' + Qu = 0 \tag{3.2.19}$$

with the boundary conditions

$$u_0(0) = 0, \quad u_0'(0) = 1, \quad u_1(1) = 0, \quad u_1'(1) = -1 \tag{3.2.20}$$

and because of (3.2.18),

$$u_0 > 0, \quad u_1 > 0, \quad u_0' > 0, \quad u_1' < 0 \quad \text{on } (0,1) \tag{3.2.21}$$

Let

$$W(x) = u_0'(x)u_1(x) - u_0(x)u_1'(x) > 0 \tag{3.2.22}$$

which is a constant, call it W_0, since $\frac{dW}{dx} = u_0''u_1 - u_0u_1'' = Q(u_0u_1 - u_0u_1) = 0$.

Define the *Green's function*, $G(x,y)$, by

$$G(x,y) = W_0^{-1}\big[u_0(x_<)u_1(x_>)\big] \tag{3.2.23}$$

where

$$x_< = \min(x,y), \qquad x_> = \max(x,y) \tag{3.2.24}$$

Notice that G is unchanged if u_0 and/or u_1 are multiplied by a nonzero constant. Thus the conditions on u' in (3.2.20) are merely for convenience. The critical conditions are $u_0(0) = u_1(1) = 0$ and the u_j is not identically zero. By (3.2.21), (3.2.22), and continuity of u_0 and u_1, we have $G > 0$ and G is bounded on $[0,1] \times [0,1]$, and so L^2 and thus,

$$(Ag)(x) = \int G(x,y)g(y)\,dy \tag{3.2.25}$$

defines a Hilbert–Schmidt operator, which is thus compact (see Theorem 3.1.16). Moreover, by (3.1.14), since $G(x,y) = \overline{G(y,x)}$, A is self-adjoint.

Proposition 3.2.8. *Let $g \in L^2([0,1],dx)$ and*

$$f = Ag \tag{3.2.26}$$

given by (3.2.25). *Then*

(a) *f is continuous on $[0,1]$ and obeys*

$$f(0) = f(1) = 0 \tag{3.2.27}$$

(b *If g is continuous, then f is C^2 and*

$$-f''(x) + Q(x)f(x) = g(x) \tag{3.2.28}$$

Proof. (a) Let

$$h_0(x) = W_0^{-1}u_1(x)g(x), \qquad h_1(x) = W_0^{-1}u_0(x)g(x) \tag{3.2.29}$$

Then

$$f(x) = u_0(x)\int_x^1 h_0(y)\,dy + u_1(x)\int_0^x h_1(y)\,dy \tag{3.2.30}$$

If $g \in L^2$, then $h_0, h_1 \in L^2$, so the integrals are continuous, showing f is also, and since $u_0(0) = 0$ and $\int_0^0 h(y)\,dy = 0$, we have $f(0) = 0$. Similarly, $f(1) = 0$.

(b) By (3.2.30) and (to kill the term from differentiating the integrals)

$$-u_0(x)h_0(x) + u_1(x)h_1(x) = 0 \tag{3.2.31}$$

we have that f is C^1 and

$$f'(x) = u_0'(x)\int_x^1 h_0(y)\,dy + u_1'(x)\int_0^x h(y)\,dy \tag{3.2.32}$$

Since u_0, u_1 are C^2 and the integrals are C^1, f' is C^1, so f is C^2. When we differentiate, we first use $u_j'' = Qu_j$ to get one term, which is Qf. There is a second from differentiating the integrals:

$$f''(x) = Q(x)f(x) + W_0^{-1}\big[-u_0'(x)u_1(x) + u_0(x)u_1'(x)\big]g(x) \tag{3.2.33}$$

$$= Q(x)f(x) \quad g(x) \tag{3.2.34}$$

by (3.2.22). This proves (3.2.28). □

Proposition 3.2.9. (a) *If f_1, f_2 are C^2 with*

$$f_1(0) = f_2(0) = f_1(1) = f_2(1) = 0 \tag{3.2.35}$$

then

$$\int_0^1 f_1(x)[-f_2''(x) + Q(x)f_2(x)]\,dx = \int_0^1 [f_1'(x)f_2'(x) + Q(x)f_1(x)f_2(x)]\,dx \tag{3.2.36}$$

(b) *If f is real-valued and C^2 and obeys* (3.2.27) *and $-f'' + Qf = 0$, then $f = 0$.*

(c) *If f is C^2 and obeys* (3.2.27) *and $g = -f'' + Qf$, then* (3.2.26) *holds.*

(d) $A \geq 0$. (3.2.37)

(e) $\text{Ker}(A) = 0$. (3.2.38)

Remarks. 1. Recall that we are assuming $Q(x) \geq 0$ for all $x \in [0,1]$.

2. (d), (e) imply, once we have Theorem 3.2.7, that $B \geq \lambda_1 \mathbb{1} > 0$.

Proof. (a) is a simple integration by parts. By (3.2.35), the boundary terms vanish.

(b) By (3.2.36),

$$\int_0^1 (|f'(x)|^2 + Q(x)|f(x)|^2)\,dx = 0 \tag{3.2.39}$$

Since $Q > 0$, $f = 0$ for a.e. x, and then, by continuity, for all x.

(c) Let $Ag = \widetilde{f}$. By Proposition 3.2.8, both f and $\widetilde{f}$ obey $-u'' + Qu = g$, so if $f^\sharp = f - \widetilde{f}$, then $-f^\sharp + Qf^\sharp = 0$. Since $f^\sharp(0) = f^\sharp(1) = 0$. $f^\sharp = 0$, that is, $f = \widetilde{f}$.

(d) Let g be continuous and $f = Ag$. Then, by Proposition 3.2.8, f is C^2 and obeys (3.2.27) and

$$\langle g, Ag\rangle = \langle -f'' + Qf, f\rangle = \int (|f'(x)|^2 + Q(x)|f(x)|^2)\,dx \tag{3.2.40}$$

by (3.2.36). Since $Q \geq 0$, $\langle g, Ag\rangle \geq 0$. Since the continuous functions are dense, $A \geq 0$.

(e) It suffices to prove that $\operatorname{Ran}(A)$ is dense by (2.2.13). Let f be C^2 obeying (3.2.38). If $g = -f'' + Qf$, by (3.2.26), $f = Ag$, so $f \in \operatorname{Ran}(A)$. Since $C_0^\infty((0,1))$ is dense in $L^2([0,1], dx)$, $\operatorname{Ran}(A)$ is dense. □

Proof of Theorem 3.2.7. Since A is self-adjoint, positive, and Hilbert–Schmidt, hence compact, and since $\operatorname{Ker}(A) = 0$, Theorem 3.2.1 implies that there exist $u_1, u_2, \dots$ in L^2 and $\gamma_1 \geq \gamma_2 \geq \cdots > 0$, so $\{u_j\}_{j=1}^\infty$ is a basis and

$$Au_j = \gamma_j u_j \tag{3.2.41}$$

By Proposition 3.2.8, $u_j = \gamma_j^{-1} Au_j$ is first continuous and then is C^2, obeys $u_j(0) = u_j(1) = 0$, and if $\lambda_j = \gamma_j^{-1}$, then

$$-u_j'' + Qu_j = \lambda_j u_j \tag{3.2.42}$$

Since $\gamma_j \to 0$, $\lambda_j \to \infty$. □

Notes and Historical Remarks. As we noted in the Notes to the last section, Schmidt was a student of Hilbert and generalized results of Hilbert [**317**] and simplified their proofs in his thesis published in [**601**]. The result is what we call the Hilbert–Schmidt theorem.

Erhard Schmidt (1876–1959) was born in what is now Estonia (then part of the Russian empire) where his father was a professor of medical biology. He was a student of Hilbert and in turn his students include Bochner, E. Hopf, and H. Hopf (the Hopfs were unrelated). In connection with his work on integral equations, he invented ℓ^2 and many of the most basic geometric results in Hilbert space in the context of ℓ^2.

After brief appointments in Zurich, Erlangen, and Breslau, he moved in 1917 to Berlin, filling the chair vacated by the retirement of Schwarz. Under Weierstrass, Kummer, and Kronecker, Berlin had been the center of analysis in Germany, a tradition continued under the leadership of Schwarz and Frobenius. Shortly after Schmidt's arrival, Carathéodory, who had replaced Frobenius, left and Schmidt became and remained the key figure in Berlin for many years. He was dismayed by the dismissal of Schur and von Mises for anti-Semitic reasons in the 1930s. While as the head of the department, he was involved in the persecution of Jews, there were complaints that he was not sufficiently anti-Semitic, so his reputation remained intact after the war when he was a major figure in the attempt to rehabilitate German mathematics.

Variational methods for eigenvalues go back to Rayleigh [**546**] and Ritz [**571**] who realized the top and bottom eigenvalues as a max and a min and intermediate as saddle points. The actual min-max principle (which is sometimes called the Courant–Fischer theorem) was found by Fischer in 1905 [**202**] and raised to high art by Courant [**133**].

The Sturm–Liouville theory goes back to their famous paper of 1837 [**675**]. As noted, our presentation relies, in essence, on writing down the inverse to B, where B is an unbounded self-adjoint operator. We didn't need to discuss B explicitly as such because we had an explicit formula for B^{-1}. When we don't have an explicit formula, e.g., dimension larger than one, we need to rely on unbounded operator theory and a compactness criterion. This is a subject we return to in Sections 3.16, 7.4, and 7.5.

Problems

1. This problem will provide an alternate proof of Theorem 3.2.1 using the spectral theorem.

   (a) Let $A \geq 0$. For some $0 < a < b$, suppose $\text{Ran}\big(P_{(a,b)}(A)\big)$ has infinite dimension. Prove that A is not compact. (*Hint*: If $\varphi \in \text{Ran}\big(P_{(a,b)}(A)\big)$, then $\|A\varphi\| \geq a\|\varphi\|$. Thus, for an orthonormal family in $\text{Ran}\big(P_{(a,b)}(A)\big)$, $A\varphi$ does not go to zero in norm.)

   (b) If $A \geq 0$ and $\dim(\text{Ran}(P_{(a,b)}(A)))$ is finite, prove $A \restriction \text{Ran}(P_{(a,b)}(A))$ has a complete set of eigenvectors.

   (c) Prove the Hilbert–Schmidt theorem.

2. This will prove Theorem 3.2.3 via the polar decomposition. Let A be normal.

   (a) If $A = U|A|$ is the polar decomposition, prove that $[U, |A|] = 0$.

(b) Fix λ an eigenvalue of $|A|$. Let $\mathcal{H}_\lambda = \{\varphi \mid |A|\varphi = \lambda\varphi\}$. Prove that $U \restriction \mathcal{H}_\lambda$ is unitary and has a complete set of eigenvectors. Conclude that $\text{Ker}(A)^\perp$ has a complete set of simultaneous eigenvectors of $|A|$ and U.

(c) Prove Theorem 3.2.3.

3. This will prove Theorem 3.2.3 by following the proof of Theorem 3.2.1. Let A be normal.

(a) If A is compact, prove there is φ_1 so that $\|\varphi_1\| = 1$ and $|\langle \varphi_1, A\varphi_1\rangle| = \sup_{\psi, \|\psi\|=1} |\langle \psi, A\psi\rangle|$.

(b) If $\langle \varphi_1, A\varphi_1\rangle = re^{i\theta}$ with $r > 0$, by replacing A by $e^{-i\theta}A$, suppose $e^{i\theta} = 1$. Show that if $A = B + iC$ with B and C self-adjoint, then $B\varphi_1 = r\varphi_1$.

(c) Let $\mathcal{H}^{(1)} = \{\psi \mid B\psi = r\psi\}$. Show that $\mathcal{H}^{(1)}$ is finite-dimensional and $C\colon \mathcal{H}^{(1)} \to \mathcal{H}^{(1)}$.

(d) If $C \restriction \mathcal{H}_1$ has a nonzero eigenvalue, prove that $\sup_{\psi, |\psi|=1} |\langle \psi, A\psi\rangle| > r$ and conclude that $C \restriction \mathcal{H}^{(1)} \equiv 0$.

(e) Prove that φ_1 is an eigenvector of A.

(f) Complete the proof of Theorem 3.2.3.

4. In (3.2.14), prove that sup and inf are actually max and min.

5. This problem will extend Theorem 3.2.7 to the case

$$-(Pu')' + Qu = \lambda u \tag{3.2.43}$$

where P, Q are continuous with $P > 0$, $Q > 0$ on $[0, 1]$.

(a) Extend Theorem 2.2.14 to this case.

(b) If $W(x) = P(x)\big(u_0'(x)u_1(x) - u_0(x)u_1'(x)\big)$, prove $W' = 0$ if u_0, u_1 solve (3.2.43) with (3.2.20) boundary conditions.

(c) With this change in W, show the rest of the proof of Theorem 3.2.7 extends.

6. Prove the analog of Theorem 3.2.7 in the case of $u'(0) = u'(1) = 0$ boundary conditions. (*Hint*: Define u_0, u_1 with $u_0(0) = u_1(1) = 1$, $u_0'(0) = u_1'(1) = 0$.)

7. (a) For fixed λ_n, prove that any two nonzero solutions of (3.2.17) must be multiples of one another. (*Hint*: If $-u'' + Qu = \lambda u$ and $u(0) = u'(0) = 0$, then $u(x) = 0$ for all x.)

(b) In Theorem 3.2.7, prove that $\lambda_1 < \lambda_2 < \dots$.

(c) If $-u'' + Qu = \lambda u$, $-v'' + Qv = \mu v$, $\lambda \neq \mu$, and $u(0) = v(0) = u(1) = v(1) = 0$, prove that $\int u(x)v(x)\,dx = 0$.

(d) Prove that the basis in Theorem 3.2.7 includes all solutions of (3.2.17).

3.3. The Riesz–Schauder Theorem

This section will focus on the fundamental theorem of the spectral theory of general compact operators on Banach spaces:

Theorem 3.3.1 (Riesz–Schauder Theorem). *Let A be a compact operator on a Banach space, X. Then $\sigma(A)\setminus\{0\}$ is a discrete subset of $\mathbb{C}\setminus\{0\}$, that is, a set $\{\Lambda_n\}_{n=1}^N$ of distinct points with $\Lambda_n \to 0$ if N is infinite. Each Λ_n is an eigenvalue of finite (algebraic) multiplicity. Moreover, the algebraic and geometric multiplicities of Λ_n as an eigenvalue of A^{t} are the same as those as an eigenvalue of A.*

Remarks. 1. We use Λ_n for the distinct eigenvalues and $\{\lambda_n\}_{n=1}^\infty$ for a counting with each Λ_n repeated a number of times equal to the algebraic multiplicity.

2. Our proof here makes heavy use of the Riesz geometric lemma (Lemma 5.1.6 of Part 1) that says if $Y \subset X$ is a closed, proper subspace of a Banach space, then for all $\varepsilon > 0$, there is $x \in X$ with $\|x\| = 1$ and $\operatorname{dist}(x, Y) \geq 1-\varepsilon$. For a different proof in the Hilbert space case, using what we'll call the analytic Fredholm theorem (which only works if $A \in \mathrm{FA}(X)$), see Theorem 3.14.3. That will directly prove $(A-z)^{-1}$ is meromorphic on $\mathbb{C}\setminus\{0\}$ with finite rank residues.

3. We'll use the analysis of discrete eigenvalues in Section 2.3 (Corollary 2.3.6); see Problem 3 which instead uses the Riesz lemma.

Notice that if Λ_n is an eigenvalue, not only is $\operatorname{Ker}(A-\Lambda_n) \neq \{0\}$, but since $\operatorname{Ker}(A^{\mathrm{t}}-\Lambda_n) \neq \{0\}$, we have $\operatorname{Ran}(A-\Lambda_n)$ is not all of X. Thus, Theorem 3.3.1 implies (to be traditional, we consider the case $\Lambda_n = 1$ which, as we'll see, is general for many purposes, and talk about solving $y = x + Ay$ instead of $(\mathbb{1}-A)^{-1}$; of course, $y = (\mathbb{1}-A)^{-1}x$):

Theorem 3.3.2 (Fredholm Alternative). *Let $A \in \mathrm{Com}(X)$. Then either*

(i) *For every $x \in X$, there is a unique $y \in X$ obeying*

$$y = x + Ay \tag{3.3.1}$$

or

(ii) *There is $y \neq 0$ obeying*

$$y = Ay \tag{3.3.2}$$

These are exclusive, that is, if (3.3.2) has a solution $y \neq 0$, then there is $x \in X$ so that (3.3.1) has no solution.

Remarks. 1. To be explicit, if (3.3.2) has nonzero solutions, $A^{\mathrm{t}}\ell = \ell$ has nonzero solutions, and that means if (3.3.1) has a solution, then $\ell(x)$

must be 0. Indeed (Problem 1), $\operatorname{Ran}(\mathbb{1} - A)$ is precisely $\{x \mid \ell(x) = 0$, all $\ell \in \operatorname{Ker}(\mathbb{1} - A^t)\}$.

2. Notice that if (3.3.2) has a nonzero solution, then (3.3.1) has multiple solutions for some x, for example, $x \equiv 0$ (in fact, if (3.3.1) has any solutions, it has multiple solutions). Therefore, $\sim$(ii) $\Rightarrow$ (i) can be paraphrased as: Uniqueness of possible solutions implies existence!

Since $(A - \lambda)^{-1} = \lambda^{-1}(\lambda^{-1}A - \mathbb{1})$ and if $\lambda \neq 0$, then $\lambda^{-1}A$ is compact if A is, we'll begin by considering $\lambda = 1$ for simplicity of notation.

Lemma 3.3.3. *If A is compact and* $\operatorname{Ker}(\mathbb{1} - A) = \{0\}$, *then there is $\varepsilon > 0$, so that*

$$\|(\mathbb{1} - A)x\| \geq \varepsilon\|x\| \tag{3.3.3}$$

for all x.

Proof. If not, we can find $x_n \in X$ with $\|x_n\| = 1$, so that

$$\|(\mathbb{1} - A)x_n\| \leq \frac{1}{n} \tag{3.3.4}$$

and, in particular,

$$\|(\mathbb{1} - A)x_n\| \to 0 \tag{3.3.5}$$

Since A is compact, find $n(j)$ so $Ax_{n(j)}$ has limit x_∞. By (3.3.5), $x_{n(j)} = (\mathbb{1} - A)x_{n(j)} + Ax_{n(j)}$ converges to x_∞ also, so $\|x_\infty\| = 1$. Thus,

$$\|(\mathbb{1} - A)x_\infty\| = \lim_j \|(\mathbb{1} - A)x_{n(j)}\| = 0$$

so $x_\infty \neq 0$ is in $\operatorname{Ker}(\mathbb{1} - A)$, contradicting the hypothesis. $\square$

The Riesz lemma (Lemma 5.1.6 of Part 1) says if $Y \subsetneqq Z$ are closed subspaces, there is $z \in Z$ so $\|z\| = 1$ and $\operatorname{dist}(z, Y) \geq \frac{1}{2}$. An increasing (respectively, decreasing) *tower of subspaces* is a sequence $\{Y_j\}_{j=1}^\infty$ of closed subspaces with $Y_j \subsetneqq Y_{j+1}$ (respectively, $Y_{j+1} \subsetneqq Y_j$). The Riesz lemma and an easy induction (Problem 2) show that

Proposition 3.3.4. (a) *If Y_j is an increasing tower of subspaces, there exists $\{y_j\}_{j=2}^\infty$ so that*

$$y_j \in Y_j, \quad \|y_j\| = 1, \quad \operatorname{dist}(y_j, Y_{j-1}) \geq \tfrac{1}{2} \quad \textit{for } j = 2, 3, \dots \tag{3.3.6}$$

(b) *If Y_j is a decreasing tower of subspaces, there exists $\{y_j\}_{j=1}^\infty$ so that*

$$y_j \in Y_j, \quad \|y_j\| = 1, \quad \operatorname{dist}(y_j, Y_{j+1}) \geq \tfrac{1}{2} \quad \textit{for } j = 1, 2, \dots \tag{3.3.7}$$

Remark. Of course, by the inclusion, $\operatorname{dist}(y_j, Y_k) \geq \frac{1}{2}$ for all $j > k$ in case (a) and for all $j < k$ in case (b).

Lemma 3.3.5. *Let $C \in \mathcal{L}(X)$ for a Banach space, X. If* $\operatorname{Ker}(C) = \{0\}$ *and* $\operatorname{Ran}(C) \neq X$, *then for all n,* $\operatorname{Ran}(C^{n+1}) \subsetneqq \operatorname{Ran}(C^n)$.

Proof. If $u = C^{n+1}y$, then $u = C^n(Cy)$, so $\operatorname{Ran}(C^{n+1}) \subset \operatorname{Ran}(C^n)$ is always true.

If $\operatorname{Ran}(C) \neq X$, pick $u \notin \operatorname{Ran}(C)$, $u \neq 0$. Clearly, $C^n u \in \operatorname{Ran}(C^n)$. If $C^n u = C^{n+1}w$ for some w, since C is one–one, $C^{n-1}u = C^n w$, and then $C^{n-2}u = C^{n-1}w, \dots, u = Cw$, contradicting $u \notin \operatorname{Ran}(C)$. Thus, $C^n u \notin \operatorname{Ran}(C^{n+1})$, so $\operatorname{Ran}(C^{n+1}) \neq \operatorname{Ran}(C^n)$. $\square$

Proposition 3.3.6. *Let A be compact. If* $\operatorname{Ker}(\mathbb{1} - A) = \{0\}$, *then* $\operatorname{Ran}(\mathbb{1} - A) = X$. *Moreover,* $1 \in \rho(A)$.

Remark. This is the "or" but not "exclusive or" part of the Fredholm alternative.

Proof. By the first conclusion, $\mathbb{1} - A$ is a bijection and then, by (3.3.2), the inverse is bounded, so $1 \in \rho(A)$. Thus, we only need to prove the first conclusion.

Let $C = \mathbb{1} - A$ and suppose $\operatorname{Ran}(C) \neq X$. By (3.3.3), $Y_n = \operatorname{Ran}(C^n)$ is a closed subspace for each n. By Lemma 3.3.5, $Y_{n+1} \subsetneqq Y_n$, so $\{Y_n\}_{n=1}^\infty$ is a decreasing tower. Thus, by Proposition 3.3.4, we can find $\{y_n\}_{n=1}^\infty$ so

$$\|y_n\| = 1, \quad y_n \in \operatorname{Ran}(C^n), \quad \operatorname{dist}(y_n, \operatorname{Ran}(C^m)) \geq \tfrac{1}{2}, \quad \text{all } m > n \tag{3.3.8}$$

Since $C = \mathbb{1} - A$,

$$Ay_n = y_n - Cy_n \tag{3.3.9}$$

If $m > n$, $Ay_m \in \operatorname{Ran}(C^m) \subset \operatorname{Ran}(C^{n+1})$ since $[A, C^m] = 0$. Thus,

$$Ay_n - Ay_m = y_n - Cy_n - Ay_m \in y_n - \operatorname{Ran}(C^{n+1})$$

We conclude $\|Ay_n - Ay_m\| \geq \frac{1}{2}$ for all $n \neq m$. Thus, $\{Ay_n\}_{n=1}^\infty$ has no convergent subsequence, contradicting compactness of A. We conclude that $\operatorname{Ran}(\mathbb{1} - A) = X$. $\square$

Proposition 3.3.7. *Let A be compact. Then for all $\varepsilon > 0$, $\sigma(A) \cap \{z \mid |z| \geq \varepsilon\}$ is finite.*

Proof. If not, pick $\lambda_1, \lambda_2, \dots$ a sequence of distinct points in $\sigma(A)$ with

$$|\lambda_j| \geq \varepsilon \tag{3.3.10}$$

Since $\lambda_j \in \sigma(A)$, Proposition 3.3.6 (applied to $\lambda_j^{-1}A$) implies there exist $x_n \in X$ so $x_n \neq 0$ and

$$(A - \lambda_n)x_n = 0 \tag{3.3.11}$$

If $\sum_{j=1}^n \alpha_j x_j = 0$, then $\prod_{j=1}^{n-1}(A - \lambda_j)(\sum_{j=1}^n \alpha_j x_j) = \alpha_n \prod_{j=1}^{n-1}(\lambda_n - \lambda_j)x_n$, so $\alpha_n = 0$, that is, by induction, $\{x_n\}_{n=1}^\infty$ are independent.

Thus, if Y_n is the span of $\{x_j\}_{j=1}^n$, $\{Y_n\}$ is an increasing tower. (They are closed by Corollary 5.1.5 of Part 1.) So, by Proposition 3.3.4, we can find $\{y_n\}_{n=2}^\infty$ so

$$\|y_n\| = 1, \qquad y_n \in Y_n, \qquad \text{dist}(y_n, Y_{n-1}) \geq \tfrac{1}{2} \tag{3.3.12}$$

By construction, $(A - \lambda_n)[Y_n] \subset Y_{n-1}$, which implies $A[Y_m] \subset Y_m$. If $m < n$, we conclude that

$$Ay_n - Ay_m = \lambda_n y_n + (A - \lambda_n)y_n - Ay_m \in \lambda_n(y_n + Y_{n-1}) \tag{3.3.13}$$

so

$$\|Ay_n - Ay_m\| \geq \tfrac{1}{2}\,\varepsilon \tag{3.3.14}$$

Thus, no subsequence of $\{Ay_n\}_{n=1}^\infty$ can converge, contradicting compactness. We conclude that $\sigma(A) \cap \{z \mid |z| \geq \varepsilon\}$ is finite. □

Thus, each $\Lambda \in \sigma(A) \setminus \{0\}$ is an isolated point and we can pick $\varepsilon > 0$, so

$$\sigma(A) \cap \{z \mid 0 < |z - \Lambda| \leq \varepsilon\} = \emptyset \tag{3.3.15}$$

and define

$$P_\Lambda = (2\pi i)^{-1} \oint_{|z-\Lambda|=\varepsilon} (z - A)^{-1}\, dz \tag{3.3.16}$$

as in (2.3.21).

Proposition 3.3.8. *For each* $\Lambda \in \sigma(A) \setminus \{0\}$*, the projection in* (3.3.16) *is compact. Thus,* $\dim(\text{Ran}(P_\Lambda)) < \infty$.

Proof. Since $0 \in \sigma(A)$, (3.3.15) implies $\varepsilon < |\Lambda|$. Thus, $(2\pi i)^{-1} \oint_{|z-\Lambda|=\varepsilon} z^{-1}\, dz = 0$. Since

$$(z - A)^{-1} - z^{-1} = A[z^{-1}(z - A)^{-1}] \tag{3.3.17}$$

if

$$B_\Lambda = (2\pi i)^{-1} \oint_{|z-\Lambda|=\varepsilon} z^{-1}(z - A)^{-1}\, dz \tag{3.3.18}$$

we have

$$P_\Lambda = AB_\Lambda \tag{3.3.19}$$

so P_Λ is compact.

Thus, the unit ball of $\text{Ran}(P_\Lambda)$ is compact. So, by Theorem 5.1.7 of Part 1, $\dim(\text{Ran}(P_\Lambda)) < \infty$. □

Proof of Theorem 3.3.1. We've proven (Proposition 3.3.6) that each $\Lambda \in \sigma(A) \setminus \{0\}$ is an eigenvalue and that (Proposition 3.3.7) the only possible limit of eigenvalues is $\{0\}$. Thus, all eigenvalues in $\sigma(A) \setminus \{0\}$ are discrete. So, by Proposition 3.3.8, they have finite algebraic multiplicity.

All that remains to show are the relations to A^{t}. Since $\sigma(A) = \sigma(A^{\text{t}})$ (see Theorem 2.2.4) and if $z \in \rho(A)$, $(z - A^{\text{t}})^{-1} = [(z - A)^{-1}]^{\text{t}}$, we see that (with N given by Theorem 2.3.5)

$$P_\Lambda(A^{\text{t}}) = P_\Lambda(A)^{\text{t}}, \qquad N_\Lambda(A^{\text{t}}) = N_\Lambda(A)^{\text{t}} \tag{3.3.20}$$

Since $\dim(\operatorname{Ran}(B^{t})) = \dim(\operatorname{Ran}(B))$ (see (2.1.9)) and

$$\text{algebraic mult. of } \Lambda \text{ for } A = \dim(\operatorname{Ran}(P_\Lambda)) \tag{3.3.21}$$

$$\text{geometric mult. of } \Lambda \text{ for } A = \dim(\operatorname{Ran}(P_\Lambda)) - \dim(\operatorname{Ran}(N_\Lambda)) \tag{3.3.22}$$

(by (2.3.43)), we have the claimed equality of multiplicities. □

As an application of Theorem 3.3.1, we note that it provides another proof (Problem 4) of the Hilbert–Schmidt theorem. As a second application, we provide a proof of the solvability of the Dirichlet problem for bounded regions $\Omega \subset \mathbb{R}^n$ with smooth boundary. The Dirichlet problem is discussed in Sections 3.4 and 3.6 of Part 3.

In particular, in Section 3.6 of Part 3, we discussed dipole layers. If Ω is a bounded open region in $\mathbb{R}^\nu$, $\nu \geq 2$, with smooth (C^∞) boundary, the dipole potential at $y \in \partial\Omega$ is defined for $x \in \mathbb{R}^\nu$ by

$$K_0(x, y) = \frac{(x - y, n_y)}{\sigma_\nu |x - y|^\nu} \tag{3.3.23}$$

where n_y is the outward pointing normal at $y \in \partial\Omega$ and (see Problem 5 or 6 of Section 4.11 of Part 1)

$$\sigma_\nu = \frac{2(\sqrt{\pi})^\nu}{\Gamma(\frac{\nu}{2})} \tag{3.3.24}$$

is the area of the unit sphere in $\mathbb{R}^\nu$.

For any $f \in C(\partial\Omega)$ and $x \in \mathbb{R}^\nu$, define

$$(\mathcal{D}_\Omega f)(x) = \int_{\partial\Omega} K_0(x, y) f(y)\, dS(y) \tag{3.3.25}$$

where dS is surface measure. Because $\partial\Omega$ is smooth, $(x-y, n_y) = O(|x-y|^2)$ as $x - y \to 0$ uniformly in $x, y \in \partial\Omega$, so the integral in (3.3.25) converges even if $x \in \partial\Omega$. Since $K(\,\cdot\,, y)$ is harmonic away from y, $(\mathcal{D}f)(x)$ is harmonic on $\mathbb{R}^\nu \setminus \partial\Omega$.

Of course, if $x \notin \partial\Omega$, the integrand is bounded as y varies, but not uniformly as $x \to \partial\Omega$. Indeed, one can prove that if $u_\pm(x) = (\mathcal{D}f)(x)$ restricted to x inside/outside $\partial\Omega$ (call them $\Omega_\pm$), then if $u_\pm$ are extended to $\partial\Omega$ by

$$u_\pm(x) = \mp f(x) + (\mathcal{D}f)(x), \qquad x \in \partial\Omega \tag{3.3.26}$$

then $u_\pm$ are continuous on $\overline{\Omega}_\pm$. Moreover, at any point, y, on $\partial\Omega$,

$$\frac{\partial u_+}{\partial n_y}(y) = \frac{\partial u_-}{\partial n_y}(y) \tag{3.3.27}$$

That is, $\frac{\partial u}{\partial n}$ is continuous across $\partial\Omega$, but if $f(y) \neq 0$, $u_\pm$ jump as y is crossed.

Define $A\colon C(\partial\Omega) \to C(\partial\Omega)$ by

$$(Af)(x) = (\mathcal{D}f)(x), \qquad x \in \partial\Omega \tag{3.3.28}$$

To solve the Dirichlet problem (i.e., u harmonic in Ω, continuous on $\overline{\Omega}_+$, and $u \restriction \partial\Omega = g$, a given function), we need to solve

$$g = -f + Af \tag{3.3.29}$$

which is exactly a Fredholm integral equation. We have the following:

Theorem 3.3.9. (a) *A is a compact operator of $C(\partial\Omega)$ to itself.*

(b) *The only solution of $f = Af$ is $f = 0$.*

(c) *The Dirichlet problem can be solved for any $g \in C(\partial\Omega)$.*

Remark. Since we need to assume $\partial\Omega$ is C^∞, this method is more limited than some of the other methods discussed in Sections 3.4 and 3.6 of Part 3, but it nicely illustrates the power of the Fredholm alternative and was historically important.

Proof. (a) For $\delta > 0$, define

$$K_\delta(x,y) = \frac{(x-y, n_y)}{[\sigma_\nu(|x-y|^\nu + \delta)]} \tag{3.3.30}$$

and

$$(A_\delta f)(x) = \int K_\delta(x,y) f(y)\, dS(y) \tag{3.3.31}$$

Then K_δ is a continuous kernel, so by Example 3.1.6, A_δ is a compact operator. Since

$$|(Af)(x) - A_\delta f(x)| \le \|f\|_\infty \int_{\partial\Omega} |K_\delta(x,y) - K_0(x,y)|\, dS(y) \tag{3.3.32}$$

we have

$$\|A - A_\delta\| \le \sup_x \int_{\partial\Omega} |K_\delta(x,y) - K_0(x,y)|\, dS(y) \tag{3.3.33}$$

which the reader will prove (in Problem 5) goes to zero. Thus, A is a norm limit of compact operators, and so compact.

(b) Let $u_\pm = \mathcal{D}f$ restricted to $\Omega_\pm$. By $f = Af$ and (3.3.26), u_+ set equal to 0 on $\partial\Omega$ is continuous on Ω, so by the maximum principle (see Corollary 3.1.5 of Part 3), $u_+ \equiv 0$. Thus, by (3.3.27), $\frac{\partial u_-}{\partial n} = 0$ on $\partial\Omega$.

Since

$$\operatorname{div}(u_-(\operatorname{grad} u_-)) = u_- \Delta u_- + |\operatorname{grad} u_-|^2 = |\operatorname{grad} u_-|^2 \tag{3.3.34}$$

then for $R > \sup_{x\in\partial\Omega}|x|$, we have that, by Gauss's theorem, with $\Omega_-^R = \Omega_- \cap \{|x| < R\}$,

$$\begin{aligned}&\int_{\Omega_-^R} |\text{grad}\, u_-(x)|^2 \, d^\nu x \\ &\quad = -\int_{\partial\Omega} u_-(y)\, \frac{\partial u_-}{\partial n}(y)\, dS(y) + \int_{|y|=R} u_-(y)\, \frac{\partial u_-}{\partial n}\, dS(y)\end{aligned} \tag{3.3.35}$$

Since $\frac{\partial u_-}{\partial n} = 0$ on $\partial\Omega$, the first integral on the right is zero, and since (Problem 6), uniformly in direction, $u_-(y) = O(|y|^{-(\nu-1)})$, $\frac{\partial u_-}{\partial n} = O(|y|^{-\nu})$, the second integral is $O(R^{-2\nu+1}R^{\nu-1}) \to 0$. Thus, $\int_\Omega |\text{grad}\, u_-|^2 \, d^\nu y = 0$, that is, u_- is constant.

Since $u_-(y) \to 0$ as $y \to \infty$, $u_- \equiv 0$. Thus, the jump $u_+ - u_- = 2f$ is zero, that is, $f = 0$ as claimed.

(c) Immediate from (a), (b), and the Fredholm alternative (Theorem 3.3.2). □

We end this section by presenting a partially alternate proof of what we called the weak Lomonosov theorem (Corollary 5.12.29 of Part 1) and, in particular, one that avoids the use of the Schauder-Tychonoff fixed point theorem. Recall that if X is a Banach space and $A \in \mathcal{L}(X)$, we say that $Y \subset X$, a closed subspace, is an *invariant subspace* for A if $A[Y] \subset Y$. If $\mathcal{A} = \{A\}' = \{B \in \mathcal{L}(X) \mid BA = AB\}$, we say that Y is a *hyperinvariant subspace* for A if it is invariant for all $B \in \{A\}'$. Recall that such a subspace is called *nontrivial* if $Y \neq 0, Y \neq X$. We'll prove

Theorem 3.3.10 (Weak Lomonosov Theorem). *Let K be a nonzero compact operator on an infinite-dimensional Banach space, X. Then K has a nontrivial hyperinvariant subspace.*

Remark. If $\dim(X) < \infty$ and $K \neq c\mathbb{1}$ for any c, then K also has a nontrivial hyperinvariant subspace, for if $\lambda \in \sigma(K)$, $Y = \{x \mid Kx = \lambda x\}$ is hyperinvariant and $Y \neq \{0\}$ since $\lambda \in \sigma K$, and $Y \neq X$ since $K \neq \lambda\mathbb{1}$.

Proof. If there is $\lambda \in \sigma(K)\setminus\{0\}$, then the associated eigenspace is hyperinvariant and, by the Riesz-Schauder theorem not $\{0\}$ and finite-dimensional, so not all of X. Thus, we can restrict to the case where K is quasinilpotent, i.e., as $n \to \infty$

$$\|K^n\|^{1/n} \to 0 \tag{3.3.36}$$

Without loss we can suppose $\|K\| = 1$ and for some x_0 with $\|x_0\| > 1$, we have that $\|Kx_0\| > 1$. Thus, since $\|K\| = 1$, $0 \notin L \equiv \overline{K([\overline{B_1(x_0)}])}$ and, of course $0 \notin \overline{B_1(x_0)}$ since $\|x_0\| > 1$.

As in Section 5.12 of Part 1, if K has no hyperinvariant spaces, $\mathcal{A} = \{K\}'$ and $y \neq 0$, we have that $\{Ay \mid A \in \mathcal{A}\}$ is dense in X since its closure is hyperinvariant for K and is not $\{0\}$ since $y = \mathbb{1}y$ is in its closure.

Define, for $T \in \mathcal{A}$,

$$U_T = \{y \in X \mid \|Ty - x_0\| < 1\} \tag{3.3.37}$$

By the density result, any $y \neq 0$ lies in some U_T and since $\|x_0\| > 1$, $0 \notin U_T$ for all T. Thus $\bigcup_{T \in \mathcal{A}} U_T = X \setminus \{0\}$ so since L is compact and in $X \setminus \{0\}$, we can find $T_1, \dots, T_n \in \mathcal{A}$ so

$$L \subset \bigcup_{j=1}^{n} U_{T_j} \tag{3.3.38}$$

Define $j_1, j_2, \dots$ inductively so $x_m \equiv T_{j_m} K \cdots K T_{j_1} K x_0 \in \overline{B_1(x_0)}$. For $Kx_0 \in L$ is in some $U_{T_{j_1}}$ and then by induction $Kx_{m-1} \in L$ so in some $U_{T_{j_m}}$. Let $\varepsilon = \min\{\|y\| \mid y \in \overline{B_1(x_0)}\} = \|x_0\| - 1 > 0$. Let $C = \max_{j=1,\dots,n}\{\|T_j\|\}$. Since K commutes with all T_j, $x_m = K^m T_{j_m} \cdots T_{j_1} x_0$ so

$$\varepsilon < \|K^m\| \, C^m \, \|x_0\| \tag{3.3.39}$$

Taking m-th roots and using (3.3.36), we see that

$$\varepsilon^{1/m} \to 0 \tag{3.3.40}$$

violating $\varepsilon > 0$. This contradiction proves K must have a hyperinvariant subspace. □

Notes and Historical Remarks. The Riesz–Schauder theorem is named after a 1916 paper of Riesz [**566**] and a 1930 paper of Schauder [**598**]. Riesz had most of the theory using the Riesz lemma as we do in most of our development (he used the argument in Problem 3 in place of the use of spectral projections). Schauder added the missing piece of the theory involving the dual space.

If f is the function on $\sigma(A)$ equal to 1 very near Λ and 0 elsewhere, g the function equal to Λ^{-1} very near Λ and 0 elsewhere, then $zg(z) = f(z)$. So, by the functional calculus, $Ag(A) = f(A)$, which is precisely (3.3.19).

The reduction of the Dirichlet problem to an integral equation by using dipole layers goes back to Neumann [**500**]. The analysis using Fredholm theory was mentioned in Fredholm and developed by Hilbert and his school. Our discussion follows Courant–Hilbert [**134, 135**].

As we explained in the Notes to Section 5.12 of Part 1, Theorem 3.3.10 is due to Lomonosov [**455**]. The proof here is due to H. M. Hilden (unpublished). He found it in response to a query of L. J. Wallen who said it would be good to have a proof where the Schauder–Tychonoff theorem were replaced by the contraction mapping theorem and instead Hilden found a proof with no use of any fixed point theorem. Wallen then presented to talk at a conference and it appeared in print in Michaels [**476**], Pearcy–Shields [**514**], and Radjavi–Rosenthal [**542**], all with attribution to Hilden.

Problems

1. (a) Let A be compact, $y_n = (\mathbb{1}-A)x_n$, and $y_n \to y_\infty$. Prove there is x_∞ so $y_\infty = (\mathbb{1}-A)x_\infty$, that is, $\mathrm{Ran}(\mathbb{1}-A)$ is closed. (*Hint*: Let $\widetilde{X} = \mathrm{Ker}(\mathbb{1}-A)$ and consider $X/\widetilde{X} = X^\sharp$. Prove there is $A^\sharp\colon X^\sharp \to X^\sharp$ with $A^\sharp[x] = [Ax]$, that $A^\sharp$ is compact, and $\mathrm{Ker}(\mathbb{1}-A^\sharp) = \{0\}$, so $\|(\mathbb{1}-A^\sharp)[x]\| \ge \varepsilon\|[x]\|$.)

   (b) Prove $\mathrm{Ran}(\mathbb{1}-A) = \{x \mid \ell(x) = 0 \text{ for all } \ell \in \mathrm{Ker}(\mathbb{1}-A^t)\}$.

2. Prove Proposition 3.3.4.

3. (a) If A is compact, prove that $\mathrm{Ker}(\mathbb{1}-A)$ is finite-dimensional. (*Hint*: Consider $\mathrm{Ker}(\mathbb{1}-A) \cap A[X_1]$.)

   (b) Prove that $\mathbb{K}_m \equiv \mathrm{Ker}((\mathbb{1}-A)^m)$ is finite-dimensional for all m. (*Hint*: Prove that $\mathbb{1}-(\mathbb{1}-A)^m$ is compact.)

   (c) Prove that either for some $m < \infty$, $\mathbb{K}_n = \mathbb{K}_m$ for all $n \ge m$ or else $\mathbb{K}_n \subsetneq \mathbb{K}_{n+1}$ for all n.

   (d) If $x_n \in \mathbb{K}_n$ and $x_m \in \mathbb{K}_m$ and $n > m$, prove that $Ax_n - Ax_m \in x_n + \mathbb{K}_{n-1}$.

   (e) Use the Riesz lemma to conclude eventually $\mathbb{K}_n = \mathbb{K}_m$ for some m and conclude 1 has finite algebraic multiplicity.

4. (a) Let A be a self-adjoint operator on a Hilbert space, $\mathcal{H}$. For $x \in \mathbb{R}$, prove that $\|(A-x-i\varepsilon)\varphi\| \ge |\varepsilon|\|\varphi\|$ and conclude that $\|(A-x-i\varepsilon)^{-1}\| \le |\varepsilon|^{-1}$.

   (b) Let $\lambda_0 \in \mathbb{R}$ be an isolated point of $\sigma(A)$. Prove that $(A-\lambda)^{-1}$ has a simple pole at $\lambda = \lambda_0$ and conclude that if P_{λ_0} is the spectral projection, (3.1.5), then $\mathrm{Ran}(P_{\lambda_0}) \equiv \mathrm{Ker}(A-\lambda_0)$.

   (c) If $\lambda_0 \ne \mu_0 \in \mathbb{R}\setminus\{0\}$ and $A\varphi = \lambda_0\varphi$, $A\psi = \mu_0\psi$, prove that $\langle\varphi,\psi\rangle = 0$.

   (d) Use the Riesz–Schauder theorem (Theorem 3.3.1) to prove the Hilbert–Schmidt theorem (Theorem 3.2.1).

5. This problem will prove the sup in (3.3.33) goes to zero as $\delta \downarrow 0$.

   (a) For each $\varepsilon > 0$, prove that $K_\delta(x,y) - K_0(x,y) \to 0$ uniformly in $|x-y| \ge \varepsilon$.

   (b) Prove that

$$\lim_{\varepsilon\downarrow 0}\ \sup_{x\in\partial\Omega} \int_{|x-y|\le\varepsilon} |K(x,y)|\, dS(y) = 0$$

(c) Prove that the sup in (3.3.33) goes to zero as $\delta \downarrow 0$.

6. If $\mathcal{D}_\Omega f$ is given by (3.3.25) with $f \in C(\partial\Omega)$, prove that in $\{x \mid |x| \geq 1 + \sup_{y \in \partial\Omega} |y|\}$, we have

$$|(\mathcal{D}_\Omega f)(x)| \leq C|x|^{-\nu+1}$$
$$|\vec{\nabla}(\mathcal{D}_\Omega f)(x)| \leq C|x|^{-\nu}$$

3.4. Ringrose Structure Theorems

The Riesz–Schauder theorem provides a kind of analog for compact operators of the Jordan normal form for finite matrices. But there are two limitations. It says nothing about a compact operator, A, on the rest of the space outside the span of the spectral projections, P_λ, for $\lambda \in \sigma(A) \setminus \{0\}$. And it says nothing in the quasinilpotent case where $\sigma(A) = \{0\}$. In this section, we'll discuss two results that provide some information. The first is limited to the Hilbert space case:

Theorem 3.4.1 (Ringrose–West Decomposition). *Let A be a compact operator on a Hilbert space, $\mathcal{H}$. Then*

$$A = N + Q \tag{3.4.1}$$

where N is a compact normal operator with

$$\sigma(A) \setminus \{0\} = \sigma(N) \setminus \{0\} \tag{3.4.2}$$

and with algebraic multiplicities of $\lambda \neq 0$ in $\sigma(A) \setminus \{0\}$ the same as those for N and where Q is quasinilpotent, i.e.,

$$\sigma(Q) = \{0\}, \quad \|Q^n\|^{1/n} \to 0 \text{ as } n \to \infty \tag{3.4.3}$$

To describe the other result, we need some terminology.

Definition. A *nest* of subspaces of a Banach space, X, is a collection, $\mathcal{F}$, of closed subspaces which is totally ordered under inclusion, i.e., if $Y, Z \in \mathcal{F}$, then either $Y \subset Z$ or $Z \subset Y$. A nest, $\mathcal{F}$, is called *complete* if $\{0\}, X \in \mathcal{F}$ and if for any subfamily $\mathcal{F}_0 \subset \mathcal{F}$

$$\sup \mathcal{F}_0 \equiv \overline{\bigcup_{L \in \mathcal{F}_0} L} \text{ and } \inf \mathcal{F}_0 = \bigcap_{L \in \mathcal{F}_0} L \tag{3.4.4}$$

lie in $\mathcal{F}$.

If $\mathcal{F}$ is a complete nest, for any $M \in \mathcal{F}$, $M \neq \emptyset$, we define

$$M_- = \sup \{L \in \mathcal{F} \mid L \subset M, L \neq M\} \tag{3.4.5}$$

Definition. A nest, $\mathcal{F}$, is called *simple* if it is complete and for any M, either $M_- = M$ or $\dim(M/M_-) = 1$. Given $A \in \mathcal{L}(X)$, a nest is called a simple *invariant nest* if it is simple and, for each $M \in \mathcal{F}$, $A[M] \subset M$. We say that $\mathcal{F}$ is discontinuous at M if $M_- \neq M$.

If $\mathcal{F}$ is a simple invariant nest, and $\mathcal{F}$ is discontinuous at M, there is a unique $\alpha_M \in \mathbb{K}$ so that for any $x \in M$, $Ax - \alpha_M x \in M_-$ (for $\widetilde{A}\colon M/M_- \to M/M_-$ by $\widetilde{A}[x] = [Ax]$ is multiplication by some α_M). α_M is called the *eigenjump* for M. Here is the second main theorem of this section.

Theorem 3.4.2 (Ringrose Structure Theorem). *Let A be a compact operator on a separable Banach space, X. Then there is a simple invariant nest, $\mathcal{F}$, for A. For any such nest, the set of points, M, of discontinuity is countable and the eigenjumps, $\alpha_M \neq 0$, are exactly the points λ of $\sigma(A)\setminus\{0\}$ and the multiplicity of $\lambda \neq 0$ as an eigenjump is its algebraic multiplicity as a point of $\sigma(A)$.*

Example 3.4.3 (Volterra Nest). One might think that the only nests are discrete, i.e., have only jumps and countably many of them. Here is an example of a maximal nest, invariant for a certain compact operator (of necessity, by Theorem 3.4.2, an operator with $\sigma(A) = \{0\}$) that has no points of discontinuity.

Let S be a Volterra operator

$$(Sf)(x) = \int_0^x f(y)\,dy \tag{3.4.6}$$

on any of Banach spaces, $C([0,1])$, $L^p([0,1],dx)$ with $1 \leq p < \infty$. In Example 2.2.13, we proved that S is quasinilpotent on $C([0,1])$ but it is easy to see (Problem 2) that it is compact and quasinilpotent on each L^p and on $C([0,1])$.

Let $0 \leq c \leq 1$. Define

$$Y_c = \{f \mid f(x) = 0 \text{ for } x \in [0,c]\} \tag{3.4.7}$$

It is trivial to see that each Y_c is invariant for A and if $c > d$, then $Y_c \subset Y_d$. In L^p, $\{Y_c\}_{0\leq c\leq 1}$ is a maximal invariant nest (Problem 2) and in $C([0,1])$ we get a maximal invariant nest if we add $C([0,1])$ (For L^p, $Y_{c=0} = L^p$ but not for $C([0,1])$). Moreover (see the Notes and Problem 5) for L^p, $C([0,1])$ the $\{Y_c\}_{0\leq c\leq 1}$ (with perhaps $\{0\}$ added for $C([0,1])$ are the only (closed) invariant subspaces for S. Of course, for any $T \subset [0,1]$, $\inf\{Y_c \mid c \in T\} = Y_{\inf\{c|c\in T\}}$ and $\sup\{Y_c \mid c \in T\} = Y_{\sup\{c|c\in T\}}$, so this is a continuous nest. It is called the *Volterra nest.*

We'll provide two proofs of Theorem 3.4.1. The first we will do with our bare hands and, for the second, we will use the machinery of simple invariant nests.

First Proof of Theorem 3.4.1. Let A be a compact operator on the Hilbert space, $\mathcal{H}$. Let $\{\Lambda_k\}_{k=1}^K$ be a labeling of the distinct points of $\sigma(A)\setminus\{0\}$ and $M_k = \dim(\operatorname{Ran} P_k)$ where P_k is the spectral projection for Λ_k. By picking a Jordan basis for each $A \restriction \operatorname{Ran} P_k$, we find an independent set of vectors $\{f_j\}_{j=1}^J$ with $J = \sum_{k=1}^K M_k$ with $Af_j = \lambda_j f_j + \sum_{q=1}^{j-1} \tilde{a}_{qj} f_q$. Then apply Gram–Schmidt to the f's and so get an orthonormal set (called a *Schur basis*), $\{e_j\}_{j=1}^J$, so that

$$Ae_j = \lambda_j e_j + \sum_{q=1}^{j-1} a_{qj} e_q \tag{3.4.8}$$

Let $\mathcal{H}_1$ be the closure of the span of the e_j's. Define $N \in \mathcal{L}(\mathcal{H})$ by

$$Ne_j = \lambda_j e_j \tag{3.4.9}$$

$$N\varphi = 0 \quad \text{if} \quad \varphi \in \mathcal{H}_1^\perp \tag{3.4.10}$$

Since N has a diagonal matrix and $|\lambda_j| \to 0$ as $j \to \infty$, N is a compact normal operator. Clearly $Q \equiv A - N$ is then compact. We only need to prove that Q is quasinilpotent.

By (3.4.8) and (3.4.9), $\mathcal{H}_1$ is an invariant subspace for A and for N and so for Q. Let $Q_1 = Q \restriction \mathcal{H}_1$. We begin by proving Q_1 is quasinilpotent. If not, let λ_1 be a nonzero eigenvalue of the compact operator Q_1. Let $\sigma_1 = \sigma(Q_1)\setminus\{\lambda_1\}$ and $z_0 = 0$ and apply Proposition 2.3.7. If $\varphi = \sum_{\ell=1}^L a_\ell e_\ell$ ($L < \infty, \{a_\ell\}_{\ell=1}^L \subset \mathbb{C}^L$) $Q_1^L \varphi = 0$, so by that proposition, $P_{\lambda_1}\varphi = 0$ (since the $\varphi \in \operatorname{Ran} P = \operatorname{Ker} P_{\lambda_1}$). Thus P_{λ_1} vanishes on a dense set of $\mathcal{H}_1$ and so on $\mathcal{H}_1$, i.e., λ_1 is not an eigenvalue after all. This shows that Q_1 is quasinilpotent.

Suppose Q is not quasinilpotent. Since Q is compact, there is $\varphi \in \mathcal{H}$ and $\mu \neq 0$ so that $\varphi \neq 0$ and

$$Q\varphi = \mu\varphi \tag{3.4.11}$$

Since Q_1 is quasinilpotent, $\varphi \notin \mathcal{H}_1$, i.e.,

$$\varphi = \psi + \eta, \quad \psi \in \mathcal{H}_1^\perp, \eta \in \mathcal{H}_1, \psi \neq 0 \tag{3.4.12}$$

By (3.4.11) and $N \restriction \mathcal{H}_1^\perp = 0$

$$(A - \mu)\psi = -(Q - \mu)\eta \in \mathcal{H}_1 \tag{3.4.13}$$

since $\mathcal{H}_1$ is invariant for Q.

Let $\mathcal{H}_2$ be the span of $\mathcal{H}_1$ and ψ, i.e., $\mathcal{H}_1 \oplus [\psi]$. By (3.4.13), $\mathcal{H}_2$ is an invariant subspace of A. Let $A_2 = A \restriction \mathcal{H}_2$. By a simple argument

(Problem 1), $\sigma(A_2) \subset \sigma(A)$ and for $z \notin \sigma(A)$, $\mathcal{H}_2$ is invariant for $(A-z)^{-1}$ and $(A_2-z)^{-1} = (A-z)^{-1} \restriction \mathcal{H}_2$. This implies that if $\lambda \in \sigma(A_2)$, with $\lambda \neq 0$,

$$P_\lambda(A_2) = P_\lambda(A) \restriction \mathcal{H}_2 \tag{3.4.14}$$

If $\mu \notin \sigma(A)$, $(A_2-\mu)$ is invertible and by (3.4.13), $\psi = (A-\mu)^{-1}\left[-(Q-\mu)\eta\right] \in \mathcal{H}_1$ since $(A-\mu)^{-1}$ leaves $\mathcal{H}_1$ invariant. This contradicts $\psi \in \mathcal{H}_1^\perp$, $\psi \neq 0$. Thus $\mu \in \sigma(A)$.

By Theorem 2.3.8 and (3.4.14)

$$\big(1 - P_\mu(A_2)\big)\,\psi \subset \operatorname{Ran}(A_2 - \mu) \subset \mathcal{H}_1 \tag{3.4.15}$$

By (3.4.14), $P_\mu(A_2)\,\psi = P_\mu(A)\,\psi \in \mathcal{H}_1$ since the range of $P_\mu(A)$ is the span of a finite number of f_j's. But then $\psi \in \mathcal{H}_1$ which violates $\psi \in \mathcal{H}_1^\perp \setminus \{0\}$. This contradiction implies that Q is quasinilpotent. □

We turn towards the proof of the Ringrose structure theorem. We begin with the existence of simple invariant nests.

Proposition 3.4.4. *Let A be a compact operator on a seperable Banach space, X. Then an invariant nest $\mathcal{F}$ is simple if and only if it is maximal among all invariant nests.*

Remark. This result is also true (by a similar proof) if "invariant" is dropped in both places. But this alternate result does not immediately imply this one. If A were not compact, it could be that only $\{0\}$ and X are invariant (see the Notes to Section 5.12 of Part 1). Then the unique invariant nest is not simple.

Proof. Suppose $\mathcal{F}$ is a simple invariant nest and $\mathcal{F}_0$ is an invariant nest containing $\mathcal{F}$. Let $M \in \mathcal{F}_0 \setminus \mathcal{F}$. Let $\mathcal{F}_- = \{L \in \mathcal{F} \mid L \subset M\}$, $\mathcal{F}_+ = \{L \in \mathcal{F} \mid M \subset L\}$,

$$L_+ = \inf_{L \in \mathcal{F}_+} L, \quad L_- = \sup_{L \in \mathcal{F}_-} L \tag{3.4.16}$$

Clearly, L_+, $L_- \in \mathcal{F}$ since $\mathcal{F}$ is complete and by construction

$$L_- \subset M \subset L_+ \tag{3.4.17}$$

Let $N \in \mathcal{F}$ with $N \subset L_+$, $N \neq L_+$. Since $N \neq L_+$, $N \notin \mathcal{F}_+$ so it is false that $M \subset N$. Since $\mathcal{F}_0$ is a nest, $N \subset M$, i.e., $N \in \mathcal{F}_-$ and thus

$$\{N \in \mathcal{F} \mid N \subset L_+, N \neq L_+\} = \mathcal{F}_- \tag{3.4.18}$$

so

$$(L_+)_- = L_-$$

and $\dim(L_+/L_-) \leq 1$ by the simplicity hypothesis. But that means any M obeying (3.4.17) must have $M = L_+$ or $M = L_-$, i.e., $M \in \mathcal{F}$ after all. This contradiction shows that $\mathcal{F}$ is maximal.

Conversely, suppose $\mathcal{F}$ is a maximal invariant nest. If $\mathcal{F}_1 \subset \mathcal{F}$ is a subset, $\sup(\mathcal{F}_1)$ and $\inf(\mathcal{F}_1)$ are clearly invariant for A. If $M \in \mathcal{F}$, either $N \subset M$ for all $N \in \mathcal{F}_1$ in which case $N \subset \inf(\mathcal{F}_1)$, or for some M, $M \subset N$ in which case $\inf(\mathcal{F}_1) \subset N$. Thus we can add $\inf(\mathcal{F}_1)$ to $\mathcal{F}$ and still get an invariant nest, so, by maximality, $\inf(\mathcal{F}_1) \in \mathcal{F}$. Similarly, $\sup(\mathcal{F}_1) \in \mathcal{F}$. Thus $\mathcal{F}$ is complete.

Let $M \subset \mathcal{F}$ and suppose $\dim(M/M_-) > 1$. $\widetilde{A}\colon M/M_- \to M/M_-$ by $\widetilde{A}[x] = [Ax]$ is well-defined since M_- is invariant and compact by a simple argument on quotients. By the Aronszajn–Smith theorem (see Corollary 5.12.30 of Part 1), $\widetilde{A}$ has a nontrivial invariant subspace, S, $\{x \in X \mid [x] \in S\}$ is then an invariant subspace for A lying strictly between M_- and M violating maximality. Thus, $\dim(M/M_-) \leq 1$ and $\mathcal{F}$ is simple. □

By a simple Zornification, maximal invariant nests exist—thus, we have the first part of Theorem 3.4.2:

Corollary 3.4.5. *Let A be a compact operator on a Banach space X. Then A has simple invariant nests.*

We now turn to the connection of eigenvalues and eigenjumps. We need a technical preliminary:

Lemma 3.4.6. *Let $\mathcal{F}$ be a simple invariant nest for a compact operator, $A \in \mathcal{L}(X)$. Let $M \in \mathcal{F}$ have $M_- = M$. Then for any $\delta > 0$, there is $L \subset M$, $L \neq M$ so that for all $x \in M$*

$$d(Ax, L) \leq \delta \|x\| \tag{3.4.19}$$

Proof. Fix $\delta > 0$. Since $A\bigl[\{x \in M \mid \|x\| \leq 1\}\bigr] \equiv Q$ is precompact, it is totally bounded, i.e., we can find $\{x_j\}_{j=1}^k$ so that

$$Q \subset \bigcup_{j=1}^k \{y \mid \|y - Ax_j\| \leq \frac{\delta}{2}\} \tag{3.4.20}$$

Since $M = \overline{(\bigcup_{L \in M, L \neq M, L \in \mathcal{F}} L)}$, we can find L_j and then $\ell_j \in L_j$ so that $\|\ell_j - x_j\| \leq \frac{\delta}{2\|A\|}$. By renumbering we can suppose that $L_1 \subset L_2 \subset \dots \subset L_k \subset M$ with $L_k \neq M$.

Since $\|A\ell_j - Ax_j\| \leq \frac{\delta}{2}$, if $y_j = A\ell_j$, then

$$Q \subset \bigcup_{j=1}^k \{y \mid \|y - y_j\| < \delta\} \tag{3.4.21}$$

Let $L = L_k$. (3.4.21) implies for $x \in M$ with $\|x\| \leq 1$, $d(Ax, L) \leq \delta$ which is (3.4.19). □

Proposition 3.4.7. *Let $\mathcal{F}$ be a simple invariant nest for a compact operator A. Suppose $x \neq 0$ and $Ax = \lambda x$ for $\lambda \in \mathbb{C}$, $\lambda \neq 0$. Let*

$$M = \inf \{N \in \mathcal{F} \mid x \in N\} \tag{3.4.22}$$

Then $\mathcal{F}$ is discontinuous at M and $\alpha_M = \lambda$.

Proof. Define M by (3.4.22), Since $M = \bigcap_{x \in N} N$, we have $x \in M$. If $M_- = M$, by Lemma 3.4.6 with $\delta = \frac{|\lambda|}{2}$, we have L obeying (3.4.19), $L \neq M$ so $x \notin L$ and $d(x, L) > 0$.

Since L is invariant for A, for any $y \in A$, $d(Ax, L) = d\big(A(x+y), L\big)$ so for all $y \in L$

$$d(Ax, L) \leq \delta \|x + y\| \tag{3.4.23}$$

and thus taking inf over $y \in L$

$$d(Ax, L) \leq \delta d(x, L) \tag{3.4.24}$$

Since $Ax = \lambda x$, we see that

$$|\lambda|\, d(x, L) \leq \delta d(x, L) \tag{3.4.25}$$

This is a contradiction given that $\delta = \frac{1}{2}|\lambda|$ and $d(x, L) > 0$. Thus $M_- \neq M$.

By definition of α_M

$$(\lambda - \alpha_M)x = (A - \alpha_M)x \in M_- \tag{3.4.26}$$

Since $x \notin M_-$, we must have $\lambda - \alpha_M = 0$. □

Proposition 3.4.8. *Let $\mathcal{F}$ be a simple invariant nest for a compact operator, A, and $M \in \mathcal{F}$ a point of discontinuity. Then $\alpha_M \in \sigma(A)$. If $\alpha_M \neq 0$ and has index one, then there is $x \in M \setminus M_-$ with $Ax = \alpha_M x$.*

Remark. Recall (see Section 2.3) that index one means $\mathrm{Ker}(A - \alpha_M)^2 = \mathrm{Ker}(A - \alpha_M)$.

Proof. Let $B = A \restriction M$ and $B_- = B \restriction M_-$. If $\alpha_M \notin \sigma(A)$, by Problem 1, $\alpha_M \notin \sigma(B) \cup \sigma(B_-)$ and $(B - \alpha_M)^{-1} \restriction M_- = (B_- - \alpha_M)^{-1}$. Pick any $x \in M \setminus M_-$ and let $y = (B - \alpha_M)x$. Then $y \in M_-$, so

$$x = (B - \alpha_M)^{-1} y = (B_1 - \alpha_M)^{-1} y \in M_- \tag{3.4.27}$$

contrary to the choice of x. Thus $\alpha_M \in \sigma(A)$.

If now $\alpha_M \neq 0$ and has index one, pick $x \in M \setminus M_-$. By Corollary 2.3.9, we can write

$$x = k + y, \quad k \in \mathrm{Ker}(B - \lambda_0),\ y \in \mathrm{Ran}(B - \lambda_0)$$

But $B - \lambda_0$ maps M to M_- so $y \in M_-$. Thus $k \in M \setminus M_-$ also and is the required eigenvector. □

Now let $A \in \mathcal{L}(X)$ be compact and $\mathcal{F}$ a simple invariant nest. Let $\lambda \in \sigma(A) \setminus \{0\}$. Let m be its algebraic multiplicity and d its multiplicity as an eigenjump. If we prove $d = m$, then we see that the number of points of discontinuity is countable (since $\sigma(A)$ is countable) so we'll have Theorem 3.4.2. We'll prove $d = m$ when λ has index one and then reduce the general case to this case.

Proposition 3.4.9. *Let $\lambda \in \sigma(A)$ have index one. Then $d \le m$.*

Proof. Suppose $d > m$. Then we can find distinct $M_1 \subset M_2 \subset \dots \subset M_{m+1}$ all points of discontinuity with $\alpha_{M_j} = \lambda$. Since λ has index one, by Proposition 3.4.8, we can find $x_j \in M_j \setminus M_{j-1}$ for $j = 2, \dots, m+1$ and $x_1 \in M_1 \setminus \{0\}$ so that $(A - \lambda_0)x_j = 0$ and thus, $x_j \in \operatorname{Ran}(P_\lambda)$.

We claim that the x_j are independent violating the fact that $\dim(P_\lambda) = m$. For suppose $\sum_{j=1}^{m+1} \alpha_j x_j = 0$ with $(\alpha_1, \dots, \alpha_{m+1}) \ne 0$. Pick $J = \max_j(\alpha_j \ne 0)$. Then $x_J = \alpha_J^{-1}(-\sum_{j=1}^{J-1} \alpha_{j-1} x_j) \subset M_{J-1}$ violating the condition that $x_J \notin M_{J-1}$. □

Proposition 3.4.10. *Let $\lambda \in \sigma(A)$ have index one. Then $m \le d$.*

Proof. If not, $m > d$. We pick $\{M_j\}_{j=1}^d \subset \mathcal{F}$ so $M_1 \subset M_2 \subset \dots \subset M_d$, all distinct with $(M_j)_- \ne M_j$ and $\alpha_{M_j} = \lambda$ for $j = 1, \dots, n$. By Proposition 3.4.8, pick $x_j \in M_j$ with $Ax_j = \lambda x_j$. Define ψ_j on $M_j = \{\alpha x_j + y \mid \alpha \in \mathbb{C}, y \in (M_j)_-\}$ by

$$\psi_j(\alpha x_j + y) = \alpha \tag{3.4.28}$$

Since $\operatorname{Ker}\psi_j = (M_j)_-$ is closed, $\psi_j \in M_j^*$. By the Hahn–Banach theorem, we can extend ψ_j to $\varphi_j \in X^*$.

Let $K = \operatorname{Ker}(A - \lambda)$ which, by hypothesis has dimension $m > d$. Let $\eta_j = \varphi_j \restriction K$. Since $\varphi_j(x_j) = 1, \varphi_j \not\equiv 0$ and thus $\operatorname{Ker}\varphi_j$ has dimension $m - 1$ and $\bigcap_{j=1}^d \operatorname{Ker}\varphi_j$ has dimension at least $m - d > 0$ so we can find $x_0 \in \bigcap_{j=1}^d \operatorname{Ker}\varphi_j$.

By Proposition 3.4.7, there is M_0, a point of discontinuity with $\alpha_{M_0} = \lambda$ and $x_0 \in M_0 \setminus (M_0)_-$. By the counting of M_j, M_0 must be an M_j but then $\varphi_j(x_0) \ne 1$ contrary to $x_0 \in \operatorname{Ker}\varphi_j$. This contradiction proves that $m \le d$. □

Proof of Theorem 3.4.2. Fix $\lambda_0 \in \sigma(A)$, $\lambda_0 \ne 0$. We want to prove that $m = d$ for this λ_0. If λ_0 has index one for A, we know that $m = d$ by Propositions 3.4.9 and 3.4.10. If not, there is a minimal $k \ge 1$ so that for $m = 1, 2, \dots$, $\operatorname{Ker}(A - \lambda_0)^{k+m} = \operatorname{Ran} P_{\lambda_0} \ne \operatorname{Ker}(A - \lambda_0)^k$. Define

$\mu_0 \equiv -(-\lambda_0)^{k+1}$.

$$B = (A - \lambda_0)^{k+1} - (-\lambda_0)^{k+1} = \left[\sum_{j=1}^{k+1} (-\lambda_0)^{k+1-j} A^{j-1}\right] A \tag{3.4.29}$$

Then B is compact and

$$(B - \mu_0)^\ell = (A - \lambda_0)^{\ell(k+1)} \tag{3.4.30}$$

Thus $P_{\mu_0}(B) = P_{\lambda_0}(A)$, so μ_0 has the same algebraic multiplicity for B as λ_0 has for A.

If M is an invariant subspace for A, it is also for B and thus $\mathcal{F}$ is a simple invariant nest for B. Moreover, a simple calculation (Problem 7) shows that if M is a discontinuity point of $\mathcal{F}$ and β_M is the eigenjump for B, then

$$\beta_M = \mu_0 - (\alpha_M - \lambda_0)^{k+1} \tag{3.4.31}$$

so $\beta_M = \mu_0 \Leftrightarrow \alpha_m = \lambda_0$. Thus, the m, d for (A, λ_0) are the same as for (B, μ_0). By (3.4.30), μ_0 has index one for B, so by Propositions 3.4.9 and 3.4.10, $m = d$.

This implies the set of nonzero jumps is countable (actually, so are the zero jumps; see Problem 8 for the Hilbert space case) and the multiplicities are equal. $\square$

Second Proof of Theorem 3.4.1. Pick a simple invariant nest, $\mathcal{F}$, for A. For each jump, $M \in \mathcal{F}$, with eigenjump, $\alpha_M \neq 0$, pick $\varphi_M \in M \cap (M_-)^\perp$ with $\|\varphi_M\| = 1$. If $P \subset M_-$ is another jump, $\varphi_P \in M_-$ so $\varphi_P \perp \varphi_M$. Thus, if $\mathcal{N} \subset \mathcal{F}$ is the set of jumps with $\alpha_M \neq 0$, $\{\varphi_P\}_{P\in\mathcal{N}}$ is an orthonormal set.

Define N by

$$N\varphi_P = \alpha_P \varphi_P, \quad P \in \mathcal{N} \tag{3.4.32}$$

$$N\psi = 0 \quad \text{if} \quad \psi \in \{\varphi_P\}_{P\in\mathcal{N}}^\perp \tag{3.4.33}$$

Then N has an orthonormal set of eigenvectors and so is normal. Since $\{P \mid |\alpha_P| > \delta\}$ is finite, N is compact.

For any $R \in \mathcal{F}$, R is spanned by $\{\varphi_P\}_{P\in\mathcal{N}, P\subset R}$ plus some orthonormal vectors in $\{\varphi_P\}_{P\in\mathcal{N}}^\perp$. It follows that R is also invariant for N and so for $Q \equiv A - N$.

For any jump, P, in $\mathcal{F}$, the eigenjump for Q is either $\alpha_P - \alpha_P$ if $P \in \mathcal{N}$ or $0 - 0$ if $\alpha_P = 0$. Either way Q has only 0 eigenjumps. Q is compact as the difference of compact operators. By the Ringrose Structure Theorem, $\sigma(Q) = \{0\}$, i.e., Q is quasinilpotent. $\square$

Notes and Historical Remarks.

The notion of invariant nests, Theorem 3.4.2, and the ideas in its proof are taken from the seminal paper of Ringrose [**569**] and its exposition in his book [**570**]. Theorem 3.4.1 proven by analyzing maximal nests in a Hilbert space (by associating orthogonal projections onto each $M \in \mathcal{F}$ and so getting a kind of projectional-valued measure) is implicit in Ringrose [**569, 570**]. Our approach, as well as the explicit statement, is from West [**740**].

Further developments include the work of Erdos [**187, 188**] who, in particular, used the theory to prove that a trace class quasinilpotent operator is a trace norm limit of finite rank quasinilpotent operators. This immediately implies the trace of a trace class quasinilpotent operator is zero. In turn, this plus Theorem 3.4.1 implies Lidskii's theorem, that the trace of a trace class operator is the sum of its eigenvalues counting algebraic multiplicity (one needs the theorem, see Theorem 3.9.1, that A trace class implies $\sum_{n=1}^{K}|\lambda_n(A)| < \infty$ for this implies that in the decomposition (3.4.1), C is trace class and thus $Q = A - C$ is trace class; in fact, the proof of this uses the fact that if $\{\varphi_n\}_{n=1}^{K}$ is the orthonormal set obtained from Gram–Schmidt on $\cup \operatorname{Ran}(P_{\lambda_j})$ then $\sum_{n=1}^{K}|\lambda_n(A)| = \sum_{n=1}^{K}|\langle\varphi_n, A\varphi_n\rangle| \leq \operatorname{Tr}(|A|)$). See the Notes to Section 3.12 for the Dunford–Schwarz proof of Lidskii's Theorem which is not unrelated to this argument.

Another important tool in the further study of nests (including Erdos' work) is that every continuous simple nest on a Hilbert space is unitarily equivalent to the Volterra nest of Example 3.4.3—a result implicitly in Kadison–Singer [**363**].

The remainder of these notes focus on the result about the Volterra operator, S, of (3.4.6), that the only invariant subspaces of S are the Y_c. In 1938, Gel'fand [**225**] asked the question of which $f \in L^2([0,1], dx)$ has $\{S^n f\}_{n=0}^{\infty}$ total in L^2. This was first answered by Agmon [**5**] in 1949 although much of the later literature seemed unaware of this paper. Using complex variable techniques, Agmon proved this set is total if and only if $0 \in \operatorname{supp}(f)$. It is not hard (Problem 5) to see that this result is equivalent to the following invariant subspaces result (in that either leads to the other in a few steps).

Theorem 3.4.11. *The only subspaces of $L^p([0,1], dx)$ invariant for the operator, S, of* (3.4.6) *are the space Y_c of* (3.4.7), $0 \leq c \leq 1$. *For $C([0,1])$ the only other invariant space is all of $C([0,1])$.*

In 1957, Brodskiĭ [**90**] and Donoghue [**165**] explicitly proved the invariant subspace result. Brodskiĭ used operator theory and Donoghue used Titchmarsh's theorem as we will explain shortly. Theorem 3.4.11 might seem related to Theorem 5.10.1 of Part 3—Beurling's theorem which specifies the invariant subspaces of the left shift on $\ell^2(\mathbb{Z}_+)$. Indeed, Sarason [**588**] has proven Theorem 3.4.11 from Beurling's theorem.

There is a close connection between Theorem 3.4.11 and a theorem of Titchmarsh some state as follows. If f is continuous on $[0,\infty)$ and $f \not\equiv 0$, one defines

$$I(f) = \inf\big(\operatorname{supp}(f)\big) = \inf\big(\{x \mid f(x) \neq 0\}\big) \tag{3.4.34}$$

Theorem 3.4.12 (Titchmarsh's Theorem, First Form). *Let f and $g \in C([0,\infty))$ with neither identically zero. Then*

$$I(f * g) = I(f) * I(g) \tag{3.4.35}$$

Remarks. 1. $(f*g)(x) = \int_0^x f(y)g(x-y)\,dy$ so the integral converges for all $x \in [0,\infty)$ and only depends on $f \restriction [0,x]$, $g \restriction [0,x]$.

2. $I(f+g) \geq I(f) + I(g)$ is trivial. The subtle point is that there can't be cancellations. It is the kind of result that seems obviously true but upon some thought perhaps less so. While there are proofs that are not long (the one in Yoshida [**768**] is two pages), they are not especially intuitive. Some proofs use some rather advanced complex variable results.

Donoghue's proof of Theorem 3.4.11 deduced it from Titchmarsh's theorem. In the other direction, Kalisch [**367**] noted that Theorem 3.4.11 in turn implies Titchmarsh's theorem. The reader will derive these implications in Problems 4, 5, and 6.

The reader will look at equivalent forms of Titchmarsh's theorem in Problem 3 including

Theorem 3.4.13 (Titchmarsh's Theorem, Second Form). *If $f, g \in C\big([0,1]\big)$ and $I(f) = 0$, $f * g = 0$, then $g = 0$.*

Titchmarsh proved his theorem using complex functions in 1926 [**700**]. Other complex-variable methods are Crum [**138**] and Dufresnoy [**170**]. A real-variable proof is Mikusinski [**477**], the basis of the proof in Yoshida [**768**]. Lions [**451**] has the following ν-dimensional analog.

Theorem 3.4.14 (Lions' Theorem). *If $S \in \mathcal{S}'(\mathbb{R}^\nu)$ is a distribution of compact support, let $c(S)$ be the closed convex hull of the support of S. Then, for any two distributions of compact support, S and T,*

$$c(S * T) = c(S) + c(T) \tag{3.4.36}$$

Remarks. 1. In one dimension $c(S) = [i(S), m(S)]$ where m is the $\sup\{x \mid x \in \operatorname{supp}(S)\}$ so $c(S)+c(T) = [i(S)+i(T), m(S)+m(T)]$ so Lions' Theorem is equivalent to Titchmarsh (given that under reflection $m(S)$ and $i(S)$ are interchanged).

2. One can determine $c(S)$ from the support of each one-dimensional distribution obtained by integrating out $\nu - 1$ variables. One can thus deduce Lions' theorem from Titchmarsh's.

Problems

1. Let X be a Banach space and $A \in \mathcal{L}(X)$. Let $Y \subset X$ be a (closed) invariant subspace for A.

   (a) If λ is in the unbounded component of $\sigma(A)$, prove that Y is invariant for $(A-\lambda)^{-1}$ and that $(A \restriction Y - \lambda)^{-1} = (A-\lambda)^{-1} \restriction Y$. (*Hint*: By Corollary 5.5.11 of Part 1, it suffices to prove $f_{\ell,y}(z) = \ell\big((A-z)^{-1}y\big)$ vanishes for each $y \in Y$ and $\ell \in Y^\perp$ and z in the unbounded component. Show this is so first for $|z| > \|A\|$.)

   (b) For A compact, prove that $\sigma(A \restriction Y) \subset \sigma(A)$.

   (c) Find an example where $\sigma(A \restriction Y) \neq \sigma(A)$.

   (d) Find a noncompact example where $\sigma(A \restriction Y)$ is strictly larger than $\sigma(A)$. (*Hint*: Right shift.)

2. Let A be given by (3.4.6)

   (a) For any $p \in [1,\infty)$, prove that $\text{Ran}(A) \subset C\big([0,1]\big)$.

   (b) Knowing that A is compact on $C\big([0,1]\big)$, conclude that it is compact on $L^p\big([0,1],dx\big)$.

   (c) Prove that $\|A^{n+1}\|_{L^p} \leq \|A^n\|_{C([0,1])}$.

   (d) Prove that A, as an operator on L^p, is quasinilpotent.

   (e) With Y_c given by (3.4.7), prove that for L^p, $\{Y_c\}_{0\leq c\leq 1}$ is a maximal nest and so a maximal invariant nest.

   (f) Prove the same for $C\big([0,1]\big)$ if $C\big([0,1]\big)$ is added.

3. (a) If $I(f)$ is defined for any f with $f \not\equiv 0$ and $\text{supp}(f)$ bounded and if $\big(T_a f\big)(x) = f(x-a)$, prove that $I\big(T_a f\big) = I(f) - a$ and $\big(T_a f\big) * \big(T_b f\big) = T_{a+b}(f * g)$.

   (b) Prove that it suffices to prove Titchmarsh's theorem in the special case $I(f) = I(g) = 0$.

   (c) Let $f, g \in C\big([0,1]\big)$ with $I(f) = 0$. Assuming Titchmarsh's theorem, prove $f * g = 0$ for $x \in [0,1] \Rightarrow g = 0$. Thus, the first form implies the second form.

   (d) Let $f, g \in C\big([0,1]\big)$. If $f * g = 0$ on $[0,1]$, using that $I(f) + I(g) \geq 1$, show that $I(f) = a$, $f * g \restriction [0,1] = 0 \Rightarrow I(g) \geq 1 - a$.

   (e) Suppose $f, g \in C\big([0,\infty)\big)$ has $I(f) = I(g) = 0$ and that contrary to the first form of Titchmarsh's theorem, $I(f * g) = a > 0$. By scaling, find $\widetilde{f}$, $\widetilde{g}$ on $[\,\cdot\,,1]$ so $I(\widetilde{f}) = I(\widetilde{g}) = 0$. Let $\widetilde{f} * \widetilde{g} \restriction [0,1] = 0$. Thus, the second form implies the first.

4. This problem will show that Theorem 3.4.11 proves Theorem 3.4.13.

(a) Let $u(x) = \chi_{[0,1]}$. Prove that for $x \in [0,1]$

$$Sf(x) = \big(u * f\big)(x) \tag{3.4.37}$$

and that

$$\big(S^n f\big) * g = (u * \cdots * u) * (f * g) \tag{3.4.38}$$

(b) Show that Theorem 3.4.11 for $L^1\big([0,1], dx\big)$ implies that the L^1-span of $\{S^n f\}_{n=0}^\infty$ is all of L^1 if $I(f) = 0$. (This is Agmon's result.)

(c) If $I(f) = 0$ and $f * g = 0$ as functions in $L^1\big([0,1], dx\big)$, prove that $h * g = 0$ for all $h \in L^1\big([0,1], dx\big)$.

(d) Picking h as an approximate identity, prove that $f * g = 0$ plus $I(f) = 0 \Rightarrow g = 0$.

5. This problem will prove that Theorem 3.4.11 is equivalent to the result of Agmon that if $f \in L^1\big([0,1], dx\big)$ and $I(f) = 0$, then $\{S^n f\}_{n=0}^\infty$ is total in L^1.

(a) Assume Agmon's result. Let Y_c be given by (3.4.7). If $I(f) = c$, prove that the span of $\{S^n f\}_{n=0}^\infty$ is Y_c.

(b) Assume Agmon's result. Let Z be a subspace of $L^1\big([0,1], dx\big)$ invariant for S. Let $c = \inf\{I(g) \mid g \in Z\}$. Prove that $Z = Y_c$ so Agmon's result implies Theorem 3.4.11.

Note. The other half of the equivalence is (b) of the last problem.

6. This problem will show that Theorem 3.4.13 for $L^2\big([0,1], dx\big)$ implies Theorem 3.4.11. It will use Problem 5 that it suffices to show $I(f) = 0 \Rightarrow \{S^n f\}_{n=0}^\infty$ is total in L^2.

(a) With $u = \chi_{[0,1]}$, prove that $\big[u * \cdots * u\,(n \text{ times})\big](x) = \frac{x^n}{n!}$

(b) If $\bar{g} \perp \{S^n f\}_{n=0}^\infty$, prove that

$$\langle \bar{g}, p * f \rangle = 0$$

for any polynomial p.

(c) Let $\widetilde{h}(x) = h(1-x)$. Prove that

$$\begin{aligned}\int_{\substack{0 \le x \le 1 \\ 0 \le y \le x}} g(x)p(y)f(x-y)\,dx\,dy &= \int_{\substack{0 \le y \le 1 \\ 0 \le x \le y}} \widetilde{g}(x)\widetilde{p}(y)f(y-x)\,dx\,dy \\ &= \langle \overline{\widetilde{p}}, \widetilde{g} * f \rangle\end{aligned}$$

(d) If $\bar{g} \perp \{S^n f\}_{n=0}^\infty$, prove that $\widetilde{g} * f = 0$ so if $I(f) = 0$ and we assume Theorem 3.4.12, then $g = 0$.

(e) Conclude that Theorem 3.4.12 $\Rightarrow$ Theorem 3.4.11.

Remark. See the Notes for the history of the ideas in Problems 3–6.

7. Let A be compact and B given by (3.4.29). Let $\mathcal{F}$ be a simple invariant nest for A and so for B. Let M be a discontinuity point of $\mathcal{F}$ and α_M (respectively, β_M), the eigenjumps to A and B. Prove that (3.4.31) holds.

8. Let $\mathcal{F}$ be a simple nest in a Hilbert space, $\mathcal{H}$. Let $\{M_\alpha\}_{\alpha\in I}$ be a family of discontinuity points of $\mathcal{F}$. Pick $\varphi_\alpha \in (M_\alpha)\cap\big[(M_\alpha)_-\big]^\perp$ with $\|\varphi_\alpha\| = 1$. Prove that $\{\varphi_\alpha\}_{\alpha\in I}$ is an orthonormal family and conclude that I is countable

3.5. Singular Values and the Canonical Decomposition

In this section, we'll show any compact operator, A, on a Hilbert space, $\mathcal{H}$, has a natural decomposition

$$A = \sum_{j=1}^{N(A)} \mu_j(A)\langle\varphi_j, \cdot\rangle\psi_j \tag{3.5.1}$$

where $N(A) = \operatorname{rank}(A)$, $\mu_1 \geq \mu_2 \geq \cdots \geq \mu_N$ (or $\geq \dots$ if $N = \infty$), and $\{\varphi_j\}_{j=1}^N$, $\{\psi_j\}_{j=1}^N$ are each orthonormal families. The φ's and ψ's are "essentially unique" and the μ_j must be the eigenvalues of $|A|$, since (3.5.1) implies

$$A^*A = \sum_{j=1}^{N(A)} \mu_j(A)^2\langle\varphi, \cdot\rangle\varphi_j \tag{3.5.2}$$

Thus, we have the following

Definition. If A is compact, its *singular values* $\{\mu_j(A)\}_{j=1}^{N(A)}$ (where $N(A) = \operatorname{rank}(A)$) are the eigenvalues of $|A|$, written in decreasing order (counting multiplicities). If $N(A) < \infty$, we will sometimes write $\mu_j(A) = 0$ for $j > N(A)$. By the argument in Problem 16 of Section 2.4, $|A|$ and $|A^*|$ have the same nonzero eigenvalues, so $\{\mu_j(A)\}_{j=1}^N$ are also the eigenvalues of $|A^*|$.

Theorem 3.5.1. *Any compact operator has an expansion* (3.5.1) *where* $\mu_j(A)$ *are the singular values,* $\{\varphi_j\}_{j=1}^{N(A)}$ *are orthonormal, and obey*

$$|A|\varphi_j = \mu_j(A)\varphi_j, \qquad A\varphi_j = \mu_j(A)\psi_j \tag{3.5.3}$$

Remarks. 1. (3.5.1) is called the *canonical decomposition.*

2. If $\mu_1 > \mu_2 > \dots$, the φ_j's are unique up to a phase factor. If some $\mu_j = \mu_{j+1}$, there are degeneracies that destroy uniqueness. ψ_j is always determined by $U\varphi_j = \psi_j$, where U is the partial isometry in the polar decomposition of A.

3. As we'll see, this essentially encodes Hilbert–Schmidt theorem & polar decomposition.

Proof. Let $\{\varphi_j\}_{j=1}^{N(A)}$ be an orthonormal basis for $\operatorname{Ker}(A)^\perp = \operatorname{Ker}(|A|)^\perp$ of eigenvectors for $|A|$. Then the Fourier expansion plus $|A|\varphi_j = \mu_j(A)\varphi_j$ implies

$$|A| = \sum_{j=1}^{N} \mu_j(A)\langle\varphi_j, \cdot\rangle\varphi_j \tag{3.5.4}$$

Since U_A is an isometry on $\operatorname{Ker}(A)^\perp$, it takes the orthonormal set $\{\varphi_j\}_{j=1}^{N(A)}$ to an orthonormal set $\{\psi_j\}_{j=1}^{N}$, where

$$\psi_j = U\varphi_j \tag{3.5.5}$$

The polar decomposition, $A = U_A|A|$, then implies (3.5.1). □

In the rest of this section, we'll explore some of the equalities and inequalities for $\{\mu_j(A)\}_{j=1}^{\infty}$. By the Hilbert–Schmidt theorem,

$$\lim_{j\to\infty} \mu_j(A) = 0 \tag{3.5.6}$$

There is a simple min-max characterization of the μ_j.

Theorem 3.5.2. *For any compact operator,*

$$\mu_j(A) = \inf_{\psi_1,\dots,\psi_{j-1}} \Bigl(\sup_{\substack{\varphi\perp\psi_1,\dots,\psi_{j-1} \\ |\varphi|=1}} \|A\varphi\| \Bigr) \tag{3.5.7}$$

Remarks. 1. One can define $\mu_j(A)$ by (3.5.7) for any A and then show (Problem 1) that A is compact if and only if (3.5.6) holds.

2. In particular,

$$\mu_1(A) = \|A\| \tag{3.5.8}$$

Proof. By (3.2.14),

$$\begin{aligned} \mu_j(A)^2 = \lambda_j(A^*A) &= \inf_{\psi_1,\dots,\psi_{j-1}} \Bigl(\sup_{\substack{\varphi\perp\psi_1,\dots,\psi_{j-1} \\ \|\varphi\|=1}} \langle\varphi, A^*A\varphi\rangle \Bigr) \\ &= (\text{RHS of } (3.5.7))^2 \end{aligned}$$

since $\langle\varphi, A^*A\varphi\rangle = \|A\varphi\|^2$. □

Theorem 3.5.3. *For any A compact and $B \in \mathcal{L}(H)$,*

(a) $\mu_j(A) = \mu_j(A^*)$ (3.5.9)

(b) $\mu_j(BA) \le \|B\|\mu_j(A)$ (3.5.10)

(c) $\mu_j(AB) \le \|B\|\mu_j(A)$

Remark. (a) says A^*A and AA^* have the same nonzero eigenvalues, so there is an alternate proof of (a) using Problem 5 of Section 2.2.

Proof. (a) By (3.5.1),

$$A^* = \sum_{j=1}^{N} \mu_j(A) \langle \psi_j, \cdot \rangle \varphi_j \tag{3.5.11}$$

so

$$AA^* = \sum_{j=1}^{N} \mu_j(A)^2 \langle \psi_j, \cdot \rangle \psi_j \tag{3.5.12}$$

implying

$$|A^*| = \sum_{j=1}^{N} \mu_j(A) \langle \psi_j, \cdot \rangle \psi_j \tag{3.5.13}$$

so $\mu_j(A)$ are also the eigenvalues of $|A^*|$ (as we noted above).

(b) is immediate from $\|BA\varphi\| \le \|B\|\|A\varphi\|$ and (3.5.7).

(c) By (a) and (b), we have that

$$\begin{aligned} \mu_j(AB) &= \mu_j((AB)^*) = \mu_j(B^*A^*) \\ &\le \|B^*\|\mu_j(A^*) = \|B\|\mu_j(A) \end{aligned} \tag{3.5.14}$$

□

Notes and Historical Remarks. The name "singular value" goes back to Picard [**522**] in 1910, but in some quarters took a while to catch on. As late as 1949, Weyl [**751**] referred to "two kinds of eigenvalues" rather than eigenvalues and singular values. But by now it is fairly standard, although *s-number* is also used.

Besides the inequalities of Theorem 3.5.3 and Problems 2 and 3, there are loads of other inequalities on singular values. Goh'berg–Krein [**257**] has a whole chapter on them and Simon [**650**] several sections. A key tool is rearrangement inequalities (see Problems 24–26 of Section 5.3 of Part 1)—for example, Horn's inequality (Problem 3) and those ideas imply that for any $r > 0$,

$$\sum_{n=1}^{\infty} \mu_n(AB)^r \le \sum_{n=1}^{\infty} [\mu_n(A)\mu_n(B)]^r \tag{3.5.15}$$

and this plus Hölder's inequality for sums implies Hölder's inequality of operator ideals, that is, another proof of Theorem 3.7.6.

While I follow my usage of forty years in calling (3.5.1) the canonical expansion, it is not the only name. Both Goh'berg–Krein [**257**] and Retherford [**559**] call it the *Schmidt expansion* after his 1907 paper [**601**].

The canonical expansion for finite matrices is essentially equivalent to what is called the *singular value decomposition* (svd). This says that if A is an $n \times m$ real matrix or rank k ($k \le \min(n, m)$, of course), then A can be written

$$A = U\Sigma V^{\mathrm{t}} \tag{3.5.16}$$

where Σ is a $k \times k$ diagonal matrix with diagonal matrix elements $\mu_1 \ge \mu_2 \ge \dots \ge \mu_k$, U is an $n \times k$ matrix with $U^{\mathrm{t}}U = \mathbb{1}$, V an $m \times k$ matrix with $V^{\mathrm{t}}V = \mathbb{1}$. In terms of (3.5.1), U is the matrix with rows $\psi_1, \dots, \psi_{N(A)}$ and V with rows $\varphi_1, \dots, \varphi_{N(A)}$, so one easily goes from (3.5.1) to (3.5.16) and vice-versa.

The svd for square matrices goes back to Beltrami [**49**] and Jordan [**352, 353**] in 1873–74. It was rediscovered by Sylvester [**678, 679, 680**]. Its extension to the nonsquare case is from the 1930s; see Williamson [**760**], Eckart–Young [**180, 181**]. For the early history, see Stewart [**665**].

The svd is heavily used in numerical analysis [**259, 260**] and data analysis [**158, 481, 710**].

Problems

1. If $\mu_j(A)$, defined by (3.5.7), has $\mu_j(A) \to 0$, show that A is compact. (*Hint*: Look at the proof of Theorem 3.1.13.)

2. This problem will prove the Ky Fan inequalities [**192**]:

$$\mu_{n+m+1}(A+B) \le \mu_{n+1}(A) + \mu_{m+1}(B) \tag{3.5.17}$$

(a) If

$$Q(A; \psi_1, \dots, \psi_\ell) = \sup_{\substack{\varphi \perp \psi_1, \dots, \psi_\ell \\ \|\varphi\|=1}} \|A\varphi\|$$

prove that

$$Q(A+B; \psi_1, \dots, \psi_{n+m}) \le Q(A; \psi_1, \dots, \psi_n) + Q(B; \psi_{n+1}, \dots, \psi_{n+m})$$

(b) Prove (3.5.17).

(c) Why doesn't this argument work to prove $\mu_{n+1}(A+B) \le \mu_{n+1}(A) + \mu_{n+1}(B)$? Is this inequality true or false in general?

3. (a) If $\wedge^k$ is the antisymmetric product (see Section 3.8 of Part 2A), prove that

$$\|\wedge^k(A)\| = \prod_{j=1}^k \mu_j(A) \tag{3.5.18}$$

(b) Prove *Horn's inequality* that

$$\prod_{j=1}^k \mu_j(AB) \le \prod_{j=1}^k \mu_j(A)\mu_j(B) \tag{3.5.19}$$

4. Prove that (due to Allahverdiev [**13**])
$$\mu_k(A) = \inf\{\|A - C\| \mid \text{rank}(C) \le k-1\} \tag{3.5.20}$$

5. Prove that (this inf is called the k-th Gel'fand number)
$$\mu_k(A) = \inf\{\|AP\| \mid P \text{ a projection with } \dim\big(\text{Ran}(1-P)\big) \le k-1\}$$

6. For each n, prove that $|\mu_k(A) - \mu_k(B)| \le \|A - B\|$. (*Hint*: Use (3.5.7) or (3.5.17) or (3.5.20).)

7. (a) Prove that
$$\sum_{k=1}^{n} \mu_k(A) = \sup\{|\text{Tr}(AC)| \mid \text{rank}(C) = n,\ \|C\| \le 1\}$$

   Remark. AC is finite rank, so trace can be defined as for finite-dimensional matrices.

   (b) Prove that
$$\sum_{k=1}^{n} \mu_k(A+B) \le \sum_{k=1}^{n} \mu_k(A) + \sum_{k=1}^{n} \mu_k(B)$$

   (c) Is it always true that $\mu_2(A+B) \le \mu_2(A) + \mu_2(B)$?

8. Let $\mathcal{H}_1, \mathcal{H}_2$ be two Hilbert spaces. Develop a canonical expansion for any compact operator $A \in \mathcal{L}(\mathcal{H}_1, \mathcal{H}_2)$.

   Remark. See Problem 15 of Section 2.4 and Problems 5–8 of Section 3.1.

3.6. The Trace and Trace Class

In this section, we continue our study of compact operators on a Hilbert space. We begin by showing if $A \ge 0$,
$$S(A; \{\varphi_n\}_{n=1}^{\infty}) = \sum_{n=1}^{\infty} \langle \varphi_n, A\varphi_n \rangle \tag{3.6.1}$$
is the same for all ON bases $\{\varphi_n\}_{n=1}^{\infty}$ and then define the trace class, $\mathcal{I}_1$, to be those A for which $S(|A|; \{\varphi_n\}_{n=1}^{\infty}) < \infty$. We'll then show any $A \in \mathcal{I}_1$ is compact and has $\sum_{n=1}^{\infty} \mu_n(A) < \infty$ and the fact that for such A,
$$\text{Tr}(A) = \sum_{n=1}^{\infty} \langle \varphi_n, A\varphi_n \rangle \tag{3.6.2}$$
is an absolutely convergent sum and the same for all ON bases $\{\varphi_n\}_{n=1}^{\infty}$. On $\mathcal{I}_1$, we define $\|A\|_1$ by
$$\|A\|_1 = \text{Tr}(|A|) \tag{3.6.3}$$

We'll prove in $\|\cdot\|_1$, $\mathcal{I}_1$ is a Banach space and that $\mathcal{I}_1$ is a $*$-ideal in $\mathcal{L}(\mathcal{H})$. We'll also prove that if $A \in \mathcal{I}_1$, $B \in \mathcal{L}(\mathcal{H})$, then

$$\operatorname{Tr}(AB) = \operatorname{Tr}(BA) \tag{3.6.4}$$

Eventually (Section 3.12), we prove Lidskii's theorem, a main theorem of this chapter, that if $A \in \mathcal{I}_1$ and $\{\lambda_n(A)\}_{n=1}^N$ is a listing of its nonzero eigenvalues (counted up to algebraic multiplicity), then

$$\operatorname{Tr}(A) = \sum_{n=1}^{\infty} \lambda_n(A) \tag{3.6.5}$$

Since AB and BA have the same nonzero eigenvalues (see Problem 5 of Section 2.2), Lidskii's theorem also shows that (3.6.4) is valid whenever $A, B \in \mathcal{L}(\mathcal{H})$ with AB and BA both in $\mathcal{I}_1$.

Theorem 3.6.1. *For any $B \in \mathcal{L}(\mathcal{H})$ and orthonormal basis $\{\varphi_n\}_{n=1}^\infty$, we define*

$$S_2(B; \{\varphi_n\}_{n=1}^\infty) = \sum_{n=1}^{\infty} \|B\varphi_n\|^2 \tag{3.6.6}$$

This sum is independent of basis and

$$S_2(B; \{\varphi_n\}_{n=1}^\infty) = S_2(B^*; \{\varphi_n\}_{n=1}^\infty) \tag{3.6.7}$$

Remarks. 1. Later (see Section 3.8), we'll define those B with $S_2(B; \{\varphi_n\}_{n=1}^\infty) < \infty$ to be the abstract Hilbert–Schmidt ideal.

2. Both sides of (3.6.7) and (3.6.9) below may be infinite. Because summands are positive, the sums below may be interchanged.

Proof. Let $\{\varphi_n\}_{n=1}^\infty$ and $\{\psi_m\}_{m=1}^\infty$ be two ON bases. By Parseval's relation, (Theorem 3.4.1 of Part 1),

$$\sum_{n=1}^{\infty} \|B\varphi_n\|^2 = \sum_{m,n} |\langle \psi_m, B\varphi_n\rangle|^2 = \sum_{m,n} |\langle \varphi_n, B^*\psi_n\rangle|^2 = \sum_m \|B^*\psi_m\|^2 \tag{3.6.8}$$

Taking $\varphi = \psi$, we get (3.6.7) and given that, (3.6.8) implies

$$S_2(B; \{\varphi_n\}_{n=1}^\infty) = S_2(B; \{\psi_m\}_{m=1}^\infty) \tag{3.6.9}$$

□

If $A \geq 0$, $\langle \varphi, A\varphi\rangle = \|\sqrt{A}\,\varphi\|^2$, so if we define S by (3.6.1), we immediately have

Theorem 3.6.2. *Let $A \geq 0$. Then for any two ON bases $\{\varphi_n\}_{n=1}^\infty$ and $\{\psi_n\}_{n=1}^\infty$, we have*

$$S(A; \{\varphi_n\}_{n=1}^\infty) = S(A; \{\psi_n\}_{n=1}^\infty) \tag{3.6.10}$$

Definition. Let $A \in \mathcal{L}(\mathcal{H})$. We say $A \in \mathcal{I}_1$, the *trace class*, if and only if for one and hence for all ON bases $\{\varphi_n\}$, we have

$$S(|A|; \{\varphi_n\}_{n=1}^{\infty}) < \infty \tag{3.6.11}$$

We let $\mathcal{B}$ denote the set of all orthonormal sets $\{\varphi_n\}_{n=1}^{\infty}$ (may not be a basis).

Proposition 3.6.3. *Let $A \in \mathcal{L}(\mathcal{H})$. Suppose for each $\{\varphi_n\}_{n=1}^{\infty}, \{\psi_n\}_{n=1}^{\infty} \in \mathcal{B}$, we have as $n \to \infty$ that $\langle \psi_n, A\varphi_n \rangle \to 0$. Then A is a compact operator.*

Proof. This will follow from Proposition 3.1.14. Let $\{\varphi_n\}_{n=1}^{\infty}$ be an orthonormal set in $\mathrm{Ker}(A)^{\perp}$. Let $A = U|A|$ be the polar decomposition. Since U is an isometry on $\mathrm{Ker}(A)^{\perp}$, $\psi_n = U\varphi_n$ is also in $\mathcal{B}$. Thus, since A takes $\mathrm{Ker}(A)^{\perp}$ to $\mathrm{Ker}(A)^{\perp}$,

$$\langle \psi_n, A\varphi_n \rangle = \langle U\varphi_n, U|A|\varphi_n \rangle = \langle \varphi_n, |A|\varphi_n \rangle \to 0$$

Since $\mathrm{Ker}(|A|) = \mathrm{Ker}(A) = \mathrm{Ker}(|A|^{\frac{1}{2}})$ (see Problem 1), $|A|^{\frac{1}{2}}$ obeys the hypotheses of Proposition 3.1.14. Thus, $|A|^{\frac{1}{2}}$ is compact, so $A = (U|A|^{\frac{1}{2}})|A|^{\frac{1}{2}}$ is compact. □

Lemma 3.6.4. *If $\varphi, \psi, \gamma, \kappa \in \mathcal{B}$, then*

$$b_{nm} = \langle \psi_n, \kappa_m \rangle \langle \gamma_m, \varphi_n \rangle \tag{3.6.12}$$

obeys

$$\forall m, \quad \sum_{n=1}^{\infty} |b_{nm}| \le 1 \tag{3.6.13}$$

$$\forall n, \quad \sum_{m=1}^{\infty} |b_{nm}| \le 1 \tag{3.6.14}$$

Remark. A matrix obeying (3.6.13)/(3.6.14) is called *doubly substochastic* (dss) for reasons we discuss in the Notes.

Proof. We'll prove (3.6.13); the other is similar. By the Schwarz inequality for ℓ^2,

$$\begin{aligned}\left(\sum_{n=1}^{\infty} |b_{nm}|\right)^2 &\le \sum_{n=1}^{\infty} |\langle \psi_n, \kappa_m \rangle|^2 \sum_{n=1}^{\infty} |\langle \gamma_m, \varphi_n \rangle|^2 \\ &\le \|\kappa_m\|^2 \|\gamma_n\|^2 = 1 \end{aligned} \tag{3.6.15}$$

by Bessel's inequality. □

Proposition 3.6.5. *For any $A \in \mathcal{L}(\mathcal{H})$,*

$$S(|A|; \{\varphi_n\}_{n=1}^{\infty}) = \sup_{\{\varphi\},\{\psi\} \in \mathcal{B}} \sum_{n=1}^{\infty} |\langle \psi_n, A\varphi_n \rangle| \tag{3.6.16}$$

Moreover, if these quantities are finite, then A is compact and they equal

$$\|A\|_1 \equiv \sum_{n=1}^{\infty} \mu_n(A) \tag{3.6.17}$$

Proof. We'll first show that if A is compact, then both sides of (3.6.16) are equal and also equal (3.6.17) (all may be infinite). Then we'll show if either side of (3.6.16) is finite, then A is compact. Finally, we'll show these facts imply the claims in the proposition.

If A is compact, it has a canonical decomposition

$$A = \sum_{m=1}^{\infty} \mu_m(A)\langle \gamma_m, \cdot\,\rangle \kappa_m, \qquad |A| = \sum_{m=1}^{\infty} \mu_m(A)\langle \gamma_m, \cdot\,\rangle \gamma_m \tag{3.6.18}$$

Then

$$S(|A|; \{\varphi_n\}_{n=1}^{\infty}) = \sum_{n,m=1}^{\infty} \mu_m(A)|\langle \gamma_m, \varphi_n\rangle|^2 = \sum_{m=1}^{\infty} \mu_m(A) \tag{3.6.19}$$

where positivity allows rearrangement of sums.

Also, if b_{nm} is given by (3.6.12), we have

$$\langle \psi_n, A\varphi_n\rangle = \sum_{m=1}^{\infty} \mu_m(A) b_{nm} \tag{3.6.20}$$

Since $\sum_m |b_{nm}| < 1$, the sum is absolutely convergent. This implies

$$\sum_n |\langle \psi_n, A\varphi_n\rangle| \le \sum_{n,m=1}^{\infty} \mu_m(A)|b_{nm}| \le \sum_{m=1}^{\infty} \mu_m(A) \tag{3.6.21}$$

by (3.6.13). If $\psi = \kappa$ and $\varphi = \gamma$, then $b_{nm} = \delta_{nm}$ and $\langle \psi_n, A\varphi_n\rangle = \mu_n(A)$. So we have equality in (3.6.20). Thus,

$$\sup_{\varphi,\psi\in\mathcal{B}} \sum_{n=1}^{\infty} |\langle \psi_n, A\varphi_n\rangle| = \sum_{m=1}^{\infty} \mu_m(A) \tag{3.6.22}$$

This completes the proof of the claim that if A is compact, (3.6.16) and (3.6.17) hold.

If $S(A; \{\varphi_n\}_{n=1}^{\infty})$ is finite for one and so for all $\{\varphi_n\}_{n=1}^{\infty}$, we get $\sum_{n=1}^{\infty} \langle \varphi_n, |A|\varphi_n\rangle = \sum_{n=1}^{\infty} \|\sqrt{|A|}\varphi_n\|^2 < \infty$, and thus, $\|\sqrt{A}\varphi_n\| \to 0$ for all ON bases and so for all ON sets. Proposition 3.1.14 implies $\sqrt{|A|}$, and so A, are compact.

If the right side of (3.6.16) is finite, Proposition 3.6.3 implies A is compact.

Thus, if either side of (3.6.16) is finite, A is compact, and so (3.6.16) holds and equals (3.6.17). If neither is finite, both are infinite, and again equality holds in (3.6.16). □

Theorem 3.6.6. *The trace class, $\mathcal{I}_1$, is a two-sided $*$-ideal and $\|\cdot\|_1$ is a norm on it. Moreover, for $A \in \mathcal{I}_1$ and $B \in \mathcal{L}(\mathcal{H})$,*

$$\|A^*\|_1 = \|A\|_1, \quad \|BA\|_1 \le \|B\|\|A\|_1, \quad \|AB\|_1 \le \|B\|\|A\|_1 \tag{3.6.23}$$

Remark. We'll see below (Theorem 3.6.8) that $\mathcal{I}_1$ is complete in $\|\cdot\|_1$.

Proof. (3.6.23) and the fact that $A \in \mathcal{I}_1$, $B \in \mathcal{L}(\mathcal{H}) \Rightarrow A^* \in \mathcal{I}_1$, $AB \in \mathcal{I}_1$, $BA \in \mathcal{I}_1$ are immediate from Theorem 3.5.3 (i.e., $\mu_n(A^*) = \mu_n|A|$, $\mu_n(BA) \le \|B\|\mu_n(A)$, $\mu_n(AB) \le \|B\|\mu_n(A)$) and (3.6.17) implies that $A \in \mathcal{I}_1 \Leftrightarrow \|A\|_1 < \infty$.

We see that if $F(A)$ is the right side of (3.6.16), then

$$F(A+B) \le F(A) + F(B) \tag{3.6.24}$$

This plus $A \in \mathcal{I}_1 \Leftrightarrow F(A) < \infty$, and in that case, $F(A) = \|A\|_1$ implies that $A, B \in \mathcal{I}_1 \Rightarrow A + B \in \mathcal{I}_1$ and $\|A+B\|_1 \le \|A\|_1 + \|B\|_1$. □

Next, we define the trace!

Theorem 3.6.7. *For any $A \in \mathcal{I}_1$ and any* ON *basis, $\{\varphi_n\}$,*

$$\mathrm{Tr}_\varphi(A) = \sum_{n=1}^\infty \langle \varphi_n, A\varphi_n \rangle \tag{3.6.25}$$

is absolutely convergent and has the same value, $\mathrm{Tr}(A)$, *for all φ. In fact, for any $B \in \mathcal{L}(\mathcal{H})$ and A given by (3.5.1),*

$$\mathrm{Tr}(AB) = \mathrm{Tr}(BA) = \sum_{n=1}^{N(A)} \mu_n(A)\langle \varphi_n, B\psi_n \rangle \tag{3.6.26}$$

Remarks. 1. $\mathrm{Tr}(A)$ is called the *trace* of A.

2. $\mathrm{Tr}(AB) = \mathrm{Tr}(BA)$ implies that for any unitary U, $\mathrm{Tr}(UAU^{-1}) = \mathrm{Tr}(A)$ and is equivalent since (see Corollary 2.4.5) every B is a linear combination of at most four unitaries.

3. $\mathrm{Tr}(AB) = \mathrm{Tr}(BA)$ has other proofs. That $\mathrm{Tr}_\varphi(A)$ is independent of φ implies unitary invariance which, as we've seen, implies $\mathrm{Tr}(AB) \setminus \{0\} = \mathrm{Tr}(BA) \setminus \{0\}$. Once one has $\mathrm{Tr}(A) = \sum_{n=1}^\infty \lambda_n(A)$, we also can get this formula from (2.4.48) and its implication that $\sigma(AB) = \sigma(BA)$ with identical algebraic multiplicities.

Proof. We begin by establishing (3.6.26). That is, for any ON basis $\{\eta_m\}_{m=1}^\infty$ with A given by (3.5.1),

$$\sum_{m=1}^\infty \langle \eta_m, BA\eta_m\rangle = \sum_{\substack{m=1,\dots \\ n=1,\dots,N(A)}} \mu_n(A)\langle \eta_m, B\psi_n\rangle\langle \varphi_n, \eta_m\rangle \tag{3.6.27}$$

$$= \sum_{n=1}^{N(A)} \mu_n(A)\langle \varphi_n, B\psi_n\rangle \tag{3.6.28}$$

where (3.6.28) comes from summing the Fourier coefficients in the basis $\{\eta_m\}_{m=1}^\infty$. In this calculation, we use the Schwarz inequality

$$\sum_{m=1}^\infty |\langle \eta_m, B\psi_n\rangle||\langle \varphi_n, \eta_m\rangle| \le \|B\psi_n\|\|\varphi_n\| \le \|B\|$$

plus $\sum_{n=1}^{N(A)} \mu_n(A) < \infty$ to get absolute summability of the double sum in (3.6.27), justifying the interchange of iterated sums.

The same computation for $\sum_{m=1}^\infty \langle \eta_m, AB\eta_m\rangle$ yields (3.6.26). The rest of the theorem follows from taking $B = 1$ and noting (3.6.28) is independent of choice of basis $\{\eta_m\}_{m=1}^\infty$. □

We close the section by showing $\mathrm{Tr}(\,\cdot\,)$ can be used to describe some dualities. We use $\mathcal{I}_\infty$ for the compact operators on a Hilbert space, $\mathcal{H}$.

Theorem 3.6.8. *We have*

$$\mathcal{I}_\infty^* = \mathcal{I}_1, \qquad \mathcal{I}_1^* = \mathcal{L}(\mathcal{H}) \tag{3.6.29}$$

In the explicit sense that if

$$\ell(A, B) = \mathrm{Tr}(AB) \tag{3.6.30}$$

then $\{\ell(A, \,\cdot\,)\}_{A\in\mathcal{I}_1}$ *describes all* BLT's *on* $\mathcal{I}_\infty$ *with*

$$\|\ell(A, \,\cdot\,)\|_{\mathcal{I}_\infty^*} = \|A\|_1 \tag{3.6.31}$$

and $\{\ell(\,\cdot\,, B)\}_{B\in\mathcal{L}(\mathcal{H})}$ *describes all* BLT's *on* $\mathcal{I}_1$ *with*

$$\|\ell(\,\cdot\,, B)\|_{\mathcal{I}_1^*} = \|B\|_{\mathcal{L}(\mathcal{H})} \tag{3.6.32}$$

Remarks. 1. This implies $\mathcal{I}_1$ is complete in $\|\cdot\|_1$.

2. This is reminiscent of Example 5.11.3 of Part 1 that $c_0^* = \ell_1$, $\ell_1^* = \ell_\infty$.

Proof. By (3.6.28), we have that

$$|\ell(A, B)| \le \sum_{n=1}^{N(A)} \mu_n(A)\|B\| = \|B\|_{\mathcal{L}(\mathcal{H})}\|A\|_1 \tag{3.6.33}$$

which implies each $\ell(A,\,\cdot\,) \in \mathcal{I}_\infty^*$ and $\ell(\,\cdot\,,B) \in \mathcal{I}_1^*$ and

$$\|(\ell(A,\,\cdot\,)\|_{\mathcal{I}_\infty^*} \le \|A\|_1, \qquad \|\ell(\,\cdot\,,B)\|_{\mathcal{I}_1^*} \le \|B\|_{\mathcal{L}(\mathcal{H})} \tag{3.6.34}$$

We thus must prove equality and that these are all the element of $\mathcal{I}_\infty^*$ and $\mathcal{I}_1^*$.

Given $A \in \mathcal{I}_1$, let (3.5.1) be its canonical expansion and for each $m \le N(A)$, $m < \infty$, let

$$B_m = \sum_{j=1}^m \langle \psi_j,\,\cdot\,\rangle \varphi_j \tag{3.6.35}$$

Then (Problem 2(a)), $\|B_m\| = 1$, B_m is finite rank and so in $\mathcal{I}_\infty$. By (3.6.26) and $B_m\psi_j = \varphi_j$,

$$\mathrm{Tr}(AB_m) = \sum_{j=1}^m \mu_j(A) \tag{3.6.36}$$

so

$$\sup_{\substack{B\in\mathcal{I}_\infty \\ \|B\|=1}} |\mathrm{Tr}(AB)| \ge \|A\|_1 \tag{3.6.37}$$

proving (3.6.32).

Given $B \in \mathcal{L}(\mathcal{H})$ and unit vectors φ and ψ, let $A_{\varphi,\psi} = \langle \psi,\,\cdot\,\rangle\varphi$. Then (Problem 2(b), $\|A\|_1 = 1$ and

$$\mathrm{Tr}(AB) = \langle \psi, B\varphi\rangle \tag{3.6.38}$$

so taking the sup over unit vectors ψ and φ,

$$\sup_{\|A\|_1=1} |\ell(A,B)| \ge \|B\| \tag{3.6.39}$$

proving (3.6.6).

Let $\lambda \in \mathcal{I}_\infty^*$. Define

$$\beta(\varphi,\psi) = \lambda(\langle \varphi,\,\cdot\,\rangle\psi) \tag{3.6.40}$$

which is linear in ψ and antilinear in φ. Moreover,

$$|\beta(\varphi,\psi)| \le \|\lambda\|_{\mathcal{I}_\infty^*} \|\langle \varphi,\,\cdot\,\rangle\varphi\|_1 = \|\lambda\|_{\mathcal{I}_\infty^*}\|\varphi\|\|\psi\| \tag{3.6.41}$$

so there exists $A \in \mathcal{L}(\mathcal{H})$ so that

$$\beta(\varphi,\psi) = \langle \varphi, A\psi\rangle \tag{3.6.42}$$

For any $m < \infty$ and $\{\varphi_j\}_{j=1}^m$, $\{\psi_j\}_{j=1}^m$, let

$$B(\{\varphi\}_{j=1}^m, \{\psi\}_{j=1}^m) = \sum_{j=1}^m \langle \varphi_j,\,\cdot\,\rangle\psi_j \tag{3.6.43}$$

which lies in $\mathcal{I}_\infty$ (since it is finite rank) and (Problem 2(a))

$$\{\varphi_j\}_{j=1}^m \text{ ON and } \{\psi_j\}_{j=1}^m \text{ ON} \Rightarrow \|B\| = 1 \tag{3.6.44}$$

Given ON sets $\{\varphi_n\}_{n=1}^\infty, \{\psi_n\}_{n=1}^\infty$, let $\langle \varphi_n, A\psi_n\rangle = e^{i\theta_n}|\langle \varphi_n, A\psi_n\rangle|$ and $\eta_n = e^{i\theta_n}\varphi_n$, so

$$\langle \eta_n, A\psi_n\rangle = e^{-i\theta_n}\langle \varphi_n, A\psi_n\rangle = |\langle \varphi_n, A\psi_n\rangle| \tag{3.6.45}$$

Then

$$\lambda((\{\eta_m\}_{m=1}^k, \{\psi_m\}_{m=1}^k)) = \sum_{m=1}^k |\langle \varphi_m, A\psi_m\rangle| \leq \|\lambda\| \tag{3.6.46}$$

Thus, RHS of (3.6.16) $\leq \|\ell\|$, so by Proposition 3.6.5, $A \in \mathcal{I}_1$. Thus, λ is an $\ell(A, \cdot\,)$.

Now let $\lambda \in \mathcal{I}_1^*$. Define β by (3.6.40). Since $\|\langle \varphi, \cdot\,\psi\rangle\|_1 = \|\varphi\|\|\psi\|$, β obeys

$$|\beta(\varphi, \psi)| \leq \|\lambda\|_{\mathcal{I}_1^*}\|\varphi\|\|\psi\| \tag{3.6.47}$$

so there is a bounded operator, B, with

$$\beta(\varphi, \psi) = \langle \varphi, B\psi\rangle \tag{3.6.48}$$

Thus, if $A \in \mathcal{I}_1$ obeys (3.5.1),

$$\lambda(A) = \lim_{M\to\infty} \lambda\biggl(\sum_{n=1}^M \mu_n(A)\langle \varphi_n, \cdot\,\rangle\psi_n\biggr) \tag{3.6.49}$$

$$= \lim_{M\to\infty} \sum_{n=1}^M \mu_n(A)\langle \varphi_n, B\psi_n\rangle \tag{3.6.50}$$

$$= \sum_{n=1}^\infty \mu_n(A)\langle \varphi_n, B\psi_n\rangle \tag{3.6.51}$$

$$= \operatorname{Tr}(AB) = \ell(A, B) \tag{3.6.52}$$

Thus, λ is an $\ell(\,\cdot\,, B)$.

In the above, (3.6.49) comes from continuity of λ and (3.6.50) from linearity (3.6.40) and (3.6.48). Since $A \in \mathcal{I}_1$, $\mu_n(A) \in \ell_1$, so we get (3.6.51). We get (3.6.52) from (3.6.26). □

Notes and Historical Remarks. The calculation (3.6.8) central to the basis independence of trace in a Hilbert space goes back to von Neumann's great 1932 book [**723**] on the foundations of quantum mechanics. He needed the trace to discuss quantum statistical mechanics. The formal definition of trace class is due to Schatten–von Neumann [**593**] who define a trace class operator as a product of two Hilbert–Schmidt operators (see the Notes to Section 3.8 for the earlier history of that class).

Where we use $\operatorname{Tr}(\,\cdot\,)$, some use $\operatorname{tr}(\,\cdot\,)$ or $\operatorname{sp}(\,\cdot\,)$ (sp is short for spur, the German word for trace).

Grothendieck [**268**] defined a trace and "trace class" in a general Banach space as follows. Any finite rank operator, A, on a Banach space, X, can be written

$$A = \sum_{i=1}^{n} \ell_i(\,\cdot\,)x_i \tag{3.6.53}$$

where $\{x_i\}_{i=1}^n \subset X$ and $\{\ell_i\}_{i=1}^n \subset X^*$ (just let the x_i be a basis for the finite-dimensional space $\operatorname{Ran}(A)$). One defines

$$N_1(A) = \inf\left\{ \sum_{i=1}^{n} \|\ell_1\|_{X^*} \|x_i\|_X \,\middle|\, A \text{ obeys } (3.6.53) \right\} \tag{3.6.54}$$

where the inf is over all representations as (3.6.53).

One shows (Problem 3) that

$$\operatorname{Tr}(A) = \sum_{i=1}^{n} \ell_i(x_i) \tag{3.6.55}$$

is independent of representations defining $\operatorname{Tr}(\,\cdot\,)$ on the finite rank operators. Clearly,

$$|\operatorname{Tr}(A)| \le N_1(A) \tag{3.6.56}$$

Also, we have for any $y \in X$ that

$$\|Ay\| \le N_1(A)\|y\|$$

so if A_n is Cauchy in N_1-norm, it is Cauchy in $\|\cdot\|$. Thus, the completion of the finite rank operators in N_1 is a family of operators called the *nuclear operators* on X. By (3.6.56), $\operatorname{Tr}(\,\cdot\,)$ extends to a linear functional on the nuclear operators. The reader will prove (Problem 4) that $\mathcal{I}_1(\mathcal{H})$ are exactly the nuclear operators on $\mathcal{H}$ in the Hilbert space case and, indeed, $N_1(A) = \|A\|_1$.

If B is an $n \times n$ matrix with $b_{ij} \ge 0$ and, for all j, $\sum_{i=1}^n b_{ij} = 1$, then B can be viewed as the generator of a simple Markov chain (see Section 7.5 of Part 1). Such matrices are called *stochastic matrices.* If also B^* has the property, then $(1\dots1)$ is an invariant state. Such matrices are thus called *doubly stochastic matrices.* Replacing the equality by an inequality leads to the name doubly substochastic .

Problems

1. (a) For any $A \in \mathcal{L}(\mathcal{H})$, prove that $\operatorname{Ker}(A) = \operatorname{Ker}(|A|)$.

   (b) For any $B \in \mathcal{L}(\mathcal{H})$, prove that $\operatorname{Ker}(B) = \operatorname{Ker}(B^*B)$. (*Hint:* $\|B\varphi\|^2 = \langle\varphi, B^*B\varphi\rangle$.)

2. (a) Prove B_m of (3.6.35) is a partial isometry with $\|B_m\| = 1$.

   (b) If $A_{\varphi,\psi} = \langle\psi,\,\cdot\,\rangle\varphi$, prove that $\|A_{\varphi,\psi}\| = \|\psi\|\|\varphi\|$.

3. In a general Banach space, if $\sum_{i=1}^n \ell_i(\cdot)x_i = \sum_{j=1}^m \lambda_j(\cdot)y_j$, prove that

$$\sum_{i=1}^n \ell_i(x_i) = \sum_{j=1}^m \lambda_j(y_j)$$

(*Hint*: Invariance of trace on finite-dimensional spaces.)

4. This problem will prove that $N_1 = \mathcal{I}_1$ on a Hilbert space.

(a) Prove that it suffices to prove

$$N_1(A) = \|A\|_1 \tag{3.6.57}$$

for finite rank A.

(b) Prove that $\|A_1\| \le N_1(A)$. (*Hint*: Canonical expansion.)

(c) Prove that $N_1(A) \le \|A\|_1$. (*Hint*: If A has the form (3.6.53), estimate $\text{Tr}(|A|)$.)

3.7. Bonus Section: Trace Ideals

In the last section, we saw $\mathcal{I}_1$ was defined as those compact A's with

$$\sum_{n=1}^\infty \mu_n(A) < \infty \tag{3.7.1}$$

and $\mathcal{I}_\infty$ is all compact operators, that is, compact A's with

$$\sup_n \mu_n(A) < \infty \tag{3.7.2}$$

In the next section, we'll discuss $\mathcal{I}_2$, the abstract Hilbert–Schmidt operators, those with

$$\text{Tr}(A^*A) = \sum_{n=1}^\infty \mu_n(A)^2 < \infty \tag{3.7.3}$$

In this section, we'll discuss for $1 \le p < \infty$ the class, $\mathcal{I}_p$, of compact A's, those with

$$\sum_{n=1}^\infty \mu_n(A)^p < \infty \tag{3.7.4}$$

called the *trace ideals* or *Schatten classes* with the natural norm

$$\|A\|_p = \left(\sum_{n=1}^\infty \mu_n(A)^p\right)^{1/p} \tag{3.7.5}$$

(It is not obvious that $\mathcal{I}_p$ is an ideal nor that $\|\cdot\|_p$ is a norm; these are things we will prove below.) These objects will be mentioned in remarks later in the chapter, but it is $\mathcal{I}_1$ and $\mathcal{I}_2$ that are central—which is why this is a bonus section. Just as one can look at L^p for $0 < p < 1$ as metric spaces (but not NLS) and at L^p_w, there are analogs of these spaces among the ideals of compact operators.

We'll first prove that $\mathcal{I}_p$ is a $*$-ideal and $\|\cdot\|_p$ a norm, and then Hölder's inequality for $\mathcal{I}_p$, namely, if $1 \le p, q, r \le \infty$ so $\frac{1}{p} + \frac{1}{q} = \frac{1}{r}$, then we'll show that $\mathcal{I}_p^* = \mathcal{I}_q$ for $1 < p < \infty$ if $q = p/(p-1)$, proving each $\mathcal{I}_p$ is complete in $\|\cdot\|_p$. Finally, we prove $A \in \mathcal{I}_p, B \in \mathcal{I}_q \Rightarrow AB \in \mathcal{I}_r$,

$$\|AB\|_r \le \|A\|_p \|B\|_q \tag{3.7.6}$$

The reader will notice that $\mathcal{I}_p$ looks a lot like ℓ^p or L^p, so the theory of $\mathcal{I}_p$ (in a more general context) is sometimes called *noncommutative integration theory.*

Recall a matrix $\{b_{nm}\}_{n,m=1}^\infty$ is called doubly stochastic (dss) if and only if (3.6.13)/(3.6.14) hold. The following will be useful:

Theorem 3.7.1. *If $B = (b_{nm})_{n,m=1}^\infty$ is a dss matrix,*

$$\alpha \mapsto (\mathcal{B}\alpha), \qquad (\mathcal{B}\alpha)_n = \sum_{m=1}^\infty b_{nm}\alpha_m \tag{3.7.7}$$

obeys

$$\|\mathcal{B}\alpha\|_p \le \|\alpha\|_p \tag{3.7.8}$$

for all $\alpha \in \ell^p$.

Remarks. 1. Our proof actually shows the sum in (3.7.7) converges for all n and all $\alpha \in \ell_p$. For by (3.7.11) with $\beta_n \in \ell^q$, $\beta_n \ne 0$ for all n, the left side of (3.7.9) is finite, so the sum over m is finite.

2. A simple but less elementary way to understand this result is to note that (3.6.12) implies (3.7.8) for $p = 1$ and (3.6.14) for $p = \infty$. The general p result then follows by the Riesz–Thorin theorem (see Section 5.2 of Part 2A).

Proof. As noted, the result is easy if $p = 1$ or $p = \infty$, so suppose $1 < p < \infty$. Let $\alpha \in \ell^p$ and $\beta \in \ell^q$, where $q = p/(p-1)$ is the Hölder dual index. Then

$$\sum_{m,n=1}^\infty |b_{nm}||\alpha_m||\beta_n| = \sum_{m,n=1}^\infty (|b_{nm}|^{1/p}|\alpha_m|)(|b_{nm}|^{1/q}|\beta_m|) \tag{3.7.9}$$

$$\le \biggl(\sum_{m,n=1}^\infty |b_{nm}||\alpha_m|^p\biggr)^{1/p}\biggl(\sum_{m,n=1}^\infty |b_{nm}||\beta_n|^q\biggr)^{1/q} \tag{3.7.10}$$

$$\le \biggl(\sum_{m=1}^\infty |\alpha_m|^p\biggr)^{1/p}\biggl(\sum_{n=1}^\infty |\beta_n|^q\biggr)^{1/q} \tag{3.7.11}$$

We get (3.7.9) from $p^{-1} + q^{-1} = 1$, (3.7.10) from Hölder's inequality, and (3.7.11) from (3.6.13)/(3.6.14).

By ℓ^p duality,

$$\|\gamma\|_p = \sup_{\substack{\beta \in \ell^q \\ \|\beta\|_q = 1}} \sum_{n=1}^{\infty} |\gamma_n \beta_n| \tag{3.7.12}$$

so (3.7.8) is proven. □

$\mathcal{I}_p$ is defined by (3.7.4). The following will be useful. Recall $\mathcal{B}$ is the family of all orthonormal sets.

Proposition 3.7.2. *Let $1 < p < \infty$ and $A \in \mathcal{L}(\mathcal{H})$. Then A is in $\mathcal{I}_p$ if and only if the right side of*

$$\|A\|_p^p = \sup_{\varphi,\psi \in \mathcal{B}} \sum_{n=1}^{\infty} |\langle \psi_n, A\varphi_n \rangle|^p \tag{3.7.13}$$

is finite and then, as indicated, the $\mathcal{I}_p$-norm is given via the sup.

Proof. If the right side of (3.7.13) is finite, for all $\varphi, \psi \in \mathcal{B}$, $\langle \psi_n, A\varphi_n \rangle \to 0$, so by Proposition 3.6.3, A is compact. Thus,

$$A - \sum_{m=1}^{\infty} \mu_m(A) \langle \gamma_n, \cdot \rangle \kappa_m \tag{3.7.14}$$

So with b_{nm} given by (3.6.12),

$$\langle \psi_n, A\varphi_n \rangle = \sum_n b_{nm} \mu_m(A) \tag{3.7.15}$$

Therefore, by Lemma 3.6.4 and Theorem 3.7.1,

$$[\text{RHS of (3.7.11)}]^{1/p} \le \left(\sum_{m=1}^{\infty} \mu_m(A)^p \right)^{1/p} = \|A\|_p$$

proving LHS of (3.7.13) $\ge$ RHS of (3.7.13).

Choosing $\varphi_n = \gamma_n$, $\kappa_m = \psi_m$, we have

$$\langle \psi_n, A\varphi_n \rangle = \mu_n(A) \tag{3.7.16}$$

showing equality. □

Theorem 3.7.3. *$\mathcal{I}_p$ is a $*$-ideal and $\|\cdot\|_p$ a norm of it. Specifically,*

$$A \in \mathcal{I}^p \Rightarrow A^* \in \mathcal{I}^p \quad \text{and} \quad \|A\|_p = \|A^*\|_p \tag{3.7.17}$$

$$A \in \mathcal{I}^p, B \in \mathcal{L}(\mathcal{H}) \Rightarrow AB, BA \in \mathcal{I}^p \quad \text{and}$$

$$\|BA\|_p \le \|B\|\|A\|_p, \ \|AB\|_p \le \|B\|\|A\|_p \tag{3.7.18}$$

$$A, B \in \mathcal{I}_p \Rightarrow A + B \in \mathcal{I}_p \quad \text{and} \quad \|A + B\|_p \le \|A\|_p + \|B\|_p \tag{3.7.19}$$

Remark. We'll see below (see Theorem 3.7.5) that $\mathcal{I}_p$ is complete in $\|\cdot\|_p$.

Proof. (3.7.17) is immediate from $\mu_n(A^*) = \mu_n(A)$ (see (3.5.9)) and (3.7.18) from $\mu_n(AB) \le \|B\|\mu_n(A)$, $\mu_n(BA) \le \|B\|\mu_n(A)$ (see (3.5.10)). (3.7.19) follows from (3.7.13),

$$|\langle\psi_n(A+B)\varphi_n\rangle| \le |\langle\psi_n, A\varphi_n\rangle| + |\langle\psi_n, B\varphi_n\rangle| \tag{3.7.20}$$

and thus, $\|\cdot\|_p$ is a norm on $\mathcal{I}_p$. $\square$

Theorem 3.7.4. *Let $1 < p < \infty$ and $q = p/(p-1)$. Then for any $A \in \mathcal{I}_p$, $B \in \mathcal{I}_q$, we have $AB \in \mathcal{I}_1$ and*

$$|\mathrm{Tr}(AB)| \le \|AB\|_1 \le \|A\|_p \, \|B\|_q \tag{3.7.21}$$

We also have that

$$\mathrm{Tr}(AB) = \mathrm{Tr}(BA) \tag{3.7.22}$$

Moreover, for any $A \in \mathcal{I}_p$,

$$\|A\|_p = \sup\{|\mathrm{Tr}(AB)| \mid B \in \mathcal{I}_q,\ \|B\|_q = 1\} \tag{3.7.23}$$

Proof. Let the canonical decompositions for A and B be

$$A = \sum_{m=1}^{\infty} \mu_m(A)\langle\gamma_n, \cdot\,\rangle\kappa_m$$

$$B = \sum_{n=1}^{\infty} \mu_n(B)\langle\psi_n, \cdot\,\rangle\varphi_n$$

Suppose first that $AB \ge 0$. Then

$$\begin{aligned}\sum_{m=1}^{\infty}\langle\kappa_n, AB\kappa_m\rangle &= \sum_{m=1}^{\infty}\mu_m(A)\langle\gamma_m, B\kappa_m\rangle \\ &= \sum_{m,n=1}^{\infty}\mu_m(A)\mu_n(B)b_{nm}\end{aligned} \tag{3.7.24}$$

(where b_{nm} is given by (3.6.12)). By Theorem 3.7.1, (3.7.24) $\le \|A\|_p\|B\|_q$, proving (3.7.21) in this case. The double sum (3.7.24) is absolutely convergent, justifying the change from single to double sum.

For general AB, let $C \equiv AB$, $C = U_C|C|$, its polar decomposition, so $|C| = U_C^* AB \equiv \widetilde{A}B$, where $\widetilde{A} = U_C^* A$. By the case already treated, since $\widetilde{A}B = |C| \ge 0$, $|C| \in \mathcal{I}_1$ and

$$\||C|\|_1 \le \|\widetilde{A}\|_p\|B\|_q \tag{3.7.25}$$

Thus, $C \in \mathcal{I}_1$ and since $\|\widetilde{A}\|_p \le \|U_C^*\|\|A\|_p \le \|A\|_p$, so we have (3.7.21) in general.

By (3.7.21), $(A, B) \mapsto \mathrm{Tr}(AB)$ is continuous from $\mathcal{I}_p \times \mathcal{I}_q$ to $\mathbb{C}$. Since the truncated canonical expansion for A converges in $\|\cdot\|_p$ to A, it suffices to prove (3.7.22) for A finite rank. But in that case, $A \in \mathcal{I}_1$ and (3.7.22) follow from Theorem 3.6.7.

By (3.7.21), if $\|B\|_q = 1$, then

$$|\mathrm{Tr}(AB)| \le \|AB\|_1 \le \|A\|_p \tag{3.7.26}$$

so RHS of (3.7.23) $\le \|A\|_p$.

Let $C = |A|^{p-1}U_A^*$. Then $C \in \mathcal{I}_q$ with

$$\|C\|_q = \|A\|_p^{p/q} \tag{3.7.27}$$

since $q(p-1) = p$, and thus,

$$(|A|^{p-1})^q = |A|^p \Rightarrow \||A|^{p-1}\|_q^q = \|A\|_p^p \tag{3.7.28}$$

and $C^* = U_A^*|A|^{p-1} \Rightarrow |C^*| = |A|^{p-1} \Rightarrow \|C\|_q = \|C^*\|_q = \||A|^{p-1}\|_q$.

Let $B = C/\|C\|_q$. Then

$$\mathrm{Tr}(AB) = \mathrm{Tr}(BA) = \frac{\mathrm{Tr}(|A|^p)}{\|C\|_q} = \|A\|_p^{p-p/q} = \|A\|_p \tag{3.7.29}$$

since $p - p/q = p(1 - \frac{1}{q}) = 1$. Therefore, equality holds in (3.7.23). $\square$

Theorem 3.7.5 (Duality for $\mathcal{I}_p$). *Let $1 < p < \infty$. Then*

$$\mathcal{I}_p^* = \mathcal{I}_q \tag{3.7.30}$$

in the sense that any $\ell \in \mathcal{I}_p^*$ has the form

$$\ell(A) = \mathrm{Tr}(AB) \equiv \ell_B(A) \tag{3.7.31}$$

for some $B \in \mathcal{I}_q$ and

$$\|\ell_B\|_{\mathcal{I}_p^*} = \|B\|_q \tag{3.7.32}$$

In particular, each $\mathcal{I}_p$, $1 < p < \infty$, is complete and is reflexive.

Proof. By the previous theorem, if ℓ_B is defined by (3.7.31), then (3.7.32) holds, so it suffices to prove that every ℓ is any ℓ_B for some $B \in \mathcal{I}_q$.

Since $\mathcal{I}_1 \subset \mathcal{I}_p$, with $\|A\|_p \le \|A\|_1$ (since $\ell^1 \subset \ell^p$ for $\sum_{n=1}^\infty |a_n|^p \le (\max|a_n|^{p-1}) \sum_{n=1}^\infty |a_n| \le (\sum_{n=1}^\infty |a_n|^p)$), if $\ell \in \mathcal{I}_p^*$, $A \mapsto \ell(A)$ defines a map in $\mathcal{I}_1^*$, so $\ell(A) = \mathrm{Tr}(AB)$ for a unique $B \in \mathcal{L}(\mathcal{H})$, so we need only show that $B \in \mathcal{I}_q$.

Let $\varphi, \psi \in \mathcal{B}$. Suppose it is false that $\langle \varphi_n, B\psi_n \rangle \to 0$. Then, by passing to a subsequence (a subsequence of $\varphi \in \mathcal{B}$ is still in $\mathcal{B}$) and multiplying ψ_n by a phase, we can suppose $\langle \varphi_n, B\psi_n \rangle \ge 0$ and

$$\langle \varphi_n, B\psi_n \rangle \to a > 0 \tag{3.7.33}$$

Pick β so $0 < \beta < 1$ but $\beta p > 1$. Let

$$A_N = \sum_{n=1}^{N} n^{-\beta} \langle \varphi_n, \cdot \rangle \psi_n \tag{3.7.34}$$

Then

$$\|A_N\|_p^p = \sum_{n=1}^{N} n^{-\beta p} < \sum_{n=1}^{\infty} n^{-\beta p} < \infty \tag{3.7.35}$$

since $\beta p > 1$, while

$$\ell(A_N) = \operatorname{Tr}(A_N B) = \sum_{n=1}^{N} n^{-\beta} \langle \varphi_n, B\psi_n \rangle \to \infty \tag{3.7.36}$$

by (3.7.33) and $\sum_{n=1}^{\infty} n^{-\beta} = \infty$ by $\beta < 1$.

It follows from (3.7.35) and (3.7.36) that $|\ell(A_N)| \leq C\|A_N\|_p$ cannot hold and this contradiction shows that for all $\varphi, \psi \in \mathcal{B}$, $\langle \varphi_n, B\psi_n \rangle \to 0$. Thus, by Proposition 3.6.3, B is compact.

Write

$$B = \sum_{n=1}^{\infty} \mu_n(B) \langle \varphi_n, \cdot \rangle \psi_n \tag{3.7.37}$$

Thus, for any $N < \infty$,

$$\ell\biggl(\sum_{n=1}^{N} a_n \langle \psi_n, \cdot \rangle \varphi_n \biggr) = \sum_{n=1}^{N} a_n \mu_n(B) \tag{3.7.38}$$

so $|\ell(A)| \leq C\|A\|_p^p$ implies that

$$\biggl| \sum_{n=1}^{N} a_n \mu_n(B) \biggr| \leq C \biggl(\sum_{n=1}^{N} |a_n|^p \biggr)^{1/p} \tag{3.7.39}$$

Picking $a_n = \mu_n(B)^{q-1}$, we see that

$$\sum_{n=1}^{N} \mu_n(B)^q \leq C \biggl(\sum_{n=1}^{N} \mu_n(B)^q \biggr)^{1/p} \tag{3.7.40}$$

or

$$\biggl(\sum_{n=1}^{N} \mu_n(B)^q \biggr)^{1/q} \leq C \tag{3.7.41}$$

so $B \in \mathcal{I}_q$. □

Finally, we want to prove Hölder's inequality for trace ideals.

Theorem 3.7.6 (Hölder's Inequality for Trace Ideals). *Let $p, q, r \in [1, \infty)$ so $p^{-1} + q^{-1} = r^{-1}$, and if $A \in \mathcal{I}_p$, $B \in \mathcal{I}_q$, then $AB \in \mathcal{I}_r$ and*

$$\|AB\|_r \leq \|A\|_p \|B\|_q \tag{3.7.42}$$

Remarks. 1. The elementary reduction to $r = 1$ for L^p-space which uses $|fg|^r = |f|^r|g|^r$ and $\frac{r}{p} + \frac{r}{q} = 1$ doesn't work since $|AB|^r$ is not $|A|^r|B|^r$ because of the noncommutativity.

2. This proof is patterned after the proof of the Riesz–Thorin theorem (see Section 5.2 of Part 2A).

Proof. If p or q is ∞, this is just (3.7.18), so we'll suppose $p < \infty$, $q < \infty$, which means $r < \infty$ also. We can also suppose $r > 1$ since $r = 1$ is Theorem 3.7.4. Let $r' = (r/r - 1)$. We claim it suffices to show for A, B, C finite rank, we have that

$$|\text{Tr}(ABC)| \le \|A\|_p \|B\|_q \|C\|_{r'} \tag{3.7.43}$$

for then, by (3.7.23) and a density argument, we get (3.7.41) for A, B finite rank and then, by another density argument for the general case (Problem 1). By replacing A by $A/\|A\|_p$, etc., we can suppose $\|A\|_p = \|B\|_q = \|C\|_{r'} = 1$.

Thus, we let

$$A = \sum_{n=1}^{N} \mu_n(A) \langle \varphi_n, \cdot \rangle \psi_n, \qquad B = \sum_{m=1}^{M} \mu_m(B) \langle \gamma_m, \cdot \rangle \kappa_m \tag{3.7.44}$$

$$\sum_{n=1}^{N} \mu_n(A)^p = 1 = \sum_{m=1}^{M} \mu_m(B)^q = \|C\|_{r'} \tag{3.7.45}$$

For $z \in \mathbb{C}$, define

$$A(z) = \sum_{n=1}^{N} \mu_n(A)^{pz/r} \langle \varphi_n, \cdot \rangle \psi_n, \qquad B(z) = \sum_{m=1}^{M} \mu_m(B)^{q(1-z)/r} \langle \gamma_n, \cdot \rangle \kappa_m \tag{3.7.46}$$

$$f(z) = \text{Tr}\big((B(z) C A(z)\big) \tag{3.7.47}$$

If $\text{Re}\, z = 0$, $A(z)$ is a partial isometry since $|\mu_n(A)|^{pz/r} = 1$ and $\|C\|_{r'} = 1$, $\|B(z)\|_r = 1$, since $\mu(B(z)) = \mu_m(B)^{q/r}$. Thus, by (3.7.21) and $\|CB(z)\|_1 \le \|C\|_{r'} \|B(z)\|_r \le \|C\|_{r'} \|B(z)\|_r$,

$$\text{Re}\, z = 0 \Rightarrow |f(z)| \le 1 \tag{3.7.48}$$

Similarly, if $\text{Re}\, z = 1$, $B(z)$ is the partial isometry and $\|A(z)\|_r = 1$ and

$$\text{Re}\, z = 1 \Rightarrow |f(z)| \le 1 \tag{3.7.49}$$

Since all operators are finite rank and uniformly bounded on $\{z \mid 0 \le \text{Re}\, z \le 1\}$, $f(z)$ is bounded there. So (3.7.48)/(3.7.49) and the maximum principle (see Theorem 5.1.9 of Part 2A) imply that $|f(z)| \le 1$ for all z with $\text{Re}\, z \in [0, 1]$.

$A(\frac{r}{p}) = A$ and $B(\frac{r}{p}) = B$ since $(1-\frac{r}{p})\frac{q}{r} = 1$ so

$$f\left(\frac{r}{p}\right) = \operatorname{Tr}(BCA) = \operatorname{Tr}(ABC) \tag{3.7.50}$$

by cyclicity of trace for trace class operators. Thus, $|\operatorname{Tr}(ABC)| \leq 1$ as required. $\square$

Notes and Historical Remarks. As we'll see in the Notes to the next section, $\mathcal{I}_2$ goes back to the first decade of the twentieth century, and we've seen that the trace was studied in the early 1930s. General $\mathcal{I}_p$ is a product mainly of the 1940s. In an earlier paper related to the subject, von Neumann [**727**] considered unitary invariant norms on finite matrices. Closely related to $\mathcal{I}_p$ itself is a 1941 classic of Calkin [**96**] on ideals of operators that we'll discuss below. The subject proper came into its own in three papers of Schatten (two with von Neumann) [**590, 593, 594**] and a monograph [**591**]. Three books are Schatten [**592**], Goh'berg–Krein [**257**], and Simon [**650**].

Fundamental to Schatten's approach is the idea of cross norm. This began as a study of norms on $\mathcal{H}_1 \otimes_{\text{alg}} \mathcal{H}_2$ (finite combination of $\varphi_j \otimes \psi_j$) invariant under applying unitaries (i.e., $\|(U \otimes V)\eta\| = \|\eta\|$ for U and V unitary). By thinking of $\mathcal{H}_1 \otimes_{\text{alg}} \mathcal{H}_2$ as finite rank operators from $\mathcal{H}_1^*$ to $\mathcal{H}_2$, we get, by completeness, a family of operators from $\mathcal{H}_1^*$ to $\mathcal{H}_2$ invariant under right and left multiplication.

In Calkin [**96**], there is a classification of all ideals (not necessarily $\|\cdot\|$-closed) in $\mathcal{L}(\mathcal{H})$. The analysis starts with part of what the reader did in Problem 1 of Section 3.1. One notes that the following does not depend on norm closure: the proof that if an ideal contains a noncompact A, it is all of $\mathcal{L}(\mathcal{H})$; and if an ideal contains a nonzero A, then it contains all finite rank operators. Thus, any nontrivial ideal contains only compact operators. If $\mu_n(A) = \mu_n(B)$ for all n, there are partial isometries U and V, so $B = UAV$ (by the canonical decomposition), so $\mathcal{I}$ is determined by its set of sequences of possible singular values, or even better, by the eigenvalues of its positive operators (which can have any order). Also, if $A \in \mathcal{I}$ and $\mu_n(B) \leq \mu_n(A)$, it is again easy to see that $B = CAD$ for bounded C, D. A *Calkin space* is a space, s, of sequences going to zero, invariant under arbitrary permutations, and so that if $a \in s$ and $|b_n| \leq |a_n|$ for all n, then $b \in s$. Calkin proved there is a one–one correspondence between Calkin spaces, s, and ideals, $\mathcal{I}$, in $\mathcal{L}(\mathcal{H})$ via $A \in \mathcal{I} \Leftrightarrow \mu_n(A) \in s$. For a proof and further developments, see Goh'berg–Krein [**257**, Ch. III] and Simon [**650**, Sect. 2].

Problems

1. Prove that (3.7.43) for finite rank A, B, C implies the result for all $A \in \mathcal{I}_p$, $B \in \mathcal{I}_q$, $C \in \mathcal{I}_r$ (including that ABC is trace class).

2. Prove the following result of Goh'berg–Krein [**257**]: If $\{P_j\}_{j=1}^J$ is a finite family of mutually orthogonal self-adjoint projections, then for any $A \in \mathcal{I}_p$, $1 \le p < \infty$,

$$\sum_{j=1}^{J} \|P_j A P_j\|_p^p \le \|A\|_p^p \tag{3.7.51}$$

(*Hint*: Let $B = \sum_{j=1}^J P_j A P_j$; show that the singular values of B are related to A via a dss matrix.)

3. Suppose $\{A_n\}_{n=1}^\infty$, $A, B \in \mathcal{L}(\mathcal{H})$, with $B \ge 0$ and $B \in \mathcal{I}_p$, $1 \le p < \infty$,

$$|A_n| \le B, \qquad |A_n^*| \le B \tag{3.7.52}$$

and $A_n \to A$ in the weak operator topology. Prove that $\{A_n\}_{n=1}^\infty$, $A \in \mathcal{I}_p$, and $\|A_n - A\|_p \to 0$ ([**650**, Thm. 2.1.6]).

4. Let $1 \le p < \infty$.

(a) If $A_n \to A$, $|A_n| \to |A|$, $|A_n^*| \to |A^*|$ all in the weak operator topology and $\|A_n\|_p \to \|A\|_p$, prove that $\|A_n - A\|_p \to 0$. (*Hint*: You'll need Problem 3.)

(b) If $A_n \to A$ and $A_n^* \to A^*$ in the strong operator topology and $\|A_n\|_p \to \|A\|_p$, prove that $\|A_n - A\|_p \to 0$.

Remarks. 1. (b) is due to Grümm [**274**].

2. If $p > 1$, $\mathcal{I}_p$ is uniformly convex, and strong convergence can be replaced by weak convergence; see [**650**, Sect. 2].

5. Let $A \in \mathcal{L}(\mathcal{H})$ and $\{\psi_m\}_{m=1}^\infty$ an orthonormal basis. This problem will prove that if $p \le 2$, then

$$\|A\|_p^p = \sum_{n=1}^\infty \mu_n(A)^p \le \sum_{n=1}^\infty \|A\psi_m\|^p \tag{3.7.53}$$

in the sense that if the right side is finite, $A \in \mathcal{I}_p$ and (3.7.53) holds.

(a) If $\sum_{n=1}^\infty \|A\psi_m\|^p < \infty$, prove that A is compact. (*Hint*: Prove A is Hilbert–Schmidt.)

(b) If $\sum_{n=1}^\infty \beta_n \le 1$ with $\beta_n \ge 0$, prove for any sequence $\{\alpha_n\}_{n=1}^\infty$ with $\alpha_n \ge 0$, we have that

$$\Big(\sum \alpha_n^p \beta_n\Big)^{\frac{1}{p}} \le \Big(\sum \alpha_n^2 \beta_n\Big)^{\frac{1}{2}} \tag{3.7.54}$$

(c) If $A = \sum_{n=1}^\infty \mu_n(A)\langle\varphi_n, \cdot\rangle\eta_n$ is the canonical decomposition, prove that

$$\sum_{n=1}^\infty \mu_n^p |\langle\varphi_n, \psi_m\rangle|^2 \le \|A\psi_m\|^p \tag{3.7.55}$$

Prove (3.7.53).

6. By abstracting, the argument in the proof of Theorem 3.7.6, prove the following:

   If $f : \{z \mid 0 \le \operatorname{Re} z \le 1\} \to \mathcal{I}_\infty$ is analytic on the interval and continuous on the whole region (in operator norm) and $f(iy) \in \mathcal{I}_{p_0}$ and $f(1+iy) \in \mathcal{I}_{p_1}$ with $\sup_y \big[\|f(iy)\|_{p_0} + \|f(1+iy)\|_{p_1}\big] < \infty$, then for $t \in (0,1)$, $f(t) \in \mathcal{I}_{p_t}$ where $p_t^{-1} \equiv tp_1^{-1} + (1-t)p_0^{-1}$. (Hint: To bound $\operatorname{Tr}\big(Bf(t)\big)$ for B finite rank, pick $g(z)$ carefully, for z in the strip, so that $g(t) = B$ and consider $\operatorname{Tr}\big(g(z)f(z)\big)$).

3.8. Hilbert–Schmidt Operators

In this section, we study the trace ideal, $\mathcal{I}_2(\mathcal{H})$, for a Hilbert space, $\mathcal{H}$—it will soon be clear why we want to specify the underlying space. As for general p, we can define $\mathcal{I}_2(\mathcal{H})$ in terms of singular values:

Definition. The (abstract) *Hilbert–Schmidt operators*, $\mathcal{I}_2(\mathcal{H})$, are all compact operators on $\mathcal{H}$ with

$$\|A\|_2^2 = \sum_{n=1}^\infty \mu_n(A)^2 < \infty \tag{3.8.1}$$

In Example 3.1.15, we defined a class of integral operators we called "concrete Hilbert–Schmidt operators," which is why we use "abstract" above—we will see shortly the notions are "essentially" the same.

We saw in Theorem 3.6.1 that for any ON basis, $\{\varphi_n\}_{n=1}^\infty$, $\sum_{n=1}^\infty \|A\varphi_n\|^2$ is independent of basis, and if it is finite, A is compact (by Proposition 3.1.14). Picking φ_n to be a basis of eigenvectors of $|A|$ and using $\|A\varphi_n\| = \||A|\varphi_n\|$, we see this sum is exactly $\|A\|_2^2$. That is,

Theorem 3.8.1. *$A \in \mathcal{I}_2$ if and only if for any* ON *basis, $\{\varphi_n\}$, $\sum_{n=1}^\infty \|A\varphi_n\|^2 < \infty$ and*

$$\|A\|_2^2 = \sum_{n=1}^\infty \|A\varphi_n\|^2 \tag{3.8.2}$$

Our main goal here is to show the equality of $\mathcal{I}_2(\mathcal{H})$, as defined by (3.8.1) or (3.8.2), with two other objects we've studied: first, $\mathcal{I}_2(\mathcal{H})$ with the tensor product $\mathcal{H}^* \otimes \mathcal{H}$ and second, $\mathcal{I}_2(L^2(\Omega, d\mu))$ with the concrete

Hilbert–Schmidt integral operators on $L^2(\Omega, d\mu)$. Along the way, we'll see that $\mathcal{I}_2(\mathcal{H})$ is a Hilbert space and find direct proofs that $\mathcal{I}_2$ is a $*$-ideal and of the Hölder inequality $\|AB\|_1 \le \|A\|_2 \|B\|_2$.

Theorem 3.8.2. *The family of operators obeying* $\|A\|_2$, *defined by* (3.8.2), *finite is a vector space and* $*$*-ideal with (for all* $B \in \mathcal{L}(\mathcal{H})$*)*

$$\|A^*\|_2 = \|A\|_2, \qquad \|BA\|_2 \le \|B\| \|A\|_2, \qquad \|AB\|_2 \le \|B\| \|A\|_2 \tag{3.8.3}$$

Proof. Since

$$\|(A+B)\varphi\|^2 \le (\|A\varphi\| + \|B\varphi\|)^2 \le 2(\|A\varphi\|^2 + \|B\varphi\|^2) \tag{3.8.4}$$

(since $(x+y)^2 \le (x+y)^2 + (x-y)^2 = 2x^2 + 2y^2$), $\mathcal{I}_2$ is a vector space. $\|A\|_2 = \|A^*\|_2$ is just (3.6.7). $\|BA\|_2 \le \|B\| \|A\|_2$ is immediate from $\|BA\varphi\| \le \|B\| \|A\varphi\|$ and the final inequality in (3.8.3) from $\|C\| \equiv \|C^*\|$ and the first two inequalities. □

Theorem 3.8.3. (a) *If* $A \in \mathcal{I}_2$, *then* $A^*A \in \mathcal{I}_1$ *and*

$$\operatorname{Tr}(A^*A) = \|A\|_2^2 \tag{3.8.5}$$

(b) *If* $A, B \in \mathcal{I}_2$, *then* $AB \in \mathcal{I}_1$.

(c) $A, B \mapsto \operatorname{Tr}(A^*B)$ *defines an inner product in* $\mathcal{I}_2$ *and* $\|A\|_2$ *is the inner product space norm and, in particular, we have the triangle inequality*

$$\|A+B\|_2 \le \|A\|_2 + \|B\|_2 \tag{3.8.6}$$

(d) $\mathcal{I}_2$ *is complete in* $\|\cdot\|_2$, *so* $\mathcal{I}_2$ *is a Hilbert space in the inner product*

$$\langle A, B\rangle_{\mathcal{I}_2} = \operatorname{Tr}(A^*B) \tag{3.8.7}$$

(e) *For all* $A, B \in \mathcal{I}_2$, *we have*

$$\|AB\|_1 \le \|A\|_2 \|B\|_2 \tag{3.8.8}$$

Proof. (a) $A^*A \ge 0$ and

$$\sum_{n=1}^\infty \langle \varphi_n, A^*A\varphi_n\rangle = \sum_{n=1}^\infty \|A\varphi_n\|^2 \tag{3.8.9}$$

showing A^*A is trace class and (3.8.5) holds.

(b) Since $\mathcal{I}_2$ is a vector space, $A, B \in \mathcal{I}_2 \Rightarrow A, B, A+B, A+iB \in \mathcal{I}_2$, so A^*A, B^*B, $(A+B)^*(A+B)$, $(A+iB)^*(A+B)$ are all in $\mathcal{I}_1$. Since $\mathcal{I}_1$ is a vector space,

$$A^*B + B^*A = (A+B)^*(A+B) - A^*A - B^*B \in \mathcal{I}_1 \tag{3.8.10}$$

$$i(A^*B - B^*A) = (A+iB)^*(A+iB) - A^*A - B^*B \in \mathcal{I}_1 \tag{3.8.11}$$

so $A^*B \in \mathcal{I}_1$.

(c) If $A, B \in \mathcal{I}_2$, so by (b), we can define $\operatorname{Tr}(A^*B)$, which is obviously an inner product, and the norm is $\|\cdot\|_2$ by (3.8.5).

(d) The basic idea is that if A_m is Cauchy in $\|\cdot\|_2$, $A_m\varphi$ is Cauchy in $\mathcal{H}$ since $\varphi/\|\varphi\|$ is part of a basis. So A_m has a limit in $\mathcal{L}(\mathcal{H})$, which one can show is in $\mathcal{I}_2$ and is the $\mathcal{I}_2$ limit. The details are left to the reader (Problem 1).

(e) By (b), $C = AB \in \mathcal{I}_1$, so with $C = U|C|$ its polar decomposition,

$$\begin{aligned}\|AB\|_1 &= \operatorname{Tr}(U^*AB) = \langle A^*U, B\rangle \\ &\le \|A^*U\|_2\|B\|_2 \le \|A^*\|_2\|U\|\|B\|_2 \\ &= \|A\|_2\|B\|_2\end{aligned}$$

□

Theorem 3.8.4. *Map $\mathcal{H}^* \times \mathcal{H} \to \mathcal{L}(\mathcal{H})$ by*

$$\mathcal{A}_{\ell,\varphi}(\psi) = \ell(\psi)\varphi \tag{3.8.12}$$

Then the map $\mathcal{A}\colon (\ell,\varphi) \in \mathcal{H}^* \times \mathcal{H} \to \mathcal{L}(\mathcal{H})$ has a unique extension $\widetilde{\mathcal{A}}\colon \mathcal{H}^* \otimes \mathcal{H} \to \mathcal{L}(\mathcal{H})$ with

$$\widetilde{\mathcal{A}}(\ell \otimes \varphi) = \mathcal{A}_{\ell,\varphi} \tag{3.8.13}$$

$\widetilde{\mathcal{A}}$ is a Hilbert space isomorphism of $\mathcal{H}^* \otimes \mathcal{H}$ and $\mathcal{I}_2(\mathcal{H})$.

Proof. Clearly, $\mathcal{A}_{\ell,\varphi}$ is a rank-one operator and so in $\mathcal{I}_2$. If $\{\ell_j\}_{j=1}^N, \{\varphi_j\}_{j=1}^N$ are finite orthonormal families in $\mathcal{H}^*$ and $\mathcal{H}$, a simple calculation (Problem 2) shows that for $a_{ij} \in \mathbb{C}$,

$$\Big\|\sum a_{ij}\mathcal{A}_{\ell_i,\varphi_j}\Big\|_2^2 = \sum_{i,j} |a_{ij}|^2 \tag{3.8.14}$$

so $\mathcal{A}$ has an extension $\widetilde{\mathcal{A}}$ to a dense set of $\mathcal{H}^* \otimes \mathcal{H}$ which is an isometry.

All that remains is to check $\operatorname{Ran}(\widetilde{\mathcal{A}})$ is all of $\mathcal{I}_2$. But if $A \in \mathcal{I}_2$ and

$$A = \sum_{n=1}^{N(A)} \mu_n(A)\langle\varphi_n, \cdot\,\rangle\psi_n \tag{3.8.15}$$

and $\ell_n = \langle\varphi_n, \cdot\,\rangle \in \mathcal{H}^*$, then $\{\ell_n \otimes \psi_n\}_{n=1}^\infty$ are orthonormal, so since $\sum_n \mu_n(A)^2 < \infty$, $\sum \mu_n(A)\ell_n \otimes \psi_n \in \mathcal{H}^* \otimes \mathcal{H}$ and, by (3.8.15),

$$\widetilde{\mathcal{A}}\Big(\sum_{n=1}^{N(A)} \mu_n(A)\ell_n \otimes \psi_n\Big) = A \tag{3.8.16}$$

□

Next, we see abstract Hilbert–Schmidt ≡ concrete Hilbert–Schmidt:

Theorem 3.8.5. *Let $\mathcal{H} = L^2(\Omega, d\mu)$, where (Ω, Σ, μ) is a σ-finite (separable) measure space. Then any $A \in \mathcal{I}_2$ has an integral kernel, $K_A(\omega, \omega')$, in $L^2(\Omega \times \Omega, d\mu \otimes d\mu)$,*

$$(Af)(\omega) = \int K_A(\omega, \omega') f(\omega')\, d\mu(\omega') \tag{3.8.17}$$

Moreover,

$$\|A\|_2 = \|K_A\|_{L^2(\Omega\times\Omega, d\mu\otimes d\mu)} \tag{3.8.18}$$

Conversely, any integral operator of the form (3.8.17) with $K_A \in L^2(\Omega \times \Omega, d\mu \otimes d\mu)$ is an $\mathcal{I}_2$ operator obeying (3.8.18). Thus, (3.8.17) sets up an isometric isomorphism of $\mathcal{I}_2(L^2(\Omega, d\mu))$ and $L^2(\Omega \times \Omega, d\mu \otimes d\mu)$.

Remark. Our requirement that (Ω, Σ, μ) be separable means that $L^2(\Omega, d\mu)$ is separable, so by our definition (see Section 3.1 in Part 1), L^2 is a Hilbert space. It is not hard to extend the theory of $\mathcal{I}_2$ to nonseparable Hilbert spaces (in any event, if $A \in \mathcal{I}_2$, $\overline{\text{Ran}(A)}$ will then be separable) and then to extend this theorem to nonseparable cases.

Proof. Let $A \in \mathcal{I}_2$ and let (3.8.15) be its canonical expansion. In $L^2(\Omega \times \Omega, d\mu \otimes d\mu)$, $\{\overline{\varphi_n(\omega')}\, \psi_n(\omega)\}_{n=1}^{N(A)}$ is an orthonormal family. Since $\sum \mu_n(A)^2 < \infty$, we can define K_A in $L^2(\Omega \otimes \Omega, d\mu \otimes d\mu)$ by

$$K_A(\omega, \omega') = \sum_{n=1}^{N(A)} \mu_n(A)\, \overline{\varphi_n(\omega')}\, \psi_n(\omega) \tag{3.8.19}$$

By Fubini's theorem, this converges in $L^2(\Omega, d\mu)$ for a.e. fixed ω, so

$$\int K_A(\omega, \omega') f(\omega')\, d\mu(\omega') = \sum_{n=1}^{N(A)} \mu_n(A) \langle \varphi_n, f \rangle \psi_n(\omega) = (Af)(\omega) \tag{3.8.20}$$

proving (3.8.17). (3.8.18) holds since both sides are $\sum_{n=1}^{N(A)} \mu_n(A)^2$.

For the converse: given $K \in L^2(\Omega \otimes \Omega, d\mu \otimes d\mu)$, let $\{\eta_n(\omega)\}_{n=1}^\infty$ be an orthonormal basis for $L^2(\Omega, d\mu)$. Then (see Theorem 4.11.8 of Part 1), $\{\eta_n(\omega)\, \overline{\eta_m(\omega')}\}_{n,m=1}^\infty$ is a basis of $L^2(\Omega \otimes \Omega, d\mu \otimes d\mu)$, so

$$K = \sum_{n,m=1} \alpha_{nm} \bar{\eta}_m \otimes \eta_n \tag{3.8.21}$$

with $\{\alpha_{nm}\} \in \ell^2$.

In $\mathcal{I}_2$, if $A_{mn} = \langle \eta_m, \cdot \rangle \eta_n$, then A_{mn} are orthonormal in $\mathcal{I}_2$, so $\sum \alpha_{nm} A_{nm}$ converges in L^2 to an operator A that obeys (3.8.17). □

The above gives necessary and sufficient conditions for an integral operator to be in $\mathcal{I}_2$ but, in general, there are no such conditions for the trace class, $\mathcal{I}_1$ (if $A \geq 0$ and has a jointly continuous kernel, there is; see Theorem 3.11.9). There is a special but often occurring class of A's for which there is an effective sufficient condition.

On $\mathcal{H} = L^2(\mathbb{R}^\nu, d^\nu x)$, if f, g are measurable functions on $\mathbb{R}^\nu$, we define $f(X)$ and $g(P)$ by

$$(f(X)\varphi)(x) = f(x)\varphi(x), \qquad (g(P)\varphi)(x) = \widecheck{[g(k)\widehat{\varphi}]}(x) \tag{3.8.22}$$

In general, $f(X)\varphi$ and $g(P)\varphi$ are not defined on all of L^2 but on

$$\mathcal{D}_X(f) = \{\varphi \in L^2 \mid f\varphi \in L^2\}, \qquad \mathcal{D}_P(X) = \{\varphi \in L^2 \mid g\widehat{\varphi} \in L^2\} \tag{3.8.23}$$

which are dense in $\mathcal{H}$.

To say $f(X)g(P)$ is trace class is shorthand for saying there is a trace class operator, A, so that for all $\varphi \in \mathcal{D}_P(g)$, $\psi \in \mathcal{D}_X(\overline{f})$, we have

$$\langle \varphi, A\psi \rangle = \langle \overline{f}(X)\varphi, g(P)\psi \rangle \tag{3.8.24}$$

For $1 \leq p, q < \infty$, we define the *Birman–Solomyak space*, $\ell^q(L^p)$, by letting χ_α be the characteristic function of the unit cube in $\mathbb{R}^\nu$ centered at $\alpha \in \mathbb{Z}^\nu$ and defining

$$\|f\|_{p,q} = \biggl(\sum_{\alpha \in \mathbb{Z}^\nu} \|f\chi_\alpha\|_p^q\biggr)^{1/q} \tag{3.8.25}$$

where $\|\cdot\|_p$ is $L^p(\mathbb{R}^\nu, d^\nu x)$-norm. $\ell^q(L^p)$ is the set of f with $\|f\|_{p,q}$ finite. We also define for $\delta > 0$,

$$L^2_\delta(\mathbb{R}^\nu) = \{f \mid (1 + |\cdot|^2)^{\delta/2} f \in L^2\} \tag{3.8.26}$$

$$\|f\|_{\delta,2} = \|(1 + |\cdot|^2)^{\delta/2} f\|_2 \tag{3.8.27}$$

In Problem 3, the reader will show that if $\delta > \nu/2$,

$$L^2_\delta(\mathbb{R}^\nu) \subset \ell_1(L^2) \subset L^1(\mathbb{R}^\nu) \cap L^2(\mathbb{R}^\nu) \tag{3.8.28}$$

We are heading towards the following:

Theorem 3.8.6. (a) *$f(X)g(P) \in \mathcal{I}_2$ if and only if $f, g \in L^2(\mathbb{R}^\nu)$.*

(b) *$f(X)g(P) \in \mathcal{I}_1$ if and only if $f, g \in \ell_1(L^2)$.*

Remarks. 1. There are implicit bounds in the proof, for example,

$$\|f(X)g(P)\|_{\mathcal{I}_1} \leq c\|f\|_{2;1}\|g\|_{2;1} \tag{3.8.29}$$

2. Of course, to conclude $g \in L^2$ (or $\ell^1(L^2)$), one needs $f \neq 0$.

3. We'll leave the necessity for $f(X)g(P)$ to be in $\mathcal{I}_1$ or $\mathcal{I}_2$ to the reader (Problem 4).

Proof (First Part). If $g \in L^2$, then $L^1(\mathbb{R}^\nu) \cap L^2(\mathbb{R}^\nu) \subset \mathcal{D}_g$, and for $\varphi \in L^1 \cap L^2$, we have

$$(g(P)\varphi)(x) = (2\pi)^{-\nu/2} \int \check{g}(x-y)\varphi(y)\, d^\nu y \tag{3.8.30}$$

Thus, $f(X)g(P)$ has the integral kernel

$$K_{f(X)g(P)}(x,y) = (2\pi)^{-\nu/2}\check{g}(x-y)f(x) \tag{3.8.31}$$

Since $g \in L^2$, we have $\check{g} \in L^2$, so $K \in L^2(\mathbb{R}^\nu \times \mathbb{R}^\nu, d^{2\nu}x)$. Thus, $f(X)g(P) \in \mathcal{I}_2$. □

The following special case of (b) will be needed for its proof; it is most often used in applications of (b).

Lemma 3.8.7. *Let $\delta > \nu/2$. If $f,g \in L^2_\delta(\mathbb{R}^\nu)$, then $f(X)g(P) \in \mathcal{I}_1$ and*

$$\|f(X)g(P)\| \le c_{\delta,\nu}\|f\|_\delta\|g\|_\delta \tag{3.8.32}$$

Proof. Write

$$\begin{aligned} f(X)g(P) = AB; \qquad A &= f(X)(1+P^2)^{-\delta/2}(1+X^2)^{\delta/2}; \\ B &= (1+X^2)^{-\delta/2}(1+P^2)^{\delta/2}g(P) \end{aligned} \tag{3.8.33}$$

We need only prove $A, B \in \mathcal{I}_2$ with suitable bounds and then use $\|AB\|_1 \le \|A\|_2\|B\|_2$.

Since $(1+|\cdot|^2)^{-\delta/2}$, $(1+|\cdot|^2)^{\delta/2}g \in L^2$, $B \in \mathcal{I}_2$ by part (a),

Let h be $(2\pi)^{\nu/2}$ times the inverse Fourier transform of $(1+k^2)^{-\delta/2}$. Since $\widehat{h}$ is analytic in a strip and L^2 along lines, the suitable Paley–Wiener theorem (see Theorem 11.1.1 of Part 2A) implies for any $|a| < 1$,

$$e^{a|\cdot|}h \in L^2 \tag{3.8.34}$$

A has integral kernel

$$f(x)h(x-y)(1+y^2)^{\delta/2} \tag{3.8.35}$$

Using

$$(1+|y|^2)^{\delta/2} \le C(1+|x|^2)^{\delta/2}(1+|x-y|^2)^{\delta/2} \tag{3.8.36}$$

and (3.8.34), we see the kernel in (3.8.35) is $L^2(\mathbb{R}^\nu \times \mathbb{R}^\nu, d^{2\nu}x)$, so $A \in \mathcal{I}_2$ also. □

Proof of Theorem 3.8.6 (Conclusion). Let $\{\chi_\alpha\}_{\alpha\in\mathbb{Z}^\nu}$ be as in the definition of $\ell^1(L^2)$. Since, for any $\delta > 0$,

$$\|f\chi_0\|_\delta \le C_1\|f\chi_0\|_2 \tag{3.8.37}$$

the lemma implies

$$\|(f\chi_0)(X)(g\chi_0)(P)\|_1 \le C_2\|f\chi_0\|_2\|g\chi_0\|_2 \tag{3.8.38}$$

If

$$(U_{\alpha,\beta}\varphi)(x) = e^{i\beta\cdot x}\varphi(x+\alpha) \tag{3.8.39}$$

then

$$U_{\alpha,\beta}\chi_0(X)\chi_0(P)U_{\alpha,\beta}^{-1} = \chi_\alpha(X)\chi_\beta(P) \tag{3.8.40}$$

we see, by (3.8.38), that

$$\|(f\chi_\alpha)(X)(g\chi_\beta)(P)\|_1 \le C_2\|f\chi_\alpha\|_2\,\|g\chi_\beta\|_2 \tag{3.8.41}$$

Thus,

$$\begin{aligned}\|f(X)g(P)\|_1 &\le \sum_{\alpha,\beta}\|(f\chi_\alpha)(X)(g\chi_\beta)(P)\|_1\\ &\le C_2\sum_{\alpha,\beta}\|f\chi_\alpha\|_2\,\|g\chi_\beta\|_2\\ &= C_2\|f\|_{2;1}\,\|g\|_{2;1}\end{aligned} \tag{3.8.42}$$

□

There is also a general compactness criterion.

Definition. $f \in L^\infty_\varepsilon(\mathbb{R}^\nu, d^\nu x)$ if and only if $f \in L^\infty$ and for all ε, there is an R so

$$|x| > R \Rightarrow |f(x)| < \varepsilon \tag{3.8.43}$$

Theorem 3.8.8. *If $f, g \in L^\infty_\varepsilon$, then $f(X)g(P)$ is compact.*

Proof. Pick R so (3.8.43) holds for f and g. Let $f_{\le R}, f_{\ge R}$ be $f\chi_{\{x\mid |x|\le R\}}, f\chi_{\{x\mid |x|>R\}}$, and similarly for g. Then

$$\|f(X)g(P) - f_{\le R}(X)g_{\le R}(P)\| \le \varepsilon^2 + \varepsilon\|f\|_\infty + \varepsilon\|g\|_\infty \tag{3.8.44}$$

so $f(X)g(P)$ is the operator norm limit of $f_{\le R}(X)g_{\le R}(X)$ as $R \to \infty$. Since $f_{\le R}, g_{\le R} \in L^2(\mathbb{R}^\nu, d^\nu x)$, $f_{\le R}(X)g_{\le R}(P)$ is Hilbert–Schmidt, so compact. It follows that $f(X)g(P)$ is compact. □

Notes and Historical Remarks. The study of concrete Hilbert–Schmidt operators goes back to the work of Hilbert and Schmidt mentioned in the Notes to Section 3.1. An important later development is the 1921 paper of Carleman [**101**] to which we return in the Notes to Sections 3.11 and 3.13. In his monograph on Hilbert space, Stone [**670**] considered operators for which $\sum_{n=1}^\infty \|A\varphi_n\|^2 < \infty$ for an orthonormal basis and noted those operators with inner product $\langle A, B\rangle_{\mathcal{I}_2} = \sum_{n=1}^\infty \langle A\varphi_n, B\varphi_n\rangle$ were a Hilbert space. That $\mathcal{I}_2$ is a Hilbert space tensor product is connected to the work on cross norms mentioned in the previous section.

The Birman–Solomyak spaces were introduced exactly for the context here in their paper [**64**].

Problems

1. Suppose A_m is Cauchy in $\mathcal{I}_2$ in the sense that for all ON bases, $\{\varphi_n\}_{n=1}^\infty$, $\forall\varepsilon$, $\exists M(\varepsilon)$,
$$k,m \geq M \Rightarrow \sum_{n=1}^{\infty} \|(A_k - A_m)\varphi_n\|^2 \leq \varepsilon$$
   (a) For every φ_0, prove that $\lim_{n\to\infty} A_n\varphi_0 \equiv A\varphi_0$ exists and defines $A \in \mathcal{L}(\mathcal{H})$.
   (b) For each N and $m \geq M(\varepsilon)$, prove that $\sum_{n=1}^{N} \|(A - A_m)\varphi_n\| \leq \varepsilon$.
   (c) Prove that $A \in \mathcal{I}_2$ and $A_m - A \to 0$ in $\mathcal{I}_2$-norm.

2. Let $\mathcal{A}_{\ell,\varphi}$ be given by (3.8.13). Prove that
$$\operatorname{Tr}(\mathcal{A}^*_{\ell_1,\varphi_1}\mathcal{A}_{\ell_2,\varphi_2}) = \langle \ell_1,\ell_2\rangle\langle\varphi_1,\varphi_2\rangle$$
   and conclude that (3.8.14) holds.

3. Prove that for $\delta > \nu/2$, (3.8.28) holds. (*Hint*: Use $L^2(\Delta_\alpha, d^\nu x) \subset L^1(\Delta_\alpha, d^\nu x)$ and $\ell^1 \subset \ell^2$.)

4. Prove the necessity parts of Theorem 3.8.6. (*Hint*: Cut off f, g by replacing f by $\min(n, f)$ and multiplying by $\chi_{\{x \mid |x|<R\}}$, obtain bounds on the cutoff, f, g, and take the cutoffs away. You'll use the fact that if $|\widetilde{f}(x)| \leq |f(x)|$ pointwise, then $\widetilde{f}(X) = Cf(X)$, where C is a contraction.)

5. If $f, h \in L^2(\mathbb{R}^\nu)$ and $g \in L^1(\mathbb{R}^\nu)$, prove that $f(X)g(P)h(X)$ is trace class with
$$\|f(X)\,g(P)\,h(X)\|_1 \leq C\|f\|_2\,\|h\|_2\,\|g\|_1 \tag{3.8.45}$$
   where C is independent of f, g, h but may depend on ν.

6. Let $2 < p < \infty$. If $f, g \in L^p(\mathbb{R}^\nu)$, prove that $f(X)g(P) \in \mathcal{I}_p$. (*Hint*: Interpolate between $p = 2$ and $p = \infty$ using Problem 6 of Section 3.7.)

   Remark. This is a result of Seiler–Simon [**619**]. Cwikel [**143**] proved that for the same range of p, $f(X)g(P)$ is in a weak trace ideal if $f \in L^p$ and g in weak-L^p. This was a conjecture of Simon [**634**]. For a proof, see [**650**].

3.9. Schur Bases and the Schur–Lalesco–Weyl Inequality

Let A be a compact operator on a Hilbert space, $\mathcal{H}$, $\{\Lambda_j(A)\}_{j=1}^{M(A)}$ a listing of its distinct nonzero eigenvalues ordered so that $|\Lambda_n(A)| \geq |\Lambda_{n+1}(A)|$. Let $\{\lambda_j(A)\}_{j=1}^{K(A)}$ be a listing of eigenvalues counted up to algebraic multiplicity, that is, if Λ_n has multiplicity m_n, then

$$\lambda_j(A) = \Lambda_\ell(A), \qquad m_1 + \cdots + m_{\ell-1} < j \le m_1 \cdots + m_\ell \tag{3.9.1}$$

Let $\{\mu_n(A)\}_{n=1}^{N(A)}$ be its singular values. In this section, our goal is to prove:

Theorem 3.9.1 (Schur–Lalesco–Weyl Inequality). *Let $1 \le p < \infty$. If $A \in \mathcal{I}_p$, then $\sum_{j=1}^{K(A)} |\lambda_j(A)|^p < \infty$ and*

$$\sum_{j=1}^{K(A)} |\lambda_j(A)|^p \le \sum_{n=1}^{N(A)} \mu_n(A)^p \tag{3.9.2}$$

Remark. We will need this especially for $p = 1$ and care most that in that case, $\sum_{j=1}^{K(A)} |\lambda_j(A)| < \infty$. For $p = 1$, the proof below needs only the easy fact that if $\{b_{nm}\}$ is a dss matrix, the associated operator is bounded on ℓ^1. The more subtle Theorem 3.7.1 is needed only for $p > 1$.

The key to the proof is a simple construction of Schur:

Theorem 3.9.2 (Schur Basis). *For any compact operator, A, there is an orthonormal set, $\{\eta_j\}_{j=1}^{K(A)}$, so that for all j,*

$$\langle \eta_j, A\eta_j \rangle = \lambda_j(A) \tag{3.9.3}$$

Moreover, A is upper triangular in the sense that

$$\langle \eta_k, A\eta_j \rangle = 0 \qquad \text{if } k > j \tag{3.9.4}$$

Remarks. 1. The $\{\eta_j\}$ may not be a complete ON set, so it need not be a basis. Despite that (because it was originally introduced in the finite-dimensional case with $\{\lambda_j\}$ all eigenvalues in which $\{\eta_j\}$ are a basis), this is called a *Schur basis.*

2. We will only use (3.9.3), not (3.9.4).

Proof. Let P_ℓ be the spectral projection (see Theorem 2.3.5 and Corollary 2.3.6) for Λ_ℓ. By Corollary 2.3.6,

$$AP_\ell = \Lambda_\ell P_\ell + N_\ell \tag{3.9.5}$$

where N_ℓ is nilpotent, so $\text{Ran}(P_\ell)$ has a basis (*not* necessarily orthonormal), $\kappa_j^{(\ell)}$, so $AP_\ell \restriction \text{Ran}(P_\ell)$ has the form

$$A\kappa_j^{(\ell)} = \Lambda_\ell \kappa_j^{(\ell)} + \alpha_j^{(\ell)} \kappa_{j-1}^{(\ell)} \tag{3.9.6}$$

where each $\alpha_j^{(\ell)}$ is 0 or 1 (Jordan normal form).

Putting the κ's in order $(\kappa_1^{(1)}, \dots, \kappa_{n_1}^{(1)}, \kappa_1^{(2)}, \dots) \equiv (\kappa_1, \kappa_2, \dots)$, we get an independent set, $\{\kappa_j\}_{j=1}^{K(A)}$, so

$$A\kappa_j = \lambda_j(A)\kappa_j + \alpha_j \kappa_{j-1} \tag{3.9.7}$$

where α_j is 0 or 1.

Now apply Gram–Schmidt to $\{\kappa_j\}_{j=1}^{K(A)}$. We get an orthonormal set, $\{\eta_j\}_{j=1}^{K(A)}$, so that for upper-triangular matrices, $\{a_{ij}\}_{1\le i\le j}$, $\{b_{ij}\}_{1\le i\le j}$, with $a_{ii} > 0$, $b_{ii} > 0$, we have that

$$\eta_j = \sum_{k=1}^{j} a_{kj}\kappa_k \tag{3.9.8}$$

$$\kappa_j = \sum_{k=1}^{j} b_{kj}\eta_k \tag{3.9.9}$$

It follows that (Problem 1) for suitable $\{c_{ij}\}_{1\le i<j}$, we have

$$A\eta_j = \lambda_j(A)\eta_j + \sum_{k=1}^{j-1} c_{kj}\eta_k \tag{3.9.10}$$

which implies (3.9.3) and (3.9.4). $\square$

Proof of Theorem 3.9.1. Write the canonical decomposition

$$A = \sum_{k=1}^{N(A)} \mu_k(A)\langle \varphi_k, \cdot\rangle \psi_k \tag{3.9.11}$$

Then, by (3.9.3),

$$\lambda_j(A) = \sum_{k=1}^{N(A)} b_{jk}\mu_k(A) \tag{3.9.12}$$

where

$$b_{jk} = \langle \varphi_k, \eta_j\rangle\langle \eta_j, \psi_k\rangle \tag{3.9.13}$$

Since φ, ψ, η are orthonormal families, by Lemma 3.6.4, $\{b_{jk}\}$ is dss. So, by Theorem 3.7.1, we get (3.9.2). $\square$

Notes and Historical Remarks. Theorem 3.9.1 for $p = 2$ and the invention of Schur bases to prove it are due to I. Schur [**608**] in 1909. The case $p = 1$ is due to Lalesco [**420**] in 1915. This was long before the trace class was invented—Lalesco proved a result on the sum of the absolute values of the eigenvalues of the operators A of the form $A = BC$ where B and C are Hilbert–Schmidt and his bound had $\|B\|_2\,\|C\|_2$, not $\|A\|_1$.

The general p result is due to Weyl [**751**] in 1949 who had a stronger result and different method of proof. He considered functions φ, nonnegative and monotone on $[0,\infty)$, so that $t \mapsto \varphi(e^t)$ is convex (e.g., $\varphi(s) = s^p$ for $p \in (0,\infty)$, not just $p \geq 1$) and proved that $\sum_n \varphi(|\lambda_n(A)|) \leq \sum_n \varphi(\mu_n(A))$. He obtained this by starting with (see Problem 5 of Section 3.10)

$$|\lambda_1(A)\cdots\lambda_n(A)| \leq \mu_1(A)\cdots\mu_n(A) \tag{3.9.14}$$

and then using the result in Problem 26 of Section 5.3 of Part 1.

Problems

1. Prove (3.9.10).

3.10. Determinants and Fredholm Theory

So far, we extended the trace from finite dimensions to infinite but the determinant is as fundamental as the trace. The trace involves the sum of matrix elements which we required to be summable and, in particular, go to zero. Bearing in mind the relation of convergence of $\prod_{n=1}^\infty w_n$ to convergence of $\sum_{n=1}^\infty |w_n - 1|$ (see Lemma 9.1.2 of Part 2A), it is natural to try to define $\det(B)$ if $B - \mathbb{1}$ is trace class. That is what we'll succeed in doing: defining $\det(1+A)$ for $A \in \mathcal{I}_1$.

One use we'll make of $\det(1+A)$ will be to prove Lidskii's theorem, that $\text{Tr}(A)$ is $\sum_{n=1}^{N(A)} \lambda_j(A)$, in Section 3.12.

Another use concerns Cramer's rule. In finite dimensions, Cramer's rule writes matrix elements of B^{-1} as a ratio

$$(B^{-1})_{ij} = \frac{\widetilde{D}_{ij}(B)}{\det(B)} \tag{3.10.1}$$

where $\widetilde{D}_{ij}(B)$ is the *minor*: $(-1)^{i+j}$ times the determinant of the matrix obtained from removing row i and column j from B. A key way of understanding minors is to note that for $n \times n$ matrices,

$$\det(B) = \sum_{j=1}^n b_{ij}\widetilde{D}_{ij}(B) \tag{3.10.2}$$

for any i so that (since $\widetilde{D}_{i\ell}(B)$ is independent of each b_{ij})

$$\widetilde{D}_{ij}(B) = \frac{\partial \det(B)}{\partial(b_{ij})} \tag{3.10.3}$$

So a second goal will be to obtain a formula for $(1+A)^{-1}$ as a ratio of a derivative of a determinant and a determinant—that is, an infinite-dimensional analog of Cramer's rule.

Our definition of determinants will rely on the use of alternating algebra, $\wedge^k(A)$, so the reader might want to review Section 1.3 of this part or Section 3.8 of Part 1. Consider first B on $\mathbb{C}^\ell$ with eigenvalues $\{\lambda_j(B)\}_{j=1}^\ell$, counting up to algebraic multiplicity. If B has a Schur basis (see Theorem 3.9.2) $\{e_j\}_{j=1}^\ell$, then it is easy to see that

$$\langle (e_{j_1} \wedge \cdots \wedge e_{j_k}), B(e_{j_1} \wedge \cdots \wedge e_{j_k})\rangle = \prod_{m=1}^k \lambda_{j_m}(B) \tag{3.10.4}$$

and thus,

$$\operatorname{Tr}(\wedge^k(B)) = \sum_{1\le j_1<\cdots<j_k\le\ell} \lambda_{j_1}(B)\dots\lambda_{j_k}(B) \tag{3.10.5}$$

Therefore, if A is an $n \times n$ matrix,

$$\begin{aligned}\det(1+A) &= \prod_{j=1}^n \lambda_j(1+A) \\ &= \prod_{k=1}^n (1+\lambda_j(A)) \\ &= 1 + \sum_{k=1}^n \biggl(\sum_{j_1<\cdots<j_k} \lambda_{j_1}(A)\dots\lambda_{j_k}(A)\biggr) \\ &= \sum_{k=0}^n \operatorname{Tr}(\wedge^k(A)) \end{aligned} \tag{3.10.6}$$

This suggests we try to define

$$\det(1+A) = \sum_{k=0}^\infty \operatorname{Tr}(\wedge^k(A)) \tag{3.10.7}$$

Of course, we need to prove the convergence of this sum, which is one purpose of

Proposition 3.10.1. *Let $A \in \mathcal{I}_1(\mathcal{H})$. Then for all $k = 0, 1, 2, \dots,$*

(a) $\wedge^k(A) \in \mathcal{I}_1(\wedge^k(\mathcal{H}))$

(b) $$\|\wedge^k(A)\|_1 = \sum_{j_1<\cdots<j_k} \mu_{j_1}(A)\dots\mu_{j_k}(A) \tag{3.10.8}$$

(c) $$\|\wedge^k(A)\|_1 \le \frac{\|A\|_1^k}{k!} \tag{3.10.9}$$

Proof. Since $\sqrt{\wedge^k(B)} = \wedge^k(\sqrt{B})$ for $B \ge 0$ (by uniqueness of square root), we have

$$|\wedge^k(A)| = \wedge^k(|A|) \tag{3.10.10}$$

so it suffices to consider the case $A = |A|$, that is, $A \ge 0$.

Let $\{\varphi_j\}_{j=1}^{N(A)}$ be an orthonormal basis of $\overline{\operatorname{Ran}(A)}$ of eigenvectors with

$$A\varphi_j = \mu_j(A)\varphi_j \tag{3.10.11}$$

Then

$$\wedge^k(A)(\varphi_{j_1} \wedge \cdots \wedge \varphi_{j_k}) = \biggl(\prod_{m=1}^{k} \mu_{j_m}(A)\biggr)(\varphi_{j_1} \wedge \cdots \wedge \varphi_{j_k})$$

and $\{\varphi_{j_1} \wedge \cdots \wedge \varphi_{j_k}\}_{j_1<\cdots<j_k}$ is an ON basis of $\operatorname{Ran}(\wedge^k(A))$, so $\{\mu_{j_1}(A)\dots\mu_{j_k}(A)\}_{j_1<\cdots<j_k}$ is the set of eigenvalues of this $A \geq 0$, proving (3.10.8) where, for now, both sides might be infinite.

$$\begin{aligned}\|A\|_1^k = \biggl(\sum_{j=1}^{N(A)} \mu_j(A)\biggr)^k &= \sum_{j_1,\dots,j_k} \mu_{j_1}(A)\dots\mu_{j_k}(A) \\ &\geq k! \sum_{j_1<\cdots<j_k} \mu_{j_1}(A)\dots\mu_{j_k}(A) \end{aligned}\tag{3.10.12}$$

by dropping terms with some $j_a = j_b$ $(a \neq b)$ and using the fact that any $j_1 < \cdots < j_k$ occurs $k!$ times as the product. This proves (3.10.9) and incidentally that $\wedge^k(A) \in \mathcal{I}_1(\wedge^k(\mathcal{H}))$. $\square$

Lemma 3.10.2. *Let $A_1,\dots,A_\ell \in \mathcal{I}_1$. Fix k and define for $(z_1,\dots,z_\ell) \in \mathbb{C}^\ell$,*

$$F(z_1,\dots,z_\ell) = \operatorname{Tr}\biggl(\wedge^k\biggl(\sum_{m=1}^{\ell} z_m A_m\biggr)\biggr)$$

Then F is a polynomial of total degree at most k in $z_1,\dots,z_\ell$.

Proof. Evaluate $\operatorname{Tr}(\wedge^k(\,\cdot\,))$ in a basis $e_{j_1}\wedge\cdots\wedge e_{j_k}$ with $\{e_j\}_{j=1}^n$ an arbitrary ON basis for $\mathcal{H}$. Each $\langle e_{j_1} \wedge \cdots \wedge e_{j_k}, \wedge^k(\sum_{m=1}^\ell z_m A_m)(e_{j_1} \wedge \cdots \wedge e_{j_k})\rangle$ is a polynomial of degree at most k. The trace class estimate of the last theorem implies the sum converges uniformly on compacts and so to a polynomial. $\square$

Lemma 3.10.3. *Fix $R > 0$. Let f be analytic in a neighborhood of $\{z \mid |z| \leq R + \frac{1}{2}\}$. Then*

$$|f(\tfrac{1}{2}) - f(-\tfrac{1}{2})| \leq R^{-1} \sup_{|w|=R+\frac{1}{2}} |f(w)| \tag{3.10.13}$$

Proof. Let $t \in [-\frac{1}{2}, \frac{1}{2}]$. By a Cauchy estimate (see Theorem 3.1.8 of Part 2A),

$$|f'(t)| \leq R^{-1} \sup_{|w-t|\leq R} |f(w)| \leq R^{-1} \sup_{|w|=R+\frac{1}{2}} |f(w)| \tag{3.10.14}$$

by the maximum principle. Now use

$$|f(\tfrac{1}{2}) - f(-\tfrac{1}{2})| = \left(\int_{-1/2}^{1/2} f'(t)\,dt\right) \le \int_{-1/2}^{1/2} |f'(t)|\,dt \qquad \square$$

Here are the most basic properties of the determinant:

Theorem 3.10.4. *Let $A, A_1, \dots, A_\ell, B \in \mathcal{I}_1$. Then the sum (3.10.7) is absolutely convergent. Moreover,*

(a) $|\det(1+A)| \le \prod_{\ell=1}^{\infty}(1+\mu_\ell(A)) = \det\bigl(1+|A|\bigr)$ (3.10.15)

(b) $|\det(1+A)| \le \exp(\|A\|_1)$ (3.10.16)

(c) $(z_1, \dots, z_\ell) \mapsto \det(1+\sum_{j=1}^{\ell} z_j A_j)$ *is an entire function on* $\mathbb{C}^\ell$.

(d) $|\det(1+A) - \det(1+B)| \le \|A-B\|_1 \exp(\|A\|_1 + \|B\|_1 + 1)$ (3.10.17)

(e) *If $\mathcal{H} = \mathcal{F} \oplus \mathcal{F}^\perp$ with $\mathcal{F}$ finite-dimensional and $A \restriction \mathcal{F}^\perp = 0$, $A[\mathcal{F}] \subset \mathcal{F}$, then*

$$\det(1+A) = \det\nolimits_{\mathcal{F}}(1 + A \restriction \mathcal{F}) \tag{3.10.18}$$

where $\det_{\mathcal{F}}$ *is the finite-dimensional determinant.*

(f) $\det(1+A+B+AB) = \det(1+A)\det(1+B)$ (3.10.19)

Proof. (a) $|\mathrm{Tr}(\wedge^k(A))| \le \|\wedge^k(A)\|_1 \le$ RHS of (3.10.8) and

$$1 + \sum_{k=1}^{\infty} \text{RHS of (3.10.8)} = \prod_{j=1}^{\infty}(1+\mu_j(A))$$

so we get absolute convergence and (3.10.15).

(b) This follows from (a) and

$$1 + \mu_j(A) \le e^{\mu_j(A)} \tag{3.10.20}$$

and $\|A\|_1 = \sum_j \mu_j(A)$.

(c) On compact subsets of $\mathbb{C}^\ell$, we have, by the proof of (a), uniformly convergent series of polynomials.

(d) If $\|A-B\|_1 = 0$, there is nothing to prove. Otherwise, let $R = \|A-B\|_1^{-1}$ and apply Lemma 3.10.3 to $f(z) = \det(1 + \frac{1}{2}(A+B) + z(A-B))$ to see

$$\begin{aligned} |f(\tfrac{1}{2}) - f(-\tfrac{1}{2})| &\le \|A-B\|_1 \exp(\tfrac{1}{2}\,\|A\| + \tfrac{1}{2}\,\|B\| + (R+\tfrac{1}{2})\|A-B\|_1) \\ &\le \text{RHS of (3.10.17)} \end{aligned}$$

(e) We have

$$\operatorname{Tr}(\wedge^k(A)) = \begin{cases} \operatorname{Tr}_{\mathcal{F}}(\wedge^k(A) \restriction \mathcal{F}), & k \le \dim(\mathcal{F}) \\ 0, & k > \dim(\mathcal{F}) \end{cases} \tag{3.10.21}$$

so (3.10.18) follows from (3.10.6) for finite matrices.

(f) If A and B are both finite rank, we can find a single finite-dimensional space $\mathcal{F}$ so $A \restriction \mathcal{F}^\perp = B \restriction \mathcal{F}^\perp = 0$ and $A[\mathcal{F}] \subset \mathcal{F}$, $B[\mathcal{F}] \subset \mathcal{F}$ (e.g., $\mathcal{F}$ is the subspace spanned by $\operatorname{Ran}(A)$, $\operatorname{Ker}(A)^\perp$, $\operatorname{Ran}(B)$, and $\operatorname{Ker}(B)^\perp$). Thus, the result follows by $\det_{\mathcal{F}}(CD) = \det_{\mathcal{F}}(C)\det_{\mathcal{F}}(D)$ (see Theorem 1.3.6). Since $A \mapsto \det(1+A)$ is continuous by (d) and the finite rank operators are dense in $\mathcal{I}_1$, we get the result for all A, B. □

det is important because $\det(B) = 0$ is equivalent to noninvertibility.

Theorem 3.10.5. *Let $A \in \mathcal{I}_1$.*

(a) *If $1 + A$ is invertible, $\det(1 + A) \neq 0$.*
(b) *If $1 + A$ is not invertible, $\det(1 + A) = 0$. In fact, $f(z) = \det(1 + zA)$ has a zero order ℓ at $z = 1$, where ℓ is the algebraic multiplicity of -1 as an eigenvalue of A.*

Remark. We use the fact that if $1 + A$ is not invertible, then -1 is an eigenvalue of A (see Theorem 3.3.1).

Proof. (a) If $1 + A$ is invertible, then with

$$B = -A(1 + A)^{-1} \tag{3.10.22}$$

we have that $B \in \mathcal{I}_1$ and

$$(1 + A)(1 + B) = 1 + A - (1 + A)A(1 + A)^{-1} = 1 \tag{3.10.23}$$

so by (3.10.19),

$$\det(1 + A)\det(1 + B) = \det(1) = 1 \tag{3.10.24}$$

and thus, $\det(1 + A) \neq 0$.

(b) If $1+A$ is not invertible, $-1 \in \sigma(A)$, and so an isolated point of the spectrum. Let P be the associated spectral projection which is finite-dimensional by Theorem 3.3.1. Since $[P, A] = 0$ and $P(1 - P) = 0$, we have

$$1 + zA = (1 + zAP)(1 + zA(1 - P)) \tag{3.10.25}$$

and thus, by (3.10.19),

$$\det(1 + zA) = \det(1 + zAP)\det(1 + zA(1 - P)) \tag{3.10.26}$$

By (2.3.29), $-1 \notin \sigma(A(1 - P))$, so

$$\det(1 + A(1 - P)) \neq 0 \tag{3.10.27}$$

by (a).

Since $\mathrm{Ran}(P) < \infty$, AP is finite rank. If $\mathcal{F}$ is such that $AP \restriction \mathcal{F}^\perp = 0$ and $A[\mathcal{F}] \subset \mathcal{F}$ (e.g., $\mathcal{F}$ is the span of $\mathrm{Ker}(AP)^\perp$ and $\mathrm{Ran}(AP)$), then $AP \restriction \mathcal{F}$ has eigenvalue -1, ℓ times where $\ell = \text{alg mult}(-1)$ and otherwise 0. By (3.10.18), we see that

$$\det(1 + zAP) = (1 - z)^\ell \tag{3.10.28}$$

(3.10.26), (3.10.27), and (3.10.28) show that $\det(1 + zA)$ has a zero of order ℓ at $z = 1$. $\square$

Recall (see Section 5.1 of Part 1) that $f\colon X \to Y$, Banach spaces, are called (Fréchet) differentiable with (Fréchet) derivative, Df_{x_0}, a bounded linear map of X to Y, if and only if

$$f(x_0 + x) - f(x_0) - Df_{x_0}(x) = o(\|x\|) \tag{3.10.29}$$

If $Y = \mathbb{C}$ (and X is a complex Banach space), if $Df_{x_0}(x)$ exists, it is an element of X^*. If $X = \mathcal{I}_1$, then $X^* = \mathcal{L}(\mathcal{H})$, as we saw in Theorem 3.6.8.

Theorem 3.10.6. *Let $f(A) = \det(1+A)$ as a function from $\mathcal{I}_1$ to $\mathbb{C}$. Then f is differentiable at any $A \in \mathcal{I}_1$, with $-1 \notin \sigma(A)$, and its derivative is*

$$D_A \equiv Df_A = [\det(1 + A)](1 + A)^{-1} \tag{3.10.30}$$

Remark. It is not hard to see (Problem 1) that f is differentiable also at A with $-1 \in \sigma(A)$ although (3.10.30) fails.

Proof. Let $A, B \in \mathcal{I}_1$ with $1 + A$ invertible. Then

$$1 + A + B = (1 + A)(1 + (1 + A)^{-1}B) \tag{3.10.31}$$

Thus,

$$f(A + B) = f(A)f((1 + A)^{-1}B) \tag{3.10.32}$$

By looking at the estimates (3.10.9) and series (3.10.7), we see that for any $C \in \mathcal{I}_1$,

$$\det(1 + C) = 1 + \mathrm{Tr}(C) + O(\|C\|_1^2) \tag{3.10.33}$$

Thus, (3.10.32) implies

$$f(A + B) - f(A) - f(A)\mathrm{Tr}((1 + A)^{-1}B) + O(\|B\|_1^2) \tag{3.10.34}$$

Thus,

$$Df_A(B) = f(A)\,\mathrm{Tr}((1 + A)^{-1}B) \tag{3.10.35}$$

Taking into account the explicit formula (3.6.17) for $\mathcal{I}_1^* = \mathcal{L}(\mathcal{H})$, we have (3.10.30). $\square$

Corollary 3.10.7 (Cramer's Rule for 1+Trace Class). *If $A \in \mathcal{I}_1$ with $1 + A$ invertible, then*

$$(1 + A)^{-1} = \frac{Df_A}{\det(1 + A)} \tag{3.10.36}$$

Here is a calculation that makes Df_A look more like the usual minor. We note that, by polarization, $\langle\varphi,(Df_A)\varphi\rangle$ for all φ determines $\langle\varphi,(Df_A)\psi\rangle$ for all φ,ψ.

Proposition 3.10.8. *Let $A\in\mathcal{I}_1$, $\varphi\in\mathcal{H}$ with $\|\varphi\|=1$. Suppose Q is the projection onto $\{\eta\mid\langle\eta,\varphi\rangle=0\}$. Then*

$$\langle\varphi,(Df_A)\varphi\rangle=\det\nolimits_{Q\mathcal{H}}(1+QAQ) \tag{3.10.37}$$

where $\det_{Q\mathcal{H}}(\,\cdot\,)$ means as an operator on $Q\mathcal{H}$.

Proof. Problem 3 has a proof manipulating derivatives of $\mathrm{Tr}(\wedge^*(A))$. Instead we prove it as follows: $A\mapsto D_Af$ is continuous in A (Problem 1), so it suffices to prove (3.10.37) for A finite rank. In that case, we pick a basis in which φ is the first vector and use the classical Cramer's rule. □

There is a final set of formulae we should mention, although we leave their proofs to the Problems (Problem 4). Let $A\in\mathcal{I}_1$ and

$$\det(1+zA)=\sum_{m=0}^\infty\frac{z^m\alpha_m(A)}{m!} \tag{3.10.38}$$

$$D_{zA}=\det(1+zA)\mathbb{1}+\sum_{m=1}^\infty\frac{z^m\beta_m(A)}{(m-1)!} \tag{3.10.39}$$

Then (*Plemelj–Smithies formulae*; $m\times m$ determinants),

$$\alpha_m(A)=\begin{vmatrix}\mathrm{Tr}(A) & m-1 & 0 & \dots & 0\\ \mathrm{Tr}(A^2) & \mathrm{Tr}(A) & m-2 & \dots & 0\\ \mathrm{Tr}(A^3) & \mathrm{Tr}(A^2) & \mathrm{Tr}(A) & \dots & 0\\ \vdots & \vdots & \vdots & \ddots & \vdots\\ \mathrm{Tr}(A^m) & \mathrm{Tr}(A^{m-1}) & \mathrm{Tr}(A^{m-2}) & \dots & \mathrm{Tr}(A)\end{vmatrix} \tag{3.10.40}$$

$$\beta_m(A)=\begin{vmatrix}A & m-1 & 0 & \dots & 0\\ A^2 & \mathrm{Tr}(A) & m-2 & \dots & 0\\ A^3 & \mathrm{Tr}(A^2) & \mathrm{Tr}(A) & \dots & 0\\ \vdots & \vdots & \vdots & \ddots & \vdots\\ A^m & \mathrm{Tr}(A^{m-1}) & \mathrm{Tr}(A^{m-2}) & \dots & \mathrm{Tr}(A)\end{vmatrix} \tag{3.10.41}$$

In particular, (3.10.40) implies $\alpha_m(A)$ is a polynomial in $\{\mathrm{Tr}(A^j)\}_{j=1}^m$. Once we have Lidskii's theorem, we'll have for $|z|<\|A\|_1^{-1}$ that

$$\log(\det(1+zA))=\sum_{j=1}^{N(A)}\log(1+z\lambda_j(A))$$

$$= \sum_{k=1}^{\infty} \frac{(-1)^k}{k} z^k \sum_{j=1}^{N(A)} (\lambda_j(A))^k \tag{3.10.42}$$

$$= \sum_{k=1}^{\infty} \frac{(-1)^k}{k} z^k \mathrm{Tr}(A^k) \tag{3.10.43}$$

so for z small,

$$\det(1 + zA) = \exp\biggl(\sum_{k=1}^{\infty} \frac{(-1)^k}{k} z^k \mathrm{Tr}(A^k)\biggr) \tag{3.10.44}$$

This, in principle, leads to a formula for $\alpha_m(A)$ as polynomials in $\{\mathrm{Tr}(A^j)\}_{j=1}^m$. While we'll prove (3.10.40) from (3.10.44), it is useful for later purposes to know that any two expressions for $\alpha_m(A)$ as polynomials in $\{\mathrm{Tr}(A^j)\}_{j=1}^m$ must agree.

Proposition 3.10.9. *Let $\mathcal{H}$ be a Hilbert space of dimension $d \geq m$. Let $f\colon \mathcal{I}_1(\mathcal{H}) \to \mathbb{C}^m$ by*

$$f(A) = (\mathrm{Tr}(A), \dots, \mathrm{Tr}(A^m)) \tag{3.10.45}$$

Then $\mathrm{Ran}(f)$ *has nonempty interior.*

Proof. By looking at f restricted to the rank at most m operators, we see that we can suppose $\dim(\mathcal{H}) = m$. Then $f(A)$ only depends on $\{\lambda_j(A)\}_{j=1}^m$, so it suffices to show that

$$\tilde{f}(\lambda_1, \dots, \lambda_m) = \biggl(\sum_{j=1}^m \lambda_j, \sum_{j=1}^m \lambda_j^2, \dots, \sum_{j=1}^m \lambda_j^m\biggr)$$

from $\mathbb{C}^m$ to $\mathbb{C}^m$ has a range with nonempty interior. Clearly,

$$\frac{\partial \widetilde{f}_j}{\partial \lambda_\ell} = j\lambda_\ell^{j-1} \tag{3.10.46}$$

so

$$\det\biggl(\frac{\partial \widetilde{f}_j}{\partial \lambda_\ell}\biggr) = m! \det|\lambda_\ell^{j-1}| = m! \prod_{1 \leq i < j \leq m} (\lambda_i - \lambda_j) \tag{3.10.47}$$

recognizing the Vandermonde determinant (see Problem 23 of Section 6.9 of Part 1). Thus, $\widetilde{f}_j$ is a local homeomorphism at any point with all $\lambda_i \neq \lambda_j$ $(i \neq j)$. Thus, $\mathrm{Ran}(f)$ has nonempty interior. □

Remark. On $\mathbb{C}^m$, $|\mathrm{Tr}(A)|^2 \leq m\mathrm{Tr}(A^2)$ by the Schwarz inequality with equality if $A = \mathbb{1}$. Thus, if $d = m$ and $y \in \mathrm{Ran}(f)$, $y_1^2 \leq my_2$, so $\mathrm{Ran}(f)$ is a proper subset of $\mathbb{C}^m$.

Because the range has a nonempty interior, any two polynomials in $\{\mathrm{Tr}(A^j)\}_{j=1}^m$ that are equal as functions are the same polynomial.

Notes and Historical Remarks. Most of this section is an abstract version of the concrete Fredholm theory that solved $f(x) = g(x) + \int_a^b K(x,y)f(y)\,dy$ for f. We discuss this concrete theory in Section 3.11. The idea of proving a bound like (3.10.17) using Lemma 3.10.3 is from Seiler–Simon [**618**]. The Plemelj–Smithies formulae were first found in 1904 by Plemelj [**528**] and rediscovered and popularized by Smithies [**658**].

Problems

1. Let X be a complex Banach space and $f\colon X \to \mathbb{C}$ a function so that for all $A, B, C \in X$ and z, w, ζ, we have $(z, w, \zeta) \to f(zA + wB + \zeta C)$ is entire in the three variables. Suppose also

$$|f(A)| \leq G(\|A\|) \tag{3.10.48}$$

for $G\colon [0,\infty) \to [0,\infty)$ and monotone. For all $A, B \in X$, define

$$g_A(B) = \frac{d}{dz} f(A + zB)\bigg|_{z=0} \tag{3.10.49}$$

This problem will show that $g_A(B)$ is linear in B, norm-continuous in A, and the Fréchet derivative of f.

(a) Prove that $f(A)$ is norm-Lipschitz in A.

(b) Prove that uniformly in $\{B \mid \|B\| \leq 1\}$ and $\{A \mid \|A\| \leq R\}$, $A \mapsto g_A(B)$ is norm-continuous in A.

(c) For $|z| \leq \frac{1}{2}$, prove that

$$|f(A + zB) - f(A) - zg_A(B)| \leq 2z^2 G(\|A\| + \|B\|) \tag{3.10.50}$$

(d) For $\|C\| \leq \frac{1}{2}$, prove that

$$\|f(A + C) - f(A) - g_A(C)\| \leq 2\|C\|^2 G(\|A\| + 1) \tag{3.10.51}$$

(e) For each A, prove that $B \mapsto g_A(B)$ is linear in B.

(f) Prove that f is Fréchet differentiable at each A in X and $A \mapsto D_A f$ is a norm-continuous map of X to X^*.

(g) Prove that $\det(1 + A)$ is Fréchet differentiable at $A \in \mathcal{I}_1$ even at points where $-1 \in \sigma(A)$.

2. Let $-1 \in \sigma(A)$ for $A \in \mathcal{I}_1$ and let P be the corresponding spectral projection. Let $f(B) = \det(1 + B)$.

(a) If the algebraic multiplicity of -1 for A is 1, prove that

$$D_A f = \big[\det\nolimits_{(1-P)\mathcal{H}}(1 + (1-P)A)\big] P \tag{3.10.52}$$

(b) If the algebraic multiplicity is $\ell \geq 2$ and $AP = -P + D$, prove that

$$D_A f = \big[\det\nolimits_{(1-P)\mathcal{H}}(1 + (1-P)A)\big] D^{\ell-1} \tag{3.10.53}$$

3. (a) Let $g_\ell(A) = \text{Tr}(\wedge^\ell(A))$ for $\ell = 1, 2, \dots$ and $A \in \mathcal{I}_1$. Using the notation of Proposition 3.10.8, prove that

$$\langle \varphi, (Dg_\ell)_A \varphi \rangle = \text{Tr}_{Q\mathcal{H}}(\wedge^{\ell-1}(QAQ)) \tag{3.10.54}$$

(b) Prove (3.10.37).

4. (a) Let f, g be analytic near $z = 0$ with $f(0) = 1$, $g(0) = 0$, $f(z) = \sum_{m=0}^\infty a_n z^n/n!$, $g(z) = \sum_{n=1}^\infty (-1)^{n+1} b_n z^n/n$ and $f(z) = \exp(g(z))$. Prove that ($a_0 = 1$)

$$a_n = \sum_{j=1}^n (-1)^{j+1} b_j a_{n-j} \frac{(n-1)!}{(n-j)!} \tag{3.10.55}$$

(*Hint*: $f' = fg'$.)

(b) Prove that ($n \times n$ determinant)

$$a_n = \begin{vmatrix} b_1 & n-1 & 0 & \dots & 0 \\ b_2 & b_1 & n-2 & \dots & 0 \\ b_3 & b_2 & b_1 & \dots & 0 \\ \vdots & \vdots & \vdots & \ddots & \vdots \\ b_n & b_{n-1} & b_{n-2} & \dots & b_1 \end{vmatrix} \tag{3.10.56}$$

(c) For z small, prove that

$$\det(1 + zA) = \exp\left[\sum_{m=1}^\infty \frac{(-1)^{m+1}}{m} z^m \text{Tr}(A^m)\right] \tag{3.10.57}$$

(*Hint*: Prove it for A finite rank using det as a product of eigenvalues and take limits. Alternatively, use Lidskii's theorem from Section 3.12.)

(d) Prove (3.10.40).

(e) For z small, prove that

$$(1 + zA)^{-1} \det(1 + zA) = (1 - zA + z^2 A \dots) \det(1 + zA)$$

(f) Deduce the formula

$$\frac{\beta_{m+1}(A)}{m!} = \left[A \frac{\alpha_m(A)}{m!} - A^2 \frac{\alpha_{m-1}(A)}{(m-1)!} + \dots\right]$$

(g) Prove (3.10.41).

5. (a) For any compact operator A on a Hilbert space, $\mathcal{H}$, prove that

$$\|\wedge^k(A)\| = \mu_1(A) \cdots \mu_k(A) \tag{3.10.58}$$

(b) If $\lambda_{i_1}(A), \dots, \lambda_{i_k}(A)$ are eigenvalues of A, prove that $\lambda_{i_1}(A) \cdots \lambda_{i_k}(A)$ is an eigenvalue of $\wedge^k(A)$ and conclude that

$$|\lambda_{i_1}(A) \cdots \lambda_{i_k}(A)| \le \mu_1(A) \cdots \mu_k(A) \tag{3.10.59}$$

3.11. Operators with Continuous Integral Kernels

Let X be a compact metric space with μ a probability measure on X. Let $K\colon X \times X \to \mathbb{C}$ be a continuous function. In this section, we'll consider the operator, A_K, given by

$$(A_K f)(x) = \int K(x,y) f(y)\, d\mu(y) \tag{3.11.1}$$

Throughout most of the section, f lies in $L^2(X, d\mu)$ and we think of A_K as an operator on L^2, but later we'll deduce information on A_K as an operator on $C(X)$. In particular, we look at solutions of

$$f = g - A_K f \tag{3.11.2}$$

At first sight, it would appear this fits exactly into the framework of the last section. Since $\int |K(x,x)|\, d\mu(x) < \infty$, one might guess that A_K is trace class. Sometimes that's true, but as we'll see, it may not be. Even for the trace class case, we'll get integral formulae for $\operatorname{Tr}(\wedge^\ell(A_K))$ that are of interest. Remarkably, even when A_K is not trace class, these formulae continue to hold, although we'll need a different method (going back to Fredholm!) for proving convergence of power series.

We'll begin by presenting the Fredholm formulae for his determinant and minor, and then show they can always be used to solve (3.11.2). If A_K is trace class, we'll prove that the Fredholm formula agree with those of the last section. Finally, we'll consider the special case when $A_K \ge 0$. We'll see that, then, A_K is trace class and the Hilbert–Schmidt eigenfunction expansion converges uniformly (Mercer's theorem).

The n-fold *Fredholm kernel* is defined by

$$K_n \begin{pmatrix} x_1 \dots x_n \\ y_1 \dots y_n \end{pmatrix} = \det(K(x_i, y_j))_{1 \le i,j \le n} \tag{3.11.3}$$

The *Fredholm determinant* is defined by

$$d_K(z) = \sum_{n=0}^\infty \frac{z^n}{n!} \int \cdots \int K_n \begin{pmatrix} x_1 \dots x_n \\ x_1 \dots x_n \end{pmatrix} d\mu(x_1) \dots d\mu(x_n) \tag{3.11.4}$$

where the $n = 0$ term is interpreted as 1. Of course, it is not obvious this sum converges—we'll soon prove that below, using Hadamard's inequality that $|\det(a_{ij})_{1\le i,j\le n}| \le n^{n/2}(\sup\|a_{ij}\|)^n$. The *Fredholm minor* is the kernel defined by

$$M_{K,a}(x,y) = \sum_{n=1}^{\infty} \frac{z^n}{(n-1)!} \int \cdots \int K_n \begin{pmatrix} x & x_1 \dots x_{n-1} \\ y & x_1 \dots x_{n-1} \end{pmatrix} d\mu(x_1)\dots d\mu(x_{n-1}) \tag{3.11.5}$$

Example 3.11.1. Let $X = \partial\mathbb{D}$, with $d\mu = \frac{d\theta}{2\pi}$. Let f be a continuous function on $\partial\mathbb{D}$ and let

$$K(e^{i\theta}, e^{i\varphi}) = f(e^{i(\theta-\varphi)}) \tag{3.11.6}$$

A_K has $\{e^{in\theta}\}_{n=-\infty}^{\infty}$ as an orthonormal basis with eigenvalues $f_n^\sharp = \int f(e^{i\theta})e^{-in\theta}\frac{d\theta}{2\pi}$. A_K is normal with an orthonormal basis of eigenfunctions so A_K is trace class if and only if

$$\sum_{n=-\infty}^{\infty} |f_n^\sharp| < \infty \tag{3.11.7}$$

In Problem 10 of Section 5.4 of Part 1 (see also Example 6.7.8 of that part), we showed there exists $f \in C(\partial\mathbb{D})$ so (3.11.7) fails. Thus, there are A_K with continuous kernels which are not trace class. Worse, $\sum_{n=-N}^{N} f_n^\sharp e^{in\theta}$ does not converge absolutely, so the Hilbert–Schmidt expansion of Theorem 3.2.1 for A_K does not converge uniformly. □

Given X and μ, we'll construct a natural family of finite rank projections, P_n, so that

$$\text{s-lim}\, P_n = \mathbb{1} \tag{3.11.8}$$

Cover X with open balls of radius $1/n$ and use compactness to pick a finite cover $U_1^{(n)}, \dots, U_{\ell_n}^{(n)}$. Choose them so for all j, $\mu(\bigcup_{k\ne j} U_k^{(n)}) < 1$. This can be arranged if we drop some U's that may make the remaining ones only cover X up to a set of μ-measure 0. Let

$$V_1^{(n)} = U_1^{(n)}, \qquad V_j^{(n)} = U_j^{(n)} \Big\backslash \Big[\bigcup_{k=1}^{j-1} U_k^{(n)}\Big] \tag{3.11.9}$$

Let $\mathcal{H}_n$ be the functions constant on each $V_j^{(n)}$ and P_n the orthogonal projection on $\mathcal{H}_n$. Since each $V_j^{(n)}$ is contained in a ball of radius $1/n$, we have

$$\|f - P_n f\|_\infty \le \sup_{\rho(x,y)\le n^{-1}} |f(x) - f(y)| \tag{3.11.10}$$

so $P_n f \to f$ in L^∞ and so in L^2. Thus, (3.11.8) holds.

Throwing out those $V_j^{(n)}$ with $\mu(V_j^{(n)}) = 0$, we see that

$$\psi_j^{(n)} = \mu(V_j^{(n)}))^{-1/2} \chi_{V_j^{(n)}} \tag{3.11.11}$$

is an orthonormal basis for $\mathcal{H}_n$. Note that

$$\langle \psi_j^{(n)}, A_K\psi_j^{(n)}\rangle = \mu(V_j^{(n)})^{-1} \int_{x,y\in V_j^{(n)}} K(x,y)\, d\mu(x)d\mu(y) \tag{3.11.12}$$

In particular,

$$\left| \langle \psi_j^{(n)}, A_K\psi_j^{(n)}\rangle - \int_{x\in V_j^{(n)}} K(x,x)\, d\mu(x)\right| \le \mu(V_j^{(n)})\varepsilon_n \tag{3.11.13}$$

where

$$\varepsilon_n = \sup_{x;\rho(y,y')\le \frac{1}{n}} |K(x,y) - K(x,y')| \to 0 \tag{3.11.14}$$

as $n\to\infty$. We therefore have:

Proposition 3.11.2. *For any continuous function, K, we have*

$$\mathrm{Tr}(P_nA_KP_n) \to \int K(x,x)\, d\mu(x) \tag{3.11.15}$$

If A_K is trace class, then

$$\mathrm{Tr}(A_K) = \int K(x,x)\, d\mu(x) \tag{3.11.16}$$

Proof. Since $\{\psi_j^{(n)}\}$ is a basis for $\mathcal{H}_n$ and $P_nA_KP_n$ is trace class (since it is finite rank),

$$\mathrm{Tr}(P_nA_KP_n) = \sum_j \langle \psi_j^{(n)}, A_K\psi_j^{(n)}\rangle \tag{3.11.17}$$

By (3.11.13) and (3.11.14), as $n\to\infty$,

$$\left| \mathrm{Tr}(P_nA_KP_n) - \int K(x,x)\, d\mu(x)\right| \le \varepsilon_n \to 0 \tag{3.11.18}$$

If A_K is trace class and $\sum_{m=1}^M \mu_m(A_K)\langle \varphi_m, \cdot\rangle\psi_m$ its canonical expansion, then

$$\mathrm{Tr}(P_nA_KP_n) = \mathrm{Tr}(A_KP_n) = \sum_{m=1}^M \mu_m(A_K)\langle \varphi_m, P_n\psi_m\rangle \tag{3.11.19}$$

$$\to \sum_{m=1}^M \mu_m(A_K)\langle \varphi_m, \psi_m\rangle = \mathrm{Tr}(A) \tag{3.11.20}$$

as $n\to\infty$, by a use of the dominated convergence theorem for sums since $|\langle \varphi_m, P_n\psi_m\rangle| \le 1$. □

Recall (see Section 3.8 of Part 1 and Section 1.3 of this part) $\wedge^k(\mathcal{H})$ is constructed by defining σ_π, for $\pi\in\Sigma_n$, on $\otimes^n\mathcal{H}$ by requiring

$$\sigma_\pi(\psi_1\otimes\cdots\otimes\psi_n) = \psi_{\pi(1)}\otimes\cdots\otimes\psi_{\pi(n)} \tag{3.11.21}$$

and

$$\mathcal{A}_n = \frac{1}{n!} \sum_{\pi \in \Sigma_n} (-1)^\pi \sigma_\pi \tag{3.11.22}$$

We define $\wedge^n(\mathcal{H})$ to be $\mathrm{Ran}(\mathcal{A}_n)$. If $\mathcal{H} = L^2(X, d\mu)$, we can associate $\otimes^n \mathcal{H}$ with $L^2(X \times \cdots \times X, d\mu \otimes \cdots \otimes d\mu)$ and σ_π is then

$$(\sigma_\pi f)(x_1, \dots, x_n) = f(x_{\pi^{-1}(1)}, \dots, x_{\pi^{-1}(n)}) \tag{3.11.23}$$

Proposition 3.11.3. *Viewed as an operator on* $L^2(\times^n X, \otimes^n d\mu)$, $\wedge^n(A_K)\mathcal{A}_n$ *has* $\frac{1}{n!}K_n\binom{x_1 \dots x_n}{y_1 \dots y_n}$ *as integral kernel.*

Proof. $\wedge^n(A_K)\mathcal{A}_n = \otimes^n(A_K)\mathcal{A}_n$ and $\otimes^n A_K$ has $\prod_{j=1}^n K(x_j, y_j)$ as integral kernel. Thus, $\wedge^n(A_K)\mathcal{A}_n$ has kernel

$$\frac{1}{n!}\sum_\pi (-1)^n \prod_{j=1}^n K(x_j, y_{\pi(j)}) = (n!)^{-1} K_n \begin{pmatrix} x_1 \dots x_n \\ y_1 \dots y_n \end{pmatrix} \tag{3.11.24}$$

□

Theorem 3.11.4 (Hadamard's Inequality). *Let* $B = (b_{ij})_{1 \le i,j \le n}$ *be an* $n \times n$ *matrix. Then*

$$|\det(B)| \le \prod_{j-1}^n \biggl(\sum_{i-1}^n |b_{ij}|^2 \biggr)^{1/2} \tag{3.11.25}$$

$$\le n^{n/2} \prod_{j-1}^n \bigl[\sup_{i=i,\dots,n} |b_{ij}| \bigr] \tag{3.11.26}$$

$$\le n^{n/2} \bigl[\sup_{i,j} |b_{ij}| \bigr]^n \tag{3.11.27}$$

Proof. Let $\delta_1, \dots, \delta_n$ be the standard basis of $\mathbb{C}^n$ and $b_j = B\delta_j$. By the alternating algebra definition of determinant (see Section 1.3), $\wedge^n(B)$ is multiplication by $\det(B)$, so

$$b_1 \wedge \cdots \wedge b_n = \det(B)(\delta_1 \wedge \cdots \wedge \delta_n)$$

and thus, by (3.8.41) of Part 1,

$$\det(B) = \|b_1 \wedge \cdots \wedge b_n\| \le \prod_{j=1}^n \|b_j\| \tag{3.11.28}$$

Since b_j has components b_{ij}, this is (3.11.25). The others are obvious. □

Theorem 3.11.5. *For any continuous kernel, K, the series* (3.11.4) *for* $d_K(z)$ *converges to an entire function. Moreover,*

(a) $d_K(z) = \lim_{m \to \infty} \det(1 + zP_m A_K P_m)$ (3.11.29)

uniformly on compact subsets of $\mathbb{C}$.

(b) *If A is trace class, $d_K(z) = \det(1 + zA)$.*
(c) *$1+zA$ is invertible as an operator on $L^2(X, d\mu)$ if and only if $d_K(z) \neq 0$.*

Proof. By Hadamard's inequality, the n-th term in the series for $d_K(z)$ is bounded by

$$n^{n/2}|z|^n \frac{\|K\|_\infty^n}{n!} \tag{3.11.30}$$

Since $n! \sim n^n$ to leading order (we don't need the full Stirling approximation; indeed, we could use the trivial $n! \geq (\frac{n}{3})^{2n/3}$), these bounds are summable in n uniformly as z runs through compact subsets of $\mathbb{C}$.

(a) $P_m A_K P_m$, as an operator on L^2, has an integral kernel bounded $\|K\|_\infty$ so the terms of $\mathrm{Tr}(\wedge^m(P_m A_K P_m))$ are also bounded by (3.11.30). We thus need convergence of the terms of the Taylor series which follows from Proposition 3.11.2 (applied to kernels on X^n).

(b) If A is trace class, $P_n A P_n \to A$ in trace class, so $\det(1 + zP_n A P_n) \to \det(1 + zA)$. (Alternatively, one can apply Proposition 3.11.2.)

(c) $P_n A P_n$ converges to A in Hilbert–Schmidt norm, so by Problem 4 of Section 2.3, the points in $\sigma_d(A)$ are precisely limits of points in $\sigma_d(P_n A P_n)$. By Hurwitz's theorem, Theorem 6.4.1 of Part 2A, and (a), the zeros of $d_K(z)$ are limits of zeros of $\det(1 + zP_n A P_n)$. □

Remark. The proof of (c) can be refined to show if $d_K(z_0) = 0$, the order of the zero is the algebraic multiplicity of $-z_0^{-1}$ as an eigenvalue of A_K (see Problem 1).

Theorem 3.11.6. *For any continuous kernel, K, and for each R, the series (3.11.5) for $M_{K,z}(x, y)$ converges uniformly in x, y and $|z| < R$. Moreover,*
(a) *If $M_K(z)$ is the operator with integral kernel, $M_{K,z}(x, y)$, then*

$$d_K(z)[zA_K(1 + zA_K)^{-1}] = M_K(z) \tag{3.11.31}$$

(b) *If $d_K(1) \neq 0$, (3.11.2) is solved by*

$$f = g + d_K(1)^{-1}[M_K(1)]g \tag{3.11.32}$$

Remark. If $d_K(z)$ has a simple zero at z_0, it is easy to show (Problem 2) that $M_K(z_0)$ is rank 1 and its range is the solutions of $(1 + z_0 A_K)f = 0$. Higher-order zeros are more complicated but the Taylor coefficients of $M_K(z)$ at z_0 determine eigenfunctions and generalized eigenfunctions.

Proof. As in the last proof, Hadamard's inequality implies the convergence of $M_{K,z}(x, y)$ uniformly in x, y and $|z| < R$ and it shows $M_{P_m K P_m}(z) \to M_K(z)$ in Hilbert–Schmidt norm (in fact, the kernels converge uniformly in x, y and $|z| < R$).

(a) By the limit results above and in that theorem, it suffices to prove this for the finite rank $P_m K P_m$. In fact, we'll prove it when A_K is trace class. By replacing K by zK, we can take $z = 1$. (3.11.31) is then equivalent to

$$\det(1 + A_K)[(1 + A^n)^{-1} - 1] = M_K(z) \tag{3.11.33}$$

This in turn follows from (3.10.30) and some manipulations (Problem 3).

(b) This follows from (3.11.33). □

Theorem 3.11.7. *Let K be a continuous integral kernel on a compact metric space, X. Then A_K given by* (3.11.1) *is a bounded linear transformation of $C(X)$ to itself. $1 + A_K$ is invertible on $C(X)$ if and only if $d_K(1) \neq 0$.*

Proof. It is easy to see that A_K maps $C(X)$ to $C(X)$ with norm $\|K\|_\infty$. Indeed, it maps $L^2(X, d\mu)$ to $C(X)$ with that norm. If $d_K(1) = 0$, there is a nonzero f in L^2 with $A_K f = -f$. As noted, A_K maps L^2 to $C(X)$ so $f \in C(X)$. Thus, $\mathrm{Ker}_{C(X)}(1 + A_X) \neq 0$ and $1 + A_K$ is not invertible in $C(X)$.

If $d_K(1) \neq 0$, $M_{K,z=1}(x, y)$ is continuous, so $M_K(1)$ maps $C(X)$ to itself, and for $y \in C(X)$, the solution f of (3.11.32) is in $C(X)$, so $1 + A_K$ is onto $C(X)$. If $A_K f = -f$ with f in $C(X)$, $f \in L^2$ so $d_K(1) = 0$. Thus, $\mathrm{Ker}_{C(X)}(1 + A_K) = \{0\}$ and $(1 + A_K)$ is invertible in $C(X)$. □

We know that the integral $\alpha_n(A)$ on the right side of (3.11.4) is a polynomial in $\{\mathrm{Tr}(A_K^j)\}_{j=1}^n$ if A_K is trace class. By the argument at the end of Section 3.10, the polynomials are the same for all such polynomial expressions. Note that if A_K is trace class,

$$\mathrm{Tr}(A^k) = \int \cdots \int_{X^k} K(x_1, x_2) \dots K(x_{k-1}, x_k) K(x_k, x_1)\, d\mu(x_1) \dots d\mu(x_k) \tag{3.11.34}$$

Every permutation, π, of $\{1, \dots, n\}$ has a decomposition into cycles of length $\ell_1, \dots, \ell_m$ with $\sum_{j=1}^m \ell_j = n$. By (3.11.34),

$$\int \cdots \int_{X^n} \prod_{j=1}^n K(x_j, x_{\pi(j)})\, d\mu(x_j) = \prod_{j=1}^n \mathrm{Tr}(A^{\ell_j}) \tag{3.11.35}$$

The sign of such a π is $(-1)^{n-m}$. If $N(\ell_1, \dots, \ell_m)$ is the number of such permutations (see the Notes), then

$$\alpha_n(A) = \sum_{\substack{\ell_1 + \cdots \ell_m = n \\ \ell_j \geq 1}} N(\ell_1, \dots, \ell_m)(-1)^{n-m} \prod_{j=1}^m \mathrm{Tr}(A^{\ell_j}) \tag{3.11.36}$$

This completes what we have to say about the general continuous integral kernel. We turn finally to the positive definite case. The following is left to the reader (Problem 4):

Proposition 3.11.8. *Let K be a continuous kernel on a compact metric space, X. Then $A_K \geq 0$ as an operator on $L^2(X, d\mu)$ if and only if for all $\{x_j\}_{j=1}^n$ in $\mathrm{supp}(d\mu)$, $\{K(x_i, x_j)\}_{1\leq i,j\leq n}$ is a positive matrix.*

In this case, we say K is a *positive definite kernel.* This is a notation we've seen already in Problems 4 and 5 of Section 3.3 of Part 1.

Here is the key result for positive definite kernels.

Theorem 3.11.9 (Mercer's Theorem). *Let K be a positive definite continuous kernel on a compact metric space, X, and μ a positive Baire measure on X. Let A_K be given by (3.11.1). Then*

(a) *A_K is trace class.*

(b) $$\mathrm{Tr}(A_K) = \int K(x,x)\, d\mu(x) \tag{3.11.37}$$

(c) *The nonzero eigenvalues $\{\mu_j(A)\}_{j=1}^N$ have continuous eigenfunctions $\{\varphi_j(x)\}_{j=1}^N$ and, as $J \to \infty$,*

$$\sum_{j=1}^{J} \mu_j(A)\, \overline{\varphi_j(y)}\, \varphi_j(x) \tag{3.11.38}$$

converges uniformly to $K(x,y)$ for $x, y \in \mathrm{supp}(d\mu)$.

As a preliminary, we note K is Hilbert–Schmidt, so as an L^2-operator,

$$A_K = \sum_{j=1}^{N} \mu_j(A)(\varphi_j, \cdot\,)\varphi_j \tag{3.11.39}$$

If $\{\psi_\ell\}_{\ell=1}^M$ is an orthonormal basis for $\mathrm{Ker}(A_K)$, we have $\{\varphi_j\}_{j=1}^N \cup \{\psi_\ell\}_{\ell=1}^M$ is an orthonormal basis for L^2.

As noted already, A_K maps L^2 to $C(X)$, so we can define $(A_K f)(x)$ for all $x \in \mathrm{supp}(d\mu)$ by (3.11.1) and it is continuous in x. We have

Lemma 3.11.10. (a) *For fixed $x_0 \in X$, we have that as $J \to N$*

$$\sum_{n=1}^{J} \mu_n(A)\, \overline{\varphi_n(\cdot)}\, \varphi_n(x_0) \to K(x_0, \cdot\,) \tag{3.11.40}$$

as functions in $L^2(X, d\mu)$.

(b) *For each* $x_0 \in \operatorname{supp}(d\mu)$,

$$\sum_{n=1}^{N} \mu_n(A)|\varphi_n(x_0)|^2 \leq K(x_0, x_0) \tag{3.11.41}$$

(c) *For each* $x_0 \in \operatorname{supp}(d\mu)$, *the convergence in* (3.11.40) *is in* $\|\cdot\|_\infty$.
(d) *Equality holds in* (3.11.41) *for each* x_0.
(e) *The convergence of the sum in* (3.11.41) *in* x_0 *is uniform in* $x_0 \in \operatorname{supp}(d\mu)$.

Remark. $K(x,y)$ is irrelevant for the definition of A_K on $(X \times X) \setminus (\operatorname{supp}(d\mu) \times \operatorname{supp}(d\mu))$, so its values there are arbitrary and need not be related to $\varphi_n(x)\varphi_n(y)$ there. That explains our various references to $x \in \operatorname{supp}(d\mu)$. It would be simpler if we supposed $\operatorname{supp}(d\mu) = X$, which is no loss since we can restrict to the compact set $\operatorname{supp}(d\mu)$.

Proof. (a) $K_{x_0} \equiv K(x_0, \cdot\,) \in L^2(X, d\mu)$, so with L^2-convergence,

$$K_{x_0}(\,\cdot\,) = \sum_{n=1}^{N} \langle \bar{\varphi}_n, K_{x_0}\rangle \bar{\varphi}_n(\,\cdot\,) \tag{3.11.42}$$

If $\psi \in \operatorname{Ker}(A_K)$, $\int K(x,y)\psi(y)\, d\mu(y) = 0$ for a.e. $x \in \operatorname{supp}(d\mu)$ and so, by continuity, for all such x. Thus, $\langle \bar{\psi}, K_{x_0}\rangle = 0$, which is why we have no ψ terms in the Fourier expansion (3.11.42).

By (3.11.39) and (3.11.42) and the fact that, as noted, (3.11.1) holds for all $x \in \operatorname{supp}(d\mu)$,

$$\langle \bar{\varphi}_n, K_{x_0}\rangle = \int K(x_0, y)\varphi_n(y)\, d\mu(y) = (A_K\varphi_n)(x_0) = \mu_n(A)\varphi_n(x_0) \tag{3.11.43}$$

so (3.11.42) is (3.11.40).

(b) Fix $x_0 \in \operatorname{supp}(d\mu)$. Let $U_m = \{y \mid \operatorname{dist}(x_0, y) < \frac{1}{m}\}$. Since $x_0 \in \operatorname{supp}(d\mu)$, $\mu(U_m) \neq 0$ and we can define $\eta_m = \mu(U_m)^{-1}\chi_{U_m}$. By continuity of K and φ_j,

$$\lim_{n\to\infty} \langle \eta_m, A_K\eta_m\rangle = K(x_0, x_0), \qquad \lim_{n\to\infty} \langle \eta_m, \varphi_j\rangle = \varphi_j(x_0) \tag{3.11.44}$$

For each $J < \infty$, by (3.11.39),

$$\langle \eta_m, A_K\eta_m\rangle \geq \sum_{n=1}^{J} \mu_n(A)|\langle \eta_m, \varphi_n\rangle|^2 \tag{3.11.45}$$

By (3.11.41), taking $m \to \infty$, we get (3.11.41).

(c) Call the left side of (3.11.40) $K_J(x_0, \cdot\,)$. Then by the Schwarz inequality,

$$\|K(x_0, \cdot\,) - K_J(x_0, \cdot\,)\|_\infty \le \biggl(\sum_{n=J+1}^{\infty} \mu_n(A)|\varphi_n(x_0)|^2 \biggr)^{1/2} \|K\|_\infty \qquad (3.11.46)$$

since

$$\sum_{n=J+1}^{\infty} \mu_n(A)|\varphi_n(y)|^2 \le K(y,y) \le \|K\|_\infty \qquad (3.11.47)$$

Thus, $\|K(x_0, \cdot\,) - K_J(x_0, \cdot\,)\|_\infty \to 0$.

(d) This is immediate from (c) for $\cdot = x_0$.

(e) This is essentially Dini's theorem (see Problem 6 of Section 4.4 of Part 1) since $K_J(x_0, x_0)$ is monotone in J. Here is the argument. Given x_0 and ε, first find N_{x_0} so $|K_{N_{x_0}}(x_0,x_0) - K(x_0,x_0)| < \varepsilon/2$. $G_N(x) \equiv K_N(x,x) - K(x,x)$ is continuous in x, so find V_{x_0} so that for $y \in V_{x_0}$, $|G_{N_x}(y) - G_{N_x}(x_0)| < \varepsilon/2$. Thus, in V_x, $|G_n(x_0)| < \varepsilon$ for all $n \ge N_x$ (by monotonicity of $G_N(x)$ in N). By compactness, there is N_ε so $|G_n(x)| < \varepsilon$ for all $n \ge N_\varepsilon$ and all x. □

Proof of Theorem 3.11.9. (a), (b) By Lemma 3.11.10(d), for each x,

$$K(x_0, x_0) = \sum_{n=1}^{N} \mu_n(A_K)|\varphi_n(x_0)|^2 \qquad (3.11.48)$$

Integrating $d\mu(x_0)$ and using positivity to justify exchanges of sum and integral, we conclude

$$\sum_{n=1}^{N} \mu_n(A_K) = \int K(x_0,x_0)\, d\mu(x_0) < \infty \qquad (3.11.49)$$

Thus, A_K is trace class and the trace is given by (3.11.37).

(c) We've seen all eigenfunctions with nonzero eigenvalue are continuous. Uniform convergence of $K - K_J$ is immediate by (3.11.46) and the uniform convergence of $K(x_0,x_0) - K_J(x_0,x_0)$. □

A_K depends, of course, on K and μ. By Problem 5 of Section 3.3 of Part 1, K is a reproducing kernel. By Problem 6 of Section 3.3 of Part 1, K is an L^2 reproducing kernel if and only if there is a μ so that $\sigma(A_K) = \{0,1\}$, that is, so that $A_K^2 = A_K$.

Notes and Historical Remarks. Most of this section realizing the abstract theory when kernels are continuous is folk wisdom, although the basic formulae (3.11.3)/(3.11.4) go back to Fredholm's famous paper [**206**]. Mercer's theorem was proven by James Mercer (1883–1932) in 1909 [**473**].

To write down the formula for the $N(\ell_j, \dots, \ell_m)$ of (3.11.36), it is convenient to parametrize in a different way. Pick $k_1, \dots, k_n$ so

$$\sum_{J=1}^{n} jk_j = n \tag{3.11.50}$$

and associate to $k_1, \dots, k_n$, k_1 ℓ's of length 1, k_2 ℓ's of length 2, ... Then (see, e.g., [**645**, Sec. VI.1]) define

$$\widetilde{N}(k_1, \dots, k_n) = \frac{n!}{\prod_{j=1}^{n} (k_j!)(j)^{k_j}} \tag{3.11.51}$$

and (3.11.36) becomes

$$\alpha_n(A) = \sum_{\{k_j \text{ obeying } (3.11.50)\}} (-1)^{n-\sum_{j=1}^{n} k_j} \widetilde{N}(k_1, \dots, k_n) \prod_{j=1}^{n} \operatorname{Tr}(A^j)^{k_j} \tag{3.11.52}$$

There is considerable literature on doing Fredholm theory in general Banach spaces, summarized in the monograph of Ruston [**586**].

Problems

1. This will refine the arguments in the proof of Theorem 3.11.5 to show if $-z_0^{-1}$ is an eigenvalue of A, then the order of the zeros of $d_K(z)$ is the algebraic multiplicity of that eigenvalue.

   (a) Let A_n, A be compact and $\|A_n - A\| \to 0$. Let $\lambda_0 \neq 0$ with $\lambda_0 \in \sigma(A)$. Prove that if n is large and the eigenvalues of A_n near λ_0 are $\{\lambda_j^{(n)}\}_{j=1}^{N_n}$, then the sum of the algebraic multiplicities of the $\lambda_j^{(n)}$ is, for n large, the algebraic multiplicity of λ_0.

   (b) Prove the order of zeros of $d_K(A)$ at $-z_0^{-1}$ is the algebraic multiplicity. (*Hint*: Hurwitz's theorem or the argument principle.)

2. (a) Let A be a compact operator and $z_0 \neq 0$. If $-z_0^{-1}$ is an eigenvalue of algebraic multiplicity 1, prove that $\lim_{z \to z_0}(z - z_0)(1 + zA)^{-1}$ is rank 1 and its range is the set of solutions of $(1 + z_0 A)f = 0$.

   (b) If $d_K(z)$ has a simple zero at $z = z_0$, prove that $M_K(z_0)$ is rank 1 and its range is the solutions of $(1 + z_0 A_K)f = 0$.

3. Complete the proof of (3.11.33).

4. This will prove Proposition 3.11.8.

   (a) If K is a continuous kernel on a compact metric space and $A_K \geq 0$ as an operator, prove for $\{x_j\}_{j=1}^n \subset \operatorname{supp}(d\mu)$ and $\{z_j\}_{j=1}^n \subset \mathbb{C}$, we have that

$$\sum \bar{z}_i z_j K(x_i, x_j) \geq 0 \tag{3.11.53}$$

(*Hint*: Let $U_j^{(m)} = \{x \mid \rho(x, x_j) \leq \frac{1}{m}\}$, $f_j^{(m)} = \mu(U_j^{(m)})\chi_{U_j^{(m)}}$. Prove that, as $m \to \infty$, the left side of (3.11.53) is $\langle \sum_{i=1}^n z_i f_j^{(m)}, A_K(\sum_{j=1}^n z_j f_j^{(m)})\rangle$.)

(b) If K is a continuous kernel and (3.11.53) holds, prove A_K is a positive operator. (*Hint*: Prove first that $\int K(x,y)\, d\eta(x) d\eta(y) \geq 0$ for any complex measure, $d\eta$, on $\mathrm{supp}(d\mu)$.)

3.12. Lidskii's Theorem

Our main goal in this section is to prove Lidskii's theorem that $A \in \mathcal{I}_1$ implies (each λ_j is repeated to count its algebraic multiplicity)

$$\mathrm{Tr}(A) = \sum_{j=1}^{N(A)} \lambda_j(A) \tag{3.12.1}$$

We recall the following sharp Hadamard factorization theorem (Theorem 9.10.9 of Part 2A, specialized to order 1): If f is an entire function with zeros (counting multiplicity), $\{z_j\}_{j=1}^\infty$, $f(0) \neq 0$, and

(i) $\forall \varepsilon > 0, \exists C_\varepsilon$ so $|f(z)| \leq C_\varepsilon \exp(\varepsilon|z|)$

$$\text{(ii)} \ \sum_j |z_j|^{-1} < \infty \tag{3.12.2}$$

then

$$f(z) = f(0) \prod_{j=1}^\infty \left(1 - \frac{z}{z_j}\right) \tag{3.12.3}$$

Lemma 3.12.1. *Let $A \in \mathcal{I}_1$ and $f(z) = \det(1 + zA)$. Then for any ε, there is a C_ε so that*

$$|f(z)| \leq C_\varepsilon \exp(\varepsilon|z|) \tag{3.12.4}$$

Proof. Fix $\varepsilon > 0$. Since $\sum_{k=1}^\infty \mu_k(A) < \infty$, pick N so

$$\sum_{k=N+1}^\infty \mu_k(A) < \frac{\varepsilon}{2} \tag{3.12.5}$$

By (3.10.15) and $1 + y \leq e^y$ for $y \geq 0$,

$$|f(z)| \leq \prod_{j=1}^\infty (1 + |z|\mu_k(A)) \tag{3.12.6}$$

$$\leq \left[\prod_{j=1}^N (1 + |z|\mu_k(A))\right] e^{\varepsilon|z|/2} \tag{3.12.7}$$

The first factor in (3.12.7) is a polynomial in $|z|$, so it times $e^{-\varepsilon|z|/2}$ goes to zero as $|z| \to \infty$. Thus, this product has a finite sup, C_ε, over all $|z|$ in $[0, \infty)$. □

Theorem 3.12.2. *Let $A \in \mathcal{I}_1$. Let $\{\lambda_k(A)\}_{k=1}^{N(A)}$ be the eigenvalues of A (counted up to algebraic multiplicity). Then*

$$\det(1 + zA) = \prod_{k=1}^{N(A)} (1 + z\lambda_k(A)) \tag{3.12.8}$$

Proof. Let $f(z) = \det(1 + zA)$. By Theorem 3.10.5, $f(z)$ vanishes only at z's for which $1+zA$ has a nonzero kernel, that is, A has $-z^{-1}$ as an eigenvalue and the multiplicity of the zero is the multiplicity of the eigenvalue, that is, $z_k = -\lambda_k(A)^{-1}$ are the zeros of f. By the Schur–Lalesco–Weyl inequality (Theorem 3.9.1), we have $\sum_k |z_k|^{-1} < \infty$.

Since $f(0) = 1$ and we have (3.12.4), the refined Hadamard theorem applies. Thus, (3.12.3) is (3.12.8). □

Corollary 3.12.3 (Lidskii's Theorem). *If $A \in \mathcal{I}_1$, then* (3.12.1) *holds.*

Proof. We have

$$\text{LHS of } (3.12.8) = 1 + z\,\mathrm{Tr}(A) + O(z^2) \tag{3.12.9}$$

$$\text{RHS of } (3.12.8) = 1 + z\sum_{k=1}^{N(A)} \lambda_k(A) + O(z^2) \tag{3.12.10}$$

so (3.12.8) implies (3.12.1). □

Notes and Historical Remarks. Let $n = 2, 3, \dots$ and $A \in \mathcal{I}_n$. Then $A^n \in \mathcal{I}_1$, so Lidskii's theorem implies

$$A \in \mathcal{I}_n, \quad n = 2, \dots \quad \Rightarrow \quad \sum_{k=1}^{N(A)} \lambda_k(A)^n = \mathrm{Tr}(A^n) \tag{3.12.11}$$

However, this result does not imply the $\mathcal{I}_1$ result, for while any $A \in \mathcal{I}_1$ is a product of two $\mathcal{I}_2$ operators, it is not necessarily the square of a single $\mathcal{I}_2$ operator (e.g., $\left(\begin{smallmatrix} 0 & 1 \\ 0 & 0 \end{smallmatrix}\right)$ is not a square).

One might think (3.12.8) is much stronger than Lidskii's theorem because Lidskii's theorem comes from first-order Taylor coefficients. There are infinitely many Taylor coefficients, but the n-th coefficient equality is just (3.12.11)! Thus, Lidskii's theorem implies (3.12.8).

As first sight, it appears that Lidskii's theorem should follow from the existence of Schur bases since they have $(e_n, Ae_n) = \lambda_n(A)$, and $\mathrm{Tr}(A)$ should be $\sum_n (e_n, Ae_n)$. This doesn't work because, despite its name, a

Schur basis might not be a basis! Indeed, the hard case of Lidskii's theorem is to show that quasinilpotent trace class A's (i.e., with $\sigma(A) = \{0\}$) have zero trace.

That Lidskii's theorem is subtle is shown by an example of Grothendieck [**268**]. As discussed in the Notes to Section 3.6, any Banach space has a natural class of operators, the nuclear operators, N_1, which have a trace. Moreover, $N_1 = \mathcal{I}_1$ on a Hilbert space (Problem 4 of Section 3.5). So one can ask if Lidskii's theorem extends to arbitrary Banach spaces. What Grothendieck does is find $A \in N_1(\ell^1)$ so $\text{Tr}(A) = 1$ but $A^2 = 0$ (so the only eigenvalue is 0). Lidskii's theorem does not extend to ℓ^1!

In 1921, Carleman [**101**] proved for a Hilbert–Schmidt kernel (which we now know is just $\mathcal{I}_2$), one has a "Fredholm determinant," which in modern language is $\det_2(1+A) \equiv \det((1+A)e^{-A})$ (see the discussion in Section 3.13), so that

$$\det((1+zA)e^{-zA}) = \prod_{j=1}^{N(A)} (1+z\lambda_j)e^{-z\lambda_j} \tag{3.12.12}$$

While Carleman defined the left side of (3.12.12) via a Fredholm series, he has

$$\det\big((1+zA)e^{-zA}\big) = 1 - \frac{z^2}{2}\,\text{Tr}(A^2) + O(z^3)$$

So implicit in Carleman's (3.12.12) is (3.12.11) for $n = 2$.

The full Lidskii theorem appeared first in Grothendieck [**269**] in 1956 by a method close to the one we use here. Because this paper was on Banach space theory and Lidskii's theorem an aside, his work on this result was not widely known to those working on Hilbert space operator theory. In 1959, Lidskii [**445**], unaware of Grothendieck, rediscovered the theorem, and in a case of Arnol'd's principle, got the theorem named after him.

Dunford–Schwartz [**175**] (who seemed not to know of either Grothendieck or Lidskii) have a proof of Lidskii's theorem that proves it first for quasinilpotent operators by a proof close to our general proof here and then use a Schur basis argument to obtain the full result. To be explicit, they look at the closure of the span of $\bigcup_{\lambda\neq 0,\, \lambda\in\sigma(A)} P_\lambda$, the space we called $\mathcal{H}_1$ in the proof of Theorem 3.4.1. Let $\{e_j\}_{j=1}^J$ be a Schur basis for $A \restriction \mathcal{H}_1$ and $\{f_k\}_{k=1}^K$ an orthonormal basis for $\mathcal{H}_1^\perp$. By construction of the Schur basis

$$\sum_{j=1}^J \langle e_j, Ae_j\rangle = \sum_{j=1}^J \lambda_j(A) \tag{3.12.13}$$

Next they note that since $A\colon \mathcal{H}_1 \to \mathcal{H}_1$, $A^*\colon \mathcal{H}_1^\perp \to \mathcal{H}_1^\perp$. Moreover, since $P_\lambda(A^*) = P_{\bar\lambda}(A)^*$, if $A^*g = \lambda g$ for $\lambda \in \sigma(A^*)$ and $g \in \mathcal{H}_1^\perp$, then

$$\|g\|^2 = \langle g, P_\lambda(A^*)g\rangle = \langle P_{\bar\lambda}(A)g, g\rangle = 0$$

since $\operatorname{Ran} P_{\bar\lambda}(A) \in \mathcal{H}_1$ and $g \in \mathcal{H}_1^\perp$. Thus $A^* \restriction \mathcal{H}_1^\perp$ is quasinilpotent. Moreover, since A^* is trace class, so is $A^* \restriction \mathcal{H}_1^\perp$. If one has Lidskii's theorem for such quasinilpotent operators,

$$\begin{aligned} 0 = \overline{\operatorname{Tr}(A^*)} &= \sum_{k=1}^{K} \langle A^* f_k, f_k\rangle \\ &= \sum_{k=1}^{K} \langle f_k, Af_k\rangle \end{aligned}$$

so by (3.12.13) and completeness of $\{e_j\}_{j=1}^J \cup \{f_k\}_{k=1}^K$, we get Lidskii's theorem for A.

The point is that once one has $\det(1+B)$ for trace class B, if A is quasinilpotent, $\det(1+zA)$ is entire and nonvanishing, so $g(z) = \log\left[\det(1+zA)\right]$ is entire and, by (3.12.4), $\limsup |z|^{-1}\log(z) \le 0$ so $g'(z) = 0$ (by an improved Cauchy estimate), i.e., $\det(1+zA) = 0 \Rightarrow \operatorname{Tr}(A) = 0$.

We note that in place of the argument showing $A^* \restriction \mathcal{H}_1^\perp$, we can use Theorem 3.4.1 and so get a simple proof of Lidskii's theorem.

Simon [**635**] rediscovered Grothendieck's proof. Leiterer Pietsch have a proof via approximation by finite rank operators—they didn't publish it but it is given in the books by König [**401**] and Garling [**221**]. Erdos [**188**] has a proof that relies on nests of invariant subspaces which he developed in [**187**]. There is a large literature on extending Lidskii's theorem to suitable Banach spaces. Pisier [**525**] is an earlier paper on this subject and Reinov [**553**] is a current one with many references to earlier works on this subject. Goh'berg et al. [**258**] discuss determinants and trace on fairly general Banach spaces.

3.13. Bonus Section: Regularized Determinants

$\det(1+zA)$ is only defined for $A \in \mathcal{I}_1$ and is generally infinite for $A \in \mathcal{I}_\infty \setminus \mathcal{I}_1$ with $A \ge 0$, $z > 0$ (Problem 1). In this section, we'll define a set of regularized determinants, $\det_n(1+zA)$, $n = 2, 3, \dots$, defined for $A \in \mathcal{I}_n$ (and so for $A \in \mathcal{I}_p$, $1 \le p \le n$) with the critical property $1 + A$ is invertible $\Leftrightarrow \det_n(1+A) \ne 0$. But we'll lose the product rule. Here are the related ways of understanding what we'll do.

Lidskii's theorem says for $A \in \mathcal{I}_1$, we have that

$$\det(1+zA) = \prod_{j=1}^{N(A)} (1+z\lambda_j(A)) \tag{3.13.1}$$

which we realize as a canonical product with zeros exactly at $\{z_1 \equiv -\lambda_j^{-1}(A)\}_{j=1}^{N(A)}$. Since $\sum_{j=1}^{N(A)} |z_j|^{-1} < \infty$, this product converges. If $\sum_{j=1}^{N(A)} |z_j|^{-1} \equiv \infty$ but $\sum_{j=1}^{N(A)} |z_j|^{-n} < \infty$, we know we should use a product of Weierstrass factors. This is precisely what $\det_n$ will do. While we'll define it differently, we'll prove that

$$\det_n(1+zA) = \prod_{j=1}^{N(A)} (1+z\lambda_j(A)) e^{\sum_{k=1}^{n-1}(-1)^k \lambda_j(A)^k z^k/k} \tag{3.13.2}$$

Another way of understanding $\det_n$ is in terms of (3.10.44), which says for A small,

$$\det(1+A) = \exp\biggl(\frac{\sum_{k=1}^{\infty}(-1)^{k+1}\mathrm{Tr}(A^k)}{k}\biggr) \tag{3.13.3}$$

If $A \in \mathcal{I}_n$, in general $\{\mathrm{Tr}(A^k) \mid k = 1, \dots, n-1\}$ are problematic, so divide them out (which won't affect vanishing if $A \in \mathcal{I}_1$, since the exponential is never zero). We'll prove

Theorem 3.13.1. *Let $n = 2, 3, \dots$. If $\det_n$ is defined on $\mathcal{I}_1$ by*

$$\det_n(1+A) = \det(1+A)\exp\biggl(\frac{\sum_{k=1}^{n-1}(-1)^k\mathrm{Tr}(A^k)}{k}\biggr) \tag{3.13.4}$$

then $\det_n$ has a continuous extension to $\mathcal{I}_n$ and $1+A$ is invertible if and only if $\det_n(1+A) \neq 0$.

We won't define $\det_n$ by (3.13.4) but rather derive this from our actual definitions, but it explains our intuition.

Definition. Let $n = 2, 3, \dots$. Given a bounded operator, A, on a Hilbert space, $\mathcal{H}$, define $\mathcal{R}_n(A)$ by

$$\mathcal{R}_n(A) = \biggl[(\mathbb{1}+A)\exp\biggl(\frac{\sum_{k=1}^{n-1}(-1)^k A^k}{k}\biggr)\biggr] - \mathbb{1} \tag{3.13.5}$$

Lemma 3.13.2. (a) *If $A \in \mathcal{I}_n$, then $\mathcal{R}_n(A) \in \mathcal{I}_1$.*

(b) *For any $K \in (0,\infty)$, there is $C_K^{(n)}$ so that*

$$\|A\| \le K, \quad A \in \mathcal{I}_n \Rightarrow \|\mathcal{R}_n(A)\|_1 \le C_K^{(n)} \|A\|_n^n \tag{3.13.6}$$

(c) *For $A_1, \dots, A_\ell \in \mathcal{I}_n$, the map $F_\ell(z_1, \dots, z_\ell) = \mathcal{R}_n(\sum_{k=1}^{\ell} z_j A_j)$ is an entire analytic function of $\mathbb{C}^\ell$ to $\mathcal{I}_1$.*

(d) *For* $\|A\| \le K$, $\|B\| \le K$, $A, B \in \mathcal{I}_n$, *we have*

$$\|\mathcal{R}_n(A) - \mathcal{R}_n(B)\|_1 \le \|A - B\|_n C_{2K+1}^{(n)}(\|A\|_n + \|B\|_n + 1)^n \tag{3.13.7}$$

Remark. f_n below is closely related to the Weierstrass factor, E_{n-1}, of (9.4.2) of Part 2A; the factorization in (3.13.9) was used in Proposition 9.4.2 of Part 2A.

Proof. (a) Let

$$f_n(z) = (1+z)\exp\biggl(\frac{\sum_{k=1}^{n-1}(-1)^k z^k}{k}\biggr) - 1 \tag{3.13.8}$$

f_n is clearly an entire function with $f_n(0) = 0$. Moreover,

$$\begin{aligned} f_n'(z) &= \bigl[1 + (1+z)(-1 + z + \dots + (-1)^{n-1} z^{n-2})\bigr] \exp\biggl(\frac{\sum_{k=1}^{n-1}(-1)^k z^k}{k}\biggr) \\ &= (-1)^{n-1} z^{n-1} \exp\biggl(\frac{\sum_{k=1}^{n-1}(-1)^k z^k}{k}\biggr) \end{aligned} \tag{3.13.9}$$

by telescoping. Thus, there is an entire function, $h_n(z)$, so

$$f_n(z) = z^n h_n(z) \tag{3.13.10}$$

It follows that $\mathcal{R}_n(A) = A^n h_n(A)$ so, by Theorem 3.7.6, $\mathcal{R}_n(A) \subset \mathcal{I}_n$, and by (3.7.42),

$$\|\mathcal{R}_n(A)\|_1 \le \|A\|_n^n \|h_n(A)\| \tag{3.13.11}$$

(b) Let $h_n(A) = \sum_{\ell=0}^\infty a_\ell^{(n)} z^\ell$ and

$$C_K^n = \sum_{\ell=0}^\infty |a_\ell^{(n)}| K^n \tag{3.13.12}$$

(3.13.6) follows from (3.13.11).

(c) Let $f_n^{(L)}$ be (3.13.8) with $\exp(\,\cdot\,)$ replaced by $\sum_{j=0}^L \frac{1}{j!}(\cdot)^j$. If $\mathcal{R}_n^{(L)}$ is defined with f_n replaced by $f_n^{(L)}$, then $F_\ell^{(L)}(z_1, \dots, z_\ell) \equiv f_n^{(L)}(\sum_{j=1}^\ell z_j A_j)$ is a polynomial in the A's and so analytic. By the bounds in (3.13.6), $F_\ell^{(L)}(z_1, \dots, z_\ell)$ is bounded uniformly in L and in the z's in compacts of $\mathbb{C}^\ell$ and converges for each fixed $(z_1, \dots, z_\ell)$ as $L \to \infty$, so we get the analyticity result.

(d) A straightforward application of Lemma 3.10.3 (Problem 2). □

Thus, we define for $A \in \mathcal{I}_n$,

$$\det_n(1 + A) = \det(1 + \mathcal{R}_n(A)) \tag{3.13.13}$$

Theorem 3.13.3. $\det_n$ *has the following properties for* $A, B \in \mathcal{I}_n$.

(a) (3.13.2) *holds.*

(b) *For a constant* Γ_n, *we have*

$$|\det_n(1+A)| \le \exp(\Gamma_n \|A\|_n^n) \tag{3.13.14}$$

(c) $|\det_n(1+A) - \det_n(1+B)| \le \|A-B\|_n \exp(\Gamma_n(\|A\|_n + \|B\|_n + 1)^n)$ (3.13.15)

(d) *For* $A \in \mathcal{I}_1$, (3.13.4) *holds.*

(e) $1+A$ *is invertible if and only if* $\det_n(1+A) \neq 0$. *If* $1+A$ *is not invertible, then* $\det_n(1+zA)$ *has a zero at* $z=1$ *of order equal to the algebraic multiplicity of* -1 *as an eigenvalue of* A.

Remarks. 1. (d), (c) prove Theorem 3.13.1.

2. We lose the analog of $\det(1+A)\det(1+B) = \det(1+A+B+AB)$, but there are replacements (Problems 3 and 4), for example,

$$\det_2(1+A)\det_2(1+B) = \det_2(1+A+B+AB)e^{\mathrm{Tr}(AB)} \tag{3.13.16}$$

for $A, B \in \mathcal{I}_2$.

Proof. (a) By the spectral mapping theorem for entire functions, the eigenvalues of $\mathcal{R}_n(zA)$ are $f_n(\lambda_j(A))$ with f_n given by (3.13.8), so (3.13.2) follows from Lidskii's theorem in the form (3.12.8).

(b) $1+f_n(z)$ defined by (3.13.8) clearly obeys

$$|1+f_n(z)| \le \exp\biggl(\frac{|z| + \sum_{k=1}^{n-1}|z|^k}{k}\biggr) \le \exp(\Gamma_n^{(1)}|z|^n) \tag{3.13.17}$$

for $|z| \ge 1$. By (3.13.10), for $|z| \le 1$,

$$|1+f_n(z)| \le 1 + \Gamma_n^{(2)}|z|^n \le \exp(\Gamma_n^{(2)}|z|^n) \tag{3.13.18}$$

so for all z and $\Gamma_n = \max(\Gamma_n^{(1)}, \Gamma_n^{(2)})$,

$$|1+f_n(z)| \le \exp(\Gamma_n |z|^n) \tag{3.13.19}$$

Therefore, by (3.13.2),

$$|\det_n(1+A)| \le \exp\biggl(\Gamma_n \sum_{j=1}^{N(A)} |\lambda_j(A)|^n\biggr) \tag{3.13.20}$$

$$\le \exp(\Gamma_n \|A\|_n^n) \tag{3.13.21}$$

by (3.9.2).

(c) is a direct application of Lemma 3.10.3 (Problem 5).

(d) is an immediate consequence of (3.13.2) and Lidskii's theorem.

(e) By (3.13.2), $\det(1+A)$ is zero if and only if $1+\lambda_j(A)=0$ for some j if and only if $1+A$ is not invertible (since A is compact). The order of the zero of $\det(1+zA)$ is the number of j's for which $\lambda_j(A)$ is -1. $\square$

If $A \in \mathcal{I}_1$, $\det_n(1+zA)$ for small z is just $\exp(\sum_{k=n}^\infty (-1)^{k+1}\operatorname{Tr}(A^k)/k)$. Its Taylor coefficients are thus the same polynomials in $\{\operatorname{Tr}(A^j)\}_{j=1}^m$ as for $\det(1+zA)$, but with each $\operatorname{Tr}(A^j)$, $j=1,\dots,n-1$ replaced by zero. As we saw (see Proposition 3.10.9), any two polynomial expressions are the same. Thus, we have the following pair of results:

Theorem 3.13.4 (Plemelj–Smithies Formulae for $\det_n(1+zA)$)**.** *We have that*

$$\det_n(1+zA) = \sum_{m=0}^\infty \frac{z^m \alpha_m^{(n)}(A)}{m!} \tag{3.13.22}$$

where $\alpha_m^{(n)}(A)$ is given by (3.10.40), except that $\{\operatorname{Tr}(A^j)\}_{j=1}^{n-1}$ is replaced by 0. In particular,

$$\alpha_m^{(n)}(A) = 0 \quad \textit{for } m \le n-1 \tag{3.13.23}$$

Theorem 3.13.5 (Fredholm Formulae for $\det_n(1+zA)$)**.** *If K is a continuous integral kernel on a compact metric space, X, with Borel measure μ, then $\det_n(1+zA)$ is given by the series (3.11.4), except that $K\binom{x_1\dots x_m}{x_1\dots x_m}$ is replaced by the sum (with $(-1)^\pi$) over only those permutations with no $1,2,\dots,n-1$ cycles. In particular, $\det_2$ is given by $\det(K(x_i,x_j))_{1\le i\le j\le m}$ with $K(x_j,x_j)$ set to 0.*

Corollary 3.13.6 (Hilbert–Fredholm Kernel)**.** *Let K be a Hilbert–Schmidt kernel on $L^2(M,d\mu)$. Then $\det_2(1+A_K)$ is given by the series (3.11.4), where*

$$K_m\begin{pmatrix} x_1\dots x_m \\ x_1\dots x_m\end{pmatrix} = \begin{vmatrix} 0 & K(x_1,x_2) & \dots & K(x_1,x_m) \\ K(x_2,x_1) & 0 & \dots & K_2(x_2,x_m) \\ \vdots & \vdots & \ddots & \vdots \\ K(x_m,x_1) & K(x_1,x_2) & \dots & 0 \end{vmatrix} \tag{3.13.24}$$

Remark. $K(x,y)$ is only determined for $d\mu\otimes d\mu$ a.e. x. Thus, $K(x,x)$ is not well-defined, even for a.e. x, but $K(x,y)K(y,x)$ is.

Proof. $\det_2(1+A_K)$ is given by a sum of polynomials in $\{\operatorname{Tr}(A^j)\}_{j=2}^\infty$. These are exactly the ones from Fredholm theory, except all $\operatorname{Tr}(A)$ terms are dropped. These are precisely the $K(x,x)$ terms, that is, we just set the diagonal to zero. $\square$

Notes and Historical Remarks. From the earliest days of understanding and extending Fredholm theory, it was clear that one wanted to go beyond continuous kernels. In applications, kernels were often infinite on the diagonal due to a power law divergence, as in a power of Green's function. But they were usually Hilbert–Schmidt. In the first part of his series on integral equations, Hilbert [**317**] discovered Fredholm's formula still gave a solution if $K(x_j, x_j)$ is replaced by 0. Even though he didn't realize the general $\det_2$ format, the resulting formulae are called Hilbert–Fredholm formulae.

Poincaré [**533**] had an approach to A with singular integral kernels where A^2 has a continuous kernel based on $(1-A)^{-1} = (1+A)(1-A^2)^{-1}$. It has a problem of extraneous zeros of $\det(1-A^2) = 0$ if $1-A$ is invertible but $1+A$ is not. In some sense, the $\mathcal{R}_n(A)$ approach of this section can be viewed as a variant of Poincaré without extraneous zeros!

A major advance in understanding Fredholm theory for L^2-kernels is Carleman's great 1921 paper [**101**] where he defined a modified determinant using Hilbert–Fredholm formulae and proved that it obeyed (3.12.2), thereby foretelling Lidskii's theorem!

Other important contributions to Fredholm theory for nontrace class operators are works of Hille–Tamarkin [**321**], Smithies [**658**], Dunford–Schwartz [**175**], and Goh'berg–Krein [**257**].

The approach we follow using $\mathcal{R}_n(A)$ is due to Seiler [**617**], with follow-up by Seiler–Simon [**618**] and Simon [**635, 650**]. The two Seiler papers involved application of the theory to construct certain quantum fields in two space-time dimensions and the extra $e^{-\mathrm{Tr}(A)}$ terms are related to quantum renormalization!

Problems

1. Let A be a compact operator on a Hilbert space, $\mathcal{H}$. Suppose $A \geq 0$, P_N is the projection onto the space of the eigenvectors for $\{\lambda_j(A)\}_{j=1}^N$ and $A_N = P_N A P_N$. For all $z > 0$, prove that $\lim_{N\to\infty} \det(1 + zA_N) = \infty$ if $A \notin \mathcal{I}_1$.

2. Prove (3.13.7) using Lemma 3.10.3.

3. Prove (3.13.16). (*Hint*: Prove it first for $A, B \in \mathcal{I}_1$.)

4. Find the analog of (3.13.16) for $\det_n$.

5. Prove (3.13.15) using Lemma 3.10.3.

3.14. Bonus Section: Weyl's Invariance Theorem

Definition. Let A be an operator on a Banach space, X. We define $\sigma_d(A)$, the *discrete spectrum* of A, to be the isolated points, λ_0, of $\sigma(A)$, where the spectral projection, $P = (2\pi i)^{-1} \oint_{|\lambda-\lambda_0|=\varepsilon} (\lambda - A)^{-1}\, d\lambda$ is finite rank. The *essential spectrum*, $\sigma_{\text{ess}}(A)$, is defined to be $\sigma(A) \setminus \sigma_d(A)$.

Remarks. 1. We already discussed the discrete spectrum in Section 2.3.

2. There are distinct definitions of $\sigma_{\text{ess}}(A)$ in the literature; see the Notes. For the main result of this section, what is critical is that all definitions of $\sigma_{\text{ess}}(A)$ agree if A is a bounded self-adjoint operator on a Hilbert space.

3. For infinite-dimensional Banach spaces, $\sigma_{\text{ess}}(A)$, as we defined it, is always nonempty (Problem 1).

A main result of this section is

Theorem 3.14.1 (Weyl's Invariance Theorem). *Let A be a bounded self-adjoint operator on a Hilbert space, $\mathcal{H}$, and let C be a self-adjoint compact operator. Then*

$$\sigma_{\text{ess}}(A + C) = \sigma_{\text{ess}}(A) \tag{3.14.1}$$

Remarks. 1. That A is self-adjoint is not critical; what is are the two conditions (interior in $\mathbb{C}$):

$$\sigma_{\text{ess}}(A)^{\text{int}} = \emptyset, \qquad \mathbb{C} \setminus \sigma_{\text{ess}}(A) \text{ is connected} \tag{3.14.2}$$

These hold automatically if $\sigma_{\text{ess}}(A)$ is a bounded subset of $\mathbb{R}$.

2. See Section 7.8 for extensions to unbounded self-adjoint A.

Example 3.14.2. This example shows what can happen if (3.14.2) fails. Let $\mathcal{H} = \ell^2(\mathbb{Z})$, square summable sequences indexed by $\mathbb{Z}$. Let $\{\delta_n\}_{n\in\mathbb{Z}}$ be the standard basis. Let A be a unitary operator,

$$A\varphi_n = \varphi_{n+1} \tag{3.14.3}$$

It is easy to see (Problem 2) that

$$\sigma(A) = \sigma_{\text{ess}}(A) = \partial\mathbb{D} \tag{3.14.4}$$

We say $\sigma(A) = \sigma_{\text{ess}}(A)$ since $\sigma(A)$ has no isolated points. Define B by

$$B\varphi_n = \begin{cases} 0 & \text{if } n = -1 \\ \varphi_{n+1} & \text{if } n \neq -1 \end{cases} \tag{3.14.5}$$

Then under the direct sum decomposition $\ell^2(\mathbb{Z}) = \ell^2(\{-1, -2, \dots\}) \oplus \ell^2(\{0, 1, 2, \dots\})$, we see $B \cong \mathrm{R} \oplus \mathrm{L}$, where R (respectively, L) is right (respectively, left) shift. As we saw in Example 2.2.5,

$$\sigma(\mathrm{L}) = \sigma(\mathrm{R}) = \mathbb{D} \tag{3.14.6}$$

Since this has no isolated points,

$$\sigma(B) = \sigma_{\text{ess}}(B) = \mathbb{D} \tag{3.14.7}$$

$C = A - B$ is rank 1, hence compact. (3.14.1) obviously fails. $\mathbb{C} \setminus \sigma_{\text{ess}}(A)$ is not connected while $\sigma_{\text{ess}}(B)^{\text{int}} \neq \emptyset$. Thus, one can have $\sigma_{\text{ess}}(A+C) \neq \sigma_{\text{ess}}(A)$ and $\sigma_{\text{ess}}(B - C) \neq \sigma_{\text{ess}}(B)$. □

The key to our proof of Weyl's invariance theorem will be the following result of independent interest. (See Problems 5 and 6 for another approach to Weyl's theorem.)

Theorem 3.14.3 (Analytic Fredholm Theorem). *Let Ω be a region in $\mathbb{C}$. Let f be an analytic function on Ω with values in the compact operators on a Hilbert space, $\mathcal{H}$. Then either*

(a) $1 - f(z)$ *is not invertible for any $z \in \Omega$, and for any $z_0 \in \Omega$, there is $\psi \neq 0$ so that*

$$f(z_0)\psi = \psi \tag{3.14.8}$$

(b) *There is a set, $S \subset \Omega$, which is discrete (i.e., no limit points in Ω) so that $1 - f(z)$ is invertible if and only if $z \notin S$. $g(z) = (1 - f(z))^{-1}$ is analytic for $z \in \Omega \setminus S$. The points in S are poles of finite order with finite rank residues.*

Remarks. 1. The same proof works in a Banach space, X, if we assume $f(z)$ takes its values in $\text{FA}(X)$, the operators which are norm limits of finite rank operators.

2. Picking $f(z) = zA$ and $\Omega = \mathbb{C}$, this result yields another proof of the Riesz–Schauder theorem (Theorem 3.3.1) in the Hilbert space case (or for operators in $\text{FA}(X)$ for any Banach space). Using the Banach–Mazur theorem as mentioned in the Notes to Section 3.1, this allows an alternate proof even for $\text{Com}(X)$ when $\text{Com}(X) \neq \text{FA}(X)$.

Proof. By a simple connectedness argument (Problem 3), it suffices to show for any $z_0 \in \Omega$, there is a neighborhood N_{z_0} of z_0 so (a) or (b) holds in N_{z_0}. We can pick N_{z_0} so that for $z \in N_{z_0}$,

$$\|f(z) - f(z_0)\| \leq \tfrac{1}{3} \tag{3.14.9}$$

Since $f(z_0)$ is compact, find A_0 finite rank so that $\|f(z_0) - A\| \leq \frac{1}{3}$ and thus,

$$\|f(z) - A\| \leq \tfrac{2}{3} \tag{3.14.10}$$

By writing a geometric series, we know $1 - (f(z) - A)$ invertible. Thus, we can write

$$\begin{aligned} 1 - f(z) &= 1 - (f(z) - A) - A \\ &= \big[1 - A(1 - (f(z) - A))^{-1}\big]\{1 - (f(z) - A)\} \end{aligned} \tag{3.14.11}$$

Therefore, if

$$g(z) \equiv A\big(1 - (f(z) - A)\big)^{-1} \tag{3.14.12}$$

we see

$$1 - f(z) \text{ is invertible} \Leftrightarrow 1 - g(z) \text{ is invertible} \tag{3.14.13}$$

(3.14.11) and the invertibility of $1-(f(z)-A)$ also implies $\mathrm{Ker}(1-f(z)) \neq \{0\} \Leftrightarrow \mathrm{Ker}(1-g(z)) \neq \{0\}$, that is,

$$f(z)\varphi = \varphi \text{ has solutions} \Leftrightarrow g(z)\psi = \psi \text{ has solutions} \tag{3.14.14}$$

We are thus reduced to the case where each $g(z)$ is finite rank with $\mathrm{Ran}(g(z)) \subset F$, a fixed finite-dimensional space. Thus,

$$g(z) = \sum_{n=1}^{N} \langle \psi_n(z), \cdot \rangle \varphi_n \tag{3.14.15}$$

where $\{\varphi_n\}_{n=1}^N$ is a fixed orthonormal set and $\psi_n(z)$ are antianalytic functions.

Notice if $g(z_0)\varphi = \varphi$, then $\varphi = \sum_{n=1}^N \alpha_n \varphi_n$ and if

$$B_{nm}(z) - \langle \psi_n(z), \varphi_m \rangle, \qquad n, m = 1, \dots, N \tag{3.14.16}$$

(which are analytic in z), then $g(z_0)\varphi = \varphi$ if and only if

$$\sum_{m=1}^{N} B_{nm}(z_0)\alpha_m - \alpha_m \tag{3.14.17}$$

so we have proven that if ($N \times N$ determinant)

$$d(z) = \det(\mathbb{1} - B(z)) \tag{3.14.18}$$

then

$$g(z)\varphi = \varphi \text{ has nonzero solutions} \Leftrightarrow d(z) = 0 \tag{3.14.19}$$

By a similar argument, one sees (Problem 4)

$$d(z) \neq 0 \Leftrightarrow 1 - g(z) \text{ is invertible} \tag{3.14.20}$$

and that $1 - g(z)$ can be expressed in terms of $(\mathbb{1} - B(z))^{-1}$.

Since d is analytic, either:

(a) $d(z) \equiv 0$ in N_{z_0}. Then $1 - g(z)$ and so $1 - f(z)$ is never invertible for $z \in N_{z_0}$ and $g(z)\varphi = \varphi$ (and so, $f(z)\varphi = \varphi$) has solutions for all $z \in N_{z_0}$.

(b) $d(z)$ has isolated zeros in N_{z_0} so $(1 - g(z))^{-1}$ and so $(1 - f(z))^{-1}$ are analytic except at these zeros and the Laurent series at these zeros can be built from the Laurent series of $(1 - B(z_0))^{-1}$ which only uses finite-order poles (by Cramer's rule) with finite rank residues.

□

Proof of Theorem 3.14.1. It suffices to prove $\sigma_{\text{ess}}(A+C) \subset \sigma_{\text{ess}}(A)$ for $A = (A+C) - C$, so the same result then implies $\sigma_{\text{ess}}(A) \subset \sigma_{\text{ess}}(A+C)$.

We can also suppose $\sigma_{\text{ess}}(A) = \sigma(A)$. For let $\{\lambda_j\}_{j=1}^N$ be a counting of the points in $\sigma_d(A)$, including multiplicity, and $\{\lambda_j\}_{j=1}^N$ an orthonormal set with

$$A\varphi_j = \lambda_j \varphi_j \tag{3.14.21}$$

and P_j the orthogonal projection onto φ_j.

$\sigma_{\text{ess}}(A)$ is closed and nonempty, so we can find $\mu_j \in \sigma_{\text{ess}}(A)$ so $|\lambda_j - \mu_j| = \text{dist}(\lambda_j, \sigma_{\text{ess}}(A))$. Since the only limit points of $\{\lambda_j\}_{j=1}^\infty$ are in $\sigma_{\text{ess}}(A)$, we see $|\lambda_j - \mu_j| \to 0$. Let

$$K = \sum_{j=1}^N (\mu_j - \lambda_j)P_j, \qquad \widetilde{A} = A + K \tag{3.14.22}$$

Thus, $\sigma(\widetilde{A}) = \sigma_{\text{ess}}(A)$ since the eigenvalues in $\sigma_d(A)$ have been moved to $\sigma_{\text{ess}}(A)$. Moreover, $\sigma_{\text{ess}}(\widetilde{A}) = \sigma_{\text{ess}}(A)$ since nonisolated points remain nonisolated and multiplicities only go up. $A + C = \widetilde{A} + (C - K)$ is a compact perturbation of $\widetilde{A}$, so we are reduced to the situation $\sigma(A) = \sigma_{\text{ess}}(A)$.

In that case, let $\Omega = \mathbb{C} \setminus \sigma_{\text{ess}}(A)$. For $z \in \Omega$, write

$$A + C - z = (A - z)(\mathbb{1} + (A - z)^{-1}C) \tag{3.14.23}$$

Thus, $z \in \Omega$ has $A + C - z$ invertible if and only if $\mathbb{1} + F(z)$ is invertible, where

$$F(z) = (A - z)^{-1}C \tag{3.14.24}$$

and if $1 + F(z)$ is invertible, then

$$(A + C - z)^{-1} = (\mathbb{1} + F(z))^{-1}(A - z)^{-1} \tag{3.14.25}$$

By (2.2.15), if $x, y \in \mathbb{R}$, $\|A - (x+iy)\|^{-1} \le |y|^{-1}$ so $\lim_{y\to\infty} F(x+iy) = 0$ and $1 + F(z)$ is invertible for $|y|$ large. By the analytic Fredholm theorem, $1 + F(z)$ is invertible except for a discrete subset of Ω, that is, points of $\sigma(A+C)$ in Ω are isolated. By the same theorem and (3.14.25), at points in $\sigma(A+C) \cap \Omega$, $(A+C-z)^{-1}$ has finite-order poles with finite-rank residues, so the multiplicities of such points are finite, that is, $\sigma(A+C) \cap \Omega \subset \sigma_d(A)$, so $\sigma_{\text{ess}}(A+C) \subset \sigma_{\text{ess}}(A)$. □

Example 3.14.4. Let J be the Jacobi matrix (see (4.1.13)) with Jacobi parameters $\{a_n, b_n\}_{n=1}^\infty$ obeying $a_n \to 1$, $b_n \to 0$. Then $\sigma_{\text{ess}}(J) = [-2, 2]$. For if J_0 has $a_n \equiv 1$, $b_n \equiv 0$, $J - J_0$ is compact. □

We end with a discussion of the general min-max principle. If A is a (bounded) self-adjoint operator on an infinite-dimensional Hilbert space, then a simple argument (Problem 15) shows that $\sigma_{\text{ess}}(A) \neq \emptyset$. Let

$$\Sigma^{\pm}(A) = \sup_{\inf} \{\lambda \mid \lambda \in \sigma_{\text{ess}}(A)\} \tag{3.14.26}$$

Let $\lambda_1^-(A) \leq \lambda_2^-(A), \dots$ be the points in $\sigma_d(A) \cap (-\infty, \Sigma^-(A))$ counting multiplicity. This set may be finite, in which case if $\lambda_N^-(A)$ is the last point in $\sigma_d(A) \cap (-\infty, \Sigma^-(A))$, we set $\lambda_j^-(A) = \Sigma^-(A)$ for $j \geq N+1$. $\lambda_1^+(A) \geq \lambda_2^+(A) \geq \dots$ is defined similarly.

Theorem 3.14.5 (Min-Max Principle). *If A is a general (bounded) self-adjoint operator on an infinite-dimensional Hilbert space, $\mathcal{H}$, we have*

$$\lambda_n^+(A) = \inf_{\psi_1,\dots,\psi_{n-1}} \Bigg(\sup_{\substack{\varphi \perp \psi_1,\dots,\psi_{n-1} \\ \|\varphi\|=1}} \langle \varphi, A\varphi \rangle \Bigg) \tag{3.14.27}$$

$$\lambda_n^-(A) = \sup_{\psi_1,\dots,\psi_{n-1}} \Bigg(\inf_{\substack{\varphi \perp \psi_1,\dots,\psi_{n-1} \\ \|\varphi\|=1}} \langle \varphi, A\varphi \rangle \Bigg) \tag{3.14.28}$$

$$\lim_{n\to\infty} \lambda_n^{\pm}(A) = \Sigma^{\pm}(A) \tag{3.14.29}$$

Proof. We prove the + case. By the same argument in the proof of Theorem 3.2.4, if $\lambda_n^+(A) > \Sigma^+(A)$, then for all $\psi_1, \dots, \psi_{n-1}$,

$$\sup_{\substack{\varphi \perp \psi_1,\dots,\psi_{n-1} \\ \|\varphi\|=1}} \langle \varphi, A\varphi \rangle \geq \lambda_n^+(A) \tag{3.14.30}$$

with equality if $\psi_1, \dots, \psi_{n-1}$ are the eigenvectors associated to $\{\lambda_j^+(A)\}_{j=1}^{n-1}$.

That only leaves the case $\lambda_n^+(A) = \Sigma^+(A)$. In that case $\dim \operatorname{Ran}(P_{(\Sigma^+-\varepsilon,\Sigma^+]}(A) = \infty$, so we can find $\varphi \perp \psi_1, \dots, \psi_{n-1}$ in that range, that is,

$$\langle \varphi, A\varphi \rangle \geq \Sigma^+ - \varepsilon \tag{3.14.31}$$

Since ε is arbitrary, (3.14.30) holds.

If $\psi_1, \dots, \psi_{n-1}$ include all the eigenvectors associated to eigenvalues in (Σ_+, ∞), we have equality in (3.14.30).

(3.14.29) is an easy consequence of the fact that either $\lambda_n^+(A) = \Sigma^+(A)$ for all large n or else λ_n^+ is decreasing to a point clearly in $\sigma_{\text{ess}}(A)$. □

Notes and Historical Remarks. Weyl's invariance theorem goes back to his 1909 paper [**741**]. His approach is outlined in Problems 5 and 6 and some instructors may wish to use that proof instead. We use the proof here because we want to introduce the analytic Fredholm theorem, which is of independent interest (see, e.g., Problem 8).

There is a converse to Weyl's theorem due to Weyl and von Neumann (see Section 5.9 where we prove the theorem and discuss its history): If A and

B are two bounded self-adjoint operators on $\mathcal{H}_A$ and $\mathcal{H}_B$ respectively, with $\sigma_{\text{ess}}(A) = \sigma_{\text{ess}}(B)$, then there is a unitary $U: \mathcal{H}_A \to \mathcal{H}_B$ so that $UAU^{-1} - B$ is a compact operator on $\mathcal{H}_B$.

Many of the applications of Weyl's theorem are to unbounded self-adjoint operators, especially to positive ones. One still defines $\sigma_d(A)$ and $\sigma_{\text{ess}}(A)$ by requiring $\lambda \in \sigma_d(A)$ to be an isolated point of $\sigma(A)$ which is an eigenvalue of finite multiplicity and $\sigma_{\text{ess}}(A) = \sigma(A) \setminus \sigma_d(A)$. Then, if $A \geq 0$,

$$\sigma_{\text{ess}}(A) = \{z \mid (z+1)^{-1} \in \sigma_{\text{ess}}((A+1)^{-1})\} \tag{3.14.32}$$

Formally,

$$(A+C+1)^{-1} = (A+1)^{-1} - (A+C+1)^{-1}C(A+1)^{-1} \tag{3.14.33}$$

Thus, for example, if $C(A+1)^{-1}$ is compact (we say then the C is *relatively compact*), we get, by (3.14.1) and (3.14.33), that $\sigma_{\text{ess}}(A+C) = \sigma_{\text{ess}}(A)$. This applies, for example, if $A = -\Delta$ on $L^2(\mathbb{R}^\nu, d^\nu x)$ and C is multiplication by $V(x)$, a continuous function with $|V(x)| \to 0$ as $|x| \to \infty$. Multiplication by V is not compact, but $V(-\Delta + 1)^{-1}$ is. This is discussed further in Example 7.8.5.

For non–self-adjoint operators, the situation is complicated, as shown by Example 3.14.2. If C is compact and both A and $A+C$ are normal, one still has (3.14.1) (in Example 3.14.2, A is normal but $A+C$ is not), as the reader will prove in Problem 7.

Otherwise, the truth of (3.14.1) depends on the meaning of σ_{ess}. For example, Gustafson–Weidmann [**276**] and the book of Edmunds–Evans [**182**] discuss a variety of alternate definitions. For example, $\bigcap_{C \text{ compact}} \sigma(A+C)$ is obviously invariant under compact perturbations and can be seen (Problem 11) to agree with our definition if A is self-adjoint.

Another definition, $\tilde{\sigma}_{\text{ess}}(A)$, is as follows. We say B is *invertible modulo compacts* (one also says D is a pseudoinverse of B) if there is D so both $BD - \mathbb{1}$ and $DB - \mathbb{1}$ are compact. It can be seen (see Section 3.15) that this is true if and only if (i) $\text{Ran}(B)$ is closed; (ii) $\text{Ran}(B)^\perp$ is finite-dimensional; and (iii) $\text{Ker}(B)$ is finite-dimensional. One then defines the *index* of B to be $i(B) = \dim(\text{Ker}(B)) - \dim(\text{Ran}(B)^\perp)$. Then

$$\tilde{\sigma}_{\text{ess}}(A) = \{z \mid A - z \text{ is not invertible modulo compacts}\} \tag{3.14.34}$$

For A bounded and self-adjoint, one has (Problem 12) that $\sigma_{\text{ess}}(A) = \tilde{\sigma}_{\text{ess}}(A)$.

Notice that if $BD - \mathbb{1}$ and $DB - \mathbb{1}$ are compact and K is compact, then $(B+K)D - \mathbb{1}$ and $D(B+K) - \mathbb{1}$ are compact, so trivially $\tilde{\sigma}_{\text{ess}}(B+K) = \tilde{\sigma}_{\text{ess}}(B)$. We discuss this further in Section 3.15.

One way to rephrase this is in terms of the *Calkin algebra*, an object named after work of Calkin [**96**]. $\mathcal{I}_\infty$, the compact operators, are a closed two-sided ideal so $\mathcal{C} = \mathcal{L}(\mathcal{H})/\mathcal{I}_\infty$ is a Banach algebra (indeed, in the language of Sections 6.4 and 6.7, a C^*-algebra, albeit one that can only be realized as a space of operators on a nonseparable Hilbert space!). Let $\pi\colon \mathcal{L}(\mathcal{H}) \to \mathcal{C}$ by taking B to its equivalence class. Then $\tilde{\sigma}_{\text{ess}}(B)$ is exactly $\sigma(\pi(B))$, the spectrum of $\pi(B)$ as an element of $\mathcal{C}$.

Brown–Douglas–Fillmore (BDF) [**92, 93**] have found a complete set of unitary invariants for operators mod compacts. If $z \in \mathbb{C} \setminus \tilde{\sigma}_{\text{ess}}(A)$, then by the above, $A - z$ has an index $i_A(z)$. Fredholm theory (see Section 3.16) says $i_A(z)$ is continuous on each connected component of $\mathbb{C} \setminus \tilde{\sigma}_{\text{ess}}(A)$. The BDF theorem says that a pair of operators, A, B, obey $UAU^{-1} - B$ is compact for some U if and only if $\tilde{\sigma}_{\text{ess}}(A) = \tilde{\sigma}_{\text{ess}}(B)$ and $i_A(z) = i_B(z)$ for all $z \in \mathbb{C} \setminus \tilde{\sigma}_{\text{ess}}(A)$. Thus, unitary classes mod compact operators are described by $\tilde{\sigma}_{\text{ess}}(A)$ and an integer on each connected component of $\mathbb{C} \setminus \tilde{\sigma}_{\text{ess}}(A)$ (and all sets of integers are possible for a given $\tilde{\sigma}_{\text{ess}}(A)$).

Problems

1. Let X be a Banach space. Let $A \in \mathcal{L}(X)$. Suppose that $\sigma(A)$ is finitely many points, each an eigenvalue of finite multiplicity. Prove $\dim(X) < \infty$. Conclude that if $\dim(X) = \infty$, any A has $\sigma_{\text{ess}}(A) \neq \emptyset$. (*Hint*: For R large, prove $(2\pi i)^{-1} \int_{|z|=\mathrm{R}} (z - A)^{-1}\, dz = \mathbb{1}$.)

2. Let A be the operator on $\ell^2(\mathbb{Z})$ given by (3.14.4). This problem will prove $\sigma(A) = \partial\mathbb{D}$.

   (a) For any unitary U, prove that $\sigma(U) \subset \partial\mathbb{D}$. (*Hint*: For $|z| > 1$, find an explicit series in U and z^{-1} converging to $(U - z)^{-1}$ and if $|z| < 1$ in U^* and z.)

   (b) Prove $\partial\mathbb{D} \subset \sigma(A)$. (*Hint*: For any $z \in \partial\mathbb{D}$, find φ_m with $\|\varphi_m\| = 1$ so that $\|(A - z)\varphi_m\| \to 0$.)

3. If one has (a) or (b) of Theorem 3.14.3 in a neighborhood of any point of $z \in \Omega$, prove (a) or (b) holds globally. (*Hint*: Show $\{z_0 \mid 1 - f(z)$ is not invertible on a neighborhood of $z_0\}$ is open and closed.)

4. Let $g(z)$ have the form (3.14.15). Prove that φ solves $\varphi = \eta + g(z)\varphi$ if and only if $\varphi = \eta + \sum_{n=1}^N \alpha_n \varphi_n$ where α_n solves

$$\alpha_n = \langle \psi_n, \eta \rangle + \sum_{n=1}^{\infty} B_{nm}(z)\alpha_m \tag{3.14.35}$$

   Conclude that $1 - g(z)$ is invertible $\Leftrightarrow d(z) \neq 0$, where $d(z)$ is given by (3.14.18).

5. This problem assumes the spectral theorem (see Theorem 2.4.13 or Chapter 5). Let A be a bounded self-adjoint operator.

(a) Prove $\lambda \in \sigma_{\text{ess}}(A)$ if and only if for all $\varepsilon > 0$, $\dim(P_{(\lambda-\varepsilon,\lambda+\varepsilon)}(A)) = \infty$.

(b) If $\lambda \in \sigma_{\text{ess}}(A)$, find φ orthonormal so that $\|(A-\lambda)\varphi_n\| \to 0$. (*Hint*: Use (a).)

(c) If $\lambda \notin \sigma_{\text{ess}}(A)$ and $\varphi_n \xrightarrow{w} 0$ with $\|\varphi_n\| = 1$, prove that $\|(A-\lambda)\varphi_n\| \nrightarrow 0$.

(d) Conclude that

$$\lambda \in \sigma_{\text{ess}}(A) \Leftrightarrow \exists \varphi_n,\ \|\varphi_n\| = 1,\ \varphi_n \xrightarrow{w} 0,\ \|(A-\lambda)\varphi_n\| \to 0 \tag{3.14.36}$$

Remark. A sequence φ_n, as in (d), is called a *Weyl sequence* for $A - \lambda$.

6. Let A, C be self-adjoint with C compact. Prove that φ_n is a Weyl sequence for $A - \lambda$ if and only if it is a Weyl sequence for $A + C - \lambda$. Use this to provide another proof of Weyl's invariance theorem.

7. By following the proof in Problems 5 and 6 and using the spectral theorem for normal operators, prove that if A and $A + C$ are both normal and C is compact, then (3.14.1) holds.

8. This problem will lead you through a proof of the Riesz–Schauder theorem in the Hilbert space case (indeed, for $A \in \text{FA}(X)$ for any Banach space, X).

   (a) By applying the analytic Fredholm theorem to $F(z) = zA$, prove that if A is compact, $\{z \mid z^{-1} \in \sigma(A)\}$ is a discrete subset of $\mathbb{C}$.

   (b) Prove that each point in $\sigma(A) \setminus \{0\}$ has a finite-dimensional spectral projection.

9. Let Ω be a region in $\mathbb{C}$. Let F be meromorphic on Ω so that if D is the discrete set of poles of F, that $F(z)$ is compact for $z \in \Omega \setminus D$ and the residues at points $z_0 \in D$ are all finite rank. Prove an extension of the analytic Fredholm theorem to $\mathbb{1} - F(z)$.

   Remark. This *meromorphic Fredholm theorem* seems to have first appeared explicitly in Ribarič–Vidav [**560**].

10. Our proof of Theorem 3.14.1 began by eliminating the points in $\sigma_d(A)$. Using the meromorphic Fredholm theorem of Problem 9, find a proof that doesn't need this preliminary step.

11. If A is self-adjoint and C is a perhaps not self-adjoint compact operator, prove that $\sigma_{\text{ess}}(A) \subset \sigma(A + C)$ (*Hint*: For $\lambda \in \sigma_{\text{ess}}(A)$, find φ_n so $(A + C - \lambda)\varphi_n \to 0$) and conclude that $\sigma_{\text{ess}}(A) = \bigcap_{C \text{ compact}} \sigma(A + C)$.

12. This problem will prove $\sigma_{\text{ess}}(A) = \tilde{\sigma}_{\text{ess}}(A)$ for self-adjoint A. Assume below that A is self-adjoint.

(a) Let $\lambda \in \sigma_d(A)$. Let $B = f(A)$ where

$$f(x) = \begin{cases} \frac{1}{x-\lambda}, & x \neq \lambda \\ 0, & x = \lambda \end{cases}$$

Prove that B is bounded and $B(A-\lambda) - \mathbb{1} = (A-\lambda)B - \mathbb{1}$ is finite rank.

(b) Suppose $K = (A-\lambda)B - \mathbb{1}$ is compact. For each ε, prove that

$$\|P_{[\lambda-\varepsilon,\lambda+\varepsilon]}(A) - P_{[1-\varepsilon,1+\varepsilon]}(K)\| \leq \varepsilon\|B\| \tag{3.14.37}$$

(c) If $\varepsilon < \|B\|^{-1}$, prove that $P_{[\lambda-\varepsilon,\lambda+\varepsilon]}(A)$ is finite rank and conclude that $\lambda \in \sigma_d(A)$.

(d) Show $\sigma_{\text{ess}}(A) = \tilde{\sigma}_{\text{ess}}(A)$.

13. (a) Let A be an arbitrary bounded operator on a Hilbert space, $\mathcal{H}$. Let Ω be the unbounded component of $\mathbb{C} \setminus \sigma_{\text{ess}}(A)$. Let C be compact. Prove that any $\lambda \in \Omega \cap \sigma(A+C)$ is in $\sigma_d(A+C)$. (*Hint*: Analytic Fredholm theorem.)

    (b) Conclude that $\sigma_{\text{ess}}(A+C) \subset \mathbb{C} \setminus \Omega$.

    (c) Prove that the outer boundaries of $\sigma_{\text{ess}}(A)$ and $\sigma_{\text{ess}}(A+C)$ agree.

14. Let A be an arbitrary bounded operator on a Hilbert space, $\mathcal{H}$. Let Ω be a bounded component of $\mathbb{C} \setminus \sigma_{\text{ess}}(A)$. Let C be compact. Prove that either $\Omega \subset \sigma(A+C)$ or else $\Omega \cap \sigma_{\text{ess}}(A+C) = \emptyset$.

15. Let A be an arbitrary bounded self-adjoint operator on a Hilbert space, $\mathcal{H}$. If $\sigma_{\text{ess}}(A) = \emptyset$, prove that $\dim(\mathcal{H}) < \infty$. (*Hint*: Show A has a complete finite ON set of eigenvectors.)

3.15. Bonus Section: Fredholm Operators and Their Index

One of the striking properties of compact operators, C, is that $\text{Ker}(\mathbb{1} + C)$ is finite-dimensional, $\text{Ran}(\mathbb{1} + C)$ is closed and has finite codimension, and these two finite numbers are equal. In this section, we discuss a class of important operators, A, from X to Y, two Banach spaces, where we ask $\dim(\text{Ker}(A))$ and $\text{codim}(\text{Ran}(A))$ are both finite, but no longer demand equality—rather we define their difference to be the index of A. This class is especially important because this index often has a geometric or topological significance and is also often given by some kind of integral formula. We'll present the basic definitions and properties here and discuss some simple examples that illustrate the topological context. Since compact operators will play a major role, we discuss this topic in this chapter. We begin by recalling and developing the notion of finite codimension.

If $Z \subset Y$ is a subspace, the *annihilator*, $\mathcal{A}(Z)$, is the set of linear functionals ℓ on Y with $\ell \restriction Z = 0$. Usually, we are interested in functionals in Y^*, but we'll sometimes also care about not necessarily continuous functionals, in which case we'll refer to the *algebraic annihilator*, $\mathcal{A}_{\text{alg}}(Z)$. If $T\colon X \to Y$ is a linear map, we define the *cokernel* of T, $\text{Coker}(T)$, to be the algebraic annihilator of $\text{Ran}(T)$.

Recall (see Section 2.1 or see Section 5.4 of Part 1) a *complement*, $\widetilde{Z}$, of $Z \subset Y$ is a subspace with

$$\widetilde{Z} \cap Z = \{0\}, \qquad \widetilde{Z} + Z = Y \tag{3.15.1}$$

that is, any $y \in Y$ is uniquely $y = z + \tilde{z}$ with $z \in Z$, $\tilde{z} \in \widetilde{Z}$. In that case, we write

$$Y = \widetilde{Z} \dotplus Z \tag{3.15.2}$$

We can map $Y \mapsto Y$ by $Py = z$ so $(1-P)y = \tilde{z}$. P is a projection, that is, $P^2 = P$ and $Q \equiv 1 - P$ is also a projection. $P \in \mathcal{L}(Y)$ if and only if Z and $\widetilde{Z}$ are closed, as we saw in Proposition 2.1.2. We then call them topological conjugates.

We say $\{y_1, \dots, y_n\} \subset Y$ are independent over $Z \subset Y$ if and only if $\sum_{j=1}^n \alpha_j y_j \in Z$ with $\alpha_j \in \mathbb{K}$ ($= \mathbb{R}$ or $\mathbb{C}$, the field over which X, Y are vector spaces) implies $\alpha_1 = \dots = \alpha_n = 0$. The following includes Theorem 5.5.12 of Part 1 and is left to the reader (Problem 1).

Proposition 3.15.1. *Let $Z \subset Y$ be a subspace of a vector space. The following are equivalent:*

(1) *$\mathcal{A}_{\text{alg}}(Z)$ has finite dimension.*
(2) *Z has a finite-dimensional complement, $\widetilde{Z}$.*
(3) *There are finitely many $y_1, \dots, y_\ell \in Y$ so*

$$\left\{ z + \sum_{j=1}^n \alpha_j y_j \;\middle|\; z \in Z,\, \alpha_j \in K \right\} = Y \tag{3.15.3}$$

(4) *Y/Z, the quotient space, is finite-dimensional.*

If these hold, $\dim(Y/Z)$, $\dim(\mathcal{A}_{\text{alg}}(Y))$, $\dim(\widetilde{Z})$, *and the minimal number of y's so* (3.15.3) *holds are all equal and called the* codimension *of Z. We say Z has* finite codimension *in this case. A minimal set of y's with* (3.15.3) *are linearly independent over Z.*

Moreover, if Y is a Banach space, then Z is closed if and only if all $\ell \in \mathcal{A}_{\text{alg}}(Y)$ are continuous, and in that case, $\widetilde{Z}$ is also closed.

A kind of dual statement is (Problem 2):

Proposition 3.15.2. *Let Y be a Banach space with a subspace $Z \subset Y$. If Z is finite-dimensional, it is always closed and it has a closed complement, $\widetilde{Z}$, whose codimension is* $\dim(Z)$.

The following is important to allow an "algebraic" definition of Fredholm operators.

Theorem 3.15.3. *Let $T\colon X \to Y$ be a bounded linear transformation between Banach spaces. If* $\text{Ran}(T)$ *has finite codimension (i.e., if* $\text{Coker}(T)$ *is finite-dimensional), then* $\text{Ran}(T)$ *is closed. Moreover, any $\ell \in \mathcal{A}_{\text{alg}}(\text{Ran}(T))$ is in Y^*.*

Remarks. 1. If $T\colon L^2([-1,1],dx) \to L^2([1,1],dx)$ by $Tf(x) = xf(x)$, then $\text{Coker}(T)$ is infinite-dimensional (Problem 3), which might seem surprising at first sight!

2. As a warning, we note that in the Hilbert space case, we have $\overline{\text{Ran}(T)} = \text{Ker}(T^*)^\perp$. If $\text{Ran}(T)$ is closed, it has finite codimension if and only if $\text{Ker}(T^*)$ is finite-dimensional. But as Remark 1 shows, it can happen that $\text{Ker}(T^*) = \{0\}$ but $\text{Ran}(T)$ is not closed. Thus, $\text{Ran}(T)$ having finite codimension is not the same as $\text{Ker}(T^*)$ being finite-dimensional, and in this theorem, we mean algebraic codimension, not the codimension of $\overline{\text{Ran}(T)}$.

Proof. Let $\widetilde{Z}$ be a finite-dimensional complement to $\text{Ran}(T)$. Let $\widetilde{X} = X \oplus \widetilde{Z}$ with the direct sum topology. Let $\widetilde{T}\colon \widetilde{X} \to Y$ by $\widetilde{T}(x,\tilde{z}) = \tilde{z} + Tx$. $\widetilde{T}$ is clearly continuous and $\text{Ran}(\widetilde{T}) = Y$. Thus, by the open mapping theorem (Theorem 5.4.11 of Part 1), $\widetilde{T}$ is open. Therefore, $\widetilde{T}[\{\tilde{x} \mid \|\tilde{x}\| < 1\}]$ contains a ball of radius ε about 0, that is,

$$\forall y \in Y, \quad \exists \tilde{x} \in \widetilde{X},\ \|\tilde{x}\| \le \varepsilon^{-1}\|y\| \text{ and } \widetilde{T}\tilde{x} = y \tag{3.15.4}$$

Since $\widetilde{T}(x,\tilde{z}) = Tx + \tilde{z}$, we see if $y \in \text{Ran}(T)$ the $\tilde{x}$ of (3.15.4) lies in X (i.e., $\tilde{z} = 0$).

Suppose $y_\infty \in \overline{\text{Ran}(T)}$. Then there exists $x_n \in X$ so $Tx_n \to y_\infty$ and $\|Tx_{n+1} - Tx_n\| \le 2^{-n}$. By (3.15.4), there $x_n^\sharp \in X$ so that $Tx_n = Tx_n^\sharp$ and $\|x_{n+1}^\sharp - x_n^\sharp\| \le 2^{-n}\varepsilon^{-1}$. Thus, $\{x_n^\sharp\}_{n=1}^\infty$ are Cauchy and so have a limit $x_\infty^\sharp$. Thus, $y_\infty = \lim Tx_n^\sharp = Tx_\infty^\sharp$, so $y_\infty \in \text{Ran}(T)$, that is, $\text{Ran}(T)$ is closed.

If $\ell \in \mathcal{A}_{\text{alg}}(\text{Ran}(T))$ and $\tilde{z}_1, \dots, \tilde{z}_\ell$ a basis for $\widetilde{Z}$, any $y \in Y$ is, by the above, $y = Tx + \sum_{j=1}^d \alpha_j \tilde{z}_j$ and $y \mapsto \alpha_j$ is continuous because $\text{Ran}(T)$ is closed. Thus, $\ell(y) = \sum_{j=1}^d \alpha_j(y)\ell(\tilde{z}_j)$ is continuous. $\square$

Definition. Let X, Y be Banach spaces. $T\colon X \to Y$, a bounded linear map, is called a *Fredholm operator* if and only if T has finite-dimensional kernel and cokernel. The *index* of T is

$$\operatorname{ind}(T) = \dim\bigl(\operatorname{Ker}(T)\bigr) - \operatorname{codim}\bigl(\operatorname{Ran}(T)\bigr) \tag{3.15.5}$$

By Theorem 3.15.3, if T is Fredholm, $\operatorname{Ran}(T)$ is closed.

Proposition 3.15.4. *Let X, Y, Z, W be Banach spaces. $T \in \mathcal{L}(X,Y)$, $V\colon Z \to X$, and $U\colon Y \to W$ invertible maps (bounded with bounded inverses). Then*

(a) *If $Y_1 \subset Y$ has finite dimension, then so does $U[Y_1]$ and*

$$\dim(U[Y_1]) = \dim(Y_1) \tag{3.15.6}$$

(b) *If $Y_2 \subset Y$ has finite codimension, then so does $U[Y_2]$ and*

$$\operatorname{codim}(U[Y_2]) = \operatorname{codim}(Y_2) \tag{3.15.7}$$

(c) *If T is Fredholm, so are VT and TU and*

$$\operatorname{ind}(VT) = \operatorname{ind}(T) = \operatorname{ind}(TU) \tag{3.15.8}$$

Proof. (a) If $y_1 \dots, y_\ell$ are a basis for Y_1, $\{Uy_j\}_{j=1}^{\ell}$ is a basis for $U[Y_1]$.

(b) By Proposition 3.15.1, Y_2 has a complement, Y_1, with

$$\operatorname{codim}(Y_2) = \dim(Y_1) \tag{3.15.9}$$

It is immediate that $U[Y_1]$ is a complement to $U[Y_2]$ so (3.15.7) is implied by (3.15.6) and (3.15.9).

(c) We have

$$\operatorname{Ker}(VT) = \operatorname{Ker}(T), \qquad \operatorname{Ran}(VT) = V[\operatorname{Ran}(T)] \tag{3.15.10}$$

$$\operatorname{Ker}(TU) = U^{-1}[\operatorname{Ker}(T)], \quad \operatorname{Ran}(TU) = \operatorname{Ran}(T) \tag{3.15.11}$$

Thus, (a) & (b) $\Rightarrow$ (c). □

Proposition 3.15.5. *Let X, Y be finite-dimensional Banach spaces. Then any $T \in \mathcal{L}(X,Y)$ is Fredholm and*

$$\operatorname{ind}(T) = \dim(X) - \dim(Y) \tag{3.15.12}$$

Remark. This helps explain why index is natural. In this finite-dimensional case, $\dim(\operatorname{Ker}(T))$ is T-dependent but index is not.

Proof. By standard linear algebra, if $X_1 = \operatorname{Ker}(T)$ and X_2 is a complement, then T is a linear bijection of X_2 and $\operatorname{Ran}(T)$, so

$$\begin{aligned} \dim(\operatorname{Ker}(T)) &= \dim(X) - \dim(X_2) \\ &= \dim(X) - \dim(\operatorname{Ran}(T)) \\ &= \dim(X) - \dim(Y) + \operatorname{codim}(\operatorname{Ran}(T)) \end{aligned}$$

which is (3.15.12). □

We'll eventually prove the set of Fredholm operators is open and $\mathrm{ind}(T)$ is invariant under small changes in operator norm. This will imply compact perturbations preserve index.

Example 3.15.6 (Example 2.2.5, revisited)**.** Recall R and L on $\mathcal{H} = \ell^2(\mathbb{Z}_+)$ by

$$R(a_1, a_2, \dots) = (0, a_1, a_2, \dots) \tag{3.15.13}$$

$$L(a_1, a_2, \dots) = (a_2, a_3, \dots) \tag{3.15.14}$$

$\mathrm{Ran}(L) = \mathcal{H}$, so $\dim \mathrm{Coker}(L) = 0$. $\mathrm{Ker}(L) = \{(\alpha, 0, \dots) \mid \alpha \in \mathbb{C}\}$ so $\dim(\mathrm{Ker}(L)) = 1$. Similarly, $\mathrm{Ker}(R) = 0$, but $\mathrm{Ker}(R^*) = (\mathrm{Ran}(R))^\perp = \{(\alpha, 0, 0, \dots) \mid \alpha \in \mathbb{C}\}$, so $\dim \mathrm{Coker}(R) = 1$. Thus, L and R are both Fredholm and

$$\mathrm{ind}(L) = 1, \qquad \mathrm{ind}(R) = -1 \tag{3.15.15}$$

An easy extension of these calculations shows

$$\mathrm{ind}(L^n) = n, \qquad \mathrm{ind}(R^n) = -n \tag{3.15.16}$$

including $n = 0$ (i.e., $L^0 = \mathbb{1}$). Thus, any $n \in \mathbb{Z}$ can be the value of an index. □

Definition. Let T be a bounded operator from X to Y. $S \in \mathcal{L}(Y, X)$ is called a *pseudoinverse* to T if and only if

$$ST = \mathbb{1} + K, \qquad TS = \mathbb{1} + L \tag{3.15.17}$$

with K a compact operator on X and L a compact operator on Y. If only the first equation is known to hold, we say S is a *left pseudoinverse* and only the second S is a *right pseudoinverse.*

Proposition 3.15.7. *If S_1 is a left pseudoinverse to T and T has a right pseudoinverse, then S_1 is a pseudoinverse to T.*

Proof. If

$$S_1 T = \mathbb{1} + K, \qquad T S_2 = \mathbb{1} + L \tag{3.15.18}$$

then

$$S_1 T S_2 = S_1 + S_1 L = S_2 + K S_2 \tag{3.15.19}$$

so

$$S_2 - S_1 = S_1 L - K S_2 \tag{3.15.20}$$

is compact. Thus,

$$T S_1 = T S_2 + T(S_1 - S_2) = \mathbb{1} + L + T(S_1 - S_2) \tag{3.15.21}$$

is $\mathbb{1}+$ compact. □

Theorem 3.15.8 (Atkinson's Theorem). *Let $T \in \mathcal{L}(X,Y)$. Then the following are equivalent:*

(1) *T is Fredholm.*

(2) *T has a pseudoinverse.*

(3) *There exists T so* (3.15.17) *holds where K and L are finite-rank operators.*

Moreover, in that case, the index is alg. $\dim(K = -1)-$ alg. $\dim(L = -1)$. *In particular, if* $\mathrm{Ker}((\mathbb{1}+K)^2) = \mathrm{Ker}(\mathbb{1}+K)$ *and* $\mathrm{Ker}((\mathbb{1}+L)^2) = \mathrm{Ker}(\mathbb{1}+L)$*, then*

$$\mathrm{ind}(T) = \dim(\mathrm{Ker}(\mathbb{1}+K)) - \dim(\mathrm{Ker}(\mathbb{1}+L)) \qquad (3.15.22)$$

Remarks. 1. In fact, we'll find a pseudoinverse so $-K$ is a projection onto $\mathrm{Ker}(T)$ and $\mathbb{1}+L$ a projection onto $\mathrm{Ran}(T)$.

2. There is more than one S and so more than one (K,L). (3.15.17) says the combination on the right is the same for different choices of S.

3. For simplicity, we'll only consider the case where $\mathrm{Ker}((\mathbb{1}+K)^2) = \mathrm{Ker}(\mathbb{1}+K)$ and $\mathrm{Ker}((\mathbb{1}+L)^2) = \mathrm{Ker}(\mathbb{1}+L)$.

Proof. We'll first show (1) $\Rightarrow$ (3) $\Rightarrow$ (2) $\Rightarrow$ (1).

(1) $\Rightarrow$ (3). Let $Z \subset X$ be a complementary subspace to $\mathrm{Ker}(T)$ and P the projection onto $\mathrm{Ker}(T)$ with $\mathrm{Ran}(1-P) = Z$. Then $T \restriction Z$ is a continuous map of Z to $\mathrm{Ran}(T)$ which is a bijection from Z to $\mathrm{Ran}(T)$. Since $\mathrm{Ran}(T)$ is closed, the inverse mapping theorem (Theorem 5.4.14 of Part 1) implies there is a continuous inverse $\widetilde{S}\colon \mathrm{Ran}(T) \to Z$. Let W be a complementary subspace to $\mathrm{Ran}(T)$ and Q the projection onto W with $\mathrm{Ran}(1-Q) = \mathrm{Ran}(T)$. Let $S\colon Y \to X$ by

$$S\eta = \widetilde{S}(1-Q)\eta \qquad (3.15.23)$$

For $\varphi \in \mathrm{Ker}(T)$, $ST\varphi = 0$. For $\varphi \in Z$, $ST\varphi = \varphi$. Thus, $ST = \mathbb{1} - P$. If $\varphi \in \mathrm{Ran}(T)$, $TS\varphi = \varphi$, and if $\varphi \in W$, $TS\varphi = 0$ since $(\mathbb{1}-Q)\varphi = 0$. Thus, $TS = \mathbb{1} - Q$.

(3) $\Rightarrow$ (2). This is trivial.

(2) $\Rightarrow$ (1). Since $ST = \mathbb{1}+K$, $T\varphi = 0 \Rightarrow (\mathbb{1}+K)\varphi = 0$. Thus, $\mathrm{Ker}(T) \subset \mathrm{Ker}(\mathbb{1}+K)$ is finite-dimensional. Since $TS = \mathbb{1}+L$, $\mathrm{Ran}(T) \supset \mathrm{Ran}(\mathbb{1}+L)$, so any functional $\ell \in X^*$ that annihilates $\mathrm{Ran}(T)$ annihilates $\mathrm{Ran}(\mathbb{1}+L)$, that is, $L^t\ell = -\ell$. Thus, the annihilators of $\mathrm{Ran}(T)$ lie in the finite-dimensional space $\mathrm{Ker}(\mathbb{1}+L^t)$.

Thus, $\mathrm{Ker}(\mathbb{1}+K) = \mathrm{Ker}(\mathbb{1}-P) = \mathrm{Ran}(P) = \mathrm{Ker}(T)$ and $\mathrm{Ker}(\mathbb{1}+L) = \mathrm{Ker}(\mathbb{1}-Q) = W$, which has dimension $\dim(\mathrm{Coker}(T))$. Thus, (3.15.22) holds for this S. We just need to show (3.15.22) holds for any other choice of S. We'll leave this result, which is straightforward but lengthy, to Problem 4. □

Corollary 3.15.9. *If $T\colon X \to Y$ is a Fredholm operator, so is $T^{\mathrm{t}}\colon Y^* \to X^*$ and*

$$\operatorname{ind}(T^{\mathrm{t}}) = -\operatorname{ind}(T) \tag{3.15.24}$$

Remark. One can show (Problem 5) that $\dim(\operatorname{Ker}(T^{\mathrm{t}})) = \operatorname{codim}(\operatorname{Ran}(T))$ and $\operatorname{codim}(\operatorname{Ran}(T^{\mathrm{t}})) = \dim(\operatorname{Ker}(T))$.

Proof. Let S be a pseudoinverse for T. Then it is easy see that S^t is a pseudoinverse for T^{t}. Indeed, if (3.15.17) holds, $T^{\mathrm{t}}S^{\mathrm{t}} = \mathbb{1} + K^{\mathrm{t}}$, $S^{\mathrm{t}}T^{\mathrm{t}} = \mathbb{1} + L^{\mathrm{t}}$. By the equality of the algebraic multiplicities of K and K^{t} and (3.15.22), we have (3.15.24). □

Corollary 3.15.10. *Let $T\colon X \to Y$ be a Fredholm operator and S a pseudoinverse. Then S is a Fredholm operator and*

$$\operatorname{ind}(S) = \quad \operatorname{ind}(T) \tag{3.15.25}$$

Proof. Immediate from (3.15.17) and (3.15.22). □

We turn next to several stability results for the index. We begin by considering 2×2 direct sum operators. Let X, Y be Banach spaces and X_1, X_2 (respectively, Y_1, Y_2) a pair of closed complementary subspaces in X (respectively, Y). Then (see Corollary 5.4.16 of Section 5.4 of Part 1), $X \cong X_1 \oplus X_2$, $Y \cong Y_1 \oplus Y_2$ in the sense of equivalent norms, and so any $\mathcal{L}(X,Y)$ can be written as a matrix

$$T = \begin{pmatrix} A & B \\ C & D \end{pmatrix} \tag{3.15.26}$$

with $A \in \mathcal{L}(X_1, Y_1)$, $B \in \mathcal{L}(X_2, Y_1)$, etc.

Proposition 3.15.11. *Let $X = X_1 \oplus X_2$, $Y = Y_1 \oplus Y_2$, and $T = \mathcal{L}(X,Y)$ be written in the form (3.15.26). Then*

(a) *If $T = A \oplus D$ (i.e., $B = C = 0$), T is Fredholm if and only if A and D are Fredholm and*

$$\operatorname{ind}(T) = \operatorname{ind}(A) + \operatorname{ind}(D) \tag{3.15.27}$$

(b) *If D is invertible, then T is Fredholm if and only if $A - BD^{-1}C$ is Fredholm and*

$$\operatorname{ind}(T) = \operatorname{ind}(A - BD^{-1}C) \tag{3.15.28}$$

(c) *If D is invertible and $\dim(X_1) < \infty$, $\dim(Y_1) < \infty$, then T is Fredholm and*

$$\operatorname{ind}(T) = \dim(X_1) - \dim(Y_1) \tag{3.15.29}$$

Remark. Since $B\colon X_2 \to Y_1$, $D^{-1}\colon Y_2 \to X_2$, and $C\colon X_1 \to Y_2$, $BD^{-1}C$, like A, maps X_1 to Y_1.

Proof. (a) Since $\operatorname{Ker}(A \oplus D) = \operatorname{Ker}(A) \oplus \operatorname{Ker}(D)$ and $\operatorname{Ran}(A \oplus D) = \operatorname{Ran}(A) \oplus \operatorname{Ran}(D)$, this is immediate.

(b) Let $U \in \mathcal{L}(Y)$, $V \in \mathcal{L}(X)$ by

$$U = \begin{pmatrix} \mathbb{1}_{Y_1} & -BD^{-1} \\ 0 & \mathbb{1}_{Y_2} \end{pmatrix}, \qquad V = \begin{pmatrix} \mathbb{1}_{X_1} & 0 \\ -D^{-1}C & \mathbb{1}_{X_2} \end{pmatrix} \tag{3.15.30}$$

Then U, V are invertible, for example, $U^{-1} = \left(\begin{smallmatrix} \mathbb{1} & BD^{-1} \\ 0 & \mathbb{1} \end{smallmatrix}\right)$, and by a direct calculation,

$$UTV = \begin{pmatrix} A - BD^{-1}C & 0 \\ 0 & D \end{pmatrix} \tag{3.15.31}$$

By Proposition 3.15.4(c), T is Fredholm if and only if UTV is and $\operatorname{ind}(T) = \operatorname{ind}(UTV)$. Since D invertible means D is Fredholm with $\operatorname{ind}(D) = 0$, (a) implies (b).

(c) This is immediate from (b) and Proposition 3.15.5. □

Remark. The calculation (3.15.31) is due to Schur [**609**] and $A - BD^{-1}C$ is called the *Schur complement* (originally by Haynsworth [**305**]). Zhang [**773**] is an entire book of articles on it!

Theorem 3.15.12 (Dieudonné's Theorem). *Let T be a Fredholm operator for X to Y. Then there is $\varepsilon > 0$ so that $\|T' - T\| < \varepsilon \Rightarrow T'$ is a Fredholm operator and*

$$\operatorname{ind}(T') = \operatorname{ind}(T) \tag{3.15.32}$$

Proof. As in the proof of Atkinson's theorem, if $X_1 = \operatorname{Ker}(T)$, $Y_2 = \operatorname{Ran}(T)$, and X_2, Y_1 closed complements of X_1, Y_2, respectively, then $X = X_1 \oplus X_2$, $Y = Y_1 \oplus Y_2$, and $T = \left(\begin{smallmatrix} 0 & 0 \\ 0 & D \end{smallmatrix}\right)$, where D is a bijection in $\mathcal{L}(X_2, Y_2)$ and so invertible (by the inverse mapping theorem). Write $T' = \left(\begin{smallmatrix} A' & B' \\ C' & D' \end{smallmatrix}\right)$. Clearly, $\|D - D'\| \le \|T - T'\|$, so if $\|T - T'\| < \|D^{-1}\|^{-1} = \varepsilon$, then writing $D' = [(D' - D)D^{-1} + \mathbb{1}]D$, we see D' is invertible. By Proposition 3.15.11, D' is invertible, so since X_1 and Y_1 are finite-dimensional, T' is Fredholm and

$$\operatorname{ind}(T') = \dim(X_1) - \dim(Y_1) \tag{3.15.33}$$

Since $\dim(X_1) = \dim(\operatorname{Ker}(T))$, $\dim(Y_2) = \operatorname{codim}(\operatorname{Ran}(T))$, this implies (3.15.32). □

Remark. Since $\dim(\operatorname{Ker}(UT'V)) = \dim(\operatorname{Ker}(T'))$ where U and V are invertible, we see that

$$\dim(\operatorname{Ker}(T')) = \dim(\operatorname{Ker}(A' - B'(D')^{-1}C'))$$

which can be strictly smaller than $\dim(X_1)$ but never larger. Thus, if T is Fredholm and $\|T - T'\|$ is small, we see $\dim(\mathrm{Ker}(T')) \le \dim(\mathrm{Ker}(T))$ with inequality possible.

Corollary 3.15.13. *In $\mathcal{L}(X,Y)$ with the $\|\cdot\|$-topology, the Fredholm operators are open and* $\mathrm{ind}(\,\cdot\,)$ *is constant on each connected component.*

Corollary 3.15.14 (Homotopy Invariance of Index)**.** *If $\gamma\colon [0,1] \to \mathcal{L}(X,Y)$ is norm-continuous and $\gamma(t)$ is Fredholm for all $t \in [0,1]$, then*

$$\mathrm{ind}(\gamma(0)) = \mathrm{ind}(\gamma(1)) \tag{3.15.34}$$

Proof. A simple compactness and connectedness argument (Problem 7). □

Corollary 3.15.15. *If T is a Fredholm operator from X to Y and C is compact from X to Y (i.e., $C[\{x \mid \|x\| < 1\}]$ has compact closure in Y), then $T + C$ is Fredholm and*

$$\mathrm{ind}(T + C) = \mathrm{ind}(T) \tag{3.15.35}$$

Proof. If S is a pseudoinverse for T, $S(T+C) = \mathbb{1}+K+SC$ is $\mathbb{1}+$ compact, so S is a pseudoinverse for $T+C$. By Atkinson's theorem, $T+C$ is Fredholm.

Let $\gamma(t) = T+tC$. By the above, it is Fredholm for each t, so $\mathrm{ind}(\gamma(0)) = \mathrm{ind}(\gamma(1))$, which is (3.15.37). □

The next general result we want to consider is additivity of index, that is, if $T\colon X \to Y$ is Fredholm and so is $R\colon Y \to W$, then so is RT and

$$\mathrm{ind}(RT) = \mathrm{ind}(R) + \mathrm{ind}(T) \tag{3.15.36}$$

As a preliminary, we need

Lemma 3.15.16. *Let $T \in \mathcal{L}(X,Y)$ for a pair of Banach spaces and let $\alpha, \beta \in \mathbb{C} \setminus \{0\}$. Then the map $A : Y \bigoplus X \to Y \bigoplus X$ given by*

$$A = \begin{pmatrix} \alpha \mathbb{1}_Y & T \\ 0 & \beta \mathbb{1}_X \end{pmatrix} \tag{3.15.37}$$

is invertible.

Proof. It is easy to check that

$$B = \begin{pmatrix} \alpha^{-1} \mathbb{1}_Y & -(\alpha\beta)^{-1} T \\ 0 & \beta^{-1} \mathbb{1}_X \end{pmatrix} \tag{3.15.38}$$

is a two-sided inverse for A. □

Theorem 3.15.17. *Let $T\colon X \to Y$ and $R\colon Y \to Z$ be Fredholm operators. Then so is RT and* (3.15.36) *holds.*

Proof. Define $A \in \mathcal{L}(Z\oplus Y, Y\oplus Z)$, $B \in \mathcal{L}(Z\oplus Y, Z\oplus Y)$, $C \in \mathcal{L}(Y\oplus X, Z\oplus Y)$, $D \in \mathcal{L}(Y\oplus X, Y\oplus X)$, $E \in \mathcal{L}(Y\oplus X, Y\oplus Z)$ by

$$A = \begin{pmatrix} 0 & \mathbb{1}_Y \\ \mathbb{1}_Z & 0 \end{pmatrix}, \ B = \begin{pmatrix} \mathbb{1}_Z & -\varepsilon^{-1}R \\ 0 & \mathbb{1}_Y \end{pmatrix}, \ C = \begin{pmatrix} R & 0 \\ \varepsilon\mathbb{1}_Y & T \end{pmatrix} \tag{3.15.39}$$

$$D = \begin{pmatrix} \varepsilon^{-1}\mathbb{1}_Y & T \\ 0 & -\varepsilon\mathbb{1}_X \end{pmatrix}, \ E = \begin{pmatrix} \mathbb{1}_Y & 0 \\ 0 & RT \end{pmatrix} \tag{3.15.40}$$

where $\varepsilon \neq 0$ will be chosen shortly. By a direct calculation

$$ABCD = E \tag{3.15.41}$$

By the lemma, for any choice of $\varepsilon \neq 0$, B and D are invertible. Since $\left(\begin{smallmatrix} 0 & \mathbb{1}_Z \\ \mathbb{1}_Y & 0 \end{smallmatrix}\right)$ is a two-sided inverse for A, A is invertible. By Proposition 3.15.11(a), $\left(\begin{smallmatrix} R & 0 \\ 0 & T \end{smallmatrix}\right)$ is Fredholm, so by Theorem 3.15.12, for ε small, C is Fredholm and

$$\operatorname{ind}(C) = \operatorname{ind}(R) + \operatorname{ind}(T) \tag{3.15.42}$$

where we used Proposition 3.15.11(a) again.

By Proposition 3.15.4(c) and (3.15.41), E is Fredholm and

$$\operatorname{ind}(E) = \operatorname{ind}(C) \tag{3.15.43}$$

Finally, by Proposition 3.15.11(a), E Fredholm implies RT is Fredholm and

$$\operatorname{ind}(E) = \operatorname{ind}(\mathbb{1}_Y) + \operatorname{ind}(RT) = \operatorname{ind}(RT) \tag{3.15.44}$$

By the last theorem and the ind equalities, (3.15.36) holds. $\square$

As a general application of the theory, we recall we defined $\tilde{\sigma}_{\text{ess}}(A)$ in (3.14.34) by

$$\mathbb{C} \setminus \tilde{\sigma}_{\text{ess}}(A) = \{z \mid A - z \text{ is a Fredholm operator}\} \tag{3.15.45}$$

See the discussion in the Notes to Section 3.14.

Finally, we turn to some examples.

Example 3.15.18 (Self-adjoint Operators)**.** Let $A \in \mathcal{L}(\mathcal{H})$ be a self-adjoint operator on a Hilbert space, $\mathcal{H}$. We claim A is a Fredholm operator if and only if 0 is an isolated point of $\sigma(A)$ and the corresponding eigenspace is finite-dimensional; see Problems 8 and 9. If A is Fredholm, $\operatorname{ind}(A) = -\operatorname{ind}(A^*)$ (see (3.15.24)), so $\operatorname{ind}(A) = 0$. $\square$

Example 3.15.19 (Index of a Pair of Projections)**.** Let P, Q be two self-adjoint projections on a Hilbert space, $\mathcal{H}$, so that $P - Q$ is compact. We want to define their *relative index*, $\operatorname{ind}(P, Q)$, to be the Fredholm index of QP. This requires some care in specifying where P and Q act! In most cases,

QP is not Fredholm as an operator on $\mathcal{H}$—indeed, it is (Problem 10(a)) if and only if $1-P$ and $1-Q$ are finite rank. Moreover (Problem 10(b)), if they are, $\text{ind}(PQ)=0$.

However, we can view QP as an operator from $X=\text{Ran}(P)$ to $Y=\text{Ran}(Q)$, that is, $QP\varphi=Q\varphi$ for $\varphi\in\text{Ran}(P)$. We claim PQ as a map from Y to X is a pseudoinverse for $1_X-PQP=P(P-Q)P$ as a map of X to X and this is compact. Thus, $QP\colon X\to Y$ is Fredholm. $\text{Ker}(QP)=\{\varphi\mid P\varphi=\varphi, Q\varphi=0\}$ and $\text{Ran}(QP)^\perp=\text{Ker}(PQ)=\{\varphi\mid Q\varphi=\psi, P\varphi=0\}$. Thus,

$$\begin{aligned}\text{ind}(P,Q)\equiv\text{ind}(QP\colon X\to Y)&=\dim(\text{Ran}(P)\cap(\text{Ran}(Q))^\perp)\\&\quad-\dim(\text{Ran}(Q)\cap(\text{Ran}(P)^\perp))\end{aligned}\tag{3.15.46}$$

Suppose now that $P-Q$ and $Q-R$ are compact. Then $(RQ)(QP)=R(Q-P)P+RP$ and $R(Q-P)P$ is compact as a map from $\text{Ran}(P)$ to $\text{Ran}(R)$. Thus, $\text{ind}((RQ)(QP))=\text{ind}(RP)$, so (3.15.35) implies

$$\text{ind}(P,R)=\text{ind}(Q,R)+\text{ind}(P,Q)\tag{3.15.47}$$

Here is a dangling question. Since $P-Q$ is trace class, in the finite-dimensional case, $\text{Tr}(P-Q)$ is the difference of dimensions, so an integer. Is this true in general, and if so, how is this integer related to their relative index?

The answer involves a beautiful symmetry argument. Let

$$A=P-Q,\qquad B=\mathbb{1}-P-Q\tag{3.15.48}$$

Then $P^2=P$, $Q^2=Q$ implies (Problem 13)

$$A^2+B^2=\mathbb{1}\tag{3.15.49}$$

$$\{A,B\}\equiv AB+BA=0\tag{3.15.50}$$

Clearly, $P-Q$ commutes with A^2 and $P+Q$ with $B^2=\mathbb{1}-A^2$. Thus, P and Q commute with A^2.

For $\lambda\in\mathbb{R}$, let $\mathcal{H}_\lambda=\{\varphi\in\mathcal{H}\mid A\varphi=\lambda\varphi\}$. By (3.15.50), if $\varphi\in\mathcal{H}_\lambda$,

$$AB\varphi=-BA\varphi=-\lambda B\varphi\tag{3.15.51}$$

$$B^2\varphi=(\mathbb{1}-A^2)\varphi=(1-\lambda^2)\varphi\tag{3.15.52}$$

Thus, B maps $\mathcal{H}_\lambda$ to $\mathcal{H}_{-\lambda}$, and if $\lambda\neq\pm1$, $(1-\lambda^2)^{-1}B$ is an inverse to this map, that is, $\lambda\neq\pm1\Rightarrow\dim(\mathcal{H}_\lambda)=\dim(\mathcal{H}_{-\lambda})$. In general, (3.15.49) implies $\|A\|\le1$, so $\mathcal{H}_\lambda\neq0\Rightarrow|\lambda|\le1$. In particular, if $A\in\mathcal{I}_{2k+1}$ $(k=0,1,2,\dots)$, then

$$\text{Tr}((P-Q)^{2k+1})=\sum_{\lambda\mid\mathcal{H}_\lambda\neq\{0\}}\dim(\mathcal{H}_\lambda)\lambda^{2k+1}=\dim(\mathcal{H}_{+1})-\dim(\mathcal{H}_{-1})\tag{3.15.53}$$

This is an integer, is independent of k, and noting $A\varphi = \varphi \Leftrightarrow P\varphi = \varphi$, $Q\varphi = 0$ (since $\langle \varphi, A\varphi\rangle = \langle \varphi, P\varphi\rangle - \langle \varphi, Q\varphi\rangle$ and $\langle \varphi, P\varphi\rangle \in [0,1]$), and similarly, $A\varphi = -\varphi \Leftrightarrow P\varphi = 0$, $Q\varphi = \varphi$, we conclude the right side of (3.15.53) is $\operatorname{ind}(P,Q)$. We summarize in the theorem below. $\square$

Theorem 3.15.20. *Let P, Q be orthogonal projections on a Hilbert space, $\mathcal{H}$, so that $P - Q$ is compact. Then QP as an operator from* $\operatorname{Ran}(P)$ *to* $\operatorname{Ran}(Q)$ *is Fredholm. If* $\operatorname{ind}(P,Q)$ *is its index, then whenever $P - Q$ and $Q - R$ are compact,* (3.15.47) *holds. Moreover, if $P - Q \in \mathcal{I}_{2k+1}$ for $k \in \{0, 1, \dots\}$, we have*

$$\operatorname{Tr}((P-Q)^{2k+1}) = \operatorname{ind}(P,Q) \tag{3.15.54}$$

Example 3.15.21 (Toeplitz Operator with Continuous Symbols). A *Toeplitz matrix* is an infinite matrix, T, indexed by $n, m \in \mathbb{Z}_+$ with

$$t_{n,m} = \varphi_{n-m} \tag{3.15.55}$$

for a sequence $\{\varphi_k\}_{k=-\infty}^{\infty}$. Thus, T has the form

$$T = \begin{pmatrix} \varphi_0 & \varphi_{-1} & \varphi_{-2} & \cdots \\ \varphi_{+1} & \varphi_0 & \varphi_{-1} & \cdots \\ \varphi_{+2} & \varphi_{+1} & \varphi_0 & \cdots \\ \vdots & \vdots & \vdots & \ddots \end{pmatrix} \tag{3.15.56}$$

We'll be interested in such matrices where T defines a bounded operator on $\ell^2(\mathbb{Z}_+)$, that is, $\sum_{k=-n}^{\infty} |\varphi_k|^2 < \infty$ for all n (so we can define Ta for any $a \in \ell^2$) and for some C,

$$\|Ta\|_2 \le C\|a\|_2 \tag{3.15.57}$$

Here is one way to get such bounded Toeplitz matrices (the only way, as we'll see). Let $f \in L^\infty(\partial\mathbb{D}, \frac{d\theta}{2\pi})$. Let $u \in L^2(\partial\mathbb{D}, \frac{d\theta}{2\pi})$. Then if $M_f g = fg$,

$$\|M_f u\|_2 \le \|f\|_\infty \|u\|_2 \tag{3.15.58}$$

In the standard Fourier basis, $\{e^{in\theta}\}_{n=-\infty}^{\infty}$, we have

$$\langle e^{in\theta}, M_f e^{im\theta}\rangle = \int f(e^{i\theta}) e^{i(m-n)\theta} \, \frac{d\theta}{2\pi} \tag{3.15.59}$$

$$= f_{n-m}^\sharp \tag{3.15.60}$$

where, as usual (see Section 3.5 of Part 1),

$$f_k^\sharp = \int_0^{2\pi} e^{-ik\theta} f(e^{i\theta}) \, \frac{d\theta}{2\pi} \tag{3.15.61}$$

Of course, M_f in this basis looks like (3.15.56) but as a doubly infinite matrix. What we've, of course, done here is realize T as a truncated convolution of sequences.

If $H^2(\partial\mathbb{D}, \frac{d\theta}{2\pi})$ is the boundary values of $H^2(\mathbb{D})$ functions (see Section 5.2 of Part 3 for a discussion of H^2), they are precisely the subset of $L^2(\partial\mathbb{D}, \frac{d\theta}{2\pi})$ of those f's with $f_n^\sharp = 0$ for $n < 0$. If P_+ is the projection onto $H^2(\partial\mathbb{D}, \frac{d\theta}{2\pi})$, then

$$T_f u \equiv P_+ M_f u \tag{3.15.62}$$

For any $f \in L^\infty(\partial\mathbb{D}, \frac{d\theta}{2\pi})$, T_f is a bounded operator whose matrix element in $\{e^{im\theta}\}_{m=0}^\infty$ basis is the Toeplitz matrix with $\varphi_k = f_k^\sharp$. T_f given by (3.15.64) is called the Toeplitz operator with symbol, f. In Problem 14, the reader will prove that any bounded Toeplitz matrix arises in this way.

We want to focus here on Toeplitz matrices whose symbol f is continuous. In Problem 15, the reader will prove that the C^*-algebra generated by the operator R of (3.15.13) (which is T_f for $f(e^{i\theta}) = e^{i\theta}$) is precisely all operators of the form $T_f + K$, where K is compact and f is continuous.

We want to see which T_f, $f \in C(\partial\mathbb{D})$, are Fredholm operators and compute their index and then compute the spectra of T_f's when $f \in C(\partial\mathbb{D})$. As a preliminary, we look at $P_- M_f P_+$, where $P_- = 1 - P_+$ on $L^2(\partial\mathbb{D}, \frac{d\theta}{2\pi})$, that is, $u \in \operatorname{Ran}(P_-)$ if and only if $u_n^\sharp = 0$ for $n \geq 0$. If

$$f_m(e^{i\theta}) = e^{im\theta} \tag{3.15.63}$$

then $P_- M_{f_m} P_+ = 0$ if $m \geq 0$ and rank $|m|$ if $m < 0$, since $P_- M_f P_+ u = 0$ if $u_k^\sharp = 0$ for $k = 0, 1, 2, \dots, |m|-1$; equivalently, $\operatorname{Ran}(P_- M_f P_+) = \{u \mid u_k^\sharp = 0$ if $k \notin \{m, m+1, \dots, -1\}\}$. Thus, $P_- M_f P_+$ is compact if f is a trigonometric polynomial. Since

$$\|M_f - M_g\| \leq \|f - g\|_\infty \tag{3.15.64}$$

by the Weierstrass approximation theorem, $P_- M P_+$ is compact for any $f \in C(\partial\mathbb{D})$.

Since $M_f M_g = M_{fg}$ and $T_f = P_+ M_f P_+$, we see that

$$T_f T_g - T_{fg} = P_+ M_f P_+ M_g P_+ - P_+ M_f M_g P_+ = -P_+ M_f P_- M_g P_+ \tag{3.15.65}$$

is compact for any $f, g \in C(\partial\mathbb{D})$.

Suppose f is everywhere nonvanishing on $\partial\mathbb{D}$ and $g = 1/f$. Since $T_{\mathbb{1}} = \mathbb{1}$ on $\operatorname{Ran}(P_+)$, we see T_g is a pseudoinverse for T_f, that is,

$$f \text{ nonvanishing on all of } \partial\mathbb{D} \Rightarrow T_f \text{ is Fredholm} \tag{3.15.66}$$

In Problem 16, the reader will show that if f does vanish, then T_f is not Fredholm.

A nonvanishing function f is equivalent to a closed curve in $\mathbb{C} \setminus \{0\}$. Every such curve has a winding number, $W(f)$, as we discussed in the Notes to Section 3.3 of Part 2A; see also Problem 17. If $W(f) = n$, there is a function $F\colon \partial\mathbb{D} \times [0,1] \to \mathbb{C} \setminus \{0\}$ so that $F(e^{i\theta}, 0) = f(e^{i\theta})$, $F(e^{i\theta}, 1) = e^{in\theta}$. Since $t \mapsto f_t \equiv F(\,\cdot\,, t)$ is continuous in $\|\cdot\|_\infty$, $t \mapsto T_{f_t}$ is continuous in operator norm and each T_{f_t} is Fredholm. By the homotopy invariance of index, $\operatorname{ind}(T_f) = \operatorname{ind}(T_{f_m})$ with f_m given by (3.15.63) and $m = W(f)$.

But T_{f_m} is R^m for $m \geq 0$ and L^{-m} if $m \leq 0$, so by (3.15.15) and $\operatorname{ind}(A^m) = m \operatorname{ind}(A)$, we see that $\operatorname{ind}(T_{f^m}) = -m$. Thus, for any $f \in C(\partial\mathbb{D})$, we have proven that

$$f \text{ everywhere nonvanishing} \Rightarrow \operatorname{ind}(T_f) = -W(f) \tag{3.15.67}$$

Finally, in studying Toeplitz operators with continuous symbols, we note that if $W(f) = 0$, one can prove that (Problems 18, 19)

$$W(f) = 0 \Rightarrow \operatorname{Ker}(T_f) = 0 = \operatorname{Ker}(T_f^*) \tag{3.15.68}$$

In particular,

$$W(f) = 0 \Rightarrow T_f \text{ is invertible} \tag{3.15.69}$$

This in turn lets us find the spectrum of any T_f, $T_f - z = T_{f-z}$. T_{f-z} is Fredholm if and only if $z \notin \operatorname{Ran}(f)$. $\operatorname{ind}(T_{f-z})$ is continuous on each connected component of $\mathbb{C} \setminus \operatorname{Ran}(f)$. If T_{f-z} is invertible, its index is 0, and by (3.15.69), if the index is 0, it is invertible. Thus,

$$\sigma(T_f) = \operatorname{Ran}(f) \cup \{\text{components of } \mathbb{C} \setminus \operatorname{Ran}(f) \mid \operatorname{ind}(T_f) \neq 0\} \tag{3.15.70}$$

□

We summarize in a theorem:

Theorem 3.15.22. *Let T_f be the Toeplitz operator with continuous symbol, f. Then*

(a) *T_f is Fredholm if and only if $f(e^{i\theta}) \neq 0$ for all $e^{i\theta} \in \partial\mathbb{D}$.*
(b) *If T_f is Fredholm, $\operatorname{ind}(T_f) = -W(f)$, the winding number of f.*
(c) *$\sigma(T_f)$ is given by*(3.15.70).

Example 3.15.23 (Aharonov–Casher Theorem)**.** Our final example involves physical poetry rather than careful mathematics. It also involves an unbounded operator which is not Fredholm, but where there is still an interesting index theorem-like quality. In two dimensions, $p_1 = \frac{1}{i}\frac{\partial}{\partial x_1}$, $p_2 = \frac{1}{i}\frac{\partial}{\partial x_2}$. If $\mathbf{a} = (a_1, a_2)$, a vector potential, is a smooth function then

$$B(x) = \left(\frac{\partial}{\partial x_1} a_2 - \frac{\partial}{\partial x_2} a_1\right)(x) \tag{3.15.71}$$

represents the magnetic field. We'll be interested in situations where B is C^∞ of compact support, but where

$$\phi_0 = \int B(x)\, d^2x \tag{3.15.72}$$

the flux, is not zero. Of necessity, a will not have compact support since, by Stoke's theorem, if B is supported inside $\{x \mid |x| < R\}$, then

$$\oint_{|x|=R} \mathbf{a} \cdot d\mathbf{x} = \phi_0 \tag{3.15.73}$$

A convenient choice for a will be to define

$$\phi(x) = \frac{1}{2\pi} \int \log(|x - x'|) B(x')\, d^2x' \tag{3.15.74}$$

so $\Delta\phi = -B$, and so

$$a \equiv \left(\frac{\partial\phi}{\partial y}, -\frac{\partial\phi}{\partial x} \right) \tag{3.15.75}$$

will obey (3.15.71).

The *Pauli equation* describes the quantum energy operator of an electron in magnetic field, B. The wave function is a function from $\mathbb{R}^2$ to $\mathbb{C}^2$ and

$$H = (p_1 - a_1)^2 + (p_2 - a_2)^2 + \sigma_3 B(x) \tag{3.15.76}$$

where $\sigma_3 = \left(\begin{smallmatrix} 1 & 0 \\ 0 & -1 \end{smallmatrix}\right)$. The connection between the first two terms in (3.15.76) and the spin term, $\sigma_3 B$, is made transparent by defining

$$\sigma_1 = \begin{pmatrix} 0 & 1 \\ 1 & 0 \end{pmatrix}, \qquad \sigma_2 = \begin{pmatrix} 0 & -i \\ i & 0 \end{pmatrix} \tag{3.15.77}$$

If for $\mathbf{c} = (c_1, c_2)$, we define $\not{c} = c_1\sigma_1 + c_2\sigma_2$, then

$$H = (\not{p} - \not{a})^2 \tag{3.15.78}$$

Among other things, this shows that $H \geq 0$, something that isn't obvious from (3.15.76). Indeed, if

$$Q = (p_1 - a_1) - i(p_2 - a_2) \tag{3.15.79}$$

(3.15.78) says

$$Q^*Q = (p_1 - a_1)^2 + (p_2 - a_2)^2 + B, \quad QQ^* = (p_1 - a_1)^2 + (p_2 - a_2)^2 - B \tag{3.15.80}$$

A suitable variant of $\sigma(xy) \setminus \{0\} = \sigma(yx) \setminus \{0\}$ (see (2.2.64) and Problem 5 of Section 2.2) says that the operators with $+B$ and $-B$ have the same spectrum except perhaps for zero.

Moreover, $Q^*Q\varphi = 0 \Leftrightarrow Q\varphi = 0$ and $QQ^*\varphi = 0 \Leftrightarrow Q^*\varphi = 0$ so that a measure of the difference between spin up and spin down is $\dim(\text{Ker}(Q)) - \dim(\text{Ker}(Q^*))$, which is a kind of index of Q (even though $\text{Ran}(Q)$ is not closed and Q is *not* Fredholm in this $\mathbb{R}^2$ case).

Next, notice that

$$Q\psi = 0 \Leftrightarrow \left(\frac{\partial}{\partial x_1} - i\,\frac{\partial}{\partial x_2}\right) e^{\phi(x_1,x_2)}\psi = 0 \tag{3.15.81}$$

$$Q^*\psi = 0 \Leftrightarrow \left(\frac{\partial}{\partial x_1} + i\,\frac{\partial}{\partial x_2}\right) e^{-\phi(x_1,x_2)}\psi = 0 \tag{3.15.82}$$

Thus, by the Cauchy–Riemann equations (for $z = x_1 + ix_2$),

$$\begin{aligned} Q^*\psi = 0 &\Leftrightarrow e^{-\phi}\psi \text{ is analytic} \\ Q\psi = 0 &\Leftrightarrow e^{\phi}\psi \text{ is antianalytic} \end{aligned} \tag{3.15.83}$$

We also want L^2-solutions (i.e., true eigenvalues). By (3.15.74),

$$\phi(x) = \frac{\phi_0}{2\pi}\log|x| + O(|x|^{-1}) \tag{3.15.84}$$

If $\phi_0 > 0$, then $e^{\pm\phi} = |z|^{\pm\phi_0/2\pi}(1 + O(|x|^{-1}))$. If $[\phi_0/2\pi]$ is the biggest integer less than $\phi_0/2\pi$, then $e^{+\phi}\psi$ is never L^2 and antianalytic. And $e^{-\phi}\psi$ is L^2 and analytic if and only if ψ is a polynomial of degree at most $[\phi_0/2\pi]-1$, that is,

$$\phi_0 \ge 0 \Rightarrow \text{Ker}(Q) = \{0\}, \qquad \dim(\text{Ker}(Q^*)) = \left[\frac{\phi_0}{2\pi}\right] \tag{3.15.85}$$

Similarly,

$$\phi_0 \le 0 \Rightarrow \text{Ker}(Q^*) = \{0\}, \qquad \dim(\text{Ker}(Q)) = \left[\frac{|\phi_0|}{2\pi}\right] \tag{3.15.86}$$

Thus, the "index" of Q is not given by an integral, but only the integral part of the integral. As we'll explain in the Notes, if one replaces $\mathbb{R}^2$ by $\mathbb{S}^2$ (the two-sphere), the flux, ϕ_0, is quantized, the analog of Q is Fredholm, and its index is $-\phi_0$. $\square$

Notes and Historical Remarks. In one sense, the name "Fredholm operator" is a misnomer since Fredholm's operators were all of the form $\mathbb{1} - K$ with K compact, so of index 0, as established by Riesz in his analysis discussed in Section 3.3. The phenomenon of nonzero index was first discovered by Fritz Noether (1884–1941) who found in 1921 [**506**] that certain singular integral equations related to the Hilbert transform could have nonzero index and even expressed the index as an integral, the first example of an index theorem.

Fritz Noether was Emmy Noether's brother and, like her, was dismissed in 1934 from his "permanent" professorship in Breslau because he was Jewish. Emmy accepted a position in Bryn Mawr in the U.S. (and died of complications connected to surgery for cancer a year later), but Fritz made the fateful decision of going to the Soviet Union, accepting a professorship in Tomsk. In 1937, he was arrested by the NKVD and sentenced for being a German spy. He was then accused of anti-Soviet propaganda and shot in 1941. In 1988, the Supreme Court of the Soviet Union exonerated him!

In 1943, in connection with the Bourbaki book on topological vector spaces, Dieudonné [**160**] noted the invariance under small perturbations of the index (Theorem 3.15.12). Because of the war, it was some time before this work was taken up, but in 1951, three mathematicians—Atkinson [**37**] (he was British but published his paper in a Russian journal), Goh'berg [**255**], and Yood [**767**]—published independent works on Fredholm operators. All three noted the invariance under compact operators; Atkinson introduced the notion of pseudoinverse and proved Theorem 3.15.8. Atkinson and Goh'berg proved that the composition of Fredholm operators is Fredholm and that (3.15.36) holds.

Lax [**432**] and Arveson [**31**] present approaches to the basics of Fredholm theory that first show $\text{ind}(ST) = \text{ind}(S) + \text{ind}(T)$ and use that to prove Dieudonné's theorem and invariance under compact perturbations. Our proof that $\text{ind}(ST) = \text{ind}(S) + \text{ind}(T)$ is taken from Kirillov–Grishiani [**390**]. Problems 11/12 have a third proof which is not as slick as these other proofs but may be more illuminating as to why the result is true. For another book reference, see Schechter [**599**].

It was, of course, the Atiyah–Singer index theorem that captured the imagination of mathematicians and raised the theory of Fredholm operators to an important place in the canon of analysis. In 1960, Gel'fand [**229**] suggested that the index of elliptic differential operators acting on vector bundles over compact manifolds should have a topological expression—he reasoned that changes in the leading term that preserved ellipticity and all changes of lower order were relatively compact, and so the index should be independent of the details of the operator, and thus intrinsic to the manifold. In 1963, Atiyah–Singer [**35**] announced their results, with a detailed paper appearing in 1968 [**36**]. In 1973, Atiyah–Bott–Patodi [**34**], following up on an idea of Patodi (that we would now call supersymmetry), gave a heat kernel proof. Getzler [**244**] found a simpler version of this proof. For a textbook version of this approach, see the book of Gilkey [**246**]. For the argument specialized to the case of the Gauss–Bonnet theorem, see Cycon et al. [**144**].

The extra element of the index theorem is that, often, the index has a formula as an integral—something we saw in some of the examples at the end of this section.

Our discussion of the index of a pair of projections follows Avron–Seiler–Simon [**40, 41**]. Earlier, Kato [**374**] had introduced the operator B and noted that $A^2 + B^2 = 1$. The supersymmetry relation $\{A, B\} = 0$ appeared first in [**41**] with the various consequences noted here. A complementary view of pairs of projections is in the paper of Davis [**149**]. That $P - Q \in \mathcal{I}_1 \Rightarrow \text{Tr}(P - Q) \in \mathbb{Z}$ was proven earlier by Effros [**183**]; see Problem 20. We'll provide another proof of this fact in Problem 1 of Section 5.9. Amrein–Sinha [**17**] have a different proof that $\text{Tr}(P - Q) \in \mathbb{Z}$ using, in part, an analysis of pairs of projections due to Halmos [**283**]. Kelton [**386**] has a proof that extends to a large class of Banach spaces, and in particular, to non–self-adjoint projections on a Hilbert space.

There is an enormous literature on Toeplitz operators (Example 3.15.21) because of their role in the theory of orthogonal polynomials [**648**] and to the exact solution of the two-dimensional Ising model; see, for example, Deift–Its–Krasovsky [**153**]. There is a close connection (both by analogy and via conformal mapping) to Wiener–Hopf operators. The connection of index of T_f and winding numbers for $f \in C(\partial\mathbb{D})$ is due to Goh'berg [**256**]. The result (3.15.70) on $\sigma(T_f)$ is due to Krein [**413**]. For comprehensive books on the subject, see Böttcher–Silbermann [**84**] and Douglas [**169**].

Matrices with $h_{n,m} = \varphi_{n+m}$ (+ rather than the − of Toeplitz matrices) are called *Hankel matrices.* Just as bounded Toeplitz matrices have the form $P_+ M_f P_+$ for $f \in L^\infty$, bounded Hankel matrices have the form $P_+ M_f P_-$ for $f \in L^\infty$. This result, Nehari's theorem, was proven in Problem 6 of Section 5.11 of Part 3 and Hartman's theorem on when Hankel matrices are compact was Problem 7 there. For a comprehensive book on Hankel matrices and operators, see Peller [**517**].

The Aharonov–Casher theorem (Example 3.15.23) appeared first in Aharonov–Casher [**7**] without rigor. Its analog for S^2 rather than $\mathbb{R}^2$, which is a strict use of a Fredholm operator, is due to Avron–Tomares [**39**]. $\mathbb{C}^2$ in that case is a bundle and the total flux is its Chern integer; see [**144**] for details of that case. Erdős–Vougalter [**189**] discuss the extension to B's with finite flux but not compact support.

Problems.

1. This problem will prove Proposition 3.15.1. Y is a vector space over $\mathbb{K} = \mathbb{R}$ or $\mathbb{C}$ (no topology is given).

   (a) Using Zorn's lemma, prove that if $Z \subset Y$ is a subspace, Z always has a complement (i.e., W so $W \cap Z = \{0\}$, $W + Z = Y$).

   (b) If $\widetilde{Z}$ is a complement and $y_1, \dots, y_k$ are in $\widetilde{Z}$ and independent, prove there are $\ell_1, \dots, \ell_k$ in $\mathcal{A}_{\text{alg}}(Z)$ so $\ell_j(y_k) = \delta_{jk}$ and conclude that $\dim(\mathcal{A}_{\text{alg}}(Z)) \geq k$.

   (c) If $\ell_1, \dots, \ell_k$ in $\mathcal{A}_{\text{alg}}(Z)$ are linearly independent, prove there must be $y_1, \dots, y_k \in \widetilde{Z}$ so the matrix $\ell_j(y_k)$ is invertible and conclude

$\dim(\widetilde{Z}) \geq k$. Thus, you've proven (1) $\Leftrightarrow$ (2) in the proposition and $\dim(\widetilde{Z}) = \dim(Y/Z)$.

(d) Prove (3.15.3) is a rewriting of $\dim(\widetilde{Z}) = \ell < \infty$.

(e) If $\pi\colon Y \to Y/Z$ is the canonical projection, since that π is a bijection of $\widetilde{Z}$ and Y/Z, conclude (2) $\Leftrightarrow$ (4) in Proposition 3.15.1 and $\dim(Y/Z) = \dim(\widetilde{Z})$.

2. Prove Proposition 3.15.2. (*Hint*: See Corollary 5.1.5 of Part 1 and use the Hahn–Banach theorem.)

3. (a) Let $f \in L^2([-1,1],dx)$ with $f \geq 0$. Prove there is $g \in L^2([-1,1],dx)$ so that for a.e. x, $g(x) \geq f(x)$ and $\lim_{x\to 0} f(x)/g(x) = 0$.

(b) Use (a) to prove $\{xf \mid f \in L^2([-1,1],dx)\}$ does not have finite codimension.

4. This problem will prove (3.15.22). Suppose that $T\colon X \to Y$ is a Fredholm operator $S\colon Y \to X$ such that $ST = \mathbb{1} + K$, $TS = \mathbb{1} + L$ with K, L compact. P is the projection onto $\bigl[\text{Ker}(\mathbb{1}+K)^k\bigr]$ and Q the projection onto $\text{Ker}\bigl[(\mathbb{1}+L)^k\bigr]$ for large k (i.e., the algebraic eigenspaces).

(a) Show it suffices to prove $X = X_0 \dotplus X_1 \dotplus X_2$, $Y = Y_0 \dotplus Y_1 \dotplus Y_2$ with $\text{Ran}(P) = X_1 \dotplus X_2$, $\text{Ran}(Q) = Y_1 \dotplus Y_2$, $\text{Ker}(T) = X_2$, $\text{Ran}(T) = Y_0 + Y_1$, and T is a bijection of X_1 onto Y_1.

(b) Show there exist bounded invertible maps $B\colon X \to X$ and $C\colon Y \to Y$ so

$$BST = STB = \mathbb{1} - P, \qquad CTS = TSC = \mathbb{1} - Q \tag{3.15.87}$$

(*Hint*: Functional calculus.)

(c) Prove $X_2 \equiv \text{Ker}(T) \subset \text{Ran}(P)$. Let X_1 be a complementary subspace to X_2 in $\text{Ran}(P)$ and $X_0 = \text{Ran}(\mathbb{1} - P)$. Show that $X = X_0 \dotplus X_1 \dotplus X_2$.

(d) Let $Y_0 = T[X_0]$, $Y_1 = T[X_1]$ and Y_2 a complement of $\text{Ran}(T) \cap \text{Ran}(Q)$ in $\text{Ran}(Q)$.

(e) Prove that $\text{Ran}(\mathbb{1} - Q) \subset \text{Ran}(T)$ and show that $Y = X_0 \dotplus Y_1 \dotplus Y_2$.

(f) Let $y_0 = Tx_0$ for $x_0 \in X_0$ so $y_0 \in Y_0$. Prove that

$$(1 - Q)y_0 \neq 0 \tag{3.15.88}$$

(*Hint*: First show $Sy_0 \neq 0$ since B is invertible and $BSTx_0 = x_0$. Then show $STSy_0 = B^{-1}STx_0 \neq 0$ so $TSy_0 \neq 0$. Then show $CTSy_0 \neq 0$ and use $CTS = \mathbb{1} - Q$.)

(g) If $(1 - Q)[y_0 + y_1 + y_2]$ with $y_a \in Y_a$, show $y_0 = 0$ and conclude $\text{Ran}(Q) = Y_1 \dotplus Y_2$.

(h) Check that this completes the proof.

5. (a) Prove $\ell \in \mathrm{Ker}(T^t)$ if and only if $\ell \restriction \mathrm{Ran}(T) = 0$ and conclude that

$$\dim(\mathrm{Ker}(T^t)) = \mathrm{codim}(\mathrm{Ran}(T)) \tag{3.15.89}$$

(b) Prove that

$$\dim(\mathrm{Ker}(T)) = \mathrm{codim}(\mathrm{Ran}(T^t)) \tag{3.15.90}$$

(*Hint*: Use (3.15.24).)

6. Suppose $A \in \mathcal{L}(\mathcal{H})$ is such that there exists $\{\varphi_n\}_{n=1}^\infty \in \mathcal{H}$ with $\varphi_n \xrightarrow{w} 0$, $\|\varphi_n\| = 1$, and $\|A\varphi\|_n \to 0$. Prove that A is not Fredholm. (*Hint*: If B is a pseudoinverse, use $1 = \|(BA - K)\varphi_n\|$ for K compact.)

7. Let $f\colon [0,1] \to \mathbb{R}$ so that for all $x \in [0,1]$, $\exists \varepsilon > 0$ so that f is constant on $(x-\varepsilon, x+\varepsilon) \cap [0,1]$. Prove that $f(\gamma(0)) = f(\gamma(1))$. (*Hint*: Cover $[0,1]$ with finitely many such intervals and show you can order them to overlap.)

8. Let A be a self-adjoint operator on a Hilbert space, $\mathcal{H}$.

   (a) If 0 is not an isolated point of $\sigma(A)$, prove that $\mathrm{Ran}(A)$ is not closed.

   (b) Prove A is Fredholm if and only if 0 is not in $\sigma(A)$ or is in $\sigma_d(A)$.

9. If A is self-adjoint and 0 is not an isolated point of $\sigma(A)$, find A_n so $\|A - A_n\| \to 0$ and so $\mathrm{Ker}(A_n)$ is infinite-dimensional. Why does this show that A is not Fredholm?

10. (a) Let $P - Q$ be compact. If $\mathrm{Ker}(QP)$ is finite-dimensional, prove $\mathrm{Ran}(1-P)$ is finite-dimensional and thus, so is $\mathrm{Ran}(1-Q)$.

    (b) If $\mathrm{Ran}(1-P)$ and $\mathrm{Ran}(1-Q)$ are finite-dimensional, prove that QP as an operator on $\mathcal{L}(\mathcal{H})$ is Fredholm and $\mathrm{ind}(PQ) = 0$. (*Hint*: $Q - QP$ is compact.)

11. Let T in $\mathcal{L}(X,Y)$ be Fredholm. You can use the invariance of index under compact perturbation but not that $\mathrm{ind}(RT) = \mathrm{ind}(R) + \mathrm{ind}(T)$.

    (a) Prove $\mathrm{Ker}(T) = \{0\} \Leftrightarrow \exists\, S \in \mathcal{L}(Y,X)$ so that $ST = \mathbb{1}_X$.

    (b) If $\mathrm{ind}(T) \le 0$, prove there is a compact C so that $\mathrm{Ker}(T+C) = \{0\}$ and conversely, if there is such a C, then $\mathrm{ind}(T) \le 0$.

    (c) Prove $\mathrm{Ran}(T) = Y \Leftrightarrow \exists\, S \in \mathcal{L}(Y,X)$ so that $TS = \mathbb{1}_Y$.

    (d) If $\mathrm{ind}(T) \ge 0$, prove there is a compact C so that $\mathrm{Ran}(T+C) = Y$ and conversely, if there is such a C, then $\mathrm{ind}(T) \ge 0$.

    (e) Prove there is a compact C with $T + C$ invertible if and only if $\mathrm{ind}(T) = \{0\}$.

12. This problem will lead the reader through an alternate proof of Theorem 3.15.17.

    (a) If $R \in \mathcal{L}(Y, Z)$ and $T \in \mathcal{L}(X, Y)$ are Fredholm, and S is a pseudoinverse for T and Q for R, prove that SQ is a pseudoinverse for RT and conclude that RT is Fredholm.

    (b) Prove that

$$\operatorname{ind}(RT) = \operatorname{ind}(R) + \operatorname{ind}(T) \tag{3.15.91}$$

    if $\operatorname{Ker} R = 0$, $\operatorname{Ker} T = 0$.

    (c) Prove (3.15.91) holds if $\operatorname{ind}(R) \leq 0$, $\operatorname{ind}(T) \leq 0$ (Hint: Use Problem 11 (b))

    (d) As in (b), (c), prove (3.15.91) holds if $\operatorname{ind}(R) \geq 0$, $\operatorname{ind}(T) \geq 0$.

    (e) Suppose $\operatorname{ind}(T) \geq 0$, $\operatorname{ind}(R) < 0$, and $\operatorname{ind}(RT) \leq 0$. Let S be a pseudoinverse for T. Prove that $\widetilde{R} \equiv (RT)(S)$ has $\operatorname{ind}(\widetilde{R}) = \operatorname{ind}(R)$ and then that $\operatorname{ind}(RT)S = \operatorname{ind}(RT) + \operatorname{ind}(S)$ and conclude (3.15.91) holds.

    (f) Part (e) discusses one case where $\operatorname{ind}(T)$ and $\operatorname{ind}(R)$ have opposite signs. Do the other three cases.

13. Verify (3.15.49)/(3.15.50).

14. This problem will prove that any Toeplitz matrix is the matrix of a Toeplitz operator with L^∞-symbol.

    (a) If T is a matrix of the form (3.15.56) with $\|Tu\| \leq C\|u\|$, prove that $\sum_{n=-\infty}^{\infty} |\varphi_j|^2 < \infty$.

    (b) Prove that if f is the L^2-function with $f_k^\sharp = \varphi_k$, then (3.15.62) holds for n a trigonometric polynomial in $\operatorname{Ran}(P_+)$.

    (c) If $g \in L^2$, prove $\|g\|_2 = \lim_{n\to\infty} \|P_+(z^n g)\|_2$ and conclude that if $\|T_f u\|_2 \leq C\|u\|_2$ for all trigonometric polynomials in H, then $\|M_f u\|_2 \leq C\|u\|_2$ for all trigonometric polynomials, u, in L^2.

    (d) Prove that T is a T_f for some $f \in L^\infty$.

15. Let R be the operator (3.15.13). This problem will prove that the smallest C^*-algebra containing R is $\{T_f + K \mid f \in C^\infty(\partial\mathbb{D}),\ K \text{ compact}\} \equiv \mathfrak{A}_0$.

    (a) Prove that $\|T_f\| = \|f\|_\infty$ for $f \in C(\partial\mathbb{D})$.

    (b) Prove that $\mathfrak{A}_0$ is norm-closed.

    (c) Prove that $\mathfrak{A}_0$ is an algebra.

    (d) Prove that $T_f^* = T_{\bar{f}}$ and conclude that $\mathfrak{A}_0$ is a C^*-algebra.

    (e) Compute $F = R^*R - RR^*$ and then $R^k F (R^*)^\ell$ and conclude that the C^*-algebra generated by R contains all compact operators.

(f) Prove every T_f lies in the C^*-algebra generated by R. Conclude $\mathfrak{A}_0$ is the C^*-algebra generated by R.

16. Let $f \in C(\partial\mathbb{D})$ with $f(e^{i\theta_0}) = 0$. This problem will show that T_f is not Fredholm.

(a) Find g_n so $\|g_n\|_2 = 1$ and $\|fg_n\| \to 0$ as $n \to \infty$.

(b) By picking $k_n \to \infty$, suitably find g_n so $\|P_+ z^{k_n} g_n\| \to 1$ and conclude there is h_n with $\|h_n\|_2 = 1$, $h_n \in \operatorname{Ran}(P_+)$, $h_n \to 0$ weakly, and $\|T_f h_n\| \to 0$.

(c) Conclude the T_f is not Fredholm. (*Hint*: See Problem 6.)

17. (a) Let $\gamma\colon [0,1] \to \partial\mathbb{D}$. Prove there is $\tilde{\gamma}\colon [0,1] \to \mathbb{R}$ so that $\gamma(t) = \exp(2\pi i\tilde{\gamma}(t))$. If $\gamma(0) = \gamma(1)$, prove that $\tilde{\gamma}(1) = \tilde{\gamma}(0) + W(\gamma)$ for an integer, $W(\gamma)$.

(b) Let $\gamma, \tilde{\gamma}\colon [0,1] \to \partial\mathbb{D}$ with $\gamma(0) = \gamma(1) = \tilde{\gamma}(0) = \tilde{\gamma}(1) = 1$. Prove there exists $\Gamma\colon [0,1]\times[0,1] \to \partial\mathbb{D}$ with $\gamma(0,s) = \Gamma(1,s)$ and $\gamma(t,0) = \gamma(t)$, $\gamma(t,1) = \tilde{\gamma}(t)$ if and only if $W(\gamma) = W(\tilde{\gamma})$. In particular, for any γ, if $\tilde{\gamma}(t) = \exp[2\pi i t W(\gamma)]$, then there exists such a Γ.

(c) Let $f \in C(\partial\mathbb{D})$ with $f(e^{i\theta}) \neq 0$ for all θ. Let $m = W(f(e^{i\theta})/|f(e^{i\theta})|)$. Prove $\operatorname{ind}(T_f) = -m$.

18. (a) Let $f \in C(\partial\mathbb{D})$ with $f(e^{i\theta}) \neq 0$ for all θ and $W(f) = 0$. Prove there is $\ell \in C(\partial\mathbb{D})$ so $f(e^{i\theta}) = \exp(i\ell(\theta))$.

(b) If f is C^1, prove $\ell = \ell_+ + \ell_-$, where $\ell_\pm$ are in $C(\partial\mathbb{D})$ and $P_\pm \ell_\pm = \ell_\pm$. Conclude $f = f_- f_+$ with $f_\pm \in C(\partial\mathbb{D})$ nonvanishing and $f_+, f_+^{-1} \in H_+$, $f_-, f_-^{-1} \in H_+^\perp$.

(c) If $T_f u = 0$ for $u \in H_+$, show $f_+ u \in H_+^\perp$ and conclude that $f_+ u = 0$ and then that $u = 0$, that is, conclude that $\operatorname{Ker}(T_f) = 0$.

(d) Using $T_f^* = T_{\bar{f}}$ and $W(\bar{f}) = 0$, prove $\operatorname{Ran}(T_f) = H_+$ and prove that T_f is invertible.

(e) If now f is not assumed C^1 but only has $W(f) = 0$, pick g C^1 with $W(g) = 0$ and $\|fg^{-1} - 1\|_\infty < 1$. Prove that with $g = g_+ g_-$ as in (b) above,

$$T_{fg^{-1}} = T_{g_-^{-1}} T_f T_{g_+^{-1}}$$

(f) Knowing that $T_{g_\pm^{-1}}$ are invertible, prove that T_f is invertible, that is, you have proven that $f \in C(\partial\mathbb{D})$, $W(f) = 0 \Rightarrow T_f$ is invertible.

19. (a) If $W(f) \geq 0$, prove that $\operatorname{Ker}(T_f) = 0$. (*Hint*: If $\operatorname{Ker}(T) = \operatorname{Ker}(S) = \{0\}$, then $\operatorname{Ker}(TS) = 0$.)

(b) If $W(f) \leq 0$, prove that $\operatorname{Ran}(T_f) = H_+$.

20. This will lead the reader through Effros' proof that $\operatorname{Tr}(P-Q) \in \mathbb{Z}$ if P, Q are self-adjoint projections with $P - Q \in \mathcal{I}_1$.

(a) Let $S = (P-Q)^2$. Prove that $[P, S] = [Q, S] = 0$ (either by direct calculation or using $A^2 + B^2 = 1$).

(b) Let $S = \sum_{j=0}^{M} \alpha_j R_j$ be the Hilbert–Schmidt expansion for S where R_j are projections and $R_\infty =$ projection onto $\operatorname{Ker}(S)$. Prove that $\operatorname{Tr}(P - Q) = \operatorname{Tr}((P-Q)R_\infty) + \sum_{j=1}^{M} \operatorname{Tr}((P-Q)R_j)$ where the sum is absolutely convergent.

(c) Prove that $\operatorname{Tr}((P-Q)R_\infty) = 0$ and $\operatorname{Tr}((P-Q)R_j) \in \mathbb{Z}$.

(d) Conclude that $\operatorname{Tr}(P-Q) \in \mathbb{Z}$.

3.16. Bonus Section: M. Riesz's Criterion

To see if an operator is compact on $L^p(\mathbb{R}^\nu, d^\nu x)$, it is useful to know when subsets are compact. Remarkably, there is a simple necessary and sufficient condition for $S \subset L^p(\mathbb{R}^\nu, d^\nu x)$ to be $\|\cdot\|_p$-compact. Clearly, if S is compact, it must lie in a ball since $\|\cdot\|_p$ is continuous, so we'll state it for $S \subset [L^p(\mathbb{R}^\nu, d^\nu x)]_1$, the unit ball.

Theorem 3.16.1 (M. Riesz's Criterion). *Let $1 \le p < \infty$. Let $S \subset [L^p(\mathbb{R}^\nu, d^\nu x)]_1$. Then $\overline{S}$ is compact if and only if*

(i) *S is uniformly small at infinity, that is, for all $\varepsilon > 0$, $\exists R_\varepsilon$ so*

$$\sup_{f \in S} \int_{|x| \ge R_\varepsilon} |f(y)|^p \, d^\nu y \le \varepsilon^p \tag{3.16.1}$$

(ii) *S is uniformly L^p-continuous, that is, for all $\varepsilon > 0$, $\exists \delta$ so that*

$$\sup_{f \in S} \sup_{\substack{y \in \mathbb{R}^\nu \\ |y| < \delta}} \int |f(x-y) - f(x)|^p \, d^\nu x \le \varepsilon^p \tag{3.16.2}$$

Proof. Suppose first $\overline{S}$ is compact. Given ε, find $f_1, \dots, f_n$ in S so that the $\varepsilon/3$ balls about f_j cover $\overline{S}$. For each j,

$$\lim_{R \to \infty} \int_{|x| \ge R} |f_j(x)|^p \, d^\nu x = 0 \tag{3.16.3}$$

so we can find R so that the integral is less than $(2\varepsilon/3)^p$ for each j. With $\tilde{\chi}_R$ the characteristic function of $\{x \mid |x| \ge R\}$, we have for $f \in B_{\varepsilon/3}(f_j)$,

$$\|\tilde{\chi}_R f\|_p \le \|\tilde{\chi}_R f_j\|_p + \|f - f_j\|_p \le \varepsilon \tag{3.16.4}$$

Similarly, if

$$(\tau_y f)(x) = f(x-y) \tag{3.16.5}$$

then $y \mapsto \tau_y g$ is L^p-norm continuous for each $g \in L^p$ (first for C_0^∞-functions and then, by density, for all g). Thus, find δ so that

$$\sup_{j=1,\dots,n} \sup_{|y|<\delta} \|f_j - \tau_y f_j\|_p < \frac{\varepsilon}{3} \tag{3.16.6}$$

If $f \in B_{\varepsilon/3}(f_j)$, then

$$\|f - \tau_y f\|_p \le 2\|f - f_j\|_p + \|f_j - \tau_y f_j\| < \varepsilon \tag{3.16.7}$$

if $|y| < \delta$. Therefore, compactness of $\overline{S}$ implies (i), (ii).

Conversely, suppose S obeys (i), (ii). Let $Q_{\alpha,\beta,\gamma}$ be defined as

$$Q_{\alpha,\beta,\gamma} = \{f \in C(\overline{B_\alpha(0)}) \mid \|f\|_\infty \le \beta,\ |f(x) - f(y)| \le \gamma|x - y|\} \tag{3.16.8}$$

By the Ascoli–Arzelà theorem (Theorem 2.3.14 of Part 1), $Q_{\alpha,\beta,\gamma}$ is compact in $C(\overline{B_\alpha(0)})$ and so, since $\|f\|_p \le \|f\|_\infty |B(0)|^{1/p}$ (which means the identity map of $C(\overline{B_\alpha(0)})$ to $L^p(\mathbb{R}^\nu)$ is continuous), $Q_{\alpha,\beta,\gamma}$ is compact in $L^p(\mathbb{R}^\nu)$.

Suppose we prove that for each ε, there is α, β, γ so that

$$S \subset \{g \mid \inf_{f \in Q_{\alpha,\beta,\gamma}} \|f - g\|_p \le \varepsilon\} \tag{3.16.9}$$

Then $\overline{S}$ is compact. For given $\{g_n\}_{n=1}^\infty \subset S$ and $m = 1, 2, \dots$, find $Q_{\alpha_m,\beta_m,\gamma_m}$ so (3.16.9) holds with $\varepsilon = 1/2m$. Find $f_n^{(m)}$ so $\|f_n^{(m)} - g_m\| \le 1/m$ and a convergent subsequence of the $f_n^{(m)}$. Then $\|g_{n(j)} - g_{n(k)}\| \le 2/m$ if $j, k \ge N_m$. Doing this inductively so each subsequence is a subsubsequence of the one before and using the diagonalization trick, we get a Cauchy and so convergent subsequence of $\{g_n\}_{n=1}^\infty$. Therefore, (3.16.9) implies compactness of $\overline{S}$.

Given ε, pick R so (3.16.1) holds for $(\varepsilon/4)^p$ and δ so that (3.16.2) holds for $(\varepsilon/4)^p$. Pick $\eta \in C_0^\infty$ supported in $B_\delta(0)$ with $\eta \ge 0$ and $\int \eta(x)\, d^\nu x = 1$. Let (with q the dual index to p)

$$\alpha = R + \delta, \qquad \beta = \|\eta\|_q, \qquad \gamma = \|\nabla\eta\|_q \tag{3.16.10}$$

We'll prove that (3.16.9) holds.

Let χ_R be the characteristic function of $\overline{B_R(0)}$. Since $\|\tau_y(\chi_R f) - \chi_R f\|_p \le \|\tau_y f - f\| + 2\|(1 - \chi_R)f\|$, we see that

$$\sup_{|y|\le\delta} \sup_{f\in S} \|\tau_y(\chi_R f) - \chi_R f\| \le \frac{3\varepsilon}{4} \tag{3.16.11}$$

Let $g = \eta * (\chi_R f)$, the convolution of $\chi_R f$ and η. Then g is C^∞ and, by Hölder's inequality,

$$\|g\|_\infty \le \beta, \qquad \|\nabla g\|_\infty \le \gamma \tag{3.16.12}$$

Clearly, g is supported on $\overline{B_\alpha(0)}$. Thus, $g \in Q_{\alpha,\beta,\gamma}$ and

$$\|g - f\|_p \le \|g - \chi_R f\|_p + \|(1-\chi_R)f\|_p \le \frac{3\varepsilon}{4} + \frac{\varepsilon}{4} \tag{3.16.13}$$

on account of (3.16.11). □

As an immediate corollary, we have

Theorem 3.16.2 (Rellich's Criterion). *Let F, G be two continuous functions on $\mathbb{R}^\nu$ with*

$$\forall y,\ |F(y)| \ge 1, \quad |G(y)| \ge 1, \qquad \lim_{|y|\to\infty} |F(y)| = \infty = \lim_{|y|\to\infty} |G(y)| \tag{3.16.14}$$

Fix $a, b \ge 0$. Let $S \subset L^2(\mathbb{R}^\nu, d^\nu x)$ be given by

$$S = \left\{ \varphi \;\middle|\; \int F(x)|\varphi(x)|^2\, d^\nu x \le a,\ \int G(k)|\widehat{\varphi}(k)|^2\, d^\nu x \le b \right\} \tag{3.16.15}$$

Then S is compact.

Remark. Unlike Riesz's criterion, this one is only sufficient, not necessary, but it covers many applications.

Proof. Given ε, pick R_ε so that $F(x) \ge a/\varepsilon^2$ if $|x| \ge R_\varepsilon$. Thus, by the first inequality in (3.16.4),

$$\int_{|x|\ge R_\varepsilon} |\varphi(x)|^2\, d^\nu x \le \frac{\varepsilon^2}{a} \int F(x)|\varphi(x)|^2\, d^\nu x \le \varepsilon^2 \tag{3.16.16}$$

proving (3.16.1) for S.

In the same way, we can find K_ε so that for $\varphi \in S$,

$$\int_{|k|\ge K_\varepsilon} |\widehat{\varphi}(k)|^2\, d^\nu x \le \left(\frac{\varepsilon}{3}\right)^2 \tag{3.16.17}$$

Let $\delta = \varepsilon/3K_\varepsilon$. Then $|e^{i\alpha} - 1| \le |\alpha|$ (for α real) implies

$$\sup_{\substack{|k|\le K_\varepsilon \\ |y|\le\delta}} |e^{ik\cdot y} - 1|^2 \le \left(\frac{\varepsilon}{3}\right)^2 \tag{3.16.18}$$

It follows that for $\varphi \in S$,

$$\begin{aligned}\|\tau_y\varphi - \varphi\|_2 &= \|\widehat{\tau_y\varphi} - \widehat{\varphi}\|_2 \\ &\le \frac{2\varepsilon}{3} + \|\chi_{|k|\le K_\varepsilon}(e^{ik\cdot y} - 1)\widehat{\varphi}\|_2 \\ &\le \frac{2\varepsilon}{3} + \frac{\varepsilon}{3} = \varepsilon\end{aligned} \tag{3.16.19}$$

proving (3.16.2). Thus, by Riesz's criterion, S is compact. □

Some of the more significant applications of Rellich's criterion involve unbounded operators on $L^2(\mathbb{R}^\nu)$. We'll quote here two results discussed later:

(1) (See Section 7.2) There is a class of densely defined operators called self-adjoint so if B is such an operator, we can define $f(B)$ for any Borel function $f\colon \mathbb{R} \to \mathbb{R}$ and $f(B)$ is also self-adjoint. If f is bounded, so is $f(B)$. If $\langle \psi, B\psi \rangle \geq 0$ for all ψ in the domain of B (we write $B \geq 0$), f need only be defined on $[0, \infty)$. In particular, if $a > 0$, $(B+a)^{-1}$ is well-defined. Moreover, formulae like

$$(B+1)^{-1/2}(B-1)(B+1)^{-1/2} = 1 \tag{3.16.20}$$

that are true for bounded self-adjoint operators continue to hold.

(2) (See Section 7.5) Let $D \subset \mathcal{H}$ be a dense subset of a Hilbert space, $\mathcal{H}$, and $A\colon D \to \mathcal{H}$ a linear operator with $\langle \varphi, A\varphi \rangle \geq 0$ for all $\varphi \in D$. Then there is a (perhaps unbounded) self-adjoint operator, A_F, the Friedrichs extension, so that for $\varphi \in D$, $A_F\varphi = A\varphi$, and $A_F \geq 0$, and for any $\psi \in D(A_F^{1/2})$, there is $\{\varphi_n\} \subset D$, so $\|\varphi - \varphi_n\| + \|A_F^{1/2}(\varphi - \varphi_n)\| \to 0$.

These two facts imply the following: In the context of (2), if

$$\{\varphi \in D \mid \langle \varphi, (A+1)\varphi \rangle \leq 1\} \text{ has compact closure} \tag{3.16.21}$$

then $(A_F+1)^{-1}$ is a compact operator. For, by (2),

$$\{(A_F+1)^{-1/2}\eta \mid \|\eta\| \leq 1,\ (A_F+1)^{-1/2}\eta \in D\} \equiv X \tag{3.16.22}$$

is dense in $(A_F+1)^{-1/2}[\{\eta \in \mathcal{H} \mid \|\eta\| = 1\}]$, so it suffices to prove X has compact closure. But, by (3.16.20), if $\varphi \in X$,

$$\langle \varphi, (A+1)\varphi \rangle = \langle \eta, (A_F+1)^{-1/2}(A_F+1)(A_F+1)^{-1/2}\eta \rangle = \|\eta\| \leq 1$$

so the compactness criterion (3.16.21) implies $(A_F+1)^{-1/2}$ is compact, which implies $(A_F+1)^{-1} = [(A_F+1)^{-1/2}]^2$ is compact. By the Hilbert–Schmidt theorem, this implies in turn that A_F has a complete set of eigenvectors, that is, there is an orthonormal basis of vector $\{\varphi_n\}_{n=1}^\infty$ in $D(A_F)$ so $A_F\varphi_n = \lambda_n\varphi_n$ with $\varphi_n \geq 0$ and $\lambda_n \to \infty$.

Example 3.16.3 (Dirichlet Laplacians). Let Ω be a bounded region (connected open set) in $\mathbb{R}^\nu$. Let $D = C_0^\infty(\Omega)$ viewed initially as a subset of $L^2(\mathbb{R}^\nu)$. We claim

$$S = \left\{\varphi \in D \,\middle|\, \int (|\nabla\varphi|^2 + |\varphi|^2)\, d^\nu x \leq 1\right\} \tag{3.16.23}$$

is a compact subset of $L^2(\mathbb{R}^\nu)$. For let $F(y) = G(y) = |y|^2 + 1$. Then $\varphi \in S \Rightarrow \int F(x)|\varphi(x)|^2\, d^\nu x \leq (1+b^2)$, where $b = \sup_{y\in\Omega}|y|$ and $\int G(k)|\widehat{\varphi}(k)|^2\, d^\nu k \leq 1$. By Rellich's criterion, S is contained in a compact set and so compact. Since $\mathcal{H}_\Omega \equiv L^2(\Omega, d^\nu x) \to L^2(\mathbb{R}^\nu, d^\nu x)$ a homeomorphism onto its range, S is also compact in $\mathcal{H}_\Omega$. D is dense in $\mathcal{H}_\Omega$ (but not in $L^2(\mathbb{R}^\nu, d^\nu x)$, of course). If $A\varphi = -\Delta\varphi$, we have

$$\langle \varphi, A\varphi \rangle = \|\nabla\varphi\|^2 \tag{3.16.24}$$

Its Friedrichs extension defines an operator called the *Dirichlet Laplacian* for Ω, $H_0^{(\Omega)}$. When Ω has a smooth boundary, there is a sense in which this is a Laplacian with Dirichlet boundary conditions, that is, $\varphi = 0$ on $\partial\Omega$. This will be discussed in Section 7.5.

By the above analysis, $(H_0^\Omega + 1)^{-1}$ is a compact operator and it has a complete set of eigenfunctions. □

Example 3.16.4 (Higher-Dimensional Sturm–Liouville Theory). Let Ω be a bounded region as above, V a nonnegative L^2-function on Ω. If $D = C_0^\infty(\Omega)$ and $A\varphi = -\Delta\varphi + V\varphi$, we have

$$\langle \varphi, A\varphi \rangle = \|\nabla\varphi\|^2 + \langle \varphi, V\varphi \rangle \geq \|\nabla\varphi\|^2 \tag{3.16.25}$$

so, as above, the Friedrichs extension has a compact resolvent. For $\Omega = (a,b) \subset \mathbb{R}$, this specializes to a Sturm–Liouville operator, so the $\mathbb{R}^\nu$-result generalizes Theorem 3.2.7. As we'll see in Sections 7.5 and 7.6, one can handle $-\Delta + V$ with $V \in L^1(\Omega, d^\nu x), V \geq 0$ with "quadratic form methods." □

Example 3.16.5 (Trapping Schrödinger Operators). Let V be continuous on $\mathbb{R}^\nu$ with

$$\lim_{|x|\to\infty} V(x) = \infty \tag{3.16.26}$$

Let $D = C_0^\infty(\mathbb{R}^\nu)$ and $A\varphi = -\Delta\varphi + V\varphi$. Then

$$\langle \varphi, (A+1)\varphi \rangle = \|\nabla\varphi\|^2 + \|\varphi\|^2 + \int V(x)|\varphi(x)|^2\, d^\nu x \tag{3.16.27}$$

By Rellich's criterion with $F(y) = 1 + V(y)$, $G(y) = 1 + |y|^2$, $\{\varphi \in D \mid \langle \varphi, (A+1)\varphi \rangle \leq 1\}$ lies in a compact subset, so A_F has compact resolvent and a complete set of eigenfunctions. □

Example 3.16.6. In $\mathbb{R}^2$, let

$$V(x_1, x_2) = x_1^2 x_2^2 \tag{3.16.28}$$

We'll show $-\Delta + V$ has compact resolvent even though (3.16.26) fails, showing that condition is not necessary.

In Section 6.4 of Part 1, we proved there is an orthonormal basis on $L^2(\mathbb{R}, dx)$ of eigenfunctions, $\{\varphi_n\}_{n=0}^\infty$, of $H_0 = \frac{1}{2}(-\frac{d^2}{dx^2} + x^2 - 1)$ with $H_0\varphi_n = n\varphi_n$. In particular, if $f \in \mathcal{S}$,

$$\|\nabla f\|^2 + \|xf\|^2 \geq \|f\|^2 \tag{3.16.29}$$

By scaling $x \to \omega^{1/2}x$,

$$-\frac{d^2}{dx^2} + \omega^2 x^2 \geq |\omega| \tag{3.16.30}$$

Thus, in $\mathbb{R}^2$,

$$-\frac{\partial^2}{\partial x_1^2} + x_1^2 x_2^2 \geq |x_2| \tag{3.16.31}$$

Multiplying by $\frac{1}{2}$ and using $x_1 \leftrightarrow x_2$ symmetry, we see

$$A_1 \equiv -\Delta + x_1^2 x_2^2 \geq -\tfrac{1}{2}\,\Delta + \tfrac{1}{2}\,(|x_1| + |x_2|) \equiv A_2 \tag{3.16.32}$$

Since $|x_1| + |x_2| \to \infty$ as $|x| \to \infty$, $\{\varphi \mid \langle \varphi, A_2\varphi\rangle \leq 1\}$ has compact closure, so by (3.16.32), $\{\varphi \mid \langle \varphi, A_1\varphi\rangle \leq 1\}$ has compact closure, that is, $(A_1)_F$ has compact resolvent. □

Notes and Historical Remarks. $L^p(\mathbb{R}^\nu, d^\nu x)$-compactness criteria were heavily studied in the early 1930s. In 1931, Kolmogorov [**399**] found necessary and sufficient conditions for $S \subset L^p([a,b], d^\nu x)$ to be compact: $\sup_{f\in S}\|f\|_p < \infty$ and $\forall\varepsilon$, $\exists\delta$ so that $|y| \leq \delta \Rightarrow \|A_y f - f\|_p \leq \varepsilon$, where $(A_y f)(x) = |2y|^{-1}\int_{x-y}^{x+y} f(w)\,dw$. This has caused some authors to call Theorem 3.16.1 the "Kolmogorov compactness criteria." Theorem 3.16.1 first appeared in 1933 in M. Riesz [**568**]. Theorem 3.16.2 (Rellich's criterion) is from Rellich's 1930 paper [**555**].

We will have a lot more to say about Dirichlet Laplacians in Section 7.5. The argument in Example 3.16.6 is taken from Simon [**643**], which has five proofs that the operator A_1 in (3.16.32) has compact resolvent! For more on this and related phenomena, see Rellich [**557**], Robert [**574**], Simon [**642, 652**], Solomyak [**660**] and Tamura [**695**]. For a more general context, see Fefferman–Phong [**196**] and Maz'ya [**471**].

Chapter 4

Orthogonal Polynomials

Big Notions and Theorems: Orthogonal Polynomials, OPRL, OPUC, Recursion Relations, Jacobi Parameters, Verblunsky Coefficients, Favard's Theorem, Real Roots for OPRL, Interlacing Roots for OPRL, Classical OPs, Hermite Polynomials, Laguerre Polynomials, Jacobi Polynomials, Bessel Polynomials, Bochner–Brenke Theorem, L^2-Variational Principle, Chebyshev Polynomials, Alternation Principle, Faber–Fekete–Szegő Theorem, Fekete Points, Szegő Recursion, Zeros in Unit Disk, Verblunsky's Theorem, Geronimus–Wendroff Theorem, Bernstein–Szegő Approximation, Szegő's Theorem, Szegő Condition, Szegő Asymptotics

Let μ be a probability measure on $\mathbb{C}$ with finite moments, that is,

$$\int |z|^n \, d\mu(z) < \infty \tag{4.0.1}$$

for all $n = 0, 1, 2, \dots$. Let μ be *nontrivial* in the sense that it is not supported at a finite set of points. Then $\{z^n\}_{n=0}^{\infty}$ are in $L^2(\mathbb{C}, d\mu)$ (by (4.0.1)) and are independent (by nontriviality, if $\int |f(z)|^2 \, d\mu = 0$, then f vanishes at infinitely many points). Thus, by Gram–Schmidt (see Section 3.4 of Part 1), we can form *monic orthogonal polynomials*, $\Phi_n(z)$, that is,

$$\Phi_n(z) = z^n + \text{lower order} \tag{4.0.2}$$

$$\int \overline{\Phi_n(z)}\, \Phi_m(z) \, d\mu(z) = 0, \qquad n \neq m \tag{4.0.3}$$

and *orthonormal polynomials*,

$$\varphi_n(z) = \frac{\Phi_n(z)}{\|\Phi_n\|_{L^2(\mathbb{C}, d\mu)}} \tag{4.0.4}$$

Two special cases which concern us in this chapter are orthogonal polynomials on the real line (OPRL) and on the unit circle (OPUC) where $\operatorname{supp}(d\mu) \subset \mathbb{R}$ and $\operatorname{supp}(d\mu) \subset \partial\mathbb{D}$. They are special because they are associated with three-term recurrence relations, as we'll see.

This chapter will present the basics and some of the more advanced features of the theory of OPs. In some sense, this whole chapter is a bonus chapter, although we'll regard Section 4.1 as nonbonus for two reasons: we believe a comprehensive real analysis course should at least expose students to the three-term recurrence relation for OPRL. Secondly, as we'll explain shortly, one basic result for OPRL is a variant of the spectral theorem and will be our "official" proof of that theorem, although we'll have several other proofs.

The orthonormal polynomials for OPRL (denoted p_n, not φ_n) will obey a three-term recurrence relation ($p_{-1} = 0$)

$$x p_n(x) = a_{n+1} p_{n+1}(x) + b_{n+1} p_n(x) + a_n p_{n-1}(x) \tag{4.0.5}$$

The $a_n \in (0,\infty)$, $b_n \in \mathbb{R}$ are called *Jacobi parameters* and

$$\sup_n (|a_n| + |b_n|) < \infty \Leftrightarrow \operatorname{supp}(\mu) \text{ is bounded} \tag{4.0.6}$$

A basic result we'll prove in Section 4.1 is Favard's theorem that any set of numbers $\{a_n, b_n\}_{n=0}^\infty$, $a_n \in (0,\infty)$, $b_n \in \mathbb{R}$ obeying the right side of (4.0.6) are the Jacobi parameters of a unique measure. (In Section 7.7, we'll see if (4.0.6) is dropped, one still has existence of μ but uniqueness may fail.) This will imply the spectral theorem (and, in turn, could be proven from the spectral theorem) as follows: If A is a bounded self-adjoint operator and $\varphi \in \mathcal{H}$ is such that $\{A^n\varphi\}_{n=0}^\infty$ are independent, then one can apply Gram–Schmidt to these vectors and get $\psi_n = p_n(A)\varphi_n$ are orthonormal with $\operatorname{span}[\varphi, A\varphi, \dots, A^n\varphi] = \operatorname{span}[\psi_0, \psi_1, \dots, \psi_n]$. One proves the $p_n(A)$ obey a three-term recurrence relation and then, by Favard's theorem, its Jacobi parameters are associated to a measure, $d\mu$. One shows $A \restriction \operatorname{span}\{A^n\varphi\}_{n=0}^\infty$ is unitarily equivalent to multiplication by x on $L^2(\mathbb{R}, d\mu)$.

The theory of OPs is roughly divided in two: the analytic theory which discusses general relations between measures $d\mu$ and the recursion parameters, the zeros of p_n, and the asymptotics of $p_n(x)$ as $n \to \infty$; and the algebraic theory which discusses the many specific, special cases of OPs, which are often crucial elements in the study of problems in analysis (or, more generally, in mathematics and applied science). Section 4.2 will illustrate the algebraic theory by providing a theorem that classifies all OPRL that also obey a second-order differential equation in x.

One central tool of the theory of OPs is the use of L^2-variational principles—Section 4.3 discusses solutions of L^∞-variational principles and

Section 4.4 includes a result on asymptotics of $\|\varphi_n\|$ for OPUC obtained via a variational principle. More generally, Section 4.4 discusses OPUC which also obey a three-term recurrence relation, called the *Szegő recurrence relation*,

$$z\Phi_n(z) = \Phi_{n+1}(z) - \bar{\alpha}_n z^n \,\overline{\Phi_n(1/\bar{z})} \tag{4.0.7}$$

The parameters $\{\alpha_n\}_{n=0}^\infty$ lie in $\mathbb{D}$ and are called *Verblunsky coefficients.* We'll prove an analog of Favard's theorem called Verblunsky's theorem and a result on $\lim_{n\to\infty}\|\Phi_n\|_{L^2(\mathbb{C},d\mu)}$.

Notes and Historical Remarks. OPRL were developed slowly in the hundred years after 1790, initially as special cases (Legendre polynomials, then Hermite polynomials by Laplace, ...). OPUC were developed initially by Szegő around 1920, especially in a two-part paper [**683**].

Szegő's book [**686**] is a classic in both the analytic and algebraic sides of the subject. Monographs on the algebraic side include Andrews et al. [**19**], Beals–Wong [**48**], and especially Ismail [**334**]. Monographs on the analytic side include Geronimus [**240**], Chihara [**123**], Nikishin–Sorokin [**504**], Stahl–Totik [**662**], and three books by Simon [**648, 649, 653**] (the first two on OPUC, the third mainly on OPRL).

4.1. Orthogonal Polynomials on the Real Line and Favard's Theorem

In this section, we'll focus on the case where the measure, μ, on $\mathbb{C}$ with

$$\int |z|^n \, d\mu(z) < \infty \tag{4.1.1}$$

for all n is supported on $\mathbb{R}$—thus, *orthogonal polynomials on the real line* or OPRL for short. We use $P_n(x)$ and $p_n(x)$ for which the general theory uses Φ_n and φ_n. Thus, $P_n(x)$ is the unique degree-n polynomial with

$$P_n(x) = x^n + \text{ lower order}, \qquad \int P_n(x)x^j \, d\mu(x) = 0, \quad j = 0,1,\dots,n-1 \tag{4.1.2}$$

The uniqueness implies P_n is real, that is,

$$\overline{P_n(z)} = P_n(\bar{z}) \tag{4.1.3}$$

equivalently, all coefficients are real. We also define

$$p_n(x) = \frac{P_n(x)}{\|P_n\|} \tag{4.1.4}$$

where $\|\cdot\|$ is $L^2(\mathbb{R}, d\mu)$-norm.

For P_n, p_n to exist for all n, we need that $\{x^j\}_{j=0}^\infty$ are linearly independent in $L^2(\mathbb{R}, d\mu)$ so we can apply Gram–Schmidt. μ is called *trivial* if

$\text{supp}(\mu)$ is a finite set of points $\{x_k\}_{k=1}^N$ and *nontrivial* if it is not trivial. If $\int |\sum_{j=0}^k a_j x^j|^2 \, d\mu(x) = 0$, then $Q(x) = \sum_{j=0}^k a_j x^j$ must be a.e. zero, which can only happen if μ is supported on the zeros of Q. Conversely, if μ is supported on $\{x_k\}_{k=1}^N$, then

$$P_N(x) = \prod_{j=1}^N (x - x_j) \tag{4.1.5}$$

has

$$\int |P_N(x)|^2 \, d\mu(x) = 0 \tag{4.1.6}$$

That is, $\{x^j\}_{j=0}^\infty$ are linearly independent if and only if μ is nontrivial. Equivalently, if μ is nontrivial, we can define $\{P_n\}_{j=0}^\infty$, $\{p_n\}_{n=0}^\infty$; if $\text{supp}(\mu) = \{x_j\}_{j=0}^N$, then $L^2(\mathbb{R}, d\mu)$ is N-dimensional, and we can define $\{P_j\}_{j=0}^N$ and $\{p_j\}_{j=0}^{N-1}$ with P_N given by (4.1.5). By (4.1.6), p_N cannot be defined.

If μ is trivial, $\{p_j\}_{j=0}^{N-1}$ are an ON basis of $L^2(\mathbb{R}, d\mu)$. If μ is nontrivial and $\text{supp}(\mu)$ is a bounded set, by the Weierstrass approximation theorem, $\{p_j\}_{j=0}^\infty$ is an ON basis. If $\text{supp}(d\mu)$ is unbounded, it can be proven (see Section 7.7) that $\{p_j\}_{j=0}^\infty$ is an ON basis if the moment problem is determinate (in the sense of Section 5.6 of Part 1) and for some but not all μ's if the moment problem is indeterminate.

We will mainly discuss the nontrivial case, although we'll see that the trivial case is useful tool. For now, unless we state otherwise, we'll suppose μ is nontrivial.

Our main initial result is that in this case (and not more general cases of measures on $\mathbb{C}$), the p_n and P_n obey a three-term recursion relation,

$$xP_n(x) = P_{n+1}(x) + b_{n+1}P_n(x) + a_n^2 P_{n-1}(x) \tag{4.1.7}$$

with $b_n \in \mathbb{R}$ (and $P_{-1} \equiv 0$),

$$a_n = \frac{\|P_n\|}{\|P_{n-1}\|} > 0 \tag{4.1.8}$$

This implies that

$$\|P_n\| = a_1 \dots a_n \mu(\mathbb{R})^{1/2} \tag{4.1.9}$$

and

$$xp_n(x) = a_{n+1}p_{n+1}(x) + b_{n+1}p_n(x) + a_n p_{n-1}(x) \tag{4.1.10}$$

$\{a_n, b_n\}_{n=1}^\infty$ are called the Jacobi parameters of μ. In particular, (4.1.9) implies that if $\mu(\mathbb{R}) = 1$, then

$$p_n(x) = \frac{1}{a_1 \dots a_n} x^n + \text{lower order} \tag{4.1.11}$$

Once we have (4.1.10), we'll prove that

$$\sup_n(|a_n| + |b_n|) < \infty \Leftrightarrow \text{supp}(d\mu) \text{ is bounded} \tag{4.1.12}$$

and we'll focus only on that case. If A is the self-adjoint operator of multiplication by x, then the conditions in (4.1.12) imply that A is bounded and its matrix in the ON basis $\{p_j\}_{j=0}^\infty$ is the *Jacobi matrix*

$$J = \begin{pmatrix} b_1 & a_1 & 0 & 0 & \cdots \\ a_1 & b_2 & a_2 & 0 & \cdots \\ 0 & a_2 & b_3 & a_3 & \cdots \\ \vdots & \vdots & \vdots & \vdots & \ddots \end{pmatrix} \tag{4.1.13}$$

We have thus seen that any μ of bounded support leads to a bounded matrix of the form (4.1.13). The converse, namely, any matrix of the form J with (4.1.12) and $b_j \in \mathbb{R}$, $a_j > 0$ is the Jacobi matrix of some measure, $d\mu$, is known as *Favard's theorem* and will be a major goal for us. We especially care about it since it is equivalent to the spectral theorem for bounded self-adjoint operators. Indeed, it will be the basis for our proof of that theorem in Section 5.3. We now turn to the details.

Theorem 4.1.1 (Three-Term Recurrence for OPRL). *Let μ be a nontrivial measure on $\mathbb{R}$ with* (4.1.1). *Then there exist $\{b_n\}_{n=0}^\infty \subset \mathbb{R}$ so that* (4.1.7) *holds, with a_n given by* (4.1.8). *Moreover,* (4.1.9) *and* (4.1.10) *hold.*

If μ is trivial with $\text{supp}(\mu) = \{x_j\}_{j=1}^N$, *$N$ distinct points,* (4.1.7)–(4.1.10) *hold for $n = 0, 1, 2, \dots, N-1$, where P_N is given by* (4.1.5) *and $p_N = 0$.*

Remark. In the nontrivial case, (4.1.7), (4.1.10) hold for all x and n (not just as functions in L^2). In the trivial case, (4.1.7), (4.1.10) hold for all x and $j = 0, \dots, N-2$, and (4.1.7) for all x and $j = N-1$ (with P_n given by (4.1.5)). But for $j = N-1$, (4.1.10) holds only as a statement of L^2-functions with $p_N = 0$. This is so since $\|P_N\| = 0$, so either a_{N+1} is infinite or undefined.

Remark. $\text{supp}(\mu) \subset \mathbb{R}$ enters via (4.1.15) below; equivalently, because it implies multiplication by x is self-adjoint.

Proof. By a simple induction, since $P_j(x)$ is monic, $\{P_j(x)\}_{j=0}^{n+1}$ is a basis for the polynomials of degree at most $n+1$. Since these functions are orthogonal in L^2, we have for $\deg(Q) \le n+1$,

$$Q = \sum_{j=0}^{n+1} \frac{\langle P_j, Q\rangle}{\langle P_j, P_j\rangle} P_j \tag{4.1.14}$$

Since $\deg(xP_j) = j+1$, if $j+1 < n$ (i.e., $j < n-1$), we have

$$\langle P_j, xP_n\rangle = \langle xP_j, P_n\rangle = 0 \tag{4.1.15}$$

Thus, by (4.1.14),

$$xP_n(x) = \alpha P_{n+1}(x) + \beta P_n(x) + \gamma P_{n-1}(x) \tag{4.1.16}$$

Since xP_n and P_{n+1} are both monic and P_n, P_{n-1} of lower degree, $\alpha = 1$. In particular, by (4.1.14) for all n,

$$\langle xP_n, P_{n+1}\rangle = \|P_{n+1}\|^2 \tag{4.1.17}$$

Replacing n by $n-1$,

$$\langle xP_{n-1}, P_n\rangle = \langle P_{n-1}, xP_n\rangle = \|P_n\|^2 \tag{4.1.18}$$

so the γ in (4.1.16) is $\|P_n\|^2/\|P_{n-1}\|^2$, that is, we have (4.1.8).

Since

$$b_{n+1} = \frac{\langle xP_n, P_n\rangle}{\langle P_n, P_n\rangle} \tag{4.1.19}$$

we see that b_{n+1} is real. We have thus proven (4.1.7)/(4.1.8) and that

$$\|P_n\| = \biggl(\prod_{j=1}^n \frac{\|P_j\|}{\|P_{j-1}\|}\biggr)\|P_0\| = \mu(\mathbb{R})^{1/2} a_1 \dots a_n$$

that is, (4.1.9).

Dividing (4.1.19) by $\|P_n\|$, we get

$$xp_n(x) = \biggl(\frac{\|P_{n+1}\|}{\|P_n\|}\biggr)p_{n+1}(x) + b_{n+1}p_n(x) + a_n^2\biggl(\frac{\|P_{n-1}\|}{\|P_n\|}\biggr)p_{n+1}(x) \tag{4.1.20}$$

which, given (4.1.8), is (4.1.10). □

Henceforth, we suppose $\mu(\mathbb{R}) = 1$.

Proposition 4.1.2. *Given a measure, μ, on $\mathbb{R}$ obeying* (4.1.1), *let*

$$\eta = 2\sup_n|a_n| + \sup_n|b_n| \tag{4.1.21}$$

Then

(a) *If η is finite,*

$$\operatorname{supp}(\mu) \subset [-\eta, \eta] \tag{4.1.22}$$

(b) *If*

$$\operatorname{supp}(\mu) \subset [-R, R] \tag{4.1.23}$$

then

$$\eta \le 3R \tag{4.1.24}$$

In particular, supp(μ) *is bounded if and only if η is finite.*

Proof. (a) If $\eta < \infty$, then the matrix J is the matrix of multiplication by x in $\{p_n\}_{n=0}^\infty$ basis (even if the polynomials aren't dense, they are left invariant by multiplication by x) and J is the matrix of this operator on polynomials.

A matrix with nonzero elements along a single diagonal clearly has norm the sup of the elements on that diagonal. So writing J as a sum of three such matrices, we get $\|J\| \le \eta$. In particular,

$$\left| \int x^n \, d\mu(x) \right| = |\langle \mathbb{1}, J^n \mathbb{1} \rangle| \le \eta^n \tag{4.1.25}$$

which it is easy to see (Problem 1) implies (4.1.22).

(b) For any f, g with $\|f\|_{L^2} = \|g\|_{L^2} = 1$, we have

$$\left| \int x |f(x) g(x)| \, d\mu(x) \right| \le R \tag{4.1.26}$$

since $|fg| \le \frac{1}{2}|f|^2 + \frac{1}{2}|g|^2$. Moreover,

$$b_n = \langle x p_n, p_n \rangle, \qquad a_n = \langle x p_n, p_{n+1} \rangle \tag{4.1.27}$$

which leads from (4.1.26) to (4.1.24). $\square$

We have seen that any measure nontrivial on $\mathbb{R}$ of bounded support leads to an infinite Jacobi matrix with $\eta < \infty$. That this has a converse is called Favard's theorem:

Theorem 4.1.3 (Favard's Theorem). *Let $\{a_n\}_{n=1}^\infty, \{b_n\}_{n=1}^\infty$ be a set of Jacobi parameters (i.e., $a_n > 0$, $b_n \in \mathbb{R}$) with η, given by (4.1.21), finite. Then there exists a unique probability measure, μ, with the given Jacobi parameters.*

Remarks. 1. There is a result also if $\eta = \infty$, which is more subtle and also called Favard's theorem. We'll prove it in the bonus chapter on unbounded self-adjoint operators; see Section 7.7.

2. Uniqueness is easy. For let μ, ν be two probability measures with the same Jacobi parameters as above. By Proposition 4.1.2, since $\eta < \infty$, they both have bounded support. So by the Weierstrass approximation theorem (see Theorem 2.4.1 of Part 1), it suffices to prove for all n that

$$\int x^n \, d\mu = \int x^n \, d\nu \tag{4.1.28}$$

(see Section 4.17, especially Theorem 4.17.1, of Part 1). This holds for $n = 0$. Suppose it holds for $n = 0, 1, 2, \dots, k-1$. Since the a_n's and b_n's determine $P_k(x)$, we have

$$\int P_k(x) \, d\mu = \int P_k(x) \, d\nu \tag{4.1.29}$$

by orthogonality of P_k to 1. This expresses the k-th moment in terms of the n-th moments for $n = 0, \dots, k-1$, so by the induction hypothesis, (4.1.28) holds for $n = k$.

The full proof of the existence part of Theorem 4.1.3 requires some preliminaries. Essentially, we will first prove the spectral theorem and analog of Favard's theorem for finite Jacobi matrices and then get Favard's theorem by approximating J by its finite approximations:

$$J_{n;F} = \begin{pmatrix} b_1 & a_1 & \dots & \dots & 0 \\ a_1 & b_1 & \ddots & \ddots & \vdots \\ \vdots & \ddots & \ddots & b_{n-1} & a_{n-1} \\ 0 & \dots & \dots & a_{n-1} & b_n \end{pmatrix} \tag{4.1.30}$$

Eventually (see Theorem 4.1.5), we'll prove that

$$\det(x - J_{n;F}) = P_n(x) \tag{4.1.31}$$

so zeros of P_n as eigenvalues of $J_{n;F}$ are important. We begin with

Proposition 4.1.4. *Let $J \equiv J_{n;F}$ be an $n \times n$ Jacobi matrix as in* (4.1.30). *Let P_n be the monic polynomial obtained from these Jacobi parameters and* (4.1.7). *Then*

(a) *All roots of P_n are real.*
(b) *All roots of P_n are simple.*
(c) *The roots of P_n and P_{n-1} strictly interlace.*

Remarks. 1. The proof below is an adaptation of a standard argument that proves the same thing for orthogonal polynomials.

2. To say the zeros of two polynomials, P and Q, *strictly interlace* means they are simple for each and the zeros of P and Q are disjoint sets, and between any two zeros P, there are zeros of Q and vice-versa.

3. See Problem 5 for proofs using the theory of rank-one perturbations of finite matrices discussed in Section 1.3. Problem 6 has an extended notion of interlacing.

Proof. Let Q be a monic polynomial of degree $n-1$ and P a monic polynomial of degree n. Suppose Q has all its zeros on $\mathbb{R}$ as $n-1$ distinct points $y_1^Q < \dots < y_{n-1}^Q$ and P is nonvanishing with sign $(-1)^j$ at y_{n-j}^Q. Then P has zeros $y_1^P < \dots < y_n^P$, all on $\mathbb{R}$, with

$$y_1^P < y_1^Q < y_2^P < y_2^Q < \dots < y_{n-1}^Q < y_n^P \tag{4.1.32}$$

For since P changes sign from y_k^Q to y_{k+1}^Q, P has to have at least one zero between them. Since $P \to \infty$ as $x \to \infty$ and is negative at y_{n-1}^Q, P has to have a zero in (y_{n-1}^Q,∞). Similarly, $(-1)^n P \to \infty$ as $x \to -\infty$ and $(-1)^n P(y_1^Q) < 0$. P has a zero in $(-\infty, y_1^Q)$. So we accounted for all zeros of P and get (4.1.32).

We now prove (a)–(c) inductively by proving inductively that P_n is negative at the largest zero of P_{n-1} and has alternating signs at successive zeros. The key fact is implied by (4.1.7):

$$\begin{aligned} P_n(y) = 0 &\Rightarrow P_{n+1}(y) + a_n^2 P_{n-1}(y) = 0 \\ &\Rightarrow P_n \text{ and } P_{n-1} \text{ have opposite signs at } y \end{aligned}$$

We'll leave the details of the induction from here to the reader (Problem 2). □

Here is the main finite-dimensional result that concludes both the spectral theorem for $J_{n;F}$ and the analog of Favard's theorem:

Theorem 4.1.5. *Let $J \equiv J_{n;F}$ be an $n \times n$ Jacobi matrix of the form (4.1.30). Let $\{P_j\}_{j=1}^n$ and $\{p_j\}_{j=1}^{n-1}$ be the polynomials defined by (4.1.7)/(4.1.8) inductively, with $p_{-1} = P_{-1} = 0$, $p_0 = P_0 = 1$. Then*

(a) *All the zeros of $P_n(x)$ are real and simple.*

(b) *If y_j, $j = 1, \dots, n$, is a zero of P_n, then it is an eigenvalue of J. Indeed, if φ^j is the n-component vector with components*

$$(\varphi^j)_k = \frac{p_{k-1}(y_j)}{(\sum_{\ell=1}^n |p_{\ell-1}(y_j)|^2)^{1/2}} \tag{4.1.33}$$

$k = 1, \dots, n$, then

$$(J - y_j)\varphi^j = 0 \tag{4.1.34}$$

(c) *For $j \neq k$ ($\langle \cdot , \cdot \rangle$ = normal $\mathbb{C}^n$ inner product),*

$$\langle \varphi^j, \varphi^k \rangle = 0 \tag{4.1.35}$$

so $\{\varphi^j\}_{j=1}^n$ is an orthonormal basis for $\mathbb{C}^n$, that is, J has an orthonormal basis of eigenvectors.

(d) *Let μ be the measure (δ_y = point mass at y)*

$$\mu = \sum_{j=1}^n |(\varphi^j)_1|^2 \delta_{y_j} \tag{4.1.36}$$

then μ is a probability measure with monic OPs *$\{P_j\}_{j=1}^n$ and orthonormal* OPs *$\{p_j\}_{j=1}^{n-1}$. The Jacobi parameters for μ are $\{a_j\}_{j=1}^{n-1}$, $\{b_j\}_{j=1}^n$.*

(e) (4.1.31) *holds.*

Remarks. 1. (b) + (c) is essentially the spectral theorem for J. In Sections 5.2 and 5.3, we'll see any finite-dimensional operator is a direct sum of finite Jacobi matrices in some orthonormal basis, so this implies the full finite-dimensional spectral theorem.

2. (d) is an explicit form of the analog of the Favard theorem for the finite Jacobi parameters.

3. Since P_n is monic and its zeros are the eigenvalues of $J_{n;F}$, we have (4.1.31).

Proof. (a) This is essentially a restatement of Proposition 4.1.4.

(b) Define vectors $\nu(x)$ in $\mathbb{C}^n$ by $(\nu(x))_k = p_{k-1}(x)$. Then

$$\big[(J-x)\nu(x)\big]_j = -(a_1 \dots a_{n-1})^{-1}\delta_{jn}P_n(x) \tag{4.1.37}$$

This follows from (4.1.10) and, for $j=n$, from (4.1.7), (4.1.9). It follows that if $P_n(y)=0$, then y is an eigenvalue of J and $\varphi^j = \nu(y_j)/N_j$ is that eigenvalue normalized so $\langle \varphi^j, \varphi^j\rangle = 1$. Since there are n distinct zeros, we have n eigenvectors.

(c) If $j \neq k$, by self-adjointness of J,

$$0 = \langle J\varphi^j, \varphi^k\rangle - \langle \varphi^j, J\varphi^k\rangle = (y_j - y_k)\langle \varphi^j, \varphi^k\rangle$$

so since $y_j - y_k \neq 0$, $\langle \varphi^j, \varphi^k\rangle = 0$. By counting dimensions, $\{\varphi^j\}_{j=1}^n$ is an orthonormal basis.

(d) Since φ^j is an orthonormal basis, the matrix

$$U_{\ell j} = (\varphi^j)_\ell \tag{4.1.38}$$

has

$$\sum_{\ell=1}^n \overline{U}_{\ell j} U_{\ell k} = \delta_{jk} \tag{4.1.39}$$

that is, $U^*U = \mathbb{1}$. Since we are in finite dimensions, $UU^* = \mathbb{1}$, that is,

$$\sum_{j=1}^n \overline{U}_{\ell j} U_{mj} = \delta_{\ell m} \tag{4.1.40}$$

Since $(\varphi^j)_\ell = p_{\ell-1}(y_j)(\varphi^j)_1$, (4.1.40) says

$$\sum_{j=1}^n |(\varphi^j)_1|^2\, \overline{p_{\ell-1}(y_j)}\, p_{m-1}(y_j) = \delta_{\ell m} \tag{4.1.41}$$

that is, the p's are the orthogonal polynomials for the measure (4.1.36). Thus, the a's and b's are this measure's Jacobi parameters.

(e) Since the zeros and eigenvalues are both simple, this is an immediate consequence of the fact that $\det(B)$ is the product of eigenvalues of B. □

To handle the infinite case, we need to extend the argument in Remark 2 after Theorem 4.1.3 to handle partial uniqueness.

Proposition 4.1.6. *Fix n among $1, 2, \ldots$. Let $d\rho, d\mu$ be two probability measures on $\mathbb{R}$, each of which has a support with at least n points. Then their first $2n-1$ Jacobi parameters $\{a_j\}_{j=1}^{n-1} \cup \{b_j\}_{j=1}^{n}$ are equal if and only if*

$$\int x^k \, d\rho = \int x^k \, d\mu \tag{4.1.42}$$

for $k = 0, 1, \ldots, 2n-1$.

Remark. Since $k = 0$ is assured by the assumption of their being probability measures, (4.1.42) has $2n-1$ conditions, the same number as the $n-1$ a_j's and n b_j's.

Proof. If (4.1.42) holds, for any polynomials P and Q of degree at most $n-1$,

$$\begin{aligned} \int P(x)Q(x) \, d\rho &= \int P(x)Q(x) \, d\mu \\ \int xP(x)Q(x) \, d\rho &= \int xP(x)Q(x) \, d\mu \end{aligned} \tag{4.1.43}$$

Thus, by Gram–Schmidt, $\{p_j(x)\}_{j=0}^{n-1}$ are determined with the same $\{a_j, b_j\}_{j=1}^{n-1}$. b_j is given by $\int x p_{j-1}(x)^2 \, d\mu(x)$, and so the second equation in (4.1.43) proves the b_n's are equal.

Conversely, if the a's and b's are equal for the given values by (4.1.7), $\{P_j(x)\}_{j=1}^{n}$ are equal. The conditions

$$\int P_j(x) x^\ell \, d\mu(x) = 0, \quad j = 1, \ldots, n; \ \ell = 0, 1, \ldots, j-1 \tag{4.1.44}$$

inductively determine $\int x^k \, d\mu(x)$ for $k = 0, 1, \ldots, 2n-1$ (Problem 3). $\square$

Finally, we can prove Favard's theorem.

Proof of Theorem 4.1.3. Given a Jacobi matrix with $\eta < \infty$, let $J_{n;F}$ be the truncated matrices in (4.1.30). By Theorem 4.1.5, there are measures μ_n supported by n points for which $J_{n;F}$ is the associated finite Jacobi matrix. Since $\|J_{n;F}\| \le \eta$, eigenvalues lie in $[-\eta, \eta]$, so all measures, $d\mu_n$, are supported in $[-\eta, \eta]$.

By Proposition 4.1.6, if $n \le m$,

$$\int x^\ell \, d\mu_n = \int x^\ell \, d\mu_m \tag{4.1.45}$$

for $\ell \le 2n-1$. Thus, $\int x^\ell \, d\mu_n$ is n-independent once $n \ge \frac{1}{2}(\ell+1)$, so $\lim \int x^\ell \, d\mu_n$ exists for each fixed ℓ. Thus, $\lim \int f(x) \, d\mu_n$ exists for each polynomial, f, and so, by the Weierstrass approximation theorem for all $f \in C([-\eta, \eta])$.

The limit function is a positive functional on $C([-\eta,\eta])$ and so (by Section 4.4 of Part 1) defines a measure $d\mu$ on $[-\eta,\eta]$. By construction for $\ell \le 2n-1$,

$$\int x^\ell \, d\mu_n = \int x^\ell \, d\mu \tag{4.1.46}$$

By Proposition 4.1.6 in the opposite direction, μ and μ_n have the same $\{a_j\}_{j=1}^{n-1} \cup \{b_j\}_{j=1}^n$ Jacobi parameters, so μ has all the required Jacobi parameters. □

Notes and Historical Remarks.

> The classical orthogonal polynomials are mostly attributed to someone other than the person who introduced them. Szegő refers to Abel and Lagrange and Tschebyscheff for work on the Laguerre Polynomials.
>
> —*R. Askey* [**688**][1]

OPRL had its start via the special examples mentioned in the introductory notes and which star in the next section. The general theory came out of work on the moment problem. If μ is a measure on $\mathbb{R}$ with finite moments,

$$F(z) = \int \frac{d\mu(x)}{x-z} \tag{4.1.47}$$

then $F(z)$ has a formal continued fraction expansion at infinity

$$F(z) = \cfrac{1}{-z+b_1-\frac{a_1^2}{-z+b_2-a_2^2\cdots}} \tag{4.1.48}$$

and one way of defining the Jacobi parameters is from this expansion.

As far as OPRL is concerned, the point is, as we'll see in Section 7.7, if one truncates by setting $a_n = 0$, the truncated continued fraction is

$$F_n(z) = -Q_n(z)/P_n(z) \tag{4.1.49}$$

where P_n is the n-th monic orthogonal polynomial (and Q is a polynomial of degree $n-1$). In Theorem 7.5.1 of Part 2A, we saw that numeric truncated continued fractions are related by a three-term recursion relation called the *Euler-Wallis equations.* Their derivation was purely algebraic and so applies here. The resulting recursion relation for the P_n's is exactly (4.1.7)!

One can go further (see e.g., Theorem 7.7.22) and prove that if $\{a_n, b_n\}_{n=1}^\infty$ are bounded, then F_n converges on $\mathbb{C}\setminus[\alpha,\beta]$ (for α, β defined in terms of $\sup(a_n)$ and $\sup(b_n)$) to an $F(z)$ for which there is a measure $d\mu$ obeying (4.1.47). The orthogonality of the P_n's and the relation to the moments of μ are, in this approach, an afterthought.

[1] in Notes on Szegő's paper, *An outline of the history of orthogonal polynomials.*

The above, while differing in technical details is essentially historically how the process of going from Jacobi parameters to measure was discussed! In 1848, Jacobi [**345**], in trying to diagonalize a general real quadratic form, was lead to finite Jacobi matrices, found the finite continued fraction (4.1.48), and showed that the zeros of the denominator were the eigenvalues of his matrix (or rather the coefficients in an eventual diagonalization of the form), so in essence he had (4.1.31).

In [**119**], Chebyshev looked at the moment problem as part of his attempts to prove the central limit theorem. He asked how to go from moments to $\mu(c,d)$ for c,d in the support of μ. He computed the weight of the finite approximations one gets by truncating the continued fraction and claimed, but did not prove, inequalities that implied convergence as $N \to \infty$. These were proven independently by Markov [**467**] and Stieltjes [**666**]. Stieltjes was unaware of the work of either Chebyshev or Markov but once he was, he wrote a note agreeing to their priority.

All this work started with $d\mu(x) = w(x)dx$ and did not address the issue of going from the moments of the Jacobi parametrization to the measure without knowing a priori that $d\mu$ existed. Indeed, without the Stieltjes integral, they couldn't study the general moment problem.

It was Stieltjes [**667**] who invented this integral and then implicitly had what we have called Favard's theorem already in 1895 (and its analog for positive, even unbounded, Jacobi matrices).

Favard's theorem appeared in Favard's 1935 paper [**195**] and this name is used extensively within the OP literature. Not only was the result already implicit in Stieltjes, but explicitly in Stone's 1932 book [**670**] which had a whole chapter on Jacobi matrices! We also note the OPUC analog predated Favard (see the Notes to Section 4.4).

Problems

1. Let μ be a Baire measure on $\mathbb{R}$ with finite moments.

   (a) If $\operatorname{supp}(\mu)$ is not in $[-R,R]$, prove that for a constant C_R and all n that
$$\int x^{2n}\,d\mu \geq C_R R^{2n}$$

   (b) If (4.1.25) holds, prove that (4.1.22) is true.

2. Complete the proof of Proposition 4.1.4 by induction in n.

3. Prove that (4.1.44) for $\ell = 0$ determines $c_k \equiv \int x^k\,d\mu(x)$ for $k = 0,1,\dots,n$, then for $\ell = 1$, we get c_{n+1} until for $\ell = n-1$, we get c_{2n-1}.

4. (a) Suppose $p_n(x)$ has a zero at $\zeta+i\kappa$ with $\kappa \neq 0$. Show this implies that

$$\int p_n(x)q(x)d\mu(x) > 0, \quad q(x) = p_n(x)/|x-\zeta-i\kappa|^2 \tag{4.1.50}$$

and conclude there cannot be such zeros.

(b) Let $[\alpha,\beta]$ be an interval with $\operatorname{supp}(\mu) \subset [\alpha,\beta]$. By considering $p_n(x)/(x-\zeta)$ if $\zeta < \alpha$ or $p_n(x)/(\zeta - x)$ if $\zeta > \beta$, prove that p_n has all its zeros in $[\alpha,\beta]$.

(c) Prove that $p_n(x)$ cannot have a real zero at ζ of multiplicity ≥ 2 by considering $p_n(x)/(x-\zeta)^2$.

(d) If $(\gamma,\lambda) \subset [\alpha,\beta]$ so $\operatorname{supp}(\mu) \subset [\alpha,\gamma] \cup [\lambda,\beta]$, prove that $p_n(x)$ cannot have two zeros $\zeta,\kappa \in (\gamma,\lambda)$ by considering $p_n(x)/(x-\zeta)(x-\kappa)$.

(e) Find a μ so $p_n(\lambda)$ has a zero in (γ,λ). (*Hint*: Pick μ so that $p_n(x) = (-1)^n p_n(-x)$.)

5. This will reprove Proposition 4.1.4 by using the theory of rank-one perturbations of finite matrices (see Section 1.3).

(a) Let $D_n(z) = \det(z - J_{n;F})$. By expanding in minors prove that

$$D_n(z) = (z-b_n)\, D_{n-1}(z) - a_n^2\, D_{n-2}(z) \tag{4.1.51}$$

(b) Conclude by induction that (4.1.31) holds.

(c) Prove that the zeros of P_n are real using (4.1.31).

(d) Prove that the zeros are simple.

(e) Prove that the zeros of P_n and P_{n-1} interlace by using Theorem 1.3.5.

6. (a) Let $J_{n;F}^{(j)}(\alpha)$ be the Jacobi matrix $J_{n;F}$ with b_j replaced by $b_j+\alpha$. Show that as $\alpha \to \infty$, the zeros of $P_n(x)$ converge to the zeros of $P_j(x)$ together with the zeros of the orthogonal polynomial $Q_{n-j-1}(x)$ with Jacobi parameters $\{a_{j+\ell}, b_{j+\ell}\}_{\ell=1}^{n-j}$. (*Hint*: See Problem 2 of Section 1.3.)

(b) Prove *Stieltjes interlacing*: If $j < n$ and P_j and P_n have no common zeros, then each interval between successive zeros of P_n has exactly one or exactly zero eigenvalues of P_j and all eigenvalues of P_j occur in those intervals.

Remark. This approach to Stieltjes interlacing is due to de Boor–Saff [**150**].

4.2. The Bochner–Brenke Theorem

Note. In this section, we will use the *Pochhammer symbol*, $(a)_n$, defined for $n = 0, 1, 2, \ldots,$ and $a \in \mathbb{C}$ by

$$(a)_n = a(a+1)\ldots(a+n-1) \tag{4.2.1}$$

(with $(a)_0 = 1$). If $a \notin \{0, -1, -2, \dots\}$, we have

$$(a)_n = \frac{\Gamma(n+a)}{\Gamma(a)} \tag{4.2.2}$$

in terms of the Gamma function. In particular, $(1)_n = n!$.

As discussed in this chapter's introduction, there is an algebraic side of the theory of OPs. This section will describe one aspect of that side. We've seen that OPRL, $\{p_n(x)\}_{n=0}^\infty$, obey a three-term recurrence relation in n. Here we'll ask which OPs also obey a particular equation in x, namely, we'd like them to all be eigenfunctions of the same second-order differential equation, that is, for there to be functions, $f(x), g(x), h(x)$, and complex numbers, $\{\lambda_n\}_{n=0}^\infty$, so that

$$f(x)\,\frac{d^2}{dx^2}\,p_n(x) + g(x)\,\frac{d}{dx}\,p_n(x) + h(x)p_n(x) = \lambda_n p_n(x) \tag{4.2.3}$$

Our goal will be to prove a remarkable theorem of Bochner and Brenke that classified all such OPs.

We begin by describing them—all have appeared already in Chapter 14 of Part 2B.

Definition. The *Hermite polynomials* are defined by

$$H_n(x) = (-1)^n e^{x^2} \left(\frac{d}{dx}\right)^n (e^{-x^2}) \tag{4.2.4}$$

The *Laguerre polynomials* are defined (depending on complex parameter α) by

$$L_n^{(\alpha)}(x) = (n!)^{-1} x^{-\alpha} e^x \left(\frac{d}{dx}\right)^n (e^{-x} x^{n+\alpha}) \tag{4.2.5}$$

The *Jacobi polynomials* are defined (depending on the complex parameters α and β) by

$$P_n^{(\alpha,\beta)}(x) = \frac{(-1)^n}{n!2^n}\,(1-x)^{-\alpha}(1+x)^{-\beta}\left(\frac{d}{dx}\right)^n \left[(1-x)^{n+\alpha}(1+x)^{n+\beta}\right] \tag{4.2.6}$$

The $H_n, L_n^{(\alpha)}$ with $\alpha > -1$, $P_n^{(\alpha,\beta)}$ with $\alpha, \beta > -1$ are called the *classical orthogonal polynomials.*

Remarks. 1. The parameters are often restricted to real values with $\alpha > -1$ for Laguerre and to $\alpha, \beta > -1$ for Jacobi. We will often do this below but will sometimes want to consider general complex α, β.

2. These are the conventional normalizations—they are neither monic nor normalized! Indeed, the leading term for $L_n^{(\alpha)}$ is $\frac{(-1)^n}{n!} x^n + O(x^{n-1})$, which is not even always positive! For $P_n^{(\alpha,\beta)}$ and H_n, it is at least positive for the classical values of α and β.

3. These are known as Rodrigues formulae. We'll use them as definitions in this section, although there are often other definitions and these are then derived formulae.

4. The Hermite and Laguerre polynomials are constants times hypergeometric functions of z^2 and z, respectively. The Jacobi polynomials are ${}_2F_1$ hypergeometric functions of $\frac{1}{2}(1-z)$; see Examples 14.4.16 and 14.4.17 of Part 2B.

5. $L_n^{(\alpha=0)}$ are sometimes called Laguerre polynomials and $L_n^{(\alpha)}$ *generalized Laguerre polynomials.*

6. Special cases of Jacobi polynomials include Legendre and Chebyshev polynomials; see Section 14.4 of Part 2B.

7. The definition (4.2.6) is intended near $x = 0$ taking the branch with $(1 \pm x)^\gamma = 1$. Once we know it is a polynomial, it is defined for all x!

Theorem 4.2.1. (a) $H_n, L_n^{(\alpha)}, P_n^{(\alpha,\beta)}$ *are polynomials in x of degree n for any complex values of α, β.*

(b) *The $H_n(x)$ are orthogonal with weight $e^{-x^2}\,dx$ on $\mathbb{R}$, that is,*

$$\int_{-\infty}^{\infty} H_m(x)H_n(x)e^{-x^2}\,dx = 0 \qquad (n \neq m) \tag{4.2.7}$$

For $\alpha \in (-1,\infty)$, the $L_n^{(\alpha)}$ are orthogonal with weight $x^\alpha e^{-x}$ on $(0,\infty)$, that is,

$$\int_0^{\infty} L_n^{(\alpha)}(x)L_m^{(\alpha)}(x)x^\alpha e^{-x}\,dx = 0 \qquad (n \neq m) \tag{4.2.8}$$

For $\alpha, \beta \in (-1,\infty)$, the $P_n^{(\alpha,\beta)}$ are orthogonal on $[-1,1]$ with weight $(1-x)^\alpha(1+x)^\beta$, that is,

$$\int_{-1}^{1} (1-x)^\alpha(1+x)^\beta P_n^{(\alpha,\beta)}(x)P_m^{(\alpha,\beta)}(x)\,dx = 0 \qquad (n \neq m) \tag{4.2.9}$$

(c) *H_n obeys the differential equations*

$$\left(-\frac{d^2}{dx^2} + 2x\,\frac{d}{dx}\right)H_n(x) = 2nH_n(x) \tag{4.2.10}$$

$L_n^{(\alpha)}$ obeys the differential equations

$$\left(-x\,\frac{d^2}{dx^2} + (x+1-\alpha)\,\frac{d}{dx}\right)L_n^{(\alpha)}(x) = nL_n^{(\alpha)}(x) \tag{4.2.11}$$

$P_n^{(\alpha,\beta)}(x)$ obeys the differential equation

$$\left(-(1-x^2)\frac{d}{dx^2}+[(\alpha+\beta+2)x+\alpha-\beta]\frac{d}{dx}\right)P_n^{(\alpha,\beta)}(x) \\ = n(n+\alpha+\beta+1)P_n^{(\alpha,\beta)}(x) \tag{4.2.12}$$

Remarks. 1. One can prove the normalization formulae (see the Notes)

$$\int_{-\infty}^{\infty} H_n(x)^2 e^{-x^2}\,dx = \sqrt{\pi}\,n!2^n \tag{4.2.13}$$

$$\int_0^\infty x^\alpha e^{-x} L_n^{(\alpha)}(x)^2\,dx = \frac{\Gamma(n+\alpha+1)}{n!} \tag{4.2.14}$$

$$\int_{-1}^{1} [P_n^{(\alpha,\beta)}(x)]^2(1-x)^\alpha(1+x)^\beta\,dx = \frac{2^{\alpha+\beta+1}\Gamma(n+\alpha+1)\Gamma(n+\beta+1)}{(2n+\alpha+\beta+1)\Gamma(n+\alpha+\beta+1)n!} \tag{4.2.15}$$

2. These polynomials have degree exactly n in the Hermite and Laguerre cases, and in the Jacobi case if $\alpha+\beta+2$ is not in $\{0,-1,-2,\dots\}$ and, in particular, if $\alpha,\beta\in(-1,\infty)$.

Proof. (a) For H_n, note that as operators

$$e^{+x^2}\frac{d}{dx}e^{-x^2} = \left(\frac{d}{dx}-2x\right) \tag{4.2.16}$$

so

$$H_n(x) = \left(2x-\frac{d}{dx}\right)^n 1 \tag{4.2.17}$$

is clearly a polynomial of degree n with

$$H_n(x) = 2^n x^n + O(x^{n-1}) \tag{4.2.18}$$

For $L_n^{(\alpha)}$, note that as operators

$$e^x\frac{d}{dx}e^{-x} = \left(\frac{d}{dx}-1\right) \tag{4.2.19}$$

so

$$n!L_n^{(\alpha)}(x) = x^{-\alpha}\left(\frac{d}{dx}-1\right)^n (x^{n+\alpha}) \tag{4.2.20}$$

and

$$x^{-\alpha}\left(\frac{d}{dx}\right)^k x^{n+\alpha} = (n+\alpha)(n+\alpha-1)\dots(n+\alpha-k+1)x^{n-k} \tag{4.2.21}$$

Thus, we see that $L_n^{(\alpha)}(x)$ is a polynomial of degree n with

$$L_n^{(\alpha)}(x) = \frac{(-1)^n}{n!}x^n + O(x^{n-1}) \tag{4.2.22}$$

For $P_n^{(\alpha,\beta)}(x)$, note that

$$\left(\frac{d}{dx}\right)^n (1-x)^{n+\alpha}(1+x)^{n+\beta} = \sum_{j=0}^{n} \binom{n}{j} \left[\left(\frac{d}{dx}\right)^j (1-x)^{n+\alpha}\right]\left[\left(\frac{d}{dx}\right)^{n-j} (1-x)^{n+\beta}\right] \tag{4.2.23}$$

and that

$$\left(\frac{d}{dx}\right)^\ell (1\pm x)^{n+\gamma} = (\pm)^\ell (n+\gamma)(n+\gamma-1)\dots(n+\gamma-\ell+1)(1\pm x)^{n+\gamma-\ell} \tag{4.2.24}$$

to see $P_n^{(\alpha,\beta)}(x)$ is a polynomial of degree at most n with (Problem 1)

$$\frac{(\alpha+\beta+n+1)_n}{2^n n!}\, x^n + O(x^{n-1}) \tag{4.2.25}$$

(b) Let $n < m$. All these have the form

$$\int Q_n(x)w(x)w^{-1}(x)\left(\frac{d}{dx}\right)^m f_m(x)\,dx \tag{4.2.26}$$

where Q_n is a polynomial, w the weight, and f_m a suitable function. We can use $ww^{-1} = 1$ and then integrate by parts $n+1$ times to get 0. One needs to show in each case that there is no boundary term where one integrates by parts, and that is easy to check if $\alpha, \beta \in (-1,\infty)$.

(c) This is left to the reader (Problem 2). □

We can now state the main result of this section.

Theorem 4.2.2 (Bochner–Brenke Theorem). *Let $\{p_n\}_{n=0}^\infty$ be a family of orthonormal* OPRL *that obey a differential equation of the form* (4.2.3) (*where only λ_n is n-dependent*). *Then up to a change of variables $x \mapsto \alpha x + \beta$, ($\alpha \in \mathbb{R}\setminus\{0\}$, $\beta \in \mathbb{R}$), p_n is one of the classical orthogonal polynomials.*

We begin with a simple algebraic analysis of the possibilities for (4.2.3). $p_0(x)$ is a constant and $\frac{d}{dx}p_0 = \frac{d^2}{dx}p_0 = 0$, so h is a constant. By subtracting it, without loss we can and will suppose that

$$h(x) = \lambda_0 = 0 \tag{4.2.27}$$

Theorem 4.2.3. *Suppose*

$$f(x)\,\frac{d^2}{dx^2}\,p_n(x) + g(x)\,\frac{d}{dx}\,p_n(x) = \lambda_n p_n(x) \tag{4.2.28}$$

has a polynomial solution, $p_n(x)$, of exact degree n for each $n = 1, 2, \dots$. Then up to an $x \to \zeta x + \omega$ change of variables (where ζ and ω are complex), we have one of the following:

(a) (Jacobi polynomials)

$$f(x) = (1 - x^2), \qquad g(x) = \beta - \alpha - x(\alpha + \beta + 2) \tag{4.2.29}$$

$$p_n(x) = P_n^{(\alpha,\beta)}(x), \qquad \lambda_n = -n(n + \alpha + \beta + 1) \tag{4.2.30}$$

(b) (Bessel polynomials; see below)

$$f(x) = x^2, \qquad g(x) = ax + 1 \tag{4.2.31}$$

$$p_n(x) = y_n(x; a, 1), \qquad \lambda_n = n(n + a - 1) \tag{4.2.32}$$

(c) (Pure power)

$$f(x) = x^2, \qquad g(x) = ax \tag{4.2.33}$$

$$p_n(x) = x^n, \qquad \lambda_n = n(n + a - 1) \tag{4.2.34}$$

(d) (Laguerre polynomials)

$$f(x) = x, \qquad g(x) = 1 + \alpha - x \tag{4.2.35}$$

$$p_n(x) = L_n^{(\alpha)}(x), \qquad \lambda_n = -n \tag{4.2.36}$$

(e) (Hermite polynomials)

$$f(x) = 1, \qquad g(x) = -2x \tag{4.2.37}$$

$$p_n(x) = H_n(x), \qquad \lambda_n = -2n \tag{4.2.38}$$

Remarks. 1. Since we require that p_n have exact degree n, in (a) we must add $\alpha + \beta + 2 \neq 0, -1, -2, \dots$.

2. α, β in (4.2.29) can be complex in (a), a complex in (b)/(c), and α complex in (d).

3. In case (c), if $1 - a$ is an integer greater than n, $x^n + cx^{1-a-n}$ is also allowed. But for a fixed and n large, $p(x) = x^n$.

Proof. Putting in p_1, we see that $g(x)p_1'(0) = \lambda_1 p_1(x)$ so g is a polynomial with $\deg g = 1$. Similarly, once we know this, by putting in p_2, $p_2''(0)f(x) = \lambda_2 p_2(x) - g(x)p_2'(x)$ is a polynomial with $\deg f \leq 2$, that is, f, g polynomials, $\deg f \leq 2$, $\deg g = 1$.

<u>Case 1: $\deg f = 0$.</u> By $\zeta x + \omega$ change of variables, we can suppose $g(0) = 0$. By scaling x and multiplying the whole of (4.2.28) by a constant, we can arrange $f(x) = 1$, $g'(0) = -2$, that is, this case is (e).

<u>Case 2: $\deg f = 1$.</u> By $\zeta x + \omega$ change of variables, we can suppose $f(0) = 0$, $f'(0) = 1$, that is, $f(x) = x$. By scaling plus multiplication by a constant, we can arrange $f(x) = x$, $g'(0) = -1$. Thus, we are in case (d).

Case 3: $\deg f = 2$, f has distinct roots. By $\zeta x + \omega$ transformation, we can suppose that the roots are ± 1, in which case after multiplication by a constant, we can suppose $f(x) = 1 - x^2$. Any degree-1 polynomial, g, can be written as $\beta - \alpha - x(\alpha + \beta + 2)$ for some α, β, so we are in case (a).

Case 4: $\deg f = 2$, f has a multiple root. We can translate so the root is at $x = 0$, so $f(x) = x^2$. If $g(0) = 0$, then $g(x) = ax$. Equating the $O(x^n)$ terms on both sides leads to $\lambda_n = n(n + a - 1)$, and then equating the x^k, $k = 0, 1, \dots, n-1$, terms on both sides shows all these terms are zero unless $1-a$ is an integer greater than n, in which case $k = 1-a-n$ can also occur. (The two roots of $r(r + a - 1) = n(n + 1 - a)$ sum to $1 - a$ so they are n and $1 - a - n$.) This is case (c).

If $g(0) \neq 0$, by scaling and multiplication by a constant, we can arrange $g(x) = ax + 1$. This is case (b). $\square$

To get Theorem 4.2.2, we have to eliminate (b)/(c) as yielding OPRL, fix the $\zeta x + \omega$ transformations in the other cases to ζ, ω real, and eliminate $\alpha \notin (-1, \infty)$ and/or $\beta \notin (-1, \infty)$ in case (a) and $\alpha \notin (-1, \infty)$ in case (d). Eliminating (c) is easy. For n large, $p_n(x) = x^n$ which has multiple zeros and we know, by Proposition 4.1.4 that for OPRL, the roots of p_n are simple, so (c) cannot lead to OPRLs. To handle case (b), we need to study its solutions.

For any a, b, the differential equation

$$z^2 y_n'' + (az + b)y_n' = n(n + a - 1)y_n \tag{4.2.39}$$

has solutions which are polynomials. Normalized by $y_n(0) = 1$, we denote them $y_n(z; a, b)$. They can be written in terms of hypergeometric functions as

$$y_n(z; a, b) = {}_2F_0\left(\begin{matrix} -n & n+a-1 \\ - & \end{matrix} ; -\frac{z}{b}\right) \tag{4.2.40}$$

$$= \sum_{j=0}^{n} \binom{n}{j} (n + a - 1)_j \left(\frac{z}{b}\right)^j \tag{4.2.41}$$

The coefficient, $-n$, guarantees that in the power series for ${}_2F_0$ (see (14.4.38) of Part 2B), all coefficients vanish for z^k, $k > n$. These are called *Bessel polynomials* because in terms of Bessel functions of imaginary argument, $K_\alpha(z)$, (see (14.5.37) of Part 2B), one has

$$y_n\left(\frac{1}{z}; 2, 2\right) = e^z z^{1/2} \frac{K_{n+\frac{1}{2}}(z)}{\sqrt{\pi}} \tag{4.2.42}$$

In our analysis below, the formulae (4.2.42) is not significant. We include only to explain where the name comes from. We will, however, need (4.2.41).

That it solves

$$z^2 \frac{d^2}{dz^2} f + (az+1) \frac{d}{dz} f = n(n+a-1)f \tag{4.2.43}$$

is left to Problem 3 as well as the fact that it is the unique polynomial solution.

We also need the following fact:

Proposition 4.2.4. *$y_n(z;a,1) \equiv y_n(z)$ obeys the recursion relation*

$$(2n+a-2)_3 z y_n(z) = A_n y_{n+1}(z) + B_n y_n(z) + C_n y_{n-1}(z) \tag{4.2.44}$$

$$\begin{aligned} A_n &= (n+a-1)(2n+a-2) \\ B_n &= -(a-2)(2n+a-1) \\ C_n &= -n(2n+a) \end{aligned} \tag{4.2.45}$$

Proof. A straightforward but tedious calculation of coefficients of the polynomials (Problem 4); one can get additional insight from the fact that the y_n are orthogonal with respect to a complex measure on $\partial\mathbb{D}$ (Problem 5). □

Proposition 4.2.5. *Let $\zeta, \omega, a \in \mathbb{C}$ with $\zeta \neq 0$ and let*

$$Q_n(x) = y_n(\zeta x + \omega; a, 1) \tag{4.2.46}$$

Suppose each Q_n is a complex multiple of a polynomial, real when x is real. Then ζ, ω, and a are all real.

Proof. A polynomial is real if and only if all its coefficients are real, so $p_n(x)$ is a constant multiple of a real polynomial if and only if the ratios of any two of its coefficients are real.

From (4.2.41), one sees (Problem 6) that

$$Q_n(z) = a_{nn} x^n + a_{n\,n-1} x^{n-1} + \dots \tag{4.2.47}$$

with

$$\frac{a_{n\,n-1}}{a_{nn}} = \frac{n\omega}{\zeta} + \frac{n}{\zeta(2n+a-2)} \tag{4.2.48}$$

Taking $n \to \infty$, we see ω/ζ is real and then that $\zeta(2 + \frac{a-2}{n})$ is real. Taking $n \to \infty$, we first see ζ is real and then that $a-2$ is real. □

Corollary 4.2.6. *For no $\zeta, \omega, a \in \mathbb{C}$ with $\zeta \neq 0$ are $\{y_n(x)\}_{n=0}^\infty$ given by (4.2.46) a set of* OPRL *(up to complex multiples).*

Proof. OPRL are real, so by Proposition 4.2.5, ζ, ω, a are all real. Since real affine transformations map OPRL to OPRL, we can suppose $\zeta = 1$, $\omega = 0$, that is, we need to show that for no real a are $\{y_n(x;a,1)\}_{n=0}^\infty$ a set of OPRL.

Fix $a \in \mathbb{R}$. For n large, all coefficients in (4.2.41) are positive and $(2n+a-2)_3 > 0$, A_n, $-C_n$ of (4.2.45) are positive. Thus, the normalized OPs are positive multiples of the y_n.

For OPRL, the p_n's are linearly independent, so the recursion relation coefficients are uniquely determined and would require $C_n/A_n > 0$. But for n large, $C_n/A_n < 0$. This contradiction proves that the $\{y_n(x;a,1)\}_{n=0}^\infty$ are not OPRL. □

Proposition 4.2.7. *Let $\alpha, \beta, \zeta, \omega \in \mathbb{C}$ with $\zeta \neq 0$. Then*

(a) *If*

$$Q_n(x) = L_n^{(\alpha)}(\zeta x + \omega) \tag{4.2.49}$$

is a complex multiple of a real polynomial for each n, then α, ζ, ω are all real.

(b) *If*

$$Q_n(x) = P_n^{(\alpha,\beta)}(\zeta x + \omega) \tag{4.2.50}$$

is a complex multiple of a real polynomial for each n, then either $\alpha, \beta, \zeta, \omega$ are all real or $\beta = \bar{\alpha}$, $\zeta, \omega \in i\mathbb{R}$.

Remark. In the Jacobi case,

$$P_n^{(\alpha,\beta)}(-x) = (-1)^n P_n^{(\beta,\alpha)}(x) \tag{4.2.51}$$

$$P_n^{(\bar{\alpha},\bar{\beta})}(\bar{x}) = \overline{P^{(\alpha,\beta)}(x)} \tag{4.2.52}$$

so if $\beta = \bar{\alpha}$, then $(-i)^n P_n^{(\alpha,\beta)}(ix)$ is a real polynomial and Q_n is real if $\zeta, \omega \in i\mathbb{R}$.

Proof. (a) From (4.2.20),

$$(-1)^n n! L_n^{(\alpha)}(z) = \sum_{j=0}^{n} (-1)^j \binom{n}{j} (n+\alpha-j+1)_j z^{n-j} \tag{4.2.53}$$

Thus,

$$(-1)^n n! Q_n(x) = \sum_{j=0}^{n} (-1)^j \binom{n}{j} (n+\alpha-j+1)_j \sum_{k=0}^{n-j} \binom{n-j}{k} \omega^k (\zeta x)^{n-j-k} \tag{4.2.54}$$

Letting $j+k=\ell$, we see

$$(-1)^n n! Q_n(x) = \sum_{\ell=0}^{n} \zeta^{n-\ell} x^{n-\ell} \biggl(\sum_{j=0}^{\ell} (-1)^j \binom{n}{j} \binom{n-j}{\ell-j} (n+\alpha-j+1)_j \omega^{\ell-j} \biggr) \tag{4.2.55}$$

The $\ell = 0$ term is ζ^n, so dividing the $(n-\ell)$-th coefficient by n-th, we see that

$$\zeta^{-\ell} \sum_{j=0}^{\ell} \frac{(n-\ell+1)_\ell (n+\alpha-j+1)_j}{j!(\ell-j)!} \, \omega^{\ell-j}$$

must be real for all $n \geq \ell$. Since this is a polynomial in n, it is real for all real n (see Problem 7).

Since $(n-\ell+1)_\ell$ is j-independent, we can divide out by it. Thus, we find for $\ell = 1, 2$,

$$\zeta^{-1}[\omega + n - \alpha] \tag{4.2.56}$$

$$\tfrac{1}{2}\,\zeta^{-2}[\omega^2 + 2(n-\alpha)\omega + (n-\alpha)(n-\alpha+1)] \tag{4.2.57}$$

are real for all real n.

By (4.2.56), ζ^{-1} and $\zeta^{-1}[\omega - \alpha]$ are real, so ζ^{-1} and $\omega - \alpha$ are real. The n^0 term in (4.2.57) is

$$\tfrac{1}{2}\,\zeta^{-2}[\omega^2 - 2\alpha\omega + \alpha^2 - \alpha] = \tfrac{1}{2}\,\zeta^{-1}[(\omega-\alpha)^2 - \alpha]$$

so α is real. Thus, ζ, ω, and α are all real.

(b) We will again use the reality of multiples of the $Q_n(x)$. It will be useful to consider the Jacobi polynomials renormalized so the leading term is 2^{-n}:

$$\widetilde{P}_n^{(\alpha,\beta)}(z) = \frac{n!}{(\alpha+\beta+n+1)_n}\, P_n^{(\alpha,\beta)}(z) \tag{4.2.58}$$

Writing $P_n^{(\alpha,\beta)}$ in terms of hypergeometric functions (Problem 8), one finds

$$\widetilde{P}_n^{(\alpha,\beta)}(z) = \left(\frac{z}{2}\right)^n + b_n(\alpha,\beta)\left(\frac{z}{2}\right)^{n-1} + c_n(\alpha,\beta)\left(\frac{z}{2}\right)^{n-2} + O(z^{n-3}) \tag{4.2.59}$$

where

$$b_n(\alpha,\beta) = \frac{n(\alpha-\beta)}{2[\alpha+\beta+2n]} \tag{4.2.60}$$

$$c_n(\alpha,\beta) = \frac{n}{8}\left(1 + O\left(\frac{1}{n}\right)\right) \tag{4.2.61}$$

Define $\tilde{\zeta}, \widetilde{\omega}$ by

$$\tilde{\zeta} x + \widetilde{\omega} = \tfrac{1}{2}\,(\zeta x + \omega) \tag{4.2.62}$$

Then if

$$\widetilde{Q}_n(x) = \tilde{\zeta}^{-n}\, \widetilde{P}_n^{(\alpha,\beta)}(\tilde{\zeta} x + \widetilde{\omega}) \tag{4.2.63}$$

we have that

$$\widetilde{Q}_n(x) = x^n + \left[\frac{\widetilde{\omega} n}{\tilde{\zeta}} + \tilde{\zeta}^{-1} b_n(\alpha,\beta)\right] x^{n-1}$$
$$+ \left[\left(\frac{\widetilde{\omega}}{\tilde{\zeta}}\right)^2 \binom{n}{2} + (n-1)\frac{\widetilde{\omega}}{\tilde{\zeta}}\tilde{\zeta}^{-1} b_n(\alpha,\beta) + \tilde{\zeta}^{-2} c_n(\alpha,\beta)\right] x^{n-2} + O(x^{n-3}) \tag{4.2.64}$$

Each coefficient must be real. For the x^{n-1} term, we see

$$\frac{\widetilde{\omega} n}{\tilde{\zeta}} + \frac{n(\alpha-\beta)}{2\tilde{\zeta}(\alpha+\beta+n)} \tag{4.2.65}$$

is real. The $O(n)$ term is $\frac{\widetilde{\omega}}{\tilde{\zeta}} n$ so $\frac{\widetilde{\omega}}{\tilde{\zeta}}$ is real. Thus, the second term of (4.2.65) is real, so taking $n \to \infty$, $(\alpha-\beta)/\tilde{\zeta}$ is real. Thus, $\alpha+\beta+n$ is real. We conclude that

$$\widetilde{\omega}\tilde{\zeta}^{-1}, \qquad (\alpha-\beta)\tilde{\zeta}^{-1}, \qquad \alpha+\beta \in \mathbb{R} \tag{4.2.66}$$

This implies $\tilde{\zeta}^{-2} c_n$ is real. Since the leading order of c_n is real, $\tilde{\zeta}^{-2}$ is real, so either $\tilde{\zeta} \in \mathbb{R}$ or $\tilde{\zeta} \in i\mathbb{R}$.

If $\tilde{\zeta} \in \mathbb{R}$, then $\alpha - \beta \in \mathbb{R}$, so $\widetilde{\omega}, \tilde{\zeta}, \alpha, \beta \in \mathbb{R}$ and thus, $\omega, \zeta\, \alpha, \beta \in \mathbb{R}$. If $\tilde{\zeta} \in i\mathbb{R}$, then $\alpha - \beta \in i\mathbb{R}$, that is, $\operatorname{Im}(\alpha+\beta) = 0 = \operatorname{Re}(\alpha-\beta) \Rightarrow \beta = \bar{\alpha}$. Moreover, $\widetilde{\omega} \in i\mathbb{R}$, so $\zeta, \omega \in i\mathbb{R}$, that is, $\beta = \bar{\alpha}, \zeta, \omega \in i\mathbb{R}$. □

We'll need two facts about Jacobi polynomials, $P_n^{(\alpha,\beta)}(x)$, below. First, as we've seen, their leading coefficient is

$$\frac{(\alpha+\beta+n+1)_n}{n!} \tag{4.2.67}$$

and this is positive for all large n if $\alpha+\beta \in \mathbb{R}$. Second, their recursion relation (Problem 9) is

$$K_n z P_n^{(\alpha,\beta)}(z) = A_n P_{n+1}^{(\alpha,\beta)}(z) + B_n P_n^{(\alpha,\beta)}(z) + C_n P_n^{(\alpha,\beta)}(z) \tag{4.2.68}$$

$$K_n = (2n+\alpha+\beta)_3, \quad A_n = (2n+2)(2n+\alpha+\beta)(n+1+\alpha+\beta) \tag{4.2.69}$$

$$B_n = (\alpha^2-\beta^2)(2n+1+\alpha+\beta), \quad C_n = 2(n+\alpha)(n+\beta)(2n+2+\alpha+\beta) \tag{4.2.70}$$

Here we'll need

$$\beta = \bar{\alpha}, \quad n \text{ large} \Rightarrow K_n > 0, \quad A_n > 0, \quad C_n > 0, \quad B_n \in i\mathbb{R} \tag{4.2.71}$$

Proof of Theorem 4.2.2. We first eliminate $P_n^{(\alpha,\bar{\alpha})}(iz)$ as candidates for multiples of OPRL using the same strategy we used for Bessel polynomials. If

$$Q_n(z) = i^{-n} P_n^{(\alpha,\bar{\alpha})}(iz) \tag{4.2.72}$$

then, by (4.2.67), for all large n, Q_n has positive leading coefficient. Moreover, Q_n obeys the recursion relation

$$K_n z Q_n(z) = A_n Q_{n+1}(z) - iB_n Q_n(z) - C_n Q_{n+1}(z) \qquad (4.2.73)$$

Since $C_n > 0$, $A_n > 0$, this is not an appropriate recursion relation for OPRL.

Given what we have already, we need only show if $L_n^{(\alpha)}$ or $P_n^{(\alpha,\beta)}$ are OPs, then $\alpha > -1$ or $\alpha, \beta > -1$. If $Q_n(x) = \sum_{j=0}^n d_{n,j} x^j$ is a monic OP for a measure μ, then

$$c_n \equiv \int x^n \, d\mu(x) = -\sum_{j=0}^{n-1} d_{n,j} c_j \qquad (4.2.74)$$

Thus, inductively, c_n is given by a polynomial in the $\{d_{k,j}\}_{k=1,\dots,n}^{j=0,\dots,n-1}$. If the Q_n are instead only OPs (not monic), c_n is given by a rational function of the d's. It follows that if $L_n^{(\alpha)}$ or $P_n^{(\alpha,\beta)}$ are multiples of OPs, then the moments of the corresponding measures are given by explicit rational functions of α (or α and β).

For the Laguerre case, if $\alpha > -1$, there is a probability measure

$$d\mu_\alpha(x) = \Gamma(1+\alpha)^{-1} x^\alpha e^{-x} \chi_{[0,\infty)}(x) \, dx \qquad (4.2.75)$$

Note if $\alpha > -1$ and $k = 0, 1, 2, \dots$, then since

$$\begin{aligned} d\mu_{\alpha+k} &= \Gamma(1+\alpha)\Gamma(1+\alpha+k)^{-1} x^k \, d\mu_\alpha \\ &= (1+\alpha)^{-1} \dots (1+\alpha+k-1)^{-1} x^k \, d\mu_\alpha \end{aligned} \qquad (4.2.76)$$

we have that

$$c_n(\alpha+k) = (1+\alpha)^{-1} \dots (1+\alpha+k-1)^{-1} c_{n+k}(\alpha) \qquad (4.2.77)$$

Suppose $\alpha_0 \in (-\infty, -1]$ is such that $\{L_n^{(\alpha_0)}\}_{n=0}^\infty$ are OPs for a measure $d\mu$. Pick k an even integer with

$$k + \alpha_0 > -1 \qquad (4.2.78)$$

Then $c_k(\alpha) = \infty$ if α_0 is a negative integer, that is, the putative formula for $\int x^k \, d\mu$ has a pole at α_0, which means $d\mu$ cannot exist.

If α_0 is not a negative integer, (4.2.77) implies $x^k \, d\mu$ has the same moments as $d\mu_{\alpha+k}$ up to an overall constant. Since the moments of $d\mu_{\alpha+k}$ are bounded by $C^n n!$, there is a unique measure with those moments (see Theorem 5.6.6. of Part 1 or Section 7.7). Thus,

$$x^k \, d\mu = (1+\alpha) \dots (1+\alpha+k-1) \, d\mu_{\alpha+k} \qquad (4.2.79)$$

so

$$d\mu = c\,\delta_0 + \Gamma(1+\alpha)^{-1} x^\alpha e^{-x} \chi_{[0,\infty)}(x) \, dx \qquad (4.2.80)$$

But $\int_0^\infty x^\alpha e^{-x} \, dx = \infty$, so there is no such measure.

The argument for the Jacobi case is similar. Instead of (4.2.77), one uses

$$c_n(\alpha+k,\beta+k) = N_k(\alpha,\beta) \sum_{j=0}^{k} (-1)^j \binom{k}{j} c_{n+2j}(\alpha,\beta) \tag{4.2.81}$$

since $(1-x)^k(1+x)^k = \sum_{j=0}^{k}(-1)^j x^{2j}$ and $N(\alpha,\beta)$ is the ratio of integrals of $\int_{-1}^{1}(1-x)^\alpha(1+x)^\beta\,dx$ to $\int_{-1}^{1}(1-x)^{\alpha+k}(1+x)^{\beta+k}\,dx$. The argument is now identical. If $k+\alpha, k+\beta > -1$ for $k \in \{1,2,\dots\}$, then any measure, $d\mu$, for which $P_n^{(\alpha,\beta)}$ are the OPs has $(1-x^2)^k\,d\mu = N_k(\alpha,\beta)^{-1}\,d\mu_{\alpha+k,\beta+k}$. If $\alpha \le -1$ or $\beta \le -1$, the positive measure is not integrable, so no such measure exists. □

Notes and Historical Remarks. The arguments in this section are typical of the algebraic theory of OPs. There are some analytic aspects but even more arguments that depend on detailed analysis of specific polynomial coefficients and/or zeros.

We have named this section and its main theorem, Theorem 4.2.2, after a 1929 paper of Bochner [**74**] and a 1930 paper of Brenke [**87**]. It is often called by Bochner's name only, although he only looked at the situation of and proved the result Theorem 4.2.3 which does not mention orthogonality! Earlier, in 1885, Routh [**580**] had looked at a related question but made stronger hypotheses than Theorem 4.2.3.

Despite the fact that "Bochner's Theorem" is often informally quoted as the result we prove in Theorem 4.2.2, I have not found a proof of this in the literature—all the results that I could find make extra hypotheses on the putative measure; for example, Brenke supposed the measure has the form $w(x)\,dx$ where $w(x) > 0$ on an interval and continuous. In finding this proof, input from Mourad Ismail, Andrei Martínez Finkelshtein, and Ed Saff was invaluable.

There is some discussion of the three classes of classical OPs and their history in Section 14.4 of Part 2A. Proofs of the normalization formulae (4.2.13), (4.2.14), (4.2.15) can be found in the book of Andrews et al. [**19**].

The Bessel polynomials first appeared in Bochner's paper and reappeared in Hahn [**280**]. Their connection to Bessel functions, their name, and many of their basic properties appear in a 1948 paper of Krall–Frink [**408**].

Bessel polynomials are OP-like in several ways. They obey a Rodrigues formula with a "weight" $x^a e^{-b/x}$ (with no reasonable integrability properties on $\mathbb{R}$) and an orthogonality relation on a curve in $\mathbb{C}$ (e.g., $\partial\mathbb{D}$) with a complex, not positive measure (see Problem 5). As we have seen, they obey a three-term recursion relation

$$zy_n = \sum_{j=-1}^{1} A_{n,n+j} y_{n+j}$$

but with $A_{n,n+1}/A_{n,n-1}$ negative for a real and n large (unlike the OP case).

One reason for interest in Bessel polynomials is that $y_n(1/x)$ are related to continued fraction expansions of e^x; see Barnes [**46**]. Of course, $\{z^n\}_{n=0}^{\infty}$ are also orthogonal (no complex conjugate) for Lebesgue measure on $\partial\mathbb{D}$.

There are a number of Bochner–Brenke-like theorems in the literature. Marcellán et al. [**465**] find all sets of polynomials obeying a differential equation and orthogonal in a set of moments without establishing when the moments come from a positive measure. Császár [**139**] and Aczél [**2**] study when a set of polynomials obey a second-order differential equation for which there is a positive weight, w, so that the differential operator is formally self-adjoint with respect to $w(x)\,dx$. The same result can be found in the book of Beals and Wong [**48**]. Hahn [**280**] answers the question of when P_n are a family of monic OPs so the $P_{n+1}'/(n+1)$ are also a family of monic OPs (different weight). Grünbaum–Haine [**275**] study all double-sided, not necessarily symmetric Jacobi matrices whose eigenfunctions obey a differential equation in x.

A recent direction has been to consider situations where only a subset of polynomials are the eigenfunctions of a second-order differential operator, for example, starting with $n = 1$ rather than $n = 0$. There are examples beyond these of the Bochner–Brenke theorem known as *exceptional orthogonal polynomials*. The first examples were found by Gomez-Ullate, Kamran, and Milson [**261, 262**]. For further developments, see [**539, 540, 508, 176, 177**].

There have been several explorations of the theme of functions in two parameters obeying equations in each. Polynomials in x which obey recursion relations in the degree and difference (as opposed to differential) equations in x are studied in the books of Ismail [**334**, Ch. 20] and Lesky [**438**] and references therein. Duistermaat–Grünbaum [**172**] study the situation where the difference equation in n is replaced by a differential equation where n is replaced by a continuous parameter. There has also been work on matrix analogs of Bochner's theorem (i.e., where the polynomials and OP measures are matrix-valued; see, for example, Durán–Grünbaum [**178**]).

Problems

1. From (4.2.6), (4.2.23), and (4.2.34), derive (4.2.25).

2. Verify the differential equations (4.2.10), (4.2.11), (4.2.12).

3. (a) Let y_n be defined by (4.2.43). Verify that (4.2.43) holds.

(b) By using the theory of singular points of ODEs (see Chapter 14 of Part 2B), verify that any linearly independent solution of (4.2.43) is not a polynomial.

4. Starting from (4.2.41), verify (4.2.44)/(4.2.45).

5. Let $\rho(z;a) = {}_1F_1(1, a-1, -1/x)$. Prove that for $n \neq m$,

$$\oint_C \rho(z;a) y_n(z;a,1) y_m(z;b,1)\, dz = 0$$

for any curve C surrounding 0.

6. Verify (4.2.47)/(4.2.48).

7. Let P be a polynomial in z. Prove that $P(z)$ is real for $z \in \{\ell, \ell+1, \dots\}$ (for some $\ell \in \mathbb{Z}$) if and only if all the coefficients of P are real. (*Hint*: Look at $\overline{P(\bar{z})}$.)

8. (a) Prove that (${}_2F_1$ is defined and discussed in Section 14.4 of Part 2B)

$$P_n^{(\alpha,\beta)}(z) = \frac{(\alpha+1)_n}{n!}\, {}_2F_1\biggl(-n, 1+\alpha+\beta; \alpha+1; \frac{1-z}{2}\biggr)$$

(b) Using relations of ${}_2F_1$ at its singular points, prove that $P_n^{(\alpha,\beta)}(-z) = P_n^{(\beta,\alpha)}(z)$.

(c) Prove that ($\widetilde{P}$ given by (4.2.58))

$$\widetilde{P}_n^{(\alpha,\beta)}(z) = \sum_{j=0}^{n} \frac{(\alpha+n-j+1)_j}{(\alpha+\beta+2n+1-j)_j} \binom{n}{j} \left(\frac{z-1}{2}\right)^{n-j} \tag{4.2.82}$$

(d) Verify (4.2.59)–(4.2.61).

9. Verify (4.2.68)–(4.2.70) by using (4.2.82).

4.3. L^2- and L^∞-Variational Principles: Chebyshev Polynomials

> I assume that I am not the only one who does not understand the interest in and significance of these strange problems on maxima and minima studied by Chebysbev in memoirs whose titles often begin with, "On functions deviating least from zero ... ." Could it be that one must have a Slavic soul to understand the great Russian Scholar?
>
> —*H. Lebesgue*[2]

We begin with a basic variational property of OPs; other variational aspects of OPs can be found in Problem 2.

[2]in report on the work of S. Bernstein dated June 17, 1926 as quoted in [**659**]

Theorem 4.3.1. *Let μ be a nontrivial measure on $\mathbb{C}$ with $\int |z|^n \, d\mu(z) < \infty$ for some $n \in \mathbb{Z}_+$. Let $\Phi_n(z)$ be the monic* OP *of degree n. Then for any monic polynomial, Q, of degree n, with $\|\cdot\|_2$ the $L^2(\mathbb{C}, d\mu)$-norm, we have*

$$\|\Phi_n\|_2 \le \|Q\|_2 \tag{4.3.1}$$

with equality only if $Q = \Phi_n$.

Proof. If

$$R = Q - \Phi_n \tag{4.3.2}$$

then $R \perp \Phi_n$ so

$$\|Q\|_2^2 = \|\Phi_n\|_2^2 + \|R\|_2^2 \tag{4.3.3}$$

This implies (4.3.1). One has equality if and only if R is zero in L^2. Since μ is nontrivial, this implies that R is the zero polynomial (since otherwise R vanishes at $n-1$ or fewer points). $\square$

In the remainder of this section, we'll study the analogous question when L^2 is replaced by L^∞. Since polynomials are continuous, the L^∞-norm is the same as the sup over the support, $\mathfrak{e}$, of μ. If $\mathfrak{e}$ is unbounded, this sup is always ∞, so we restrict $\mathfrak{e}$ to be a compact set. We always suppose that $\mathfrak{e}$ is an infinite set of points.

Definition. Let $\mathfrak{e}$ be a compact subset of $\mathbb{C}$. Define

$$\|f\|_\mathfrak{e} = \sup_{z \in \mathfrak{e}} |f(z)| \tag{4.3.4}$$

Definition. Let $\mathfrak{e}$ be a compact subset of $\mathbb{C}$. A monic polynomial, T_n, of degree n is called a *Chebyshev polynomial for* $\mathfrak{e}$ of degree n if for any other monic polynomial, Q_n, of degree n, we have

$$\|T_n\|_\mathfrak{e} \le \|Q_n\|_\mathfrak{e} \tag{4.3.5}$$

It is not hard to see (Problem 3) that if $Q_n(z) = z^n + \sum_{j=0}^{n-1} b_j z^j$, then $\|Q_n\|_\mathfrak{e} \to \infty$ as $\sum_{j=0}^{n-1} |b_j|^2 \to \infty$, so

$$\{Q_n \mid \|Q_n\|_\mathfrak{e} \le 2 \inf_{P_n} \|P_n\|_\mathfrak{e}\}$$

is compact. Thus, a polynomial minimizing $\|\cdot\|_\mathfrak{e}$ exists. While the theory can be developed for general $\mathfrak{e} \subset \mathbb{C}$, we'll only discuss the case where $\mathfrak{e} \subset \mathbb{R}$. We'll show that T_n is unique in this case and that $\|T_n\|_\mathfrak{e}^{1/n} \to C(\mathfrak{e})$, the logarithmic capacity of $\mathfrak{e}$ (see Section 3.6 of Part 3).

Example 4.3.2. Recall (see Problem 8 of Section 3.1 of Part 2A) the classical Chebyshev polynomials are defined by

$$p_n(\cos\theta) = \cos(n\theta) \tag{4.3.6}$$

Since $\cos\theta = (e^{i\theta}+e^{-i\theta})/2$, $\cos(n\theta) = (e^{in\theta}+e^{-in\theta})/2$, we see

$$T_n(x) = 2^{-(n-1)} p_n(x) \tag{4.3.7}$$

is a monic polynomial. We claim it is a Chebyshev polynomial for $[-1,1]$. For

$$\|T_n\|_{[-1,1]} = 2^{-(n-1)} \tag{4.3.8}$$

since $\cos\theta \in [-1,1] \Leftrightarrow \theta$ is real, and then $\cos(n\theta) \in [-1,1]$.

As θ runs from $-\pi$ to 0, $\cos\theta$ runs from -1 to 1 and $|\cos(n\theta)|$ takes the value 1, $n+1$ times. Thus, there exists

$$-1 = x_0^{(n)} < x_1^{(n)} < x_2^{(n)} < \dots < x_n^{(n)} = 1 \tag{4.3.9}$$

so that

$$T_n(x_j^{(n)}) = (-1)^{n-j} 2^{-(n-1)} \tag{4.3.10}$$

Suppose Q_n is a monic polynomial with

$$\|Q_n\|_{[-1,1]} < 2^{-(n-1)} \tag{4.3.11}$$

Then $(-1)^{n-j}[T_n(x_j^{(n)}) - Q_n(x_j^{(n)})] > 0$ for $j = 0,\dots,n$ and thus, $T_n - Q_n$ has to have a zero in each interval $(x_j^{(n)}, x_{j+1}^{(n)})$, $j = 0,\dots,n-1$ (since $T_n - Q_n$ has opposite signs at the two ends of the interval). Thus, $T_n - Q_n$ has at least n zeros. But since both are monic, $\deg(T_n - Q_n) \le n-1$. Thus, $T_n - Q_n = 0$, violating (4.3.11). It follows that T_n is a Chebyshev polynomial for $[-1,1]$. □

Definition. Let P_n be a real degree-n monic polynomial and $\mathfrak{e} \subset \mathbb{R}$ a compact set. We say that P_n has an *alternating set* in $\mathfrak{e}$ if there exists $\{x_j\}_{j=0}^n \subset \mathfrak{e}$ with

$$x_0 < x_1 < \ldots < x_n \tag{4.3.12}$$

and so that

$$P_n(x_j) = (-1)^{n-j} \|P_n\|_{\mathfrak{e}} \tag{4.3.13}$$

Remark. Since P_n changes sign in (x_j, x_{j+1}), there must be n zeros in (x_0, x_n). It follows that for $y > x_n$, $P_n(y) > P(x_n)$ and for $y < x_0$, $(-1)^n P_n(y) > (-1)^n P_n(x_0)$. This implies that $\mathrm{cvh}(\mathfrak{e})$ must be $[x_0, x_n]$.

We will prove the following

Theorem 4.3.3 (Alternation Principle). *Let $\mathfrak{e} \subset \mathbb{R}$ be compact and P_n a monic polynomial. Then the following are equivalent.*

(1) *P_n is real and has an alternating set.*
(2) *P_n is a Chebyshev polynomial.*

Remark. If one merely has points with sign alternation, one can obtain lower bounds on $\|T_n\|_{\mathfrak{e}}$; see Problem 1.

The argument that (4.3.7) are Chebyshev polynomials for $[-1,1]$ works also to show (1) $\Rightarrow$ (2) in Theorem 4.3.3 so we turn towards (2) $\Rightarrow$ (1) by looking at moving zeros of a real polynomial by a small amount.

Lemma 4.3.4. *Let $q_m(x)$ be a real polynomial. Then*

(a) *For any fixed a and any infinite compact set $K_1 \subset (-\infty, a]$ and $K_2 \subset [a,\infty)$, we have for all small $\varepsilon > 0$, that if*

$$p_{m+1}^{(\varepsilon)}(x) = \big(x-(a-\varepsilon)\big)q_m(x) \tag{4.3.14}$$

then

$$\sup_{x\in K_1} |p_{m+1}^{(\varepsilon)}(x)| < \sup_{x\in K_1} |p_{m+1}^{(0)}(x)|, \quad \sup_{x\in K_2} |p_{m+1}^{(\varepsilon)}(x)| \le \sup_{x\in K_2} |p_{m+1}^{(0)}| + O(\varepsilon) \tag{4.3.15}$$

(b) *For any fixed a and any infinite compact set $K \subset \mathbb{R}$ for all small $\varepsilon > 0$, we have if*

$$r_{m+2}^{(\varepsilon)}(x) = \big(x-(a+\varepsilon)\big)\,\big(x-(a-\varepsilon)\big)q_m(x) \tag{4.3.16}$$

then

$$\sup_{x\in K} |r_{m+2}^{(\varepsilon)}(x)| < \sup_{x\in K} |r_{m+2}^{(0)}(x)| \tag{4.3.17}$$

(c) *For any fixed $a<b$ and any infinite compact set, $K_1 \subset \mathbb{R}\setminus(a,b)$ and $K_2 \subset [a,b]$, we have for all small $\varepsilon > 0$, that if*

$$s_{m+2}^{(\varepsilon)}(x) = \big(x-(a-\varepsilon)\big)\,\big(x-(b+\varepsilon)\big)q_m(x) \tag{4.3.18}$$

then

$$\sup_{x\in K_1} |s_{m+2}^{(\varepsilon)}(x)| < \sup_{x\in K_1} |s_{m+2}^{(0)}(x)|; \quad \sup_{x\in K_2} |s_{m+2}^{(\varepsilon)}(x)| \le \sup_{x\in K_2} |s_{m+2}^{(0)}(x)| + O(\varepsilon) \tag{4.3.19}$$

Remarks. 1. Rather than get bogged down in algebra, the reader should think what is being stated. For example, (c) says if a pair of zeros are moved symmetrically outwards by a small amount, the sup on K, outside $[a,b]$ goes strictly down and the sup inside can't increase by more than $O(\varepsilon)$.

2. We only require K infinite to be sure $q_n(y) \neq 0$ for some $y \in K$ so the sup isn't near the zeros that are moving.

Proof. (a) Since K_1 is infinite, $\sup_{x\in K_1} |p_{m+1}^{(0)}(x)|$ is taken at some point $y<a$. Thus for ε small, the sup is taken on $K_1 \cap (-\infty, a-\varepsilon)$. On this set, $|p_{m+1}^{(\varepsilon)}(x)|$ is strictly monotone decreasing in ε. Thus the first inequality in (4.3.15) is immediate. Since $\varepsilon\, q_m$ is bounded on K_2, the second inequality is trivial.

(b) Without loss, we suppose $a=0$. For small ε, the sup occurs on $K\setminus [-2\varepsilon_0, 2\varepsilon_0]$ when $\varepsilon<\varepsilon_0$ and $(x^2-\varepsilon^2)$ is strictly monotone there.

(c) Without loss, we suppose $a = -b$, $b > 0$. Then $|s_{m+2}^{(\varepsilon)}| = \big(x^2 - (b+\varepsilon)^2\big)\,|q_m(x)|$ is strictly monotone in $\mathbb{R}\setminus[-b-2\varepsilon_0, b+2\varepsilon_0]$ for $\varepsilon < \varepsilon_0$ so we get the first inequality in (4.3.19). Since $\|s_{m+2}^{(\varepsilon)} - s_{m+2}^{(0)}\|_{K_2} = O(\varepsilon)$, the second is trivial $\square$

Theorem 4.3.5. *Let $\mathfrak{e} \subset \mathbb{R}$ and let T_n be a Chebyshev polynomial for $\mathfrak{e}$. Then*

(a) *All zeros of T_n are real and thus T_n is real.*
(b) *All zeros of T_n are simple.*
(c) *All zeros of T_n lie in* $\operatorname{cvh}(\mathfrak{e})$, *the convex hull of $\mathfrak{e}$.*
(d) *Let $y_1 < y_2$ be two zeros of T_n. Then there is a point, w, of $\mathfrak{e}$ in (y_1, y_2) so that*
$$|T_n(w)| = \|T_n\|_{\mathfrak{e}} \tag{4.3.20}$$
(e) *Let y_1 be the smallest zero of T_n. Then there is a point, w, of $\mathfrak{e}$ in $(-\infty, y_1)$, so that* (4.3.20) *holds.*
(f) *T_n has an alternating set.*

Proof. (a) Let $y = a + ib$ be a zero with $b \neq 0$. Write $T_n(x) = (x - a - ib)q_{n-1}(x)$ and note that
$$|T_n(x)| > |(x-a)q_{n-1}(x)| \tag{4.3.21}$$
at the non-zeros of q_{n-1} so $\|(x-a)q_{n-1}\|_{\mathfrak{e}} < \|T_n\|_{\mathfrak{e}}$ (since $\mathfrak{e}$ is infinite). This contradicts the minimality of $\|T_n\|_{\mathfrak{e}}$. Thus all zeros are real and so T_n is real since it is monic.

(b) If a is a double or higher zero, $T_n(x) = (x-a)^2 q_{n-2}$ so, by (b) of the lemma, for ε small, $\|r_n^{(\varepsilon)}\|_{\mathfrak{e}} < \|T_n\|_{\mathfrak{e}}$ violating minimal. Hence all zeros are simple.

(c) Let $[\alpha, \beta] = \operatorname{cvh}(\mathfrak{e})$. If $a > \beta$ is a zero, (a) of the lemma lets us decrease $\|p_n^{(\varepsilon)}\|_{\mathfrak{e}}$ by moving the zero down. Similarly, there is no zero in $(-\infty, \alpha]$.

(d) If $\sup_{\mathfrak{e}\cap[y_1,y_2]} |T_n(x)| < \|T_n\|_{\mathfrak{e}}$, we can move the zeros outwards symmetrically and use (c) of the lemma to get a contradiction.

(e) If $\sup_{\mathfrak{e}\cap(-\infty,y_1]} |T_n(x)| < \|T_n\|_{\mathfrak{e}}$, we can move the zero at y_1 up slightly and use (a) of the lemma to get a contradiction.

(f) Let $y_1 < y_2 < \dots < y_n$ be the zeros of T_n. By (d), (e), we can find $w_0 < y_1 < w_1 < y_2 < \dots < y_n < w$ so $|T_n(w_j)| = \|T_n\|_{\mathfrak{e}}$ with $\{w_0, \dots, w_n\} \subset \mathfrak{e}$. Clearly since $T_n(x) = \prod_{j=1}^n (x - y_j)$, we see $(-1)^{n-j}T_n(w_j) > 0$. Thus we have an alternating set. $\square$

Corollary 4.3.6. *Let $\mathfrak{e} \subset \mathbb{R}$ be compact. The Chebyshev polynomial, T_n, of degree n is unique.*

Proof. Suppose $A = \inf\{\|P_n\|_{\mathfrak{e}} \mid P_n \text{ monic of degree } n\}$ and there are two distinct polynomials, P_n, Q_n, with

$$\|P_n\|_{\mathfrak{e}} = \|Q_n\|_{\mathfrak{e}} = A \tag{4.3.22}$$

Let

$$T_n = \tfrac{1}{2}(P_n + Q_n) \tag{4.3.23}$$

Since $\|T_n\|_{\mathfrak{e}} \geq A$ and $\|T_n\|_{\mathfrak{e}} \leq \frac{1}{2}(\|P_n\|_{\mathfrak{e}} + \|Q_n\|_{\mathfrak{e}}) = A$, we see T_n is also a Chebyshev polynomial.

By Theorem 4.3.5, we have $w_0 < w_1 < \cdots < w_n$, so

$$|T_n(w_j)| = \|T_n\|_{\mathfrak{e}} \tag{4.3.24}$$

Since $|P_n(w_j)|, |Q_n(w_j)| \leq \|T_n\|_{\mathfrak{e}}$, we conclude

$$T_n(w_j) = P_n(w_j) = Q_n(w_j) \tag{4.3.25}$$

In particular, for $j = 0, \dots, n$,

$$P_n(w_j) - Q_n(w_j) = 0 \tag{4.3.26}$$

But $\deg(P_n - Q_n) \leq n - 1$ and $P_n - Q_n \neq 0$, so $P_n - Q_n$ cannot have more than $n - 1$ zeros (rather than $n + 1$). This contradiction shows there is not more than one minimizer. $\square$

We are heading towards a proof that

$$\lim_{n\to\infty} \|T_n\|_{\mathfrak{e}}^{1/n} = C(\mathfrak{e}) \tag{4.3.27}$$

the logarithmic capacity of $\mathfrak{e}$. We will need the Bernstein–Walsh lemma (see Theorem 3.7.1 of Part 3) which says that if $Q_n(z)$ is any polynomial of degree n and $G_{\mathfrak{e}}(z)$ is the Green's function for $\mathfrak{e} \subset \mathbb{R}$, then

$$|q_n(z)| \leq \|q_n\|_{\mathfrak{e}} \exp(nG_{\mathfrak{e}}(z)) \tag{4.3.28}$$

We also need that $G_{\mathfrak{e}}(z)$ obeys (see Theorem 3.6.8 of Part 3)

$$G_{\mathfrak{e}}(z) - \log(|z|) \to -\log(C(\mathfrak{e})) \tag{4.3.29}$$

as $|z| \to \infty$.

Theorem 4.3.7. *For any $\mathfrak{e} \subset \mathbb{R}$, one has*

$$\|T_n\|_{\mathfrak{e}} \geq C(\mathfrak{e})^n \tag{4.3.30}$$

Proof. By (4.3.28),

$$|T_n(z)||z|^{-n} \leq \|T_n\|_{\mathfrak{e}} \exp(n[G_{\mathfrak{e}}(z) - \log(|z|)]) \tag{4.3.31}$$

Take $|z| \to \infty$. Since T_n is monic, the left side of (4.3.31) goes to 1. Thus, by (4.3.29),

$$1 \leq \|T_n\|_{\mathfrak{e}} \exp(-n \log(C(\mathfrak{e}))) \tag{4.3.32}$$

which is (4.3.30). $\square$

In fact, one can improve (4.3.30) by a factor of 2 and find many examples where $\|T_n\|_{\mathfrak{e}} = 2C(\mathfrak{e})^n$.

Theorem 4.3.8. (a) *Let $y_1 < \dots < y_n$ be the zeros of $T_n(x)$. Then $T_n'(x)$ has exactly one simple zero, q_j, in each interval (y_j, y_{j+1}) and*

$$|T_n(q_j)| \geq \|T_n\|_{\mathfrak{e}} \tag{4.3.33}$$

(b) *The following is a subset of $\mathbb{R}$:*

$$\mathfrak{e}_n \equiv T_n^{-1}\left[\left[-\|T_n\|_{\mathfrak{e}}\,\|T_n\|_{\mathfrak{e}}\right]\right] \tag{4.3.34}$$

(c) *$\mathfrak{e} \subset \mathfrak{e}_n$.*
(d) *T_n is also the n-th Chebyshev polynomial of $\mathfrak{e}_n$.*
(e) *The Green's function for $\mathfrak{e}_n$ is given by*

$$G_{\mathfrak{e}_n}(z) = \frac{1}{n}\log\left|\frac{T_n(z)}{\|T_n\|_{\mathfrak{e}}} + \sqrt{\frac{T_n(z)}{\|T_n\|_{\mathfrak{e}}}^2 - 1}\right| \tag{4.3.35}$$

where the branch of $\sqrt{\ }$ is taken with it equal to $z^n + O(z^{n-1})$ near $z = \infty$.

(f) $\|T_n\|_{\mathfrak{e}} = 2C(\mathfrak{e}_n)^n$. (4.3.36)

(g) (Schiefermayr's Theorem). $\|T_n\|_{\mathfrak{e}} \geq 2C(\mathfrak{e})^n$. (4.3.37)

Remarks. 1. Given $\mathfrak{e}$, (d) and (f) say that for $\mathfrak{e} = \mathfrak{e}_n$, we have equality in (4.3.37), so equality holds for many $\mathfrak{e}$'s. We'll explain exactly which $\mathfrak{e}$'s in the Notes.

2. The square root in (4.3.35) has branch points at z_0 with $T_n(z_0) = \pm\|T_n\|_{\mathfrak{e}}$. We draw branch cuts on the sets of $\mathfrak{e}$. The square root is discontinuous across the cuts but the $|\cdot|$ is not so the right side of (4.3.35) is continuous and well defined on all of $\mathbb{C}$.

Proof. (a) By Snell's theorem, T_n' has at least one zero in each interval (y_j, y_{j+1}), $j = 1, \dots, n-1$. Since T_n' only has $n-1$ zeros counting multiplicity, each interval has precisely one single zero. Since $|T_n(w_j)| \in [0, |T_n(q_j)|]$, and $|T_n(w_j)| = \|T_n\|_{\mathfrak{e}}$, (4.3.33) is immediate.

(b) Suppose first that n is even. Then T_n maps $[y_n, \sup_{x\in\mathfrak{e}} x]$ and $[\inf_{x\in\mathfrak{e}} x, y_1]$ to $[0, \|T_n\|_{\mathfrak{e}}]$. By (4.3.33), each $\gamma \in [0, \|T_n\|_{\mathfrak{e}})$ gets taken twice in $(y_2, y_3), (y_4, y_5), \dots (y_{n-2}, y_{n-1})$ so in total n times and similarly for $\gamma \in (-\|T_n\|_{\mathfrak{e}}, 0)$. Thus

$$T_n^{-1}\big[\big(-\|T_n\|_{\mathfrak{e}}, \|T_n\|_{\mathfrak{e}}\big)\big] \subset \operatorname{cvh}(\mathfrak{e}) \subset \mathbb{R} \tag{4.3.38}$$

since $T_n(z) = \gamma$ has at most n solutions. By the open mapping theorem, we see $\mathfrak{e}_n \subset \mathbb{R}$.

(c) Immediate from $\|T_n\|_{\mathfrak{e}} = \sup_{x\in\mathfrak{e}} |T_n(x)|$ and T_n real.

(d) By definition, $\|T_n\|_{\mathfrak{e}_n} = \|T_n\|_{\mathfrak{e}}$ so the alternating set for $\mathfrak{e}$ is one for $\mathfrak{e}_n$. Thus, T_n is a Chebyshev polynomial for $\mathfrak{e}$.

(e), (f) The expression inside $|\cdot|$ of $Q(z)$, the right side of (4.3.35), is analytic on $\mathbb{C} \setminus \mathfrak{e}_n$ so $Q(z)$ is harmonic on $\mathbb{C} \setminus \mathfrak{e}_n$. As noted in Remark 2, $Q(z)$ is continuous across $\mathfrak{e}_n$ and since $|\cdots| = 1$ on $\mathfrak{e}_n$, we see $Q \restriction \mathfrak{e}_n \equiv 0$. Near ∞, $T_n(z) + \sqrt{T_n(z)^2 - \|T_n\|_n^2} = 2z^n + O(z^{n-1})$, so near ∞

$$Q(z) = \log|z| - \frac{1}{n}\log\left(\frac{\|T_n\|_{\mathfrak{e}}}{2}\right) + o(1) \tag{4.3.39}$$

Therefore $Q(z) = G_{\mathfrak{e}_n}(z)$ and

$$\log\big(C(\mathfrak{e}_n)\big) = \frac{1}{n}\left(\frac{\|T_n\|_{\mathfrak{e}}}{2}\right) \tag{4.3.40}$$

which is (4.3.36).

(g) Immediate from (4.3.36) and $\mathfrak{e} \subset \mathfrak{e}_n \Rightarrow C(\mathfrak{e}) \le C(\mathfrak{e}_n)$. □

To get the upper bound on $\|T_n\|_{\mathfrak{e}}$, we need a new notion:

Definition. Given $x_1, \dots, x_n \in \mathbb{R}$, define

$$c_n(x_1, \dots, x_n) = \prod_{i \neq j} |x_i - x_j| \tag{4.3.41}$$

the product over the $n(n-1)$ distinct pairs. This is a continuous function on $\mathfrak{e} \times \cdots \times \mathfrak{e}$, so the maximum is achieved. An n-point *Fekete set for* $\mathfrak{e}$ is $x_1^{(0)}, \dots, x_n^{(0)} \in \mathfrak{e}$ so that

$$c_n(x_1^{(0)}, \dots, x_n^{(0)}) = \sup_{x_j \in \mathfrak{e}} c_n(x_1, \dots, x_n) \tag{4.3.42}$$

We define

$$\zeta_n(\mathfrak{e}) = c_n(x_1^{(0)}, \dots, x_n^{(0)})^{1/n(n-1)} \tag{4.3.43}$$

the maximum multiplicative average distance between points in $\mathfrak{e}$.

Theorem 4.3.9.

(a) $\zeta_{n+1}(\mathfrak{e}) \le \zeta_n(\mathfrak{e})$ (4.3.44)

(b) *For any* n,

$$\|T_n\|_{\mathfrak{e}}^{1/n} \le \zeta_{n+1}(\mathfrak{e}) \tag{4.3.45}$$

Proof. (a) Let $x_1^{(0)}, \dots, x_{n+1}^{(0)}$ be an $n+1$-point Fekete set. By definition of $\zeta_n(\mathfrak{e})$, for $j = 1, \dots, n+1$,

$$c_n(x_1^{(0)}, \dots, \widehat{x}_j^{(0)}, \dots, x_{n+1}^{(0)}) \le \zeta_n(\mathfrak{e})^{n(n-1)} \tag{4.3.46}$$

(where the symbol on the left means the value at the n point set with $x_j^{(0)}$ removed).

Take the product over the $n+1$ values of j. Each pair $|x_k - x_\ell|$ occurs on the left side $n-1$ times (for $j \neq k, \ell$). Thus,

$$c_{n+1}(x_1^{(0)}, \dots, x_{n+1}^{(0)})^{n-1} \le \zeta_n(\mathfrak{e})^{n(n-1)(n+1)} \tag{4.3.47}$$

which is (4.3.44).

(b) Let $x_1^{(0)}, \dots, x_{n+1}^{(0)}$ be an $n+1$-point Fekete set. For $j = 1, \dots, n+1$, let p_j be the monic polynomial

$$p_j(x) = \prod_{k \neq j} (x - x_k^{(0)}) \tag{4.3.48}$$

We have that

$$p_j(x) \prod_{\substack{k,\ell \neq j \\ k \neq \ell}} (x_k^{(0)} - x_\ell^{(0)}) \tag{4.3.49}$$

is maximized for $x \in \mathfrak{e}$ when $x = x_j^{(0)}$. Thus,

$$\|p_j\|_{\mathfrak{e}} = \prod_{k \neq j} |x_j^{(0)} - x_k^{(0)}| \tag{4.3.50}$$

(j is fixed and the product is over n values of k).

Therefore,

$$\prod_{j=1}^{n+1} \|p_j\|_{\mathfrak{e}} = c_{n+1}(x_1^{(0)}, \dots, x_{n+1}^{(0)}) = \zeta_{n+1}(\mathfrak{e})^{(n+1)n} \tag{4.3.51}$$

Since T_n minimizes $\|p\|_{\mathfrak{e}}$, we see $\|T_n\|_{\mathfrak{e}} \le \|p_j\|_{\mathfrak{e}}$ for each j, so the left side of (4.3.51) $\ge \|T_n\|_{\mathfrak{e}}^{n+1}$ which yields (4.3.45). □

Since $\zeta_n(\mathfrak{e})$ is monotone, its limit

$$\zeta_\infty(\mathfrak{e}) = \lim_{n\to\infty} \zeta_n(\mathfrak{e}) \tag{4.3.52}$$

exists. It is called the *transfinite diameter*. We have that

$$C(\mathfrak{e}) \le \liminf \|T_n\|_{\mathfrak{e}}^{1/n} \le \limsup \|T_n\|_{\mathfrak{e}}^{1/n} \le \zeta_\infty(\mathfrak{e}) \tag{4.3.53}$$

by (4.3.43) and (4.3.45).

Theorem 4.3.10 (Faber–Fekete–Szegő Theorem). *For any compact set $\mathfrak{e} \subset \mathbb{R}$,*

$$\zeta_\infty(\mathfrak{e}) = C(\mathfrak{e}) \tag{4.3.54}$$

In particular,

$$\lim_{n\to\infty} \|T_n\|_{\mathfrak{e}}^{1/n} = C(\mathfrak{e}) = \zeta_\infty(\mathfrak{e}) \tag{4.3.55}$$

Remark. (4.3.54) has a Coulomb energy interpretation.

$$-\log(\zeta_n(\mathfrak{e})) \text{ is the minimum of } \frac{1}{n(n-1)}\Bigg[\sum_{1\le j\ne k\le n} -\log(|x_j - x_k|)\Bigg] \tag{4.3.56}$$

the Coulomb energy of n particles. So as $n \to \infty$, we expect this to go the minimum of the continuum Coulomb energy of a unit charge spread out, that is, the $-\log(C(\mathfrak{e}))$.

In taking a limit, we need to include self-energies with a cutoff. Since there are only n self-energies and we divide by n^2, the cutoff goes away in the limit.

Proof. (4.3.54) plus (4.3.53) implies (4.3.55) so we need only prove (4.3.54). Since (4.3.53) says $C(\mathfrak{e}) \le \zeta_\infty(\mathfrak{e})$, we need only prove

$$\zeta_\infty(\mathfrak{e}) \le C(\mathfrak{e}) \tag{4.3.57}$$

Define the cutoff Coulomb energy of a probability measure, ν, by

$$\mathcal{E}_a(\nu) = \int \log(\min(a, |x-y|^{-1}))\, d\nu(x)d\nu(y) \tag{4.3.58}$$

where $a \in (0,\infty)$. By the monotone convergence theorem, for any ν,

$$\lim_{a\to\infty} \mathcal{E}_a(\nu) = \mathcal{E}(\nu) \equiv \int \log(|x-y|^{-1})\, d\nu(x)d\nu(y) \tag{4.3.59}$$

the Coulomb energy (see Section 5.9 of Part 1).

For each n, pick an n-point Fekete set and let ν_n be the normalized counting measure $\frac{1}{n}\sum_{j=1}^n \delta_{x_j^{(0)}}$. By passing to a subsequence, we can suppose $\nu_{n(j)} \to \nu_\infty$, a limiting measure.

Since

$$|x-y| \le \min(|x-y|, a^{-1}) = \exp\big(-\log(a, |x-y|^{-1})\big) \tag{4.3.60}$$

we have for each $a < \infty$ that

$$a^{-1/n} \prod_{j\ne k} |x_j - x_k|^{1/n^2} \le \exp[-\mathcal{E}_a(\nu_{n(j)})] \tag{4.3.61}$$

Take $n \to \infty$ and use that $a^{-1/n} \to 1$ to see

$$\zeta_\infty(\mathfrak{e}) \le \exp(-\mathcal{E}_a(\nu_\infty)) \tag{4.3.62}$$

Now take $a \to \infty$ and use (4.3.54) to see

$$\zeta_\infty(\mathfrak{e}) \le \exp(-\mathcal{E}(\nu_\infty)) \tag{4.3.63}$$

By (4.3.53), this says $\mathcal{E}(\nu_\infty) \le \log(C(\mathfrak{e}^{-1})) = \mathcal{E}(\rho_\mathfrak{e})$, where $\rho_\mathfrak{e}$ is the minimizing measure for potential theory (see Theorem 5.9.3 of Part 1). (If $C(\mathfrak{e}) = 0$, $\mathcal{E}(\nu_\infty) = 0$ and (4.3.63) says $\zeta_\infty(\mathfrak{e}) = 0$, so (4.3.57) holds.)

Thus, $\nu_\infty = \rho_\mathfrak{e}$ and therefore, ν_n converges to $\rho_\mathfrak{e}$ and (4.3.63) is (4.3.57). □

In many cases, this argument can be extended to prove the zero counting measure for the T_n's converges to $\rho_\mathfrak{e}$.

Corollary 4.3.11. *Let μ be a measure with compact support $\mathfrak{e} \subset \mathbb{R}$. Let P_n be the monic* OPRL. *Then*

$$\limsup \|P_n\|_{L^2(d\mu)}^{1/n} \le C(\mathfrak{e}) \tag{4.3.64}$$

Proof. By Theorem 4.3.1,

$$\|P_n\|_{L^2(d\mu)} \le \|T_n\|_{L^2(d\mu)} \le \|T_n\|_\mathfrak{e} \mu(\mathbb{R})^{1/2} \tag{4.3.65}$$

so taking roots and letting $n \to \infty$, we get (4.3.64). □

μ is called *regular* if $\lim \|P_n\|_{L^2(d\mu)}^{1/n} = C(\mathfrak{e})$. We discuss this in Section 3.7 of Part 3.

Notes and Historical Remarks.

> Chebyshev polynomials are everywhere dense in numerical analysis.
> —quoted in *Mason-Handscomb* [**469**] [3]

Chebyshev (whose capsule biography can be found in Section 7.2 of Part 1) introduced and exploited Chebyshev polynomials on $[-1, 1]$ in a series of papers starting in 1854 [**116, 117, 118**]. He never noted the definition that many modern presentations use: $T_n(\cos\theta) = \cos n\theta$ but focused instead on the min-max property (4.3.5). His first paper involved applications to building explicit mechanical devices: there is a website [**398**] devoted to his devices as well as models in the Moscow Polytechnical Museum and in the Paris Musée des Arts. There are several books [**204, 469, 573**] on the subject of these $[-1, 1]$ T_n polynomials. One should think of T_n as answering the question, which polynomial, q_{n-1} of degree $n-1$ minimizes $\|x^n - q_{n-1}\|_\mathfrak{e}$ and more generally, there is an alteration principal for the minimizer of $\|f - q_{n-1}\|_\mathfrak{e}$ explaining why the subject is central to approximation theory and to numerical analysis.

The general theory of Chebyshev polynomials was developed in Borel [**82**] in 1905, including existence and the use of the principle of alternation—a version of that had appeared earlier in the thesis of Kirchberger [**388**], a

[3] They attribute it to "George Forsythe (and maybe Philip Davis)."

student of Hilbert. The general theory also appeared in 1905 lectures of Markov formally only published in 1948 [**466**]. See Christiansen et al. [**124**] for his proof of the alternation theorem. Further significant developments of the general theory are due to Tonelli [**703**], de la Vallée Poussin [**154**], Jackson [**338**], and Haar [**278**].

Schiefermayr's theorem appeared in his paper [**600**]. From (4.3.35) and the fact that harmonic measure of $\mathfrak{e}$ in case $\mathfrak{e}$ contains an interval is given by $\frac{1}{\pi}\frac{\partial G_{\mathfrak{e}}(x+iy)}{\partial y}\big|_{y=0}$ (see the proof of Theorem 3.7.12 of Part 3), we see that for $\mathfrak{e}_n$ and $x \in \mathfrak{e}_n^{\text{int}}$

$$\frac{d\rho_{\mathfrak{e}_n}}{dx}(x) = \frac{1}{n\pi}\frac{T_n'(x)}{\sqrt{\|T_n\|_{\mathfrak{e}}^2 - T_n(x)^2}} \tag{4.3.66}$$

$$= \frac{1}{n\pi}\left|\frac{d}{dx}\arccos\left(\frac{T_n(x)}{\|T_n\|_{\mathfrak{e}}}\right)\right| \tag{4.3.67}$$

It follows that if $y_1 < \dots < y_n$ are the zeros of T_n and $w_1^- < y_1 < w_1^+ \le w_2^- < y_2 < \dots < y_n < w_n^+$ (with $w_j^+ = w_j^-$ if $\frac{dT_n}{dx}(w_j^+) = 0$) where $|T_n(w_j^\pm)| = \|T_n\|_{\mathfrak{e}}$ then

$$\rho_{\mathfrak{e}_n}\big((w_j^-, y_j)\big) = \rho_{\mathfrak{e}_n}\big((y_j, w_j^+)\big) = \frac{1}{2n} \tag{4.3.68}$$

since $\arccos^{-1}(y)$ runs from 0 to $\frac{\pi}{2}$ as y runs from 1 to 0.

$\mathfrak{e}_n = \bigcup_{j=1}^n [w_j^-, w_j^+]$ has at most n connected components with each component of the form $[w_j^-, w_\ell^+]$ for $\ell \ge j$ (usually $\ell = j$ unless $w_{j+1}^- = w_j^+$). From (4.3.68), we conclude the harmonic measure is $\frac{\ell+1-j}{n}$, i.e., we have shown that

Theorem 4.3.12. *Each component of $\mathfrak{e}_n$ has harmonic measure $\frac{j}{n}$ with $0 < j \le n$.*

There is a converse of this: if $\mathfrak{e} \subset \mathbb{R}$ is a compact set with finitely many connected components and all have rational harmonic measure, say $\frac{j_1}{n}, \dots, \frac{j_k}{n}$, then the Chebyshev polynomial T_n for $\mathfrak{e}$ has $\mathfrak{e} = \mathfrak{e}_n$ and so $\|T_n\|_{\mathfrak{e}} = 2C(\mathfrak{e})^n$. Since for $\ell = 2, 3, \dots$, $\frac{j_m}{n} = \frac{j_m \ell}{\ell n}$, we see that $\mathfrak{e}$ also has $\|T_{n\ell}\|_{\mathfrak{e}} = 2C(\mathfrak{e})^{n\ell}$ for $\ell = 1, 2, \dots$. This can be proven using the connection of this whole subject to periodic Jacobi matrices, see Chapter 5 of [**653**].

Any book on approximation theory is likely to have a chapter on general Chebyshev polynomials asking for the best $\|\cdot\|_{\mathfrak{e}}$-approximation of x^n by polynomials of degree $n-1$; see [**10, 120, 121, 405, 475, 572**]. Trefethen [**704**] deals with their use in numeric approximation and Steffens has a book [**663**] on the history of the subject.

Theorem 4.3.10 appeared first in Szegő [**685**]—earlier Fekete [**199**] had that ζ_n has a limit. Faber [**191**] had asymptotics of Chebyshev polynomials and their norms in the special case of an analytic Jordan region. In that case, he proved the stronger result that $\lim \|T_n\|_e C(e)^{-n} = 1$.

Problems

1. Let $\mathfrak{e}$ be a compact subset of $\mathbb{R}$. Let q_n be a monic polynomial of degree n so there exist points $\{x_j\}_{j=0}^n \subset \mathfrak{e}$ so that $x_0 < x_1 < \dots < x_n$ and $(-1)^{n-j} q_n(x_j) > 0$. Prove that

$$\|T_n\|_{\mathfrak{e}} \geq \min_{j=0,\dots,n} |q_n(x_j)| \tag{4.3.69}$$

Remark. This is a theorem of de la Vallée Poussin [**154**].

2. Let μ be a nontrivial probability measure on $[a, b] \subset \mathbb{R}$. Let $\mathbb{P}_{n-1}$ be the set of nonzero polynomials of degree at most $n-1$ and let P_n be the n-th monic OP for μ. Prove that $\sup_{Q \in \mathbb{P}_{n-1}} \left[\int x Q_n(x)^2 d\mu\right] \left[\int Q_n(x)^2 \, d\mu\right]^{-1}$ is the largest zero of P_n. (*Hint*: (4.1.31).)

3. Let $\mathfrak{e} \subset \mathbb{C}$ be compact and

$$F(b_0, \dots, b_{n-1}) = \sup_{z \in \mathfrak{e}} \Bigl| z^n + \sum_{j=0}^{n-1} b_j z^j \Bigr|$$

Prove that $F \to \infty$ as $\sum_{j=0}^{n-1} |b_j|^2 \to \infty$.

4.4. Orthogonal Polynomials on the Unit Circle: Verblunsky's and Szegő's Theorems

In this section, we consider the case where the measure, μ, on $\mathbb{C}$ is supported on $\partial\mathbb{D}$, the unit circle—hence, *orthogonal polynomials on the unit circle* or OPUC for short. OPRL and OPUC are special because, in exploiting orthogonality for recursion, the adjoint M_z^* of M_z, multiplication by z on $L^2(\mathbb{C}, d\mu)$, is an important object. For general $d\mu$, M_z^* has no relation to M_z, but if $\operatorname{supp}(d\mu) \subset \mathbb{R}$, we have $M_z^* = M_z$, and if $\operatorname{supp}(d\mu) \subset \partial\mathbb{D}$, we have $M_z^* = M_z^{-1}$. This is responsible for the fact that in these cases, the OPs have simple recursion relations.

We'll first derive these recursion relations. The Jacobi parameters, $\{a_n, b_n\}_{n=1}^\infty$, will be replaced by Verblunsky coefficients, $\{\alpha_n\}_{n=0}^\infty$, numbers in $\mathbb{D}$. We'll then prove Verblunsky's theorem (the analog of Favard's theorem) that every sequence $\{\alpha_n\}_{n=0}^\infty \subset \mathbb{D}$ arises as the Verblunsky coefficients of a unique probability measure, $d\mu$, on $\partial\mathbb{D}$. Finally, expanding on the theme of the last section, we'll discuss a variational proof of a celebrated theorem of Szegő: If μ, a probability measure on $\partial\mathbb{D}$, has the form

$$d\mu = w(\theta)\,\frac{d\theta}{2\pi} + d\mu_{\rm s} \tag{4.4.1}$$

with $d\mu_{\rm s}$ singular wrt $d\theta$, and $\{\alpha_n\}_{n=0}^\infty$ are its Verblunsky coefficients, then

$$\prod_{n=0}^\infty (1-|\alpha_n|^2) = \exp\biggl(\int_0^{2\pi} \log(w(\theta))\,\frac{d\theta}{2\pi}\biggr) \tag{4.4.2}$$

We use $\Phi_n(z)$ for the monic OPs and $\varphi_n = \Phi_n/\|\Phi_n\|$ for the orthonormal polynomials.

The recursion relation will use a map of polynomials, P, of degree n:

$$P_n^*(z) = z^n\,\overline{P_n(1/\bar z)} \tag{4.4.3}$$

called *reversed polynomials* This is standard, although terrible, notation since the right is n-dependent. In particular, that $(P_n^*)^*$ is P_n applied to $P(z) = z^n$ requires $1^* = z^n$. When there is a possibility of confusion, one talks about "$*$ of degree n." Here is why $*$ is important:

Proposition 4.4.1. (a) *On any $L^2(\partial\mathbb{D}, d\mu)$, $P \to P^*$ is an anti-unitary map.*

(b) $(z^j)^* = z^{n-j}$. *Thus,*

$$P(z) = \sum_{j=0}^n a_j z^j \Rightarrow P^*(z) = \sum_{j=0}^n \bar a_{n-j} z^j \tag{4.4.4}$$

(c) *Constant multiples of $\Phi_n^*(z)$ are the set of polynomials of degree at most n which are orthogonal to $z, z^2, \dots, z^n$.*

Proof. (a) For functions on $\partial\mathbb{D}$, $f(1/\bar z) = f(z)$, so $P \to \overline{P(1/\bar z)}$ is just $f \mapsto \bar f$ is anti-unitary. Multiplication by z^n is unitary.

(b) is trivial.

(c) By (a), $P \perp z, z^2, \dots, z^n \Leftrightarrow P^* \perp 1, z, \dots, z^{n-1} \Leftrightarrow P^* = c\Phi_n \Leftrightarrow P = \bar c\Phi_n^*$. □

Theorem 4.4.2. *Let $d\mu$ be a probability measure on $\partial\mathbb{D}$.*

(a) *Then there exists $\{\alpha_n\}_{n=0}^\infty \subset \mathbb{C}$ so that*

$$z\Phi_n(z) = \Phi_{n+1} - \bar\alpha_n \Phi_n^* \tag{4.4.5}$$

(b) *We have*

$$\|\Phi_{n+1}\|^2 = (1-|\alpha_n|^2)\|\Phi_n\|^2 \tag{4.4.6}$$

(c) $\alpha_n \in \mathbb{D}$ *for each n.*

(d) *All the zeros of $\Phi_n(z)$ lie in $\mathbb{D}$.*

Remarks. 1. The α_n are called the *Verblunsky parameters* of μ. (4.4.5) is the *Szegő recursion.*

2. The Notes explain the a priori strange use of $-\bar{\alpha}_n$ rather than, say, $+\alpha_n$.

3. One defines

$$\rho_n = (1 - |\alpha_n|^2)^{1/2} \tag{4.4.7}$$

so, by $\mu(\partial\mathbb{D}) = 1$, yields

$$\|\Phi_n\| = \rho_1 \dots \rho_{n-1} \tag{4.4.8}$$

which implies

$$z\varphi_n(z) = \rho_n \varphi_{n+1} - \bar{\alpha}_n \varphi_n^* \tag{4.4.9}$$

Proof. (a) If $j = 1, \dots, n$, then

$$\langle z\Phi_n, z^j \rangle = \langle \Phi_n, z^{j-1} \rangle = 0 \tag{4.4.10}$$

so $\Phi_{n+1} - z\Phi_n \perp \{z, z^2, \dots, z^n\}$. Since Φ_n is monic, $\Phi_{n+1} - z\Phi_n$ is a polynomial of degree n. Thus, by the proposition above, $\Phi_{n+1} - z\Phi_n$ is a multiple of Φ_n^*.

(b) Since $\Phi_n^* \perp \Phi_{n+1}$ and multiplication by z is unitary, we have

$$\|\Phi_n\|^2 = \|z\Phi_n\|^2 = \|\Phi_{n+1} - \bar{\alpha}_n \Phi_n^*\|^2 = \|\Phi_{n+1}\|^2 + |\alpha_n|^2 \|\Phi_n\|^2 \tag{4.4.11}$$

which is (4.4.6).

(c) Since μ is nontrivial, $\|\Phi_{n+1}\| \neq 0$, so (4.4.6) implies $|\alpha_n| < 1$.

(d) Let z_0 be a zero of $\Phi_n(z)$. Then

$$\Phi_n(z) = (z - z_0)p(z) \tag{4.4.12}$$

for a polynomial, p, of degree $n - 1$, so $p \perp \Phi_n$. (4.4.12) implies that $zp = \Phi_n + z_0 p_n$. Since $p_n \perp \Phi_n$,

$$\|p\|^2 = \|zp\|^2 = \|\Phi_n\|^2 + |z_0|^2 \|p_n\|^2 \tag{4.4.13}$$

Since $\|\Phi_n\| \neq 0$, $|z_0| < 1$. □

Remark. (d) is not directly related to (a)–(c). We include it here because its proof is an analog of the proof of (c). See the Notes for a discussion of other proofs of (d). We'll also provide a second proof later; see Proposition 4.4.7.

Our next signpost is:

Theorem 4.4.3 (Verblunsky's Theorem). *$\mu \mapsto \{\alpha_n(d\mu)\}_{n=0}^{\infty}$ sets up a one–one correspondence between all nontrivial probability measures on $\partial\mathbb{D}$ and $\mathbb{D}^\infty$ (all sequences in $\mathbb{D}$).*

This is obviously an analog of Favard's theorem, Theorem 4.1.3. As in that case, we need to show that, given any sequence $\{\alpha_n\}_{n=0}^\infty \in \mathbb{D}^\infty$, there is a probability measure with $\alpha_n(d\mu) = \alpha_n$. We do this in two steps: We first prove if $d\mu_m$ is a sequence of measures, so that for each $k = 0, 1, 2, \dots$, $\alpha_k(d\mu_m)$ is constant for m large, then $d\mu_m$ has a nontrivial weak limit, $d\mu$, whose Verblunsky coefficients are $\lim_{m\to\infty} \alpha_n(d\mu_m)$. We then prove, given any sequence $\{\alpha_n\}_{n=0}^\infty$, there is $d\mu_m$ with

$$\alpha_j(d\mu_m) = \begin{cases} \alpha_j, & j = 0, \dots, m-1 \\ 0, & j \geq m \end{cases} \tag{4.4.14}$$

Proposition 4.4.4. *Let $d\mu_m$ be a sequence of nontrivial measures on $\partial\mathbb{D}$. Suppose for each α_k, $\alpha_k(d\mu_m)$ is a constant $\alpha_k^{(\infty)}$ for all large m. Then $d\mu_m$ converges weakly to a nontrivial measure, $d\mu_\infty$, with*

$$\alpha_k(d\mu_\infty) = \alpha_k^{(\infty)} \tag{4.4.15}$$

Proof. The α's determine the Φ_j inductively by (4.4.5). Thus, for each j, $\Phi_j(z, d\mu_m) = \Phi_j^{(\infty)}(z)$ for all large m. By (4.4.5) and induction, the coefficients of $\Phi_j(z, d\mu)$ are polynomials in $\bar{\alpha}_0, \bar{\alpha}_1, \dots, \bar{\alpha}_{j-1}$. If $\Phi_n(z) = z^n + \sum_{j=1}^n b_j z^{n-j}$, then $\int \Phi_n(z)d\mu = 0$ for $n \geq 0$ implies

$$\int z^n \, d\mu = \sum_{j=1}^n b_j \int z^{n-j} \, d\mu \tag{4.4.16}$$

proving inductively the n-th moment is a polynomial in $\bar{\alpha}_0, \dots, \bar{\alpha}_{n-1}$.

Thus, the moments $\int z^n \, d\mu_m$ are constant for m large and $n \geq 0$. Since $\int z^{-n} \, d\mu_m = \overline{(\int z^n \, d\mu_m)}$, the same is true for $n < 0$ and n large.

Since the polynomials in $e^{\pm i\theta}$ are dense in $C(\partial\mathbb{D})$, we see $d\mu_m$ has a weak limit, $d\mu_\infty$. Since all moments are eventually constant, for any $z_1, \dots, z_k$, we have that $\int \prod_{j=1}^k |z - z_j|^2 \, d\mu_m$ is eventually constant and equal to the same for $d\mu_\infty$. Since $d\mu_m$ is nontrivial, this integral is nonzero for $d\mu_\infty$, that is, $d\mu_\infty$ is nontrivial.

Similarly, the Φ_m are polynomials in the moments $\{c_j\}_{j=-m}^m$ and so the $\bar{\alpha}_j$ are such polynomials. Thus, convergence of the moments implies convergence of the α_j's, proving (4.4.15). $\square$

We need some preliminaries for the explicit construction of measures with (4.4.14). Applying * for $(n+1)$ polynomials to (4.4.5), we see

$$\Phi_{n+1} = z\Phi_n + \bar{\alpha}_n \Phi_n^* \tag{4.4.17}$$

$$\Phi_{n+1}^* = \Phi_n^* + \alpha_n z \Phi_n \tag{4.4.18}$$

or

$$\begin{pmatrix} \Phi_{n+1}^* \\ \Phi_{n+1} \end{pmatrix} = \begin{pmatrix} 1 & \alpha_n \\ \bar{\alpha}_n & 1 \end{pmatrix} \begin{pmatrix} \Phi_n^* \\ z\Phi_n \end{pmatrix} \tag{4.4.19}$$

Since $\left(\begin{smallmatrix} 1 & \alpha_n \\ \bar{\alpha}_n & 1 \end{smallmatrix}\right)^{-1} = \rho_n^{-2}\left(\begin{smallmatrix} 1 & -\bar{\alpha}_n \\ -\alpha_n & 1 \end{smallmatrix}\right)$, this yields

Proposition 4.4.5 (Inverse Szegő Recursion). *If Φ_n are constructed by the Szegő recursion relations, then*

$$z\Phi_n = \rho_n^{-2}[\Phi_{n+1} + \bar{\alpha}_n \Phi_{n+1}^*] \tag{4.4.20}$$

$$\Phi_n^* = \rho_n^{-2}[\Phi_{n+1}^* + \alpha_n \Phi_{n+1}] \tag{4.4.21}$$

$$\Phi_{n+1}(0) = \bar{\alpha}_n \tag{4.4.22}$$

Proof. We showed (4.4.20)/(4.4.21) above. Since Φ_n is monic, $\Phi_n^*(0) = 1$, so Szegő recursion evaluated at 0 is (4.4.22). □

Theorem 4.4.6 (Geronimus–Wendroff Theorem). *Let $\{\widetilde{\Phi}_j\}_{j=0}^n$ be constructed from $\tilde{\alpha}_0, \dots, \tilde{\alpha}_{j-1} \in \mathbb{D}$ by Szegő recursion. If for a fixed n and some nontrivial probability measure, $d\mu$, on $\partial\mathbb{D}$, we have*

$$\Phi_n(z, d\mu) = \widetilde{\Phi}_n(z) \tag{4.4.23}$$

then

$$\alpha_j(d\mu) = \tilde{\alpha}_j, \qquad j = 0, 1, 2, \dots, n-1 \tag{4.4.24}$$

$$\Phi_j(z, d\mu) = \widetilde{\Phi}_j(z), \quad j = 0, 1, 2, \dots, n-1 \tag{4.4.25}$$

In particular, if $\Phi_n(z, d\mu) = \Phi_n(z, d\nu)$ for two measures μ, ν, then their α_j's, $j = 0, \dots, n-1$, and Φ_j's, $j = 0, \dots, n-1$, are equal.

Remarks. 1. Remarkably, this says Φ_n determines Φ_j, $j = 0, \dots, n-1$, and α_j, $j = 0, \dots, n-1$. Put differently, since Φ_n is monic, the zeros of Φ_n determine the zeros of Φ_j, $j = 0, \dots, n-1$.

2. Since Φ_n does not determine $\alpha_n, \alpha_{n+1}, \dots$, it cannot determine Φ_j for $j \geq n+1$.

Proof. By inverse Szegő recursion, Φ_n determines α_{n-1} and Φ_{n-1}. By induction on k, it determines α_{n-k}, Φ_{n-k} for $k = 1, \dots, n$. □

Next, we need to prove Szegő recursion alone implies all zeros of Φ_n lie in $\mathbb{D}$.

Proposition 4.4.7. *Let P be a monic polynomial of degree n and $\alpha \in \mathbb{D}$. Suppose all the zeros of P lie in $\mathbb{D}$. Then*

$$Q(z) = zP(z) - \bar{\alpha}P^*(z) \tag{4.4.26}$$

has all its zeros in $\mathbb{D}$. In particular, if $\widetilde{\Phi}_n$ is constructed from $\tilde{\alpha}_0, \dots, \tilde{\alpha}_{n-1} \in \mathbb{D}$ via Szegő recursion, then $\widetilde{\Phi}_n$ has all its zeros in $\mathbb{D}$.

Proof. Notice $|P^*(e^{i\theta})| = |P(e^{i\theta})|$ so $|zP(z)/P^*(z)| = 1$ on $\partial\mathbb{D}$. (Since P has all its zeros in $\mathbb{D}$, P^* is nonvanishing on $\partial\mathbb{D}$.) Then if (4.4.26) is Q_α, for all $\alpha \in \mathbb{D}$, $z \in \partial\mathbb{D}$, $Q_\alpha(z) \neq 0$. It follows, by the argument principle applied to $Q_\alpha(z)$, that the number of zeros of $Q_\alpha(z)$ inside $\mathbb{D}$ is independent of α for all $\alpha \in \mathbb{D}$. But $Q_{\alpha=0}(z) = zP(z)$ has all its zeros in $\mathbb{D}$.

The final "in particular" follows from the first assertion and induction. □

Lemma 4.4.8. *Let p_n be a degree n polynomial all of whose zeros lie in $\mathbb{D}$. Let*

$$d\mu(\theta) = \frac{d\theta}{2\pi|p_n(e^{i\theta})|^2} \tag{4.4.27}$$

Then in $L^2(\partial\mathbb{D}, d\mu(\theta))$, we have that

$$\langle z^q, z^\ell p_n(z)\rangle = 0, \qquad \ell = 0, 1, \dots, \quad q = 0, 1, \dots, \ell + n - 1 \tag{4.4.28}$$

Proof. Notice that

$$p_n(z)\, d\mu_n(\theta) = \frac{1}{\overline{p_n(1/\bar{z})}} \frac{d\theta}{2\pi} = \frac{z^n}{p_n^*(z)} \frac{d\theta}{2\pi} = \frac{1}{2\pi i} \frac{z^{n-1}}{p_n^*(z)}\, dz \tag{4.4.29}$$

since $d\theta = \frac{dz}{iz}$. Thus,

$$\text{LHS of (4.4.28)} = \frac{1}{2\pi i} \oint \frac{z^{\ell+n-1-q}\, dz}{p_n^*(z)} = 0 \tag{4.4.30}$$

since $\ell + n - 1 - q \geq 0$ and $p_n^*(z)$ has all its zeros outside $\overline{\mathbb{D}}$. □

Theorem 4.4.9 (Bernstein–Szegő Approximation). *Let $\{\alpha_n\}_{n=0}^\infty$ be a sequence in $\mathbb{D}$ and let Φ_n, φ_n be the polynomials constructed via Szegő recursion and*

$$\varphi_n = \frac{\Phi_n}{\prod_{j=1}^{n-1} \rho_j} \tag{4.4.31}$$

Then

$$d\mu_n = \frac{d\theta}{2\pi|\varphi_n(e^{i\theta})|^2} \tag{4.4.32}$$

is a nontrivial probability measure whose Verblunsky coefficients are

$$\alpha_j(d\mu_n) = \begin{cases} \alpha_j, & j = 0, \dots, n-1 \\ 0, & j \geq n \end{cases} \tag{4.4.33}$$

Proof. Let $d\tilde{\mu}_n = \frac{d\mu_n}{\mu_n(\partial\mathbb{D})}$. We'll see soon that $\mu_n(\partial\mathbb{D}) = 1$. By (4.4.28),

$$\langle z^q, z^\ell \Phi_n(z)\rangle_{d\tilde{\mu}_n} = 0, \qquad \ell = 0, \dots, q = 0, \dots, \ell + n - 1 \tag{4.4.34}$$

Thus, $z^\ell \Phi_n$ is a degree $n+\ell$ polynomial orthogonal to all lower degrees, that is,

$$\Phi_{n+\ell}(z, d\tilde{\mu}_n) = z^\ell \Phi_n(z) \tag{4.4.35}$$

By Theorem 4.4.6, we see that

$$\Phi_j(z, d\tilde{\mu}_n) = \Phi_j(z, d\mu), \quad j = 0, \dots, n \tag{4.4.36}$$

$$\alpha_j(z, d\tilde{\mu}_n) = \alpha_j, \qquad j = 0, \dots, n-1 \tag{4.4.37}$$

and so,

$$\varphi_j(z, d\tilde{\mu}_n) = \varphi_j(z) \tag{4.4.38}$$

Thus,

$$\int |\varphi_n(z)|^2 \, d\tilde{\mu}_n(\theta) = \frac{1}{\mu_n(\partial\mathbb{D})} = 1$$

so $\mu_n(\partial\mathbb{D}) = 1$ and $\tilde{\mu}_n = \mu_n$. Since $\alpha_{k+1} = \varphi_k(0)$, (4.4.35) implies $\alpha_j(z, d\tilde{\mu}_n) = 0$ for $j \geq n$. □

Proof of Theorem 4.4.3. Given $\{\alpha_n\}_{n=0}^\infty \subset \mathbb{D}^\infty$, we construct $d\mu_n$ as in the last theorem, so $\alpha_n(d\mu_m) = \alpha_n$ if $m \geq n-1$. Thus, Proposition 4.4.4 applies and we get a limiting measure, $d\mu_\infty$, with $\alpha_n(d\mu_\infty) = \alpha_n$ for all n. □

Finally, we turn to a remarkable theorem that we'll see has important spectral theory consequences, a variational interpretation, and consequences for polynomial asymptotics (and in the Notes we'll discuss its significance for Toeplitz determinants).

Theorem 4.4.10 (Szegő's Theorem). *Let μ be a probability measure on $\partial\mathbb{D}$ written as (4.4.1) with $d\mu_{\rm s} \perp \frac{d\theta}{2\pi}$. Let $\{\alpha_n\}_{n=0}^\infty$ be its Verblunsky coefficients. Then*

$$\prod_{n=0}^\infty (1 - |\alpha_n|^2) = \exp\biggl(\int_0^{2\pi} \log(w(\theta)) \, \frac{d\theta}{2\pi}\biggr) \tag{4.4.39}$$

Remarks. 1. $\prod_{n=0}^N (1 - |\alpha_n|^2)^{1/2}$ is monotone decreasing, so it has a limit (which might be zero). By Jensen's inequality and concavity of log, $\int_0^{2\pi} \log(w(\theta)) \, \frac{d\theta}{2\pi} \leq \log(\int_0^{2\pi} w(\theta) \frac{d\theta}{2\pi}) \leq 0$, so either the integral is convergent or else is $-\infty$ (put differently, $\int \log_+(w(\theta)) \frac{d\theta}{2\pi} < \infty$ always). If the integral is $-\infty$, we interpret the right side of (4.4.39) as 0.

2. (4.4.39) is sometimes called a sum rule—it may look like a product rule, but if we take logs, we get

$$\tfrac{1}{2} \sum_{n=0}^\infty \log(1 - |\alpha_n|^2) = \int_0^{2\pi} \log(w(\theta)) \, \frac{d\theta}{2\pi} \tag{4.4.40}$$

It is a precursor to the KdV sum rules (as we explain in the Notes).

3. Remarkably, $d\mu_{\rm s}$ plays no role in (4.4.39).

Before turning to the proof, we mention two corollaries and one reformulation.

Corollary 4.4.11. *Let μ be a probability measure on $\partial\mathbb{D}$ written as* (4.4.1) *with $d\mu_s \perp \frac{d\theta}{2\pi}$. Then*

$$\sum_{n=0}^{\infty} |\alpha_n|^2 < \infty \Leftrightarrow \int_0^{2\pi} \log(w(\theta))\, \frac{d\theta}{2\pi} > -\infty \tag{4.4.41}$$

Remarks. 1. There is a sense in which OPUC is a kind of nonlinear Fourier transform and (4.4.41) is a kind of Plancherel theorem (if $w(\theta) = 1 + f(\theta)$ with f small, then $-\log(w(\theta)) = f(\theta) + f(\theta)^2/2 + \dots$. Since $\int w(\theta)\, d\theta = 1$ (if $d\mu_s = 0$), we see to leading order $-\int_0^{2\pi} \log(w(\theta)) \frac{d\theta}{2\pi} = \frac{1}{2}\int f(\theta)^2 \frac{d\theta}{2\pi} +$ higher order, making the nonlinear Plancherel analogy clearer).

2. This is a result in spectral theory, that is, the study of the connection of "objects"—in this case, measures—and their defining coefficients. (4.4.41) has two consequences. First, if $\sum_{n=0}^{\infty} |\alpha_n|^2 < \infty$, then Σ_{ac}, the essential support of the a.c. part of μ, is all of $\partial\mathbb{D}$ (since $w(\theta) > 0$ for a.e. θ). This is a borderline result. There are examples (see the Notes) of $d\mu$'s with $\sum_{n=0}^{\infty} |\alpha_n|^p < \infty$ for all $p > 2$ but $d\mu_{ac} = 0$! Second, the singular part of $d\mu$ is arbitrary, that is, if $\int_0^{2\pi} w(\theta) \frac{d\theta}{2\pi} = 1 - \beta$ ($\beta > 0$) and the right side of (4.4.41) holds, then $d\mu_s$ can be any singular measure with $\mu_s(\partial\mathbb{D}) = \beta$.

Proof. If $\sum_{n=0}^{\infty} |\alpha_n|^2 < \infty$, the product $\prod_{n=0}^{\infty}(1 - |\alpha_n|^2)$ is absolutely convergent and so nonzero (see Theorem 9.1.3 of Part 2A). On the other hand, if $\beta < 1$,

$$-\log(1-\beta) = \beta + \tfrac{1}{2}\beta^2 + \dots \tag{4.4.42}$$

so

$$-\log(1 - |\alpha_n|^2) \ge |\alpha_n|^2 \tag{4.4.43}$$

and

$$\sum_{n=0}^{\infty} |\alpha_n|^2 \le -\sum_{n=0}^{\infty} \log(1 - |\alpha_n|^2) \tag{4.4.44}$$

Thus,

$$\sum_{n=0}^{\infty} |\alpha_n|^2 < \infty \Leftrightarrow \prod_{n=0}^{\infty} (1 - |\alpha_n|^2) > 0 \tag{4.4.45}$$

so (4.4.39) implies (4.4.41). □

For the second corollary, we need a definition. If

$$\int \log(w(\theta))\, \frac{d\theta}{2\pi} > -\infty \tag{4.4.46}$$

(the *Szegő condition*), we define, for $z \in \mathbb{D}$, the *Szegő function*

$$D(z) = \exp\biggl(\frac{1}{4\pi} \int \frac{e^{i\theta} + z}{e^{i\theta} - z} \log(w(\theta))\, d\theta \biggr) \tag{4.4.47}$$

Proposition 4.4.12. *Let w obey the Szegő condition. Then $D \in H^2(\mathbb{D})$, novanishing on $\mathbb{D}$, and if*

$$\lim_{r\uparrow 1} D(re^{i\theta}) \equiv D(e^{i\theta}) \tag{4.4.48}$$

then for Lebesgue a.e. θ,

$$|D(e^{i\theta})|^2 = w(\theta) \tag{4.4.49}$$

Proof. Suppose first that $w(\theta) \le c$ with $c > 0$ for all θ. Then $D(z)$ defined by (4.4.47) is less than $c^{1/2}$. For $\operatorname{Re}(\frac{e^{i\theta}+z}{e^{i\theta}-z}) \ge 0$, so Re integral in (4.4.47) $\le \frac{1}{2}\log c$. Thus, $D(z) \in H^\infty$. Moreover, since $\log(w(\theta)) \in L^1$, by the convergence result for h^1 functions (see Theorems 2.5.5 and 5.4.4 of Part 3), the real part of the integral in (4.4.47) has limit $\frac{1}{2}\log(w(\theta))$ so that the boundary value of the H^∞ function (see Theorem 5.2.1 of Part 3) has (4.4.49).

In particular, for any $r < 1$,

$$\int_0^{2\pi} |D(re^{i\theta})|^2 \, \frac{d\theta}{2\pi} \le \int_0^{2\pi} w(\theta) \, \frac{d\theta}{2\pi} \tag{4.4.50}$$

For general w obeying the Szegő condition, let D_n be the function with w replaced by $w_n \equiv \min(n, w(\theta))$. Then, by the dominated convergence theorem in the integral for D, we see $D_n(z) \to D(z)$ for all $z \in \mathbb{D}$, and for $0 < r < 1$,

$$|D(re^{i\theta})|^2 = \exp\biggl(\int P_r(\theta,\varphi)\log(w(\varphi))\,\frac{d\varphi}{2\pi}\biggr) \tag{4.4.51}$$

and by (4.4.50) and the dominated convergence theorem,

$$\int_0^{2\pi} |D(re^{i\theta})|^2 \, \frac{d\theta}{2\pi} \le \int w(\theta) \, \frac{d\theta}{2\pi} \tag{4.4.52}$$

so $D \in H^2$. By (4.4.51), we have $\lim_{r\uparrow 1}|D(re^{i\theta})|^2 = w(\theta)$ for a.e. θ as claimed. (4.4.51) also implies that $|D(re^{i\theta})| \neq 0$ for all $re^{i\theta} \in \mathbb{D}$. □

We can now state and prove the second corollary of Szegő's theorem.

Corollary 4.4.13. *Let μ be a probability measure on $\partial\mathbb{D}$ written as (4.4.1) with $d\mu_{\mathrm{s}} \perp \frac{d\theta}{2\pi}$. Suppose that μ obeys the Szegő condition. Let D be its Szegő function. Then*

(a) $$\lim_{n\to\infty}\biggl[\int |D(e^{i\theta})\varphi_n^*(e^{i\theta}) - 1|^2 \, \frac{d\theta}{2\pi} + \int |\varphi_n^*(e^{i\theta})|^2 \, d\mu_{\mathrm{s}}(\theta)\biggr] = 0 \tag{4.4.53}$$

(b) *In $H^2(\partial\mathbb{D})$, $D\varphi_n^* \to 1$.*

(c) *For all $z \in \mathbb{D}$,*

$$\varphi_n^*(z) \to D(z)^{-1} \tag{4.4.54}$$

uniformly in each $\mathbb{D}_\rho(0)$, $\rho < 1$.

(d) *For all* $z \in \mathbb{C} \setminus \overline{\mathbb{D}}$,

$$z^{-n}\varphi_n(z) \to \overline{D(1/\bar{z})}^{-1} \tag{4.4.55}$$

(e) $$\lim_{n\to\infty} \int |\varphi_n^*(e^{i\theta})|^2 \, d\mu_s(\theta) = 0 \tag{4.4.56}$$

Remarks. 1. (4.4.55) is called *Szegő asymptotics.* We'll discuss it further in the Notes.

2. Since $\overline{(\frac{e^{i\theta}+1/\bar{z}}{e^{i\theta}-1/\bar{z}})} = -(\frac{e^{i\theta}+z}{e^{i\theta}-z})$, we have for $z \in \mathbb{C} \setminus \overline{\mathbb{D}}$ that

$$\overline{D(1/\bar{z})}^{-1} = \exp\biggl(\frac{1}{4\pi} \int \frac{e^{i\theta}+z}{e^{i\theta}-z} \log(w(\theta)) \, d\theta\biggr) \tag{4.4.57}$$

The right side of (4.4.56) is sometimes called the *exterior Szegő function.*

Proof. (a) Note that since $|D(e^{i\theta})|^2 = w(\theta)$, before taking limits, the left side of (4.4.53) is

$$\|\varphi_n^*\|^2 + 1 - 2\operatorname{Re}\biggl[\int D(e^{i\theta})\varphi_n^*(e^{i\theta}) \frac{d\theta}{2\pi}\biggr] - 2 - 2\operatorname{Re} D(0)\varphi_n^*(0) \tag{4.4.58}$$

since $D\varphi_n^* \in H^2$, and for $g \in H^2$, $\int g(e^{i\theta}) \frac{d\theta}{2\pi} = g(0)$. By $\Phi_n^*(0) = 1$ (since Φ_n is monic) and $\|\Phi_n^*\| = \rho_1 \dots \rho_n$ and Szegő's theorem which says

$$D(0) = \prod_{j=1}^{\infty} \rho_j \tag{4.4.59}$$

we see

$$\text{LHS of (4.4.53)} = \lim_{n\to\infty} \biggl[2 - 2 \prod_{j=n+1}^{\infty} \rho_j\biggr] = 0$$

since the infinite product is absolutely convergent.

(b) Since both terms in (4.4.44) are nonnegative, the first goes to zero.

(c) By (b), $D\varphi_n^* \to 1$ uniformly on compact subsets of D.

(d) $z^{-n}\varphi_n(z) = \overline{\varphi_n^*(1/\bar{z})}$, so this follows from (c).

(e) The second term in (4.4.44) goes to zero. □

We want to rephrase Szegő's theorem as a variational principle. Notice that

$$\prod_{j=0}^{n-1} \rho_j = \|\Phi_n\| = \|\Phi_n^*\| \tag{4.4.60}$$

As discussed in Theorem 4.3.1,

$$\|\Phi_n\| = \min\{\|P\| \mid \deg(P) = n,\ P \text{ is monic}\} \tag{4.4.61}$$

Thus,

$$\|\Phi_n^*\| = \min\{\|P\| \mid \deg(P) \leq n,\ P(0) = 1\} \tag{4.4.62}$$

Therefore,

$$\prod_{j=0}^{\infty} \rho_j = \inf\{\|P\| \mid P \text{ a polynomial, } P(0) = 1\} \tag{4.4.63}$$

We thus have the following rephrasing of Theorem 4.4.10:

Theorem 4.4.14 (Szegő's Theorem—Variational Form). *Let μ be a positive measure on $\partial\mathbb{D}$ written as (4.4.1) with $d\mu_{\mathrm{s}} \perp \frac{d\theta}{2\pi}$. Then*

$$\inf\{\|P\|^2_{L^2(d\mu)} \mid P \text{ a polynomial, } P(0) = 1\} = \exp\biggl(\int_0^{2\pi} \log(w(\theta))\,\frac{d\theta}{2\pi}\biggr) \tag{4.4.64}$$

Remarks. 1. We've shifted from probability measures to general positive measures. Replacing $d\mu$ by $c\,d\mu$, $w \to cw$ and both sides of (4.4.64) are multiplied by c, so this is an immediate (but useful) extension.

2. It is this form we'll prove. Let $\mathrm{Sz}(d\mu)$ be the left side of (4.4.64). We'll first prove (4.4.64) if $d\mu_{\mathrm{s}} = 0$. Then we'll prove that

$$\mathrm{Sz}\biggl(w\,\frac{d\theta}{2\pi} + d\mu_{\mathrm{s}}\biggr) = \mathrm{Sz}\biggl(w\,\frac{d\theta}{2\pi}\biggr) \tag{4.4.65}$$

We begin with the lower bound:

Proposition 4.4.15. *For any polynomial, P, with $P(0) = 1$, we have that*

$$\|P\|^2_{L^2(d\mu)} \geq \exp\biggl(\int_0^{2\pi} \log(w(\theta))\,\frac{d\theta}{2\pi}\biggr) \tag{4.4.66}$$

Proof. We have

$$\|P\|^2_{L^2(d\mu)} \geq \int_0^{2\pi} |P(e^{i\theta})|^2 w(\theta)\,\frac{d\theta}{2\pi} \tag{4.4.67}$$

$$= \int_0^{2\pi} \exp(2\log|P(e^{i\theta})| + \log(w(\theta)))\,\frac{d\theta}{2\pi}$$

$$\geq \exp\biggl(\int_0^{2\pi} [2\log|P(e^{i\theta})| + \log(w(\theta))]\,\frac{d\theta}{2\pi}\biggr) \tag{4.4.68}$$

$$\geq \exp\biggl(\int_0^{2\pi} \log(w(\theta))\,\frac{d\theta}{2\pi}\biggr) \tag{4.4.69}$$

In (4.4.67), we use $d\mu_{\mathrm{s}} \geq 0$. In (4.4.68), we use Jensen's inequality (Theorem 5.3.14 of Part 1). In (4.4.69), we use

$$\int_0^{2\pi} \log|P(e^{i\theta})|\,\frac{d\theta}{2\pi} \geq \log|P(0)| = 0 \tag{4.4.70}$$

which comes from the fact that $\log|P(z)|$ is subharmonic (see Theorem 3.2.6 of Part 3) or a direct calculation (see Problem 1). $\square$

To maximize $\|P\|$, we want to try to arrange the inequalities in the above proof to be equalities. For equality in Jensen's inequality, we want $2\log|P(e^{i\theta})| + \log[w(\theta)]$ to be a constant. For equality in (4.4.70), we want $\log|P(z)|$ to be harmonic, that is, $P(z)$ should have no zeros in $\mathbb{D}$, so in an inner-outer representation, P should be outer with $\log|P(e^{i\theta})| = -\frac{1}{2}\log(w(\theta))$, that is, we want to arrange $P(z) = D(z)^{-1}$. Of course, in general, $D(z)^{-1}$ is not a polynomial, so we want to first arrange polynomial approximations.

Proposition 4.4.16. *Let $d\mu = w(\theta)\frac{d\theta}{2\pi}$ obey the Szegő condition. Then*

$$\operatorname{Sz}\left(w\,\frac{d\theta}{2\pi}\right) = \inf\left\{\int w(\theta)|f^*(e^{i\theta})|^2\,\frac{d\theta}{2\pi}\;\middle|\; f\in H^\infty(\mathbb{D}),\ f(0)=1\right\} \tag{4.4.71}$$

Remark. $f^*(e^{i\theta})$ is the a.e. in $d\theta$ limit of $f(re^{i\theta})$.

Proof. We sketch the idea, leaving the details to the reader (Problem 2). Since polynomials are in H^∞, we need only show that for any $f\in H^\infty(\mathbb{D})$, there exist polynomials, p_n, with $p_n(0)=1$,

$$\sup_{n,z\in\mathbb{D}} |p_n(z)| \le 2\|f\|_\infty \tag{4.4.72}$$

and for a.e. θ,

$$\lim_{n\to\infty} |p_n(e^{i\theta})| = |f^*(e^{i\theta})| \tag{4.4.73}$$

since dominated convergence then implies $\int w(\theta)|p_n(e^{i\theta})|^2\frac{d\theta}{2\pi} \to \int w(\theta)|f^*(e^{i\theta})|^2\frac{d\theta}{2\pi}$.

For any $n>1$, $f((1-\frac{1}{n})z)$ is analytic in $\mathbb{D}_{(1-\frac{1}{n})^{-1}}(0)$, so its Taylor series converges uniformly, so we can find polynomials, $p_n(z)$, so (4.4.72) holds and

$$\left|p_n\left(\left(1-\frac{1}{n}\right)e^{i\theta}\right) - f\left(\left(1-\frac{1}{n}\right)e^{i\theta}\right)\right| \le \frac{1}{n} \tag{4.4.74}$$

This implies (4.4.73). $\square$

Proof of Theorem 4.4.14 if $d\mu_s = 0$. For each $\varepsilon>0$, let

$$D_\varepsilon(z) = \exp\left(\frac{1}{4\pi}\int_0^{2\pi}\left(\frac{e^{i\theta}+z}{e^{i\theta}-z}\right)\log(w(\theta)+\varepsilon)\,d\theta\right) \tag{4.4.75}$$

and

$$f_\varepsilon(z) = \frac{D_\varepsilon(0)}{D_\varepsilon(z)} \tag{4.4.76}$$

Then

$$|f_\varepsilon(z)| \leq D_\varepsilon(0)\varepsilon^{-1/2} \tag{4.4.77}$$

so $f_\varepsilon \in H^\infty$. Clearly, $f_\varepsilon(0) = 1$, so we can use f_ε in (4.4.71).

We have that

$$|f_\varepsilon^*(e^{i\theta})| = D_\varepsilon(0)(w(\theta) + \varepsilon)^{-1/2} \tag{4.4.78}$$

so

$$\mathrm{Sz}\bigg(w\,\frac{d\theta}{2\pi}\bigg) \leq D_\varepsilon(0)^2 \int_0^{2\pi} \bigg[\frac{w(\theta)}{w(\theta)+\varepsilon}\bigg]\frac{d\theta}{2\pi} \tag{4.4.79}$$

As $\varepsilon \to 0$, $D_\varepsilon(0) \to D(0)$, while by monotone convergence, the integral converges to 1. Thus,

$$\mathrm{Sz}\bigg(w\,\frac{d\theta}{2\pi}\bigg) \leq \exp\bigg(\int (w(\theta))\frac{d\theta}{2\pi}\bigg) \tag{4.4.80}$$

Combined with (4.4.66), this yields (4.4.64) for when $d\mu_{\mathrm{s}} = 0$. $\square$

To complete the proof, we want to show we can use polynomials to kill $d\mu_{\mathrm{s}}$. That is the purpose of the following:

Theorem 4.4.17. *Let $d\nu$ be a positive singular measure on $\partial\mathbb{D}$. Then there exists a sequence of polynomials $\pi_n(z)$ so that*

(a) $\sup\limits_{z\in\mathbb{D},n} |\pi_n(z)| \leq 1$ (4.4.81)

(b) *For $d\theta$-a.e. $e^{i\theta}$, we have that*

$$\lim_{n\to\infty} |\pi_n(e^{i\theta})| = 0 \tag{4.4.82}$$

(c) *For each $\delta > 0$,*

$$\sup_{|z|\leq 1-\delta} |\pi_n(z)| = 0 \tag{4.4.83}$$

(d) *For $d\nu$-a.e. θ, we have that*

$$\lim_{n\to\infty} \pi_n(e^{i\theta}) = 1 \tag{4.4.84}$$

Proof. We sketch the proof, leaving the details to the reader (Problem 3). Without loss, suppose $d\nu$ is a probability measure and define $g(z)$ for $z \in \mathbb{D}$ by

$$\frac{1+g(z)}{1-g(z)} = \int \frac{e^{i\theta}+z}{e^{i\theta}-z}\,d\nu(\theta) \tag{4.4.85}$$

By Theorem 2.5.5 of Part 3 and the fact that $w \mapsto \frac{1+w}{1-w}$ maps $\mathbb{D}$ to $\mathbb{H}$ taking 1 to ∞ and 0 to 1, we see that

$$|g(z)| < 1 \text{ if } z \in \mathbb{D}, \qquad g(0) = 0 \tag{4.4.86}$$

For $d\theta$-a.e. θ, $\lim_{r\uparrow 1} g(re^{i\theta})$ exists and $\neq 1$.

If $E = \{\theta \mid \lim_{r\uparrow 1} g(re^{i\theta}) = 1\}$, then

$$\nu(\partial\mathbb{D} \setminus E) = 0 \tag{4.4.87}$$

Notice that (4.4.86) implies

$$|g(z)| \leq |z| \tag{4.4.88}$$

Let $h(z) = \frac{1}{2}(1+g(z))$ and Q_n are polynomial approximations (truncated Taylor series to $h((1-\frac{1}{n})z))$ so that

$$(1)\ |Q(z)| \leq 1 - \frac{1}{6n}, \qquad \text{all } z \in \mathbb{D} \tag{4.4.89}$$

(2) For $d\theta$-a.e. θ,

$$\sup_n |Q_n(e^{i\theta})| < 1 \tag{4.4.90}$$

$$(3)\ \lim_{n\to\infty} Q_n(e^{i\theta}) = 1, \quad e^{i\theta} \in E \tag{4.4.91}$$

By (4.4.91), we can find $k_1 < k_2 < \dots \to \infty$ inductively and integers $j_0(\theta) < \infty$, so for ν-a.e. θ, we have

$$|z - Q_n(e^{i\theta})| \leq \frac{1}{j^2} \quad \text{if } n \geq k_j,\ j \geq j_0(\theta) \tag{4.4.92}$$

Pick

$$\pi_j(z) = [Q_{k_j}]^j \tag{4.4.93}$$

By (4.4.90), we have (4.4.82), and by (4.4.89), we have (4.4.81). These two plus Poisson representation and $\pi_j(0)$ real implies (4.4.83).

Finally, $|Q_j(z)| \leq 1$ and (4.4.92) implies that for ν-a.e. $e^{i\theta}$, we have for j large that

$$|1 - \pi_j(e^{i\theta})| = |1 - Q_j(e^{i\theta})^j| \leq j|1 - Q_j(e^{i\theta})| \to 0 \tag{4.4.94}$$

□

The following result proves

$$\mathrm{Sz}\biggl(w(\theta)\,\frac{d\theta}{2\pi} + d\mu_{\mathrm{s}}\biggr) = \mathrm{Sz}\biggl(w(\theta)\,\frac{d\theta}{2\pi}\biggr) \tag{4.4.95}$$

and so complete the proof of Theorems 4.4.14 and 4.4.10.

Theorem 4.4.18. *For any $d\mu$ of the form* (4.1.2) *and any polynomial, $Q(z)$, with $Q(0) = 1$, there exist polynomials, $P_n(z)$, so that $P_n(0) = 1$ and*

$$\lim_{n\to\infty} \int |P_n(e^{i\theta})|^2\, d\mu(\theta) = \int |Q(e^{i\theta})|^2 w(\theta)\,\frac{d\theta}{2\pi} \tag{4.4.96}$$

Proof. Let $d\nu = d\mu_s$. Let π_n be as in Theorem 4.4.17. Let

$$P_n(z) = Q(z)\bigl[1 - \pi_n(z)\bigr]\,\bigl[1 - \pi_n(0)\bigr]^{-1} \tag{4.4.97}$$

Clearly, $P_n(0) = 1$ and $\sup_{n,z\in\overline{\mathbb{D}}}|P_n(z)| < \infty$. For $d\theta$-a.e. θ, $P_n(e^{i\theta}) \to Q(e^{i\theta})$, and for $d\nu$-a.e. θ, $P_n(e^{i\theta}) \to 0$. By the dominated convergence theorem,

$$\int |P_n(e^{i\theta})|^2 w(\theta)\,\frac{d\theta}{2\pi} \to \int |Q(e^{i\theta})|^2 w(\theta)\,\frac{d\theta}{2\pi}, \qquad \int |P_n(e^{i\theta})|^2\,d\nu(\theta) \to 0 \tag{4.4.98}$$

□

Notes and Historical Remarks. The subject of this section is dear to my heart—so much so that I've written three long books on the subject: two [**648, 649**] on OPUC and one on extensions of Szegő's theorem, especially to OPRL [**653**]. Other books on OPUC include Szegő [**686**] and Geronimus [**240**].

While OPRL arose slowly as the nineteenth century unfolded, OPUC is an invention of Gabor Szegő (1895–1985), initially as parts of a few papers but then in a remarkable two-part paper in 1920–21 [**683**], motivated by his work on Toeplitz determinants (see below).

Szegő, a Hungarian by birth was a student of Fejér. He was nineteen when he found the first version of his theorem which was published when he was fighting in the First World War. He was a *privatdozent* in Berlin when the 1920–21 paper was written and became a Professor in Königsberg in 1926. In 1925, he and his dear friend Pólya wrote their famous problem book [**534, 535**]. In the mid-1930's, he saw that he was likely to lose his position since he was Jewish. He came to a temporary position at Washington University in St. Louis in 1936, became chair at Stanford in 1938, and served as chair until 1966. He turned a rather pedestrian department into a world powerhouse. His *Colloquium Lectures* [**686**] has been a standard reference in the theory of orthogonal polynomials. Askey–Nevai [**33**] is a wonderful biographical note.

In the late 1930s, Szegő found the Szegő recursion and put it in his book on OPs [**686**] and Geronimus wrote several papers—it appears he discovered Szegő recursion slightly after but independently of Szegő.

Earlier, in 1935, Verblunsky wrote two remarkable papers [**714, 715**] on the subject of positive harmonic functions. He never discussed orthogonal polynomials but considered Carathéodory functions and parametrized the moments of a probability measure in terms of coefficients α_n His scheme, described in [**648**, Sect. 3.1], noted that once $c_0, \dots, c_n$ were determined, positivity required c_{n+1} to lie in a disk: α_n parameters the precise location within the disk, that is, if $r_n(c_0, \dots, c_n) > 0$ and $\zeta_n(c_0, \dots, c_n)$ are the radius and center of the disk, one has that

$$c_{n+1} = \alpha_n r_n(c_0, \dots, c_n) + \zeta_n(c_0, \dots, c_n) \tag{4.4.99}$$

He went on to prove, in terms of this definition, that all possible α_n's in $\mathbb{D}^\infty$ occurred, that is, what we have called Verblunsky's theorem. He introduced approximations to the Carathéodory functions as a ratio of polynomials—one was Φ_n^*, although he didn't know that nor did he find the recursion. He was the first person to write Szegő's theorem as a sum rule and the first to prove the theorem when $d\mu$ had a singular part. His papers, while deep and containing results only rediscovered later, were mainly ignored; occasionally, someone in the Geronimus school referred to it. At the beginning of this century, I was pointed to the papers, realized their importance, and suggested the terms Verblunsky coefficients and Verblunsky's theorem which have stuck!

Samuel Verblunsky (1906–1996) was born to poor parents of Polish-Jewish extraction, growing up in London. He won a scholarship to Magdalene College, Cambridge where he stayed on to get a Ph.D. under Littlewood's supervision. He wrote the two great papers mentioned above while at the University of Manchester and spent most of his career at Queen's College, Belfast.

The state of OPUC was advanced by two papers of Geronimus [**238, 239**]. The second was published in the *Annals of Mathematics* with submission and correspondence going from Russia to the U.S. in the middle of the Second World War! The first paper showed the Verblunsky coefficients associated to the OPUC of a measure, $d\mu$, were identical to the Schur parameters (see Section 7.5 of Part 2A) of the functions, f, defined by

$$\frac{1 + zf(z)}{1 - zf(z)} = \int \frac{e^{i\theta} + z}{e^{i\theta} - z}\, d\mu(\theta) \tag{4.4.100}$$

This is one reason that the parameters have a negative and a complex conjugate in the Szegő recursion.

The Geronimus–Wendroff theorem appeared first in Geronimus [**239**]. He even had a footnote about the OPRL result rediscovered fifteen years later by Wendroff [**738**]. Since it became known under Wendroff's name in the West, we've called this theorem by the name we do. The OPRL result says:

Theorem 4.4.19 (Geronimus–Wendroff Theorem for OPRL). *Let P_n and P_{n+1} be two monic polynomials of degree n and $n+1$ whose zeros are all real and interlace. Then there is a probability measure, μ, on $\mathbb{R}$ with finite moments so that P_n and P_{n+1} are among the monic* OPRL *for μ. $P_{n-1}, P_{n-2}, \dots$ are determined by P_n, P_{n+1}, as are $\{a_j\}_{j=1}^n \cup \{b_j\}_{j=1}^{n+1}$.*

Remarks. 1. The reader will prove this in Problem 4.

2. P_n and P_{n+1} are determined by the zeros, that is, by $2n+1$ parameters. In turn, they determine $2n+1$ Jacobi parameters.

The Bernstein–Szegő approximation (Theorem 4.4.9) is due to Geronimus [**239**]. Its name comes from the fact that Bernstein and Szegő originally studied measures of the form $\frac{dx}{|Q(x)|^2}$ or $\frac{d\theta}{|Q(e^{i\theta})|^2}$.

Verblunsky's theorem appears (with his definition of parameters) in his first paper [**714**]. [**648**] has several proofs. One of the most interesting writes a matrix representation for multiplication by z in terms of the α_n's and appeals to the spectral theorem, somewhat like our proof of Favard's theorem. This matrix representation is called the CMV matrix for Cantero, Moral, and Velázquez [**98**]. Remarkably, it only appeared in the OPUC literature in 2003 although a variant appeared somewhat earlier in the numerical linear algebra literature. Since Verblunsky's paper was ignored for many years, there were several rediscoveries and the result was sometimes called "Favard's theorem for the unit circle." This is especially ironic since Verblunsky's paper slightly predated Favard!

Szegő asymptotics has been a touchstone in the theory of OPs. Szegő [**684**] found a mapping of certain OPRL to OPUC and so extended Szegő asymptotics to OPRL—z^n is replaced (when the essential support of $d\mu$ is $[-2,2]$) by $(\frac{x}{2} = \sqrt{(\frac{x}{2})^2 - 1})^n$ (where $x \in \mathbb{C} \setminus [-2,2]$). For a modern treatment of the Szegő mapping, see [**649**, Sect. 13.1].

As we mentioned, there are examples of μ's with $\sum_{n=0}^{\infty} |\alpha_n(d\mu)|^p < \infty$ for all $p > 2$ and where $w = 0$, that is, μ is purely singular. [**648**, Sects. 2.10–2.12] has three different constructions of such $\{\alpha_n\}_{n=0}^{\infty}$. [**648**] also has higher-order Szegő theorems.

Verblunsky's form of Szegő's theorem as a sum rule, that is, (4.4.40) is an analog and precursor of KdV sum rules. This is made explicit, for example, in Killip–Simon [**387**].

Szegő's original form of his theorem did not involve OPUC but Toeplitz determinants. Given a measure, μ, on $\partial\mathbb{D}$, we define its moments by

$$c_n = \int z^n \, d\mu(z) \tag{4.4.101}$$

A Toeplitz matrix is constant along diagonals:

$$T^{(n)} = \begin{pmatrix} c_0 & c_1 & \dots & c_{n-1} \\ c_{-1} & c_0 & \dots & c_{n-2} \\ \vdots & & \ddots & \vdots \\ c_{-n+1} & & \dots & c_0 \end{pmatrix} \tag{4.4.102}$$

(i.e., $(T^{(n)})_{ij} = c_{j-i}$). The Toeplitz determinant is

$$D_n(d\mu) = \det(T^{(n+1)}) \tag{4.4.103}$$

What Szegő proved in 1914 [**682**] is that

$$\lim_{n\to\infty} D_n\biggl(w(\theta)\,\frac{d\theta}{2\pi}\biggr)^{1/n} = \exp\biggl(\int_0^{2\pi} \log(w(\theta))\,\frac{d\theta}{2\pi}\biggr) \tag{4.4.104}$$

The connection to OPUC, which Szegő found in [**683**], is that since $(T^{(n)})_{ij} = \langle z^i, z^j\rangle$ and determinants are invariant under upper-triangular monic change of basis, the determinant is the same for $\langle \Phi_i, \Phi_j\rangle = \delta_{ij}\|\Phi_j\|^2$. Thus,

$$D_n\biggl(w(\theta)\,\frac{d\theta}{2\pi}\biggr) = \|\Phi_0\|^2\,\|\Phi_1\|^2 \dots \|\Phi_n\|^2 \tag{4.4.105}$$

showing the connection of (4.4.105) to (4.4.64).

[**648**] has several other proofs of Szegő's theorem; the one we give for $\mathrm{Sz}(w\,\frac{d\theta}{2\pi})$ is essentially Szegő's proof in [**683**]; the proof we give for (4.4.95) first appeared in [**648**], although there were earlier proofs of this using results like Theorem 4.4.17 and the theory of peak sets.

Remarkably, having found the first term in the large-n behavior of $\frac{1}{n}\log(D_n(d\mu))$ in 1914, Szegő found the second term in 1952 [**687**]! This result, of importance in the study of the Ising model of statistical physics, is also sometimes called Szegő's theorem. To avoid confusion, I call it the strong Szegő theorem. [**648**, Ch. 6] has five proofs of this result.

We stated and proved the variational form of Szegő's theorem for L^2-norms and $P(0) = 1$ conditions. The same idea handles L^p-norms, $0 < p < \infty$, and $P(z_0) = 1$ conditions for $z_0 \in \mathbb{D}$. Details can be found in [**648**, Sect. 2.5].

[**648**] has several proofs of the theorem that the zeros of OPUC lie in $\mathbb{D}$. One of the most interesting is a special case of a theorem of Fejér [**198**] that if μ is a measure of compact support in $\mathbb{C}$, its OPs have zeros in the closed convex hull of the support (and not on its boundary if the set doesn't lie in line). The reader will prove this in Problem 5.

Problems

1. Given a polynomial, P, suppose that its zeros within $\mathbb{D}$ are $\{z_j\}_{j=1}^{\ell}$. Let

$$Q(z) = \frac{1-\bar{z}_j z}{z - z_j}\,P(z) \tag{4.4.106}$$

(a) Prove that

$$\int \log|P(e^{i\theta})|\,\frac{d\theta}{2\pi} = \int \log|Q(e^{i\theta})|\,\frac{d\theta}{2\pi} = \log|Q(0)| \tag{4.4.107}$$

(b) Prove $|P(0)| \le |Q(0)|$ and conclude that (4.4.70) holds.

2. Provide the details of the proof of Proposition 4.4.16.

3. Provide the details of the proof of Theorem 4.4.17.

4. This will prove Theorem 4.4.19. Suppose P_{n+1} and P_n are monic polynomials of degree $n+1$ and n whose zeros are real and interlace.

   (a) Prove there is a unique b_{n+1} so that
$$Q = P_{n+1} + (b_{n+1} - x)P_n(x) \tag{4.4.108}$$
   has degree $n-1$.

   (b) Prove that the zeros of Q interlace those of P_n.

   (c) Prove that the leading coefficient of Q is negative so there exists a unique $a_n > 0$ so that $Q = -a_n^2 P_{n-1}$.

   (d) Show that P_n, P_{n+1} determine $\{P_j\}_{j=0}^{n-1}$ and $\{a_j\}_{j=1}^{n}, \{b_j\}_{j=1}^{n+1}$.

   (e) Construct μ. (*Hint*: Extend the given Jacobi parameters by, say, taking $b_j = 0$ if $j \geq n+2$ and $a_j = 1$ if $j \geq n+1$.)

5. This will prove Fejér's theorem that if μ has support on a compact $K \subset \mathbb{C}$ and Φ_n is the degree n monic OP, then all its zeros lie in $\mathrm{cvh}(K)$, the closed convex hull of K, and if K is not contained in a line, not in the boundary of $\mathrm{cvh}(K)$.

   (a) Let $\Phi_n(z_0) = 0$ and $Q(z) = (z - z_0)^{-1}\Phi_n(z)$. Let $d\nu = |Q(z)|^2\, d\mu$. Prove that
$$z_0 = \frac{\int z\, d\nu(z)}{\int d\nu(z)} \tag{4.4.109}$$
   and conclude $z_0 \in \mathrm{cvh}(K)$.

   (b) Prove that z_0 cannot be an extreme point of $\mathrm{cvh}(K)$.

   (c) Prove that if K is not in a straight line, then $z_0 \notin \partial[\mathrm{cvh}(K)]$.

Chapter 5

The Spectral Theorem

Big Notions and Theorems: Resolution of the Identity, Spectral Theorem, Continuous Functional Calculus, Borel Functional Calculus, Multiplication Operator, Cyclic Vector, Cyclic Subspace, Multiplicity Theory, Measure Class, Commutant, Simple Spectrum, Spectral Theorem for Unitary Operators, RAGE Theorem, Spectral Theorem for Normal Operators, Rank-One Perturbations, Aronszajn–Krein Formula, Krein Spectral Shift, Kato's Finite Rank Theorem, Aronszajn–Donoghue Theorem, Simon–Wolff Criterion, Kato–Birman Theorem, Weyl–von Neumann–Kuroda Theorem

If one had to pick one result that stood out among the many in the theory of operators, especially self-adjoint operators, on a Hilbert space, it would be the spectral theorem of Hilbert (and in the unbounded case, von Neumann). This result and related issues is the subject of this chapter.

Recall the finite-dimensional spectral theorem (Theorem 1.3.3) which says that any self-adjoint matrix has a complete orthonormal basis of eigenvectors. While we saw the result as just stated extends to the self-adjoint compact operators (see Theorem 3.2.1), one needs a different format, in general, as Example 2.2.1 shows. Over the years, three different forms have been in vogue at different times. Our view is that they are equivalent and each illuminating in its own way. The three are:

(1) *Projection-valued measure.* If $\lambda_1, \dots \lambda_J$ are the distinct eigenvalues of A, a finite-dimensional self-adjoint matrix, and E_j is the orthogonal projection onto $\{\varphi \mid A\varphi = \lambda_j \varphi\}$, then the spectral theorem is equivalent to

$$A = \sum_{j=1}^{J} \lambda_j E_j \tag{5.0.1}$$

Define a projection-valued measure $dE = \sum_{j=1}^{J} \delta(\lambda - \lambda_j)E_j$. Then (5.0.1) becomes

$$A = \int t dE_t \tag{5.0.2}$$

where E_t is a Stieltjes measure. This form does generalize—all that is needed is to allow measures beyond the finite pure point measure of the finite-dimensional case.

(2) *Functional calculus.* If $A \in \mathcal{L}(X)$ for any Banach space, we can define $P(A)$ for any polynomial. In Section 2.3, we even defined $f(A)$ for certain analytic functions, f. In the finite-dimensional self-adjoint case, (5.0.1) lets us define

$$f(A) = \sum_{j=1}^{J} f(\lambda_j)E_j \tag{5.0.3}$$

for any continuous function, f, a definition that agrees with $P(A)$ if P is a polynomial. We'll see that not only does the spectral theorem in the infinite-dimensional case allow one to define $f(A)$, one can go backwards from the functional calculus to the other forms.

(3) *Multiplication operator.* Suppose A is a finite-dimensional self-adjoint operator and suppose that the spectrum is simple, i.e., each E_j is one-dimensional. Let μ be the point probability measure

$$d\mu(x) = \frac{1}{n} \sum_{j=1}^{n} \delta(x - \lambda_j) \tag{5.0.4}$$

It is easy to see that if B is the operator on $L^2(\mathbb{R}, d\mu)$,

$$(Bf)(x) = xf(x) \tag{5.0.5}$$

then A and B are unitarily equivalent, i.e., for some unitary $U\colon \mathbb{C}^n \to L^2(\mathbb{R}, d\mu)$, $UAU^{-1} = B$. To accomodate degenerate eigenvalues, we need to allow direct sums of multiplication operators, but one form of the finite-dimensional spectral theorem is that A is unitary equivalent to a direct sum of multiplications on $L^2(\mathbb{R}, d\mu_k)$ where μ_k is a pure point measure with only finitely many pure points. The corresponding form of the general spectral theorem is to say this is true in the infinite-dimensional case if we allow a general measure supported on $\big[-\|A\|, \|A\|\big]$.

In Section 5.1, we'll explore these three forms more precisely and prove their equivalence. After a technical result in Section 5.2, we'll give the "official" proof of the spectral theorem (relying on Favard's theorem) in Section 5.3. Section 5.4 answers the question of when two bounded self-adjoint operators are unitarily equivalent. Section 5.5 discusses the spectral theorem for unitary operators and gives a direct proof while Section 5.6 does general normal operators based on the self-adjoint case. Section 5.7

sketches another five proofs of the spectral theorem! The last two sections discuss perturbations from a spectral theory point of view including the Krein spectral shift. We also prove a spectacular result of Weyl and von Neumann that for any self-adjoint operator A, there is a compact self-adjoint operator C so that $A+C$ has a complete orthonormal basis of eigenvectors.

5.1. Three Versions of the Spectral Theorem: Resolutions of the Identity, the Functional Calculus, and Spectral Measures

As indicated in the introduction to this chapter, the spectral theorem has three "equivalent" forms: a representation as a Stieltjes integral (with projection-valued increments), an extension of the functional calculus from analytic functions on $\sigma(A)$ to continuous or even bounded measurable functions, and the realization of any bounded self-adjoint operator as a direct sum of multiplication operators. Our goal in this section is to make these forms precise and to show their equivalence. Several proofs appear later in Sections 5.3 and 5.7, and we provided a proof of one form already in Theorem 2.4.13.

At the end of the section, we use the spectral theorem to provide another proof of the existence of square roots, discuss the relation of the objects of this chapter to the spectrum, and discuss a spectral decomposition. We begin with a preliminary needed to define a projection-valued Stieltjes measure:

Proposition 5.1.1. *Let P, Q be two self-adjoint projections on a Hilbert space, $\mathcal{H}$, with $P \le Q$. Then*

(a) $\mathrm{Ran}(P) \subseteq \mathrm{Ran}(Q)$.

(b) *$Q - P$ is a projection onto* $\mathrm{Ran}(Q) \cap (\mathrm{Ran}(P))^\perp$.

(c) $QP = PQ$.

Proof. (a) For any self-adjoint projection, R,

$$\varphi \in \mathrm{Ran}(R) \Leftrightarrow \langle \varphi, R\varphi \rangle = \langle \varphi, \varphi \rangle \tag{5.1.1}$$

Since $\langle \varphi, Q\varphi \rangle \le \langle \varphi, \varphi \rangle$ for all φ, we see

$$P \le Q \,\&\, (\langle \varphi, P\varphi \rangle = \langle \varphi, \varphi \rangle) \Rightarrow \langle \varphi, Q\varphi \rangle = \langle \varphi, \varphi \rangle \tag{5.1.2}$$

implying $\mathrm{Ran}(P) \subseteq \mathrm{Ran}(Q)$.

(b) Let R be the orthogonal projection onto $(\mathrm{Ran}(P))^\perp \cap \mathrm{Ran}(Q)$. Then for $\varphi \in \mathrm{Ran}(Q)$, $Q\varphi = \varphi = P\varphi + R\varphi \Rightarrow R = Q - P$ is a projection onto $\mathrm{Ran}(Q - P) = \mathrm{Ran}(Q) \cap (\mathrm{Ran}(P))^\perp$.

(c) $QP = (P + R)P = P^2 = P$ and $PQ = P$. □

Definition. Let $\mathcal{H}$ be a Hilbert space. A *resolution of the identity* is a function $t \mapsto E_t$ from $(-\infty, \infty)$ to the orthogonal projections on $\mathcal{H}$ so that

$$\text{(i) } t < s \Rightarrow E_t \leq E_s \tag{5.1.3}$$

$$\text{(ii) } \operatorname*{s-lim}_{t \to -\infty} E_t = 0, \qquad \operatorname*{s-lim}_{t \to +\infty} E_t = \mathbb{1} \tag{5.1.4}$$

$$\text{(iii) } \operatorname*{s-lim}_{t \downarrow s} E_t = E_s \tag{5.1.5}$$

If $-\infty < a < b < \infty$ with $E_{a-\varepsilon} = 0$ for all $\varepsilon > 0$ and $E_b = \mathbb{1}$, we say E is *supported* in $[a, b]$.

It is not hard to see (Problem 1) that for all s, $\text{s-lim}_{t\uparrow s} E_t$ exists, and for all but countably many s, this limit is E_s. We begin with a Riemann–Stieltjes integral.

Proposition 5.1.2. *Let E_t be a resolution of the identity supported on $[-R, R]$. Let $g \in C([-R, R])$. Let*

$$\Phi_n(g) = \sum_{j=-2^n R}^{2^n R} g\left(\frac{j}{2^n}\right)\left[E_{j/2^n} - E_{(j-1)/2^n}\right] \tag{5.1.6}$$

Then $\Phi_n(g)$ converges in norm to an operator $\Phi(g)$ and

$$\Phi(g)\Phi(h) = \Phi(gh) \tag{5.1.7}$$

$$g \mapsto \Phi(g) \text{ is linear} \tag{5.1.8}$$

$$\Phi(\bar{g}) = \Phi(g)^* \tag{5.1.9}$$

$$\|\Phi(g)\| \leq \|g\|_\infty \tag{5.1.10}$$

Remark. In fact (Problem 2),

$$\|\Phi(g)\| = \sup_{\lambda \in \sigma(\Phi(\text{id}))} |g(\lambda)| \tag{5.1.11}$$

where id is the function that takes λ to λ.

Using Proposition 5.1.1, it is easy to see that Φ_n obeys (5.1.7)–(5.1.9), so once we know the limit exists, we get these results for Φ.

Proof. Let $\psi \in \mathcal{H}$. Then, by Proposition 5.1.1, for all n, j, and k, with $j \neq k$, we have

$$\operatorname{Ran}(E_{j/2^n} - E_{(j-1)/2^n}) \perp \operatorname{Ran}(E_{k/2^n} - E_{(k-1)/2^n}) \tag{5.1.12}$$

from which we conclude:

$$\begin{aligned} \|\Phi_n(g)\psi\|^2 &= \sum_j \left|g\left(\frac{j}{2^n}\right)\right|^2 \|(E_{j/2^n} - E_{(j-1)/2^n})\psi\|^2 \\ &\leq \sup_j \left|g\left(\frac{j}{2^n}\right)\right| \|\psi\|^2 \end{aligned} \tag{5.1.13}$$

since

$$\sum_j \|(E_{j/2^n} - E_{(j-1)/2^n})\psi\|^2 = \|(E_R - E_{(-R-1)/2^n})\psi\|^2 = \|\psi\|^2 \quad (5.1.14)$$

This plus the existence of the limit implies (5.1.10). It also implies for $\ell > n$,

$$\|\Phi_\ell(g) - \Phi_n(g)\| \leq \sup_{|x-y|\leq 2^{-n}} |g(x) - g(y)| \quad (5.1.15)$$

$\to 0$ as $n \to \infty$ by uniform continuity of g. □

We can now state our first form of the spectral theorem.

Theorem 5.1.3 (Spectral Theorem: Resolution of Identity Form). *Let A be a bounded self-adjoint operator on a Hilbert space, $\mathcal{H}$. Then there exists a resolution of the identity, E_t, supported on $[-\|A\|, \|A\|]$ so that*

$$A = \int t\, dE_t \quad (5.1.16)$$

Remarks. 1. We'll see soon (the remark after (5.1.38)) that E_t is uniquely determined by A. Thus, given any resolution of the identity, we can define A by (5.1.16) and see E_t is its resolution of identity. In other words, (5.1.16) sets up a one–one correspondence between bounded self-adjoint operators and resolutions of identity with bounded support.

2. Theorem 2.4.13 is essentially an equivalent form of this theorem.

The second version (in two variants) involves the functional calculus. We already have $f \mapsto f(A)$ for functions analytic in a neighborhood of $\sigma(A)$ (see Section 2.3). We'll extend this to continuous and even bounded measurable functions. An *algebra*, $\mathfrak{A}$, is a vector space with an associative product that obeys the distributive laws. If $AB = BA$, for all $A, B \in \mathfrak{A}$, it is a *commutative algebra*. An *adjoint* on $\mathfrak{A}$ obeys the usual rules (i.e., $(\lambda A)^* = \bar{\lambda}A^*$, $(AB)^* = B^*A^*$, $(A+B)^* = A^* + B^*$). An A with an adjoint is called a $*$-algebra. Chapter 6 will discuss algebras where $\mathfrak{A}$ is also a Banach space.

$\mathcal{L}(\mathcal{H})$, the bounded operators, is a $*$-algebra with $*$ given by the operator adjoint. So are $C([a,b])$, the complex-valued continuous functions on $[a,b]$ and $\mathcal{B}([a,b])$, the bounded Borel functions on $[a,b]$ with adjoint given by pointwise complex conjugate.

If $\mathfrak{A}_1, \mathfrak{A}_2$ are $*$-algebras, a *$*$-homomorphism* $\Phi\colon \mathfrak{A}_1 \to \mathfrak{A}_2$ is a map that preserves the algebraic structure, (i.e., Φ is linear and $\Phi(ab) = \Phi(a)\Phi(b)$) and obeys

$$\Phi(a^*) = \Phi(a)^* \quad (5.1.17)$$

Theorem 5.1.4. (Spectral Theorem: Continuous Functional Calculus Version). *Let A be a bounded self-adjoint operator on a Hilbert space, $\mathcal{H}$. Then there exists a $*$-homomorphism,*

$$\Phi_A\colon C\big([-\|A\|,\|A\|\,]\big)\to\mathcal{L}(\mathcal{H}) \tag{5.1.18}$$

(we write $\Phi_A(f)=f(A)$) so that

$$\mathbb{1}(x)\equiv 1\Rightarrow\Phi_A(\mathbb{1})=\mathbb{1},\qquad \mathrm{id}(x)=x\Rightarrow\Phi_A(\mathrm{id})=A \tag{5.1.19}$$

$$\|\Phi_A(f)\|\le\|f\|_\infty \tag{5.1.20}$$

Moreover, (5.1.19) *and* (5.1.20) *determine Φ_A.*

Remarks. 1. The adjoint on $C\big([-\|A\|,\|A\|\,]\big)$ is $f^*(x)=\overline{f(x)}$ and on $\mathcal{L}(\mathcal{H})$ the operator adjoint. The $\|\Phi_A(f)\|$ in (5.1.20) is the operator norm.

2. We note the proof of uniqueness now: That Φ_A is a $*$-homomorphism implies for a polynomial p in x, $\Phi_A(p)=p(A)$ by (5.1.19). Continuity of Φ_A (i.e., (5.1.20)) and the Weierstrass approximation theorem (see Theorem 2.4.1 of Part 1) show that $\Phi_A(f)$ is then determined for any continuous f.

3. (5.1.11) implies that $\Phi_A(f)$ only depends on $f\restriction\sigma(A)$ so the natural domain is $C(\sigma(A))$; see Theorem 5.1.9.

Theorem 5.1.5 (Spectral Theorem: Borel Functional Calculus). *Let A be a bounded self-adjoint operator on a Hilbert space, $\mathcal{H}$, and $\mathcal{B}\big([-\|A\|,\|A\|\,]\big)$ the bounded Borel functions on $[-\|A\|,\|A\|\,]$. Then there exists a $*$-homomorphism, Φ_A, of $\mathcal{B}\big([-\|A\|,\|A\|\,]\big)$ to $\mathcal{L}(\mathcal{H})$ obeying* (5.1.19) *and* (5.1.20) *and so that*

$$f_n(x)\to f(x)\text{ for all }x\text{ and }\sup_{x,n}|f_n(x)|<\infty\Rightarrow\text{s-lim}\,\Phi_A(f_n)=\Phi_A(f) \tag{5.1.21}$$

Remark. Problem 3 will prove the uniqueness part of the theorem.

Let $\Omega\subset\mathbb{R}$ be a Borel set and χ_Ω its characteristic function. Since $\chi_\Omega^2=\chi_\Omega=\bar\chi_\Omega$,

$$P_\Omega(A)=\chi_\Omega(A) \tag{5.1.22}$$

is a self-adjoint projection. $\{P_\Omega(A)\}$ are called the *spectral projections* for A. The strong continuity in Theorem 5.1.4 implies for $\{\Omega_j\}_{j=1}^\infty$ Borel sets,

$$\Omega_i\cap\Omega_j=\emptyset\text{ for }i\ne j\Rightarrow\underset{N\to\infty}{\text{s-lim}}\sum_{j=1}^N P_{\Omega_j}(A)=P_{\cup\Omega_j}(A) \tag{5.1.23}$$

so $\Omega\mapsto P_\Omega(A)$ is called a *projection-valued measure*. It is clearly closely related to resolutions of the identity

$$E_t=P_{(-\infty,t]}(A) \tag{5.1.24}$$

The relation is an analog of the relation of scalar measures on $[0,1]$ to the monotone functions defining the Riemann–Stieltjes integral.

The last version also comes in two variants. The first form says every self-adjoint operator is a direct sum of multiplication operators:

Theorem 5.1.6 (Spectral Theorem: Multiplication Operator Form). *Let A be a self-adjoint operator on a Hilbert space, $\mathcal{H}$. Then there exist measures $\{\mu_j\}_{j=1}^N$ (N finite or infinite) on $[-\|A\|, \|A\|\,]$ and a unitary map*

$$U\colon \mathcal{H} \to \bigoplus_{j=1}^{N} L^2(\mathbb{R}, d\mu_j) \tag{5.1.25}$$

onto the direct sum so that UAU^{-1} is multiplication by x, that is,

$$(UA\varphi)_j(x) = x(U\varphi)_j(X) \tag{5.1.26}$$

Remarks. 1. The μ_j's and U are nonunique, but there is an involved uniqueness result in Section 5.4.

2. Each μ_j has support contained in $\sigma(A)$.

For the last version, we need a pair of related notions.

Definition. Let A be a bounded operator on a Hilbert space, $\mathcal{H}$. The *cyclic subspace for A generated by $\varphi \in \mathcal{H}$, $\mathcal{H}_\varphi^{(A)}$*, is the smallest closed subspace containing φ and invariant for A and A^*. It can be constructed by taking the closure of the span of all $B_1 B_2 \dots B_k \varphi$, where $k = 0, 1, 2, \dots$ and each B_j is A or A^*. If $\mathcal{H}_\varphi^{(A)} = \mathcal{H}$, we say φ is a *cyclic vector* for A. For many authors, the cyclic subspace for A is only required to be invariant under A and this would be called the cyclic subspace for $\{A, A^*\}$. We'll find this notion more useful and, when the A is clear, will use the shorthand that even drops "for A."

Theorem 5.1.7 (Spectral Theorem: Spectral Measure Form). *Let A be a self-adjoint operator on a Hilbert space, $\mathcal{H}$, and suppose $\varphi \in \mathcal{H}$ is cyclic for A. Then there exists a measure, $d\mu_\varphi$ (we sometimes will use $d\mu_\varphi^{(A)}$) on $[-\|A\|, \|A\|\,]$ and a unitary map*

$$U\colon \mathcal{H} \to L^2(\mathbb{R}, d\mu_\varphi) \tag{5.1.27}$$

$$(UA\psi)(x) = x(U\psi)(x) \text{ for all } x \text{ and } \psi \in \mathcal{H}, \quad (U\varphi)(x) \equiv 1 \tag{5.1.28}$$

Moreover, (5.1.27) *and* (5.1.28) *determine $d\mu_\varphi$ and U uniquely.*

Remark. As in the other uniqueness results, we sketch the proof for this one. By (5.1.28),

$$\langle \varphi, A^n \varphi\rangle = \langle U\varphi, U(A^n\varphi)\rangle = \int x^n \, d\mu_\varphi(x) \tag{5.1.29}$$

Thus, φ and A determine $\int p(x)\,d\mu_\varphi(x)$ for any polynomial, and so since polynomials are dense in $C([-\|A\|,\|A\|\,])$, $\int f(x)\,d\mu_\varphi(x)$ for all continuous f. Thus, the uniqueness part of the Riesz–Markov theorem (i.e., the construction of measures in Section 4.4 of Part 1) implies A and φ determine $d\mu_\varphi$ uniquely. Moreover, by (5.1.28),

$$(UA^n\varphi)(x) = x^n \tag{5.1.30}$$

so U is determined on the span $\{A^n\varphi\}_{n=0}^\infty$, which is dense by the cyclicity hypothesis.

In the remainder of this section, we'll show that each of these versions implies the others as follows:

Resolution of the Identity (Theorem 5.1.3)
[1] ⇓
Continuous Functional Calculus (Theorem 5.1.4)
[2] ⇓
Spectral Measure Version (Theorem 5.1.7)
[3] ⇓ (see below)
Multiplication Operator Form (Theorem 5.1.6)
[4] ⇓
Borel Functional Calculus (Theorem 5.1.5)
[5] ⇓
Resolution of the Identity (Theorem 5.1.3)

Step [3] will use a result which is the main result of the next section (put there not because it is complicated, but because it is conceptually different):

Theorem 5.1.8 (≡ Theorem 5.2.1). *Let A be any operator on a Hilbert space, $\mathcal{H}$. Then there exists an orthonormal family $\{\varphi_j\}_{j=1}^N$ ($N \in \mathbb{Z}_+$ or else $N = \infty$) so that for all $j \neq k$, $\mathcal{H}_{\varphi_j}^{(A)} \perp \mathcal{H}_{\varphi_k}^{(A)}$ and*

$$\mathcal{H} = \bigoplus_{j=1}^N \mathcal{H}_{\varphi_j}^{(A)} \tag{5.1.31}$$

Here are the details of the equivalences:

[1] (Theorem 5.1.3 ⇒ Theorem 5.1.4). Given the resolution of the identity with (5.1.16), let $\Phi_A(g) = \int g(t)\,dE_t$, as defined by Proposition 5.1.2. That proposition shows that Φ_A is a $*$-homomorphism obeying (5.1.19) and $\Phi_A(\mathbb{1}) = \mathbb{1}$. (5.1.16) implies $\Phi_A(\mathrm{id}) = A$. $\square$

[2] (Theorem 5.1.4 ⇒ Theorem 5.1.7). Given a continuous functional calculus, Φ_A, and $\varphi \in \mathcal{H}$, let

$$\ell_\varphi^{(A)}(g) = \langle \varphi, \Phi_A(g)\varphi \rangle \tag{5.1.32}$$

as a linear functional on $C([-\|A\|,\|A\|])$. If $g \geq 0$, we can write $g = f^2$ with $\overline{f} = f$ (where $f(x) = \sqrt{g(x)}$). Then

$$\ell_\varphi^{(A)}(g) = \langle \varphi, \Phi_A(f)^* \Phi_A(f)\varphi\rangle = \|\Phi_A(f)\varphi\|^2 \geq 0 \tag{5.1.33}$$

so by the Riesz–Markov theorem (Theorem 4.5.4 of Part 1), there is a measure $d\mu \equiv d\mu_\varphi^{(A)}$ with

$$\langle \varphi, \Phi_A(g)\varphi\rangle = \int g(x)\,d\mu(x) \tag{5.1.34}$$

In particular,

$$\langle A^n\varphi, A^m\varphi\rangle = \int x^{n+m}\,d\mu(x) = \langle x^n, x^m\rangle_{L^2(\mathbb{R},d\mu)} \tag{5.1.35}$$

Thus, if we define $U[P(A)\varphi] = P(x)$, we get an isometry from $\mathcal{H} = \overline{\{P(A)\varphi\}}$ to $L^2(\mathbb{R}, d\mu_\varphi)$. Since polynomials are dense in $L^2([-\|A\|,\|A\|\,], d\mu_\varphi)$, the map is onto L^2, that is, U extends to a unitary map. By construction, UAU^{-1} is multiplication by x. $\square$

3 (Theorem 5.1.7 $\Rightarrow$ Theorem 5.1.6). This is immediate from Theorem 5.1.8. $\square$

4 (Theorem 5.1.6 $\Rightarrow$ Theorem 5.1.5). Let $F(x)$ be multiplication by $F(x)$ on each $L^2(\mathbb{R}, d\mu_j)$. Then

$$\Phi_A(F) = U^{-1}F(x)U \tag{5.1.36}$$

defines a map with all the required properties. The strong continuity, (5.1.21), follows from the dominated convergence theorem (Problem 4). $\square$

5 (Theorem 5.1.5 $\Rightarrow$ Theorem 5.1.3). Let χ_t be the function

$$\chi_t(x) = \begin{cases} 1, & x \leq t \\ 0, & x > t \end{cases} \tag{5.1.37}$$

Let $E_t = \Phi_A(\chi_t)$. Then by the properties in Theorem 5.1.5, E_t is a resolution of the identity. If Φ_n is given by (5.1.6) and $\mathrm{id}(x) = x$, then $\Phi_n(\mathrm{id}) = \Phi_A(q_n)$, where q_n is a function with $|q_n(x)| \leq \|A\|$ and $q_n(x) \to x$. Thus,

$$\int t\,dE_t = \Phi_A(\mathrm{id}) = A \tag{5.1.38}$$

$\square$

Remark. Given this form for E_t, we can prove uniqueness of the resolution of the identity associated to A. Given a resolution of the identity, $\widetilde{E}_t$, let $A = \widetilde{\Phi}(\mathrm{id})$. Then for any polynomial, P, $P(A) = \widetilde{\Phi}(P)$ since $\widetilde{\Phi}$ is a $*$-homomorphism, so by norm continuity, for all $g \in C([-\|A\|,\|A\|\,])$,

$$g(A) = \widetilde{\Phi}(g) \tag{5.1.39}$$

Given t, let g_n be defined by

$$g_n(s) = \begin{cases} 1, & s \le t \\ 1 - n(s-t), & t \le s \le t + n^{-1} \\ 0, & s \ge t + n^{-1} \end{cases} \tag{5.1.40}$$

By construction of $\widetilde{\Phi}$,

$$\widetilde{E}_t \le \widetilde{\Phi}(g_n) \le \widetilde{E}_{t+n^{-1}} \tag{5.1.41}$$

so, by (5.1.5),

$$\widetilde{E}_t = \operatorname*{s-lim}_{n\to\infty} \widetilde{\Phi}(g_n) \tag{5.1.42}$$

Thus, A determines $\widetilde{E}_t$, implying uniqueness.

Here are two applications of the spectral theorem.

Theorem 5.1.9. *Let A be a bounded self-adjoint operator. The following sets are the same:*

(1) *$\sigma(A)$, the spectrum of A;*
(2) *the set of points of increase of E_λ, the resolution of the identity for A, that is, $\{\lambda \mid \forall \varepsilon > 0,\, E_{\lambda+\varepsilon} \ne E_{\lambda-\varepsilon}\}$;*
(3) *the support of the spectral measure for A, that is, $\overline{\cup_n \operatorname{supp}(\mu_n)}$ if (5.1.25)/(5.1.26) hold.*

Remarks. 1. In particular, (3) implies that for every φ, $\operatorname{supp}(d\mu_\varphi^{(A)}) \subset \sigma(A)$, consistent with (5.1.11).

2. This connection to the spectrum is one reason for the name "spectral theorem."

Proof. Let $\sigma_{(2)}(A)$ (respectively, $\sigma_{(3)}$) be the set in (2) (respectively, (3)). We'll prove $\mathbb{R} \setminus \sigma_{(3)}(A) \subset \mathbb{R} \setminus \sigma(A) \subset \mathbb{R} \setminus \sigma_{(2)}(A)$ and $\sigma_{(3)}(A) \subset \sigma_{(2)}(A)$.

$\underline{\mathbb{R} \setminus \sigma_{(3)}(A) \subset \mathbb{R} \setminus \sigma(A)}$. If $\lambda \notin \sigma_{(3)}(A)$, let $d = \operatorname{dist}(\lambda, \sigma_{(3)}(A)) \ge 0$ since $\sigma_{(3)}(A)$ is closed. It is easy to see that if

$$g(x) = \begin{cases} \frac{1}{x-\lambda} & \text{if } \operatorname{dist}(x,\lambda) < \frac{d}{2} \\ \frac{2}{d} & \text{if } \operatorname{dist}(x,\lambda) \ge \frac{d}{2} \end{cases} \tag{5.1.43}$$

then $g \in \mathcal{B}(\mathbb{R})$ and $g(A)(A-\lambda) = (A-\lambda)g(A) = \mathbb{1}$.

$\underline{\mathbb{R} \setminus \sigma(A) \subset \mathbb{R} \setminus \sigma_{(2)}(A)}$. Suppose $\operatorname{dist}(\lambda, \sigma(A)) = d > 0$ and $\varepsilon < d/2$. Let

$$g(x) = \begin{cases} 1 & \text{if } |x-\lambda| < \frac{d}{2} \\ 0 & \text{if } |x-\lambda| > d \\ \text{linear,} & \text{in between} \end{cases} \tag{5.1.44}$$

Then, by (5.1.11), $g(A) = 0$. Since $0 \le E_{\lambda+\varepsilon} - E_{\lambda-\varepsilon} \le g(A)$, we see $E_{\lambda-\varepsilon} = E_{\lambda+\varepsilon}$.

$\sigma_{(3)}(A) \subset \sigma_{(2)}(A)$. If $\lambda \in \sigma_{(2)}(A)$ and $\varepsilon > 0$, we can find $\varphi \neq 0$ in some $L^2(\mathbb{R}, d\mu_n)$ with $\text{supp}(\varphi) \subset (\lambda - \varepsilon, \lambda + \varepsilon)$. It follows that $\varphi \in \text{Ran}(E_{\lambda+\varepsilon} - E_{\lambda-\varepsilon})$, so $E_{\lambda+\varepsilon} \neq E_{\lambda-\varepsilon}$. □

Theorem 5.1.10. *Let A be a bounded self-adjoint operator. Then*

(a) $A \ge 0 \Leftrightarrow \sigma(A) \subset [0, \infty)$

(b) *If $A \ge 0$, there exists $B \ge 0$ so $B^2 = A$ and so that*

$$[C, A] = 0 \Rightarrow [B, A] = 0 \tag{5.1.45}$$

Remarks. 1. If we use that $A = B^2$ with $B = B^*$, then the spectral mapping theorem (Theorem 2.2.16) implies $\sigma(A) \subset [0, \infty)$. But we want to use (a) to prove (b).

2. This is an alternate proof of the existence half of the square root lemma (Theorem 2.4.4). Our first proof of the spectral theorem (in Section 2.4) uses the square root lemma but not the other ones.

Proof. (a) If $\sigma(A)$ is not in $[0, \infty)$, there is $\lambda \in (-\infty, 0) \cap \mathbb{R}$, and so, by Theorem 5.1.9, some n and $\varphi \in L^2(\mathbb{R}, d\mu_n)$ with $\varphi \neq 0$ and φ supported in $(-\infty, \frac{\lambda}{2})$. Thus, $\langle U^{-1}\varphi, AU^{-1}\varphi\rangle \le -\frac{\lambda}{2}\|\varphi\|^2 < 0$, so A is not positive. On the other hand, if $\sigma(A) \subset [0, \infty)$, every μ_n is supported on $[0, \infty)$, so for any $\psi \in \mathcal{H}$,

$$\langle \psi, A\psi\rangle = \sum_{n=1}^{N} \int x|(U\psi)_n(x)|^2 \, d\mu_n(x) \ge 0$$

so A is nonnegative.

(b) Since each μ_n is supported on $[0, \infty)$, we can let M be multiplication by $\sqrt{x}$ and $B = U^{-1}MU$, that is, $B = f(A)$ with f the continuous real-valued function which is 0 on $(-\infty, 0]$ and $\sqrt{x}$ on $[0, \infty)$. Thus, (5.1.45) holds. □

We can extend the spectral mapping theorem to continuous functions of a self-adjoint operator:

Theorem 5.1.11 (Spectral Mapping Theorem for Self-adjoint Operators). *Let A be a bounded self-adjoint operator and f a continuous function on $\sigma(A)$. Then*

$$\sigma(f(A)) = f[\sigma(A)] = \{f(\lambda) \mid \lambda \in \sigma(A)\}$$

Proof. Let $\lambda \notin f[\sigma(A)]$. Since $\sigma(A)$ is compact and f continuous, $d = \text{dist}(\lambda, f(\sigma(A))) > 0$ so on each $L^2(\mathbb{R}, d\mu_n)$, $(f(x) - \lambda)^{-1}$ is bounded (by d^{-1}). Thus, if M is multiplication by $f - \lambda$, then $B = U^{-1}MU$ is an inverse to $f(A) - \lambda$, so $\lambda \notin \sigma(f(A))$.

Suppose $\lambda \in f[\sigma(A)]$, say $\lambda = f(\zeta_0)$ with ζ_0 in $\sigma(A)$. Since f is continuous, given ε, find δ so that

$$|\zeta - \zeta_0| \le \delta \Rightarrow |f(\zeta) - \lambda| \le \varepsilon \tag{5.1.46}$$

Since $Q_\delta \equiv E_{\zeta_0+\delta} - E_{\zeta_0-\delta} \neq 0$, find $\varphi_\varepsilon \in \mathrm{Ran}(Q_\delta)$ with $\|\varphi_\varepsilon\| = 1$. By (5.1.46), $\|(f(A)-\lambda)\varphi_\varepsilon\| \le \varepsilon\|\varphi_\varepsilon\|$, so $\lambda \in \sigma(f(A))$. □

The final result we want to note is connected with the decomposition (4.7.28) of Part 1 for any Baire measure on $\mathbb{R}$

$$d\mu = d\mu_{\mathrm{ac}} + d\mu_{\mathrm{pp}} + d\mu_{\mathrm{sc}} \tag{5.1.47}$$

as a sum of a piece absolutely continuous with respect to Lebesgue measure, a pure point piece, and a singular continuous piece.

Theorem 5.1.12. *Let A be a bounded self-adjoint operator on a Hilbert space, $\mathcal{H}$. Then $\mathcal{H} = \mathcal{H}_{\mathrm{ac}} \oplus \mathcal{H}_{\mathrm{pp}} \oplus \mathcal{H}_{\mathrm{sc}}$, where each space is invariant for A and $\varphi \in \mathcal{H}_{\mathrm{ac}}$ (respectively, $\mathcal{H}_{\mathrm{pp}}$ or $\mathcal{H}_{\mathrm{sc}}$) if and only if $d\mu_\varphi^{(A)}$ is a.c. wrt dx (respectively, pure point or singular continuous).*

Remark. $\mathcal{H}_{\mathrm{pp}}$ is the span of the eigenvectors for A.

Proof. Write $UAU^{-1} = \oplus_{j=1}^N M_{\mu_j}$, where M_μ is multiplication by x on $L^2(\mathbb{R}, d\mu)$. Since the pieces in (5.1.47) are mutually singular, $L^2(\mathbb{R}, d\mu) = L^2(\mathbb{R}, d\mu_{\mathrm{ac}}) \oplus L^2(\mathbb{R}, d\mu_{\mathrm{pp}}) \oplus L^2(\mathbb{R}, d\mu_{\mathrm{sc}})$ with each invariant for M_μ, we get $\mathcal{H}_{\mathrm{ac}}$ as $\oplus_{j=1}^N L^2(\mathbb{R}, d\mu_{\mathrm{ac}})$, etc. □

One defines the *a.c. spectrum, pure point spectrum,* and *singular continuous spectrum* of A, a self-adjoint operator, by

$$\sigma_{\mathrm{ac}}(A) = \sigma(A \restriction \mathcal{H}_{\mathrm{ac}}), \quad \sigma_{\mathrm{pp}}(A) = \sigma(A \restriction \mathcal{H}_{\mathrm{pp}}), \quad \sigma_{\mathrm{sc}}(A) = \sigma(A \restriction \mathcal{H}_{\mathrm{sc}}) \tag{5.1.48}$$

In some ways, these are rough invariants since they are closed sets. For example, $\sigma_{\mathrm{pp}}(A)$ is the closure of the set of eigenvalues and the set of eigenvalues has more information.

Example 5.1.13. Let $\{q_n\}_{n=1}^\infty$ be a counting of the rationals in $(0,1)$. Let $S_{n,\varepsilon} = (q_n - \frac{\varepsilon}{2^n}, q_n + \frac{\varepsilon}{2^n}) \cap (0,1)$. Let $S_\varepsilon = \bigcup_n S_{n,\varepsilon}$ so $|S_\varepsilon| \le \sum_{n=1}^\infty \frac{2\varepsilon}{2^n} = 2\varepsilon$ even though S_ε is a dense open set in $[0,1]$. If A is multiplication by x in $L^2([0,1], dx)$ and B in $L^2([0,1], \chi_{S_\varepsilon}(x)\,dx)$, then A and B have purely a.c. spectrum and $\sigma_{\mathrm{ac}}(A) = \sigma_{\mathrm{ac}}(B)$, but A and B are not unitarily equivalent. For example, if $P_\Omega(\,\cdot\,)$ are the spectral projections (5.1.22), then $P_{[0,1]\setminus S_\varepsilon}(B) = 0$ but $P_{[0,1]\setminus S_\varepsilon}(A) \neq 0$. □

This suggests we want a finer invariant than σ_{ac}.

Definition. A *Lebesgue measure class* is an equivalence class of Borel subsets of $\mathbb{R}$ under the relation $S \sim T \Leftrightarrow |S \Delta T| = 0$, where $|\cdot|$ is Lebesgue measure.

Definition. Let A be a bounded self-adjoint operator. Let $A_{\text{ac}} = A \restriction \mathcal{H}_{\text{ac}}$. Then $\Sigma_{\text{ac}}(A)$ is the measure class of those S with $P_{\mathbb{R}\setminus S}(A_{\text{ac}}) = 0$, $P_T(A_{\text{ac}}) \neq 0$ for all T with $|T| > 0$ and $|T \setminus S| = 0$.

It is not hard to see that

$$\sigma_{\text{ac}}(A) = \overline{\Sigma_{\text{ac}}}^{\text{ess}} \equiv \{\lambda \mid \forall \varepsilon\, |(\lambda - \varepsilon, \lambda + \varepsilon) \cap \Sigma_{\text{ac}}| > 0\} \tag{5.1.49}$$

the essential closure of Σ_{ac}. In (5.1.49), by the intersection we mean with any set in the equivalence class. Thus, Σ_{ac} determines σ_{ac}. It is also easy to see that if $d\mu$ and $d\nu$ are two measures a.c. wrt dx, then multiplication by x in $L^2(\mathbb{R}, d\mu)$ and $L^2(\mathbb{R}, d\nu)$ (call them A and B) are unitarily equivalent if and only if $\Sigma_{\text{ac}}(A) = \Sigma_{\text{ac}}(B)$.

Notes and Historical Remarks.

> It is unfortunate therefore that even the bare statement of the spectral theorem is widely regarded as somewhat mysterious and deep, and probably inaccessible to the nonspecialist ... Another reason the spectral theorem is thought to be hard is that its proof is hard.
>
> —*P. Halmos* [**284**][1]

The spectral theorem for bounded self-adjoint operators is due to Hilbert in 1906, the fourth part of his series [**317**] on the theory of operators on what we know as Hilbert space. Important reformulations and extensions are due to Schmidt [**602**], Weyl [**741**], Hellinger [**310**], and Hellinger–Toeplitz [**311**]. Except for Toeplitz (who was Hilbert's postdoc, not his student), all these authors were students of Hilbert.

All this work was done using the resolution of the identity formulation and using, not operators, but quadratic forms. It was F. Riesz [**564**] in 1913 who phrased things in the operator language now standard. Riesz also had some discussion of the functional calculus, but it appeared more prominently in the work of Stone [**670**] and von Neumann [**720**] on the unbounded case. With the advent of the theory of operator algebras with von Neumann's work in the 1930s and the Banach algebra work of Gel'fand and his school in the 1940s, the functional calculus framework became more central still.

The multiplication operator format occurred first in the context of multiplicity theory, most clearly in work of Segal [**616**]. It was Halmos, in a note [**284**] in 1963 that emphasized the simplicity of this point of view, which has become the most common approach in more recent texts although, for some,

[1] in a paper championing the multiplication operator version.

the resolution of the identity persists. While I agree that the multiplication operator format is the simplest conceptually, I feel that the multiplicity of viewpoints is not only historically correct but pedagogically illuminating.

Problems

1. (a) Let $\{P_j\}_{j=1}^{\infty}$ be a family of orthogonal projections, with $\mathcal{H}_j = \text{Ran}(P_j)$ and $P_j \le P_{j+1}$. Prove that $\text{s-lim}_{j\to\infty} P_j$ exists and is the orthogonal projection onto $\overline{\cup_j \mathcal{H}_j}$.

   (b) If $\{E_t\}$ is a resolution of the identity, prove that

$$\operatorname*{s-lim}_{s\uparrow t} E_s \equiv E_t^- \tag{5.1.50}$$

   exists and that $E_t^- \le E_t$.

   (c) If $t \neq s$, prove that $E_t - E_t^-$ is a projection onto a space orthogonal to $E_s - E_s^-$.

   (d) Prove that $\{t \mid E_t^- \neq E_t\}$ is a countable set.

2. (a) Let $\{E_t\}$ be a resolution of the identity. Define

$$\sigma(\{E_t\}) = \{t \mid \forall \varepsilon > 0,\, E_{t-\varepsilon} \neq E_{t+\varepsilon}\} \tag{5.1.51}$$

   Let $A = \int t\, dE_t$. Prove that $\sigma(A) = \sigma(\{E_t\})$.

   (b) Prove that (5.1.11) holds.

3. (a) Let U be a bounded open subset of R and

$$g_{n,U}(x) = \begin{cases} 1 & \text{if } \text{dist}(x, \mathbb{R}\setminus U) \ge n^{-1} \\ n\,\text{dist}(x, \mathbb{R}\setminus U) & \text{if } \text{dist}(x, \mathbb{R}\setminus U) \le n^{-1} \end{cases}$$

   If A is a bounded self-adjoint operator and χ_U the characteristic function of U, prove that

$$\Phi_A(\chi_U) = \operatorname*{s-lim}_{n\to\infty} \Phi_A(g_{n,U}) \tag{5.1.52}$$

   and conclude that if $\Phi_A, \widetilde{\Phi}_A$ are two $*$-homomorphisms realizing Theorem 5.1.5, then $\Phi_A(\chi_U) = \widetilde{\Phi}_A(\chi_U)$ for all open sets U.

   (b) Prove $\{B, \text{ Borel sets} \mid \Phi_A(\chi_B) = \widetilde{\Phi}_A(\chi_B)\}$ is a σ-algebra and conclude it is all Borel sets.

   (c) Prove $\Phi_A(f) = \widetilde{\Phi}_A(f)$ for all bounded Borel functions, f.

   Remark. One can also get this uniqueness from uniqueness of the spectral measures.

4. Let μ be a measure of compact support, F_n, F Borel functions on $\text{supp}(\mu)$ so $F_n \to F$ for μ a.e. x and $\sup_{x,n}|F_n(x)| < \infty$. Let φ be in $L^2(\mathbb{R}, d\mu)$. Prove that $F_n\varphi \to F\varphi$ in $L^2(\mathbb{R}, d\mu)$.

5.2. Cyclic Vectors

Recall that, given $A \in \mathcal{L}(\mathcal{H})$ with $\mathcal{H}$ a Hilbert space and $\varphi \in \mathcal{H}$, we defined the *cyclic subspace*, $\mathcal{H}_\varphi^{(A)}$, to be the closure of the span of all products of A's and A^*'s applied to φ. In this section, our goal is to prove:

Theorem 5.2.1 ($\equiv$ Theorem 5.1.8). *Let A be any operator on a Hilbert space, $\mathcal{H}$. Then there exists an orthonormal family $\{\varphi_j\}_{j=1}^N$ ($N \in \mathbb{Z}_+$ or else $N = \infty$) so that for all $j \neq k$, $\mathcal{H}_{\varphi_j}^{(A)} \perp \mathcal{H}_{\varphi_k}^{(A)}$ and*

$$\mathcal{H} = \bigoplus_{j=1}^N \mathcal{H}_{\varphi_j}^{(A)} \tag{5.2.1}$$

The key is the following lemma:

Lemma 5.2.2. *If $\psi \perp \mathcal{H}_\varphi^{(A)}$, then $\mathcal{H}_\psi^{(A)} \perp \mathcal{H}_\varphi^{(A)}$.*

Proof. If $\eta \in \mathcal{H}_\varphi^\perp$, $\zeta \in \mathcal{H}_\varphi^A$, and $A^\sharp$ is A or A^*, then

$$\langle A^\sharp \eta, \zeta \rangle = \langle \eta, (A^\sharp)^* \zeta \rangle = 0$$

since $A^{**}\zeta = A\zeta \in \mathcal{H}_\varphi^{(A)}$ and $A^*\zeta \in \mathcal{H}_\varphi^{(A)}$. Therefore, $A^\sharp \eta \perp \mathcal{H}_\varphi^{(A)}$. Thus, any product of A and A^*'s on ψ is in $(\mathcal{H}_\varphi^{(A)})^\perp$. By continuity of the inner product, $\mathcal{H}_\psi^{(A)} \perp \mathcal{H}_\varphi^{(A)}$. □

Proof of Theorem 5.2.1. Let $\{\eta_j\}_{j=1}^\infty$ be an orthonormal basis for $\mathcal{H}$. Let $\varphi_1 = \eta_1$, $\mathcal{H}_1 = \mathcal{H}_{\varphi_1}^{(A)}$, and P_1 be the projection onto $\mathcal{H}_1^\perp$. If for all j, $P_1\eta_j = 0$, then let $N = 1$. If not, let k_1 be the smallest j with $P_1\eta_j \neq 0$ and let $\varphi_2 = P_1\eta_{k_1}/\|P_1\eta_{k_1}\|$. By the lemma, $\mathcal{H}_{\varphi_2}^{(A)} \perp \mathcal{H}_{\varphi_1}^{(A)}$, and by construction, $\{\eta_j\}_{j=1}^{k_1} \subset \mathcal{H}_{\varphi_1}^{(A)} \oplus \mathcal{H}_{\varphi_2}^{(A)}$.

Proceed inductively, that is, given $\varphi_1, \dots, \varphi_{m-1}$, let P_{m-1} be the projection onto $[\oplus_{j=1}^{m-1} \mathcal{H}_{(\varphi_j)}^{(A)}]^\perp$. If $P_{m-1} = 0$, take $N = m-1$. Otherwise, let k_{m-1} be the first j with $P_{m-1}\eta_j \neq 0$. Define $\varphi_m = P_{m-1}\eta_{k_{m-1}}/\|P_{m-1}\eta_{k_{m-1}}\|$. Then by the lemma, $\mathcal{H}_{\varphi_m}^{(A)} \subset (\mathcal{H}_{\varphi_j}^{(A)})^\perp$ for $j = 1, \dots, m-1$. Since $\{\eta_j\}_{j=1}^m \subset \oplus_{q=1}^m \mathcal{H}_{\varphi_q}^{(A)}$, we see that $\oplus_{q=1}^N \mathcal{H}_{\varphi_q}^{(A)} = \mathcal{H}$. □

5.3. A Proof of the Spectral Theorem

In this section, we'll prove Theorem 5.1.7 and thereby, given our discussion in Section 5.1, all the forms of the spectral theorem (for bounded self-adjoint operators). So let A be a bounded self-adjoint operator with cyclic vector, φ. The idea will be to find an orthonormal basis for $\mathcal{H}$ in which A is a Jacobi matrix. Then we can appeal to Favard's theorem (Theorem 4.1.3).

For $n = 1, 2, \ldots$, define

$$\psi_n = A^{n-1}\varphi \tag{5.3.1}$$

and $\{\varphi_n\}_{n=1}^N$ the ON set obtained by applying Gram–Schmidt to this set. If $p(A)\varphi = 0$ for all polynomials, p, of minimal degree, N, then $\{A^j\varphi\}_{j=0}^{N-1}$ are independent (since N is the minimal degree) and, by induction (Problem 1), any $A^\ell\varphi$, $\ell \geq N$, is a linear combination of $\{A^j\varphi\}_{j=0}^{N-1}$. So either N is finite or $\{A^j\varphi\}_{j=1}^\infty$ are linearly independent.

Thus, for $k \leq N$,

$$\varphi_k = \sum_{j=1}^k \alpha_{kj}\psi_j \tag{5.3.2}$$

and

$$\psi_k = \sum_{j=1}^k \beta_{kj}\varphi_j \tag{5.3.3}$$

where

$$\alpha_{kk} > 0, \qquad \beta_{kk} > 0 \tag{5.3.4}$$

Since φ is cyclic, $\{\varphi_j\}_{j=1}^N$ is an orthonormal basis.

It follows from this and $A\psi_j = \psi_{j+1}$ that $A\varphi_k$ is in the span of $\{\varphi_j\}_{j=1}^{k+1}$. In particular, if $\ell > k+1$, $\langle \varphi_\ell, A\varphi_k\rangle = 0$. But then, $\langle \varphi_k, A\varphi_\ell\rangle = \overline{\langle \varphi_\ell, A\varphi_k\rangle} = 0$ also, so $\langle \varphi_\ell, A\varphi_k\rangle \neq 0 \Rightarrow |\ell - k| \leq 1$, that is, in the ON basis $\{\varphi_j\}_{j=1}^N$, A is tridiagonal. Moreover, (5.3.4) is easily seen (Problem 2) to imply $\langle \varphi_{k+1}, A\varphi_k\rangle > 0$ (so $\langle \varphi_k, A\varphi_{k+1}\rangle > 0$ also). We have thus proven:

Theorem 5.3.1. *Let A be a bounded self-adjoint operator on a Hilbert space, $\mathcal{H}$, with a cyclic vector, φ. Then in the basis obtained via Gram–Schmidt on $\{A^j\varphi\}_{j=0}^\infty$, A is a Jacobi matrix.*

There exists a unique measure, μ, on $[-\|A\|, \|A\|\,]$ so that

$$\int x^n \, d\mu_\varphi(x) = \langle \varphi, A^n\varphi\rangle \tag{5.3.5}$$

There exists a unitary map, $U\colon \mathcal{H} \to L^2(\mathbb{R}, d\mu)$ obeying (5.1.28).

Proof. The first paragraph is as noted above. Favard's theorem (Theorem 4.1.3) then implies there is a measure $d\mu_\varphi$ with the same Jacobi parameters as this Jacobi matrix. By Proposition 4.1.6, one has (5.3.5). As in the proof of uniqueness in the remark after Theorem 5.1.7, we have a unique unitary obeying (5.1.28). □

Remarks. 1. This is an extended form of the spectral measure form of the spectral theorem (Theorem 5.1.7).

2. If $\{p_n\}_{n=1}^N$ are the orthonormal polynomials for $d\mu_\varphi$, then the Gram–Schmidt basis $\{\varphi_n\}_{n=1}^N$ constructed above is given by

$$\varphi_n = p_n(A)\varphi \tag{5.3.6}$$

Problems

1. If $A^n\varphi = \sum_{j=0}^{n-1} \alpha_j A^j \varphi$, prove by induction that for any $\ell \geq 0$, $A^{n+\ell}$ is a linear combination of $\{A^j\varphi\}_{j=0}^{n-1}$.
2. Show that (5.3.4) implies $\langle \varphi_{n+1}, A\varphi_n \rangle > 0$.

5.4. Bonus Section: Multiplicity Theory

In this section, we'll explore two closely related issues: first, lack of uniqueness in Theorem 5.1.6 and second, the question of unitary invariants for (bounded) self-adjoint operators, that is, data to be assigned to such operators so that two such operators are unitarily equivalent if and only if their data are the same. We'll see there are canonical choices in Theorem 5.1.6 (albeit not of measures) that provide the data. As a first hint, we note that in the finite-dimensional case, unitary invariant data are the eigenvalues and their multiplicity. Our results will also allow the study of the structure of the set of all operators commuting with a bounded self-adjoint operator, A.

We start with simple operators. Given a measure, μ, on $\mathbb{R}$ with bounded support, we let M_μ be the multiplication operator on $L^2(\mathbb{R}, d\mu)$, that is,

$$(M_\mu f)(x) = xf(x) \tag{5.4.1}$$

A self-adjoint operator, A, is called *simple* if it has a cyclic vector. By picking $\varphi(x) \equiv 1$, we see every M_μ is simple and the spectral theorem in the form Theorem 5.1.7 says that every simple operator is unitarily equivalent to some M_μ.

Theorem 5.4.1. *Let μ, ν be two measures of bounded support on $\mathbb{R}$. Then M_μ is unitarily equivalent to M_ν if and only if μ and ν are mutually absolutely continuous, that is, each is a.c. wrt the other.*

Remark. μ and ν are mutually a.c. if and only if $(\mu(A) = 0 \Leftrightarrow \nu(A) = 0)$ if and only if $d\mu = f\,d\nu$ with $f(x) \neq 0$ for ν-a.e. x.

Proof. If $VM_\mu V^{-1} = M_\nu$ for a unitary V, then $M_\nu^\ell(V1) = V(M_\mu^\ell 1)$ for all ℓ, so if $V1 = g(z)$, we have

$$\int x^\ell |g(x)|^2 \, d\nu(x) = \int x^\ell \, d\mu(x) \tag{5.4.2}$$

so by the fact that moments determine bounded measures, we have

$$d\mu(x) = |g(x)|^2 \, d\nu(x) \tag{5.4.3}$$

so μ is a.c. wrt ν. By symmetry, they are mutually absolutely continuous. $\square$

Definition. Two measures are called equivalent if they are mutually absolutely continuous (written $\mu \sim \nu$). An equivalence class is called a *measure class.*

If A is simple, by the last theorem and the spectral theorem, if φ, ψ are two cyclic vectors, $d\mu_\varphi^{(A)}$ and $d\mu_\psi^{(A)}$ are equivalent, so simple A's define a unique measure class. Theorem 5.4.1 can thus be restated.

Theorem 5.4.2. *Let A and B be two simple, bounded self-adjoint operators. Then there is a unitary V so $VAV^{-1} = B$ if and only if their measure classes are the same. Moreover, every measure class of bounded support occurs.*

To turn to general operators, consider the finite-dimensional case where the natural unitary invariants are to list the eigenvalues of multiplicity one, then those of multiplicity two, ..., requiring disjointness of the list.

Thus, we need to define an operator of uniform multiplicity ℓ. By $M_\mu^{(\ell)}$, we mean the ℓ-fold direct sum $M_\mu \oplus \cdots \oplus M_\mu$ (ℓ times). $\ell = \infty$ is allowed, in which case we take infinitely many copies. Alternatively, we can talk about $M_\mu \otimes \mathbb{1}$ on $L^2(\mathbb{R}, d\mu) \otimes \mathbb{C}^\ell$ ($\mathbb{C}^\infty$ is ℓ^2) or as multiplication by x on $L^2(\mathbb{R}, d\mu; \mathbb{C}^\ell)$ (or $L^2(\mathbb{R}, d\mu; \ell^2)$) allowing $\mathbb{C}^\ell$-valued functions.

The main result of this section is:

Theorem 5.4.3 (Multiplicity Theorem: First Form)**.** *Let A be a bounded self-adjoint operator. Then there exist measures supported in $[-\|A\|, \|A\|\,]$, $\{\mu_\ell\}_{\ell=1}^\infty$ and μ_∞ so that A is unitarily equivalent to $M_{\mu_\infty}^{(\infty)} \oplus \bigoplus_{\ell=1}^\infty M_{\mu_\ell}^{(\ell)}$ and so that for any distinct j and ℓ (including ∞), we have $\mu_j \perp \mu_\ell$. The measure classes of μ_ℓ's are uniquely determined by A, and two A's are unitarily equivalent if and only if all their μ_ℓ measure classes agree.*

Remarks. 1. To accommodate empty "sets" of a given multiplicity, we allow $\mu = 0$, in which case $M_\mu = 0$ on the zero-dimensional space!

2. Recall $\mu \perp \nu$ means they are mutually singular, that is, $\exists \Omega$ so $\mu(\Omega) = 0 = \nu(\mathbb{R} \setminus \Omega)$.

3. μ_ℓ is called the *spectral part of uniform multiplicity ℓ.*

It will be convenient to first state an equivalent form, prove that it implies Theorem 5.4.3, and then turn to proving it.

Theorem 5.4.4 (Multiplicity Theorem: Second Form). *Let A be a bounded self-adjoint operator. Then there exist measures supported in $[-\|A\|, \|A\|\,]$, $\{\nu_\ell\}_{\ell=1}^\infty$, so $\nu_1 \geq \nu_2 \geq \dots$ and so that A is unitarily equivalent to $\oplus_{\ell=1}^\infty M_{\nu_\ell}$. The measure classes of the ν_ℓ are unique.*

Given a Baire measure, μ, on $\mathbb{R}$ and $\Omega \subset \mathbb{R}$, a Borel set, we define a measure μ_Ω by

$$\mu_\Omega(\Sigma) = \mu(\Sigma \cap \Omega) \tag{5.4.4}$$

Theorem 5.4.5. *For any Baire measure, μ, and Borel set, Ω, we have*

$$M_\mu = M_{\mu_\Omega} \oplus M_{\mu_{\mathbb{R}\setminus\Omega}} \tag{5.4.5}$$

Indeed, if $P = P_\Omega(M_\mu)$ is the spectral measure, then Ran(P) *is invariant for M_μ and*

$$M_\mu \restriction \operatorname{Ran}(P) \cong M_{\mu_\Omega}, \qquad M_\mu \restriction \operatorname{Ran}(1-P) \cong M_{\mu_{\mathbb{R}\setminus\Omega}}$$

Proof. Since $M_\mu P = PM_\mu$, Ran(P) is invariant for M_μ. In $L^2(\mathbb{R}, d\mu)$, $1 = \chi_\Omega + \chi_{\mathbb{R}\setminus\Omega}$. χ_Ω is cyclic for M_μ in $L^2(\mathbb{R}, d\mu_\Omega)$ and $d\mu_{\chi_\Omega}^{(M_\mu)} = d\mu_\Omega$. □

Lemma 5.4.6. *Given Baire measures η and κ on $\mathbb{R}$ with $\eta \geq \kappa$, there is a Baire set, $\Omega \subset \mathbb{R}$, so that*

$$\eta_\Omega \perp \kappa, \qquad \eta_{\mathbb{R}\setminus\Omega} \sim \kappa \tag{5.4.6}$$

Proof. By the combined Lebesgue decomposition and Radon–Nikodym theorem of von Neumann (Theorem 4.7.6 of Part 1),

$$d\eta = f\,d\kappa + d\zeta$$

where $d\zeta \perp d\kappa$ and $f \geq 1$. Since $d\zeta \perp d\kappa$, we can pick Ω so

$$\zeta(\mathbb{R}\setminus\Omega) = 0, \qquad \kappa(\Omega) = 0 \tag{5.4.7}$$

Then $\eta_\Omega = \zeta \perp \kappa$ and $d\eta_{\mathbb{R}\setminus\Omega} = f\,d\kappa \sim d\kappa$ since $f \geq 1$. □

Proof that Theorem 5.4.4 ⇒ Theorem 5.4.3. By Lemma 5.4.6, find $\widetilde{\Omega}_j$ so $(\nu_j)_{\widetilde{\Omega}_j} \perp \nu_{j+1}$ and $(\nu_j)_{\mathbb{R}\setminus\widetilde{\Omega}_j} \sim \nu_{j+1}$. It follows (Problem 1) that if

$$\Omega_j = \widetilde{\Omega}_j \cap \bigcap_{k=1}^{j-1} (\mathbb{R}\setminus\widetilde{\Omega}_k), \qquad \Omega_\infty = \mathbb{R} \setminus \bigcup_{j=1}^\infty \Omega_j \tag{5.4.8}$$

then, with $\nu_{j,\ell} = (\nu_j)_{\Omega_\ell}$,

$$\nu_j = \left(\sum_{\ell=1}^\infty \nu_{j,\ell}\right) + \nu_{j,\infty} \tag{5.4.9}$$

and

(a) $\nu_{j,\ell} = 0 \quad \text{if } \ell < j;$ (5.4.10)

(b) $\nu_{j,\ell} \sim \nu_{k,\ell} \quad \text{if } j,k \le \ell;$ (5.4.11)

(c) $\nu_{j,\ell} \perp \nu_{j,m} \quad \text{if } \ell \ne m.$ (5.4.12)

Thus, by Lemma 5.4.6 and its iteration and limits,

$$M_{\nu_j} = \bigoplus_{\ell=j}^{\infty} M_{\nu_{j,\ell}} \tag{5.4.13}$$

Since (b) holds, Theorem 5.4.1, $M_{\nu_{\ell,\ell}}$ is unitarily equivalent to $M_{\nu_{k,\ell}}$ for $k \le \ell$, so with $\cong$ meaning unitarily equivalent,

$$\bigoplus_{k=1}^{\ell} M_{\nu_{k,\ell}} \cong M_{\nu_{\ell,\ell}}^{(\ell)} \tag{5.4.14}$$

so we get Theorem 5.4.5 with $\mu_\ell = \nu_{\ell,\ell}$.

It is easy to see uniqueness of ν_ℓ implies that of the μ_ℓ (Problem 2). $\square$

We now turn to the existence half of Theorem 5.4.4.

Definition. Let A be a bounded self-adjoint operator on a Hilbert space, $\mathcal{H}$. A vector $\varphi \in \mathcal{H}$ is called *dominant* if and only if for any $\psi \in \mathcal{H}$,

$$d\mu_\psi^{(A)} \ll d\mu_\varphi^{(A)} \tag{5.4.15}$$

We call $d\mu_\varphi^{(A)}$ a *dominant measure* for A. (Recall that $d\mu << d\nu$ means that μ is a.c. wrt ν.)

Remark. It is not hard to see (Problem 3) that φ is dominant if and only if for all $f \in \mathcal{B}(\mathbb{R})$, $f(A)\varphi = 0 \Rightarrow f(A) = 0$. Because of the role of vectors obeying an analog of this for operator algebras, such vectors are also called *separating vectors.*

Lemma 5.4.7. *Let A be a bounded self-adjoint operator on a Hilbert space, $\mathcal{H}$, and $\psi \in \mathcal{H}$ a unit vector. Then there exists a dominant vector φ so that $\psi \in \mathcal{H}_\varphi^{(A)}$.*

Proof. By the construction in Section 5.2, we can find orthonormal vector $\{\psi_j\}_{j=1}^\infty$ with $\psi_1 = \psi$ and with

$$\mathcal{H}_{\psi_j}^{(A)} \perp \mathcal{H}_{\psi_k}^{(A)} \text{ if } j \ne k, \qquad \mathcal{H} = \bigoplus_{j=1}^{\infty} \mathcal{H}_{\psi_j}^{(A)} \tag{5.4.16}$$

Let $\mu = \mu_\psi^{(A)}$, $\nu = \sum_{n=2}^\infty 2^{-n} \mu_{\psi_n}^{(A)}$. Write $d\nu = g\, d\mu + d\nu_{\rm s}$ with $d\nu_{\rm s} \perp d\mu$ and find Ω so that

$$\mu(\Omega) = 0, \qquad \nu_{\rm s}(\mathbb{R} \setminus \Omega) = 0 \tag{5.4.17}$$

Pick

$$\varphi = \psi + \sum_{n=2}^{\infty} 2^{-n} P_\Omega(A)\psi_n \tag{5.4.18}$$

Then it is easy to see that φ is dominant (Problem 4) and $\psi \in \mathcal{H}_\varphi^{(A)}$. □

Lemma 5.4.8. (a) *Let μ, ν be two Baire measures on $\mathbb{R}$ with $\nu \ll \mu$. Then there exists $\nu_1 \sim \nu$ so $\nu_1 \le \mu$.*

(b) *Let μ be a Baire measure on $\mathbb{R}$, $\mathcal{H}$ a Hilbert space, $\varphi \in \mathcal{H}$, and A self-adjoint. Suppose that*

$$d\mu_\varphi^{(A)} \ll d\mu \tag{5.4.19}$$

Then there exists $\psi \in \mathcal{H}_\varphi^{(A)}$ so that

$$\mathcal{H}_\psi^{(A)} = \mathcal{H}_\varphi^{(A)}, \qquad d\mu_\psi^{(A)} \le d\mu \tag{5.4.20}$$

Proof. (a) By the Radon–Nikodym theorem, write $d\nu = f\,d\mu$. Define

$$d\nu_1(x) = \min(f(x), 1)\,d\mu(x) \tag{5.4.21}$$

Since $\min(f(x), 1) = 0 \Leftrightarrow f(x) = 0$, we see $\nu_1 \sim \nu$. Since $\min(f(x), 1) \le 1$, $d\nu_1 \le d\mu$.

(b) Let f be as in the proof of (a) when $\nu = d\mu_\varphi^{(A)}$ and $g = \sqrt{\min(f(x), 1)}$. Let $\psi = g(A)\varphi$. Then since $g(x) \neq 0$ for $d\mu_\varphi^{(A)}$ a.e. x, we have $\mathcal{H}_\psi^{(A)} = \mathcal{H}_\varphi^{(A)}$. Since $d\mu_\psi^{(A)}(x) = g(x)^2\,d\mu_\varphi^{(A)}(x)$, we have $d\mu_\varphi^{(A)} \le d\mu$. □

Proof of the Existence Half of Theorem 5.4.3. Let $\{\eta_\ell\}_{\ell=1}^\infty$ be an orthonormal basis. By Lemma 5.4.7, pick a dominant vector φ_1 so $\eta_1 \in \mathcal{H}_{\varphi_1}^{(A)}$. Let $d\nu_1 = d\mu_{\varphi_1}^{(A)}$ and P_1 the projection onto $(\mathcal{H}_{\varphi_1}^{(A)})^\perp$.

We can now pick $\varphi_1, \dots, \varphi_n$ inductively so that with $d\nu_j = d\mu_{\varphi_j}^{(A)}$, we have

$$\begin{aligned} &\mathcal{H}_{\varphi_j}^{(A)} \perp \mathcal{H}_{\varphi_{j+1}}^{(A)}, \quad j = 1, 2, \dots, n-1; \\ &\eta_\ell \in \bigoplus_{k=1}^{\ell} \mathcal{H}_{\varphi_j}^{(A)}; \qquad d\nu_j \perp d\nu_{j+1}, \quad j = 1, \dots, n-1 \end{aligned} \tag{5.4.22}$$

For given $\varphi_1, \dots, \varphi_{n-1}$ with those properties, let P_{n-1} be the projection onto $(\oplus_{\ell=1}^{n-1} \mathcal{H}_{\varphi_\ell}^{(A)})^\perp$, let $\psi_n = P_{n-1}\eta_n$ and $A_n = A \restriction \operatorname{Ran}(P_{n-1}) = \mathcal{H}_{n-1}$. By Lemma 5.4.7, pick $\kappa_1 \in \mathcal{H}$ dominant for A_n so $\psi_n \in \mathcal{H}_{\kappa_n}^{(A)}$. Then by Lemma 5.4.8, pick φ_n so $d\nu_n \equiv d\mu_{\varphi_n}^{(A)} \le d\nu_{n-1}$ and $\mathcal{H}_{\varphi_n}^{(A)} = \mathcal{H}_{\kappa_n}^{(A)}$. By construction, the first and third parts of (5.4.22) hold. Since $(1 - P_{n-1})\eta_n \in \oplus_{j=1}^{n-1} \mathcal{H}_{\varphi_j}^{(A)}$ and $P_{n-1}\eta_n \in \mathcal{H}_{\varphi_n}^{(A)}$, the second assertion holds.

Since $\eta_j \in \oplus_{\ell=1}^{\infty} \mathcal{H}_{\varphi_\ell}^{(A)}$ for all j, we see that the direct sum in $\mathcal{H}$ and $A \restriction \mathcal{H}_{\varphi_\ell}^{(A)}$ is M_{ν_ℓ}. This proves existence. $\square$

We now turn to the proof of uniqueness. If $\mathcal{F}$ is a family of operators, its *commutant*, written $\mathcal{F}'$, is defined by

$$\mathcal{F}' = \{C \in \mathcal{L}(\mathcal{H}) \mid CA = AC \text{ for all } A \in \mathcal{F}\}$$

If $\mathcal{F}^* = \mathcal{F}$, then $\mathcal{F}'$ is a weakly closed $*$-algebra. The following will be useful also for discussing general commutants:

Lemma 5.4.9. *Let μ be a Baire measure of bounded support. Then*

$$\{M_\mu\}' = \{f(M_\mu) \mid f \in \mathcal{B}(\mathbb{R})\} \tag{5.4.23}$$

In particular, each B in $\{M_\mu\}'$ is normal.

Remark. If A is self-adjoint but doesn't have a cyclic vector, that is, isn't a M_μ, we'll show (see Corollary 5.4.16) that there are nonnormal $B \in \{A\}'$.

Proof. Clearly, $f(M_\mu) \in \{M_\mu\}'$, so we need only show that if $BM_\mu = M_\mu B$, then $B = f(M_\mu)$ for some f in $\mathcal{B}(\mathbb{R})$, that is, as an operator on $L^2(\mathbb{R}, d\mu)$, B is multiplication by $f(x)$ for f in $L^\infty(\mathbb{R})$.

We know $B\mathbb{1} = f(x)$ for some $f \in L^2$. Since $B \in \{M_\mu\}'$, $Bx^n = x^n f(x)$. By the Weierstrass approximation theorem, $Bg(x) = f(x)g(x)$ for any $g \in C(\operatorname{supp}(\mu))$. By taking strong limits, this holds for any $g \in \mathcal{B}(\mathbb{R})$. Fix ε. Let $g = \chi_\Omega$, where $\Omega = \{x \mid |f(x)| > (\|B\| + \varepsilon)\}$. Then $\|Bg\| \le \|B\| \mu(\Omega)^{1/2}$, while $\|fg\| \ge (\|B\| + \varepsilon)\mu(\Omega)^{1/2}$. It follows that $\mu(\Omega) = 0$, that is, for μ a.e. x, $|f(x)| \le \|B\|$. $\square$

We introduce $A \cong B$ for operators A, B on $\mathcal{H}_A$, $\mathcal{H}_B$ if there is a unitary map U of $\mathcal{H}_A$ onto $\mathcal{H}_B$ so $UAU^{-1} = B$.

Lemma 5.4.10. *Let A, B be self-adjoint operators on $\mathcal{H}_A$, $\mathcal{H}_B$. Let $C\colon \mathcal{H}_A \to \mathcal{H}_B$ so that*

$$CA = BC \tag{5.4.24}$$

Then

(a) *A leaves* $\operatorname{Ker}(C)$ *and* $\operatorname{Ker}(C)^\perp$ *invariant. B leaves* $\operatorname{Ker}(C^*)$ *and* $\operatorname{Ker}(C^*)^\perp$ *invariant.*
(b) $A \restriction \operatorname{Ker}(C)^\perp \cong B \restriction \operatorname{Ker}(C^*)^\perp$.

Proof. (a) If $C\varphi = 0$, then $CA\varphi = BC\varphi = 0$, so $C(A\varphi) = 0$, that is, A leaves $\operatorname{Ker}(C)$ invariant. Since A is self-adjoint, it leaves $\operatorname{Ker}(C)^\perp$ invariant. Since (5.4.24) implies

$$AC^* = C^*B \tag{5.4.25}$$

we have, by the same argument, that $\operatorname{Ker}(C^*)$ and so $\operatorname{Ker}(C^*)^\perp$.

(b) By (5.4.24) and (5.4.25), we have

$$AC^*C = C^*BC = C^*CA \tag{5.4.26}$$

By (2.4.20) (C and A have reversed roles!), we have

$$A|C| = |C|A \tag{5.4.27}$$

By Lemma 2.4.10 (which also follows from the spectral theorem for $|C|$), if $P =$ projection onto $\operatorname{Ker}(|C|)^\perp = \operatorname{Ker}(C)^\perp$, we have $\text{s-lim}|C|(|C|+\varepsilon)^{-1} = P$, so if $C = U|C|$, we see

$$CA(|C|+\varepsilon)^{-1} = U|C|(|C|+\varepsilon)^{-1}A \to U(AP) \tag{5.4.28}$$

By (5.4.24),

$$CA(|C|+\varepsilon)^{-1} = BU|C|(|C|+\varepsilon)^{-1} \to BUP \tag{5.4.29}$$

If Q is the projection onto $\operatorname{Ker}(C^*)^\perp$, since $|C^*|U = U|C|$ (see (2.4.48)), we have $QU = UP$, so (5.4.28), (5.4.29) imply

$$U(AP) = (BQ)U \tag{5.4.30}$$

Since U is unitary from $\operatorname{Ran}(P)$ to $\operatorname{Ran}(Q)$, we see that

$$A \restriction \operatorname{Ran}(P) \cong B \restriction \operatorname{Ran}(Q) \tag{5.4.31}$$

$\square$

Lemma 5.4.11. *If N is normal, then*

$$\operatorname{Ker}(N) = \operatorname{Ker}(N^*), \qquad \overline{\operatorname{Ran}(N)} = \overline{\operatorname{Ran}(N^*)} \tag{5.4.32}$$

Proof.

$$\begin{aligned}\varphi \in \operatorname{Ker}(N) &\Leftrightarrow \|N\varphi\| = 0\\ &\Leftrightarrow \langle \varphi, N^*N\varphi\rangle = 0\\ &\Leftrightarrow \langle \varphi, NN^*\varphi\rangle = 0\\ &\Leftrightarrow \|N^*\varphi\| = 0\\ &\Leftrightarrow \varphi \in \operatorname{Ker}(N^*)\end{aligned}$$

proving the first assertion. This implies the second since (see (2.2.13))

$$\overline{\operatorname{Ran}(N)} = \operatorname{Ker}(N^*)^\perp = \operatorname{Ker}(N)^\perp = \overline{\operatorname{Ran}(N^*)} \tag{5.4.33}$$

$\square$

The key to uniqueness is:

Theorem 5.4.12. *Let μ be a measure on $\mathbb{R}$ of bounded support and let A, B be self-adjoint operators on Hilbert spaces $\mathcal{H}_A$ and $\mathcal{H}_B$. Then*

$$M_\mu \oplus A \cong M_\mu \oplus B \Rightarrow A \cong B \tag{5.4.34}$$

Remarks. 1. By an induction, M_μ can be replaced by an operator of fixed finite multiplicity.

2. If M_μ is replaced by an operator of infinite multiplicity, the analog of this result may be false. If $0^{(n)}$ is the zero operator on $\mathbb{C}^n$ (ℓ^2 if $n=\infty$), then $0^{(\infty)} \oplus 0^{(\infty)} \cong 0^{(\infty)} \oplus 0^{(1)}$ but $0^{(\infty)}$ is not unitarily equivalent to $0^{(1)}$.

Proof. Let $V\colon L^2(\mathbb{R}, d\mu) \oplus \mathcal{H}_A \to L^2(\mathbb{R}, d\mu) \oplus \mathcal{H}_B$ be unitary so that

$$V(M_\mu \oplus A) = (M_\mu \oplus B)V \tag{5.4.35}$$

Write

$$V = \begin{pmatrix} V_{11} & V_{12} \\ V_{21} & V_{22} \end{pmatrix} \tag{5.4.36}$$

according to the direct sum decomposition theorem. Then (5.4.35) becomes

$$V_{22}A = BV_{22}, \quad V_{12}A = M_\mu V_{12}, \quad V_{21}M_\mu = BV_{21}, \quad V_{11}M_\mu = M_\mu V_{11} \tag{5.4.37}$$

By Lemma 5.4.10, $\text{Ker}(V_{22})^\perp$ (respectively, $\text{Ker}(V_{22}^*)^\perp$) is invariant for A (respectively, B) and

$$A \restriction \text{Ker}(V_{22})^\perp \cong B \restriction \text{Ker}(V_{22}^*)^\perp \tag{5.4.38}$$

so it suffices to prove that

$$A \restriction \text{Ker}(V_{22}) \cong B \restriction \text{Ker}(V_{22}^*) \tag{5.4.39}$$

Since V is unitary, we have

$$V_{12}^*V_{12} + V_{22}^*V_{22} = \mathbb{1}_{\mathcal{H}_A}, \qquad V_{21}V_{21}^* + V_{22}V_{22}^* = \mathbb{1}_{\mathcal{H}_B} \tag{5.4.40}$$

$$V_{11}V_{11}^* + V_{12}V_{12}^* = \mathbb{1}_{L^2(\mathbb{R},d\mu)}, \quad V_{21}V_{11}^* + V_{22}V_{12}^* = 0 \tag{5.4.41}$$

Let

$$\mathcal{R}_1 = V_{12}[\text{Ker}(V_{22})], \qquad \mathcal{R}_2 = V_{21}^*[\text{Ker}(V_{22}^*)] \tag{5.4.42}$$

By the first equation in (5.4.40), if $\varphi \in \mathcal{H}_A$, then

$$\|V_{12}\varphi\|^2 + \|V_{22}\varphi\|^2 = \|\varphi\|^2 \tag{5.4.43}$$

so V_{12} is an isometry on $\text{Ker}(V_{22})$ and so a unitary map of that space to $\mathcal{R}_1$. Thus, by the second equation in (5.4.37),

$$A \restriction \text{Ker}(V_{22}) \cong M_\mu \restriction \mathcal{R}_1 \tag{5.4.44}$$

Similarly, using the second equation in (5.4.40),

$$B \restriction \text{Ker}(V_{22}^*) \cong M_\mu \restriction \mathcal{R}_2 \tag{5.4.45}$$

so (5.4.39) is implied by

$$\mathcal{R}_1 = \mathcal{R}_2 \tag{5.4.46}$$

Suppose we prove that

$$\mathcal{R}_1 = \text{Ker}(V_{11}^*), \qquad \mathcal{R}_2 = \text{Ker}(V_{11}) \tag{5.4.47}$$

By the last equation in (5.4.37), V_{11} commutes with M_μ, so by Lemma 5.4.9, V_{11} is normal, and then by Lemma 5.4.11, we get (5.4.46). Thus, we need only prove (5.4.47) to get (5.4.39). We'll prove the first assertion in (5.4.47); the second is similar.

Suppose $g \in \mathcal{R}_1$, that is, $g = V_{12}\varphi$ for some $\varphi \in \operatorname{Ker}(V_{22})$. Since $V_{12} \restriction \operatorname{Ker}(V_{22})$ is unitary,

$$\|g\| = \|\varphi\| \tag{5.4.48}$$

By the first equation in (5.4.40), $V_{12}^*V_{12}\varphi = \varphi \Rightarrow V_{12}^*g = \varphi$, so by (5.4.48),

$$\|V_{12}^*g\| = \|g\| \tag{5.4.49}$$

By the first equation in (5.4.41), for any $g \in L^2(M, d\mu)$,

$$\|V_{12}^*g\|^2 + \|V_{11}^*g\|^2 = \|g\|^2 \tag{5.4.50}$$

so (5.4.49) implies $g \in \operatorname{Ker}(V_{11}^*)$, that is, we have proven that $\mathcal{R}_1 \subset \operatorname{Ker}(V_{11}^*)$.

On the other hand, if $g \in L^2(\mathbb{R}, d\mu)$ and $V_{11}^*g = 0$, then by (5.4.41),

$$V_{22}V_{12}^*g = -V_{21}V_{11}^*g = 0 \tag{5.4.51}$$

so $\varphi \equiv V_{12}^*g \in \operatorname{Ker}(V_{22})$. But the first equation in (5.4.41) and $V_{11}^*g = 0$ then implies

$$g = V_{12}\varphi \in \mathcal{R}_1 \tag{5.4.52}$$

that is, we have proven

$$\operatorname{Ker}(V_{11}^*) \subset \mathcal{R}_1 \tag{5.4.53}$$

□

Proof of the Uniqueness Half of Theorem 5.4.4. Suppose

$$A \equiv \bigoplus_{j=1}^{\infty} M_{\nu_j} \cong \bigoplus_{j=1}^{\infty} M_{\tilde{\nu}_j} \tag{5.4.54}$$

with $\nu_j \geq \nu_{j+1}$ and $\tilde{\nu}_j \perp \tilde{\nu}_{j+1}$. Then ν_1 and $\tilde{\nu}_1$ are both dominant measures for A and so each is a.c. wrt the other, that is, $\nu_1 \sim \tilde{\nu}_1$. By Theorem 5.4.1, $M_{\tilde{\nu}_1} \cong M_{\nu_1}$, so

$$M_{\nu_1} \oplus \bigoplus_{j=2}^{\infty} M_{\nu_j} \cong M_{\nu_1} \oplus \bigoplus_{j=2}^{\infty} M_{\tilde{\nu}_j} \tag{5.4.55}$$

By Theorem 5.4.12,

$$\bigoplus_{j=2}^{\infty} M_{\nu_j} \cong \bigoplus_{j=2}^{\infty} M_{\tilde{\nu}_j} \tag{5.4.56}$$

By iteration, $\tilde{\nu}_j \sim \nu_j$. □

Finally, we turn to the calculation of $\{A\}'$ for any bounded self-adjoint A. Recall $M_\mu^{(\ell)}$ is multiplication by x on $L^2(\mathbb{R}, d\mu; \mathbb{C}^\ell) \equiv \mathcal{H}_\ell$ (where $\mathbb{C}^\infty$ is interpreted as ℓ^2). Given $Q \in L^\infty(\mathbb{R}, d\mu; \mathrm{Hom}(\mathbb{C}^\ell))$, that is, $Q(x)$ is a bounded operator of $\mathbb{C}^\ell$ to itself so $\sup_x \|Q(x)\| < \infty$, we define an operator, N_Q, by

$$(N_Q f)(x) = Q(x) f(x) \tag{5.4.57}$$

Theorem 5.4.13.

$$\{M_\mu^{(\ell)}\}' = \{N_Q \mid Q \in L^\infty(\mathbb{R}, d\mu; \mathcal{L}(\mathbb{C}^\ell))\} \tag{5.4.58}$$

Remark. Lemma 5.4.9 is the $\ell = 1$ case and the proof is similar.

Proof. Clearly, each N_Q commutes with $M_\mu^{(\ell)}$. For the converse, let $B \in \{M_\mu^{(\ell)}\}$. Let $\{\delta_j\}_{j=1}^\ell$ be the usual basis for $\mathbb{C}^\ell$ and let $\mathbb{1}_j$, $j < \ell + 1$, be the function

$$\mathbb{1}_j(x) = \delta_j \quad \text{for all } x \tag{5.4.59}$$

Define

$$Q_{kj}(x) = (B\mathbb{1}_j)_k(x) \tag{5.4.60}$$

Then if g is a polynomial in x in each component (and if $\ell = \infty$ has only finitely many nonzero components), we have

$$(Bg)(x) = (N_Q g)(x) \tag{5.4.61}$$

By taking norm and then L^2-limits, this extends to all $g \in L^\infty(\mathbb{R}, d\mu; \mathbb{C}^\ell)$. That $\|N_Q g\|_2 \le \|B\| \|g\|_2$ for all such bounded Borel functions proves that for a.e. x, $\|Q(x)\| \le \|B\|$. $\square$

Lemma 5.4.14. *Let A, C be self-adjoint operators on Hilbert spaces $\mathcal{H}_A$ and $\mathcal{H}_C$. Let $B\colon \mathcal{H}_A \to \mathcal{H}_C$ obey*

$$BA = CB \tag{5.4.62}$$

Let μ_A, μ_C be dominant measures for A and C. Then

$$\mu_A \perp \mu_C \Rightarrow B = 0 \tag{5.4.63}$$

Proof. If $\mathrm{Ker}(B)^\perp \neq \{0\}$, by Lemma 5.4.10, there is $\varphi \in \mathcal{H}_A$, $\varphi \neq 0$, and $U\varphi \neq 0$, where $B = U|B|$ is the polar decomposition of B, so

$$d\mu_\varphi^{(A)} = d\mu_{U\varphi}^{(C)} \neq 0 \tag{5.4.64}$$

Since $d\mu_\varphi^{(A)} \le d\mu_A$ and $d\mu_\varphi^{(C)} \le d\mu_C$, $d\mu_A \perp d\mu_C \Rightarrow d\mu_\varphi^{(A)} = 0$, contradicting (5.4.64). $\square$

Theorem 5.4.15. *Let A be a self-adjoint operator with $A \cong \oplus_{j=1}^{\infty} M_{\mu_j}^{(j)}$ with $\mu_j \perp \mu_k$ for $j \neq k$. Then $\{A\}'$ is described by*

$$\left\{ \bigoplus_{j=1}^{\infty} M_{\mu_j}^{(j)} \right\}' = \bigoplus_{j=1}^{\infty} L^{\infty}(\mathbb{R}, d\mu_j; \mathcal{L}(\mathbb{C}^j)) \tag{5.4.65}$$

Remark. The direct sum on the right of (5.4.65) is the L^∞-direct sum, that is, $\{Q_j\}_{j=1}^{\infty}$ with $Q_j(x) \in \mathcal{L}(\mathbb{C}^j)$ and

$$\|\{Q_j\}\| \equiv \sup_j \| \|Q_j(x)\|_{\mathcal{L}(\mathbb{C}^j)} \|_{L^\infty(\mathbb{R}, d\mu_j)} < \infty \tag{5.4.66}$$

Proof. If $B \in \{\oplus_{j=1}^{\infty} M_{\mu_j}^{(j)}\}'$, then B has matrix elements $B_{kj}\colon L^2(\mathbb{R}, d\mu_j; \mathbb{C}^j) \to L^2(\mathbb{R}, d\mu_k; \mathbb{C}^k)$ with

$$M_{\mu_k}^{(k)} B_{kj} = B_{kj} M_{\mu_j}^{(j)} \tag{5.4.67}$$

Since $\mu_j \perp \mu_k$ if $j \neq k$, $B_{kj} = 0$ for such (j,k) by Lemma 5.4.14. By Theorem 5.4.13, $B_{jj} = N_{Q_{jj}}$ with $B_{jj} \in L^\infty(\mathbb{R}, d\mu_j; \mathcal{L}(\mathbb{C}^j))$. It is easy to see $\|\{Q_j\}\|$, given by (5.4.66), is given by $\|B\|$. $\square$

Corollary 5.4.16. *Let A be a bounded self-adjoint operator. Then the following are equivalent:*

(1) *A is simple.*
(2) *$\{A\}'$ is abelian.*
(3) *$\{A\}'$ has only normal operators.*

Proof. Immediate, given that $\mathcal{L}(\mathbb{C}^j)$ is nonabelian and is nonnormal for $j \geq 2$ and is abelian for $j = 1$. $\square$

Notes and Historical Remarks.

> ... multiplicity theory is a magnificent measure-theoretic tour de force ... Although the material is complicated, it is subdivided into pleasant compact packages which the reader absorbs as he proceeds on his journey. The author meanwhile is a veritable ringmaster who marshals his troupe of techniques, cracking the whip over each aspect of the theory in turn, thus keeping it at the proper pitch for its part.
>
> —*E. R. Lorch* [**458**][2]

Multiplicity theory for bounded self-adjoint operators within Hilbert's framework goes back to the 1907 dissertation of Hilbert's student, Ernst Hellinger (1883–1950) [**309**] and was brought to a fairly complete state by a second paper of Hellinger [**310**] and a paper of Hahn [**279**]. It is often called the Hahn–Hellinger theory.

Starting in the late 1930s, the theory was extended to nonseparable Hilbert spaces by Wecken [**734**], Nakano [**493**], and Plesner–Rohlin [**529**]

[2] in his review of Halmos' book on multiplicity theory [**282**].

and their results were codified in a small book of Halmos [**282**]. Many results in Hilbert space theory extend from the separable to the general case fairly easily—a Zornification normally suffices but this is not true for multiplicity theory. You now have pieces of multiplicity m for each infinite cardinal m and because the number of cardinals is not countable, you have to overcome considerable measure-theoretic difficulties.

For those who glory in pyrotechnics for their own sake, this extension is fascinating. But you need to bear in mind that while an analyst can't go out for a stroll without tripping over a separable Hilbert space, in my career, I've never come across a nonseparable Hilbert space in the wild. For me, this situation justifies my decision to (mainly) stick to the separable case. It is an unfortunate byproduct of this situation that multiplicity theory has a reputation of being very difficult, while, as we've seen, it is fairly straightforward in the separable case.

Another textbook presentation of multiplicity theory (in the separable case) is Arveson [**30**].

Traditionally, multiplicity theory focusses on the spectral resolution. The formulation in terms of measure classes and the notion that measure classes are a sort of generalized subset is due to Mackey [**460**].

This section in discussing commutants peeks at the theory of operator algebras on Hilbert spaces, a subject with vast literature; for example [**30, 71, 85, 86, 126, 127, 128, 129, 130, 145, 162, 163, 361, 362, 370, 444, 485, 516, 587, 677, 692, 775**] is a list of just some of the books on the subject.

One first result that sets the stage is the celebrated double commutant theorem of von Neumann [**722**].

Theorem 5.4.17 (von Neumann's Double Commutant Theorem). *Let $\mathfrak{A}$ be a star algebra with unit in a Hilbert space, $\mathcal{H}$ (i.e., $\mathfrak{A}$ is a set of bounded operators on $\mathcal{H}$ which is closed under $(A,B) \to A+B$, AB and $A \to A^*$, λA for all $\lambda \in \mathbb{C}$). Then the following sets are identical: $\overline{\mathfrak{A}}^s$, $\overline{\mathfrak{A}}^w$, $\mathfrak{A}''$ where the first two are the closures in the strong and weak operator topology.*

The reader will prove this in Problem 6. A stronger result, known as a Kaplansky density theorem (after Kaplansky [**369**]) states that for each $A \in \mathfrak{A}''$, not only can one find a net $\{B_\alpha\}_{\alpha\in\mathcal{I}}$ so $B_\alpha \to A$ strongly, one can find one with $\sup_\alpha \|B_\alpha\| \le \|A\|$.

Problems

1. Fill in the details of the proof of (5.4.10)–(5.4.12).

2. Prove that uniqueness of $\{\nu_j\}_{j=1}^\infty$ in Theorem 5.4.4 implies uniqueness of $\{\mu_j\}_{j=1}^\infty$ in Theorem 5.4.5. (*Hint*: Show the Ω_j are determined up to sets of ν_1 measure 0.)

3. (a) If $f \in \mathcal{B}(\mathbb{R})$, prove that $f(A)\varphi = 0 \Leftrightarrow \int |f(x)|^2\, d\mu_\varphi^{(A)} = 0$.

   (b) Prove that $f(A)\varphi = 0 \Rightarrow f(A) = 0$ if and only if φ is dominant.

4. (a) Prove that the vector φ of (5.4.18) is a dominant vector. (*Hint*: Compute $d\mu_\varphi^{(A)}$.)

   (b) If φ is given by (5.4.18), prove that $P_{\mathbb{R}\setminus\Omega}(A)\varphi = \psi$ and conclude that $\psi \in \mathcal{H}_\varphi^{(A)}$.

5. If A is self-adjoint, compute $\{A\}''$. (*Hint*: Use Theorem 5.4.15.)

6. This problem will lead the reader through a proof of *von Neumann's double commutant theorem* (aka *von Neumann's density theorem*), Theorem 5.4.17.

   (a) Prove that any set closed in the weak topology is closed in the strong topology so that $\overline{\mathfrak{A}}^s \subset \overline{\mathfrak{A}}^w$.

   (b) Prove that the double commutant $\mathfrak{A}'' = (\mathfrak{A}')'$ is weakly closed (indeed $\mathfrak{A}'$ is weakly closed) and conclude that $\overline{\mathfrak{A}}^w \subset \mathfrak{A}''$. Thus the key to proving the theorem is to show for any $A \in \mathfrak{A}''$, there is a set $\{A_\alpha\}_{\alpha\in\mathcal{I}}$ in $\mathfrak{A}$ with $A_\alpha \to A$ in the strong operator topology.

   (c) Prove it suffices to prove such a net exists for $A = A^*$.

   (d) As an intermediate step, the reader will show for a fixed $\varphi \in \mathcal{H}$ and $A \in \mathfrak{A}''$, one can find $A_m \in \mathfrak{A}$, $m = 1, 2, \ldots$ so that $A_m\varphi \to A\varphi$. Given such a φ, let $\mathcal{K} = \overline{\{B\varphi \mid B \in \mathfrak{A}\}}$ and P the orthogonal onto $\mathcal{K}$. Prove that for any $B \in \mathfrak{A}$, $BP = PBP$ and conclude that $BP = PB$ so that $P \in \mathfrak{A}'$ and that $AP = PA$.

   (e) Prove that $A\varphi \in \mathcal{K}$, so there is $A_m \in \mathfrak{A}$ with $A_m\varphi \to A\varphi$.

   (f) For each n, define $\mathfrak{A}^{(n)}$, a set of operators on $\overset{n}{\underset{j=1}{\oplus}} \mathcal{H}_j$, by $\mathfrak{A}^{(n)} = \{B \oplus \ldots \oplus B \mid B \in \mathfrak{A}\}$. If $A \in \mathfrak{A}''$, prove that $\overset{n}{\underset{j=1}{\oplus}} A \in \left(\mathfrak{A}^{(n)}\right)''$.

   (g) Prove that for any fixed $A \in \mathfrak{A}''$ and $\varphi_1, \ldots, \varphi_n \in \mathcal{H}$, there are $A_m \in \mathfrak{A}$ so $A_m\varphi_j \to A\varphi_j$ for $j = 1, \ldots, n$. (*Hint*: Apply (e) to $\mathfrak{A}^{(n)}$.)

   (h) Find the required net.

7. Extend Theorem 5.4.12 to the case where M_μ is replaced by a finite direct sum of M_{μ_j}'s (i.e., by a general operator with bounded multiplicity).

5.5. Bonus Section: The Spectral Theorem for Unitary Operators

In this section, we'll prove the spectral theorem for unitary operators. Since we'll prove the spectral theorem for normal operators in the next section and unitary operators are normal, this section is overkill—which is why it is a bonus section. After the proof, we discuss some applications relevant to ergodic theory. We'll focus on the following form:

Theorem 5.5.1. *Let U be a unitary operator on a Hilbert space, $\mathcal{H}$, with a cyclic vector, φ. Then there exists a measure, $d\mu_\varphi^{(U)}$, on $\partial\mathbb{D}$ so that for $n = 0, \pm 1, \pm 2, \dots,$*

$$\langle \varphi, U^n \varphi \rangle = \int e^{in\theta}\, d\mu_\varphi^{(U)}(\theta) \tag{5.5.1}$$

Remarks. 1. Since $U^{-1} = U^*$, $\langle \varphi, U^{-n}\varphi\rangle = \overline{\langle \varphi, U^n\varphi\rangle}$ and, of course, $\int e^{-in\theta}\, d\mu(\theta) = \overline{\int e^{in\theta}\, d\mu(\theta)}$, so (5.5.1) for $n \ge 0$ implies it for $n < 0$.

2. The existence of the measure and (5.5.1) hold even if φ is not cyclic since one can restrict to the cyclic subspace.

As in Section 5.1, once we have Theorem 5.5.1, we can construct, using cyclicity of φ, a unitary map $V\colon \mathcal{H} \to L^2(\partial\mathbb{D}, d\mu_\varphi^{(U)})$ with $VUV^{-1} =$ multiplication by $e^{i\theta}$ (and then $VU^*V^{-1} = (VUV^{-1})^* =$ multiplication by $e^{-i\theta}$). Once one has that, one can deduce analogs of the other variants of the spectral theorem: Resolutions of the identity have support $[0, 2\pi]$ and $U = \int e^{i\theta}\, dE_\theta$.

We'll prove Theorem 5.5.1 by viewing (5.5.1) as a moment problem. We'll find necessary and sufficient conditions for $\{c_n\}_{n=-\infty}^\infty$ to be the moments of a measure on $\partial\mathbb{D}$ and then verify that $\langle \varphi, U^n\varphi\rangle$ obey the conditions. Our approach to solving the moment problem on $\partial\mathbb{D}$ is related to the one for $[0,\infty)$ or $[-1,1]$ or $\mathbb{R}$ (see Sections 4.17 and 5.6 of Part 1), only simpler (compare Theorem 5.5.2 below to Lemma 5.6.4 of Part 1).

We start by defining a *Laurent polynomial* as a finite linear combination of $\{z^k\}_{k=-\infty}^\infty$ viewed as an analytic function on $\mathbb{C}\setminus\{0\}$. If

$$f(z) = \sum_{n=k}^{\ell} \alpha_n z^n \tag{5.5.2}$$

with $\alpha_\ell \neq 0 \neq \alpha_k$, we say $\ell - k$ is the *degree* of f. If

$$Q(z) = z^{-k} f(z) \tag{5.5.3}$$

then Q is a polynomial of exact degree $\ell - k$ with $p(0) \neq 0$, so f has precisely $\ell - k$ zeros (counting multiplicity) on $\mathbb{C} \setminus \{0\}$. If f is real on $\partial\mathbb{D}$, we claim

$$f\left(\frac{1}{z}\right) = \overline{f(\bar{z})} \tag{5.5.4}$$

since both sides are analytic on $\mathbb{C} \setminus \{0\}$ and they are equal on $\partial\mathbb{D}$. It follows that in (5.5.2), $\alpha_{-n} = \overline{\alpha_n}$, so $k = -\ell$ with $\ell \geq 0$. In particular, $\deg(f) = \ell - k = 2\ell$ is even.

Theorem 5.5.2 (Fejér–Riesz Theorem). *If f is a Laurent polynomial with $f(e^{i\theta}) \geq 0$ for all $e^{i\theta} \in \partial\mathbb{D}$, then there is a polynomial P nonvanishing on $\mathbb{D}$ so that*

$$f(z) = P(z)\,\overline{P(1/\bar{z})} \tag{5.5.5}$$

In particular,

$$f(e^{i\theta}) = |P(e^{i\theta})|^2 \tag{5.5.6}$$

Remarks. 1. It can be seen (Problem 1) that P is unique up to an overall multiplicative constant but only because we have limited the location of its zeros. When $f \geq 0$ on $\partial\mathbb{D}$, there is even an explicit formula for P on $\mathbb{D}$ in terms of $f(e^{i\theta})$ (Problem 2).

2. Conversely, any f obeying (5.5.6) is clearly nonnegative on $\partial\mathbb{D}$.

This theorem was also proved and used in Section 4.6 of Part 3.

Proof. By the above, f has degree 2ℓ for some ℓ, and so 2ℓ zeros in $\mathbb{C} \setminus \{0\}$. Since $f(e^{i\theta})$ is smooth and nonnegative on $\partial\mathbb{D}$, it has zeros of even order on $\partial\mathbb{D}$. It is easy to see that means Q, given by (5.5.3), has zeros of even order on $\partial\mathbb{D}$. By (5.5.4), if $z_0 \in \mathbb{C} \setminus \overline{\mathbb{D}}$ has $f(z_0) = 0$, then $f(1/\bar{z}_0) = 0$ also. Thus, the zeros of Q are $z_1, \dots, z_\ell \in \mathbb{C} \setminus \mathbb{D}$ and $\{1/\bar{z}_j\}_{j=1}^{\ell}$, that is,

$$Q(z) = c\prod_{j=1}^{\ell}(z - z_j)(z - \bar{z}_j^{-1}) = c\prod_{j=1}^{\ell}(-\bar{z}_j^{-1})\prod_{j=1}^{\ell}(z - z_j)(1 - \bar{z}_j z) \tag{5.5.7}$$

so (note that $k = -\ell$)

$$f(z) = z^{-\ell}Q(z) = c\prod_{j=1}^{\ell}(-\bar{z}_j^{-1})\prod_{j=1}^{\ell}(z - z_j)\left(\frac{1}{z} - \bar{z}_j\right) \tag{5.5.8}$$

Let

$$P(z) = \left[c\prod_{j=1}^{\ell}(-\bar{z}_j^{-1})\right]^{1/2}\prod_{j=1}^{\ell}(z - z_j) \tag{5.5.9}$$

Then

$$f(z) = e^{i\psi}P(z)\,\overline{P(1/\bar{z})} \tag{5.5.10}$$

with

$$e^{i\psi} = \left[c\prod_{j=1}^{\ell}(-\bar{z}_j)^{-1}\right]^{1/2}\left[\bar{c}\prod_{j=1}(-z_j)^{-1}\right]^{-1/2} \tag{5.5.11}$$

Since f and $P(z)\overline{P(1/\bar{z})}$ are nonnegative on $\partial\mathbb{D}$, $e^{i\psi} = 1$. $\square$

The moments of a measure on $\partial\mathbb{D}$ use its discrete Fourier transform, so the following result on the moment problem for $\partial\mathbb{D}$ can be regarded as a discrete version of Bochner's theorem (see Theorem 6.6.6 of Part 1).

Definition. A two-sided sequence $\{c_n\}_{n=-\infty}^{\infty}$ is called *positive definite* if and only if for all $\{\zeta_j\}_{j=1}^{\ell}$ in $\mathbb{C}^\ell$ and all ℓ,

$$\sum_{j,k} \bar{\zeta}_j \zeta_k c_{k-j} \geq 0 \tag{5.5.12}$$

Theorem 5.5.3 (Carathéodory–Toeplitz Theorem). *A sequence $\{c_n\}_{n=-\infty}^{\infty}$ are the moments of a measure, μ, on $\partial\mathbb{D}$, that is,*

$$c_n = \int e^{in\theta}\, d\mu(\theta) \tag{5.5.13}$$

for a positive measure, μ, on $\partial\mathbb{D}$ if and only if it is positive definite.

Proof. It is immediate that (5.5.13) implies (5.5.12), so we need only prove the converse. So suppose (5.5.12) for all $\{\zeta_j\}_{j=1}^{\ell}$. Define a linear functional, L, on the set of Laurent polynomials by

$$L\biggl(\sum_{j=k}^{\ell} \alpha_j z^j\biggr) \equiv \sum_{j=k}^{\ell} \alpha_j c_j \tag{5.5.14}$$

If $f(z)$ has the form (5.5.5) with

$$P(z) = \sum_{j=1}^{\ell} \zeta_j z^{j-1} \tag{5.5.15}$$

then

$$f(z) = \sum_{j,k} \bar{\zeta}_j \zeta_k z^{k-j} \tag{5.5.16}$$

so

$$L(f) = \sum_{j,k} \bar{\zeta}_j \zeta_k c_{k-j} \geq 0 \tag{5.5.17}$$

by (5.5.12). Thus, by the Fejér–Riesz theorem, L is a positive linear functional on the Laurent polynomials. In particular, for any Laurent polynomial which is real on $\partial\mathbb{D}$,

$$L(\|f\|_\infty \pm f) \geq 0 \Rightarrow |L(f)| \leq \|f\|_\infty c_0 \tag{5.5.18}$$

Thus, since Laurent polynomials are dense in $C(\partial\mathbb{D})$ (by Weierstrass' second approximation theorem, Theorem 2.4.2 of Part 1), L defines a positive linear functional and so a measure $d\mu$ (by the Riesz–Markov theorem, Theorem 4.5.4 of Part 1) with

$$L(f) = \int f(e^{i\theta})\, d\mu(\theta) \tag{5.5.19}$$

Taking $f(z) = z^n$, we get (5.5.13). □

Proof of Theorem 5.5.1. Let

$$c_n = \langle \varphi, U^n \varphi \rangle \tag{5.5.20}$$

Then, since $U^{-1} = U^*$,

$$\sum_{j,k} \bar{\zeta}_j \zeta_k c_{k-j} = \sum_{j,k} \bar{\zeta}_j \zeta_k \langle U^j \varphi, U^k \varphi \rangle = \left\| \sum_k \zeta_k U^k \varphi \right\|^2 \geq 0 \tag{5.5.21}$$

so, by Theorem 5.5.3, there is a measure obeying (5.5.1). □

The following is equivalent to the von Neumann ergodic theorem (Theorem 2.6.7 of Part 3):

Theorem 5.5.4. *Let U be a unitary operator and let P be the projection onto $\{\varphi \mid U\varphi = \varphi\}$. Then*

$$\operatorname*{s-lim}_{N\to\infty} \frac{1}{N} \sum_{j=0}^{N-1} U^j = P \tag{5.5.22}$$

Remarks. 1. This applies directly only in case the underlying dynamical system is invertible. One needs the same result when U is only an isometry. But it can be shown (Problem 3) that given such a U, there is W unitary on $\widetilde{\mathcal{H}}$ and a subspace $\mathcal{H}^\sharp \subset \widetilde{\mathcal{H}}$ invariant for W (but not W^*) so that $U \cong W \restriction \mathcal{H}^\sharp$. So (5.5.22) holds for isometries also.

2. This is the discrete analog of von Neumann's original proof of the ergodic theorem.

Proof. Let F_N be the function on $\partial\mathbb{D}$

$$F_N(e^{i\theta}) = \frac{1}{N} \sum_{j=0}^{N-1} e^{ij\theta} = \begin{cases} \frac{1}{N}(e^{iN\theta} - 1)/(e^{i\theta} - 1), & \theta \neq 0 \\ 1, & \theta = 0 \end{cases} \tag{5.5.23}$$

Then $\|F_N\|_\infty \leq 1$ for all N and $F_N(e^{i\theta}) \to 0$ (respectively, 1) for $\theta \neq 0$ (respectively, $\theta = 0$). (5.5.22) follows from the strong continuity of the functional calculus (equivalently, the dominated convergence theorem). □

The following completes the characterization of weak mixing in (2.6.90) of Part 3:

Theorem 5.5.5. *Let U be a unitary operator on a Hilbert space, $\mathcal{H}$, so that for a unique unit vector φ_0, $U\varphi_0 = \varphi_0$ and so that $U \restriction \{\varphi_0\}^\perp$ has no eigenvalues. Then for all $\varphi, \psi \in \mathcal{H}$,*

$$\frac{1}{N} \sum_{j=0}^{N-1} |\langle \varphi, U^j \psi \rangle - \langle \varphi, \varphi_0 \rangle \langle \varphi_0, \psi \rangle| \to 0 \tag{5.5.24}$$

Proof. By restricting U to $\{\varphi_0\}^\perp$, we need only prove that if U has no eigenvectors, then

$$\frac{1}{N} \sum_{j=0}^{N-1} |\langle \varphi, U^j \psi \rangle| \to 0 \tag{5.5.25}$$

By the Schwarz inequality (on the sum),

$$\left(\frac{1}{N} \sum_{j=0}^{N-1} |\langle \varphi, U^j \psi \rangle| \right)^2 \leq \frac{1}{N} \sum_{j=0}^{N-1} |\langle \varphi, U^j \psi \rangle|^2 \tag{5.5.26}$$

By polarization, it suffices to prove for all φ that

$$\lim_{N\to\infty} \frac{1}{N} \sum_{j=0}^{N-1} |\langle \varphi, U^j \varphi \rangle|^2 = 0 \tag{5.5.27}$$

By the spectral theorem,

$$\langle \varphi, U^j \varphi \rangle = \int e^{ij\theta} \, d\mu_\varphi^{(U)}(\theta) \tag{5.5.28}$$

$$= (\mu_\varphi^{(U)})^\sharp(-j) \tag{5.5.29}$$

in terms of the Fourier series coefficients for μ. By the discrete version of Wiener's theorem (see Problem 8 of Section 6.6 of Part 1),

$$\text{LHS of (5.5.27)} = \sum_{e^{i\theta}} |\mu_\varphi^{(U)}(\{e^{i\theta}\})|^2 \tag{5.5.30}$$

Since $U \restriction \{\varphi_0\}^\perp$ has no eigenvalues, $\mu_\varphi^{(U)}(\{e^{i\theta}\}) = 0$ for all 0. $\square$

As a final application, we prove a form of what is called the RAGE theorem:

Theorem 5.5.6 (RAGE Theorem). *Let U be a unitary operator on a Hilbert space, $\mathcal{H}$, and $P_{\text{pp}}(U)$ the projection onto the span of all the eigenvectors for U. Let C be compact. Then*

$$\lim_{N\to\infty} \frac{1}{N} \sum_{j=0}^{N-1} \|CU^j \varphi\|^2 = \|CP_{\text{pp}}(U)\varphi\|^2 \tag{5.5.31}$$

Proof. Suppose first C is of the form $\langle\psi, \cdot\rangle\psi$. By polarization, it suffices to prove the result for $\psi = \varphi$, in which case it is just (5.5.30). Since both sides of (5.5.31) are invariant under changing C to $|C|$, we can suppose $C = C^*$. The case $C = \langle\psi, \cdot\rangle\psi$ extends to any self-adjoint finite rank, and so, by taking limits, for any self-adjoint compact C. $\square$

As an example, let U be a unitary on $\mathcal{H} = \ell^2(-\infty, \infty)$ and $C_n = P_{[-n,n]}$, the projection onto φ's, supported on $[-n, n]$. If $\varphi \perp \text{Ran}(P_{\text{pp}})$, (5.5.31) says the probability that after j time steps (under the quantum mechanical evolution $j \mapsto U^j\varphi$), $U^j\varphi$ lies in $[-n, n]$ goes to zero on average. And if $\varphi \notin (\text{Ran}(P_{\text{pp}}))^\perp$, then for all large n, this probability does not go to zero.

Notes and Historical Remarks. As we'll discuss in Section 7.2, von Neumann proved the spectral theorem for unbounded self-adjoint operators by using the Cayley transform to reduce it to the spectral theorem for unitary operators. This was proven by extending the usual self-adjoint theorem to a theorem for finitely many commuting bounded self-adjoint operators, then going to bounded normal operators including unitaries.

The direct approach here is not as well-known as it should be. It relies on tools developed between 1910 and 1915 in response to a stimulating 1907 paper of Carathéodory [**99**]. He asked what finite Taylor series $\sum_{n=0}^N a_n z^n$ were the start of the Taylor series of an analytic function, F, on $\mathbb{D}$ where $\text{Re}\, F(z) > 0$ for all $z \in \mathbb{D}$. Numerous young leaders in analysis followed up and there resulted a fairly complete theory of what are now called Carathéodory functions. We've discussed this in part already in Section 7.5 of Part 2A and Section 5.4 of Part 3. For more on the subject, the reader can consult Section 1.3 of [**648**].

In particular, the Fejér–Riesz theorem is due to them [**197, 565**] in 1916 and the Carathéodory–Toeplitz theorem to those authors [**100, 702**] in 1911.

The RAGE theorem is named after work of Ruelle [**585**], Amrein–Georgescu [**16**] and Enss [**186**].

Problems

1. If P, P_1 obey the conditions of Theorem 5.5.2, show they have the same degree and the same zeros and so agree up to a constant.

2. When $f \geq 0$ on $\partial\mathbb{D}$, show that the P of Theorem 5.5.2 is given by

$$P(z) = \exp\biggl(\frac{1}{4\pi}\int \frac{e^{i\theta} + z}{e^{i\theta} - z} \log(f(e^{i\theta}))\, d\theta\biggr)$$

If T is an operator on a Hilbert space, $\mathcal{H}$, we say S is a *dilation* of T if S acts on a Hilbert space, $\mathcal{H}'$ which has $\mathcal{H}$ as a closed subspace and if

$n = 1, 2, \ldots$, we have that

$$T^n = (PS^n) \restriction \mathcal{H} \tag{5.5.32}$$

where P is the orthogonal projection from $\mathcal{H}'$ onto $\mathcal{H}$. One calls T the *compression* of S. In the next problem, the reader will prove the celebrated dilation theorem of Sz.-Nagy [**690**] that any contraction, T, on $\mathcal{H}$ (i.e., $\|T\| \le 1$) has a unitary dilation.

3. (a) If S is a dilation of T and U is a dilation of S, prove that U is a dilation of T. Conclude that it suffices to prove that every contraction has a dilation by an isometry and every isometry has a contraction by a unitary.

   (b) Let T be a contraction on $\mathcal{H}$. Let $\mathcal{H}' = \overset{\infty}{\underset{n=1}{\oplus}} \mathcal{H}$ with $P(\varphi_1, \varphi_2, \ldots) = (\varphi_1, 0, 0, \ldots)$. Let

$$S(\varphi_1, \varphi_2, \ldots) = \big(T\varphi_1, \, (\sqrt{\mathbb{1} - T^*T})\, \varphi_1, \varphi_2, \varphi_3, \ldots\big) \tag{5.5.33}$$

   Prove that S is an isometry and a dilation of T.

   (c) If S is an isometry and $Q = \sqrt{\mathbb{1} - SS^*}$, prove that $S^*Q^2S = 0$ so $QS = S^*Q = 0$.

   (d) Let S be an isometry on $\mathcal{H}$. Let $\mathcal{H}' = \mathcal{H} \oplus \mathcal{H}$ and let

$$U(\varphi_1, \varphi_2) = \big(S\varphi_1 + (\sqrt{\mathbb{1} - SS^*})\varphi_2, -S^*\varphi_2\big)$$

   Prove that U is unitary and a dilation of S and so conclude the dilation theorem.

   Remark. There is a huge literature on this subject of which two high points are the dilation theorem for strongly continuous contraction semigroups and the Foias–Nagy commutant lifting theorem. Nagy et al. [**691**] is a text on the subject.

4. This problem will lead the reader through a proof of von Neumann's contraction theorem. To state the theorem, let f be analytic in a neighborhood of $\overline{\mathbb{D}}$. Then (see (a) below), if $f(z) = \sum_{n=0}^\infty a_n \, z^n$ is the Taylor series

$$\sum_{n=0}^\infty |a_n| < \infty \tag{5.5.34}$$

   so, if $\|T\| \le 1$, one can define

$$f(T) = \sum_{n=0}^\infty a_n \, T^n \tag{5.5.35}$$

   von Neumann's theorem asserts that

$$\|f(T)\| \le \sup_{z \in \mathbb{D}} |f(z)| \tag{5.5.36}$$

This theorem is due to von Neumann [**728**]. The modern proof usually uses the Sz.-Nagy dilation theorem (see (b) below). The proof we lead the reader through in (c)–(h) below is due to Davies–Simon [**148**] (related to a proof of Nelson [**497**]).

(a) Prove that if f is analytic in a neighborhood of $\overline{\mathbb{D}}$, then (5.5.34).

(b) Assuming the dilation theorem of Problem 3, prove (5.5.36).

(c) Prove that it suffices to prove (5.5.36) when $\|T\| < 1$.

(d) Suppose now $T = U|T|$ for a unitary U and $\|T\| < 1$. Prove that

$$g(z) = U\left[\frac{z + |T|}{1 + z|T|}\right] \tag{5.5.37}$$

is an analytic $\mathcal{L}(\mathcal{H})$-valued function in a neighborhood of $\mathbb{D}$ and for all $e^{i\theta} \in \partial\mathbb{D}$, $g(e^{i\theta})$ is unitary. (*Hint*: Use the spectral theorem for $|T|$.)

(e) Prove that $\|g(z)\| \le 1$ for all $z \in \overline{\mathbb{D}}$ so that $f\bigl(g(z)\bigr)$ is well defined as a continuous function on $\overline{\mathbb{D}}$, analytic on $\mathbb{D}$.

(f) Prove that for all $e^{i\theta} \in \partial\mathbb{D}$

$$\|f\bigl(g(e^{i\theta})\bigr)\| \le \sup_{z\in\mathbb{D}} |f(z)| \tag{5.5.38}$$

(*Hint*: Use the spectral theorem for $g(e^{i\theta})$.)

(g) Prove (5.5.38) for $g(0)$ and so (5.5.36) if $T = U|T^*|$ with U unitary.

(h) For any T on $\mathcal{H}$, let $\widetilde{T} = T \oplus 0$ on $\mathcal{H} \oplus \mathcal{H}_1$, where $\mathcal{H}_1$ is an infinite-dimensional space. Prove that one can write $\widetilde{T} = U|\widetilde{T}|$ with U unitary and that $f(\widetilde{T}) = f(T) \oplus a_0$ and conclude (5.5.36) for any T with $\|T\| < 1$.

(i) Show that f analytic in a neighborhood of $\mathbb{D}$ can be replaced by f obeys (5.5.34) (i.e., f in the Wiener space).

(j) Prove a version that only requires f in $H^\infty(\mathbb{D})$ if $\|T\| < 1$ and (5.5.36) has $z \in \mathbb{D}$ rather than $z \in \overline{\mathbb{D}}$.

5.6. Commuting Self-adjoint and Normal Operators

In this section, we'll prove the spectral theorem for several bounded self-adjoint operators which commute with one another and obtain the spectral theorem for normal operators as a byproduct.

Definition. Given a family, $\mathfrak{A}$, of bounded operators on a Hilbert space, $\mathcal{H}$, and $\varphi \in \mathcal{H}$, the *cyclic subspace* for φ generated by $\mathfrak{A}$, $\mathcal{H}_\varphi^{(\mathfrak{A})}$ is the smallest closed subspace containing φ and invariant under $\{A, A^*\}_{A\in\mathfrak{A}}$. If $\mathcal{H}_\varphi^{(\mathfrak{A})} = \mathcal{H}$, we say φ is a *cyclic vector* for $\mathfrak{A}$.

Explicitly, if $\widetilde{\mathfrak{A}}$ is the set of polynomials in $A_1, \dots, A_n, A_{n+1}^*, \dots, A_{n+m}^*$, with $A_j \in \mathfrak{A}$, then $\mathcal{H}_\varphi^{(\mathfrak{A})} = \overline{\{C\varphi \mid C \in \widetilde{\mathfrak{A}}\}}$. The main result of this section is

Theorem 5.6.1. (Spectral Theorem for Commuting Operators: Spectral Measure Form). Let $\mathfrak{A} \equiv \{A_1, \dots, A_n\}$ be a finite family of bounded self-adjoint operators on a Hilbert space, $\mathcal{H}$. Suppose for all j, k,

$$A_j A_k = A_k A_j \tag{5.6.1}$$

Let $\varphi \in \mathcal{H}$ be a cyclic vector for $\mathfrak{A}$. Then there exists a measure $d\mu_\varphi$ on $\times_{j=1}^n [-\|A_j\|, \|A_j\|\,]$ and a unitary map $U\colon \mathcal{H} \to L^2(\mathbb{R}^n, d\mu_\varphi)$ so that

$$UA_j\psi(\mathbf{x}) = x_j(U\psi)(\mathbf{x}), \quad (U\varphi)(\mathbf{x}) \equiv 1 \tag{5.6.2}$$

As in the one–operator case, $d\mu_\varphi$ and U are uniquely determined. For it suffices, as in that case, to prove there is a measure, $d\mu_\varphi$, with

$$\langle \varphi, A_1^{\ell_1} \dots A_n^{\ell_n} \varphi \rangle = \int x_1^{\ell_1} \dots x_n^{\ell_n} \, d\mu_\varphi(\mathbf{x}) \tag{5.6.3}$$

and as in that case, once we have this result, we get all the other versions (the resolution of the identity form is not quite the same and will be proven as a step in proving Theorem 5.6.1).

Given a measure, μ on $X_1 \times X_2 \times \cdots \times X_n$, a product of n compact Hausdorff spaces, there are measures $\pi_j\mu$ on X_j given by

$$\int f(y) \, d(\pi_j\mu)(y) = \int f(x_j) \, d\mu(\mathbf{x}) \tag{5.6.4}$$

Since (5.6.3) implies

$$\langle \varphi, A_j^\ell \varphi \rangle = \int x^\ell \, d(\pi_j \mu_\varphi^{(\mathfrak{A})})(x) \tag{5.6.5}$$

we see that

$$\pi_j(d\mu_\varphi^{(\mathfrak{A})}) = d\mu_\varphi^{(A_j)} \tag{5.6.6}$$

but the reader should avoid the temptation of thinking $d\mu_\varphi^{(\mathfrak{A})}$ is just a product measure! Here is an example that shows this:

Example 5.6.2. Let $d\mu$ be a measure of $\mathbb{R}^2$ which we think of as $\mathbb{C}$. Let $A \in \mathcal{L}(L^2(\mathbb{C}, d\mu))$ by

$$(Af)(z) = zf(z) \tag{5.6.7}$$

if $B = \frac{1}{2}(A + A^*)$, $C = \frac{1}{2i}(A - A^*)$, then $d\mu_{\varphi\equiv 1}^{(B,C)}(x, y) = d\mu(x + iy)$, showing any measure of compact support and, in particular, nonproduct measures, can occur. This example, as we'll see, is the model for a normal operator with a cyclic vector. □

Example 5.6.3. Let $p(x)$ be a polynomial in one variable with real coefficients. Let A be a bounded self-adjoint operator and let $B = p(A)$. Let φ be cyclic for A. Then $d\mu_\varphi^{(A,B)}$ is supported on $\{(x,y) \mid y = p(x), x \in \sigma(A)\}$. Indeed (Problem 1), one can compute $d\mu_\varphi^{(A,B)}$ in terms of $d\mu_\varphi^{(A)}$. $\square$

The key behind our construction of $d\mu_\varphi$ in Theorem 5.6.1 will be the fact that while spectral measures aren't products, spectral projections are! As a preliminary, we need

Lemma 5.6.4. *Let A, B be bounded self-adjoint operators. If $[A,B] = 0$, then for all bounded Borel functions, f, g, we have $[f(A), g(B)] = 0$; in particular, the spectral projections of A and B commute.*

Proof. Without loss, we can suppose f and g are real-valued, so $f(A)$ is self-adjoint and thus, it suffices to prove

$$[f(A), B] = 0 \tag{5.6.8}$$

and then use symmetry. We do this by a by-now standard song-and-dance.

(5.6.8) holds for all polynomials in A and so, by taking norm limits, for all continuous f. Since for any interval (a,b), $\chi_{(a,b)}$ is a pointwise limit of continuous f's with $\|f\|_\infty \le 1$, by taking strong limits,

$$[P_\Omega(A), B] = 0 \tag{5.6.9}$$

for any interval Ω. It is easy to see (Problem 2) that the set of Ω for which (5.6.9) holds in a Σ-algebra. So, since it holds for all open intervals, it holds for all Borel sets. Thus, (5.6.8) holds for finite linear combinations of such P_Ω's, and so, since they are $\|\cdot\|_\infty$-dense in $\mathcal{B}(\mathbb{R})$, for all bounded Borel f's. $\square$

Theorem 5.6.5. (Spectral Theorem for Commuting Operators: Functional Calculus Form). *Let $A_1, \dots, A_n$ be commuting bounded self-adjoint operators on a Hilbert space, $\mathcal{H}$. Then there exists a $*$-homeomorphism*

$$\Phi_{A_1,\dots,A_n} \colon \bigtimes_{j=1}^{n} [-\|A_j\|, \|A_j\|\,] \to \mathcal{L}(\mathcal{H})$$

(we write $\Phi_{A_1\dots A_n}(f) = f(A_1, \dots, A_n)$) so that

$$\mathbb{1}(x) \equiv 1 \Rightarrow \Phi_{A_1,\dots,A_n}(\mathbb{1}) = \mathbb{1}, \qquad id_j(\mathbf{x}) \equiv x_j \Rightarrow \Phi_{A_1\dots A_n}(\mathrm{id}_j) = A_j \tag{5.6.10}$$

$$\|\Phi_{A_1,\dots,A_n}(f)\| \le \|f\|_\infty \tag{5.6.11}$$

Moreover, (5.6.10) *and* (5.6.11) *determine $\Phi_{A_1,\dots,A_n}$.*

Proof. Uniqueness is the same as in one variable: (5.6.10) determines $\Phi_{A_1\dots A_n}(f)$ when f is a polynomial in $x_1, \dots, x_n$ and then the Stone–Weierstrass theorem plus (5.6.11) implies uniqueness for all f.

So we turn to existence. By scaling, we can suppose that $\|A_j\| < 1$ for $j = 1, \dots, n$. For $m = 1, 2, \dots$ and $\ell = 0, \pm 1, \dots, \pm 2^m - 1, -2^m$, define

$$P_{m,\ell}^{(j)} = P_{[\frac{\ell}{2^m}, \frac{\ell+1}{2^m})}(A_j) \tag{5.6.12}$$

and

$$\Phi_{A_1,\dots,A_n}^{(m)}(f) = \sum_{\ell_1,\dots,\ell_n} f\left(\frac{\ell_1}{2^m}, \dots, \frac{\ell_n}{2^m}\right) P_{m,\ell_1}^{(1)} P_{m,\ell_2}^{(2)} \dots P_{m,\ell_n}^{(n)} \tag{5.6.13}$$

Since $\sum_\ell (P_{m,\ell}^{(j)})^2 = \mathbb{1}$ and the P's commute, we have

$$\|\Phi_{A_1\dots A_n}^{(m)}(f)\varphi\|^2 = \sum_{\ell_1\dots\ell_n} \left| f\left(\frac{\ell_1}{2^m}, \dots, \frac{\ell_m}{2^m}\right)\right|^2 \|P_{m,\ell_1}^{(1)} \dots P_{m,\ell_n}^{(n)}\varphi\|^2 \tag{5.6.14}$$

$$\leq \|f\|_\infty^2 \sum_{\ell_1\dots\ell_m} \|P_{m,\ell_1}^{(1)} \dots P_{m,\ell_n}^{(n)}\varphi\|^2 \tag{5.6.15}$$

$$= \|f\|_\infty^2 \|\varphi\|^2 \tag{5.6.16}$$

There are no cross-terms in (5.6.14) since the P's commute (by the lemma) and $P_{m,\ell}^{(j)} P_{m,\ell'}^{(j)} = \delta_{\ell\ell'} P_{m,\ell}^{(j)}$.

As in the one-variable case, (5.6.15) implies for $\tilde{m} > m$,

$$\|\Phi_{A_1\dots A_n}^{(m)}(f) - \Phi_{A_1\dots A_n}^{(\tilde{m})}(f)\| \leq \sup_{\substack{|x_1-y_1|\leq 2^{-m} \\ \vdots \\ |x_n-y_n|\leq 2^{-m}}} |f(x) - f(y)| \tag{5.6.17}$$

$\to 0$ as $m \to \infty$. Thus, $\Phi = \lim \Phi^{(m)}$ exists. If is easy to see (Problem 3) that it has the required properties. □

Proof of Theorem 5.6.1. Identical to step $\boxed{2}$ in Section 5.1. □

Corollary 5.6.6 (Spectral Theorem for Normal Operators). *Let A be a bounded normal operator (i.e., $AA^* = A^*A$) on a Hilbert space, $\mathcal{H}$. Let φ be a cyclic vector for A. Then there is a unique measure, μ_φ, on $\{z \mid |\mathrm{Re}\, z| \leq \|A\|, |\mathrm{Im}\, z| \leq \|A\|\}$ and with*

$$\langle \varphi, (A^*)^k (A)^\ell \varphi \rangle = \int \bar{z}^k z^\ell \, d\mu(z) \tag{5.6.18}$$

Remark. We'll see below that $\mathrm{supp}(\mu)$ is equal to $\sigma(A)$.

Proof. Immediate from Theorem 5.6.1 using $A = B + iC$ with $B = B^*$, $C = C^*$, and $A^*A = AA^* \Leftrightarrow [B, C] = 0$. □

Given n commuting self-adjoint operators $A_1, \dots, A_n$ by Theorems 5.6.1 and 5.2.1, we can find $\{\mu^{(k)}\}_{k=1}^K$ measures on $\mathbb{R}^n$ (indeed, on $\times_{j=1}^n [-\|A_j\|, \|A_j\|\,]$) and $U\colon \mathcal{H} \to \oplus_{k=1}^K L^2(\mathbb{R}^n, d\mu^{(k)})$ so that

$$(UA_jU^{-1}f)(x_1, \dots, x_n) = x_j f(x_1, \dots, x_n) \tag{5.6.19}$$

We define

$$\operatorname{supp}(\{\mu^{(k)}\}_{k=1}^K) = \overline{\bigcup_{k=1}^K \operatorname{supp}(\mu^{(k)})} \tag{5.6.20}$$

(the closure is only needed if K is infinite), called their *joint spectrum.*

Theorem 5.6.7. *Let $\{A_j\}_{j=1}^n$ be n commuting bounded self-adjoint operators. Let F be a continuous function on $\times_{j=1}^n [-\|A_j\|, \|A_j\|\,]$ and $B = \Phi_{A_1 \dots A_n}(F)$. Then*

$$\sigma(B) = F\bigl[\operatorname{supp}(\{\mu^{(k)}\}_{k=1}^K)\bigr] \tag{5.6.21}$$

Proof. Let $z \notin F[\operatorname{supp}(\{\mu^{(k)}\}_{k=1}^K)] = Q$. Since the supp is compact, Q is compact, so $d = \operatorname{dist}(z, Q) > 0$. On each $L^2(\mathbb{R}^n, d\mu^{(k)})$, $|F(w) - z|^{-1} \le d^{-1}$, so multiplication by $(F(w) - z)^{-1}$ is an operator on L^2 bounded by d^{-1}. It follows that $B - z$ has a two-sided bounded inverse, so $z \notin \sigma(B)$.

Conversely, if $z \in F[\operatorname{supp}(\{\mu^{(k)}\}_{k=1}^K)]$, for every $\varepsilon > 0$, there is a k so $\mu^{(k)}(\{\mathbf{x} \mid |F(x) - z| < \varepsilon\}) > 0$. Picking $\varphi_\varepsilon \in L^2(\mathbb{R}^n, d\mu^{(k)})$ supported in this set with $\|\varphi_\varepsilon\|_2 = 1$, we find $\|(B - z)\varphi_\varepsilon\| \le \varepsilon \|\varphi_\varepsilon\|_2$, so $z \in \sigma(B)$. □

Corollary 5.6.8. *Let $A_1, \dots, A_n$ be commuting, bounded self-adjoint operators. Let $G\colon \mathbb{R}^n \to \mathbb{R}$ be continuous so that $G(A_1, \dots, A_n) = 0$. Then the joint spectrum of $(A_1, \dots, A_n)$ lies in $\{\mathbf{x} \in \mathbb{R}^n \mid G(\mathbf{x}) = 0\}$.*

Remark. If $U = A + iB$ is unitary with A, B self-adjoint, then $U^*U = UU^* = \mathbb{1}$ implies $[A, B] = 0$ and $A^2 + B^2 - \mathbb{1} = 0$, that is, the spectral measures are supported on $\{(x + iy) \mid x^2 + y^2 = 1\}$. Thus, we recover the spectral theorem for unitary operators.

Proof. If $B = G(A_1, \dots, A_n)$, then $\sigma(B) = 0$, so by (5.6.21), if $(x_1, \dots, x_n)$ is in the joint spectrum, then $G(x) = 0$. □

Problems

1. (a) Let A be a bounded self-adjoint operator on $\mathcal{H}$, $B = p(A)$, where p is a polynomial real on $\mathbb{R}$, and let $\varphi \in \mathcal{H}$. If $F \in C([-\|A\|, \|A\|\,] \times [-\|B\|, \|B\|\,])$, prove that

$$\int F(x, y)\, d\mu_\varphi^{(A,B)} = \int F(x, p(x))\, d\mu_\varphi^{(A)} \tag{5.6.22}$$

(b) Prove that $\operatorname{supp}(d\mu_\varphi^{(A,B)}) = \{(x, p(x)) \mid x \in \operatorname{supp}(d\mu_\varphi^{(A)})\}$.

2. Prove that the set of Ω for which (5.6.9) holds is a σ-algebra.
3. In the context of the proof of Theorem 5.6.5, prove that $\Phi = \lim \Phi^{(m)}$ obeys (5.6.10) and (5.6.11).

5.7. Bonus Section: Other Proofs of the Spectral Theorem

As one might expect for one of the signature results of twentieth-century analysis and a theorem which is the starting point for so much, the spectral theorem has attracted many proofs. Besides our "official" proof in Section 5.3, we had a proof in Section 2.4 (as Theorem 2.4.13), using the polar decomposition, and it will follow from the theory of commutative Banach algebras of Chapter 6. Here we summarize various proofs.

5.7.1. Polar Decomposition Proof of Riesz. This is the proof of Section 2.4 that uses the fact that the partial isometry for a self-adjoint operator, B, has spectrum contained in $\{0, \pm 1\}$, and so we can define projections onto the set where $B > 0$. Using $B = a - A$, we get spectral projections $P_{(-\infty,a)}(A)$ and so a resolution of the identity.

5.7.2. Favard Theorem Proof. This is the proof in Section 5.3. We reduce the spectral theorem to the case of Jacobi matrices by looking at cyclic vectors and then use finite approximations and the spectral theorem for finite matrices.

5.7.3. Moment Problem Proof. In Sections 4.17 and 5.6 of Part 1, we studied the question of when $\{a_n\}_{n=0}^\infty$ is the set of moments of a measure, that is, there exists a measure μ on $\mathbb{R}$ (or on $[0,1]$ or on $[0,\infty)$) so that

$$a_n = \int x^n \, d\mu(x) \tag{5.7.1}$$

One way to prove the spectral measure version of the spectral theorem, that is, Theorem 5.1.7, is to show that

$$a_n = \langle \varphi, A^n \varphi \rangle \tag{5.7.2}$$

obeys the conditions for there to be a measure of bounded support obeying (5.7.1). This is, of course, the same strategy used in Section 5.5 to prove the spectral theorem for unitary operators, but for reasons we'll come to, it is slightly more involved. There are two approaches: using the Hamburger moment problem or using the Hausdorff moment problem.

For the Hamburger moment problem approach, one uses the fact that if $\{\zeta_j\}_{j=1}^n \in \mathbb{C}^n$, then if $A = A^*$ and if a_n is given by (5.7.2), then

$$\sum_{j,k=1}^{n} \bar{\zeta}_j \zeta_k a_{j+k-2} = \sum_{j,k=1}^{n} \langle \zeta_j A^{j-1}\varphi, \zeta_k A^{k-1}\varphi\rangle = \biggl\| \sum_{j=1}^{n} \zeta_j A^{j-1}\varphi \biggr\|^2 \geq 0 \tag{5.7.3}$$

so, by Theorem 5.6.1 of Part 1, there is a measure obeying (5.7.1). So far, we just follow what we did in Section 5.5, but now we have to show $\operatorname{supp}(d\mu) \subset [-\|A\|, \|A\|\,]$ and this requires an extra argument (Problem 1).

For the Hausdorff moment problem approach, we note that

$$\widetilde{A} = \frac{A + \|A\|}{2\|A\|} \tag{5.7.4}$$

has $0 \leq \widetilde{A} \leq \mathbb{1}$ (Problem 2), so without loss, we can suppose $0 \leq A \leq \mathbb{1}$ and prove $d\mu$ exists with $\operatorname{supp}(d\mu) \subset [0,1]$. By Theorem 4.17.4 of Part 1, there exists such a μ if and only if $(\delta^k a)_n \geq 0$ for all $k, n = 0, 1, 2, \dots$, where $(\delta b)_n = b_n - b_{n-1}$. With a_n given by (5.7.2), this is, by the analysis in the proof of Proposition 4.17.5 of Part 1, equivalent to proving for all n, k that

$$\langle \varphi, A^n(\mathbb{1} - A)^k \varphi\rangle \geq 0 \tag{5.7.5}$$

By the square root lemma (Theorem 2.4.4), if $0 \leq A \leq \mathbb{1}$, there are self-adjoint B and C so $[B, C] = 0$ and

$$A = B^2, \qquad \mathbb{1} - A = C^2 \tag{5.7.6}$$

Thus,

$$\langle \varphi, A^n(\mathbb{1} - A)^k \varphi\rangle = \|B^n C^k \varphi\|^2 \geq 0 \tag{5.7.7}$$

Here the extra element beyond the unitary case is the need for the square root lemma.

5.7.4. Bochner Theorem Proof. One can define

$$e^{itA} = \sum_{n=0}^{\infty} \frac{(itA)^n}{n!} \tag{5.7.8}$$

and show that $f(t) = \langle \varphi, e^{itA}\varphi\rangle$ is positive definite in the sense of Section 6.6 of Part 1, so by Bochner's theorem (Theorem 6.6.6 of Part 1), for a measure $d\mu_\varphi^A$,

$$\langle \varphi, e^{itA}\varphi\rangle = \int e^{itx}\, d\mu_\varphi^A(x) \tag{5.7.9}$$

and from this deduce the spectral measure for the spectral theorem.

5.7.5. Spectral Radius Formula to Get Functional Calculus. This proof uses the spectral radius formula for normal operators (Theorem 2.2.11) and the spectral mapping theorem (Theorem 2.2.16). Let p be a polynomial in x. Then $p(A)$ is normal, so

$$\begin{aligned}\|p(A)\| &= \sup_{\lambda\in\sigma(p(A))} |\lambda| &\quad (5.7.10)\\ &= \sup_{\lambda\in p[\sigma(A)]} |\lambda| &\quad (5.7.11)\\ &= \sup_{\lambda\in\sigma(A)} |p(\lambda)| &\\ &= \|p\|_{C(\sigma(A))} &\quad (5.7.12)\end{aligned}$$

Here (5.7.10) uses the spectral radius formula and (5.7.11) the spectral mapping theorem. In (5.7.12), $\|\cdot\|_{C(\sigma(A))}$ means the $\|\cdot\|_\infty$ in $C(\sigma(A))$, the continuous complex-valued functions on $\sigma(A)$.

Let $\mathcal{P}(\sigma(A))$ be the polynomials viewed as a subspace of $C(\sigma(A))$. Then $p \mapsto p(A)$ is a norm isometry of $\mathcal{P}(\sigma(A))$ to $\mathcal{L}(\mathcal{H})$ and so, by the Weierstrass approximation theorem, of $C(\sigma(A))$ to $\mathcal{L}(\mathcal{H})$. Since $p \mapsto p(A)$ is a $*$-homeomorphism, so is its extension. We have thus constructed the continuous functional calculus.

5.7.6. Cayley Transform Proof. When we studied fractional linear transformations in Section 7.3 of Part 2A, the map

$$f(z) = \frac{i-z}{i+z} \tag{5.7.13}$$

played a special role. If $z = \tan\theta$ ($\theta \in (-\frac{\pi}{2}, \frac{\pi}{2})$), then

$$f(\tan\theta) = e^{2i\theta} \tag{5.7.14}$$

so f maps $\mathbb{R}$ to $\partial\mathbb{D} \setminus \{-1\}$ and $\mathbb{R} \cup \{\infty\}$ to $\partial\mathbb{D}$. The functional inverse of f is

$$g(z) = i\left(\frac{1-z}{1+z}\right) \tag{5.7.15}$$

A natural analog for operators, given a bounded self-adjoint operator, A, is

$$U \equiv f(A) = \frac{i-A}{i+A} \tag{5.7.16}$$

$$A \equiv g(U) = i\left(\frac{1-U}{1+U}\right) \tag{5.7.17}$$

It is easy to see (Problem 3) that for each $\Theta \in (0, \frac{\pi}{2})$, $A \mapsto U$ is a bijection of $\{A \mid A = A^*; \|A\| \le \tan\Theta\}$ and $\{U \mid U^*U = UU^* = \mathbb{1},\ \sigma(U) \subset \{e^{2i\theta} \mid$

$|\theta| \leq \Theta\}\}$. U is called the *Cayley transform* of A. Knowing that there is a spectral theorem for U implies one for A (Problem 4).

But what about U with $-1 \in \sigma(A)$? At least if -1 is not an eigenvalue, it turns out that one can use (5.7.17) to define an unbounded self-adjoint operator. Put differently, an unbounded self-adjoint operator defines a unitary via (5.7.16) (as we'll see in Section 7.2)—this provides one approach to the spectral theorem for unbounded self-adjoint operators.

5.7.7. Herglotz Representation Proof. In Theorem 5.9.1 of Part 3, we proved that if F is analytic in $\mathbb{C}_+$ with $\operatorname{Im} F(z) > 0$ there, then there is a measure μ on $\mathbb{R}$ with

$$\int \frac{d\mu(x)}{1+x^2} < \infty \tag{5.7.18}$$

$\alpha > 0$ and $\beta \in \mathbb{R}$ so that

$$F(z) = \alpha z + \beta + \int \left[\frac{1}{x-z} - \frac{x}{1+x^2}\right] d\mu(x) \tag{5.7.19}$$

If F is analytic across $\mathbb{R}\setminus[a,b]$, then $\operatorname{supp}(d\mu)$ is in $[a,b]$, so $\int d\mu(x) < \infty$ and the $x/(1+x^2)$ term can be dropped and absorbed in β, that is,

$$F(z) = \alpha z + \beta + \int \frac{1}{x-z}\, d\mu(x) \tag{5.7.20}$$

We then have

$$\alpha = \lim_{y\to\infty} \frac{[\operatorname{Im} F(iy)]}{y}, \qquad \beta = \lim_{y\to\infty} \operatorname{Re} F(iy) \tag{5.7.21}$$

If A is a bounded self-adjoint operator on a Hilbert space, $\mathcal{H}$, and $\varphi \in \mathcal{H}$ and $z \in \mathbb{C}\setminus\sigma(A)$, we can define

$$F(z) = \left\langle \varphi, \frac{1}{A-z}\varphi \right\rangle \tag{5.7.22}$$

which is analytic there. Since

$$\frac{1}{A-z} - \frac{1}{A-\bar{z}} = \frac{2(\operatorname{Im} z)}{|A-z|^2} \tag{5.7.23}$$

we see on $\mathbb{C}\setminus\mathbb{R}$,

$$\frac{\operatorname{Im} F(z)}{\operatorname{Im} z} > 0 \tag{5.7.24}$$

Thus, $\operatorname{Im} F(z) > 0$ if $z \in \mathbb{C}_+$. Moreover, $F(iy) \to 0$ as $y \to \infty$, so in (5.7.20), $\alpha = \beta = 0$. Thus, for a measure μ on $[-\|A\|, \|A\|\,]$,

$$\left\langle \varphi, \frac{1}{A-z}\varphi \right\rangle = \int \frac{d\mu(x)}{x-z} \tag{5.7.25}$$

Using for $|x| \le \|A\|$ and $|z| > \|A\|$,

$$(x-z)^{-1} = -\sum_{n=0}^{\infty} x^n z^{-n-1}, \quad (A-z)^{-1} = -\sum_{n=0}^{\infty} A^n z^{-n-1} \tag{5.7.26}$$

shows that

$$\langle \varphi, A^n \varphi \rangle = \int x^n \, d\mu(x) \tag{5.7.27}$$

proving once more the spectral theorem. One advantage of this proof is its focus on (5.7.25). This provides a tool for the study of spectral measures using the ideas in Theorem 2.5.4 of Part 3.

There is a close connection between the resolvent, $(A-z)^{-1}$, and the resolution of the identity. For $a, b \in \mathbb{R}$ and, for $\varepsilon > 0$, let

$$f_\varepsilon(x) = \frac{1}{2\pi} \int_a^b \operatorname{Im}(x-y-i\varepsilon)^{-1} \, dy \tag{5.7.28}$$

Then a simple calculation (Problem 5) shows that

$$\begin{aligned} f(x) &\equiv \lim_{\varepsilon \downarrow 0} f_\varepsilon(x) \\ &= \begin{cases} 1, & a < x < b \\ \frac{1}{2}, & x = a \text{ or } x = b \\ 0, & x < a < \text{ or } x > b \end{cases} \end{aligned} \tag{5.7.29}$$

Moreover, $\sup_{x,\varepsilon<1} |f_\varepsilon(x)| < \infty$. It follows that

$$\begin{aligned} \lim_{\varepsilon \downarrow 0} \frac{1}{4\pi} \int_a^b \langle \varphi, [(A-y-i\varepsilon)^{-1} - (A-y+i\varepsilon)^{-1}]\varphi \rangle \\ = \langle \varphi, \tfrac{1}{2}\,(P_{(a,b)}(A) - P_{[a,b]}(A))\varphi \rangle \end{aligned} \tag{5.7.30}$$

a formula known as *Stone's formula.*

5.7.8. Gel'fand Theory Proof. In the next chapter, we'll see (Theorems 6.4.4 and 6.4.7) that if $\mathfrak{A}$ is any norm-closed commutative $*$-subalgebra of $\mathcal{L}(\mathcal{H})$, then there is a natural isomorphism (Gel'fand isomorphism) of $\mathfrak{A}$ and $C(\sigma(\mathfrak{A}))$, where $\sigma(\mathfrak{A})$ is a natural compact Hausdorff space. In case A is a bounded self-adjoint operator and $\mathfrak{A}$ is the norm closure of polynomials in A, we'll see $\sigma(\mathfrak{A})$ is just $\sigma(A)$, so the inverse of the Gel'fand isomorphism defines the functional calculus.

Problems

1. (a) Suppose μ is a measure on $\mathbb{R}$ so that for $n = 1, 2, \dots,$

$$\int x^{2n} \, d\mu(x) \le C^{2n} \tag{5.7.31}$$

Prove that $\operatorname{supp}(d\mu) \subset [-C, C]$.

(b) Complete the proof of the spectral theorem using the Hamburger moment problem given in Subsection 5.7.3.

2. Prove that $\widetilde{A}$ given by (5.7.4) has $0 \leq \widetilde{A} \leq \mathbb{1}$.

3. If A and U are related by (5.7.16)/(5.7.17), prove that $\|A\| \leq \tan\Theta$ if and only if $\sigma(U) \subset \{e^{2i\theta} \mid |\theta| \leq \Theta\}$.

4. Complete the proof of the spectral measures form of the spectral theorem for A bounded and self-adjoint, knowing that theorem for unitaries.

5. Verify (5.7.29) by computing $f_\varepsilon(x)$ directly in terms of arctan.

5.8. Rank-One Perturbations

In some ways, this section and the next that study $A = B + C$, where B and C are self-adjoint and C rank-one (or compact in the next section), belong in Chapter 3—they are here because the spectral theorem is central to our analysis.

Throughout this section, B is a bounded self-adjoint operator on a Hilbert space, $\mathcal{H}$, and $\varphi \in \mathcal{H}$ a vector of norm 1. Define for $\alpha \in \mathbb{R}$,

$$A_\alpha = B + \alpha\langle\varphi, \cdot\,\rangle\varphi \tag{5.8.1}$$

The reader should compare the results with the finite-dimensional case discussed in Section 1.3. Those results extend to isolated eigenvalues of A. If $\mathcal{H}_\varphi$ is the cyclic subspace for φ and B (i.e., closure of the span of $\{B^j\varphi\}_{j=0}^\infty$), it is easy to see (Problem 1) that $\mathcal{H}_\varphi$ is also the cyclic subspace of φ and each A_α. Thus, A_α and B leave $\mathcal{H}_\varphi^\perp$ invariant, and clearly, $A_\alpha \restriction \mathcal{H}_\varphi^\perp = B \restriction \mathcal{H}_\varphi^\perp$ for all α. Thus, to understand the relation of A_α to B, we can focus on their restrictions to $\mathcal{H}_\varphi$. Put differently, we can suppose φ is cyclic—equivalently, if $d\mu_\alpha$ is the spectral measure of A_α and φ and $d\mu \equiv d\mu_{\alpha=0}$, we can suppose

$$\mathcal{H} = L^2(\mathbb{R}, d\mu), \qquad \varphi \equiv 1, \qquad B = \text{multiplication by } x \tag{5.8.2}$$

B bounded means $d\mu$ has compact support. Here are the main results we'll prove:

(a) For all α, $\Sigma_{\text{ac}}(A_\alpha) = \Sigma_{\text{ac}}(B)$, that is, since the spectrum is simple, the a.c. parts of A_α and B are unitarily equivalent.

(b) There exist Borel sets, S_α, of Lebesgue measure 0 so that

$$\mu_{\alpha,\text{s}}(\mathbb{R} \setminus S_\alpha) = 0, \qquad S_\alpha \cap S_\beta = \emptyset \text{ for } \alpha \neq \beta \tag{5.8.3}$$

that is, the singular parts of the measures $d\mu_\alpha$ are mutually singular.

(c) $\displaystyle\int d\alpha[d\mu_\alpha(x)] = dx$ (5.8.4)

In the sense that if f is a continuous function on $\mathbb{R}$ with compact support, then

$$\int \biggl[\int f(x)\,d\mu_\alpha(x)\biggr]\,d\alpha = \int f(x)\,dx \tag{5.8.5}$$

(d) $x_0 \in \mathbb{R}$ is a pure point of some $d\mu_\alpha$, $\alpha \neq 0$, if and only if

$$G_\mu(x_0) \equiv \int \frac{d\mu(x)}{(x-x_0)^2} < \infty, \qquad \int \frac{d\mu(x)}{x-x_0} = -\alpha^{-1} \tag{5.8.6}$$

(*Note*: If the first integral in (5.8.6) is finite, then the second integrand is absolutely integrable and so, real.)

(e) The set of x_0 for which $G_\mu(x_0) = \infty$ is a dense G_δ in $\text{supp}(d\mu)$. More is true—the set of α for which (5.8.6) holds for some $x_0 \in \mathbb{R}$ is always a nowhere dense F_σ; we won't prove this here, but see the Notes.

(f) We'll also analyze the Krein spectral shift for this case.

In proving these results, we'll study the Stieltjes transforms on $\mathbb{C}_+$,

$$F_\alpha(z) = \int \frac{d\mu_\alpha(x)}{x-z} \tag{5.8.7}$$

We'll need the following properties of Stieltjes transforms:

$$F^{(\nu)}(z) = \int \frac{d\nu(x)}{x-z} \tag{5.8.8}$$

all proven in Part 3 (some by noting that $\exp(iF^{(\nu)}(z))$ is bounded:

(1) For Lebesgue a.e. $x \in \mathbb{R}$, $\lim_{\varepsilon\downarrow 0} F^{(\nu)}(x+i\varepsilon) \equiv F^{(\nu)}(x+i0)$ exists (see (2.5.20) and Theorem 5.2.1 of Part 3).

(2) If

$$d\nu(x) = f(x)\,dx + d\nu_{\text{s}} \tag{5.8.9}$$

with $d\nu_{\text{s}}$ singular wrt dx, then

$$f(x) = \frac{1}{\pi}\,\text{Im}\,F^{(\nu)}(x+i0) \tag{5.8.10}$$

(see Theorem 2.5.4 of Part 3).

(3) If

$$S^{(\nu)} = \bigl\{x \in \mathbb{R} \bigm| \lim_{\varepsilon\downarrow 0} \text{Im}\,F^{(\nu)}(x+i\varepsilon) = \infty\bigr\} \tag{5.8.11}$$

then

$$|S^{(\nu)}| = 0, \qquad \nu_{\text{s}}(\mathbb{R}\setminus S^{(\nu)}) = 0 \tag{5.8.12}$$

(see Theorem 2.5.4 of Part 3).

(4) For all $x_0 \in \mathbb{R}$,

$$\mu(\{x_0\}) = \lim_{\varepsilon \downarrow 0} \varepsilon \operatorname{Im} F^{(\nu)}(x_0 + i\varepsilon) = \lim_{\varepsilon \downarrow 0} i^{-1} \varepsilon F^{(\nu)}(x_0 + i\varepsilon) \tag{5.8.13}$$

(see Theorem 2.5.4 of Part 3).

(5) For any $w \in \mathbb{C}$, $\{x \in \mathbb{R} \mid F(x + i0) = w\}$ has Lebesgue measure zero (see Theorem 5.2.6 of Part 3).

Here is the simple result that is central to the analysis:

Theorem 5.8.1. *In the setup of* (5.8.2) (*where* μ *has bounded support*), *with* F_α *given by* (5.8.7), *we have*

(a) (Aronszajn–Krein Formula)

$$F_\alpha(z) = \frac{F_0(z)}{1 + \alpha F_0(z)} \tag{5.8.14}$$

(b) *For* $z \in \mathbb{C} \setminus \mathbb{R}$,

$$\operatorname{Tr}((B - z)^{-1} - (A_\alpha - z)^{-1}) = \frac{d}{dz} \log(1 + \alpha F_0(z)) \tag{5.8.15}$$

(c) *For* $z \in \mathbb{C} \setminus \mathbb{R}$,

$$\det((A_\alpha - z)(B - z)^{-1}) = 1 + \alpha F_0(z) \tag{5.8.16}$$

(d) (Spectral Averaging) $\int d\alpha \big[d\mu_\alpha(x)\big] - dx$.

(e) *For each* α *and a.e.* x,

$$\operatorname{Im} F_\alpha(x + i0) = \frac{\operatorname{Im} F_0(x + i0)}{|1 + \alpha F_0(x + i0)|^2} \tag{5.8.17}$$

(f) *If*

$$S_\alpha = \{x \mid F(x + i\varepsilon) = -\alpha^{-1}\} \tag{5.8.18}$$

then

$$|S_\alpha| = 0, \qquad \mu_{\alpha,\mathrm{s}}(\mathbb{R} \setminus S_\alpha) = 0 \tag{5.8.19}$$

(g) *If* G *is given by* (5.8.6), *then*

$$G(x_0) < \infty \Rightarrow \lim_{\varepsilon \downarrow 0} F(x_0 + i\varepsilon) \text{ exists and is real} \tag{5.8.20}$$

(h) *Fix* $\alpha \in \mathbb{R} \setminus \{0\}$. *Let*

$$P_\alpha = \{x \in \mathbb{R} \mid F(x + i0) = -\alpha^{-1},\ G(x) < \infty\} \tag{5.8.21}$$

Then P_α *is countable and*

$$(d\mu_\alpha)_{\mathrm{pp}}(x) = \sum_{x_n \in P_\alpha} [\alpha^2 G(x_n)]^{-1} \delta_{x_n} \tag{5.8.22}$$

Remark. At the end of the section, we'll explore what (5.8.22) implies about eigenvalues of A_α in gaps of $\sigma(B)$ and of $\sigma_{\mathrm{ess}}(B)$.

Proof. (a) Let

$$P_\varphi = \langle \varphi, \cdot \rangle \varphi \tag{5.8.23}$$

Then

$$(A_\alpha - z)^{-1} = (B - z)^{-1} - \alpha (B - z)^{-1} P_\varphi (A_\alpha - z)^{-1} \tag{5.8.24}$$

since $\alpha P_\varphi = (A_\alpha - z) - (B - z)$. Taking expectation values in $\langle \varphi, \cdot\, \varphi \rangle$, we get

$$F_\alpha(z) = F_0(z) - \alpha F_0(z) F_\alpha(z) \tag{5.8.25}$$

which yields (5.8.14).

(b) Applying (5.8.24) to φ, we get

$$(A_\alpha - z)^{-1}\varphi = (1 - \alpha F_\alpha(z))(B - z)^{-1}\varphi \tag{5.8.26}$$

Noting that, by (5.8.25), $1 - \alpha F_\alpha(z) = F_\alpha(z)/F_0(z) = (1 + \alpha F_0(z))^{-1}$ by (5.8.14), we see (5.8.24) becomes

$$(A_\alpha - z)^{-1} - (B - z)^{-1} = -\frac{\alpha}{1 + \alpha F_0(z)} \langle (B - \bar z)^{-1}\varphi, \cdot \rangle (B - z)^{-1}\varphi$$

Thus,

$$\operatorname{Tr}((B - z)^{-1} - (A_\alpha - z)^{-1}) = \frac{\alpha}{1 + \alpha F_0(z)} \langle \varphi, (B - z)^{-2}\varphi \rangle \tag{5.8.27}$$

Since

$$\frac{d}{dz} F_0(z) = \int \frac{d\mu(x)}{(x - z)^2} = \langle \varphi, (B - z)^{-2}\varphi \rangle \tag{5.8.28}$$

we get (5.8.15).

(c) $(A_\alpha - z)(B - z)^{-1} = 1 + \alpha P_\varphi (B - z)^{-1}$. $P_\varphi (B - z)^{-1}$ is a rank-one operator whose only nonzero eigenvalue is its trace, that is, $F_0(z)$. Thus,

$$\det(1 + \alpha P_\varphi (B - z)^{-1}) = \prod_{\substack{\text{eigenvalues, } \lambda, \\ \text{of } P_\varphi (B - z)^{-1}}} (1 + \alpha\lambda) = 1 + \alpha F_0(z) \tag{5.8.29}$$

(d) Suppose first for $z \in \mathbb{C} \setminus \mathbb{R}$, we try

$$f(x) = (x - z)^{-1} - (x + i)^{-1} \tag{5.8.30}$$

Then, by closing the contour in the upper half-plane, we get

$$\int_{-\infty}^{\infty} f(x)\, dx = \begin{cases} 0 & \text{if } \operatorname{Im} z < 0 \\ 2\pi i & \text{if } \operatorname{Im} z > 0 \end{cases} \tag{5.8.31}$$

On the other hand, by (5.8.14),

$$\begin{aligned} \int f(x)\, d\mu_\alpha(E) &= F_\alpha(z) - F_\alpha(-i) \\ &= \frac{1}{\alpha + F_0(z)^{-1}} - \frac{1}{\alpha + F_0(-i)^{-1}} \end{aligned} \tag{5.8.32}$$

If $\operatorname{Im} z < 0$, the right side of (5.8.32) has both poles in α in the lower half-plane and so $\int_{-\infty}^{\infty} \dots d\alpha = 0$. If it has one pole in each half-plane, then closing the contour in the upper half-plane, we get $2\pi i$. Thus, we've proven (5.8.5) for f of the form (5.8.30). A standard density argument (Problem 2) proves it for general f's.

(e) This follows from (5.8.14) given

$$\begin{aligned} \operatorname{Im} F_\alpha(z) &= \frac{1}{2i}\,[F_\alpha(z) - \overline{F_\alpha(z)}] \\ &= \frac{1}{2i}\left[\frac{F_0}{1+\alpha F_0} - \frac{\overline{F_0}}{1+\alpha\overline{F_0}}\right] = \frac{1}{2i}\,\frac{[F_0 - \overline{F_0}]}{|1+\alpha F_0|^2} \end{aligned} \tag{5.8.33}$$

(f) By (e) of the properties of $F^{(\nu)}$, $|S_\alpha| = 0$. If $\operatorname{Im} F_\alpha(x+i\varepsilon) \to \infty$, then $1/|F_\alpha| \to \infty$, so $|F_0^{-1} + \alpha| \to 0$ by (5.8.14) and thus, $x \in S_\alpha$. That is, $S^{(\mu_\alpha)} \subset S_\alpha$, so by (5.8.12), we have $\mu_{\alpha,\mathrm{s}}(\mathbb{R} \setminus S_\alpha) = 0$.

(g) If $G(x_0) < \infty$, $|x - x_0|^{-1}$ is integrable, so by the dominated convergence theorem, $\lim_{\varepsilon\downarrow 0} F(x_0 + i\varepsilon)$ exists and equals $\int (x - x_0)^{-1}\, d\mu(x)$ which is real.

(h) Let $\mathcal{H} = L^2(\mathbb{R}, d\mu_0)$, $\varphi(x) \equiv 1$. Then A_α acts on $\mathcal{H}$ by

$$(A_\alpha\psi)(x) = x\psi(x) + \alpha c(\psi), \qquad c(\psi) \equiv \int \psi(x)\, d\mu(x) \tag{5.8.34}$$

We claim that $A_\alpha\psi_n = x_n\psi_n(x)$ with $\|\psi_n\| = 1$ if and only if (the first since $\psi_n \in L^2$)

$$G(x_n) < \infty, \qquad c^2\alpha^2 G(x_n) = 1, \qquad -\alpha F(x_n + i0) = 1 \tag{5.8.35}$$

for $\psi = -\alpha c(\psi)(x - x_n)^{-1}$ and integrating $d\mu(x)$ yields $-\alpha F(x_n + i0) = 1$. For the converse, if the first and third conditions hold, $\psi \in L^2$ and obeys (5.8.34) and the middle condition is equivalent to $\|\psi_n\| = 1$.

We have thus proven that x_n is an eigenvalue of A_α if and only if $x_n \in P_\alpha$, which implies P_α is countable, and since

$$|\langle\varphi, \psi_n\rangle|^2 = c(\psi_n)^2 = [\alpha^2 G(x_n)]^{-1} \tag{5.8.36}$$

we see $[\alpha^2 G(x_n)]^{-1}$ is the weight of the point mass at x_n in $d\mu_\alpha$. □

Corollary 5.8.2 (Kato's Finite Rank Theorem). *In the rank-one setup of* (5.8.2), *we have for all α that*

$$\Sigma_{\mathrm{ac}}(A_\alpha) = \Sigma_{\mathrm{ac}}(B) \tag{5.8.37}$$

Remark. Since there is a cyclic vector, the operators are simple, so there is a single spectral measure. Thus, the a.c. parts of A_α and B are unitarily equivalent.

Proof. Fix α. By (5) of our list of properties of Stieltjes transforms, for Lebesgue a.e. x, $1+\alpha F(x+i0)\neq 0$. Thus, (5.8.17) implies up to sets of Lebesgue measure 0 that

$$\{x \mid \operatorname{Im} F_\alpha(x+i0)\neq 0\} = \{x \mid \operatorname{Im} F_0(x+i0)\neq 0\} \tag{5.8.38}$$

By (5.8.10), this implies (5.8.37). □

Corollary 5.8.3 (Aronszajn–Donoghue Theorem). *For every $\alpha\neq\beta$, $d\mu_{\alpha,\mathrm{s}}$ and $d\mu_{\beta,\mathrm{s}}$ are mutually singular.*

Remark. Aronszajn–Donoghue also proved (5.8.37), so it is often included as part of the statement of their theorem.

Proof. By (5.8.19), $d\mu_{\alpha,\mathrm{s}}$ is supported on the set S_α of (5.8.18). Since $S_\alpha\cap S_\beta=\emptyset$, we have the mutual singularity. □

Corollary 5.8.4 (Simon–Wolff Criterion). *Let $E\subset\mathbb{R}$ have positive Lebesgue measure. Then the following are equivalent:*

(1) *For Lebesgue a.e. α, the spectrum of A_α in E is pure point (i.e., $\mu_{\alpha,\mathrm{sc}}(E)=\mu_{\alpha,\mathrm{ac}}(E)=0$).*
(2) *For Lebesgue a.e. $x\in E$, $G(x)<\infty$.*

Proof. (1) $\Rightarrow$ (2). Let

$$E_1=\{x\in E \mid G(x)<\infty\} \tag{5.8.39}$$

By hypothesis and (5.8.22), for a.e. α, μ_α is supported on E_1, that is, $\mu_\alpha(E\setminus E_1)=0$. By (5.8.4), $|E\setminus E_1|=0$, that is, $G(x)<\infty$ for a.e. $x\in E$.

(2) $\Rightarrow$ (1). Since $G(x)<\infty$ implies $\operatorname{Im} F_0(x+i0)=0$, we see $\operatorname{Im} F_0(x+i0)=0$ for a.e. $x\in E$, so $\mu_{\alpha,\mathrm{ac}}(E)=0$ for all α by Corollary 5.8.2. Let E_1 be given by (5.8.39). By $|E\setminus E_1|$ and (5.8.4), for a.e. α, $\mu_\alpha(E\setminus E_1)=0$. Thus, μ_α is supported by a countable set (by (5.8.19) and the fact that P_α is countable), so $\mu_{\alpha,\mathrm{sc}}(E)=0$. □

Theorem 5.8.5. *For any measure μ, $\{x \mid G(x)=\infty\}$ is a dense G_δ in $\operatorname{supp}(\mu)$.*

Proof. Let $G^{(\varepsilon)}(x_0)=\int(|x-x_0|+\varepsilon)^{-2}\,d\mu(x)$ which is continuous. Clearly, $G(x_0)\leq\alpha \Leftrightarrow \forall n\ G^{(1/n)}(x_0)\leq\alpha$ since $G^{(\varepsilon)}\to G$ monotonically. Thus, for each $\alpha<\infty$, $\{x_0 \mid G(x_0)\leq\alpha\}$ is an intersection of closed sets, so closed, and $\{x_0 \mid G(x_0)<\infty\}$ is an F_σ. Thus, $\{x \mid G(x)=\infty\}$ is a G_δ.

Let $x_0\in\operatorname{supp}(\mu)$. We want to find $x_n\in[x_0-\frac{1}{n},x_0+\frac{1}{n}]$ with $G(x_n)=\infty$, proving the set where G is infinite is dense in $\operatorname{supp}(\mu)$. Since $x_0\in\operatorname{supp}(\mu)$, $\mu([x_0-\frac{1}{n},x_0+\frac{1}{n}])=C>0$. Define I_m inductively. I_1 is the closed left or right half of $[x_0-\frac{1}{n},x_0+\frac{1}{n}]$ so that $\mu(I_1)\geq\frac{1}{2}C$. I_{n+1} is the closed left or right half of I_n, so $\mu(I_{m+1})\geq\frac{1}{2^{m+1}}C$.

Since $I_{m+1} \subset I_m$ and they are all closed by the finite intersection property, $\bigcap_m I_m$ is nonempty. It is a single point x_n. By construction,

$$\mu(\{y \mid |y - x_n| \le 2^{-m+1} n^{-1}\}) \ge \frac{C}{2^m} \tag{5.8.40}$$

This implies $\int |x - x_n|^{-1}\, d\mu = \infty$, so certainly, $G(x_n) = \infty$. □

This shows that if $E \subset \operatorname{supp}(d\mu)$ has $G(x) < \infty$ for a.e. x (and E is closed), then there is still a big set with $G(x) = \infty$. A more detailed analysis (see the Notes) implies that while there is then pure point spectrum for Lebesgue a.e. α, there is purely singular spectrum for a dense G_δ of α's.

Example 5.8.6. Let $\{x_n\}_{n=1}^\infty$ be a countable dense set in $[0,1]$. Let μ be a measure with pure points exactly at $\{x_n\}_{n=1}^\infty$ with

$$\mu(\{x_n\}) = 2^{-n} \tag{5.8.41}$$

Then $G(x) < \infty$ for Lebesgue a.e. x. For let

$$S_m = \{x \mid |x - x_n| \ge 2^{-m} 2^{-n/4} \text{ for all } n\} \tag{5.8.42}$$

Then $S_m \subset S_{m+1}$,

$$\mu([0,1] \setminus S_m) \le \sum_{n=1}^\infty 2^{-m+1} 2^{-n/4} = 2^{-m+3/4}(1 - 2^{-1/4}) \tag{5.8.43}$$

goes to 0 as $m \to \infty$. If $x \in S_m$, then

$$G(x) \le \sum_{n=1}^\infty 2^{2m} 2^{-n} 2^{n/2} < \infty \tag{5.8.44}$$

Thus, $G(x) < \infty$ for $x \in \bigcup_m S_m$ and, by (5.8.43), $\lim_{m\to\infty} \mu([0,1] \setminus S_m) = 0$. Thus, for a.e. α, μ_α has pure points only, but for a dense G_δ of α, $d\mu_\alpha$ is singular continuous. □

Example 5.8.7. Let $d\mu_n = \frac{1}{2^n} \sum_{j=1}^{2^n} \delta_{j/2^n}$. Then for all x_0 in $[0,1]$,

$$\int |x - x_0|^{-2}\, d\mu_n \ge (2^{-n})^{-2} 2^{-n} = 2^n \tag{5.8.45}$$

Let $a_n = C 2^{-n/2}$, where C is such that $\sum_{n=1}^\infty a_n = 1$. Let $d\mu = \sum_{n=1}^\infty a_n\, d\mu_n$. By (5.8.45) for all x_0,

$$G_\mu(x_0) \ge \sum_{n=1}^\infty a_n 2^n = \infty \tag{5.8.46}$$

Thus, $d\mu$ is a pure point measure, so for all $\alpha \ne 0$, $d\mu_\alpha$ is purely s.c. □

Next, we want to discuss the Krein spectral shift in this rank-one case. This is of intrinsic interest, but will also be a key tool in the next section. We'll define the *Krein spectral shift*, $\xi_\alpha(x)$, by

$$\xi_\alpha(x) = \frac{1}{\pi} \arg(1 + \alpha F_0(x + i0)) \tag{5.8.47}$$

For a.e. x, $F_0(x+i0)$ exists and is not $-\alpha^{-1}$, so $1+\alpha F_0(x+i0)$ is a number in $\mathbb{C}_+ \setminus \{0\}$ if $\alpha > 0$ or $\mathbb{C}_- \setminus \{0\}$ if $\alpha < 0$. In the first case, we pick arg in $[0, \pi]$ and in the latter case, in $[-\pi, 0]$. Here are the remarkable properties of ξ_α:

Theorem 5.8.8. *$\xi_\alpha(x)$ has the following properties:*

(a) $0 \le \pm\xi_\alpha(x) \le 1$ *if* $0 < \pm\alpha$.

(b) $\xi_\alpha(x) = 0$ *if* $x \le \min(\sigma(A_\alpha) \cup \sigma(B))$ *or* $x \ge \max(\sigma(A_\alpha) \cup \sigma(B))$.

(c) (Exponential Herglotz Representation) *We have that for* $z \in \mathbb{C}_+$,

$$1 + \alpha F_\alpha(z) = \exp\biggl(\int_{-\infty}^{\infty} (x-z)^{-1} \xi_\alpha(x)\, dx\biggr) \tag{5.8.48}$$

(d) $$\int |\xi_\alpha(x)|\, dx = |\alpha| \tag{5.8.49}$$

(e) $$\mathrm{Tr}((B-z)^{-1} - (A_\alpha - z)^{-1}) = \int (x-z)^{-2} \xi_\alpha(x)\, dx \tag{5.8.50}$$
for $z \in \mathbb{C}_+$.

(f) *For Lebesgue a.e.* $x \notin \Sigma_{\mathrm{ac}}(B)$, *we have* $\xi_\alpha(x) \in \{0, \pm 1\}$.

(g) *Let f obey*

$$\int (1 + |k|)|\widehat{f}(k)|\, dk < \infty \tag{5.8.51}$$

Then $f(A_\alpha) - f(B)$ is trace class and

$$\mathrm{Tr}(f(A_\alpha) - f(B)) = \int f'(x) \xi_\alpha(x)\, dx \tag{5.8.52}$$

Proof. (a) is immediate from $\arg(z) \in [0, \pi]$ (respectively, $[-\pi, 0]$) if $z \in \bar{\mathbb{C}}_+ \setminus \{0\}$ (respectively, $\bar{\mathbb{C}}_- \setminus \{0\}$).

(b) We begin by noting

$$F_0(z) = -\frac{1}{z} + O(z^{-2}) \tag{5.8.53}$$

For $F_0(z)$ is analytic at ∞ with

$$F_0(z) = -\sum_{n=0}^{\infty} z^{-n-1} \int x^n\, d\mu(x) \tag{5.8.54}$$

In particular, $F_0(x) \to 0$ as $|x| \to \infty$, which means $1 + \alpha F_0(x) \to 1$, so $\xi_\alpha(x) = 0$ for x near $\pm\infty$. Since F_0 is analytic on $\mathbb{C} \setminus \sigma(B)$ and real on $\mathbb{R} \setminus \sigma(B)$, we have $\xi_\alpha(x) = 0$ unless there is a point, x_0, where $1 + \alpha F_0(x_0) = 0$. But then $F_\alpha(z)$ has a pole there, so μ_α has a pure point so $x_0 \in \sigma(A_\alpha)$. Therefore, $\xi_\alpha(x)$ =0 on $[\mathbb{R}\setminus\sigma(B)]\cup\{x \mid x > \max(\sigma(A_n))$ or $x < \min(\sigma(A_n))\}$.

(c), (d) For $\alpha > 0$, define for $z \in \mathbb{C}_+ \cup \{x \in \mathbb{R}$ given by (b)$\}$,

$$L_\alpha(z) = \log(1 + \alpha F_\alpha(z)) \tag{5.8.55}$$

with the branch of log which has 0 imaginary part near infinity. $\operatorname{Im} L_\alpha(z) \in [0, \pi]$ if $z \in \mathbb{C}_+$. By (5.8.53), near infinity,

$$L_\alpha(z) = -\frac{\alpha}{z} + O(z^{-2}) \tag{5.8.56}$$

We thus have a Herglotz representation,

$$L_\alpha(z) = \int \frac{d\nu_\alpha(x)}{x - z} \tag{5.8.57}$$

where, by (5.8.56),

$$\int d\nu_\alpha(x) = \alpha \tag{5.8.58}$$

and

$$d\nu_\alpha(x) = \operatorname*{w-lim}_{\varepsilon\downarrow 0} \frac{1}{\pi} \operatorname{Im} L_\alpha(x + i\varepsilon)\, dx \tag{5.8.59}$$

Since $\operatorname{Im} L_\alpha$ is bounded, the limit measure is purely absolutely continuous and the weight is $\frac{1}{\pi}\operatorname{Im} L_\alpha(x + i0) = \xi_\alpha(x)$. Thus, (5.8.57) says

$$L_\alpha(z) = \int \frac{\xi_\alpha(x)\, dx}{x - z} \tag{5.8.60}$$

which is (5.8.48). (5.8.58) is (5.8.49). The proof for $\alpha < 0$ is similar.

(e) This is immediate from (5.8.60) and (5.8.15).

(f) $\Sigma_{\text{ac}}(B) = \{x \mid \operatorname{Im} F_0(x + i0) \neq 0\} = \{x \mid \operatorname{Im}(1 + \alpha F_0(x + i0)) \neq 0\}$, so for a.e. $x \notin \Sigma_{\text{ac}}(B)$, $1 + \alpha F_0(x + i\varepsilon)$ is real, so π^{-1} times its argument is 0 or 1 (if $\alpha > 0$) and 0 or -1 (if $\alpha < 0$).

(g) Writing $B^n - A_\alpha^n = \sum_{j=0}^{n-1} B^j(B - A_\alpha)A_\alpha^{n-1-j}$, we see $B^n - A_\alpha^n$ is trace class and

$$\|B^n - A_\alpha^n\|_1 \le n \sup(\|A_\alpha\|, \|B\|)^{n-1}\|B - A_\alpha\|_1 \tag{5.8.61}$$

This implies that for any $z \in \mathbb{C}$, $e^{zA_n} - e^{zB}$ is trace class and

$$e^{zA_n} - e^{zB} = \sum_{n=1}^\infty \frac{z^n(A_\alpha^n - B^n)}{n!} \tag{5.8.62}$$

converging in $\|\cdot\|_1$.

We also see that $(B - z)^{-1} - (A_\alpha - z)^{-1}$ is analytic in z near $z = \infty$ with

$$(B - z)^{-1} - (A_\alpha - z)^{-1} = \sum_{n=1}^\infty z^{-n-1}(A_\alpha^n - B^n) \tag{5.8.63}$$

Taking derivatives in (5.8.57) and using (5.8.50) shows that

$$\operatorname{Tr}((B-z)^{-1}-(A_\alpha-z)^{-1})=\sum_{n=1}^\infty nz^{-n-1}\int x^{n-1}\xi_\alpha(x)\,dx \tag{5.8.64}$$

Given (5.8.64), we conclude that

$$\operatorname{Tr}(B^n-A_\alpha^n)=-\int nx^{n-1}\xi_\alpha(x)\,dx \tag{5.8.65}$$

Using (5.8.62), this implies that

$$\operatorname{Tr}(e^{zB}-e^{zA_\alpha})=-z\int e^{zx}\xi_\alpha(x)\,dx \tag{5.8.66}$$

which is (5.8.52) for $f(x)=e^{zx}$.

We have (in trace class using the estimate (5.8.61))

$$\begin{aligned} e^{ikA_\alpha}-e^{ikB}&=[e^{ikA_\alpha}e^{-ikB}-1]e^{ikB} \\ &=ik\int_0^1 e^{ik\beta A_\alpha}(A_\alpha-B)e^{ik(1-\beta)B}\,d\beta \end{aligned} \tag{5.8.67}$$

Since for k real, $\|e^{ikA_\alpha}\|=\|e^{ikB}\|=1$, we see

$$\|e^{ikA_\alpha}-e^{ikB}\|_1\le |k|\|A_\alpha-B\|_1 \tag{5.8.68}$$

Thus, in

$$f(A_\alpha)-f(B)=(2\pi)^{-1/2}\int \widehat{f}(k)(e^{ikA_\alpha}-e^{ikB})\,dk$$

the integral converges in $\mathcal{I}_1$-norm if $(1+|k|)\widehat{f}\in L^1$. We can thus interchange $\operatorname{Tr}(\,\cdot\,)$ and the integral and use (5.8.66) for $z=ik$ to get

$$\operatorname{Tr}(f(A_\alpha)-f(B))=(2\pi)^{-1/2}\iint (ik)\widehat{f}(k)e^{ikx}\xi_\alpha(x)\,dxdk=\int f'(x)\xi_\alpha(x)\,dx$$

proving (5.8.56). □

Finally, we discuss eigenvalues in gaps under rank-one perturbations.

Theorem 5.8.9. *Let B be a self-adjoint operator with A_α given by (5.8.1), where φ is a cyclic vector for B. Suppose a and b are finite and $(a,b)\cap\sigma(B)=\emptyset$. Then*

(a) *For each α, $\sigma(A_\alpha)\cap(a,b)$ is either a single point, $\lambda(\alpha)$, or is empty.*
(b) *If $\mathcal{A}=\{\alpha^{-1}\mid\sigma(A_\alpha)\cap(a,b)=\emptyset\}$, then $\mathcal{A}$ is an interval $(\alpha_0^{-1},\alpha_1^{-1})$ or is $(\alpha_0^{-1},\alpha_1^{-1})\setminus\{0\}$, where α_0^{-1} may be $-\infty$ and/or α_1^{-1} may be $+\infty$.*
(c) *In (α_1,α_0), $\lambda(\alpha)$ is a strictly monotone function of α.*
(d) *If a and b are eigenvalues of B, then $\mathcal{A}=(-\infty,\infty)$.*

In addition, if $(a,b)\cap\sigma_{\text{ess}}(B)=\emptyset$, for each α, eigenvalues of A_α and B in (a,b) strictly interlace, and if $\#((a,b)\cap\sigma(B))<\infty$, then

$$|\#((a,b)\cap\sigma(A_\alpha))-\#((a,b)\cap\sigma(B))|\le 1 \tag{5.8.69}$$

Proof. Suppose $(a,b)\cap\sigma(B)=\emptyset$. Then for $\lambda\in(a,b)$,

$$F(\lambda)=\int\frac{d\mu(x)}{x-\lambda} \tag{5.8.70}$$

$$F'(\lambda)=\int\frac{d\mu(x)}{(x-\lambda)^2}=G(x)<\infty \tag{5.8.71}$$

(a) $F'(\lambda)>0$, so F' is strictly monotone on (a,b). Thus, $-\alpha^{-1}=F(\lambda)$ has either no solution or one solution, $\lambda(\alpha)$ in (a,b), and by (5.8.14), $F_\alpha(\lambda+i0)$ is finite for $\lambda\neq\lambda(\alpha)$, so $\sigma(A_\alpha)\cap(a,b)$ is $\emptyset$ or $\{\lambda(\alpha)\}$.

(b) Since $F(\lambda)$ is monotone on (a,b), $F_+=\lim_{\lambda\uparrow b}F(\lambda)$, $F_-=\lim_{\lambda\downarrow a}F(\lambda)$ exist with $F_-<F_+$. Define $\alpha_1=-F_-^{-1}$, $\alpha_0=-F_+^{-1}$. Then $\{-F(\lambda)\mid\lambda\in(a,b)\}=(\alpha_0^{-1},\alpha_1^{-1})$. Taking into account that $F(\lambda)=0$ corresponds to $\lambda=\infty$, we conclude $-\alpha^{\ 1}=F(\lambda)$ has a solution exactly in the interval claimed.

(c) This follows from the strict monotonicity of F and the monotonicity of $y\mapsto -y^{-1}$. Alternatively, $\alpha F(\lambda(\alpha))=-1$ implies $F(\lambda(\alpha))+\lambda'(\alpha)\alpha F'(\lambda(\alpha))$, so (using $\alpha F(\lambda(\alpha))=-1$ and $F'(\lambda)=G(\lambda)$)

$$\lambda'(\alpha)=[\alpha^2G(\lambda(\alpha))]^{-1} \tag{5.8.72}$$

(d) If a and b are eigenvalues, $F_+=\infty$, $F_-=-\infty$, so $\alpha_1^{-1}=-F_-^{-1}=\infty$, $\alpha_0^{-1}=-F_+=-\infty$.

σ_{ess} results. Let λ_0,λ_1 be two successive eigenvalues of B. Then, by (d), A_α has an eigenvalue in (λ_0,λ_1). Since $F_\alpha(\lambda_j)\neq\infty$, λ_0,λ_1 are not eigenvalues of A_α. On the other hand, by (a), there must be a point of $\sigma(B)$ between any two discrete eigenvalues of A_α. This proves interlacing, which in turn proves (5.8.69). □

Remark. It may seem surprising that, by (5.8.72), $\lambda'(\alpha)$ is exactly the weight in (5.8.69). But this weight is just $|\langle\varphi,\psi\rangle|^2$, where ψ is a normalized solution of $A_\alpha\psi=\lambda(\alpha)\psi$. On the other hand, by first-order perturbation theory, $\lambda'(\alpha)=\langle\psi,P_\varphi\psi\rangle=|\langle\psi,\varphi\rangle|^2$.

Notes and Historical Remarks. The Aronszajn–Krein formula is named after Krein [**412**] and Aronszajn [**24, 25, 26**]. It is related to earlier work of Weinstein [**736**]. The basic properties of rank-one perturbations were developed in independent 1957 works of Kato [**376**], Aronszajn [**26**], and a later paper of Donoghue [**166**]. Kato was motivated by scattering theory (see the next section) and his results included Corollary 5.8.2. Aronszajn–Donoghue had this result by different means (close to what we do here) and also Corollary 5.8.3 and (5.8.22).

Spectral averaging, (5.8.4), was found by Simon–Wolff [**655**] who also had Corollary 5.8.4 and Examples 5.8.6 and 5.8.7. For the special case of boundary condition variation of Sturm–Liouville operators (which can be put into this rank-one perturbation framework; see below), Javrjan [**347**] had spectral averaging earlier.

The instability of dense point spectrum expressed by Theorem 5.8.5 and by the stronger result below are due to Gordon [**263, 264**] and del Rio et al. [**155, 156**]. In this regard, [**156**] proved that if μ is a measure of compact support on $\mathbb{R}$, $\{\alpha \equiv -F(x+i0)^{-1} \mid x \in \text{supp}(d\mu),\, G(x) < \infty\}$ is a nowhere dense F_σ.

We'll discuss the history of the Krein spectral shift in the next section.

The theory of $A_\alpha = B + \alpha(\varphi, \cdot\,)\varphi$ most naturally extends to $B \geq 0$, which are self-adjoint but may be unbounded. φ need not be in $\mathcal{H}$, but need only be in the completion, $\mathcal{H}_{-1}$, of $\mathcal{H}$ in the norm $\|\varphi\|_{-1}^2 = \langle \varphi, (B+1)^{-1}\varphi\rangle$ (see Section 7.5).

Our presentation here owes a lot to Simon [**650**] and to Gesztesy–Pushnitski–Simon [**242**].

There is a parallel theory of rank-one perturbations of unitary operators, U. In the self-adjoint case, we wanted to preserve self-adjointness, which forced $A_\alpha - B = \alpha(\varphi, \cdot\,)\varphi$, while in this case, we want to preserve unitarity. Thus, for $\|\varphi\| = 1$, we want $V = U + \langle \varphi, \cdot\,\rangle\psi$, $V = U$ on $\{\varphi\}^\perp$, and so is unitary there. $U[\{\varphi\}^\perp] = \{U\varphi\}^\perp$ since $\langle \eta, \varphi\rangle = 0 \Rightarrow \langle U\eta, U\varphi\rangle = 0$. Thus, $V\varphi \in [\{U\varphi\}^\perp]^\perp = \{\gamma U\varphi \mid \gamma \in \mathbb{C}\}$. For $\|V\varphi\| = \|U\varphi\|$, we need $V\varphi = e^{i\alpha}U\varphi$ or

$$V_\alpha = U + (e^{i\alpha} - 1)\langle \varphi, \cdot\,\rangle U\varphi \tag{5.8.73}$$

This is equivalent to

$$V = Ue^{i\alpha P}, \qquad P = \langle \varphi, \cdot\,\rangle\varphi \tag{5.8.74}$$

One finds that with

$$F_\alpha(z) = \left\langle \varphi, \frac{V_\alpha + z}{V_\alpha - z}\varphi \right\rangle \tag{5.8.75}$$

the Carathéodory function associated to the spectral measure for V_α (see Theorem 2.5.5 of Part 3), one has

$$F_\alpha(z) = \frac{(1 - e^{-i\alpha}) + (1 + e^{-i\alpha})F_0(z)}{(1 + e^{-i\alpha}) + (1 - e^{-i\alpha})F_0(z)} \tag{5.8.76}$$

For some background on the resulting theory, see Simon [**648**].

Problems

1. Let φ be cyclic for a self-adjoint B and let $A_\alpha = B + \alpha\langle\varphi, \cdot\rangle\varphi$. Prove that φ is cyclic for A_α. (*Hint*: If $\psi \perp \varphi$ and $A_\alpha\varphi \Rightarrow \psi \perp B\varphi$ and then inductively handle $B^\ell\varphi$.)

2. Let X be the closure of the continuous functions of compact support in the norm $\|f\|_X = \sup\|(1+|x|)^{3/2}f(x)\|$. Let Y be the closure of the span of the functions $(x-z)^{-1} - (x-i)^{-1}$ for $z \in \mathbb{C}\setminus\mathbb{R}$. Prove that Y includes polynomials, P, in $(x \pm i)^{-1}$ of the form $\sum_{n,m}(x+i)^{-m}(x-i)^{-n}$, where $m+n \geq 2$, and then that it includes the continuous functions of compact support. Verify (5.8.5) from the special case $f(x) = (x-z)^{-1} - (x-i)^{-1}$, all $z \in \mathbb{C}\setminus\mathbb{R}$.

5.9. Trace Class and Hilbert–Schmidt Perturbations

In the last section, we studied the spectral theory of

$$A = B + C \tag{5.9.1}$$

where C is rank-one, and it is easy to then inductively get some information for finite rank C. In this section, we'll consider C compact (we already know C compact $\Rightarrow \sigma_{\text{ess}}(A) = \sigma_{\text{ess}}(B)$; see Theorem 3.14.1). The interesting fact is that $C \in \mathcal{I}_1$ vs. $C \in \mathcal{I}_2$ can be very, very different, as can be seen by the two main theorems of this section:

Theorem 5.9.1. *Let A, B, C be related by* (5.9.1)*, where A, B are bounded self-adjoint operators and C is trace class. Then*

$$\Sigma_{\text{ac}}(A) = \Sigma_{\text{ac}}(B) \tag{5.9.2}$$

Remark. This is a weak form of the Kato–Birman theorem (discussed in the Notes), which implies $A \restriction \mathcal{H}_{\text{ac}}(A)$ is unitarily equivalent to $B \restriction \mathcal{H}_{\text{ac}}(B)$ (i.e., not only are the essential supports the same but their multiplicities agree).

Theorem 5.9.2 (Weyl–von Neumann–Kuroda Theorem). *Let B be any bounded self-adjoint operator. Then there exists C in every $\mathcal{I}_p$, $p > 1$, so that A given by* (5.9.1) *has only point spectrum (i.e., an orthonormal basis of eigenvectors). Moreover, for any $p_0 > 1$ and any $\varepsilon_0 > 0$, we can choose C so that $\|C\|_{p_0} \leq \varepsilon_0$.*

Remark. Since $p_1 > p_0 \Rightarrow \|C\|_{p_1} \leq \|C\|_{p_0}$, we can restrict to $p \in (1, 2]$, which we henceforth do.

These two theorems are dramatically different! For in Theorem 5.9.2, $\Sigma_{\text{ac}}(A) = \emptyset$ and, of course, $\Sigma_{\text{ac}}(B)$ is arbitrary. To prove Theorem 5.9.1, we'll develop a general theory of the Krein spectral shift.

Once we have Theorem 5.9.2, we'll prove there is a unitary U so $UAU^{-1} = B + C$ with C compact if and only if $\sigma_{\text{ess}}(A) = \sigma_{\text{ess}}(B)$ (multiplicity doesn't count!).

We begin with the extension to the Krein spectral shift. In (5.8.47), we defined $\xi_{A,B}(x)$ whenever A, B are bounded self-adjoint operators with $A - B$ rank-one so that

$$\int |\xi_{A,B}(x)|\, dx \leq \|A - B\|_1 \tag{5.9.3}$$

Now let $A - B$ be trace class:

$$A - B = \sum_{n=1}^{N} \alpha_n \langle \varphi_n, \cdot \rangle \varphi_n, \quad \langle \varphi_n, \varphi_m \rangle = \delta_{nm}, \quad \sum_{n=1}^{N} |\alpha_n| \equiv \|A - B\|_1 < \infty \tag{5.9.4}$$

Define

$$\xi_{A,B}(x) = \sum_{n=1}^{\infty} \xi_{A_n, A_{n-1}}(x) \tag{5.9.5}$$

where ($n = 0$ means the sum is 0)

$$A_n = B + \sum_{j=1}^{n} \alpha_j \langle \varphi_j, \cdot \rangle \varphi_j \tag{5.9.6}$$

Since $A_n - A_{n-1}$ is rank-one, $\xi_{A_n,A_{n-1}}$ is defined by (5.8.47) and by (5.9.3) for the rank-one case,

$$\sum_{n=1}^{\infty} \int |\xi_{A_n,A_{n-1}}(x)|\, dx < \infty \tag{5.9.7}$$

so the sum converges in $L^1(\mathbb{R}, dx)$, which is complete. This shows the definition (5.9.5) yields an L^1-function.

Theorem 5.9.3. *Let A, B, C be bounded self-adjoint operators with $A - B$, $B - C$ trace class. $\xi_{A,B}$ obeys the following:*

(a) $$\int |\xi_{A,B}(x)|\, dx \leq \|A - B\|_1 \tag{5.9.8}$$

(b) *For any f obeying (5.8.51), $f(A) - f(B)$ is trace class and*

$$\operatorname{Tr}(f(A) - f(B)) = \int f'(x) \xi_{A,B}(x)\, dx \tag{5.9.9}$$

(c) *For $z \in \mathbb{C}_+$,*

$$\det((A - z)(B - z)^{-1}) = \exp\biggl(\int_{-\infty}^{\infty} (x - z)^{-1} \xi_{A,B}(x)\, dx\biggr) \tag{5.9.10}$$

(d) $\xi_{A,B}(x) = 0$ *if* $|x| \geq \|B\| + \|A - B\|_1$ (5.9.11)

(e) $\xi_{A,B}$ *is uniquely determined by* (5.9.9) *and* $\xi_{A,B}(x) = 0$ *for x near $\pm\infty$.*

(f) $\xi_{A,B}(x) + \xi_{B,C}(x) = \xi_{A,C}(x)$ *for a.e.* x.

(g) *For a.e.* $x \notin \Sigma_{\text{ac}}(B)$,

$$\xi_{A,B}(x) \in \mathbb{Z} \tag{5.9.12}$$

(h) *If* $x \notin \sigma(A) \cup \sigma(B)$, $E_{(-\infty,x)}(A) - E_{(-\infty,x)}(B)$ *is trace class and*

$$\operatorname{Tr}(E_{(-\infty,x)}(A) - E_{(-\infty,x)}(B)) = \xi_{A,B}(x) \in \mathbb{Z} \tag{5.9.13}$$

(i) *For a.e.* $x \in \mathbb{R}$,

$$\lim_{\varepsilon\downarrow 0} \det((A-x-i\varepsilon)(B-x-i\varepsilon)^{-1}) \equiv \det((A-x-i0)(B-x-i0)^{-1}) \tag{5.9.14}$$

exists, and for a.e. $x \notin \Sigma_{\text{ac}}(B)$, *it is real.*

Remarks. 1. If $x \notin \sigma(A) \cup \sigma(B)$, $(x-\varepsilon, x+\varepsilon) \cap \big[\sigma(A) \cup \sigma(B)\big] \neq \emptyset$ for some $\varepsilon > 0$. We'll prove $\xi_{A,B}(y)$ is a.e. constant on that interval, so we can define $\xi_{A,B}(y)$ to be that constant everywhere on the interval. That's the sense in which we mean (5.9.13).

2. In Theorem 3.15.20, we proved that if P, Q are projections and $P-Q \in \mathcal{I}_1$, then its trace is an integer. We won't use that in proving (5.9.13), so (5.9.13) provides a second proof (Problem 1).

Proof. (a) This is (5.9.3) which follows from

$$\|\xi_{A,B}\|_1 \le \sum_n \|\xi_{A_n,A_{n-1}}\|_1 = \sum_n |\alpha_n| = \|A-B\|_1 \tag{5.9.15}$$

$\|\xi_{A_n,A_{n-1}}\|_1 = |\alpha_n|$ is just (5.8.49).

(b) By (5.8.67) and (5.8.68),

$$\|f(A_n) - f(A_{n-1})\|_1 \le |\alpha_n| (2\pi)^{-1/2} \int |k| |\widehat{f}(k)|\, dk \tag{5.9.16}$$

It follows that $f(A) - f(B)$ is trace class and its trace is

$$\lim_{n\to\infty} \int f'(x) \bigg(\sum_{j=1}^n \xi_{A_j,A_{j-1}}(x) \bigg)\, dx = \text{RHS of (5.9.9)}$$

(c) For each $z \in \mathbb{C}_+$,

$$(A-z)(B-z)^{-1} - 1 = (A-B)(A-z)^{-1} \tag{5.9.17}$$

is trace class, so

$$\det((A-z)(B-z)^{-1}) = \lim_{n\to\infty} \det((A_n - z)(B-z)^{-1})$$

$$= \lim_{n\to\infty} \exp\bigg(\int_{-\infty}^{\infty} (x-z)^{-1} \xi_{A_n,B}(x)\, dx \bigg) \tag{5.9.18}$$

$$= \text{RHS of (5.9.10)} \tag{5.9.19}$$

(d) If $|x| \geq \|B\| + \|A - B\|_1$, for all n, $x \geq \max(\sigma(A_n), \sigma(A_{n-1}))$ or $x \leq \min(\sigma(A_n), \sigma(A_{n-1}))$. Thus, $\xi_{A_n,A_{n-1}}(x) = 0$, implying $\xi_{A,B} = \sum_n \xi_{A_n,A_{n-1}}$ is zero.

(e) If ξ_1, ξ_2 are two L^1-functions obeying (5.9.9), as distributions $(\xi_1 - \xi_2)' = 0$, so $\xi_1 - \xi_2$ is constant. Thus, if both are zero near infinity, they are equal.

(f) This is immediate from uniqueness of solutions of (5.9.9).

(g) Since rank-one perturbations preserve Σ_{ac}, we have $x \notin \Sigma_{\text{ac}}(B) \Rightarrow x \notin \Sigma_{\text{ac}}(A_{n-1})$ for all n. By Theorem 5.8.8(f), $\xi_{A_n,A_{n-1}}(x) \in \mathbb{Z}$, so $\lim(\sum_{j=1}^n \xi_{A_j,A_{j-1}}) \in \mathbb{Z}$ (for the limit is given pointwise since $\sum_{j=1}^\infty \int |\xi_{A_j,A_{j-1}}(x)|\, dx < \infty$).

(h) Pick an open interval J with $x \in J$ and $J \cap [\sigma(A) \cup \sigma(B)] = 0$. Then $\int f'(x)\xi_{A,B}(x) = 0$ for any f supported in J. It follows $\xi_{A,B}$ is a.e. constant on J, so constant. Since $\xi_{A,B}(x) \in \mathbb{Z}$ for a.e. $x \in J$, it is in $\mathbb{Z}$ for all $x \in J$.

(i) The existence of the limit (5.9.14) follows from (5.9.10) and existence of $\lim_{\varepsilon\downarrow 0} \int (y - x - i\varepsilon)^{-1}\xi_{A,B}(y)\, dy$. For a.e. $x \notin \Sigma_{\text{ac}}(B)$, by (g), $\xi_{A,B}(x)$ is in $\mathbb{Z}$, so $\text{Im}[\lim \int \dots] \in \mathbb{Z}\pi$, that is, $\exp(\lim \int \dots)$ is real. (5.9.10) implies the claim. □

We need one more preliminary before proving Theorem 5.9.1.

Lemma 5.9.4. *Let A, B be bounded self-adjoint operators with $A - B$ trace class. Let φ have $\|\varphi\| = 1$ and $P_\varphi = \langle \varphi, \cdot \rangle \varphi$. Then for $z \in \mathbb{C}_+$,*

$$1 - \langle \varphi, (A - z)^{-1}\varphi \rangle = \frac{\det((A + P_\varphi - z)(B - z)^{-1})}{\det((A - z)(B - z)^{-1}} \tag{5.9.20}$$

Proof. Clearly,

$$[(A - z)(B - z)^{-1}][(B - z)(C - z)^{-1}] = (A - z)(C - z)^{-1} \tag{5.9.21}$$

and all operators are $1 + \mathcal{I}_1$, so the product of determinants is the determinant of $(A - z)(C - z)^{-1}$. In particular, if $C = A$, we see

$$\frac{1}{\det((A - z)(B - z)^{-1})} = \det((B - z)(A - z)^{-1}) \tag{5.9.22}$$

Using this and (5.9.21) for $C = A$ and $A = A + B$, we get

$$\begin{aligned} \text{RHS of (5.9.20)} &= \det((A + P_\varphi - z)(A - z)^{-1}) \\ &= \det(1 + P_\varphi(A - z)^{-1}) \\ &= 1 + \langle \varphi, (A - z)^{-1}\varphi \rangle \end{aligned}$$

since $P_\varphi(A - z)^{-1}$ is rank-one and its trace is $\langle \varphi, (A - z)^{-1}\varphi \rangle$. □

Proof of Theorem 5.9.1. Let $\varphi \in \mathcal{H}$. For a.e. real $x_0 \notin \Sigma_{\rm ac}(B)$, the limit as $\varepsilon \downarrow 0$ of $z = x_0 + i\varepsilon$ on the right side of (5.9.20) is real, so by (5.9.20), $\operatorname{Im}\langle \varphi, (A - x_0 + i0)^{-1}\varphi\rangle = 0$ for a.e. $x_0 \in \Sigma_{\rm ac}(B)$. Thus, $\mu_{\varphi,A,{\rm ac}}(\mathbb{R}\setminus\Sigma_{\rm ac}(B)) = 0$ for each φ, so $\mathbb{R}\setminus\Sigma_{\rm ac}(B) \subset \mathbb{R}\setminus\Sigma_{\rm ac}(A)$, that is, $\Sigma_{\rm ac}(A) \subset \Sigma_{\rm ac}(B)$. By $A \leftrightarrow B$ symmetry, $\Sigma_{\rm ac}(B) \subset \Sigma_{\rm ac}(A)$. $\square$

We turn next to the Weyl–von Neumann–Kuroda theorem and its consequences. One key to the proof will be Problem 5 of Section 3.7, which showed for $1 \le p \le 2$ (recall we are restricting to $p \in (1,2]$ in our discussion of Theorem 5.9.2),

$$\|C\|_p^p \le \sum_{n=1}^\infty \|C\psi_m\|^p \tag{5.9.23}$$

for any orthonormal basis $\{\psi_m\}_{m=1}^\infty$. Suppose we find such a basis and $\{\alpha_m\} \subset \mathbb{R}$ so that

$$\sum_{n=1}^\infty \|(A - \alpha_m)\psi_m\|^p \le \varepsilon^p \tag{5.9.24}$$

Then, if $B\psi_m = \alpha\psi_m$, B has only point spectrum and $\|A - B\|_p \le \varepsilon$. For $\|(A - \alpha_m)\psi_m\|$ to be small, we want ψ_m to have little spread in a spectral representation. Thus, the following analog of Haar bases (see Example 4.6.1 of Part 3) will be critical.

Proposition 5.9.5 (Modified Haar Basis). *Let μ be a positive measure on $[0,1]$ with no point masses. Then for any $N_0 = 1, 2, \dots$, there is an orthonormal basis $\{\varphi_{m,j}\}_{m=N_0; j=1,\dots,Q_m}^\infty$, where $Q_{N_0} \le 2\,2^{N_0}$, and for $m > N_0$, $Q_m \le 2^m$ and so that $\varphi_{m,j}$ is supported on an interval of the form $[\frac{k_{m,j}}{2^m}, \frac{k_{m,j}+1}{2^m}]$.*

Proof. At level N_0, look at those intervals $I_{N_0,k} \equiv [\frac{k}{2^{N_0}}, \frac{k+1}{2^{N_0}}]$ so $\mu(I_{N_0,k_0}) \neq 0$ and take $\varphi_{N_0,j}$ of the form $\mu(I_{N_0,k_0})^{-1/2}\chi_{I_{N,k_0}}$. That gives at most 2^{N_0} such $\varphi_{N_0,j}$. If $\mu(I_{N_0,k_0}) \neq 0$, and if also

$$\mu(I_{N_0+1,2k}) \neq 0 \neq \mu(I_{N_0+1,2k+1}) \tag{5.9.25}$$

take a $\varphi_{N_0,j} = \alpha\chi_{I_{N_0+1,2k}} + \beta\chi_{I_{N_0+1,2k+1}}$, where $\alpha > 0$,

$$\alpha^2\mu(I_{N_0+1,2k}) + \beta^2\mu(I_{N_0+1,2k+1}) = 1, \ \alpha\mu(I_{N_0+1,2k}) + \beta\mu(I_{N_0+1,2k+1}) = 0 \tag{5.9.26}$$

This gives an additional at most 2^{N_0} vectors. All these vectors are orthonormal, their span includes all $\chi_{I_{N_0+1,k}}$, and each is supported in some $I_{N_0,k}$.

Having constructed the φ's for $m = N_0, \dots, N_0 + q - 1$, inductively, we have all $\chi_{I_{N_0+q,k}}$ in the span of the constructed φ's. If $\mu(I_{N_0+q,k}) \neq 0$, we'll construct a $\varphi_{N_0+q,j}$ so long as

$$\mu(I_{N_0+q+1,2k}) \neq 0 \neq \mu(I_{N_0+q+1,2k+1}) \tag{5.9.27}$$

of the form $\alpha\chi_{N_0+q+1,2k} + \beta\chi_{N_0+q+1,2k+1}$ with α, β obeying analogs of (5.9.26).

This constructs an orthonormal set so that every $\chi_{I_{m,k}}$ is in its span with $m \geq N_0$. If f is orthogonal to the φ's then

$$\int_{I_{m,k}} f(x)\, d\mu(x) = 0 \tag{5.9.28}$$

so

$$\int_0^y f(x)\, d\mu(x) = 0 \tag{5.9.29}$$

for all dyadic rational y. By continuity of $d\mu$, (5.9.29) holds for all y so f is orthogonal to all continuous g's and so is 0. Thus, $\{\varphi_{m,j}\}$ is an orthonormal basis. □

Proof of Theorem 5.9.2. By replacing B by $\alpha B + \beta$ for suitable α, β, we can suppose that

$$0 \leq B \leq 1 \tag{5.9.30}$$

Let's also suppose for the moment that B has a cyclic vector, φ, that is, B is multiplication by x on $L^2([0,1], d\mu(x))$ for some probability measure, μ. If μ has pure points, we can write $\mu = \mu_{\rm c} + \mu_{\rm pp}$ where $d\mu_{\rm c}$ is continuous. $\mathcal{H}_{\rm pp} = B \restriction L^2([0,1], d\mu_{\rm pp})$ is spanned by eigenvectors, so take $C = 0$ on $\mathcal{H}_{\rm pp}$, that is, we need only prove the result when μ has no pure points.

Let N_0 be fixed and let $\{\varphi_{m,j}\}_{m=N_0;j=1,\dots,P_m}^{\infty}$ be the basis of Proposition 5.9.5. Let $\alpha_{m,j}$ be the center of the dyadic interval, $I_{m,j}$, on which $\varphi_{m,j}$ lives. Then

$$\|(B - \alpha_{m,j})\varphi_{m,j}\| \leq 2^{-m-1} \tag{5.9.31}$$

Thus, if A is the operator on $L^2([0,1], d\mu(x))$ with

$$A\varphi_{m,j} = \alpha_{m,j}\varphi_{m,j} \tag{5.9.32}$$

by (5.9.23),

$$\|A - B\|_p^p \leq 2^{N_0} 2^{-(N_0+1)p} + \sum_{m=N_0}^{\infty} 2^m 2^{-(m+1)p} \tag{5.9.33}$$

$$= \varepsilon(N_0, p) \tag{5.9.34}$$

If $p > 1$, $\varepsilon < \infty$ by summing the geometric series, and for each $p > 1$, $\lim_{N_0 \to \infty} \varepsilon(N_0, p) = 0$. Notice if $p = 1$, the sum diverges as it must if we are to avoid contradicting Theorem 5.9.1.

Since A has point spectrum, we've proven the result when B is self-adjoint and has a cyclic vector. In general, $\mathcal{H}$ is a direct sum of cyclic subspaces for any self-adjoint B (see Theorem 5.2.1). Given $p_0 \in (1,2]$, pick $p_1, p_2, \dots$ so $p_0 \geq p_1 \geq p_2 \geq \cdots > 1$ with $p_n \downarrow 1$. On the k-th direct summand, find a pure point A_k so $\|A_k - B_k\|_{p_k} \leq \varepsilon_0 2^{-k}$ via the fact that

$\varepsilon(N_0, p_k) \to 0$ as $N_0 \to \infty$. Then with $A = \oplus A_k$, $\|B - A\|_{p_0} \le \varepsilon_0$ and $\|B - A\|_p < \infty$ for all $p > 1$. $\square$

Finally, we want to prove

Theorem 5.9.6. *Let A, B be two bounded self-adjoint operators on a Hilbert space, $\mathcal{H}$. Then there exists a compact operator, C, and unitary operator $U\colon \mathcal{H} \to \mathcal{H}$ so that*

$$A = U(B + C)U^{-1} \tag{5.9.35}$$

if and only if

$$\sigma_{\text{ess}}(A) = \sigma_{\text{ess}}(B) \tag{5.9.36}$$

We begin with some preliminary remarks:

(1) Weyl's invariance principle and invariance of spectra under unitaries says that (5.9.35) implies (5.9.36), so we need focus only on the converse.
(2) Since $A + C_1 = U(B + C_2)U^{-1} \Leftrightarrow A = U(B - U^{-1}C_1U + C_2)U^{-1}$ we can, if convenient, talk about making compact perturbations to both A and to B.
(3) By Theorem 5.9.2, we can make preliminary compact perturbations so that A and B both have point spectrum.
(4) We'll show below that we can also make a preliminary compact perturbation to eliminate all discrete spectra.
(5) Thus, the key will be to pair eigenvalues of A and B. The unitary will then adjust the actual eigenvectors. One might think multiplicity should count, but a small change in one eigenvalue of a degenerate pair will turn it into a nondegenerate pair. Thus, multiplicity shouldn't—and doesn't—matter.

Definition. We call a sequence $\{x_n\}_{n=1}^\infty$ *perfect* if and only if for all n and $\varepsilon > 0$, there are infinitely many different j's with $|x_n - x_j| < \varepsilon$.

Remark. $\overline{\{x_n\}_{n=1}^\infty}$ can have isolated points but only if for some y, $x_n = y$ for infinitely many n's.

Lemma 5.9.7. *Let $\{x_n\}_{n=1}^\infty, \{y_n\}_{n=1}^\infty$ be two perfect sequences with the same limit points. Then there is a bijection $N\colon \{1, 2, \dots\} \to \{1, 2, \dots\}$ so that $\sum_{n=1}^\infty |x_{N(n)} - y_n| < \infty$.*

Proof. Suppose we find two bijections, S, T, so $\sum_{n=1}^\infty |x_{S(n)} - y_{T(n)}| < \infty$. Then picking $N = ST^{-1}$ does what we want. Define S, T inductively as follows. $S(1) = 1$, $T(1)$ is some j so $|y_j - x_{S(1)}| \le 2^{-1}$ (possible since there are y_j's arbitrarily close to $x_{S(1)}$). Pick $T(2\ell) = \ell$ if $T(1) \ne \ell, \dots, T(2\ell - 1) \ne \ell$; otherwise, pick $T(2\ell)$ to be any previously unpicked j. Then pick $S(2\ell)$ so $|x_{S(2\ell)} - y_{T(2\ell)}| \le 2^{-2\ell}$. Then pick $S(2\ell + 1) = \ell$ if $S(1) \ne \ell, \dots, S(2\ell) \ne \ell$; otherwise, any index not previous an $S(k)$ and $T(2\ell + 1)$ so $|x_{S(2\ell+1)} -$

$y_{T(2\ell+1)}| \leq 2^{-(2\ell+1)}$. It is easy to check S and T are bijections, and clearly, $\sum_{n=1}^{\infty} |x_{S(n)} - y_{T(n)}| < \infty$. □

Lemma 5.9.8. *Let B be any bounded self-adjoint operator. Then there is a compact perturbation, C, so $A = B + C$ has only point spectrum and so that $\sigma_d(A) = \emptyset$.*

Proof. By a first compact, C_1, find $A_1 = B + C_1$ with point spectrum (use Theorem 5.9.2). There are at most countably many eigenvalues x_n with $x_n \in \sigma_d(A_1)$, say $A_1\varphi_n = x_n\varphi_n$. Any limit point of x_n's is not an isolated point of $\sigma(A_1)$, so $\lim_{n\to\infty} \operatorname{dist}(x_n, \sigma_{\text{ess}}(A_1)) = 0$. For each n, pick $y_n \in \sigma_{\text{ess}}(A_1)$ so $\operatorname{dist}(x_n, \sigma_{\text{ess}}(A_1)) = |x_n - y_n|$. Let $C_2\varphi_n = (y_n - x_n)\varphi_n$. Since $|x_n - y_n| \to 0$, C_2 is compact, and if $A = A_1 + C_2$, then $A\varphi_n = y_n\varphi_n$ and $\sigma_d(A) = \emptyset$. □

Proof of Theorem 5.9.6. As noted, we need only show that (5.9.36) implies (5.9.35). By Theorem 5.9.2 and Lemma 5.9.8, we can find preliminary C's changing A and B to operators with only point spectrum and only essential spectrum. That is, there are orthonormal bases, $\{\varphi_n\}_{n=1}^{\infty}$ and $\{\psi_n\}_{n=1}^{\infty}$, and real numbers, $\{x_n\}_{n=1}^{\infty}, \{y_n\}_{n=1}^{\infty}$, so

$$A\varphi_n = x_n\varphi_n, \qquad B\psi_n = y_n\psi_n \tag{5.9.37}$$

Moreover,

$$\forall n, \qquad x_n \in \sigma_{\text{ess}}(A), \quad y_n \in \sigma_{\text{ess}}(B) \tag{5.9.38}$$

By (5.9.36), the set of limit points of the x_n and y_n are the same, so by Lemma 5.9.7, there is a bijection, N, of $\{1, \dots, n, \dots\}$ to itself so that $\sum_{n=1}^{\infty} |x_{N(n)} - y_n| < \infty$. Let

$$U\psi_n = \varphi_{N(n)}, \quad C\psi_n = (x_{N(n)} - y_n)\psi_n \tag{5.9.39}$$

Since $|x_{N(n)} - y_n| \to 0$, C is compact. Clearly,

$$AU\psi_n = A\varphi_{N(n)} = x_{N(n)}\varphi_{N(n)} = U[(B + C)]\psi_n \tag{5.9.40}$$

so (5.9.35) holds. □

Notes and Historical Remarks. Our proof of Theorem 5.9.1 follows Gesztesy–Pushnitski–Simon [**242**]. The theorem itself is a special case of a much older theorem of Kato [**377**]. Kato proved that if $A - B$ is trace class and A, B are bounded self-adjoint operators, then $A \restriction \mathcal{H}_{\text{ac}}(A)$ is unitarily equivalent to $B \restriction \mathcal{H}_{\text{ac}}(B)$, that is, (5.9.2) plus an equality of multiplicities.

He did this by proving something of additional interest. Namely, if $P_{\text{ac}}(C)$ is the orthogonal projection onto $\mathcal{H}_{\text{ac}}(C)$, Kato proved that if A, B are self-adjoint and $A - B$ is trace class, then

$$\Omega^{\pm}(A, B) = \operatorname*{s-lim}_{t\to\mp\infty} e^{itA}e^{-itB}P_{\text{ac}}(B) \tag{5.9.41}$$

exists and is called the wave operator. It is not hard to see (since $\lim_{t\to\infty} = \lim_{t+s\to\infty}$) that

$$e^{isA}\Omega^{\pm}(A,B) = \Omega^{\pm}(A,B)e^{isB} \tag{5.9.42}$$

and that since $\Omega^{\pm}(B,A)$ also exists, we have $\mathrm{Ran}(\Omega^{\pm}(A,B)) = \mathcal{H}_{\mathrm{ac}}(A)$. These together imply $A \restriction \mathcal{H}_{\mathrm{ac}}(A)$ is unitarily equivalent to $B \restriction \mathcal{H}_{\mathrm{ac}}(B)$, with either Ω^+ or Ω^- restricted to $\mathcal{H}_{\mathrm{ac}}(A)$ being the unitary map.

The background for (5.9.41) involves scattering theory and the physics is discussed in Reed–Simon [**550**, Sec. IX.3]. The theorem on $A - B \in \mathcal{I}_1$ implying the wave operators existing is sometimes called the Kato–Birman theorem because of various extensions of Birman (discussed below). Earlier than [**377**], Kato [**376**] had proved the result when $A-B$ is finite rank, and then Rosenblum [**578**] proved it for $A - B$ trace class, assuming A and B only have a.c. spectral measures. As mentioned, the general result is from Kato [**377**].

deBranges [**151**] has a proof that doesn't use wave operators that A_{ac} and B_{ac} are unitarily equivalent if $A - B$ is trace class.

Theorem 5.9.2 for C compact goes back to Weyl [**741**] in 1909. von Neumann [**725**] in 1935 handled the case C Hilbert–Schmidt and introduced the method of proof we use here. Kuroda [**417**] did the general $\mathcal{I}_p$, $p > 1$, and indeed, he handled any symmetrically normed ideal bigger than $\mathcal{I}_1$.

The spectral shift function was introduced nonrigorously by the physicist, Lifschitz [**449**], and then carefully by Krein [**412**] in 1953. Krein used an approach, much like ours here, of doing rank-one, then finite rank, and finally, trace class.

The realization of a Herglotz function as the exponential of a Herglotz function whose measure is a.c. was developed by Aronszajn–Donoghue [**27**] in 1956 and in a second paper [**28**] in 1964.

Among papers with further developments and reviews of the Krein spectral shift are Birman–Solomyak [**65**], Birman–Yafaev [**66**], Jonas [**349, 350**], Krein [**414**], Krein–Yavryan [**415**], Sinha–Mohapatra [**657**], and Yafaev [**766**].

There are variants of the Krein spectral shift when $A - B$ is Hilbert–Schmidt, going back to Koplienko [**403**] with further developments by Chattopadhyay–Sinha [**114**], Gesztesy–Pushnitski–Simon [**242**], Neidhardt [**495**], and Peller [**518**].

Problems

1. Let P, Q be two projections so $P - Q$ is trace class and $\xi(x)$ its spectral shift.

   (a) Prove $\xi(x) = 0$ for $x < 0$ or $x > 1$ and is an integer n for $0 < x < 1$.

(b) Let f be a C_0^∞-function equal to 1 in a neighborhood of 0 supported in $(-\frac{1}{2}, \frac{1}{2})$. Prove that $f(P) = 1 - P$, $f(Q) = 1 - Q$.

(c) Prove that $\mathrm{Tr}(P - Q) = -n$ is an integer.

Remarks. 1. Section 3.15 has two other proofs that $\mathrm{Tr}(P - Q)$ is an integer, one in Example 3.15.19 and one in Problem 20.

2. Amrein–Sinha [**17**] also discuss the Krein spectral shift of P, Q when P, Q are projections with $P - Q \in \mathcal{I}_1$.

Chapter 6

Banach Algebras

> It is hardly possible to believe what economy of thought, as Mach used to say, can be effected by a well-chosen term. I think I have already said somewhere that mathematics is the art of giving the same name to different things. It is enough that these things, though differing in matter, should be similar in form, to permit of their being, so to speak, run in the same mould. When language has been well chosen, one is astonished to find that all demonstrations made for a known object apply immediately to many new objects: nothing requires to be changed, not even the terms, since the names have become the same.
>
> —*H. Poincaré* [**532**]

Big Notions and Theorems: Banach Algebra Isomorphic to Operator Algebra, Add Identities, Invariance of Outer Boundary of the Spectrum, Invariance of Spectrum, Maximal Ideals, Radical, Semisimple Banach Algebra, Function Algebra, Uniform Algebra, Multiplicative Linear Functional, Mazur's Theorem, Gel'fand Spectrum, Gel'fand Topology, Gel'fand Transform, Spectra of Many Examples, Spectra of Singly Generated Algebras, Character, Involutions on a Banach Algebra, Norm-Compatible Involution, Positive Functionals, Bochner–Raikov Theorem, B^*-Algebra, C^*-Identity, Commutative Gel'fand–Naimark Theorem, B^*-Algebra Proof of Spectral Theorem, Compact Morphism, Compactification, Strict Compactification, Stone–Čech Compactification, Almost Periodic, Bochner Almost Periodic, Bohr Almost Periodic, Bohr Compactification, Besicovitch Almost Periodic, Noncommutative Gel'fand Naimark Theorem, State, GNS Construction, Kadison Positive, Kadison State, $\{x^*x\}$ Is a Convex Cone, W^*-Algebra, Group Representation, Irredicible Representation, Gel'fand Raikov Theorem, Peter–Weyl Theorem, Schur's Lemma, Commutant and Extreme Points, Orthogonality Relations, Bochner–Weil Theorem, Fourier Inversion for LCA Groups, Plancherel Theorem for LCA Groups, Wiener Algebra as Convolutions of L^2, Pontryagin Duality, Shilov Boundary, Function Algebra, Hartogs–Rosenthal Theorem, Lebesgue–Walsh Theorem, Lavrentiev's Theorem, Mergelyan's Theorem, Boundary, Choquet Boundary, Cauchy Transform, Swiss Cheese, Dirichlet Algebra, Gleason Parts, Wiener Tauberian

Theorem for $L^1(\mathbb{R})$, Pitt's Extension of Wiener's Theorem, Slowly Oscillating Functions, Almost Monotone Functions, Ingham's Tauberian Theorem, Ingham's Proof of the Prime Number Theorem

In this chapter, we see the fireworks that result if a product is added to a Banach space. Examples of Banach spaces with products are the bounded operators, $\mathcal{L}(X)$, on a Banach space, X, and its subalgebras. This link to operator theory is even stronger since we'll prove in Section 6.1 (see Theorem 6.1.2) that every Banach algebra with identity, $\mathfrak{A}$, is isometrically isomorphic to an algebra of operators on $\mathcal{L}(\mathfrak{A})$.

Besides operator compositions, the other natural products that will recur are pointwise products of functions and convolutions. Section 6.1 will provide lots of examples, including $C(X)$, C^k-functions, many algebras of analytic functions, and convolution algebras on locally compact groups and semigroups. We'll also show how to add an identity to a Banach algebra without one, recall some basic spectral theory, and begin the discussion of ideals in Banach algebras.

Section 6.2, the central one of the chapter, is on the Gel'fand theory of abelian Banach algebras. Associated to any such algebra, $\mathfrak{A}$, is locally compact space, $\widehat{\mathfrak{A}}$, which is compact if $\mathfrak{A}$ has an identity. $\widehat{\mathfrak{A}}$ will be the set of nonzero bounded linear functionals on $\mathfrak{A}$ which are multiplicative, that is, $\ell(xy) = \ell(x)\ell(y)$. There will be a natural map, the Gel'fand transform $x \mapsto \widehat{x} \in C(\widehat{\mathfrak{A}})$ taking x to a continuous function on $\widehat{\mathfrak{A}}$ and, in the case where $\widehat{\mathfrak{A}}$ is not compact, to a function vanishing at ∞. Remarkably, both the Fourier transform on $L^1(\mathbb{R}^\nu)$ and the inverse of the continuous functional calculus for normal operators are special cases of the Gel'fand transform! We'll also prove a Tauberian theorem of Wiener that if $\{b_n\}_{n=0}^\infty$ is in ℓ^∞ and $\{a_n\}_{n=0}^\infty \in \ell^1$ with nowhere nonvanishing Fourier sum and if $(a * b)_n \to \alpha$ as $n \to \infty$, then $b_n \to \alpha/(\sum_{n=-\infty}^\infty a_n)$ as $n \to \infty$.

Sections 6.3 and 6.4 explore involutions on Banach algebras, that is, maps, $x \to x^*$, with the usual properties of adjoint and complex conjugation. We'll be especially interested in involutions on abelian Banach algebras that obey

$$\widehat{x^*} = \overline{\widehat{x}}, \qquad \|x^*\| = \|x\| \tag{6.0.1}$$

They will have a notion of positivity and will provide an extension of Bochner's theorem to abelian Banach algebras. Section 6.4 will study abelian Banach algebras with identity and involution obeying

$$\|x^*x\| = \|x\|^2 \tag{6.0.2}$$

which will imply (6.0.1) and, more importantly, that $\widehat{\ }$ is an isometric isomorphism of $\mathfrak{A}$ and $C(\widehat{\mathfrak{A}})$ (the commutative Gel'fand–Naimark theorem).

Sections 6.5 and 6.6 discuss applications of the Gel'fand theory to two special situations: the classification of compactifications of a locally compact space and the theory of almost periodic functions on a locally compact abelian group.

Section 6.7 proves the noncommutative Gel'fand–Naimark theorem that a general (i.e., not necessarily abelian) Banach algebra with identity and involution obeying (6.0.2) is isometrically isomorphic to an algebra of operators on a Hilbert space.

The chapter ends with five bonus sections: two on applications to the representation theory of groups, including Fourier analysis on a general locally compact Abelian group; one on function algebras; and two on Tauberian theorems (one extends the ℓ^1 Wiener Tauberian theorem to L^1 and the second applies it to a proof of the prime number theorem).

Notes and Historical Remarks. Banach algebras were defined in a 1936 paper of Naguno [**488**], but except for an important paper of Mazur [**470**], the subject was dominated in the 1940s by I. M. Gel'fand and his protégés, Naimark (of the Gel'fand–Naimark theorem), Raikov (of the Bochner–Raikov theorem), and Shilov (of the Shilov boundary)—detailed references to their work appear later.

After the war, various aspects of the theory became major subjects, including rephrasing and expanding Fourier analysis on groups, using the theory in the study of operator algebras on a Hilbert space, and the analysis of function algebras.

A beautiful, vast subject like Banach algebra has spawned many books starting with the early review article of Gel'fand, Raikov, and Shilov, published as a book [**236**], which had considerable impact. For other texts, see [**80, 322, 368, 428, 482, 492, 511, 562, 772**]. There are also well-written introductions as chapters in Katznelson [**384**] and Rudin [**584**].

6.1. Banach Algebra: Basics and Examples

We repeat the definitions from Section 2.2.

Definition. A *Banach algebra*, $\mathfrak{A}$, is a complex Banach space together with a product, that is, a map $(x, y) \mapsto xy$ obeying

(a) *distributivity*, that is, for each fixed x, $y \mapsto xy$ is a linear map, and for each fixed y, $x \mapsto xy$ is a linear map;
(b) *associativity*, that is, for each $x, y, z \in \mathfrak{A}$, $(xy)z = x(yz)$;
(c) *Banach algebra property*

$$\|xy\| \leq \|x\|\|y\| \tag{6.1.1}$$

Definition. A Banach algebra is said to have an *identity* if and only if there is $e \in \mathfrak{A}$ with $\|e\| = 1$ so that for all $x \in \mathfrak{A}$,

$$xe = ex = x \tag{6.1.2}$$

In Problem 1, you'll show if $\mathfrak{A}$ is a real Banach space with these properties, you can complexify to get a complex algebra. Below we'll see that if $\mathfrak{A}$ has an identity and (6.1.1) is replaced by separate continuity of $(x,y) \to xy$, there is an equivalent norm on $\mathfrak{A}$ in which (6.1.1) holds. We'll also see how to add an identity to $\mathfrak{A}$ if it doesn't have one.

Definition. $\mathfrak{A}$ is called an *abelian Banach algebra* (or commutative) if for all $x, y \in \mathfrak{A}$,

$$xy = yx \tag{6.1.3}$$

Example 6.1.1 (Operator Algebras). If X is a Banach space, $\mathcal{L}(X)$, the bounded operators in the norm

$$\|A\| = \sup_{\substack{x \in X \\ \|x\|=1}} \|Ax\| \tag{6.1.4}$$

is a Banach algebra. Closed subalgebras of any Banach algebra are again Banach algebras. $\mathcal{L}(X)$ has $\mathbb{1}$ as identity. Subalgebras may or may not. If $\mathcal{H}$ is a Hilbert space, a norm-closed subalgebra, $\mathfrak{A}$ with $\mathbb{1} \in \mathfrak{A}$ and

$$A \in \mathfrak{A} \Rightarrow A^* \in \mathfrak{A} \tag{6.1.5}$$

is called a *C^*-algebra.* If $\mathfrak{A}$ is also closed in the weak operator topology, it is called a *W^*-algebra* or a *von Neumann algebra.*

If $\{A_1, \dots, A_n\} \subset \mathcal{L}(X)$, $\mathfrak{A}_{A_1,\dots,A_n}$ is the smallest subalgebra of $\mathcal{L}(X)$ containing $\{A_1, \dots, A_n\}$. It can be constructed by taking finite linear combinations of arbitrary products of the A_j's (in any order with repeated A's allowed) and taking the norm closure. In a Hilbert space, $\mathcal{H}$, $\mathfrak{A}_{A,A^*}$ is called the *C^*-algebra generated by A.* If A is normal, it is an abelian Banach algebra. □

Theorem 6.1.2. *Any Banach algebra with identity is isometrically isomorphic to an algebra of operators.*

Proof. For each $x \in \mathfrak{A}$, let $L_x \in \mathcal{L}(\mathfrak{A})$ be left multiplication, that is,

$$L_x(y) = xy \tag{6.1.6}$$

Then

$$\|L_x\| = \sup_{\|y\|=1} \|xy\| \le \|x\| \tag{6.1.7}$$

while

$$\|L_x\| \ge \|L_x e\| = \|x\| \tag{6.1.8}$$

so $x \mapsto L_x$ is an isometry.

Since multiplication is distributive, $x \mapsto L_x$ is linear, and since it is associative, $L_{xy} = L_x L_y$. Thus, $x \mapsto L_x$ is an algebraic isomorphism. $\square$

The same idea implies

Theorem 6.1.3. *Let $\mathfrak{A}$ be a Banach space with an associative, distributive product. Suppose $\mathfrak{A}$ has an identity (so (6.1.2) holds) and that for each fixed x, $y \mapsto xy$ and $y \mapsto yx$ are continuous. But suppose neither $\|xy\| \le \|x\|\|y\|$ nor $\|e\| = 1$ is necessarily true. Then $\mathfrak{A}$ has an equivalent norm, $|\!|\!| \cdot |\!|\!|$, in which $\mathfrak{A}$ is a Banach algebra with identity (i.e., so that $|\!|\!|xy|\!|\!| \le |\!|\!|x|\!|\!|\,|\!|\!|y|\!|\!|$ and $|\!|\!|e|\!|\!| = 1$).*

Remark. We'll get $|\!|\!|x|\!|\!| \le C\|x\|$ by using the inverse mapping theorem; one can instead use the uniform boundedness principle (Problem 2).

Proof. Define L_x by (6.1.6). Since $y \mapsto xy$ is continuous, each $L_x \in \mathcal{L}(\mathfrak{A})$. Let $\widetilde{\mathfrak{A}} \subset \mathcal{L}(\mathfrak{A})$ be the family of $\{L_x\}_{x\in\mathfrak{A}}$.

We claim $\widetilde{\mathfrak{A}}$ is closed in $\mathcal{L}(\mathfrak{A})$. For if $L_{x_n} \to A$ in $\mathcal{L}(\mathfrak{A})$, then $L_{x_n}e = x_n \to Ae$ in $\mathfrak{A}$ and then, by continuity of $x \mapsto xy$, we have $Ay = \lim L_{x_n}y = L_{Ae}y$, so $A = L_{Ae} \in \widetilde{\mathfrak{A}}$. Thus, $\widetilde{\mathfrak{A}}$ is a Banach algebra which obeys $\|\mathbb{1}\| = 1$ and $\|L_x L_y\| \le \|L_x\|\|L_y\|$.

Let $Q\colon \widetilde{\mathfrak{A}} \to \mathfrak{A}$ by

$$QA = Ae \tag{6.1.9}$$

Since

$$\|QA\| = \|Ae\| \le \|e\|\,\|A\| \tag{6.1.10}$$

Q is a bounded linear map of $\widetilde{\mathfrak{A}}$ to $\mathfrak{A}$. Since $L_{QA} = A$ and $QL_x = x$, we see that Q is a bijection. Therefore, by the inverse mapping theorem (Theorem 5.4.14 of Part 1), L is bounded, that is, if

$$|\!|\!|x|\!|\!| = \|L_x\| \tag{6.1.11}$$

then for some C_1,

$$|\!|\!|x|\!|\!| \le C_1\|x\| \tag{6.1.12}$$

while, by (6.1.10) with $A = L_x$,

$$\|x\| \le \|e\|\,|\!|\!|x|\!|\!| \tag{6.1.13}$$

Therefore, $|\!|\!| \cdot |\!|\!|$ is an equivalent norm to $\|\cdot\|$ in which $|\!|\!|e|\!|\!| = 1$ and $|\!|\!|xy|\!|\!| \le |\!|\!|x|\!|\!|\,|\!|\!|y|\!|\!|$. $\square$

Example 6.1.4 ($C(X)$ and its Subalgebras)**.** Let X be a topological space and $C(X)$ the bounded, continuous, complex-valued functions on X. This is a Banach algebra with identity when one uses pointwise operations and

$\|\cdot\|_\infty$ as the norm. We'll be most interested in the case where X is a compact Hausdorff space, but in the theory of compactifications, locally compact spaces will also arise.

A *function algebra* (aka *uniform algebra*) is a closed subalgebra, $\mathfrak{A}$, of $C(X)$, X a compact Hausdorff space, with $1 \in \mathfrak{A}$, using $\|\cdot\|_\infty$, and so that $\mathfrak{A}$ separates points, that is, $\forall x, y \in X$, $\exists f \in \mathfrak{A}$ so $f(x) \neq f(y)$. Lest the reader think that means $\mathfrak{A} = C(X)$ because of the Stone–Weierstrass theorem, we do not assume that $f \in \mathfrak{A} \Rightarrow \bar{f} \in \mathfrak{A}$ so $\mathfrak{A}$ need not be all of $C(X)$. The canonical example is $\mathbb{A}(\overline{\mathbb{D}})$, the functions continuous on $\overline{\mathbb{D}}$ and analytic on $\mathbb{D}$. We'll say more about $\mathbb{A}(\overline{\Omega})$ in Example 6.1.6 and more about function algebras in Section 6.10. Some authors use "function algebras" to be algebras of functions even if the norm is not $\|\cdot\|_\infty$, for example, $C^k([0,1])$ and $\mathbb{W}(\overline{\mathbb{D}})$ below, and they reserve the name "uniform algebra" if the norm is $\|\cdot\|_\infty$. □

Example 6.1.5 (C^k-functions). Let $C^k([0,1])$ be the C^k-functions on $(0,1)$ so each derivative $f^{(j)}$, $j = 0, 1, 2, \dots, k$, has a continuous extension to $[0,1]$. With

$$\|f\| = \sum_{j=0}^{k} (j!)^{-1} \sup_{x\in[0,1]} |f^{(j)}(x)| \tag{6.1.14}$$

this is a Banach algebra (where $\|fg\| \le \|f\|\|g\|$ needs Leibniz's rule; see Problem 4). □

Example 6.1.6 (Algebras of Analytic Functions). Let Ω be a bounded region in $\mathbb{C}^n$ with $\overline{\Omega}^{\text{int}} = \Omega$. We are interested in four families of analytic functions in Ω, all with the $\|\cdot\|_\infty$-norm: $H^\infty(\Omega)$ is the set of bounded analytic functions on Ω. We studied $H^\infty(\mathbb{D})$ extensively in Chapter 5 of Part 3. $\mathbb{A}(\overline{\Omega})$ is the set of functions continuous on $\overline{\Omega}$ and analytic on Ω. $\mathbb{P}(\overline{\Omega})$ is closure in $\mathbb{A}(\overline{\Omega})$ of the set of polynomials in z and $\mathbb{R}(\overline{\Omega})$ the closure in $\mathbb{A}(\overline{\Omega})$ of the set of rational functions, P/Q, where P and Q are polynomials and Q is nonvanishing on $\overline{\Omega}$. We clearly have

$$\mathbb{P}(\overline{\Omega}) \subset \mathbb{R}(\overline{\Omega}) \subset \mathbb{A}(\overline{\Omega}) \subset H^\infty(\Omega) \tag{6.1.15}$$

These are all function algebras (H^∞ isn't given as a function algebra since Ω isn't compact but since $\|f^n\|_\infty = \|f\|_\infty^n$, $\widehat{\ }$ is an isometric isomorphism so H^∞ is a function algebra on $\widehat{H^\infty}$, the Gel'fand spectrum of the next section). More generally still, one considers for $U \subset K$ with U open and K compact, $\mathbb{A}(K;U)$ of functions continuous on K and analytic in U which allows for Ω's like $\mathbb{D} \setminus [0,1)$ with $\overline{\Omega}^{\text{int}}$ bigger than Ω.

Another algebra of analytic functions but with a different norm than $\|\cdot\|_\infty$ is $\mathbb{W}(\overline{\mathbb{D}})$, the *Wiener algebra* of all functions analytic in $\mathbb{D}$, so that

$$f(z) = \sum_{n=0}^\infty a_n z^n, \qquad \|f\|_{\mathbb{W}} = \sum_{n=0}^\infty |a_n| < \infty \tag{6.1.16}$$

The product is pointwise in z. Clearly, the power series converges on all of $\overline{\mathbb{D}}$, so as sets $\mathbb{W}(\overline{\mathbb{D}}) \subset \mathbb{A}(\overline{\mathbb{D}})$ but the norms are different, and the spaces are not equal (see Problem 9). That $\|fg\|_{\mathbb{W}} \le \|f\|_{\mathbb{W}}\|g\|_{\mathbb{W}}$ is easy to check (Problem 7). □

Example 6.1.7 (Group Algebras)**.** Let G be a group and let $\ell^1(G)$ be all formal sums of the form

$$f = \sum_{n=1}^\infty \alpha_n \delta_{x_n} \tag{6.1.17}$$

where $\{x_n\}_{n=1}^\infty \subset G$ and $\alpha_n \in \mathbb{C}$ with

$$\sum_{n=1}^\infty |\alpha_n| < \infty \tag{6.1.18}$$

$\ell^1(G)$ is a vector space; indeed, a Banach space with norm $\|f\| = \sum_{n=1}^\infty |\alpha_n|$. If we set $\delta_x * \delta_y = \delta_{xy}$, $\ell^1(G)$ gets a natural product structure, making it into an algebra. If we think of $\ell^1(G)$ as point measures, that is, for bounded maps $F\colon G \to \mathbb{C}$, we have for f given by (6.1.17),

$$\mu_f(F) = \sum_{n=1}^\infty \alpha_n F(x_n) \tag{6.1.19}$$

Then

$$\mu_{f*g}(F) = \int F(xy)\, d\mu_f(x)\, d\mu_g(y) \tag{6.1.20}$$

Put differently, if f is given by (6.1.17) and we let

$$f(x) = \begin{cases} \alpha_n & \text{if } x = x_n \\ 0, & \text{otherwise} \end{cases} \tag{6.1.21}$$

so $f = \sum_x f(x)\delta_x$ and $g = \sum_g g(y)\delta_y$, then

$$f * g = \sum_x p(x)\delta_x \tag{6.1.22}$$

where

$$p(x) = \sum_y f(y) g(y^{-1}x) \tag{6.1.23}$$

Now suppose G has a topology making it into a locally compact group. Let $\mathcal{M}(G)$ be the finite complex Baire measures on G. Then if G has a metric topology, $\ell^1(G)$ viewed as a family of point measures is a weakly dense subspace and $*$ has a unique weakly continuous extension to $\mathcal{M}(G)$ given by

$$\int f(x)\,d(\mu*\nu)(x) = \int f(xy)\,d\mu(x)\,d\mu(y) \tag{6.1.24}$$

$\mathcal{M}$ with this product is a Banach algebra with identity.

G has a left-invariant Haar measure, $d\mu$, that obeys

$$\int f(xy)\,d\mu(y) = \int f(y)\,d\mu(y) \tag{6.1.25}$$

for all $x \in G$ and continuous f's of compact support. Inside $\mathcal{M}(G)$ is the subalgebra, $L^1(G, d\mu)$, of those measures a.c. wrt μ. It is a subalgebra because a simple calculation yields (Problem 10)

$$f\,d\mu * g\,d\mu = (f*g)\,d\mu \tag{6.1.26}$$

where

$$(f*g)(y) = \int f(x)g(x^{-1}y)\,d\mu(x) \tag{6.1.27}$$

$$= \int f(yz)g(z^{-1})\,d\mu(z) \tag{6.1.28}$$

The identity in $\mathcal{M}(G)$ is δ_e, the unit mass at the identity. If G is not discrete, this is not in L^1, so $L^1(G, d\mu)$ is a Banach algebra without identity—one of the most significant examples.

(6.1.24) only involves the product and not inverses. Thus, if G is a locally compact semigroup, one can define $\mathcal{M}(G)$, and if there is a Haar-type measure, also $L^1(G, d\mu)$. This happens if G is a locally compact group with Haar measure μ, if $H \subset G$ has $x, y \in H \Rightarrow xy \in H$ and $e \in H$, and if $\mu(H) > 0$. Then $\{f \in L^1(G) \mid f = 0 \text{ on } G \setminus H\}$ is a subalgebra of $L^1(G, d\mu)$. An example is $G \equiv \mathbb{R}$, $H = [0, \infty)$ in which case, for $x > 0$,

$$(f*g)(x) = \int_0^x f(y)g(x-y)\,dy \tag{6.1.29}$$

Problem 3 provides a generalization of $L^1(G, d\mu)$ to $L^1(G, w\,d\mu)$ for certain *Beurling weights*, w, on G (positive measurable functions on G with $w(xy) \le w(x)w(y)$). □

To add an identity to $L^1(G, d\mu)$, we need only take

$$\mathbb{V}(G) = \{f\,d\mu + \lambda\delta_e \mid f \in L^1,\ \lambda \in G\} \subset \mathcal{M}(G) \tag{6.1.30}$$

This adding-one-dimension idea actually works in general:

Theorem 6.1.8. *Let $\mathfrak{A}$ be a Banach algebra. Let $\mathfrak{A}_e = \mathfrak{A} \oplus \mathbb{C}$ with norm*

$$\|(x, \lambda)\| = \|x\| + |\lambda| \tag{6.1.31}$$

and product

$$(x, \lambda)(y, \mu) = (xy + \lambda y + \mu x, \lambda\mu) \tag{6.1.32}$$

Then $\mathfrak{A}_e$ is a Banach algebra with identity $e = (0, 1)$.

Remark. This works even if $\mathfrak{A}$ has an identity.

Proof. Everything is straightforward to check. □

Recall the following spectral results obtained in Chapter 2 in a general Banach algebra with identity:

(1) (Theorems 2.2.8 and 2.2.9) If the *resolvent set* and *spectrum* are defined by

$$\rho(x) = \{\lambda \in \mathbb{C} \mid x - \lambda e \text{ has a two-sided inverse}\},\ \sigma(x) = \mathbb{C} \setminus \rho(x) \tag{6.1.33}$$

then $\sigma(x)$ is closed and

$$\sigma(x) \neq \emptyset \tag{6.1.34}$$

(2) (Theorem 2.2.10) For any x, the limit in

$$\operatorname{spr}(x) = \lim_{n\to\infty} \|x^n\|^{1/n} \tag{6.1.35}$$

exists and equals $\operatorname{spr}(x) = \sup\{|\lambda| \mid \lambda \in \sigma(x)\}$.

(3) (Theorem 2.3.1) Let $\mathcal{F}(x)$ be the family of equivalence classes of functions analytic in a neighborhood of $\sigma(x)$ with $f \equiv g$ if there is a neighborhood, N, of $\sigma(x)$ so $f \restriction N = g \restriction N$, then $\mathcal{F}(x)$ is an algebra in an obvious way. For $f \in \mathcal{F}(x)$, one can define $f(x)$ so that $f(z) \equiv z$ for all $z \in \mathbb{C} \Rightarrow f(x) = x$ and $f \mapsto f(x)$ is an algebra morphism. f is defined via (2.3.4).

(4) (Theorem 2.3.2) One has the spectral mapping theorem

$$\sigma(f(x)) = f[\sigma(x)] \tag{6.1.36}$$

There is a subtlety about spectrum we need to mention. Suppose $\mathfrak{A}_1 \subset \mathfrak{A}_2$ for two Banach algebras with identity. Let $x \in \mathfrak{A}_1$. If $(x - \lambda e)$ has an inverse in $\mathfrak{A}_1$, obviously it has one in $\mathfrak{A}_2$. But the converse might not apply, that is, using $\sigma_{\mathfrak{A}}(x)$ for the spectrum as an element of $\mathfrak{A}$, we have

$$x \in \mathfrak{A}_1 \subset \mathfrak{A}_2 \Rightarrow \rho_{\mathfrak{A}_1}(x) \subset \rho_{\mathfrak{A}_2}(x) \Rightarrow \sigma_{\mathfrak{A}_2}(x) \subset \sigma_{\mathfrak{A}_1}(x) \tag{6.1.37}$$

Example 6.1.9. Let $\mathcal{H} = \ell^2(\mathbb{Z})$, that is, two-sided sequences. Let $U \in \mathcal{L}(\mathcal{H})$ be given by

$$U(a)_n = a_{n-1} \tag{6.1.38}$$

Then $\mathfrak{A}_1 \equiv \mathfrak{A}_U \subset \mathfrak{A}_{U,U^*} \equiv \mathfrak{A}_2$. Since $\|U\| = 1$, for $j = 1, 2$,

$$\sigma_{\mathfrak{A}_j}(U) \subset \overline{\mathbb{D}} \tag{6.1.39}$$

Since $U^*U = UU^* = \mathbb{1}$, $0 \notin \sigma_{\mathfrak{A}_2}(U)$. But it is easy to see (Problem 5) that if V is a polynomial in U, then $\|U^* - V\| \geq 1$. Thus, $U^* \in \mathfrak{A}_2 \setminus \mathfrak{A}_1$, so $0 \in \sigma_{\mathfrak{A}_1}(U)$, that is, the containment in (6.1.37) can be strict. In fact (Problem 6),

$$\sigma_{\mathfrak{A}_1}(U) = \overline{\mathbb{D}}, \qquad \sigma_{\mathfrak{A}_2}(U) = \partial\mathbb{D} \tag{6.1.40}$$

□

Example 6.1.10. We have $\mathbb{A}(\overline{\mathbb{D}}) \subset \mathbb{C}(\partial\mathbb{D})$ by restriction. That its norms are the same is the maximum principle for analytic functions (see Theorem 3.6.3 in Part 2A). Let z be the function $f(z) = z$. Clearly, $0 \in \sigma_{\mathbb{A}(\overline{\mathbb{D}})}(z)$ and $0 \notin \sigma_{C(\partial\mathbb{D})}(z)$ (for $z(e^{i\theta})\,\overline{z(e^{i\theta})} = 1$). Indeed, we have that for any function $f \in \mathbb{A}(\mathbb{D})$ that

$$\sigma_{\mathbb{A}(\overline{\mathbb{D}})}(f) = f[\overline{\mathbb{D}}], \qquad \sigma_{C(\partial\mathbb{D})}(f) = f[\partial\mathbb{D}]$$

$f[\overline{\mathbb{D}}]$ contains an open set if f is not constant while, typically, $f[\partial\mathbb{D}]$ has empty interior. □

Definition. Let $K \subset \mathbb{C}$ be compact. Let U be the unbounded component of $\mathbb{C} \setminus K$. ∂U is called the *outer boundary* of K and denoted $O\partial(K)$; it is contained in ∂K.

There is one general result:

Theorem 6.1.11. *Let $\mathfrak{A}_2$ be a Banach algebra with identity and $\mathfrak{A}_1$ a Banach subalgebra with $e \in \mathfrak{A}_1$. Let $x \in \mathfrak{A}_1$ and $S = \sigma_{\mathfrak{A}_2}(x)$. Then*

(a) *For every connected component K of $\mathbb{C} \setminus S$, either $K \subset \sigma_{\mathfrak{A}_1}(x)$ or $K \subset \mathbb{C} \setminus \sigma_{\mathfrak{A}_1}(x)$.*
(b) *The unbounded component of $\mathbb{C} \setminus S$ is in $\mathbb{C} \setminus \sigma_{\mathfrak{A}_1}(x)$.*
(c) *$O\partial(\sigma_{\mathfrak{A}_1}(x)) = O\partial(\sigma_{\mathfrak{A}_2}(x))$.*
(d) *$\partial\sigma_{\mathfrak{A}_1}(x) \subset \sigma_{\mathfrak{A}_2}(x)$*
(e) *If $\sigma_{\mathfrak{A}_1}(x) \subset \mathbb{R}$ or $\sigma_{\mathfrak{A}_2}(x) \subset \mathbb{R}$, then $\sigma_{\mathfrak{A}_2}(x) = \sigma_{\mathfrak{A}_1}(x)$.*

Remark. (d) is sometimes called *permanence of spectrum.*

Proof. (a) Let L be a component of $\mathbb{C} \setminus S$. Let $\widetilde{L} = \{\lambda \in L \mid (x - \lambda e)^{-1} \in \mathfrak{A}_1\}$. Since $(x - \lambda e)^{-1}$ is analytic as an $\mathfrak{A}_2$-valued function on L, it is continuous, so $\widetilde{L}$ is closed in L. By Theorem 2.2.6 (with x replaced by $x - \lambda_0 e$ and $y = (\lambda - \lambda_0)e$), $\widetilde{L}$ is open.

(b) Any λ with $\lambda > \|x\|$ is in $\sigma_{\mathfrak{A}_1}(x)$, so the unbounded components have nonempty intersection and so are equal by (a).

(c) is immediate from (b) and the definition of outer boundary.

(d) If $\lambda \in \partial\sigma_{\mathfrak{A}_1}(x)$ but $\lambda \notin \sigma_{\mathfrak{A}_2}(x)$, then there is a disk $D_r(\lambda) \subset \rho_{\mathfrak{A}_2}(x)$ for some $r > 0$. Since $D_r(\lambda)$ is in a single connected component of $\rho_{\mathfrak{A}_2}(x)$, every point is either in $\sigma_{\mathfrak{A}_1}(x)$ or every point is in $\rho_{\mathfrak{A}_1}(x)$. In neither case can λ be in $\partial\sigma_{\mathfrak{A}_1}(x)$.

(e) Since $\sigma_{\mathfrak{A}_2}(x) \subset \sigma_{\mathfrak{A}_1}(x)$, we need only consider the case $\sigma_{\mathfrak{A}_2}(x) \subset \mathbb{R}$. If $K \subset \mathbb{R}$, $K = O\partial(K)$, so by (c), $O\partial\sigma_{\mathfrak{A}_1}(x) \subset \mathbb{R}$ and thus, both $\sigma_{\mathfrak{A}_1}(x)$ and $\sigma_{\mathfrak{A}_2}(x)$ are their outer boundaries, and so they are equal by (c). □

Abelian subalgebras are useful. The awkwardness of impermanence of spectra is ameliorated by the fact that if $\mathfrak{A}_1 \subset \mathfrak{A}$ is abelian, there is an abelian $\mathfrak{A}_2$ so $\mathfrak{A}_1 \subset \mathfrak{A}_2 \subset \mathfrak{A}$, and for all $x \in \mathfrak{A}_1$ (see Problem 11),

$$\sigma_{\mathfrak{A}_2}(x) = \sigma_{\mathfrak{A}}(x) \tag{6.1.41}$$

We end this section with some preliminaries on the theory of ideals. For the rest of this section, $\mathfrak{A}$ will be an abelian Banach algebra with identity.

Definition. An *ideal* in an abelian Banach algebra with identity, $\mathfrak{A}$, is a subalgebra $\mathfrak{i} \subset \mathfrak{A}$ so that

$$x \in \mathfrak{i},\ y \in \mathfrak{A} \Rightarrow xy = yx \in \mathfrak{i} \tag{6.1.42}$$

An ideal is *proper* if $\mathfrak{i} \neq \mathfrak{A}$ and *maximal* if it is maximal among proper ideals, that is, $\mathfrak{i} \subset \mathfrak{j} \neq \mathfrak{A}$ and $\mathfrak{j}$ a proper ideal implies $\mathfrak{i} = \mathfrak{j}$. Given $x \in \mathfrak{A}$, the *principal ideal* generated by x, (x) is given by

$$(x) = \{xy \mid y \in \mathfrak{A}\} \tag{6.1.43}$$

Proposition 6.1.12. *Let $\mathfrak{A}$ be an abelian Banach algebra with identity.*

(a) *(x) is an ideal. It is proper if and only if x is not invertible, that is, if $0 \in \sigma(x)$.*
(b) *The closure of an ideal is closed. The closure of a proper ideal is proper. Every maximal ideal is closed.*
(c) *If $0 \in \sigma(x)$, x is contained in a maximal ideal.*
(d) *If $\mathfrak{m}$ is a maximal ideal and $x \notin \mathfrak{m}$, there exists $y \in \mathfrak{A}$ and $z \in \mathfrak{m}$ so that*

$$xy + z = e \tag{6.1.44}$$

Proof. (a) It is easy to check that (x) is an ideal. If x is invertible and $y \in \mathfrak{A}$, then $y = x(x^{-1}y) \in (x)$, so $y \in (x)$, that is, $(x) = \mathfrak{A}$. If x is not invertible, then $e \notin (x)$, so (x) is proper.

(b) It is easy to confirm that $\bar{\mathfrak{i}}$ is an ideal if $\mathfrak{i}$ is. If $\mathfrak{i}$ is proper, then $e \notin \mathfrak{i}$ (since otherwise, all $x = xe \in \mathfrak{i}$), and if $\|y\| < 1$, then $e - y$ is invertible by Theorem 2.2.6, so $(e-y) \notin \mathfrak{i}$ (since if it were, $(e-y)(e-y)^{-1} = e$ would be in $\mathfrak{i}$). Thus, $\operatorname{dist}(e, \mathfrak{i}) = 1$, so $e \notin \bar{\mathfrak{i}}$. If $\mathfrak{m}$ is a maximal ideal, $\bar{\mathfrak{m}}$ is an ideal, proper (by the above), and $\mathfrak{m} \subset \bar{\mathfrak{m}}$, so $\mathfrak{m} = \bar{\mathfrak{m}}$.

(c) By a simple Zornification (see Problem 12), every proper ideal is contained in a maximal ideal. If x is not invertible, $x \in (x) \subset \mathfrak{m}$, some maximal ideal.

(d) Let $x \notin \mathfrak{m}$, a maximal ideal. $\mathcal{B} = \{xy + z \mid y \in \mathfrak{A},\, x \in \mathfrak{m}\}$ is an ideal by an easy argument. Since $x = xe + 0 \in \mathcal{B}$ and $x \notin \mathfrak{m}$, $\mathfrak{m} \subsetneq \mathcal{B}$. Since $\mathfrak{m}$ is maximal, $\mathcal{B} = \mathfrak{A}$, that is, $e \in \mathcal{B}$. □

The following ideal will be important. We'll eventually prove it is the intersection of all maximal ideals.

Definition. Let $\mathfrak{A}$ be an abelian Banach algebra with identity. The *radical* of $\mathfrak{A}$, $\operatorname{rad}(\mathfrak{A}) = \mathfrak{r}$ is the set of all $x \in \mathfrak{A}$ with $\sigma(x) = \{0\}$. If $\mathfrak{r} = \{0\}$, we say $\mathfrak{A}$ is *semisimple.*

Theorem 6.1.13. $\operatorname{rad}(\mathfrak{A})$ *is an ideal.*

Proof. $x \in \mathfrak{r} \Leftrightarrow \lim\|x^n\|^{1/n} = 0$. Since

$$\|(xy)^n\|^{1/n} \le \|x^n\|^{1/n}\|y^n\|^{1/n} \le \|y\|\|x^n\|^{1/n} \tag{6.1.45}$$

we see $x \in \mathfrak{r} \Rightarrow xy \in \mathfrak{r}$ for all y in $\mathfrak{A}$.

If $x, y \in \mathfrak{r}$, given ε, there is C so

$$\|x^n\| \le C\varepsilon^n, \qquad \|y^n\| \le C\varepsilon^n \tag{6.1.46}$$

Thus,

$$\|(x+y)^n\| \le \sum_{j=0}^{n} \binom{n}{j} \|x^k y^{n-j}\| \le C^2 2^n \varepsilon^n \tag{6.1.47}$$

so

$$\lim\|(x+y)^n\|^{1/n} \le 2\varepsilon \tag{6.1.48}$$

Since ε is arbitrary, $x + y \in \mathfrak{r}$. □

Example 6.1.14. Let X be a Banach space and $A \in \mathcal{L}(X)$ quasinilpotent. Examples 2.2.12 and 2.2.13 present quasinilpotent operators on ℓ^2 and $C([0,1])$, respectively. Let $\mathfrak{A} = \mathfrak{A}_A$. $\mathfrak{A}$ is $\mathbb{1} \oplus (A)$ and A, and so any element of (A), is in $\operatorname{rad}(\mathfrak{A})$. Thus, $\operatorname{rad}(\mathfrak{A})$ can be large, so large that $\operatorname{codim}(\operatorname{rad}(\mathfrak{A})) = 1$. On the other hand, in the next section and in the next theorem, we'll see many Banach algebras of interest are semisimple. □

Theorem 6.1.15. *Every function algebra is semisimple.*

Remark. This proof depends on the use of $\|\cdot\|_\infty$. Once we have the Gel'fand theory, the result will extend to an algebra of functions where all point evaluations are multiplicative linear functions, for example, $\mathbb{W}(\overline{\mathbb{D}})$ or $C^k[0,1]$; see Corollary 6.2.6.

Proof. Clearly, $\|f^2\|_\infty = \|f\|_\infty^2$, so $\|f^{2^n}\|_\infty^{1/2^n} = \|f\|_\infty$ and $\operatorname{spr}(f) = \|f\|_\infty$. Thus, $\operatorname{spr}(f) = 0 \Rightarrow f = 0$. □

Notes and Historical Remarks. During the 1930s, four disparate strands were developed and then brought together by thinking of a general theory of Banach algebras (explicit references appear in the Notes to later sections): the work of Stone on algebraic aspects of $C(X)$, the work of von Neumann on algebras of operators on a Hilbert space, the work of Pontryagin, von Kampen, and Weil on Fourier analysis on general locally compact abelian groups, and the work of Bohr, Bochner, and Besicovitch on almost periodic functions.

The existence of elements in $\mathbb{A}(\overline{\mathbb{D}})$ whose Taylor coefficients are not in L^1 (i.e., functions in $\mathbb{A}[\overline{\mathbb{D}}] \setminus \mathbb{W}(\overline{\mathbb{D}}))$ is somewhat subtle. If $a_n \geq 0$ and $\sum_{n=0}^\infty a_n = \infty$, then $\lim_{x\uparrow 1} \sum_{n=0}^\infty a_n x^n = \infty$ by the monotone convergence theorem, so $\sum_{n=0}^\infty a_n z^n$ cannot lie in $\mathbb{A}(\overline{\mathbb{D}})$. Thus, the phases of a_n have to vary to get an $f \in \mathbb{A}(\overline{\mathbb{D}}) \setminus \mathbb{W}(\overline{\mathbb{D}})$ and the variation can't be too regular—for example, if $(-1)^n a_n \geq 0$, $\sum_{n=0}^\infty |a_n| = \infty$ implies $\lim_{x\downarrow -1} \sum_{n=0}^\infty a_n x^n < \infty$. One tends to think that random or close to random phases are best.

The simplest explicit example goes back to Hardy–Littlewood [**293**] and is discussed in Zygmund [**777**, Sect. 5.4]: If $\alpha \in (0, \frac{1}{2})$ and $c > 0$,

$$f(z) = \sum_{n=1}^\infty e^{icn \log n} n^{-\frac{1}{2}-\alpha} z^n \tag{6.1.49}$$

is in $\mathbb{A}(\overline{\mathbb{D}})$ but obviously not in $\mathbb{W}(\overline{\mathbb{D}})$.

Paley–Zygmund [**510**] (and discussed in Zygmund [**777**, Sect. 5.8]) prove that if $\varepsilon > 0$ and $\sum_{k=1}^\infty |c_k|^2 [\log k]^{1+\varepsilon} < \infty$, then for almost every $\sigma \in \{-1, 1\}^{\mathbb{N}}$ (with $\mathbb{P}(\sigma_j = +1) = \frac{1}{2}$), one has that $\sum_{k=1}^K \sigma_k c_k z^k$ converges in $\mathbb{A}(\overline{\mathbb{D}})$ as $K \to \infty$. If, for example, $c_k = k^{-\beta}$ ($\frac{1}{2} < \beta < 1$), then this limit is not in $\mathbb{W}(\overline{\mathbb{D}})$.

Problem 9 has still another example following Katznelson [**384**].

Beurling weights were introduced by Beurling [**54**] who noted that they could be used to define convolution weights and were the natural setting for extending Wiener's Tauberian theorem.

Problems

1. Let $\mathfrak{A}$ have all the properties of a Banach algebra, except it is a real Banach space. Let $\mathfrak{A}_\mathbb{C}$ be the complexification of $\mathfrak{A}$, that is, $\mathfrak{A} \oplus \mathfrak{A}$ with $i(x, y) = (-y, x)$ and norm such as $\|(x, y)\|_1 = \|x\| + \|y\|$. Show $\mathfrak{A}_\mathbb{C}$ can be given an equivalent norm and product making it into a Banach algebra with norm equivalent to $\|\cdot\|_1$. (*Hint*: If necessary, add an identity and then use Theorem 6.1.3.)

2. Let $\mathfrak{A}$ be a Banach space with an associative, distributive, separately continuous product. This problem will have you prove that for some C,

$$\|xy\| \leq C\|x\|\|y\| \tag{6.1.50}$$

(a) Let L_x be given by $L_x y = xy$. Prove that each $L_x \in \mathcal{L}(\mathfrak{A})$.

(b) For each y, prove that $\sup_{\|x\|\leq 1}\|L_x(y)\| < \infty$.

(c) Prove (6.1.50). (*Hint*: Uniform boundedness principle.)

3. Let G be a locally compact group with Haar measure, $d\mu$. A continuous function $w\colon G \to (0,\infty)$ is called a *Beurling weight* if for all $x, y \in G$,

$$w(xy) \leq w(x)w(y) \tag{6.1.51}$$

(a) If $\mathfrak{A} = L^1(G, w\,d\mu)$, prove that $f, g \in \mathfrak{A} \Rightarrow f * g \in \mathfrak{A}$ and $\|f * g\|_{L^1(w\,d\mu)} \leq \|f\|_{L^1(w\,d\mu)}\|g\|_{L^1(w\,d\mu)}$.

(b) For $G = \mathbb{R}^\nu$ with addition, prove that for any $\alpha > 0$, $(1+|x|^2)^\alpha$ and $e^{\alpha|x|}$ are Beurling weights.

4. Prove that the norm (6.1.14) on $C^k([0,1])$ has $\|fg\| \leq \|f\|\|g\|$ and $\|\mathbb{1}\| = 1$.

5. Let U be the operator (6.1.38) on $\ell^2(\mathbb{Z})$ and V a polynomial in U. With δ_0 the vector with 1 at $n = 0$ and 0 elsewhere, prove that $\|(U^* - V)\delta_0\| \geq 1$ and conclude that $U^* \notin \mathfrak{A}_U$.

6. Let U be the operator (6.1.38) on $\ell^2(\mathbb{Z})$.

(a) For $|z| > 1$, prove that $-\sum_{n=0}^\infty z^{-n-1}U^n$ is an inverse for $(U - z)$.

(b) For $|z| < 1$, prove that $\sum_{n=0}^\infty z^n (U^*)^{n+1}$ is an inverse for $(U - z)$.

(c) For $\omega \in \partial\mathbb{D}$, let $\varphi_n(\omega) = \frac{1}{\sqrt{n}}\sum_{j=0}^{n-1} \bar\omega^j \delta_j$ (with $(\delta_j)_k = \delta_{jk}$). Prove that $\|(U-\omega)\varphi_n\| \to 0$ as $n \to \infty$ and conclude that $\omega \in \sigma_{\mathfrak{A}_U}(U)$ and $\omega \in \sigma_{\mathfrak{A}_{U,U^*}}(U)$.

(d) Prove $\sigma_{\mathfrak{A}_{U,U^*}}(U) = \partial\mathbb{D}$.

(e) Prove that $\sigma_{\mathfrak{A}_U}(U) = \overline{\mathbb{D}}$. (*Hint*: Theorem 6.1.11 and Problem 5.)

7. Let $\mathbb{W}$ be the Wiener algebra on $\mathbb{D}$. Prove that

$$\|fg\|_{\mathbb{W}} \leq \|f\|_{\mathbb{W}}\|g\|_{\mathbb{W}} \tag{6.1.52}$$

8. (a) In $\mathbb{A}(\overline{\Omega})$, prove that $\|f\|_{\mathbb{A}(\Omega)} = \sup_{z\in\partial\Omega}|f(z)|$.

(b) In $\mathbb{A}(\overline{\mathbb{D}})$, suppose $f_n \to f_\infty$ and $f_n(z) = \sum_{j=0}^\infty a_j^{(n)} z^j$. Prove that $a_j^{(n)} \to a_j^{(\infty)}$ as $n \to \infty$ for each j.

9. This problem will construct a function $f \in \mathbb{A}(\overline{\mathbb{D}})$ but not in $\mathbb{W}(\overline{\mathbb{D}})$. The *Rudin–Shapiro polynomials* of degree $2^n - 1$ are defined inductively by

$$\begin{aligned} P_0(z) = Q_0(z) = 1, \qquad P_{n+1}(z) &= P_n(z) + z^{2^n} Q_n(z), \\ Q_{n+1}(z) &= P_n(z) - z^{2^n} Q_n(z) \end{aligned} \tag{6.1.53}$$

(a) For $e^{i\theta} \in \partial\mathbb{D}$, prove that

$$|P_{n+1}(e^{i\theta})|^2 + |Q_{n+1}(e^{i\theta})|^2 = 2(|P_n(e^{i\theta})|^2 + |Q_n(e^{i\theta})|^2) \tag{6.1.54}$$

so that

$$|P_n(e^{i\theta})|^2 + |Q_n(e^{i\theta})|^2 = 2^{n+1} \tag{6.1.55}$$

and

$$\|P_n\|_{\mathbb{A}(\overline{\mathbb{D}})} \le 2^{(n+1)/2}, \qquad \|Q_n\|_{\mathbb{A}(\overline{\mathbb{D}})} \le 2^{(n+1)/2} \tag{6.1.56}$$

(b) Prove there are $\{\sigma_j\}_{j=0}^\infty$ each ± 1 so that

$$P_n(z) = \sum_{j=0}^{2^{n-1}} \sigma_j z^j \tag{6.1.57}$$

(c) Let

$$f_N = \sum_{j=1}^N 2^j (P_j - P_{j-1}) = \sum_{j=0}^N 2^{-j-1} z^{2^j} Q_j(z)$$

Prove that f_N is Cauchy in $\mathbb{A}(\overline{\mathbb{D}})$ and the limit is

$$f_\infty(z) = \sum_{k=1}^\infty 2^{-k-1} \sum_{j=2^k}^{2^{k+1}-1} \sigma_j z^j \tag{6.1.58}$$

(d) Show $f_\infty \in \mathbb{A}(\overline{\mathbb{D}})$ but $f_\infty \notin \mathbb{W}(\mathbb{D})$.

Remarks. 1. The Rudin–Shapiro polynomials were found by Shapiro [**621**] and rediscovered and published by Rudin [**581**]. If R_d is a degree $(d-1)$ polynomial with coefficients ± 1, $\int_0^{2\pi} |R_d(e^{i\theta})|^2 \frac{d\theta}{2\pi} = d$, so $\|R_d\|_\infty \ge \sqrt{d}$. The RS polynomial has $\|R_d\|_\infty = \sqrt{2}\sqrt{d}$ almost optimal, given the bound, which is why they were constructed.

2. As noted by Katznelson [**384**, Sect. 1.6], the f_∞ we constructed restricted to $\partial\mathbb{D}$ is Hölder-continuous of order $\frac{1}{2}$. It is known that if $g \in \mathbb{A}[\overline{\mathbb{D}}]$ restricted to $\partial\mathbb{D}$ is Hölder-continuous of order greater than $\frac{1}{2}$ (see Problem 3 in Section 6.11), then $g \in \mathbb{W}(\mathbb{D})$.

3. The sequence $\{\sigma_j\}_{j=0}^\infty$ is sometimes called the *Golay–Rudin–Shapiro sequence*. As we explained in the Notes, the phases of the Taylor coefficients of a function in $\mathbb{A}(\overline{\mathbb{D}}) \setminus \mathbb{W}(\mathbb{D})$ have to be random-like, and this sequence is a paradigm of a pseudorandom sequence. For example, the

asymptotic fraction of any combination of k successive $+/-$ signs is 2^{-k}. That said, it is missing some aspects of truly random sequences. The sums of the σ's, $\sum_{j=0}^{n-1} \sigma_j = \tau_n$ are $O(\sqrt{n})$, but definitely *not* Gaussian distributed! For example, $\tau_n > 0$ for all n (!) and the set of limit points of $\tau_n/\sqrt{n}$ is the interval $[\sqrt{3/5}, \sqrt{6}]$ (!!). See the papers of Brillhart–Morton [**88, 89**].

10. Verify (6.1.26) where $*$ on the left is given by (6.1.24) and on the right by (6.1.27).

11. This will prove that if $\mathfrak{A}$ is a Banach algebra with identity $\mathfrak{A}_1 \subset \mathfrak{A}$ with $\mathfrak{A}_1$ abelian, then there is $\mathfrak{A}_2 \subset \mathfrak{A}$ with $\mathfrak{A}_2$ abelian and $\mathfrak{A}_1 \subset \mathfrak{A}_2$, so $x \in \mathfrak{A}_1 \Rightarrow$ (6.1.41).

    (a) Let $\mathfrak{A}_2$ be the algebra generated by $\{(x-\lambda)^{-1} \mid x \in \mathfrak{A}_1, \lambda \in p(x)\}$. Prove $\mathfrak{A}_2$ has the required properties.

    (b) For any $\mathcal{B} \subset \mathfrak{A}$, let $\mathcal{B}' = \{y \in \mathfrak{A} \mid xy = yx \text{ for all } x \in \mathcal{B}\}$, the *commutant* of $\mathcal{B}$. Prove if $\mathcal{B}$ is abelian, so is $\mathcal{B}''$ and that $\mathfrak{A}_1'' = \mathfrak{A}_2$ has the required properties. (*Note*: These two $\mathfrak{A}_2$'s may not be the same.)

12. Use Zorn's lemma to prove every proper ideal of an abelian Banach algebra with identity is contained in a maximal ideal.

6.2. The Gel'fand Spectrum and Gel'fand Transform

In this section, we'll associate to every abelian Banach algebra with identity, $\mathfrak{A}$, a compact Hausdorff space, $\widehat{\mathfrak{A}}$. In all our results in this section, $\mathfrak{A}$ will be assumed to be abelian unless explicitly stated otherwise. We'll start by showing $\widehat{\mathfrak{A}}$ has two equivalent descriptions: as multiplicative linear functionals (mlf), $\ell \in \mathfrak{A}^*$, with

$$\ell(xy) = \ell(x)\ell(y), \qquad \ell \not\equiv 0 \tag{6.2.1}$$

or as maximal ideals, $\mathfrak{m}$, with the association

$$\mathfrak{m} = \operatorname{Ker}(\ell) \tag{6.2.2}$$

Duality will then define $\widehat{\ }: \mathfrak{A} \to C(\widehat{\mathfrak{A}})$ and we'll see

$$\|\widehat{x}\|_\infty = \operatorname{spr}(x), \qquad \operatorname{Ran}(\widehat{x}) = \sigma(x) \tag{6.2.3}$$

We'll also show that

$$\operatorname{Ker}(\widehat{\ }) = \mathfrak{r}(\mathfrak{A}) \tag{6.2.4}$$

Next we'll define $\widehat{\mathfrak{A}}$ when $\mathfrak{A}$ does not have an identity and see it is a locally compact space with (6.2.3). We'll then turn to identifying $\widehat{\mathfrak{A}}$ in many of the examples of the last section. Finally, we'll discuss an important subclass of Banach algebras: the regular ones where for any $K \subset \widehat{\mathfrak{A}}$ closed and $\ell_0 \notin K$, there is $x \in \mathfrak{A}$ with $\widehat{x} \restriction K = 0$ and $\widehat{x}(\ell_0) = 1$.

The key to the association of maximal ideals and mlf's is:

Theorem 6.2.1 (Mazur's Theorem). *Let $\mathfrak{A}$ be a Banach field, that is, an abelian Banach algebra with identity so that any $y \neq 0$ has an inverse. Then $\mathfrak{A} \cong \mathbb{C}$ via $\lambda \in \mathbb{C} \leftrightarrow \lambda e \in \mathfrak{A}$.*

Remark. That $\mathfrak{A}$ is abelian plays no role. Lest the reader think the quaternions are a counterexample, we note that their multiplication is not complex linear.

Proof. Let $x \in \mathfrak{A}$ and $\lambda \in \sigma(x)$ which is nonempty. Since $x - \lambda e$ is not invertible, we must have that $x - \lambda e = 0$, that is, $x = \lambda e$. $\square$

Definition. A *multiplicative linear functional* (mlf), ℓ, on a Banach space with identity is a linear map $\ell\colon \mathfrak{A} \to \mathbb{C}$ (not a priori bounded) so that (6.2.1) holds. The set of all mlf's is called the *Gel'fand spectrum* of $\mathfrak{A}$ and denoted $\widehat{\mathfrak{A}}$.

Theorem 6.2.2. (a) *Every $\ell \in \widehat{\mathfrak{A}}$ obeys $\ell(e) = 1$.*

(b) *Every $\ell \in \widehat{\mathfrak{A}}$ is a bounded linear functional.*

(c) *If $\ell \in \widehat{\mathfrak{A}}$ and $x \in \mathfrak{A}$, then*

$$\ell(x) \in \sigma(x) \tag{6.2.5}$$

(d) *$\|\ell\| = 1$ for all $\ell \in \widehat{\mathfrak{A}}$.*

(e) *$\widehat{\mathfrak{A}}$ is a compact Hausdorff space in the weak-* topology, $\sigma(\widehat{\mathfrak{A}}, \mathfrak{A})$ topology.*

(f) *If $\ell \in \widehat{\mathfrak{A}}$, $\mathfrak{m}$ given by (6.2.2) is a maximal ideal. Conversely, if $\mathfrak{m}$ is a maximal ideal, there is a unique $\ell \in \widehat{\mathfrak{A}}$ with (6.2.2).*

Remarks. 1. (b) also follows from (c), (d), but the maximal ideal link provides insight into why (b) holds. This automatic continuity of mlf's implies $\widehat{\mathfrak{A}}$ is independent of norm.

2. Because of (f), $\widehat{\mathfrak{A}}$ is somctimcs called the *maximal ideal space* of $\mathfrak{A}$.

3. The $\sigma(\mathfrak{A}^*, \mathfrak{A})$-topology on $\widehat{\mathfrak{A}}$ is called the *Gel'fand topology.*

Proof. (a) Since ℓ is an mlf, $\ell(e) = \ell(e^2) = \ell(e)^2$, so $\ell(e)$ is 0 or 1. If $\ell(e) = 0$, then for any x, $\ell(x) = \ell(xe) = \ell(x)\ell(e) = 0$, so $\ell \equiv 0$. Thus, $\ell \not\equiv 0$ implies $\ell(e) = 1$.

(b) If $x \in \operatorname{Ker}(\ell)$, $y \in \mathfrak{A}$, then $\ell(xy) = \ell(x)\ell(y) = 0$, so $\operatorname{Ker}(\ell)$ is a proper ideal. Since $\operatorname{codim}(\operatorname{Ker}(\ell)) = 1$ (for $z = \ell(x)y - \ell(y)x \in \operatorname{Ker}(\ell)$, so $\ell(x) \neq 0 \Rightarrow y = \ell(x)^{-1}(z + \ell(y)x) \Rightarrow \operatorname{codim}(\operatorname{Ker}(\ell)) = 1$), $\operatorname{Ker}(\ell)$ is a maximal proper ideal. Such ideals are closed by Proposition 6.1.12(b), so $\operatorname{Ker}(\ell)$ is closed. By Theorem 3.3.2 of Part 1, ℓ is continuous, and so bounded.

(c) $\ell(x - \ell(x)e) = 0$, so $x - \ell(x)e$ is not invertible, since $yz = e$ implies $\ell(y)\ell(z) = 1$ implies $\ell(y) \neq 0$. Thus, $\ell(x) \in \sigma(x)$. For if $yz = e$, $\ell(y)\ell(z) = 1$. Thus, $\ell(y) = 0$ implies y is not invertible.

(d) $|\ell(x)| \leq \operatorname{spr}(x) \leq \|x\|$.

(e) $\widehat{\mathfrak{A}}$ is a subset of the unit ball of $\mathfrak{A}^*$, so by the Banach–Alaoglu theorem, it suffices to show $\widehat{\mathfrak{A}}$ is closed in $\mathfrak{A}^*$. Any weak limit, ℓ, of a net $\ell_\alpha \in \widehat{\mathfrak{A}}$ has $\ell(xy) = \ell(x)\ell(y)$ and $\ell(e) = 1$ by taking limits, so $\ell \in \widehat{\mathfrak{A}}$.

(f) We proved in (b) above that $\mathfrak{m}$ is a maximal ideal. Since, as also noted in (b), $\operatorname{Ker}(\ell)$ plus $\ell(x)$ for one $x \notin \operatorname{Ker}(\ell)$ determine ℓ and $\ell(e) = 1$ for any $\ell \in \widehat{\mathfrak{A}}$, we see that, given $\mathfrak{m}$, there is at most one ℓ with (6.2.2).

Thus, we must show any maximal ideal $\mathfrak{m}$, is the kernel of an mlf. We know that $\mathfrak{m}$ is closed, so we can form $X = \mathfrak{A}/\mathfrak{m}$, the quotient Banach space, with $\pi\colon x \to [x]$ the quotient map. Since $\mathfrak{m}$ is an ideal, $[x][y]$ is the coset $[xy]$. Thus, X is a Banach algebra. By Proposition 6.1.12(d), if $[x] \neq [0]$, $[x]$ is invertible, that is, X is a Banach field. By Mazur's theorem, $X \cong \mathbb{C}$ and π is a linear functional, ℓ, which is multiplicative since $[xy] = [x][y]$. Clearly, $\operatorname{Ker}(\ell) = \mathfrak{m}$. $\square$

Definition. Let $\mathfrak{A}$ be an abelian Banach algebra with identity. We define the *Gel'fand transform*, $\widehat{x}$, for $x \in \mathfrak{A}$ as the function on $\widehat{\mathfrak{A}}$ given by

$$\widehat{x}(\ell) = \ell(x) \tag{6.2.6}$$

Theorem 6.2.3. (a) *$\widehat{x}$ is continuous on $\widehat{\mathfrak{A}}$.*

(b) $\operatorname{Ran}(\widehat{x}) = \operatorname{spr}(x)$.

(c) $\|\widehat{x}\|_\infty = \operatorname{spr}(x)$.

(d) $\widehat{\ }: x \mapsto \widehat{x}$ *is a Banach algebra homomorphism.*

(e) $\operatorname{Ker}(\widehat{\ }) = \mathfrak{r}(\mathfrak{A})$, *the radial of* $\mathfrak{A}$.

(f) $\{\widehat{x}\}_{x\in\mathfrak{A}}$ *separates points in* $\widehat{\mathfrak{A}}$.

(g) $\mathfrak{r}(\mathfrak{A})$ *is the intersection of all maximal ideals in* $\mathfrak{A}$.

Proof. (a) Immediate from the definition of the weak-* topology.

(b) We already showed $\operatorname{Ran}(\widehat{x}) \subset \sigma(x)$. For the converse, suppose $\lambda \in \sigma(x)$. $0 \in \sigma(x - \lambda e)$, so by Proposition 6.1.12(c), $x - \lambda e \in \mathfrak{m}$, some maximal ideal. If $\ell \in \widehat{\mathfrak{A}}$ has $\operatorname{Ker}(\ell) = \mathfrak{m}$, then $\ell(x) = \lambda$, so $\widehat{x}(\ell) = \lambda$, that is, $\sigma(x) \subset \operatorname{Ran}(\widehat{x})$.

(c) By (b), $\|\widehat{x}\|_\infty = \sup\{|\lambda| \mid \lambda \in \sigma(x)\} = \operatorname{spr}(x)$.

(d) Since $\operatorname{spr}(x) \leq \|x\|$, we have

$$\|\widehat{x}\|_\infty \leq \|x\| \tag{6.2.7}$$

so $\widehat{\ }$ is continuous. That it preserves algebraic operations is immediate.

(e) $\operatorname{Ker}(\widehat{\ }) = \{x \mid \|\widehat{x}\|_\infty = 0\} = \{x \mid \operatorname{spr}(x) = 0\} = \mathfrak{r}(\mathfrak{A})$.

(f) If $\ell_1 \neq \ell_2$, then there is x so $\ell_1(x) \neq \ell_2(x)$. For this x, $\widehat{x}(\ell_1) \neq \widehat{x}(\ell_2)$.

(g) $x \in \mathfrak{r}(\mathfrak{A}) \Leftrightarrow \|\widehat{x}\|_\infty = 0 \Leftrightarrow \widehat{x}(\ell) = 0$ for all $\ell \Leftrightarrow x \in$ every maximal ideal. $\square$

Corollary 6.2.4. *Let $\mathfrak{A}$ be a Banach algebra with identity.*

(a) *If $\mathfrak{A}$ is abelian, for all $x, y \in \mathfrak{A}$,*

$$\sigma(x+y) \subset \sigma(x) + \sigma(y) \equiv \{\lambda + \mu \mid \lambda \in \sigma(x),\, \mu \in \sigma(y)\} \tag{6.2.8}$$

(b) *In general (even if $\mathfrak{A}$ is not abelian), if $xy = yx$, then* (6.2.8) *holds.*

Proof. (a) If $\kappa \in \sigma(x+y)$, there exists $\ell \in \widehat{\mathfrak{A}}$, so that $\ell(x+y) = \kappa$. Then $\lambda = \ell(x) \in \sigma(x)$, $\mu = \ell(y) \in \sigma(y)$, so $\kappa = \lambda + \mu \in \sigma(x) + \sigma(y)$.

(b) By Problem 11 of Section 6.1 (with $\mathfrak{A}_1$ the abelian algebra generated by x and y), there is an abelian algebra $\mathfrak{A}_2$ so $\sigma_{\mathfrak{A}_2}(z) = \sigma_{\mathfrak{A}}(z)$ for all $z \in \mathfrak{A}_1$. Thus, (a) for $\mathfrak{A}_2$ implies (b). $\square$

Example 6.2.5. (i) If $y = -x$, $\sigma(x+y) = \{0\}$, while $\sigma(x) + \sigma(y) = \sigma(x) - \sigma(x)$ is typically much larger than $\{0\}$. Thus, the containment in (6.2.8) is often strict (but say, if $y = 0$, it must be equality).

(ii) In $\mathfrak{A} = \mathrm{Hom}(\mathbb{C}^2)$, $x = \left(\begin{smallmatrix} 0 & 1 \\ 0 & 0 \end{smallmatrix}\right)$, $y = \left(\begin{smallmatrix} 0 & 0 \\ 1 & 0 \end{smallmatrix}\right)$, then $\sigma(x) = \sigma(y) = \{0\}$, while $\sigma(x+y) = \{-1, 1\}$. Thus, (6.2.8) can fail if $xy \neq yx$.

(iii) To show the phenomenon of (ii) doesn't depend on nonsymmetric matrices, if $x = \left(\begin{smallmatrix} 1 & 0 \\ 0 & -1 \end{smallmatrix}\right)$, $y = \left(\begin{smallmatrix} 0 & 1 \\ 1 & 0 \end{smallmatrix}\right)$, then $\sigma(x) = \sigma(y) = \{-1, 1\}$, $\sigma(x+y) = \{-\sqrt{2}, \sqrt{2}\}$, so again (6.2.8) fails. $\square$

Corollary 6.2.6. *Let $\mathfrak{A}$ be a Banach algebra of functions, $f\colon X \to \mathbb{C}$ for some set X so that $\mathbb{1} \in \mathfrak{A}$ and the operations are pointwise. Then $\mathfrak{A}$ is semisimple.*

Proof. For each $x \in X$, $\delta_x\colon \mathfrak{A} \to \mathbb{C}$ by $\delta_x(f) = f(x)$ is an mlf. Thus, if $f \in \mathfrak{r}(\mathfrak{A})$, $\delta_x(f) = f(x) = 0$ for all x, so $f \equiv 0$. $\square$

Thus far, we've considered the case where $\mathfrak{A}$ is abelian with identity. If $\mathfrak{A}$ is abelian but has no identity, one can form $\mathfrak{A}_e$ as in Theorem 6.1.8. Of course, the theory above applies to $\mathfrak{A}_e$. If $\tilde{\ell} \in \widehat{\mathfrak{A}}_e$, then

$$\tilde{\ell}(x \oplus \lambda e) = \lambda + \ell(x) \tag{6.2.9}$$

where ℓ is an mlf on $\mathfrak{A}$ and vice-versa, except $\ell \equiv 0$ does not imply that $\tilde{\ell}$ is 0. Put differently, $\mathfrak{A}$ is a maximal ideal in $\mathfrak{A}_e$ corresponding to $\ell \equiv 0$. Notice automatic continuity of $\tilde{\ell}$ implies it for ℓ, that is, automatic continuity holds in abelian Banach algebras without identity.

We define $\widehat{\mathfrak{A}}$, the *Gel'fand spectrum* of $\mathfrak{A}$, to be $\{\ell \in \mathfrak{A}^* \mid \tilde{\ell}$ given by (6.2.9) $\in \widehat{\mathfrak{A}}_e;\ \ell \not\equiv 0\} = \{\ell \in \mathfrak{A}^* \setminus \{0\} \mid \forall x, y \in \mathfrak{A}, \ell(xy) = \ell(x)\ell(y)\}$. This is a locally compact space with $\widehat{\mathfrak{A}}_e$, in general, the one-point compactification (unless $\widehat{\mathfrak{A}}$ is already compact). We summarize in:

Theorem 6.2.7. *Let $\mathfrak{A}$ be an abelian Banach algebra. Then $\widehat{\mathfrak{A}}$, the set of ℓ in $\mathfrak{A}^*$ obeying* (6.2.1), *is a locally compact space in the $\sigma(\mathfrak{A}^*, \mathfrak{A})$-topology. We have:*

(a) *Any ℓ obeying* (6.2.1) *is continuous on $\mathfrak{A}$.*
(b) *$\widehat{x}$, given by* (6.2.6), *lies in $C_0(\widehat{\mathfrak{A}})$.*
(c) *$\widehat{\ }$ is a Banach algebra homeomorphism of $\mathfrak{A}$ into $C_0(\widehat{\mathfrak{A}})$.*
(d) $\sigma_{\mathfrak{A}_e}(x) = \operatorname{Ran}(\widehat{x}) \cup \{0\}$.
(e) $$\|\widehat{x}\|_\infty = \lim_{n\to\infty} \|x^n\|^{1/n}. \tag{6.2.10}$$
(f) $\operatorname{Ker}(\widehat{\ }) = \mathfrak{r}(\mathfrak{A})$.

It can happen, if $\mathfrak{A}$ doesn't have an identity, that $\mathfrak{r}(\mathfrak{A}) = \mathfrak{A}$ (e.g., $\mathfrak{A} = \{P(T) \mid P \text{ polynomial } P(0) = 0\}$ for $T \in \mathcal{L}(X)$ with $\lim_{n\to\infty}\|T^n\|^{1/n} = 0$), so $\widehat{\mathfrak{A}}$ is empty.

Corollary 6.2.8. *Let $\mathfrak{A}$ be a semisimple abelian Banach algebra with norm $\|\cdot\|_1$. Suppose $\|\cdot\|_2$ is another norm on $\mathfrak{A}$ making it into a Banach algebra. Then $\|\cdot\|_1$ and $\|\cdot\|_2$ are equivalent norms.*

Proof. Maximal ideals, and so semisimplicity (given Theorem 6.2.3(g)), are algebraic and thus, $(\mathfrak{A}, \|\cdot\|_2)$ is also semisimple. So, by symmetry, we need only show that the identity map id: $(\mathfrak{A}, \|\cdot\|_1) \to (\mathfrak{A}, \|\cdot\|_2)$ is continuous.

By the closed graph theorem (Theorem 5.4.17 of Part 1), we need only show that if $\|x_n - x\|_1 \to 0$ and $\|x_n - y\|_2 \to 0$, then $x = y$. Let $\ell \in \widehat{\mathfrak{A}}$ which is norm-independent by automatic continuity. Since ℓ is continuous in each norm, $\ell(x_n) \to \ell(x)$ and $\ell(x_n) \to \ell(y)$.

Thus, for all $\ell \in \widehat{\mathfrak{A}}$, $\ell(x - y) = 0$. By semisimplicity, $x = y$. □

Before turning to examples, one more notion. If $x_1, \dots, x_n \in \mathfrak{A}$, we define $\widehat{x}_1 \otimes \cdots \otimes \widehat{x}_n$ on $\widehat{\mathfrak{A}}$ as the $\mathbb{C}^n$-valued function

$$(\widehat{x}_1 \otimes \cdots \otimes \widehat{x}_n)(\ell) = (\widehat{x}_1(\ell), \dots, \widehat{x}_n(\ell)) = (\ell(x_1), \dots, \ell(x_n)) \tag{6.2.11}$$

and the *joint spectrum* of $(x_1, \dots, x_n)$, $j\sigma(x_1, \dots, x_n)$, by

$$j\sigma(x_1, \dots, x_n) = \operatorname{Ran}(\widehat{x}_1 \otimes \cdots \otimes \widehat{x}_n) \tag{6.2.12}$$

There are two wide-ranging methods in computing $\widehat{\mathfrak{A}}$. The first works for some algebras of functions.

Example 6.2.9. ($\widehat{C(X)}$, with X a Compact Hausdorff Space; Example 6.1.4 (revisited)). Clearly, if $x \in X$, δ_x defined by

$$\delta_x(f) = f(x) \tag{6.2.13}$$

lies in $\widehat{C(X)}$. Now let $\mathfrak{m}$ be a maximal ideal. For some x_0, if $\mathfrak{m} \neq \operatorname{Ker}(\delta_{x_0})$, then since $\mathfrak{m}$ is maximal, $\mathfrak{m}$ is not contained in $\operatorname{Ker}(\delta_{x_0})$, that is, there is $f_{x_0} \in \mathfrak{m}$ with $f_{x_0}(x_0) \neq 0$. Suppose for all x_0, $\mathfrak{m} \neq \operatorname{Ker}(\delta_{x_0})$. Define

$$U_{x_0} = \{y \mid f_{x_0}(y) \neq 0\} \tag{6.2.14}$$

Then $\{U_{x_0}\}_{x_0 \in X}$ is an open cover of X, so we can find $x_1, \dots, x_n$ with

$$U_{x_1} \cup U_{x_2} \cup \cdots \cup U_{x_n} = X \tag{6.2.15}$$

Let

$$g = \sum_{j=1}^{n} \bar{f}_{x_j} f_{x_j} \tag{6.2.16}$$

Since $\mathfrak{m}$ is an ideal, $g \in \mathfrak{m}$. But by (6.2.14)/(6.2.15), g is never zero, so by compactness, $\min|g(x)| > 0$, that is, g is invertible. Thus, $\mathfrak{m}$ is not proper!

We conclude that $\mathfrak{m} = \operatorname{Ker}(\delta_{x_0})$ for some x_0. It follows that $\widehat{C(X)} = X$ as a set. By the definition of the weak-* topology, the map $x \mapsto \delta_x$ of X into $\widehat{C(X)}$ is continuous when X is given the original topology and $\widehat{C(X)}$ the Gel'fand topology. Thus, the map is a homeomorphism by Theorem 2.3.11 of Part 1. Notice $\operatorname{Ker}(\widehat{\ })= \{0\}$, so $C(X)$ is semisimple. We summarize in the theorem below.

If X is locally compact and $\mathfrak{A} = C_\infty(X)$ the functions vanishing at ∞, then (unless X is compact), $\mathfrak{A}$ has no identity and $\mathfrak{A}_e = C(X_\infty)$, the continuous functions on the one-point compactifications. Thus, $\widehat{\mathfrak{A}}_e = X_\infty$, the point with $\operatorname{Ker}(\ell) = \mathfrak{A}$ is ∞, and so $\widehat{\mathfrak{A}}$ is X with the original topology. If we instead take $\mathfrak{A} = C(X)$, the set of all bounded continuous functions, $\widehat{\mathfrak{A}}$ is huge—it is a compact space, usually not separable, called the Stone–Čech compactification. We'll discuss it in Section 6.5 (see Theorem 6.5.8). □

Theorem 6.2.10. *Let X be a compact Hausdorff space. Then $\widehat{C(X)}$ is X with the original topology and $\widehat{f}$ is f.*

Example 6.2.11 ($\widehat{C^k[0,1]}$; Example 6.1.5 (revisited)). The same analysis as in the last example shows that $\widehat{C^k[0,1]} = [0,1]$ and $\widehat{f} = f$, so $C^k[0,1]$ is semisimple. □

Example 6.2.12 (General Function Algebras). Let $\mathfrak{A}$ be a function algebra on a compact Hausdorff space, X. It can happen there is another compact Hausdorff space, Y, so that $\mathfrak{A}$ is isometrically isomorphic to a function algebra on Y. For example, let $\mathfrak{A} = \mathbb{A}(\overline{\mathbb{D}})$, the analytic functions on $\mathbb{D}$ with a continuous extension to $\overline{\mathbb{D}}$. By the maximum principle, the map $\alpha\colon \mathfrak{A} \to C(\partial\mathbb{D})$ by $\alpha(f) = f \restriction \partial\mathbb{D}$ is an isometry, and so an isometric isomorphism of $\mathfrak{A}$ and $\operatorname{Ran}(\alpha)$, a function algebra on $\partial\mathbb{D}$. Problem 4 of Section 6.10 will identify $\operatorname{Ran}(\alpha)$ intrinsically. In the general case, $\widehat{\ }\colon \mathfrak{A} \to C(\widehat{\mathfrak{A}})$

is an isometry in this case since $\|f^n\|_\infty = \|f\|_\infty^n$ so $\|\widehat{f}\| = \text{spr}(f) = \|f\|$ and $\mathfrak{A}$ can be realized as a function algebra on $\widehat{\mathfrak{A}}$ (e.g., if we look at the $\mathfrak{A}$ above as a function algebra on $\partial\mathbb{D}$, this recovers $\mathbb{D}$!). If $\mathfrak{A}$ is a function algebra on Y and $\tau_y(f) = f(y)$, then $\tau_y \in \widehat{\mathfrak{A}}$ and $y \mapsto \tau_y$ is one–one (since $\mathfrak{A}$ separates points) and continuous, so $Y \subset \widehat{\mathfrak{A}}$. Thus, $\widehat{\mathfrak{A}}$ can be viewed as the maximal space on which $\mathfrak{A}$ can be realized as a function algebra. We will return to this theme in Section 6.10. □

The other major tool in identifying $\widehat{\mathfrak{A}}$ involves the following:

Definition. Let $\mathfrak{A}$ be an abelian Banach algebra with identity. We say $x_1, \dots, x_n$ *generate* $\mathfrak{A}$ if the polynomials in $x_1, \dots, x_n$ are dense in $\mathfrak{A}$. We say they *weakly generate* $\mathfrak{A}$ if polynomials in $\{x_1, \dots, x_n\} \cup \{(x_j - \lambda e)^{-1}, \lambda \notin \sigma(x_j)\}$ are dense in $\mathfrak{A}$.

Theorem 6.2.13. *Let $\{x_1, \dots, x_n\}$ generate or weakly generate $\mathfrak{A}$, an abelian Banach algebra with identity. Then*

$$\widehat{x}_1 \otimes \cdots \otimes \widehat{x}_n \colon \widehat{\mathfrak{A}} \to j\sigma(x_1, \dots, x_n) \subset \mathbb{C}^n \tag{6.2.17}$$

is a homeomorphism.

Proof. $\widehat{x}_1 \otimes \cdots \otimes \widehat{x}_n$ is a continuous map from $\widehat{\mathfrak{A}}$ onto $j\sigma(x_1, \dots, x_n)$. If we can show it is one–one, then it is a homeomorphism by Theorem 2.3.11 of Part 1.

If $\ell(x_j) = \tilde{\ell}(x_j)$ for $j = 1, \dots, n$ and some $\ell, \tilde{\ell} \in \mathfrak{A}$, then since $\ell((x_j - \lambda e)^{-1}) = (\ell(x_j) - \lambda)^{-1}$, we see ℓ and $\tilde{\ell}$ agree on polynomials in $\{x_1, \dots, x_n\} \cup \{(x_j - \lambda e)^{-1}, \lambda \in \sigma(x_j)\}$. Since they are dense in $\mathfrak{A}$, ℓ and $\tilde{\ell}$ agree on $\mathfrak{A}$. We have thus proven $\widehat{x}_1 \otimes \cdots \otimes \widehat{x}_n$ is one–one. □

Example 6.2.14 (Single Operator Algebra). Let $A \in \mathcal{L}(X)$ for a Banach space, X. Let $\mathfrak{A}_A$ be generated by A so $\widehat{\mathfrak{A}}_A = \sigma_{\mathfrak{A}}(A)$. This may not be the spectrum of A as an element of $\mathcal{L}(X)$ (see Example 6.1.9). Let $\mathfrak{A}_A^\sharp$ be the algebra generated by $\{A\} \cup \{(A - \lambda)^{-1} \mid \lambda \notin \sigma_{\mathcal{L}(X)}(A)\}$. Clearly, $\sigma_{\mathfrak{A}^\sharp}(A) = \sigma_{\mathcal{L}(X)}(A)$ and A weakly generates $\mathfrak{A}^\sharp$, so $\widehat{\mathfrak{A}_A^\sharp} = \sigma_{\mathcal{L}(X)}(A)$.

We will prove later (see Theorem 6.4.7) that if A is a normal operator on a Hilbert space, then $\widehat{\ } \colon \mathfrak{A}_{A,A^*}^\sharp \to C(\sigma(A))$ is an isometric isomorphism. Its inverse will be the continuous functional calculus! □

Example 6.2.15. (Singly Generated Analytic Function Algebra; Example 6.1.6 (revisited)). Let Ω be a bounded region in $\mathbb{C}$. $\mathbb{P}(\overline{\Omega})$ and $\mathbb{W}(\overline{\mathbb{D}})$ are singly generated (by the function $f(z) = z$ which we'll call z) and $\mathbb{R}(\overline{\Omega})$ is weakly generated by z. Thus, we need to determine their spectra.

$\mathbb{W}(\overline{\mathbb{D}})$ is easy. For any $z_0 \in \overline{\mathbb{D}}$, $f \mapsto f(z_0)$ is an mlf, so $\overline{\mathbb{D}} \subset \sigma_{\mathbb{W}(\overline{\mathbb{D}})}(z)$. On the other hand, if $z_0 \notin \overline{\mathbb{D}}$,

$$(z - z_0)^{-1} = -\sum_{n=0}^{\infty} \frac{z^n}{z_0^{n+1}} \tag{6.2.18}$$

Since $\sum_{n=0}^{\infty} |z_n|^{-n-1} < \infty$, this inverse is in $\mathbb{W}(\overline{\mathbb{D}})$, so $\sigma_{\mathbb{W}(\overline{\mathbb{D}})}(z) = \overline{\mathbb{D}}$. Therefore,

$$\widehat{\mathbb{W}(\overline{\mathbb{D}})} = \overline{\mathbb{D}} \tag{6.2.19}$$

and the mlf's are all evaluation maps. We'll say more about this later.

For $\mathbb{R}(\overline{\Omega})$, we note $\overline{\Omega} \subset \sigma(z)$ since for $z_0 \in \Omega$, $\delta_{z_0} \colon f \mapsto f(z_0)$ is an mlf. Moreover, if $z_0 \notin \Omega$, $(z - z_0)^{-1} \in \mathbb{R}(\overline{\Omega})$, so $z_0 \in \rho(z)$. Thus, $\sigma(z) = \overline{\Omega}$. Since $\mathbb{R}$ is weakly generated by z,

$$\widehat{\mathbb{R}(\overline{\Omega})} = \overline{\Omega} \tag{6.2.20}$$

$\mathbb{P}(\overline{\Omega})$ is a little more complicated, but not much more. Let $\widetilde{\Omega}$ be the union of $\overline{\Omega}$ and all bounded components of $\mathbb{C} \setminus \overline{\Omega}$, that is, the bounded simply connected set with $O\partial(\widetilde{\Omega}) = O\partial(\overline{\Omega})$. By the maximum principle, for any polynomial,

$$\sup_{z \in \widetilde{\Omega}} |p(z)| - \sup_{z \in \overline{\Omega}} |p(z)| \tag{6.2.21}$$

so any $f \in \mathbb{P}(\overline{\Omega})$ extends to a continuous function on $\widetilde{\Omega}$ (indeed, one analytic on $(\widetilde{\Omega})^{\text{int}}$). Thus, for any $z_0 \in \widetilde{\Omega}$, $\delta_{z_0} \colon f \mapsto f(z_0)$ is an mlf, so $\widetilde{\Omega} \subset \sigma(z)$. On the other hand, $\mathbb{P}(\overline{\Omega}) \subset C(\widetilde{\Omega})$, so by Theorem 6.1.11, $\mathbb{C} \setminus \sigma_{C(\Omega)}(z) \subset \mathbb{C} \setminus \sigma_{\mathbb{P}(\overline{\Omega})}(z)$ because $\mathbb{C} \setminus \sigma_{C(\widetilde{\Omega})}(z) = \mathbb{C} \setminus \widetilde{\Omega}$ only has an unbounded component. Thus, $\sigma(z) = \widetilde{\Omega}$ and

$$\widehat{\mathbb{P}(\overline{\Omega})} = \widetilde{\Omega} \tag{6.2.22}$$

since z generates $\mathbb{P}$.

Neither $\mathbb{A}(\overline{\Omega})$ nor $H^\infty(\Omega)$ are singly generated so that their Gel'fand spectra are harder to describe. For both, evaluation at points (of $\overline{\Omega}$ for $\mathbb{A}$ and of Ω for H^∞) are mlf's and, as algebras of functions, both are semisimple. Below we'll prove $\widehat{\mathbb{A}(\overline{\Omega})} = \overline{\Omega}$. That $\mathbb{D}$ is dense in $\widehat{H^\infty(\mathbb{D})}$ is discussed in the Notes. □

The Gel'fand theory provides effortless proofs of the following results about $\mathbb{W}(\overline{\mathbb{D}})$:

Theorem 6.2.16. (a) (Wiener Tauberian Theorem). *If $f \in \mathbb{W}(\overline{\mathbb{D}})$ and $f(z) \neq 0$ for all $z \in \overline{\mathbb{D}}$, then $f(z)^{-1} \in \mathbb{W}(\overline{\mathbb{D}})$.*

(b) (Wiener–Lévy Theorem). *Let $f \in \mathbb{W}(\overline{\mathbb{D}})$ and let G be analytic in a neighborhood of* $\mathrm{Ran}(f)$. *Then*

$$g(z) = G(f(z)) \tag{6.2.23}$$

lies in $\mathbb{W}(\overline{\mathbb{D}})$.

Remarks. 1. The point is that this shows certain functions have absolutely convergent Taylor series.

2. (b) implies (a) by taking $G(w) = w^{-1}$.

3. Wiener's original proof was complicated, so Gel'fand's proof caused a stir. Problem 1 has an elementary proof of Newman; see also Problem 2.

Proof. (a) f is not invertible if and only if $0 \in \sigma_{\mathbb{W}}(f) = \mathrm{Ran}(\widehat{f}) = f[\overline{\mathbb{D}}]$ since $\widehat{\mathbb{W}} = \overline{\mathbb{D}}$. Thus, $f(z) \neq 0$ for all $z \in \overline{\mathbb{D}}$ implies f is invertible as an element of $\mathbb{W}(\overline{\mathbb{D}})$.

(b) Since $\widehat{\mathbb{W}} = \overline{\mathbb{D}}$, $\mathrm{Ran}(f)$ is $\sigma(f)$, and the claim follows from the analytic functional calculus of Section 2.3. $\square$

Tauberian theorems are converses of easy limit theorems that require an extra hypothesis. If $a \in \ell^\infty(\mathbb{N})$ and $a_n \to \alpha$ as $n \to \infty$, then it is easy to see $(b * a)_n \to \alpha(\sum_{m=1}^\infty b_m)$ as $n \to \infty$ for any $b \in \ell^1(\mathbb{N})$. The following result shows the Tauberian nature of the Wiener Tauberian theorem:

Corollary 6.2.17. *Let $a \in \ell^\infty(\mathbb{N})$, $b \in \ell^1(\mathbb{N})$ so that $(a * b)_n \to \beta$ as $n \to \infty$. Suppose b is such that for all $z \in \overline{\mathbb{D}}$, $B(z) = \sum_{n=0}^\infty b_n z^n \neq 0$. Then as $n \to \infty$,*

$$a_n \to \frac{\beta}{\sum_{n=1}^\infty b_n} \tag{6.2.24}$$

Remarks. 1. Since $B(1) = \sum b_n$, by hypothesis, $\sum_{n=1}^\infty b_n \neq 0$.

2. Section 6.11, discusses analogs of this for $L^1(\mathbb{R})$ rather than $\ell^1(\mathbb{N})$.

Proof. The map $\Phi\colon \ell^1 \to \mathbb{W}(\overline{\mathbb{D}})$

$$\Phi(a)(z) = \sum_{n=0}^\infty a_n z^n \tag{6.2.25}$$

is an isometric isomorphism of algebras if ℓ^1 is given convolution as product and $\mathbb{W}(\overline{\mathbb{D}})$ pointwise products. By the hypothesis $B = \Phi(b)$ nonvanishing on $\overline{\mathbb{D}}$ and the Wiener Tauberian theorem, there is $C = \Phi(c)$ with $C(z)^{-1} = B(z)$, that is,

$$c * b = \delta_0 \tag{6.2.26}$$

Thus, by the abelian result mentioned above,

$$a_n = [(c * b) * a]_n = [c * (b * a)]_n \to \beta \sum_{m=1}^{\infty} c_m \tag{6.2.27}$$

Since $B(1)C(1) = 1$, we see $\sum_{m=1}^{\infty} c_m = (\sum_{m=1}^{\infty} b_m)^{-1}$, so (6.2.27) is (6.2.24). $\square$

There is an analog for $\mathbb{W}(\partial\mathbb{D})$, which is the set of $f\colon \partial\mathbb{D} \to \mathbb{C}$, so $\sum_{n=-\infty}^{\infty} |f_n^\sharp| < \infty$. This is isomorphic to $\ell^1(\mathbb{Z})$ and is an algebra under convolution on the ℓ^1 side and pointwise multiplication on the $\mathbb{W}(\partial\mathbb{D})$ side. If $f \in \mathbb{W}(\partial\mathbb{D})$, since $f_n^\sharp \in \ell^1$, $\sum_{n=-N}^{N} f_n^\sharp e^{in\theta} \to f$ in $\|\cdot\|_{\mathbb{W}}$. Thus, $e^{\pm i\theta}$ generate $\mathbb{W}$ since $e^{-i\theta} = (e^{i\theta})^{-1}$, $e^{i\theta}$ weakly generates $\mathbb{W}$. Evaluation shows $\partial\mathbb{D} \subset \sigma_{\mathbb{W}(\partial\mathbb{D})}(e^{i\theta})$ while explicit formulae for $(e^{i\theta} - z_0)^{-1}$ in terms of geometric power series in z_0 for $|z_0| < 1$ and geometric Laurent series for $|z_0| > 1$ show that $\mathbb{C} \setminus \mathbb{D} \subset \rho_{\mathbb{W}(\partial\mathbb{D})}(e^{i\theta})$. Therefore,

$$\sigma_{\mathbb{W}(\partial\mathbb{D})}(e^{i\theta}) = \partial\mathbb{D} \tag{6.2.28}$$

and the same proof as Theorem 6.2.16(a) shows that:

Theorem 6.2.18 (Wiener's Tauberian Theorem for $\mathbb{W}(\partial\mathbb{D})$). *$f \in \mathbb{W}(\partial\mathbb{D})$ is invertible if and only if for all $z \in \partial\mathbb{D}$, $f(z) \neq 0$.*

This has an interesting reformulation on the ℓ^1 side:

Theorem 6.2.19 (Wiener's Tauberian Theorem for $\ell^1(\mathbb{Z})$). *Let $a \in \ell^1(\mathbb{Z})$ and let $\tau_m(a)$ be defined by*

$$(\tau_m(a))_n = a_{n+m} \tag{6.2.29}$$

Then the space, T_a, spanned by $\{\tau_m(a)\}_{m=-\infty}^{\infty}$ is dense in $\ell^1(\mathbb{Z})$ if and only if for all $e^{i\theta} \in \partial\mathbb{D}$, $\sum_{n=-\infty}^{\infty} a_n e^{in\theta} \neq 0$.

Proof. If b has finite support, $b * a \in T_a$, so for any $b \in \ell^1$, $b * a \in \overline{T}_a$. If a is invertible in ℓ^1 (with convolution as product), $(c * a^{-1}) * a = c \in \overline{T}_a$ for any $c \in \ell^1$. If a is not invertible, since $b * a$ is not invertible for any b, $\|e - b * a\|_1 \geq 1$ for all b, so $e \notin \overline{T}_a$ and $\overline{T}_a$ is not all of ℓ^1. The criteria for invertibility in $\mathbb{W}(\partial\mathbb{D})$ completes the proof. $\square$

Remark. This reformulation is important because, while Theorem 6.2.18 requires an identity in $\ell^1(\mathbb{Z})$, Theorem 6.2.19 does not. $L^1(\mathbb{R}, dx)$ doesn't have an identity, so it will be the analog of Theorem 6.2.19 that we'll prove in Section 6.11.

Before leaving the subject of algebras of analytic functions, we want to turn to determining $\widehat{\mathbb{A}(\overline{\Omega})}$. Let z be the function taking value z_0 at $z_0 \in \overline{\Omega}$.

If $z_1 \notin \overline{\Omega}$, $(z-z_1)^{-1} \in \mathbb{A}(\overline{\Omega})$ so $\sigma_{\mathbb{A}(\overline{\Omega})}(z) \subset \overline{\Omega}$. On the other hand, for each $z \in \overline{\Omega}$, $\delta_z \colon f \mapsto f(z)$ is an mlf, so $\overline{\Omega} \subset \sigma_{\mathbb{A}(\Omega)}(z)$, that is, we have proven that

$$\sigma_{\mathbb{A}(\overline{\Omega})}(z) = \overline{\Omega} \tag{6.2.30}$$

Since, in general, z does not generate $\mathbb{A}(\overline{\Omega})$, this is not enough to conclude that $\widehat{\mathbb{A}(\overline{\Omega})} = \overline{\Omega}$ (which we will prove below by other means). But it is an important first step for it implies that if $\ell \in \widehat{\mathbb{A}(\overline{\Omega})}$, then $z_0 \equiv \ell(z) \in \sigma(z) = \overline{\Omega}$. Suppose $z_0 \in \Omega$. Then every $f \in \mathbb{A}(\overline{\Omega})$ has

$$g(z) = \frac{f(z) - f(z_0)}{z - z_0} \in \mathbb{A}(\overline{\Omega}) \tag{6.2.31}$$

so $f = f(z_0) + (z - z_0)g \Rightarrow \ell(f) = f(z_0) + (\ell(z) - z_0)\ell(y) = f(z_0)$, showing $\ell = \delta_{z_0}$.

The same argument would work when $z_0 \in \partial\Omega$ if (6.2.31) held for all f's in $\mathbb{A}(\overline{\Omega})$. It is easy to see it does not. But it suffices this is true for a dense set of f's. The following lemma is thus critical:

Lemma 6.2.20. *Let $f \in \mathbb{A}(\overline{\Omega})$ and $z_0 \in \overline{\Omega}$. Then there exists $f_n \in \mathbb{A}(\overline{\Omega})$ so that each f_n has an analytic continuation to a neighborhood of z_0 and $\|f - f_n\|_\infty \to 0$.*

Sketch. We'll sketch the construction and leave the details to Problem 5. Without loss (by translating), we can suppose $z_0 = 0$.

(1) As a preliminary, we note that for any δ,

$$h_\delta(z) = \int_{|w|\le\delta} |w - z|^{-1}\, d^2w \tag{6.2.32}$$

is maximized at $z = 0$ (this follows, e.g., from Problem 17 of Section 2.2 of Part 3), where the value is $2\pi\delta$, that is,

$$|h_\delta(z)| \le 2\pi\delta \tag{6.2.33}$$

(2) It follows that if h has compact support in $\mathbb{C}$ and is bounded and measurable that

$$Q(h)(z) = \int \frac{1}{w - z}\, h(w)\, d^2w \tag{6.2.34}$$

is continuous on $\widehat{\mathbb{C}} = \mathbb{C} \cup \{\infty\}$ (by cutting off $(w-z)^{-1}$ and using (6.2.33) to prove uniform convergence as the cutoff goes away) and is analytic in z on $(\mathbb{C} \cup \{\infty\}) \setminus \operatorname{supp}(h)$ by Morera's theorem.

(3) As another preliminary, we note, using $\Delta(\frac{1}{2\pi}\log|z|) = \delta(z)$, $\Delta = 4\bar\partial\partial$ (where $\partial = \frac{1}{2}(\frac{\partial}{\partial x} - i\frac{\partial}{\partial y})$, $\bar\partial = \frac{1}{2}(\frac{\partial}{\partial x} + i\frac{\partial}{\partial y})$), $\frac{1}{2\pi}\log|z| = \frac{1}{4\pi}[(\log z) + (\log \bar z)]$,

and $\partial \log z = 1/z$, that as distributions,

$$\frac{\partial}{\partial \bar{w}} \left[\frac{1}{\pi} (w-z)^{-1} \right] = \delta(w-z) \tag{6.2.35}$$

that is, if $g \in C_0^\infty(\mathbb{C})$, then

$$\frac{1}{\pi} \int \frac{1}{w-z} \frac{\partial g}{\partial \bar{w}} (w) \, d^2 w = -g(z) \tag{6.2.36}$$

(4) For $f \in \mathbb{A}(\overline{\Omega})$ extended continuously to $\widehat{\mathbb{C}}$ (using the Tietze extension theorem (Theorem 2.2.5 of Part 1)) and $g \in C_0^\infty(\mathbb{C})$, define

$$F(f,g)(z) = \int \frac{f(w)-f(z)}{w-z} \frac{\partial g}{\partial \bar{w}} \, d^2 w \tag{6.2.37}$$

We claim first since, for $z \in \Omega$, $(f(w)-f(z))/(w-z)$ is analytic in w on Ω, $F(f,g)(z)$ is analytic in z on Ω (by Morera's theorem). Moreover, by using (6.2.36) on the $-f(z)$ term, we see

$$F(f,g)(z) = f(z)g(z) + Q\biggl(f \frac{\partial q}{\partial \bar{w}} \biggr) \tag{6.2.38}$$

(5) Pick $g_n(w)$ in C_0^∞ supported in $\{w \mid |w| \le \frac{2}{n}\}$ so that $g_n(w) = 1$ on $\{w \mid |w| \le \frac{1}{n}\}$, and so that, for a constant, D,

$$\left\| \frac{\partial g_n}{\partial \bar{w}} \right\|_\infty \le Dn \tag{6.2.39}$$

and define on $\widehat{\mathbb{C}}$,

$$f_n(z) = f(z) - F(f, g_n) \tag{6.2.40}$$

As noted, $F(f,g_n)$ is analytic in Ω and by (6.2.38) and continuity of Q, continuous on $\widehat{\mathbb{C}}$, so $f_n \in \mathbb{A}(\overline{\Omega})$. On $\{z \mid |z| < \frac{1}{n}\}$, by (6.2.38) and $g_n(z) = 1$ there, we see that $f_n(z) = -Q(f\frac{\partial g}{\partial \bar{w}})(z)$ which is analytic there since Q is analytic off $\operatorname{supp}(f\frac{\partial g}{\partial \bar{w}}) \subset \{w \mid \frac{1}{n} \le |w| \le \frac{2}{n}\}$.

(6) For $|z| > \frac{2}{n}$, by (6.2.38), $(f - f_n) = Q(f\frac{\partial g}{\partial \bar{w}})$ is analytic, including at ∞ on account of (2). Thus, by the maximum principle,

$$\sup_{z \in \widehat{\mathbb{C}}} |f(z) - f_n(z)| = \sup_{|z| \le \frac{2}{n}} |f(z) - f_n(z)| \tag{6.2.41}$$

$$\le Dn \sup_{\substack{|z| \le \frac{2}{n} \\ |w| \le \frac{2}{n}}} |f(w) - f(z)| (2\pi) \left(\frac{2}{n} \right) \tag{6.2.42}$$

where, crudely estimated, $F(f,g_n)$ is given by (6.2.37) by bounding $|\frac{\partial g}{\partial \bar{w}}|$ by (6.2.39), and $\int_{|w| \le \frac{2}{n}} |w-z|^{-1} \, d^2 w$ by (6.2.33). In (6.2.42), the n and $\frac{1}{n}$ cancel and the sup goes to zero as $n \to \infty$ by continuity of f at $z_0 = 0$. Thus, in $\mathbb{A}(\overline{\Omega})$, $f - f_n \to 0$. $\square$

Theorem 6.2.21 (Arens' Theorem). $\widehat{\mathbb{A}(\overline{\Omega})} = \overline{\Omega}$ *in that every mlf is of the form* $\ell(f) = f(z_0)$ *for a* $z_0 \in \overline{\Omega}$.

Proof. Let $\ell \in \widehat{\mathbb{A}(\overline{\Omega})}$. As noted above, $\ell(z) \equiv z_0 \in \overline{\Omega}$. Given $f \in \mathbb{A}(\overline{\Omega})$, use the lemma to find $f_n \in \mathbb{A}(\overline{\Omega})$ so that

$$\|f_n - f\|_\infty \to 0, \quad f_n \text{ analytic in a neighborhood of } z_0 \tag{6.2.43}$$

Let

$$g_n(z) = \frac{f_n(z) - f_n(z_0)}{z - z_0} \tag{6.2.44}$$

which is in $\mathbb{A}(\overline{\Omega})$ if we define it to be $g_n'(z_0)$ at $z = z_0$.

Thus,

$$\begin{aligned} \ell(f) &= \lim \ell(f_n) \\ &= \lim \ell(f_n(z_0) + (z - z_0)g_n) \\ &= \lim\bigl[f_n(z_0) + \ell(g_n)(\ell(z - z_0))\bigr] \\ &= \lim f_n(z_0) = f(z_0) \end{aligned} \tag{6.2.45}$$

proving that ℓ is evaluation at z_0. $\square$

That completes our discussion of $\widehat{\mathfrak{A}}$ when $\mathfrak{A}$ is an algebra of analytic functions.

Example 6.2.22 ($L^\infty(X, d\mu)$). Let μ be a Baire measure on X, a locally compact, σ-compact space. Let $\mathfrak{A} = L^\infty(X, d\mu)$. Then (Problem 6) $\widehat{\ }$ is an isometry of L^∞ onto $C(\widehat{\mathfrak{A}})$ and $\widehat{\mathfrak{A}}$ is totally disconnected, that is, the only connected subsets are single points.

Next we turn to $\mathfrak{A} = L^1(G, d\mu)$, where G is a locally compact abelian (henceforth LCA) group and μ is Haar measure. We use $(x, y) \mapsto x + y$ for the group product, minus for the inverse, and 0 for the group identity.

Definition. A *character* on an LCA group is a continuous function $\chi\colon G \to \partial\mathbb{D} = \{z \in \mathbb{C} \mid |z| = 1\}$ obeying

$$\chi(x + y) = \chi(x)\chi(y), \quad \chi(0) = 1 \tag{6.2.46}$$

Given a character χ, we define ℓ_χ on L^1 by

$$\ell_\chi(f) = \int f(x)\,\overline{\chi(x)}\,d\mu(x) \tag{6.2.47}$$

Theorem 6.2.23. *Every ℓ_χ is an element of $\widehat{L^1(G,d\mu)}$, and conversely, for every $\ell \in \widehat{L^1(G,d\mu)}$, there is a unique character, χ, so $\ell = \ell_\chi$.*

Remark. Our proof that ℓ_χ is an mlf only requires χ be measurable and obey (6.2.46), but we prove that every mlf is an ℓ_χ with χ continuous. Thus, any measurable character is a.e. a continuous character and if we demand (6.2.46) for all x, y, one can see (Problem 7) that they are equal everywhere, not just a.e. This is a special case of a general result we'll prove for group representation in Theorem 6.8.2.

Proof. If ℓ_χ is given by (6.2.47), we have

$$\begin{aligned}\ell_\chi(f*g) &= \int f(x-y)g(y)\,\overline{\chi(x)}\,d\mu(x)d\mu(y) \\ &= \int f(w)g(y)\,\overline{\chi(w+y)}\,d\mu(w)d\mu(y) \\ &= \int f(w)\,\overline{\chi(w)}\,g(y)\,\overline{\chi(y)}\,d\mu(w)d\mu(y) \\ &= \ell_\chi(f)\ell_\chi(g) \end{aligned}\tag{6.2.48}$$

so ℓ_χ is an mlf.

Uniqueness of χ follows from uniqueness of L^∞-representatives of $\ell \in (L^1)^*$.

Conversely, suppose ℓ is an mlf on L^1 not identically 0. Pick $f \in L^1(G,d\mu)$ with $\ell(f) \neq 0$. Define for each $x \in G$ and $h \in L^1$, $\tau_x(h)$ by

$$\tau_x(h)(y) = h(y-x) \tag{6.2.49}$$

and define $\chi(x)$ by

$$\chi(x) = \frac{\ell(\tau_x(f))}{\ell(f)} \tag{6.2.50}$$

It is easy to see (Problem 8) that $x \mapsto \tau_x f$ is continuous in L^1, so χ is continuous. Since $\|\tau_x(f)\|_1 = \|f\|_1$ and ℓ is bounded, $\chi \in L^\infty$.

It is easy to see that τ_x commutes with convolution, that is,

$$\tau_x(g*h) = (\tau_x g)*h = g*(\tau_x h) \tag{6.2.51}$$

In particular,

$$\tau_x f * \tau_{-x} f = \tau_x\tau_{-x}(f*f) = f*f \tag{6.2.52}$$

so applying ℓ to this and using $\ell(g*h) = \ell(g)\ell(h)$, we see

$$\chi(x)\chi(-x) = 1 \tag{6.2.53}$$

Similarly,

$$\tau_x f * \tau_y f = \tau_{x+y} f * f \tag{6.2.54}$$

which implies

$$\chi(x+y) = \chi(x)\chi(y) \tag{6.2.55}$$

In particular, $\chi(nx) = \chi(x)^n$, so $\chi \in L^\infty$ implies $|\chi(x)| \leq 1$. Since $|\chi(-x)| \leq 1$, we have by (6.2.53) that

$$|\chi(x)| = 1, \qquad \chi(-x) = \overline{\chi(x)} \tag{6.2.56}$$

Convolution can be rewritten

$$f * g = \int g(x)(\tau_{-x}(f))\, d\mu(x) \tag{6.2.57}$$

first pointwise, but then multiplying by $h(y) \in L^\infty$ and integrating and using Fubini's theorem,

$$L(f * g) = \int g(x) L(\tau_{-x}(f))\, d\mu(x) \tag{6.2.58}$$

for any linear functional L. In particular, picking $\ell = L$ and using $\ell(f * g) = \ell(f)\ell(g)$ and $\ell(f) \neq 0$, we see that

$$\ell(g) = \int g(x)\chi(-x)\, d\mu(x) = \int g(x)\, \overline{\chi(x)}\, d\mu(x) \tag{6.2.59}$$

so ℓ is an ℓ_χ. $\square$

In Section 6.9, we'll study $\widehat{L^1(G, d\mu)}$ further, but we mention several results now:

(1) Using the notion of positive functionals that we'll discuss in the next section, we'll prove there are enough characters so that $f \mapsto \widehat{f}$ is injective, that is, $L^1(G, d\mu)$ is semisimple.

(2) We'll show the Gel'fand topology on $\widehat{L^1}$ is equivalent to pointwise convergence, that is, $\ell_{\chi_\alpha} \to \ell_\chi$ in $\sigma((L^1)^*, L^1)$ if and only if $\chi_\alpha(x) \to \chi(x)$ for each x. In this topology, the characters form a topological group under pointwise multiplication which is also an LCA group, called $\widehat{G}$. There is a natural induced normalization of Haar measure on $\widehat{G}$ so that the Fourier inversion and Plancherel theorems hold.

Example 6.2.24 (Characters for $\mathbb{R}^\nu, \mathbb{T}^\nu, \mathbb{Z}^\nu$). By a simple argument (Problem 9),

$$\widehat{G_1 \times G_2} = \widehat{G_1} \times \widehat{G_2}$$

that, is, any character on $G_1 \times G_2$ has the form $(x_1, x_2) \mapsto \chi_1(x_1)\chi_2(x_2)$ for χ_j a character on G_j. Thus, we need only look at $\mathbb{R}, \mathbb{T}, \mathbb{Z}$.

Consider $\mathbb{R}$ first. If $f \in C_0^\infty(\mathbb{R})$, $x \mapsto \tau_x f$ is a C^∞-function of $\mathbb{R}$ to L^1. Since $C_0^\infty(\mathbb{R})$ is dense in L^1, for any $\ell \in (L^1)^*$, we can pick $f \in C_0^\infty(\mathbb{R})$ with $\ell(f) \neq 0$. This shows that $\chi \in \widehat{\mathbb{R}}$ is a C^∞-function. Letting $\beta = \chi'(0)$ and differentiating $\chi(x+y) = \chi(x)\chi(y)$ at $y = 0$ leads to $\chi'(x) = \beta\chi(x)$. Thus, since $\chi(0) = 1$, $\chi(x) = e^{\beta x}$. $|\chi(x)| = 1$ for all x implies $\beta \in i\mathbb{R}$. It is easy to see every such function is in $\widehat{\mathbb{R}}$. Thus, if

$$\chi_k(x) = e^{ikx}, \qquad k \in \mathbb{R} \tag{6.2.60}$$

then

$$\widehat{\mathbb{R}} = \{\chi_k \mid k \in \mathbb{R}\} \cong \mathbb{R} \tag{6.2.61}$$

Therefore, the Gel'fand transform on $L^1(\mathbb{R}^\nu, d^\nu x)$ is

$$\widehat{f}(\mathbf{k}) = \int e^{-i\mathbf{k}\cdot\mathbf{x}} f(x)\, d^\nu x \tag{6.2.62}$$

which is essentially the Fourier transform (except for the pesky $(2\pi)^{-\nu/2}$).

For $\mathbb{T}$, again χ is C^∞ in θ (as a function of $e^{i\theta}$), so $\chi(e^{i\theta}) = e^{ik\theta}$, but the requirement of single-valuedness and continuity on $\mathbb{T}$ (i.e., 2π-periodicity in θ) implies $k \in \mathbb{Z}$. Thus,

$$\widehat{\mathbb{T}^\nu} = \mathbb{Z}^\nu \tag{6.2.63}$$

and the Gel'fand transform maps $L^1(\mathbb{T}^\nu, \frac{d^\nu\theta}{(2\pi)^\nu})$ to $\ell_0^\infty(\mathbb{Z}^\nu)$ (sequences going to zero at ∞) and is just $f \mapsto f^\sharp$ in the language of Fourier series (see Section 3.5 of Part 1).

For $\mathbb{Z}$, $\chi(1) \in \partial\mathbb{D} \cong \mathbb{T}$ and $\chi(n) = \chi(1)^n$, that is, $\chi(n) = e^{in\theta}$. Thus,

$$\widehat{\mathbb{Z}^\nu} \cong \mathbb{T}^\nu \tag{6.2.64}$$

and the Gel'fand transform is the inverse of Fourier series. Thus, Fourier analysis is a special case of the Gel'fand transform!

Notice also $\mathbb{Z}^\nu$ and $\mathbb{T}^\nu$ have the fact that $\widehat{\widehat{G}} = G$. This illustrates Pontryagin duality which says $\widehat{\widehat{G}} = G$ in general. It is discussed in Section 6.9. $\square$

Given the spectral radius formula and the fact that the Gel'fand transform is the Fourier transform (given our normalization, only up to $(2\pi)^{-\nu/2}$), we obtain the following theorem of Beurling (proven before the Gel'fand theory):

Theorem 6.2.25 (Beurling [**54**]). *If $f \in L^1(\mathbb{R}^\nu)$, then*

$$\|\widehat{f}\|_\infty = \lim_{n\to\infty} (2\pi)^{-\nu/2} \|\underbrace{f * \cdots * f}_{n}\|_1^{1/n} \tag{6.2.65}$$

Example 6.2.26 (Characters for $\mathbb{R}_+$ and Laplace Transforms). The same argument that shows mlf's on $L^1(G, d\mu)$ are associated to characters works for abelian semigroups with an invariant measure. In particular, for $\mathbb{R}_+ = [0, \infty)$, we want χ's obeying (6.2.46), but since there aren't additive inverses, we can't argue $|\chi(x)| = 1$ but only $|\chi(x)| \le 1$. Thus, $\chi_\alpha(x) = e^{-\alpha x}$ with $\operatorname{Re}\alpha \ge 0$, that is,

$$\widehat{L^1([0,\infty), dx)} = \mathbb{H}_+ \equiv \{\alpha \mid \operatorname{Re}\alpha \ge 0\}$$

The Gel'fand transform is

$$\widehat{f}(\alpha) = \int_0^\infty e^{-\alpha x} f(x)\, dx \tag{6.2.66}$$

which is just the Laplace transform. □

As a final topic, we discuss regular abelian Banach algebras. In the Notes to Section 2.2 of Part 1, we defined a completely regular space as a topological space where points are closed and points can be separated from closed sets by continuous functions. This is connected with the following

Definition. An abelian Banach algebra is called *regular* if for all $\ell \in \widehat{\mathfrak{A}}$ and closed sets $K \subset \widehat{\mathfrak{A}}$ with $\ell \notin K$, there exists $f \in \mathfrak{A}$ so that $\widehat{f}(\ell) = 1$ and $\widehat{f} \restriction K \equiv 0$.

By Urysohn's lemma, $C(X)$ is regular, and for a normal operator, the spectral theorem (see Section 6.4) implies $\mathfrak{A}_{A,A^*}$ is regular. For spaces of analytic functions where, typically, $\widehat{f}$ is analytic on suitable closed sets, regularity does not hold. Using C_0^∞-functions (and the fact that $\mathcal{S} \subset (\operatorname{Ran}(\widehat{\ }) : L^1(\mathbb{R}^\nu) \to C_\infty(\mathbb{R}^\nu)))$, it is clear that $L^1(\mathbb{R}^\nu, d^\nu x)$ is regular. It is a deeper result that for all LCA groups, $L^1(G, d\mu)$ is regular.

Given a set S in $\widehat{\mathfrak{A}}$, one defines its *kernel* as

$$k(S) = \{f \in \mathfrak{A} \mid \widehat{f} \restriction S = 0\} \tag{6.2.67}$$

Notice that if S is thought of as a set of maximal ideals, $k(S)$ is the intersection of the ideals in S. Given a set $T \subset \mathfrak{A}$, we define its *hull* by

$$h(T) = \{\ell \in \widehat{\mathfrak{A}} \mid \ell(f) = 0 \text{ for all } f \in T\} \tag{6.2.68}$$

Notice $h(T)$ is the set of maximal ideals containing T.

In Problem 10, the reader will prove that if $\mathfrak{A}$ is a regular algebra, then for any $S \subset \widehat{\mathfrak{A}}$,

$$h(k(S)) = \overline{S} \tag{6.2.69}$$

the closure in the Gel'fand topology. In general, $S \mapsto h(k(S))$ can be shown to be a closure operation and so it generates a topology known as the *hull-kernel* topology. Since any hull is closed in the Gel'fand topology, this topology is weaker than the Gel'fand topology. It is known (see the Notes) that the hull-kernel topology is Hausdorff if and only if $\mathfrak{A}$ is regular if and only if the Gel'fand and hull-kernel topologies agree.

Notes and Historical Remarks.

> Vladimir Arnold, who with Gel'fand was one of Kolmogorov's most brilliant pupils, suggested that, if both men arrived in a mountainous country, Kolmogorov would start scaling the highest peaks, while Gel'fand would start building roads.
>
> —*London Times* obituary of I. M. Gel'fand[1]

What we call Mazur's theorem is due to Mazur [**470**]. Mazur actually stated the more subtle theorem that the only Banach algebras over $\mathbb{R}$ (not $\mathbb{C}$) that are division rings are $\mathbb{R}$, $\mathbb{C}$, and the quaternions. Mazur's paper has no proof but is only an announcement (it is claimed because the editor insisted he shorten an already simple proof). Mazur's proof first appeared in a 1973 book [**772**] written by a member of his school. It began with the complex case using analytic functions and the use of harmonic functions to handle the real case.

Mazur's result is an infinite-dimensional analog of an 1878 theorem of Frobenius [**211**] that the only finite-dimensional algebras over $\mathbb{R}$ which are division rings are $\mathbb{R}$, $\mathbb{C}$, and the quaternions. Gel'fand [**228**] found the complex version using the proof we do here so that some refer to this result as the Gel'fand–Mazur theorem. This paper first had the spectral radius formula used in this proof. Stone [**672**] has a proof of Mazur's theorem that avoids analytic function theory. Bogdan [**76**] has an interesting pedagogical presentation of the full division algebra over $\mathbb{R}$ theorem. In particular, he has a version of Frobenius' theorem that replaces the conditions of existence of identity and inverses with the weaker condition of no zero divisors.

Gel'fand announced his results in 1939 [**226**] and published the details in 1941 [**228**]. Shortly after the announcement in 1940, Gel'fand, Raikov, and Shilov wrote a long review [**235**], only published in 1946 because of the war. This appeared in English translation as a book [**236**].

Israel Moissevich Gel'fand (1913–2009) was a Jewish Soviet mathematician born in a small town in the southern Ukraine. Because his father had been a mill owner, he was expelled from high school, so he went to Moscow at age 16 and taught elementary mathematics in evening school while attending lectures at Moscow University. At age 19, he was admitted as a graduate student (with no elemenary degrees!) and did a thesis in functional analysis under Kolmogorov. His initial publication in the theory of normed rings in 1938 was his second thesis.

He was a professor in Moscow State University from 1941 until 1990 when he moved to the US (at age 76) and became a professor at Rutgers. In 1953, he was made an associate member of the Russian Academy, but despite being the most active Russian mathematician, because of antisemitism, he

[1] `http://www-history.mcs.st-andrews.ac.uk/Obits2/Gelfand_Times.html`

only became a full member in 1984, several years after winning the Wolf Prize.

Gel'fand was a machine builder rather than a solver of specific problems. Besides Banach algebra, he was a key figure in the development of distribution theory (with a series of five books shortly after Schwartz' work), inverse scattering theory (Gel'fand–Levitan equation), and representation and cohomology theory of Lie groups. He was also instrumental in the development that led to the Atiyah–Singer index theorem.

His students include Berezin, Bernstein, Kirillov, Shilov, and Szemerédi. His weekly seminar, which ran for many years, was a famous fixture of the Moscow mathematics scene. Around 1960, Gel'fand got interested in cell biology and became an important researcher in control systems in complex multicell systems. He also developed a famous program for gifted high-school students, something he repeated after he came to the US.

The uniqueness of the norm theorem (Corollary 6.2.8) is due to Johnson [**348**].

An important precursor to Gel'fand's work is Stone's work in 1937 [**670**], where he defined his version of the Stone–Čech compactification as the maximal ideal space (he used the term "divisorless ideals") of $C(X)$ with X a completely regular space. He did not have a connection to mlf's, so no duality and no Gel'fand transform. Also, he didn't have the Gel'fand topology but rather used what we call the hull-kernel topology (and others call the Stone topology and still others the Jacobson topology)—in the case of $C(X)$, it agrees with Gel'fand topology, as we have seen. While Stone's work is only a partial first step towards the Gel'fand theory, given that Gel'fand–Kolmogorov have a paper [**230**] extending Stone's work, it seems likely it was a factor in Gel'fand's thinking.

The Wiener Tauberian theorem was proven by Wiener in 1932 [**757**]. He also had a version for $L^1(\mathbb{R}, dx)$: If $f \in L^1$, then $\{\tau_x f\}_{x\in\mathbb{R}}$ is dense in L^1 if and only if $\widehat{f}(k) \neq 0$ for all k. This in turn implies that if $h \in L^\infty$ and $\lim_{x\to\infty}(h * f)(x) = \alpha$ for some f with $\widehat{f}(k) \neq 0$ for all k, then $\lim_{x\to\infty}(h * g)(x) = \alpha(\int g(x)\,dx / \int f(x)\,dx)$. We discuss Wiener's result in Section 6.11. The Wiener–Lévy theorem is an extension by Lévy [**443**].

Because Wiener's proof was so complicated, Gel'fand's Banach algebra proof in [**227**] caused quite a stir. Regular algebras were introduced to extend the result from the discrete ℓ^1 or ℓ^1_+ case to L^1. As noted, Problem 1 has a simple proof of Newman for the ℓ^1 case which doesn't extend to L^1.

There was, for a time, a famous open question about the maximal ideal space of $H^\infty(\mathbb{D})$. Clearly this space contains $\mathbb{D}$ via evaluating, i.e., $\ell_z(f) = f(z)$. The Corona conjecture asserted that $\mathbb{D}$ was dense in $H^\infty(\mathbb{D})$, i.e., every maximal ideal was defined via limits of nets in $\mathbb{D}$. The name comes from the fact that this says $\mathbb{D}$, unlike the sum, doesn't have a corona! It was solved affirmatively by Carleson [**103**] in 1962. A slick proof due to Tom Wolff [2] can be found in Garnett [**223**], Koosis [**402**], and Narasimhan–Nievergelt [**494**]. Gamelin [**220**] has a particularly elementary variant. For extensions of the Corona theorem to infinitely connected subsets of $\widehat{\mathbb{C}}$, see Carleson [**105, 103**] and Jones–Marshall [**351**].

Arens' theorem is due to Arens [**22**] in 1958.

That the Gel'fand spectrum of an L^1-algebra is $\widehat{G}$ is basic to the approach of Fourier analysis on LCA groups, and we discuss its history in the Notes to Section 6.9.

The notion of regular algebra goes back to Shilov [**623, 624**]. For a discussion of the relation of regularity to equality of the Gel'fand and hull-kernel topology and to the Hausdorff nature of the hull-kernel topology, see the monograph of Kaniuth [**368**].

Problems

1. This problem will lead you through an "elementary" proof of the Wiener Tauberian theorem for $\ell^1(\mathbb{Z})$. Let $\mathbb{W}(\partial\mathbb{D})$ be the set of functions, f, on $\partial\mathbb{D}$ with $\|f\|_{\mathbb{W}} \equiv \sum_{n=-\infty}^{\infty} |f_n^\sharp| < \infty$, where

$$f_n^\sharp = \int_0^{2\pi} e^{-in\theta} f(e^{i\theta}) \, \frac{d\theta}{2\pi} \tag{6.2.70}$$

(a) If f is C^1, prove that $f \in \mathbb{W}(\partial\mathbb{D})$ and for a universal constant, C,

$$\|f\|_{\mathbb{W}} \leq C(\|f\|_1 + \|f'\|_1) \tag{6.2.71}$$

(*Hint*: $\sum_{n\neq 0} \frac{1}{n^2} < \infty$.)

(b) If g is C^1 with $|g(e^{i\theta})| \geq \varepsilon > 0$ for all θ, prove that for a g-dependent constant, D, and all $k \geq 0$, that

$$\|g^{-k}\|_{\mathbb{W}} \leq Dk\varepsilon^{-k} \tag{6.2.72}$$

(c) Prove the C^1-functions are dense in $\mathbb{W}$. (*Hint*: Truncate the Fourier series!)

(d) If $x, y \in \mathfrak{A}$, an abelian Banach algebra with identity, with y invertible, prove that

$$x \sum_{n=0}^{N} \frac{(y-x)^n}{y^{n+1}} = 1 - \frac{(y-x)^{N+1}}{y^{N+1}} \tag{6.2.73}$$

---

[2] Wolff never published his proof because after he found it, he lectured on it. Garnett and Koosis heard about the proof and with permission included it in their books with attribution. At that point, Wolff didn't see any need to publish a paper about it!

and conclude that if

$$\sum_{n=0}^{\infty} \left\| \frac{(y-x)^n}{y^{n+1}} \right\| < \infty \tag{6.2.74}$$

then x is invertible.

(e) If $f \in \mathbb{W}(\partial\mathbb{D})$ has $\alpha \equiv \inf_\theta |f(e^{i\theta})| > 0$, prove that f has an inverse in $\mathbb{W}(\partial\mathbb{D})$. (*Hint*: Find $g \in C^1$ so that $\|f - g\| < \frac{1}{2}\alpha$.)

Remarks. 1. This proof is due to Newman [**503**] but is close to one in Zygmund's classic book [**777**].

2. Since this relies on strict positivity and an identity, this proof does not extend to $L^1(\mathbb{R})$. Since it uses derivatives, it doesn't extend to general discrete LCA groups.

2. (a) Extend the argument of Problem 1 to $\mathbb{W}(\overline{\mathbb{D}})$.

   (b) Use the argument of Problem 1 to prove the Lévy–Wiener theorem.

3. Find a Banach algebra proof of an analog of the Wiener Tauberian theorem for $\ell^1(G)$ with G a discrete abelian group.

4. Let $\mathfrak{A}$ be an abelian Banach algebra with identity, e.

   (a) Let $x_1, \dots, x_n \in \mathfrak{A}$. Prove that exactly one of the following holds:
   (i) $\exists \ell \in \widehat{\mathfrak{A}}$ so that $\widehat{x}_j(\ell) = 0$ for $j = 1, \dots, n$.
   (ii) $\exists y_1, \dots, y_n \in \mathfrak{A}$ so that $\sum_{j=1}^n x_j y_j = e$
   (*Hint*: Consider the set $\{\sum_{j=1}^n x_j y_j\}$.)

   (b) Prove the following theorem of Wiener: If $f_1, \dots, f_n \in \mathbb{W}(\overline{\mathbb{D}})$ with no common zero in $\overline{\mathbb{D}}$, then there exists $g_1, \dots, g_n \in \mathbb{W}(\mathbb{D})$ so that $\sum_{j=1}^n f_j g_j \equiv 1$.

5. This will provide the details of the proof of Lemma 6.2.20.

   (a) Prove that h_δ given by (6.2.32) is maximized at $z = 0$.

   (b) Let $Q_\delta(h)$ be (6.2.34) modified for the integral to go over $\{w \mid |w-z| \geq \delta\}$. Prove $Q_\delta(h)(z)$ is continuous on $\widehat{\mathbb{C}} = \mathbb{C} \cup \{\infty\}$ (with $Q_\delta(h)(\infty) = 0$) and that $\|Q_\delta - Q\|_\infty \to 0$. Conclude that $Q(h)$ is continuous on $\widehat{\mathbb{C}}$.

   (c) Prove $Q(h)$ is analytic on $\widehat{\mathbb{C}} \setminus \operatorname{supp}(h)$. (*Hint*: Morera.)

   (d) Verify (6.2.36) for $g \in C_0^\infty(\mathbb{C})$.

   (e) If $f \in \mathbb{A}(\overline{\Omega})$, prove f is the restriction of a continuous function on $\widehat{\mathbb{C}}$ (which we'll call f).

   (f) Define F by (6.2.37) and prove it is analytic in Ω.

   (g) Prove (6.2.38).

(h) Prove g_n can be picked obeying (6.2.39) with $g_n(z) = 1$ if $|z| \leq 1/n$ and $= 0$ if $|z| \geq 1/n$.

(i) Prove that f_n given by (6.2.40) is in $\mathbb{A}(\overline{\Omega})$ and analytic in $\{z \mid |z| < 1/n\}$.

(j) Prove that $\|f - f_n\|_{\infty,\widehat{\mathbb{C}}} \to 0$.

6. Let $\mathfrak{A} = L^\infty(X, d\mu)$, where μ is a Baire measure on X, a locally compact, σ-compact space (actually, the arguments below work for any σ-finite measure space). This will show $\widehat{\ }$ is an isometry of $\mathfrak{A}$ to $C(\widehat{\mathfrak{A}})$ and that $\widehat{\mathfrak{A}}$ is totally disconnected.

   (a) Prove $\mathfrak{A}$ is a Banach algebra under $\|\cdot\|_\infty$ and pointwise operations.

   (b) Prove $\|\widehat{f}\| = \|f\|$. (*Hint*: $\|f^n\|_\infty = \|f\|_\infty^n$.)

   (c) If f is real-valued, prove that $\widehat{f}$ is real-valued. (*Hint*: If $\lambda \in \mathbb{C} \setminus \mathbb{R}$, prove $\lambda \notin \sigma(f)$.)

   (d) Prove that $\widehat{\overline{f}} = \overline{\widehat{f}}$.

   (e) Prove $\text{Ran}(\widehat{\ }) = C(\widehat{\mathfrak{A}})$ and conclude $\widehat{\ }$ is an isometric isomorphism of $\mathfrak{A}$ and $C(\widehat{\mathfrak{A}})$.

   (f) Let $A \subset X$ be a Baire set. Prove that $\widehat{\chi}_A$ is the characteristic function of a set $\widehat{A} \in \widehat{\mathfrak{A}}$ with $\widehat{A}$ both open and closed.

   (g) Given distinct $\ell_1, \ell_2 \in \widehat{\mathfrak{A}}$, prove there is f which is a finite sum of $\sum_{j=1}^N \alpha_j \chi_{A_j}$, where $\alpha_j \in \mathbb{C}$ and A_j are disjoint Baire sets so that $\ell_1(f) \neq \ell_2(f)$.

   (h) Prove there is a Baire set, A, so $\ell_1 \in \widehat{A}$, $\ell_2 \notin \widehat{A}$ and conclude there is no connected set $B \subset \mathfrak{A}$ so $\ell_1, \ell_2 \in B$.

   (i) Prove that $\widehat{\mathfrak{A}}$ is totally disconnected.

7. Let G be a locally compact abelian group, $\chi\colon G \to \partial\mathbb{D}$ a measurable function, so that (6.2.46) holds for all $x, y \in G$. Suppose χ is not a.e. 0. Prove that χ is a continuous function. (*Hint*: Prove χ is given by (6.2.50).)

8. Let G be a metrizable locally compact group.

   (a) If $f \in C_0(G)$, the continuous function of compact support, prove that $x \mapsto \tau_x(f)$ is continuous in x as an L^1-valued function. (*Hint*: Dominated convergence.)

   (b) For any $f \in L^1(G, d\mu)$, prove that $x \mapsto \tau_x(f)$ is continuous in x as an L^1-valued function.

   (c) Extend this to L^p for $1 < p < \infty$.

9. (a) If G_1, G_2 are two LCA groups, prove that $\widehat{G_1 \times G_2} = \widehat{G_1} \times \widehat{G_2}$, where $\widehat{G_j}$ is the Gel'fand spectrum of $L^1(G, d\mu)$.

   (b) State and prove an analog of this for general abelian Banach algebras.

10. Let $\mathfrak{A}$ be an abelian Banach algebra.

    (a) For any $S \subset \widehat{\mathfrak{A}}$, prove that

$$S \subset h(k(S)) \tag{6.2.75}$$

    (b) For any $T \subset \mathfrak{A}$, prove that $h(T)$ is closed in the Gel'fand topology.

    (c) If $\mathfrak{A}$ is regular, prove that (6.2.70) holds, and conversely, if (6.2.70) holds, then $\mathfrak{A}$ is regular.

11. This problem will lead you through the proof of the following theorem of Gleason [**249**] and Kahane–Żelazko [**364**]: A codimension 1 subspace, X, of an abelian Banach algebra, $\mathfrak{A}$, with identity is a maximal ideal if and only if all $x \in X$ are noninvertible. One direction is immediate, so suppose X has codimension 1 and no invertible elements.

    (a) Prove there exists a continuous $\ell\colon \mathfrak{A} \to \mathbb{C}$ with $\ell(e) = 1$ and $\operatorname{Ker} \ell = X$. (*Hint*: Use the fact that the noninvertible elements are closed.)

    (b) Fix $x \in X$. Prove that $f(z) = \ell(e^{zx})$ is an entire function with no zeros (here $e^{zx} \equiv \sum_{n=0}^\infty (zx)^n/n!$).

    (c) Prove that $|f(z)| \le \|\ell\| \exp(|z|\|x\|)$ and conclude that for some α, $f(z) = e^{\alpha z}$. (*Hint*: The improved Cauchy estimate of Section 3.2 of Part 2A.)

    (d) Prove that $\ell(x^2) = \ell(x)^2$.

    (e) Prove that $\ell(xy) = \ell(x)\ell(y)$. (*Hint*: $2xy = (x+y)^2 - x^2 - y^2$.)

    (f) Conclude that X is a maximal ideal.

    (g) In the real Banach algebra, $C_{\mathbb{R}}([0,1])$, let $X = \{f \mid \int_0^1 f(x)\,dx = 0\}$. Prove X only contains noninvertible elements, is not an ideal, and has codimension 1 (so this theorem fails for real Banach algebras).

6.3. Symmetric Involutions

$C(X)$ with $f \mapsto \bar{f}$ and $\mathcal{L}(\mathcal{H})$, $\mathcal{H}$ a Hilbert space, with $A \mapsto A^*$, have extra structure motivating the following:

Definition. An *involution* on a Banach algebra, $\mathfrak{A}$, is a map $x \mapsto x^*$ of $\mathfrak{A} \to \mathfrak{A}$ obeying

(i) * is antilinear;
(ii) * is involutive, that is, $(x^*)^* = x$;
(iii) $(xy)^* = y^* x^*$ for all $x, y \in \mathfrak{A}$.

Our goal in this section is to single out a subclass in the abelian case and prove that in that case (when $\mathfrak{A}$ has an identity), there is a one–one correspondence between those $\varphi \in \mathfrak{A}^*$ with those $\varphi(x^*x) \geq 0$ for all x and all $\varphi(e) = 1$, and $\mu \in \mathcal{M}_{+,1}(\widehat{\mathfrak{A}})$ via

$$\varphi(x) = \int \widehat{x}(\ell)\, d\mu_\varphi(\ell) \tag{6.3.1}$$

This generalizes Bochner's theorem (Theorem 6.6.6 of Part 1) and will play a major role in Section 6.9.

Proposition 6.3.1. *If $\mathfrak{A}$ is a Banach algebra with identity, e, and involution, *, then*

$$e^* = e \tag{6.3.2}$$

Proof. $e = (e^*)^* = (e^*e)^* = e^*e^{**} = e^*e = e^*$. □

We have not explicitly required that * be continuous. In Problem 1, the reader will show that, for an abelian semisimple Banach algebra, all involutions are bounded. If

$$\|x^*\| = \|x\| \tag{6.3.3}$$

we say the involution is *norm-compatible.*

Definition. Let $\mathfrak{A}$ be an (even nonabelian) Banach algebra with involution, *. $x \in \mathfrak{A}$ is called

(i) *symmetric* if $x = x^*$;
(ii) *positive* if $x = y^*y$ for some $y \in \mathfrak{A}$;
(iii) *unitary* if $x^*x = xx^* = e$.

Theorem 6.3.2. *Let $\mathfrak{A}$ be an abelian Banach algebra with identity and with involution *. Then the following are equivalent:*

(1) *For all $\ell \in \widehat{\mathfrak{A}}$ and $x \in \mathfrak{A}$,*

$$\ell(x^*) = \overline{\ell(x)} \tag{6.3.4}$$

(2) *For all $x \in \mathfrak{A}$, $\widehat{x^*} = \overline{\widehat{x}}$.*
(3) *Every maximal ideal, $\mathcal{M}$, has $\mathcal{M}^* = \mathcal{M}$ in the sense that $y \in \mathcal{M} \Rightarrow y^* \in \mathcal{M}$.*
(4) *$\ell(x)$ is real for every symmetric x and $\ell \in \widehat{\mathfrak{A}}$.*
(5) *$\ell(x)$ is positive for each positive $x \in \mathfrak{A}$ and $\ell \in \widehat{\mathfrak{A}}$.*
(6) *For all $x \in \mathfrak{A}$, $e + x^*x$ is invertible.*
(7) *$|\ell(x)| = 1$ for each unitary x and $\ell \in \widehat{\mathfrak{A}}$.*

Remark. (6) is unique in that it doesn't use $\widehat{\mathfrak{A}}$ and so makes sense in the nonabelian case. Indeed, some authors define an involution to be symmetric if (6) holds. We'll see it recur in Corollary 6.7.10.

Proof. We will show (1) $\Leftrightarrow$ (2) $\Leftrightarrow$ (3) $\Rightarrow$ (7) $\Rightarrow$ (4) $\Rightarrow$ (1) $\Rightarrow$ (5) $\Rightarrow$ (6) $\Rightarrow$ (4).

<u>(1) $\Leftrightarrow$ (2).</u> Immediate from $\widehat{x}(\ell) = \ell(x)$.

<u>(1) $\Leftrightarrow$ (3).</u> If (6.3.4) holds, $\ell(y) = 0 \Rightarrow \ell(y^*) = \overline{\ell(y)} = 0$. If (3) holds, $y - \ell(y)e \in \mathcal{M} \Rightarrow y^* - \overline{\ell(y)}\, e \in \mathcal{M} \Rightarrow \ell(y^*) = \overline{\ell(y)}$.

<u>(1) $\Rightarrow$ (7).</u> If $y^*y = e$, then $1 = \ell(y^*y) = \ell(y^*)\ell(y) = |\ell(y)|^2$ if (1) holds, so $|\ell(y)| = 1$.

<u>(7) $\Rightarrow$ (4).</u> Let x be symmetric and $y = e^{ix}$ defined by the power series

$$y = \sum_{n=0}^{\infty} \frac{i^n}{n!}\, x^n \tag{6.3.5}$$

By the usual power series manipulation using $(x + w)^n = \sum_{j=1}^{n} \binom{n}{j} x^j w^{n-j}$, one has for all $x, w \in \mathfrak{A}$ that

$$e^{i(x+w)} = e^{ix} e^{iw} \tag{6.3.6}$$

Since $x = x^*$, y^* has the same form as (6.3.5) with x replaced by $-x$. Thus, $y^*y = e$ so y is unitary. By (7), $|\ell(y)| = 1$. By (6.3.5),

$$\ell(y) = \exp(i\ell(x))$$

so $|\ell(y)| = 1$ implies $\operatorname{Im} \ell(x) = 0$ so $\ell(x) \in \mathbb{R}$.

<u>(4) $\Rightarrow$ (1).</u> Any $x = w + iy$ with $w = \frac{1}{2}(x + x^*)$, $y = (2i)^{-1}(x - x^*)$, and w, y symmetric. If $\ell(w)$ and $\ell(y)$ are in $\mathbb{R}$, clearly $\ell(x^*) = \ell(w) - i\ell(y) = \overline{\ell(x)}$.

<u>(1) $\Rightarrow$ (5).</u> $\ell(y^*y) = |\ell(y)|^2$ if $\overline{\ell(y)} = \ell(y^*)$.

<u>(5) $\Rightarrow$ (6).</u> $\ell(x^*x) \geq 0 \Rightarrow \ell(e + x^*x) > 0 \Rightarrow 0 \notin \sigma(e + x^*x)$, so $(e + x^*x)$ is invertible.

<u>(6) $\Rightarrow$ (4).</u> If (4) fails, there is $x = x^*$, so $\ell(x) = \alpha + i\beta$ with $\beta \neq 0$. Let $y = \beta^{-1}(x - \alpha e)$. Then $y = y^*$ and $\ell(y) = i$ and $\ell(e + y^*y) = 1 + \ell(y)^2 = 0$, so $e + y^*y$ is not invertible. $\square$

Definition. Let $\mathfrak{A}$ be an abelian Banach algebra with identity and involution, *. If one and hence all conditions in Theorem 6.3.2 hold, we say that *, the involution, is *symmetric*. If $\mathfrak{A}$ doesn't have an identity, we say the involution is symmetric if (6.3.4) holds. If we extend * to $\mathfrak{A}$, by setting $e^* = e$, then (1)–(6) hold for $\mathfrak{A}_e \Leftrightarrow$ (6.3.4) holds for $\mathfrak{A}$.

Notice that if * is a symmetric involution, $\operatorname{Ran}(\widehat{\ }) = \{\widehat{x} \mid x \in \mathfrak{A}\}$ is an algebra of continuous functions on $\mathfrak{A}$ with $\overline{\widehat{x}} = \widehat{x^*} \in \operatorname{Ran}(\widehat{\ })$, so $\operatorname{Ran}(\widehat{\ })$ is dense in $C(\mathfrak{A})$ by the Stone–Weierstrass theorem.

Theorem 6.3.3. *Let $\mathfrak{A}$ be an abelian Banach algebra with identity and symmetric involution. If $\{x_1, \dots, x_n, x_1^*, \dots, x_n^*\}$ weakly generate $\mathfrak{A}$, then $\widehat{x}_1 \otimes \cdots \otimes \widehat{x}_n \colon \widehat{\mathfrak{A}} \to j\sigma(x_1, \dots, x_n)$ is a homeomorphism.*

Proof. As in the proof of Theorem 6.2.13, it suffices to show that $(\ell(x_1), \dots, \ell(x_n))$ determines ℓ. Since $\ell(x_j^*) = \overline{\ell(x_j)}$, $(\ell(x_1), \dots, \ell(x_n))$ determines $(\ell(x_1), \dots, \ell(x_n), \ell(x_1^*), \dots, \ell(x_n^*))$ and that determines ℓ by the proof of Theorem 6.2.13. □

Example 6.3.4. Two obvious examples are $f \mapsto \bar{f}$ on $C(X)$, which is norm-compatible and symmetric, and $A \mapsto A^*$ on $\mathcal{L}(\mathcal{H})$, which is norm-compatible. If $\mathfrak{A}$ is an abelian subalgebra with identity of $\mathcal{L}(\mathcal{H})$ and U is unitary, since $\|U^*\| = \|U\| = 1$, we have that $|\ell(U)| \le 1$, $|\ell(U^*)| \le 1$ for all $\ell \in \widehat{\mathfrak{A}}$. Since $\ell(U)\ell(U^*) = \ell(U^*U) = 1$, we conclude $|\ell(U)| = 1$. Thus, for abelian C^*-algebras, the involution is symmetric.

On $\mathbb{A}(\overline{\mathbb{D}})$, define * by

$$f^*(z) = \overline{f(\bar{z})} \tag{6.3.7}$$

This is a norm-compatible involution but is not a symmetric involution since, if $\mathcal{M}_{z_0} = \{f \mid f(z_0) = 0\}$, then $\mathcal{M}_{z_0}^* = \mathcal{M}_{\bar{z}_0}$, not $\mathcal{M}_{z_0}$.

In $L^1(G)$, define

$$f^*(x) = \overline{f(x^{-1})}\, \Delta(x^{-1}) \tag{6.3.8}$$

where Δ is the modular function for G (see Section 4.19 of Part 1). This is a norm-compatible involution (Problem 2). If G is abelian, $\Delta \equiv 1$, and using additive notation for products in G,

$$f^*(x) = \overline{f(-x)} \tag{6.3.9}$$

For χ a character on G, $\overline{\chi(x)} = \chi(-x)$, so

$$\begin{aligned} \widehat{f^*}(\chi) &= \int \overline{f(-x)}\,\overline{\chi(x)}\, d\mu(x) \\ &= \int \overline{f(x)}\,\overline{\chi(-x)}\, d\mu(x) \\ &= \int \overline{f(x)}\,\chi(x)\, d\mu(x) \\ &= \overline{\widehat{f}(\chi)} \end{aligned} \tag{6.3.10}$$

Thus, this involution is symmetric. □

Definition. Let $\mathfrak{A}$ be a Banach algebra with involution. A linear functional, φ, on $\mathfrak{A}$ is called *positive* (or *positive definite*) if and only if for all $x \in \mathfrak{A}$,

$$\varphi(x^*x) \ge 0$$

If $\mathfrak{A}$ has an identity and $\varphi(e) = 1$, we say φ is a *normalized positive function.*

Proposition 6.3.5. *Let * be a norm-compatible involution in a Banach algebra with identity. Every positive functional obeys*

$$\varphi(x^*) = \overline{\varphi(x)} \tag{6.3.11}$$

and

$$|\varphi(x)| \leq \|x\|\varphi(e) \tag{6.3.12}$$

In particular, every positive functional is continuous.

Proof. $\varphi\big((\alpha x + \beta e)^*(\alpha x + \beta e)\big) \geq 0$ implies the matrix

$$\begin{pmatrix} \varphi(e) & \varphi(x) \\ \varphi(x^*) & \varphi(x^* x) \end{pmatrix}$$

is positive. It is thus symmetric, implying (6.3.11) and obeys

$$|\varphi(x)|^2 \leq \varphi(e)\varphi(x^* x) \tag{6.3.13}$$

since the matrix has a positive determinant.

Consider the power series

$$(1+z)^{1/2} = \sum_{n=0}^{\infty} c_n z^n$$

which has radius of convergence 1. It follows if x is symmetric with $\|x\| < 1$, then one can use the power series to define

$$y_\pm = (1 \pm x)^{1/2}$$

which obeys

$$y_\pm^* = y_\pm \qquad y_\pm^2 = 1 \pm x$$

since the c_n are real and $(\sum_{n=0}^\infty c_n z^n)^2 = 1 + z$ if $|z| < 1$. Thus, $\varphi(1 \pm x) = \varphi(y_\pm^* y_\pm) \geq 0$ so if $\|x\| < 1$, $|\varphi(x)| \leq \varphi(e)$. For an arbitrary symmetric x, we have $\varphi(x/[\|x\|(1+\varepsilon)]) \leq \varphi(e)$ for any $\varepsilon > 0$, so (6.3.12) holds for x symmetric.

By (6.3.13),

$$|\varphi(x)|^2 \leq \varphi(e)\varphi(x^* x) \leq \varphi(e)^2 \|x^* x\| \leq \varphi(e)^2 \|x\|\,\|x^*\|$$

implying (6.3.12) in general since $\|x^*\| = \|x\|$. □

Remark. A Banach algebra, $\mathfrak{A}$, with involution, *, is said to have an *approximate identity* if there is a net $\{a_\alpha\}_{\alpha \in I}$ in $\mathfrak{A}$ with $\|a_\alpha\|\,\|a_\alpha^*\| = 1$ and $\lim_\alpha \|a_\alpha x - x\| = 0$ for all $x \in \mathfrak{A}$. If $\mathfrak{A}$ is not abelian, one also requires $\lim_\alpha \|x a_\alpha - x\| = 0$. It is easy to see that if φ is a positive functional on a Banach algebra without identity but with approximate identity and involution, *, and φ is continuous, then φ extended to $\mathfrak{A}_e$ by setting $\varphi(e) = \|\varphi\|$ is positive on $\mathfrak{A}_e$ (Problem 3). This allows the following theorem to be extended to $\mathfrak{A}$ without identity (but with approximate identity), if $\varphi(e) = 1$ is replaced by $\|\varphi\| = 1$ and $\widehat{\mathfrak{A}}$ by $\widehat{\mathfrak{A}}_e$.

Theorem 6.3.6 (Bochner–Raikov Theorem). *Let $\mathfrak{A}$ be an abelian Banach algebra with identity and with symmetric involution and let φ be a positive functional on $\mathfrak{A}$ with $\varphi(e) = 1$. Then there exists a unique measure $\mu \in \mathcal{M}_{+,1}(\widehat{\mathfrak{A}})$, so that*

$$\varphi(x) = \int_{\widehat{\mathfrak{A}}} \widehat{x}(\ell)\, d\mu(\ell) \tag{6.3.14}$$

Conversely, every functional of the form (6.3.14) *is a positive functional.*

Remark. In particular, any $\ell \in \widehat{\mathfrak{A}}$ is a positive functional, which is easy to see directly. Indeed, this theorem implies that the mlf's are exactly the extreme points in the compact, convex set of normalized positive functions.

Proof. If φ has the form (6.3.14), then

$$\varphi(x^*x) = \int |\widehat{x}(\ell)|^2\, d\mu(\ell) \geq 0 \tag{6.3.15}$$

since $\widehat{x^*x} = \widehat{x^*}\widehat{x} = |\widehat{x}|^2$ on account of the symmetry of the involution.

Conversely, let φ be a positive functional. Then

$$\langle x, y\rangle = \varphi(x^*y) \tag{6.3.16}$$

defines a positive semidefinite sesquilinear form on $\mathfrak{A}$, so, in the usual way, the Schwarz inequality holds, that is,

$$|\varphi(x^*y)|^2 \leq \varphi(x^*x)\varphi(y^*y) \tag{6.3.17}$$

In particular, taking $x = e$,

$$|\varphi(y)|^2 \leq \varphi(e)\varphi(y^*y) \tag{6.3.18}$$

By induction

$$\begin{aligned} |\varphi(y)|^{2^n} &\leq \varphi(e)^{2^n-1}\varphi((y^*y)^{2^{n-1}}) \\ &\leq \varphi(e)^{2^n-1}\|\varphi\|\, \|(y^*)^{2^{n-1}}y^{2^{n-1}}\| \\ &\leq \varphi(e)^{2^n-1}\|\varphi\|\, \|(y^*)^{2^{n-1}}\|\, \|y^{2^{n-1}}\| \end{aligned} \tag{6.3.19}$$

Taking 2^n roots and letting $n \to \infty$ shows that

$$\begin{aligned} |\varphi(y)| &\leq \varphi(e)\,\mathrm{spr}(y)^{1/2}\,\mathrm{spr}(y^*)^{1/2} \\ &= \varphi(e)\,\mathrm{spr}(y) \\ &= \varphi(e)\|\widehat{y}\|_\infty \end{aligned} \tag{6.3.20}$$

since $\sigma(y^*) = \overline{\sigma(y)}$ by symmetry of the involution, and therefore, $\mathrm{spr}(y) = \mathrm{spr}(y^*)$.

For $f \in \mathrm{Ran}(\widehat{\ })$, pick y with $f = \widehat{y}$ and define

$$\widehat{\varphi}(f) = \varphi(y) \tag{6.3.21}$$

$\widehat{\varphi}$ is well-defined since $f = \widehat{y} = \widehat{w}$ implies $\widehat{y - w} = 0$, so by (6.3.20), $\varphi(y) = \varphi(w)$. Also by (6.3.20) and $\varphi(e) = 1$,

$$|\widehat{\varphi}(f)| \le \|f\|_\infty \tag{6.3.22}$$

for $f \in \text{Ran}(\widehat{\ })$. Therefore, φ extends to a linear map on the closure of $\text{Ran}(\widehat{\ })$ in $\|\cdot\|_\infty$. By the Stone–Weierstrass theorem, $\hat{x}\hat{y} = \widehat{xy}$ and symmetry of the involution, this is all of $C(\widehat{\mathfrak{A}})$. Therefore,

$$\widehat{\varphi}(f) = \int_{\widehat{\mathfrak{A}}} f(\ell)\, d\mu(\ell) \tag{6.3.23}$$

for some $\mu \in \mathcal{M}(\widehat{\mathfrak{A}})$. μ is real since $\widehat{\varphi}(\bar{f}) = \overline{\widehat{\varphi}(f)}$ if $f = \widehat{y}$ and then for general f by taking limits.

Since $\widehat{\varphi}(1) = \varphi(e) = 1$, (6.3.22) implies $\|\widehat{\varphi}\|_{\mathcal{M}(\widehat{\mathfrak{A}})} \le 1$, which implies that μ is a positive measure, for if $\mu = \mu_+ - \mu_-$ is the Hahn decomposition of the real measure, μ, then $\mu_+(\widehat{\mathfrak{A}}) + \mu_-(\widehat{\mathfrak{A}}) = \|\mu\|=1$, while $\mu_+(\mathfrak{A}) - \mu_-(\mathfrak{A}) = \int 1\, d\mu = \varphi(e) = 1$, so $\mu_- = 0$.

Uniqueness follows from the fact that if both μ and ν represent ℓ, then $\mu(f) = \nu(f)$ for f in the dense set $\{\widehat{x} \mid x \in \mathfrak{A}\}$, and so for all f. □

If one specializes to $L^1(\mathbb{R}^\nu)$, then $\varphi \in L^\infty(\mathbb{R}^\nu) = L^1(\mathbb{R}^\nu)^*$ is positive if and only if

$$0 \le \int \varphi(x)(f * f^*)(x)\, dx = \int \varphi(x) f(x-y)\, \overline{f(-y)}\, dx\, dy$$

$$= \int \varphi(x) f(x+y)\, \overline{f(y)}\, dx\, dy \tag{6.3.24}$$

$$= \int \varphi(w-y) f(w)\, \overline{f(y)}\, dw\, dy \tag{6.3.25}$$

Thus, φ is positive as a functional means φ is equal a.e. to a positive definite function in Bochner sense (see Problem 6 of Section 6.6 of Part 1). Moreover, $\varphi \in L^\infty(\mathbb{R}^\nu)$ implies $\|\varphi\| < \infty$ and $L^1(\mathbb{R}^\nu)$ has approximate identities, so the extension of Theorem 6.3.6, noted in the remark before that theorem, is applicable. The Bochner–Raikov theorem is thus a generalization of Bochner's theorem.

Remarkably, even if an involution is not symmetric, positive functionals on a Banach algebra with identity and norm-compatible involution have a representation for $\mu \in \mathcal{M}_{+,1}(\widehat{\mathfrak{A}})$,

$$\varphi(x) = \int_{\mathcal{P}} \widehat{x}(\ell)\, d\mu(\ell)$$

where $\mathcal{P} = \{\ell \in \widehat{\mathfrak{A}} \mid \ell(x^*) = \overline{\ell(x)} \text{ for all } x \in \mathfrak{A}\}$. The reader will prove this in Problem 4.

Notes and Historical Remarks. Gel'fand–Naimark [**231**] first considered involutions on Banach algebras in the context of B^*-algebras as discussed in Sections 6.4 and 6.7—that is, only with the condition $\|x^*x\| = \|x\|^2$. The general theory, including the notion of symmetric involution and Theorem 6.3.6, is due to Raikov [**543, 545**]. The theory has become a standard part of the subject in the many texts mentioned in the Notes to Section 6.0.

Problems

1. Let $\mathfrak{A}$ be a semisimple abelian Banach algebra with involution, *. Define $\|\cdot\|_2$ on $\mathfrak{A}$ by
$$\|x\|_2 = \|x^*\| \tag{6.3.26}$$
(a) Prove that $\mathfrak{A}$ is complete in $\|\cdot\|_2$.
(b) Prove that * is continuous. (*Hint*: Corollary 6.2.8.)

2. Prove the map given by (6.3.8) is a norm-compatible involution. *Hint*: You'll need (4.19.8) of Part 1 which says for (left) Haar measure, μ, that
$$\int f(x)\,d\mu(x) = \int f(x^{-1})\Delta(x^{-1})\,d\mu(x) \tag{6.3.27}$$

3. Let φ be a positive, continuous functional on a Banach algebra, $\mathfrak{A}$, with continuous involution and approximate identity (but not assumed to be abelian or to have an identity). Define $\widetilde{\varphi}$ on $\mathfrak{A}_e$ by
$$\widetilde{\varphi}((x,\lambda)) = \varphi(x) + \lambda\|\varphi\| \tag{6.3.28}$$
Extend * to $\mathfrak{A}_e$ by $(x,\lambda)^* = (x^*,\bar\lambda)$.
(a) If $\{a_\alpha\}_{\alpha\in I}$ is an approximate identity, prove that
$$\varphi(a_\alpha^* a_\alpha) \le \|\varphi\| \tag{6.3.29}$$
(b) For any x and λ, prove that $\widetilde{\varphi}((x,\lambda)^*(x,\lambda)) \ge 0$. (*Hint*: Consider $\varphi((x+\lambda a_\alpha)^*(x+\lambda a_\alpha))$.)

4. Let $\mathfrak{A}$ be a Banach algebra with identity and norm-compatible involution, but which might not be symmetric. Let
$$\mathcal{P} = \{\ell \in \widehat{\mathfrak{A}} \mid \ell(x^*) = \overline{\ell(x)} \text{ for all } x \in \mathfrak{A}\} \tag{6.3.30}$$
This problem will prove that for positive functionals, φ, there is a measure, μ, on $\widehat{\mathfrak{A}}$ with $\mu(\widehat{\mathfrak{A}}\setminus\mathcal{P}) = 0$ so that
$$\varphi(x) = \int_{\mathcal{P}} \widehat{x}(\ell)\,d\mu(\ell) \tag{6.3.31}$$
Without loss, suppose henceforth that $\varphi(e) = 1$. The problem will require Theorem 6.4.3 from the next section (or at least the fact that involutions obeying (6.4.1) are symmetric involutions). For technical simplicity, you

should assume that $\widehat{\mathfrak{A}}$ is metrizable in the Gel'fand topology (e.g., if $\mathfrak{A}$ is norm-separable), so Baire and Borel sets agree.

(a) Prove that $\mathcal{P}$ is closed in the Gel'fand topology and so a Baire set.

(b) Define

$$\|x\|_\varphi = \sup\{|\varphi(z^* xw)| \mid \varphi(z^* z) = \varphi(w^* w) = 1\} \tag{6.3.32}$$

Prove that $\|\cdot\|_\varphi$ is a seminorm and that it has all the properties of a Banach algebra norm (except for completeness and strict positivity) and that

$$\|e\|_\varphi = 1, \qquad \|x^* x\|_\varphi = \|x\|_\varphi^2, \qquad \|x^*\|_\varphi = \|x\|_\varphi \tag{6.3.33}$$

(c) Let $\mathfrak{A}_0 = \{x \in \mathfrak{A} \mid \|x\|_\varphi = 0\}$. Form $\mathfrak{A}_1$, the completion of $\mathfrak{A}/\mathfrak{A}_0$ in $\|\cdot\|_\varphi$. Prove that $\mathfrak{A}_1$ is a B^*-algebra (in the involution inherited by $\mathfrak{A}_1$) so that * on $\mathfrak{A}_1$ is a symmetric involution.

(d) Prove that for a probability measure, μ, on $\widehat{\mathfrak{A}}_1$ and π the natural projection of $\mathfrak{A}$ to $\mathfrak{A}_1$ (by $\pi(x) = [x]$), one has

$$\varphi(x) = \int_{\widehat{\mathfrak{A}}_1} \ell(\pi(x))\, d\tilde{\mu}(\ell) \tag{6.3.34}$$

(e) Map $\widehat{\mathfrak{A}}_1$ to $\widehat{\mathfrak{A}}$ by π^*, that is, $\pi^*(\ell)(x) = \ell(\pi(x))$. Prove that π^* is one–one and $\operatorname{Ran}(\pi^*)$ is closed in $\widehat{\mathfrak{A}}$ and lies in $\mathcal{P}$.

(f) Define μ on $\widehat{\mathfrak{A}}$ by

$$\mu(A) = \tilde{\mu}((\pi^*)^{-1}[A]) \tag{6.3.35}$$

Prove $\mu(\mathfrak{A} \setminus \mathcal{P}) = 0$ and that (6.3.31) holds.

Remark. This argument is clarified by the GNS construction of Section 6.7. φ defines a map $\pi_\varphi \colon \mathfrak{A} \to \mathcal{L}(\mathcal{H})$ for a Hilbert space. $\|x\|_\varphi = \|\pi_\varphi(x)\|$ and $\mathfrak{A}_1$ can be realized as $\overline{\operatorname{Ran}(\pi_\varphi)}^{\|\cdot\|}$ in $\mathcal{L}(\mathcal{H})$.

6.4. Commutative Gel'fand–Naimark Theorem and the Spectral Theorem for Bounded Normal Operators

Definition. A *B^*-algebra* is a Banach algebra, $\mathfrak{A}$, with involution obeying the *C^*-identity*

$$\|x^* x\| = \|x\|^2 \tag{6.4.1}$$

for all $x \in \mathfrak{A}$.

Example 6.4.1 (C^*-algebras). If $\mathfrak{A}$ is a norm-closed subalgebra of $\mathcal{L}(\mathcal{H})$, $\mathcal{H}$ a Hilbert space, with $A \in \mathfrak{A} \Rightarrow A^* \in \mathfrak{A}$, then $\mathfrak{A}$ is a B^*-algebra by Theorem 2.1.8. □

Example 6.4.2 ($C(X)$)**.** If X is a normal topological space, $\mathfrak{A} = C(X)$ with $f^* = \bar{f}$ is a B^*-algebra since $|\bar{f}f(x)| = |f(x)|^2$ and $\sup(|a|^2) = (\sup(|a|))^2$. □

A central goal in this section is to prove the following and to show it provides yet another proof of the spectral theorem:

Theorem 6.4.3 (Commutative Gel'fand–Naimark Theorem)**.** *Every B^*-algebra with identity is isometrically isomorphic to $C(\widehat{\mathfrak{A}})$ under the Gel'fand transform.*

Remarks. 1. The name has "commutative" in it because later (see Theorem 6.7.1) we'll have a "noncommutative Gel'fand–Naimark theorem," which says every C^*-algebra with identity is isometrically isomorphic to a C^*-algebra of operators on a Hilbert space, $\mathcal{H}$.

2. We'll have a version below for $\mathfrak{A}$ a B^*-algebra without identity.

3. Many authors use the term *C^*-algebra* distinguishing what we call B^*-algebras and algebras of operators on $\mathcal{H}$ by sometimes using "abstract C^*-algebra" and "concrete C^*-algebra." Because of the noncommutative Gel'fand–Naimark theorem, this naming isn't terrible, but until we prove that theorem, it is awkward. This is why we prefer "B^*-algebra," a name once quite common, now less so.

The following shows the power of the C^*-identity:

Theorem 6.4.4. *Let $\mathfrak{A}$ be a B^*-algebra. Then*

(a) * *is norm-compatible, that is,*

$$\|x^*\| = \|x\| \tag{6.4.2}$$

(b) *If $y = y^*$, then*

$$\operatorname{spr}(y) = \|y\| \tag{6.4.3}$$

(c) *If $\mathfrak{A}$ is abelian, for any $x \in \mathfrak{A}$,*

$$\operatorname{spr}(x) = \|x\| \tag{6.4.4}$$

(d) *If $\mathfrak{A}$ has an identity and $\psi \in \mathfrak{A}^*$ with $\psi(e) = \|\psi\| = 1$, then $\psi(x^*) = \overline{\psi(x)}$.*

(e) *If $\mathfrak{A}$ is abelian with identity, then * is a symmetric involution, that is, for all $\ell \in \widehat{\mathfrak{A}}$,*

$$\ell(x^*) = \overline{\ell(x)} \tag{6.4.5}$$

(f) *If $\mathfrak{A}$ has an identity, e, then $\|e\| = 1$.*

Remark. There are other proofs of (e) (see Problems 1 and 2). We'll need (d) in Section 6.7, so we use this proof of (e) via (d).

Proof. (a) Since $0^* = 0$, we can suppose $x \neq 0$. Then

$$\|x\|^2 = \|x^*x\| \leq \|x^*\|\|x\| \tag{6.4.6}$$

Since $\|x\| \neq 0$, we have

$$\|x\| \leq \|x^*\| \tag{6.4.7}$$

Replacing x by x^* using $(x^*)^* = x$, we get the opposite inequality, and so (6.4.2).

(b) $\|y^2\| = \|y^*y\| = \|y\|^2$ (6.4.8)

Replacing y by y^{2^n}, we get

$$\|y^{2^{n+1}}\| = \|y^{2^n}\|^2 \tag{6.4.9}$$

so inductively, we get that

$$\|y^{2^n}\|^{1/2^n} = \|y\| \tag{6.4.10}$$

Since $\lim_{m\to\infty}\|y^m\|^{1/m}$ exists, we can take $m = 2^n$ and still get the limit. Thus, (6.4.3) holds.

(c) $\|x\|^4 = \|x^*x\|^2 = \|(x^*x)^*(x^*x)\| = \|(x^2)^*x^2\| = \|x^2\|^2$ (6.4.11)

where (6.4.11) used that $\mathfrak{A}$ is abelian. Thus, (6.4.8) holds for x and the argument in (b) then implies (6.4.4).

(d) Since any $x = y + iz$ with y, z symmetric, it suffices to prove

$$y = y^* \Rightarrow \operatorname{Im}\psi(y) = 0 \tag{6.4.12}$$

Since $y = y^*$, $(1 + i\varepsilon y)^*(1 + i\varepsilon y) = 1 + \varepsilon^2 y^*y$ and thus,

$$\|1 + i\varepsilon y\|^2 \leq 1 + \varepsilon^2\|y\|^2 \tag{6.4.13}$$

Since $|\psi(1 + i\varepsilon y)|^2 = (1 - \varepsilon\operatorname{Im}\psi(y))^2 + \varepsilon^2(\operatorname{Re}\psi(y))^2 = 1 + \varepsilon^2|\psi(y)|^2 - 2\varepsilon\operatorname{Im}\psi(y)$, $\|\psi\| = 1$ implies that

$$-2\varepsilon\operatorname{Im}\psi(y) \leq \varepsilon^2(\|y\|^2 - |\psi(y)|^2) \tag{6.4.14}$$

If $\operatorname{Im}\psi(y) \neq 0$, taking $|\varepsilon|$ small so $\varepsilon\operatorname{Im}\psi(y) > 0$, we get a contradiction. Thus, $\operatorname{Im}\psi(y) = 0$, that is, (6.4.12) holds.

(e) $\ell \in \mathfrak{A}^* \Rightarrow \|\ell\| = \ell(e) = 1$. Now use (d).

(f) By Proposition 6.3.1, $e^* = e$ so

$$\|e\| = \|e^*\| = \|e^*e\| = \|e\|^2 \tag{6.4.15}$$

Thus, $\|e\| = 1$ or 0. $\|e\| = 0 \Rightarrow \mathfrak{A} = \{0\}$, so $\|e\| = 1$.

□

Proof of Theorem 6.4.3. Since

$$\|\widehat{x}\|_\infty = \text{spr}(x) = \|x\| \tag{6.4.16}$$

$\widehat{\ }: \mathfrak{A} \to C(\widehat{\mathfrak{A}})$ is an isometry. By the Stone–Weierstrass theorem, Ran($\widehat{\ }$) is dense, but since $\widehat{\ }$ is an isometry, Ran($\widehat{\ }$) is complete, and thus closed. It follows that $\widehat{\ }$ is a bijection. □

An interesting question raised by this theorem is $C(X)$ when X is not compact, say $X = \mathbb{R}$ (or $X = \mathbb{Z}$, where $C(X) = \ell^\infty$). $C(\mathbb{R})$ is isometrically isomorphic to $C(\widehat{\mathfrak{A}})$ with $\widehat{\mathfrak{A}}$ compact. So what is $\widehat{C(\mathbb{R})}$? It turns out to be huge—so huge that its topology is not metrizable. We'll explore this issue in Section 6.5.

To extend this theorem to the case where $\mathfrak{A}$ does not have an identity, we need the following, which we emphasize does not require that $\mathfrak{A}$ be abelian:

Lemma 6.4.5. *Let $\mathfrak{A}$ be a (possibly nonabelian) B^*-algebra without identity. Let $\mathfrak{A}_e = \mathfrak{A} \oplus \mathbb{C}$ with norm*

$$\|(x, \lambda)\| = \|x\| + |\lambda| \tag{6.4.17}$$

Let

$$(x, \lambda)^* = (x^*, \lambda) \tag{6.4.18}$$

and product (6.1.35) as in Theorem 6.1.8. There is an equivalent norm, $\|\cdot\|_1$, on $\mathfrak{A}_e$ making it into a B^*-algebra and so that for $x \in \mathfrak{A}$,

$$\|x\|_1 = \|x\| \tag{6.4.19}$$

Proof. Define $\widetilde{L}_{(x,\lambda)}$ as an operator on $\mathfrak{A}$ (*not* $\mathfrak{A}_e$) by

$$\widetilde{L}_{(x,\lambda)} y = xy + \lambda y \tag{6.4.20}$$

(i.e., $L_{(x,\lambda)}$ restricted to $\mathfrak{A}$ viewed as a subspace of $\mathfrak{A}_e$).

Since for $\alpha, \beta \in \mathfrak{A}_e$, we have $\widetilde{L}_{\alpha\beta} = \widetilde{L}_\alpha \widetilde{L}_\beta$, if we define

$$\|\alpha\|_1 = \|\widetilde{L}_\alpha\|_{\mathcal{L}(\mathfrak{A})} \tag{6.4.21}$$

then

$$\|\alpha\beta\|_1 \le \|\alpha\|_1 \|\beta\|_1 \tag{6.4.22}$$

Since $L_e y = L_{(0,1)} y = y$, we have

$$\|e\|_1 = 1 \tag{6.4.23}$$

Since $\|\widetilde{L}_{(x,\lambda)} y\| \le \|(x, \lambda)\| \, \|y\|$, we see that

$$\|\alpha\|_1 \le \|\alpha\| \tag{6.4.24}$$

If $\|(x, \lambda)\|_1 = 0$, then for all y, $xy = -\lambda y$, so $z = -\lambda^{-1} x$ is a left inverse. But then $z^* = zz^* = (zz^*)^* = z$, and for all $y \in \mathfrak{A}$, $yz = yz^* = (zy^*)^* =$

$(y^*)^* = y$. So z is a two-sided inverse, violating the fact that $\mathfrak{A}$ has no identity. We have thus proven that $\|\cdot\|_1$ is a norm.

If $x \in \mathfrak{A}$, we have $\|L_x x^*\| = \|xx^*\| = \|x^*\|^2 = \|x\|\|x^*\|$, so $\|x\|_1 = \|L_x\| \geq \|x\|$. Combining with (6.4.24), we see that (6.4.19) holds.

Since $\|\cdot\|_1 = \|\cdot\|$ on $\mathfrak{A}$, $\mathfrak{A}$ is complete in $\|\cdot\|_1$ and so closed. It follows that the Banach quotient space, $\mathfrak{A}_e/\mathfrak{A}$, is $\mathbb{C}$ with

$$\rho(\lambda) = \inf\{\|(x,\lambda)\|_1 \mid x \in \mathfrak{A}\}$$

a norm on $\mathbb{C}$. Thus, if (x_n, λ_n) is Cauchy, λ_n is Cauchy in ρ, so it converges. So since $\mathfrak{A}$ is complete, (x_n, λ_n) has a limit, that is, $\mathfrak{A}_e$ is complete in $\|\cdot\|_1$. By the inverse mapping theorem and (6.4.24), $\|\cdot\|_1$ is equivalent to $\|\cdot\|$.

Finally, to check the C^*-identity, we note first that if $\alpha = (x, \lambda) \in \mathfrak{A}_e$, then for $y \in \mathfrak{A}$, $L_\alpha y \in xy + \alpha y \in \mathfrak{A}$, so

$$\begin{aligned} \|L_\alpha y\|^2 &= \|(L_\alpha y)^*(L_\alpha y)\| = \|y^*(\alpha^*\alpha y)\| \\ &\leq \|y^*\|\|y\|\|L_{\alpha^*\alpha}\| = \|y\|^2\|L_{\alpha^*\alpha}\| \end{aligned} \tag{6.4.25}$$

Therefore,

$$\|\alpha\|_1^2 \leq \|\alpha^*\alpha\|_1 \leq \|\alpha\|_1\|\alpha^*\|_1 \tag{6.4.26}$$

This in turn implies first that $\|\alpha\|_1 \leq \|\alpha^*\|_1$, so by symmetry and $(\alpha^*)^* = \alpha$, $\|\alpha\|_1 = \|\alpha^*\|_1$, and then using (6.4.26) again, $\|\alpha^*\alpha\|_1 = \|\alpha\|_1^2$. □

Here is the version for algebras which might not have an identity:

Theorem 6.4.6. *Let $\mathfrak{A}$ be an abelian B^*-algebra which might not have identity. Then $\widehat{\ }$ is an isometric isomorphism of $\mathfrak{A}$ onto $C_\infty(\widehat{\mathfrak{A}})$, the continuous functions on $\widehat{\mathfrak{A}}$ vanishing at ∞.*

Proof. If $\mathfrak{A}$ has an identity, this is just Theorem 6.4.1. If $\mathfrak{A}$ does not have an identity, put the C^*-norm on $\mathfrak{A}_e$. We saw that $\widehat{\mathfrak{A}}_e = \widehat{\mathfrak{A}} \cup \{\infty\}$ and any $x \in \mathfrak{A}$ has $\widehat{x}(\infty) = 0$. Thus, $\widehat{\ }$ is a map of $\widehat{\mathfrak{A}}$ onto $C_\infty(\widehat{\mathfrak{A}})$. By restriction of the result for $\mathfrak{A}_e$, the map is an isometry, and so one–one and Ran($\widehat{\ }$) is closed. Ran($\widehat{\ }$) is an algebra, invariant under complex conjugations and separates points. By Theorem 2.5.5 of Part 1, Ran($\widehat{\ }$) $= C_\infty(\mathfrak{A})$. □

Remark. Even if $\mathfrak{A}$ has an identity, one can form $\mathfrak{A}_e$ and $\widehat{\mathfrak{A}}_e$. If $\tilde{e}$ is the identity in $\mathfrak{A}$, $\tilde{e}(\ell) = 1$ ($\ell \in \widehat{\mathfrak{A}}$), $= 0$ if $\ell = \{\infty\}$. Thus, $\{\infty\}$ is isolated and $\widehat{\mathfrak{A}}$ is already compact. Conversely, if $\widehat{\mathfrak{A}}$ is compact, that is, if $\{\infty\}$ is isolated in $\widehat{\mathfrak{A}}_e$, that is, the nonzero mlf's on $\mathfrak{A}$ are weak-* closed, then the $\tilde{e}$ with $\widehat{\tilde{e}} = 1$ on $\widehat{\mathfrak{A}}$ is an identity.

We turn next to the proof this gives us of the spectral theorem for a bounded normal operator, A, on $\mathcal{H}$, a Hilbert space. Given such an operator, let $\mathfrak{A}$ be the algebra generated by $\{A, A^*\} \cup \{(A-\lambda)^{-1}, (A^*-\bar{\lambda})^{-1} \mid \lambda \notin$

$\sigma_{\mathcal{H}}(A)\}$, where $\sigma_{\mathcal{H}}(A)$ is the spectrum of A as an operator on $\mathcal{H}$. $\mathfrak{A}$ is an abelian B^*-algebra. By Theorem 6.2.3, $\widehat{A}\colon \widehat{\mathfrak{A}} \to \mathbb{C}$ is a homeomorphism. By including $(A-\lambda)^{-1}$ in $\widehat{\mathfrak{A}}$, we see $\sigma_{\mathfrak{A}}(A) \subset \sigma_{\mathcal{H}}(A)$, and since $\mathfrak{A} \subset \mathcal{H}$, $\sigma_{\mathcal{H}}(A) \subset \sigma_{\mathfrak{A}}(A)$, so $\widehat{\mathfrak{A}} \cong \sigma_{\mathcal{H}}(A)$. By the commutative Gel'fand–Naimark theorem, $\widehat{\ }: \mathfrak{A} \to C(\sigma_{\mathcal{H}}(A))$ is an isometric isomorphism, so its inverse, Φ, is also. We have thus proven:

Theorem 6.4.7 (Spectral Theorem for Normal Operators). *Let A be a normal operator on a Hilbert space, $\mathcal{H}$. Then there exists a map $\Phi\colon C(\sigma_{\mathcal{H}}(A)) \to \mathcal{L}(\mathcal{H})$ so that Φ is an algebra homomorphism, $\Phi(\bar{f}) = \Phi(f)^*$, and $\Phi(z) = A$, where z is the function mapping $z_0 \in \sigma_{\mathcal{H}}(A)$ to z_0.*

The reader will notice this is precisely the continuous functional calculus form of the spectral theorem. We notice something else. By the Stone–Weierstrass theorem, polynomials in z and $\bar{z}$ are dense in $C(\sigma_{\mathcal{H}}(A))$, so since Φ is an isometric isomorphism, $\mathfrak{A}$ is generated by A and A^*, that is, we have proven:

Theorem 6.4.8. *Let A be a normal operator on a Hilbert space, $\mathcal{H}$. Then for any $z \in \rho_{\mathcal{H}}(A)$, $(A-z)^{-1} \in \mathfrak{A}_{A,A^*}$.*

We recall Example 6.1.9 showing this may be false for $\mathfrak{A}_A$.

Notes and Historical Remarks. In their 1943 paper, Gel'fand–Naimark [**231**] focused on the noncommutative Gel'fand–Naimark theorem that any B^*-algebra is isometrically isomorphic to a C^*-algebra (see Section 6.7). The commutative Gel'fand–Naimark theorem appears as a lemma in their argument.

Their original definition of B^*-algebra included the condition that $e+x^*x$ is invertible for all $x \in \mathfrak{A}$, a condition it took ten years to show (by others; see the Notes to Section 6.7) was unnecessary. But they didn't require it for their lemma. The name B^*-algebra is due to Rickart [**561**].

The subtlest part of the proof of the commutative Gel'fand–Naimark theorem is that * is a symmetric involution. Their original proof relied on the existence of a Shilov boundary (see Problem 1)—they first proved that if $S \subset \widehat{\mathfrak{A}}$ is the Shilov boundary for $\mathrm{Ran}(\widehat{\ })$, then all $\ell \in S$ have $\ell(x^*) = \overline{\ell(x)}$. They next need to show that $\widehat{\ }$ is an isometric isomorphism of $\mathfrak{A}$ to $C(S)$. Since $\widehat{C(S)} = S$, they concluded $S = \widehat{\mathfrak{A}}$ and so all $\ell \in \widehat{\mathfrak{A}}$ are symmetric. This is a rather indirect argument.

The first direct proof that * is symmetric is due to Arens [**21**] and is essentially the proof we give (see Problem 2 for another simple proof). There is an amusing piece of historical jujitsu associated to the Arens paper. He said he wrote it because the Gel'fand–Naimark paper used the existence of the Shilov boundary quoted in a paper, namely [**235**], that was difficult to get. By the time [**235**] actually appeared, it included, with attribution, Arens' proof!

That a C^*-algebra without identity can be embedded in one with identity is an unpublished observation of Kaplansky.

Problems

1. This will lead you through the Gel'fand–Naimark [**231**] proof that in the abelian B^*-algebra with identity, $\mathfrak{A}$, the involution is symmetric. It will rely on the fact that there exists a closed set $S \subset \widehat{\mathfrak{A}}$ (the Shilov boundary) so that for every $x \in \mathfrak{A}$,
$$\sup_{\ell\in S}|\widehat{x}(\ell)| = \sup_{\ell\in\widehat{\mathfrak{A}}}|\widehat{x}(\ell)| \tag{6.4.27}$$
and which is minimal among all such sets. Minimality implies that if N is a neighborhood of $\ell_0 \in S$, there exists x so that
$$\sup_{\ell\in\mathfrak{A}\setminus N}|\widehat{x}(\ell)| < \sup_{\ell\in\mathfrak{A}}|\widehat{x}(\ell)| \tag{6.4.28}$$
We'll prove existence and uniqueness of S in Theorem 6.10.7. You may also suppose you've proven (6.4.2) and (6.4.4), that is, that
$$\|\widehat{x}\|_\infty = \|x\| \tag{6.4.29}$$
For each $\ell \in \widehat{\mathfrak{A}}$, define $\ell^* \in \widehat{\mathfrak{A}}$ by
$$\ell^*(x) \equiv \ell(x^*) \tag{6.4.30}$$
(a) Prove that * is continuous as the map of $\widehat{\mathfrak{A}} \to \widehat{\mathfrak{A}}$.

(b) Prove $\ell \in S \Rightarrow \ell^* \in S$. (*Hint*: If N, x are as in (6.4.28), prove that $N^* \cap S \neq \emptyset$ since $\widehat{x^*}$ only takes its maximum value on N^*.)

(c) If $\ell \in S$ and $\ell \neq \ell^*$, prove there is a neighborhood, N, of ℓ so $N \cap N^* = \emptyset$.

(d) If N is as in (c) and x obeys (6.4.28), prove that
$$\|\widehat{x^*x}\|_\infty < \|\widehat{x}\|_\infty\|\widehat{x^*}\|_\infty \tag{6.4.31}$$
Conclude, by (6.4.29), that $\ell \in S \Rightarrow \ell = \ell^*$.

(e) Define $\tilde{}\colon \mathfrak{A} \to C(S)$ by $\tilde{x} = \widehat{x} \restriction S$. Prove that $\tilde{}$ is an isometric isomorphism of $\mathfrak{A}$ and $C(S)$.

(f) Conclude $S = \widehat{\mathfrak{A}}$ and then * is a symmetric involution.

2. This will provide another proof that an abelian C^*-algebra with identity is symmetric. It is essentially the proof we gave in Example 6.3.4.

(a) If u is unitary, prove that $\|u\| = \|u^*\| = 1$ and conclude that for $\ell \in \widehat{\mathfrak{A}}$, $|\ell(u)| \leq 1$ and $|\ell(u^*)| \leq 1$.

(b) Prove that $|\ell(u)||\ell(u^*)| = 1$ and conclude that $|\ell(u)| = 1$.

(c) Conclude that * is symmetric.

6.5. Compactifications

In this section, we'll present the theory of compactifications of X, a locally compact space. All spaces are Hausdorff. Much of the theory extends to completely regular spaces (see Problem 1 and the Notes), but we'll restrict to this case for simplicity of exposition.

Definition. A *compact morphism* of X is a compact space, Y, and a continuous map, $f\colon X \to Y$, so that $\mathrm{Ran}(f)$ is dense in Y. If f is one–one, we say (Y, f) is a *compactification*. If f is one–one and if, when $\mathrm{Ran}(f)$ is given the relative topology, f is a homeomorphism, then we say (Y, f) is a *strict compactification*.

Definition. Let (Y_1, f_1) and (Y_2, f_2) be two compact morphisms. We say (Y_2, f_2) *covers* (Y_1, f_2) if there is a continuous map, $g\colon Y_2 \to Y_1$ so that $g \circ f_2 = f_1$.

Note that since $\mathrm{Ran}(g)$ is closed and $\mathrm{Ran}(f_j)$ is dense in X_j, a covering map is always onto. g is called a *covering map*. If g is a homeomorphism, we say that the compact morphisms are *equivalent*.

Definition. A *symmetric subalgebra* of $C(X)$ is a norm-closed subalgebra, $\mathfrak{A}$, with $1 \in \mathfrak{A}$ and $f \in \mathfrak{A} \Rightarrow \bar{f} \in \mathfrak{A}$.

Our goal here will be to prove there is a one–one correspondence between equivalence classes of compact morphisms and symmetric subalgebras where, given $\mathfrak{A}$, Y will be $\widehat{\mathfrak{A}}$ and f the map that takes $x \in \mathfrak{A}$ into $\delta_x \in \widehat{\mathfrak{A}}$ where

$$\delta_x(f) = f(x) \tag{6.5.1}$$

We'll also see that being a compactification or strict compactification can be expressed in terms of properties of $\mathfrak{A}$ and that so can the existence of covering maps. Also, we'll find a compactification so large that it covers all compact morphisms.

As a warmup, we illustrate covering maps. The instructive straightforward argument is left to the reader.

Recall that X has a one-point compactification $X_\infty = X \cup \{\infty\}$ topologized by setting the open sets to be the open sets of X and the complements in X_∞ of compact sets in X. The identity map $i\colon X \to X_\infty$ by $i(x) = x$ is a strict compactification (Problem 2).

Proposition 6.5.1. *Let (Y, f) be a strict compactification. Let $g\colon Y \to X_\infty$ by*

$$g(y) = \begin{cases} i(f^{-1}(y)) & \text{if } y \in \operatorname{Ran}(f) \\ \infty & \text{if } y \notin \operatorname{Ran}(f) \end{cases} \tag{6.5.2}$$

Then g is a covering map.

Example 6.5.2. Let $X = [0, \infty)$. If $Y = \partial\mathbb{D}$ and $f(\theta) = e^{i\theta}$, then (Y, f) is a compact morphism which is not one–one. If $Y = \mathbb{T}^2 = \partial\mathbb{D} \times \partial\mathbb{D}$ by $f(\theta) = (e^{i\theta}, e^{i\alpha\theta})$ for a fixed irrational α, then (Y, f) is a compactification since f is one–one but not a strict one, since if $U \subset \pi^2$ is open, $f^{-1}[U]$ is always unbounded. Thus, $f[(a, b)]$ is not relatively open in $\operatorname{Ran}(f)$. Of course, the one- and two-point compactifications are familiar strict compactifications. To see there are others, we'll use the simple construction below. $\square$

Proposition 6.5.3. *Let (Y, f) be a compact morphism of X. Let $\widetilde{f}\colon X \to X_\infty \times Y$ by*

$$\widetilde{f}(x) = (x, f(x)) \tag{6.5.3}$$

Let $\widetilde{Y}$ be the closure of $\operatorname{Ran}(\widetilde{f})$. *Then $(\widetilde{Y}, \widetilde{f})$ is a strict compactification.*

Proof. By Tychonoff's theorem (Theorem 2.7.1 of Part 1), $\widetilde{Y}$ is compact and, by construction, $\widetilde{f}$ is continuous and dense in $\widetilde{Y}$.

So we need only show that if $U \subset X$ is open, then $f[U]$ is relatively open in $X_\infty \times Y$. Since

$$f[U] = (U \times Y) \cap \operatorname{Ran}(f) \tag{6.5.4}$$

this is obvious. $\square$

Example 6.5.2 (continued). Consider $f\colon X \to \partial\mathbb{D}$ by $f(x) = e^{ix}$. One can realize $X_\infty \equiv [0, \infty]$ as $[0, 1]$ (by using, say, $h(x) = \frac{2}{\pi}\arctan(x)$) and then $X_\infty \times \partial\mathbb{D}$ is $\overline{\mathbb{D}}$ and $\widetilde{f}(x) = h(x)e^{ix}$. $\widetilde{Y}$ is a spiral with a circle of points added at infinity (see Figure 6.5.1). Similarly, the winding line on the torus adds a "torus" at infinity.

Remarks. 1. If $\pi\colon X_\infty \times Y \to Y$ by $\pi(x, y) = y$, then $\pi \restriction \widetilde{Y}$ is a covering map (see Problem 4).

2. If $f^{-1}[U]$ is unbounded (i.e., its closure in X is not compact) for all open U in Y, then it is easy to see (Problem 5) that $\widetilde{Y} \setminus \operatorname{Ran}(\widetilde{f})$ is homeomorphic to Y as in the above examples.

3. It is easy to see (Problem 6) that if (Y, f) is a strict compactification, then $(\widetilde{Y}, \widetilde{f})$ is equivalent to (Y, f); indeed, the π of Remark 1 is an equivalence.

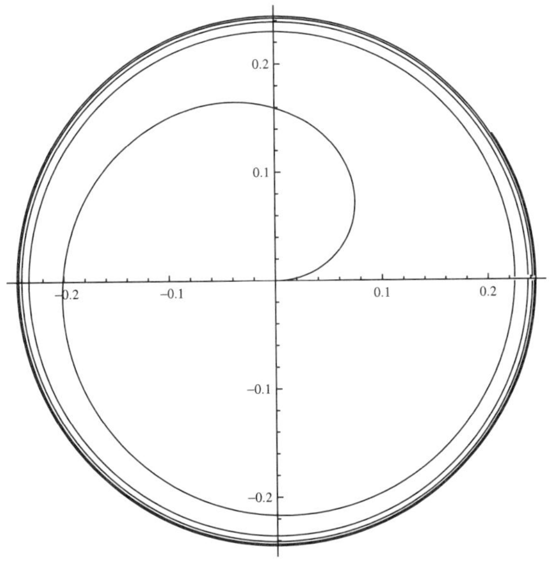

Figure 6.5.1. A winding line compactification that adds a circle at infinity.

Recall that $C(X)$ are the bounded continuous functions of X to $\mathbb{R}$. Given a compact morphism (Y, f) of X, let $G_{Y,f}\colon C(Y) \to C(X)$ by

$$G_{Y,f} h = h \circ f \tag{6.5.5}$$

Define

$$C_{Y,f}(X) = \operatorname{Ran}(G_{Y,f}) \tag{6.5.6}$$

Theorem 6.5.4. (a) *$C_{Y,f}(X)$ is a symmetric subalgebra of $C(X)$.*

(b) *(Y, f) is a compactification of X if and only if $C_{Y,f}(X)$ separates points (i.e., for all $x_1, x_2 \in X$, there is g in $C_{Y,f}$ so that $g(x_1) \neq g(x_2)$).*

(c) *(Y, f) is a strict compactification of X if and only if $C_\infty(X) \subset C_{Y,f}(x)$.*

(d) *In the construction of Proposition* 6.5.3,

$$C_{\widetilde{Y},\widetilde{f}}(X) = \overline{C_{Y,f}(X) + C_\infty(X)} \tag{6.5.7}$$

where $^-$ is $\|\cdot\|_\infty$-closure.

Remark. In many, perhaps all, cases, the sum on the right side of (6.5.7) is already norm-closed.

Proof. (a) Since $\operatorname{Ran}(f)$ is dense in Y,

$$\|G_{Y,f}(h)\|_{C(X)} = \|h\|_{C(Y)} \tag{6.5.8}$$

Since $G_{Y,f}$ is an algebraic homeomorphism with $\overline{G_{Y,f}(h)} = G_{Y,f}(\bar{h})$ and an isometry, $\operatorname{Ran}(G)$, is norm-complete, so a norm-closed subalgebra of $C(X)$ also closed under complex conjugation.

(b) If $f(x_1) = f(x_2)$, clearly $(h \circ f)(x_1) = (h \circ f)(x_2)$, so $C_{Y,f}$ does not separate points. Conversely, if $f(x_1) \neq f(x_2)$ by Urysohn's lemma, there is $h \in C(Y)$ with $h(f(x_1)) \neq h(f(x_2))$, so take $g = h \circ f$.

(c) If (Y, f) is strict, it covers (X_∞, i) by Proposition 6.5.1. Let g be the map (6.5.2). Let $h \in C_\infty(X)$. Then $h \circ g = h$ so $C_\infty(X) \subset C_{Y,f}(X)$.

For the converse, suppose $C_\infty(X) \subset C_{Y,f}(X)$. Since C_∞ separates points (by Urysohn's lemma on X_∞), f is one–one, so we need to show for a net $\{x_\alpha\}_{\alpha \in I}$, $f(x_\alpha) \to f(x_\infty)$ implies $x_\alpha \to x_\infty$ (since the converse follows by continuity). If $f(x_\alpha) \to f(x_\infty)$ is false by passing to a subnet and using that X_∞ is compact, we can suppose $x_\alpha \to y \in X_\infty$ with $y \neq x_\infty$. Find $g \in C_\infty(X)$ so $g(x_\infty) = 1$, $g(y) = 0$. Since $C_\infty(X) \subset C_{Y,f}(X)$, we have $h \in C(Y)$ so $g = h \circ f$. Thus, $0 = g(y) = \lim g(x_\alpha) = \lim h(f(x_\alpha)) = h(f(x_\infty)) = g(x_\infty) = 1$. This construction proves that $x_\alpha \to x_\infty$.

(d) By the Tietze extension theorem (Theorem 2.2.5 of Part 1), any continuous function on $\widetilde{Y} \equiv \overline{\mathrm{Ran}(\widetilde{f})} \subset X_\infty \times Y$ is the restriction to $\widetilde{Y}$ of an $H \in C(X_\infty \times Y)$. By the Stone–Weierstrass theorem, we get all H by taking norm limits of finite sums of $g(x)h(y)$ with $g \in C(X_\infty)$, $h \in C(Y)$. But

$$G_{\widetilde{Y},\widetilde{f}}(g(x)h(y)) = g(x)h^\sharp(x) \tag{6.5.9}$$

with $h^\sharp = G_{Y,f}(h)$.

We can write

$$g = \alpha + g_0, \qquad \alpha = g(\infty), \qquad g_0 \in C_\infty(X) \tag{6.5.10}$$

If we note that $g_0 h \in C_\infty(X)$ and $\alpha h \in C_{Y,f}(X)$, so $gh^\sharp \in C_{Y,f}(X) + C_\infty(X)$. Moreover, every such $h^\sharp + q \in C_{y,f}(X) + C_\infty(X)$ is $G_{\widetilde{Y},f}(h(y) + q(x))$.

It follows that $C_{\widetilde{Y},\widetilde{f}}(X)$ is the norm-closure algebra generated by $C_{Y,f}(X) + C_\infty(X)$. But this is already an algebra since $h^\sharp \in C_{Y,f}(X)$, $g \in C_\infty(X)$ has $h^\sharp g \in C_\infty(X)$. □

The next natural question is which $\mathfrak{A} \subset C(X)$ are a $C_{Y,f}(X)$—and it is here where the Gel'fand theory enters and provides a one–one correspondence.

Theorem 6.5.5. *Let X be a locally compact space and $\mathfrak{A}$ a symmetric subalgebra. For $x \in X$, let $\delta_x \in \widehat{\mathfrak{A}}$ be given by (6.5.1). Then $(\widehat{\mathfrak{A}}, \delta)$ is a compact morphism of X. Moreover,*

$$C_{\widehat{\mathfrak{A}},\delta}(X) = \mathfrak{A} \tag{6.5.11}$$

Conversely, if (Y, f) is a compact morphism and $\mathfrak{A} = C_{Y,f}(X)$, then $(\widehat{\mathfrak{A}}, \delta)$ is equivalent to (Y, f).

Proof. $\widehat{\mathfrak{A}}$ is compact and δ continuous, so we need only prove $\mathrm{Ran}(\delta)$ is dense in $\widehat{\mathfrak{A}}$.

Since $\mathfrak{A}$ is a B^*-algebra, the commutative Gel'fand–Naimark theorem (Theorem 6.4.1) implies $\widehat{\ }$ is a bijection of $\mathfrak{A}$ and $C(\widehat{\mathfrak{A}})$, that is, if $g \in C(\widehat{\mathfrak{A}})$, there is $f \in \mathfrak{A}$ with $\widehat{f} = g$, and so

$$g(\delta_x) = \widehat{f}(\delta_x) = \delta_x(f) = f(x) \tag{6.5.12}$$

If $\overline{\mathrm{Ran}(\delta)}$ is not all $\widehat{\mathfrak{A}}$, there is $\ell \in \widehat{\mathfrak{A}} \setminus \overline{\mathrm{Ran}(\delta)}$ and so, by Urysohn's lemma, $g \in C(\widehat{\mathfrak{A}})$ with $g \restriction \mathrm{Ran}(\delta) = 0$ and $g(\ell) = 1$. But then if $f \in \mathfrak{A}$ has $\widehat{f} = g$, we have, by (6.5.12) and $g \restriction \mathrm{Ran}(\delta) = 0$, that $f = 0$, so $\widehat{g}(\ell) = \widehat{f}(\ell) = \ell(f) = 0$. This is a contradiction and proves $(\widehat{\mathfrak{A}}, \delta)$ is a compact morphism.

If $g \in C(\widehat{\mathfrak{A}})$, then $g = \widehat{f}$ for some f in $\mathfrak{A}$ and $(g \circ \delta)(x) = g(\delta_x) = f(x)$ by (6.5.12), that is, $C_{\widehat{\mathfrak{A}},\delta} \subset \mathfrak{A}$. Conversely, given f, let $g = \widehat{f}$. The above calculation shows $g \circ \delta = f$, that is, $\mathfrak{A} \subset C_{\widehat{\mathfrak{A}},\delta}$.

The final assertion is left to Problem 7. □

We've thus established a bijection between $\mathfrak{A}$'s and equivalence classes of compact morphisms. Thus, we can refer to $C_{Y,f}$ or to (Y, f).

Theorem 6.5.6. *(Y_2, f_2) covers (Y_1, f_1) if and only if*

$$C_{Y_1,f_1}(X) \subset C_{Y_2,f_2}(X) \tag{6.5.13}$$

Proof. Let $g\colon Y_2 \to Y_1$ be a covering map. If $h \in C(Y_1)$, then

$$G_{Y,f}(h) = h \circ f_1 = h \circ g \circ f_2 = G_{Y_2,f_2}(h \circ g) \tag{6.5.14}$$

proving (6.5.13).

Conversely, if (6.5.13) holds, since $Y_j = \widehat{C_{Y_j,f_j}(X)}$, we can define $g\colon Y_2 \to Y_1$ by letting $\ell \in \widehat{C_{Y_2,f_2}(X)}$ and then

$$g(\ell) = \ell \restriction C_{Y_1,f_2}(X) \tag{6.5.15}$$

Since $f_j(x)$ is δ_x (given by (6.5.1)) under the association of $\widehat{C_{Y_j,f_j}(X)}$ to Y_j, we see $g \circ f_2 = f_1$. □

Corollary 6.5.7. *If (Y_2, f_2) covers (Y_1, f_1) and (Y_1, f_1) covers (Y_2, f_2), then they are equivalent.*

Example 6.5.2 (continued). If $f\colon \mathbb{R} \to \partial\mathbb{D}$ by $f(\theta) = e^{i\theta}$, then $C_{\partial\mathbb{D},f}$ is exactly the periodic functions. $C_{\partial\mathbb{D},f} + C_\infty$ is already closed and is the asymptotically periodic functions. Similarly, the torus examples involve some asymptotically quasiperiodic functions (see Section 6.6). □

The compactification associated to $C(X)$ is universal in that it covers every other compact morphism. It is not hard to see (Theorem 2.3.7 of Part 1) that for a B^*-algebra, $\widehat{\mathfrak{A}}$ is second countable if and only if $\mathfrak{A}$ is separable. In $C(\mathbb{R})$, $\{e^{i\lambda x}\}_{\lambda\in\mathbb{R}}$ are all distance 2 from each other, so $C(\mathbb{R})$ is not separable and $\widehat{C(\mathbb{R})}$ is not second countable. It is huge. It is called the *Stone–Čech compactification.* We summarize in:

Theorem 6.5.8. *For every locally compact space, X, there is a strict compactification, $\widetilde{X}$, with the property that every $f \in C(X)$ has a continuous extension to $\widetilde{X}$. This Stone–Čech compactification covers every other compact morphism.*

Notes and Historical Remarks. Recall that a completely regular space is one in which points are closed and for any point, $x \notin A$, a closed set, there is a continuous f with $f(x) = 1$, $f \restriction A = \theta$. In 1930, Tychonoff [**707**], in a followup to his proof of Tychonoff's theorem, constructed compactifications of general completely regular spaces (also called Tychonoff spaces!). He formed

$$Y = \bigtimes_{f} [-\|f\|, \|f\|]$$

over a closure of f's and mapped $X \xrightarrow{\varphi} Y$ by $\varphi(x)_f = f(x)$; see Problem 1.

In 1937, Čech [**112**] and Stone [**671**] independently proved the existence of the universal compactification that bears their names and developed a theory of compactifications. Čech used Tychonoff's product construct. Stone instead formed the spaces of maximal ideals, then put the hull-kernel topology on it.

Three books on the subject of compactifications are [**113, 247, 730**].

Problems

1. Let X be a topological space and $C(X)$ the set of bounded continuous functions from X to $\mathbb{R}$. Let

$$\widetilde{X} = \bigtimes_{g\in C(X)} [-\|g\|, \|g\|] \tag{6.5.16}$$

Define $f\colon X \to \widetilde{X}$ by $f(x)_g = g(x)$.

(a) Prove that f is continuous and that if $Y = \overline{\text{Ran}(f)}$, then (f, Y) is a compact morphism.

(b) If $C(X)$ separates points, that is, $\forall x, y \in X$, $\exists f \in C(X)$ so $f(x) \neq f(y)$, prove that (Y, f) is a compactification.

(c) If X is completely regular (i.e., all closed sets A and $x \notin A$, there is $g \in C(X)$ so $g \restriction A = 0$, $g(x) = 1$), prove that (Y, f) is a strict compactification. (*Hint*: Let A be closed and $g_{A,x}$ a continuous function with $g \restriction A = 1$, $g(x) = 0$. Prove that $f[A] = \bigcap_{x \notin A}\{\tilde{x} \in \text{Ran}(f) \mid \tilde{x}_{G_{A,x}} \in [f_2, \|G_{A,x}\|]\}$.)

2. Let X be locally compact and X_∞ the one-point compactification. Prove that $f\colon X \to X_\infty$ by $f(x) = x$ is a strict compactification.

3. Let (Y, f) be a strict compactification and $g\colon X \to X_\infty$ given by (6.5.2). Prove that g is a covering map by direct argument. Then prove the result using Theorem 6.5.6.

4. If (Y, f) is a compact morphism, $(\widetilde{Y}, \widetilde{f})$ is the strict compactification of Proposition 6.5.3, and $\pi\colon X_\infty \times Y \to Y$ by $\pi(x, y) = y$, prove that $\pi \restriction \widetilde{Y}$ is a covering map of $(\widetilde{Y}, \widetilde{f})$ onto (Y, f).

5. Let (Y, f) be a compact morphism of X so that $f^{-1}[U]$ is unbounded (i.e., $\overline{f^{-1}[U]}$ is not compact) for all nonempty, open U (give an example of this), prove that the $(\widetilde{Y}, \widetilde{f})$ construction of Proposition 6.5.3 yields a strict compactification.

6. If (Y, f) is a strict compactification of X, prove that the $(\widetilde{Y}, \widetilde{f})$ construction of Proposition 6.5.3 yields a compact morphism equivalent to the original (Y, f). Do this directly and also using the $C_{Y,f}(X)$ formalism.

7. Let (Y, f) be a compact morphism of X, $\mathfrak{A} = C_{Y,f}(X)$, and $(\widehat{\mathfrak{A}}, \delta)$ the morphism given by (6.5.1). Prove $(\widehat{\mathfrak{A}}, \delta)$ is equivalent to (Y, f).

6.6. Almost Periodic Functions

A *locally compact abelian group* (aka LCA group), G, has a natural set of maps $\{\tau_g\}_{g \in G}$ on $C(G)$, the bounded, complex-valued continuous functions on G, given by

$$(\tau_g f)(h) = f(g + h) \tag{6.6.1}$$

Since

$$\|\tau_g(f)\|_\infty = \|f\|_\infty \tag{6.6.2}$$

each τ_g is a continuous map of $C(G)$ to $C(G)$ and

$$\tau_g \tau_h = \tau_{g+h}, \quad \tau_e = \mathbb{1} \tag{6.6.3}$$

$g \mapsto \tau_g$ is not continuous if G is not compact; indeed, a function like $f(x) = \sin(x^2)$ on $\mathbb{R}$ shows that $g \mapsto \tau_g f$ may not even be continuous. We will focus in this section on a class of f's for which (we'll show) not only is $g \mapsto \tau_g f$ continuous but more is true.

Definition. A function $f \in C(G)$ is called *almost periodic* (*in Bochner sense*) if and only if $\{\tau_g f\}_{g \in G}$ has compact closure. Its closure is called the *hull*, H_f, of f. The set of all almost periodic functions will be denoted $\mathrm{AP}(G)$.

Example 6.6.1. Let $G = \mathbb{R}$ and

$$f(x) = e^{i\alpha x} + e^{i\beta x} \tag{6.6.4}$$

If α/β is rational, say $\beta = (p/q)\alpha$, then f is periodic (i.e., $f(x+\sigma) = f(x)$ where $\sigma = 2\pi q\alpha^{-1}$ (so $e^{i\alpha\sigma} = e^{2\pi i q} = 1$ and $e^{i\beta\sigma} = e^{2\pi i p} = 1$)) and $\{f_\tau(x)\}$ is a circle in $C(\mathbb{R})$ and so compact. For general α, β, f is not periodic, but

$$f_\tau(x) = e^{i\alpha\tau} e^{i\alpha x} + e^{i\beta\tau} e^{i\beta x} \tag{6.6.5}$$

so $\{f_\tau\}$ lies in the torus of $\{\omega_1 e^{i\alpha\cdot} + \omega_2 e^{i\beta\cdot} \mid (\omega_1, \omega_2) \in \partial\mathbb{D} \times \partial\mathbb{D}\}$ and again is compact. Thus, f is almost periodic in the above sense. It is also "almost" periodic in that if $q|\frac{\beta}{\alpha} - \frac{p}{q}|$ is very small (and we know there are such rational approximations; see Theorem 7.5.2 of Part 2A), then (Problem 1)

$$\left\| f\left(\cdot + \frac{2\pi q}{\alpha}\right) - f(\cdot) \right\|_\infty \le 2\pi q \left| \frac{\beta}{\alpha} - \frac{p}{q} \right| \tag{6.6.6}$$

will be small. □

Notice that $x \mapsto e^{i\alpha x}$ is a character on $\mathbb{R}$. One of our first main results will be to prove the equivalence of Bochner almost periodic, being close to a finite linear combination of characters and being close to periodic in that there are large τ's so $\|f_\tau - f\|_\infty$ is small. This subject is clearly related to the last section. The set of all almost periodic functions is a Banach algebra, as we'll prove shortly, and since characters separate points (we'll prove this under a separability hypothesis on G in Section 6.9), so by Theorem 6.5.4, the almost periodic functions will be associated to a compactification, H, of G called the *Bohr compactification*, which we'll prove has a topological group structure.

Definition. A function, f, on G is called *Bohr almost periodic* if and only if for all ε, there is a compact set K so that for all $g \in G$, there is $h \in g + K$ so

$$\|f_h - f\|_\infty \le \varepsilon \tag{6.6.7}$$

We'll see soon that this is equivalent to the Bochner definition.

The reader should note that this implies there are arbitrary large h's, so capturing the intuitive notion of close to periodic.

Definition. f on G is called *uniformly continuous* if for all $\varepsilon > 0$, there exists a neighborhood, N_ε, of e so that

$$x - y \in N_\varepsilon \Rightarrow |f(x) - f(y)| < \varepsilon \tag{6.6.8}$$

As a warmup:

Theorem 6.6.2. (a) $x \mapsto \tau_x f$ *as a function from* G *to* $C(G)$ *is continuous if and only if* f *is uniformly continuous.*

(b) *Any Bohr almost periodic function,* f*, is uniformly continuous.*

(c) *Any (Bochner) almost periodic function,* f*, is uniformly continuous.*

(d) *The Bochner almost periodic functions are a Banach subalgebra of* $C(G)$*.*

Proof. (a) If $x \mapsto \tau_x f$ is continuous, given ε, there is a neighborhood N_ε of e so $u \in N_\varepsilon \Rightarrow \|\tau_u f - f\|_\infty < \varepsilon$ which means for all z, $|f(u+z) - f(z)| < \varepsilon$, so if x and y have $x - y \in N_\varepsilon$, then setting $z = y$ and $u = x - y$, we have $|f(x) - f(y)| < \varepsilon$, that is, (6.6.8) holds. Conversely, if (6.6.8) holds and ε, x are given, then for all $y \in x - N$ and all $z \in G$, $x + z - y + z = x - y \in N$ so $|f(x+z) - f(y+z)| < \varepsilon$, that is, $\|\tau_x f - \tau_y f\|_\infty \le \varepsilon$, proving continuity at x.

(b) For each y, $\tau_y f$ is continuous at e, so there is a neighborhood N_y of e so $w \in N_y \to |\tau_y f(w) - \tau_y f(e)| < \varepsilon/4$. Thus, $w, w' \in N_y \Rightarrow |\tau_y f(w) - \tau_y f(w')| < \varepsilon/2$. By continuity of addition, find M_y a neighborhood of e, so $M_y + M_y \subset N_y$. Since $w, w'' \in M_y \Rightarrow w + w'' \in N_y$, we see

$$w, w', w'' \in M_y \Rightarrow |(\tau_{y+w''} f)(w) - (\tau_{y+w''})(w')| < \frac{\varepsilon}{2} \tag{6.6.9}$$

Let K be chosen so (6.6.7) holds with $\varepsilon/4$ in place of ε. Since $K \subset \bigcup_{y\in K}(y + M_y)$, find $y_1, \dots, y_\ell$ so $K \subset \bigcup_{j-1}^{\ell}(y_j + M_{y_j})$ and let $M = \bigcap_{j-1}^{\ell} M_{y_j}$. By (6.6.9), we have

$$y \in K,\ w, w' \in M \Rightarrow |(\tau_y f)(w) - (\tau_y f)(w')| < \frac{\varepsilon}{2} \tag{6.6.10}$$

Now suppose x, y are such that $x - y \in M$. By Bohr almost periodicity, there is $h \in K$ so $\|\tau_{h-y} f - f\|_\infty < \varepsilon/4$, so

$$\|\tau_h f - \tau_y f\|_\infty < \frac{\varepsilon}{4} \tag{6.6.11}$$

By (6.6.10),

$$|(\tau_h f)(e) - (\tau_h f)(x-y)| < \frac{\varepsilon}{2} \tag{6.6.12}$$

so by (6.6.10),

$$|f(y) - f(x)| = |(\tau_y f)(e) - (\tau_y f)(x-y)| < \varepsilon \tag{6.6.13}$$

Thus, f is uniformly continuous.

(c) Given ε, pick $y_1, \dots, y_\ell$ so $\{\tau_y f\} \subset \bigcup_{j=1}^{\ell}\{g \mid \|g - \tau_{y_j}\| < \varepsilon/3\}$, which we can do by compactness. By continuity of $x \to \tau_{y_f} f(x)$, pick N_j so

$$u \in N_j \Rightarrow |\tau_{y_j} f(u) - \tau_{y_j} f(e)| < \frac{\varepsilon}{3} \tag{6.6.14}$$

Let $N = \bigcap_{j=1}^{\ell} N_j$. Suppose $x - y \in N$. Pick y_j so $\|\tau_y f - \tau_{y_j} f\|_\infty < \varepsilon/3$. This plus (6.6.14) implies

$$u \in N \Rightarrow |\tau_y f(u) - \tau_y f(e)| < \varepsilon \tag{6.6.15}$$

Taking $u = x - y$ yields $|f(x) - f(y)| < \varepsilon$, that is, f is uniformly continuous.

(d) Continuity of addition and multiplication implies if K_1 and K_2 are two compact sets in $C(G)$, then so are $\{f + g \mid f \in K_1, g \in K_2\}$ and $\{fg \mid f \in K_1, g \in K_2\}$. This shows the (Bochner) almost periodic functions are an algebra. That they are norm-closed is a simple limit argument (Problem 2). □

By T^∞, we mean the countable infinite product of circles, $\partial\mathbb{D}$. In the next proof, we'll need the fact that if H is a compact separable LCA group, the set of its characters separates points. We'll actually prove this twice later in this chapter. In Section 6.8, we'll prove the matrix elements of the finite-dimensional irreducible representations of any compact group separate points (and for abelian groups, we'll prove irreducible representations are one-dimensional). And in Section 6.9, we'll prove any LCA with a certain separability condition has characters that separate points (see Theorem 6.9.5). Here is the central theorem in the theory of almost periodic functions:

Theorem 6.6.3. *Let G be a metrizable, separable LCA group and $f \in C(G)$. The following are equivalent:*

(1) *f is (Bochner) almost periodic.*
(2) *f is Bohr almost periodic.*
(3) *f is a uniform limit of finite linear combinations of elements $\chi \in \widehat{G}$.*
(4) *There is a continuous homomorphism $\varphi\colon G \to \mathbb{T}^\infty$ and $F \in C(T^\infty)$ so that*

$$f(x) = F(\varphi(x)) \tag{6.6.16}$$

Remark. Our proof of (1) $\Leftrightarrow$ (2) below will show the K can be taken finite. Thus, a corollary of the proof of (2) $\Rightarrow$ (1) $\Rightarrow$ (2) is that for a Bohr almost periodic function, the K can be taken finite.

Proof. We'll show (4) $\Rightarrow$ (3) $\Rightarrow$ (1) $\Rightarrow$ (2) $\Rightarrow$ (1) $\Rightarrow$ (4).

<u>(4) $\Rightarrow$ (3).</u> Let $\{z_j\}_{j=1}^\infty$ be the coordinate functions on $\mathbb{T}^\infty$. By the Stone–Weierstrass theorem, Laurent polynomials in the z_j are dense in $C(\mathbb{T}^\infty)$ so F is a limit of such polynomials. Since φ is a homomorphism, $(\prod_{j=1}^J z_j^{\ell_j}) \circ \varphi$ is a character on G, so $F \circ \varphi$ is a limit of a linear combination of characters of G.

(3) $\Rightarrow$ (1). Since the Bochner almost periodic functions are a norm-closed algebra, we only need to know that the translates of any character are compact. But if χ is a character,

$$(\tau_g \chi) = \chi(g)\chi \tag{6.6.17}$$

so $\{\tau_g \chi\} \subset \{e^{i\theta}\chi \mid e^{i\theta} \in \partial\mathbb{D}\}$ which is compact.

(1) $\Rightarrow$ (2). Given ε, by compactness pick $y_1, \dots, y_\ell \in G$ so every $\tau_y f$ is within ε of some $\tau_{y_j} f$. Let $K = \{-y_1, \dots, -y_\ell\}$ which is finite, and so compact. Given $x \in G$, pick y_j so $\|\tau_x f - \tau_{y_j} f\| < \varepsilon$. Then $\|f - \tau_{x-y_j} f\| < \varepsilon$ and $x - y_j \in x + K$.

(2) $\Rightarrow$ (1). Given ε, we need only find finitely many y_j so every $\tau_y f$ is within ε of some y_j. Let K be such that (6.6.7) with ε replaced by $\varepsilon/2$. By Theorem 6.6.2, $x \mapsto \tau_x f$ is continuous, so $\{\tau_x f\}_{x \in K}$ is compact. Thus, find $y_1, \dots, y_i \in K$ so every $\tau_x f$, $x \in K$, is within $\varepsilon/2$ of some $\tau_{y_j} f$.

By Bohr almost periodicity, for any $y \in G$, we can find $x \in K$ so $\|\tau_{-y+x} f - f\|_\infty < \varepsilon/2$. Thus,

$$\|\tau_y f - \tau_{y_j} f\| \le \|\tau_y(\tau_{-y+x} f - f)\| + \|\tau_x f - \tau_{y_j} f\| < \varepsilon$$

(1) $\Rightarrow$ (4). This is the only subtle part of the proof. We'll show H, the compact hull of f, is a compact group in such a way that there is $\varphi_1 \colon G \to H$, a group homomorphism, with

$$\varphi_1(x) = \tau_x f \tag{6.6.18}$$

So define φ_1 by (6.6.18). Since f is uniformly continuous, φ_1 is continuous when H is given the $\|\cdot\|_\infty$ topology. Clearly,

$$\begin{aligned} \|\varphi_1(x+y) - \varphi_1(x'+y')\| &= \|(\tau_x \tau_y f - \tau_x \tau_{y'} f) + (\tau_{y'} \tau_x f - \tau_{y'} \tau_x f)\| \\ &\le \|\varphi_1(y) - \varphi_1(y')\| + \|\varphi_1(x) - \varphi_1(x')\| \end{aligned} \tag{6.6.19}$$

By picking x_n, y_n so $\varphi(x_n) \to h$, $\varphi(y_n) \to h'$, this lets us define $h + h'$ for any $h, h' \in H$. It is easy to see (Problem 3) this makes H into a group so that $\varphi_1(x+y) = \varphi_1(x) + \varphi_1(y)$.

Thus, H is a compact metrizable group, so separable, and so we know $\widehat{H}$ separates points. By separability of $C(H)$ (see Problem 4), we can find a countable subset $\{\chi_j\}_{j=1}^\infty$ of $\widehat{H}$ which separates points. Let $\varphi_2 \colon H \to \mathbb{T}^\infty$ by

$$[\varphi_2(h)]_j = \chi_j(h) \tag{6.6.20}$$

Since $\{\chi_j\}_{j=1}^\infty$ separate points, φ_2 is a bijection, so $\varphi_2[H]$ is closed in $\mathbb{T}^\infty$ and, by compactness, φ_2 is a homeomorphism of H and $\varphi_2[H]$.

Define G on H by $G(h) = h(e)$ and $\widetilde{F}$ on $\varphi_2[H]$ as $G \circ \varphi_2^{-1}$. $\widetilde{F}$ is continuous, so by the Tietze extension theorem (Theorem 2.2.5 of Part 1), there is $F\colon \mathbb{T}^\infty \to \mathbb{C}$, so $F \restriction \varphi_2[H] = \widetilde{F}$. If $\varphi = \varphi_2\varphi_1$, φ is a homomorphism of G to $\mathbb{T}^\infty$ and, by construction, $(F\circ\varphi)(x) = G(\tau_x f) = (\tau_x f)(e) = f(x)$. $\square$

Now let $\mathcal{B}(G)$ be the Gel'fand spectrum of the Banach algebra of all almost periodic functions. The same argument that let us define a group structure on the hull of a single f lets us define it on $\mathcal{B}(G)$ and so get a natural compact group, $\mathcal{B}(G)$, the Bohr compactification, and continuous homomorphism $\Phi\colon G \to \mathcal{B}(G)$. Another way to understand $\mathcal{B}(G)$ is the following. Let $\mathbb{T}^{\widehat{G}}$ be all functions from $\widehat{G}$, the set of all characters on G, to $\partial\mathbb{D}$. $\mathbb{T}^{\widehat{G}}$ is compact by Tychonoff's theorem. Let $\widetilde{\Phi}(x)(\chi) = \chi(x)$. $\overline{\text{Ran}(\widetilde{\Phi})}$ is a compact subgroup of $\mathbb{T}^{\widehat{G}}$, and by construction, Φ, the map from G to $\overline{\text{Ran}(\widetilde{\Phi})}$, is continuous. This range provides an explicit realization of $\mathcal{B}(G)$ and shows that $\mathcal{B}(G)$ is typically nonseparable.

Finally, we want to discuss a natural "integral" on AP(G) and the associated Fourier expansion. We have seen that for any f, H_f, its hull, is a compact group in the topology induced by $\|\cdot\|_\infty$. It has a Haar probability measure, $d\mu_f$. Thus, we can define

$$\mathbb{A}(f) \equiv \int g(e)\, d\mu_f(g) \tag{6.6.21}$$

It can be shown (Problem 5) that $\mathbb{A}(f)$ is a linear functional on AP(G) (indeed, since AP$(G) = C(\mathcal{B}(G))$, it defines a measure on $\mathcal{B}(G)$, which is just Haar measure on $\mathcal{B}(G)$). Since $H_{\tau_x f} = H_f$ as functions in $C(G)$ and $\tau_y(\tau_x f) = \tau_{x+y} f$, so that the association is just translation, and since $d\mu_f$ is invariant under translation, we see

$$\mathbb{A}(\tau_x f) = \mathbb{A}(f) \tag{6.6.22}$$

Of course, $H_{\mathbb{1}}$ is a single point, so

$$\mathbb{A}(\mathbb{1}) = 1 \tag{6.6.23}$$

Since $\tau_x\chi = \chi(x)\chi$, (6.6.22) shows that if $\chi \not\equiv \mathbb{1}$,

$$\mathbb{A}(\chi) = 0, \quad \chi \not\equiv 1 \tag{6.6.24}$$

Next, if $g \in H_f$, $\|g\|_\infty \le \|f\|_\infty$, so (6.6.18) implies

$$|\mathbb{A}(f)| \le \|f\|_\infty \tag{6.6.25}$$

Since any $f \in$ AP(G) is a uniform limit of linear combinations of characters, the last three equations uniquely determine $\mathbb{A}$.

For any $\chi \in \widehat{G}$ and $f \in$ AP(G), we define its Fourier coefficients to be

$$\widehat{f}(\chi) = \mathbb{A}(\bar{\chi} f) \tag{6.6.26}$$

By approximating any f by finite linear combinations of characters, it is easy to see (Problem 6(a)) that

$$\sum_{\chi\in\widehat{G}} |\widehat{f}(\chi)|^2 = \mathbb{A}(|f|^2) \tag{6.6.27}$$

In particular, the set of χ for which $\widehat{f}(\chi) \neq 0$, called the *spectrum* of f, is countable. The subgroup (*not* closed subgroup, just finite products of the χ's and $\bar{\chi}$'s) generated by the spectrum is called the *frequency module*, FM(f), of f. If FM(f) is finitely generated (equivalently, see Problem 7, if there is a finite-dimensional torus, $\mathbb{T}^\ell$, and $\varphi\colon G \to \mathbb{T}^\ell$ and $F\colon \mathbb{T}^\ell \to \mathbb{C}$ so $f(g) = F(\varphi(g))$), we say f is *quasiperiodic*.

Since $\{\chi \mid \widehat{f}(\chi) \neq 0\}$ is countable, let $\{\chi_j\}_{j=1}^\infty$ be a counting. Then, it can be seen (Problem 6(b)) that as $N \to \infty$,

$$\mathbb{A}\left(\left|f - \sum_{j=1}^N \widehat{f}(\chi_j)\chi_j\right|^2\right) \to 0 \tag{6.6.28}$$

Notes and Historical Remarks. The definition of (Bohr) almost periodic on functions on $\mathbb{R}$ and some of their properties is due to Bohr [**77, 78**], while the Bochner definition approach is due to Bochner [**73, 75**].

On $\mathbb{R}$, one can show (Problem 8) for $f \in \text{AP}(\mathbb{R})$ that

$$\mathbb{A}(f) = \lim_{T\to\infty} \frac{1}{2T}\int_{-T}^T f(x)\,dx \tag{6.6.29}$$

A function f on $\mathbb{R}$ is called *Besicovitch almost periodic* if there are f_n finite sums of e^{ikx} so that

$$\lim_{n\to\infty}\left[\limsup_{T\to\infty}\frac{1}{2T}\int_{-T}^T |f(x) - f_n(x)|^2\,dx\right] = 0 \tag{6.6.30}$$

This is a larger class than AP($\mathbb{R}$). Thus, the smaller class is sometimes called "uniformly almost periodic."

For book treatments of the theory of uniform and/or Besicovitch almost periodic functions, see [**53, 79, 132, 442**].

In terms of the general theory of dual groups, one can construct $\mathcal{B}(G)$ as follows: Let $\widehat{G}_{\text{disc}}$ be $\widehat{G}$ with the discrete topology. Then $(\widehat{G}_{\text{disc}})\widehat{\ } = \mathcal{B}(G)$. This is discussed further in the Notes to Section 6.9. (6.6.27)/(6.6.28) can then be viewed as standard Fourier analysis on the compact group $\mathcal{B}(G)$ whose dual is $\widehat{G}_{\text{disc}}$.

Problems

1. Let $f\colon \mathbb{R} \to \mathbb{C}$ be given by (6.6.4). Prove (6.6.6).

2. (a) Prove f is Bochner almost periodic if and only if for all $\varepsilon > 0$, there exists $x_1, \dots, x_n$ so $\sup_{y\in G} \min_{j=1,\dots,n} \|\tau_y f - \tau_{x_j} f\| < \varepsilon$.

   (b) Prove that a uniform limit of Bochner almost periodic functions is Bochner almost periodic.

3. Let $f \in \mathrm{AP}(G)$. Using (6.6.19), confirm that H_f has a topological group structure (with $\|\cdot\|_\infty$ topology) so that φ_1 is a continuous group homomorphism.

4. (a) Let X be a separable metric space. Prove that $C(X)$ is separable. (*Hint*: If ρ is the metric and $\{x_j\}_{j=1}^\infty$ a dense set, consider polynomials in $\rho(x, x_j)$ with rational coefficients.)

   (b) Let $\{f_\alpha\}_{\alpha\in I}$ be a family of functions including $\mathbb{1}$ that separates points. Prove there is a countable subfamily that separates points. (*Hint*: First show that polynomials in the f_α are dense in $C(X)$. Then for a countable dense set g_n, find a countable subfamily so polynomials in that subfamily approximate the g_n's well, and then conclude this countable family must separate points.)

5. (a) Given $f, g \in \mathrm{AP}(G)$, let $\mathfrak{A}_{f,g}$ be the subalgebra of $C(G)$ generated by f and g. Show that $\widehat{\mathfrak{A}}_{f,g}$ has a group structure so $x \mapsto \delta_x$ is a group homomorphism of G to $\widehat{\mathfrak{A}}_{f,g}$.

   (b) If η is normalized Haar measure on $\widehat{\mathfrak{A}}_{f,g}$, prove that $\mathbb{A}(f) = \int \widehat{f}(\ell)\, d\eta(\ell)$ and similarly for g, and conclude that $\mathbb{A}$ is a linear functional.

6. (a) Prove (6.6.27) if f is a finite linear combination of characters and deduce it for all $f \in \mathrm{AP}(G)$.

   (b) Verify (6.6.28) for f a finite linear combination of characters and deduce it for all $f \in \mathrm{AP}(G)$.

7. (a) Prove that any $f \in \mathrm{AP}(G)$ is a uniform limit of polynomials in the χ's with $\widehat{f}(\chi) \neq 0$.

   (b) If $\mathrm{FM}(f)$ is finitely generated, show there is a finite ℓ, a group homomorphism $\varphi\colon G \to \mathbb{T}^\ell$, and continuous $f\colon \mathbb{T}^\ell \to \mathbb{C}$ so $f = F \circ \varphi$.

   (c) If f has the form $F \circ \varphi$ in (b), show that f is almost periodic and $\mathrm{FM}(f)$ is finitely generated.

8. (a) Let $f\colon \mathbb{R} \to \mathbb{C}$ be a finite linear combination of characters. Prove (6.6.29).

   (b) Prove (6.6.29) for all $f \in \mathrm{AP}(\mathbb{R})$.

   (c) Find analogs of (6.6.29) for $\mathbb{R}^\nu$, $\mathbb{Z}$, and $\mathbb{Z}^\nu$.

6.7. The GNS Construction and the Noncommutative Gel'fand–Naimark Theorem

The bulk of this chapter discusses abelian Banach algebras although some of the early results were stated in nonabelian generality. This section and the next will discuss some not necessarily abelian cases—this section will prove a structure theorem for general B^*-algebras and the next some simple results about representations of general locally compact groups. Both depend critically on the theory of representations of Banach algebras with norm-compatible involution, which is the first topic of this section.

Recall (see Section 6.4) that a B^*-algebra is a Banach algebra with involution, *, obeying the C^*-identity

$$\|x^*x\| = \|x\|^2 \tag{6.7.1}$$

The main theorem of this section is:

Theorem 6.7.1 (Noncommutative Gel'fand–Naimark Theorem). *Every B^*-algebra, $\mathfrak{A}$, is isometrically isomorphic to an algebra of operators on a not necessarily separable Hilbert space under an isomorphism π with*

$$\pi(x^*) = \pi(x)^* \tag{6.7.2}$$

(the * on the right is operator adjoint). If $\mathfrak{A}$ is separable, the Hilbert space may be taken to be separable.

Remarks. 1. Recall (see Section 3.1 of Part 1) our definition of "Hilbert space" includes separability and we use "not necessarily separable Hilbert space" if that axiom is dropped.

2. A norm-closed algebra, $\mathfrak{A}$, of operators on a Hilbert space with $A \in \mathfrak{A} \Rightarrow A^* \in \mathfrak{A}$ is called a C^*-algebra. This theorem is sometimes paraphrased as "every B^*-algebra is a C^*-algebra" or as "every abstract C^*-algebra is a concrete C^*-algebra."

Note that Lemma 6.4.5 states that every B^*-algebra without identity can be realized as a codimension 1 subalgebra of a B^*-algebra with one, so in proving Theorem 6.7.1, we can and will suppose $\mathfrak{A}$ has an identity (one can also use approximate identities; see the Notes and Problem 3). The key to both the proof of this theorem and our discussion of group representations in the next section will be positive linear functionals and their relation to representations of *-algebras.

Definition. Let $\mathfrak{A}$ be a Banach algebra with identity, e, and norm-compatible involution, *. We call this a *-algebra for short. A *representation*, π, of $\mathfrak{A}$ is a map $\pi\colon \mathfrak{A} \to \mathcal{L}(\mathcal{H})$, the bounded operators on a not necessarily separable Hilbert space, $\mathcal{H}$, so that

$$\pi(xy) = \pi(x)\pi(y), \qquad \pi(e) = \mathbb{1}, \qquad \pi(x^*) = \pi(x)^* \tag{6.7.3}$$

Definition. A *generalized state*, φ, on a *-algebra, $\mathfrak{A}$, is a function $\varphi \in \mathfrak{A}^*$ so that for all $x \in \mathfrak{A}$,

$$\varphi(x^*x) \geq 0 \tag{6.7.4}$$

$\mathcal{S}_+(\mathfrak{A})$ will denote the set of generalized states on $\mathfrak{A}$. A *state* is a generalized state with

$$\varphi(e) = 1 \tag{6.7.5}$$

The set of all states will be denoted $\mathcal{S}_{+,1}(\mathfrak{A})$.

Recall Proposition 6.3.5 proved that any $\varphi \in \mathcal{S}_+(\mathfrak{A})$ obeys

$$|\varphi(x)| \leq \|x\|\varphi(e) \tag{6.7.6}$$

This explains why we didn't require continuity in our definition of representation because it is automatic:

Proposition 6.7.2. *For any representation, π, of a *-algebra, $\mathfrak{A}$, we have*

$$\|\pi(x)\| \leq \|x\| \tag{6.7.7}$$

In particular, π is $\|\cdot\|$-continuous.

Proof. For $\psi \in \mathcal{H}$, define

$$\varphi_\psi^{(\pi)}(x) = \langle \psi, \pi(x)\psi \rangle \tag{6.7.8}$$

Since

$$\varphi_\psi^{(\pi)}(x^*x) = \|\pi(x)\psi\|^2 \tag{6.7.9}$$

we see that $\varphi_\psi^{(\pi)}$ is a generalized state and is a state if $\|\psi\| = 1$. Since $\varphi(e) = \|\psi\|^2$, (6.7.6) implies that

$$|\varphi_\psi^{(\pi)}(x)| \leq \|\psi\|^2\|x\| \tag{6.7.10}$$

In particular, by (6.7.9),

$$\|\pi(x)\psi\|^2 \leq \|x^*x\|\|\psi\|^2 \leq \|x\|^2\|\psi\|^2 \tag{6.7.11}$$

which implies (6.7.7). $\square$

Thus, representations are associated to states. A key point is that a strong converse is true. We need some definitions.

Definition. A *cyclic representation* of a *-algebra, $\mathfrak{A}$, is a representation $\pi\colon \mathfrak{A} \to \mathcal{L}(\mathcal{H})$ and distinguished unit vector, ψ, so that

$$\mathcal{H} = \overline{\{\pi(x)\psi \mid x \in \mathfrak{A}\}} \tag{6.7.12}$$

Two cyclic representations, $(\pi, \mathcal{H}, \psi)$ and $(\widetilde{\pi}, \widetilde{\mathcal{H}}, \widetilde{\psi})$, are called *equivalent* if there exists a unitary $V\colon \mathcal{H} \to \widetilde{\mathcal{H}}$ so that

$$V\pi(x)V^{-1} = \widetilde{\pi}(x), \qquad V\psi = \widetilde{\psi} \tag{6.7.13}$$

The *state of* $(\pi, \mathcal{H}, \psi)$ is

$$\varphi_\psi^{(\pi)} = \langle \psi, \pi(x)\psi\rangle \tag{6.7.14}$$

The same argument that led to Theorem 5.2.1 (except to handle that in the nonseparable case, one needs a Zornification) implies

Proposition 6.7.3. *Any representation is a direct sum of cyclic representations.*

Theorem 6.7.4 (GNS Construction). *Any state is the state of a cyclic representation. Two cyclic representations are equivalent if and only if their states are equal. If* $\mathfrak{A}$ *is separable, every cyclic representation is on a separable Hilbert space.*

Remark. GNS stands for "Gel'fand–Naimark–Segal" (see the Notes).

Proof. Let φ be a state. Define $\langle\ ,\ \rangle_\varphi$ on $\mathfrak{A}$ by

$$\langle x, y\rangle_\varphi = \varphi(x^*y) \tag{6.7.15}$$

This is a sesquilinear form with $\langle x, x\rangle_\varphi \geq 0$. It may not, however, be strictly positive, so we define

$$\mathcal{I}_\varphi = \{x \mid \langle x, x\rangle_\varphi = \varphi(x^*x) = 0\} \tag{6.7.16}$$

Since φ is continuous, $\mathcal{I}_\varphi$ is closed. The Schwarz inequality for $\langle\ ,\ \rangle_\varphi$ says

$$|\varphi(z^*w)|^2 \leq \varphi(z^*z)\varphi(w^*w) \tag{6.7.17}$$

In particular, if $x \in \mathcal{I}_\varphi$, $y \in \mathfrak{A}$, and $z = x^*y^*y$,

$$\varphi((yx)^*(yx)) \leq \varphi(zz^*)\varphi(x^*x) = 0$$

so $x \in \mathcal{I}_\varphi \Rightarrow yx \in \mathcal{I}_\varphi$, that is,

$$\mathcal{I}_\varphi \text{ is a closed left ideal in } \mathfrak{A} \tag{6.7.18}$$

Let $\mathcal{H}_0 = \mathfrak{A}/\mathcal{I}_\varphi$ and define $\langle\ ,\ \rangle$ on $\mathcal{H}_0$ by

$$\langle [x], [y]\rangle = \langle x, y\rangle_\varphi \tag{6.7.19}$$

which is easily seen to be well-defined with a positive definite inner product, so its completion, $\mathcal{H}$, is a Hilbert space.

Define $\pi_0\colon \mathfrak{A} \to \mathcal{L}(\mathcal{H}_0)$ by

$$\pi_0(x)[y] = [xy] \tag{6.7.20}$$

By (6.7.18), this is well-defined. By the same argument as in Proposition 6.7.2,

$$\|\pi_0(x)[y]\|_{\mathcal{H}} \le \|x\|_{\mathfrak{A}} \|[y]\|_{\mathcal{H}} \tag{6.7.21}$$

so $\pi_0(x)$ extends to a map, $\pi(x)$, in $\mathcal{L}(\mathcal{H})$. It is easy to check that π is a representation of $\mathfrak{A}$. Notice if $\psi = [e]$,

$$\langle \psi, \pi(x)\psi \rangle = \langle [e], [xe] \rangle = \varphi(exe) = \varphi(x)$$

so φ is the associated state. Since $\{\pi(x)\psi\} = \{[x]\}$, ψ is cyclic. So we have the promised cyclic representation.

The other parts of the theorem are easy to check (Problem 1). □

The key to proving the noncommutative Gel'fand–Naimark theorem is the following, whose proof we defer:

Theorem 6.7.5. *Let $\mathfrak{A}$ be a B^*-algebra with identity. For any $x \in \mathfrak{A}$, there is a state, φ, with*

$$\varphi(x^*x) = \|x\|_{\mathfrak{A}}^2 \tag{6.7.22}$$

Proof of Theorem 6.7.1 given Theorem 6.7.5. Let π_φ be the cyclic representation generated by φ via Theorem 6.7.4. Let

$$\pi = \bigoplus_{\varphi \in \mathcal{S}_{+,1}} \pi_\varphi \tag{6.7.23}$$

on $\oplus \mathcal{H}_\varphi$. Since $\|\pi(x)\| \le \|x\|$ on every $\mathcal{H}_\varphi$ and if $\varphi(x^*x) = \|x\|^2$, then $\|\pi_\varphi(x)[e]\|_{\mathcal{H}} = \|[x]\|_{\mathcal{H}}^2 = \|x\|^2$, we see $\|\pi(x)\| = \|x\|$. Thus, π is an isometric isomorphism on $\mathrm{Ran}(\pi)$.

If $\mathfrak{A}$ is separable, the weak topology on $\mathfrak{A}^*$ is metrizable and $\mathcal{S}_{+,1}$ is separable (Problem 2(a)). Let $\{\varphi_n\}_{n=1}^\infty$ be a dense subset of $\mathcal{S}_{+,1}$. It is easy to see (Problem 2(b))

$$\sup_n \varphi_n(x^*x) = \|x\|_{\mathfrak{A}}^2 \tag{6.7.24}$$

so that $\oplus_n \pi_{\varphi_n}$ is an isometry and is on a separable space since each $\mathcal{H}_{\varphi_n}$ is. □

The key to proving Theorem 6.7.5 will be to show if $\varphi \in \mathfrak{A}^*$ and $\varphi(e) = 1 = \|\varphi\|$, then $\varphi \in \mathcal{S}_{+,1}$. For it will be easy to construct such a φ with $\varphi(x^*x) = \|x\|^2$ on the two-dimensional space $\{\alpha e + \beta x^*x \mid \alpha, \beta \in \mathbb{C}\}$ and then use the Hahn–Banach theorem to get a φ with $\varphi(e) = 1 = \|\varphi\|$. We need some preliminary temporary notions:

Definition. Let $\mathfrak{A}$ be a B^*-algebra with identity. An $x \in \mathfrak{A}$ with $x = x^*$ and $\sigma(x) \subset [0, \infty)$ will be called *Kadison positive.* A *Kadison state* is a $\varphi \in \mathfrak{A}^*$ with

$$\varphi(e) = 1, \qquad x \text{ Kadison positive} \Rightarrow \varphi(x) \ge 0 \tag{6.7.25}$$

Remark. We'll eventually prove that x is Kadison positive if and only if $x = y^*y$ for some y, so Kadison positive is the same as positive and Kadison state is the same as state.

Proposition 6.7.6. *If $\varphi \in \mathfrak{A}^*$ has*

$$\varphi(e) = 1 = \|\varphi\| \tag{6.7.26}$$

then φ is a Kadison state.

Remark. For now we don't know Kadison states have $\|\varphi\| = 1$, but we will see Kadison states are states (see Theorem 6.7.8) so that (6.7.26) is equivalent to being a Kadison state.

Proof. By Theorem 6.4.4(d), if $x = x^*$, and (6.7.26) holds, then $\varphi(x)$ is real so $\varphi(x)$ is real for any Kadison positive x.

By $\sigma(x) \subset [0,\infty)$ and $\sigma(x) \subset \{\lambda \mid |\lambda| \leq \|x\|\}$, we know that $\sigma(x) \subset [0,\|x\|]$. If $y = \|x\|e - x$, then $\sigma(y) \subset \|x\| - [0,\|x\|] = [0,\|x\|]$. By Theorem 6.4.4(c), we conclude

$$\|y\| \leq \|x\| \tag{6.7.27}$$

Since $\|\varphi\| = 1$, this implies $|\varphi(y)| \leq \|x\|$, that is,

$$\big|\|x\| - \varphi(x)\big| \leq \|x\| \tag{6.7.28}$$

Since this is false if $\varphi(x) < 0$, we see $\varphi(x) \geq 0$. □

Let x be symmetric and $\mathfrak{A}(x)$ the closure of the polynomials in x. $\mathfrak{A}(x)$ is an abelian B^*-algebra, so by Theorem 6.4.3, it is isomorphic to $C(\widehat{\mathfrak{A}(x)}) = C(\sigma(x))$ (since x generates $\mathfrak{A}(x)$), and we have a continuous functional calculus. Here is what we get from this:

Proposition 6.7.7. (a) *If x is symmetric, $\sigma(x) \subset \mathbb{R}$.*

(b) *If x is Kadison positive, then $x = y^2$ for a symmetric y.*

(c) *If y is symmetric and $x = y^2$, then x is Kadison positive.*

(d) *For any symmetric $x \in \mathfrak{A}$ and $\lambda \in \sigma(x)$, there exists a Kadison state, φ, with*

$$\varphi(x) = \lambda \tag{6.7.29}$$

(e) *For any symmetric $x \in \mathfrak{A}$, we can find symmetric $x_\pm$ Kadison positive so that*

$$x = x_+ - x_-, \qquad x_+x_- = x_-x_+ = 0 \tag{6.7.30}$$

Proof. (a) Since * is a symmetric involution on $\mathfrak{A}(x)$ (see Theorem 6.4.4(e)), $\widehat{x}$ is real on $\widehat{\mathfrak{A}(x)}$, so $\sigma_{\mathfrak{A}(x)}(x) \subset \mathbb{R}$. By Theorem 6.1.11(e),

$$\sigma_{\mathfrak{A}}(x) = \sigma_{\mathfrak{A}(x)}(x) \subset \mathbb{R} \tag{6.7.31}$$

proving (a).

(b) $\mathfrak{A}(x) \cong C(\sigma(x))$. Since $\sigma(x) \subset [0, \|x\|]$ and, under this isomorphism, $\widehat{x}(\lambda) = \lambda$, we can find $y \in \mathfrak{A}(x)$ with $\widehat{y}(\lambda) = \sqrt{\lambda}$. Thus, $\widehat{y^2} = (\widehat{y})^2 = \widehat{x}$ so $y^2 = x$ since $\widehat{\ }$ is an isomorphism.

(c) By the spectral mapping theorem (Theorem 2.2.16), $\sigma(x) = \sigma(y)^2$, so $\sigma(y) \subset \mathbb{R} \Rightarrow \sigma(x) \subset [0, \infty)$.

(d) Let $\ell \in \widehat{\mathfrak{A}(x)}$ have $\ell(x) = \lambda$. As a functional on $\mathfrak{A}(x)$, ℓ obeys

$$\ell(e) = \|\ell\| = 1 \tag{6.7.32}$$

By the Hahn–Banach theorem (Theorem 5.5.5 of Part 1), there is $\varphi \in \mathfrak{A}^*$ so

$$\|\varphi\| = 1, \qquad \varphi \restriction \mathfrak{A}(x) = \ell \tag{6.7.33}$$

Since $\varphi(e) = \ell(e) = 1$, φ is Kadison state by Proposition 6.7.6.

(e) Since $\widehat{\ }$ is an isomorphism of $\mathfrak{A}(x)$ and $C(\sigma(x))$, we can find $x_\pm \in \mathfrak{A}(x)$ with

$$\widehat{x}_+(\lambda) = \max(\widehat{x}(\lambda), 0), \qquad \widehat{x}_-(\lambda) = \max(-\widehat{x}(\lambda), 0) \tag{6.7.34}$$

Clearly, $\widehat{x} = \widehat{x}_+ - \widehat{x}_-$, $\widehat{x}_\pm \geq 0$, $\overline{\widehat{x}}_\pm = \widehat{x}_\pm$ and $\widehat{x}_+\widehat{x}_- = \widehat{x}_-\widehat{x}_+ = 0$. Since $\widehat{\ }$ is an isomorphism, these translate to $x_\pm$ Kadison positive and (6.7.30). □

Theorem 6.7.8. *Let $\mathfrak{A}$ be a B^*-algebra with identity.*

(a) *The Kadison positive elements are a convex cone, that is, x, y Kadison positive $\Rightarrow x + y$ is Kadison positive (λx Kadison positive for $\lambda \geq 0$ is trivial).*
(b) *Every positive element of $\mathfrak{A}$ (i.e., x of the form y^*y) is Kadison positive, so positive $\Leftrightarrow$ Kadison positive and the Kadison states and states are the same.*

Proof. (a) Let x, y be Kadison positive. Then $x + y$ is symmetric and so it has $\sigma(x + y) \subset \mathbb{R}$. Let $\lambda \in \sigma(x + y)$. By Proposition 6.7.7(d), there is a Kadison state φ so

$$\varphi(x + y) = \lambda \tag{6.7.35}$$

Since φ is a Kadison state, $\varphi(x) \geq 0$ and $\varphi(y) \geq 0$, so $\lambda \geq 0$, that is, $\sigma(x + y) \subset [0, \infty)$, i.e., $x + y$ is Kadison positive.

(b) Let $x = y^*y = x_+ - x_-$. We need to show $x_- = 0$. Let $z = yx_-$. Then

$$z^*z = x_-(x_+ - x_-)x_- = -x_-^3 \tag{6.7.36}$$

so by the spectral mapping theorem, $-z^*z$ is Kadison positive. Since $\sigma(-zz^*) \subset \sigma(-z^*z) \cup \{0\}$ (By Problem 5 of Section 2.2), $-z^*z - zz^* \equiv -w$ is Kadison positive by (a), that is,

$$\sigma(w) \subset (-\infty, 0] \tag{6.7.37}$$

Write $z = a + ib$ with a, b symmetric. Then

$$w = 2(a^2 + b^2) \tag{6.7.38}$$

Since a^2, b^2 are Kadison positive, w is also by (a). Thus,

$$\sigma(w) \subset [0, \infty) \tag{6.7.39}$$

Therefore, $\sigma(w) = \{0\}$. By (6.4.3),

$$(w = w^* + \sigma(w) = \{0\}) \Rightarrow \|w\| = 0 \Rightarrow w = 0 \tag{6.7.40}$$

Thus, by (6.7.38), $a^2 = -b^2$, so $\sigma(a^2) \subset [0, \infty]$ and $\sigma(b^2) \subset [0, \infty) \Rightarrow \sigma(a^2) \subset [0, \infty) \cap (-\infty, 0] = \{0\}$. By (6.7.40) for a, $a = 0$, and then $b = 0$, so $z = 0$.

By (6.7.37), $x_-^3 = 0$. Thus, by the spectral mapping theorem, $\sigma(x_-)^3 = \sigma(x_-^3) = \{0\} \Rightarrow \sigma(x_-) = 0 \Rightarrow x_- = 0$ by (6.7.40) again. $\square$

Corollary 6.7.9. *In any B^*-algebra, $\mathfrak{A}$, with identity, $\{x^*x \mid x \in \mathfrak{A}\}$ is a convex cone.*

Proof. This set is the same as $\{y \mid \sigma(y) \subset [0, \infty)\}$ which we saw was a convex cone. $\square$

Corollary 6.7.10. *In any B^*-algebra, $\mathfrak{A}$, with identity, $e + x^*x$ is invertible for all $x \in \mathfrak{A}$.*

Remarks. 1. As we noted in Theorem 6.3.2, this is one criterion for an involution to be symmetric in the abelian case.

2. In their original paper, Gel'fand–Naimark included that $e + x^*x$ is invertible in their axioms for a B^*-algebra. This corollary shows this axiom is implied by the others.

Proof. $\sigma(x^*x) \subset [0, \infty)$ so $-1 \notin \sigma(x^*x)$. $\square$

Theorem 6.7.11. *Let $\mathfrak{A}$ be a B^*-algebra with identity. Let $\varphi \in \mathfrak{A}^*$ obey*

$$\varphi(e) = 1 = \|\varphi\| \tag{6.7.41}$$

Then $\varphi \in \mathcal{S}_{+,1}(\mathfrak{A})$.

Proof. Immediate from Proposition 6.7.6 and Theorem 6.7.8. $\square$

Proof of Theorem 6.7.5. (This completes the proof of Theorem 6.7.1.) Since $\sigma(x^*x) \subset [0, \infty)$ and, by (6.4.3),

$$\text{spr}(x^*x) = \|x^*x\| \tag{6.7.42}$$

we see $\|x^*x\| \in \sigma(x^*x)$. Thus, Proposition 6.7.7(d) (and the fact that Kadison states are the same as states) implies the result. $\square$

Notes and Historical Remarks. The noncommutative Gel'fand–Naimark theorem (or rather, something slightly weaker since they had extra axioms for B^*-algebras; see below) comes from their seminal 1943 paper [**231**]. Considerations of things close to C^*-algebras arose in the 1930s in attempts to axiomatize quantum mechanics, and von Neumann worked with weakly-closed subalgebras of $\mathcal{L}(\mathcal{H})$ (called W^*-algebras). But it was the paper of Gel'fand–Naimark that, by emphasizing the algebraic aspects of the theory, gave birth to the general study of C^*-algebras—so much so that the AMS had a fifty-year anniversary conference and book [**168**] in 1993.

The GNS construction is named after this paper and work of Segal [**615**]. The GN paper had an explicit example of the construction in the B^*-case; Segal, realizing that he had found something similar in the case of group algebras, formalized the general construction, emphasizing the structure of states (and showed that irreducible representations are associated to extreme points; see the next section). The name "state" is from Segal, who was motivated by its use in quantum theory (density matrices and mixed states).

Segal dealt with representations of $\|\cdot\|$-closed *-algebras of $\mathcal{L}(\mathcal{H})$ rather than B^*-algebras.

Gel'fand–Naimark had six axioms for B^*-algebras:

(i)–(iii) said * was an involution.
(iv) said $\|x^*x\| = \|x\|\|x^*\|$.
(v) said $e + x^*x$ is invertible.
(vi) said $\|x\| = \|x^*\|$.

They conjectured (i)–(iv) $\Rightarrow$ (v), (vi). We use

(vii) $\|x^*x\| = \|x\|^2$

which we showed (see (6.4.2)) imply (iv), (vi), but it is a fact proven by Glimm–Kadison [**252**] in 1960 that (i)–(iv) $\Rightarrow$ (vi); see also Kadison [**360**].

That (i)–(iii) + (vii) $\Rightarrow$ (v) is intimately connected to the fact that $\{x^*x \mid x \in \mathfrak{A}\}$ is a cone. That the cone condition follows from (i)–(iv) is a result of Kelley–Vaught [**385**] and Fukamiya [**216**] in 1952. That this in turn implies every x^*x has nonnegative spectra (and so establishes axiom (v)) is an unpublished remark of Kaplansky that appeared in Schatz's MathSciNet review of Fukamiya [**595**]! In particular, this has the x_-^3 argument we used in the proof of Theorem 6.7.8(b) and also states that Kaplansky had a way to embed a B^*-algebra without identity in one with identity.

Our elegant approach to proving the cone result is due to Kadison [**360**] after whom we have named Kadison positive and Kadison state.

For most purposes, it suffices that B^*-algebras without identity can be embedded in one with identity, but it is sometimes useful to know they have approximate identities—this was first observed by Segal [**615**]; see Problem 3.

A W^**-algebra* is a weakly closed C^*-algebra, $\mathfrak{A}$, in $\mathcal{L}(\mathcal{H})$. We saw that $B \to \ell_B$ (with $\ell_B(A) = \text{Tr}(AB)$) is a map of $\mathcal{L}(\mathcal{H})$ isometrically onto $\mathcal{I}_1^*$, the dual of the trace class. It is not hard to see if $h(\mathfrak{A}) = \{A \in \mathcal{I}_1 \mid \text{Tr}(AB) = 0 \text{ for all } B \in \mathfrak{A}\}$, then $\mathfrak{A}$ is the dual of the Banach space $\mathcal{I}_1/h(\mathfrak{A})$, precisely because $\mathfrak{A}$ is weakly closed (these are called ultra-weakly continuous functionals); that is, any W^*-algebra is the dual of a Banach space. It is a theorem of Sakai [**587**] that any B^*-algebra that is also the dual of some Banach space is isometrically isomorphic to a W^*-algebra.

Problems

1. (a) Prove that equivalent cyclic representations have the same state.

   (b) If $(\tilde{\pi}, \widetilde{\mathcal{H}}, \widetilde{\psi})$ is a cyclic representation, show that $\tilde{\pi}$ is equivalent to the π generated by the GNS representation from its state and conclude that two cyclic representations with the same state are equivalent.

   (c) If $\mathfrak{A}$ is separable, prove that any cyclic representation is on a (separable) Hilbert space.

   (d) Find a nonseparable B^*-algebra with a cyclic injective representation on a (separable) Hilbert space.

2. (a) If $\mathfrak{A}$ is separable, prove that $\mathcal{S}_{+,1}(\mathfrak{A})$ is metrizable in the weak (i.e., $\sigma(\mathfrak{A}^*, \mathfrak{A})$) topology.

   (b) Supposing Theorem 6.7.5, prove (6.7.24) for any countable dense set.

   (c) If $\mathcal{S}_{+,1}(\mathfrak{A})$ is metrizable, prove it has a countable dense set. (*Hint*: Banach–Alaoglu theorem.)

3. This problem will prove that any B^* algebra, $\mathfrak{A}$, without identity has an approximate identity following Segal [**615**]. $\mathfrak{A}_e$ will be the algebra with identity added in the B^*-norm (see Lemma 6.4.5). Let $\varepsilon_n = 1/n$. Given $(x_1, \dots, x_n) \in \mathfrak{A}$, define

$$b_n(x_1, \dots, x_n) = \sum_{j=1}^{n} x_j^* x_j \tag{6.7.43}$$

   and (which lies in $\mathfrak{A}$ although $\mathfrak{A}_e$ is used to define it)

$$u_n(x_1, \dots, x_n) = (b_n + \varepsilon_n e)^{-1} b_n \tag{6.7.44}$$

   Think of $u_n(x_1 \dots, x_n)$ as a net indexed by $(x_1, \dots, x_n)$ with $(x_1, \dots, x_\ell) \rhd (\tilde{x}_1, \dots, \tilde{x}_n)$ if $\{\tilde{x}_1, \dots, \tilde{x}_n\} \subset \{x_1, \dots, x_\ell\}$. This problem will lead you through a proof that $\{u_n\}$ is an approximate identity.

(a) Prove $(b_n + \varepsilon_n e)$ is invertible and that

$$u_n - e = -\varepsilon_n(b_n + \varepsilon_n e)^{-1} \tag{6.7.45}$$

(b) Prove that

$$\sum_{j=1}^{n} [x_j(u_n - e)]^*[x_j(u_n - e)] = \varepsilon_n^2 b_n[(b_n + \varepsilon_n e)^{-1}]^2 \tag{6.7.46}$$

(c) Prove $\sigma(b_n) \subset [0, \infty)$ and, by the spectral mapping theorem and that $\|x\| = \sup_{\lambda\in\sigma(x)}|\lambda|$ if x is Hermitian, prove that

$$\|\varepsilon_n^2 b_n(b_n + \varepsilon_n e)^{-1}\| \le \tfrac{1}{2}\,\varepsilon_n \tag{6.7.47}$$

(d) For $j = 1, \dots, n$, prove that

$$\|x_j u_n - x_j\| \le \frac{1}{2n} \tag{6.7.48}$$

Conclude that $\{u_n\}$ is an approximate right identity.

(e) Prove that a symmetric approximate right identity is a two-sided approximate identity so $\{u_n\}$ is one.

(f) If $\mathfrak{A}$ is separable, find a sequential approximate identity.

6.8. Bonus Section: Representations of Locally Compact Groups

> You boil it in sawdust: you salt it in glue
> You condense it with locusts and tape
> Still keeping one principal object in view
> To preserve its symmetrical shape.
>
> —*Lewis Carroll* (*C. L. Dodgson*) [**107**]

Let G be a locally compact group. A *representation* of G is a map $U\colon G \to \mathcal{L}(\mathcal{H})$, the bounded operators on a not necessarily separable Hilbert space, $\mathcal{H}$, so that

(i) $U(xy) = U(x)U(y), \quad U(e) = \mathbb{1}$ (6.8.1)

(ii) $U(x^{-1}) = U(x)^*$ (6.8.2)

(iii) $x \mapsto U(x)$ is strongly continuous.

Notice that (6.8.1)/(6.8.2) imply that each $U(x)$ is unitary so, in particular, $\|U(x)\| = 1$. Thus, by Theorem 3.6.2 of Part 1, strong continuity is equivalent to weak continuity (and, we'll see shortly, to weak measurability if $\mathcal{H}$ is separable).

Definition. A representation, U, on a Hilbert space, $\mathcal{H}$, is called *irreducible* if and only if $\mathcal{H}' \subset \mathcal{H}$, a subspace with $U(g)\mathcal{H}' \subset \mathcal{H}'$ for all g implies $\mathcal{H}' = \{0\}$ or $\mathcal{H}' = \mathcal{H}$.

The idea is that *irreps* (our shorthand for "irreducible representations") are building blocks. In the finite-dimensional case, every representation is a direct sum of irreps (see Proposition 6.8.12 below) and there are senses in which one can usually write an infinite-dimensional representation as some kind of "direct integral" of irreps. Our goal in this section is to prove two results:

Theorem 6.8.1 (Gel'fand–Raikov Theorem). *For any locally compact group, G, the irreps separate points, that is, for any $x, y \in G$, there is an irrep, U, with $U(x) \neq U(y)$.*

Theorem 6.8.2 (Peter–Weyl Theorem). *For any compact group, the finite-dimensional irreps separate points and any representation is a direct sum of finite-dimensional irreps. In particular, any irrep is finite-dimensional.*

We'll prove that there is a one–one correspondence between representations of G and $*$-representations of $L^1(G)$ and then between cyclic representations of $L^1(G)$ and suitably defined states on $L^1(G)$. We'll then show a cyclic representation of $L^1(G)$ is an irrep if and only if its corresponding state is an extreme point of the set of states, and then prove Theorem 6.8.1 from the Krein–Milman theorem. Theorem 6.8.2 will rely on the fact that when G is compact, convolution operators are Hilbert–Schmidt and the eigenspaces provide finite-dimensional representations. As a preliminary,

Theorem 6.8.3. *Let $U\colon G \to \mathcal{L}(\mathcal{H})$ obey* (6.8.1) *and* (6.8.2) *and suppose $x \mapsto U(x)$ is weakly measurable, that is, for all $\varphi, \psi, x \mapsto \langle \varphi, U(x), \psi\rangle$ is a Baire function. Suppose $\mathcal{H}$ is separable. Then U is strongly continuous.*

Proof. For $f \in L^1(G)$ and $\psi \in \mathcal{H}$, define

$$\psi_f = \int f(x) U(x)\psi\, d\mu(x) \tag{6.8.3}$$

This is a weak integral, that is, the unique vector with

$$\langle \eta, \psi_f\rangle = \int f(x)\langle \eta, U(x)\psi\rangle\, d\mu(x) \tag{6.8.4}$$

for all $\eta \in \mathcal{H}$. By the left invariance of Haar measure,

$$U(y)\psi_f = \int f(x) U(yx)\psi\, d\mu(x) \tag{6.8.5}$$

$$= \psi_{\tau_y f} \tag{6.8.6}$$

where

$$(\tau_y f)(x) = f(y^{-1}x) \tag{6.8.7}$$

Noting that

$$\|\psi_f - \psi_g\|_{\mathcal{H}} \le \|f - g\|_{L^1}\|\psi\| \tag{6.8.8}$$

and that (by a density argument starting with continuous $f's$ of compact support), $y \mapsto \tau_y f$ is $\|\cdot\|_1$-continuous, we see that $y \mapsto U(y)\psi_f$ is continuous. So it suffices to show $\{\psi_f \mid f \in L^1, \psi \in \mathcal{H}\}$ is dense in $\mathcal{H}$.

Let φ be such that $\langle \varphi, \psi_f \rangle = 0$ for all $\psi \in \mathcal{H}$, $f \in L^1$. Then for each ψ, $\langle \varphi, U(x)\psi \rangle = 0$ for a.e. x. Let $\{\psi_n\}_{n=1}^\infty$ be an orthonormal basis for $\mathcal{H}$. Since this is countable, we can find x so that $\langle \varphi, U(x)\psi_n \rangle = 0$ for all n. It follows that $U(x)^*\varphi = 0$ so $\varphi = U(x)U^*(x)\varphi = 0$. □

Example 6.8.4. Let $\mathcal{H} = \ell^2(\mathbb{R})$, that is, $f\colon \mathbb{R} \to \mathbb{C}$ so that $\sum_{x\in\mathbb{R}}|f(x)|^2 < \infty$ (so $f(x) = 0$ for all but countable x). It is easy to see this is a non-separable Hilbert space. Define for $y \in \mathbb{R}$,

$$(U(y)f)(x) = f(x + y) \tag{6.8.9}$$

Then $\langle g, U(y)f \rangle$ is zero except for countably many y, and so $U(y)$ is weakly measurable. It is not strongly continuous. Indeed, each ψ_f is zero! This shows that the separability hypothesis is essential. □

If $y \mapsto U(y)$ is a representation of a locally compact group, G, then one can define $\pi_U\colon L^1(G) \to \mathcal{L}(\mathcal{H})$ by

$$\pi_U(f) = \int f(x)U(x)\, d\mu(y) \tag{6.8.10}$$

With the involution on $L^1(G)$ given by (6.3.8), it is easy to see (Problem 1) that π_U is a representation of the $*$-algebra $L^1(G)$, that is, obeys (6.7.3). We are heading towards a converse. As a preliminary, pick a neighborhood base $\{U_\alpha\}_{\alpha\in I}$ of compact neighborhoods of e indexed and ordered by inverse inclusion, that is, $\alpha \rhd \beta \Leftrightarrow U_\alpha \subset U_\beta$. Define

$$g_\alpha(x) = \mu(U_\alpha)^{-1} \begin{cases} 1, & x \in U_\alpha \\ 0, & x \notin U_\alpha \end{cases}$$

Note that $\{g_\alpha\}_\alpha$ is an approximate identity.

Theorem 6.8.5. *Let π be a $*$-representation of $L^1(G, d\mu)$ obeying*

$$\|\pi(f)\| \le \|f\|_1 \tag{6.8.11}$$

Then there is a unitary representation, U, of G so $\pi = \pi_U$. Moreover,

(a) $\operatorname{s-lim}_\alpha \pi(g_\alpha) = \mathbb{1}$. (6.8.12)

(b) *$\psi \in \mathcal{H}$ is cyclic for π if and only if ψ is cyclic for U.*

Proof. Fix $\psi \in \mathcal{H}$ and define $\varphi_\psi^{(\pi)}\colon L^1(G) \to \mathbb{C}$ by (6.7.8). We'll use φ for simplicity. Let $\psi_\alpha = \pi(g_\alpha)\psi$. Then, since g_α is an approximate identity for any $f \in L^1$,

$$\langle \pi(f)\psi, \psi_\alpha \rangle = \varphi(f^* * g_\alpha) \to \varphi(f^*) = \langle \pi(f)\psi, \psi \rangle \tag{6.8.13}$$

It follows that $\psi_\alpha \to \psi$ weakly since

$$\sup_\alpha \|\psi_\alpha\| = \|\pi(g_\alpha)\| \|\psi\| \le \sup_\alpha \|g_\alpha\|_1 \|\psi\| \le \|\psi\| \tag{6.8.14}$$

by (6.8.11) and $\langle \eta, \psi_\alpha \rangle \equiv 0$ for $\eta \perp \{\pi(f)\psi\}$.

By (6.8.14),

$$\|\psi - \psi_\alpha\|^2 = \|\psi\|^2 + \|\psi_\alpha\|^2 - 2\operatorname{Re}\langle \psi_\alpha, \psi \rangle \le 2\|\psi\|^2 - 2\operatorname{Re}\langle \psi_\alpha, \psi \rangle \to 0$$

so $\psi_\alpha \to \psi$ strongly, proving (6.8.12).

Fix $\psi \in \mathcal{H}$. For any $f \in L^1$ and $y \in G$, define

$$U(y)(\pi(f)\psi) = \pi(\tau_y f)\psi \tag{6.8.15}$$

This defines $U(y)$ uniquely on the closed span of $\{\pi(f)\psi\}$ and is a representation since $\tau_y\tau_z = \tau_{yz}$. Since $\tau_{-y} f * \tau_y g = f * g$, $U(y)$ is unitary, and since

$$\int f(x)(\tau_x g)(y)\, d\mu(x) = (f * g)(y) \tag{6.8.16}$$

We see that (6.8.3) holds.

By a standard cyclic vector construction (see Section 5.2), we can vary ψ to define $U(y)$ on all of $\mathcal{H}$. By construction, the statement about cyclic vectors is immediate. □

Theorem 6.8.6. *If $(\pi, \mathcal{H}, \psi)$ is a cyclic representation of $L^1(G)$, then $\varphi \equiv \varphi_\psi^{(\pi)}$ of* (6.7.8) *obeys*

(i) $\varphi \in (L^1)^*$, $\|\varphi\| = 1$, $\varphi(f^*) = \overline{\varphi(f)}$;
(ii) $\varphi(f^* * f) \ge 0$ *for all* $f \in L^1$;
(iii) $\lim_\alpha \varphi(g_\alpha) = 1$.

Conversely, if φ obeys (i)–(ii), *there is a unique (up to unitary equivalence) cyclic representation $(\pi, \mathcal{H}, \psi)$ of $L^1(G)$ with φ given by* (6.7.8). *It obeys* (iii).

Remarks. 1. This is close to Theorem 6.7.4, but since $\mathfrak{A} = L^1(G, d\mu)$ does not have an identity, there is a difference.

2. One can go further (Problem 2) and prove that $\varphi(f) = \int \varphi(x) f(x)\, dx$ with φ continuous, $\{\varphi(x_i^{-1}x_j)\}_{1 \le i,j \le n}$ positive definite, and $\|\varphi\|_\infty = \varphi(0)$.

Proof. That φ obeys (i) and (ii) is immediate. ($\|\varphi\| \leq 1$ follows from $\|\psi\| = 1$. That $\|\varphi\| = 1$ then follows from (iii).) We proved (iii) in the proof of Theorem 6.8.5.

For the converse, use the same (GNS) construction as in the proof of Theorem 6.7.4. We wind up with an inner product space whose elements are equivalence classes, $[f]$, of f's in L^1 ($f \sim h \Leftrightarrow \langle f-h, f-h\rangle_\varphi = 0$) with inner product

$$\langle [f], [h]\rangle_\varphi = \varphi(f^* * h) \tag{6.8.17}$$

and $\mathcal{H}$ is the completion of this inner product space. We define

$$\pi(f)[h] = [f * h] \tag{6.8.18}$$

This implies that (by the Schwarz inequality for φ)

$$\|\pi(f)[h]\|_\varphi^{2^n} = \varphi(h^* * f^* * f * h)^{2^n} \tag{6.8.19}$$

$$\leq \|[h]\|_\varphi^{2^n-1} \varphi(h^* * (f^* * f)^2 h)^{2^{n-1}} \cdots \leq \|[h]\|_\varphi^{2^n-2} \|f\|_1^{2^n} \|h\|_1^2 \tag{6.8.20}$$

Taking 2^n-th roots and taking n to infinity, we get

$$\|\pi(f)[h]\|_\varphi \leq \|f\|_1 \|[h]\|_\varphi \tag{6.8.21}$$

or

$$\|\pi(f)\| \leq \|f\|_1 \tag{6.8.22}$$

For any f,

$$\langle [f], [g_\alpha]\rangle_\varphi = \varphi(f^* * g_\alpha) \to \varphi(f^*) \tag{6.8.23}$$

and

$$\|[g_\alpha]\|_\varphi^2 = \varphi(g_\alpha^* * g_\alpha) \leq \|g_\alpha\|_1^2 = 1 \tag{6.8.24}$$

so $[g_\alpha]$ converges weakly to some $\psi \in \mathcal{H}$. By $\varphi(f^*) = \overline{\varphi(f)}$, we have that

$$\varphi(f) = \langle \psi, [f]\rangle \tag{6.8.25}$$

By (6.8.18),

$$\pi(f)\psi = \text{w-lim}[f * g_\alpha] = \psi \tag{6.8.26}$$

Thus, by (6.8.25),

$$\varphi(f) = \langle \psi, \pi(f)\psi\rangle \tag{6.8.27}$$

which implies

$$\|\varphi\| \leq \|\psi\|^2 \tag{6.8.28}$$

This means $\|\psi\| \geq 1$, so by (6.8.24), $\|\psi\| = 1$ and $[g_\alpha]$ converges to ψ in norm. Thus, $\varphi(g_\alpha) = \langle \psi, [g_\alpha]\rangle \to \|\psi\|^2 = 1$, proving (iii). $\square$

The *states*, $\mathcal{S}_{+,1}(G)$, on $L^1(G, d\mu)$ are the functionals $\varphi \in (L^1)^*$ obeying (i)–(iii) of Theorem 6.8.6. This is a convex set. The norm of a convex combination in $\mathcal{S}_{+,1}(G)$ still has norm 1 because of (iii). In case G is abelian, Theorem 6.3.6 says $\mathcal{S}_{+,1}(G)$ are the probability measures on $\widehat{G}$. If $\widehat{G}$ is only locally compact, this set is not closed and so not compact in the weak topology. The closure is measures of total mass at most 1. We thus define $\mathcal{S}_{+,\le 1}(G)$ as the $\varphi \in (L^1(G))^*$ which obey (i) and (ii) with $\|\varphi\| = 1$ replaced by $\|\varphi\| \le 1$. We have

Theorem 6.8.7. *$\mathcal{S}_{+,\le 1}$ is a compact convex set in the $\sigma(\mathcal{S}, L^1)$ topology. Its extreme points consist of $\{0\}$ and the extreme points of $\mathcal{S}_{+,1}$.*

Proof. All conditions defining $\mathcal{S}_{+,\le 1}$ are closed under convex combinations and weak limits. If φ is an extreme point with $0 < \|\varphi\| < 1$, then

$$\varphi = \|\varphi\| \left(\frac{\varphi}{\|\varphi\|} \right) + (1 - \|\varphi\|)0$$

so φ is not extreme after all. Thus, any extreme point has $\|\varphi\| = 0$ or $\|\varphi\| = 1$. □

It is easy to see that, in general, for any irrep, all nonzero vectors are cyclic (Problem 3). The converse, i.e., that cyclic representations are irreducible, is not in general true in the nonabelian case, as seen by

Example 6.8.8. Let $G = \mathbb{SU}(2, \mathbb{C})$. Let

$$\varphi(f) = \int f(A) \operatorname{Tr}(A)\, d\mu(A) \tag{6.8.29}$$

The associated cyclic representation acts on $\operatorname{Hom}(\mathbb{C}^2)$ (i.e., 2×2 matrices) by

$$U_A B = AB \tag{6.8.30}$$

with

$$\langle B, C \rangle = \operatorname{Tr}(B^* C) \tag{6.8.31}$$

$B = \mathbb{1}$ is a cyclic vector. But for any $\varphi \neq 0$ in $\mathbb{C}^2$, $\{B \in \operatorname{Hom}(\mathbb{C}^2) \mid B\varphi = 0\}$ is an invariant subspace for $\{U_A\}$. It is two-dimensional (while $\operatorname{Hom}(\mathbb{C}^2)$ is four-dimensional), so this cyclic representation is not an irrep. □

We are heading towards a proof that the cyclic representations that are irreps are exactly those whose associated φ is a nonzero extreme point of $\mathcal{S}_{+,1}$. Given a family, $\mathcal{F}$, of operators on a Hilbert space, $\mathcal{H}$, we define its *commutant* by

$$\mathcal{F}' = \{C \in \mathcal{L}(\mathcal{H}) \mid CA = AC \text{ for all } A \in \mathcal{F}\} \tag{6.8.32}$$

It is easy to see ($\mathcal{F}^* = \{C^* \mid C \in \mathcal{F}\}$)

$$\mathcal{F}^* = \mathcal{F} \Rightarrow (\mathcal{F}')^* = \mathcal{F}' \tag{6.8.33}$$

Theorem 6.8.9 (Schur's Lemma). *Let $\{U_g\}_{g\in G}$ be a representation of a group, G. Then U is an irrep if and only if*

$$\{U_g\}'_{g\in G} = \{c\mathbb{1} \mid c \in \mathbb{C}\} \tag{6.8.34}$$

Proof. If U is not an irrep, there is a nontrivial invariant subspace, $\mathcal{H}'$. Let P be the orthogonal projection onto $\mathcal{H}'$. By Theorem 2.1.4, for all g,

$$PU_gP = U_gP \tag{6.8.35}$$

Replacing g by g^{-1}, using $U_g^* = U_g^{-1}$, and taking adjoints,

$$PU_gP = PU_g \tag{6.8.36}$$

Thus, $P \in \{U_g\}'$, so (6.8.34) fails.

Conversely, suppose (6.8.34) fails, that is, there is $B \neq c\mathbb{1}$ in $\{U_g\}'$. By (6.8.33), $B^* \in \{U_g\}'$ also, so either $B + B^*$ or $(B - B^*)/i$ is in $\{U_g\}'$ and not $c\mathbb{1}$. Thus, we can suppose $B = B^*$ and is not $c\mathbb{1}$. By the spectral theorem, B has a nontrivial spectral projection, Q. Since $U_xBU_x^{-1} = B$ for all x, the uniqueness or the functional calculus shows $U_xQU_x^{-1} = Q$. Since $0 \neq Q \neq 1$, $\text{Ran}(Q)$ is a nontrivial invariant subspace for U, that is, U is not an irrep. □

Proposition 6.8.10. *Let $\varphi \in \mathcal{S}_{+,1}(G)$ be associated with a cyclic representation, $(U, \mathcal{H}, \psi)$, of G and representation, π, of $L^1(G)$. If $B \in \{U_x\}'_{x\in G}$ with $0 \leq B \leq \mathbb{1}$, then*

$$\eta(f) = \langle \psi, B\pi(f)\psi\rangle \tag{6.8.37}$$

lies in $\mathcal{S}_{+,\leq 1}(G)$ and obeys $\psi - \eta \in \mathcal{S}_{+,\leq 1}(G)$. Conversely, if $\eta, \psi - \eta \in \mathcal{S}_{+,\leq 1}$, then there exists B obeying (6.8.37) so that $0 \leq B \leq 1$.

Proof. Since $B\pi(g) = \pi(g)B$,

$$\eta(f^*f) = \langle \psi, B\pi(f)^*\pi(f)\psi\rangle = \|B^{1/2}\pi(f)\psi\|^2 \tag{6.8.38}$$

and

$$(\varphi - \eta)(f^*f) = \|(1 - B)^{1/2}\pi(f)\psi\|^2 \tag{6.8.39}$$

so $\eta, \varphi - \eta \in \mathcal{S}_{+,\leq 1}(G)$.

Conversely, if $\eta, \psi - \eta \in \mathcal{S}_{+,\leq 1}(G)$, define a sesquilinear form on $\mathcal{H}$ by

$$b([f],[g]) = \eta(f^* * g) \tag{6.8.40}$$

If $[f] = 0$, then

$$|\eta(f^*g)| \leq \eta(f^*f)^{1/2}\eta(g^*g)^{1/2} \leq \psi(f^*f)^{1/2}\psi(g^*g)^{1/2} = 0 \tag{6.8.41}$$

so b is well-defined on $\mathcal{H}$ and, by the same calculation,

$$|b([f],[g])| \leq \|f\|_{\mathcal{H}}\|g\|_{\mathcal{H}} \tag{6.8.42}$$

Thus, there is a bounded operator on $\mathcal{H}$ with

$$b([f],[g]) = ([f], B[g]) \tag{6.8.43}$$

It follows easily that $0 \leq B \leq \mathbb{1}$. Since

$$\eta(f^* * (h * g)) = \eta((h^* * f)^* * g) \tag{6.8.44}$$

we have $\langle [f], [\pi(h)B - B\pi(h)][g] \rangle = 0$ so $B \in (\pi(f))' = \{U_x\}'$. $\square$

Theorem 6.8.11. *Let $(U, \mathcal{H}, \psi)$ be the cyclic representation associated to $\varphi \in \mathcal{S}_{+,1}(G)$. Then U is an irrep if and only if φ is an extreme point in $\mathcal{S}_{+,\leq 1}(G)$.*

Proof. It is easy to see (Problem 4) that $\varphi \in \mathcal{S}_{+,1}$ is an extreme point of $\mathcal{S}_{+,\leq 1}$ if and only if

$$\psi, \varphi - \psi \in \mathcal{S}_{+,\leq 1} \Rightarrow \psi = t\varphi \text{ for some } t \in [0,1]$$

Thus, by Proposition 6.8.10, φ is an extreme point if and only if

$$\{B \in \{U_x\}' \mid 0 \leq B \leq \mathbb{1}\} = \{t\mathbb{1} \mid 0 \leq t \leq 1\} \tag{6.8.45}$$

As we saw, if $\{U_x\}'$ has a $B \neq \mathbb{1}$, there is a nontrivial projection in $\{U_x\}'$, so in that case, (6.8.45) fails.

Thus, φ is an extreme point if and only if $\{U_x\}' = \{c\mathbb{1}\}$. By Schur's lemma, this happens if and only if U is an irrep. $\square$

Proof of Theorem 6.8.1. Define L_x on $L^2(G, d\mu)$ by

$$(L_x h)(y) = h(x^{-1}y) \tag{6.8.46}$$

Then L is a representation (called the *left regular representation*). It is easy to see (Problem 5) for all $f, g \in L^1$, $f \neq g \Rightarrow \pi_L(f) \neq \pi_L(g)$. Fix $f \neq g$, both in $L^1(G)$. Then, by looking at $\varphi(f) = \langle \eta, \pi_L(f)\eta \rangle$, for some $\varphi \in \mathcal{S}_{+,1}(G)$, $\varphi(f) \neq \varphi(g)$.

By the Krein–Milman theorem (Theorem 5.11.1 of Part 1), there is an extreme point, φ_0, necessarily nonzero, with $\varphi_0(f) \neq \varphi_0(g)$. Given $x, y \in G$, let $z = x^{-1}y$. Pick f supported in a neighborhood, N, of e so $f(w)$ is zero in zN. Let $g = \tau_z f$. If $U_z = \mathbb{1}$, then $\pi_U(\tau_z f) = \pi_U(f)$ for all $f \in L^1(G)$. Thus, $\varphi_0(\tau_z f) \neq \varphi_0(f)$ implies in the associated irrep, $U_z \neq \mathbb{1}$, so $U_x \neq U_y$. $\square$

We finally turn to the compact case. Rather than use extreme points to identify irreps, we use the following which will let us reduce everything to finding finite-dimensional representations.

Proposition 6.8.12. *Any finite-dimensional representation of a locally compact group is a direct sum of irreps.*

Proof. Let $(U, \mathcal{H})$ be a representation and $d = \dim(\mathcal{H})$. We'll prove this proposition by induction on d. If $d = 1$, U is an irrep, so a direct sum of irreps. Thus, suppose we have the result for $d < d_0$ and $\dim(\mathcal{H}) = d_0$.

If U is not an irrep, there is an invariant subspace, $\mathcal{H}_1$, with $1 \leq \dim(\mathcal{H}_1) < d_0$. If P is the orthogonal projection onto $\mathcal{H}_1$, $PU(x)P = U(x)P$ for all x, so as above, taking adjoints using $U(x)^* = U(x^{-1})$, we get $PU(x) = U(x)P$ so $(1-P)U(x) = U(x)(1-P)$, which implies $\mathcal{H}_1^\perp$ is also invariant.

Thus, U is the direct sum of $U \restriction \mathcal{H}_1$ and $U \restriction \mathcal{H}_1^\perp$. Since $\dim(\mathcal{H}_1) < d$ and $\dim(\mathcal{H}_1^\perp) = d - d_1 < d$, we conclude by induction that U is a direct sum of irreps. □

Let $\widehat{G}$ be the set of equivalence classes of finite-dimensional irreps of a compact group, G, where we say $(U, \mathcal{H}) \cong (V, \widetilde{\mathcal{H}})$ if there is a unitary $W\colon \mathcal{H} \to \widetilde{\mathcal{H}}$ so that $WU(x)W^{-1} = V(x)$ for all x. For each $\alpha \in \widehat{G}$, pick a fixed representative and fixed basis so $U(x)$ has a matrix $D_{ij}^{(\alpha)}(x)$. We'll eventually prove their linear combinations are dense in $C(G)$ in $\|\cdot\|_\infty$, and that if d_α is the dimension of $D(\alpha)$, then $\{d_\alpha^{-1/2} D_{ij}^{(\alpha)}\}_{\alpha \in \widehat{G}, 1 \leq i,j \leq d_n}$ is an orthonormal basis for $L^2(G)$. As a start, we note

Theorem 6.8.13 (Orthogonality Relations). *If μ is normalized Haar measure on G (i.e., $\mu(G) = 1$), then*

$$\int \overline{D_{ij}^{(\alpha)}(x)}\, D_{k\ell}^{(\beta)}(x)\, d\mu(x) = d_\alpha^{-1} \delta_{\alpha\beta} \delta_{ik} \delta_{j\ell} \tag{6.8.47}$$

Moreover,

$$D_{ij}^{(\alpha)} * D_{k\ell}^{(\beta)} = d_\alpha^{-1} \delta_{\alpha\beta} \delta_{jk} D_{i\ell}^{(\alpha)} \tag{6.8.48}$$

Proof. Given $C\colon \mathcal{H}_\beta \to \mathcal{H}_\alpha$, let

$$\widetilde{C} = \int D^{(\alpha)}(x)^{-1} C D^{(\beta)}(x)\, d\mu(x) \tag{6.8.49}$$

Since μ is right invariant (see Problem 3 of Section 4.19 of Part 1), it is easy to see that for all y,

$$\widetilde{C} D^{(\beta)}(y) = D^{(\alpha)}(y) \widetilde{C} \tag{6.8.50}$$

This shows that $\operatorname{Ker}(\widetilde{C})$ is an invariant subspace for $D^{(\beta)}$ and $\operatorname{Ran}(\widetilde{C})$ an invariant subspace for $D^{(\alpha)}$. It follows that if $\widetilde{C} \neq 0$, then $\widetilde{C}$ is invertible. Thus, if $|\widetilde{C}| = \sqrt{\widetilde{C}^* \widetilde{C}}$, we see that this is invertible and commutes with $D^{(\alpha)}$ (Problem 6(a)). Thus, $\widetilde{C} = W|\widetilde{C}|$ with W unitary and

$$W D^{(\beta)}(y) W^{-1} = D^{(\alpha)}(y) \tag{6.8.51}$$

We conclude that if $\alpha \neq \beta$, then $\widetilde{C} = 0$ and, by Schur's lemma, if $\alpha = \beta$, then $\widetilde{C}$ is constant. Noting $\text{Tr}(\widetilde{C}) = \text{Tr}(C)$, we conclude that if $\alpha = \beta$, then

$$\widetilde{C} = d_\alpha^{-1}[\text{Tr}(C)]\mathbb{1} \tag{6.8.52}$$

Letting C run though suitable rank-one operators yields (6.8.47) (Problem 6(c)). Using

$$D_{ij}^{(\alpha)}(xy^{-1}) = \sum_m D_{im}^{(\alpha)}(x)\,\overline{D_{jm}^{(\alpha)}(y)} \tag{6.8.53}$$

one gets (6.8.48) from (6.8.47) (Problem 6(d)). □

The idea behind the proof of Theorem 6.8.2 will be to use the left regular representation $\{L_x\}_{x\in G}$ given by (6.8.46). We'll show for any $g \in L^1$, the operator $T_g(f) = f * g$ of right convolution on L^2 commutes with each L_x. Moreover, if $g(x) = \overline{g(x^{-1})}$, T_g will be self-adjoint and Hilbert–Schmidt. Thus, its eigenspaces will be finite-dimensional invariant subspaces for L. Therefore, the following will be critical.

Lemma 6.8.14. (a) *Let $B \subset C(G)$ be a finite-dimensional subspace of functions left invariant by $\{L_x\}_{x\in G}$. Then B lies in the span of a finite number of $\{D_{ij}^{(\alpha)}(x)\}$.*

(b) *Let $T\colon L^2(G) \to L^2(G)$ be a bounded operator. Suppose for all $x \in G$, $L_xT = TL_x$. Then for any $\lambda \in \mathbb{C}$, $\{f \mid Tf = \lambda f\}$ is left invariant by all $\{L_x\}_{x\in G}$.*

Proof. (a) $L_x \restriction B$ is finite-dimensional, so a direct sum of $D^{(\alpha)}$'s, some perhaps multiple times, say m_α times. That means B has a basis $\{\psi_{\alpha;k;i}\}_{1\le k\le m_\alpha;1\le i\le d_\alpha}$ so that

$$\psi_{\alpha;k;i}(x^{-1}y) = L_x\psi_{\alpha;k;i}(y) = \sum_j D_{ji}^{(\alpha)}(x)\psi_{\alpha;k;j}(y) \tag{6.8.54}$$

Since the ψ's are continuous, this formula holds for all x, y, not just for a.e. y. Picking $y = e$, replacing x by x^{-1}, and using $D_{ji}^{(\alpha)}(x^{-1}) = \overline{D_{ij}^{(\alpha)}(x)}$, we see

$$\psi_{\alpha;k;i}(x) = \sum_j \psi_{\alpha;k;j}(e)\,\overline{D_{ij}^{(\alpha)}(x)} \tag{6.8.55}$$

$\overline{D^{(\alpha)}}$ is also a finite-dimensional irrep, so it is unitarily equivalent to one of the $D^{(\beta)}$'s, which implies $\overline{D_{ij}^{(\alpha)}(x)}$ is a linear combination of $D_{pq}^{(\beta)}$'s.

(b) Immediate, since $Tf = \lambda f \Rightarrow TL_xf = L_xTf = \lambda L_xf$. □

Remarks. 1. The reader might be surprised that (6.8.54) has D_{ji} and not D_{ij}. That's because the usual matrix definitions involve how components transform and the ψ's are basis vectors. Put differently, if M is a matrix, $M\begin{pmatrix}1\\0\\\vdots\\0\end{pmatrix}$ is the first column of M, that is, M_{j1}, not M_{1j}.

2. For each α, $\overline{D^{(\alpha)}}$ is an irrep. If it is not equivalent to $D^{(\alpha)}$, we can choose $\beta = \bar{\alpha}$ so $D_{ij}^{(\beta)}(x) = \overline{D_{ij}^{(\alpha)}(\lambda)}$. But if they are equivalent, it can happen $D_{ij}^{(\alpha)}$ cannot be picked all real (so-called *quaternionic irreps*) and we need linear combinations.

The final preparatory result is:

Proposition 6.8.15. *For $g \in C(G)$, define $T_g\colon L^2 \to L^2$ by*

$$[T_g f](x) = \int f(y)g(y^{-1}x)d\mu(y) \quad (\equiv (f * g)(x)) \tag{6.8.56}$$

Then

(a) *T_g maps L^2 to $C(G)$ with*

$$\|T_g f\|_\infty \le \|g\|_\infty \|f\|_2 \tag{6.8.57}$$

(b) *Each T_g is a Hilbert–Schmidt operator with*

$$\|T_g\|_2 \le \|g\|_\infty \tag{6.8.58}$$

(c) *If*

$$g(x^{-1}) = \overline{g(x)} \tag{6.8.59}$$

then T_g is self-adjoint.

(d) *If g obeys (6.8.59), any $f \in \mathrm{Ran}(T_g)$ is in the L^2-limits of finite linear combinations of $\{D_{ij}^{(\alpha)}(x)\}$.*

(e) *If g_1, g_2 obey (6.8.59), any $f \in \mathrm{Ran}(T_{g_1}T_{g_2})$ is a uniform limit of finite linear combinations of $\{D_{ij}^{(\alpha)}(x)\}$.*

Proof. (a) Using $d\mu(x^{-1}) = d\mu(x)$,

$$T_g f(x) = \int f(xy^{-1})g(y)\,d\mu(y) \tag{6.8.60}$$

so, by the triangle inequality for L^2,

$$\|T_g f\|_\infty \le \|g\|_\infty \int \|f(\,\cdot\, y^{-1})\|_2\, d\mu(y) = \|g\|_\infty \|f\|_2 \tag{6.8.61}$$

since $d\mu$ is right invariant and $\mu(G) = 1$. Since $T_g f \in C(G)$ if $f \in C(G)$, by a density argument, $\mathrm{Ran}(T_g) \subset C(G)$.

(b) T_g has integral kernel bounded by $\|g\|_\infty$, so its $L^2(G \times G, d\mu \otimes d\mu)$-norm is bounded by $\|g\|_\infty$.

(c) If $t(x, y) = g(y^{-1}x)$, then (6.8.59) implies that $t(y, x) = \overline{t(x, y)}$.

(d) By the Hilbert–Schmidt theorem (Theorem 3.2.1) extended to not necessarily positive self-adjoint operators, any $f \in \operatorname{Ran}(T_g)$ is an L^2-limit of finite sums of eigenfunctions with nonzero eigenvalues. By a direct calculation (Problem 7), T_g commutes with each L_x, so by Lemma 6.8.14 and the fact that eigenspaces for nonzero eigenvalues are finite-dimensional, we get (d).

(e) Since $T_{g_2} f$ is an L^2-limit of finite linear combinations of $D_{ij}^{(\alpha)}$, by (a), $T_{g_1}(T_{g_2} f)$ is a uniform limit of finite sums of $T_{g_1}(D_{ij}^{(\alpha)})$. By a simple calculation (Problem 8), $T_{g_1}(D_{ij}^{(\alpha)})$ is a finite combination of $D_{k\ell}^{(\alpha)}$'s. $\square$

Here is the result that includes the first half of Theorem 6.8.2.

Theorem 6.8.16. *Finite linear combinations of $D_{ij}^{(\alpha)}(x)$ are $\|\cdot\|_\infty$-dense in $C(G)$ and so $\{D_{ij}^{(\alpha)}\}$ separate points. $\{\sqrt{d_\alpha} D_{ij}^{(\alpha)}\}$ is an orthonormal basis for $L^2(G)$.*

Proof. If g_β is a continuous approximate identity, for any $f \in C(G)$, $T_{g_\beta} f \to f$ uniformly, so $T_{g_\beta} T_{g_\gamma} f$ does also, first taking β and γ to the limit. Thus, any f is a uniform limit of finite sums by (e) of the proposition. The rest is immediate. $\square$

The following completes the proof of the Peter–Weyl theorem (Theorem 6.8.2).

Theorem 6.8.17. *Any unitary representation of a compact group, G, is a direct sum of finite-dimensional irreps.*

Proof. Let $(U, \mathcal{H})$ be a representation of G. Let

$$W_{ij}^{(\alpha)} = \pi_U(d_\alpha \overline{D_{ij}^{(\alpha)}}) \tag{6.8.62}$$

$$= d_\alpha \int \overline{D_{ij}^{(\alpha)}(x)}\, U(x)\, d\mu(x) \tag{6.8.63}$$

(weak integral). Since π_U is a $*$-representation of $L^1(G)$ and we have (6.8.47), we find

$$W_{ij}^{(\alpha)} W_{k\ell}^{(\beta)} = \delta_{\alpha\beta} \delta_{jk} W_{i\ell}^{(\alpha)} \tag{6.8.64}$$

Since $D_{ij}^{(\alpha)}(x^{-1}) = \overline{D_{ji}^{(\alpha)}(x)}$, we see

$$(W_{ij}^{(\alpha)})^* = W_{ji}^{(\alpha)} \tag{6.8.65}$$

Using $U(x)U(y) = U(xy)$ and the invariance of $d\mu$, we have

$$U(x)W_{jj}^{(\alpha)} = \sum_{k=1}^{d_\alpha} D_{ki}^{(\alpha)}(x)W_{kj}^{(\alpha)} \tag{6.8.66}$$

We also define

$$P_{\alpha;j} = W_{jj}^{(\alpha)}, \qquad P_\alpha = \sum_{j=1}^{d_\alpha} P_{\alpha;j} \tag{6.8.67}$$

By (6.8.64)/(6.8.65), we have

$$P_{\alpha;j}^2 = P_{\alpha;j}, \quad P_{\alpha;j}^* = P_{\alpha;j}, \quad P_{\alpha;j}P_{\beta;k} = \delta_{\alpha\beta}\delta_{jk}P_{\alpha;j} \tag{6.8.68}$$

so the P's are mutually orthogonal self-adjoint projections. This means

$$P_\alpha = P_\alpha^*, \qquad P_\alpha P_\beta = \delta_{\alpha\beta}P_\alpha \tag{6.8.69}$$

(6.8.64)/(6.8.65) also imply

$$(W_{ij}^{(\alpha)})^* W_{ij}^{(\alpha)} = P_{\alpha;j}, \qquad W_{ij}^{(\alpha)}(W_{ij}^{(\alpha)})^* = P_{\alpha;i} \tag{6.8.70}$$

so $W_{ij}^{(\alpha)}$ is a partial isometry with initial space $P_{\alpha;j}$ and final space $P_{\alpha;i}$.

Suppose $\psi \in \mathcal{H}$ is orthogonal to each $P_{\alpha;j}$, that is, for all α, j,

$$P_{\alpha;j}\psi = 0 \tag{6.8.71}$$

By (6.8.70), $\|W_{ij}^{(\alpha)}\psi\|^2 = \|P_{\alpha;j}\psi\|^2 = 0$, so $\langle\psi, W_{ij}^{(\alpha)}\psi\rangle = 0$ for all α, i, j. By (6.8.56) and the fact that $\{D_{ij}^{(\alpha)}\}$ is an orthonormal basis, we conclude $\langle\psi, U(x)\psi\rangle = 0$, first as an L^2-function and then, by continuity, for all x. Taking $x = e$, $\|\psi\| = 0$, that is, we have proven that

$$\mathcal{H}_\alpha \equiv \text{Ran}(P_\alpha), \qquad \mathcal{H} = \oplus\mathcal{H}_\alpha \tag{6.8.72}$$

By (6.8.66), each $\mathcal{H}_\alpha$ is an invariant space for $U(x)$. Pick an orthonormal basis, $\psi_{\alpha;k;1}$ for $\text{Ran}(P_{\alpha;1})$. Define

$$\psi_{\alpha;k;j} = W_{j1}^{(\alpha)}\psi_{\alpha;k;1} \tag{6.8.73}$$

Since the W's are partial isometries, $\{\psi_{\alpha;k;j}\}$ is an orthonormal basis for $\mathcal{H}_\alpha$. By (6.8.66),

$$U(x)\psi_{\alpha;k;j} = \sum_{\ell=1}^{d_\alpha} D_{\ell j}^{(\alpha)}(x)\psi_{\alpha;k;\ell} \tag{6.8.74}$$

showing $U \restriction \mathcal{H}_\alpha$ is a direct sum of copies of $D^{(\alpha)}$. By (6.8.72), U is a direct sum of $\{D^{(\alpha)}\}$. □

A finite group is a special case of a compact group. Since $L^2(G, d\mu)$ in that case has finite dimension equal to $o(G)$, the order of the group, the fact that the $\{D_{ij}^{(\alpha)}\}_{\alpha\in\widehat{G},1\le i,j\le d_\alpha}$ are a basis implies

Theorem 6.8.18. *If G is a finite group and $\widehat{G}$ its equivalence classes of irreps with corresponding dimension d_α, then*

$$\sum_{\alpha\in\widehat{G}} d_\alpha^2 = o(G) \tag{6.8.75}$$

It is easy to see (Problem 9) that if $d_\alpha = 1$ for all α, then G is abelian so nonabelian groups must have irreps with $d_\alpha \ge 2$. For small $o(G)$, this fact and (6.8.75) determine the d_α's, e.g., if $o(G) = 6$ and G is nonabelian, one must have one $d_\alpha = 2$ and two $d_\alpha = 1$. It is also known (see the Notes) that any d_α must divide $o(G)$ and the number of one-dimensional irreps must also divide $o(G)$ so if G is nonabelian and $o(G) = 10$, one must have two $d_\alpha = 2$ and two $d_\alpha = 1$.

Notes and Historical Remarks.

> It may then be asked why ... a considerable space is devoted to substitution groups [permutation groups] while other particular modes of representation such as groups of linear transformations are not even referred to. My answer to this question is that ... it would be difficult to find a result that could be most directly obtained by the consideration of groups of linear transformations.
>
> —*Burnside* **[94]**[3]

Group representation theory is a major and unifying theme of mathematics with many applications, so the number of books is enormous [**4, 142, 159, 217, 346, 316, 325, 335, 395, 461, 478, 620, 645, 664, 737, 750, 749, 758**]. This eclectic list includes books on the general theory, on specific groups, on applications to physics or statistics, and ones of historical importance. Of course, my favorite is Simon [**645**]. There are even several books [**81, 141, 303, 462**] that focus on the history of group representation (sometimes as part of the history of Lie groups). Knapp [**393, 394**] and Mackey [**462**] discuss the history in relation to harmonic analysis.

The group concept arose in number theory (modular arithmetic and Dirichlet's characters) and geometry, especially projective geometry. As early as 1854, Cayley tried to define abstract groups [**110**]. By the 1890s, there was an explosion of interest in different aspects of groups from the work

[3]dismissing the idea of studying representations of groups by matrices. Ironically, once he learned of Frobenius' work, representation theory became a major tool for studying finite groups in Burnside's hands. By the 1911 second edition he admitted in its preface that "the reason given in the original preface for omitting any account of it [linear substitutions] no longer holds good."

of Klein and Poincaré on Fuchsian groups, Klein's Erlanger program, the work of Lie and others on continuous groups, and the start of classification of finite groups under Jordan, Sylow, Frobenius, and Burnside.

The basics of representation theory for finite groups was set in place by Georg Frobenius (1849–1917) in three papers in 1896 [**213, 214, 212**] and one in 1897 [**215**]. In fact, on the basis of his correspondence with Richard Dedekind (1831–1916), we know that most of the research in the three 1896 papers fell into place in a six-day period from April 12 to April 17.

Dedekind was a key player in this saga. He was Gauss' last student and returned to Göttingen as a privatdozent and attended Dirichlet's lectures on number theory; Dirichlet had been hired to take Gauss' chair, but died not long afterwards, and one of Dedekind's earliest claims to fame was as the editor and organizer of Dirichlet's published lectures. By the time he retired in 1894, he was a giant of German algebra, especially number theory. We've seen him as the discoverer of Dedekind cuts, and he was a valuable correspondent and supporter of Cantor. He made significant contributions to algebraic number theory and ring theory, including the formal definition of ideals. He spent most of his career as a professor at the Braunschweig Technical School preferring to live in his town of birth with his sister to a position at a more prestigious institution.

Dedekind had serious correspondence with several mathematicians; in the Notes to Section 1.6 of Part 1, we saw his correspondence with Cantor impacted the latter's work. His correspondence with Frobenius (in 1882–86, 1895–98, and 1901) runs to over 300 pages.

In presenting Dirichlet's work on his prime progression theorem (see Section 13.4 of Part 2B), Dedekind had formalized the notion of characters of abelian groups, i.e., homomorphism of G to $\partial\mathbb{D}$. In this context, Dedekind had gotten interested in what he called *group determinants*. For each $g \in G$, a finite group of order $o(G)$, one introduces a commuting variable, X_g, and defines a polynomial in these $o(G)$ variables

$$D_G\bigl(X_{g_1}, \dots, X_{g_{o(G)}}\bigr) = \det\bigl(X_{gh^{-1}}\bigr) \tag{6.8.76}$$

For example, if $G = \{0, 1\}$ is the group with two elements

$$D(X_0, X_1) = \det\begin{pmatrix} X_0 & X_1 \\ X_1 & X_0 \end{pmatrix} = X_0^2 - X_1^2 \tag{6.8.77}$$

For abelian groups, Dedekind could prove that

$$D_G(X) = \prod_{\chi\in\widehat{G}} \Bigl(\sum_{g\in G} \chi(g) X_g\Bigr) \tag{6.8.78}$$

(we'll prove this shortly but with ideas that Dedekind didn't have). In case (6.8.77), this is $X_0^2 - X_1^2 = (X_0 + X_1)(X_0 - X_1)$.

Dedekind computed D_G for some simple nonabelian groups of orders 6, 8, and 10 and factored by hand to find irreducible (in the algebraic sense) factors of order 2 unlike the only linear factors of (6.8.78). Dedekind worked on this for almost 15 years never publishing anything on the subject. In 1896, Dedekind was 65 and retired and suggested to Frobenius that he look into this question.

Using the modern theory, we easily solve the factorization problem. If U is a representation of G on a finite-dimensional space, define

$$D_G^{(U)}(X) = \det\Bigl(\sum_{g\in G} X_g U(g)\Bigr) \tag{6.8.79}$$

It is easy to see that D_G is just $D_G^{(L)}$ for the left regular representation and that $D_G^{(U_1\oplus U_2)} = D_G^{(U_1)} D_G^{(U_2)}$. Thus, the decomposition of L into the sum of irreducible $D^{(\alpha)}$, $\alpha \in \widehat{G}$, each d_α term implies a factorization

$$D_G(X) = \prod_{\alpha\in\widehat{G}} \bigl[D_G^{(D^{(\alpha)})}(X)\bigr]^{d_\alpha} \tag{6.8.80}$$

One can further show that each $D_G^{D(\alpha)}$ is an irreducible polynomial (the origin of the name irreducible in group representation theory!) and a polynomial of degree, d_α (the origin of the name degree for the dimension of an irrep). Of course, Frobenius didn't have the idea of a representation in 1896 but was still able to prove a factorization like (6.8.80) where the degree of the irreducible factors was the number of times they occurred (this final fact was only proven later in 1896 than the April burst).

As an important tool in his proofs, Frobenius had the idea of making G close to abelian by replacing the $o(G)$ variables, $X_1, \dots, X_{o(G)}$ by $\sharp(\mathcal{C})$ variables, one for each conjugancy class in G, i.e., if $x_1 = hx_2h^{-1}$ for some h, we set $X_1 = X_2 = Y$ in D. He discovered (connected to Problem 10) that the new polynomial again factors into linear factors up to normalization $\sum_{C\in\mathcal{C}} \chi(C)Y_C$ where χ were his new "characters." He proved orthogonality relations for these characters and even realized their relation implied $o(G)/d_\alpha$ was an algebraic integer and so an integer.

Quite remarkably by the end of his three 1896 papers, Frobenius had almost all the basic general facts about representations of finite groups stated in terms of factorization of the group determinant: that the number of distinct irreducible factors is the number of classes, that $\Sigma d_\alpha^2 = o(G)$, that $o(G)/d_\alpha$ is an integer, and the group characters and their orthogonality relations. What was missing, of course, is the connection to representations via the definition of irrep in terms of homomorphisms of G into unitary operators. It is this, following a suggestion of Dedekind, that is the main theme of the 1897 paper [**215**]. We note that there were precursors to Frobenius by Cartan [**108**] and Molien [**480**] in terms of the structure of the group algebra, but Frobenius seems to have been unaware of this work. Hawkins [**300, 301, 302**] has three papers on this history and Curtis [**140**] and Lam [**421**] have more popular accounts. Hawkins has a book [**304**] on Frobenius' life and work, including, of course, this work on group representations.

Frobenius' proofs were difficult although there was important progress by Burnside and by Frobenius, who, in particular, found the irreps of the symmetric group. The situation was changed by a 1905 paper of I. Schur [**607**] who used what is now called Schur's lemma (although he noted it was in Burnside's work) to revamp the subject, generalize the orthogonality relations (see Theorem 6.8.13), and present a digestable theory.

Ferdinand Georg Frobenius was born in a suburb of Berlin. Like Euler, his father was a pastor. He spent a year studying in Göttingen before going to Berlin, where in 1870, he earned a Ph.D. supervised by Weierstrass. He taught secondary school for four years, in Berlin for one, and at the ETH in Zurich for 17 years before returning to Berlin, filling the chair vacated by Kronecker's death. He was then the dominant figure in Berlin for twenty-five years. This period is marked by bad relations with Göttingen which, under Klein and Hilbert, surplanted Berlin as the internationally recognized center of mathematics. He objected to the Göttingen style, at one point describing it as [**141**] one "in which one amuses oneself more with rosy images than hard ideas."

Besides the theory of group representation, Frobenius was a pioneer in the general theory of finite groups with the first proof of Sylow theorems for general (as operator permutations) groups. As we saw in Section 14.3 of Part 2B, Frobenius was also a pioneer in the theory of complex linear ODEs. We've also seen him in the Perron–Frobenius theorem (see Section 7.5 of Part 1) and for the first proof of the Cayley–Hamilton theorem (see Section 1.4). He is also known for contributions to number theory (Frobenius automorphisms) and for conditions of vector fields being integrable (the so-called Frobenius theorem was first proven by Clebsch and Deahna but Frobenius applied it in a way that gives his name to it). His students include Landau and Schur.

Earlier, in his 1901 thesis under Frobenius [**606**], Schur found connections between polynomial invariants for $\mathbb{SL}(n, \mathbb{C})$ and the representations of the permutation group (now sometimes called Schur–Weyl duality—see the discussion in Section IX.11 of Simon [**645**]) but he didn't seem to realize the connection to unitary representations of $\mathbb{SU}(n)$ until 1924 [**610**] when spurred by queries from Weyl, he reexamined the subject. The modern subject of representations of the compact Lie groups was founded by Weyl in 1925–26 who, in a remarkable three-part paper [**748**], like Frobenius with finite groups, established not only the basics but much more. Any kind of understanding of group representations should include this topic and not just the results we present here intended to illustrate the connection to the theory of Banach algebras.

The approach to representation theory in terms of Banach algebra representations of $L^1(G, d\mu)$ goes back to Segal in 1941 [**613**]. The Gel'fand–Raikov theorem was presented by them in 1943 [**234**]; our proof is closer to the later version of Segal [**615**] and Gel'fand–Naimark [**232**].

The Peter–Weyl theorem is from their 1927 paper [**519**].

Problems

1. Prove that π_U given by (6.8.10) is a $*$-rep of $L^1(G)$ if U is a representation of G.

2. While it is a standard abuse of notation to use the same symbol for a functional on $L^1(G)$ and the L^∞-function that generates it, in this problem, we'll use distinct symbols to avoid any possible confusion. Thus $(\pi, \mathcal{H}, \psi)$ is a cyclic representation of $L^1(G)$, Φ on $L^1(G)$ is defined by
$$\Phi(f) = \langle \psi, \pi(f)\psi \rangle \tag{6.8.81}$$
and $(U, \mathcal{H}, \psi)$ the corresponding group representation (of Theorem 6.8.5) with $\pi = \pi_U$.
   (a) Define $\varphi : G \to \mathbb{C}$ by
$$\langle \psi, U(x)\psi \rangle = \varphi(x) \tag{6.8.82}$$
   Prove that
$$\Phi(f) = \int f(x)\, \varphi(x)\, d\mu(x) \tag{6.8.83}$$
   (b) Prove that φ is continuous and (if $\|\psi\| = 1$),
$$\varphi(0) = 1, \quad \|\varphi\|_\infty = 1 \tag{6.8.84}$$
   (c) For any $x_1, \dots, x_n \in G$, prove that $\{\varphi(x_i^{-1}x_j)\}_{1\le i,j\le n}$ is a positive definite matrix.

3. If U is any representation of a group, prove that for any fixed $\psi \in \mathcal{H}$, the closure of $\{\text{span of } U(x)\psi \mid x \in G\}$ is an invariant subspace. Conclude that if U is an irrep, any $\psi \neq 0$ is a cyclic vector.

4. Let C be a convex subset of a locally convex tvs with $0 \notin C$. Let $\widetilde{C} = \{\alpha x \mid 0 \le \alpha \le 1, x \in C\}$ and suppose that $\widetilde{C}$ is a compact convex set. Prove that $x \in C$ is an extreme point of $\widetilde{C}$ if and only if
$$y \in \widetilde{C} \text{ so } x - y \in \widetilde{C} \Rightarrow y = \alpha x \text{ for some } \alpha \in [0,1] \tag{6.8.85}$$

5. (a) Let $f \in L^1(G)$ so that for all $g \in L^2(G)$, $f * g = 0$. Prove that $f = 0$. (*Hint*: Approximate identity.)

   (b) If L is a left regular representation of G, $f \in L^1(G)$, and $\pi_L(f) = 0$, prove that $f = 0$.

   (c) If $f \neq g$ in $L^1(G)$, prove $\pi_L(f) \neq \pi_L(g)$.

6. Let (6.8.50) hold.

   (a) Prove that $\widetilde{C}^*\widetilde{C}$ commutes with $D^{(\alpha)}$ and conclude that so does $|\widetilde{C}|$.

   (b) If $\widetilde{C} \neq 0$, conclude that $|\widetilde{C}| = 1$ and that $\widetilde{C}$ is unitary.

   (c) Obtain (6.8.47).

   (d) Obtain (6.8.48).

7. Prove that T_g given by (6.8.56) commutes with each L_x.

8. Compute $T_g(D_{ij}^{(\alpha)})$ and prove that it is a sum of $D_{kl}^{(\alpha)}$'s. (*Hint*: See (6.8.48).)

9. If every irrep of a locally compact group is one-dimensional, prove that G is abelian. (*Hint*: $U(xyx^{-1}y^{-1}) \neq \mathbb{1}$ for some irrep if $xyx^{-1}y^{-1} \neq e$.)

10. (a) If U is an irrep of a compact group, prove that for any y

$$\int U(xyx^{-1})\,d\mu(x)$$

   is a constant multiple of $\mathbb{1}$.

   (b) If $\mu(G) = 1$, prove that this constant multiple is

$$\dim(U)^{-1}\chi_U(y)$$

6.9. Bonus Section: Fourier Analysis on LCA Groups

In this section, we'll study G, a locally compact abelian group with Haar measure, $d\mu$. This study began in Theorem 6.2.23 and Example 6.3.4. We'll suppose $G \cup \{\infty\}$, the one-point compactification, is metrizable (and so, also separable). This is easily seen to be equivalent to G being second countable and σ-compact or to $C_\infty(G)$ being separable (see Theorem 2.3.7 of Part 1). It implies (Problem 1) that each $L^p(G, d\mu)$ is separable, $1 \le p < \infty$, and thus it implies the metrizability of the unit ball of $L^\infty(G, d\mu)$ in the weak-$*$ topology. Since $\widehat{\mathfrak{A}} \cup \{\infty\}$ lies in this space, this implies $\widehat{G} \cup \{\infty\}$ is also metrizable. These countability assumptions are convenient because they allow us to use sequences, and so the dominated convergence theorem (which holds for sequences but not for nets).

As we've seen (see Theorem 6.2.23), the Gel'fand spectrum of $L^1(G, d\mu)$, which we denoted $\widehat{G}$ (instead of $\widehat{L^1(G)}$) is naturally associated with the family of characters, that is, continuous maps of G to $\partial\mathbb{D}$ obeying

$$\chi(x+y) = \chi(x)\chi(y) \tag{6.9.1}$$

These define mlf's on $L^1(G, d\mu)$ (with product given by convolution) via

$$\ell_\chi(f) = \int \overline{\chi}(x) f(x)\, d\mu(x) \equiv \widehat{f}(\chi) \tag{6.9.2}$$

A central result of this section is the Fourier inversion formula: $\widehat{G}$ is also an LCA and there is a normalization of Haar measure on $\widehat{G}$ so that for suitable f's,

$$f(x) = \int_{\widehat{G}} \chi(x) \widehat{f}(x)\, d\mu_{\widehat{G}}(\chi) \tag{6.9.3}$$

Recall we defined an involution on $L^1(G)$ by

$$f^*(x) = \overline{f(-x)} \tag{6.9.4}$$

and showed (see Example 6.3.4) that

$$\widehat{f^*}(\chi) = \overline{\widehat{f}(\chi)} \tag{6.9.5}$$

This in turn implies $\{\widehat{f} \mid f \in L^1\}$ is $\|\cdot\|_\infty$-dense in $C_\infty(\widehat{G})$, the continuous functions vanishing at ∞.

A main technical tool will be positive definite functions (pdf) which we'll show have the form (Bochner–Raikov theorem)

$$\varphi(x) = \int \chi(x)\, d\nu_\varphi(\chi) \tag{6.9.6}$$

for a measure, $d\nu_\varphi$, on $\widehat{G}$. By finding a rich set of pdf's, we'll prove that the χ's separate points of G, that $L^1(G)$ is semisimple and, eventually, that $L^1(G)$ is regular. It will also provide functions in $L^1(G, d\mu)$ so that $\widehat{f}$ is in $L^1(\widehat{G}, d\mu_{\widehat{G}})$ since we'll show that if $f \in L^1(G, d\mu)$ and is also positive definite, then $\widehat{f} \in L^1(\widehat{G}, d\mu_{\widehat{G}})$. And we'll first prove (6.9.3) for such f's.

Once we have established (6.9.3) for the span of such f's, we'll get the Plancherel theorem and then describe $\mathbb{W}(G)$ (i.e., the inverse Fourier transform of g's in $L^1(\widehat{G}, d\mu_{\widehat{G}})$) as exactly the convolutions of two $L^2(G, d\mu)$-functions. These results will then allow a proof of *Pontryagin duality*

$$K \equiv \widehat{G} \Rightarrow \widehat{K} \cong G \tag{6.9.7}$$

We begin with several classes of functions that will be useful.

Definition. A function, $\varphi\colon G \to \mathbb{C}$, is called a *positive definite function* (pdf) if and only if

(i) φ is continuous.
(ii) For all $x_1, \dots, x_n \in G$, the $n \times n$ matrix

$$C_{k\ell} = \varphi(x_k - x_\ell) \tag{6.9.8}$$

is a positive definite matrix. The set of all pdf's will be denoted $\mathcal{B}_+$. A function φ in L^∞ is called *weakly positive definite* if for all $f \in L^1$,

$$\int \overline{f(x)}\, \varphi(x-y) f(y)\, d\mu(x) d\mu(y) \geq 0 \tag{6.9.9}$$

We also define

$$\mathcal{P}_+ = \mathcal{B}_+ \cap L^1(G, d\mu) \tag{6.9.10}$$

and $\mathcal{P}$ (respectively, $\mathcal{B}$) to be the complex vector space generated by $\mathcal{P}_+$ (respectively, $\mathcal{B}_+$).

Remarks. 1. This agrees with the notion on $\mathbb{R}^\nu$ discussed in Section 6.6 of Part 1.

2. $\mathcal{P}_+$ is clearly a convex cone (i.e., closed under sums and positive multiples), so any $\widetilde{\varphi} \in \mathcal{P}$ can be written

$$\widetilde{\varphi} = \varphi_1 - \varphi_2 + i\varphi_1 - i\varphi_4 \tag{6.9.11}$$

for some $\varphi_j \in \mathcal{P}_+$.

3. As in the case of $\mathbb{R}^\nu$, $\left(\begin{smallmatrix} \varphi(0) & \varphi(x) \\ \varphi(-x) & \varphi(0) \end{smallmatrix}\right)$ positive definite implies

$$\varphi \in \mathcal{B}_+ \Rightarrow \|\varphi\|_\infty \leq \varphi(0), \quad \varphi(-x) = \overline{\varphi(x)} \tag{6.9.12}$$

Thus, any pdf lies in L^∞.

There is an analog of Bochner's theorem (Theorem 6.6.6 of Part 1) to general LCA groups—it is essentially a translation of the Bochner–Raikov theorem (Theorem 6.3.6) to this setting. We need one preliminary.

Recall that an abelian Banach algebra (perhaps without identity), $\mathfrak{A}$, is said to have an approximate identity if there exists a net, $\{a_\alpha\}_{\alpha\in I} \in \mathfrak{A}$, so $\|a_\alpha x - x\| \to 0$ for all $x \in \mathfrak{A}$.

Proposition 6.9.1. *Let $\mathfrak{A} = L^1(G, d\mu)$ for an LCA group, G. Let $\mathcal{U}$ be the net of all neighborhoods of 0 with compact closure, with $U \triangleright V$ if $U \subset V$, and $\{f_U\}_{U\in\mathcal{U}}$, a collection of nonnegative L^1-functions with $f_U(x) = 0$ if $x \notin U$ and*

$$\int f_U(x)\, d\mu(x) = 1 \tag{6.9.13}$$

for all U. Then $\{f_U\}_{U\in\mathcal{U}}$ is an approximate identity. In particular, $L^1(G, d\mu)$ has approximate identities.

Remark. Since G is metrizable, we can take a countable neighborhood base of U's, that is, U_n with $U_{n+1} \subset U_n$.

Proof. Suppose first g is a continuous function of compact support, N_0. By a standard uniform continuity argument (Problem 2), there is for each $\varepsilon > 0$, $U_\varepsilon \in \mathcal{U}$ so

$$x - y \in U_\varepsilon \Rightarrow |g(x) - g(y)| < \varepsilon \tag{6.9.14}$$

Thus, by (6.9.13), if $W \subset U_\varepsilon$,

$$|f_W * g(x) - g(x)| \leq \int f_W(x-y)\, |g(y) - g(x)|\, d\mu(x) < \varepsilon \tag{6.9.15}$$

If $W \subset U_1$ also, then $\operatorname{supp}(f_W * g) \subset W + N \subset U_1 + N$. This set is the continuous image of $W \times N$, so also compact, and thus, of finite μ measure. It follows that

$$W \subset U_1 \cap U_\varepsilon \Rightarrow \|f_W * g - g\|_1 \leq \varepsilon\mu(U_1 + N) \tag{6.9.16}$$

Therefore, $\|f_W * g - g\|_1 \to 0$. Since the set of such g's is dense in L^1, $\{f_U\}_{U \in \mathcal{U}}$ is an approximate identity. $\square$

Theorem 6.9.2 (Bochner–Weil Theorem). *The following are equivalent for* $\varphi \in L^\infty(G)$:

(1) φ *is in* $\mathcal{B}_+$ *(after a possible change on a set of measure* 0*).*
(2) φ *is weakly positive definite.*
(3) *There exists a finite Baire measure,* ν_φ*, on* $\widehat{G}$ *so for a.e.* $x \in G$,

$$\varphi(x) = \int \chi(x)\, d\nu_\varphi(\chi) \tag{6.9.17}$$

Remarks. 1. In particular, if (6.9.17) holds, then for all $f \in L^1$,

$$\int \overline{\varphi(x)}\, f(x)\, d\mu(x) = \int \widehat{f}(\chi)\, d\nu_\varphi(\chi) \tag{6.9.18}$$

Equivalently,

$$\int \varphi(x) f(x)\, d\mu(x) = \int \widehat{f}(\overline{\chi}_1)\, d\nu_\varphi(\chi_1) \tag{6.9.19}$$

Replacing f by $f\overline{\chi}$ and noting that

$$\widehat{f\overline{\chi}}(\overline{\chi}_1) = \widehat{f}(\chi\overline{\chi}_1) \tag{6.9.20}$$

we see that ($\varphi \in L^\infty$, $f \in L^1 \Rightarrow \varphi f \in L^1$)

$$\widehat{\varphi f}(\chi) = \int \widehat{f}(\chi\overline{\chi}_1)\, d\nu_\varphi(\chi_1) \tag{6.9.21}$$

Once we have that $\varphi \in \mathcal{B}_+ \cap L^1 \Rightarrow d\nu_\varphi(\chi) = \widehat{\varphi}(\chi)\, d\mu_{\widehat{G}}(\chi)$, this will imply that $\widehat{\varphi f} = \widehat{\varphi} * \widehat{f}$.

2. The measure, $d\nu_\varphi$, is unique because (6.9.18) holds and, as noted, $\{\widehat{f} \mid f \in L^1\}$ is $\|\cdot\|_\infty$ dense in $C_\infty(\widehat{G})$.

Proof. (1) $\Rightarrow$ (2). For a complex measure of finite $\mathcal{M}(G)$-norm, $d\eta$, on G and $\varphi \in L^\infty$, define

$$Q_\varphi(\eta) = \int \varphi(x-y)\,\overline{d\eta(x)}\,d\eta(y) \tag{6.9.22}$$

φ positive definite implies $Q_\varphi(\eta)$ is nonnegative for every η which is a complex point measure with finitely many points.

If $N \subset G$ is compact, any probability measure on N is a weak limit of point probability measures with finitely many points (see, e.g., Theorem 5.11.8 in Part 1), so any complex measure, η, of compact support and total norm, m, is a limit point (in the $\sigma(\mathcal{M}(G), C(G))$-topology) of finite point measures of norm at most m with points in $\mathrm{supp}(\eta)$. So suppose η is a measure of compact support, N, and η_α a net of finite point measures with support in N so $\eta_\alpha \to \eta$ (weakly). Then, since sums of product functions are dense in all continuous functions on the product, $\bar{\eta}_\alpha \otimes \eta_\alpha \to \bar{\eta} \otimes \eta$ and so $Q_\varphi(\eta) \geq 0$ for η of compact support.

Therefore, (6.9.9) holds for $f \in L^1$ with compact support (using $d\eta = f\,d\mu$), and thus for all $f \in L^1$ by a limiting argument.

(2) $\Rightarrow$ (3). Let

$$\Phi(f) = \int \varphi(x) f(x)\,d\mu(x) \tag{6.9.23}$$

Then (6.9.9) says

$$\begin{aligned}\Phi(f^* * f) &= \int \varphi(x)\,\overline{f(-(x-y))}\,f(y)\,d\mu(x)d\mu(y)\\ &= \int \varphi(y-w)\,\overline{f(w)}\,f(y)\,d\mu(w)d\mu(y) \geq 0\end{aligned} \tag{6.9.24}$$

(changing variables from (x,y) to $(w = y-x, y)$ and using $d\mu(x) = d\mu(-x) = d\mu(y-x)$).

Therefore, Φ is a positive functional (in the sense of Section 6.3) on $L^1(G, d\mu)$. Since L^1 has an approximate identity (by Proposition 6.9.1), by Problem 3 of Section 6.3, Φ has a positive extension, $\widetilde{\Phi}$, to $\widetilde{\mathfrak{A}}(G) = L^1(G, d\mu) \oplus \mathbb{C}$. By the Bochner–Raikov theorem (Theorem 6.3.6), there is a measure, ν_Φ, on $\widehat{G} \cup \{\infty\}$ so

$$\widetilde{\Phi}((f,\lambda)) = \int \widehat{(f,\lambda)}(\chi)\,d\nu_\Phi(\chi) + \lambda\nu_\Phi(\{\infty\}) \tag{6.9.25}$$

In particular, taking $\lambda = 0$ and setting $d\nu_\varphi(\chi) = [d\nu_\Phi \restriction \widehat{G}](\overline{\chi})$,

$$\int f(x)\varphi(x)\,d\mu(x) = \int \widehat{f}(\overline{\chi})\,d\nu_\varphi(\chi)$$

$$= \int \biggl(\int f(x)\chi(x)\,d\mu(x) \biggr)\,d\nu_\varphi(\chi) \tag{6.9.26}$$

$$= \int \biggl[\int \chi(x)\,d\nu_\varphi(\chi) \biggr] f(x)\,d\mu(x) \tag{6.9.27}$$

where we used Fubini's theorem in the last step. Since f is an arbitrary L^1-function, (6.9.17) holds for μ-a.e. x.

(3) $\Rightarrow$ (1). By the continuity of each $\chi(\cdot)$, $\|\chi\|_\infty \le 1$, and the dominated convergence theorem, the function φ defined by (6.9.17) is continuous. Since

$$\chi(x_i - x_j) = \chi(x_i)\overline{\chi(x_j)} \tag{6.9.28}$$

we have for any $x_1, \dots, x_n \in G$, $\zeta_1, \dots, \zeta_n \in \mathbb{C}$,

$$\sum_{i,j=1}^n \bar{\zeta}_i \chi(x_i - x_j)\zeta_j = \biggl| \sum_{j=1}^n \zeta_j\,\overline{\chi(x_j)} \biggr|^2 \ge 0 \tag{6.9.29}$$

Thus,

$$\sum_{i,j=1}^n \bar{\zeta}_i\,\varphi(x_i - x_j)\zeta_j = \int \biggl| \sum_{i=1}^n \zeta_j \overline{\chi(x_j)} \biggr|^2 d\nu(\chi) \ge 0 \tag{6.9.30}$$

□

Remarks. 1. In particular, every weakly positive definite function is equal a.e. to a continuous function.

2. The canonical choice of Φ with $\Phi((0,1)) = \|\varphi\| = \varphi(0)$ has $\nu_\Phi(\{\infty\}) = 0$ (Problem 3).

3. We can use dominated convergence because G is metrizable.

Positive definite functions are useful because one can find lots of them:

Proposition 6.9.3. *For any $f \in L^1(G, d\mu)$, $\varphi \equiv f^* * f$ is a pdf and $\lim_{x\to\infty} \varphi(x) = 0$.*

Remark. $\lim_{x\to\infty} g(x) = 0$ means for any ε, there is a compact K with $|g(x)| \le \varepsilon$ on $G \setminus K$.

Proof. It is easy to see (Problem 4), if $f \in L^p(G, d\mu)$, $h \in L^q(G, d\mu)$, where $1 < p < \infty$ and $q = (1 - p^{-1})^{-1}$, then $f * h$ is a continuous function vanishing at infinity. Thus, φ is continuous and vanishes at infinity.

Moreover, by translation invariance of Haar measure,

$$\begin{aligned}\varphi(x_i - x_j) &= \int \overline{f(y - x_i + x_j)}\, f(y)\, d\mu(y) \\ &= \int \overline{f(y + x_j)}\, f(y + x_i)\, d\mu(y) \qquad (6.9.31)\end{aligned}$$

so

$$\sum \bar{\zeta}_i \zeta_j \varphi(x_i - x_j) = \Big\| \sum \zeta_j \overline{f(\cdot + x_j)} \Big\|_{L^2}^2 \geq 0 \qquad (6.9.32)$$

proving that φ is positive definite. □

Note that by a variant of the polarization identity ((3.1.8) of Part 1), if $\varphi_f = f^* * f$, then

$$f^* * g = \tfrac{1}{4}\left(\varphi_{f+g} - \varphi_{f-g} - i\varphi_{f+ig} + i\varphi_{f-ig}\right) \qquad (6.9.33)$$

so the complex span of the φ_f's for all $f \in L^2(G, d\mu)$ includes every $f * g$ with $f, g \in L^2 G, d\mu)$.

Definition. $\mathbb{W}_+(G)$ is the set of $f^* * f$ with $f \in L^2(G, d\mu)$. $\mathbb{W}(G)$ is the complex linear span of $\mathbb{W}_+(G)$. $\mathcal{C}_+(G)$ is the linear space of $f^* * f$, where f is an L^2-function of compact support and $\mathcal{C}(G)$ is its complex linear span.

Remarks. 1. We, of course, have

$$\mathcal{C}_+(G) \subset \mathbb{W}_+(G) \subset \mathcal{B}_+ \quad \text{and} \quad \mathcal{C}_+(G) \subset \mathcal{P}_+ \qquad (6.9.34)$$

2. For $\mathbb{Z}, \mathbb{T}, \mathbb{R}$, we defined $\mathbb{W}$, the Wiener algebra, as the Fourier transform of L^1-functions on the dual group. We'll show later (once we have the Plancherel theorem) that this is true here. This will also show (Theorem 6.9.21) $\mathbb{W}_+(G)$ is a convex cone and the linear span, $\mathbb{W}(G)$, is already $\{f * g \mid f, g \in L^2\}$, that is, the set of such convolutions is a vector space. Moreover, $\mathbb{W}(G)$ is an algebra under pointwise multiplication.

Proposition 6.9.4. (a) *If N and M are disjoint closed subsets of G and N is compact, there exists $h \in \mathcal{C}(G)$ with $h \restriction N \equiv 1$, $h \restriction M \equiv 0$.*

(b) *$\mathcal{C}(G)$ is norm-dense in $C_\infty(G)$ and in each $L^p(G, d\mu)$, $1 \leq p < \infty$.*

(c) *$\mathcal{C}(G)$ is weak-$*$ dense in $L^\infty(G, d\mu)$, that is, for $h \in L^\infty(G, d\mu)$, there exists a net, $h_\alpha \in \mathcal{C}(G)$, so that for each $f \in L^1(G, d\mu)$,*

$$\int f(x) h_\alpha(x)\, d\mu(x) \to \int f(x) h(x)\, d\mu \qquad (6.9.35)$$

and $\|h_\alpha\|_\infty \leq \|h\|_\infty$.

Proof. (a) As noted, by polarization, if f, g are continuous functions of compact support, $f * g \in \mathcal{C}(G)$. First note that if U is an open neighborhood of 0, there is V a neighborhood of 0 with $V = -V$, $V + V \subset U$, and $\overline{V}$ compact.

Next, find for each $x \in N$, a neighborhood U_x of 0, so

$$(x + U_x) \cap M = \emptyset \tag{6.9.36}$$

obviously possible since $G \setminus M$ is open. Now pick V_x with the above properties for U_x. $\bigcup_{x \in N}(x + V_x) \supset N$, so find $x_1, \dots, x_m$ with $N \subset \bigcup_{j=1}^n (x_j + V_{x_j})$. Let $V = \bigcap_{j=1}^n V_{x_j}$. If $x \in x_j + V_{x_j}$, then $x + V \subset x_j + V_{x_j} + V \subset x_j + U_{x_j}$, and therefore, by (6.9.36),

$$(N + V) \cap M = \emptyset \tag{6.9.37}$$

Now find W related to V as V is to U. Since W is open, $\mu(W) > 0$, and since $\overline{W}$ is compact, $\mu(W) < \infty$.

Define

$$h(x) = \mu(W)^{-1}(\chi_{N+W} * \chi_W)(x) = \mu(W)^{-1} \int_{y \in W} \chi_{N+W}(x+y)\, d\mu(y) \tag{6.9.38}$$

Then $h \in \mathcal{C}(G)$, $h \restriction M \equiv 0$ by (6.9.37) and $W - W \subset V$ and $h \restriction N \equiv 1$ since $x \in N$, $y \in W \Rightarrow x + y \in N + W$.

(b) If $\mathcal{U}$ is the set of symmetric neighborhoods of 0 with compact closure, $\{|U|^{-1}\chi_U\}_{u \in \mathcal{U}}$ is an approximate identity, so if g is a continuous function of compact support, $g * |U|^{-1}\chi_U \to g$ in $\|\cdot\|_\infty$ and in each L^p. Since these functions are dense in $C_\infty(G)$ in $\|\cdot\|_\infty$ and in each L^p, $1 \le p < \infty$, we get the required density result.

(c) If it easy to see (Problem 5) that the L^∞-functions of compact support are weak-$*$ dense in L^∞. If g is such a function and $f \in L^1$ and $\{U^{-1}\chi_U\}_{U \in \mathcal{U}}$, the above approximate identity, then

$$\int ((U^{-1}\chi_U) * g)(x) f(x)\, d\mu(x) = \int g(x)((U^{-1}\chi_U) * f)(x)\, d\mu(x)$$

converges to $\int g(x) f(x)\, d\mu(x)$, so $\{|U|^{-1}\chi_U * g \mid g \in L^\infty$ with compact support$\}$ are dense in L^∞. Each $U^{-1}\chi_U * g \in \mathcal{C}(G)$. □

This lets us prove there are lots of elements in $\widehat{G}$.

Theorem 6.9.5. (a) *$L^1(G, d\mu)$ is semisimple.*

(b) *For any $x, y \in G$, there is $\chi \in \widehat{G}$ with $\chi(x) \neq \chi(y)$.*

Proof. (a) Let $f \in L^1$ with $\widehat{f} \equiv 0$. Let $\varphi \in \mathcal{B}_+$. By (6.9.7) and Fubini's theorem,

$$\begin{aligned} \int \varphi(x)\, \overline{f(x)}\, d\mu(x) &= \int \chi(x) \overline{f(x)}\, d\mu(x) d\nu_\varphi(\chi) \\ &= \int \overline{\widehat{f}(\chi)}\, d\nu_\varphi(\chi) = 0 \end{aligned} \tag{6.9.39}$$

Since $\mathcal{C}_+ \subset \mathcal{B}_+$, this is true for all $\varphi \in \mathcal{C}_+$ and so for all finite combinations, that is, all $\varphi \in \mathcal{C}(G)$. Since $\bar{f} \in L^1$ and $\mathcal{C}(G)$ is weak-$*$ dense in L^∞, this is true for all $\varphi \in L^\infty$. Picking

$$\varphi(x) = \begin{cases} 0 & \text{if } f(x) = 0 \\ f(x)/|f(x)| & \text{if } f(x) \neq 0 \end{cases} \tag{6.9.40}$$

we conclude that $\|f\|_1 = 0$ so $f = 0$. Thus, $\widehat{f} = 0 \Rightarrow f = 0$.

(b) If not, picking $z = xy^{-1}$, we find $z \neq 0$, so for all $\chi \in \widehat{G}$, $\chi(z) = \chi(0) = 1$ and thus,

$$\forall x \in G,\, \chi \in \widehat{G}, \quad \chi(z + x) = \chi(x) \tag{6.9.41}$$

Pick U a neighborhood of zero with $z \notin U$ and then V, a symmetric neighborhood of zero with compact closure, so that

$$V + V \subset U \tag{6.9.42}$$

This implies

$$V \cap (V + z) = \emptyset \tag{6.9.43}$$

Let $f = \chi_V - \chi_{V+z}$. Since $\mu(V) < \infty$, $f \in L^1$ and, by (6.9.43), $\|f\|_1 = 2\mu(V) > 0$. But, by (6.9.41), $\widehat{f}(\chi) = 0$ for all χ. It follows that $f = 0$. This contradiction proves that there is no such $z \neq 0$. $\square$

Remark. Since characters separate point and $\chi \in \widehat{G} \Rightarrow \overline{\chi} \in \widehat{G}$ and $\chi_1, \chi_2 \in \widehat{G} \Rightarrow \chi_1\chi_2 \in \widehat{G}$, the Stone–Weierstrass theorem implies that finite linear combinations of characters are dense in each $C(Q)$, $Q \subset G$ compact.

We turn next to the topology on $\widehat{G}$.

Theorem 6.9.6. *Let $\{\chi_n\}_{n=1}^\infty$ be a sequence in $\widehat{G}$ and let $\chi \in \widehat{G}$. Then the following are equivalent:*

(1) *$\chi_n \to \chi$ in the Gel'fand topology on $\widehat{G}$.*
(2) *$\chi_n(x) \to \chi(x)$ uniformly for $x \in K$, an arbitrary compact subset of G.*
(3) *$\chi_n(x) \to \chi(x)$ for each $x \in G$.*

Proof. (1) $\Rightarrow$ (2). Let ℓ_n be the multiplicative linear functional associated to χ_n and ℓ to χ. Pick f so $\ell(f) \neq 0$ and then since $\ell_n(f) \to \ell(f)$ (by (1)), pick n_0 so $n > n_0$ implies

$$|\ell_n(f)| \geq \tfrac{1}{2}\,|\ell(f)| \tag{6.9.44}$$

By (6.2.50) and Theorem 6.2.2(d),

$$\begin{aligned} |\chi_n(x) - \chi(x)| &\leq 2|\ell(f)|^{-2}[\,|\ell_n(f)|\,|\ell(\tau_x f) - \ell_n(\tau_x f)| + |\ell(\tau_x f)|\,|\ell(f) - \ell_n(f)|\,] \\ &\leq 2|\ell(f)|^{-2}\|f\|_1[\,|\ell_n(\tau_x f) - \ell(\tau_x f)| + |\ell(f) - \ell_n(f)|\,] \end{aligned}$$

so we need only show

$$\sup_{x\in K} |\ell_n(\tau_x f) - \ell(\tau_x f)| = 0 \tag{6.9.45}$$

Since $\|\ell_n\| = 1 = \|\ell\|$ and $x \mapsto \tau_x f$ is continuous, $\{\tau_x f\}_{x\in K}$ is compact, so for any ε, we can find $x_1, \dots, x_m$ so that for each $x \in K$, $\inf_j \|\tau_x f - \tau_{x_j} f\| < \varepsilon$, and thus

$$\sup_{x\in K} |\ell_n(\tau_x f) - \ell(\tau_x f)| \le 2\varepsilon + \sup_{j=1,\dots,m} |\ell_n(\tau_{x_j} f) - \ell(\tau_x f)| \tag{6.9.46}$$

Since each $\ell_n(\tau_{x_j} f) \to \ell(\tau_{x_j} f)$, we conclude (6.9.45) holds.

(2) $\Rightarrow$ (3). This is immediate.

(3) $\Rightarrow$ (1). $f\overline{\chi}_n \to f\overline{\chi}$ pointwise, so for $f \in L^1$, by the dominated convergence theorem, $\widehat{f}(\chi_n) \to \widehat{f}(\chi)$ so $\chi_n \to \chi$ in the Gel'fand topology. □

Theorem 6.9.7. *$\widehat{G}$ is a topological group in the Gel'fand topology.*

Proof. Clearly, once we know the topology is given by pointwise convergence, products and inverses (given by complex conjugation) are continuous. □

Theorem 6.9.8. (a) *If G is compact, $\widehat{G}$ is discrete.*

(b) *If G is discrete, $\widehat{G}$ is compact.*

Proof. (a) Since G is compact, $\mu(G) < \infty$, so every $\chi \in \widehat{G}$ lies in $L^1(G, d\mu)$. Normalize μ as $\mu(G) = 1$. Let $\chi \in \widehat{G}$ with

$$\int \chi(x)\, d\mu(x) \neq 0 \tag{6.9.47}$$

Then for any y,

$$\chi(y) \int \chi(x)\, d\mu(x) = \int \chi(x+y)\, d\mu(x) = \int \chi(x)\, d\mu(x) \tag{6.9.48}$$

by translation invariance of μ. Thus,

$$(6.9.47) \Rightarrow \chi \equiv 1 \tag{6.9.49}$$

so

$$\chi \not\equiv 1 \Rightarrow \int \chi(x)\, d\mu = 0 \tag{6.9.50}$$

In $L^\infty(G) \equiv L^1(G)^*$, since $1 \in L^1$,

$$U = \left\{ f \in L^\infty \;\middle|\; \left| \int f(x)\, d\mu(x) - 1 \right| < \tfrac{1}{2} \right\}$$

is a $\sigma(L^\infty, L^1)$-open set. $1 \in U$ but, by (6.9.50), no other χ lies in U. This shows 1 is an isolated point, so by translation invariance of the topology, any point is open, so the topology is discrete.

(b) δ_e is an identity in L^1, so $\widehat{L^1}$ is compact. $\square$

Remark. Since $\overline{\chi}_1\chi_2 \in \widehat{G}$ if $\chi_1, \chi_2 \in \widehat{G}$ and it is not the identity if $\chi_1 \neq \chi_2$, we have $\int \overline{\chi}_1(x)\chi_2(x)\, d\mu(x) = 0$. Thus, if $\mu(G) = 1$, then $\{\chi\}_{\chi\in\widehat{G}}$ are an orthonormal set. Since $f \in L^2 \Rightarrow f \in L^1 \Rightarrow (\widehat{f}(\chi) = 0 \Rightarrow f = 0)$, we see it is a basis.

Example 6.2.24 (revisited). We saw that as groups $\widehat{\mathbb{T}} = \mathbb{Z}$, $\widehat{\mathbb{Z}} = \mathbb{T}$, $\widehat{\mathbb{R}} = \mathbb{R}$. We can now confirm the Gel'fand topology in these cases agrees with the usual ones. Since $\mathbb{T}$ is compact, $\widehat{\mathbb{T}}$ is discrete, and so $\mathbb{Z}$ has its usual topology. For $\widehat{\mathbb{Z}}$, the topology is the weak one defined by the functions, $\{e^{in\theta}\}_{n\in\mathbb{Z}}$. In the usual topology, each such function is continuous, and if $e^{i\theta}$ is continuous, then the topology is stronger than the usual one, so the topology on $\widehat{\mathbb{Z}}$ is the usual on $\mathbb{T}$. It is easy to see (Problem 6) that for $\{x_n\}_{n=1}^\infty, x_\infty \in \mathbb{R}$, $e^{ikx_n} \to e^{ikx_\infty}$ for each k if and only if $x_n \to x_\infty$. $\square$

We now turn to the central result of this section: the Fourier inversion formula, (6.9.3). We'll show that for a suitable normalization, $d\mu_{\widehat{G}}$, of Haar measure on $\widehat{G}$, we have for $\varphi \in \mathcal{P}_+$ that

$$d\nu_\varphi(\chi) = \widehat{\varphi}(\chi)\, d\mu_{\widehat{G}}(\chi) \tag{6.9.51}$$

in which case, (6.9.17) is exactly (6.9.3). A key preliminary to proving (6.9.51) is to show that if $\varphi, \psi \in \mathcal{P}_+$, then

$$\widehat{\varphi}(\chi)\, d\nu_\psi(\chi) = \widehat{\psi}(\chi)\, d\nu_\varphi(\chi) \tag{6.9.52}$$

We will then define $d\mu$ locally as $\widehat{\varphi}(\chi)^{-1}\, d\nu_\varphi(\chi)$ on sets where $\widehat{\varphi}$ is nonvanishing and prove this is a Haar measure. For this to work, we need to know $\varphi \in \mathcal{B}_+ \Rightarrow \widehat{\varphi} \geq 0$ and that there are φ's positive at each χ. Thus, the following preliminary is important:

Proposition 6.9.9. (a) *If $\varphi \in \mathcal{C}_+(G)$, then for all χ,*

$$\widehat{\varphi}(\chi) \geq 0 \tag{6.9.53}$$

(b) *For any $\chi_0 \in \widehat{G}$, there is $\varphi \in \mathcal{C}_+(G)$ with*

$$\widehat{\varphi}(\chi_0) > 0 \tag{6.9.54}$$

(c) *For any compact, Q, in $\widehat{G}$, there is $\varphi \in \mathcal{P}_+(G)$ so that*

$$\inf_{\chi_0\in Q} \varphi(\chi_0) > 0 \tag{6.9.55}$$

Remark. We'll eventually prove that $\mathcal{C}_+(G)$ is a convex cone, but since we don't know that yet, we use $\mathcal{P}_+(G)$ in (c).

Proof. (a) If $\varphi = f^* * f$, then, since $\widehat{f^*} = \overline{\widehat{f}}$, we have

$$\widehat{\varphi}(\chi) = |\widehat{f}(\chi)|^2 \geq 0 \tag{6.9.56}$$

(b) By (6.9.56), we need only find $f \in L^2$ of compact support, so $\widehat{f}(\chi_0) \neq 0$. Let $g \in L^2$ of compact support with g nonnegative and $g \not\equiv 0$. Let

$$f(x) = \chi_0(x)g(x) \tag{6.9.57}$$

Then

$$\widehat{f}(\chi_0) = \int g(x)\, d\mu(x) > 0 \tag{6.9.58}$$

(c) For any $\chi_0 \in Q$, find $\varphi_{\chi_0} \in \mathcal{C}_+(G)$ obeying (6.9.54) so, since $\widehat{\varphi}$ is continuous, for some neighborhood, U_{χ_0} of χ_0, we have $\widehat{\varphi}(\chi) > 0$ on U_{χ_0}. By compactness, find $\chi_1, \dots, \chi_m$ so

$$Q \subset \bigcup_{j=1}^{m} U_{\chi_j} \tag{6.9.59}$$

Then $\varphi = \sum_{j=1}^m \varphi_{\chi_j} \in \mathcal{P}_+$ (which is a convex cone) and, by definition of U_{χ_j} and (6.9.53), we have $\widehat{\varphi} > 0$ on Q. By compactness, the inf is strictly positive. $\square$

Proposition 6.9.10. *Let $\varphi, \psi \in \mathcal{P}_+$. Then*

$$\widehat{\psi}(\chi)\, d\nu_\varphi(\chi) = \widehat{\varphi}(\chi)\, d\nu_\psi(\chi) \tag{6.9.60}$$

Remarks. 1. Since $\psi \in L^1$, $\widehat{\psi}$ is bounded, so $\widehat{\psi}\, d\nu_\varphi$ is a priori a complex measure which we'll see is a positive one.

2. Formally, (6.9.61) below says $(h * \psi * \varphi)(0) = (h * \varphi * \psi)(0)$, but since commutativity of convolution is true only for a.e. x, we write out integrals below.

Proof. Since $\widehat{\psi}$ and $\widehat{\varphi}$ are bounded functions (since $\psi, \varphi \in L^1$), both sides of (6.9.60) are complex but bounded measures. Thus, it suffices to prove equality of the integrals of any function in $C_\infty(\widehat{G})$. Since $\{\widehat{h} \mid h \in L^1(G, d\mu)\}$ is $\|\cdot\|_\infty$-dense in $C_\infty(\widehat{G})$, it suffices to prove that

$$\int \widehat{h}\widehat{\psi}\, d\nu_\varphi = \int \widehat{h}\widehat{\varphi}\, d\nu_\psi \tag{6.9.61}$$

Since $h, \psi \in L^1$, we have $\widehat{h}\widehat{\psi} = \widehat{h * \psi}$, so

$$\begin{aligned}\int \widehat{h}\widehat{\psi}\, d\nu_\varphi(\chi) &= \int \widehat{h * \psi}(\chi)\, d\nu_\varphi(\chi) \\ &= \int (h * \psi)(-x)\chi(x)\, d\nu_\varphi(\chi)\, d\mu(x)\end{aligned}$$

$$= \int (h * \psi)(-x)\varphi(x)\, d\mu(x)$$

$$= \int h(-x-y)\psi(y)\varphi(x)\, d\mu(x)d\mu(y) \tag{6.9.62}$$

where we used Fubini's theorem. (6.9.62) is symmetric in φ and ψ, so (6.9.61) holds. □

Corollary 6.9.11. *Let $\varphi \in \mathcal{P}_+$ with $\widehat{\varphi}(\chi_0) > 0$. Then $\chi_0 \in \operatorname{supp}(d\nu_\varphi)$.*

Proof. Let $\psi \in \mathcal{P}_+$ with $\psi \not\equiv 0$. Then for any χ_1,

$$(\chi_1\psi)(x) = \int (\chi_1\chi)(x)\, d\nu_\psi(\chi) = \int \chi(x)\, d\nu_\psi(\chi_1^{-1}\chi) \tag{6.9.63}$$

we see $\chi_1\psi \in \mathcal{P}_+$ and

$$d\nu_{\chi_1\psi}(\chi) = d\nu_\psi(\chi_1^{-1}\chi) \tag{6.9.64}$$

Since $\psi \not\equiv 0$, $\operatorname{supp}(d\nu_\psi) \neq \emptyset$, so by replacing ψ by $\chi_1\psi$ for suitable χ_1, we can suppose we've picked ψ so

$$\chi_0 \in \operatorname{supp}(d\nu_\psi) \tag{6.9.65}$$

Suppose $h \in C_\infty(\widehat{G})$ is nonnegative, $h(\chi_0) > 0$, and $\operatorname{supp}(h) \subset \{\chi \mid \widehat{\varphi}(\chi) > 0\}$. Then $h\widehat{\varphi} \geq 0$, $(h\widehat{\varphi})(\chi_0) > 0$, $\chi_0 \in \operatorname{supp}(d\nu_\psi)$, we see, by (6.9.60), that

$$0 < \int (h\widehat{\varphi})(\chi)\, d\nu_\psi(\chi) = \int h(\chi)\widehat{\psi}(\chi)\, d\nu_\varphi(\chi) \tag{6.9.66}$$

If $\chi_0 \notin \operatorname{supp}(d\nu_\varphi)$, then we can find h with the required properties so the right side of (6.9.66) is 0, contradicting its positivity. We conclude that $\chi_0 \in \operatorname{supp}(d\nu_\varphi)$. □

Corollary 6.9.12. *For all $\varphi \in \mathcal{P}_+$, $\widehat{\varphi}(\chi) \geq 0$ for all χ.*

Proof. Suppose that $\widehat{\varphi}(\chi_0) < 0$. Pick any $f \neq 0$ in $L^2(G, d\mu)$ with $\int f(x)\, d\mu(x) > 0$ and let $\psi = \chi_0(f^* * f)$ so

$$\widehat{\psi}(\chi_0) = \left| \int f(x)\, d\mu(x) \right|^2 > 0 \tag{6.9.67}$$

Thus, by Corollary 6.9.11, $\chi_0 \in \operatorname{supp}(d\nu_\psi)$.

By continuity, $\widehat{\varphi}(\chi_0) < 0$ in a neighborhood of χ_0, so we can find $h \in C(\widehat{G})$ supported in a neighborhood, U, of χ_0 with compact closure so $h \geq 0$, $h(\chi_0) > 0$, $\nu_\psi(U) > 0$, and $\widehat{\varphi} < 0$ on U. Thus,

$$\int h(\chi)\widehat{\varphi}\, d\nu_\psi(\chi) < 0 \tag{6.9.68}$$

But $\widehat{\psi}(\chi) = |\widehat{f}(\chi\bar{\chi}_0)|^2 \geq 0$, so

$$\int h(\chi)\widehat{\psi}(\chi)\, d\nu_\varphi(\chi) \geq 0 \tag{6.9.69}$$

This contradiction shows that $\widehat{\varphi} \geq 0$. $\square$

Theorem 6.9.13. *There is a Haar measure, $d\mu_{\widehat{G}}$, on $\widehat{G}$ so that for all $\varphi \in \mathcal{P}_+$,*

$$d\nu_\varphi(\chi) = \widehat{\varphi}(\chi)\, d\mu_{\widehat{G}}(\chi) \tag{6.9.70}$$

Proof. Let g be a continuous function with $\text{supp}(g) = Q$ compact. Let φ_1, φ_2 be two functions in $\mathcal{P}_+$ obeying (6.9.55) (some exist by Proposition 6.9.9). Then $g/\widehat{\varphi}_1\widehat{\varphi}_2 \equiv h$ is continuous of compact support, so

$$\int \frac{g(\chi)}{\widehat{\varphi}_1(\chi)\widehat{\varphi}_2(\chi)}\, \widehat{\varphi}_2(\chi)\, d\nu_{\varphi_1}(\chi) = \int \frac{g(\chi)}{\widehat{\varphi}_1(\chi)\widehat{\varphi}_2(\chi)}\, \widehat{\varphi}_1(\chi)\, d\nu_{\varphi_2}(\chi) \tag{6.9.71}$$

Thus,

$$\ell(g) \equiv \int g(\chi)\widehat{\varphi}_1(\chi)^{-1}\, d\nu_{\varphi_1}(\chi) \tag{6.9.72}$$

is independent of the choice of φ_1 obeying (6.9.55).

Since $\widehat{\varphi}_1 > 0$, $\ell(g) \geq 0$ if $g \geq 0$, and by the uniqueness (using $\widehat{\varphi}_1 > 0$ on $\text{supp}(g_1) \cup \text{supp}(g_2)$), we see ℓ is linear, so there is a Baire measure which we'll call $d\mu_{\widehat{G}}$ so that

$$\ell(g) = \int g(\chi)\, d\mu_{\widehat{G}} \tag{6.9.73}$$

Given h of continuous compact support, Q, in $\widehat{G}$, first pick $\varphi_1 \in \mathcal{P}_+$ so $\varphi_1 > 0$ on Q and then $\chi_0 \in \widehat{G}$. Then, by (6.9.64),

$$d\nu_{\overline{\chi}_0\varphi_1}(\chi) = d\nu_{\varphi_1}(\chi_0\chi) \tag{6.9.74}$$

Clearly,

$$\widehat{\overline{\chi}_0\varphi_1}\,(\chi) = \widehat{\varphi}_1(\chi\chi_0) \tag{6.9.75}$$

so if $\varphi_2 \equiv \overline{\chi}_0\varphi_1$, then

$$(\varphi_2^{-1}\, d\nu_{\varphi_2})(\chi) = (\varphi_1^{-1}\, d\nu_{\varphi_1})(\chi\chi_0) \tag{6.9.76}$$

as measures on functions g supported on $\chi_0^{-1}Q$. Picking $g(\chi) = h(\chi_0\chi)$, we see that

$$\begin{aligned}\ell(g) &= \int h(\chi_0\chi)(\varphi_2^{-1}\, d\nu_2)(\chi) = \int h(\chi_0\chi)(\varphi_1^{-1}\, d\nu_{\varphi_1})(\chi_0\chi) \\ &= \ell(h)\end{aligned} \tag{6.9.77}$$

proving that $d\mu_G$ is translation invariant, that is, a choice of Haar measure (only normalization is up to us to choose).

Finally, given $\varphi \in \mathcal{P}_+$ and a continuous function, h, of support on $\widehat{G}$, pick $\varphi_1 \in \mathcal{P}_+$ strictly positive on $\operatorname{supp}(h)$ and note

$$\begin{aligned}\int h(\chi)\widehat{\varphi}(\chi)\,d\mu_G(\chi) &= \int h(\chi)\widehat{\varphi}(\chi)\widehat{\varphi}_1(\chi)^{-1}\,d\nu_{\varphi_1}(\chi)\\ &= \int h(\chi)\widehat{\varphi}_1(\chi)[\widehat{\varphi}_1(\chi)]^{-1}\,d\nu_\varphi(\chi)\\ &= \int h(\chi)\,d\nu_\varphi(\chi)\end{aligned}$$

proving (6.9.70). □

We define $\check{}$ on $L^1(\widehat{G}, d\mu_{\widehat{G}})$ as a function on G by

$$\check{g}(x) = \int \chi(x)g(\chi)\,d\mu_{\widehat{G}}(\chi) \tag{6.9.78}$$

Theorem 6.9.14 (Fourier Inversion Formula). (a) $g \in L^1(\widehat{G}, d\mu_{\widehat{G}}) \Rightarrow \check{g} \in \mathcal{B}(G)$.

(b) *For all* $f \in L^1(G, d\mu)$, $g \in L^1(\widehat{G}, d\mu_{\widehat{G}})$,

$$\int \overline{f(x)}\,\check{g}(x)\,d\mu_G(x) = \int \overline{\widehat{f}(\chi)}\,g(\chi)\,d\mu_{\widehat{G}}(\chi) \tag{6.9.79}$$

(c) $g \in L^1(\widehat{G}, d\mu_{\widehat{G}})$ *and* $\check{g}(x) = 0$ *for all* $x \Rightarrow g = 0$.

(d) $f \in \mathcal{P}(G) \Rightarrow \widehat{f} \in L^1(G, d\mu_{\widehat{G}})$ *and*

$$(\widehat{f})\check{} = f \qquad \text{for } f \in \mathcal{P}(G) \tag{6.9.80}$$

(e) $g \in L^1(\widehat{G}, d\mu_{\widehat{G}})$ *and* $\check{g} \in L^1(G, d\mu) \Rightarrow$

$$\widehat{\check{g}} = g \qquad \text{for } g \in L^1(\widehat{G}),\ \check{g} \in L^1(G) \tag{6.9.81}$$

Proof. (a) Since both $L^1(\widehat{G}, d\mu_{\widehat{G}})$ and $\mathcal{C}(G)$ are finite linear combinations of positive objects, we need only note that if $g \geq 0$, $\check{g}$ has the form (6.9.17) for $d\nu_{\check{g}} = g\,d\mu_{\widehat{G}}(\chi)$, so $\check{g} \in \mathcal{B}_+$.

(b) Both sides of (6.9.79) equal

$$\iint \overline{f(x)}\,g(\chi)\chi(x)\,d\mu(x)d\mu_{\widehat{G}}(\chi) \tag{6.9.82}$$

so (6.9.79) holds by Fubini's theorem.

(c) By (6.9.79), if $\check{g} = 0$, then (since $\overline{\widehat{f}} = \widehat{f^*}$)

$$\int \widehat{h}(\chi)\,d\eta(\chi) = 0 \tag{6.9.83}$$

for all $h \in L^1$, where $d\eta = g(\chi)\,d\mu_{\widehat{G}}(\chi)$. Since $\{\widehat{h} \mid h \in L^1\}$ is $\|\cdot\|_\infty$-dense in $C_\infty(\widehat{G})$, $d\eta$ is the zero measure, so $g = 0$.

(d) By linearity, we can suppose $f \in \mathcal{P}_+(G)$. Since Theorem 6.9.13 says

$$d\nu_f(\chi) = \widehat{f}(\chi)\, d\mu_{\widehat{G}}(\chi) \tag{6.9.84}$$

and $d\nu_f$ is a finite measure, $\widehat{f} \in L^1$. By (6.9.84), (6.9.17) is exactly (6.9.80).

(e) By (a), if $\widehat{g} \in L^1$, then $\check{g} \in \mathcal{P}$. Thus, if $h = \widehat{\check{g}}$, we have $h \in L^1$ and, by (d), $\check{h} = \check{g}$, that is, $(g-h)\check{} = 0$ so $h = g$ by (c). Thus, $\widehat{\check{g}} = g$. $\square$

We do the bulk of the proof of the Plancherel theorem in four lemmas (the difficulty will be showing that $\widehat{L^2}$ is all of L^2).

Lemma 6.9.15. *If $f \in \mathcal{C}(G)$, then $\widehat{f} \in L^2(\widehat{G}, d\mu_G)$ and*

$$\|\widehat{f}\|_{L^2(\widehat{G})} = \|f\|_{L^2(G)} \tag{6.9.85}$$

Proof. $\mathcal{C}(G) \subset \mathcal{P}(G)$ so $\widehat{f} \in L^1 \cap L^\infty \subset L^2$. If $g = \widehat{f}$, then $\check{g} = f$ (by (6.9.80)), so (6.9.79) is (6.9.85). $\square$

Lemma 6.9.16. *$\widehat{\ }$ defined on $L^1(G) \cap L^2(G)$ extends to an isometry of $L^2(G, d\mu)$ into $L^2(\widehat{G}, d\widehat{\mu})$. Moreover, for all $f, g \in L^2(G, d\mu)$,*

$$\langle \widehat{f}, \widehat{g} \rangle_{L^2(G)} = \langle f, g \rangle_{L^2(G)} \tag{6.9.86}$$

Proof. Since $f \in L^1 \cap L^2$ can be approximated by $f_n \in \mathcal{C}(G)$ so that one has convergence in both L^1- and L^2-norm, a simple argument (Problem 7) shows $\widehat{\ }$ defined by (6.9.2) has $\widehat{f} \in L^2(\widehat{G})$ and (6.9.85) holds. By a density argument, one can extend $\widehat{\ }$ uniquely to an isometry. (6.9.86) holds by polarization. $\square$

Lemma 6.9.17. *If $g \in L^1(\widehat{G}, d\mu_{\widehat{G}}) \cap L^2(\widehat{G}, d\mu_{\widehat{G}})$, then $\check{g} \in L^2$ and*

$$\|\check{g}\|_{L^2(G,d\mu)} \le \|g\|_{L^2(\widehat{G},d\mu_{\widehat{G}})} \tag{6.9.87}$$

Proof. Let $U = \widehat{\ }$ on $L^2(G, d\mu)$. Then (6.9.85) says $U^*U = 1$, so U^* obeys $\|U^*g\| \le \|g\|$ since UU^* is the projection onto $\text{Ran}(U)$. By using (6.9.79) for $f \in L^1(G, d\mu) \cap L^2(G, d\mu)$ and $g \in L^1(\widehat{G}) \cap L^2(\widehat{G})$, we see for $g \in L^1(\widehat{G}) \cap L^2(\widehat{G})$, $U^*g = \check{g}$. Thus, (6.9.87) holds for such g. $\square$

Lemma 6.9.18. *If $h, k \in L^1(\widehat{G}, d\mu_{\widehat{G}}) \cap L^2(\widehat{G}, d\mu_{\widehat{G}})$, then*

$$f = \check{h}\check{k} \in L^1(G, d\mu) \cap L^\infty(G, d\mu) \tag{6.9.88}$$

and

$$\widehat{f} = h * k \in L^1(\widehat{G}, d\mu_{\widehat{G}}) \cap C_\infty(\widehat{G}) \tag{6.9.89}$$

Proof. By Theorem 6.9.14(a) and Lemma 6.9.17,

$$\check{h},\check{k}\in\mathcal{B}(G)\cap L^2(G,d\mu)\subset L^\infty(G)\cap L^2(G,d\mu)$$

so (6.9.88) holds.

Moreover,

$$\begin{aligned}(h*k)^\vee(x)&=\int\chi(x)h(\chi_1)k(\chi\bar{\chi}_1)\,d\mu_{\widehat{G}}(\chi_1)d\mu_{\widehat{G}}(\chi)\\&=\check{h}(x)\check{k}(x)\equiv f(x)\end{aligned}\tag{6.9.90}$$

since $\chi=(\chi\bar{\chi}_1)\chi_1$ and $d\mu(\chi)=d\mu(\chi\bar{\chi}_1)$. This implies $f\in\mathcal{B}(G)$ by Theorem 6.9.14(a) and, as noted, $f\in L^1(G,d\mu)$, so $f\in\mathcal{P}(G)$.

Thus, $\widehat{f}\in L^1$ and $(\widehat{f})^\vee=f$. By (6.9.90) and Theorem 6.9.14(c), we have (6.9.89). □

Define $\mathcal{C}(\widehat{G})$, $\mathbb{W}(\widehat{G})$ analogously to G. Proposition 6.9.4 extends to $\widehat{G}$ since it only depends on the group structure.

Theorem 6.9.19 (Plancherel Theorem). $\widehat{\ }$ *defined initially on* $L^1(G,d\mu)\cap L^2(G,d\mu)$ *extends to a unitary map of* $L^2(G,d\mu)$ *onto all of* $L^2(\widehat{G},d\mu_{\widehat{G}})$. $\check{\ }$ *defined originally on* $L^1(\widehat{G},d\mu_{\widehat{G}})\cap L^2(\widehat{G},d\mu_{\widehat{G}})$ *extends to a unitary map in the other direction and is the two-sided inverse to* $\widehat{\ }$.

Proof. We've already seen that $\widehat{\ }$ is an isometry. If $q=h*k$ with $h,k\in L^2(\widehat{G},d\mu_{\widehat{G}})$ with compact support, then $h,k\in L^1(\widehat{G})\cap L^2(\widehat{G})$. So by Lemma 6.9.18, there is $f\in L^1(G)\cap L^2(G)$ so $\widehat{f}=h*k$, that is, $q\in\text{Ran}(\widehat{\ }\restriction L^2)$. Since $\mathcal{C}(G)$ is the linear span of such q's and is dense in $L^2(\widehat{G},d\mu_{\widehat{G}})$, Ran($\widehat{\ }$) is dense, and so all of $L^2(\widehat{G},d\mu_{\widehat{G}})$.

Since $\widehat{\ }$ is onto, it is unitary and its adjoint is its two-sided inverse. We've already seen that this adjoint on $L^1(\widehat{G})\cap L^2(\widehat{G})$ is $\check{\ }$. □

As a corollary, we can extend Lemma 6.9.18 and prove a dual version:

Lemma 6.9.20. *Let* $f,g\in L^2(G,d\mu)$. *Then*

$$\widehat{fg}=\widehat{f}*\widehat{g}\tag{6.9.91}$$

where $\widehat{\ }$ *on the left is the* L^1 *Fourier transform and* $\widehat{\ }$ *on the right is the* L^2 *Fourier transform. Let* $h,k\in L^2(\widehat{G},d\mu_{\widehat{G}})$. *Then*

$$(hk)^\vee=\check{h}*\check{k}\tag{6.9.92}$$

Proof. We'll prove (6.9.92). The proof of (6.9.91) is similar. Let $f=\check{h}$, $g=\check{k}$. Suppose $f,g\in L^1(G)\cap L^2(G)$. Then

$$\widehat{f*g}=\widehat{f}\widehat{g}\tag{6.9.93}$$

where all $\widehat{\ }$'s are L^1. But all functions are in $L^2(G)$ or $L^2(\widehat{G})$, so using $L^2 - \check{\ }$ and that $\widehat{\ }$ and $\check{\ }$ are L^2 inverse,

$$\check{h} * \check{k} \equiv f * g = (hk)\check{\ } \tag{6.9.94}$$

Since $L^1 \cap L^2$ is dense in L^1, we get the result for $f, g \in L^2$, and so $h, k \in L^2$. $\square$

The following is both an important consequence and links our definition of $\mathbb{W}(G)$ here to the earlier definition for the case $G = \mathbb{R}$ or $\mathbb{T}$.

Theorem 6.9.21 (Krein's Factorization Theorem). *We have*

(a) $\mathbb{W}(\widehat{G}) = \{\widehat{f} \mid f \in L^1(G, d\mu)\}$ (6.9.95)

(b) $\mathbb{W}(G) = \{\check{g} \mid g \in L^1(\widehat{G}, d\mu_{\widehat{G}})\}$ (6.9.96)

Remarks. 1. Once we have Pontryagin duality, these will be equivalent, but we'll use both halves in the proof of this duality result.

2. The same argument (Problem 8) shows that $\mathbb{W}_+(G) = \{\check{g} \mid g \in L^1,\ g \geq 0\}$.

Proof. $f \in L^1(G) \Leftrightarrow f = gq$ with $g, q \in L^2(G)$. But then, by (6.9.91), $\widehat{f} = \widehat{g} * \widehat{q} \in \mathbb{W}(\widehat{G})$. Conversely, if $h, k \in L^2(\widehat{G})$, $f = \check{h}\check{k}$ has $\widehat{f} = h * k$, so $h * k \in \{\widehat{f} \mid f \in L^1\}$. Thus, $\mathbb{W}(\widehat{G})$, which is generated by $\{h * k \mid h, k \in L^2(\widehat{G})\}$ is in $\{\widehat{f} \mid f \in L^1\}$. This proves (6.9.95). The proof for (6.9.96) is similar, using (6.9.92) in place of (6.9.91). $\square$

Corollary 6.9.22. *Any $g \in \mathbb{W}(\widehat{G})$ can be written*

$$g = k * h = \widehat{f} \tag{6.9.97}$$

where $f \in L^1(G)$, $h, k \in L^2(\widehat{G})$, and

$$\|\widehat{f}\|_{L^1(G)} = \|k\|^2_{L^2(\widehat{G})} \|h\|^2_{L^2(\widehat{G})} \tag{6.9.98}$$

and similarly for $L^1(\widehat{G})\mathbb{W}(G)$.

Remark. In particular, this shows $\{k * h \mid k, h \in L^2(\widehat{G})\}$ is a vector space.

Proof. By the theorem, $g = \widehat{f}$ with $f \in L^1(G)$. Let $k = \widehat{|f|^{1/2}}$ and $h = (f/|f|^{1/2})\widehat{\ }$ and use (6.9.91). $\square$

Corollary 6.9.23. *If $g \in L^1(\widehat{G}, d\mu_{\widehat{G}})$, then $\check{g} \in C_\infty(G)$.*

Remark. It is easy to see $\check{g}$ is a bounded continuous function. The new element is that it vanishes at ∞.

Proof. By the theorem, $\check{g} \in \mathbb{W}(G)$ is $f * g$ for $f, g \in L^2$. By approximating f, g by L^2-functions of compact support, we see $\check{g}$ is a $\|\cdot\|_\infty$ of continuous functions of compact support. $\square$

Corollary 6.9.24. *If M, N are disjoint and closed sets in $\widehat{G}$ and N is compact, there is $f \in L^1(G, d\mu)$ so $\widehat{f} \restriction N \equiv 1$, $\widehat{f} \restriction M = 0$. In particular, $L^1(G, d\mu)$ is a regular Banach algebra. A similar result is true for G with $\widehat{\ }$ replaced by $\check{\ }$.*

Proof. We proved this for $\mathbb{W}(G)$ in Proposition 6.9.4 and the proof of $\mathbb{W}(\widehat{G})$ is identical. Since $\mathbb{W}(\widehat{G}) = \{\widehat{f} \mid f \in L^1(G)\}$, we are done. $\square$

We are heading towards a proof of Pontryagin duality that $\widehat{\widehat{G}} = G$ as topological groups (under a natural embedding of G in $\widehat{\widehat{G}}$). To avoid too many $\widehat{\widehat{G}}$ and to emphasize the distinction between G related to $\widehat{G}$ and $\widehat{\widehat{G}}$ related to $\widehat{G}$, we'll denote $K = \widehat{G}$, in which case we are heading towards proving that $G = \widehat{K}$. We need one preliminary:

Lemma 6.9.25. *$x_n \to x$ in $G \Leftrightarrow \chi(x_n) \to \chi(x)$ for all $\chi \in \widehat{G}$.*

Proof. $\Rightarrow$ is just continuity of each $\chi \in \widehat{G}$.

$\Leftarrow$. Suppose for $\{x_n\}_{n=1}^\infty$, $x \in G$ and $\chi(x_n) \to \chi(x)$ for every $\chi \in \widehat{G}$. By the dominated convergence theorem, for every $k \in L^1(\widehat{G}, d\mu_{\widehat{G}})$, $\check{k}(x_n) \to \check{k}(x)$. By Corollary 6.9.24, for any open neighborhood U of x, there is $\check{k}$ so $\check{k}(y) = 0$ if $y \notin U$ and $\check{k}(x) = 1$. Since $\check{k}(x_n) \to \check{k}(x) = 1$ for large n, $|\check{k}(x_n) - 1| < \frac{1}{2} \Rightarrow x_n \in U$ for n large. Since U is arbitrary, $x_n \to x$. $\square$

Let $K = \widehat{G}$ and $\alpha\colon G \to \widehat{K}$ by

$$\alpha(x)(\chi) = \chi(x) \tag{6.9.99}$$

Theorem 6.9.26 (Pontryagin Duality). *α is a homomorphism and homeomorphism of G onto $\widehat{K}$, that is,*

$$\widehat{K} \cong G \tag{6.9.100}$$

Proof. α is obviously a homomorphism. By the lemma and Theorem 6.9.6, $x_n \to x \Leftrightarrow \alpha(x_n) \to \alpha(x)$, so we need only show α is a bijection. By Theorem 6.9.5(b), α is one–one.

We first claim $\operatorname{Ran}(\alpha)$ is closed in $\widehat{K}$. For suppose $\alpha(x_n) \to y \in \widehat{K}$. Find $k \in L^1(K, d\mu_K)$, so $\widehat{k}(y) = 1$. Since

$$\widehat{k}(\alpha(x)) = \check{k}(-x) \tag{6.9.101}$$

we know that $\check{k}(-x_n) = \widehat{k}(\alpha(x_n)) \to \widehat{k}(y) = 1$, so for n large, $x_n \in Q = \{x \mid |\check{k}(-x) - 1| \le \frac{1}{2}\}$ which is compact by Corollary 6.9.23. Thus, there is a subsequence and $x_\infty \in G$ so $x_{n(i)} \to x_\infty$. Thus, since α is continuous, $\alpha(x_{n(i)}) \to \alpha(x_\infty)$, so $y = \alpha(x_\infty)$, that is, $y \in \mathrm{Ran}(\alpha)$.

Finally, suppose that $\mathrm{Ran}(\alpha)$ is not all of $\widehat{K}$. Pick $y \in \widehat{K} \setminus \mathrm{Ran}(\alpha)$. By Corollary 6.9.24, there is $k \in L^1(K, d\mu_K)$ so that $\widehat{k} \restriction \mathrm{Ran}(\alpha) = 0$, but $\widehat{k}(y) = 1$.

By (6.9.101), $\check{k} \equiv 0$, so by Theorem 6.9.14(c), $k = 0$. Thus, $\widehat{k} = 0$ so $\widehat{k}(y) = 0$. This contradiction shows that $\mathrm{Ran}(\alpha) = \widehat{K}$, so α is a bijection. $\square$

Notes and Historical Remarks. The theory of Fourier analysis on LCA groups was developed in the 1930s prior to the theory of Banach algebras, but then absorbed into the Gel'fand framework. Key figures in the early phase were van Kampen, Pontryagin, Weil, and Krein, and in the later phase Raikov and Godement.

Our approach here of developing pdf's first, then applying that to the Fourier inversion and Plancherel theorems, and finally getting Pontryagin duality is almost the exact opposite of the historical where the duality theory based on structure theorems came first, then Plancherel and Fourier inversion usually together, and finally pdf's almost as an afterthought. During the Banach algebra consolidation though, the central role of the Bochner–Weil theorem in an abstract approach came to the fore.

After an early short exploration of characters on general abelian topological groups by Paley–Wiener [**509**], the general theory was developed in papers by van Kampen [**712**] and Pontryagin [**536**] in 1935. They put topologies on $\widehat{G}$ (Pontryagin used the topology, now called the *Pontryagin topology*, of uniform convergence on compact subsets of G; van Kampen used an equivalent topology) and proved what is usually called Pontryagin duality (but, by some, Pontryagin–van Kampen duality). That the Pontryagin topology is the same as pointwise convergence is a result of Yosida–Iwamura [**769**] and that it is the same as the Gel'fand topology a result of Raikov [**543**].

General Fourier analysis was then developed by Weil [**735**] in a book that also had the first construction of Haar measure on nonseparable groups and by Krein [**409**]. Weil used the structure theory of LCA groups while Krein exploited Gel'fand's then new theory of abelian Banach algebras and pdf's (see also Gel'fand–Raikov [**233**]). In particular, Krein's paper has what we call Krein's factorization (Theorem 6.9.21) and Weil's book what we call the Bochner–Weil theorem (also found by Povzner [**537**] and

Raikov [**543**]). Development of the theory from a Banach algebra perspective was accomplished by Gel'fand–Raikov–Shilov [**236**], Godement [**254**], and Loomis [**456**].

Our proof of Pontryagin duality follows Raikov [**544**] with one exception. In place of the proof that $\mathrm{Ran}(\alpha)$ is closed, Raikov uses the argument in Problem 10.

Among book treatments of Fourier analysis on LCA groups are [**456, 554, 583, 316**].

One consequence of the Pontryagin duality theorem is that for a compact abelian subgroup, G, the relation of the irreducible representation under tensor product determines the group (since it determines the dual group whose topology is discrete). There is a generalization of this to arbitrary compact groups called Tannaka–Krein duality after [**696, 411**]; see the discussion in Hewitt–Ross [**316**, Vol. 2].

If G is an LCA group, let $\widehat{G}_{\text{disc}}$ be $\widehat{G}$ given the discrete topology. Then $(\widehat{G}_{\text{disc}})\widehat{\ }$ is a compact group containing G since any $x \in G$ defines a character on $\widehat{G}_{\text{disc}}$ by $x(\chi) = \chi(x)$ which is continuous since all functions are continuous in the discrete topology. Since the identity map of $\widehat{G}_{\text{disc}}$ to $\widehat{G}$ is continuous, by duality the induced map of $G = (\widehat{G})\widehat{\ }$ into $(\widehat{G}_{\text{disc}})\widehat{\ }$ is continuous. This is, of course, another construction of the Bohr compactification, $\mathcal{B}(G)$, discussed in Section 6.6.

Problems

1. If G, an LCA group, is σ-compact and second countable, prove that $L^p(G, d\mu)$ is separable for each $p < \infty$.

2. Let f be a continuous function of compact support on G, an LCA group.

   (a) For each $x \in G$, show there is a neighborhood, U_x, of 0 so that $y - x \in U_x \Rightarrow |f(x) - f(y)| < \varepsilon/2$.

   (b) Let $\{U_{x_j}\}_{j=1}^n$ cover $\mathrm{supp}(f)$. Let $V = \bigcap_{j=1}^n U_{x_j}$. Prove that $y - z \in U \Rightarrow |f(z) - f(y)| < \varepsilon$.

3. If Φ is given by (6.9.25) and $\Phi((0,1)) = \|\varphi\|_\infty$, prove that $\nu_\Phi(\{\infty\}) = 0$.

4. Let $1 < p < \infty$ and q the dual index. Prove that if G is an LCA group, $f \in L^p(G, d\mu)$ and $h \in L^q(G, d\mu)$, then $f * h \in C_\infty(G)$. (*Hint*: Use density of the continuous functions of compact support.)

5. Let G be a σ-compact LCA group. Let $K_j \subset K_{j+1}^{\text{int}}$ with K_j compact and $\cup K_j = G$. If $f \in L^\infty$ and $f_j = f\chi_{K_j}$ and if $g \in L^1$, prove that $\int g(x)[f(x) - f_j(x)]\, d\mu(x) \to 0$.

6. Let $x_n \in \mathbb{R}$ so that $e^{ikx_n} \to 1$ for all $k \in \mathbb{R}$. Prove that $x_n \to 0$. (*Hint*: Prove that for f in $\mathcal{S}(\mathbb{R})$, $\widehat{f}(x_n) \to \widehat{f}(0)$ and conclude that x_n lies in a compact set.)

7. Fill in the details of the proof of Lemma 6.9.16.

8. Prove that $\mathbb{W}_+(G) = \{f^* * f \mid f \in L^2(G, d\mu)\}$ is the same as $\{\check{g} \mid g \in L^1(\widehat{G}, d\mu_{\widehat{G}}),\, g \geq 0\}$.

9. This problem will show a metrizable LCA group where $L^1(G, d\mu)$ has an identity is a discrete group. Suppose e is the identity and $\{f_{U_n}\}_{n=1}^\infty$ an approximate identity with $\text{supp}(f_{U_n}) \subset U_n$, where U_n is a neighborhood of 0 with $U_{n+1} \subset U_n$ and $\bigcap_{n=1}^\infty U_n = \{0\}$.

   (a) Prove that $\lim_{n\to\infty} \|e - f_{U_n}\|_1 = 0$. (*Hint*: $f_{U_n} = f_{U_n} * e$.)

   (b) Prove that if $\mu(\{0\}) = 0$, then $\|e\|_1 = 0$, so conclude $\mu(\{0\}) > 0$.

   (c) Prove that G is discrete.

10. This problem will provide a partially alternate proof that the map α in Theorem 6.9.26 has $\text{Ran}(\alpha) = \widehat{K}$ without proving $\text{Ran}(\alpha)$ is closed; the argument follows Raikov [**544**].

    (a) Use the argument in the third paragraph of the proof of Theorem 6.9.26 to show if $V \subset \widehat{K}$ is nonempty and open, then $V \cap \text{Ran}(\alpha) \neq \emptyset$. (*Hint*: If $V \cap \text{Ran}(\alpha) = \emptyset$, find $f \in L^1(K)$ and $\widehat{f}(x_0) = 1$ for some $x_0 \in V$ and $\widehat{f} \restriction \widehat{K} \setminus V = 0$.)

    (b) Pick $U \subset G$ and a neighborhood of 0 so $\overline{U}$ is compact. Show there is $V \subset \widehat{K}$ open so $V \cap \alpha[G] = \alpha[U]$.

    (c) Prove that any point in V lies in $\overline{\alpha[U]}$. (*Hint*: Use (a).)

    (d) Prove $\overline{\alpha[U]} = \alpha[\overline{U}]$ and conclude $V \subset \text{Ran}(\alpha)$.

    (e) Fix $x \in \widehat{K}$. Find $y \in \text{Ran}(\alpha)$ so $y \in x + V$ (*Hint*: Use (a).)

    (f) Conclude $x \in \text{Ran}(\alpha)$ so that $\widehat{K} = \text{Ran}(\alpha)$.

6.10. Bonus Section: Introduction to Function Algebras

Recall that a function algebra is a Banach algebra, $\mathfrak{A}$, of functions on a compact Hausdorff space, X, with $\mathbb{1} \in \mathfrak{A}$, with $\|\cdot\|_\infty$-norm and which separates points of X. For many of the most studied examples, X is a subset of $\mathbb{C}$, including $\mathbb{P}(X)$, the closure of the polynomials in z, $\mathbb{R}(X)$, the closure of the rational functions with poles in $\mathbb{C} \setminus X$, and $\mathbb{A}(X)$, the functions continuous on X and analytic in X^{int}.

Our first goal will be to study boundaries of general abelian Banach algebras, $\mathfrak{A}$, subsets of $\widehat{\mathfrak{A}}$, where every $\widehat{f}$ takes its maximum modulus. If

$\mathfrak{A}$ is a function algebra and Y is a closed boundary, then $f \mapsto \widehat{f} \restriction Y$ is an isometric isomorphism of $\mathfrak{A}$ to a function algebra on Y. A key result will be the existence of a smallest closed boundary called the Shilov boundary.

Many of the most significant results in the theory involve density results where the key tool will be the existence of annihilating measures when density fails, for if $Q \subset \mathfrak{A}$ and Q is not dense, there will be $\ell \in \mathfrak{A}^*$, $\ell \neq 0$, so $\ell \restriction Q = 0$. By the Hahn–Banach theorem, ℓ will extend to $C(Y)^*$ for any boundary and so define a measure. We'll illustrate some of the techniques by proving four theorems:

Theorem 6.10.1 (Hartogs–Rosenthal Theorem). *If $X \subset \mathbb{C}$ is a compact set of zero (two-dimensional) Lebesgue measure, then*

$$\mathbb{R}(X) = C(X) \tag{6.10.1}$$

Theorem 6.10.2 (Lebesgue–Walsh Theorem). *Let X be a compact subset of $\mathbb{C}$ so that $\mathbb{C} \setminus X$ has m bounded components (m finite), $G_1, \dots, G_m$. Pick $a_j \in G_j$. Then any real-valued continuous function on ∂X is a uniform limit of u's of the form $\operatorname{Re} f + \sum_{j=1}^m \beta_j \log|z - a_j|$, where f is a rational function with poles only at the a_j. In particular, if $m = 0$ (i.e., X is simply connected), then $\operatorname{Re}[\mathbb{P}(X)]$ is dense in $\mathbb{C}_{\mathbb{R}}(X)$.*

This theorem has consequences for solvability of the Dirichlet problem for X^{int}, as we'll see.

Theorem 6.10.3 (Lavrentiev's Theorem). *Let X be a compact subset of $\mathbb{C}$. Then*

$$\mathbb{P}(X) = C(X) \tag{6.10.2}$$

if and only if X has empty interior and $\mathbb{C} \setminus X$ is simply connected.

Theorem 6.10.4 (Mergelyan's Theorem). *Let $K \subset \mathbb{C}$ be a compact set so that $\mathbb{C} \setminus K$ is connected. Then $\mathbb{P}(K) = \mathbb{A}(K)$, that is, any function f continuous in K and analytic in K^{int} is a uniform limit (on K) of polynomials in z.*

Remark. The "if" part of Theorem 6.10.3 is a special case of this result. We prove that result separately as a warmup for Theorem 6.10.4 which is, by far, the most subtle of these four theorems.

We start with the study of boundaries:

Definition. Let $\mathfrak{A}$ be an abelian Banach algebra with identity. A *boundary* for $\mathfrak{A}$ is a subset $K \subset \widehat{\mathfrak{A}}$ so that for all $f \in \mathfrak{A}$, there is $\ell \in K$ with

$$|\widehat{f}(\ell)| = \sup_{k \in \widehat{\mathfrak{A}}} |\widehat{f}(k)| \tag{6.10.3}$$

Notice that $\overline{\text{Ran}(\widehat{\ })}$ is a function algebra, $\mathfrak{A}_1$, and that (Problem 1) if K is a boundary for $\mathfrak{A}_1$, it is one for $\mathfrak{A}$; and that if K is a closed boundary for $\mathfrak{A}$, it is a boundary for $\mathfrak{A}_1$. Thus, we can and will focus on function algebras. For a function algebra, $\|f^n\| = \|f\|^n$, so by the spectral radius formula, in a function algebra,

$$\|\widehat{f}\|_\infty = \|f\| \tag{6.10.4}$$

and so (6.10.3) says, for a function algebra, that

$$|\widehat{f}(\ell)| = \|f\| \tag{6.10.5}$$

Example 6.10.5. Let $K \subset \mathbb{C}$ be compact. By the maximum principle, ∂K is a boundary for $\mathbb{A}(K)$. Moreover, for each $z_0 \in \partial K$, by considering $z_n \in \mathbb{C} \setminus K$ with $z_n \to z$ and $f_n(z) = (z - z_n)^{-1}$, one sees (Problem 2) that z_0 lies in any closed boundary. So $\mathbb{A}(K)$ has a smallest closed boundary, namely ∂K, something we'll see holds generally. This example also explains the reason for the name "boundary." If $\mathbb{A}(\overline{\mathbb{D}} \times \overline{\mathbb{D}})$ is the space of functions continuous on $\overline{\mathbb{D}} \times \overline{\mathbb{D}}$ and analytic in $\mathbb{D} \times \mathbb{D}$, it can be shown (Problem 3) that $\partial\mathbb{D} \times \partial\mathbb{D}$, which is much smaller than the topological boundary of $\overline{\mathbb{D}} \times \overline{\mathbb{D}}$ (which is $(\partial\mathbb{D} \times \overline{\mathbb{D}}) \cup (\overline{\mathbb{D}} \times \partial\mathbb{D})$), is the minimal closed boundary. □

One reason boundaries are important is:

Proposition 6.10.6. *Let K be a closed boundary for a function algebra, $\mathfrak{A}$. Then $f \mapsto \widehat{f} \restriction K$ is an isometric isomorphism of $\mathfrak{A}$ and a function algebra on K.*

Proof. As noted in (6.10.5), $\|\widehat{f} \restriction K\|_\infty = \|f\|$. □

Example 6.10.5 (revisited). Thus, $\mathbb{A}(\overline{\mathbb{D}})$ can be viewed as a function algebra on $\partial\mathbb{D}$. Indeed (Problem 4), it is those continuous f's with $f_n^\sharp = 0$ for $n < 0$. □

Theorem 6.10.7. *Let $\mathfrak{A}$ be an abelian Banach algebra with identity. Then there is a closed boundary contained in any other closed boundary.*

Remarks. 1. This minimum closed boundary is called the *Shilov boundary.*

2. As noted, the closed boundary for $\mathfrak{A}$ and $\overline{(\text{Ran}(\widehat{\ }))}$ are the same, so we can and will restrict ourselves in the proof to function algebras.

3. We'll give two proofs that illustrate different aspects of the set. The first will be "hands on" and the second relies on the structure of certain compact subsets of $\mathfrak{A}^*$.

First Proof. We'll first show that minimal closed boundaries exist by a Zornification and then that this minimal boundary lies in every closed boundary, so is the intersection of all closed boundaries.

If $\{Y_\alpha\}_{\alpha\in I}$ is a chain of closed boundaries, that is, for all α,β, either $Y_\alpha \subset Y_\beta$ or $Y_\beta \subset Y_\alpha$, then $\bigcap_{\alpha\in I} Y_\alpha$ is a closed boundary. For if $f \in \mathfrak{A}$, then $y_\alpha \in Y_\alpha$ with $f(y_\alpha) = \|f\|_\infty$. Order I so $\alpha \rhd \beta$ if $Y_\alpha \subset Y_\beta$. The net $\{y_\alpha\}_{\alpha\in I}$ has a limit point y_∞. By continuity, $f(y_\infty) = \|f\|_\infty$ and $y_\infty \in \bigcap_{\alpha\in I} Y_\alpha$. Thus, $\bigcap_{\alpha\in I} Y_\alpha$ is a closed boundary. By Zorn's lemma (see Section 1.5 of Part 1), there are minimal closed boundaries. Let Y_∞ be such a minimal closed boundary.

Let $y \in Y_\infty$ and let U be an open set of the form

$$U = \{\ell \in \widehat{\mathfrak{A}} \mid |\widehat{f_1}(\ell)| < \varepsilon, \dots, |\widehat{f_n}(\ell)| < \varepsilon\} \tag{6.10.6}$$

for some $\varepsilon > 0$ and $f_1,\dots,f_n \in \mathfrak{A}$. Suppose $y \in U$. Since Y_∞ is minimal, $Y_\infty \setminus U$ is not a boundary. Thus, there is $f \in \mathfrak{A}$ so $\|\widehat{f}\|_\infty = 1$, but

$$\sup_{\ell\in Y_\infty\setminus U} |\widehat{f}(\ell)| < 1 \tag{6.10.7}$$

By replacing f by f^n, we can arrange for the sup in (6.10.7) to be as small as we want. So we can pick f so $\|\widehat{f}\|_\infty = 1$, but

$$\sup_{\ell\in Y_\infty\setminus U} |\widehat{f}(\ell)| \le \varepsilon/\bigl(2 \max_{j=1,\dots,n} (\|\widehat{f_j}\|)\bigr) \tag{6.10.8}$$

Let $g_j = f_j f$. By (6.10.8), on $Y_\infty \setminus U$, $|\widehat{g_j}| \le \frac{1}{2}\varepsilon$. By (6.10.6) and $\|\widehat{f}\|_\infty = 1$, $|\widehat{g_j}| < \varepsilon$ on U. Thus, since Y_∞ is a boundary, $\|\widehat{g_j}\|_\infty < \varepsilon$.

Let $\widetilde{Y}$ be any boundary and let $\tilde{y} \in \widetilde{Y}$ be such that $\widehat{f}(\tilde{y}) = 1$. Then $|\widehat{f_j}(\tilde{y})| = |\widehat{g_j}(\tilde{y})| < \varepsilon$ so $\tilde{y} \in U$. Thus, $\widetilde{Y}$ intersects U.

The set of U's obeying (6.10.6) and containing y is a neighborhood base for y, so $y \in \overline{\widetilde{Y}}$. Thus, Y_∞ is in any closed boundary. □

For the second proof, we consider the following:

$$\mathcal{R}(\mathfrak{A}) = \{\varphi \in \mathfrak{A}^* \mid \|\varphi\| = \varphi(e) = 1\} \tag{6.10.9}$$

The reader will recognize these conditions; for $\mathfrak{A} = C(X)$, $\mathcal{R}(C(X))$ is the probability measures on X. For any C^*-algebra, it is the set of states which played such an important role in Section 6.7.

Proposition 6.10.8. (a) $\widehat{\mathfrak{A}} \subset \mathcal{R}(\mathfrak{A})$

(b) *$\mathcal{R}(\mathfrak{A})$ is a convex set, compact in the $\sigma(\mathfrak{A}^*,\mathfrak{A})$-topology.*

(c) *If $\mathfrak{A}$ is a function algebra on X and φ_0 is an extreme point of $\mathcal{R}(\mathfrak{A})$, then for a unique $x_0 \in X$,*

$$\varphi_0(f) = f(x_0) \tag{6.10.10}$$

that is, every extreme point is a point evaluation.

Remarks. 1. The proof of (c) is essentially identical to the case $\mathfrak{A} = C(X)$, that is, that extreme points of $\mathcal{M}_{+,1}(X)$ are point measures; see Problem 4 of Section 5.11 of Part 1.

2. A similar argument to that in (c) shows any $\varphi \in \mathcal{R}(\mathfrak{A})$ is of the form

$$\varphi(f) = \int_{\widehat{\mathfrak{A}}} \widehat{f}(x)\, d\mu(x) \tag{6.10.11}$$

for some probability measure μ on $\widehat{\mathfrak{A}}$. It is not hard to see (Problem 5) another characterization of the extreme points of $\mathcal{R}(\mathfrak{A})$ are those φ for which the measure μ in (6.10.11) is unique. This is another definition of the Choquet boundary.

3. Evaluation is a mlf so (c) implies that extreme points of $\mathcal{R}(\mathfrak{A})$ are elements of $\widehat{\mathfrak{A}}$.

Proof. (a) If $\ell \in \widehat{\mathfrak{A}}$, $\ell(\mathfrak{e}) = 1$ is immediate and $\|\ell\| = 1$ follows from $\|\widehat{f}\|_\infty = \|f\|$.

(b) If $\varphi_\alpha \to \varphi$ with $\varphi_\alpha \in \mathcal{R}(\mathfrak{A})$, then $\|\varphi\| \le 1$. Since $\varphi(\mathfrak{e}) = 1$, we see $\|\varphi\| = 1$, showing $\mathcal{R}(\mathfrak{A})$ is a closed subset of the unit ball which is compact. Similarly, if $\varphi = \theta\varphi_1 + (1-\theta)\varphi_2$ with $\theta \in [0,1]$, $\varphi_1, \varphi_2 \in \mathcal{R}(\mathfrak{A})$, then $\|\varphi\| \le 1$, $\varphi(\mathfrak{e}) = 1$, so $\|\varphi\| = 1$ and $\varphi \in \mathcal{R}(\mathfrak{A})$.

(c) By the Hahn–Banach theorem, φ_0 can be extended to $\widetilde{\varphi}_0 \in C(X)^*$, so $\widetilde{\varphi}_0(\mathbb{1}) = \|\widetilde{\varphi}_0\| = 1$. $\widetilde{\varphi}_0$ is thus given by a measure, that is, there is a probability measure μ_0 on X so that

$$\varphi_0(f) = \int_X f(x)\, d\mu_0(x) \tag{6.10.12}$$

Pick $x_0 \in \operatorname{supp}(d\mu)$, so $\mu(A) > 0$ for all open F_σ sets, A, containing x_0. We claim for all such A,

$$\varphi_0(f) = \frac{1}{\mu(A)} \int_A f(x)\, d\mu_0 \tag{6.10.13}$$

If $\mu(A) = 1$, this is (6.10.12). If $\mu(A) < 1$, let $\widetilde{\varphi}_0^{(A)}$ be the right side of (6.10.13) and note

$$\varphi_0 = \mu(A)\widetilde{\varphi}_0^{(A)} + (1-\mu(A))\widetilde{\varphi}_0^{(X\setminus A)} \tag{6.10.14}$$

Since φ_0 is an extreme point $\varphi_0 = \widetilde{\varphi}_0^{(A)}$, that is, (6.10.13) holds.

By continuity, for any f and any ε, we can find A so $|f(x) - f(x_0)| < \varepsilon$ on A. Thus, by (6.10.13),

$$|\varphi_0(f) - f(x_0)| < \varepsilon \tag{6.10.15}$$

Since ε is arbitrary, (6.10.10) holds. Since $\mathfrak{A}$ separates points of X, x_0 is unique. $\square$

Definition. Let $\mathfrak{A}$ be a function algebra on X. The set of $x_0 \in X$ so that φ_0 given by (6.10.10) is an extreme point of $\mathcal{R}(\mathfrak{A})$ is called the *Choquet boundary* of $\mathfrak{A}$ and denoted $\mathrm{Ch}(\mathfrak{A})$.

Theorem 6.10.9. (a) *Let $\mathfrak{A}$ be a function algebra.* $\mathrm{Ch}(\mathfrak{A})$ *is a boundary for* $\mathfrak{A}$.

(b) $\mathrm{Ch}(\mathfrak{A})$ *is a subset of any closed boundary of* $\mathfrak{A}$.

Remark. Boundaries are subsets of $\widehat{\mathfrak{A}}$, while the Choquet boundary is a subset of X. We associate X with $\widehat{\mathfrak{A}}$ via (6.10.10) and so view the Choquet boundary as a subset of $\widehat{\mathfrak{A}}$.

Proof. (a) Let $f \in \mathfrak{A}$. Pick $\lambda \in \partial\mathbb{D}$, so $\exists x_1 \in X$ so that $\lambda f(x_1) = \|f\|_\infty$. Let

$$\mathcal{F}_f = \{\varphi \in \mathcal{R}(\mathfrak{A}) \mid \lambda\varphi(f) = \|f\|_\infty\} \tag{6.10.16}$$

Then $\mathcal{F}_f$ is nonempty since evaluation at x_1 is in $\mathcal{R}(\mathfrak{A})$ and obeys $\lambda\varphi(f) = \|f\|_\infty$. Since $\varphi \mapsto \mathrm{Re}[\lambda\varphi(f)]$ is linear and continuous and $\mathrm{Re}(\lambda\varphi(f)) \le \|f\|_\infty$ for all φ, we see $\mathcal{F}_f$ is a face of $\mathcal{R}(\mathfrak{A})$. Thus, it contains an extreme point of $\mathcal{R}(\mathfrak{A})$ by Proposition 5.11.5 of Part 1. Therefore, for some x in $\mathrm{Ch}(\mathfrak{A})$, $\lambda f(x) = \|f\|_\infty$ and $|f(x)| = \|f\|_\infty$.

(b) Let Y be a closed boundary for $\mathfrak{A}$. By Proposition 6.10.6, $\mathfrak{A}$ can be viewed as a function algebra on Y. Thus, by Proposition 6.10.8, any extreme point in $\mathcal{R}(\mathfrak{A})$ is evaluation by some $y \in Y$. Thus, elements of $\mathrm{Ch}(\mathfrak{A})$ are points in Y, that is, $\mathrm{Ch}(\mathfrak{A}) \subset Y$. □

Second Proof of Theorem 6.10.7. $\overline{\mathrm{Ch}(\mathfrak{A})}$ is a closed boundary contained in any closed boundary, that is, the Shilov boundary is $\overline{\mathrm{Ch}(\mathfrak{A})}$. □

Example 6.10.10. This will present a function algebra, $\mathfrak{A}$, for which $\mathrm{Ch}(\mathfrak{A})$ is not closed, so the Choquet and Shilov boundaries can differ. Start with disk algebra $\mathbb{A}(\overline{\mathbb{D}})$ of functions continuous on $\overline{\mathbb{D}}$ and analytic on $\mathbb{D}$. Let

$$\mathfrak{A} = \{f \in \mathbb{A}(\overline{\mathbb{D}}) \mid f(0) = f(1)\} \tag{6.10.17}$$

which has codimension 1 in $\mathbb{A}(\overline{\mathbb{D}})$ and which is obviously an algebra and norm-closed. $\mathfrak{A}$ can be viewed as an algebra on $\partial\mathbb{D}$ since, by the maximum principle, $f \restriction \partial\mathbb{D}$ determines f and is norm-preserving. Indeed, as a function algebra on $\partial\mathbb{D}$, $\mathfrak{A}$ is given by (Problem 7)

$$\mathfrak{A} = \left\{ f \in C(\partial\mathbb{D}) \,\middle|\, \int e^{-in\theta} f(\theta)\,\frac{d\theta}{2\pi} = 0 \text{ if } n < 0, = f(1) \text{ if } n = 0 \right\} \tag{6.10.18}$$

It is easy to see (Problem 8) that if $y \in \partial\mathbb{D} \setminus \{1\}$, there exists $f \in \mathfrak{A}$ with $f(y) > f(x)$ for all $x \in \partial\mathbb{D} \setminus \{y\}$ (y is called a *peak point*), which implies

$y \in \mathrm{Ch}(\mathfrak{A})$ (Problem 6). On the other hand, $1 \notin \mathrm{Ch}(\mathfrak{A})$ since

$$f(1) = \int_0^{2\pi} f(e^{i\theta})\, \frac{d\theta}{2\pi} \tag{6.10.19}$$

implies δ_1 is not an extreme point (Problem 8). Thus, $\mathrm{Ch}(\mathfrak{A}) = \partial\mathbb{D} \setminus \{1\}$ is not closed and its closure, $\partial\mathbb{D}$, is the Shilov boundary. $\square$

Let $\mathfrak{A}$ be a function algebra, $\ell \in \widehat{\mathfrak{A}}$, and S the Shilov boundary. Then

$$|\ell(f)| = |\widehat{f}(\ell)| \le \|f\|_{\mathfrak{A}} = \sup\|\widehat{f}\|_S \tag{6.10.20}$$

Thus, ℓ is a well-defined linear functional on $\{\widehat{f} \mid f \in \mathfrak{A}\}$ which is a closed subspace of $C(S)$. Since $\ell(1) = \|\ell\| = 1$ as a functional on $\{\widehat{f} \mid f \in \mathfrak{A}\}$, by the Hahn–Banach theorem, there is an extension to $C(S)$ with norm 1. Thus, there is a probability measure, μ_ℓ, on S with

$$\ell(f) = \int_S \widehat{f}(q)\, d\mu_\ell(q) \tag{6.10.21}$$

It is called a *representing measure* for ℓ. In particular, on $\mathbb{A}(\mathbb{D})$, $S = \partial\mathbb{D}$, so if ℓ is evaluation at some $re^{i\varphi} \in \mathbb{D}$, then $d\mu_\ell(\theta) = P_r(\theta, \varphi)\frac{d\theta}{2\pi}$ with P_r the Poisson kernel, so representing measures extend the notion of the Poisson representation.

The above didn't use that $\mathfrak{A}$ was an algebra. Indeed, if one has a subspace, $Y \subset C(X)$ with $1 \in Y$ and $\ell\colon Y \to \mathbb{C}$ with $\ell(1) = \|\ell\| = 1$, then by the Hahn–Banach theorem, there is a measure, μ_ℓ, on X so that

$$\ell(f) = \int f(q)\, d\mu_\ell(q) \tag{6.10.22}$$

In particular, if K is compact in $\mathbb{C}$ and $\widetilde{Y}$ is the set of real-valued $u\colon K \to \mathbb{R}$ continuous on K and harmonic on K^{int}, then u is determined by $u \restriction \partial K$ by the maximum principle. Letting $X = \partial K$ and $Y \subset C_{\mathbb{R}}(X)$, the set of $u \restriction X$, then by the maximum principle, if $z \in K$, $\ell(u) = u(x)$ obeys $\|\ell\| = 1$ and, of course, since $u \equiv 1$ is harmonic, $\ell(1) = 1$. Thus, there is a probability measure, $d\mu_z$ on ∂K so that for $u \in \widetilde{Y}$,

$$u(z) = \int u(w)\, d\mu_z(w) \tag{6.10.23}$$

It is called *harmonic measure*, studied in Chapter 3 of Part 3.

We now turn to the various density theorems for algebras on subsets of $\mathbb{C}$. The key idea is the following: If $Y \subset Z \subset C(X)$, where Y, Z are subspaces with Z closed, then Y is dense in Z if and only if for all $\mu \in C(X)^*$, $\mu \restriction Y = 0 \Rightarrow \mu \restriction Z = 0$. For, if Y is not dense, the Hahn–Banach theorem implies there is $\varphi \in Z^*$, $\varphi \neq 0$, so that $\varphi \restriction Y = 0$. By the Hahn–Banach theorem again, φ has an extension to $\mu \in C(X)^*$. Thus, we'll need to prove certain measures are 0 or at least obey $\int f(z)\, d\mu(z) = 0$ for $f \in Z$. The tools we'll use rely on two functions defined by a finite (complex) Borel measure μ on $\mathbb{C}$.

Definition. The *Cauchy transform*, $\widehat{\mu}$, of a finite measure on $\mathbb{C}$ is defined by

$$\widehat{\mu}(z) = \int \frac{d\mu(w)}{w-z} \tag{6.10.24}$$

and the *potential* of μ by

$$\check{\mu}(z) = \int \log(|z-w|)\, d\mu(w) \tag{6.10.25}$$

Remarks. 1. $\widehat{\ }$ and $\check{\ }$ differ radically from our use for Fourier transform and will only be used in this section. $\check{\mu}$ is, up to a factor of 2π, what we called Φ_μ in Section 3.1 of Part 3.

2. $\widehat{\mu}$ and $\check{\mu}$ are only defined on the set where the integrals are absolutely convergent, for example, $\widehat{\mu}$ where

$$\int \frac{d|\mu|(w)}{|z-w|} < \infty \tag{6.10.26}$$

We'll shortly show this is true for Lebesgue a.e. z.

Proposition 6.10.11. *Let μ be a finite measure of compact support in $\mathbb{C}$. Then*

(a) *$\widehat{\mu}$ and $\check{\mu}$ are defined for d^2z-a.e. z and are locally $L^1(\mathbb{C}, d^2z)$-functions. Thus, they define distributions.*
(b) *We have, as distributions,*

$$\widehat{\mu}(z) = -2\partial(\check{\mu})(z) \tag{6.10.27}$$

$$-\bar{\partial}(\widehat{\mu})(z) = \tfrac{1}{2}\,(\Delta\check{\mu})(z) = \pi\mu \tag{6.10.28}$$

(c) *On $\mathbb{C}\setminus\operatorname{supp}(\mu)$, $\widehat{\mu}$ and $\check{\mu}$ are defined, $\check{\mu}$ is harmonic, and $\widehat{\mu}$ is analytic and* (6.10.27) *holds in the classical sense.*
(d) *If $\check{\mu}$ or $\widehat{\mu}$ is zero for a.e. z, then μ is zero.*

Proof. (a) Let $M = \operatorname{supp}(\mu)$. Fix R. Then

$$\sup_{w\in M} \int_{|z|\le R} \frac{d^2z}{|z-w|} = C < \infty \tag{6.10.29}$$

Thus,

$$\int_{|z|\le R} \frac{d^2z\, d\mu(w)}{|z-w|} < \infty \tag{6.10.30}$$

so (6.10.26) holds for d^2z-a.e. z with $|z| < R$, and so varying R for a.e. z. Moreover, (6.10.30) implies $\widehat{\mu}$ is L^1 over $|z| < R$. The argument for $\check{\mu}$ is similar.

(b) As noted [(6.2.35)], we have as distributions that

$$\partial \log|z|^{-1} = -\tfrac{1}{2}\, z^{-1}, \qquad \bar{\partial} z^{-1} = \pi\delta(z) \tag{6.10.31}$$

which implies (using $\int g(z) f(z-w)\, d\mu(w)$ is uniformly convergent for $g \in C_0^\infty(\mathbb{C})$ and f one of the two kernels here) (6.10.27)/(6.10.28).

(c) Analyticity of $\widehat{\mu}$ follows from the usual Morera's theorem argument (see Theorem 3.1.6 of Part 2A). Outside $\text{supp}(d\mu)$, it is easy to see (6.10.27) holds classically, so $\Delta\check{\mu} = -2\bar{\partial}\widehat{\mu}(z) = 0$ since $\widehat{\mu}$ is analytic.

(d) Since they are L^1_{loc}-functions, being zero implies $\check{\mu}$ or $\widehat{\mu}$ are zero as distributions, so their distributional derivatives are zero. □

Proof of Theorem 6.10.1. If $\mathbb{R}(X) \neq C(X)$, there is a complex measure μ on X, the compact set in $\mathbb{C}$, so $\mu \neq 0$ but

$$\int f(w)\, d\mu(w) = 0 \tag{6.10.32}$$

for all rational functions with poles in $\mathbb{C} \setminus X$. Since this includes $f(w) = 1/(w-z)$ for all $z \in \mathbb{C} \setminus X$, we see $\widehat{\mu}(z) = 0$ for all $z \in \mathbb{C} \setminus X$. Since $|X| = 0$, $\widehat{\mu}(z)$ is 0 for a.e. z in $\mathbb{C}$, so $\mu = 0$ by (d) of Proposition 6.10.11. This contradiction proves that the rational functions are dense in $C(X)$. □

Example 6.10.12 (The Swiss Cheese). Some thought suggests the last theorem may be too weak—it might be that the right condition is not that $|X| = 0$ but that $X^{\text{int}} = \emptyset$ (as is true for simply connected X by Theorem 6.10.3). We'll dispel this idea with a set $X \subset \overline{\mathbb{D}}$ with empty interior but $\mathbb{R}(X) \neq C(X)$!

Let $\{z_n\}_{n=1}^\infty$ be a countable dense set in $\mathbb{D}$ and α_n a decreasing sequence with

$$\sum_{n=1}^\infty |\alpha_n| < \infty \tag{6.10.33}$$

Let $\rho_n = \min(1-|z_n|, \alpha_n)$ and pick r_n and U_n inductively as follows: $r_1 = \rho_1$ and $U_1 = \mathbb{D}_{r_1}(z_1)$. Once $r_1, \dots, r_n$ and $U_1 \subseteq U_2 \subseteq U_3 \subseteq \cdots \subseteq U_n$ are picked, pick r_{n+1} as follows. If $z_{n+1} \in \overline{U}_n$, set $r_{n+1} = 0$ and $U_{n+1} = U_n$. Otherwise, pick

$$r_{n+1} = \min(\rho_{n+1}, \tfrac{1}{2}\,\text{dist}(z_{n+1}, \overline{U}_n)), \qquad U_{n+1} = U_n \cup \mathbb{D}_{r_{n+1}}(z_{n+1})$$

We'll let X be the complement of $\bigcup_{n=1}^\infty U_n$. It is easy to see that infinitely many r_n are nonzero and that X contains $\partial\mathbb{D}_{r_n}(z_n)$ for any $r_n \neq 0$.

Moreover, we claim $X^{\text{int}} = \emptyset$, for if V is open and nonempty, some $\mathbb{D}_\eta(z_n)$, $\eta \neq 0$, is contained in V. But either $z_n \in U_{n+1}$ or $z_n \in \overline{U}_n$, so any $\mathbb{D}_\eta(z_n)$ intersects U_n, that is, $V \not\subset X$.

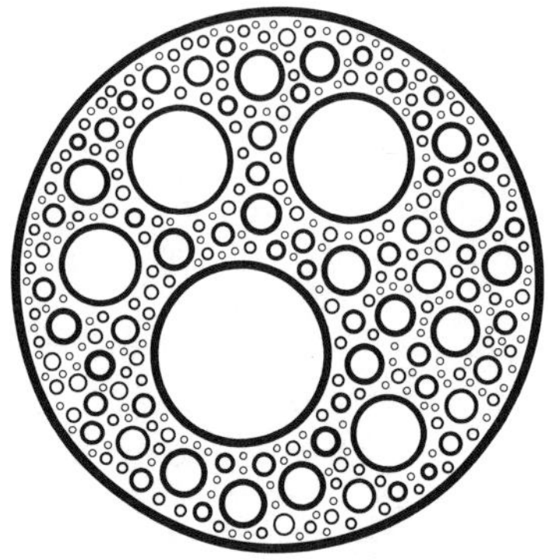

Figure 6.10.1. The Swiss cheese.

Relabel so all $r_j \neq 0$ (by dropping the zero r_j's) and use $\mathbb{D}_j$ for $\mathbb{D}_{r_j}(z_j)$. The z_j are no longer dense, but since α_j is decreasing, $r_j \leq \alpha_j$ and so, by (6.10.33),

$$\sum_{n=1}^{\infty} r_j < \infty \tag{6.10.34}$$

By construction, $\overline{\mathbb{D}_j} \cap \overline{\mathbb{D}_k} = \emptyset$ for $j \neq k$ (see Figure 6.10.1).

Let $\{\mu_j\}_{j=0}^{\infty}$ be the measure dz along the curve $\partial\mathbb{D}$ if $j = 0$, the curve $\partial\mathbb{D}_j$ if $j \neq 0$. If $w \in \mathbb{D}_j$,

$$\frac{1}{2\pi i} \int \frac{d\mu_k(z)}{z - w} = \begin{cases} 0, & k \neq 0, j \\ 1, & k = 0, j \end{cases} \tag{6.10.35}$$

Since (6.10.34) holds and $\|\mu_j\| = 2\pi r_j$ ($r_0 \neq 1$),

$$\mu = \mu_0 - \sum_{j=1}^{\infty} \mu_j \tag{6.10.36}$$

is a finite complex measure on X. By (6.1.34) for any $w \notin X$, $\int \frac{d\mu(z)}{z-w} = 0$, so by a partial fraction expansion, $\int f(z)\, d\mu(z) = 0$ for any rational function with poles in $\mathbb{C} \setminus X$ (of course, if p is a polynomial, $\int p(z)\, d\mu_j = 0$ so $\int p(z)\, d\mu = 0$). We conclude $\mathbb{R}(X) \neq C(X)$ even though $X^{\text{int}} = \emptyset$.

By Theorem 6.10.1, $|X| \neq 0$, although that may not be obvious. In fact, there is a direct argument (Problem 9) that $|X| > 0$. □

Next, we turn to Theorem 6.10.2 where we'll use the potential $\check{\mu}$. We'll need to go from $\check{\mu}$ vanishing on $\mathbb{C} \setminus X$ to most points on ∂X.

Proposition 6.10.13. *Let U be a connected open set in $\mathbb{C}$ and μ a measure of compact support in $\mathbb{C} \setminus U$. If $\check{\mu}(z) = 0$ for $z \in U$, then*

$$w \in \partial U + \widecheck{|\mu|}(w) < \infty \Rightarrow \check{\mu}(w) = 0 \tag{6.10.37}$$

Proof. Suppose $w \in \partial U$ with $\breve{|\mu|}(w) < \infty$. For simplicity of notation, suppose $w = 0$. Since 0 is a limit of points in U, find $\rho_0 \in (0, \frac{1}{2})$ and $z_0 \in U$ with $|z_0| = \rho_0$. Since U is connected and U contains points arbitrarily close to 0, $U_r \equiv U \cap \partial\mathbb{D}_r(0) \neq \emptyset$ for all $r \in (0, \rho_0)$, and since U_r is relatively open, $|U_r| > 0$ (where $|\cdot|$ is $\frac{d\theta}{2\pi}$ measure). For $0 < \delta < \rho_0$, let σ_δ be the measure

$$\int g(z)\, d\sigma_\delta(z) = \int_0^\delta \left[\frac{1}{|U_r|}\left(\int_{U_r} g(z)\,\frac{d\theta}{2\pi}\right)\right] dr \tag{6.10.38}$$

$\sigma_\delta \geq 0$ has two critical properties:

$$\sigma_\delta(\mathbb{C}\setminus U) = 0, \quad \sigma_\delta(\{z \mid r_1 < |z| < r_2\}) = r_2 - r_1 \tag{6.10.39}$$

Fix $0 < \rho < \rho_0$ and vary $\delta \in (0, \rho)$. Let μ_+ be $\mu \restriction \{z \mid |z| \geq \rho\}$ and $\mu_- = \mu - \mu_+$. Uniformly in $|w| \geq \rho$, since $\int d\sigma_\delta(z) = \delta$,

$$\lim_{\delta\downarrow 0} \frac{1}{\delta} \int \log \frac{1}{|z-w|}\, d\sigma_\delta(z) = \log \frac{1}{|w|} \tag{6.10.40}$$

so

$$\lim_{\delta\downarrow 0} \frac{1}{\delta} \int \breve{\mu}_+(z)\, d\sigma_\delta(z) \to \breve{\mu}_+(0) \tag{6.10.41}$$

(Note that since $|\mu_+| \leq |\mu|$, $\breve{|\mu_+|}(z) < \infty$ for $z \in U$ and since μ_+ is supported away from 0, $\breve{|\mu_+|}(0) < \infty$, so we can talk about $\breve{\mu}_+(0)$.)

On the other hand, by the triangle inequality,

$$|z-w| \geq ||z| - |w|| \Rightarrow \log\left(\frac{1}{|z-w|}\right) \leq \log\left(\left|\frac{1}{|z|-|w|}\right|\right) \tag{6.10.42}$$

This implies that for any $w \neq 0$,

$$\begin{aligned}
\frac{1}{\delta}\int \log\left(\frac{1}{|z-w|}\right) d\sigma_\delta(z) &\leq \frac{1}{\delta}\int \log\left(\left|\frac{1}{|z|-|w|}\right|\right) d\sigma_\delta(z) \\
&= \frac{1}{\delta}\int_0^\delta \log\left(\frac{1}{|r-|w||}\right) dr \\
&= \log\left(\frac{1}{|w|}\right) + \frac{1}{\delta}\int_0^\delta \log\left(\frac{1}{|1-r/|w||}\right) dr \\
&= \log\left(\frac{1}{|w|}\right) + \frac{1}{\delta/|w|}\int_0^{\delta/|w|} \log\left(\frac{1}{|1-z|}\right) dt \\
&\leq \log\left(\frac{1}{|w|}\right) + C
\end{aligned} \tag{6.10.43}$$

where

$$C = \sup_{S>0} \frac{1}{S}\int_0^S \log\left(\frac{1}{|1-t|}\right) dt \tag{6.10.44}$$

since the integral is always finite $O(S^2)$ at $S = 0$ and negative for S large. Thus,

$$\frac{1}{\delta}\int |(\check{\mu}_-)(z)|\,d\sigma_\delta(z) \le \frac{1}{\delta}\int \widecheck{|\mu_-|}(z)\,d\sigma_\delta(z) \le \int_{|w|\le\rho}\left[\log\left(\frac{1}{|w|}\right)+C\right]d|\mu|(w) \tag{6.10.45}$$

Since $\check{\mu}_+(z) + \check{\mu}_-(z) = \check{\mu}(z) = 0$ on U and $\sigma_\delta(\mathbb{C}\setminus U) = 0$,

$$\frac{1}{\delta}\int \check{\mu}_+(z)\,d\sigma_\delta(t) = -\frac{1}{\delta}\int \check{\mu}_-(z)\,d\sigma_\delta(t) \tag{6.10.46}$$

By (6.10.41) and (6.10.45), we see that

$$\left|\int_{|w|\ge\rho}\log\left(\frac{1}{|w|}\right)d\mu(w)\right| \le \int_{|w|\le\rho}\left[\log\left(\frac{1}{|w|}\right)+C\right]d|\mu|(w) \tag{6.10.47}$$

The condition $\widecheck{|\mu|}(0) < \infty$ implies the integral on the right of (6.10.47) goes to 0 as $\rho \downarrow 0$ and the integral on the right goes to $\check{\mu}(0)$. Thus, $\check{\mu}(0) = 0$ as claimed. □

Lemma 6.10.14. *Let X be a compact set where $\mathbb{C}\setminus X$ has finitely many components. Let $z \in X^{\text{int}}$ and $d\mu_z$ harmonic measure on ∂X for functions continuous on X, harmonic on X^{int}. Then for each $w \in \partial X$,*

$$\int \log(|w-y|^{-1})\,d\mu_z(y) = \log(|z-w|^{-1}) \tag{6.10.48}$$

In particular, the integral on the left is finite.

Proof. Since $\mathbb{C}\setminus X$ has finitely many components, w is a limit of points, w_n, in one component, U, so $w \in \partial U$. Let $\alpha = \mu_z - \delta_z$. Since $\log(|y-w_n|^{-1})$ is continuous on X and harmonic in X^{int}, $\check{\alpha}(y) = 0$ on U. In particular, $\check{\alpha}(w_n) = 0$, that is,

$$\int \log(|y-w_n|^{-1})\,d\mu_z(y) = \log(|z-w_n|^{-1}) \tag{6.10.49}$$

Since μ_z is a positive measure, $\check{\mu}_z$ as a sup of increasing continuous functions (replace $\log(|y|^{-1})$ by $\min(\log(|y|^{-1}, n))$ and take $n \to \infty$), $\check{\mu}_z$ is lower semicontinuous. It follows that

$$\int \log(|w-y|^{-1})\,d\mu_z(y) \le \log(|z-w|^{-1}) < \infty \tag{6.10.50}$$

so $\widecheck{|\alpha|}(w) < \infty$. Thus, by Proposition 6.10.13, $\check{\alpha}(w) = 0$ which is (6.10.48). □

Proof of Theorem 6.10.2. Let α be a real measure on ∂X with

$$\int \operatorname{Re} f(w)\,d\alpha(w) = 0 = \int \log|w-a_j|\,d\alpha(w) \tag{6.10.51}$$

for $j = 1, \dots, m$ and for any f which is a polynomial in w and $\{(w - a_j)^{-1}\}_{j=1}^m$. $\breve{\alpha}(z)$ is harmonic in each G_j, $j = 1, \dots, m$, and in

$$G_0 = \mathbb{C} \Big\backslash \Big(X \cup \bigcup_{j=1}^m G_j \Big) \tag{6.10.52}$$

the unbounded component of $\mathbb{C} \setminus X$.

By (6.10.51), $\breve{\alpha}(a_j) = 0$. If $|z - a_j|$ is small, uniformly for all $w \in \partial X$,

$$\log|w-z| - \log|w-a_j| = \log\left|1 - \frac{z-a_j}{w-a_j}\right| = \operatorname{Re}\left[-\sum_{n=1}^\infty \frac{(z-a_j)^n}{(w-a_j)^n}\right] \tag{6.10.53}$$

so by (6.10.51), $\breve{\alpha}(a_j) = 0$ for $|z - a_j|$ small. Since $\breve{\alpha}$ is harmonic in G_j, $\breve{\alpha} = 0$ in G_j. Similarly, for z large, uniformly in $w \in \partial X$,

$$\log|w-z| - \log|z| = \log\left|1 - \frac{w}{z}\right| = \operatorname{Re}\left[-\sum_{n=1}^\infty \frac{w^n}{z^n}\right] \tag{6.10.54}$$

Since $\int \log|z| \, d\alpha(w) = \log|z| \int d\alpha(w) = 0$ (taking $f = 1$ in (6.10.51)) and $\operatorname{Re} \int w^n \, d\alpha(w) = 0$, we conclude $\breve{\alpha} = 0$ in G_0 also.

Now let $z \in X^{\text{int}}$ and μ_z harmonic measure on ∂X. Integrating (6.10.48) with $d|\alpha|(w)$, we see

$$\int \breve{|\alpha|}(y) \, d\mu_z(y) = \breve{|\alpha|}(z) < \infty \tag{6.10.55}$$

since $z \notin \operatorname{supp}(d|\alpha|)$. Thus, for $d\mu_z$ a.e. y, $\breve{|\alpha|}(y) < \infty$. By Proposition 6.10.13 (and the fact that any $x \in \partial X$ is in ∂U for an open U in $\mathbb{C} \setminus X$), $\breve{\alpha}(y) = 0$ for $d\mu_z$ a.e. y.

It follows from (6.10.48) that

$$\breve{\alpha}(z) = \int \breve{\alpha}(y) \, d\mu_z(y) = 0 \tag{6.10.56}$$

Also, since $|\breve{\alpha}|(y) < \infty$ for d^2z a.e. $y \in \partial X$ (should $|\partial X| > 0$), we see $\breve{\alpha}(y)$ is 0 for a.e. $y \in \partial X$. Since it is zero on $\mathbb{C} \setminus X$ and X^{int}, $\breve{\alpha}(z)$ is zero for a.e. $z \in \mathbb{C}$. Thus, $\alpha = 0$ by Proposition 6.10.11.

By the Hahn–Banach theorem, if the claimed set were not dense in $C(\partial X)$, there would exist a real measure α on ∂X obeying (6.10.51). Thus, the claimed set is dense. $\square$

If $f \in C(X)$ and $\operatorname{Re}(p_n) \to f$ uniformly on ∂X, by the maximum principle applied to $\operatorname{Re}(p_n - p_m)$, $\operatorname{Re} p_n$ is Cauchy in $\|\cdot\|_\infty$ on X. Its limit, u, is harmonic on X^{int} and $u \restriction \partial X = f$. We thus have:

Corollary 6.10.15. *Let X be compact in $\mathbb{C}$ so $\mathbb{C} \setminus X$ has finitely many components. Then the Dirichlet problem is solvable for X^{int}, that is, given $f \in C(\partial X^{\text{int}})$, there is u harmonic in X^{int} and continuous on X^{int} so that $u \restriction \partial X = f$. If $\mathbb{C} \setminus X$ is connected, u is a uniform limit of the real part of polynomials.*

If X is bigger than $\overline{X^{\text{int}}}$, we extend f from ∂X^{int} to ∂X using the Tietze extension theorem. Using Kelvin transforms (see Problem 15 of Section 3.1 of Part 3), one can replace X compact by X closed and proper in $\mathbb{C}$ with finitely many components in $\mathbb{C} \setminus X$. One needs to require that u is continuous at infinity (also to get uniqueness). The one limitation of this result is that it only applies to region Ω where Ω is the interior of a closed set, for example, if $K \subset \mathbb{R}$ is compact, $\Omega \setminus K$ is not the interior of a closed set.

For Theorems 6.10.3 and 6.10.4, we need an analog of Proposition 6.10.13 for Cauchy transforms rather than potentials.

Proposition 6.10.16. *Let $X \subset \mathbb{C}$ be compact so $\mathbb{C} \setminus X$ is connected. Let μ be a complex measure on ∂X so $\widehat{\mu}(z) = 0$ for $z \in \mathbb{C} \setminus X$. Then $\widehat{\mu}(z_0) = 0$ for any point $z_0 \in \partial X$ where* (6.10.26) *holds.*

Remark. There is an idea here that will return later in a more complicated guise and which we'll call the exponential trick. Theorem 6.10.2 lets us approximate certain real-valued f by $\operatorname{Re} p_n$ for polynomials p_n. We want to control something involving the imaginary part of p_n. If we look at $g = e^{p_n}$ or $g = e^{-p_n}$, we can control $|g|$ so, for example, if $\operatorname{Re} p_n \to \infty$, we know $|e^{-p_n}| \to 0$.

Proof. We begin by noting that for $|z|$ large,

$$\widehat{\mu}(z) = \int d\mu(w)(w-z)^{-1} = \sum_{n=0}^{\infty} z^{-n-1} \int w^m \, d\mu(w) \tag{6.10.57}$$

so $\widehat{\mu}(z) = 0$ for large $|z|$ (which, by analyticity, is equivalent to $\widehat{\mu}(z) = 0$ on the connected component of $\mathbb{C} \setminus X$) is equivalent to

$$\int w^n \, d\mu(w) = 0, \qquad n = 0, 1, 2, \dots \tag{6.10.58}$$

Since $\operatorname{supp}(\mu)$ is bounded, we get

$$h \text{ entire} \Rightarrow \int h(w) \, d\mu(w) = 0 \tag{6.10.59}$$

By Theorem 6.10.2, there exist polynomials, $p_m(z)$, so

$$|\operatorname{Re} p_m(z) - |z - z_0|| \le \frac{1}{m}, \qquad z \in \partial X \tag{6.10.60}$$

and with $p_m(z_0) = 0$ (for pick an initial $\tilde{p}_m$ obeying (6.10.60) and let $p_m = \tilde{p}_{2m} - \tilde{p}_{2m}(z_0)$). Let

$$h_m(z) = (z - z_0)^{-1}[1 - e^{-mp_m(z)}] \tag{6.10.61}$$

Since $p_m(z_0) = 0$, h_m is entire so

$$\int h_m(w)\, d\mu(w) = 0 \tag{6.10.62}$$

(6.10.60) implies that for $z \in \partial X$,

$$m \operatorname{Re} p_m(z) \geq -1 + m|z - z_0| \tag{6.10.63}$$

so if $z \neq z_0$, $e^{-mp_m(z_0)} \to 0$, that is,

$$z \in \partial X \setminus \{z_0\} \Rightarrow h_m(z) \to (z - z_0)^{-1} \tag{6.10.64}$$

On the other hand, (6.10.63) says $|e^{-mp_m(z)}| \leq e$, so

$$|h_m(z)| \leq (1 + e)|z - z_0|^{-1} \tag{6.10.65}$$

If (6.10.26) holds, $|\mu|(\{z_0\}) = 0$, so $h_m(z) \to (z - z_0)^{-1}$ for $d|\mu|$-a.e. z. (6.10.65) and (6.10.26) implies $|h_m|$ is dominated by an $L^1(\partial X, d|\mu|)$-function. Thus, by the dominated convergence theorem,

$$\check{\mu}(z_0) = \int (w - z_0)^{-1}\, d\mu(w) - \lim_m \int h_m(w)\, d\mu(w) - 0 \tag{6.10.66}$$

$\square$

Proof of Theorem 6.10.3. If X is not simply connected, then by (6.2.22), $\widehat{\mathbb{P}(X)}$ is strictly bigger than $\widehat{C(X)} \cong X$, so (6.10.2) fails. If $X^{\text{int}} \neq \emptyset$ and $x_0 \in X^{\text{int}}$, then $|x - x_0| \in C(X) \setminus \mathbb{P}(X)$ since it is not analytic at x_0. Thus, the conditions are necessary for (6.10.2) and we need only prove them sufficient.

If $\mathbb{P}(X) \neq C(X)$, then there exists a nonzero measure μ on $\partial X = X$ with $\mu(x^\ell) = 0$ for all ℓ. Since $(x - z)^{-1} = -\sum_{\ell=0}^\infty x^\ell z^{-\ell-1}$ uniformly in $x \in X$ and $|z|$ large; this means $\widehat{\mu}(z) = 0$ for large z, and so by analyticity and connectedness of $\mathbb{C} \setminus X$, $\widehat{\mu}(z) = 0$ for $z \in \mathbb{C} \setminus X$.

By Proposition 6.10.16, $\widehat{\mu}(z) = 0$ for those z in $X = \partial X$ where (6.10.26) holds. Since d^2z-a.e. z has (6.10.26), $\widehat{\mu}(z) = 0$ for a.e. z and so, by Proposition 6.10.11(d), $\mu = 0$. This contraction shows $\mathbb{P}(X) = C(X)$. $\square$

We turn finally to the proof of Theorem 6.10.4, the subtlest of these approximation results. As a warmup, we consider the case $X = \overline{\mathbb{D}}$. There is an easy proof using the fact that $f(rz) \to f(z)$ uniformly as $r \uparrow 1$ and that Taylor approximations uniformly converge to $f(rz)$. But this is not available in general, so let's find an argument using the ideas of this section.

By the Hahn–Banach theorem and the fact that $f \in \mathbb{A}(X)$ is determined by its values on ∂X,

$$\mathbb{P}(X) = \{f \mid \forall \alpha \, \forall \ell \int z^\ell \, d\alpha(z) = 0 \Rightarrow \int f(z)\, d\alpha(z)\} \tag{6.10.67}$$

We already see the proof will be more subtle since we'll need to show not that $\alpha = 0$, just that $\alpha \restriction \mathbb{A}(X) = 0$. Given $f \in \mathbb{A}(X)$, by Theorem 6.10.2, we know there are polynomials p_n so that $\|\text{Re}\, p_n - \text{Re}\, f\|_\infty \to 0$. We can pick p_n so that $\text{Im}\, p_n(0) = \text{Im}\, f(0)$. Since $\int \text{Re}((p_n - f)^2)\, \frac{d\theta}{2\pi} = \text{Re}([(p_n - f_n)(0)]^2) = [\text{Re}(p_n - f)(0)]^2 > 0$, we see (as noted in Section 5.2 of Part 3; see equations (5.2.34) and (5.2.36) of that section) that

$$\int \big[\text{Im}(p_n - f)\big]^2 (e^{i\theta})\, \frac{d\theta}{2\pi} \le \int \big[\text{Re}(p_n - f)\big]^2 (e^{i\theta})\, \frac{d\theta}{2\pi} \tag{6.10.68}$$

so by passing to a subsequence, $p_n(e^{i\theta}) \to f(e^{i\theta})$ for a.e. θ. If $d\alpha$ is purely a.c., then this implies $\int f(e^{i\theta})\, d\alpha(\theta) = 0$. (Actually, we also need to be able to choose p_n bounded and for that we'll need to use the exponential trick.) But this tells us nothing about the singular part. For this, the F. and M. Riesz theorem comes to the rescue—it says $\int z^\ell \, d\alpha(z) = 0 \Rightarrow \alpha_\text{s} = 0$! This suggests we need to be prepared in the general case to look at a Lebesgue decomposition of $d\alpha$ relative to a representing measure and to separately prove $\alpha_\text{s} = 0$. Because X^int may have many components, we actually need to do an iteration and interpret α_s as singular with respect to all representing measures. With this background to guide us, we begin with a reformulation of Harnack's inequality:

Proposition 6.10.17. *Let X be a compact subset of $\mathbb{C}$ for which the Dirichlet problem is solvable and $d\mu_z$ harmonic measure on ∂X for $z \in X^\text{int}$. For z and w in the same component of X^int, $d\mu_z$ and $d\mu_w$ are mutually a.c.; in fact, for some constant $c \in (0, \infty)$ (depending on z and w),*

$$c^{-1}\, d\mu_w \le d\mu_z \le c\, d\mu_w \tag{6.10.69}$$

Remark. We defined the Dirichlet problem being solvable for open sets. We mean here solvable for X^int, that is, for any continuous f on ∂X, there is a continuous u on X so u is harmonic on X^int and $u \restriction \partial X = f$. This implies $d\mu_w$ is unique for each w. By Theorem 6.10.2, it holds if $\mathbb{C} \setminus X$ is connected.

Proof. By Harnack's inequality (see Theorem 3.1.29 of Part 3), since w, z lie in the same open set, A, there is $c \in (0, \infty)$ so that for all u harmonic and positive in A, we have

$$c^{-1} u(w) \le u(z) \le c u(w) \tag{6.10.70}$$

u harmonic in $X^{\rm int}$, continuous in X with $u \restriction \partial X = f$ has $u > 0$ on $X^{\rm int}$ if and only if $f \geq 0$. Thus, for any positive f,

$$c^{-1} \int f(q)\, d\mu_w(q) \leq \int f(q)\, d\mu_z(q) \leq c \int f(q)\, d\mu_w(q) \tag{6.10.71}$$

which is precisely (6.10.69). □

Write $X^{\rm int} = \bigcup_{j=1}^{J} U_j$ (J finite or infinite) for the components of $X^{\rm int}$ and pick $a_j \in U_j$ and let $d\mu_j = d\mu_{a_j}$.

Lemma 6.10.18. *Fix j and a compact X with $\mathbb{C}\backslash X$ connected. Let $S \subset \partial X$ be closed with*

$$d\mu_j(S) = 0 \tag{6.10.72}$$

For any $\varepsilon > 0$, there is a polynomial P so that

$$1 + \varepsilon \geq \operatorname{Re} P(z) \geq 1 \text{ if } z \in S, \quad 1 + \varepsilon \geq \operatorname{Re} P(z) \geq 0 \text{ for } z \in \partial X \tag{6.10.73}$$

$$\int [\operatorname{Re} P(z)]^2\, d\mu_j(z) \leq \varepsilon \tag{6.10.74}$$

Proof. By decreasing ε, suppose $\varepsilon < 1$. (6.10.72) and the fact that Baire measures of closed sets on a compact space can be approximated from above by continuous functions (apply (4.4.28) of Part 1 to $\partial X \setminus S$) implies we can find $f \in C_{\mathbb{R}}(\partial X)$ so that

$$1 \geq f \geq 0, \quad f \geq \chi_S, \qquad \int f(x)\, d\mu_j(z) \leq \frac{\varepsilon}{4} \tag{6.10.75}$$

Since $f + \frac{\varepsilon}{8} \in C_{\mathbb{R}}(\partial X)$, also by Theorem 6.10.2, we can find a polynomial, P, so that $\|\operatorname{Re} P - (f + \frac{\varepsilon}{8})\|_\infty \leq \frac{\varepsilon}{8}$. Thus, for all $x \in \partial X$,

$$f \leq \operatorname{Re} P \leq f + \frac{\varepsilon}{4} \tag{6.10.76}$$

(6.10.75) then implies (6.10.73) and

$$\int \operatorname{Re} P(z)\, d\mu_j(z) \leq \frac{\varepsilon}{2} \tag{6.10.77}$$

Since $\varepsilon < 1$, $\operatorname{Re} P \leq 2$, so (6.10.77) $\Rightarrow$ (6.10.74). □

Proposition 6.10.19. *Fix j and a compact $X \subset \mathbb{C}$ with $\mathbb{C} \setminus X$ connected. Let σ be a measure singular with respect to μ_j. Then there exist polynomials P_n so that*

(a) $e^{-P_n(z)} \to 1$ *for $d\mu_j$-a.e. z in ∂X* (6.10.78)

(b) $e^{-P_n(z)} \to 0$ *for σ-a.e. z in ∂X* (6.10.79)

(c) $\operatorname{Re} P_n(z) \geq 0$ *on ∂X* (6.10.80)

(d) $\displaystyle\int |P_n(z)|^2\, d\mu_j(z) \to 0$ (6.10.81)

Proof. Let S be a Baire set with $\mu_j(S) = 0$ and $\sigma(\partial X \setminus S) = 0$. Since σ is inner regular (see Theorem 4.5.6 of Part 1), we can find closed sets $\{S_n\}_{n=1}^\infty$, $S_1 \subset S_2 \subset \cdots \subset S$ so that $\sigma(S \setminus \bigcup_{j=1}^\infty S_n) = 0$. By picking $S = S_n$ and $\varepsilon = 2^{-4n}$ in Lemma 6.10.18 and multiplying that P (which is already n-dependent) by 2^n, we get polynomials P_n so that

$$2^{n+1} \geq \operatorname{Re} P_n(z) \geq 2^n, \quad z \in S_n, \quad \operatorname{Re} P_n(z) \geq 0, \quad z \in \partial X \tag{6.10.82}$$

$$\int [\operatorname{Re} P_n(z)]^2 \, d\mu_j(z) \leq 2^{-2n}, \quad \operatorname{Im} P_n(a_j) = 0 \tag{6.10.83}$$

(and we can add an imaginary constant to P_n to arrange the last).

By (6.10.82) and $S_n \subset S_{n+1}$, we see (b) holds for $z \in \bigcup_j S_j$ which is σ-a.e. z. Since P_n^2 is analytic, we have

$$\operatorname{Re}(P_n)^2(a_j) = \int \operatorname{Re}((P_n)^2(z)) \, d\mu_j(z) \tag{6.10.84}$$

Since $\operatorname{Im} P_n(a_j) = 0$, $\operatorname{Re}(P_n)^2(a) = (\operatorname{Re} P_n(a))^2 \geq 0$, so (6.10.84) says

$$\int (\operatorname{Im} P_n(z))^2 \, d\mu_j(z) \leq \int (\operatorname{Re} P_n(z))^2 \, d\mu_j(z) \tag{6.10.85}$$

Thus,

$$\sum_{n=0}^\infty \int |P_n(z)|^2 \, d\mu_j(z) < \infty \tag{6.10.86}$$

so $\sum_{n=0}^\infty |P_n(z)|^2 < \infty$ for μ_j-a.e. z, which implies $|P_n(z)| \to 0$ which implies (a). (6.10.86) also implies (d). $\square$

Lemma 6.10.20. *Fix j and a compact $X \subset \mathbb{C}$ with $\mathbb{C} \setminus X$ connected. Let β be a measure which is a.c. wrt $d\mu_j$ for some j. Suppose*

$$\int z^\ell \, d\beta(z) = 0, \qquad \ell = 0, 1, 2, \dots \tag{6.10.87}$$

Let f be continuous on X and analytic on X^{int}. Then

$$\int f(z) \, d\beta(z) = 0 \tag{6.10.88}$$

Remark. We use $y = \log(f)$ below because we want to use the exponential trick since we have no control on $\operatorname{Im} P_n$.

Proof. We first note that since any polynomial P is bounded on ∂X, $\exp(P(z))$ is a uniform limit of polynomials, so (6.10.87) implies that for every polynomial P,

$$\int e^{P(z)} \, d\beta(z) = 0 \tag{6.10.89}$$

Since (6.10.87) includes $\ell = 0$, we can add a constant to f without affecting whether (6.10.88) is true. We can also multiply by a constant. Thus, without loss, we can suppose that for $z \in X$,

$$\operatorname{Re} f(z) > 0, \qquad |f(z)| \le 1, \qquad \operatorname{Im} f(a_j) = 0 \tag{6.10.90}$$

Let

$$g(z) = \log(f(z)) \tag{6.10.91}$$

taking the branch with $|\operatorname{Im} \log(f(z))| < \pi/2$.

By Theorem 6.10.2, we can find polynomials P_n so

$$\sup_{z \in \partial X} |\operatorname{Re}(g(z) - P_n(z))| \le 2^{-n}, \quad \operatorname{Im} P_n(a_j) = 0 \tag{6.10.92}$$

As in the last proof for $d\mu_j$-a.e. z, we have $P_n(z) \to g(z)$, so

$$e^{P_n(z)} \to f(z)\, d\mu_j\text{-a.e. } z \tag{6.10.93}$$

Since $\operatorname{Re} g(z) < 0$ and (6.10.92) holds, for $n \ge 1$, $\operatorname{Re} P_n(z) \le \frac{1}{2}$, so

$$\sup_{z \in \partial X, n} |e^{P_n(z)}| \le \sqrt{e} \tag{6.10.94}$$

Thus, by the dominated convergence theorem and the fact that, since β is $d\mu_j$-a.e., (6.10.93) holds for $d\beta$-a.e. z, we see, using (6.10.89), that (6.10.88) holds. □

The approximation in Proposition 6.10.19 gets us

Proposition 6.10.21. *Fix j and a compact X with $\mathbb{C} \setminus X$ connected. Let $d\alpha$ be a measure on ∂X obeying*

$$\int z^\ell \, d\alpha(z) = 0, \qquad \ell = 0, 1, 2, \dots \tag{6.10.95}$$

Write

$$d\alpha = d\beta + d\alpha_{\text{s}} \tag{6.10.96}$$

where β is $d\mu_j$-a.c. and $d\alpha_{\text{s}}$ is $d\mu_j$-singular. Then

(a) $$\int z^\ell \, d\beta(z) = \int z^\ell \, d\alpha_{\text{s}}(z) = 0, \quad \ell = 0, 1, 2, \dots \tag{6.10.97}$$

(b) *For $z \in \mathbb{C} \setminus \bar{U}_j$,*

$$\int \frac{d\beta(w)}{w - z} = 0 \tag{6.10.98}$$

(c) *For $z \in U_j$,*

$$\int \frac{d\alpha(w)}{w - z} = \int \frac{d\beta(w)}{w - z}, \qquad \int \frac{d\alpha_{\text{s}}(w)}{w - z} = 0 \tag{6.10.99}$$

Proof. Let P_n be as in Proposition 6.10.19. Since $|e^{-P_n(z)}| \le 1$, the dominated convergence theorem shows for $g \in C(\partial X)$,

$$\lim_{n\to\infty} \int g(z)e^{-P_n(z)}\, d\alpha = \int g(z)\, d\beta \tag{6.10.100}$$

As in the last few proofs, (6.10.95) $\Rightarrow \int z^\ell e^{-P_n(z)}\, d\alpha(z) = 0$, so (6.10.100) implies $\int z^\ell\, d\beta(z) = 0$, and then $\alpha = \beta + \alpha_{\rm s}$ implies (6.10.97).

Since $\mathbb{C}\setminus X$ is connected, so is $\mathbb{C}\setminus \overline{U}_j$, so (6.10.97) and $\mathrm{supp}(d\beta) \subset \partial U_j$ implies (b). Clearly, the first statement in (6.10.99) implies the second, so we can focus on the first.

Since $P_n(z)$ has real and imaginary parts that are harmonic, we have

$$|P_n(a_j)|^2 = \left|\int P_n(w)\, d\mu_j(w)\right|^2 \le \int |P_n(w)|^2\, d\mu_j(w) \to 0 \tag{6.10.101}$$

by (6.10.81). Write

$$\begin{aligned} e^{-P_n(a_j)} \int \frac{d\alpha(w)}{w-a_j} &= \int \frac{e^{-P_n(\alpha_j)} - e^{-P_n(w)}}{w-a_j}\, d\alpha(w) \\ &\quad + \int \left[\frac{1}{w-a_j}\right] e^{-P_n(w)}\, d\alpha(w) \end{aligned} \tag{6.10.102}$$

Since $(w-a_j)^{-1}[e^{-P_n(a_j)} - e^{P_n(w)}]$ is entire, the first integral is 0. As above, the second integral converges to $\int (w-a_j)^{-1}\, d\beta$. Since $P_n(a_j)\to 0$, the left side goes to $\int (w-a_j)^{-1}\, d\alpha$. This proves the first assertion in (6.10.99) at $z = a_j$.

Since $d\mu_z$ is $d\mu_j$ mutually equivalent for $z \in U_j$ (by Proposition 6.10.17), the same argument shows $P_n(z) \to 0$, so by an analog of (6.10.102), we get the first assertion in (6.10.99) for all $z \in U_j$. □

Proof of Theorem 6.10.4. Given a measure $d\alpha$ obeying (6.10.95), expand it by $d\alpha = d\beta_1 + d\alpha_{{\rm s},1}$, where $d\beta_1$ is $d\mu_1$-a.c. and $d\alpha_{{\rm s},1}$ is $d\mu_1$-singular. Note that with $\|\cdot\|$, the $C(\partial X)^*$-norm, by the general theory of Lebesgue decompositions (see Section 4.7 of Part 1),

$$\|\alpha\| = \|\beta_1\| + \|\alpha_{{\rm s},1}\| \tag{6.10.103}$$

Next, write $d\alpha_{{\rm s},1} = d\beta_2 + d\alpha_{{\rm s},2}$ and repeat. If J, the number of components of $X^{\rm int}$, is finite, stop after J steps. Otherwise, continue indefinitely. One gets, for each $k<\infty$,

$$d\alpha = \sum_{j=1}^k d\beta_j + d\alpha_{{\rm s},k} \tag{6.10.104}$$

where $d\beta_j$ is $d\mu_j$-a.c. and $d\alpha_{s,k}$ is singular wrt $\{d\mu_j\}_{j=1}^k$. We have

$$\|\alpha\| = \sum_{j=1}^{k} \|\beta_j\| + \|\alpha_{s,j}\| \tag{6.10.105}$$

This implies (if $J = \infty$) that $\alpha_{s,k}$ is $\|\cdot\|$-Cauchy and so has a limit α_s and

$$d\alpha = d\alpha_s + \sum_{j=1}^{J} d\beta_j \tag{6.10.106}$$

By Lemma 6.10.20, if $f \in C(X)$ and analytic in X^{int}, then

$$\int f(z)\, d\beta_j(z) = 0, \qquad j = 1, 2, \dots, J \text{ (or } 1, 2, \dots) \tag{6.10.107}$$

At Step 1, $\widehat{\sigma}_{1,s}(z) = 0$ if $z \in U_1 \cup \mathbb{C} \setminus X$. Since $\int z^n\, d\beta_2(z) = 0$ and $\mathbb{C} \setminus U_2$ is connected, we have $\widehat{\beta}_2(z) = 0$ on $\mathbb{C} \setminus U_2$, and so on U_1. Thus, on U_1, $\widehat{\sigma}_{2,s}(z) = \widehat{\sigma}_{1,s}(z) - \beta_2(z) = 0$, so $\widehat{\sigma}_{2,s}(z) = 0$ on $U_1 \cup U_2$. Inductively, $\widehat{\sigma}_{j,s}(z) = 0$ on $\bigcup_{k=1}^{j} U_k$.

All measures are supported on ∂X, $(\,\cdot\, - z)^{-1}$ is in $C(\partial X)$ if $z \in U_j$, and $\sigma_{j,s} \to \sigma_s$ in norm. We conclude that $\widehat{\sigma}_s(z) = 0$ on X^{int}. Since $\int z^\ell d\widehat{\sigma}_s = 0$, $\ell = 0, \dots,$ we conclude that $\widehat{\sigma}_s(z)$ is 0 on $\mathbb{C} \setminus X$. By Proposition 6.10.16, $\widehat{\sigma}_s(z) = 0$ for d^2z-a.e. $z \in \partial X$. Thus, by Proposition 6.10.11(d), $\sigma = 0$, so (6.10.106) and (6.10.107) imply (6.10.88). □

Notes and Historical Remarks. The four density theorems that are stated at the start of the section predate the extensive development of function algebras and originally had classical proofs. The Hartogs–Rosenthal theorem (Theorem 6.10.1) is from their joint 1931 paper [**296**]; the Lebesgue–Walsh theorem (Theorem 6.10.2) is named after 1907 work of Lebesgue [**434**] and 1929 work of Walsh [**731**]; Lavrentiev's theorem (Theorem 6.10.3) is from his 1936 paper [**431**]; and Mergelyan's theorem (Theorem 6.10.4) is from his 1952 paper [**474**].

Function algebras were heavily studied in the five years before and after 1960. Besides the major players we'll mention soon, important work on problems other than the approximation theory results we focus on was done by Arens, Helson, Lumer, and Singer. Books on this subject include [**91, 218, 219, 323, 436, 520, 673, 739**].

The Shilov boundary attributed to him appeared first in the review of Gel'fand–Raikov–Shilov [**235**]. Earlier, in 1931, Bergman [**52**] has noted that for several complex variables on a region $\Omega \subset \mathbb{C}^n$, a small subset of the topological boundary of Ω suffices for the maximum principle (e.g., $\partial\mathbb{D} \times \partial\mathbb{D}$

for $\mathbb{D} \times \mathbb{D}$). He called this the distinguished boundary. For this reason, some authors refer to the Bergman–Shilov boundary.

What is called the Choquet boundary was introduced in a seminal paper of Bishop [**68**]. He showed functional algebras have a minimal boundary (not merely a minimal closed boundary) and presented Example 6.10.10 to show that this minimal boundary need not be closed. In the metrizable case, he defined the boundary as those x for which there is $f \in \mathfrak{A}$ with $|f(x)| > |f(y)|$ for all $y \in X \setminus \{x\}$, that is, what are called *peak points*. He also showed these were exactly the points with unique measure μ_0 on X obeying (6.10.2). He referred to future work on representing measures supported on $\text{Ch}(\mathfrak{A})$ and mentioned the relevance of Choquet's work. He then joined forces with de Leeuw [**70**] to extend Choquet theory to nonmetrizable X's. They introduced the name "Choquet boundary" although "Bishop boundary" would have made more sense.

Thus, the representation (6.10.21) can be viewed as a case of the strong Krein–Milman theorem (Theorem 5.11.10 of Part 1).

Bishop was also an important contributor to the functional analytic approach to the four approximation theorems introducing the use of the Cauchy transform [**67, 69**] and finding a functional analytic approach to parts of Mergelyan's theorem [**67, 68, 69**]. The first complex function algebra approach is due to Glicksberg and Wermer [**251**]. A streamlining of their proof is due to Carleson [**104**] whose paper we largely follow in this section.

The Swiss cheese example (Example 6.10.12) is named not only because of its holes, but because it was discovered by the Swiss mathematician, Alice Roth [**579**].

Mergelyan also proved that if X is compact so that $\mathbb{C} \setminus X$ has finitely many components, then $\mathbb{R}(X) = \mathbb{A}(X)$. Function algebra proofs of this can be found in [**8, 222, 250, 397**].

Among the central ideas in further studies of function algebras are two ideas in a significant paper of Gleason [**248**]: first, the definition of Dirichlet algebras, $\mathfrak{A}$'s, in $C(X)$ so $\{\text{Re} f \mid f \in \mathfrak{A}\}$ is dense in $C(X)$. Thus, Theorem 6.10.2 in case $\mathbb{C} \setminus X$ is connected can be rephrased as $\mathbb{P}(X)$ is a Dirichlet algebra.

The second idea involves Gleason parts. Gleason showed for a function algebra, $\mathfrak{A}$, to say $\ell, q \in \widehat{\mathfrak{A}}$ are equivalent if and only if $\|\ell - q\| < 2$ is an equivalence relation. Equivalence classes are called *Gleason parts*. It can be seen the bound $\|\ell - q\| < 2$ is equivalent to a Harnack inequality on real parts of $\{\widehat{f}\}$ (see Problems 10 and 11). The idea is that in some cases, one can embed disks into $\widehat{\mathfrak{A}}$ when parts have more than one point so that $\widehat{f}$ (for $f \in \mathfrak{A}$) is analytic on the disk. Gamelin's book [**218**] discusses this.

Problems

1. Let $\mathfrak{A}$ be an abelian Banach algebra with identity. Let $\mathfrak{A}_0 = \{\widehat{x} \mid x \in \mathfrak{A}\} \subset C(\widehat{\mathfrak{A}})$ and $\mathfrak{A}_1 = \overline{\mathfrak{A}_0}$, the $\|\cdot\|_\infty$-closure.

   (a) If $\ell \in \widehat{\mathfrak{A}}_1$, prove that $\ell \in \widehat{\mathfrak{A}}$ (*Hint*: You'll need automatic continuity) and conclude that $\widehat{\mathfrak{A}}_1 = \widehat{\mathfrak{A}}$ (including topologies).

   (b) Let K be a boundary for $\mathfrak{A}_1$. Prove that K is a boundary for $\mathfrak{A}$.

   (c) Let K be a closed boundary for $\mathfrak{A}$. Prove K is a closed boundary for $\mathfrak{A}_1$.

2. Let $K \subset \mathbb{C}$ be compact and S a closed boundary for $\mathbb{A}(K)$. Let $z_0 \in \partial K$ and $z_n \in \mathbb{C} \setminus K$ with $z_n \to z$.

   (a) For any n, define $f_n(z) = (z - z_n)^{-1}$ and use it to conclude that there is $w_n \in S$ with $|w_n - z_0| \le 2|z_n - z_0|$. Conclude that $S \supset \partial K$.

   (b) Prove ∂K is a boundary and conclude that ∂K is a unique smallest closed boundary for $\mathbb{A}(K)$.

3. Prove that $\partial\mathbb{D} \times \partial\mathbb{D}$ is a boundary for $\mathbb{A}(\overline{\mathbb{D}} \times \overline{\mathbb{D}})$.

4. (a) Let $f \in \mathbb{A}(\mathbb{D})$. If $g = f \restriction \partial\mathbb{D}$, prove that $g_n^\sharp \equiv (2\pi)^{-1}\int_0^{2\pi} e^{-in\theta} g(e^{i\theta})\, d\theta$ is 0 for all $n < 0$.

   (b) Let $g \in C(\partial\mathbb{D})$, obey $g_n^\sharp = 0$ for $n < 0$. Prove there is a family P_m of polynomials in z, so $\sup_{e^{i\theta} \in \partial\mathbb{D}} |g(e^{i\theta}) - P_m(e^{i\theta})| \to 0$ and conclude there is $f \in \mathbb{A}(\mathbb{D})$ so that $f \restriction \partial\mathbb{D} = g$. (*Hint*: Fejér's theorem.)

5. Let $\mathfrak{A}$ be a function algebra on a compact metrizable space. Let $\varphi \in \mathcal{R}(\mathfrak{A})$.

   (a) Let μ be an arbitrary measure on $\widehat{\mathfrak{A}}$ and $f \in \mathfrak{A}$ so that for every $A \in \widehat{\mathfrak{A}}$, $\int_A \widehat{f}(x)\, d\mu(x) = \mu(A) \int_{\widehat{\mathfrak{A}}} \widehat{f}(x)\, d\mu(x)$. Prove that $\widehat{f}$ is constant on the support of μ.

   (b) If φ has a representation of the form (6.10.11) with $\operatorname{supp}(\mu)$ not a single point, prove that φ is not an extreme point of $\mathcal{R}(A)$.

   (c) Conclude that if there is more than one μ obeying (6.10.11) (for the given φ), then φ is not an extreme point of $\mathcal{R}(A)$.

   (d) Suppose $\varphi(f) = f(x)$ and φ is not an extreme point of $\mathcal{R}(A)$. Prove there are multiple μ's obeying (6.10.11).

   (e) Conclude that $\operatorname{ch}(\mathfrak{A}) \subset \widehat{\mathfrak{A}}$ is the set of these x's in $\widehat{\mathfrak{A}}$ for which there is a unique measure μ on $\widehat{\mathfrak{A}}$ with

$$\widehat{f}(x) = \int \widehat{f}(y)\, d\mu(y) \tag{6.10.108}$$

6. Let $\mathfrak{A}$ be a function algebra on Y. A point $y \in Y$ is called a *peak point* for $\mathfrak{A}$ if and only if there exists $f \in \mathfrak{A}$ so that for all $x \neq y$, $f(y) > |f(x)|$. Prove that any peak point lies in any boundary of $\mathfrak{A}$ and so in $\text{ch}(\mathfrak{A})$.

Remarks. 1. Bishop [**68**] has proven that if Y is metrizable, then $\text{ch}(\mathfrak{A})$ is exactly the peak points of $\mathfrak{A}$.

2. In this case, $\text{ch}(\mathfrak{A})$ is contained in every boundary for $\mathfrak{A}$ and so is a minimal boundary (that, as we've seen, may be smaller than the Shilov boundary).

7. Verify that if $\mathfrak{A}$ is given by (6.10.17), then $\{f \restriction \partial\mathbb{D} \mid f \in \mathfrak{A}\}$ is given by (6.10.18).

8. Fix any arc $[\varepsilon, 2\pi - \varepsilon] = \{e^{i\theta} \in \partial\mathbb{D} \mid \varepsilon < \theta < 2\pi - \varepsilon\} \equiv I_\varepsilon \subset \partial\mathbb{D}$ for some small ε. On $\mathbb{D}$, define

$$f_{\theta,\delta}(z) = z(1-z)[z - (1+\delta)e^{i\theta}]^{-1} \tag{6.10.109}$$

which lies in the algebra $\mathfrak{A}$ of (6.10.17).

(a) Prove that for each $\varepsilon > 0$, there is δ_0 so that for $\delta < \delta_0$, and $\theta \in I_\varepsilon$, $|f_{\theta,\delta}(z)|$ takes its maximum at unique point $e^{i\Theta(\theta,\delta)}$ on $\partial\mathbb{D}$.

(b) As $\delta \downarrow 0$, prove that $\Theta(\varepsilon, \delta) \to \varepsilon$.

(c) Prove that any $e^{i\theta} \in I_\varepsilon$ is a peak point in the sense of Problem 6.

(d) Conclude that $\text{ch}(\mathfrak{A}) \supset \partial\mathbb{D} \setminus \{1\}$. (*Hint*: Use Problem 6.)

(e) Prove that $1 \notin \text{ch}(\mathfrak{A})$. (*Hint*: Prove $f \mapsto \int_0^\pi f(e^{\pm i\theta})\, \frac{d\theta}{2\pi}$ are distinct for $\pm$ and show δ_1 is not an extreme point.)

(f) Conclude $\text{Ch}(\mathfrak{A}) = \partial\mathbb{D} \setminus \{1\}$ and that it is not closed.

9. We saw in Example 6.10.12 (Swiss cheese) that $\overline{\mathbb{D}} \setminus X$ has positive d^2z measure indirectly since if the measure were 0, $\mathbb{R}(X)$ would be $C(X)$! This problems leads you through a direct proof of W. Allard. We'll consider slices $\text{Im}\, z = y$ constant.

(a) Let $n_j(y) = \#[\{\text{Im}\, z = y\} \cap \partial\mathbb{D}_j]$ which is $0, 1$, or 2. Prove that $\int_{-1}^1 n_j(y)\, dy = 2r_j$.

(b) Prove that $\int_{-1}^1 [\sum_j n_j(y)]\, dy < \infty$.

(c) Conclude that for a.e. $y \in (-1, 1)$, $\#\{y \mid \{\text{Im}\, z = y\} \cap \mathbb{D}_j \neq \emptyset\}$ is finite.

(d) If this number is finite for a given y, prove that $|\{\text{Im}\, z = y\} \cap X| > 0$, where $|\cdot|$ is one-dimensional Lebesgue measure. (*Hint*: $\overline{\mathbb{D}}_j \cap \overline{\mathbb{D}}_k = \emptyset$ if $j \neq k$.)

(e) Prove that the d^2z measure of X is positive.

10. Let $\mathfrak{A}$ be a function algebra with $\mathbb{1} \in \mathfrak{A}$. Let $\ell, q \in \widehat{\mathfrak{A}}$. We say q *dominates* ℓ, written $\ell \ll q$, if there is a C so that for all $f \in \mathfrak{A}$ with $\operatorname{Re} \widehat{f} > 0$ on $\widehat{\mathfrak{A}}$, we have

$$\operatorname{Re} \widehat{f}(\ell) \le C \operatorname{Re} \widehat{f}(q) \tag{6.10.110}$$

We say ℓ, q are *Harnack related* if $\ell \ll q$ and $q \ll \ell$. Recall that if $\mathbb{H}_+ = \{z \mid \operatorname{Re} z > 0\}$ and $F(z) = (1-z)/(1+z)$, then

$$F\colon \mathbb{D} \to \mathbb{H}_+, \qquad F\colon \mathbb{H}_+ \to \mathbb{D} \tag{6.10.111}$$

(a) If $\ell \ll q$, prove that $q \ll \ell$. (*Hint*: If $\operatorname{Re} \widehat{f} > 0$ on $\widehat{\mathfrak{A}}$, prove f is invertible. Also, if $\operatorname{Im} \widehat{f}(\ell) = 0$, prove $[\operatorname{Re} \widehat{f}(\ell)]^{-1} \le C|\widehat{f}(q)|^{-1}$.)

(b) Prove that being Harnack related is an equivalence relation.

(c) Suppose there exists $f_n \in \mathfrak{A}$ with $\|f_n\| = 1$ and $\widehat{f_n}(q) \to 1$, $\widehat{f}_n(\ell) \to -1$ and if $g_n = F((1-\frac{1}{n})f_n)$, then $\operatorname{Re} \widehat{g}_n > 0$, but $\operatorname{Re} \widehat{g}_n(\ell)/\operatorname{Re} \widehat{g}_n(q) \to \infty$. Conclude that if $\ell \ll q$, then $\|\ell - q\| < 2$.

(d) If $\ell \ll q$ fails, show there exists $g_n \in \mathfrak{A}$ so that $\operatorname{Re} \widehat{g}_n > 0$, $\widehat{g}_n(\ell) \to 0$, $|\widehat{g}_n(q)| \to \infty$. Let $f_n = F(g_n)$. Prove $\|f_n\|_\infty \le 1$, $\widehat{f}_n(\ell) \to 1$, $\widehat{f}_n(q) \to -1$, and conclude that $\|\ell - q\| = 2$. Conclude that $\|\ell - q\| < 2 \Leftrightarrow \ell \ll q \Leftrightarrow \ell \ll q$ and $q \ll \ell$.

(e) Conclude that $\ell \sim q \Leftrightarrow \|\ell - q\| < 2$ is an equivalence relation (ℓ and q lie in the same Gleason part).

11. (Assumes Problem 10) (a) In $\mathbb{A}(\overline{\mathbb{D}})$, prove if $y \in \partial\mathbb{D}$ and $w \in \overline{\mathbb{D}} \setminus \{y\}$, then $\|\delta_y - \delta_w\| = 2$. (*Hint*: $f_n(z) = \frac{1}{n} + \frac{1}{1+z\bar{y}}\frac{z\bar{y}}{}$.)

(b) If $y, w \in \mathbb{D}$, prove that $\|\delta_y - \delta_w\| < 2$. (*Hint*: Harnack's inequality.)

6.11. Bonus Section: The $L^1(\mathbb{R})$ Wiener and Ingham Tauberian Theorems

Our first goal in this section is to prove:

Theorem 6.11.1 (Wiener Tauberian Theorem for $L^1(\mathbb{R})$). *Let $h \in L^1(\mathbb{R}, dx)$. Let*

$$(\tau_x h)(y) = h(y-x) \tag{6.11.1}$$

Then the linear span of $\{\tau_x h \mid x \in \mathbb{R}\}$ is dense in L^1 if and only if for all $k \in \mathbb{R}$,

$$\widehat{h}(k) \ne 0 \tag{6.11.2}$$

The proof will not rely directly on the theory of Banach algebras but will indirectly, since we'll prove it using the Wiener Tauberian theorem for $\ell^1(\mathbb{Z})$. As an application, we'll prove a specialized-looking Tauberian theorem of

Ingham that, as we'll show in the next section, implies the prime number theorem!

Define $\mathbb{W}(\mathbb{R})$, the Wiener algebra for $\mathbb{R}$, to be set of bounded continuous f on $\mathbb{R}$ so that $\widehat{f} \in L^1$ with

$$\|f\|_{\mathbb{W}(\mathbb{R})} = (2\pi)^{-1/2}\|\widehat{f}\|_1 \tag{6.11.3}$$

Since pointwise product is, up to a factor of $\sqrt{2\pi}$, convolution of the Fourier transform, $\mathbb{W}(\mathbb{R})$ is a Banach algebra equivalent to $L^1(\mathbb{R}, dx)$ under the map $f \mapsto \widehat{f}$.

Note that if $h \in L^1$ and $\widehat{h}(k_0) = 0$, then $\widehat{\tau_x h}(k_0) = 0$ for all x, and so, for all q in the closure of the linear span, $\widehat{q}(k_0) = 0$ (since $q \mapsto \widehat{q}(k_0)$ is continuous). Thus, one-half of Theorem 6.11.1 is easy. For the deeper result, we note that if h is continuous of compact support, it is easy to approximate $h * g$ by finite sums of $\tau_x g$'s (replace the Riemann integral $\int h(y)(\tau_y g)\, dy$ by Riemann sums) and then, since C_0^∞ is dense in L^1, any $h * g$. Shifting to Fourier transforms, we see Theorem 6.11.1 is implied by the following, which is what we'll directly prove:

Theorem 6.11.2 (Wiener Tauberian Theorem for $\mathbb{W}(\mathbb{R})$). *Let $f \in \mathbb{W}(\mathbb{R})$ be everywhere nonvanishing. Then $\{gf \mid g \in \mathbb{W}(\mathbb{R})\}$ is dense in $\mathbb{W}(\mathbb{R})$.*

As a preliminary, we note C^1-functions on $\partial\mathbb{D}$ (respectively, $\mathbb{R}$) are in $\mathbb{W}(\partial\mathbb{D})$ (respectively, $\mathbb{W}(\mathbb{R})$ if f also has compact support). For in the $\partial\mathbb{D}$ case, we have, by the Schwartz inequality, that

$$\begin{aligned}\sum_n |f_n^\sharp| &\le \Big(\sum_n (1+n^2)^{-1}\Big)^{1/2}\Big(\sum_n (1+n^2)|f_n^\sharp|^2\Big)^{1/2} \\ &\le C(\|f\|_{L^2}^2 + \|f'\|_{L^2}^2)^{1/2}\end{aligned} \tag{6.11.4}$$

Similarly, using $\int (1+x^2)^{-1}\, dx < \infty$, we see that $\int |\widehat{f}(k)|\, dk$ is bounded by $C(\|f\|_2 + \|f'\|_2)$. We summarize in (using $\|q\|_{L^2} \le \|(1+x^2)^{-1}\|_1^{1/2}\|(1+x^2)^{1/2}q\|_\infty$):

Theorem 6.11.3. (a) *If f is C^1 on $\partial\mathbb{D}$, then $f \in \mathbb{W}(\partial\mathbb{D})$ and*

$$\|f\|_{\mathbb{W}(\partial\mathbb{D})} \le C(\|f\|_2 + \|f'\|_2) \tag{6.11.5}$$

(b) *If f is C^1 on $\mathbb{R}$ and xf and xf' are bounded, then $f \in \mathbb{W}(\mathbb{R})$ and*

$$\|f\|_{\mathbb{W}(\mathbb{R})} \le C(\|(1+x^2)^{1/2}f\|_\infty + \|(1+x^2)^{1/2}f'\|_\infty) \tag{6.11.6}$$

We now make a brief aside to prove a special case of Theorem 6.11.1 where we also suppose that

$$\int_{-\infty}^{\infty} |x||h(x)|\, dx < \infty \tag{6.11.7}$$

The proof is quite easy and this is the case we need for Theorem 6.11.9 below (which implies the prime number theorem): in that case, $f = K_1$ (given by (6.11.66)/(6.11.67)) and we prove that $|K_1(x)| \le Ce^{-|x|}$ so (6.11.7) holds.

Proof of Theorem 6.11.1 if (6.11.7) **holds (Kac).** Let $\mathbb{K}(\mathbb{R})$ be the set of C^1-functions of compact support. By (6.11.6), $\mathbb{K}(\mathbb{R})$ lies in $\mathbb{W}(\mathbb{R})$. By (6.11.7), $\widehat{h}(k)$ lies in $C^1(\mathbb{R})$ and is everywhere nonvanishing.

If $g \in \mathbb{K}(\mathbb{R})$, $\inf_{k \in \operatorname{supp}(g)} |\widehat{h}(k)| > 0$ since $\widehat{h}$ is C^1 and $\operatorname{supp}(g)$ is compact. Thus, $f = (\widehat{h})^{-1}g$ is also C^1, so $f \in \mathbb{K}(\mathbb{R})$, and so, $\check{g} = (2\pi)^{-1/2}(h * \check{f}) \in L^1$, that is, $\check{\mathbb{K}}(\mathbb{R}) \subset \{h * q \mid q \in L^1\}$.

Therefore, it suffices to prove $\check{\mathbb{K}}(\mathbb{R})$ is dense in L^1. If not, by the Hahn–Banach theorem, there exists $\varphi \in L^\infty(\mathbb{R})$, $\varphi \ne 0$, so that for all $g \in \mathbb{K}(\mathbb{R})$,

$$\int \varphi(x)\check{g}(x)\, dx = 0 \tag{6.11.8}$$

Pick any nonzero $g \in \mathbb{K}(\mathbb{R})$. Then for all $q \in \mathbb{R}$,

$$g_q(k) \equiv g(k-q) \tag{6.11.9}$$

is in $\mathbb{K}(\mathbb{R})$ and

$$\check{g}_q(x) = e^{iqx}\check{g}(x) \tag{6.11.10}$$

so by (6.11.8),

$$\widehat{(\varphi\check{g})} = 0 \tag{6.11.11}$$

so

$$(\varphi\check{g})(x) = 0 \text{ for a.e. } x \tag{6.11.12}$$

Since g has compact support, $\check{g}$ is analytic, so it has only countably many zeros on $\mathbb{R}$. Thus, (6.11.12) implies $\varphi(x) = 0$ for a.e. x, that is, $\check{\mathbb{K}}(\mathbb{R})$ is dense. $\square$

As a second preliminary to the general proof, we note that $C_0^\infty(\mathbb{R})$ is dense in $\mathbb{W}(\mathbb{R})$, for by (6.11.6), convergence in $\mathcal{S}(\mathbb{R})$ implies convergence in $\mathbb{W}(\mathbb{R})$, and since $\mathcal{S}(\mathbb{R})$ is dense in $L^1(\mathbb{R}, dx)$ and $\mathcal{S}(\mathbb{R})$ is left invariant by Fourier transform, we see $\mathcal{S}(\mathbb{R})$ is dense in $\mathbb{W}(\mathbb{R})$.

If $g \in C_0^\infty(\mathbb{R}^\nu)$ and j is C_0^∞ and equal to 1 on $\operatorname{supp}(g)$, then $g = fh \Leftrightarrow g = jfh$ so that we see it will suffice to show that g, f of bounded support in $\mathbb{W}(\mathbb{R})$ and $f \ne 0$ on $\operatorname{supp}(g)$ implies $g = hf$ for some $h \in \mathbb{W}(\mathbb{R})$. This reduces results on $\mathbb{W}(\mathbb{R})$ to ones only involving functions in a fixed interval, $[a, b]$, which we can map to a proper subinterval of $(-\pi, \pi)$ and thus think of the functions as ones on $\partial\mathbb{D}$ supported on $[-\pi + \varepsilon, \pi - \varepsilon]$. Such an f is associated to $\widetilde{f}$ on $\partial\mathbb{D}$ by

$$\widetilde{f}(e^{i\theta}) = f(\theta) \text{ for } -\pi \le \theta < \pi \tag{6.11.13}$$

We'll prove $f \in \mathbb{W}(\mathbb{R})$ if and only if $\widetilde{f} \in \mathbb{W}(\partial\mathbb{D})$ with comparable norms. We'll obtain the necessary $\mathbb{W}(\partial\mathbb{D})$ result by using the Wiener Tauberian theorem for $\mathbb{W}(\partial\mathbb{D})$ and localization with C^1 partitions of unity. We begin with the $f, \widetilde{f}$ result:

Theorem 6.11.4. *Fix $\varepsilon \in (0, \pi)$. Let f be a function on $\mathbb{R}$ supported on $[-\pi+\varepsilon, \pi-\varepsilon]$. Define $\widetilde{f}$ on $\partial\mathbb{D}$ by (6.11.13). Then $f \in \mathbb{W}(\mathbb{R}) \Leftrightarrow \widetilde{f} \in \mathbb{W}(\partial\mathbb{D})$ and there are constants $C_1, C_2 \in (0, \infty)$ (depending only on ε) so that*

$$C_1 \|\widetilde{f}\|_{\mathbb{W}(\partial\mathbb{D})} \le \|f\|_{\mathbb{W}(\mathbb{R})} \le C_2 \|\widetilde{f}\|_{\mathbb{W}(\partial\mathbb{D})} \tag{6.11.14}$$

Proof. Fix $\alpha \in \mathbb{R}$. Let $e^{-i\alpha\theta}$ be the function on $\partial\mathbb{D}$ which has that value for $-\pi \le \theta < \pi$. So long as f, and so $\widetilde{f}$, is L^2, we have

$$(e^{-i\alpha\theta}\widetilde{f})^\sharp_n = \big(\sqrt{2\pi}\,\big)^{-1} \widehat{f}(n+\alpha) \tag{6.11.15}$$

Suppose first $\widetilde{f} \in \mathbb{W}(\partial\mathbb{D})$. Let $\psi \in C_0^\infty(\partial\mathbb{D})$ so that $\psi \equiv 1$ on $[-\pi + \varepsilon, \pi - \varepsilon]$ and $\text{supp}(\psi) \subset [-\pi + \frac{1}{2}\varepsilon, \pi - \frac{1}{2}\varepsilon]$. Then $e^{i\alpha\theta}\psi$ is in C_0^∞ and its derivative is continuous in α, so by (6.11.5),

$$C_2 = (2\pi)^{1/2} \sup_{0 \le \alpha < 1} \|e^{i\alpha\theta}\psi\|_{\mathbb{W}(\partial\mathbb{D})} < \infty \tag{6.11.16}$$

Since $\widetilde{f}\psi = \widetilde{f}$ and $\|\cdot\|$ is a Banach algebra norm,

$$\|e^{i\alpha\theta}\widetilde{f}\|_{\mathbb{W}(\partial\mathbb{D})} \le (2\pi)^{-1/2} C_2 \|f\|_{\mathbb{W}(\partial\mathbb{D})} \tag{6.11.17}$$

Thus, by (6.11.15),

$$\int_{-\infty}^{\infty} |\widehat{f}(k)|\, dk = \sum_{n=-\infty}^{\infty} \int_0^1 |\widehat{f}(n+\alpha)|\, d\alpha \tag{6.11.18}$$

$$= (2\pi)^{1/2} \int_0^1 \|e^{-i\alpha\theta}\widetilde{f}\|_{\mathbb{W}(\partial\mathbb{D})}\, d\alpha \tag{6.11.19}$$

$$\le C_2 \|\widetilde{f}\|_{\mathbb{W}(\partial\mathbb{D})}$$

proving $f \in \mathbb{W}(\mathbb{R})$ and the second inequality in (6.11.14) holds.

Suppose next $f \in \mathbb{W}(\mathbb{R})$. Since $\text{supp}(f) \subset [-\pi + \varepsilon, \pi + \varepsilon]$, we can define a C_0^∞-function g so

$$g(x) f(x) = x f(x) \tag{6.11.20}$$

for all x. Since $\widehat{xf} = i(\widehat{f})'$ and $\|fg\|_{\mathbb{W}(\mathbb{R})} \le \|f\|_{\mathbb{W}(\mathbb{R})} \|g\|_{\mathbb{W}(\mathbb{R})}$, we conclude for a constant C_3, we have

$$\int |(\widehat{f})'(k)|\, dk \le C_3 \|f\|_{\mathbb{W}(\mathbb{R})} \tag{6.11.21}$$

Note that

$$|y - n| \le \tfrac{1}{2} \Rightarrow |\widehat{f}(y) - \widehat{f}(n)| \le \tfrac{1}{2} \int_{n-\frac{1}{2}}^{n+\frac{1}{2}} |(\widehat{f})'(k)|\, dk \tag{6.11.22}$$

to conclude

$$|\widehat{f}(n)| \le \int_{n-\frac{1}{2}}^{n+\frac{1}{2}} [|\widehat{f}(k)| + \tfrac{1}{2}\,|(\widehat{f})'(k)|]\, dk \tag{6.11.23}$$

Using (6.11.15) with $\alpha = 0$,

$$\begin{aligned} \|\widehat{f}\|_{\mathbb{W}(\partial\mathbb{D})} &\le (\sqrt{2\pi}\,)^{-1} \int_{-\infty}^{\infty} [|\widehat{f}(k)| + \tfrac{1}{2}\,|(\widehat{f})'(k)|]\, dk \\ &\le (\sqrt{2\pi}\,)^{-1}(1 + \tfrac{1}{2}\, C_3)\|f\|_{\mathbb{W}(\mathbb{R})} \end{aligned} \tag{6.11.24}$$

proving $\widetilde{f} \in \mathbb{W}(\partial\mathbb{D})$ and the first inequality in (6.11.14) with $C_1 = [(\sqrt{2\pi}\,)^{-1}(1+\frac{1}{2}C_3)]^{-1}$. □

Let I, J be two proper open intervals on $\partial\mathbb{D}$ with $I \cup J = \partial\mathbb{D}$. In the usual way (see Theorem 1.4.6 of Part 2A), we can find j_I, j_J, C_0^∞ with values in $[0,1]$ so $\operatorname{supp}(j_I) \subset I$, $\operatorname{supp}(j_J) \subset J$, and

$$j_I^2 + j_J^2 = 1 \tag{6.11.25}$$

In particular,

$$e^{i\theta} \in I \setminus J \Rightarrow j_I(e^{i\theta}) = 1 \tag{6.11.26}$$

Proposition 6.11.5. *Let $f, f_1 \in \mathbb{W}(\partial\mathbb{D})$ with* $\operatorname{supp}(f_1)$ *contained in a proper interval $I_1 \subset \partial\mathbb{D}$ and suppose*

$$|f(e^{i\theta})| > 0 \quad \text{on } I_1 \tag{6.11.27}$$

Then there exists $g \in \mathbb{W}(\partial\mathbb{D})$ with $\operatorname{supp}(g) \subset \operatorname{supp}(f_1)$ *so that*

$$f_1 = fg \tag{6.11.28}$$

Remark. This is connected with a notion of localizing functions in $\mathbb{W}(\partial\mathbb{D})$. See Problem 1.

Proof. By (6.11.27) and continuity of f, there is slightly larger interval I containing I_1 and $1 \ge \varepsilon > 0$ so that

$$|f(e^{i\theta})| > \varepsilon \quad \text{on } \bar{I} \tag{6.11.29}$$

Let $J = \partial\mathbb{D} \setminus I_1$ so $J \cup I = \partial\mathbb{D}$. Let j_I, j_J be the functions above obeying (6.11.25), so by (6.11.29), $\bar{f}f \ge \varepsilon^2 j_I^2$, so if

$$h = \bar{f}f + j_J^2 \tag{6.11.30}$$

then $h \ge \varepsilon^2$, and so h is everywhere nonvanishing. By the Wiener Tauberian theorem for $\mathbb{W}(\partial\mathbb{D})$, $h^{-1} \in \mathbb{W}(\partial\mathbb{D})$, so

$$g = f_1\bar{f}h^{-1} \tag{6.11.31}$$

is in $\mathbb{W}(\partial\mathbb{D})$ also. Since $j_J = 0$ on $\operatorname{supp}(f_1)$, $g = f_1\bar{f}(\bar{f}f)^{-1} = f_1 f^{-1}$ (with $0 \cdot \infty = 0$) and (6.11.28) holds. □

Theorem 6.11.4 extends the above result to $f_1, f \in \mathbb{W}(\mathbb{R})$ with supports in $[-\pi+\varepsilon, \pi-\varepsilon]$ and then by translation and scaling any interval. We have thus proven:

Proposition 6.11.6. *Let $f, f_1 \in \mathbb{W}(\mathbb{R})$ so $\text{supp}(f_1) \subset [a,b]$, $f > 0$ on $[a,b]$ and f of compact support. Then there exists $g \in \mathbb{W}(\mathbb{R})$ so that $\text{supp}(g) \subset \text{supp}(f_1)$ and*

$$f_1 = fg \tag{6.11.32}$$

Proof of Theorem 6.11.1. As noted, we need only prove for any $f_1 \in \mathbb{W}(\mathbb{R})$ of compact support, we have (6.11.32) for some $g \in \mathbb{W}(\mathbb{R})$. Pick $[a,b]$ with $\text{supp}(f_1) \subset [a,b]$ and $j \in C_0^\infty$ with $j \equiv 1$ on $[a,b]$. Let $f_2 = jf$ and g so $\text{supp}(g) \subset \text{supp}(f_1) \subset [a,b]$ and $f_1 = gf_2 = gjf = gf$. $\square$

We can explore the Tauberian aspect of Theorem 6.11.1.

Definition. A function q on $\mathbb{R}$ is called *slowly oscillating* if and only if $\forall \varepsilon > 0$, $\exists \delta$ and R so that

$$x > R, \quad |x-y| < \delta \Rightarrow |q(x) - q(y)| < \varepsilon \tag{6.11.33}$$

We say q is *almost monotone* if and only if $\forall \varepsilon > 0$, $\exists \delta$ and R so that $R < x < y < x + \delta \Rightarrow q(x) \le q(y) + \varepsilon$.

Theorem 6.11.7. *Let $f \in L^1(\mathbb{R})$ with $\widehat{f}$ everywhere nonvanishing and let $q \in L^\infty(\mathbb{R})$. Suppose that*

$$\lim_{x\to\infty} \int f(x-y)q(y)\,dy = \alpha \int f(x)\,dx \tag{6.11.34}$$

Then

(a) *For all $g \in L^1$,*

$$\lim_{x\to\infty} \int g(x-y)q(y)\,dy = \alpha \int g(x)\,dx \tag{6.11.35}$$

(b) *If also q is slowly oscillating, then*

$$\lim_{x\to\infty} q(x) = \alpha \tag{6.11.36}$$

(c) (6.11.36) *holds if q is almost monotone.*

Remark. (a) is sometimes called the Wiener Tauberian theorem. (b) is sometimes called *Pitt's Tauberian theorem*.

Proof. (a) Since for x_0 fixed, $x \to \infty$ if and only if $x - x_0 \to \infty$, (6.11.34) holds for translates of f, so by taking linear combinations and limits (using $\|g * q - h * q\|_\infty \le \|g-h\|_1 \|q\|_\infty$), (6.11.35) holds for all g on account of Theorem 6.11.1.

(b) Fix $\varepsilon > 0$. Let g be the characteristic function of $(0, \delta)$. Then, if $|x| > R$,

$$|\delta q(x) - (g * q)(x)| = \int_x^{x+\delta} |q(x) - q(y)|\, dy \le \varepsilon\delta \tag{6.11.37}$$

Since $(g * q)(z) \to \alpha\delta$ by (a), we see that

$$\limsup_{x\to\infty} q(x) \le \alpha + \varepsilon, \qquad \liminf_{x\to\infty} q(x) \ge \alpha - \varepsilon \tag{6.11.38}$$

Since ε is arbitrary, we get (6.11.36).

(c) By (a), for any δ,

$$\frac{1}{\delta}\int_x^{x+\delta} q(y)\, dy \to \alpha \tag{6.11.39}$$

Given ε, pick δ, R as in the definition of almost monotone. Then for $x > R$ and $y \in [x, x + \delta]$,

$$q(x) \le \varepsilon + q(y) \le 2\varepsilon + q(x + \delta) \tag{6.11.40}$$

so

$$q(x) \le \varepsilon + \frac{1}{\delta}\int_x^{x+\delta} q(y)\, dy \le 2\varepsilon + q(x + \delta) \tag{6.11.41}$$

Going through subsequences where $q(x_n) \to \limsup q(x)$ and another where $q(x_n + \delta) \to \liminf q(x)$, we conclude

$$\limsup q(x) \le \alpha + \varepsilon \le 2\varepsilon + \liminf q(x) \tag{6.11.42}$$

Since ε is arbitrary, we have (6.11.36). □

Example 6.11.8. Let

$$q(x) = \sin x^2 \tag{6.11.43}$$

Then $q = \frac{d}{dx} q_1 + q_2$, where

$$q_1(x) = -(2x)^{-1} \cos x^2, \qquad q_2(x) = -(2x^2)^{-1} \cos x^2 \tag{6.11.44}$$

Therefore, if $f \in C_0^\infty$, by integrating by parts for the q_1 term and using $q_j(x) \to 0$, we have $q * f \to 0$. Since C_0^∞ is dense in $L^1(\mathbb{R})$ and $q \in L^\infty$, $q * f \to 0$ for all f in L^1. But $q(x)$ does not go to zero pointwise. Thus, we see some kind of additional condition, like slowly oscillating, is needed and even why "slow oscillation" is the right name. □

As a final topic in this section, we will prove the following:

Theorem 6.11.9 (Ingham's Tauberian Theorem). *Let $f(x)$ be a positive increasing function on $(1, \infty)$ and define*

$$F(x) = \sum_{\substack{n \le x \\ n \in \mathbb{Z}}} f\left(\frac{x}{n}\right) \tag{6.11.45}$$

Suppose for $a > 0$ and b real, as $x \to \infty$,

$$F(x) = ax \log x + bx + o(x) \tag{6.11.46}$$

Then

$$\lim_{x\to\infty} x^{-1} f(x) = a \tag{6.11.47}$$

Remarks. 1. For each x, the sum in (6.11.45) is finite with $[x]$ terms.

2. As we discuss in the Notes, there are two senses of Tauberian theorem. In the Abelian/Tauberian pair sense, this does not fit since the conclusion only involves a, but there is a stronger conclusion which we won't prove (or need) which gives such a pair; see the Notes. In the other sense of going from a summation result to a direct limit, this does fit the mold.

As we'll see, this theorem implies the prime number theorem fairly directly without any further use of complex analysis. That theorem typically relies on the nonvanishing of the Riemann zeta function, $\zeta(s)$, (defined below) on the line $\operatorname{Re} s = 1$. We'll use this in the proof of Ingham's theorem because a Fourier transform will be "essentially" $\zeta(1+is)$ and that will allow us to apply the almost monotone extension of the Wiener Tauberian theorem! So we begin with a summary of some facts about the function, $\zeta(s)$, discussed in detail in Chapter 13 of Part 2B. ζ is defined for $\operatorname{Re} s > 1$ by

$$\zeta(x) = \sum_{n=1}^{\infty} n^{-s} \tag{6.11.48}$$

Noting that

$$s \int_n^{n+1} [x] x^{-s-1} = n[n^{-s} - (n+1)^{-s}] \tag{6.11.49}$$

and that

$$\sum_{n=1}^{N} n(b_n - b_{n+1}) = \sum_{n=1}^{N} b_n[n - (n-1)] - Nb_{N+1} \tag{6.11.50}$$

we see that if $\operatorname{Re} s > 1$ (so $N(N+1)^{-s} \to 0$), then

$$\operatorname{Re} s > 1 \Rightarrow \zeta(s) = s \int_1^{\infty} [x] x^{-s-1}\, dx \tag{6.11.51}$$

Noting that $s \int_1^\infty x^{-s}\, dx = s/(s-1)$ and that $\int_1^\infty |[x]-x| x^{-\operatorname{Re} s - 1}\, dx < \infty$ if $\operatorname{Re} s > 0$, we see directly that

$$\zeta(s) - \frac{1}{s-1} \text{ has analytic continuation into } \operatorname{Re} s > 0 \tag{6.11.52}$$

Besides the formula (6.11.51) and this analyticity, we'll need that

$$y \in \mathbb{R} \setminus \{0\} \Rightarrow \zeta(1+iy) \neq 0 \tag{6.11.53}$$

a deep fact proven in Theorem 13.3.5 of Part 2B. We now return to the analysis of f, F in Theorem 6.11.9.

Lemma 6.11.10. *If f is extended to $[0,\infty)$ by setting it to 0 on $[0,1)$ and F, f are related by (6.11.45) and F obeys (6.11.46), then for some $K > 0$ and all $x > 0$,*

$$f(x) \le Kx \tag{6.11.54}$$

Remark. Ingham notes his proof of this, which we give below, was motivated by Chebyshev's argument that we gave in Proposition 13.5.3 of Part 2B.

Proof. Because we have set f to be zero on $[0,1]$, we can make the sums formally infinite so

$$\begin{aligned} F(x) - 2F(\tfrac{1}{2}x) &= \sum_{n=1}^{\infty} f\left(\frac{x}{2n-1}\right) + \sum_{n=1}^{\infty} f\left(\frac{x}{2n}\right) - 2f\left(\frac{x}{2n}\right) \\ &= \sum_{n=1}^{\infty}\left[f\left(\frac{x}{2n-1}\right) - f\left(\frac{x}{2n}\right)\right] \ge f(x) - f(\tfrac{1}{2}x) \end{aligned} \tag{6.11.55}$$

since f is monotone on $[0,\infty)$ by the assumptions on the original f. On the other hand, by (6.11.46),

$$F(x) - 2F(\tfrac{1}{2}x) = ax\log 2 + o(x) \le \tfrac{1}{2}Kx \tag{6.11.56}$$

for some $K > 0$ and all x in $[0,\infty)$.

Since $f(\frac{1}{2^\ell}x) = 0$ for large ℓ, using (6.11.55) and (6.11.56),

$$f(x) = \sum_{\ell=0}^{\infty}\left[f\left(\frac{1}{2^\ell}x\right) - f\left(\frac{1}{2^{\ell+1}}x\right)\right] \le \tfrac{1}{2}Kx[1+\tfrac{1}{2}+\tfrac{1}{4}+\dots] = Kx \tag{6.11.57}$$

$\square$

Next, we note that, by (6.11.50),

$$\begin{aligned} \int_1^x \frac{F(u)}{u}\,du &= \int_1^x (a\log u + b + o(1))\,du \\ &= ax\log x + (b-a)x + o(x) \end{aligned} \tag{6.11.58}$$

On the other hand,

$$\int_1^x \frac{F(u)}{u}\,du = \int_1^x \sum_{n\le u} f\left(\frac{u}{n}\right)\frac{du}{u} \tag{6.11.59}$$

$$= \sum_{n\le x}\int_n^x f\left(\frac{u}{n}\right)\frac{du}{u} \tag{6.11.60}$$

$$= \sum_{n \le x} \int_1^{x/n} f(y)\, \frac{dy}{y} \tag{6.11.61}$$

$$= \int_1^x \frac{f(y)}{y} \left[\frac{x}{y}\right] dy \tag{6.11.62}$$

$$= \int_0^\infty \frac{f(y)}{y} \left[\frac{x}{y}\right] dy \tag{6.11.63}$$

where we extend the region of integration from 0 to 1 since $f(y)$ is 0 there and from x to ∞ since $[\frac{x}{y}] = 0$ there.

Thus, we have that

$$\frac{1}{x} \int_0^\infty \frac{f(y)}{y} \left[\frac{x}{y}\right] dy = a \log x + (b-a) + o(1) \tag{6.11.64}$$

This is a multiplication convolution! If we look at $(0,\infty)$ as a group under multiplication, $\frac{dy}{y}$ is Haar measure and $[\frac{x}{y}]$ is xy^{-1}. To make it look like an L^1-convolution, we change variables exponentially, that is, set $x = e^u$, $y = e^v$, so $\frac{dy}{y} = dv$ and the $\frac{1}{x}$ in front of (6.11.65) is $e^{-u} = e^{-v}e^{-(u-v)}$. So (6.11.64) becomes

$$\int_{-\infty}^\infty \phi(v) K_0(u-v)\, dv = au + (b-a) + o(1) \tag{6.11.65}$$

$$\phi(v) = e^{-v} f(e^v), \qquad K_0(u) = e^{-u}[e^u] \tag{6.11.66}$$

Notice since $\phi(v) = 0$ if $v \le 0$, we have, by (6.11.54), that $\phi \in L^\infty$. The theorem is reduced to showing (6.11.65) $\Rightarrow \lim_{v\to\infty} \phi(v) = a$.

This is almost the framework for using the almost monotone extension of Wiener's theorem. There isn't a limit as $u \to \infty$ (since we have au, not a) and, more importantly, K_0 is *not* in L^1 since $K_0(u) \to 1$ at ∞. The reason the integral exists in (6.11.65) is that $\phi(v) = 0$ if $v < 0$ and $K_0(u) = 0$ if $u < 0$ (so the integral is really from $v = 0$ to $v = u$). The key is to pick $\mu, \lambda > 0$ (to be specified shortly) and pick

$$K_1(u) = 2K_0(u) - K_0(u-\mu) - K_0(u-\lambda) \tag{6.11.67}$$

in which case (6.11.65) becomes

$$\int_{-\infty}^\infty \phi(v) K_1(u-v)\, dv = a(\mu+\lambda) \tag{6.11.68}$$

The two key lemmas are:

Lemma 6.11.11. *ϕ is almost monotone.*

Lemma 6.11.12. *$K_1 \in L^1$ and if μ/λ is irrational, then $\widehat{K}_1(k) \neq 0$ for all k. Moreover,*

$$\int K_1(u)\, du = (\mu + \lambda) \tag{6.11.69}$$

Given these two lemmas, we have

Proof of Theorem 6.11.9 given Lemmas 6.11.11 and 6.11.12. Pick μ, λ so μ/λ is irrational. By (6.11.69), the right side of (6.11.68) is $a \int K_1(u)\, du$. By Lemma 6.11.12, the Wiener theorem applies, so by Lemma 6.11.11 and the almost monotone extension, $\phi(u) \to a$. As noted, this implies $x^{-1} f(x) \to a$. $\square$

Proof of Lemma 6.11.11. If $x < y$, then $f(e^x) \leq f(e^y)$, so $e^{-x} f(e^x) \leq e^{y-x}[e^{-y} f(e^y)]$. Thus, $x < y < x+\delta \Rightarrow \varphi(x) \leq e^\delta \varphi(y) \leq \varphi(y) + (e^\delta - 1)\varphi(y)$.

Recall $\varphi(y) \leq K$, so given ε, pick δ so $(e^\delta - 1)K < \varepsilon$ and get $x < y < x + \delta \Rightarrow \varphi(x) \leq \varphi(y) + \varepsilon$. $\square$

Proof of Lemma 6.11.12. We start by noting that

$$x - 1 < [x] \leq x \Rightarrow e^u - 1 < [e^u] \leq e^u \Rightarrow 1 - e^{-u} \leq e^{-u}[e^u] \leq 1$$

so

$$|K_0(u) - 1| \leq e^{-u} \tag{6.11.70}$$

so

$$|K_1(u)| \leq (1 + e^\mu + e^\lambda) e^{-u} \tag{6.11.71}$$

lies in $L^1(\mathbb{R})$ (remember $K_0(u) = 0$ for $u < 0$, so $K_1(u) = 0$ for $u < 0$).

Now let $\operatorname{Re} s > 0$. Letting $y = e^u$, we find

$$\int_{-\infty}^{\infty} K_0(u) e^{-us}\, du = \int_0^\infty \frac{[y]}{y^{2+s}}\, dy \tag{6.11.72}$$

$$= \frac{\zeta(1+s)}{1+s} \tag{6.11.73}$$

on account of (6.11.51).

Translating to get $K_0(u-\mu)$ and $K_0(u-\lambda)$ and then, because K_1 is L^1, taking the limit as $\operatorname{Re} s \downarrow 0$, we find

$$\widehat{K}_1(k) = (2\pi)^{-1/2}(2 - e^{-ik\lambda} - e^{-ik\mu}) \frac{\zeta(1+ik)}{1+ik} \tag{6.11.74}$$

Since $\lim_{k\downarrow 0}[\zeta(1+ik)/(1+ik)](ik) = 1$, we conclude that (6.11.69) holds. Moreover, $\operatorname{Re}(1 - e^{-ik\alpha}) \geq 0$ and is 0 only if $k\alpha \in 2\pi\mathbb{Z}$, we see that μ/λ irrational implies $(2 - e^{-ik\lambda} - e^{-ik\mu}) \neq 0$ for all $k \in \mathbb{R}$ so that (6.11.53) and (6.11.74) imply that $\widehat{K}_1(k) \neq 0$ for all $k \in \mathbb{R}$. $\square$

Notes and Historical Remarks. The L^1 Wiener Tauberian theorem is from Wiener [**757**]. Pitt's extension is from [**527**] and Ingham's Tauberian theorem from Ingham [**333**].

The proof of the special case of Wiener's theorem on $L^1(\mathbb{R})$, where $(1+|x|)f \in L^1$, is due to Kac [**358**]. Actually, Kac required $(1+|x|^2)f \in L^1$ and uses C^2-functions of compact support where we use C^1. By the Riemann–Lebesgue lemma if $g \in L^1$, g is C^2 and $g'' \in L^1$, then $|\widehat{g}(k)| \leq C(1+k^2)^{-1}$, and so in L^1, which is what Kac uses. We instead use the slightly more subtle fact that, by using the Plancherel theorem, g C^1 with $g, g' \in L^2$ implies $\widehat{g} \in L^1$, so we only need $(1+|x|)f \in L^1$. Actually, by using Hölder continuous functions, one only needs $(1+|x|^\alpha)f \in L^1$ for some $\alpha > \frac{1}{2}$ (see Problem 3).

The proof we give of the full Wiener $L^1(\mathbb{R})$ theorem is from Gel'fand–Raikov–Shilov [**236**].

The Wiener theorem holds for general LCA groups, that is, the result that $f \in L^1(G, d\mu)$ and its translates generate a dense subset of all of L^1 if and only if $\widehat{f}(\chi) \neq 0$ for all $\chi \in \widehat{G}$. This is proven, for example, in Rudin [**583**]. The general $L^1(G, d\mu)$ result is due to Godement [**253**] and Segal [**614**].

The Wiener theorem can be translated to a statement about ideals in $\mathfrak{A} = L^1(G, d\mu)$, namely, the closure of the principal ideal $\{f * g \mid g \in L^1\}$ generated by f is either contained in a maximal proper ideal or is all of $\mathfrak{A}$. It is true and not too hard to extend this to the statement that every closed proper ideal in $\mathfrak{A}$ is contained in a maximal proper ideal (the analog for algebras with identity is easy).

There is a further generalization: Recall that the hull, $h(\mathcal{I})$, of an ideal $\mathcal{I} \subset \mathfrak{A} = L^1(G, d\mu)$ is $\{\chi \in \widehat{G} \mid \widehat{f}(\chi) = 0 \text{ for all } f \in \mathcal{I}\}$. If $f \in \mathfrak{A}$, we set $Z(f)$, the *zero set* of f, to be the set of zeros of $\widehat{f}$. Thus, $f \in \mathcal{I} \Rightarrow h(\mathcal{I}) \subset Z(f)$.

The general Wiener theorem (which some authors call the Wiener–Shilov theorem) says that if $\mathcal{I}$ is a closed ideal and (i) $h(\mathcal{I}) \subset Z(f)$; (ii) $\partial(Z(f)) \cap \partial(h(\mathcal{I}))$ contains no nonempty perfect set (recall perfect means closed with no isolated points), then $f \in \mathcal{I}$. Note that if $\mathcal{I}$ is the ideal generated by $g \in L^1$, where $\widehat{g}$ is everywhere nonvanishing, then $h(\mathcal{I}) = \emptyset$ and (i), (ii) hold for all f, so $\mathcal{I} = \mathfrak{A}$. That is, this does generalize Wiener's theorem. The theorem in this generality is due to Loomis [**456**] and Naimark [**490**]. For a detailed exposition, see the books of Rudin [**583**] or Hewitt–Ross [**316**, Vol. II]. We note that there were seminal papers by Ditkin [**161**] and Shilov [**624**] along the way—see [**316**, Sect. 39 Notes] for many more references.

While on the subject of ideals in abelian Banach algebras, especially L^1, we should mention the issue of *spectral synthesis*, a subject related to the Wiener theorem and heavily studied in the middle of the twentieth century. Let $E \subset \widehat{\mathfrak{A}}$ be a closed set. If $\mathfrak{A}$ is regular, we've seen in Section 6.2 that $E = h(k(E))$ so E is the hull of a closed ideal in $\mathfrak{A}$. One says E is an S-set (S for synthesis) if there is a unique closed ideal, $\mathcal{I}$, with $E = h(\mathcal{I})$. Wiener's theorem says that the empty set, $\emptyset$, is an S-set. The books of Rudin [**583**], Hewitt–Ross [**316**], and Benedetto [**50**] discuss spectral synthesis. (The name comes from the following: Associated to any ideal, $\mathcal{I} \subset L^1$, is the translation invariant subspace $S \subset L^\infty$ of those $\varphi \in L^\infty$ with $\int \varphi(x) f(x)\, d\mu(x) = 0$ for all $f \in \mathcal{I}$. $h(\mathcal{I})$ is precisely the set of $\chi \in S$ and is called the spectrum of S. If the spectrum uniquely determines S, we can "synthesize" S from its spectrum.)

There are two slightly different meanings to "Tauberian": one as an "Abelian–Tauberian" pair and one as conditions under which convergence of a summability method implies convergence of the original sum. The Abelian–Tauberian pairing is named after the first example. In 1816, Abel [**1**] proved that if

$$\sum_{n=0}^{N} a_n \to \alpha \tag{6.11.75}$$

then

$$\lim_{x\uparrow 1} \sum_{n=0}^{\infty} a_n x^n = \alpha \tag{6.11.76}$$

Notice an example like $a_n = (-1)^n$ has (6.11.76) with $\alpha = \frac{1}{2}$ but not (6.11.75). Thus, (6.11.76) is called a *summability method* (indeed, it is abelian summation) and Abel's theorem says that this method is a *regular summability method.* Hardy's book, *Divergent Series*, [**290**] has an extensive discussion of summability methods and their history.

In 1897, Tauber [**697**] (the Notes to Section 13.0 of Part 2B have a little about his life) proved a kind of partial converse to Abel's theorem, namely,

$$na_n \to 0 + (6.11.76) \Rightarrow (6.11.75) \tag{6.11.77}$$

The name Tauberian theorem thus sometimes means there is a paired Abelian converse result where the conclusion implies that the asymptotic result in the hypotheses holds. But the term Tauberian theorem has also been used to describe a theorem that, under extra hypotheses, goes from a complicated asymptotic result to a simpler one.

As we stated it, Ingham's Tauberian theorem (Theorem 6.11.9) is in the second category of Tauberian theorem. Since the hypothesis involves both a and b in (6.11.46) but the conclusion (6.11.47) involves only a, one cannot hope to go backwards. But Ingham's paper has a stronger result, namely, he also concludes that

$$\lim_{R\to\infty}\int_1^R \frac{f(x)-ax}{x^2}\,dx = b - a\gamma \tag{6.11.78}$$

(where γ is the Euler–Mascheroni constant) and (6.11.47) plus (6.11.78) do imply (6.11.46), providing an Abelian companion to Ingham's theorem.

We conclude with a brief summary of the highlights of the history of Tauberian theorems. For much more, see the monumental monograph of Korevaar [**404**] (which exposes the theory itself) or the book review of Borwein [**83**]. In 1910, Hardy [**288**] proved that Cesàro summability plus a bound $|a_n| \le C/n$ implies summability and asked if the same was true for abelian summability. This was answered affirmatively by Littlewood [**453**] in 1911, that is, he proved

$$|a_n| \le \frac{C}{n} + \lim_{x\uparrow 1}\sum_{n=0}^{\infty} a_n x^n = \alpha \Rightarrow \lim_{N\to\infty}\sum_{n=0}^{N} a_n = \alpha \tag{6.11.79}$$

Littlewood's proof was quite complicated and this was regarded as a very difficult result for twenty years until a big breakthrough in Tauberian theory by Karamata [**371**]. (Problems 4, 6, and 7 have a proof of (6.11.79) using a variant of Karamata's proof due to Wielandt [**755**].) It also turns out that (6.11.79) also follows fairly easily from Pitt's extension of the Wiener Tauberian theorem (see Problem 5).

Littlewood's paper led to a long collaboration with Hardy (see the Notes to Section 2.3 of Part 3 for more on them and their collaboration), including a paper where they named "Tauberian" [**291**], and several others on the subject [**292, 294, 295**]. In his great 1932 paper, Wiener said: "I feel it would be far more appropriate to term these theorems Hardy–Littlewood theorems, were it not that usage has sanctioned the other appellation." In particular, Hardy and Littlewood showed how some of their theorems implied the prime number theorem (PNT) [**295**], but their proof of this Tauberian theorem alas required results stronger than the PNT.

Ingham and Schmidt encouraged Wiener to look at the Tauberian theorems in his attempts to compare asymptotics, resulting in a 1928 breakthrough [**756**], a proof of the PNT by Wiener's student, Ikehara [**332**], and then his great 1932 seminal paper [**757**] that eliminated restrictions he had earlier.

Problems.

1. A function, f, on $\partial\mathbb{D}$ (respectively, $\mathbb{R}$) is said to be locally in $\mathbb{W}(\partial\mathbb{D})$ (respectively, $\mathbb{W}(\mathbb{R})$) at z_0 if there is $g \in \mathbb{W}(\partial\mathbb{D})$ (respectively, $\mathbb{W}(\mathbb{R})$) so that $f(z) = g(z)$ in a neighborhood of z_0. In the case of $\mathbb{R}$, it is said to be locally in $\mathbb{W}(\mathbb{R})$ near infinity if there is $g \in \mathbb{W}(\mathbb{R})$ so that for some

$R > 0$, $f(x) = g(x)$ for $|x| > R$. Prove that if f is locally in $\mathbb{W}(\partial\mathbb{D})$ (respectively, $\mathbb{W}(\mathbb{R})$) at every $z_0 \in \partial\mathbb{D}$ (respectively, $z_0 \in \mathbb{R} \cup \{\infty\}$), then f is in $\mathbb{W}(\partial\mathbb{D})$ (respectively, $\mathbb{W}(\mathbb{R})$). (*Hint*: Partition of unity.)

2. Prove Proposition 6.11.5 using the result in Problem 1.

3. Fix $\alpha \in (\frac{1}{2}, 1)$.

   (a) If
$$\int (1 + |x|^\alpha)|f(x)|\, dx < \infty \tag{6.11.80}$$
   prove that $\widehat{f}$ is uniformly Hölder-continuous of order α, that is, for some C and all k, ℓ with $|k - \ell| \le 1$,
$$|\widehat{f}(k) - \widehat{f}(\ell)| \le C|k - \ell|^\alpha \tag{6.11.81}$$
   (b) If $\widehat{f}$ is a continuous function of compact support and obeys (6.11.81), prove that $f \in L^1(\mathbb{R})$.

   (c) Extend the proof of Theorem 6.11.1 to the case where (6.11.80) holds.

4. This problem will lead the reader through a proof of Littlewood's Tauberian theorem, (6.11.79), due to Wielandt (using ideas of Karamata; see the Notes). Without loss, by replacing a_0 by $a_0 - \alpha$, one can and we will suppose that $\alpha = 0$. Let $\mathcal{F}$ be the set of real-valued functions, f, on $[0, 1]$ so that $|f(x)| \le D|x|$ for some D and
$$\lim_{x \uparrow 1} S(f, x) = 0, \quad S(f, x) = \sum_{n=0}^\infty a_n f(x^n) \tag{6.11.82}$$
   (a) Show that if the characteristic function, χ, of $[\frac{1}{2}, 1]$ lies in $\mathcal{F}$, then (6.11.79) holds. (*Hint*: For each x, $\sum_{n=0}^\infty a_n \chi(x^n)$ is a partial sum.)

   (b) Prove that any polynomial, p, lies in $\mathcal{F}$, assuming $f(x) = x$ is.

   (c) Prove that for each ε, there exist polynomials, p_1 and p_2, so that
$$\begin{gathered} p_1(x) \le \chi(x) \le p_2(x) \text{ for } x \in [0, 1] \text{ and} \\ \int_0^1 \frac{p_2(x) - p_1(x)}{x(1 - x)}\, dx < \varepsilon \end{gathered} \tag{6.11.83}$$
   (d) Let
$$q(x) = \frac{p_2(x) - p_1(x)}{x(1 - x)} \tag{6.11.84}$$
   Prove that (where $C = \sup n|a_n|$)
$$|S(\chi, x) - S(p_1, x)| \le C \sum_{n=0}^\infty \frac{x^n(1 - x^n)}{n}\, q(x^n) \tag{6.11.85}$$

(e) Prove that

$$(1-x)\sum_{n=0}^{\infty} x^n q(x^n) \to \int_0^1 q(s)\,ds \tag{6.11.86}$$

for any polynomial q. (*Hint*: Check $q(x) = x^\ell$.)

(f) Prove that $n^{-1}(1-x^n) \le (1-x)$ and conclude that

$$\limsup_{x\uparrow 1} |S(\chi, x)| \le \varepsilon C \tag{6.11.87}$$

so that $\chi \in \mathcal{F}$.

5. This problem will prove Littlewood's Tauberian theorem, (6.11.79), using Pitt's extension of Wiener's Tauberian theorem—indeed, only the special case of Kac (since the function, K, below is $O(e^{-|x|})$ at infinity). We define

$$S_n = \sum_{j=0}^{n} a_j, \quad F(x) = \sum_{n=0}^{\infty} a_n x^n \tag{6.11.88}$$

(a) For $0 < x < 1$, prove that

$$|S_n - F(x)| \le (1-x)\sum_{j=0}^{n} |ja_j| + \frac{1}{n}\sum_{j=n+1}^{\infty} |ja_j| x^j \tag{6.11.89}$$

(*Hint*: $(1-x^n)/(1-x) \le n$.)

(b) Prove that

$$\left| S_n - F\left(1 - \frac{1}{n}\right)\right| \le 2\sup_j |ja_j| \tag{6.11.90}$$

so that $\sup|ja_j| < \infty$ and $\lim_{x\uparrow} F(x) = 0$ implies S_n is bounded.

(c) Define

$$s(x) = \sum_{n\le x} a_n \tag{6.11.91}$$

and show that

$$F(e^{-t}) = t\int_0^\infty e^{-tx} s(x)\,dx \tag{6.11.92}$$

(d) If

$$K(y) = \exp(-y - e^{-y}),\ \widetilde{F}(u) = F(\exp(-e^{-u})),\ \varphi(y) = s(e^y) \tag{6.11.93}$$

prove that (6.11.92) says that

$$\widetilde{F}(u) = \int_{-\infty}^{\infty} K(u-y)\varphi(y)\,dy \tag{6.11.94}$$

(*Hint*: $x = e^j$, $t = e^{-u}$.)

(e) Prove that

$$\int K(x)e^{-isx}\,dx = \int_0^\infty v^{is}e^{-v}\,dv = \Gamma(1+is) \tag{6.11.95}$$

(*Hint*: $v = e^{-x}$.)

(f) Using $\sup_j|ja_j| < \infty$, prove that φ is slowly oscillating.

(g) Prove that $\lim_{y\to\infty}\varphi(y) = 0$ and conclude that (6.11.79) holds.

The next two problems involve an alternate way to provide the technicalities of the Karamata approach to proving the Littlewood Tauberian theorem. It still has the Karamata idea of using polynomial approximation (but here, polynomials in e^{-x}) and avoids the Wiener Tauberian theorem, but uses some soft measure theory in place of the sometimes tedious estimates of Problem 4. This is an idea we learned from Aizenman and reported on in [**647**] in a somewhat different context. We need some preliminaries from Problem 5. Namely, we suppose

$$\lim_{t\downarrow 0} t\int_0^\infty e^{-tx}s(x)\,dx = 0 \tag{6.11.96}$$

and need to prove that

$$\lim_{x\to\infty} s(x) = 0 \tag{6.11.97}$$

given two facts:

(1) $\|s\|_\infty < \infty$ (6.11.98)

(2) s is slowly oscillating on exponential scale

$$\forall\varepsilon\,\exists\delta, R \text{ s.t. } R < x < y < x(1+\delta) \Rightarrow |s(y) - s(x)| < \varepsilon \tag{6.11.99}$$

(which is (f) of Problem 5 and requires $\sup_j|ja_j| < \infty$). Problem 6 will show abelian limit + Littlewood bound $\Rightarrow$ Cesàro limit, explicitly

$$(6.11.96) + (6.11.98) \Rightarrow t^{-1}\int_0^t S(x)\,dx \Rightarrow 0 \tag{6.11.100}$$

and Problem 7, a result of Hardy that predated and motivated Littlewood, that Cesàro limit + Littlewood bound $\Rightarrow$ summability, explicitly

$$(6.11.99) + t^{-1}\int_0^t s(x)\,dx \to 0 \Rightarrow (6.11.97)$$

6. Suppose s obeys (6.11.98) and (6.11.96). Let

$$d\mu_t(y) = e^{-y}s\left(\frac{y}{t}\right)dy \tag{6.11.101}$$

(a) Prove that for any $n = 0, 1, 2, \dots,$

$$\lim_{t\to 0}\int_0^\infty e^{-ny}\,d\mu_t(y) = 0 \tag{6.11.102}$$

(b) Prove that for any g in $C_\infty([0,\infty))$ the continuous functions vanishing at ∞, we have

$$\lim_{t\to\infty}\int_0^\infty g(y)\,d\mu_t(y)=0 \tag{6.11.103}$$

(c) Let $h(y)=e^y\chi_{[0,1]}$. For any ε, show there is g continuous so that

$$\sup_{t<1}\int_0^\infty |h(y)-g(y)|\,d|\mu_t|(y)\le\varepsilon \tag{6.11.104}$$

(d) Conclude that

$$\lim_{T\to\infty}\frac{1}{T}\int_0^T s(y)\,dy=0 \tag{6.11.105}$$

7. Suppose s obeys (6.11.105) and (6.11.99).

(a) For any $\delta>0$, prove that

$$\lim_{T\to\infty}\frac{1}{\delta T}\int_T^{T+\delta T} s(y)\,dy=0 \tag{6.11.106}$$

(b) Prove that

$$\lim_{y\to\infty} s(y)=0 \tag{6.11.107}$$

(*Hint*: $|s(T)-\frac{1}{\delta T}\int_T^{T+\delta T} s(y)\,dy|\le\sup_{T\le y\le T+\delta T}|s(y)-s(T)|$.)

6.12. The Prime Number Theorem via Tauberian Theorems

Recall that for any $x\in(0,\infty)$, $\pi(x)$ is the number of primes, p, with $p<x$. The celebrated prime number theorem is the assertion that

$$\lim_{x\to\infty}\frac{\pi(x)}{x/\log x}=1 \tag{6.12.1}$$

In Section 13.5 of Part 2B, we proved this using complex variable methods. In this section, we'll see it follows form Ingham's Tauberian theorem.

Our proof in Part 2B used the function ($\mathcal{P}$ = set of primes)

$$\theta(x)=\sum_{\substack{p\le x\\ p\in\mathcal{P}}}\log p \tag{6.12.2}$$

Our proof here will use the closely related

$$\psi(x)=\sum_{\substack{p^n\le x\\ p\in\mathcal{P},\,n=1,2,\dots}}\log p \tag{6.12.3}$$

Our proof will rely on three facts; we'll show

(1) $$\lim_{x\to\infty} \frac{\pi(x)}{x/\log x} = 1 \Leftrightarrow \lim_{x\to\infty} \frac{\psi(x)}{x} = 1 \tag{6.12.4}$$

(2) if

$$F(x) = \sum_{n\le x} \psi\left(\frac{x}{n}\right) \tag{6.12.5}$$

then for $n \in \mathbb{Z}_+$,

$$F(n) = \log n! \tag{6.12.6}$$

(3) that by Stirling's approximation

$$\log n! = n\log n - n + O(\log n) \tag{6.12.7}$$

(2) + (3) plus monotonicity of F will show (6.11.46) holds with $a = 1$, $b = -1$ and then Theorem 6.11.9 will imply $\frac{\psi(x)}{x} \to 1$ so that, by (6.12.4), we have (6.12.1). We begin with Step (1):

Lemma 6.12.1. (6.12.4) *holds.*

Remark. The argument is essentially the same as for Proposition 13.5.4 of Part 2B (where ψ is replaced by θ).

Proof. For each prime $p \le x$, $\log p$ occurs n times on ψ where n is the largest integer with $p^n \le x$, that is, $n = [\frac{\log x}{\log p}]$ so

$$\psi(x) = \sum_{p\le x} \left[\frac{\log x}{\log p}\right] \log p \tag{6.12.8}$$

Since $[y] \le x$, we conclude

$$\psi(x) \le \pi(x)\log x \tag{6.12.9}$$

Thus (and similarly for lim sup's),

$$\liminf \frac{\psi(x)}{x} \le \liminf \frac{\pi(x)}{x/\log x} \tag{6.12.10}$$

On the other hand, if $0 < \alpha < 1$ and $p \ge x^\alpha$, then $\log p \ge \alpha \log x$ and $[\frac{\log x}{\log p}]\log p \ge \log p$, so

$$\psi(x) \ge \alpha[\pi(x) - \pi(x^\alpha)][\log x] \ge \alpha[\pi(x) - \alpha x^\alpha][\log x] \tag{6.12.11}$$

since $\pi(x) \le x$.

Since $\log x \frac{x^\alpha}{x} \to 0$, (6.12.11) implies

$$\liminf \frac{\psi(x)}{x} \ge \alpha \liminf \frac{\pi(x)}{x/\log x} \tag{6.12.12}$$

and similarly for lim sup (6.12.10)/(6.12.12) and the lim sup analogs imply (6.12.4). $\square$

Lemma 6.12.2. *Let F be given by* (6.12.5). *Then $F(1) = 0$ and*

$$F(n) - F(n-1) = \log n \tag{6.12.13}$$

so (6.12.6) *holds.*

Proof. Since $\psi(1) = 0$, $F(1) = 0$ is immediate. Note that $\psi(x)$ is monotone in x and has jumps exactly at $x = p^n$ with

$$\psi(p^n + 0) = \psi(p^n) = \log p + \psi(p^n - 0) \tag{6.12.14}$$

Define Λ on the integrals by

$$\Lambda(m) = \begin{cases} \log p & \text{if } m = p^\ell,\ \ell = 1, 2, \dots \\ 0, & \text{otherwise} \end{cases}$$

Then if n and m are integers, since $\frac{n}{m}$ and $\frac{n-1}{m}$ lie on opposite sides of a jump of ψ if $\frac{n}{m} \in \mathbb{Z}$ and the size of the jump is $\Lambda(\frac{n}{m})$.

Thus ($m \mid n$ says m divides n),

$$F(n) - F(n-1) = \sum_m \biggl[\psi\Bigl(\frac{n}{m}\Bigr) - \psi\Bigl(\frac{n-1}{m}\Bigr) = \sum_{m|n} \Lambda\Bigl(\frac{n}{m}\Bigr) \tag{6.12.15}$$

If $n = p_1^{k_1} \dots p_\ell^{k_\ell}$, then $\Lambda(\frac{n}{m}) \neq 0$ only if $\frac{n}{m} = p_j^{q_j}$, $q_j = 1, \dots, k_j$. Thus,

$$\sum_{m|n} \Lambda\Bigl(\frac{n}{m}\Bigr) = \sum_j k_j \log p_j = \log n \tag{6.12.16}$$

proving (6.12.13).

Since $F(1) = 0$,

$$\begin{aligned} F(n) &= \sum_{j=0}^{n-2} [F(n-j) - F(n-j-1)] \\ &= \log n + \log(n-1) + \dots + \log 2 = \log n! \end{aligned}$$

$\square$

For completeness, we repeat the elementary argument for the leading $\log n!$ behavior.

Lemma 6.12.3. (6.12.7) *holds.*

Proof. Since $\log x$ is monotone,

$$\log j \le \int_j^{j+1} \log x\, dx \le \log(j+1) \tag{6.12.17}$$

Thus,

$$\log n! \le \int_1^n \log x\, dx + \log n \le \log n! + \log n$$

so

$$\log n! = \int_1^n \log x\, dx + O(\log n) \tag{6.12.18}$$

But $\int_1^n \log x\, dx = x\log x - x \int_1^n = n\log n - n$, so (6.2.18) is (6.12.7). $\square$

Theorem 6.12.4 (Prime Number Theorem)**.** (6.12.1) *holds.*

Proof. Since ψ is monotone, so is F given by (6.12.5). Thus, if $n \le x < n+1$, we have

$$\log n! \le F(x) \le \log n! + \log(n+1) \tag{6.12.19}$$

so

$$F(x) = F([x]) + O(\log x) \tag{6.12.20}$$

Similarly,

$$x\log x - x = \int_1^x \log y\, dy = n\log n - n + O(\log n) \tag{6.12.21}$$

Therefore, by (6.12.6) and (6.12.7),

$$F(x) = x\log x - x + O(\log x) \tag{6.12.22}$$

Since ψ is monotone, Ingham's theorem, Theorem 6.11.9, applies. So since the coefficient of $x\log x$ is 1, we have $\lim_{x\to\infty} \frac{\psi(x)}{x} = 1$.

By Lemma 6.12.1, we conclude that (6.12.1) holds. $\square$

Notes and Historical Remarks. Since the history of the Tauberian number theorems and Tauberian theorems are intertwined, we discussed this history in the last section. The proof in this section is from Ingham's 1945 paper [**333**]. Ingham's theorem is a strengthening of results of much earlier results of Landau [**422, 423**]. Motivated by Landau but not knowing of Ingham, arguments close to Ingham's were rediscovered by Levinson [**440**].

Chapter 7

Bonus Chapter: Unbounded Self-adjoint Operators

> The aim of science is to seek the simplest explanations of complex facts. We are apt to fall into the error of thinking that the facts are simple because simplicity is the goal of our quest. The guiding motto in the life of every natural philosopher should be, "Seek simplicity and distrust it."
>
> —*A. N. Whitehead* (1861-1947) [**753**]

Big Notions and Theorems: Unbounded Operator, Domain of an Operator, Graph of an Operator, Hermitian Operator, Positive Operator, Closed Operator, Closable Operator, Adjoint, Extension, Self-adjoint, Essential Self-adjoint, Core, Closable $\Leftrightarrow$ A^* Densely Defined, $\overline{A} = A^{**}$, Resolvent, Spectrum, Deficiency Index, Fundamental Criterion for Self-adjointness, von Neumann Conjugation Result, A-Bounded, Relatively Bounded, Kato–Rellich Theorem, Kato's Theorem on Atomic Hamiltonians, Spectral Theorem for Unbounded Self-adjoint Operators, Cayley Transform, Norm Resolvent Convergence, Strong Resolvent Convergence, Continuity of the Functional Calculus, Stone's Theorem, von Neumann Classification of Self-adjoint Extensions, Limit Point, Limit Circle, Weyl's Limit Point/Limit Circle Theorem, Quadratic Form, Nonnegative Sesquilinear Form and Correspondence to Quadratic Forms, Closed Quadratic Form, One–One Correspondence of Closed Quadratic Forms and Positive Self-adjoint Operators, Form Core, KLMN Theorem, Factorizable Perturbations, Ordering of Positive Self-adjoint Operators, Regular Part of a Quadratic Form, Closable Quadratic Form, Monotone Convergence for Forms, Friedrichs Extension, Krein's Extension, Classification of Positive Self-adjoint Extensions of Positive Hamiltonian Operators, Dirichlet–Neumann Bracketing, Weyl Eigenvalue Counting Theorem, Quasiclassical Limit of Schrödinger

Operators, Kato's $L^2_{\rm loc}$-Theorem, Simon's $L^1_{\rm loc}$-Theorem, Positivity Preserving Operators, Beurling–Deny Criterion, Kato's Inequality, Second Beurling–Deny Criterion, Diamagnetic Inequalities, Chernoff's Theorem, Trotter Product Formula, Feynman–Kac Formula, Stieltjes and Hamburger Moment Problem, Hamburger and Stieltjes Moment Conditions, Determinate and Indeterminate Problems, von Neumann Solution, Nevanlinna Parameterization Theorem, Classification of Solutions to Indeterminate Stieltjes Problems, Krein Solution and Friedrichs Solutions, Rational Approximation to Krein and Friedrichs Resolvents, Spectral Mappng Theorem, Relatively Compact Perturbation, Essential Spectrum, Invariance of Essential Spectrum, Trace Class Invariance of A.C. Spectrum, Birman–Schwinger Principle, Birman–Schwinger Bound, Lieb–Thirring Inequality, Lieb–Thirring Kinectic Energy Bound, Bargmann Bound, Small Coupling Schrödinger Operators in Dimensions 1 and 2

The spectral theorem is so successful at helping us understand bounded self-adjoint operators that it is natural to see wider venues for its application. Moreover, the center of analysis is the differential and integral calculus. While integral operators have occurred earlier, differential operators are conspicuous in their absence. In this chapter, we give up boundedness of operators and develop a theory that applies to some important class of ordinary and partial differential operators. This introductory section will not be precise because it will explain the precise definitions we eventually use.

The first conceptual leap we face concerns domains of definition. The Hellinger–Toeplitz theorem (Theorem 5.4.18 of Part 1) says that if $A\colon \mathcal{H} \to \mathcal{H}$ is a linear map on Hilbert space with

$$\langle \varphi, A\psi \rangle = \langle A\varphi, \psi \rangle \tag{7.0.1}$$

for all $\varphi, \psi \in \mathcal{H}$, then A is bounded. Since we want (7.0.1) to hold in some sense to consider A to be self-adjoint, we will need to define A not on all of $\mathcal{H}$ but only on a dense subspace. Of course, this shouldn't be surprising. If $\varphi \in L^2(\mathbb{R}^\nu, d^\nu x)$ and we wish to make sense of $\Delta\varphi$, we already know we'll want $\Delta\varphi$ to be the distributional derivative. But there is no reason that this distribution derivative lies in L^2. In fact, we'll eventually show Δ on $\{\varphi \in L^2 \mid \Delta\varphi \in L^2 \text{ (distributional derivative)}\}$ is a self-adjoint operator and that this domain is only a subset of L^2.

The lodestone in deciding what self-adjoint should mean is the spectral theorem. Let μ be a measure on $\mathbb{R}$ whose support may not bounded. Thus, $\varphi \in L^2(\mathbb{R}, d\mu)$ may not imply $\int x^2 |\varphi(x)|^2\, d\mu(x) < \infty$. The "natural" domain for x is

$$D(x) \equiv \left\{ \varphi \;\middle|\; \int (1+|x|^2)|\varphi(x)|^2\, d\mu(x) < \infty \right\} \tag{7.0.2}$$

Thus, the "right" definition of self-adjoint is an operator, A, unitarily equivalent to a direct sum of such multiplication operators (of course, we'll have to decide how to define the domain of such a direct sum). Giving this definition and proving the spectral theorem will be one major high point of the chapter.

There are two different subtleties connected to choice of domain. $D(x)$ in (7.0.2) is a kind of maximal domain on which x can be defined. But one might as naturally take $D(x)$ to be $C_0^\infty(\mathbb{R})$, the C^∞-functions of compact support, or $\mathcal{S}(\mathbb{R})$ or various smaller domains than $D(x)$ which, intuitively, one might think are just as good. This issue will be addressed by the notion of operator closure. Given an operator, A, obeying (7.0.1) for all $\varphi, \psi \in D(A)$, there will be a natural "closed" operator, B, which extends A to a bigger domain. For multiplication by x on $C_0^\infty(\mathbb{R})$ or $\mathcal{S}(\mathbb{R})$ in the above $L^2(\mathbb{R}, d\mu)$ setup, the closure will be multiplication x on $D(x)$. Once we have a proper definition for "self-adjoint," we'll call A "essentially self-adjoint" if the closure of A is self-adjoint.

There is a second subtlety involving boundary conditions, illustrated most dramatically in the following. As a preliminary, we note that whatever the spectrum of an unbounded operator is, we expect the spectrum of multiplication by x on $L^2(\mathbb{R}, d\mu(x))$ to be $\mathrm{supp}(d\mu)$, that is, a nonempty closed subset of $\mathbb{R}$.

Example 7.0.1. We want to look at $\frac{1}{i}\frac{d}{dx}$ on $L^2([0,1], dx)$ with various boundary conditions. We let $D_{\max} = \{f \text{ on } [0,1] \mid f(x) = a + \int_0^x g(y)\,dy$ for some $g \in L^2([0,1], dx)\}$. We define $A_{\max} f = -ig$. We'll let $A_0 = A_{\max} \restriction \{f \in D_{\max} \mid f(0) = 0\}$, that is, $D(A_0) = \{f \in D_{\max} \mid f(0) = 0\}$. Here is the shocker: $\sigma(A_{\max}) = \mathbb{C}$, $\sigma(A_0) = \emptyset$!

For given $\lambda \in \mathbb{C}$, $f(x) = e^{i\lambda x}$ lies in $D_{\max}$ and $A_{\max} f = \lambda f$, that is, every $\lambda \in \mathbb{C}$ is an eigenvector. Whatever precise definition of $\sigma(A)$ we take, eigenvalues must be in the spectrum, so $\sigma(A_{\max}) = \mathbb{C}$. On the other hand, given $\lambda \in \mathbb{C}$, let

$$(K_\lambda g)(x) = i \int_0^x e^{i\lambda(x-y)} g(y)\,dy \tag{7.0.3}$$

Then clearly, $K_\lambda g \in D(A_0)$, and by an easy calculation, $(A_0 - \lambda)K_\lambda g = g$. By an integration by parts, if $g \in D(A_0)$, then $K_\lambda(A_0 - \lambda)g = g$. Thus, $A_0 - \lambda$ has a two-sided inverse so, as claimed, $\sigma(A_0) = \emptyset$!

Of course, neither $A_{\max}$ nor A_0 obeys (7.0.1) for all φ, ψ in their domain. We eventually see that associated to $A_{\max}$ are a one-parameter family of self-adjoint operators, A_θ, with

$$D(A_\theta) = \{f \in D_{\max} \mid f(1) = e^{i\theta} f(0)\}$$

and $A_\theta = A_{\max} \restriction D(A_\theta)$. We also note that if A_{00} has $D(A_{00}) = \{f \in D_{\max} \mid f(0) = f(1) = 0\}$, then A_{00} has (7.0.1) for all φ, ψ in their domain, but it will not obey the condition we need for self-adjointness. Self-adjointness will require a juggling act with boundary conditions: too few as with $A_{\max}$ and (7.0.1) fails. Too many, as with A_{00}, and we don't get the spectral theorem. Indeed, $\sigma(A_{00}) = \mathbb{C}$ also, as we'll eventually see (Example 7.1.6). □

Section 7.1 will spell out the basic definitions—graphs of operators will be a central notion. Operators obeying (7.0.1) for all φ, ψ in their domain will be called Hermitian. Every such operator, A, will have an adjoint, A^*, suitably defined. $D(A) \subset D(A^*)$ and $A^* \restriction D(A) = A$. The self-adjoint operators will be those with $A = A^*$. We'll prove that A is self-adjoint if and only if $\operatorname{Ran}(A \pm i) = \mathcal{H}$—this fundamental criterion will be central to the remainder of the chapter.

Section 7.2 will prove the spectral theorem for self-adjoint operators. Once we know a self-adjoint operator, A, is a multiplication operator, we can define $F(A)$ for any bounded Borel function $F\colon \mathbb{R} \to \mathbb{C}$ and $F(A)$ will be a bounded operator. In particular, we can define $U_t = e^{-itA}$ for all $t \in \mathbb{R}$ and it will be a family of unitaries obeying $U_{t+s} = U_t U_s$, $U_0 = \mathbb{1}$, and $t \mapsto U_t$ is strongly continuous. Stone's theorem, the goal of Section 7.3, will prove the converse: every such family of unitaries defines a self-adjoint operator.

Section 7.4 will look at which Hermitian operators have self-adjoint extensions and classify them, while Sections 7.5 and 7.6 will discuss methods of proving certain operators, especially differential operators, are self-adjoint. Sections 7.7–7.9 discuss further developments.

7.1. Basic Definitions and the Fundamental Criterion for Self-adjointness

In this section, we'll define closed operators and operator closure and then self-adjoint and essentially self-adjoint. We'll state the fundamental criterion for self-adjointness and use it to prove a basic perturbation result, the Kato–Rellich theorem. As an example, we prove Kato's theorem on the essential self-adjointness of atomic Hamiltonians in quantum mechanics, one of the signature results of modern mathematical physics.

Definition. An *unbounded operator*, A, on a Hilbert space, $\mathcal{H}$, is a dense subspace, $D(A)$, the *domain* of A, and a linear map $A\colon D(A) \to \mathcal{H}$. The *graph*, $\Gamma(A)$, is the subspace of $\mathcal{H} \times \mathcal{H}$

$$\Gamma(A) = \{(\varphi, A\varphi) \mid \varphi \in D(A)\} \tag{7.1.1}$$

Remarks. 1. In (7.1.1), do not confuse $(\varphi, A\varphi)$, the ordered pair in $\mathcal{H} \times \mathcal{H}$ with $\langle \varphi, A\varphi \rangle$, the inner product.

2. A can be bounded, so we should really say "perhaps unbounded operator." We don't want to just use "operator" since we've earlier implicitly used it for bounded operators.

3. $\Gamma(A)$ is a subspace, S, of $\mathcal{H} \times \mathcal{H}$ with two properties, $\pi_1\colon (\varphi, \psi) \mapsto \varphi$:

(a) $\pi_1[S]$ is dense (7.1.2)

(b) $\pi_1(s_1) = \pi_1(s_2) \Rightarrow s_1 = s_1$, i.e., $\pi_1 \restriction S$ is injective (7.1.3)

A subset, S, is called the *graph of an operator* if it obeys (a), (b). It is then $\Gamma(B)$ for a unique operator B.

4. At times, we will drop the requirement that $D(B)$ is dense. We'll then refer to "perhaps not densely defined operator." In such a case, it may be useful to think of B as a densely defined operator on the Hilbert (sub)space, $\overline{D(B)}$.

Definition. An (unbounded) operator, A, is called *Hermitian* (aka *symmetric*) if and only if

$$\langle \varphi, A\psi \rangle = \langle A\varphi, \psi \rangle \tag{7.1.4}$$

for all $\varphi, \psi \in D(A)$.

Since our Hilbert spaces are complex, polarization shows

$$A \text{ Hermitian} \Leftrightarrow \langle \varphi, A\varphi \rangle \in \mathbb{R} \text{ for all } \varphi \in D(A) \tag{7.1.5}$$

A Hermitian implies, for $x, y \in \mathbb{R}$ and $\varphi \in D(A)$,

$$\langle iy\varphi, (A-x)\varphi \rangle + \langle (A-x)\varphi, iy\varphi \rangle = 0 \tag{7.1.6}$$

so

$$\begin{aligned} &A \text{ Hermitian}, \ \varphi \in D(A), \ z = x + iy \in \mathbb{C} \\ &\Rightarrow \|[A - (x+iy)]\varphi\|^2 - \|(A-x)\varphi\|^2 + |y|^2 \, \|\varphi\|^2 \end{aligned} \tag{7.1.7}$$

Definition. An operator, A, is called *positive* (written $A \geq 0$) if

$$\varphi \in D(A) \Rightarrow \langle \varphi, A\varphi \rangle \geq 0 \tag{7.1.8}$$

The number

$$\inf_{\substack{\varphi \in D(A) \\ \|\varphi\|=1}} \langle \varphi, A\varphi \rangle \equiv \gamma(A) \tag{7.1.9}$$

is called the *lower bound* of A. If $A + c \geq 0$ for some $c \in (0, \infty)$, we say A is *bounded below* or *semibounded.* γ is still defined by (7.1.9) but now can be negative.

By polarization (and the fact the our Hilbert spaces are complex), positive $\Rightarrow$ Hermitian. If $x \leq 0$ and $A \geq 0$,

$$\begin{aligned} \|(A-x)\varphi\|^2 &= \|A\varphi\|^2 + |x|^2 \|\varphi\|^2 - 2x\langle \varphi, A\varphi \rangle \\ &\geq \|A\varphi\|^2 + |x|^2 \|\varphi\|^2 \end{aligned} \tag{7.1.10}$$

Thus,

$$A \geq 0, \ \varphi \in D(A), \ \operatorname{Re} z \leq 0 \Rightarrow \|(A-z)\varphi\|^2 \geq \|A\varphi\|^2 + |z|^2 \|\varphi\|^2 \tag{7.1.11}$$

Definition. An (unbounded) operator, A, is called *closed* if and only if $\Gamma(A)$ is closed. A is called *closable* if and only if $\overline{\Gamma(A)}$ is the graph of an operator, then called its closure, $\bar{A}$, that is,

$$\Gamma(\bar{A}) = \overline{\Gamma(A)} \tag{7.1.12}$$

If A is a closed operator, $D_0 \subset D(A)$ and $\overline{A \restriction D_0} = A$, we say D_0 is a *core* for A.

For a bounded operator, A, its adjoint is defined by

$$\langle A^*\eta, \varphi\rangle = \langle \eta, A\varphi\rangle \tag{7.1.13}$$

We'll define adjoint by requiring (7.1.13) for the maximal set of η's.

Definition. Let A be an (unbounded) operator. A^*, the *adjoint* of A, is the perhaps not densely defined operator with

$$D(A^*) = \{\eta \mid \exists c \text{ so that } |\langle \eta, A\varphi\rangle| \le c\|\varphi\| \text{ all } \varphi \in D(A)\} \tag{7.1.14}$$

If $\eta \in D(A^*)$, $A^*\eta$ is the unique vector so that (7.1.13) holds.

Definition. If A, B are two (unbounded) operators, we say B is an *extension* of A, written $A \subseteq B$ if and only if

$$D(A) \subset D(B) \quad \text{and} \quad B \restriction D(A) = A \tag{7.1.15}$$

Equivalently,

$$A \subseteq B \Leftrightarrow \Gamma(A) \subseteq \Gamma(B) \quad \text{(subspace containment)} \tag{7.1.16}$$

Notice

$$A \text{ is Hermitian} \Leftrightarrow A \subset A^* \tag{7.1.17}$$

Definition. A is called *self-adjoint* if and only if $A = A^*$. A is called *essentially self-adjoint* (esa) if and only if A is closable and $\bar{A}$ is self-adjoint. If B is self-adjoint, $D(A) \subset D(B)$ so $\overline{B \restriction D(A)} = B$, that is, $D(A)$ is a *core* for B; we also call it a *domain of essential self-adjointness.*

There is a deep connection between adjoints and closability of operators. It depends on its map $J\colon \mathcal{H} \times \mathcal{H} \to \mathcal{H} \times \mathcal{H}$ by

$$J(\eta, \psi) = (\psi, -\eta) \tag{7.1.18}$$

Theorem 7.1.1. *Let A be an (unbounded) operator on a Hilbert space, $\mathcal{H}$. Then*

(a) $\Gamma(A^*) = J[\Gamma(A)^\perp] = [J(\Gamma(A))]^\perp$ (7.1.19)

(b) *Let A be a perhaps not densely defined operator. Then $\pi_1 \restriction J[\Gamma(A)^\perp]$ is one–one if and only if $D(A)$ is dense.*

(c) *A is closable if and only if A^* is densely defined, and in that case, A^* is closed and*

$$\bar{A} = (A^*)^* \tag{7.1.20}$$

In particular, any Hermitian operator is closable.

(d) *If A is closable, then*

$$A^* = A^{***} = A^{5*} = \dots, \qquad A^{**} = A^{4*} = \dots \tag{7.1.21}$$

Remark. Thus,

$$A \text{ esa} \Leftrightarrow A^* = A^{**} \tag{7.1.22}$$

Proof. (a) Since J is unitary, we have the second equality in (7.1.19). We note that

$$\begin{aligned} J(\eta,\psi) \in \Gamma(A)^\perp &\Leftrightarrow \forall\varphi \in D(A),\ \langle\psi,\varphi\rangle + \langle -\eta, A\varphi\rangle = 0 \\ &\Leftrightarrow \eta \in D(A^*) \text{ and } A^*\eta = \psi \\ &\Leftrightarrow (\eta,\psi) \in \Gamma(A^*) \end{aligned}$$

(b) As we've just seen,

$$(\eta,\psi) \in J[\Gamma(A)]^\perp \Leftrightarrow \forall\varphi \in D(A),\ \langle\eta, A\varphi\rangle = \langle\psi,\varphi\rangle$$

Thus, η determines a unique ψ ($\eta = 0$ is always possible) if and only if $D(A)^\perp = \{0\}$ if and only if $D(A)$ is dense.

(c) Since $J^2 = -\mathbb{1}$ leaves subspaces fixed,

$$\overline{\Gamma(A)} = \Gamma(A)^{\perp\perp} = J[J[\Gamma(A)^\perp]]^\perp = J[\Gamma(A^*)]^\perp \tag{7.1.23}$$

By (b), $\overline{\Gamma(A)}$ obeys (7.1.22) if and only if $D(A^*)$ is dense. Since $\Gamma(A) \subset \overline{\Gamma(A)}$, $D(A) = \pi_1[\Gamma(A)] \subset \pi_1[\overline{\Gamma(A)}]$. $\overline{\Gamma(A)}$ always obeys (7.1.2).

Since $\Gamma(A^*) = [J\Gamma(A)]^\perp$, A^* is always closed (if densely defined) and, by (7.1.23), we have (7.1.20). The final assertion follows from $D(A) \subset D(A^*)$, so $D(A^*)$ is dense.

(d) We have just seen that if B is closable, $B = B^{**}$, which implies (7.1.7) given (c). □

Thus, A Hermitian implies

$$A \subset \bar{A} = A^{**} \subset A^* \tag{7.1.24}$$

We'll later see (see Section 7.4) that if A has self-adjoint extensions, then $D(\bar{A})$ has even codimension, 2ℓ, in $D(A^*)$ (ℓ may be ∞), and the self-adjoint extensions, B, lie "exactly" in between in that $D(B)/D(\bar{A})$ and $D(A^*)/D(B)$ are each of dimension ℓ. Since $A \subset B$, we have that $B^* \subset A^*$, it can happen that A has self-adjoint extensions, indeed,

$$B \text{ a self-adjoint extension of } A \Rightarrow A \subset \bar{A} \subset B = B^* \subset A^* \tag{7.1.25}$$

Since $(A^*)^* = \bar{A}$, the definition of * provides the following characterization of $\bar{A}$:

Theorem 7.1.2. *Let A be a Hermitian operator. Then $\varphi \in D(\bar{A})$ if and only if $\varphi \in D(A^*)$ and*

$$\forall \psi \in D(A^*)\colon \langle \varphi, A^*\psi \rangle = \langle A^*\varphi, \psi \rangle \tag{7.1.26}$$

Corollary 7.1.3. *Let A be a closed Hermitian operator. Let $\varphi, \psi \in D(A^*)$ so that* (7.1.26) *fails. Then $\varphi, \psi \notin D(A)$. If also $\langle \psi, A^*\psi \rangle$ and $\langle \varphi, A^*\varphi \rangle$ are real, then φ, ψ are linearly independent in $D(A^*)/D(A)$.*

Proof. If $\varphi \in D(A)$, (7.1.26) holds so we see that if (7.1.26) fails then $\varphi \notin D(A)$. Since (7.1.26) is symmetric in φ and ψ, $\psi \notin D(A)$ also. If φ, ψ are linearly dependent in $D(A^*)/D(A)$, then there are $c, d \in \mathbb{C}$, $(c, d) \neq (0, 0)$ and $\eta \in D(A)$ so $c\varphi + d\psi + \eta = 0$. Without loss, one can suppose $c = 1$ or $d = 1$. If $c = 1$, $\varphi = -d\psi - \eta$, so

$$\langle \varphi, A^*\psi \rangle - \langle A^*\varphi, \psi \rangle = -\bar{d}[\langle \psi, A^*\psi \rangle - \langle A^*\psi, \psi \rangle] = 0$$

so (7.1.26) holds. The case for $d = 1$ is similar. □

Example 7.1.4 (Nonclosable Operator)**.** Let ℓ be a densely defined not bounded linear functional on $\mathcal{H}$, for example, if $\{\varphi_n\}_{n=1}^\infty$ is an orthonormal basis, define ℓ with

$$D(\ell) = \biggl\{ \sum_{n=1}^\infty \alpha_n \varphi_n \biggm| \sum_{n=1}^\infty |\alpha_n| < \infty \biggr\} \tag{7.1.27}$$

with

$$\ell\biggl(\sum_{n=1}^\infty \alpha_n \varphi_n \biggr) = \sum_{n=1}^\infty \alpha_n \tag{7.1.28}$$

Define A with $D(A) = D(\ell)$ and

$$A\varphi = \ell(\varphi)\varphi_1 \tag{7.1.29}$$

Then

$$\langle \eta, A\varphi \rangle = \eta_1 \ell(\varphi) \tag{7.1.30}$$

$\eta \in D(A^*) \Leftrightarrow \varphi \mapsto \eta_1 \ell(\varphi)$ is bounded. If $\eta_1 \neq 0$, ℓ is not bounded. Thus,

$$D(A^*) = \{\eta \mid \eta_1 = 0\} = \{\eta \mid \langle \varphi_1, \eta \rangle = 0\}$$

is not dense and A is not closable.

By the algebraic Hahn–Banach theorem (see Problem 11 of Section 5.5 of Part 1), ℓ has an extension to a linear functional on all of $\mathcal{H}$, so $D(A)$ can be $\mathcal{H}$, while $D(A^*)$ is not dense and $\Gamma(A)$ is not closed. (By the closed graph theorem, see Theorem 5.4.17 of Part 1, A defined on $\mathcal{H}$ is closed if and only if A is bounded.) □

Example 7.1.5 (Multiplication Operators). Let $\mathcal{H} = L^2(M, d\nu)$ for some (σ-finite) separable, measurable space. Let $F\colon M \to \mathbb{R}$ be a measurable function which is not in L^∞. Define

$$D(A) = \{\varphi \in \mathcal{H} \mid F\varphi \in L^2\}, \quad (A\varphi)(x) = F(x)\varphi(x) \tag{7.1.31}$$

It is easy to see that $D(A)$ is dense, since $\bigcup_{k=1}^\infty \{x \mid |F(x)| \le k\} = M$. Suppose $(\eta, \psi) \in \Gamma(A^*)$. Then

$$\left| \int \overline{\eta(x)}\, F(x)\varphi(x)\, d\mu(x) \right| \le \|\psi\| \|\varphi\| \tag{7.1.32}$$

It follows that $\varphi \mapsto \int \bar{\eta} F \varphi \, d\mu$ extends to a unique (bounded) linear functional on $\mathcal{H}$ (since $D(A)$ is dense), that is, $F\bar{\eta} \in L^2$ and $\bar{\psi} = F\bar{\eta}$. Therefore, $(\eta, \psi) \in \Gamma(A^*) \Leftrightarrow \eta \in D(A)$ and $\psi = F\eta$. Thus, $A = A^*$ and these maximally defined multiplication operators are self-adjoint.

Of course, an example of special interest, given our expectations for the spectral theorem, is $M = \mathbb{R}$, $\nu = \mu$, and $F(x) = x$. But there are others.

For example, $\psi(\mathbf{k}) \mapsto k^2 \eta(\mathbf{k})$ defines a self-adjoint operator. Thus, since Fourier transform is unitary and the distributional derivative of ψ, $\widehat{\Delta\psi} = k^2 \widehat{\psi}$, if

$$D(A) = \{\psi \in L^2 \mid \Delta\psi \in L^2\}, \quad A\psi = -\Delta\psi \tag{7.1.33}$$

then A is self-adjoint. The same argument works for any constant coefficient partial differential operators, $P(\partial)$, which is formally self-adjoint, that is, $P(i\mathbf{k})$ is real for all k.

It is easy to show (Problem 1) that $C_0^\infty(\mathbb{R}^\nu)$ is a core for any such $P(\partial)$. □

Example 7.1.6 ($i^{-1}\frac{d}{dx}$ on $L^2([0,1], dx)$). Let $D(A) = C_0^\infty([0,1])$, that is, C^∞-functions supported in $(0,1)$, and $Af = -if'$. Then one can show (Problem 4(a)) that if

$$Q = \{f \in L^2([0,1], dx) \mid f' \in L^2(\text{distributional derivative})\} \tag{7.1.34}$$

then any $f \in Q$ is a continuous function on $[0,1]$ and (Problem 4(b)) A is Hermitian and so closable with

$$D(\bar{A}) = \{f \in Q \mid f(0) = f(1) = 0\} \tag{7.1.35}$$

and $\bar{A}f = -if'$. Moreover (Problem 4(c)),

$$D(A^*) = Q, \qquad A^*f = -if' \tag{7.1.36}$$

For each $\beta \in \partial\mathbb{D}$, let

$$D(A_\beta) = \{f \in D(A^*) \mid f(1) = \beta f(0)\} \tag{7.1.37}$$

Then (Problem 4(d)), $A^* \restriction D(A_\beta)$ is self-adjoint. In this example, we see $\bar{A}$ has many self-adjoint extensions. We'll see later (Theorem 7.4.1) this is the general rule: if a Hermitian operator A is not essentially self-adjoint, it either has no self-adjoint extension or uncountably many. We'll also see (Example 7.4.4) that in this example, $\{A_\beta\}_{\beta\in\partial\mathbb{D}}$ describes all self-adjoint extensions of A. $\square$

The following shows why closed operators are natural and why (7.1.7) plays an important role in the theory of Hermitian self-adjoint operators (take $B = A - z$ with $D(B) \equiv D(A)$, A Hermitian and $\operatorname{Im} z \neq 0$).

Proposition 7.1.7. *Let B be a densely defined operator on a Hilbert space, $\mathcal{H}$, so that for some $\varepsilon > 0$ and all $\varphi \in D(B)$,*

$$\|B\varphi\| \geq \varepsilon\|\varphi\| \tag{7.1.38}$$

Then B is closed if and only if $\operatorname{Ran}(B)$ *is closed.*

Proof. Let $\pi_2\colon \mathcal{H}\times\mathcal{H}\to\mathcal{H}$ by $\pi_2((\eta,\psi)) = \psi$. Then for all $(\varphi,\psi)\in\Gamma(B)$, we have, by (7.1.38),

$$\|\psi\| \leq \|(\varphi,\psi)\|_{\mathcal{H}\times\mathcal{H}} \leq (1+\varepsilon^{-1})\|\psi\| \tag{7.1.39}$$

since $(\varphi,\psi)\in\Gamma(B) \Rightarrow \|\psi\|\geq\varepsilon\|\varphi\| \Rightarrow \|\varphi\|\leq\varepsilon^{-1}\|\psi\|$. Thus, π_2 is a bijection of $\Gamma(B)$ and $\operatorname{Ran}(B)$ and $\|\pi_2(\,\cdot\,)\|_{\mathcal{H}}$ and $\|\cdot\|_{\mathcal{H}\times\mathcal{H}}$ are equivalent norms. $\Gamma(B)$ is complete if and only if $\operatorname{Ran}(B)$ is complete. And completeness of a subspace of a Hilbert space is equivalent to that subspace being closed. $\square$

Definition. Let B be a closed operator on a Hilbert space, $\mathcal{H}$. We say $z\in\rho(B)$, the *resolvent* set of B, if and only if there is a *bounded* operator C so that $\operatorname{Ran}(C)\subset D(B)$ and

$$\varphi\in\mathcal{H}\Rightarrow(B-z)C\varphi=\varphi,\qquad \varphi\in D(B)\Rightarrow C(B-z)\varphi=\varphi \tag{7.1.40}$$

$\mathbb{C}\setminus\rho(B)$ is called the *spectrum* of B. Thus, $B-z$ is a bijection of $D(B)$ and $\mathcal{H}$ and C is its inverse.

If $z,w\in\rho(B)$, it is easy to see that

$$(B-z)^{-1}-(B-w)^{-1}=(z-w)(B-z)^{-1}(B-w)^{-1} \tag{7.1.41}$$

often called the *first resolvent equation* (for $\operatorname{Ran}(B-w)^{-1}\subset D(B)$ so we can insert B's in between the two $(\dots)^{-1}$ on the right so $(z-w)=[(B-w)-(B-z)]$).

Notice that $(B-z)C\varphi=\varphi$ for all φ implies $\operatorname{Ran}(B-z)=\mathcal{H}$ if $z\in\rho(B)$. Notice also that, by (7.1.7) and (7.1.11), we have

$$A \text{ Hermitian},\quad \operatorname{Im} z\neq 0\Rightarrow \operatorname{Ran}(A-z)\text{ is closed} \tag{7.1.42}$$

$$A\geq 0,\quad \operatorname{Re} z<0\Rightarrow \operatorname{Ran}(A-z)\text{ is closed} \tag{7.1.43}$$

Now let A be Hermitian. For $z \in \mathbb{C} \setminus \mathbb{R}$ or $z \in \mathbb{C} \setminus [0, \infty)$ when also $A \geq 0$, we define the *deficiency index*

$$d(z) = \dim[\operatorname{Ran}(A - z)]^\perp = \dim[\operatorname{Ker}(A^* - \bar{z})] \tag{7.1.44}$$

The second equality follows from

$$\begin{aligned} \forall \varphi \in D(A), \langle \psi, (A - z)\varphi \rangle = 0 &\Leftrightarrow \psi \in D(A^*) \text{ and } \langle (A^* - \bar{z})\psi, \varphi \rangle = 0 \\ &\Leftrightarrow \psi \in D(A^*) \text{ and } (A^* - \bar{z})\psi = 0 \end{aligned} \tag{7.1.45}$$

since $D(A)$ is dense. We are heading towards the next signpost:

Theorem 7.1.8. *Let A be Hermitian. There are two (but not necessarily different) nonnegative integers $d_\pm$ so*

$$\pm \operatorname{Im} z > 0 \Rightarrow d(z) = d_\pm \tag{7.1.46}$$

If also $A \geq 0$, then $d_+ = d_-$ and $d(-x) = d_+$ if $x > 0$.

To prove this, we use

Lemma 7.1.9. (a) *If V, W are two closed subspaces with $V \cap W = \{0\}$, then*

$$\dim(V) \leq \dim(W^\perp) \tag{7.1.47}$$

(b) *If A is Hermitian,* $\operatorname{Im} z \neq 0$, *then*

$$|z - w| < \operatorname{Im} z \Rightarrow d(w) \leq d(z) \tag{7.1.48}$$

(c) *If $A \geq 0$,* $\operatorname{Re} z < 0$, *then*

$$|z - w| < |z| \Rightarrow d(w) \leq d(z) \tag{7.1.49}$$

Proof. (a) If $\dim(W^\perp)$ is infinite, there is nothing to prove. So suppose $\dim(W^\perp) = \ell < \infty$ and let $\psi_1, \dots, \psi_\ell$ be an orthonormal basis. If $\dim(V) > \ell$, the map $T \colon V \to \mathbb{C}^\ell$ by $T(\varphi_j) = \langle \psi_j, \varphi \rangle$ has a nonzero kernel, that is, $\exists \varphi \in V \cap (W^\perp)^\perp = V \cap W$ with $\varphi \neq 0$, violating the assumption that $V \cap W = \{0\}$.

(b) By (a), it suffices to show that if

$$W = \operatorname{Ran}(A - z), \qquad V = \operatorname{Ker}(A^* - \bar{w}) \tag{7.1.50}$$

then $W \cap V = \{0\}$. Let $\eta \in W \cap V$. Then for some $\varphi \in D(A)$,

$$\eta = (A - z)\varphi \tag{7.1.51}$$

By (7.1.7),

$$\|\eta\| \geq |\operatorname{Im} z| \|\varphi\| \tag{7.1.52}$$

Since $(A^* - \bar{w})\eta = 0$, $(A^* - \bar{z})\eta = (\bar{w} - \bar{z})\eta$. Thus,

$$\|\eta\|^2 = \langle \eta, (A - z)\varphi \rangle = \langle (A^* - \bar{z})\eta, \varphi \rangle = (\bar{w} - \bar{z})\langle \eta, \varphi \rangle \tag{7.1.53}$$

so, by the Schwarz inequality and (7.1.6),

$$\|\eta\|^2 \leq |\bar{w} - \bar{z}| \, \|\eta\| \, \|\varphi\| \leq |w - z| \, |\operatorname{Im} z|^{-1} \, \|\eta\|^2 \tag{7.1.54}$$

This is only consistent with $|w-z| < |\operatorname{Im} z|$ if $\|\eta\| = 0$, that is, $W \cap V = \{0\}$.

(c) the same as (b), except that (7.1.7)/(7.1.52) is replaced by (7.1.11) and

$$\|\eta\| \geq |z| \|\varphi\| \tag{7.1.55}$$

□

Proof of Theorem 7.1.8. If $|w - z| < \frac{1}{2}|\operatorname{Im} z|$, then $\operatorname{Im} w \geq \frac{1}{2}|\operatorname{Im} z|$ so $|w - z| < |\operatorname{Im} w|$. Thus, by Lemma 7.1.9(b), $d(z) = d(w)$. A connectedness argument proves $d(z)$ is constant on $\mathbb{C}_+$ and on $\mathbb{C}_-$. If also $A \geq 0$, if $x < 0$, $|z - x| < \frac{1}{2}|x| \Rightarrow |z - x| < |z|$, so again $d(z)$ is constant on $D_{\frac{1}{2}|x|}(x)$ which implies $D(z)$ is constant on $\mathbb{C} \setminus [0, \infty)$. □

This implies the most important result of this section:

Theorem 7.1.10 (Fundamental Criterion for Self-adjointness)**.** *Let A be a closed Hermitian operator. Then $A = A^*$ if and only if $d_+ = d_- = 0$.*

Remark. If A is only Hermitian, this result implies that A is essentially self-adjoint $\Leftrightarrow d_+ = d_- = 0$.

Proof. Suppose first $A = A^*$. If $\varphi \in \operatorname{Ker}(A^* - z)$ with $\operatorname{Im} z \neq 0$, then $z\|\varphi\|^2 = \langle \varphi, A\varphi \rangle$ is real, so $\|\varphi\|^2 = 0$, that is, $d(z) = 0$. Therefore, $d_+ = d_- = 0$.

Conversely, suppose A is closed and Hermitian and $d_+ = d_- = 0$. Given $\varphi \in D(A^*)$, since $\operatorname{Ran}(A + i) = \mathcal{H}$ (since $d(-i) = d_- = 0$), find $\eta \in D(A)$, so

$$(A + i)\eta = (A^* + i)\varphi \tag{7.1.56}$$

But $A \subset A^*$, so

$$(A^* + i)(\eta - \varphi) = (A + i)\eta - (A^* + i)\varphi = 0 \tag{7.1.57}$$

Thus, $\eta - \varphi \in \operatorname{Ker}(A^* - \bar{i}) = \{0\}$ since $d(i) = d_+ = 0$. Therefore, $\varphi = \eta \in D(A)$. We have proven $D(A^*) \subset D(A)$, so $D(A^*) = D(A)$ and $A = A^*$. □

Example 7.1.6 (revisited). $A^*f = zf$ is the differential equation $f' = izf$ solved by $f(x) = e^{izx}$. This lies in Q for any $z \in \mathbb{C}$ and it is easy to see (Problem 5) that these are the only solutions. Thus, $d_+(A) = d_-(A) = 1$, so A is not self-adjoint.

An easy integration by parts proves each A_β is Hermitian. Moreover, if $\beta = e^{i\theta}$ and $\varphi_\beta(x) = e^{i\theta x}$, then $D(A_\beta) = D(A) \dotplus [\varphi_\beta]$ (i.e., $D(A_\beta)$ is the unique sum of $D(A)$ and multiples of φ_β). Since $\operatorname{Ker}(A_\beta + i) = 0$, $(A_\beta + i)\varphi_\beta$ must be independent of $\operatorname{Ran}(A + i)$. Since $\operatorname{Ran}(A + i)^\perp$ is one-dimensional, we conclude $\mathrm{Ran}(A_\beta + i)$ must be all of $\mathcal{H}$, that is, $d_+(A_\beta) = d_-(A_\beta) = 0$, so A_β is self-adjoint. $\square$

The same argument that led to Theorem 7.1.10 implies

Theorem 7.1.11. *Let A be a closed Hermitian operator and*

$$\mathcal{K}_\pm = \mathrm{Ker}(A^* \mp i) \tag{7.1.58}$$

Then

$$D(A^*) = D(A) \dotplus \mathcal{K}_+ \dotplus \mathcal{K}_- \tag{7.1.59}$$

in the sense that any $\varphi \in D(A^*)$ can be written uniquely as

$$\varphi = \varphi_0 + \varphi_+ + \varphi_-, \qquad \varphi_0 \in D(A), \quad \varphi_\pm \in \mathcal{K}_\pm \tag{7.1.60}$$

Moreover,

$$A^*(\varphi_0 + \varphi_+ + \varphi_-) = A\varphi_0 + i\varphi_+ - i\varphi_- \tag{7.1.61}$$

$$\mathrm{Im}\langle(\varphi_0 + \varphi_+ + \varphi_-), A^*(\varphi_0 + \varphi_+ + \varphi_-)\rangle = \|\varphi_+\|^2 - \|\varphi_-\|^2 \tag{7.1.62}$$

Proof. If $\varphi \in D(A^*)$ and $\psi = (A^* + i)\varphi$, then since $\mathrm{Ran}(A + i)$ is closed, $\psi = (A+i)\varphi_0 + \eta$, with $\eta \in \mathrm{Ran}(A+i)^\perp = \mathcal{K}_+$. Let $\varphi_+ = (2i)^{-1}\eta$. Then since $A^* = i$ on $\mathcal{K}_+$, $(A+i)(\varphi - \varphi_0 - \varphi_+) = 0$. Therefore, $\varphi_- \equiv \varphi - \varphi_0 - \varphi_+ \in \mathcal{K}_-$. Thus, any $\varphi \in D(A^*)$ is such a sum.

To get uniqueness, suppose that

$$\varphi \equiv \varphi_0 + \varphi_+ + \varphi_- = 0 \tag{7.1.63}$$

Applying $(A^* + i)$ to this, we get

$$(A + i)\varphi_0 + 2i\varphi_+ = 0 \tag{7.1.64}$$

Since $\mathrm{Ran}(A+i) \perp \mathcal{K}_+$, we conclude $\varphi_+ = (A+i)\varphi_0 = 0$. Since $\mathrm{Ker}(A+i) = 0$, $\varphi_+ = \varphi_0 = 0$ and thus, by (7.1.63), $\varphi_- = 0$, so we have uniqueness.

Since $A^* = \pm i\mathbb{1}$ on $\mathcal{K}_\pm$, (7.1.61) is immediate. (7.1.62) is a straightforward calculation (Problem 2) using $\mathrm{Re}(\langle\eta,\psi\rangle - \langle\psi,\eta\rangle) = 0$ for all $\eta, \psi \in \mathcal{H}$ and $\langle A\varphi_0, \varphi_\pm\rangle = \langle\varphi_0, \pm i\varphi_\pm\rangle$. $\square$

We'll eventually see $d_+ = d_-$ is very useful when it holds. If $A \geq 0$, we've already seen that this is true. Here is another situation when it holds.

Definition. A *conjugation* on a Hilbert space, $\mathcal{H}$, is an anti-unitary map $C\colon \mathcal{H} \to \mathcal{H}$ so that $C^2 = \mathbb{1}$. Anti-unitary maps are defined at the end of Section 3.3 of Part 1.

If $\{\varphi_n\}_{n-1}^\infty$ is an orthonormal basis, then $C(\sum_{n=1}^\infty a_n\varphi_n) = \sum_{n=1}^\infty \bar{a}_n\varphi_n$ is a complex conjugation. Every conjugation has this form (Problem 3).

Theorem 7.1.12 (von Neumann's Conjugation Result). *Let A be a closed Hermitian operator and C a complex conjugation on $\mathcal{H}$. If*

$$C[D(A)] \subset D(A), \quad A(C\varphi) = CA\varphi \tag{7.1.65}$$

then $d_+ = d_-$.

Proof. It is easy to see $C[D(A^*)] \subset D(A^*)$ and $A^*C\psi = CA^*\psi$ and then that $C[\mathrm{Ker}(A^* + i)] = \mathrm{Ker}(A^* - i)$. □

The simplest application of the fundamental criterion is to show stability of self-adjointness under suitable small perturbations.

Definition. Let A, B be densely defined operators on a Hilbert space, $\mathcal{H}$. We say B is *A-bounded* (aka *relatively bounded wrt A*) if and only if

(i) $D(A) \subset D(B)$;
(ii) For some $\alpha, \beta \in [0, \infty)$ and all $\varphi \in D(A)$, we have

$$\|B\varphi\| \le \alpha\|A\varphi\| + \beta\|\varphi\| \tag{7.1.66}$$

The inf over α for which (7.1.66) holds (β may be α-dependent and may go to ∞ as $\alpha \downarrow$ inf) is called the A-bound for B. Sometimes, if the A-bound is 0, we say B is A-infinitesimal, aka infinitesimally bounded wrt A.

It is easy to show (Problem 6) that if B is closed, it suffices to prove (7.1.66) for φ in a core for A.

Proposition 7.1.13. *Let A be an operator and let B be A-bounded with relative bound $\alpha < 1$. Then $A + B$ defined on $D(A)$ is closed if and only if A is closed and the cores for A are exactly the cores for $A + B$.*

Proof. Define two norms on $D(A)$:

$$\|\varphi\|_A = \|A\varphi\| + \|\varphi\| \tag{7.1.67}$$

$$\|\varphi\|_{A+B} = \|(A + B)\varphi\| + \|\varphi\| \tag{7.1.68}$$

It is easy to see (Problem 7) that $A + B$ is closed if and only if $D(A)$ is complete in $\|\cdot\|_{A+B}$ and cores are exactly those $D_0 \subset D(A)$ dense in $\|\cdot\|_{A+B}$ (or $\|\cdot\|_A$ for cores of A).

Suppose for some $\alpha < 1$ and $\varphi \in D(A)$,

$$\|B\varphi\| \le \alpha\|A\varphi\| + \beta\|\varphi\| \tag{7.1.69}$$

Then

$$\|(A + B)\varphi\| \le (1 + \alpha)\|A\varphi\| + \beta\|\varphi\| \tag{7.1.70}$$

so

$$\|\varphi\|_{A+B} \le [1 + \max(\alpha, \beta)]\,\|\varphi\|_A \tag{7.1.71}$$

Moreover,

$$\begin{aligned}\|A\varphi\| &\le \|(A+B)\varphi\| + \|B\varphi\| \\ &\le \|(A+B)\varphi\| + \alpha\|A\varphi\| + \beta\|\varphi\|\end{aligned} \tag{7.1.72}$$

so

$$\|A\varphi\| \le (1-\alpha)^{-1}\|(A+B)\varphi\| + (1-\alpha)^{-1}\beta\|\varphi\| \tag{7.1.73}$$

Thus,

$$\|\varphi\|_A \le \max[(1-\alpha)^{-1}, (1-\alpha)^{-1}\beta + 1]\|\varphi\|_{A+B} \tag{7.1.74}$$

Therefore, $\|\cdot\|_A$ and $\|\cdot\|_{A+B}$ are equivalent norms. $D(A)$ is complete in both or neither and the dense subspaces are the same. □

Theorem 7.1.14 (Kato–Rellich Theorem). *Let A be a self-adjoint operator and let B be A-bounded with A bound $\alpha < 1$. Suppose $B \restriction D(A)$ is Hermitian. Then $A + B$ as an operator on $D(A)$ is self-adjoint. The cores of A and $A + B$ are the same.*

Proof. We begin by noting that (for y real and $\varphi \in D(A)$)

$$\|(A+iy)\varphi\|^2 = \|A\varphi\|^2 + y^2\|\varphi\|^2 \tag{7.1.75}$$

With $\varphi = (A+iy)^{-1}\eta$, this implies

$$\|A(A+iy)^{-1}\| \le 1, \qquad \|(A+iy)^{-1}\| \le |y|^{-1} \tag{7.1.76}$$

Suppose $D(A) \subset D(B)$ and

$$\|B\varphi\| \le \alpha\|A\varphi\| + \beta\|\varphi\| \tag{7.1.77}$$

Applying to $\varphi = (A+iy)^{-1}\eta \in D(A) \subset D(B)$ and using (7.1.76), we see

$$\|B(A+iy)^{-1}\| \le \alpha + \beta|y|^{-1} \tag{7.1.78}$$

Since $\alpha < 1$, we can pick y so $\alpha + \beta|y|^{-1} < 1$. Thus,

$$C = 1 + B(A+iy)^{-1} \tag{7.1.79}$$

is invertible. In particular, $\mathrm{Ran}(1+C) = \mathcal{H}$. Since $\mathrm{Ran}(A+iy) = \mathcal{H}$, we see as a map from $D(A)$ to $\mathcal{H}$,

$$\mathrm{Ran}((1+C)(A+iy)) = \mathcal{H} \tag{7.1.80}$$

But

$$(1+C)(A+iy) = A + iy + B \tag{7.1.81}$$

so $d_{A+B}(iy) = 0$. Similarly, $d_{A+B}(-iy) = 0$. By the invariance of $d_{A+B}(z)$ in each half-plane, $A + B$ is self-adjoint as an operator on A. The core statement follows from Proposition 7.1.13. □

One other significant fact about relatively bounded perturbations with relative bound $\alpha < 1$ is that they preserve semiboundedness:

Theorem 7.1.15. *Let A, B be Hermitian operators with A self-adjoint and bounded below. If (7.1.69) holds for all $\varphi \in D(A)$ and some $\alpha < 1$, then $A + B$ is bounded below.*

Remark. Our proof shows that

$$\gamma(A+B) \geq \gamma(A) - \frac{\beta}{1-\alpha} \tag{7.1.82}$$

Proof. We use the facts that

(1) If A is self-adjoint semibounded, then

$$\sigma(A) \subset [\gamma(A), \infty) \tag{7.1.83}$$

and $\|A(A - \gamma(A) + c)^{-1}\| \leq 1$ and

$$\|(A - \gamma(A) + c)^{-1}\| \leq c^{-1} \tag{7.1.84}$$

for any $c > 0$ (see Problem 7(a) of Section 7.2).

(2) If C is self-adjoint and $(-\infty, \gamma) \subset \mathbb{C} \backslash \sigma(C)$, then C is semibounded and $\gamma(C) \geq \gamma$ (see Problem 7(b) of Section 7.2).

If $c > 0$, then by (7.1.69), (7.1.83), and (7.1.84),

$$\|B(A - \gamma(A) + c)^{-1}\| \leq \alpha + \beta c^{-1} \tag{7.1.85}$$

if $c > \beta(1-\alpha)^{-1}$, then this norm is less than 1, and as in the last proof, $D \equiv 1 + B(A - \gamma(A) + c)^{-1}$ is bounded and invertible (with inverses given by a geometric series). Thus, $(A - \gamma(A) + c)^{-1} D^{-1}$ is an inverse for $A + B - \gamma(A) + c$, so $\sigma(A+B) \subset [\gamma(A) - \beta(1-\alpha)^{-1}, \infty)$. By (2) above, $A + B$ is semibounded. □

The following is related to the Kato–Rellich theorem:

Theorem 7.1.16. *Let A be a closed symmetric operator and B a symmetric operator which is relatively A-bounded with relative bound less than 1. Then*

$$d_\pm(A+B) = d_\pm(A) \tag{7.1.86}$$

Proof. Introduce the symbol $C \prec \alpha D$ for arbitrary operators

$$C \prec \alpha D \Leftrightarrow D(D) = D(C) \;\&\; (\varphi \in D(D) \Rightarrow \|C\varphi\| \leq \alpha \|D\varphi\|) \tag{7.1.87}$$

We first claim that C, D closed plus

$$(C - D) \prec \alpha D \;\&\; \alpha < \tfrac{1}{2} \Rightarrow \operatorname{Ran}(D)^\perp \cap \operatorname{Ran}(C) = \{0\} \tag{7.1.88}$$

For suppose $\varphi \in \operatorname{Ran}(D)^\perp \cap \operatorname{Ran}(C)$ and the left side of (7.1.88) holds. Then $\varphi = C\eta$ for some $\eta \in D(D)$ and

$$\langle \varphi, \varphi \rangle = \langle \varphi, (D + C - D)\eta \rangle = \langle \varphi, (C - D)\eta \rangle \tag{7.1.89}$$

so by the Schwarz inequality,

$$\|\varphi\| \le \|(C-D)\eta\| \le \alpha\|D\eta\| \tag{7.1.90}$$

On the other hand, since $D\eta = \varphi - (C-D)\eta$,

$$\|D\eta\| \le \|\varphi\| + \|(C-D)\eta\| \le 2\|(C-D)\eta\| \le 2\alpha\|D\eta\| \tag{7.1.91}$$

Thus, $2\alpha < 1 \Rightarrow \|D\eta\| = 0 \Rightarrow \|\varphi\| = 0 \Rightarrow \varphi = 0$, proving (7.1.88).

By the argument that led to (7.1.78), if B is A-bounded with relative bound a, there is $Y \in (0,\infty)$ so $|y| > Y$ implies $B \prec a(A+iy)$. This plus (7.1.88) plus Lemma 7.1.9 shows that (if A, B are symmetric)

$$\begin{gathered} C \text{ is } A\text{-bounded and } A+C\text{-bounded} \\ \text{with relative bounds } < \tfrac{1}{2} \Rightarrow d_\pm(A+C) = d_\pm(A). \end{gathered} \tag{7.1.92}$$

By Problem 16, if B is A-bounded with relative bound $\alpha < 1$, there is γ so that for all $\lambda \in [0,1]$, B is $A + \lambda B$-bounded with relative bound $r_\lambda \le \gamma$. Pick m an integer so that $m > 2\gamma$ and let $C = \frac{B}{m}$. Then for $j = 0, \dots, m$, C is $A + \frac{j}{m}B$-bounded with relative bound less than $\frac{\gamma}{m} < \frac{1}{2}$. Thus, by (7.1.92),

$$d_\pm\biggl(A + \frac{j}{m}B\biggr) = d_\pm\biggl(A + \frac{(j-1)}{m}B\biggr), \quad j = 1, 2, \dots, m \tag{7.1.93}$$

which proves (7.1.86). □

We turn next to examples, focusing mainly on $A = -\Delta$ on $L^2(\mathbb{R}^\nu)$ and $B = V(x)$, that is, multiplication by an unbounded function of x. We should note that, often, operator bounded perturbations have relative bound zero and we'll see that here. One reason for this is that relatively compact perturbations (a notion we'll define in Section 7.8) also have relative bound zero; see Problem 4 of Section 7.8. This said, we see below if $\nu \ge 5$, $B =$ multiplication by $|x|^{-2}$ is a relatively bounded perturbation of $-\Delta$ but with nonzero relative bound.

Example 7.1.17. Recall Sobolev estimates (see Section 6.3 of Part 3 and Theorem 1.5.10) which in one form say that if

$$\begin{aligned} &q \le \infty, && \nu = 1,2,3 \\ &q < \infty, && \nu = 4 \\ &q \le \frac{2\nu}{\nu-4}, && \nu \ge 5 \end{aligned} \tag{7.1.94}$$

then there is $c_{q,\nu}$ so for all $\varphi \in \mathcal{S}(\mathbb{R}^\nu)$, we have

$$\|\varphi\|_q \le c_{q,\nu}[\|\Delta\varphi\|_2 + \|\varphi\|_2] \tag{7.1.95}$$

Define p so $q^{-1}+p^{-1}=\frac{1}{2}$, that is,

$$\begin{aligned} &p \geq 2, && \nu = 1,2,3 \\ &p > 2, && \nu = 4 \\ &p \geq \frac{\nu}{2}, && \nu \geq 5 \end{aligned} \tag{7.1.96}$$

Then, by Hölder's inequality, if $V \in L^p$,

$$\|V\varphi\|_2 \leq c_{q,\nu}\|V\|_p(\|-\Delta\varphi\|_2 + \|\varphi\|_2) \tag{7.1.97}$$

so multiplication by V is $-\Delta$-bounded.

In fact, the relative bound is zero. For if $p<\infty$ and

$$V^{(N)}(x) = \begin{cases} V(x) & \text{if } |V(x)| \leq N \\ 0 & \text{if } |V(x)| > N \end{cases} \tag{7.1.98}$$

then $\|V - V^{(N)}\|_p \to 0$ by the dominated convergence theorem, so for any ε, one can write $V = V_{1,\varepsilon} + V_{2,\varepsilon}$ with $\|V_{1,\varepsilon}\|_p \leq \varepsilon c_{q,\nu}^{-1}$ and $\|V_{2,\varepsilon}\|_\infty < \infty$, so

$$\|V\varphi\|_2 \leq \varepsilon\|-\Delta\varphi\|_2 + (\varepsilon + \|V_{2,\varepsilon}\|_\infty)\|\varphi\|_2 \tag{7.1.99}$$

In particular, if $\nu=3$, $V(x)=|x|^{-1} \in L^2 + L^\infty$, so $|x|^{-1}$ has relative bound zero, proving $-\Delta+|x|^{-1}$, the hydrogen atom Hamiltonian of elementary quantum theory, is essentially self-adjoint on $C_0^\infty(\mathbb{R}^3)$ (and self-adjoint on $D(-\Delta)$). We summarize the above below. □

Theorem 7.1.18. *Let p obey* (7.1.96). *Then multiplication by any $V \in L^p(\mathbb{R}^\nu) + L^\infty(\mathbb{R}^\nu)$ is $-\Delta$-bounded with relative bound zero and $-\Delta+V$ is self-adjoint on $D(-\Delta)$.*

Remarks. 1. The above is fine for V's going to zero at infinity but not for singular V's which are periodic or almost periodic. For example, for $\nu=3$, one would expect

$$V(x) = \sum_{v\in\mathbb{Z}^\nu} |x-v|^{-1} e^{-|x-v|} \tag{7.1.100}$$

to be $-\Delta$-bounded also. In fact, a localization argument (Problem 8) proves the result below.

2. $(-\Delta)^{-1}$ is convolution with $c_\nu|x-y|^{-(\nu-2)}$ if $\nu \geq 2$ and $(-\Delta)^{-2}$ with $c_\nu|x-y|^{-(\nu-4)}$. There is an alternate to L^p-conditions called Stummel conditions. If $\nu>4$, it is defined by

$$\limsup_{\alpha\downarrow 0}\ \sup_x \int_{|x-y|\leq\alpha} |V(y)|^2|x-y|^{-(\nu-4)}\, d^\nu y = 0 \tag{7.1.101}$$

Such V's are said to lie in the Stummel class, S_ν. V's obeying (7.1.101) are relatively $-\Delta$-bounded with relative bound zero. Conversely, if for some

$a, b > 0$ and $\delta \in (0,1)$, one has for all $\varepsilon > 0$ and $\varphi \in D(\Delta)$ that

$$\|V\varphi\|_2 \le \varepsilon \|\Delta\varphi\|_2 + a\exp(b\varepsilon^{-\delta})\|\varphi\| \tag{7.1.102}$$

then V obeys (7.1.101). This is discussed further in the Notes and Problem 9.

Here is the promised result that the reader will prove in Problem 8.

Definition. $L^p_u(\mathbb{R}^\nu)$, the *uniformly L^p-functions*, are those V with

$$\|V\|_{p,u} = \sup_{x\in\mathbb{R}^\nu} \left(\int_{|x-y|\le 1} |V(y)|^p \, d^\nu y \right)^{1/p} < \infty \tag{7.1.103}$$

Theorem 7.1.19. *Let $p_c = 2$ if $\nu \le 4$ and $\nu/2$ if $\nu \ge 5$. Then if $V \in L^p_u(\mathbb{R}^\nu)$ with $p > p_c$, then multiplication by V is $-\Delta$-bounded with relative bound zero. If $p = p_c$ and $\nu \ne 4$, multiplication by V is $-\Delta$-bounded.*

One defect of the L^p-results above is that p_c is ν-dependent. If one wants to consider multiparticle Coulomb quantum Hamiltonians as arise in atomic and molecular physics,

$$H = \sum_{j=1}^N -(2m_j)^{-1}\Delta_j + \sum_{j<k} e_j e_k |r_j - r_k|^{-1} \tag{7.1.104}$$

on $L^2(\mathbb{R}^{3N})$ with $r \in \mathbb{R}^{3N}$ is written $(r_1, \dots, r_N)$, $r_j \in \mathbb{R}^3$, and Δ_j is the Laplacian for r_j, then $|r_j - r_k|^{-1} \in L^p_u(\mathbb{R}^n)$ so long as $p < n$, so $p > p_c$ only if $3N/2 < 3$, that is, $N = 1$! Since $r_j - r_k$ is really only a three-dimensional variable, one would expect 3, not $3N$, is the relevant dimension. Tensor product ideas resolve this problem.

Theorem 7.1.20. *Let $\mathcal{H}_1, \mathcal{H}_2$ be two Hilbert spaces. Let A be self-adjoint on $\mathcal{H}$ and B Hermitian and A-bounded. Then $A \otimes \mathbb{1}$ on $\mathcal{H}_1 \otimes \mathcal{H}_2$ with domain the span of $\{\varphi \otimes \psi \mid \varphi \in D(A),\, \psi \in \mathcal{H}_2\}$ and $(A \otimes 1)(\varphi \otimes \psi) = A\varphi \otimes \psi$ is essentially self-adjoint and the closure of $B \otimes \mathbb{1}$ on $D(A)$ is relatively bounded with respect to $\overline{A \otimes 1}$ with the same relative bound.*

Remarks. 1. See Section 3.8 of Part 1 for discussion of tensor products.

2. The straightforward proof is left to the reader (Problem 11).

This result implies (Problem 12) that if $f \in L^p(\mathbb{R}^\nu)$, $p > p_c$, $T\colon \mathbb{R}^\mu \to \mathbb{R}^\nu$ a linear surjection and $V(x) = f(Tx)$, then V is a relatively bounded perturbation (with relative bound zero) of $-\Delta_{\mathbb{R}^\nu}$ acting on $L^2(\mathbb{R}^\mu)$. Since $(-\Delta_{\mathbb{R}^\nu})^2 \le (-\Delta_{\mathbb{R}^\mu})^2$, we see V is a relatively bounded perturbation of $-\Delta_{\mathbb{R}^\mu}$ with bound zero. We have thus proven:

Theorem 7.1.21 (Kato's Theorem)**.** *The Hamiltonian (7.1.104) is essentially self-adjoint on $C_0^\infty(\mathbb{R}^{3N})$.*

Example 7.1.22 ($|x|^{-2}$ as a perturbation of $-\Delta$)**.** It follows from the Stein–Weiss inequality (see Theorem 6.2.2 of Part 3) that if $\nu \geq 5$ then $|x|^{-2}(-\Delta+1)^{-2}$ (where "$|x|^{-2}$" is multiplication by $|x|^{-2}$) is bounded on L^2. It follows that $|x|^{-2}$ is Δ–bounded, that is, for some α, β and $\varphi \in \mathcal{S}(\mathbb{R}^\nu)$,

$$\||x|^{-2}\varphi\| \leq \alpha\|\Delta\varphi\| + \beta\|\varphi\| \tag{7.1.105}$$

In fact (Problem 15), if (7.1.105) holds, then $\beta = 0$ (same α) and this implies α cannot be arbitrarily small. Indeed (see Problem 10 in Section 7.4), the optimal α is known and is known as *Rellich's inequality*. Thus, $|x|^{-2}$ is Δ-bounded but the relative bound is not 0. We'll show (see Example 7.4.26) that for λ large enough, $-\Delta - \lambda|x|^{-2}$ is not essentially self-adjoint on $D(-\Delta)$. □

Notes and Historical Remarks.

> I have often wondered why it took so long for this fundamental question to be answered. As Kato remarks in his Wiener Prize acceptance, the proof is "rather easy." In modern terminology it is a combination of a Sobolev estimate and the theorem of Kato and F. Rellich on stability of self-adjointness under certain kinds of perturbations. ... One factor could have been von Neumann's attitude. V. Bargmann told me of a conversation he had with von Neumann in 1948 in which von Neumann asserted that the multiparticle result was an impossibly hard problem and even the case of hydrogen was a difficult open problem (even though the hydrogen case can be solved by separation of variables and the use of H. Weyl's 1912 theory, which von Neumann certainly knew!). Perhaps this is a case like the existence of the Haar integral, in which von Neumann's opinion stopped work by the establishment, leaving the important discovery to the isolated researcher unaware of von Neumann's opinion. Another factor surely was the Second World War. ... In his Wiener Prize acceptance, Kato remarks dryly: "During World War II, I was working in the countryside of Japan." In fact, from a conversation I had with Kato one evening at a conference, it was clear that his experiences while evacuated to the countryside and in the chaos immediately after the war were horrific. He barely escaped death several times, and he caught tuberculosis.
>
> —*Barry Simon* [**131**]

The theory of unbounded self-adjoint operators was developed by three young (all turned 26 in 1929, the central year of the birth of the theory) mathematicians, John von Neumann (1903–1954) (his central work was a long paper [**720**]), Marshall Stone (1903–1989) (he had several papers but his masterpiece was a book [**670**]), and Aurel Wintner (1903–1958) (also a book [**761**]). Von Neumann and Wintner were motivated by the development of quantum mechanics and the need to make sense of its mathematics; indeed, some of the theory appeared in von Neumann's book [**723**] on the foundations of quantum mechanics. Stone was following up on von Neumann, but his applications and clear exposition had long-lasting impact.

Somewhat earlier, Carleman in his book [**102**] on singular integral equations, had to cope with unbounded operators and had some partial progress.

Wintner's work had less impact because he used the language of matrices and the language of operators used by von Neumann and Stone was more flexible. Since the Heisenberg approach to quantum mechanics was matrix-based, von Neumann wrote a paper [**721**] on the matrix theory and its equivalence to the operator theory.

There are some who feel that Wintner deserves greater credit for foundational work on unbounded operators including the spectral theorem and results on self-adjoint extensions. He was slightly earlier than von Neumann with some of the basic results. Unfortunately, the work was hard for his contemporaries to absorb (see [**693**]). von Neumann's clarity of exposition and the scope of his work have resulted in his getting the bulk of the credit for this theory.

von Neumann's paper had the basics: the notions of closed operators and closure (in terms of sequences rather than graphs) and of self-adjoint (which he attributed to Schmidt), deficiency indices, the spectral theorem and the theory of self-adjoint extensions that we present in Section 7.4 with its consequences for positive operators (he had what we call the Krein extension in Section 7.5), and for operators commuting with a complex conjugate. It did not have the idea of using graphs to simplify the analysis—that appeared in a 1932 paper of von Neumann [**724**].

During the 1927–1930 period, von Neumann developed this theory and its applications to quantum mechanics which included his work on quantum measurement and quantum statistical mechanics. This is an impressive opus for a life's work, let alone a three-year period. But the remarkable thing is that he was doing a lot of other mathematics at the same time. Among this was his joint work with Wigner on eigenvalue crossing and on embedded bound states, the continuation of his earlier work on the foundations of mathematics (set theory), his founding of the subject of game theory and the first work on the existence of equilibria in economics, and his discovery that the Banach–Tarski paradox was connected to the nonamenability of the rotation group in three or more dimensions.

John von Neumann was born János Neumann in Budapest. His father, a wealthy Jewish banker, purchased a title when János was 10 and eventually, János added the German "von". János was Johann when he worked in Germany and after coming to the United States became John, universally called Johnny. von Neumann is not related to Carl Neumann (1832–1925), the German mathematician after whom Neumann boundary conditions and Neumann series are named (see the Notes to Section 3.6 of Part 3).

Johnny was a child prodigy and his father arranged tutoring—the stories say that both Szegő and Fekete tutored him and both were very impressed.

He was the ultimate double major. He registered as a math student in Budapest and at the same time studied chemistry, first in Berlin and later in Zürich, returning to Budapest only to ace his math exams. And he didn't hesitate to discuss math, for example, with Weyl in Zürich when he was officially a chemistry student.

In 1926, von Neumann got a diploma in chemistry from the ETH and a Ph.D. from Budapest for revolutionary work in set theory including formulating the now standard definition of infinite ordinal numbers. He then moved to be a *privatdozent* in Berlin and then Hamburg having spent a year during that period in Göttingen studying with Hilbert. In 1930, uneasy with both the limited job market in Germany and the unstable political situation, he went to the US initially as a visitor, then Professor at Princeton University and from 1933 until his death at the Institute for Advanced Study. He died of cancer at the age of only 53.

Besides an array of work done in the 1920s mentioned above, he founded the theory of operator algebras, was the initiator of modern ergodic theory, and he did work on Haar measure. He coauthored a basic book on mathematical economics, did foundational work in lattice theory, and was a founder, arguable, the founder, of modern computer science (the standard architecture of modern digital computers is named after him). He did numerical analysis (Monte Carlo method), hydrodynamics, and work on cellular automata.

He was a key figure in the development of nuclear energy, initially as an important leader in the Manhattan Project. It was his calculations that led to the decision to detonate the atomic bombs dropped on Japan at high altitude to cause maximum damage. After the war, he did joint work with Teller and Ulam that led to the nuclear physics behind the H-bomb, and, finally, as a member of the US Atomic Energy Commission, he was involved in key decisions in the years before his death.

His speed at calculation and finding proofs are legendary. Halmos [**285**] and Ulam [**708**] have readable biographical notes that attempt to capture his remarkable personality.

The Kato–Rellich theorem (Theorem 7.1.14) was found by Rellich [**556**, part III] in 1939 and rediscovered by Kato [**376**](see the further discussion of Kato's paper below). This theorem requires $\alpha < 1$—and this is not coincidental. There are examples (see Example 7.4.26) with $\alpha > 1$ where $A+B$ is not even essentially self-adjoint on $D(A)$. Wüst [**765**] (see Problem 13) has proven that if (7.1.66) holds with $\alpha = 1$, then $A + B$ is essentially self-adjoint on $D(A)$—it may not be self-adjoint as the example $B = -A$ shows (earlier, Kato [**380**] had proven essential self-adjointness under the stronger condition $\|B\varphi\|^2 \leq \|A\varphi\|^2 + b^2\|\varphi\|^2$). Problem 14 has another extension of Theorem 7.1.14.

Kato's theorem (Theorem 7.1.21 and the implicit (7.1.19)–(7.1.21)) is from his 1951 paper [**373**]. This represents one of the signal results of modern mathematical physics. Given Stone's theorem (see Theorem 7.3.1), it asserts the solvability of the time-dependent Schrödinger equation with Coulomb forces for any regular initial condition ($\varphi \in D(A)$ suffices for classical solutions, $\varphi \in L^2$ for weak solutions). The analogous question for classical Newtonian gravitation (solvability for almost every boundary condition) remains open for N-body systems with $N \geq 5$! Kato's theorem also represents the birth of a new focus of mathematical nonrelativistic quantum mechanics—before Kato, almost all the mathematical work focused on the formalism of quantum theory; since then, on the specific models of concrete Hamiltonians.

The quote at the top gives some background about this paper. Kato was trained as a physicist and knew of von Neumann's book, but not Rellich's work. He found this theorem in 1944, but in the aftermath of the war (and his need to get his Ph.D. degree, which was on convergence of eigenvalue perturbation theory), it was only in 1948 when he submitted to *Physical Review* (!), which had no idea how to handle it. Eventually, they consulted von Neumann who had it transferred to *Transactions of the AMS* where it was finally published three years after submission. Along the way, in an era before electronic versions or even xerox copies, the manuscript was lost by the journal several times!

Tosio Kato (1917–1999) was made a faculty member in physics at Tokyo University, becoming a professor in 1958. He visited the US, including NYU and Berkeley in 1954–55 and was appointed to a position at Berkeley in 1962 where he spent the rest of his career. He was fortunate that when he needed a visa to accept the Berkeley offer that mathematical physicist, Charles Dolph, was attachée in the US Embassy in Japan, since the visa was initially denied because of Kato's wartime brush with TB.

Besides his work on perturbation theory and self-adjointness that we mentioned and his work on quadratic forms and on Kato's inequality that appear later in the chapter, Kato is known for his work in nonlinear PDEs, including the KdV and Navier–Stokes equations. His book [**380**] on perturbation theory published in 1966 has been a standard reference for fifty years. His students include E. Balslev, J. Howland, T. Ikebe, and S. Kuroda.

There are two forms of implementing when multiplication operators on $L^2(\mathbb{R}^\nu)$ are $-\Delta$-bounded, just as there are two versions of "Sobolev" inequalities. One is the L^p-approach discussed here. The other exploits convolution with $|x-y|^{-(\nu-4)}$, that is, the integral kernel of $(\Delta)^{-2}$. This latter approach goes back to Stummel [**674**], after whom Stummel conditions are named. The relation to (7.1.102) goes back to Cycon et al. [**144**] who mimicked an analogous idea for form boundedness in Aizenman–Simon [**9**]; see Problem 9.

Problems

1. Prove that $C_0^\infty(\mathbb{R}^\nu)$ is a domain of essential self-adjoint for $P(i\partial)$ for any real polynomial in ν variables. (*Hint*: Use $j * (f\varphi)$ to approximate φ for suitable f and j.)

2. Prove (7.1.62).

3. Let C be a conjugation on a complex Hilbert space, $\mathcal{H}$, and let $\mathcal{H}_r$ be the real Hilbert space obtained by viewing $\mathcal{H}$ as a vector space over the reals with inner product $\langle\varphi,\psi\rangle_r = \text{Re}\langle\varphi,\psi\rangle$. Let $J\colon \mathcal{H}_r \to \mathcal{H}_r$ by $J\varphi = i\varphi$.

   (a) Prove $C^2 = 1$, $J^2 = -1$, $JC = -CJ$.

   (b) Prove $\mathcal{H}_r = \mathcal{H}_r^+ \oplus \mathcal{H}_r^-$ with $\mathcal{H}_r^\pm = \{\psi \mid C\psi = \pm\psi\}$. (*Hint*: Prove that $\frac{1}{2}(1 \pm C)$ are projections.)

   (c) Prove J is a bijection of $\mathcal{H}_r^+$ to $\mathcal{H}_r^-$.

   (d) Let $\{\varphi_j^{(r)}\}_{j=1}^N$ be a real orthogonal basis of $\mathcal{H}_r^+$. Prove that the φ_j are orthonormal in $\mathcal{H}$ and a basis.

   (e) Show $C(\sum_{j=1}^N a_j\varphi_j) = \sum_{j=1}^N \bar{a}_j\varphi_j$.

4. (a) Let $f \in L^1([0,1], dx)$ so that its distribution derivative is also in L^1. Prove that f is continuous and then conclude the same if L^1 is replaced by L^2. (*Hint*: First show that if $g' = 0$ (distributional derivative), then g is constant, and apply to $g(x) = f(x) - \int_0^x f'(y)\,dy$.)

   (b) Verify (7.1.35) is the closure of A given on C_0^∞.

   (c) Verify (7.1.36).

   (d) Check each A_β given by (7.1.37) is Hermitian.

5. Let Q be given by (7.1.34). If $f \in \ell^2$ and $f' = izf$ (distributional derivative), prove that $f(x) = ce^{izx}$. (*Hint*: If $g = e^{-izx}f$, prove that $g' = 0$.)

6. Let A be self-adjoint, B a closed operator, and $\mathcal{D} \subset D(A) \cap D(B)$, a core for A. If (7.1.66) holds for $\varphi \in \mathcal{D}$, prove $D(A) \subset D(B)$ and (7.1.66) holds for all $\varphi \in D(A)$.

7. Let A be an operator with domain $D(A)$ and define $\|\cdot\|_A$ on $D(A)$ by (7.1.67).

(a) Prove that A is a closed operator if and only if $D(A)$ is complete in $\|\cdot\|_A$.

(b) Prove that $D_0 \subset D(A)$ is a core if and only if D_0 is $\|\cdot\|_A$-dense in $D(A)$.

In several of the following problems, it will help to have the following notation: If $D(B) \subset D(A)$ and for all $\varphi \in D(A)$, we have

$$\|B\varphi\| \le \alpha\|A\varphi\| + B\|\varphi\| \tag{7.1.106}$$

we say B is $\beta_1(\alpha,\beta)$ bounded. If

$$\|B\varphi\|^2 \le \alpha^2\|A\varphi\|^2 + \beta^2\|\varphi\|^2 \tag{7.1.107}$$

we then say B is $\beta_2(\alpha,\beta)$ bounded.

8. (a) Let $k \in \mathbb{Z}^\nu$ and $\Delta_k = \{x \in \mathbb{R}^\nu \mid |x_j - k_j| \le \frac{1}{2}\}$, the unit cube centered at k. Prove there is $j \in C_0^\infty(\mathbb{R}^\nu)$, $j \ge 0$, with $\operatorname{supp}(j) \in \bigcup_{|k|\le 1}\Delta_k$, $\inf_{x\in\Delta_0} j(x) > 0$, and so if $j_k(x) = j(x-k)$, then

$$\sum_{k\in\mathbb{Z}^\nu} j_k(x)^2 = 1 \tag{7.1.108}$$

(b) Prove for every $\varepsilon > 0$, there is $\tilde\beta_\varepsilon$ so

$$\|\Delta(j\varphi)\|^2 \le (1+\varepsilon)\|j(\Delta\varphi)\|^2 + \tilde\beta_\varepsilon^2 \sum_{|k|\le 1} \|j_k\varphi\|^2 \tag{7.1.109}$$

(c) Let χ_k be the characteristic function of $\bigcup_{|j|\le 1}\Delta_{k+j}$. Suppose for some fixed α, β and all k, $V\chi_k$ is a $\beta_2(\alpha,\beta)$ bounded perturbation of $-\Delta$. Prove that for all $\varepsilon > 0$, there is $\beta^\sharp$ so that V is a $\beta_2(\alpha+\varepsilon, \beta^\sharp)$ bounded perturbation of $-\Delta$. (*Hint*: $\|V\varphi\|^2 = \sum_k \|V\chi_k j_k\varphi\|^2$.)

(d) If $p > p_c$ and $V \in L_u^p(\mathbb{R}^\nu)$, prove that V is $-\Delta$-bounded.

(e) Prove the relative bound is zero. (*Hint*: Show that if $q < p < \infty$, then $V = V_{1,\varepsilon} + V_{2,\varepsilon}$, where $V_{1,\varepsilon}$ is bounded and $\|V_{2,\varepsilon}\chi_k\|_q \le \varepsilon\|V_{2,\varepsilon}\chi_k\|_p$.)

9. (a) Prove that $V \in S_\nu$ (see (7.1.101)) if and only if

$$\lim_{E\to\infty} \|(-\Delta+E)^{-2}|V|^2\|_{\infty,\infty} = 0 \tag{7.1.110}$$

where ∞,∞ means a map from L^∞ to L^∞.

(b) If (7.1.102) holds, prove that

$$(\exp(t\Delta)|V|^2)(y) \le C(\varepsilon t^{-2} + a\exp(-b\varepsilon^{-\delta})) \tag{7.1.111}$$

(*Hint*: Pick $\varphi(x) = [\exp(t\Delta)(x,y)]^{1/2}$.)

(c) Prove that (7.1.111) implies (7.1.110) and thus, (7.1.102) implies $V \in S_\nu$. (*Hint*: Write $(-\Delta+E)^{-2}$ as an integral of $e^{t\Delta}$ over t and pick ε in (7.1.111) t-dependent.)

10. (a) If B is a $\beta_1(\alpha,\beta)$ bounded perturbation of A, prove that it is a $\beta_2(\alpha,\beta)$ bounded perturbation.

    (b) If B is a $\beta_2(\alpha,\beta)$ bounded perturbation of A, prove that, for any $\varepsilon > 0$, it is a $\beta_1(\alpha(1+\varepsilon),\beta(1+\varepsilon^{-1}))$ bounded perturbation. (*Hint*: $2uv \le u^2+v^2$.)

    Remarks. 1. The relative bounds can be defined with or without the square.

    2. This explains why Wüst's result is stronger than the one in Kato's book.

11. Prove Theorem 7.1.20. (*Hint*: If $\{\psi_j\}_{j=1}^\infty$ is an orthonormal basis for $\mathcal{H}_2$, $\|\sum \varphi_j \otimes \psi_j\|^2 = \sum \|\varphi_j\|^2$ for any $\{\varphi_j\}_{j=1}^\infty \subset \mathcal{H}_1$.)

12. If $f\colon \mathbb{R}^\mu \to \mathbb{R}$ is a real-valued function, $T\colon \mathbb{R}^\nu \to \mathbb{R}^\nu$ is linear, and $V(x) = f(Tx)$, then prove that if f is $-\Delta_\mu$-bounded, then V is $-\Delta_\nu$-bounded.

13. This will prove Wüst's theorem mentioned in the Notes. Suppose B is $\beta(1,\beta)$ bounded and $(A+B+i)^*\gamma = 0$.

    (a) Show that for $t \in (0,1)$, there exists φ_t so $\|\varphi_t\| \le \|\gamma\|$ and $(A+tB+i)\varphi_t = \gamma$.

    (b) Define $\psi_t = \gamma - (t-1)B\varphi_t$. Prove that

    $$(1-t)\|A\varphi_t\| \le \|(A+iB)\varphi_t\| + t\beta\|\varphi_t\|$$

    (c) Prove that $(1-t)\|A\varphi_t\|$, $(1-t)\|B\varphi_t\|$, and $\|\psi_t\|$ are bounded as $t \uparrow 1$.

    (d) If $\eta \in D(A)$, prove $(\psi_t - \gamma, \eta) \to 0$ as $t \uparrow 1$, and conclude γ is the weak limit of ψ_t.

    (e) Prove $|\psi_t, h| = 0$ and conclude $h = 0$.

    (f) Conclude that $A+B$ is essentially self-adjoint on $D(A)$.

14. The reader will prove that if A and C are Hermitian operators, $D \subset D(A) \cap D(C)$ and for some $\alpha < 1$, $\beta > 0$, and all $\varphi \in D$, we have

    $$\|(A-C)\varphi\| \le \alpha(\|A\varphi\| + \|C\varphi\|) + \beta\|\varphi\| \tag{7.1.112}$$

    that $A \restriction D$ is essentially self-adjoint if and only if $C \restriction D$ is essentially self-adjoint (this essentially extends the Kato–Rellich theorem).

    (a) Let $C(\lambda) = C + \lambda(A-C)$. Prove that for any $\varphi \in D$, we have for all $\lambda \in (0,1)$ that

    $$\|(A-C)\varphi\| \le \frac{2\alpha}{1-\alpha}\|C(\lambda)\varphi\| + \frac{\beta}{1-\alpha}\|\varphi\|$$

    (b) Find k and $0 = \lambda_1, \dots, \lambda_k = 1$ so that $C(\lambda_{j+1})$ is essentially self-adjoint on D, if and only if $C(\lambda_j)$ is. (*Hint*: Kato–Rellich.)

    (c) Conclude the claimed result.

15. Prove (7.1.105) for some α, β and all $\varphi \in \mathcal{S}(\mathbb{R}^\nu)$ implies (7.1.105) for the same α and β and all $\varphi \in \mathcal{S}(\mathbb{R}^\nu)$. (*Hint*: Consider $\varphi(\lambda x)$ for $\lambda \in (0, \infty)$.)

16. (a) If $\|B\varphi\| \le \alpha\|A\varphi\| + \beta\|\varphi\|$ and $0 < \alpha < 1$, prove that

$$\|A\varphi\| \le (1-\alpha)^{-1}[\|(A+B)\varphi\| + \beta\|\varphi\|]$$

(b) If B is A-bounded with relative bound $\alpha < 1$, prove that for all $\lambda \in [0,1]$, B is $(A+\lambda B)$-bounded with relative bound at most $\alpha(1-\alpha)^{-1}$.

7.2. The Spectral Theorem for Unbounded Operators

In this section, we'll prove a version of the spectral theorem for unbounded self-adjoint operators. We'll do this by considering the map (Cayley transform)

$$U = (A-i)(A+i)^{-1} \tag{7.2.1}$$

We'll prove this is unitary when A is self-adjoint and a partial isometry if A is Hermitian. The spectral theorem for unitary operators will then yield a spectral theorem for A. In Section 7.4, we use the fact that this relates self-adjoint extensions of Hermitian operators to unitary extensions of partial isometries to present von Neumann's theory of self-adjoint extensions.

Once we have the spectral theorem, we'll explore some of its most significant consequences.

As an indication towards where we are heading, we have

Theorem 7.2.1. *Let $\{\mu_j\}_{j=1}^N$ be a finite or infinite sequence of probability measures on $\mathbb{R}$. Let $\mathcal{H} = \oplus_{j=1}^N L^2(\mathbb{R}, d\mu_j)$ and let A be the operator on $\mathcal{H}$ with*

$$D(A) = \left\{ \{\varphi_j\}_{j=1}^N \,\middle|\, \sum_{j=1}^N \int (1+x^2)|\varphi_j(x)|^2\, d\mu_j(x) < \infty \right\} \tag{7.2.2}$$

$$(A\varphi_j)(x) = x\varphi_j(x) \tag{7.2.3}$$

Then A is self-adjoint.

Proof. It is easy to see A is Hermitian. If $\psi = \{\psi_j\}_{j=1}^N \in \mathcal{H}$ and $(\varphi_\pm)_j(x) = (x+i)^{-1}\psi_j(x)$, then it is immediate that $\varphi_\pm \in \mathcal{H}$, $\varphi_\pm \in D(A)$, and $(A\pm i)\varphi = \psi$. Thus, $\operatorname{Ran}(A \pm i) = \mathcal{H}$, so A is self-adjoint by Theorem 7.1.10. $\square$

Our goal is to show every self-adjoint A is unitarily equivalent to an A of the form (7.2.2)/(7.2.3).

If A is Hermitian and closed, one has for $\varphi \in D(A)$

$$\|(A+i)\varphi\|^2 = \|A\varphi\|^2 + \|\varphi\|^2 = \|(A-i)\varphi\|^2 \tag{7.2.4}$$

Thus, if $\varphi = (A+i)^{-1}\psi$, we see

$$\|\psi\| = \|(A-i)(A+i)^{-1}\psi\|^2 \tag{7.2.5}$$

This leads to the

Definition. Let A be a closed Hermitian operator. Its *Cayley transform*, U, is defined as the linear operator with

$$U\psi = \begin{cases} (A-i)(A+i)^{-1}\psi & \text{if } \psi \in \operatorname{Ran}(A+i) \\ 0 & \text{if } \psi \in \operatorname{Ran}(A+i)^{\perp} = \operatorname{Ker}(A^* - i) \end{cases}$$

Theorem 7.2.2. *Let A be a closed Hermitian operator and U its Cayley transform. Then*

(a) *U is a partial isometry with initial subspace,* $\operatorname{Ran}(A+i)$*, and final subspace,* $\operatorname{Ran}(A-i)$.

(b) *A is self-adjoint if and only if U is unitary.*

(c) $$U^*\psi = \begin{cases} (A+i)(A-i)^{-1}\psi & \text{if } \psi \in \operatorname{Ran}(A-i) \\ 0 & \text{if } \psi \perp \operatorname{Ran}(A-i) \end{cases} \tag{7.2.6}$$

(d) *φ is cyclic for U if and only if it is cyclic for $(A+i)^{-1}$.*

(e) *$U\psi = \psi$ has no solution.*

Remark. We'll see later that any unitary U obeying (e) is the Cayley transform of a self-adjoint A.

Proof. (a) (7.2.5) implies U is an isometry on $\operatorname{Ran}(A+i)$ so U is a partial isometry. Clearly, $\operatorname{Ran}(A+i)$ is the initial space. Since $(A \pm i)$ is a bijection of $D(A)$ and $\operatorname{Ran}(A \pm i)$, U maps $\operatorname{Ran}(A+i)$ onto $\operatorname{Ran}(A-i)$.

(b) This is just a restatement of the fundamental criterion that A is self-adjoint if and only if $\operatorname{Ran}(A+i) = \operatorname{Ran}(A-i) = \mathcal{H}$.

(c) If $\varphi \in \operatorname{Ran}(A+i)$, $\psi \in \operatorname{Ran}(A-i)$ and $\varphi = (A+i)\eta$, $\psi = (A-i)\gamma$, then by self-adjointness,

$$\begin{aligned} (\psi, U\varphi) &= (\psi, (A-i)\eta) \\ &= ((A-i)\gamma, (A-i)\eta) \\ &= (A\gamma, A\eta) + (\gamma, \eta) \end{aligned} \tag{7.2.7}$$

and if V is the right side of (7.2.6), the same calculation shows

$$(V\psi, \varphi) = (A\gamma, A\eta) + (\gamma, \eta) \tag{7.2.8}$$

On the other hand, if either $\varphi \in \operatorname{Ran}(A+i)^{\perp}$ or $\psi \in \operatorname{Ran}(A-i)^{\perp}$, it is easy to see that $(\psi, U\varphi) = 0 = (V\psi, \varphi)$.

(d) Immediate if we note that

$$U = \mathbb{1} - 2i(A+i)^{-1} \tag{7.2.9}$$

(e) Since $U\psi = \psi \Rightarrow \|U\psi\| = \|\psi\|$, $\psi \in \operatorname{Ran}(A+i)$, so $U^*\psi = U^*U\psi = \psi$. Thus, this is a special case of the next proposition. $\square$

Proposition 7.2.3. *Let U be a Cayley transform of a closed Hermitian operator, A, and C an arbitrary "extension" of U, that is, $C \in \mathcal{L}(\mathcal{H})$ so that $C \restriction \operatorname{Ran}(A+i) = U \restriction \operatorname{Ran}(A+i)$. Then $C^*\psi = \psi \Rightarrow \psi = 0$.*

Proof. Let $\eta \in D(A)$. Then

$$\begin{aligned} 0 &= \langle (C^* - \mathbb{1})\psi, (A+i)\eta\rangle = \langle \psi, [C(A+i) - (A+i)]\eta\rangle \\ &= \langle \psi, [(A-i) - (A+i)]\eta\rangle = -2i\langle \psi, \eta\rangle \end{aligned} \tag{7.2.10}$$

Thus, $\psi \in D(A)^\perp = \{0\}$. $\square$

The inverse map of $F(z) = (z-i)(z+i)^{-1}$ is $F^{-1}(w) = i(1+w)/(1-w)$ or $F^{-1}(e^{i\theta}) = -\cot(\theta/2)$. If μ is a measure on $\partial\mathbb{D}$ with $\mu(\{1\}) = 0$, then we define $\mu^\sharp$, a measure on $\mathbb{R}$, by

$$\mu^\sharp(E) = \mu(F[E]), \qquad \int f(x)\, d\mu^\sharp(x) = \int f\biggl(-\cot\biggl(\frac{\theta}{2}\biggr)\biggr)\, d\mu(e^{i\theta}) \tag{7.2.11}$$

Lemma 7.2.4. *Let μ be a probability measure on $\partial\mathbb{D}$ with*

$$\mu(\{1\}) = 0 \tag{7.2.12}$$

Let U be a unitary operator on $L^2(\partial\mathbb{D}, d\mu)$

$$(U\varphi)(e^{i\theta}) = e^{i\theta}\varphi(e^{i\theta}) \tag{7.2.13}$$

Let A be the self-adjoint operator on $L^2(\mathbb{R}, d\mu^\sharp)$ with $\mu^\sharp$ given by (7.2.11):

$$D(A) = \{\psi \in L^2 \mid x\psi \in L^2\}, \qquad (A\psi)(x) = x\psi(x) \tag{7.2.14}$$

Let $V\colon L^2(\mathbb{R}, \mu^\sharp) \to L^2(\partial\mathbb{D}, d\mu)$ by

$$V\psi(e^{i\theta}) = \psi\biggl(-\cot\biggl(\frac{\theta}{2}\biggr)\biggr) \tag{7.2.15}$$

Then V is unitary and the Cayley transform of VAV^{-1} is U.

Proof. That V is an isometry is immediate from (7.2.11). That V is onto all of $L^2(\partial\mathbb{D}, d\mu)$ follows from (7.2.12). The Cayley transform result follows from the fact that $(A-i)(A+i)^{-1}$ at $x = -\cot(\theta/2)$ is multiplication by $e^{i\theta}$. $\square$

Theorem 7.2.5 (Spectral Theorem—Multiplication Operator Form). *Let B be a (possibly unbounded) operator on a Hilbert space, $\widetilde{\mathcal{H}}$. Then there exist measures $\{\mu_j\}_{j=1}^N$ on $\mathbb{R}$ and $W\colon \widetilde{\mathcal{H}} \to \oplus_{j=1}^N L^2(\mathbb{R}, d\mu_j)$, $N \in \mathbb{Z}_+ \cup \{\infty\}$, unitary, so if A is multiplication by x, then $W[D(B)] = D(A)$, and if $\varphi \in D(B)$, then*

$$WB\varphi = AW\varphi \tag{7.2.16}$$

Proof. Let V be the Cayley transform of A. By the spectral theorem for unitary operators (see Theorems 5.2.1 and 5.5.1), there are measures $\{\gamma_j\}_{j=1}^N$ on $\partial\mathbb{D}$ and a unitary $\widetilde{W}\colon \mathcal{H} \to \oplus_{j=1}^N L^2(\partial\mathbb{D}, d\gamma_j)$ so

$$(\widetilde{W}V\varphi)_j(e^{i\theta}) = e^{i\theta}(V\varphi)_j(e^{i\theta}) \tag{7.2.17}$$

By Lemma 7.2.4, $\widetilde{W}V$ is the Cauchy transform of a direct sum of multiplication operators. Chasing through the various unitaries and uniqueness of Cauchy transform (by (7.2.9)), we obtain this result. □

This argument also implies:

Theorem 7.2.6. *Any unitary U with $U\psi = \psi \Rightarrow \psi = 0$ is the Cayley transform of a self-adjoint operator.*

Once we have the spectral theorem in multiplication operator form, we can do multiplicity theory as in Section 5.4. We can define $F(B)$ for F a bounded Borel function on $\sigma(B)$ (which is easily seen to be $\overline{\bigcup_{j=1}^N \operatorname{supp}(d\mu_j)}$) by

$$(WF(B)W^{-1}\varphi)_j = F(x)\varphi_j(x) \tag{7.2.18}$$

Then $F(B)$ is bounded. Indeed,

$$\|F(B)\| \leq \sup|F(x)| \tag{7.2.19}$$

In fact, it is equal to $\operatorname{ess\,sup}|F(x)|$, where essential sup means over a.e. x with respect to $\sum_{j=1}^N 2^{-j}\, d\mu_j$. Among the most important functions are χ_S, characteristic functions for Borel sets $S \subset \mathbb{R}$; then $\chi_S(B) = P_S(B)$, the spectral projections. In particular, $E_t(B) = P_{(-\infty,t]}(B)$ is a resolution of the identity. The other two forms of the spectral theorem then read:

Theorem 7.2.7 (Spectral Theorem—Resolution of Identity Form). *Let A be a (possibly unbounded) self-adjoint operator on a Hilbert space, $\mathcal{H}$. Then there is a resolution of the identity, $\{E_t\}_{t\in\mathbb{R}}$, so that*

$$A = \int t\, dE_t \tag{7.2.20}$$

in the sense that

$$D(A) = \left\{\varphi \;\middle|\; \lim_{R\to\infty}\left\|\left[\int_{-R}^R t\, dE_t\right]\varphi\right\| < \infty\right\} \tag{7.2.21}$$

and for $\varphi \in D(A)$,

$$A\varphi = \lim_{R\to\infty}\left[\int_{-R}^R t\, dE_t\right]\varphi \tag{7.2.22}$$

Remarks. 1. It is easy to see that there is a converse: Every spectral resolution defines a self-adjoint operator via (7.2.21)/(7.2.22).

2. $t \to \|E_t\varphi\|^2$ is monotone increasing and so define a Stieltjes integral and (7.2.22) can be rewritten

$$D(A) = \{\varphi \mid \int_{\mathbb{R}} (1+t^2)\, d\,\|E_t\varphi\|^2 < \infty \tag{7.2.23}$$

Theorem 7.2.8 (Spectral Theorem—Functional Calculus Form). *Let A be a possibly unbounded self-adjoint operator on a Hilbert space, $\mathcal{H}$. Let $\mathcal{B}(\mathbb{R})$ be the bounded Borel functions on A. Let*

$$id_R(x) = \begin{cases} x, & |x| \le R \\ 0, & |x| > R \end{cases} \tag{7.2.24}$$

Then there exists $\Phi_B\colon \mathcal{B}(\mathbb{R})$ to $\mathcal{L}(\mathcal{H})$ which is a $*$-homomorphism with $\Phi_B(\mathbb{1}) = \mathbb{1}$ and

$$D(B) = \{\varphi \mid \lim_{R\to\infty} \|\Phi_B(id_R)\varphi\| < \infty\} \tag{7.2.25}$$

$$B\varphi = \lim_{R\to\infty} \Phi_B(id_R)\varphi \text{ if } \varphi \in D(B) \tag{7.2.26}$$

Moreover,

$$\|\Phi_B(f)\| \le \|f\|_\infty, \quad \|f - f_n\| \to 0 \Rightarrow \|\Phi_B(f_n) - \Phi_B(f)\| \to 0 \tag{7.2.27}$$

$$f_n(x) \to f(x) \text{ for all } x \text{ and } \sup_{n,x}|f_n(x)| < \infty \Rightarrow \text{s-lim}\, \Phi_B(f_n) = \Phi_B(f) \tag{7.2.28}$$

In terms of the spectral representation of Theorem 7.2.5, one defines the *spectral measure*, $\mu_\varphi^{(A)}$, for operator A and vector φ as

$$d\mu_\varphi^{(A)}(x) = \sum_{j=1}^N |(W\varphi)_j(x)|^2\, d\mu_j(x) \tag{7.2.29}$$

For bounded self-adjoint A, this agrees with the measure $d\mu_\varphi^{(A)}$ of Theorem 5.1.7. It is related to the objects above by

$$\langle \varphi, E_t\varphi\rangle = \mu_\varphi^{(A)}((-\infty, t]) \tag{7.2.30}$$

and

$$\langle \varphi, f(A)\varphi\rangle = \int f(x)\, d\mu_\varphi^{(A)}(x) \tag{7.2.31}$$

for all bounded Borel functions, f. If f is unbounded,

$$\varphi \in D(f(A)) \Leftrightarrow \int |f(x)|^2\, d\mu_\varphi^{(A)}(x) \tag{7.2.32}$$

and then (7.2.31) still holds. Once we have e^{itA} below, we'll also have the Fourier transform of $\mu_\varphi^{(A)}$ via the special case of (7.2.31):

$$\langle \varphi, e^{itA}\varphi\rangle = \int e^{itx}\, d\mu_\varphi^{(A)}(x) \tag{7.2.33}$$

Another important class of bounded functions are $x \mapsto e^{itx}$ for $t \in \mathbb{R}$. The associated $e^{itA} \equiv U_t$ are unitary since $U_{-t}U_t = U_tU_{-t} = \mathbb{1}$ and $U_t^* = U_{-t}$.

Definition. A *one-parameter unitary group* is a map, $t \mapsto U_t$, from $\mathbb{R}$ to $\mathcal{L}(\mathcal{H})$ so that

$$U_{t+s} = U_tU_s \qquad U_0 = \mathbb{1}, \qquad U_t^* = U_{-t} \tag{7.2.34}$$

$$t \mapsto U_t \text{ is strongly continuous} \tag{7.2.35}$$

Since $U_tU_t^* = U_t^*U_t = U_{-t}U_t = U_0 = \mathbb{1}$, each U_t is unitary. By Theorem 6.8.3, since $\mathcal{H}$ is separable, weak measurability implies strong continuity.

Theorem 7.2.9. *If A is a (possibly unbounded) self-adjoint operator on a Hilbert space, $U_t \equiv e^{itA}$ is a one-parameter unitary group.*

The main result of the next section is that this theorem has a converse.

Theorem 5.1.12 has an analog for unbounded self-adjoint operators. Once one has the spectral theorem which we've just proven, the proof is identical.

As we've seen, $(f, A) \mapsto f(A)$ is continuous in f (in two senses) for A fixed. Our final topic in this section is to explore continuity in A.

Definition. Let $\{A_n\}_{n=1}^\infty$, A be possibly unbounded self-adjoint operators. We say A_n converges to A in *norm-resolvent sense*, written $A_n \xrightarrow{\text{nrs}} A$, if and only if

$$\|(A_n \pm i) - (A \pm i)^{-1}\| \to 0 \tag{7.2.36}$$

We say A_n converges to A in *strong resolvent sense*, written $A_n \xrightarrow{\text{srs}} A$, if and only if for all $\varphi \in \mathcal{H}$,

$$\|(A_n \pm i)^{-1}\varphi - (A \pm i)^{-1}\varphi\| \to 0 \tag{7.2.37}$$

Remarks. 1. Since B self-adjoint implies $[(B+i)^{-1}]^* = (B-i)^{-1}$, one only a priori needs $+$ and not $\pm$ in (7.2.36). In fact, this is also true for (7.2.37) because, remarkably, weak convergence of $(A_n + i)^{-1}$ to $(A + i)^{-1}$ implies strong convergence of $(A_n \pm i)^{-1}\varphi$ to $(A \pm i)^{-1}\varphi$ (Problem 1).

2. We'll see shortly these convergence results (for resolvents) at $\pm i$ imply the same for resolvents for all z with $\operatorname{Im} z \neq 0$.

Theorem 7.2.10. *Let A_n be (possibly unbounded) self-adjoint operators. Then*

(a) $A_n \xrightarrow{\text{nrs}} A \Rightarrow f(A_n) \xrightarrow{\text{nrs}} f(A)$ *for all $f \in C_\infty(\mathbb{R})$, the continuous functions on $\mathbb{R}$ vanishing at ∞.*

(b) $A_n \xrightarrow{\text{srs}} A \Rightarrow f(A_n) \xrightarrow{\text{srs}} f(A)$ *for all $f \in C(\mathbb{R})$, the bounded continuous functions on $\mathbb{R}$.*

Remark. Picking $f(x) = (x - z)^{-1}$, we see norm (respectively, strong) convergence of resolvents for $z \in \mathbb{C} \setminus \mathbb{R}$.

Proof. (a) Writing

$$(A\pm i)^{-m}-(A_n\pm i)^{-m} = \sum_{j=0}^{m-1}(A_n\pm i)^{-j}[(A\pm i)^{-1}-(A_n\pm i)^{-1}](A\pm i)^{-(m-j-1)} \tag{7.2.38}$$

and using $\|(A_n \pm i)^{-1}\| \le 1$, we see (7.2.36) implies $p(A_n) \to p(A)$ in $\|\cdot\|$ for any polynomial in $(x+i)^{-1}$ or $(x-i)^{-1}$. Repeated use of $(x+i)^{-1}(x-i)^{-1} = \frac{i}{2}[(x+i)^{-1} - (x-i)^{-1}]$ shows any polynomial in $(x+i)^{-1}$ and $(x-i)^{-1}$ is a sum of polynomials in $(x+i)^{-1}$ or $(x-i)^{-1}$. By the Stone–Weierstrass theorem (see Theorem 2.5.7 of Part 1), the polynomials in $(x+i)^{-1}$ and $(x-i)^{-1}$ are dense in $C_\infty(\mathbb{R})$, so we get $f(A_n) \xrightarrow{\text{nrs}} f(A)$ (Problem 2).

(b) By mimicking the argument in (a), if $f \in C_\infty(\mathbb{R})$ and $A_n \xrightarrow{\text{srs}} A$, $f(A_n) \to f(A)$ strongly (just note that for each φ and m and $j \le m-1$, $[(A\pm i)^{-1} - (A_m \pm i)^{-1}](A \pm i)^{-(m-j-1)}\varphi \to 0$ and use (7.2.38)).

Fix $f \in C(\mathbb{R})$. For each $M = 1, 2, \dots$, let g_M be the function

$$g_M(x) = \begin{cases} 1, & |x| \le M \\ M+1-|x|, & M \le |x| \le M+1 \\ 0, & |x| \ge M+1 \end{cases} \tag{7.2.39}$$

Since as $M \to \infty$, $g_M \to 1$ pointwise and $\sup_{x,M}|g_M(x)| = 1$, we have that $g_M(A)\varphi \to \varphi$ for all φ. Given ε, pick M so that

$$\|g_{M_0}(A)\varphi - \varphi\| \le \frac{\varepsilon}{4(1+\|f\|_\infty)} \tag{7.2.40}$$

Since fg_{M_0} and g_{M_0} lie in $C_\infty(\mathbb{R})$, pick N_0 so that for $n \ge N_0$,

$$\|[g_{M_0}(A_n) - g_{M_0}(A)]\varphi\| \le \frac{\varepsilon}{4(1+\|f\|_\infty)} \tag{7.2.41}$$

$$\|[(fg_M)(A_n) - (fg_{M_0})(A)]\varphi\| \le \frac{\varepsilon}{4} \tag{7.2.42}$$

By (7.2.39)/(7.2.40) and $\|f(B)\| \le \|f\|_\infty$, we see that

$$\|(fg_{M_0})(A)\varphi - f(A)\varphi\| \le \frac{\varepsilon}{4}, \quad \|fg_{M_0}(A_n)\varphi - f(A_n)\varphi\| \le \frac{\varepsilon}{2} \tag{7.2.43}$$

This plus (7.2.42) shows that

$$\|f(A)\varphi - f(A_n)\varphi\| \le \varepsilon \tag{7.2.44}$$

Since ε is arbitrary, $f(A_n)\varphi \to f(A)\varphi$. □

We leave some further results on continuity to the problems, including

(1) (Trotter) $A_n \xrightarrow{\text{srs}} A$ if and only if for all $t \in \mathbb{R}$, $e^{itA_n} \xrightarrow{s} e^{itA}$ (Problem 3). $\xrightarrow{s}$ is shorthand for convergence in the strong operator topology (later, we'll use similar notation for weak operator convergence).

(2) If $\mu \notin \sigma(A)$ and $A_n \xrightarrow{\text{nrs}} A$, then for all large n, $\mu \notin \sigma(A_n)$ (Problem 4).

(3) If $A_n \xrightarrow{\text{srs}} A$ and $\alpha < \beta$ are real, with α and β not eigenvalues of A_i, then $P_{(\alpha,\beta)}(A_n) \xrightarrow{s} P_{(\alpha,\beta)}(A)$ (Problems 5 and 6).

Here is one way to prove strong resolvent convergence:

Theorem 7.2.11. *Let $\{A_n\}_{n=1}^\infty$ and A_∞ be self-adjoint operators. Suppose that*

$$\mathcal{D} \subset D(A_\infty) \cap \bigcap_{n=1}^\infty D(A_n) \tag{7.2.45}$$

that $\mathcal{D}$ is a core for A_∞, and for all $\varphi \in \mathcal{D}$,

$$\|(A_n - A_\infty)\varphi\| \to 0 \tag{7.2.46}$$

as $n \to \infty$. Then $A_n \xrightarrow{\text{srs}} A_\infty$.

Proof. Let $\varphi \in \mathcal{D}$ and $\eta = (A_\infty + i)\varphi$. Then

$$[(A_n + i)^{-1} - (A_\infty + i)^{-1}]\eta = (A_n + i)^{-1}(A_\infty - A_n)\varphi \tag{7.2.47}$$

goes to zero in norm, since $\|(A_n + i)^{-1}\| \le 1$. Since $\mathcal{D}$ is a core for A_∞, $\operatorname{Ran}(A_\infty + i)$ is dense in $\mathcal{H}$, so (7.2.46) holds for all φ. The proof of $(A_n - i)^{-1}$ is similar. □

Notes and Historical Remarks. von Neumann realized that it was necessary to extend Hilbert's spectral theorem to unbounded self-adjoint operators in order to make mathematical sense of quantum mechanics. He did this in 1929 in [**720**]. We follow his idea of using the Cayley transform to deduce this from the spectral theorem for unitary operators (that he also had to prove).

Consideration of what we call norm and strong resolvent convergence goes back to Rellich [**556**, part III]. It was raised to an art in Kato [**380**].

Problems

1. (a) If A is self-adjoint, prove that

$$\|(A \pm i)^{-1}\varphi\|^2 = \operatorname{Im}\langle\varphi, (A + i)^{-1}\varphi\rangle \tag{7.2.48}$$

(b) Let $\{A_n\}_{n=1}^\infty$, A be self-adjoint operators. Prove that if $(A_n+i)^{-1} \xrightarrow{w} (A+1)^{-1}$, then for all φ,

$$\|(A_n \pm i)^{-1}\varphi\| \to \|(A \pm i)^{-1}\varphi\| \tag{7.2.49}$$

(c) If $(A_n + i)^{-1} \xrightarrow{w} (A+i)^{-1}$, prove that $(A_n \pm i)^{-1} \xrightarrow{s} (A \pm i)^{-1}$. (*Hint*: See Theorem 3.6.2 of Part 1.)

2. (a) Prove polynomials in $(x+i)^{-1}$ and $(x-i)^{-1}$ which vanish at ∞ are dense in $C_\infty(\mathbb{R})$.

   (b) Complete the final step in the proof of Theorem 7.2.10(a).

3. (a) If $A_n \xrightarrow{\text{srs}} A$, prove that for any $t \in \mathbb{R}$, $e^{itA_n} \xrightarrow{s} e^{itA}$.

   (b) Suppose for all t, $e^{itA_n} \xrightarrow{s} e^{itA}$. Prove $A_n \xrightarrow{\text{srs}} A$. (*Hint*: Compute $\int_0^\infty e^{-t}(e^{\pm itA}\varphi)\,dt$.)

4. (a) Let $A_n \xrightarrow{\text{nrs}} A$ and $\mu \notin \sigma(A)$. For large n, prove that $\mu \notin \sigma(A_n)$. (*Hint*: Prove that $(-\varepsilon, \varepsilon) \cap \sigma(A) = \emptyset$ if and only if $\|(A+i\varepsilon)^{-1}\| \le \varepsilon^{-1}/\sqrt{2}$.)

   (b) Find A_n, A bounded so $A_n \xrightarrow{s} A$ (which implies $A_n \xrightarrow{\text{srs}} A$), so $0 \notin \sigma(A)$ but $0 \in \sigma(A_n)$ for all n. (*Hint*: Consider infinite diagonal matrices.)

5. (a) Let P_n, P be self-adjoint projections. Prove that $P_n \xrightarrow{w} P$ if and only if $P_n \xrightarrow{s} P$. (*Hint*: See Theorem 3.6.2 of Part 1.)

   (b) Let $d\mu_\varphi^{(A)}$ be the spectral measure for A on the cyclic subspace of φ, that is, $\langle \varphi, f(A)\varphi\rangle = \int f(x)\,d\mu_\varphi^{(A)}$. If $A_n \xrightarrow{\text{srs}} A$, prove $d\mu_\varphi^{(A_n)} \to d\mu_\varphi^{(A)}$ weakly as functionals on $C_\infty(\mathbb{R})$.

   (c) Let $A_n \xrightarrow{\text{srs}} A$ for self-adjoint operators. Suppose a, b are not eigenvalues of A. Prove that $P_{(a,b)}(A_n) \to P_{(a,b)}(A)$. (*Hint*: See Theorem 4.5.7 of Part 1.)

6. Show the result of Problem 5(c) can fail if a is an eigenvalue of A.

7. Let A be a self-adjoint operator with $(-\infty, \gamma) \cap \sigma(A) = \emptyset$.

   (a) Prove (7.1.84).

   (b) Prove that $A \ge \gamma$.

7.3. Stone's Theorem

Our goal in this section is to prove:

Theorem 7.3.1 (Stone's Theorem). *Let $\{U_t\}_{t\in\mathbb{R}}$ be a one-parameter unitary group on a Hilbert space, $\mathcal{H}$. Then there is a self-adjoint operator, A,* *so that*

$$U_t = e^{itA} \tag{7.3.1}$$

for all $t \in \mathbb{R}$.

Remarks. 1. Theorem 7.2.8 is a converse to this. Stone's theorem is often stated as both sides, that is, there is a one–one correspondence between one-parameter groups and self-adjoint operators via (7.3.1).

2. A is called the *generator* of U_t.

3. As we'll explain in the Notes, this theorem is a fundamental part of quantum mechanics—quantum dynamics is given by a unitary group—its generator is the Hamiltonian operator. Other group generators in quantum theory include momentum and angular momentum.

We'll provide two proofs which illuminate different aspects. In particular, the second will describe $D(A)$ explicitly in terms of differentiability of U_t in t and will provide one way to show that $D_0 \subset D(A)$ is a core. The first proof relies on Bochner's theorem (see Theorem 6.6.6 of Part 1). Indeed, as we'll see, in a sense, Bochner's theorem and Stone's theorem are equivalent.

First Proof of Theorem 7.3.1. Given $\varphi \in \mathcal{H}$, define

$$T(t) = \langle \varphi, U_t\varphi \rangle \tag{7.3.2}$$

Then T is continuous because U is, and for $t_1, \dots, t_n \in \mathbb{R}$, $\zeta_1, \dots, \zeta_n \in \mathbb{C}$,

$$\begin{aligned} \sum_{j,k=1}^{n} \bar{\zeta}_j \zeta_k T(t_j - t_k) &= \sum_{j,k=1}^{n} \bar{\zeta}_j \zeta_k \langle U_{-t_j}\varphi, U_{-t_k}\varphi \rangle \\ &= \left\| \sum_{k=1}^{n} \zeta_k U_{-t_k}\varphi \right\|^2 \qquad (7.3.3) \\ &\geq 0 \qquad (7.3.4) \end{aligned}$$

Thus, T is positive definite. Therefore, by Bochner's theorem (Theorem 6.6.6 of Part 1), there is a measure $d\mu_\varphi(x)$ on φ so

$$\langle \varphi, U_t\varphi \rangle = \int e^{itx}\, d\mu_\varphi(x) \tag{7.3.5}$$

It follows that if

$$f_t(x) = e^{itx} \in L^2(\mathbb{R}, d\mu_\varphi) \tag{7.3.6}$$

then

$$\begin{aligned} \langle U_t\varphi, U_s\varphi \rangle &= \langle \varphi, U_{s-t}\varphi \rangle \\ &= \int e^{i(s-t)}\, d\mu_\varphi(\varphi) \qquad (7.3.7) \\ &= \langle f_t, f_s \rangle \qquad (7.3.8) \end{aligned}$$

Let $\mathcal{H}_\varphi$ be the closed subspace generated by $\{U_t\varphi\}_{t\in\mathbb{R}}$. (7.3.8) implies if for $\zeta_1,\dots,\zeta_n \in \mathbb{C}$,

$$V\biggl(\sum_{j=1}^n \zeta_j U_t\varphi\biggr) = \sum_{j=1}^n \zeta_j f_t \tag{7.3.9}$$

then V is a well-defined unitary map of $\mathcal{H}\varphi$ onto $L^2(\mathbb{R}, d\mu)$. It is easy to see that the span of the f_t is all of $L^2(\mathbb{R}, d\mu_\varphi)$, so V is a unitary map of $\mathcal{H}_\varphi$ onto $L^2(\mathbb{R}, d\mu_\varphi)$.

The same cyclic vector construction that we used in Theorem 5.2.1 lets us find an orthonormal set $\{\varphi_j\}_{j=1}^N$ so $\mathcal{H}_{\varphi_j} \perp \mathcal{H}_{\varphi_k}$ for $j \neq k$ and $\mathcal{H} = \oplus_{j=1}^N \mathcal{H}_{\varphi_j}$. By the above, extending V to that direct sum shows U_t is unitarily equivalent to multiplication by e^{itx} on $\oplus_{j=1}^N L^2(\mathbb{R}, d\mu_{\varphi_j})$. If A is the self-adjoint operator of multiplication by x with $D(A) = \{\{f_j\}_{j=1}^N \mid \sum_{j=1}^N \int (1+|x|^2)|f_j(x)|^2\, d\mu_j(x) < \infty\}$, then $U_t = e^{itA}$. $\square$

Remarks. 1. It is not hard to go in the opposite direction and prove Bochner's theorem by combining Stone's theorem and the spectral theorem.

2. Remarkably, we did not use the spectral theorem in this proof and we obtained the spectral theorem for the generator A! The only reason this doesn't prove the spectral theorem in general is that, without that theorem, we don't know how to define e^{itA}. But if A is bounded, we can define $e^{itA} = \sum_{n=0}^\infty \frac{(it)^n}{n!} A^n$ and deduce the spectral theorem from this.

To start the second proof, we define

$$D(A) = \bigl\{\varphi \in \mathcal{H} \bigm| \lim_{t\to 0} t^{-1}(U_t - \mathbb{1})\varphi \text{ exists}\bigr\} \tag{7.3.10}$$

$$A\varphi = \lim_{t\to 0} (it)^{-1}(U_t - \mathbb{1})\varphi \tag{7.3.11}$$

We note that if $A \geq 0$, in terms of the form domain language of Section 7.5, we have that

$$Q(A) = \{\varphi \in \mathcal{H} \mid \lim_{t\downarrow 0} \langle \varphi, t^{-1}(\mathbb{1} - e^{-tA})\rangle \text{ exists}\} \tag{7.3.12}$$

$$q_A(\varphi) = \lim_{t\downarrow 0} \langle \varphi, t^{-1}(\mathbb{1} - e^{-tA})\varphi\rangle \tag{7.3.13}$$

We also define for $f \in \mathcal{S}(\mathbb{R})$ and $\varphi \in \mathcal{H}$,

$$\varphi_f = \int f(s)(U_s\varphi)\, ds \tag{7.3.14}$$

Proposition 7.3.2. (a) *$\{\varphi_f \mid f \in \mathcal{S}(\mathbb{R}),\ \varphi \in \mathcal{H}\}$ is dense in $\mathcal{H}$.*

(b) *$\{\varphi_f \mid f \in \mathcal{S}(\mathbb{R}),\ \varphi \in \mathcal{H}\} \subset D(A)$ and*

$$A\varphi_f = i\varphi_{f'} \tag{7.3.15}$$

(c) *A is Hermitian.*

(d) *If $\varphi \in D(A)$, then for all s, $U_s\varphi \in D(A)$ and*

$$A(U_s\varphi) = U_s(A\varphi) \tag{7.3.16}$$

(e) *If B is self-adjoint and $U_t = e^{itB}$, then A defined by (7.3.10)/(7.3.11) is equal to B.*

Proof. (a) Since U,φ is strongly continuous and bounded, if j_n is an approximate identity (see Theorem 3.5.11 of Part 1), then $\varphi_{j_n} \to \varphi$.

(b) Let $f_t(s) = f(s-t)$. Then

$$U_t\varphi_f = \int f(s)U_{t+s}\varphi\, ds = \int f_t(s)U_s\varphi\, ds \tag{7.3.17}$$

so

$$t^{-1}[U_t\varphi_f - \varphi_f] = \int t^{-1}[f_t(s) - f(s)]U_s\varphi\, ds \to -\int f'(s)U_s\varphi\, ds \tag{7.3.18}$$

proving $\varphi_f \in D(A)$ and (7.3.15).

(c) Immediate from $U_t^* = U_{-t}$ via

$$\langle \psi, t^{-1}(U_t - \mathbb{1})\varphi\rangle = \langle t^{-1}(U_{-t} - \mathbb{1})\psi, \varphi\rangle \tag{7.3.19}$$

which implies if $\psi,\varphi \in D(A)$ that

$$\langle \psi, iA\varphi\rangle = \langle -iA\psi, \varphi\rangle \tag{7.3.20}$$

(d) Since $U_s(U_t - \mathbb{1}) = (U_t - \mathbb{1})U_s$ and U_s is bounded, this is immediate.

(e) By the spectral theorem, it suffices to prove this when B is multiplication by x on $L^2(\mathbb{R}, d\mu)$. In that case, if $\varphi \in D(B)$, we have

$$|t|^{-1}|e^{ixt} - 1| \le |t|^{-1}\left|\int (ix)e^{ixs}, ds\right| \le |x| \tag{7.3.21}$$

so $(it)^{-1}(e^{ixt} - 1)\varphi \to B\varphi$ by the dominated convergence theorem. Thus, $D(B) \subset D(A)$ and $A\varphi = B\varphi$.

Conversely, if $\varphi \in D(A)$, for any $R \in (0,\infty)$,

$$\|t^{-1}(U_t - \mathbb{1})\varphi\|^2 \ge \int_{-R}^{R} ||t|^{-1}(e^{itx} - 1)\varphi|^2\, d\mu(x) \to \int_{-R}^{R} x^2|\varphi(x)|^2\, d\mu(x)$$

by the dominated convergence theorem again. Thus, for all R,

$$\int_{-R}^{R} x^2|\varphi(x)|^2\, d\mu \le \limsup \|t^{-1}(U_t - \mathbb{1})\varphi\|^2 \tag{7.3.22}$$

so $\varphi \in D(A) \Rightarrow \varphi \in D(B)$. $\square$

Once we have Stone's theorem, the following is of independent interest.

Theorem 7.3.3. *Let $\{U_t\}_{t\in\mathbb{R}}$ be a one-parameter unitary group on a Hilbert space, $\mathcal{H}$. Define A by (7.3.10)/(7.3.11). Let $\mathcal{D} \subset D(A)$ be dense in $\mathcal{H}$ with*

$$U_t[\mathcal{D}] \subset \mathcal{D} \tag{7.3.23}$$

for all t. Then $A \restriction \mathcal{D}$ is essentially self-adjoint.

Proof. Let $C = A \restriction \mathcal{D}$. Since A is Hermitian, so is C. Suppose $C^*\psi = \pm i\psi$. Then for any $\varphi \in \mathcal{D}$ and all t, $U(t)\varphi$ is C^1 (by Proposition 7.3.2) and if

$$f(t) = \langle \psi, U(t)\varphi \rangle \tag{7.3.24}$$

then

$$\begin{aligned} f'(t) &= \langle \psi, (iA)U(t)\varphi \rangle && (7.3.25)\\ &= \langle \psi, (iC)U(t)\varphi \rangle && (7.3.26)\\ &= i\langle C^*\psi, U(t)\varphi \rangle \\ &= \mp i^2 \langle \psi, U(t)\varphi \rangle \\ &= \pm f(t) && (7.3.27) \end{aligned}$$

where (7.3.26) uses the fact that $U(t)\varphi_0 \in \mathcal{D}$. Thus,

$$f(t) = e^{\pm t} f(0) \tag{7.3.28}$$

Since $|f(t)| \le \|\varphi\|\,\|\psi\|$, (7.3.28) can only hold for all t if $f(0) = 0$, that is, if $\psi \in \mathcal{D}^\perp = \{0\}$. Thus, $\operatorname{Ker}(C^* \mp i) = \{0\}$ and C is essentially self-adjoint by the fundamental criterion (Theorem 7.1.10). $\square$

Second Proof of Theorem 7.3.1. By Theorem 7.3.3, if B is the closure of A, B is self-adjoint. Let $\varphi \in D(A)$ and

$$g(t) = e^{-itB} U_t \varphi \tag{7.3.29}$$

Since $U_t\varphi \in D(A) \subset D(B)$, $g(t)$ is C^1 and

$$\begin{aligned} g'(t) &= (-iB)e^{-itB}U(t)\varphi + e^{-itB}(iA)U(t)\varphi && (7.3.30)\\ &= [e^{-itB}(-iB + iA)U(t)]\varphi = 0 && (7.3.31) \end{aligned}$$

since $U(t)\varphi \in D(A)$ and $B \restriction D(A) = A$. Thus, since $D(A)$ is dense, $e^{-itB}U_t$ so

$$U_t = e^{-itB} \tag{7.3.32}$$

By Proposition 7.3.2(e), $B = A$ so we have proven (7.3.1) and given the domain of the generator. $\square$

Notes and Historical Remarks. Stone's theorem was announced in his 1929 paper [**668**] and proven in his 1932 paper [**669**]. It is central to both quantum mechanics and to the representations of Lie groups. In quantum theory, the dynamics provides a one-parameter unitary group and the generator is the Hamiltonian (energy operator). (7.3.11) is the Schrödinger equation.

7.4. von Neumann's Theory of Self-adjoint Extensions

In this section, we'll consider a closed Hermitian operator, A, with domain, $D(A)$, dense in a Hilbert space, $\mathcal{H}$. We define

$$\mathcal{K}_\pm = \text{Ker}(A^* \mp i) = \text{Ran}(A \pm i)^\perp, \quad d_\pm = \dim(\mathcal{K}_\pm)$$

For any closed symmetric operator, B, we'll use U_B for the Cayley transform $(B-i)(B+i)^{-1}$ extended to $\mathcal{H}$ by setting it to 0 on $\text{Ran}(B+i)^\perp$. Here is the main result:

Theorem 7.4.1 (von Neumann's Extension Theorem). *Let A be a closed Hermitian operator. There is a one–one correspondence between closed symmetric extensions, B, of A and partial isometries, V, with initial space $\mathcal{H}_I(V) \subset \mathcal{K}_+$ and final space $\mathcal{H}_F(V) \subset \mathcal{K}_-$. This correspondence is given by*

$$U_B = U_A + V \tag{7.4.1}$$

or by

$$D(B) = \{\varphi + \psi + V\psi \mid \varphi \in D(A),\, \psi \in \mathcal{H}_I(V)\}, \quad B = A^* \restriction D(B) \tag{7.4.2}$$

B is self-adjoint if and only if

$$\mathcal{H}_I(V) = \mathcal{K}_+, \qquad \mathcal{H}_F(V) = \mathcal{K}_- \tag{7.4.3}$$

In particular, A has self-adjoint extensions if and only if $d_+ = d_-$ and, in that case, $U \mapsto A_U$ from the unitary maps of $\mathcal{K}_+$ to $\mathcal{K}_-$ to the associated self-adjoint is continuous if U is given the norm (respectively, strong) topology and A the topology of norm (respectively, strong) resolvent convergence. Thus, if $d_+ < \infty$, the set of self-adjoint extensions is a d_+^2-dimensional real manifold (in the topology of norm-resolvent convergence).

Remark. A Hermitian operator, A, is called *maximal Hermitian* if it has no proper Hermitian extensions. This theorem implies A is maximal Hermitian if and only if $d_+ = 0$ or $d_- = 0$, so there are maximal Hermitian operators which are not self-adjoint (namely, if $d_\pm = 0$ but $d_\mp \neq 0$). In some literature prior to 1960, what we call self-adjoint was called *hypermaximal Hermitian*.

Proof. Given V a partial isometry from $\mathcal{K}_+$ to $\mathcal{K}_-$, define B by (7.4.2). By (7.1.62), for any $\varphi \in D(A)$, $\psi \in \mathcal{H}_I(V)$, we have

$$\operatorname{Im}\langle(\varphi + \psi + V\psi), A^*(\varphi + \psi + V\psi)\rangle = (\|\psi\|^2 - \|V\psi\|^2) = 0 \tag{7.4.4}$$

Thus, by polarization, $B = A^* \restriction D(A)$ is Hermitian. Moreover, by an easy calculation (Problem 1)), for this B, (7.4.1) holds.

On the other hand, if B is a symmetric extension of A and $\varphi = (A+i)\eta$, then $U_B\varphi = (A-i)\eta = U_A\varphi$, so $U_B \restriction \operatorname{Ran}(A+i) = U_A$. Thus, U_B on $\mathcal{K}_+$ must be a partial isometry, and to preserve the need for U_B to preserve orthogonality on $\mathcal{H}_I(U_B)$, $U_B[\mathcal{K}_+] \subset \mathcal{K}_-$. Thus, (7.4.1) holds for some partial isometry, V, with $\mathcal{H}_I(V) \in \mathcal{K}_+$, $\mathcal{H}_F(V) \in \mathcal{K}_-$. If B_V is the "B" associated to V by (7.4.2), we also have (7.4.1) so $U_B = U_{B_V}$, which implies that $B = B_V$, that is, we have the claimed one–one correspondence.

Since

$$(A^* + i)(\varphi + \psi + V\psi) = (A+i)\varphi + 2i\psi \tag{7.4.5}$$

$$(A^* - i)(\varphi + \psi + V\psi) = (A-i)\varphi - 2iV\psi \tag{7.4.6}$$

we see $\operatorname{Ran}(B+i) = \operatorname{Ran}(A+i) + \mathcal{H}_I(V)$, $\operatorname{Ran}(B-i) = \operatorname{Ran}(A-i) + \mathcal{H}_F(V)$, so these are both $\mathcal{H}$ if and only if (7.4.3) holds. Thus, by the fundamental criterion, B is self-adjoint if and only if (7.4.3) holds.

By (7.2.9), norm-resolvent convergence of self-adjoint operators is equivalent to norm convergence of their Cayley transforms. Thus, the map $V \mapsto B_V$ (given by (7.4.2)) is continuous if V is given the topology of norm convergence for unitary maps of $\mathcal{K}_+$ to $\mathcal{K}_-$ and B_V the topology of norm-resolvent convergence. In particular, if $d_+ = d_- < \infty$ and V_0 is a unitary map of $\mathcal{K}_+ \to \mathcal{K}_-$, then $V = (VV_0^{-1})V_0$ provides a homeomorphism of unitary maps of $\mathcal{K}_+$ to $\mathcal{K}_-$ and unitary maps on $\mathcal{K}_-$, that is, $\mathbb{U}(d_+)$, the group of unitary $d_+ \times d_+$ unitaries which is a manifold of dimension d_+^2. A similar analysis works for strong resolvant convergence. □

Corollary 7.4.2. *If A is a closed Hermitian operator and $A \geq 0$, then A has self-adjoint extensions.*

Remark. We'll say much more about this in Section 7.5; in particular, we'll prove there that A has positive self-adjoint extensions.

Proof. By Theorem 7.1.8, $d_+ = d_-$. □

Corollary 7.4.3. *Let A be a closed Hermitian operator so there is a conjugation obeying (7.1.65). Then A has self-adjoint extensions.*

Proof. By Theorem 7.1.12, $d_+ = d_-$. □

Example 7.4.4 (Example 7.1.6 (continued)). Let A be $-i\frac{d}{dx}$ on $C_0^\infty([0,1])$ in $L^2([0,1],dx)$. Then $\bar{A}$ is given by (7.1.35)/(7.1.36) and A^* is given by (7.1.36). $\mathcal{K}_\pm$ are multiples of the unique distributional solutions (up to constants) of $u' = \mp u$, that is, $e^{\mp x}$. We thus define

$$\varphi_+(x) = e^{1-x}, \quad \varphi_-(x) = e^x \tag{7.4.7}$$

so $\|\varphi_+\| = \|\varphi_-\|$. Self-adjoint extensions are parametrized by $e^{i\psi} \in \partial\mathbb{D}$ via

$$D(B_\psi) = \{g + w(\varphi_+ + e^{i\psi}\varphi_-) \mid g \in D(A),\, w \in \mathbb{C}\}, \quad B_\psi = A^* \restriction D(B_\psi) \tag{7.4.8}$$

Since g is continuous on $[0,1]$ and vanishes at 0 and 1 if $f \in B_\psi$, then

$$f(0) = w(e + e^{i\psi}), \quad f(1) = w(1 + e^{i\psi}e) = e^{-i\psi}f(0) \tag{7.4.9}$$

Moreover, if $f \in D(A^*)$ so f is continuous on $[0,1]$, it is easy to see that if $f(1) = e^{-i\psi}f(0)$, then $f \in D(B_\psi)$. Thus, in the notation of Example 7.1.6,

$$B_\psi = A_\theta \quad \text{if } \theta = -\psi \tag{7.4.10}$$

We have found that all the self-adjoint extensions are $\{A_\theta\}_{e^{i\theta}\in\partial\mathbb{D}}$. □

There is an alternate approach to self-adjoint extensions that depends on looking at, for $\varphi, \psi \in D(A^*)$,

$$D(\varphi,\psi) = \langle \varphi, A^*\psi\rangle - \langle A^*\varphi, \psi\rangle \tag{7.4.11}$$

$$S(\varphi,\psi) = \langle \varphi, A^*\psi\rangle \tag{7.4.12}$$

Here is a part of this approach.

Proposition 7.4.5. (a) *Let $\varphi_1,\varphi_2,\psi_1,\psi_2 \in D(A^*)$. If $\varphi_1 - \varphi_2$, $\psi_1 - \psi_2$ are in $D(A)$, then*

$$D(\varphi_1,\psi_1) = D(\varphi_2,\psi_2) \tag{7.4.13}$$

(b) *If $\varphi_1,\varphi_2 \in D(A^*)$ and $\varphi_1 - \varphi_2 \in D(A)$, then*

$$S(\varphi_2,\varphi_2) - S(\varphi_1,\varphi_1) \in \mathbb{R}$$

(c) *$D(\varphi,\varphi) = 0 \Leftrightarrow S(\varphi,\varphi)$ is real.*

Proof. (a) This follows from sesquilinearity of D, the fact that

$$D(\varphi,\psi) = -\overline{D(\psi,\varphi)} \tag{7.4.14}$$

and that if $\varphi \in D(A^*)$ and $\eta \in D(A)$, then

$$D(\varphi,\eta) = \langle \varphi, A\eta\rangle - \langle A^*\varphi, \eta\rangle = 0 \tag{7.4.15}$$

(b) Let $\varphi_2 - \varphi_1 = \eta$. Then

$$\begin{aligned} S(\varphi_2,\varphi_2) - S(\varphi_1,\varphi_1) &= S(\eta,\eta) + S(\eta,\varphi_1) + S(\varphi_1,\eta) \\ &= \langle \eta, A\eta\rangle + \langle \eta, A^*\varphi_1\rangle + \langle \varphi_1, A^*\eta\rangle \\ &= \langle \eta, A\eta\rangle + \langle A\eta, \varphi_1\rangle + \langle \varphi_1, A\eta\rangle \end{aligned}$$

is real since $\langle \eta, A\eta\rangle$ is real and $\langle A\eta, \varphi_1\rangle = \overline{\langle \varphi_1, A\eta\rangle}$.

(c) $D(\varphi,\varphi)=0 \Leftrightarrow \langle\varphi, A^*\varphi\rangle = \langle A^*\varphi,\varphi\rangle$. Since $\langle A^*\varphi,\varphi\rangle = \overline{\langle\varphi, A^*\varphi\rangle}$, this equality is equivalent to reality of $\langle\varphi, A^*\varphi\rangle$. □

Theorem 7.4.6. *Let A be a closed Hermitian operator and let B be an extension of A. Then B is symmetric if and only if*

(i) *$D(B)\subset D(A^*)$ and $B=A^*\restriction D(B)$.*
(ii) *$S(\varphi,\varphi)$ is real for all $\varphi\in D(B)$.*

For (ii) *to be true, it is sufficient that $S(\varphi,\varphi)$ is real for a subspace, $\mathcal{D}\subset D(B)\setminus D(A)$ so $\mathcal{D}+D(A)=D(B)$.*

Proof. If B is symmetric, $A\subset B=B^*$ implies $B=B^*\subset A^*$, proving (i). B symmetric clearly implies $D(\varphi,\varphi)=0$, so $S(\varphi,\varphi)$ is real.

Conversely, if (i), (ii) hold, $S(\varphi,\varphi)$ real implies $D(\varphi,\varphi)$ is real, so by polarization, B is symmetric.

The final statement follows from (b) of the last proposition. □

We've discussed this in part to be able to present the explicit form for the deficiency index $(1,1)$ case.

Theorem 7.4.7. *Let A be a closed symmetric operator with deficiency indices $(1,1)$. Let $\varphi,\psi\in D(A^*)$ be independent over $D(A)$ (i.e., not multiples of each other in $D(A^*)/D(A)$) and so that $S(\varphi,\varphi)$, $S(\psi,\psi)$, $S(\varphi,\psi)$, and $S(\psi,\varphi)$ are all real. Then*

(a) $$S(\varphi,\psi)\neq S(\psi,\varphi) \tag{7.4.16}$$

(b) *For each $t\in\mathbb{R}\cup\{\infty\}$, let $D(B_t)=D(A)+[\varphi+t\psi]$ if $t\in\mathbb{R}$, $D(B_\infty)=D(A)+[\psi]$, and $B_t=A^*\restriction D(B_t)$. Then each B_t is self-adjoint and these are all the self-adjoint extensions.*

Proof. (a) We have for $z,w\in\mathbb{C}$ that

$$\begin{aligned}S(z\varphi+w\psi,z\varphi+w\psi)&\\ &= |z|^2S(\varphi,\varphi)+|w|^2S(\psi,\psi)+(S(\varphi,\psi)+S(\psi,\varphi))\operatorname{Re}(\bar zw)\\ &\quad + i(S(\varphi,\psi)-S(\psi,\varphi))\operatorname{Im}(\bar zw)\end{aligned} \tag{7.4.17}$$

If (7.4.16) fails, $S(\eta,\eta)=0$ for all $\eta\in D(A^*)$. Since there are $\eta_0\neq 0$ with $A^*\eta_0=i\eta_0$, so $S(\eta_0,\eta_0)=i\|\eta_0\|^2$, we would have a contradiction. Thus, (7.4.16) holds.

(b) By (7.4.16) and (7.4.17),

$$S(z\varphi+w\psi,z\varphi+w\psi)\in\mathbb{R}\Leftrightarrow \operatorname{Im}(\bar zw)=0$$

Thus, either $w=tz$ with $t\in\mathbb{R}$ or $z=0$.

Since $D(A^*)=D(\eta)\dotplus[z\varphi+w\psi\mid z,w\in\mathbb{C}]$ and $D(B)/D(A)$ has dimension 1 for any self-adjoint extension, $D(B)=D(A)+[\psi]$ for some

$\eta = z\varphi + w\psi$ with $S(\eta, \eta) \in \mathbb{R}$. By the above, any such $D(B)$ is some $D(B_t)$.

Conversely, $D(B)$ is symmetric for each t by Theorem 7.4.6 and so, since the codimension is 1, B_t is self-adjoint. □

We turn now to the operator

$$D(A) = C_0^\infty(a,b), \quad A\varphi = -\varphi'' + V\varphi \tag{7.4.18}$$

where $V \in L^2_{\text{loc}}(a,b)$, that is,

$$\int_c^d |V(y)|^2 \, dy < \infty, \quad \text{all } c,d \text{ with } [c,d] \subset (a,b) \tag{7.4.19}$$

Each of a or b may be infinite or finite. What we will discuss is called the *Weyl limit point/limit circle theory* (we'll explain the name in the Notes). We remark that while we develop the Weyl limit point/limit circle theory for L^2_{loc}, it can be extended to V in L^1_{loc} without too much extra work. We begin by describing $D(A^*)$. We need

Lemma 7.4.8. *Let $\varphi \in \mathcal{D}(a,b)$, the ordinary distributions on (a,b). Suppose that φ' is a locally L^1-function. Then φ is continuous, and for any $c \in (a,b)$, we have*

$$\varphi(x) = \varphi(c) + \int_c^x \varphi'(y) \, dy \tag{7.4.20}$$

Proof. Fix $c \in (a,b)$ and let

$$\eta(x) = \int_c^x \varphi'(y) \, dy \tag{7.4.21}$$

Then it is easy to see that η is continuous and $\eta' = \varphi'$ as distributions, so by Problem 14 of Section 6.2 of Part 1, $(\varphi - \eta)$ is constant, namely, $\varphi(c) - \eta(c) = \varphi(c)$. □

Definition. A φ obeying (7.4.20) for an L^1-function φ' is called an *absolutely continuous function.* The name comes from the fact that such φ's are locally precisely the functions of bounded variation with $d\varphi$ an absolutely continuous measure.

Proposition 7.4.9. *Let $V \in L^2_{\text{loc}}(a,b)$ be real-valued and A given by* (7.4.18). *Let $\mathcal{D}^{(2)}_{\text{loc}}(a,b)$ be given by*

$$\mathcal{D}^{(2)}_{\text{loc}}(a,b) = \{\varphi \mid \varphi \text{ is } C^1 \text{ on } (a,b) \text{ and } \varphi' \text{ is absolutely continuous}\} \tag{7.4.22}$$

Then

$$D(A^*) = \{\varphi \in \mathcal{D}^{(2)}_{\text{loc}}(a,b) \cap L^2(a,b) \mid -\varphi'' + V\varphi \in L^2(a,b)\} \tag{7.4.23}$$

and

$$A^*\varphi = -\varphi'' + V\varphi \tag{7.4.24}$$

Remark. Since φ' is absolutely continuous, φ'' is a function, and since φ is continuous and $V \in L^2_{\text{loc}}$, $V\varphi \in L^2_{\text{loc}}$. Thus, $-\varphi'' + V\varphi \in L^1_{\text{loc}}$ and the condition $-\varphi'' + V\varphi \in L^2(a,b)$ make sense.

Proof. $\varphi \in D(A^*)$ if and only if $\varphi \in L^2$ and there is $\eta \in L^2$ with

$$\langle \varphi, -\psi'' + V\psi \rangle = \langle \eta, \psi \rangle \tag{7.4.25}$$

for all $\psi \in C_0^\infty(a,b)$. (7.4.25) says that as distributions, $\bar\eta = -\bar\varphi'' + \overline{V\varphi}$. Since $\varphi \in L^2$ and $V \in L^2_{\text{loc}}$, $V\varphi \in L^1_{\text{loc}}$, so $V\varphi$ is a distribution as $-\bar\varphi'' + \overline{V\varphi}$ makes sense as a distributional sum. It follows that RHS of (7.4.23) $\subset D(A^*)$ and A^* is given by (7.4.24).

For the converse, if $-\varphi'' + V\varphi \in L^2$, then $-\varphi'' \in L^1_{\text{loc}}$ so φ' is absolutely continuous, so φ is continuous given by (7.4.20) and that φ is L^1 so $\varphi \in \mathcal{D}^{(2)}_{\text{loc}}(a,b) \cap L^2$, that is, $D(A^*)$ is given by (7.4.23). $\square$

When $[a,b]$ is a bounded interval, it will be useful to define $\mathcal{D}^{(2)}(a,b)$ as those $\varphi \in \mathcal{D}^{(2)}_{\text{loc}}(a,b)$ with $\varphi'' \in L^1(a,b)$ (rather than just L^1_{loc}). By Lemma 7.4.8, φ and φ' are continuous on $[a,b]$.

Example 7.4.10. Let $[c,d]$ be a bounded interval in $\mathbb{R}$. Define

$$\begin{gathered} A_0 = -\frac{d^2}{dx^2} \\ D(A_0) = \{\varphi \in \mathcal{D}^{(2)}(c,d) \mid \varphi(c) = \varphi(d) = \varphi'(c) = \varphi'(d) = 0, \varphi'' \in L^2(a,b)\} \end{gathered} \tag{7.4.26}$$

Here we use the fact that, by Lemma 7.4.8, φ is C^1 and φ and φ' have limits as $x \downarrow c$ or $x \uparrow d$. As above, A_0 is closed and

$$A_0^* = -\frac{d^2}{dx^2}, \quad D(A_0^*) = \mathcal{D}^{(2)}(c,d) \cap \{\varphi \mid \varphi'' \in L^2(a,b)\} \tag{7.4.27}$$

As in the analysis of Example 7.4.4, one can find solutions of $(A_0 \pm i)\varphi = 0$ and conclude A_0 has deficiency indices $(2,2)$. If we let B_0 be the operator with $\varphi'(d) = 0$ dropped, then B_0 has deficiency indices $(1,1)$. $\square$

Proposition 7.4.11. *Let $[c,d]$ be a bounded interval in $\mathbb{R}$ and $V \in L^2([c,d], dx)$. Let A_0 be given by (7.4.26) and*

$$A = A_0 + V, \qquad D(A) = D(A_0) \tag{7.4.28}$$

Then $D(A^*) = D(A_0^*)$ and $A^* = A_0^* + V$. A has deficiency indices $(2,2)$. If B is the operator with $\varphi'(d) = 0$ dropped, then B has deficiency indices $(1,1)$.

Proof. By Theorem 7.1.18, V is A_0-relatively bounded with relative bound zero. The deficiency results follow from the theory in Section 7.4 which implies that if $A \subset B \subset B^*$, then $d_\pm(B) = d_\pm(A) - \dim(B \setminus A)$. $\square$

We now need some results on solutions of

$$-\varphi'' + V\varphi = z\varphi \tag{7.4.29}$$

for $z \in \mathbb{C}$, as we studied in Example 15.5.5 of Part 2B. We proved there:

(1) For any $c \in (a,b)$, (7.4.29) has, for each $(\alpha, \beta) \in \mathbb{C}^2$, unique solutions with

$$\varphi(c) = \alpha, \quad \varphi'(c) = \beta \tag{7.4.30}$$

(2) If there is one $z \in \mathbb{C}$ so that (7.4.30) has all its solutions L^2 at b, that is true for all $z \in \mathbb{C}$, in which case we say (7.4.29) is *limit circle* at b. Otherwise, we say (7.4.29) is *limit point* at b.

Here is the main result on $-\frac{d^2}{dx^2} + V$ on $L^2(a,b)$:

Theorem 7.4.12 (Weyl Limit Point/Limit Circle Theorem). *Let $V \in L^2_{\text{loc}}(a,b)$ and A given by (7.4.18). Then $\bar{A}$ has deficiency indices:*

(i) $(2,2)$ *if* (7.4.29) *is limit circle at both a and b.*

(ii) $(1,1)$ *if* (7.4.29) *is limit point at one of a, b and limit circle at the other.*

(iii) $(0,0)$ *if* (7.4.29) *is limit point at both ends.*

In particular, A is essentially self-adjoint if and only if (7.4.29) *is limit point at both ends.*

First Part of Proof. (i) If both sides are limit circle, all solutions of $-\varphi'' + V\varphi = \pm i\varphi$ are L^2 at both ends, so L^2. Thus, $\dim \operatorname{Ker}(A^* \mp i)$ is 2.

(ii) As a preliminary, we show for $z \in \mathbb{C} \setminus \mathbb{R}$, there is a nonzero solution of

$$-\varphi'' + V\varphi - z\varphi = 0 \tag{7.4.31}$$

which is L^2 at a (and also one L^2 at b). Pick $[c,d] \subset (a,b)$, a closed bounded interval in (a,b). By Proposition 7.4.11, $B \equiv -\frac{d^2}{dx^2} + V$ on $[c,d]$ with $\varphi(c) = \varphi'(c) = \varphi(d) = 0$ boundary conditions has deficiency indices $(1,1)$ so, since $C_0^\infty([c,d])$ is dense in $L^2([c,d], dx)$, there must be $\eta \in C_0^\infty$ with $\eta \notin \operatorname{Ran}(B - z)$.

Let $\widetilde{A}$ be $-\frac{d^2}{dx^2} + V$ on $L^2(a,d)$ with $D(\widetilde{A}) = \{\varphi \in \mathcal{D}^2_{\text{loc}}(a,d) \mid \varphi(x) = 0$ for $x \in (a, a+\varepsilon)$ for some $\varepsilon > 0$ and $\varphi(d) = 0\}$. Let C be a self-adjoint extension of $\widetilde{A}$ which exists since $\widetilde{A}$ commutes with $\varphi \to \bar{\varphi}$. It is easy to show (Problem 11) that $\varphi \in D(C)$ has $\varphi(d) = 0$.

Let $\psi \in D(C)$ obey $(C - z)\psi = \eta$. We claim ψ is not identically zero on (a,c). For if it is, since ψ is C^1, we have $\psi(c) = \psi'(c) = 0$ and $\psi(d) = 0$ by the above. Thus, $\widetilde{\psi} \equiv \psi \restriction [c,d] \in D(B)$ and $(B - z)\widetilde{\psi} = \eta$, contrary to the choice of $\eta \notin \operatorname{Ran}(B - z)$. Since $\psi \not\equiv 0$ on (a, c), $\psi \restriction (a, c)$ extended to $[c, b)$ by solving (7.4.31) is L^2 near a.

Since A is limit circle at b, all solutions of (7.4.31) are L^2 at b, so since there is some solution L^2 near a, (7.4.31) has a solution in $L^2(a, b)$.

Thus, $\dim \operatorname{Ker}(A^* \pm i) > 0$. Since it is not limit circle at both ends, not all solutions of (7.4.31) are L^2 so $\dim \operatorname{Ker}(A^* \pm i) < 2$. Thus, $\dim \operatorname{Ker}(A^* \pm i) = 1$. □

The case that remains is A is limit point at both a and b. There are solutions of $-\varphi'' + V\varphi = \pm i$ which are L^2 at each side and we don't a priori know they aren't the same. Instead of showing this directly, we prove that A^* is Hermitian. A main tool is the *Wronskian*

$$W(x; \varphi, \psi) = \varphi(x)\psi'(x) - \varphi'(x)\psi(x) \tag{7.4.32}$$

which is defined for $x \in (a, b)$ if $\varphi, \psi \in \mathcal{D}^{(2)}(a, b)$.

Proposition 7.4.13. *If $\varphi, \psi \in \mathcal{D}^{(2)}(a, b)$ and $[c, d] \subset (a, b)$, then*

(a) *$W(x; \bar{\varphi}, \psi)$ is absolutely continuous and*

$$W(d; \bar{\varphi}, \psi) - W(c; \bar{\varphi}, \psi) = \int_c^d [\bar{\varphi}(y)\psi''(y) - \overline{\varphi''}(y)\psi(y)] \tag{7.4.33}$$

(b) *If $\varphi, \psi \in D(A^*)$, then $\lim_{x \downarrow a} W(x; \bar{\varphi}, \psi)$ and $\lim_{x \uparrow b} W(x; \bar{\varphi}, \psi)$ exist and*

$$W(b; \bar{\varphi}, \psi) - W(a; \bar{\varphi}, \psi) = \langle \varphi, A^*\psi \rangle - \langle A^*\varphi, \psi \rangle \tag{7.4.34}$$

Proof. It is easy to see (Problem 2) that if f, g are absolutely continuous, so is fg and Leibniz's rule holds. Thus, W is absolutely continuous and $W' = \bar{\varphi}\psi'' - \bar{\varphi}''\psi$. This proves (7.4.33).

Since $V\varphi\psi \in L^2_{\text{loc}}$, we can add/substract it in (7.4.33) to see, if $\varphi, \psi \in D(A^*)$, that

$$W(d; \bar{\varphi}, \psi) - W(c; \bar{\varphi}, \psi) = -\int_c^d [\overline{A^*\varphi}\,(y)\psi(y) - \overline{\varphi(y)}\,(A^*\psi)(y)]\, dy \tag{7.4.35}$$

Since the integral lies in $L^1(a, b)$, the claimed limits exist as $d \uparrow b$ or $c \downarrow a$ and (7.4.34) holds. □

We say a is a *regular point* (and similarly for b) if and only if a is finite and $\int_a^c |V(x)|^2\, dx < \infty$ for all $c \in (a, b)$.

Proposition 7.4.14. *If a is a regular point and $\varphi \in D(A^*)$, then $\lim_{x \downarrow a} \varphi(x)$ and $\lim_{x \downarrow a} \varphi'(x)$ exist (we call them $\varphi(a)$ and $\varphi'(a)$). If $\varphi \in D(\bar{A})$, these limits are zero. A is limit circle at a.*

Remark. It can also be shown (Problem 3) that $\varphi'(a) = \lim_{\varepsilon \downarrow 0}(\varphi(a + \varepsilon) - \varphi(a))/\varepsilon$.

Proof. $V\varphi$ is L^1 near a, so $\varphi'' = -A^*\varphi + V\varphi$ is L^1. Thus, by (7.4.20) for φ', $\lim_{x\downarrow a}\varphi'(x)$ exists. Since a is finite, φ' is bounded and so L^1 near a, so by (7.4.20) for φ, $\lim_{x\downarrow 0}\varphi(x)$ exists.

The argument above shows that $\varphi \to (\varphi(a), \varphi'(a))$ is continuous in the $\|\varphi\| + \|A^*\varphi\|$ norm. Thus, since $(\varphi(a), \varphi'(a)) = (0,0)$ for φ in A, it holds for $\varphi \in D(\bar{A})$.

Since a is finite and any φ in $D(A^*)$ is bounded at a, it is L^2 there, so a is limit circle. □

Proposition 7.4.15. *Let a be a regular point and b limit point. Then A has a self-adjoint extension, B, so that for some fixed $(\alpha, \beta) \in \mathbb{R}^2 \setminus \{0\}$, all $\varphi \in D(B)$ have*

$$\alpha\varphi(a) + \beta\varphi'(a) = 0 \tag{7.4.36}$$

Remark. We'll see shortly that every self-adjoint extension has an $(\alpha, \beta) \in \mathbb{R}^2 \setminus \{0\}$.

Proof. Let $\psi \in \operatorname{Ker}(A^* + i)$. Then $\varphi_0 = \psi + \bar{\psi}$ is real, so $(\varphi_0(a), \varphi_0'(a))$ obeys (7.4.36) for some real $(\alpha, \beta) \neq (0,0)$. Since A has deficiency indices $(1,1)$, there is a self-adjoint extension with $D(B) = \{\lambda\varphi_0 + \eta \mid \eta \in D(\bar{A})\lambda \in \mathbb{C}\}$. □

Proposition 7.4.16. *If a is a regular point and b a limit point, we have*

$$W(b; \bar{\varphi}, \psi) = 0 \tag{7.4.37}$$

for all $\varphi, \psi \in D(A^*)$.

Proof. Let B be the extension of $\bar{A}$ of the last proposition. We first claim (7.4.37) holds for $\varphi, \psi \in D(B)$. For, by (7.4.36) and the fact that (α, β) are real, we have $W(a; \bar{\varphi}, \psi) = 0$. (7.4.34) and $B = A^*$ shows that

$$0 = \langle \varphi, B\psi \rangle - \langle B\varphi, \psi \rangle = W(b; \bar{\varphi}, \psi) - W(a; \bar{\varphi}, \psi) \tag{7.4.38}$$

so we have (7.4.37).

Now pick η_0 which is C^∞ on (a,b) and vanishing on $[c,b)$ for some $c < b$ so that η_0 does not obey (7.4.36). Then $\eta_0 \in D(A^*)$ but $\eta_0 \notin D(B)$. Since A has deficiency indices $(1,1)$, $\dim(D(A^*)/D(B))$ is one, that is, any $\varphi \in D(A^*)$ has the form

$$\varphi = \widetilde{\varphi} + \lambda\eta_0 \tag{7.4.39}$$

with $\lambda \in \mathbb{C}$ and $\widetilde{\varphi} \in D(B)$. Similarly,

$$\psi = \widetilde{\psi} + \mu\eta_0 \tag{7.4.40}$$

For $x > c$, $\eta_0(x) = \eta_0'(x) = 0$, so $x > c \Rightarrow W(x; \bar{\varphi}, \psi) = W(x; \overline{\widetilde{\varphi}}, \widetilde{\psi}) \to 0$ as $x \uparrow b$, proving (7.4.37) in general. □

End of Proof of Theorem 7.4.12. Suppose A is limit point at a and b. Pick $c \in (a,b)$. Then c is a regular point for $-\frac{d^2}{dx^2}+V$ on (c,b) or (a,c). If $\varphi \in D(A^*)$, then $\varphi \restriction (c,b)$ lies in $D((A^\sharp)^*)$, the adjoint of the operator on $C_0^\infty(c,b)$ (since $\varphi \restriction (c,b)$ lies in $\mathcal{D}^{(2)}(c,d)$ and

$$-\varphi'' + V\varphi \restriction (c,d) = -\varphi'' + V\varphi \tag{7.4.41}$$

as distributions on (c,d)). Since W is locally defined, we conclude, by Proposition 7.4.16, that for $\varphi, \psi \in D(A^*)$, $W(b;\bar\varphi,\psi)=0$. The same argument implies $W(a;\bar\varphi,\psi)=0$. Therefore, by (7.4.34), A^* is Hermitian. Thus, $\operatorname{Ker}(A^* \pm i)=\{0\}$, so the deficiency indices are $(0,0)$. □

Using the same ideas as we used in the discussion of Sturm–Liouville equations on $[0,1]$ in Section 3.2, we can write down an explicit formula for $(A-e)^{-1}$ as an integral operator. A will be $-\frac{d^2}{dx^2}+V(x)$ on $L^2\big((a,b)\big)$ where one of the following hold:

(i) A is limit point at a and b. We required this if both $|a|$ and $|b|$ are ∞.

(ii) A is limit point at a or b and limit circle at the other point and a boundary condition is picked at that point; say if the limit circle end is a,

$$\big(\cos(\theta_a)\,\varphi(a) + \big(\sin\theta_a\big)\,\varphi'(a) = 0 \tag{7.4.42}$$

We require a limit point at a (respectively, b) if $a=-\infty$ (respectively, $b=\infty$).

(iii) A is limit circle at both ends and a boundary condition of the form (7.4.42) is picked at each of the points (θ_a or θ_b) different.

As we'll see in case (ii), (7.4.42) describes all boundary conditions but in case (iii), there is a four-dimensional family of boundary conditions, so the ones we pick are special and are called *separated boundary conditions.*

Suppose $e \notin \sigma(A)$. We pick unique (up to a constant) functions $u^\pm$ solving

$$-(u^\pm)'' + Vu^\pm = eu^\pm \tag{7.4.43}$$

If a is limit point, u^- is the unique solution which is L^2 at a (which exists as above) or if a is limit circle, it is the solution obeying the boundary conditions at a. Similarly, u^+ is set by the boundary condition at b. Note that $W(u^+,u^-)$ is constant. If $W(u^+,u^-)=0$, $u^+=u^-$ (up to a constant) and e is an eigenvalue of A, so $e\notin\sigma(A) \Rightarrow W(u^+,u^-)=0$. We define the *Green's function* at e to be

$$G_e(x,y) = u^-(x_<)u^+(x_>)W(u^+,u^-)^{-1} \tag{7.4.44}$$

where

$$x_< = \min(x,y), \quad x_> = \max(x,y) \tag{7.4.45}$$

The calculation of Proposition 3.2.8 applies without change to get

Theorem 7.4.17. *Let $A = -\frac{d^2}{dx^2} + V(x)$ on $L^2(a,b)$ with boundary condition* (i), (ii), *or* (iii) *above. Let $e \notin \sigma(A)$ and $u^\pm$ as described above. Let $G_e(x,y)$ be given by* (7.4.44). *Then $G_e(x,y)$ is the integral kernel of $(A-e)^{-1}$ in the sense that for any $f \in L^2(a,b)$,*

$$g(x) = \int_a^b G_e(x,y)\, f(y)\, dy$$

(or, in $D(A)$ and $(A-e)g = f$).

Example 7.4.18. Let V be limit point at b and regular at a. For $\theta \in [0,\pi)$, define

$$D(A_\theta) = \{\varphi \in D(A^*) \mid (\cos\theta)\varphi(a) + (\sin\theta)\varphi'(a) = 0\},\ A_\theta = A^* \restriction D(A_\theta) \tag{7.4.46}$$

We claim each A_θ is self-adjoint and every self-adjoint extension of A is an A_θ for some $\theta \in [0,\pi)$.

If $\varphi, \psi \in D(A_\theta)$, clearly $W(a;\bar\varphi,\psi) = 0$ since $\bar\varphi$ lies in $D(A_\theta)$. Since $W(b;\bar\varphi,\psi) = 0$ by Proposition 7.4.16, we conclude, by (7.4.34), that A_θ is Hermitian. If $\varphi \in C^\infty([a,c))$ for $c < b$, $\varphi \in D(A^*)$ so there exist $\varphi \in D(A_\theta)$ with $(\varphi(a),\varphi'(a)) \neq (0,0)$. Thus, $D(A_\theta) \neq D(\bar A)$. Since A_θ is Hermitian and $\bar A$ has deficiency indices $(1,1)$, $D(A_\theta) \neq D(A^*)$. Since $\dim(D(A^*)/D(\bar A))$ is 2, A_θ is self-adjoint since its deficiency indices are $(0,0)$ by dimension counting.

Now let B be a self-adjoint extension of $\bar A$. If $(\varphi(a),\varphi'(a)) = (0,0)$ for all $\varphi \in D(B)$, then $D(B) \subset D(A_\theta)$ for all θ, so $B = A_\theta$ for all θ. This is impossible, so there is $\varphi_0 \in D(B)$ with $(\varphi_0(a),\varphi_0'(a)) \neq (0,0)$. Since $W(b;\bar\varphi,\varphi) = 0$ (by Proposition 7.4.16), we must have $W(a;\bar\varphi,\varphi) = \langle B\varphi,\varphi\rangle - \langle\varphi,B\varphi\rangle = 0$.

Either $\varphi(a) = 0$ or else

$$0 = \frac{W(a;\bar\varphi,\varphi)}{|\varphi(a)|^2} = \frac{\overline{\varphi'(a)}}{\overline{\varphi(a)}} - \frac{\varphi'(a)}{\varphi(a)} \tag{7.4.47}$$

which implies $\varphi'(a)/\varphi(a)$ is real, that is, replacing φ by $\varphi(x)/\varphi(a)$, we get a φ with $(\varphi(a),\varphi'(a)) \in \mathbb{R}^2 \setminus \{0\}$. It follows that $\varphi \in D(A_\theta)$ for some θ. Since $\dim(D(B)/D(\bar A)) = 1$, we see $D(B) \subset D(A_\theta)$, so $B = A_\theta$. □

Remarks. 1. Thus, in this case with $C\varphi = \bar\varphi$, we see every self-adjoint extension A_θ has $C[D(A_\theta)] = D(A_\theta)$. This is no coincidence. For every deficiency index $(1,1)$ operator in the general context of Corollary 7.4.3, all self-adjoint extensions, B, have $C[D(B)] = D(B)$ (Problem 4(b)). For deficiency index (d,d) with $d \geq 2$, this is no longer true (Problem 4(d)).

2. More generally, if b is limit point and a is limit circle, the self-adjoint extensions, B, are described by picking some $\varphi_0 \in D(A^*)$ with $W(a; \bar{\varphi}_0, \varphi_0) = 0$. Then $D(B) = \{\varphi \in D(A^*) \mid W(a; \bar{\varphi}, \varphi_0) = 0\}$ (Problem 5).

Theorem 7.4.19. *Let $V(x) \geq 0$ and $b = \infty$. Then $-\frac{d^2}{dx^2} + V(x)$ is limit point at ∞.*

Proof. Consider the solution of $-u'' + Vu = 0$ with $u'(c) = 1$, $u(c) = 0$. It can be seen (Problem 6) that $u \geq 0$ on $[c, \infty)$, so $u'(x) \geq 1$ there so $u(x) \geq (x - c)$ which is not L^2. □

Corollary 7.4.20. *Let $V \in L^2_{\text{loc}}(\mathbb{R})$ with $V(x) \geq 0$. Then $-\frac{d^2}{dx^2} + V(x)$ is essentially self-adjoint on $C_0^\infty(\mathbb{R})$.*

Remarks. 1. It is known (see the Notes) that $V(x) \geq 0$ can be replaced by $V(x) \geq -\alpha - \beta x^2$ for some $\alpha, \beta \in \mathbb{R}$.

2. We'll prove an extension of this corollary to $\mathbb{R}^\nu$ in Theorem 7.6.1.

3. It is known that rather than $V \geq 0$, it suffices that for all $\varphi \in C_0^\infty(\mathbb{R})$, that $\|\varphi'\|^2 + \langle \varphi, V\varphi \rangle \geq 0$ for the limit point condition to hold.

Theorem 7.4.21. *Let V be C^2 on (a, ∞) so that for some $c \in (a, \infty)$,*

(i) *$V(x) \to -\infty$ as $x \to \infty$; $V(x) \leq -1$ if $x \geq c$.*

(ii) $$\int_c^\infty \left[\frac{|V'(x)|^2}{|V(x)|^{5/2}} + \frac{|V''(x)|}{|V(x)|^{3/2}} \right] dx < \infty \tag{7.4.48}$$

Then $-\frac{d^2}{dx^2} + V$ is limit point at ∞ if and only if

$$\int_c^\infty \frac{1}{|V(x)|^{1/2}} \, dx = \infty \tag{7.4.49}$$

Remarks. 1. If the integral in (7.4.49) is finite, this says the operator is limit circle at ∞.

2. See the Notes for a quantum mechanical interpretation of (7.4.49).

3. While (7.4.48) looks complicated, if V is nonoscillatory in that $|V'(x)|/|V(x)| = O(x^{-1})$ and $|V''(x)|/|V(x)| = O(x^{-2})$, it holds.

Proof. We use Theorem 15.5.14 of Part 2B. It implies $-u'' + Vu = 0$ has a basis of solutions so that

$$\lim_{x \to \infty} \frac{|u(x)|}{|V(x)|^{-1/4}} = 1 \tag{7.4.50}$$

Thus, (7.4.49) implies there is a basis of solutions with $\int_c^\infty |u(x)|^2 \, dx = \infty$, and if the integral is finite, a basis with $\int_c^\infty |u(x)|^2 \, dx < \infty$. □

Example 7.4.22. Let $V(x) = -|x|^\alpha$. Then (7.4.49) holds at ∞ and $-\infty$ if and only if $\alpha \le 2$ and the integral is finite if $\alpha > 2$. Thus, $-\frac{d^2}{dx^2} - |x|^{-\alpha}$ is essentially self-adjoint on $C_0^\infty(\mathbb{R})$ if and only if $\alpha \le 2$. $\square$

We finally turn to the situation where a is finite but V is not L^2 at a; a prime example we'll consider is $V(x) = \beta|x|^{-2}$ which will also let us study $|x|^{-2}$ as a perturbation of $-\Delta$ in $\mathbb{R}^\nu$. We begin with

Theorem 7.4.23. *Let a be finite and for some $\varepsilon > 0$,*

$$V_1(x) \ge V_2(x) \ge 0 \quad \textit{for } x \in (a, a+\varepsilon) \tag{7.4.51}$$

Then $-\frac{d^2}{dx^2} + V_1$ is limit point at a if $-\frac{d^2}{dx^2} + V_2$ is, and $-\frac{d^2}{dx^2} + V_2$ is limit circle at a if $-\frac{d^2}{dx^2} + V_1$ is.

Proof. Let $c > 0$ and $\varphi_j^c(x)$ be the solution of $-\varphi'' + V_j\varphi = 0$ with

$$\varphi_j(a+\varepsilon) = 0, \qquad \varphi_j'(a+\varepsilon) = -c \tag{7.4.52}$$

Then a simple argument (Problem 7) shows that on $(a, a+\varepsilon)$,

$$\varphi_1^c(x) \ge \varphi_2^c(x) \ge 0 \tag{7.4.53}$$

Taking $c = 1, 2$, we get two linear independent solutions. If $-\frac{d^2}{dx^2} + V_2$ is limit point, at least one of these solutions is not L^2 so $-\frac{d^2}{dx^2} + V_1$ has a non-L^2 solution. If $-\frac{d^2}{dx^2} + V_1$ is limit circle, both these solutions are L^2 for V_1, so the same is true for V_2. $\square$

In looking at $|x|^{-2}$, we also want to study when $-\frac{d^2}{dx^2} + V \ge 0$. In this regard, we'll use two facts:

Fact 1. If u_0 solves $-u_0'' + Vu_0 = 0$ on (a, b) and $u_0(x) > 0$ for all x, then for all $\varphi \in C_0^\infty(a, b)$,

$$\langle \varphi, (-\varphi'' + V\varphi)\rangle \ge 0 \tag{7.4.54}$$

The reader will prove this in Problem 8; see also the Notes.

Fact 2. If u_0 solves $-u_0'' + Vu_0 = 0$ on (a, b) has at least two zeros in (a, b), then for some $\varphi \in C_0^\infty(a, b)$, (7.4.54) fails. This is discussed in the Notes.

Example 7.4.24 (x^{-2} on $(0, \infty)$)**.** Let $\beta \in \mathbb{R}$ and look at $-\frac{d^2}{dx^2} + V(x)$ on $(0, \infty)$ where

$$V(x) = \beta|x|^{-2} \tag{7.4.55}$$

We begin by looking at solutions of

$$u''(x) = (\beta|x|^{-2})u(x) \tag{7.4.56}$$

Since this equation is invariant under $x \mapsto \lambda x$, scaling it is natural to try solutions of the form

$$u(x) = x^\alpha \tag{7.4.57}$$

in which (7.4.56) is

$$\alpha(\alpha-1)=\beta \tag{7.4.58}$$

solved by

$$\alpha_\pm = \tfrac{1}{2}\left(1\pm\sqrt{1+4\beta}\,\right) \tag{7.4.59}$$

Two values are special. $\beta=-\frac{1}{4}$ is one. For if $\beta\geq-\frac{1}{4}$, (7.4.56) has positive solutions. If $\beta<-\frac{1}{4}$, $\alpha_\pm$ are complex, and taking $r^{\alpha_+}+r^{\alpha_-}$, we get a solution with infinitely many zeros. Thus, by the two facts above,

$$-\frac{d^2}{dx^2}+\beta x^{-2}\geq 0 \text{ on } C_0^\infty(0,\infty)\Leftrightarrow \beta\geq\tfrac{1}{4} \tag{7.4.60}$$

Equivalently, for any $\varphi\in C_0^\infty(0,\infty)$,

$$\tfrac{1}{4}\int_0^\infty |x|^{-2}|\varphi(x)|^2\,dx\leq\int_0^\infty|\varphi'(x)|^2\,dx \tag{7.4.61}$$

and this fails for suitable φ if $\frac{1}{4}$ is replaced by a larger constant. The reader will note this is exactly Hardy's inequality as discussed in Sections 4.1 and 6.1 of Part 3.

The other special value is $\beta=\frac{3}{4}$, at which $\alpha_-=-\frac{1}{2}$. If $\beta\geq\frac{3}{4}$, $\alpha\leq-\frac{1}{2}$, and $u_-\notin L^2(0,1)$. If $\beta<\frac{3}{4}$, both solutions are in L^2. Thus,

$$-\frac{d^2}{dx^2}+\beta x^{-2} \text{ is limit point at } 0\Leftrightarrow\beta\geq\tfrac{3}{4} \tag{7.4.62}$$

□

Corollary 7.4.25. *If $V(x)\geq\frac{3}{4x^2}$ for $x\in(0,\varepsilon)$, $-\frac{d^2}{dx^2}+V(x)$ is limit point at $x=0$. If $V(x)\leq(\frac{3}{4}-\delta)|x|^{-2}$ for $x\in(0,\varepsilon)$ and some $\delta>0$, then $-\frac{d^2}{dx^2}+V(x)$ is limit circle at $x=0$.*

Proof. Immediate from Theorem 7.4.23 and (7.4.62). □

Example 7.4.26 ($|x|^{-2}$ in $\mathbb{R}^\nu$; $\nu\geq 5$)**.** Let $C_{00}^\infty(\mathbb{R}^\nu)$ be the C_0^∞-functions vanishing in a neighborhood of $x=0$. It is not hard to see (Problem 9) that $C_{00}^\infty(\mathbb{R}^\nu)$ is a core for $-\Delta$ if $\nu\geq5$. By Theorem 3.5.8 of Part 3, $-\Delta+\beta|x|^{-2}$ on $C_{00}^\infty(\mathbb{R}^\nu)$ is a direct sum of operators of the form

$$-\frac{d^2}{dx^2}+\frac{\beta}{r^2}+\frac{(\nu-1)(\nu-3)}{4r^2}+\frac{d(d+\nu-2)}{r^2} \tag{7.4.63}$$

on $C_0^\infty(0,\infty)$, where $d=0,1,\dots,$ The smallest value of $\beta+\frac{1}{4}(\nu-1)(\nu-3)+d(d+\nu-2)$ occurs at $d=0$ and then,

$$-\tfrac{1}{4}(\nu-1)(\nu-3)-\tfrac{1}{4}=-\frac{(\nu-2)^2}{4},\quad -\tfrac{1}{4}(\nu-1)(\nu-3)+\tfrac{3}{4}=-\frac{\nu(\nu-4)}{4} \tag{7.4.64}$$

We conclude:

$$(1)\quad -\Delta+\frac{\beta}{x^2}\geq 0 \Leftrightarrow \beta\geq -\frac{(\nu-2)^2}{4} \tag{7.4.65}$$

This is the ν-dimensional version of Hardy's inequality (and the argument also works for $\nu=3,4$).

(2) $-\Delta+\frac{\beta}{x^2}$ is essentially self-adjoint on $C_{0\,0}^\infty(\mathbb{R}^\nu)$ if and only if $\beta\geq -\frac{\nu(\nu-4)}{4}$. We note the following inequality of Rellich (see the Notes and Problem 10)

$$\frac{\nu(\nu-4)}{4}\,\||x|^{-2}\varphi\|\leq \|\Delta\varphi\| \tag{7.4.66}$$

for $\varphi\in C_0^\infty(\mathbb{R}^\nu)$ and $\nu\geq 5$. By the Kato–Rellich theorem (Theorem 7.1.14), this implies $-\Delta+\frac{\beta}{x^2}$ is essentially self-adjoint on $C_{0\,0}(\mathbb{R}^\nu)$ (and self-adjoint on $D(-\Delta)$) if $|\beta|\leq\frac{\nu(\nu-4)}{2}$.

Since we do not have essentially self-adjointness on $C_{0\,0}(\mathbb{R}^\nu)$ if $\beta<-\frac{\nu(\nu-4)}{2}$, the Kato–Rellich theorem implies that the constant $\frac{\nu(\nu-4)}{4}$ in (7.4.66) is optimal (as can be proven directly (Problem 10)). Significantly, $B=-x^{-2}$, $A=-\Delta$ is an example where one has relative boundedness

$$\|B\varphi\|\leq\alpha\|A\varphi\|+\gamma\|\varphi\| \tag{7.4.67}$$

for some α so $A+\lambda B$ is self-adjoint on $D(A)$ for $0<\lambda<\alpha$, but not even essentially self-adjoint if $\lambda>\alpha$. □

Notes and Historical Remarks. Remarkably, the same paper [**720**] where von Neumann singled out the notion of self-adjointness and also proved the spectral theorem and fundamental criterion for unbounded self-adjoint operators also had his theory of self-adjoint extensions (Theorem 7.4.1). Our approach is basically the one he used.

As we indicated in our discussion of the deficiency $(1,1)$ case, there is a second approach to proving Theorem 7.4.1 that looks at the form, D, given by (7.4.11) on $[D(A^*)/D(A)]\times[D(A^*)/D(A)]$ and associates self-adjoint extensions with maximal Lagrangian subspaces (subspaces V so that $D\restriction V\times V\equiv 0$). This approach has the advantage of having an expression in terms of boundary conditions for ODE's, since as we've seen by an integration by parts, $D(\varphi,\psi)$ can be expressed in terms of boundary values of φ and ψ. The earliest variant of the approach is in J. W. Calkin's Ph.D. thesis [**95**] under Stone published in 1939. The modern approach is by Dunford–Schwartz in their book [**175**], also presented in Reed–Simon [**549**].

The Weyl limit point/limit circle method is due to Weyl in 1910 [**742, 743**]. He was not specifically studying self-adjoint extensions, an idea that, for unbounded operators came only 20 years later, but asymptotics. The point is if you look at solutions of $-u'' + Vu = zu$ on $[0, a]$ and $\operatorname{Im} z$ is positive, then as you run through all real boundary conditions at a, the Weyl m-function

$$m(z) = \frac{u'(0, z)}{u(0, z)} \tag{7.4.68}$$

runs through a circle in the upper half-plane, $\mathbb{C}_+$ (see Problem 12). Moreover, if $a_1 < a_2$, the circle for a_2 is contained inside the circle for a_1. Thus, as $a \to \infty$, these circles either collapse to a point (in which case, we speak of limit point) or a circle (in which case, we speak of limit circle). This is where the names come from. This is the modern approach; the original approach, in place for many years, varied boundary values at 0. We'll see an analog of this analysis in our discussion of the moment problem as a self-adjointness problem in Section 7.7.

For books on the use of Weyl limit point/limit circle in the spectral analysis of ODE's, see [**125, 175, 179, 441, 491, 701**]. In particular, Titchmarsh's books were historically important rephrasing the subject enough that some authors speak of the Weyl–Titchmarsh limit point/limit circle method and the Weyl–Titchmarsh m-function.

If $V(x) \to -\infty$ as $x \to \infty$ and $V(x) < -1$ for all $x \geq c$, then a particle starting at c at time 0 with velocity $\sqrt{-V(c)}$ at that time, by conservation of energy has $\frac{dx}{dt} = \sqrt{-V\big(x(t)\big)}$ at time t so x at time t_0 obeys

$$t_0 = \int_c^{x(t_0)} \frac{dx}{\sqrt{|V(x)|}} \tag{7.4.69}$$

Thus, the negation of (7.4.64) says that a classical particle gets to infinity in finite time where a condition is needed. Therefore, 7.4.21 can be interpreted as an agreement between classical and quantum dynamics if (7.4.48) holds, i.e., under non-oscillation conditions. Reed–Simon [**549**] has examples (following Nelson) where quantum mechanics and classical mechanics differ because of oscillations. Theorems of this genre go back to Wintner [**762**] and Levinson [**439**].

Fact 1 before Example 7.4.24 is closely connected to the intrinsic semigroup as defined in Section 6.6 of Part 3. The reason for Fact 2 is connected with Sturm oscillation theorems, a subject discussed in several of the books on ODE's quoted above.

One last fact about one-dimensional Schrödinger operators we should mention concerns the spectral multiplicity of operators, $-\frac{d^2}{dx^2} + V$ on $L^2(\mathbb{R}, dx)$ where $V \in L^1_{\text{loc}}(\mathbb{R})$ and is limit point at both $+\infty$ and $-\infty$. Since there are at most two independent solutions of $-u'' + Vu = \lambda u$, it is not hard to see the multiplicity is at most two and the example $V = 0$ shows for a.c.

spectrum, it can be of multiplicity two. For point spectrum, since we are limit point, there is at most one L^2-solution, so the multiplicity of the point spectrum is one. For singular spectrum, it is a theorem of Kac [**355, 356**] that the multiplicity is one. Gilbert [**245**] and Simon [**651**] have alternate proofs. In particular [**651**] uses the fact that adding a rank-one perturbation changes the singular continuous spectrum (the Aronszajn–Donoghue theorem, Corollary 5.8.3) and the fact that adding a $u(0) = 0$ boundary condition is a rank-one perturbation of resolvents.

Problems

1. If B and $D(B)$ are given by (7.4.2), prove that for $\varphi \in D(A)$, $\psi \in \mathcal{H}_I(V)$,
$$(B-i)(B+i)^{-1}\big[(B+i)[\varphi+\psi+V\psi]\big] = (A-i)\varphi - 2iV\psi \tag{7.4.70}$$
and conclude that (7.4.1) holds.

2. If f and g are absolutely continuous functions on (a,b), prove that so is fg and that $(fg)' = f'g + fg'$.

3. If A is $-\frac{d^2}{dx^2} + V$ on (a,b), a is a regular point and $\varphi \in D(A^*)$, prove that with $\varphi'(a) = \lim_{x\downarrow a}\varphi'(x)$, one has that $\varphi'(a) = \lim_{\varepsilon\downarrow 0}[\varphi(a+\varepsilon) - \varphi(a)]/\varepsilon$.

4. Let A have deficiency indices (d,d) and let C be a conjugate commuting with A.

   (a) Suppose $\varphi \in \mathcal{K}_+$. Prove that $C(\varphi + e^{i\theta}C\varphi)$ is a complex multiple of $\varphi + e^{i\theta}C\varphi$.

   (b) Prove that if $d = 1$, every self-adjoint extension of A commutes with C.

   (c) Let $\varphi_1, \dots, \varphi_d$ be a basis for $\mathcal{K}_+$. Given $V \colon \mathcal{K}_+ \to \mathcal{K}_-$ unitary, define a $d \times d$ matrix, v, by
$$V\varphi_j = \sum_{k=1}^{d} v_{kj} C\varphi_k \tag{7.4.71}$$
   Let B_V be the self-adjoint operator associated to V. Prove that $C(D[B_V]) = D(B_V)$ if and only if for all j,k, we have
$$v_{kj} = v_{jk} \tag{7.4.72}$$

   (d) Conclude that if $d \geq 2$, there are B_V which are not invariant under C.

5. If b is limit point and a is limit circle (but not regular), prove that all self-adjoint extensions, B, are described by picking $\varphi_0 \in D(A^*)$ with $W(a;\bar{\varphi}_0,\varphi_0) = 0$ and letting $D(B) = \{\varphi \in D(A^*) \mid W(a;\bar{\varphi},\varphi_0) = 0\}$.

6. If $V \geq 0$ and u solves $-u'' + Vu = 0$, $u(c) = 0$, $u'(c) = 1$, prove that

   (a) If $u \geq 0$ on (c,d), then $u'(x) \geq 1$ on (c,d) so $u(d) > 0$.

   (b) Conclude that $u(x) > 0$ on (c,∞).

(c) Conclude that $u'(x) \geq 1$ on (c, ∞).

(d) Prove $u(x) \geq x - c$ is not L^2 at ∞.

7. Using the ideas of Problem 6, show that when (7.4.50) holds, the solutions of $-\varphi'' + V_j\varphi = 0$ with (7.4.52) obey (7.4.53).

8. Let u_0 solve $-u_0'' + Vu_0 = 0$ on (a, b) obey $u_0 > 0$ on (a, b).

(a) Prove that

$$\left(\frac{u_0'}{u_0}\right)^2 + \left(\frac{u_0'}{u_0}\right)' = V(x) \tag{7.4.73}$$

(b) Prove that for φ real-valued, $\varphi \in C_0^\infty(a, b)$, we have that

$$\int_a^b \left[\left(\frac{\varphi}{u_0}\right)'\right]^2 u_0^2\, dx = \int [(\varphi')^2 + V\varphi^2]\, dx \tag{7.4.74}$$

and conclude that (7.4.54) holds.

(c) Find an analog for $\mathbb{R}^\nu$.

9. If $\nu \geq 5$, prove that $C_{00}(\mathbb{R}^\nu)$ is a core for $-\Delta$. (*Hint*: Let $\psi(x)$ be C^∞ and zero near $x = 0$, 1 for $|x| \geq 1$, and given φ in $D(-\Delta)$, conclude $\psi(xn)\varphi$ as $n \to \infty$.)

10. Let $\nu \geq 5$. Then (see Theorem 6.8.1 of Part 1), show that p^{-4} given by

$$(p^{-4}\varphi) = (|k|^{-4}\widehat{\varphi})^{\check{}} \tag{7.4.75}$$

is an explicit convolution operator.

(a) Prove that for $\varphi \in \mathcal{S}(\mathbb{R}^\nu)$, the integral

$$\langle \varphi, x^{-2}p^{-4}x^{-2}\varphi\rangle \equiv M(\varphi) \tag{7.4.76}$$

is finite.

(b) Prove that

$$\|x^{-2}\psi\|_{L^2} \leq \alpha\|(-\Delta)\psi\|_{L^2} \tag{7.4.77}$$

is equivalent to $M(\varphi) \leq \alpha^2\|\varphi\|^2$ so that the optimal α in (7.4.77) is

$$\alpha_0 = \sup \frac{M(\varphi)^{1/2}}{\|\varphi\|} = \|p^{-2}x^{-2}\| \tag{7.4.78}$$

(c) If $f \in L^1(\mathbb{R}^\nu)$ with $f \geq 0$, prove that the norm of convolution with f as a map of $L^2 \to L^2$ is exactly $\|f\|_1$.

(d) $p^{-2}x^{-2}$ is scale invariant. This suggests after a change of variables $|x| \to \rho = \log(|x|)$, this operator is a convolution operator. Use this observation to prove the α_0 of (7.4.78) is $4/[\nu(\nu-4)]$, that is, prove (7.4.67) and that the constant is optimal. Thus, prove *Rellich's inequality*.

Remark. This approach to optimal constants for $\|p^{-\beta}x^{-\beta}\|$ is due to Herbst [**314**].

11. Let $\widetilde{A}$ be the operator on $L^2(a,d)$ given in the proof of Theorem 7.4.12(ii). Prove that

$$D(\widetilde{A}^*) = \{\varphi \in \mathcal{D}_{\text{loc}}^{(2)}(a,d) \mid \varphi\, C^1 \text{ and } \varphi' \text{ ac near } d \text{ with } \varphi(d)=0\}$$

Conclude that $\varphi \in D(C)$ has $\varphi(d)=0$.

12. Let $V \in L^1([0,a],dx)$, $\operatorname{Im} z > 0$ and let u solve

$$-u'' + Vu = zu \tag{7.4.79}$$

(a) Prove that

$$(u'\bar{u} - \bar{u}'\,u)' = -2(\operatorname{Im} z)\,|u|^2 \tag{7.4.80}$$

(b) Prove that

$$\operatorname{Im}\bigl[u'(0)\,\bar{u}(0)\bigr] = \operatorname{Im}\bigl[u'(a)\,\bar{u}(a)\bigr] + \int_0^a (\operatorname{Im} z)\,|u(x)|^2\,dx \tag{7.4.81}$$

(c) Prove that

$$\operatorname{Im}\left[\frac{u'(a)}{u(a)}\right] > 0 \Rightarrow \operatorname{Im}\left[\frac{u'(0)}{u(0)}\right] > 0 \tag{7.4.82}$$

(*Hint*: $\operatorname{Im}\bigl[\frac{u'}{u}\bigr] = |u|^2 \operatorname{Im}[u'\,\bar{u}]$.)

(d) Prove that for z fixed, the map of $\frac{u'(a)}{u(a)}$ to $\frac{u'(0)}{u(0)}$ is a fractional linear transformation.

(e) Prove that for z and a fixed (with $\operatorname{Im} z$), the set of $\frac{u'(0)}{u(0)}$ as $\frac{u'(a)}{u(a)}$ runs through $\mathbb{R}\cup\{\infty\}$ is a circle in the upper half-plane and that as a increases, the circle shrinks.

7.5. Quadratic Form Methods

Definition. A *quadratic form* is a map $q\colon \mathcal{H} \to [0,\infty]$ (with ∞ allowed) so that q obeys the parallelogram law and is homogeneous, that is, for all $\varphi, \psi \in \mathcal{H}$ and $\alpha \in \mathbb{C}$,

$$q(\varphi+\psi) + q(\varphi-\psi) = 2q(\varphi) + 2q(\psi) \tag{7.5.1}$$

$$q(\alpha\varphi) = |\alpha|^2 q(\varphi) \tag{7.5.2}$$

A *nonnegative sesquilinear form* (we'll drop nonnegative below, although it will be implicit unless explicitly stated otherwise) is a pair, (V,Q), where $V \subset \mathcal{H}$ is a subspace and $Q\colon V\times V \mapsto \mathbb{C}$ is sesquilinear with

$$Q(\varphi,\varphi) \geq 0 \tag{7.5.3}$$

for all $\varphi \in V$.

Remarks. 1. (7.5.1)/(7.5.2) are intended with the "usual" laws of arithmetic of positive numbers, including ∞, that is,

$$\forall a,\ a+\infty=\infty, \qquad \forall a\neq 0,\ a\cdot\infty=\infty, \quad 0\cdot\infty=0 \tag{7.5.4}$$

2. It is often useful to allow values of q in $(-\infty,\infty]$ with positivity replaced by

$$q(\varphi)\geq -c\|\varphi\|^2, \qquad Q(\varphi,\varphi)\geq -c\|\varphi\|^2 \tag{7.5.5}$$

By replacing $q(\varphi)$ by $q(\varphi)+c\|\varphi\|^2$, $Q(\varphi,\psi)$ by $Q(\varphi,\psi)+c(\varphi,\psi)$, one can understand this case from the positive case.

3. The *lower bound* of q is

$$\ell_q=\inf\{q(\varphi)\mid \|\varphi\|=1\} \tag{7.5.6}$$

which also makes sense even if (7.5.5) holds.

Given a nonnegative sesquilinear form, Q, we define the associated quadratic form by

$$q_Q(\varphi)=\begin{cases} Q(\varphi,\varphi), & \varphi\subset V\\ \infty, & \varphi\notin V\end{cases} \tag{7.5.7}$$

If we note that $\varphi,\psi\in V\Leftrightarrow \varphi+\psi,\ \varphi-\psi\in V$, one sees that q_Q obeys (7.5.1) for all $\varphi,\psi\in\mathcal{H}$. Conversely, given q, define

$$V_q\equiv\{\varphi\mid q(\varphi)<\infty\}\equiv D(q) \tag{7.5.8}$$

$$Q_q(\varphi,\psi)=\tfrac14\left(q(\varphi+\psi)-q(\varphi-\psi)-iq(\varphi+i\psi)+iq(\varphi-i\psi)\right) \tag{7.5.9}$$

By (7.5.1), if $\varphi,\psi\in V_q$, $q(\varphi+\psi)<\infty$ and if $\alpha\neq\mathbb{C}$, $q(\alpha\varphi)<\infty$ by (7.5.2). Q_q is sesquilinear by the argument used in Problem 4 of Section 3.1 of Part 1. We thus have

Proposition 7.5.1. *There is a one–one correspondence between quadratic forms, q, and sesquilinear forms given by* (7.5.7)–(7.5.9).

The following introduces an analog of closed operators:

Theorem 7.5.2. *Let q be a quadratic form and (V,Q) the associated sesquilinear form. Then the following are equivalent:*

(1) *$\varphi\mapsto q(\varphi)$ is lower semicontinuous, that is,*

$$\|\varphi_n-\varphi\|\to 0\Rightarrow q(\varphi)\leq\liminf q(\varphi_n) \tag{7.5.10}$$

(2) *V is complete in the norm*

$$\|\varphi\|_{+1}=(Q(\varphi,\varphi)+\|\varphi\|^2)^{1/2} \tag{7.5.11}$$

Remark. If these equivalent conditions hold, we say q (respectively, (V,Q)) is a *closed quadratic form* (respectively, closed sesquilinear form).

Proof. (2) $\Rightarrow$ (1). Suppose that $\|\varphi_n - \varphi\| \to 0$. By passing to a subsequence, we can suppose $q(\varphi_n)$ has a limit $a \in [0, \infty]$. If $a = \infty$, $q(\varphi) \le a$ is immediate, so suppose that $a < \infty$. Then $\|\varphi_n\|_{+1}$ is bounded, and thus, by the Banach–Alaoglu theorem (Theorem 5.8.1 of Part 1), by passing to a subsequence, there is $\eta \in V$ so that $\varphi_n \to \eta$ weakly in V. Since $V \subset \mathcal{H}$ and each $\langle \psi, \cdot \rangle$ is continuous in $\|\cdot\|_{+1}$, $\langle \psi, \varphi_n \rangle \to \langle \psi, \eta \rangle$. Thus, $\eta = \varphi$. Since the norm is weakly lower semicontinuous on any Banach space (Theorem 5.9.1 of Part 1), we see that

$$\|\varphi\|_{+1}^2 \le \liminf \|\varphi_n\|_{+1}^2 \tag{7.5.12}$$

Since $\|\varphi_n\| \to \|\varphi\|$, we see that (7.5.10) holds.

$\sim$(2) $\Rightarrow$ $\sim$(1). Suppose φ_n is Cauchy in $\|\cdot\|_{+1}$ but does not converge in V. Since $\|\cdot\| \le \|\cdot\|_{+1}$, φ_n has a limit φ in $\|\cdot\|$. If $\varphi \notin V$, then $q(\varphi) = \infty$ while $\lim q(\varphi_n) = \lim(\|\varphi_n\|_{+1}^2 - \|\varphi\|^2)^{1/2} < \infty$, so (7.5.10) fails.

If $\varphi \in V$, by replacing φ_n by $\varphi_n - \varphi$, we can suppose $\varphi = 0$, that is, $\|\varphi_n\| \to 0$, but since $\|\varphi_n\|_{+1} \not\to 0$, we have

$$a = \lim_n q(\varphi_n) > 0 \tag{7.5.13}$$

Pick n_0 so

$$n, m \ge n_0 \Rightarrow \|\varphi_n - \varphi_m\|_{+1}^2 \le \frac{a}{2} \tag{7.5.14}$$

Pick $m \ge n_0$ so that

$$q(\varphi_m) > \frac{a}{2} \tag{7.5.15}$$

Let $\psi_n = \varphi_m - \varphi_n$. Since $\varphi_n \to 0$, $\psi_n \to \varphi_m$. By (7.5.14),

$$\liminf q(\psi_n) \le \liminf \|\psi_n\|_{+1}^2 \le \frac{a}{2} \tag{7.5.16}$$

Thus,

$$q(\lim \psi_n) = \varphi(\varphi_m) > \liminf q(\psi_n) \tag{7.5.17}$$

again violating (7.5.10). □

With these preliminaries out of the way, we can describe the goals of this section. First, we'll show there is a one–one correspondence between closed quadratic forms, q, and positive self-adjoint operators, A, on $\overline{D(q)}$. We'll use this to state a form analog of the Kato–Rellich theorem and a monotone convergence theorem for forms. Finally, we'll discuss all the positive self-adjoint extensions of a positive Hermitian operator, A, in terms of forms. We'll see there exist two distinguished extensions (different if $\ell_q > 0$, perhaps the same if $\ell_q = 0$), A_∞, the *Friedrichs extension*, and A_0, the *Krein extension.* The set of positive extensions is exactly those self-adjoint operators, B, with $A_0 \le B \le A_\infty$. To prove this, we'll need to figure out what $B \le C$ means for two self-adjoint operators.

We begin with describing the quadratic form of a nonnegative self-adjoint operator.

Example 7.5.3 (Form of a Self-adjoint Operator)**.** Let A be a positive self-adjoint operator, so $\sigma(A) \subset [0,\infty)$. Thus, since $x \mapsto \sqrt{x}$ is a real-valued Borel function on $\sigma(A)$, we can define $\sqrt{A}$ which is a self-adjoint operator with

$$D(\sqrt{A}) = \left\{ \varphi \in \mathcal{H} \,\middle|\, \int |x| \, d\mu_\varphi^{(A)}(x) < \infty \right\} \tag{7.5.18}$$

where $d\mu_\varphi^{(A)}$ is the spectral measure. One defines the quadratic form of A by

$$q_A(\varphi) = \begin{cases} \|\sqrt{A}\,\varphi\|^2 & \text{if } \varphi \in D(\sqrt{A}) \\ \infty & \text{if } \varphi \notin D(\sqrt{A}) \end{cases} \tag{7.5.19}$$

(i.e., $q_A(\varphi)$ is the integral in (7.5.18)), so

$$D(q_A) = D(\sqrt{A}), \quad Q(\varphi,\psi) = \langle \sqrt{A}\,\varphi, \sqrt{A}\,\psi \rangle \tag{7.5.20}$$

Thus (with perhaps both sides infinite), for all φ,

$$q_A(\varphi) = \int |x| \, d\mu_\varphi^{(A)}(x) \tag{7.5.21}$$

Notice $\|\cdot\|_{+1}$ is the graph norm for $\sqrt{A}$ on $D(\sqrt{A})$. Since $\sqrt{A}$ is a closed operator, $D(\sqrt{A})$ is complete in this norm, that is, q_A is a closed quadratic form.

We use $Q(A)$ for $D(q_A) = D(\sqrt{A})$. Occasionally, we'll want to define $\widetilde{Q}(A) = D(A)$ and

$$\tilde{q}_A(\varphi) = \begin{cases} \langle \varphi, A\varphi \rangle & \text{if } \varphi \in D(A) \\ \infty & \text{if } \varphi \notin D(A) \end{cases} \tag{7.5.22}$$

This later is defined for any positive Hermitian operator. This form may not be closed. □

The central theorem in the subject is the converse of this example. We need a preliminary:

Proposition 7.5.4. *Let q be a closed quadratic form with $D(q)$ dense in $\mathcal{H}$. Let A be a positive self-adjoint operator so that*

(i) *$D(A) \subset D(q)$, $\varphi \in D(A) \Rightarrow q(\varphi) = \langle \varphi, A\varphi \rangle$;*
(ii) *$D(A)$ is dense in $D(q)$ in $\|\cdot\|_{+1}$.*

Then $q = q_A$, the quadratic form of A.

Proof. Let $\|\cdot\|_{+1}^{(A)}$ be the norm on $D(q_A)$. By (i), $\|\varphi\|_{+1}^{(A)} = \|\varphi\|_{+1}$ for $\varphi \in D(A)$. Since $D(A)$ is dense in both $D(q)$ and $D(q_A)$, these are the same as abstract completions and the same as subsets of $\mathcal{H}$ since $\|\varphi_n - \varphi\|_{+1} \to 0 \Rightarrow \|\varphi_n - \varphi\| \to 0$. The norms are the same on the completions, so $q_A(\varphi) + \|\varphi\|^2 = q(\varphi) + \|\varphi\|^2$ for all $\varphi \in D(q) = D(q_A)$. $\square$

Theorem 7.5.5. *Let q be a closed quadratic form. Then there is a positive self-adjoint operator, A, on $\overline{D(q)}$ so that*

$$q(\varphi) = \begin{cases} q_A(\varphi) & \text{if } \varphi \in \overline{D(q)} \\ \infty & \text{if } \varphi \notin \overline{D(q)} \end{cases} \tag{7.5.23}$$

Proof. Without loss, we can suppose $\overline{D(q)} = \mathcal{H}$, for if we have that case, apply that result to $q \restriction \overline{D(q)}$ to get the full result. Use $\mathcal{H}_{+1}$ for $V = D(q)$ and $\mathcal{H}_{-1}$ for its family of bounded antilinear functionals. We suppress the bijection of $\mathcal{H}_{+1}$ and $\mathcal{H}_{-1}$, which come from the duality via $\langle\cdot\,,\,\cdot\rangle_{+1}$ and instead use the duality defined by $\langle\cdot\,,\,\cdot\rangle$, that is, for $\varphi \in \mathcal{H}$, we define $j(\varphi) \in \mathcal{H}_{-1}$ by

$$j(\varphi)(\psi) = \langle\psi, \varphi\rangle \quad \text{for } \psi \in \mathcal{H}_{+1} \tag{7.5.24}$$

Thus,

$$|j(\varphi)(\psi)| \le \|\psi\|\,\|\varphi\| \le \|\psi\|\,\|\varphi\|_{+1} \tag{7.5.25}$$

so $j(\varphi) \in \mathcal{H}_{-1}$ and

$$\|j(\varphi)\|_{-1} \le \|\varphi\| \tag{7.5.26}$$

Notice that since $j(\varphi)(\varphi) = \|\varphi\|^2$, j is an injection.

We are pedantic and let $k\colon \mathcal{H}_{+1} \to \mathcal{H}$ be the identity map viewed as a bounded map of $\mathcal{H}_{+1}$ to $\mathcal{H}$ since

$$\|k(\varphi)\| = \|\varphi\| \le \|\varphi\|_{+1} \tag{7.5.27}$$

Notice that $k^*\colon (\mathcal{H})^* \to (\mathcal{H}_{+1})^*$ is just j.

Next, define $\widehat{B}\colon \mathcal{H}_{+1} \to \mathcal{H}_{-1}$ by

$$\widehat{B}(\varphi)(\psi) = \langle\psi, \varphi\rangle_{+1} \tag{7.5.28}$$

By the Riesz representation theorem, $\widehat{B}$ is an isometric isomorphism, that is,

$$\|\widehat{B}\varphi\|_{-1} = \|\varphi\|_{+1} \tag{7.5.29}$$

Define B by

$$D(B) = k\big[(\widehat{B})^{-1}[\operatorname{Ran}(j)]\big], \quad B\varphi = j^{-1}\widehat{B}k^{-1}(\varphi) \tag{7.5.30}$$

Here we use the fact that B^{-1} is a bijection of $\mathcal{H}_{-1}$ and $D(q) \subset \mathcal{H}$ and j^{-1} is a bijection of $\operatorname{Ran}(j)$ and $\mathcal{H}$. Since $\widehat{B}$ is a bijection, $\widehat{B}k^{-1}[D(B)] = \operatorname{Ran}(j)$, so

$$\operatorname{Ran}(B) = \mathcal{H} \tag{7.5.31}$$

Moreover, if $\varphi \in D(B)$,

$$\langle B\varphi, \varphi \rangle = j(B\varphi)(\varphi) = \widehat{B}\varphi(\varphi) = \langle \varphi, \varphi \rangle_{+1} \tag{7.5.32}$$

Thus, B is Hermitian so

$$A \equiv B - \mathbb{1}, \qquad D(A) = D(B) \tag{7.5.33}$$

is also Hermitian, and by (7.5.32),

$$\langle \varphi, A\varphi \rangle = q(\varphi) \geq 0 \tag{7.5.34}$$

By (7.5.31), $\text{Ran}(A + \mathbb{1}) = \mathcal{H}$ so A is closed and A is self-adjoint since $d_+ = d_- = \dim[\text{Ran}(A + \mathbb{1})]^\perp = 0$.

Since $\text{Ker}(i) = \{0\}$, we have that $\text{Ran}(i^*) = \text{Ran}(j)$ is dense in $\mathcal{H}_{-1}$. Since $(\widehat{B})^{-1}$ is an isometric isomorphism of $\mathcal{H}_{-1}$ and $\mathcal{H}_{+1}$, we see that $D(A)$ is $\|\cdot\|_{+1}$ dense in $\mathcal{H}_{+1} = D(q)$. Thus, by (7.5.34) and Proposition 7.5.4, $q = q_A$. $\square$

The proof of Theorem 7.5.5 shows the following characterization of $D(A)$:

Proposition 7.5.6. *Let q be a closed positive quadratic form and A the associated self-adjoint operator on $\overline{D(q)}$. Then*

$$D(A) = \{\varphi \in D(q) \mid \exists \psi \in \mathcal{H} \text{ so that } \forall \eta \in D(q),\ Q_q(\eta, \varphi) = \langle \eta, \psi \rangle\} \tag{7.5.35}$$

with

$$A\varphi = \psi \tag{7.5.36}$$

It suffices that $Q_q(\eta, \varphi) = \langle \eta, \psi \rangle$ for η in a form core for q.

Definition. Let q be a closed quadratic form. A *form core* for q is a subspace, S, of $D(q)$ so that S is dense in $D(q)$ in $\|\cdot\|_{+1}$.

If $q = q_A$, it is easy to see (Problem 1) that S is a form core for q if and only if it is an operator core for $\sqrt{A}$. The following is an analog of the Kato–Rellich theorem.

Theorem 7.5.7 (KLMN Theorem)**.** *Let q be a quadratic form and P a Hermitian sesquilinear form on $D(q)$ so that for some $\alpha < 1$ and $\beta > 0$ and all $\varphi \in D(q)$, we have that*

$$|P(\varphi, \varphi)| \leq \alpha q(\varphi) + \beta \|\varphi\|^2 \tag{7.5.37}$$

Let $\tilde{q}$ be the form

$$\tilde{q}(\varphi) = \begin{cases} q(\varphi) + P(\varphi, \varphi) + \beta\|\varphi\|^2, & \varphi \in D(q) \\ \infty, & \varphi \notin D(q) \end{cases} \tag{7.5.38}$$

Then $\tilde{q}$ is a positive form. If q is closed, so is $\tilde{q}$, and any form core for q is a form core for $\tilde{q}$.

Remarks. 1. If (7.5.37) holds for some α, β (and all $\varphi \in D(q)$) (perhaps $\alpha \geq 1$), we say that P is q *form-bounded* and that the inf over allowed α's is called the *relative form bound.*

2. Thus, we can associate a self-adjoint operator with $A +$ "P" if $q = q_A$ by taking the operator associated to $\tilde{q}$ and subtracting β.

3. If $P(\varphi, \psi) = \langle \varphi, B\psi \rangle$ for an operator B which is A-bounded with relative bound α_0, then P is q_A form-bounded with relative bound, $\alpha_1 \leq \alpha_0$ (see Problem 2).

Proof. Clearly, by (7.5.37), if $\varphi \in D(q)$,

$$\tilde{q}(\varphi) \geq q(\varphi) - \alpha q(\varphi) = (1 - \alpha) q(\varphi) \geq 0 \tag{7.5.39}$$

so $\tilde{q}$ is positive. We also have

$$\tilde{q}(\varphi) \leq (1 + \alpha) q(\varphi) + 2\beta \|\varphi\|^2 \tag{7.5.40}$$

so $\|\cdot\|_{+1,q}$ and $\|\cdot\|_{+1,\tilde{q}}$ are equivalent norms which prove the first sentence. $\square$

Example 7.5.8. Let $A = -\frac{d^2}{dx^2}$ on $L^2(\mathbb{R})$ (defined via Fourier transform or as the operator closure of the operator on $C_0^\infty(\mathbb{R})$; it is self-adjoint as we've seen). For all $\varphi \in D(A)$, φ is C^1 and bounded. Thus,

$$\varphi(0) = -\int_0^\infty (e^{-y}\varphi)'(y)\, dy \tag{7.5.41}$$

so

$$\begin{aligned} |\varphi(0)|^2 &\leq \left| \int_0^\infty [|e^{-y}\varphi(y)| + |e^{-y}\varphi'(y)|]\, dy \right|^2 \\ &\leq 2 \int_0^\infty e^{-y} [|\varphi(y)|^2 + |\varphi'(y)|^2]\, dy \end{aligned} \tag{7.5.42}$$

by the Schwarz inequality since $\chi_{[0,\infty)}(y) e^{-y}\, dy$ is a probability measure. Thus, if $\|\cdot\|_{+1}$ is form norm associated to q_A, we see if $\varphi \in D(A)$ that

$$|\varphi(0)| \leq \sqrt{2}\, \|\varphi\|_{+1} \tag{7.5.43}$$

Since $D(A)$ is dense in $\mathcal{H}_{+1}$, this extends to $\mathcal{H}_{+1}$. In fact, this argument shows that any $\varphi \in \mathcal{H}_{+1}$ is continuous (Problem 3) and

$$P(\varphi, \psi) = \overline{\varphi(0)}\, \psi(0) \tag{7.5.44}$$

is a form-bounded perturbation of $-\frac{d^2}{dx^2}$ with relative bound at most $\sqrt{2}$. In addition, scaling shows that the relative bound is 0 (Problem 4).

This lets one define as a form sum self-adjoint operators

$$A_\lambda = -\frac{d^2}{dx^2} + \lambda \delta(x), \quad \lambda \in \mathbb{R} \tag{7.5.45}$$

The form domain is λ-independent. The operator domain, $D(A_\lambda)$, is given by (Problem 5) $D(A_\lambda)$, those φ which are:

(a) φ is C^1 on $[0,\infty)$ and $(-\infty,0]$.
(b) φ' is absolutely continuous on $(0,\infty)$ and $(-\infty,0)$ and $\varphi'' \in L^2(0,\infty)$ and $L^2(-\infty,0)$.
(c) $\lim_{\varepsilon\downarrow 0}\bigl[\varphi'(\varepsilon)-\varphi'(-\varepsilon)\bigr]=\lambda\varphi(0)$. □

There is an interesting class of form-bounded perturbations called *factorizable.* Let A be a positive self-adjoint operator and $\mathcal{H}_{+1}\subset\mathcal{H}\subset\mathcal{H}_{-1}$ its scale of spaces as constructed above. Let $\mathcal{K}$ be an auxiliary Hilbert space (even $\mathbb{C}^n$ is interesting), $C\in\mathcal{L}(\mathcal{H}_{+1},\mathcal{K})$, a bounded operator from $\mathcal{H}_{+1}$ to $\mathcal{K}$, and $U\in\mathcal{L}(\mathcal{K})$ are bounded operators. Then

$$B = C^*UC \tag{7.5.46}$$

defines a sesquilinear form on $\mathcal{H}_{+1}$ by

$$b(\varphi,\psi)=\langle C\varphi, UC\psi\rangle_{\mathcal{K}} \tag{7.5.47}$$

Clearly,

$$|b(\varphi,\psi)|\le \|C\|^2_{\mathcal{L}(\mathcal{H}_{+1},\mathcal{K})}\|U\|\|\varphi\|_{+1}\|\psi\|_{+1}$$

so B is A-form bounded. If $U=U^*$, then $\overline{b(\varphi,\psi)}=b(\psi,\varphi)$.

In fact, every form-bounded perturbation is factorizable.

Theorem 7.5.9. *Let $A\ge 0$ be a self-adjoint operator and B be a sesquilinear form on $\mathcal{H}_{+1}$ which is A-form bounded. Then, B can be written in the form* (7.5.46) *where $\mathcal{K}=\mathcal{H}$.*

Proof. Let $C=(A+\mathbb{1})^{1/2}$ which is an isometry of $\mathcal{H}_{+1}$ to $\mathcal{H}$ and

$$U=(A+\mathbb{1})^{-1/2}B(A+\mathbb{1})^{-1/2} \tag{7.5.48}$$

which is bounded from $\mathcal{H}$ to $\mathcal{H}$ since it is a product of maps in $\mathcal{L}(\mathcal{H},\mathcal{H}_{+1}),\mathcal{L}(\mathcal{H}_{+1},\mathcal{H}_{-1})$, and $\mathcal{L}(\mathcal{H}_{-1},\mathcal{H})$ □

Remark. In particular, if U is rank-one, we have

$$\langle\varphi,B\varphi\rangle=|\langle\psi,\varphi\rangle|^2 \tag{7.5.49}$$

for some $\psi\in\mathcal{H}_{-1}$.

Theorem 7.5.10. (a) *If U is finite rank, b has relative form bound zero.*

(b) *If U is compact, b has relative form bound zero.*

Proof. (a) Clearly, it suffices to consider the case where U has rank-one. It is enough to prove for any $\psi\in\mathcal{H}_{-1}$ and $\varepsilon>0$, there is a constant C so that for all $\varphi\in\mathcal{H}_{+1}$,

$$|\langle\psi,\varphi\rangle|\le\varepsilon\|\varphi\|_{+1}+C\|\varphi\| \tag{7.5.50}$$

Since $\mathcal{H}$ is dense in $\mathcal{H}_{-1}$, find $\eta \in \mathcal{H}$ so that $\|\psi - \eta\|_{+1} \le \varepsilon$. Then

$$|\langle \psi, \varphi\rangle| \le |\langle \psi - \eta, \varphi\rangle| + |\langle \eta, \varphi\rangle| \le \varepsilon\|\varphi\|_{+1} + \|\eta\|\|\varphi\| \tag{7.5.51}$$

(b) Write $U = F + V$ where F is finite rank and $\|V\|\,\|C\|^2 \le \varepsilon/2$. Then

$$|b(\varphi, \psi)| \le \tfrac{\varepsilon}{2}\,\|\varphi\|_{+1}\,\|\psi\|_{+1} + |f(\varphi, \psi)| \tag{7.5.52}$$

By (a), f has relative bound zero, so that is C with

$$|b(\varphi, \psi)| \le \varepsilon\|\varphi\|_{+1}\,\|\psi\|_{+1} + C\|\varphi\|\,\|\psi\| \tag{7.5.53}$$

□

Remark. Factorizable perturbations are useful even when B is a bounded operator on $\mathcal{H}$. Suppose for simplicity that $B > 0$ and $C = B^{1/2}$. Then, with

$$R_0(z) = C(A - z)^{-1}C, \quad R(z) = C(A + B - z)^{-1}C \tag{7.5.54}$$

we have that

$$R_0(z) - R(z) = R_0(z)R(z) \tag{7.5.55}$$

and

$$R(z) = R_0(z)(\mathbb{1} + R_0(z))^{-1} \tag{7.5.56}$$

The point is that even though $(A - z)^{-1}$ will be singular as $z \to \mathbb{R}$, R_0 may have nice boundary values. The Notes discuss this further.

Quite remarkably, we can define "$A + B$" for any positive self-adjoint operators (although it may not be densely defined but will be if $D(q_A) \cap D(q_B)$ is dense even if $D(A) \cap D(B) = \{0\}$!).

Theorem 7.5.11. *Let q_1, q_2 be two closed quadratic forms. Then $q_1 + q_2$ is a closed quadratic form.*

Remarks. 1. Clearly, $D(q_1 + q_2) = D(q_1) \cap D(q_2)$.

2. If A, B are densely defined, positive self-adjoint operators and $q_1 = q_A$, $q_2 = q_B$, then we have $q_1 + q_2$ is the form of a self-adjoint operator, C, which is densely defined if $D(q_1) \cap D(q_2)$ is dense. It is natural to call $C = A + B$.

Proof. A sum of lsc functions is lsc. □

Example 7.5.12. Let $V \ge 0$ be a nonnegative function on $\mathbb{R}^\nu$ so that

$$\int_{|x| \le R} |V(x)|\, d^\nu x < \infty \tag{7.5.57}$$

for all $R < \infty$ (we say $V \in L^1_{\text{loc}}(\mathbb{R}^\nu)$). Then $Q(-\Delta) \cap Q(V)$ is dense since it contains $C_0^\infty(\mathbb{R}^\nu)$. We thus get a densely defined self-adjoint operator we can think of as $-\Delta + V$. We'll see in Theorem 7.6.2 that $C_0^\infty(\mathbb{R}^\nu)$ is a form core for $-\Delta + V$. □

Positivity allows a natural notion for inequalities among operators and forms:

Definition. Let q_1, q_2 be two quadratic forms. We say $q_1 \leq q_2$ if and only if $q_1(\varphi) \leq q_2(\varphi)$ for all φ (in particular, allowing for the case $q_2(\varphi) = \infty$, $q_1(\varphi) < \infty$, $D(q_2) \subset D(q_1)$). If A and B are positive self-adjoint operators, we say $A \leq B$ if and only if $q_A \leq q_B$ (so $Q(B) \subset Q(A)$ and $\langle \varphi, A\varphi \rangle \leq \langle \varphi, B\varphi \rangle$ for all $\varphi \in Q(B)$).

Definition. Let q_1, q_2 be two quadratic forms. We say that q_1 is an *extension* of q_2 if $D(q_1) \supset D(q_2)$ and $q_1 \restriction D(q_2) = q_2$.

Notice if q_1 is an extension, $q_2 = \infty$ on $D(q_1) \setminus D(q_2)$ so $q_1 \leq q_2$. Also, if A is self-adjoint, q_A is an extension of $\tilde{q}_A$ and $q_A \leq \tilde{q}_A$ (proper, unless A is bounded, since $q_A(\varphi) < \infty$ if $\varphi \in D(\sqrt{A}) \setminus D(A)$ but $\tilde{q}_A(\varphi) = \infty$).

Example 7.5.13 (Continuation of Example 7.4.18). Let $A = -\frac{d^2}{dx^2}$ on $L^2(0,\infty)$ with domain the closure of the operator on $C_0^\infty(0,\infty)$. The self-adjoint extensions are given by (7.4.47). If $\varphi \in D(A_\theta)$, an integration by parts shows that

$$\begin{aligned} \langle \varphi, A_\theta \varphi \rangle &= \int_0^\infty -\overline{\varphi(x)}\,(\varphi''(x))\,dx \\ &= \overline{\varphi(0)}\,\varphi'(0) + \int_0^\infty |\varphi'(x)|^2\,dx && (7.5.58) \\ &= -\cot\theta|\varphi(0)|^2 + \int_0^\infty |\varphi'(x)|^2\,dx && (7.5.59) \end{aligned}$$

If $\theta = 0$, $\varphi(0) = 0$, so $-\cot\theta|\varphi(0)|^2$ is interpreted as 0.

Define

$$\begin{aligned} Q_0 &= \{\varphi \in L^2(0,\infty) \mid \varphi \text{ is absolutely continuous}, \varphi' \in L^2\} && (7.5.60) \\ Q_\infty &= \{\varphi \in Q_0 \mid \varphi(0) = 0\} && (7.5.61) \end{aligned}$$

As in Example 7.5.8, if $\varphi \in Q_0$, φ is continuous on $[0,\infty)$, so in (7.5.61), $\varphi(0)$ makes sense.

It is not hard to see (Problem 6) that

$$Q(A_{\theta=0}) = Q_\infty, \quad q(\varphi) = \int_0^\infty |\varphi(x)|^2\,dx \tag{7.5.62}$$

and for $\theta \neq 0$,

$$Q(A_\theta) = Q_0, \quad q(\varphi) = -\cot\theta|\varphi(0)|^2 + \int_0^\infty |\varphi'(x)|\,dx \tag{7.5.63}$$

(If $\theta \in [\frac{\pi}{2}, \pi) \cup \{0\}$, $A_\theta \geq 0$. If $\theta \in (0, \frac{\pi}{2})$, A_θ is not positive but it is bounded below with $\ell_{q_{A_\theta}} = -\cot\theta$.)

Remarkably, $Q(A_{\theta=0}) \subsetneqq Q(A_{\theta\neq 0})$ for any $\theta \neq 0$, and for all θ, $A_\theta \leq A_0$. As forms, A_θ is an extension of A_0! Notice that if $A = A_{\theta=\frac{\pi}{2}}$, then $\varphi \mapsto \varphi(0)$ is a continuous map of $\mathcal{H}_{+1}(A)$ to $\mathbb{C}$ (see Problem 3) so there is $\psi \in \mathcal{H}_{-1}$ with $\langle \psi, \varphi \rangle = \varphi(0)$. What (7.5.63) says is that for $0 < \theta < \pi$,

$$A_\theta = A - \cot(\theta)\, \langle \psi, \cdot \rangle\, \psi \tag{7.5.64}$$

so these are rank-one perturbations. We'll see later (Example 7.8.7 of Section 7.8) that $A_{\theta=0}$ can be viewed as this rank-one perturbation at infinite coupling. □

This example lets us emphasize some important differences between operators and forms:

(a) Every closed quadratic form is associated to a self-adjoint operator. Not every closed Hermitian operator is self-adjoint.

(b) If A and B are self-adjoint operators and B is an extension of A, then $A = B$. However, there exist distinct self-adjoint operators whose forms are extensions one of the other.

(c) Every Hermitian operator has a closed extension. We'll see shortly (Example 7.5.17) that there are quadratic forms with no closed extensions.

If C, D are bounded positive operators, $C \leq D$ means that $\langle \varphi, C\varphi \rangle \leq \langle \varphi, D\varphi \rangle$ for all $\varphi \in \mathcal{H}$. We have:

Theorem 7.5.14. *Let A, B be positive self-adjoint operators. Then*

$$A \leq B \Leftrightarrow (B + \mathbb{1})^{-1} \leq (A + \mathbb{1})^{-1} \tag{7.5.65}$$

Remarks. 1. In (7.5.65), on the left, $\leq$ means as an inequality in forms, while on the right, $\leq$ means the familiar, everywhere defined, notion.

2. We've stated this for densely defined operators, but it remains true if a closed form, q_A, is associated to a self-adjoint operator on $\overline{V}$ with V a subspace of $\mathcal{H}$. If $(A + \mathbb{1})^{-1}$ is then interpreted as 0 on $V^\perp$, we have $q_A \leq q_B \Leftrightarrow (B + \mathbb{1})^{-1} \leq (A + \mathbb{1})^{-1}$. The proof is the same.

3. The same proof shows $q_A \leq q_B$ if and only if $(A + c\mathbb{1})^{-1} \geq (B + c\mathbb{1})^{-1}$ for all c with $-c \leq \min(\ell_{q_A}, \ell_{q_B})$ and this is true if and only if it holds for one such c.

Proof. Suppose first that $q_A \leq q_B$. Let $\varphi \in \mathcal{H}$ and $\psi_A = (A + \mathbb{1})^{-1}\varphi$, $\psi_B = (B + \mathbb{1})^{-1}\varphi$. Then

$$\begin{aligned}
\langle \varphi, (B + \mathbb{1})^{-1}\varphi \rangle^2 &= \langle (A + \mathbb{1})\psi_A, \psi_B \rangle^2 \\
&= \langle \psi_A, \psi_B \rangle_{+1,A}^2 && (7.5.66)\\
&\leq \|\psi_A\|_{+1,A} \|\psi_B\|_{+1,A} && (7.5.67)\\
&\leq \|\psi_A\|_{+1,A} \|\psi_B\|_{+1,B} && (7.5.68)\\
&= \langle \varphi, (A + \mathbb{1})^{-1}\varphi \rangle \langle \varphi, (B + \mathbb{1})^{-1}\varphi \rangle && (7.5.69)
\end{aligned}$$

proving that $(B+\mathbb{1})^{-1} \le (A+\mathbb{1})^{-1}$. Since $\mathcal{H}_{+1,A} \subset \mathcal{H}_{+1,B}$, we can write (7.5.66). (7.5.67) is then the Schwarz inequality and (7.5.68) that $q_A \le q_B$.

For the converse, let A be an arbitrary positive self-adjoint operator. The spectral theorem and monotone convergence theorem imply that

$$\langle \varphi, \varphi \rangle + q_A(\varphi) = \lim_{\varepsilon \downarrow 0} \langle \varphi, ((A+\mathbb{1})^{-1} + \varepsilon)^{-1} \varphi \rangle \tag{7.5.70}$$

including the case $q_A(\varphi) = \infty$. By the first paragraph above, if $(B+\mathbb{1})^{-1} \le (A+\mathbb{1})^{-1}$, for $\varepsilon > 0$, $((A+\mathbb{1})^{-1}+\varepsilon)^{-1} \le ((B+\mathbb{1})^{-1}+\varepsilon)^{-1}$. This plus (7.5.70) implies $q_A \le q_B$. □

Theorem 7.5.15. *Let q be an arbitrary quadratic form. Then there is a closed quadratic form, q_r, so that*

$$q_r \le q \tag{7.5.71}$$

and t closed plus

$$0 \le t \le q \Rightarrow 0 \le t \le q_r \tag{7.5.72}$$

(i.e., q_r is the largest quadratic form less than q). Moreover, for any φ,

$$q_r(\varphi) = \liminf_{\psi \to \varphi} q(\psi) \tag{7.5.73}$$

in the sense that if $\psi_n \to \varphi$, then $q_r(\varphi) \le \liminf_n q(\psi_n)$, and there exists $\psi_n^{(0)} \to \varphi$ so $q_r(\varphi) = \lim_n q(\psi_n^{(0)})$.

Remarks. 1. q_r can be the 0 form! See Example 7.5.17 below.

2. If $q_s = q - q_r$ ($= \infty$ on $\mathcal{H} \setminus D(q)$), then $(q_s)_r = 0$ (since $q_r + (q_s)_r$ is closed and less than q). Thus, $q = q_r + q_s$ is a kind of form analog of the Lebesgue decomposition theorem. See also Example 7.5.17.

3. The right side of (7.5.73) is an lsc function but it is not evident that it obeys the parallelogram law. Our explicit construct will prove it does.

4. If q is closed, $q = q_r$, and by the lsc and $\psi_n^{(0)} = \varphi$, (7.5.73) holds.

5. q_r is called the *regular part* of q.

Proof. Let $\mathcal{H}_{+1}$ denote $D(q)$ with the inner product associated to $\|\cdot\|_{+1}$. At the risk of being pedantic, let $I_q\colon \mathcal{H}_{+1} \to \mathcal{H}$ be the identity map, that is, $I_q(\varphi) = \varphi$. Since $\|\cdot\| \le \|\cdot\|_{+1}$, we have

$$\|I_q(\varphi)\|_{\mathcal{H}} \le \|\varphi\|_{\mathcal{H}_{+1}} \tag{7.5.74}$$

Let $\mathcal{H}_q$ be the abstract completion of $\mathcal{H}_{+1}$ and $\widetilde{I}_q$ the continuous extension of I_q to $\mathcal{H}_q$, unique since $\mathcal{H}_{+1}$ is dense in $\mathcal{H}_q$. For clarity, use J_q for the identity map of $\mathcal{H}_{+1}$ to $\mathcal{H}_q$ so (7.5.74) becomes, for $\psi \in \mathcal{H}_q$,

$$\|\widetilde{I}_q \psi\|_{\mathcal{H}} \le \|\psi\|_{\mathcal{H}_q}, \quad I_q = \widetilde{I}_q J_q \tag{7.5.75}$$

and we have for $\varphi \in \mathcal{H}_{+1}$,

$$q(\varphi) = \|J_q\varphi\|_{+1}^2 - \|\varphi\|^2 \tag{7.5.76}$$

Let $\mathcal{L}_q = \text{Ker}(\widetilde{I}_q)^\perp$ (as we'll see, $\text{Ker}(\widetilde{I}_q)$ may be nonempty) and $\mathcal{R}_q = \text{Ran}(\widetilde{I}_q)$. Since $\widetilde{I}_q$ is nonzero on $\mathcal{L}_q$, $\widetilde{I}_q$ is a bijection of $\mathcal{L}_q$ and $\mathcal{R}_q$, so we can define

$$\widetilde{J}_q \colon \mathcal{R}_q \to \mathcal{L}_q, \quad \widetilde{I}_q\widetilde{J}_q = \mathbb{1} \tag{7.5.77}$$

Let $P_q \colon \mathcal{H}_q \to \mathcal{L}_q$ be the orthogonal projection. Thus,

$$\widetilde{I}_q P_q = \widetilde{I}_q \tag{7.5.78}$$

which implies that

$$\varphi \in \mathcal{H}_{+1} \Rightarrow \widetilde{J}_q\varphi = P_qJ_q\varphi \tag{7.5.79}$$

Now define q_r a quadratic form on $\mathcal{H}$ by

$$q_r(\varphi) = \begin{cases} \|\widetilde{J}_q\varphi\|_{+1}^2 - \|\varphi\|^2, & \varphi \in \mathcal{R}_q \\ \infty, & \varphi \notin \mathcal{R}_q \end{cases} \tag{7.5.80}$$

We first want to prove that

$$0 \le q_r \le q \tag{7.5.81}$$

and that q_r is a closed form. $q_r \le q$ since for $\varphi \notin \mathcal{H}_{+1}$, $q(\varphi) = \infty$ and there is nothing to prove, and for $\varphi \in \mathcal{H}_{+1}$,

$$\begin{aligned} q_r(\varphi) &= \|\widetilde{J}_q\varphi\|_{+1}^2 - \|\varphi\|^2 \\ &= \|P_qJ_q\varphi\|_{+1}^2 - \|\varphi\|^2 && \text{(by (7.5.79))} \\ &\le \|J_q\varphi\|_{+1}^2 - \|\varphi\| = q(\varphi) && \text{(by (7.5.76))} \end{aligned}$$

To see that $q_r \ge 0$, use (7.5.77) to note that

$$q_r(\varphi) \equiv \|\widetilde{J}_q\varphi\|_{+1}^2 - \|\widetilde{I}_q\widetilde{J}_q\varphi\|^2 \ge 0$$

by (7.5.75). $\widetilde{J}_q$ is a bijection of $\mathcal{H}_{+1,q_r} = \mathcal{R}_{q_r}$, and by (7.5.17), $\|\cdot\|_{+1,q_r}$ is just $\|\cdot\|_{+1}$ under this bijection. Since $\mathcal{L}_q$ is closed in $\mathcal{H}_q$, it is complete so q_r is a closed form.

Next, we want to prove (7.5.73). Let $\varphi_n \to \varphi$ in $\|\cdot\|_{\mathcal{H}}$. Then, since q_r is lsc,

$$q_r(\varphi) \le \liminf q_r(\varphi_n) \le \liminf q(\varphi_n) \tag{7.5.82}$$

because $q_r \le q$.

On the other hand, if $\varphi \notin D(q_r) \equiv \mathcal{R}_q$, then $q_r(\varphi) = \infty$, so $q(\varphi) = \infty$ and we can take $\varphi_n^{(0)} \equiv \varphi$ and find

$$q_r(\varphi) = \lim_{n\to\infty} q(\varphi_n^{(0)}) \tag{7.5.83}$$

If $\varphi \in D(q_r) = \mathcal{R}_q$, we can find $\varphi_n^{(0)}$ in $\mathcal{H}_{+1}$ so

$$J_q \varphi_n^{(0)} \to \widetilde{J}_q \varphi \tag{7.5.84}$$

in $\|\cdot\|_{+1}$ (since $J_q[\mathcal{H}_{+1}]$ is dense in $\mathcal{H}_q$). Thus, $\|J_q \varphi_n^{(0)}\|_{+1} \to \|\widetilde{J}_q \varphi\|_{+1}$, so

$$q(\varphi_n^{(0)}) + \|\widetilde{I}_q J_q \varphi_n\|^2 \to q_r(\varphi) + \|\widetilde{I}_q \widetilde{J}_q \varphi\|^2 \tag{7.5.85}$$

This uses $\widetilde{I}_q J_q = \mathbb{1}$ and $\widetilde{I}_q \widetilde{J}_q = \mathbb{1}$. Since $\widetilde{I}_q$ is continuous, (7.5.84) implies $\|\widetilde{I}_q J_q \varphi_n\| \to \|\widetilde{I}_q \widetilde{J}_q \varphi\|$. Thus, we have (7.5.83) and so (7.5.73).

Finally, let $0 \le s \le q$ with s closed. Since s is lsc,

$$s(\varphi) = \lim_{\psi \to \varphi} s(\psi) \le \lim_{\psi \to \varphi} q(\psi) = q_r(\varphi) \tag{7.5.86}$$

so $s \le q_r$. □

Theorem 7.5.16. *Let q be a quadratic form on $\mathcal{H}$. Then q has a closed extension if and only if q_r is an extension of q.*

Remark. If q has a closed extension, we say it is *closable* and, in that case, q_r is its *closure*.

Proof. If q_r is an extension, it is clearly a closed extension. If s is a closed extension, then $s \le q$ so

$$s \le q_r \le q \tag{7.5.87}$$

If s is an extension, $s(\varphi) = q(\varphi)$ for $\varphi \in D(q)$, so (7.5.87) implies $q(\varphi) = q_r(\varphi)$ for such φ, proving q_r is an extension of q. □

Example 7.5.17 ($\delta(x)$ as a Quadratic Form)**.** Let $\mathcal{H} = L^2(\mathbb{R})$,

$$q(\varphi) = \begin{cases} |\varphi(0)|^2 & \text{if } \varphi \in C_0^\infty(\mathbb{R}) \\ \infty & \text{if } \varphi \notin C_0^\infty(\mathbb{R}) \end{cases}$$

We claim that $q_r = 0$ on all of $\mathcal{H}$, so q is not closable. First,

$$\mathcal{H}_q = \mathbb{C} \oplus L^2(\mathbb{R}) \tag{7.5.88}$$

for on $\mathcal{H}_+$,

$$\|\varphi\|_{+1}^2 = |\varphi(0)|^2 + \|\varphi\|_{L^2}^2 \tag{7.5.89}$$

Given $\varphi \in \mathcal{H}$ and any $\lambda \in \mathbb{C}$, we can find $\varphi_n \in C_0^\infty$ so $\varphi_n \to \varphi$ and $\varphi_n(0) \to \lambda$, since C_0^∞ is dense in L^2 and there exist $\psi_n \in C_0^\infty$ so $\|\psi_n\|_2 \to 0$ and $\psi_n(0) \to 1$. This shows (7.5.88).

Clearly, $\widetilde{I}_q((\lambda, \varphi)) = \varphi$, so $\mathcal{L}_q = \{(\lambda, 0) \mid \lambda \in \mathbb{C}\}^\perp = 0 \oplus L^2$ and $\widetilde{J}_q \varphi = (0, \varphi)$, so $\|J_q \varphi\|_{+1}^2 = \|\varphi\|^2$ and $q_r(\varphi) = 0$ for all $\varphi \in \mathcal{R}_q = L^2$.

Related to this is the following, showing the relation to the Lebesgue decomposition. Let μ, ν be two probability measures on a compact Hausdorff space, X, and

$$d\nu = f\, d\mu + d\nu_s \tag{7.5.90}$$

the Lebesgue decomposition. Let $\mathcal{H} = L^2(X, d\mu)$ and

$$q(\varphi) = \begin{cases} \int |\varphi(x)|^2 \, d\nu(x), & \varphi \in C(X) \\ \infty, & \varphi \notin C(X) \end{cases} \tag{7.5.91}$$

Then it can be seen (Problem 7) that

$$q_r(\varphi) = \int f(x)|\varphi(x)|^2 \, d\mu(x), \qquad q_s = d\nu_s \text{ on } C(X) \tag{7.5.92}$$

□

Theorem 7.5.18 (Monotone Convergence for Forms). *Let $\{q_n\}_{n=1}^\infty$ be a family of closed quadratic forms on $\mathcal{H}$ and A_n the associated operators (where A_n is self-adjoint on $\overline{D(q_n)}$). Then*

(a) *If $q_{n+1} \geq q_n$, then*

$$q_\infty(\varphi) = \sup_n q_n(\varphi) = \lim q_n(\varphi) \tag{7.5.93}$$

is a closed quadratic form, and if A_∞ is the associated operator for $z \in \mathbb{C} \setminus [0, \infty)$,

$$(A_n - z)^{-1} \xrightarrow{s} (A_\infty - z)^{-1} \tag{7.5.94}$$

strongly (i.e., $A_n \to A_\infty$ in strong resolvent sense).

(b) *If $q_{n+1} \leq q_n$, then*

$$q_\infty(\varphi) = \inf_n q_n(\varphi) = \lim q_n(\varphi) \tag{7.5.95}$$

may not be closed (or even closable). If A_∞ is the operator associated to $(q_\infty)_r$, then (7.5.94) holds, that is, $A_n \to A_\infty$ in strong resolvent sense.

Proof. (a) A sup of a family of lsc functions is lsc (see Proposition 2.1.9 of Part 1) so q_∞ is lsc. Since the sup is a limit, q_∞ obeys the parallelogram law. Thus, q_∞ is a closed form.

By Theorem 7.5.14,

$$(A_{n+1} + 1)^{-1} \leq (A_n + 1)^{-1} \tag{7.5.96}$$

so for all $\varphi \in \mathcal{H}$,

$$\lim_{n\to\infty} (\varphi, (A_n + 1)^{-1}\varphi) = \inf(\varphi, (A_n + 1)^{-1}\varphi) \tag{7.5.97}$$

exists. By polarization,

$$C = \text{w-lim}(A_n + 1)^{-1} \tag{7.5.98}$$

exists. Since Theorem 7.5.14 says $(A_\infty + 1)^{-1} \leq (A_n + 1)^{-1}$, we see

$$0 \leq (A_\infty + 1)^{-1} \leq C \leq (A_n + 1)^{-1} \tag{7.5.99}$$

C is a positive, self-adjoint, bounded (since $C \le \mathbb{1}$) operator, so the spectral theorem implies there is a positive self-adjoint operator, $\widetilde{B}$, on $\mathrm{Ker}(C)^\perp$ so that $C = (\widetilde{B})^{-1}$ (in our sense that inverses of operators only defined on a closed $V \subset \mathcal{H}$ are set to 0 on $V^\perp$). $C \le 1$ implies $\widetilde{B} \ge 1$ so $B = \widetilde{B} - \mathbb{1}$ is a positive operator.

By (7.5.99) and Theorem 7.5.14, $A_n \le B \le A_\infty$. But A_∞ is the smallest closed form greater than all A_n, so $B = A_\infty$.

Weak convergence also holds for $(A-z)^{-1}$ for all $z \in (-\infty, 0)$ and then, by analyticity, for all $z \in \mathbb{C} \setminus [0,\infty)$. Thus, since weak resolvent convergence implies strong resolvent convergence (see Problem 1 of Section 7.2), $A_n \to A_\infty$ in strong resolvent sense.

(b) The argument is similar to (a). Since $\langle \varphi, (A_n+1)^{-1}\varphi\rangle \le \|\varphi\|^2$, the weak limit still exists, and so there is B (positive and self-adjoint but perhaps not densely defined), so $(A_n+1)^{-1} \xrightarrow{w} (B+1)^{-1}$.

Since $A_n \ge A_\infty$, $(A_\infty + 1)^{-1} \ge (A_n + 1)^{-1}$, and thus, $(B+1)^{-1} \le (A_\infty+1)^{-1}$. Therefore, $A_n \ge B \ge A_\infty$. Since the form of A_∞ is the largest closed form less than all q_n, $B = A_\infty$. Going from weak to strong resolvent convergence is as above. □

Example 7.5.7 (continued). As in that example, for each $n = 1, 2, \dots,$ we can define a self-adjoint operator

$$-\frac{1}{n}\frac{d^2}{dx^2} + \delta(x) \equiv A_n \tag{7.5.100}$$

as a form sum with $Q(A_n) = Q(-\frac{d^2}{dx^2})$. $A_{n+1} \le A_n$ and the form limit is $\delta(x)$ on $Q(-\frac{d^2}{dx^2})$. The regular part is 0. This shows in the monotone decreasing case that the limit many not be closed or even closable. □

Our next topic is the forms view of self-adjoint extensions. We begin with the following that also provides many examples of closable but not closed forms:

Theorem 7.5.19 (Friedrichs Extension). *Let A be a positive Hermitian operator. Let $\tilde{q}_A$ be the form*

$$\tilde{q}_A(\varphi) = \begin{cases} \langle \varphi, A\varphi\rangle, & \varphi \in D(A) \\ \infty, & \varphi \notin D(A) \end{cases} \tag{7.5.101}$$

Then $\tilde{q}_A$ is closable. Its closure, $q_\infty^{(A)}$, is the form of positive self-adjoint operator, A_∞, which is an extension of A. Moreover,

$$\ell_{\tilde{q}_A} = \ell_{q_\infty^{(A)}} \tag{7.5.102}$$

Proof. q is not closable if and only if $\text{Ker}(\widetilde{I_q}) \neq \{0\}$ if and only if there exists $\varphi_n \in D(q)$ so $q(\varphi_n - \varphi_m) \to 0$, $q(\varphi_n) = 1$, $\|\varphi\|_n \to 0$.

If $q = \tilde{q}_A$ and $\varphi_n \in D(A)$ with $\|\varphi_n\| \to 0$ and $\|\varphi_n - \varphi_m\|_{+1} \to 0$, then

$$\lim_{n\to\infty} \|\varphi_n\|_{+1}^2 = \lim_{n\to\infty} \big(\lim_{m\to\infty} \langle \varphi_n, \varphi_m \rangle_{+1}\big) = 0 \tag{7.5.103}$$

since $\lim_m \langle \varphi_n, A\varphi_m \rangle = \lim_m \langle A\varphi_n, \varphi_m \rangle = 0$ for n fixed. Thus, $q(\varphi_n) \to 0$. Thus, by the first paragraph, $\tilde{q}_A$ is closable.

Next, let $\varphi \in D(A)$. For $\psi \in D(A_\infty)$, pick $\psi_n \in D(A)$ so $\|\psi_n - \psi\|_{+1} \to 0$. Then

$$\langle A\varphi, \psi \rangle = \lim_{n\to\infty} \langle A\varphi, \psi_n \rangle = \lim_{n\to\infty} \tilde{q}_A(\varphi, \psi_n) = q_\infty^{(A)}(\varphi, \psi) = \langle \varphi, A_\infty \psi \rangle \tag{7.5.104}$$

It follows that $\varphi \in D(A_\infty^*)$ and $A_\infty^* \varphi = A\varphi$. Since $A_\infty^* = A_\infty$, $\varphi \in D(A_\infty)$ and $A_\infty \varphi = A\varphi$, that is, A_∞ is an extension of A. $\square$

We will prove the following description of all positive self-adjoint extensions of a closed positive Hermitian operator:

Theorem 7.5.20. *Let A be a closed positive Hermitian operator on a Hilbert space, $\mathcal{H}$. Then there is a distinguished self-adjoint extension, A_0, of A so that the set of all self-adjoint extensions is the set of all self-adjoint operators, B, with*

$$A_0 \leq B \leq A_\infty \tag{7.5.105}$$

Remarks. 1. We will call A_0 the *Krein extension.* It is sometimes called the *von Neumann extension* or *Krein–von Neumann extension.* When $\ell_{q_A} > 0$, von Neumann had the construction but didn't notice its extreme properties.

2. We'll prove this below after a more detailed analysis in case $\ell_{\tilde{q}_A} = \ell_{q_{A_\infty}} > 0$ by taking limits as $\varepsilon \downarrow 0$ for $A + \varepsilon$.

3. When $\ell_{q_{\tilde{A}}} = 0$, it can happen that $A_0 = A_\infty$. Surprisingly, this can happen even if A has positive deficiency indices, that is, there exist positive Hermitian A with multiple self-adjoint extensions but only one positive extension (see Example 7.5.25 and Example 7.5.10 (continued) which precedes it). It can also happen that $\ell_{q_{\tilde{A}}} = 0$ and $A_0 \neq A_\infty$ and there are multiple positive self-adjoint extensions.

We begin by considering the case where $\ell_{q_{\tilde{A}}} > 0$. Thus, by Theorem 7.1.8,

$$N = \text{Ker}(A^*) \tag{7.5.106}$$

is a space of dimension $d = d_+ = d_-$. We know that $\dim(D(A^*)/D(A))$ is $2d$. Since $A_\infty \geq \ell_{\tilde{q}_A} > 0$, A_∞ has a bounded inverse, and clearly, $D(A_\infty) \subset D(A^*)$, so some thought suggests the following proposition, whose proof is close to that of Theorem 7.1.11.

Proposition 7.5.21. *Let A be a closed positive Hermitian operator with $\ell_{\tilde{q}_A} > 0$. Let N be given by (7.5.106). Then*

$$D(A^*) = D(A) \dotplus N \dotplus A_\infty^{-1} N \tag{7.5.107}$$

with (for $\varphi \in D(A)$, $\eta_1, \eta_2 \in N$)

$$A^*(\varphi + \eta_1 + A_\infty^{-1}\eta_2) = A\varphi + \eta_2 \tag{7.5.108}$$

Moreover,

$$D(A_\infty) = D(A) \dotplus A_\infty^{-1} N \tag{7.5.109}$$

Remark. Recall that $V \dotplus W$ means the sum when $V \cap W = \{0\}$ (disjoint union).

Proof. We know by Proposition 7.1.7 that $\text{Ran}(A)$ is closed and $\text{Ran}(A)^\perp = \text{Ker}(A^*)$ so

$$\mathcal{H} = \text{Ran}(A) \oplus \text{Ker}(A^*) \tag{7.5.110}$$

Thus, for any $\psi \in D(A^*)$, there is $\varphi \in D(A)$, $\eta_2 \in N$, so

$$A^*\psi = A\varphi + \eta_2 \tag{7.5.111}$$

Since A_∞ is a bijection of $D(A_\infty) \subset D(A^*)$ and $\mathcal{H}$,

$$A^*(\varphi + A_\infty^{-1}\eta_2) = A_\infty(\varphi + A_\infty^{-1}\eta_2) = A^*\psi \tag{7.5.112}$$

so $\eta_1 = \psi - \varphi - A_\infty^{-1}\eta_2 \in \text{Ker}(A^*) = N$, proving that (7.5.107) holds with $\dotplus$ replaced by $+$. This also shows A_∞ is a bijection of $D(A) \dotplus A_\infty^{-1} N \subset D(A_\infty)$ and $\mathcal{H}$, proving (7.5.109).

To get $\dotplus$, we only need uniqueness. Thus, suppose

$$\varphi + \eta_1 + A_\infty^{-1}\eta_2 = 0 \tag{7.5.113}$$

for $\varphi \in D(A)$, $\eta_1, \eta_2 \in N$. Applying A_∞, we see $A\varphi + \eta_2 = 0$, so by (7.5.110), $\eta_2 = 0$ and $A\varphi = 0$. But then

$$0 = \langle \varphi, A\varphi \rangle \geq \ell_{\tilde{q}_A} \|\varphi\|^2 \tag{7.5.114}$$

so $\varphi = 0$. Clearly, $\varphi = \eta_2 = 0$ and (7.5.114) implies $\eta_1 = 0$. $\square$

Lemma 7.5.22. *Let $\ell_{\tilde{q}_A} > 0$. Then $Q(A_\infty) \cap N = \{0\}$.*

Proof. Note first that since $q_{A_\infty} \geq \ell_A$, $q_{A_\infty}(\varphi)^{1/2}$ and $[q_{A_\infty}(\varphi) + \|\varphi\|^2]^{1/2}$ are equivalent norms, so $D(A_\infty)$ is dense in $q_{A_\infty}(\cdot)$-norm in $Q(A_\infty)$,

Let $\varphi \in Q(A_\infty) \cap N$. Then for $\psi \in D(A) \subset D(A^*)$,

$$\langle A\psi, \varphi \rangle = \langle \psi, A^*\varphi \rangle = 0 \tag{7.5.115}$$

But

$$Q_{A_\infty}(\psi, \varphi) = \langle A\psi, \varphi \rangle \tag{7.5.116}$$

so in Q_{A_∞}-inner product, $\varphi \perp D(A)$. Since $D(A)$ is dense in q_{A_∞}-norm in $Q(A_\infty)$, $\varphi = 0$. $\square$

Now let b be a closed quadratic form on N,

$$N_b = \overline{\{\varphi \in N \mid b(\varphi) < \infty\}} \tag{7.5.117}$$

and B the self-adjoint operator on $\overline{N}_b$ corresponding to b. Define

$$q_b(\psi) = \begin{cases} q_{A_\infty}(\varphi) + b(\eta) & \text{if } \psi = \varphi + \eta,\ q \in Q(A_\infty),\ \eta \in D(b) \\ \infty, & \text{otherwise} \end{cases} \tag{7.5.118}$$

By the lemma, $\psi = \varphi + \eta$ is unique since $D(b) \cap Q(A_\infty) \subset N \cap Q(A_\infty) = \{0\}$. Thus, q_b is a quadratic form with

$$D(q_b) = Q(A_\infty) \dotplus D(b) \tag{7.5.119}$$

Theorem 7.5.23. *Let A be a positive, closed Hermitian operator with $\ell_{\tilde{q}_A} > 0$. For every closed quadratic form, b, on N, q_b is a closed quadratic form with $\overline{D(q_b)} = \mathcal{H}$. If A_b is the associated positive self-adjoint operator, then A_b is a positive self-adjoint extension of A. Conversely, every positive self-adjoint extension of A is an A_b. Moreover,*

$$D(A_b) = \{\varphi + A_\infty^{-1}(B\eta_1 + \eta_2) + \eta_1 \mid \varphi \in D(A),\ \eta_1 \in D(B),\ \eta_2 \in N_b^\perp\} \tag{7.5.120}$$

where $N_b^\perp$ is the orthogonal complement of N_b in N with

$$A_b(\varphi + A_\infty^{-1}(B\eta_1 + \eta_2) + \eta_1) = A\varphi + B\eta_1 + \eta_2 \tag{7.5.121}$$

Remarks. 1. The φ, η_1, η_2 in (7.5.120) are unique.

2. We'll see shortly that the Krein extension is the A_b corresponding to $b = 0$.

3. In case $\dim(N) < \infty$, one can extend this result to all self-adjoint extensions. b is then a self-adjoint matrix on a subspace N_b of N, $D(A_b)$ is given by (7.5.120), and A_b by (7.5.121); see the reference in the Notes.

Proof. Since $D(A) \subset Q(A_\infty)$ is dense, $\overline{D(q_b)} = \mathcal{H}$. To see q_b is closed, it suffices to prove the norm

$$\|\psi\|_{+1,q_b} = (q_b(\psi) + \|\psi\|^2)^{1/2} \tag{7.5.122}$$

on $D(q_b)$ makes it into a Hilbert space.

Since $D(q_b)$ is given by (7.5.119) and $Q(A_\infty)$ is complete in q_{A_∞}, while $D(b)$ is complete $\|\cdot\|_{+1,b}$, it suffices to prove the norm (7.5.122) is equivalent to the direct sum norm for $\varphi \in Q(A_\infty)$, $\eta \in D(q_b)$,

$$\|\varphi + \eta\|_\oplus = \left[(q_{A_\infty}(\varphi) + \|\varphi\|^2) + (b(\eta) + \|\varphi\|^2)\right]^{1/2} \tag{7.5.123}$$

Using $\|\varphi + \eta\|^2 \le 2\|\varphi\|^2 + 2\|\eta\|^2$, we see that

$$\|\psi\|_{+1,q_b} \le \sqrt{2}\,\|\psi\|_\oplus \tag{7.5.124}$$

On the other hand, since $q_b(\varphi) \geq \ell_{\tilde{q}_b}\|\varphi\|^2$, we see, for some $c > 0$,

$$\|\varphi + \eta\|_{+1,q_b} \geq c\|\varphi\| \tag{7.5.125}$$

so that

$$\|\eta\| \leq \|\varphi + \eta\| + \|\varphi\| \leq (1 + c^{-1})\|\varphi + \eta\|_{+1,q_b}$$

proving that $\|\psi\|_\oplus$ is dominated by a constant multiple of $\|\psi\|_{+1,q_b}$. Thus, q_b is closed.

If Q_b is the sesquilinear form on $Q(q_b)$ and $\varphi \in D(A)$, then since $\varphi \in D(A) \subset D(A_\infty) \subset Q(A_\infty)$,

$$Q_b(\varphi_1 + \eta_1, \varphi) = \langle \varphi_1, A_\infty \varphi \rangle = \langle \varphi_1, A\varphi \rangle = \langle \varphi_1 + \eta_1, A\varphi \rangle \tag{7.5.126}$$

since $\langle \eta_1, A\varphi \rangle = \langle A^* \eta_1, \varphi \rangle = 0$. Thus, $\varphi \in D(A_b)$ and $A_b\varphi = A\varphi$, proving that A_b is an extension of A.

Finally, let $\widetilde{A}$ be a self-adjoint extension of A. Let $\psi \in D(\widetilde{A}) \subset D(A^*)$. By Proposition 7.5.21,

$$D(A^*) = D(A_\infty) \dot{+} N \tag{7.5.127}$$

so

$$\psi = \varphi + \eta, \qquad \varphi \in D(A_\infty), \quad \eta \in N \tag{7.5.128}$$

Since $D(A) \subset Q(\widetilde{A})$ and $q_{\widetilde{A}} \restriction D(A) = q_{A_\infty} \restriction D(A)$ and $D(A)$ is dense in $Q(A_\infty)$, we see that

$$Q(A_\infty) \subset Q(\widetilde{A}) \tag{7.5.129}$$

Thus, in (7.5.128), we have $\varphi, \eta \in Q(\widetilde{A})$. Picking $\varphi_n \in D(A)$ so $\|\varphi_n - \varphi\|_{+1,A_\infty} = \|\varphi_n - \varphi\|_{+1,\widetilde{A}} \to 0$, we see that

$$\begin{aligned} Q_{\widetilde{A}}(\varphi, \eta) &= \lim_{n\to\infty} Q_{\widetilde{A}}(\varphi_n, \eta) = \lim_{n\to\infty} \langle \widetilde{A}\varphi_n, \eta \rangle = \lim_{n\to\infty} \langle A\varphi_n, \eta \rangle \\ &= \lim_{n\to\infty} \langle \varphi_n, A^*\eta \rangle = 0 \end{aligned} \tag{7.5.130}$$

so

$$q_{\widetilde{A}}(\psi) = q_\infty(\varphi) + b(\eta) \tag{7.5.131}$$

where b is defined on N as follows:

$$b(\eta) = \begin{cases} q_{\widetilde{A}}(\eta), & \eta \in Q(\widetilde{A}) \cap N \equiv N_b \\ \infty, & \eta \in N, \eta \notin Q(\widetilde{A}) \end{cases}$$

(7.5.131) says that

$$\psi \in D(\widetilde{A}) \Rightarrow q_{\widetilde{A}}(\psi) = q_b(\psi) \tag{7.5.132}$$

Since $D(\widetilde{A})$ is a form core for $q_{\widetilde{A}}$ and q_b is closed, we obtain (7.5.132) for all $\psi \in Q(\widetilde{A})$. In particular,

$$Q(\widetilde{A}) = Q(A_\infty) \dot{+} N_b \tag{7.5.133}$$

If $\psi = \varphi + \eta$ with $\varphi \in Q(A_\infty)$, $\eta \in N$, $\eta \notin N_b$, then since $Q_{A_\infty} \subset Q(\widetilde{A})$, $\eta \notin Q(\widetilde{A})$ (since $N_b = N \cap Q(\widetilde{A})$). Thus, $q_{\widetilde{A}}(\psi) = \infty = q_b(\psi)$. Finally, if $\psi \notin Q(A_\infty) + N$, by definition of q_b, we have $q_b(\psi) = \infty$, and by (7.5.133), $q_{\widetilde{A}}(\psi) = \infty$. We have thus proven that $q_{\widetilde{A}} = q_b$ so $\widetilde{A} = A_b$. □

Corollary 7.5.24. *Under the hypotheses of Theorem 7.5.23, with A_0 the operator associated to $b = 0$,*

$$\{C \text{ self-adjoint} \mid A_0 \le C \le A_\infty\} \tag{7.5.134}$$

is the set of positive self-adjoint extensions of A.

Proof. Clearly, $0 \le b \le \infty \Rightarrow$ for any b that $A_0 \le A_b \le A_\infty$, so it suffices to show that any C in the set (7.5.134) is an A_b.

Since $A_0 \le C$, $Q(C) \subset Q(A_b) = Q(A_\infty) \dotplus N$. Since $C \le A_\infty$, $Q(A_\infty) \subset Q(C)$, so $Q(C) = Q(A_\infty) \dotplus N_1$ for a subspace $N_1 \subset N$. If $\varphi \in Q(A_\infty)$,

$$q_0(\varphi) = q_\infty(\varphi) \le q_C(\varphi) \le q_\infty(\varphi) \tag{7.5.135}$$

so $q_C(\varphi) = q_\infty(\varphi)$. If $\eta \in N_1$, $\varphi + z\eta \in Q(C)$ for any $z \in \mathbb{C}$, so

$$q_\infty(\varphi) = q_0(\varphi + z\eta) \le q_\infty(\varphi) + 2\operatorname{Re}(zQ_C(\varphi, \eta)) + |z|^2 q_C(\eta) \tag{7.5.136}$$

This is only possible for all $z \in \mathbb{C}$ if

$$Q_C(\varphi, \eta) = 0 \tag{7.5.137}$$

that is,

$$q_C(\varphi) = q_b(\varphi) \tag{7.5.138}$$

where $b = q_C \restriction N_1$. Since Q_C is closed, $\mathcal{H}_{+1,C}$ is a Hilbert space, and by (7.5.137), N_1 is the orthogonal complement of $Q(A_\infty)$ in $\langle\ ,\ \rangle_{+1,C}$. Thus, b is a closed form and $C = A_b$. □

Proof of Theorem 7.5.20. For each $\varepsilon > 0$,

$$B(\varepsilon) \equiv A + \varepsilon\mathbb{1} \ge \varepsilon\mathbb{1} \tag{7.5.139}$$

so we can apply the above theorem and get $B_0(\varepsilon)$ so all positive self-adjoint extensions of $B(\varepsilon)$ are $\{D \mid B_0(\varepsilon) \le D \le A_\infty + \varepsilon\mathbb{1}\}$ (since the closure of $B(\varepsilon)$ as a form is $A_\infty + \varepsilon\mathbb{1}$). If $A_\varepsilon = B_0(\varepsilon) - \varepsilon$, we see that all self-adjoint extensions of A, with $C \ge -\varepsilon\mathbb{1}$ are

$$\mathcal{S}(\varepsilon) = \{C \mid A_\varepsilon \le C \le A_\infty\} \tag{7.5.140}$$

If $\varepsilon' < \varepsilon$, $\mathcal{S}(\varepsilon') \subset \mathcal{S}(\varepsilon)$, so $A(\varepsilon') \ge A(\varepsilon)$. Thus, by Theorem 7.5.18,

$$A_0 = \text{monotone limit of } A(\varepsilon) \tag{7.5.141}$$

exists. Clearly,

$$\bigcap_{\varepsilon>0} \mathcal{S}(\varepsilon) = \{C \mid A_0 \le C \le A_\infty\} \tag{7.5.142}$$

It is evident that $C \geq 0$ if and only if $C \geq -\varepsilon\mathbb{1}$ for all $\varepsilon > 0$, so $\bigcap_{\varepsilon>0} \mathcal{S}(\varepsilon)$ is the set of positive self-adjoint extensions. □

Remark. The proof of Theorem 7.5.23 shows that

$$A(\varepsilon) = A^* \restriction [D(A) \dotplus \operatorname{Ker}(A^* + \varepsilon)] \tag{7.5.143}$$

and A_0 is the monotone limit of these.

Here are two examples that illustrate when $A_0 \neq A_\infty$ there must be multiple positive self-adjoint extensions. Since $\ell_{A_\infty} = \ell_{\tilde{q}_A}$ and $\ell_{A_0} = 0$, clearly, if $\ell_{\tilde{q}_A} > 0$, $A_0 \neq A_\infty$ so we want examples with $\ell_{\tilde{q}_A} = 0$.

Example 7.5.10 (continued). In this case, $A_\infty = A_{\theta=0}$ is clearly the largest A_θ and, indeed, as we've seen, $Q(A_{\theta=0}) = Q_\infty$ is a strict subset of Q_0 and all $A_{\theta\neq 0}$ have $Q(A_{\theta=0}) = Q_\infty$ and $q_{A_\theta} \restriction Q_\infty = q_{A_\infty}$.

By (7.5.65), $A_\theta \geq 0$ if and only if $\theta \in [\frac{\pi}{2}, \pi] \cup \{0\}$. $A_0 = A_{\theta=\pi/2}$ is the smallest A. So this is an example with lower bound zero but $A_0 \neq A_\infty$.

Example 7.5.25. Let A be the operator on $L^2([0,1], dx)$ with

$$D(A) = \{\varphi \mid \varphi \text{ is } C^1,\ \varphi' \text{ is a.c.},\ \varphi(0) = \varphi'(0) = 0,\ \varphi'(1) = 0\} \tag{7.5.144}$$

$$A\varphi = -\frac{d^2}{dx}\varphi - \frac{\pi^2}{4}\varphi \tag{7.5.145}$$

(we'll see the reason for the $\pi^2/4$ shortly). It is easy to see (Problem 8) that A is a closed Hermitian operator with deficiency indices $(1,1)$. The self-adjoint extensions (Problem 9) are A_θ of the same formal form (7.5.145) but on

$$D(A_\theta) = \{\varphi \mid \varphi \text{ is } C_1,\ \varphi' \text{ is a.c.},\ \varphi'(1) = 0,\ \cos(\theta)\varphi'(0) + \sin(\theta)\varphi'(0) = 0\} \tag{7.5.146}$$

Each A_θ has a complete set of eigenfunctions of the form $\cos(k(x-1))$ with eigenvalues $k^2 - (\frac{\pi}{2})^2$ or $\cosh(\kappa(x-1))$ with eigenvalues $-\kappa^2 - (\frac{\pi}{2})^2$. A_∞ has an eigenfunction with $k = \frac{\pi}{2} + \ell\pi$, $\ell = 0, 1, 2, \dots$, and so $A_\infty \geq 0$. All other A_θ's have strictly negative eigenvalues (Problem 10). Even though A has multiple self-adjoint extensions, it has only one that is positive so the Krein and Friedrichs extensions agree. □

Example 7.5.26 (Dirichlet and Neumann Laplacians)**.** Let $\Omega \subset \mathbb{R}^\nu$ be a bounded open set. Take $D(q) = \{\varphi \in L^2(\Omega) \mid \nabla\varphi \in L^2(\Omega)\}$, where $\nabla\varphi$ is the distributional Laplacian and take

$$q(\varphi) = \int |\nabla\varphi(x)|^2\, d^\nu x \tag{7.5.147}$$

then it is easy to see (Problem 11) that q is closed form. The associated self-adjoint operator is called the *Neumann Laplacian.* If instead we take the closure of this form restricted to C_0^∞, the associated operator is called the *Dirichlet Laplacian.*

In nice cases, the Dirichlet Laplacian has an operator core of functions continuous up to $\partial\Omega$ with $\varphi \restriction \partial\Omega \equiv 0$ and the Neumann Laplacian an operator core of functions C^1 up to $\partial\Omega$ with $\frac{\partial\varphi}{\partial n} \restriction \partial\Omega = 0$ (this clearly requires $\partial\Omega$ to be smooth enough that $\frac{\partial}{\partial n}$ makes sense for most points on $\partial\Omega$). For example, if Ω obeys an exterior cone condition (see Section 3.4 of Part 3), one can use path integral methods to prove the set of $\varphi \in C(\overline{\Omega})$ so that φ is C^∞ in Ω and continuous on $\overline{\Omega}$, with φ, $-\Delta\varphi \in L^2(\Omega)$, and $\varphi \restriction \partial\Omega = 0$ is an operator core for the Dirichlet Laplacian.

For the Neumann boundary condition, more is needed, but it is enough if there is a function, F, C^1 in a neighborhood, N, of $\overline{\Omega}$ with $\Omega = \{x \mid F(x) > 0\}$ and $\nabla F \neq 0$ everywhere on $\partial\Omega$ (i.e., Ω is a C^1-hypersurface), the set of $\varphi \in L^2(\Omega)$, C^2 in Ω, C^1 up to $\partial\Omega$ with $\nabla\varphi \in L^2$ and $\frac{\partial\Omega}{\partial n} = 0$ on $\partial\Omega$ is an operator core for the Neumann Laplacian. The Notes provide some references for the operator core results.

As the above indicates, there are connections between path integrals and these Laplacians. In great generality (see Problem 12), one has that if $|\partial\Omega| = 0$, then for $x \in \Omega$,

$$\mathbb{E}(e^{-\frac{1}{2}tH_D^\Omega} f)(x) = \mathbb{E}(f(x + b(t)) \mid b\text{'s with } x + b(s) \in \Omega \text{ for all } s \in [0, t]) \tag{7.5.148}$$

where H_D^Ω is the Dirichlet Laplacian. When Ω is nice enough (see the Notes), one can express $e^{-\frac{1}{2}tH_N^\Omega}$ in terms of a reflecting Brownian motion.

As we discussed in Section 3.16, H_D^Ω has compact resolvent if $|\Omega| < \infty$. This is obvious from (7.5.148) since it implies $e^{-\frac{1}{2}tH_D^\Omega}(x, y)$ is bounded by $e^{t\Delta}(x, y) \le (2\pi t)^{-\nu/2}$ so $e^{-\frac{1}{2}tH_D^\Omega}$ is Hilbert–Schmidt. There are even infinite volume regions where H_D^Ω has compact resolvent (see the Notes). It is often true that Neumann Laplacians with $|\Omega| < \infty$ have compact resolvent—but not always. Indeed, there are examples where Ω is bounded and their spectrum is absolutely continuous; see the Notes. □

There are two properties of Dirichlet and Neumann Laplacians that are not only intrinsically interesting but will lead to results on asymptotic eigenvalue counting.

Theorem 7.5.27 (Dirichlet–Neumann Bracketing). *Let $\Omega_1 \subset \Omega_2$ be open sets in $\mathbb{R}^\nu$. Then*

$$H_D^{\Omega_1} \ge H_N^{\Omega_1} \tag{7.5.149}$$

$$H_D^{\Omega_1} \ge H_D^{\Omega_2} \tag{7.5.150}$$

where $H_D^{\Omega_1}$ *is viewed as the form on* $L^2(\Omega_2, d^\nu x)$ *obtained by viewing* $C_0^\infty(\Omega_1) \subset L^2(\Omega_2, d^\nu x)$ *and closing the form* (7.5.147) (*and so* $H_D^{\Omega_1}$ *is not densely defined*).

Remarks. 1. This is called bracketing since if $\Omega_1 \subset \Omega_2 \subset \Omega_3$, it implies that

$$H_D^{\Omega_1} \geq H_D^{\Omega_2} \geq H_N^{\Omega_3} \tag{7.5.151}$$

(since $H_D^{\Omega_2} \geq H_D^{\Omega_3} \geq H_N^{\Omega_3}$).

2. (7.5.150) fails in general for the N case. It must, since otherwise it would imply that every H_N^Ω with Ω bounded has a compact resolvent, contrary to the examples mentioned in the Notes.

Proof. By definition, the form of $H_N^{\Omega_1}$ is an extension of the form $H_D^{\Omega_1}$, so (7.5.149) is immediate.

Clearly, $C_0^\infty(\Omega_1) \subset C_0^\infty(\Omega_2)$ so H_N^Ω restricted to $C_0^\infty(\Omega_2)$ is an extension of the restriction to $C_0^\infty(\Omega_1)$ and the same is thus true of their closures. Thus, $H_D^{\Omega_2}$ is an extension $H_D^{\Omega_1}$ and (7.5.150) follows. □

Theorem 7.5.28 (Dirichlet–Neumann Decoupling). *Let* Ω *in* $\mathbb{R}^\nu$ *be open,* $K \subset \Omega$ *a relatively closed subset of measure zero so that*

$$\Omega \setminus K = \Omega_1 \cup \Omega_2 \tag{7.5.152}$$

are disjoint open sets (*see Figure* 7.5.1).

View $L^2(\Omega, d^\nu x)$ *as* $L^2(\Omega_1, d^\nu x) \oplus L^2(\Omega_2, d^\nu x)$ *in the obvious way. Then*

$$H_D^\Omega \leq H_D^{\Omega_1} \oplus H_D^{\Omega_2} \tag{7.5.153}$$

$$H_N^\Omega \geq H_N^{\Omega_1} \oplus H_N^{\Omega_2} \tag{7.5.154}$$

Proof. $Q(H_N^{\Omega_1} \oplus H_N^{\Omega_2})$ is those $L^2(\Omega)$-functions, φ, where there is an L^2-function, $\vec{\psi}$, so that

$$\langle \vec{\psi}, f \rangle = \langle \varphi, \vec{\nabla} f \rangle \tag{7.5.155}$$

for all f in $C_0^\infty(\Omega_1) \cup C_0^\infty(\Omega_2)$. The set where (7.5.155) holds for all $f \subset C_0^\infty(\Omega)$ is clearly a subset. Thus, $H_N^{\Omega_1} \oplus H_N^{\Omega_2}$ an extension of H_N^Ω, so we have (7.5.154).

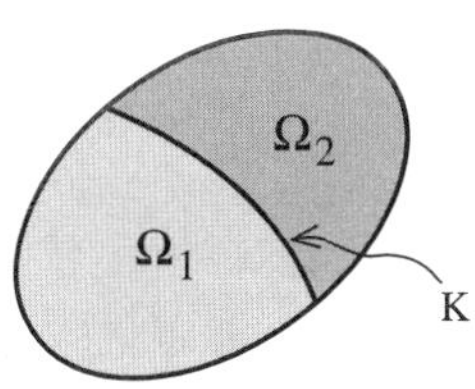

Figure 7.5.1. Dirichlet–Neumann decoupling.

Similarly, since $C_0^\infty(\Omega_1) \cup C_0^\infty(\Omega_2)$ is a subset of $C_0^\infty(\Omega_1 \cup \Omega_2)$, H_D^Ω is an extension of $H_D^{\Omega_1} \oplus H_D^{\Omega_2}$. □

Remark. If Ω_1 and Ω_2 are touching unit cubes and Ω is their union with K their common boundary, and if φ is different constants in Ω_1 and Ω_2, then $\varphi \in Q(H_N^{\Omega_1} \oplus H_N^{\Omega_2})$ but not in $Q(H_N^\Omega)$.

As a final topic, we want to use Dirichlet–Neumann bracketing to prove some results on the *quasiclassical limit*, the notion that in some kind of $\hbar \to 0$ limit, you can find the number of bound states by looking at volumes of phase space divided by $(h)^\nu = (2\pi)^\nu(\hbar)^\nu$. We'll first consider Weyl's asymptotic eigenvalue counting result.

To state the Weyl result in the form we want, we need the definition of contented set. A hyperrectangle is a set $Q = \times_{j=1}^\nu [a_j, b_j]$ with $a_j < b_j$. We say two hyperrectangles are "disjoint" if their interiors are (some authors use $[a_j, b_j)$ to make it possible to use disjoint without quotes but we prefer closed hypercubes).

The volume, $\tau(Q)$, of Q is $\prod_{j=1}^\nu |b_j - a_j|$. We let $\mathcal{C}$ be the collection of all finite sets, C, of "disjoint" hypercubes. If $C = \{Q_j\}_{j=1}^m$, then $U(C) = \bigcup_{j=1}^m Q_j$ and

$$\tau(C) = \sum_{Q_j \in C} \tau(Q_j) \tag{7.5.156}$$

Definition. Let Ω be a bounded subset of $\mathbb{R}^\nu$. The *inner content* and *outer content* of Ω is

$$\tau_{\text{inn}}(\Omega) = \sup\{\tau(C) \mid C \in \mathcal{C}, U(C) \subset \Omega\} \tag{7.5.157}$$

$$\tau_{\text{out}}(\Omega) = \inf\{\tau(C) \mid C \in \mathcal{C}, \Omega \subset U(C)\} \tag{7.5.158}$$

If $\tau_{\text{inn}}(\Omega) = \tau_{\text{out}}(\Omega)$ we say Ω is a *contented set* and set $\tau(\Omega)$ to this common value.

Remarks. 1. It is easy to see (Problem 13) that the complement of a positive measure Cantor set in $\mathbb{R}$ is not a contented set.

2. It is easy to see (Problem 14) if Ω is open and (with $|\cdot|$ = Lebesgue measure) $|\{x \mid \text{dist}(x, \partial\Omega) < \varepsilon\}| \to 0$ as $\varepsilon \downarrow 0$, then Ω is contented. The reader should keep this example in mind when looking at the next theorem.

3. It is known (see the Notes) that $\tau_{\text{inn}}(\Omega)$ is the Lebesgue measure of Ω^{int} and τ_{out} is the Lebesgue measure of $\overline{\Omega}$, so if Ω is a Borel set, it is contented if and only if $|\partial\Omega| = 0$.

The two theorems we'll prove are

Theorem 7.5.29 (Weyl's Eigenvalue Counting Theorem). *Let Ω be a bounded open contented set in $\mathbb{R}^\nu$. Let $N_\Omega(E)$ be the number (counting multiplicity) of eigenvalues of H_D^Ω less than E. Then*

$$\lim_{E\to\infty} E^{-\nu/2} N_\Omega(E) = \frac{\tau_\nu}{(2\pi)^\nu}\,|\Omega| \tag{7.5.159}$$

Theorem 7.5.30. *Let V be a continuous function of compact support on $\mathbb{R}^\nu$. Let $N(V)$ be the number of negative eigenvalues of $-\Delta + V$. Then*

$$\lim_{\lambda\to\infty} N(\lambda V)/\lambda^{\nu/2} = \frac{\tau_\nu}{(2\pi)^\nu}\int V_-(x)^{\nu/2}\, d^\nu x \tag{7.5.160}$$

Remarks. 1. τ_ν is the volume of the unit ball in $\mathbb{R}^\nu$, i.e., $\pi^{\nu/2}/\Gamma(\frac{\nu}{2}+1)$.

2. $V_-(x) = \max(-V(x), 0)$.

3. $|\{(x,p) \mid x \in \Omega, p^2 \le E\}| = \tau_\nu\, E^{\nu/2}\,|\Omega|$ so (7.5.159) says $N_\Omega(E)$ is asymptotically $(2\pi)^{-\nu}$ times a phase space volume. Similarly, (7.5.160) has a phase space volume interpretation. It's right side is $(2\pi)^{-\nu}|\{(x,p) \mid p^2 + V(x) < 0\}|$.

4. $N(\lambda V)$ is the number of negative eigenvalues of $-\lambda^{-1}\Delta + V$, so in units with $2m = 1$, $\frac{-\hbar^2}{2m}\Delta + V$ with $\hbar = \lambda^{-1/2}$; $\lambda \to \infty$ is the same as $\hbar \downarrow 0$.

5. In dimension $\nu \ge 3$, one can use the CLR inequality (Theorem 6.7.7 of Part 3) to extend (7.5.160) to all $V \in L^{\nu/2}(\mathbb{R}^\nu)$ (Problem 15).

6. One can also compute the quasiclassical value of $\sum_{E_j<0}|E_j|^p$ for any $p > 0$ and prove a limit theorem (see Problem 16).

For simplicity, we'll suppose that all of the sets in the $\mathcal{C}$'s in the definition of content are hypercubes with a common size (actually, one can see it is no loss to only take such collections). We begin the proof by counting eigenvalues of H_D^Ω and H_N^Ω when Ω is a hypercube.

Proposition 7.5.31. *Let Ω be a unit hypercube in $\mathbb{R}^\nu$. Then*

(a) *The eigenvalues of H_N^Ω are $\{\pi^2 \sum_{j=1}^\nu n_j^2 \mid n_j = 0,1,2,\dots\}$.*

(b) *The eigenvalues of H_D^Ω are $\{\pi^2 \sum_{j=1}^\nu n_j^2 \mid n_j = 1,2,\dots\}$.*

Proof. If $\nu = 1$, the eigenfunctions for $H_N^{[0,1]}$ (respectively, $H_D^{[0,1]}$) are $\cos(\pi n x)$ (respectively, $\sin \pi n x$) for $n = 0,1,\dots$ (respectively, $n = 1,2,\dots$). These are clearly eigenfunctions with eigenvalues $(\pi n)^2$ (and it is easy to prove completeness (Problem 17). Products of these are eigenfunctions for H_N^Ω (respectively, H_D^Ω) and by the $\nu = 1$ completeness are complete for general ν. $\square$

Thus we need to compute the number of lattice points in balls to count the number of eigenvalues less than some E:

Proposition 7.5.32. (a) *Let $n_L(R)$ be the number of points in $\mathbb{Z}^\nu$ with $|n| < R$. Then as $R \to \infty$*

$$n_L(R) = \tau_\nu R^\nu + O(R^{\nu-1}) \tag{7.5.161}$$

(b) *Let $n_N(R)$ (respectively, $n_D(R)$) be the points $n \in \mathbb{Z}^\nu$ with $|n| < r$ and also $n_j \geq 0$ (respectively, $n_j \geq 1$). Then as $R \to \infty$*

$$n_N(R) - n_D(R) = O(R^{\nu-1}) \tag{7.5.162}$$

$$n_D(R) = 2^{-\nu} \tau_\nu R^\nu + O(R^{\nu-1}) \tag{7.5.163}$$

Proof. (a) Let L_R be the union of hypercubes of unit side centered at points $n \in \mathbb{Z}$ with $|n| < R$. Let B_ρ be the ball $\{x \mid |x| < \rho\}$. If $x \in L_R$, there is $n \in \mathbb{Z}^\nu$ with $|x - n| \leq \frac{1}{2}\sqrt{\nu}$ (this is the distance from n to a corner) so

$$L_R \subset B_{R+\frac{1}{2}\sqrt{\nu}} \tag{7.5.164}$$

Similarly, if $|x| < R - \frac{1}{2}\sqrt{\nu}$ and x lies in the unit hypercube centered at n, then $|x - n| \leq \frac{1}{2}\sqrt{\nu}$ so $|n| < R$, i.e.,

$$B_{R-\frac{1}{2}\sqrt{\nu}} \subset L_R \tag{7.5.165}$$

Since $|B_\rho| = \tau_\nu \rho^\nu$, (7.5.120) is immediate from (7.5.164) and (7.5.165).

(b) Let $n_L^{(\nu)}(R)$ be $n_L(R)$ with the ν-dependence made explicit. By definition, for each fixed j

$$\begin{aligned} \sharp\{n \in \mathbb{Z}^\nu \mid |n| \leq R, n_j = 0\} &= n_L^{(\nu-1)}(R) \\ &= O(R^{\nu-1}) \end{aligned} \tag{7.5.166}$$

This implies (7.5.162).

By looking at $(\varepsilon_1 n_1, \dots, \varepsilon_\nu n_\nu)$ for $\varepsilon_j = \pm 1$, we see that

$$2^\nu n_D(R) \leq n_L(R) \leq 2^\nu n_N(R) \tag{7.5.167}$$

This plus (7.5.162) implies (7.5.163). □

Proposition 7.5.33. *Let $N_\Delta(E)$ be the eigenvalues of H_D^Δ for $\Delta = [0,1]^\nu$ and $\widetilde{N}_\Delta(E)$ the same for H_N^Δ. Then*

$$|\widetilde{N}_\Delta(E) - N_\Delta(E)| = O(E^{(\nu-1)/2}) \tag{7.5.168}$$

$$N_\Delta(E) = \frac{\tau_\nu}{(2\pi)^\nu} E^{\nu/2} + O(E^{(\nu-1)/2}) \tag{7.5.169}$$

Proof. $\pi^2 |n|^2 \leq E \Leftrightarrow |n| \leq E^{1/2}/\pi$ so (7.5.169) follows from Proposition 7.5.31 and (7.5.163). Similarly, (7.5.168) is immediate from that Proposition and (7.5.168). $\square$

Proof of Theorem 7.5.29. Let C_1, C_2 be two collections of hypercubes all of size L, so $U(C_1) \subset \Omega \subset U(C_2)$. Then, by (7.5.151),

$$H_N^{U(C_2)} \leq H_D^\Omega \leq H_D^{U(C_1)} \tag{7.5.170}$$

By Theorem 7.5.28, we can decouple into individual cubes

$$\bigoplus_{Q_j \in C_2} H_N^{Q_j} \leq H_D^\Omega \leq \bigoplus_{Q_\ell \in C_1} H_D^{Q_\ell} \tag{7.5.171}$$

By scaling, if Δ_L is a hypercube of size L,

$$N_{\Delta_L}(E) = N_\Delta(E/L^2), \quad \widetilde{N}_{\Delta_L}(E) = \widetilde{N}_\Delta(E/L^2) \tag{7.5.172}$$

Thus,

$$\sharp(C_1)\, N_\Delta(EL^{-2}) \leq N_\Omega(E) \leq \sharp(C_2)\, \widetilde{N}_\Delta(EL^{-2}) \tag{7.5.173}$$

Dividing by $E^{\nu/2}$ and taking $E \to \infty$

$$\frac{\tau_\nu}{(2\pi)^\nu}\, L^\nu\, \sharp(C_1) \leq \liminf E^{-\nu/2}\, N_\Omega(E)$$

$$\leq \limsup E^{-\nu/2}\, N_\Omega(E) \leq \frac{\tau_\nu}{(2\pi)^\nu}\, L^\nu \sharp(C_2) \tag{7.5.174}$$

Noting that $L^\nu \sharp(C_1) = \tau(C_1)$, and taking sup over C_1 and inf over C_2, we get the desired result. $\square$

We note that clearly

$$\sharp\{E \mid A \leq E_0\} = \sharp\{E \mid A - E_0 \leq 0\}$$

We see that looking at negative eigenvalues of $H_\Delta^D - E$ as $E \to \infty$ is the same as looking at $N_\Omega(E)$. Thus,

Proof of Theorem 7.5.30. Cover supp V by a finite family of "disjoint" hypercubes of side L, $\{\Delta_j\}_{j=1}^{m_L}$. Let $V_{(L)}^\pm$ be a function which is

$$V_{(L)}^\pm(x) = \begin{cases} \max\limits_{y \in \Delta_j} V(y) \ (\min) & \text{if } x \in \Delta_j \\ 0 & \text{if } x \in \mathbb{R}^\nu \setminus \bigcup_{j=1}^{m_L} \Delta_j \equiv X \end{cases} \tag{7.5.175}$$

Then, by Theorem 7.5.28 with $\Omega = \mathbb{R}^\nu$

$$H_D^X \oplus \bigoplus_{j=1}^{m_L} \big(H_D^{\Delta_j} + \lambda V_{(L)}^-\big) \leq -\Delta + \lambda V \leq H_N^X \oplus \bigoplus_{j=1}^{m_L} \big(H_N^{\Delta_j} + \lambda V_{(L)}^+\big) \tag{7.5.176}$$

This plus Theorem 7.5.29 for the Δ_j implies

$$\begin{aligned}\frac{\tau_\nu}{(2\pi)}\int (V^-)_-(y)\,d^\nu y \le \liminf_{\lambda\to\infty}\frac{N(\lambda V)}{\lambda^{\nu/2}} &\le \limsup\frac{N(\lambda V)}{\lambda^{\nu/2}} \\ &\le \frac{\tau_\nu}{(2\pi)^\nu}\int (V^+)_-(y)\,d^\nu y\end{aligned} \tag{7.5.177}$$

Taking L to zero, by continuity of V, the integrals converge to the right side of (7.5.160) completing the proof. □

Notes and Historical Remarks. In his original paper on self-adjointness, von Neumann [**720**] showed that if A is closed and symmetric and bounded from below and if $\alpha < \ell_q$, then $A^* \restriction [D(A)+\text{Ker}(A^*-\alpha)]$ is self-adjoint with lower bound α on its spectrum. That is, he proved there were self-adjoint extensions with lower bounds arbitrarily close to ℓ_q and he had (in case $\ell_q > 0$ and $\alpha = 0$) what we call the Krein extension (without its extremal property). For this reason, some people call this extension the *Krein–von Neumann extension.*

von Neumann conjectured that there were extensions with lower bound ℓ_q. This was proven by Stone [**670**] and Friedrichs [**208**]. Friedrichs' construction, especially, as explicated by Freudenthal [**207**], while not explained in form language, essentially closed the quadratic form and gave what we (and others) call the Friedrichs extension.

In understanding and elucidating this idea, the notion of quadratic form and the crucial representation theorem, Theorem 7.5.5, was developed in the ten years after 1954 by Kato [**375**], Lax–Milgram [**433**], Lions [**452**], and Nelson [**499**]. The KLMN theorem is named after these contributions.

The idea of factorizing operators $B = C^*D$, so to study $A+B$, one can look at $D(A-z)^{-1}C^*$, goes back to Schwartz [**611**]. In the context of the study of Schrödinger operators, while $(-\Delta - z)^{-1}V$ always blows up as an operator on L^2 as $z = x + i\varepsilon$ with $x > 0$ and $\varepsilon \downarrow 0$, if $V^{1/2} = V/|V|^{1/2}$, then $V^{1/2}(-\Delta - z)^{-1}|V|^{1/2}$ may be bounded and one-sided continuous in this limit. This becomes a powerful tool in scattering and spectral theory. Three high points in the development of this idea are Kato [**378**], Kato–Kuroda [**383**], and Agmon [**6**]. See Reed–Simon [**551**] for many examples of this notion.

In his book, Kato [**380**] noted that since uniformly bounded monotone sequences of operators have strong limits, monotone sequences of forms have strong resolvent limits. In the increasing case, he proved if the limit form is closable, it is already closed and is the form of the operator limit but, in his book, he missed that it was always closed. In the decreasing case, he noted if the limit form was closable, then $\bar{t}_\infty$ is the form of the limiting operators. Because it was raised by Kato's work, that t_∞ is closable in the increasing case was proven independently at about the same time by Davies [**147**], Kato (unpub), Robinson [**575**], and Simon [**639**].

Simon's proof used the $q_r + q_s$ decomposition (Theorem 7.5.15) which was invented in his paper. This paper also had the full monotone decreasing theorem. Kato's proof involved the discovery that a form is closed if and only if it is lsc. This theme was explored by Simon [**637**] who, in particular, proved if T_n, T_∞ are positive self-adjoint operators so that $T_n \to T_\infty$ in the strong resolvent sense, then their forms are related by

$$t_\infty(\varphi) = \liminf_{\psi_n \to \varphi} t_n(\psi_n) \tag{7.5.178}$$

Shortly before Simon, Ando [**18**] had a decomposition of operators, not unrelated but different from that of Simon. See Hassi et al. [**298**] for a discussion and development of both decompositions.

The extension theory of forms is associated with work of Birman, Krein, and Vishik. Krein [**410**] found his extension and proved that the positive extensions are exactly all operators between the Krein and Friedrichs extensions. Vishik [**716**] found the formula (7.5.120), although in terms of operators rather than forms, and Birman [**58**] codified and extended the earlier work. Grubb [**270, 271, 272**] exploited and extended the work to sectorial forms. Alonso–Simon [**14**] rephrased everything in the language of forms (rather than positive operators) and reviewed and popularized the work. We follow their approach here. In particular, they present the finite rank results (mentioned in the text) which go back to Krein. They, in turn, have an approach with much in common with Faris [**194**].

Mark Grigorievich Krein (1907–1989) was a Jewish Ukranian mathematician who we've seen many times besides his work on extensions of quadratic forms discussed here: the Kakutani–Krein Theorem (Theorem 2.5.3 of Part 1), his condition for indeterminate moment problems (Theorem 5.6.8 of Part 1), the Krein–Milman Theorem (Theorem 5.11.1 of Part 1), his book with Goh'berg [**257**] on compact operators on a Hilbert space, the Aronszajn–Krein formula ((5.8.14)), the Krein spectral shift (Theorem 5.8.8 and the Krein factorization (Theorem 6.9.21). And if we went further into spectral theory, or othogonal polynomials, his name would appear many more times.

Krein suffered from serious antisemitism throughout his career. At age seventeen, he left his hometown of Kiev to go to Odessa where he spent the rest of his life except for an interlude during the Second World War (see below). He was allowed to work for a doctor's degree under Nikolai Chebotaryov (1894–1944), and he became a faculty member at Odessa State University. Working sometimes with his friend Naum Akhiezer (1901–1980), in the 1930s, he made Odessa a center for research in functional analysis, especially the moment problem and associated theory of orthogonal polynomials (both OPRL and OPUC).

In 1941, he fled Odessa as part of an evacuation at the approach of the German army. Soon after his return to Odessa in 1944, he was accused of Jewish nationalism presumably because so many of his students in the 1930s were Jewish. He was dismissed from his university post and spent the remainder of his career at various institutes and ran a lively seminar in his home. With the help of other mathematicians, he managed to have students as far flung as Kiev, Karkov, and Leningrad.

He received the 1982 Wolf Prize. His students include V. Adamyan, Y. Berezansky, I. Glazman, I. Gohberg, D. Milman, M. Naimark, G. Popov, and L. Sakhnovich.

There is an enormous literature on Dirichlet and Neumann Laplacians. Three books that include discussion of those boundary values and often others (e.g., Robin boundary condition, i.e., $\frac{\partial u(x)}{\partial n} + h(x)u(x) = 0$ on $\partial\Omega$ where h is a function on $\partial\Omega$, and also the Krein extension) are Edmunds–Evans [**182**], Grubb [**273**], and Schmüdgen [**604**]. Ashbaugh et al. [**32**] review and extend some of the literature on the Krein extension of $-\Delta \restriction C_0^\infty(\Omega)$.

When $\partial\Omega$ is regular enough (e.g., C^2), one can prove that $\{u \in L^2(\Omega) \mid u \text{ is } C^2 \text{ with } D^\alpha u, |\alpha| \leq 2 \text{ continuous up to } \partial\Omega \text{ and } u \restriction \partial\Omega = 0\}$ is an operator core for H_D^Ω (with $Hu = -\Delta u$, of course) and also that $\{u \in L^2(\Omega) \mid u \text{ is } C^2 \text{ with } D^\alpha u, |\alpha| \leq 2 \text{ continuous up to } \partial\Omega \text{ and } \frac{\partial u}{\partial n} \restriction \partial\Omega = 0\}$ is an operator core for H_N^Ω. These results are discussed, for example, in Schmüdgen [**604**]. (7.5.148) is fundamental to the probabilistic approach to H_D^Ω; indeed, it is the definition used by probabilists. That it agrees with the form definition with no restriction on Ω is a result of Simon [**638**]; see Problem 12.

There is another equality that always holds: namely, in case Ω is bounded, we defined a Dirichlet Green's function, $G_\Omega(x,y)$, in Theorem 3.6.17 of Part 3. It obeys $-\Delta_x G_\Omega = \delta(x-y)$ and $\lim_{x_n \to x_\infty \in \partial\Omega} G_\Omega(x_n, y) = 0$ for each y and quasi-every $x_\infty \in \partial\Omega$. The result is that for any bounded Ω, $G_\Omega(x,y)$ is the integral kernel of the operator $(H_D^\Omega)^{-1}$ (since Ω is bounded, $H_D^\Omega \geq c > 0$ for some c so H_D^Ω is invertible). One can prove this as follows:

(1) Prove it for Ω's with smooth $\partial\Omega$ by classical methods;
(2) Find open $\Omega_n \subset \Omega_{n+1} \subset \dots$ with $\overline{\Omega}_n \subset \Omega_{n+1}$ and with smooth boundaries so $\Omega = \cup\Omega_n$;
(3) Use the monotone convergence theorem for forms to prove that in the strong operator topology, $\big(H_D^{\Omega_n}\big)^{-1} \to \big(H_D^\Omega\big)^{-1}$ (if we set $\big(H_D^{\Omega_n}\big)^{-1} \equiv 0$ on $L^2(\Omega \setminus \Omega_n)$);

(4) Extend the ideas of Corollary 3.8.9 of Part 3 to $\mathbb{R}^\nu$ to show $G_\Omega(x,y) = \sup_n G_{\Omega_n}(x,y)$.

When $\partial\Omega$ is nice, there is a way to define reflecting Brownian motion and prove an analog of (7.5.137) where the left side has H_N^Ω in place of H_D^Ω and the right side has reflecting Brownian motion—see Williams [**759**] and papers that he refers to. In the next section we'll discuss diamagnetic inequalities for $-\Delta + V$ on $\mathbb{R}^\nu$. For analogs on the difference of Neumann and Dirichlet semigroups, see Hundertmark–Simon [**327**].

From (7.5.148), one sees $|(e^{-\frac{1}{2}tH_D^\Omega})(x,y)| \le (2\pi t)^{-\nu/2}$ so if $|\Omega| < \infty$, $e^{-\frac{1}{2}tH_D^\Omega}$ is Hilbert–Schmidt for all t and so trace class for all t. In particular, H_D^Ω has discrete spectrum. There are some Ω's with $|\Omega| = \infty$ but still H_D^Ω has discrete spectrum because Ω is thin at infinity. The simplest example is $\{(x_1,x_2) \in \mathbb{R}^2 \mid |x_1x_2| \le 1\}$. This phenomenon was discovered by Rellich [**557**] who had this example. It is further discussed in Maz'ya [**471**], Wang–Wu [**732**], and Simon [**652**].

Remarkably, there are bounded Ω's for which H_N^Ω does not have discrete spectrum. The phenomenon called "rooms and passages," since it has larger sets connected by narrow passages (so there are infinitely many orthogonal) low-energy states living in not much more than a single room) is discussed in Courant–Hilbert [**135**] and further in Hempel et al. [**313**]. There is even a Neumann Laplacian with Ω bounded so that H_N^Ω has some absolutely continuous spectrum (!); see Simon [**644**].

The theory of Jordan content has been surplanted by the Lebesgue theory so it is primarily of historical interest. However, it turns out to be precisely what is needed for the bracketing approach to Weyl's theorem.

Jordan content goes back to Peano [**513**] and Jordan [**354**] and is therefore sometimes called Peano–Jordan measure. It was known that in one dimension, a set, S, is Jordan contented if and only if its characteristic function is Riemann integrable. The theorem that the inner content is the Lebesgue measure of S^{int} and the outer content is the Lebesgue measure of $\overline{S}$ is due to Frink [**210**]. Three books with more on Jordan content are [**136, 400, 505**]. For some discussion of the historical context of Jordan content, see Shenitzer–Steprans [**622**].

Lord Rayleigh, already in his 1877 book [**547**], noted that one could compute the asymptotics of the number of Dirichlet or Neumann Laplacians in parallelepipeds and that the leading term only depended on the volume. In 1910, Lorentz [**459**] and Sommerfeld [**661**] independently conjectured that the asymptotic number of eigenvalues of $-\Delta$ should, to leading order, only depend on the volume of Ω. They were interested in this question because of the physics of black-body radiation. Lorentz' paper was notes (taken by Max Born) of a lecture series given in the Göttingen Math Institute by Lorentz under the title *Old and New Problems of Physics.*

Interestingly enough, the lectures were funded by the Göttingen academy. They administered money set aside to fund a prize for the proof of Fermat's Last Theorem (eventually awarded to Andrew Wiles). The income from the prize fund was used for lecture series given over the years by Poincaré, Einstein, Planck, and Bohr. There is a story that Hilbert remarked that this conjecture would not be proven in his lifetime. Ironically, it was proven a year later by his student, using in part, the method of integral equations that he had pioneered.

Weyl's results were announced by Hilbert to the Göttingen academy [**744**] with several fuller papers [**745, 746, 747**]. He basically used Dirichlet decoupling (and a form of bracketing). Of course, he didn't use quadratic forms on Hilbert space but rather looked at a Dirichlet Green's function, noting the imposing of an extra Dirichlet boundary decoupled and resulted in a smaller operator (i.e., instead of $H_D^{\Omega_1} \oplus H_D^{\Omega_2} \geq H_D^{\Omega_1 \cup \Omega_2}$, he proved and used $(H_D^{\Omega_1 \cup \Omega_2})^{-1} \leq (H_D^{\Omega_1})^{-1} \oplus (H_D^{\Omega_2})^{-1}$). Interestingly enough, in reminiscences on this written forty years later, Weyl [**752**] said that the hard step was to realize and then prove that all the eigenvalues increased if one computed $A \leq B$. To do this, he discovered a min–max principle of the form that Fischer [**202**] had found in 1905.

In 1913, Weyl [**747**] conjectured for regular boundaries, one has that (for $\frac{D}{N}$ boundary conditions)

$$N(\lambda) = (2\pi)^{-\nu}\lambda^{\nu/2}|\Omega| \mp \frac{1}{n}(2\pi)^{-\nu+1}\lambda^{(\nu-1)/2} \text{ area } (\partial\Omega) + o(\lambda^{(\nu-1)}) \tag{7.5.179}$$

Modulo a technical condition, this was proven by Ivriĭ [**336, 337**] in 1984.

Other methods used to prove Weyl's leading asymptotics include heat kernel ideas. By using (7.5.148), it is easy to prove that

$$\operatorname{Tr}\big(e^{-\frac{1}{2}tH_D^\Omega}\big) = (2\pi t)^{-\nu/2}\,|\Omega|\big(1 + o(1)\big) \tag{7.5.180}$$

Tauberian theorems of the type discussed in Section 6.9 of Part 3 (see Theorem 6.9.4 of that part) then lead to a Weyl law. This is discussed in many places, for example, Simon [**647**] and Problem 18.

For the example of the Dirichlet Laplacian on $\{(x_1, x_2) \in \mathbb{R} \mid |x_1 x_2| \leq 1\}$ where $|\Omega|$ is infinite, Simon [**642**] proved that

$$N(\lambda) = \pi^{-1}\lambda^{-1}\log(\lambda)\big(1 + o(1)\big) \tag{7.5.181}$$

see also Robert [**574**] and Tamura [**695**]. For a readable account of the ideas and history of Weyl's theorem and some of its extensions, see Arendt et al. [**20**].

Weyl's law raises the general issue of what the eigenvalue spectrum of H_D^Ω says about D. Weyl's law implies that this spectrum determines $|\Omega|$. The general question was emphasized most cogently in a 1966 semi-popular article of Mark Kac [**359**] entitled *Can you hear the shape of a drum* which asked if two regions of the plane have the same spectrum, are the two regions isometric. Already before Kac, Milnor [**479**] had proven there were two sixteen-dimension non-isometric tori with the same spectrum (so you can't hear the shape of an arbitrary dimension compact manifold, possibly with boundary). Eventually, Gordon et al. [**265**], using in part ideas of Sunada [**676**], found non-isometric regions in $\mathbb{R}^2$ whose eigenvalue spectra are the same. But the question of what geometric information is encoded in this spectrum has been an enduring theme. See Benguria [**51**] for a review of the idea that if H_D^Ω has the spectrum of a ball in $\mathbb{R}^\nu$, it is a ball and Chavel [**115**] for a book on the subject.

Zworski [**776**] and Nonnenmacher et al. [**507**] have Weyl-type asymptotic results in other situations.

Weyl's result is connected to

$$N_\nu(\lambda) = \sharp\{(j_1, \dots j_\nu) \in \mathbb{Z}^\nu \mid |j| \le \lambda\} \tag{7.5.182}$$

As an object of obvious number-theoretic interest (recall Section 13.1 of Part 2B discussed $r_2(n) = \sharp\{(j_1, j_2) \in \mathbb{Z}^\nu \mid |j|^2 = n\}$ so $N_2(n^2) - \sum_{j=0}^n r_2(j)$), there has been a lot of interest in asymptotics of $N_\nu(\lambda)$, particularly $N_2(\lambda)$, beyond the leading $\tau_\nu \lambda^\nu$. This is true especially since, as originally proven by Sierpinski [**626**] in 1906, the error is less than the naive expectation of $\sigma_\nu \lambda^{\nu-1}$ (i.e., surface area); in $\nu = 2$, Sierpinski showed that the error is bounded by $\lambda^{2/3}$ (Note that much of the literature looks at $t = \lambda^2$ so the naive error is $t^{1/2}$ and Sierpinski is $t^{1/3}$).

Hardy [**289**] and Landau [**424**] showed that an $O(\lambda^{1/2})$ (aka $O(t^{1/4})$) upper bound does not hold and the belief is that the optimal upper bound is $O(t^{1/4+\varepsilon})$ for all $\varepsilon > 0$. Despite many improvements, the best current value (see Huxley [**329**]) is still over 0.3. Landau [**425**] has bounds better than $O(\lambda^{\nu-1})$ for general ν.

This refined error is *not* relevant for counting Dirichlet or Neumann boundary values since these have $j_k \ge 0$ or $j_k > 0$ restrictions and given that the error for $N_\nu(\lambda)$ is $O(\lambda^{\nu-1})$, it is easy to see that the errors with the $j_k \ge 0$ or $j_k > 0$ conditions are $O(\lambda^{\nu-1})$. However, if one looks at a torus, one is looking exactly at $N_\nu(\lambda)$.

The idea of using Weyl's ideas to study the classical limit of the number of eigenvalues of $-\Delta + V$ is due to Birman–Borzov [**63**], Martin [**468**], Robinson [**575**], and Tamura [**694**].

The extension of the polar decomposition to arbitrary closed operators, including that A closed implies A^*A is self-adjoint (as discussed in Problems 19 and 20) was in von Neumann's initial paper [**724**] on the polar decomposition.

Problems

1. (a) Let A be a positive self-adjoint operator and $\mathcal{H}_{+1}$ its form domain with the norm $\|\cdot\|_{+1}$. Let $\sqrt{A}$ be given by the functional calculus. Prove that $\mathcal{H}_{+1} = D(\sqrt{A})$ and that $\|\cdot\|_{+1}$ is equivalent to the graph norm on the graph of $\sqrt{A}$.

   (b) Prove that $\mathcal{D} \subset Q(A)$ is a form core for A if and only if it is an operator core for $\sqrt{A}$.

2. (a) Let A be a positive self-adjoint operator and B an A-bounded symmetric operator. Let $e > 0$. Prove that $B(A+e)^{-1}$ is a bounded operator and that $(A+e)^{-1}B$ can be viewed as its adjoint.

   (b) By consideration of the function $z \mapsto (A+e)^{-z} B(A+e)^{-(1-z)}$ on $\{z \mid 0 \le \operatorname{Re} z \le 1\}$ and the maximum principle for analytic functions (see Theorem 5.2.1 of Part 2A), prove that

$$\|(A+e)^{-1/2} B(A+e)^{-1/2}\| \le \|B(A+e)^{-1}\| \tag{7.5.183}$$

   (c) Prove that B is A-form bounded with relative form bound no larger than the relative operator bound.

   Remark. The same proof shows that if A and B are positive self-adjoint operators

$$0 \le B \le A, \quad 0 \le \alpha \le 1 \Rightarrow B^\alpha \le A^\alpha \tag{7.5.184}$$

   This is a special case of a result of Loewner [**454**] in 1934 who determined all f's with $0 \le B \le A \Rightarrow f(B) \le f(A)$ and showed $f(x) = \lambda^\alpha$ has this property if and only if $0 \le \alpha \le 1$. The special case (7.5.184) was rediscovered by Heinz 15 years later [**307**] so this result is sometimes called the *Heinz–Loewner theorem*. For more on Loewner's result and its proof, see Donoghue [**167**] or Simon [**654**].

3. (a) Write the analog of (7.5.41) for $\varphi(x)$.

   (b) Prove that, if $\varphi \in Q(-d^2/dx^2)$, then for $|x-y| \le 1$

$$|\varphi(x) - \varphi(y)| \le C\,\|\varphi\|_{+1}\,|x-y|^{1/2} \tag{7.5.185}$$

   and, in particular, that φ is continuous.

   Remark. This is of course a Sobolev estimate.

4. By applying (7.5.43) to $\varphi(\lambda x)$ for $\lambda > 0$, prove that $\delta(x)$ has relative bound zero.

5. Verify that the set described after (7.5.45) is the domain of the operator of (7.5.45).

6. By integration by parts, verify the form domains and formula for the forms in (7.5.62) and (7.5.63).

7. Let X be a compact Hausdorff space with probability measure μ and $d\nu = f\,d\mu + d\nu_s$ a second positive measure with indicated Lebesgue decomposition. Let q be the quadratic form with $D(q) = C(X)$ and $q(f) = \int |\varphi(x)|^2\,d\nu$ on $\mathcal{H} = L^2(X, d\mu)$.

   (a) Prove that ν is closable if and only if $\nu_s = 0$. (*Hint*: If K is compact with $\mu(K) = 0$ and $\nu_s(K) > 0$, let $0 \le \varphi_n \le 1$ with $\varphi_n \in C(X)$ and $\varphi_n \downarrow \chi_K$.)

   (b) Prove $q_r(\varphi) = \int f(x)\,|\varphi(x)|^2\,d\mu(x)$.

The next three problems deal with Example 7.5.25

8. Prove that the operator A of (7.5.145) with domain (7.5.144) is Hermitian with deficiency indices $(1,1)$.

9. Show that the self-adjoint extensions of the operator, A, of Problem 8 are given by (7.5.146).

10. Prove that for $\theta \neq \infty$, A_θ has a negative eigenvalue and conclude that A_∞ is the unique nonnegative self-adjoint extension.

11. Let $\Omega \subset \mathbb{R}^\nu$ be any open set. Define

$$D(q) = \{\varphi \in L^2(\Omega, d^\nu x) \mid \vec{\nabla}\varphi \in L^2(\Omega, d^\nu x)\} \tag{7.5.186}$$

and

$$q(\varphi) = \int |\vec{\nabla}\varphi(x)|^2\,d^\nu x \tag{7.5.187}$$

Prove that q is a closed form (here $\vec{\nabla}$ is a distributional derivative).

12. This problem which will prove (7.5.148) assumes familiarity with Brownian motion as discussed in Section 4.16 of Part 1. Let $\Omega \subset \mathbb{R}^\nu$ be open. Pick bounded open sets Ω_j with $\overline{\Omega}_j \subset \Omega_{j+1}$ and $\bigcup_{j=1}^\infty \Omega_j = \Omega$ and function $f_j \in C_0^\infty(\mathbb{R}^\nu)$ so $0 \le f_j \le 1$, $f_j \equiv 1$ on Ω_j and supp $f_j \subset \Omega_{j+1}$. Define for $x \in \Omega$

$$V(x) = \sum_{j=1}^\infty (\nabla f_j)^2 + \left[\operatorname{dist}(x, \mathbb{R}^\nu \setminus \Omega)\right]^{-3} \tag{7.5.188}$$

(a) Let $\varphi \in L^2(\Omega, d^\nu x)$, $\vec{\nabla}\varphi \in L^2(\Omega, d^\nu x)$ and $\varphi \in Q(V)$. Prove that $\varphi \in Q(H_D^\Omega)$. (*Hint*: Prove that $\nabla(f_j\varphi) \to \nabla\varphi$ in L^2.)

(b) For each $\lambda > 0$, prove that

$$q_\lambda(\varphi) = \|\vec{\nabla}\varphi\|_{L^2}^2 + \lambda \int V(x)\varphi(x) d^\nu x \tag{7.5.189}$$

on $Q(V) \cap Q(H_N^\Omega)$ is a closed quadratic form on $L^2(\Omega, d^3x)$. Let $H(\lambda)$ be the operator it defines.

(c) Prove that $C_0^\infty(\Omega) \subset Q(V) \cap Q(H_N^\Omega)$ and that $\lim_{\lambda\downarrow 0} q_{H(\lambda)}(\varphi) = q_{H_D^\Omega}(\varphi)$ for $\varphi \in C_0^\infty(\Omega)$.

(d) Prove that as $\lambda \downarrow 0$, $e^{-tH(\lambda)} \to e^{-tH_D^\Omega}$ strongly on $L^2(\Omega, d^\nu x)$. (*Hint*: Monotone convergence and (a).)

(e) Define V_N on $\mathbb{R}^\nu$ by

$$V_N(x) = \begin{cases} V(x) & \text{if } x \in \Omega \text{ and } V(x) \le N \\ N & \text{if } x \notin \Omega \text{ or } V(x) > N \end{cases}$$

Prove that for $\varphi \in L^2(\Omega, d^\nu x)$, $\exp(-t(-\Delta + \lambda V_n))\varphi \to \exp(-tH(\lambda))\varphi$ in L^2 as $N \to \infty$. (*Hint*: Monotone convergence for forms.)

(f) For each $x \in \Omega$ and a.e. Brownian path, $b(s)$, prove that $\int_0^t V(b(s))\, ds < \infty \Rightarrow b(s) \in \Omega$ for all $s \in [0, t]$. (*Hint*: b is Hölder continuous of order $1/3$ and V has the $\text{dist}(\dots)^{-3}$ term.)

(g) Prove (7.5.148).

Remark. The above proof is from Simon [**638**] (see also Simon [**647**]).

13. (a) Let C be a nowhere dense set in $[0, 1]$. Prove that its inner content is 0.

(b) If C is a closed subset of $[0, 1]$, prove that its outer content is $1 - \sum_{j=1}^J (b_j - a_j)$ where $[0, 1] \setminus C = \bigcup_{j=1}^J (a_j, b_j)$ is the decomposition to connected components.

(c) If C is a positive measure Cantor set (see Problem 4 in Section 4.2 of Part 1) show that C is not contented.

14. If Ω is any bounded set in $\mathbb{R}^\nu$ and

$$\Omega_\varepsilon^- = \{x \in \Omega \mid \text{dist}(x, \mathbb{R}^\nu \setminus \Omega) > \varepsilon\}$$
$$\Omega_\varepsilon^+ = \{x \in \mathbb{R}^\nu \mid \text{dist}(x, \Omega) < \varepsilon\}$$

prove that $\tau_{\text{inn}}(\Omega) \ge |\Omega_\varepsilon^-|$ and $\tau_{\text{out}}(\Omega) \le |\Omega_\varepsilon^+|$ (where $|\cdot|$ is Lebesgue measure). Conclude if $\lim_{\varepsilon\downarrow 0} |\Omega_\varepsilon^+ \setminus \Omega_\varepsilon^-| = 0$, then Ω is contented.

15. (a) $N(A) = \dim P_{(-\infty,0)}(A)$ for a self-adjoint operator, A, and $C \geq A+B$, prove that $n(C) \leq n(A) + n(B)$.

    (b) Knowing (7.5.160) for $V \in C_0^\infty$ and a CLR bound, prove (7.5.160) for all $V \in L^{\nu/2}$. (*Hint*: Write $-\Delta - V \geq \bigl(-\varepsilon\Delta - |V-W|\bigr) + \bigl(-(1-\varepsilon)\Delta + W\bigr)$ and reverse the role of V and W.)

    Remark. This argument is from Simon [**634**].

16. By mimicking the proof of (7.5.160), prove a quasiclassical limit result for $\lim_{\lambda\to\infty}\bigl[\sum_{j=1}^{J}|e_j\bigl(-\Delta + \lambda V\bigr)|^p\bigr]/\lambda^{p+\nu/2}$ where $e_j(\lambda)$ are the negative eigenvalues of $-\Delta + \lambda V$ and V is a continuous function of compact support.

17. Prove completeness of $\{\cos(\pi n x)\}_{n=0}^\infty$ and of $\{\sin(\pi n x)\}_{n=1}^\infty$ on $L^2\bigl([0,1], dx\bigr)$. (*Hint*: Use the fact that $\{e^{in\pi x}\}_{n=-\infty}^\infty$ is complete for $L^2\bigl([-1,1], dx\bigr)$ and consider even and odd functions.)

18. Let Ω be an arbitrary bounded open region in $\mathbb{R}^\nu$. Let $P_t^\Omega(x,y)$ be the integral kernel of $e^{-\frac{1}{2}tH_D^\Omega}$.

    (a) Prove that for all x, t

    $$P_t^\Omega(x,x) \leq (2\pi\, t)^{-\nu/2} \tag{7.5.190}$$

    (b) For each x, let Δ_x be a cube with x at its center and $\overline{\Delta}_x \subset \Omega$. Prove that as $t \downarrow 0$,

    $$(2\pi\, t)^{\nu/2}\, P_t^{\Delta_x}(x,x) \to 1 \tag{7.5.191}$$

    (*Hint*: Method of images.)

    (c) Prove that

    $$P_t^{\Delta_x}(x,x) \leq P_t^\Omega(x,x) \tag{7.5.192}$$

    (d) Conclude that $(2\pi\, t)^{\nu/2}\, P_t^\Omega(x,x) \to 1$ for each x and deduce that

    $$\lim_{t\downarrow 0}\, (2\pi\, t)^{\nu/2}\, \mathrm{Tr}\, \bigl(e^{-\frac{1}{2}tH_D^\Omega}\bigr) = 1 \tag{7.5.193}$$

    (e) Deduce Weyl's law, (7.5.158). (*Hint*: Theorem 6.9.4 of Part 3.)

    Remark. This proof doesn't require that Ω is contented and it works for unbounded Ω so long as $|\Omega| < \infty$.

19. (a) Let A be a closed densely defined operator on a Hilbert space, $\mathcal{H}$. Let q be given by

    $$q(\varphi) = \begin{cases} \|A\varphi\|^2 & \text{if } \varphi \in D(A) \\ \infty & \text{if } \varphi \notin D(A) \end{cases}$$

Prove that q is a closed quadratic form and let B be the associated positive, self-adjoint operator.

(b) By looking at the construction of B, show that

$$D(B) = \{\varphi \in D(A) \mid A\varphi \in D(A^*)\} \tag{7.5.194}$$

$$B\varphi = A^*(A\varphi) \tag{7.5.195}$$

so that A^*A can be viewed as a positive self-adjoint operator.

20. This problem provides a slick proof of the results of the last problem without using forms. A is a closed, densely defined, operator on a Hilbert space, $\mathcal{H}$.

(a) On $\mathcal{H} \oplus \mathcal{H}$, let C be the operator with $D(C) = \{(\varphi, \eta) \mid \varphi \in D(A), \eta \in D(A^*)\}$ and $C(\varphi, \eta) = (A^*\eta, A\varphi)$. In matrix language,

$$C = \begin{pmatrix} 0 & A^* \\ A & 0 \end{pmatrix} \tag{7.5.196}$$

Prove that C is self-adjoint.

(b) For any self-adjoint operator, X, prove that X^2 defined by the spectral theorem has $D(X^2) = \{\varphi \in D(X) \mid X\varphi \in D(X)\}$ and $X^2\varphi = X(X\varphi)$.

(c) Apply (b) to $X = C$ to conclude that the operator given by (7.5.194)/(7.5.195) is self-adjoint, so A^*A can be viewed as a positive self-adjoint operator.

21. (a) Let A be a closed, densely defined, operator on a Hilbert space, $\mathcal{H}$. Define $|A| = \sqrt{A^*A}$ where A^*A is defined as a positive self-adjoint operator by either of the last two problems. Prove that there is a partial isometry, U, so that

$$U\,|A| = A \tag{7.5.197}$$

(b) Prove that if we also require

$$\operatorname{Ker}(U)^\perp = \operatorname{Ker}(A)^\perp \tag{7.5.198}$$

then U and $|A|$ are uniquely determined by (7.5.197), (7.5.198) and $|A|$ is a positive self-adjoint operator. Thus, we have extended the *polar decomposition* of Theorem 2.4.8 to unbounded operators.

7.6. Pointwise Positivity and Semigroup Methods

While the analysis of some examples has relied on specifics of the underlying space, the general methods so far have involved arbitrary Hilbert spaces without structure beyond inner products. In this section, we'll discuss spaces of the form $L^2(M, d\mu)$, where (M, Σ, μ) is a σ-finite measure space. We will thus have the notion of pointwise positivity, that is, f's with $f(x) \geq 0$ for a.e. x. One of our main goals will be to prove:

Theorem 7.6.1 (Kato's L^2_{loc} Theorem). *Let $\mathcal{H} = L^2(\mathbb{R}^\nu, d^\nu x)$ and $V \in L^2_{\mathrm{loc}}(\mathbb{R}^\nu)$ with*

$$V(x) \geq 0 \tag{7.6.1}$$

for a.e. x. Then $-\Delta + V$ is essentially self-adjoint on $C_0^\infty(\mathbb{R}^\nu)$ and the domain of the closure is

$$\{\varphi \in L^2(\mathbb{R}^\nu, d^\nu x) \mid \nabla\varphi \in L^2(\mathbb{R}^\nu, d^\nu x),\ -\Delta\varphi + V\varphi \in L^2(\mathbb{R}^\nu, d^\nu x)\} \tag{7.6.2}$$

Remarks. 1. In (7.6.2), $\nabla\varphi$ and $-\Delta\varphi + V\varphi$ are intended in distributional sense, since $V \in L^2_{\mathrm{loc}}$, $\varphi \in L^2$, we have $V\varphi \in L^1_{\mathrm{loc}}$, and so it defines an ordinary distribution.

2. Positivity (or some lower bound on V) is critical. If $\nu \geq 5$, we've seen (see Example 7.4.26) that if β is large enough, $-\Delta - \beta|x|^{-2}$ is not essentially self-adjoint on $C_0^\infty(\mathbb{R}^\nu)$ but $|x|^{-2} \in L^2_{\mathrm{loc}}(\mathbb{R}^\nu)$. There are extensions that allow $V(x) \geq 0$ to be replaced, for example, by $V(x) \geq -d|x|^2$ for some d; see the Notes.

There is a form analog of Theorem 7.6.1:

Theorem 7.6.2 (Simon's L^1_{loc} Theorem). *Let $\mathcal{H} = L^2(\mathbb{R}^\nu, d^\nu x)$ and $V \in L^1_{\mathrm{loc}}(\mathbb{R}^\nu)$ with (7.6.1) for a.e. x. Let*

$$H = -\Delta + V \tag{7.6.3}$$

be the form sum, that is, the self-adjoint operator with form

$$q_H(\varphi) = \|\nabla\varphi\|^2 + \|V^{1/2}\varphi\|^2 \tag{7.6.4}$$

defined on the set of φ with $\nabla\varphi \in L^2$ and $V^{1/2}\varphi \in L^2$. Then $C_0^\infty(\mathbb{R}^\nu)$ is a form core for H.

Remarks. 1. It follows from Theorem 7.5.11 that (7.6.4) defines a closed quadratic form.

2. These theorems are analogs; the L^p_{loc} conditions are optimal if we want $-\Delta + V$ to be defined as operators (respectively, forms) on $C_0^\infty(\mathbb{R}^\nu)$.

The key to our proofs is the notion of positivity-preserving operators defined by the following:

Definition. A bounded operator, C, on an $L^2(M, d\mu)$ is called *positivity-preserving* if and only if

$$\varphi \geq 0 \Rightarrow C\varphi \geq 0 \tag{7.6.5}$$

(where ≥ 0 means pointwise a.e.).

If $\varphi \geq 0$, $\varphi \not\equiv 0 \Rightarrow C\varphi > 0$, we say C is *positivity improving.* If C has an integral kernel $c(x,y)$, $c(x,y) \geq 0$ for a.e. $x,y \Leftrightarrow C$ is positivity preserving while $c(x,y) > 0$ for a.e. $x,y \Leftrightarrow c$ is positivity improving. Below we'll focus on the positivity-preserving case although it is often useful to consider the stronger condition.

We'll first find conditions in terms of a positive self-adjoint operator, A, so that e^{-tA} is positivity-preserving for all $t > 0$ and then for two positive self-adjoint operators to obey

$$|e^{-tB}\varphi| \leq e^{-tA}|\varphi| \tag{7.6.6}$$

We'll use this to prove for $V \geq 0$ the form sum $-\Delta + V = H$ obeys (7.6.6) pointwise with $B = H$ and $A = -\Delta$. Since e^{-tA} has a bounded integral kernel, this will imply $L^\infty \cap Q(H)$ (respectively, $L^\infty \cap D(H)$) is a form (respectively, operator) core for H.

Throughout, we'll restrict to positive operators, but note that with minor changes, $A \geq 0$ can be replaced by $A + c \geq 0$ for some $c > 0$.

We'll illustrate (7.6.6) by also proving diamagnetic inequalities, that (7.6.6) holds (with $V, \vec{a}$ suitable) for $A = -\Delta + V$, $B = (i\vec{\nabla} - \vec{a})^2 + V$. This inequality can also be used to extend Theorems 7.6.1 and 7.6.2 to allow magnetic fields (i.e., $\vec{a} \neq 0$); we'll discuss this in the Notes and Problems.

Because e^{-tA} plays a major role, these are sometimes called "semigroup methods." Historically, these results and the realization of the importance of pointwise positivity was different from the path we take here. First, pointwise positivity was originally developed in probability theory and in proving that certain eigenvalues of certain operators are simple ("nondegenerate ground states")—a subject we limit to the Notes.

Second, the first proof of Theorem 7.4.1 used Kato's inequality but not via its connection to pointwise inequality and semigroups, but directly using a calculation and direct proof that $(H+1)[C_0^\infty]$ is dense in $L^2(\mathbb{R}^\nu, d^\nu x)$. We discuss this in Problems 1 and 2.

Third, the first proofs of positivity-preserving of $e^{+t\Delta}$ and (7.6.6) for $A = -\Delta$, $B = -\Delta + V$ rely on the Trotter product formula that for X, Y positive self-adjoint with suitable assumptions on X and Y, one has

$$e^{-t(X+Y)} = \operatorname*{s-lim}_{n\to\infty} (e^{-tX/n} e^{-tY/n})^n \tag{7.6.7}$$

We end this section with a proof of this fact (not the optimal result, which we discuss in the Notes) and use it to prove the Feynman–Kac formula.

We begin with a preliminary on positivity-preserving operators:

Proposition 7.6.3. *Let C be a bounded operator on $L^2(M, d\mu)$. Then C is positivity-preserving if and only if for all $\varphi \in L^2$,*

$$|C\varphi(x)| \le (C|\varphi|)(x) \tag{7.6.8}$$

Proof. If (7.6.8) holds, if $\varphi \ge 0$, then $C\varphi = C|\varphi|$ is nonnegative.

Conversely, if C is positivity-preserving, it takes real-valued functions into real-valued functions, since any real-valued function is a difference of two positive functions. Moreover, if φ is real-valued, $|\varphi| \pm \varphi \ge 0 \Rightarrow C|\varphi| \pm C\varphi \ge 0 \Rightarrow$ (7.6.8).

For general complex-valued φ and each $\omega \in \partial\mathbb{D}$,

$$|\mathrm{Re}(\omega(C\varphi))| = |C(\mathrm{Re}(\omega\varphi))| \le C|\varphi| \tag{7.6.9}$$

since $|\varphi| \pm \mathrm{Re}(\omega\varphi) \ge 0$. If $\{\omega_n\}_{n=1}^\infty$ is a countable dense set in $\partial\mathbb{D}$,

$$\sup_n |\mathrm{Re}(\omega_n \eta)| = |\eta| \tag{7.6.10}$$

so (7.6.9) for a.e. x implies (7.6.8). $\square$

One of the two main technical results of this section is:

Theorem 7.6.4. *Let $A \ge 0$ be a positive operator on $L^2(M, d\mu)$. Then the following are equivalent:*

(1) e^{-tA} *is positivity-preserving for all $t > 0$.*
(2) $(A+\gamma)^{-1}$ *is positivity-preserving for all $\gamma > 0$.*
(3) (Beurling–Deny criterion) *For all $\varphi \in L^2$,*

$$q_A(|\varphi|) \le q_A(\varphi) \tag{7.6.11}$$

(4) (Kato's inequality) *For all $\varphi \in D(A)$ and $\psi \ge 0$ in $Q(A)$, we have that $|\varphi| \in Q(A)$, and*

$$\mathrm{Re}\langle A^{1/2}\psi, A^{1/2}|\varphi|\rangle \le \mathrm{Re}\langle \overline{\mathrm{sgn}(\varphi)}\,\psi, A\varphi\rangle \tag{7.6.12}$$

where

$$\mathrm{sgn}(\varphi)(x) = \begin{cases} \overline{\varphi(x)}/|\varphi(x)| & \text{if } \varphi(x) \ne 0 \\ 0 & \text{if } \varphi(x) = 0 \end{cases} \tag{7.6.13}$$

Remarks. 1. In (7.6.12), the right side has $A\varphi$ in operator sense. The left is "$\langle \psi, A|\varphi|\rangle$" in quadratic-form sense.

2. Kato only had a special case of (7.6.12), which is often called "Kato's inequality," namely,

$$\Delta|\varphi| \ge \mathrm{Re}[\mathrm{sgn}(\varphi)\Delta\varphi] \tag{7.6.14}$$

as a distributional inequality. The reader will give a "direct" proof of (7.6.14) in Problem 1 and, in Problem 2, prove Theorem 7.6.1 in a few lines from (7.6.14).

Proof. We'll show (1) $\Leftrightarrow$ (2), (1) $\Rightarrow$ (4) $\Rightarrow$ (3) $\Rightarrow$ (2).

(1) $\Leftrightarrow$ (2). This is immediate from the pair of formulae (Problem 5)

$$(A+\gamma)^{-1} = \int_0^\infty e^{-\gamma t} e^{-tA}\,dt \tag{7.6.15}$$

$$e^{-tA} = \operatorname*{s-lim}_{n\to\infty}\biggl(1+\frac{t}{n}\,A\biggr)^{-n} \tag{7.6.16}$$

(1) $\Rightarrow$ (4). By the monotone convergence theorem in a spectral representation, for any φ and $A \geq 0$,

$$q_A(\varphi) = \lim_{t\downarrow 0}\langle \varphi, t^{-1}(\mathbb{1}-e^{-tA})\varphi\rangle \tag{7.6.17}$$

even if one side is ∞ (for $t^{-1}(1-e^{-tx}) = \int_0^x e^{-ty}\,dy$ is monotone increasing to x as $t \downarrow 0$).

By Proposition 7.6.3, (1) implies $|e^{-tA}\varphi| \leq |\varphi|$, so

$$\langle \varphi, e^{-tA}\varphi\rangle = \langle |\varphi|, |e^{-tA}\varphi|\rangle \leq \langle |\varphi|, e^{-tA}|\varphi|\rangle \tag{7.6.18}$$

This plus $\langle\varphi,\varphi\rangle = \langle |\varphi|,|\varphi|\rangle$ and (7.6.17) shows (proving (1) $\Rightarrow$ (3)!)

$$q_A(|\varphi|) \leq q_A(\varphi) \tag{7.6.19}$$

for all φ, and thus, $\varphi \in Q(A) \Rightarrow |\varphi| \in Q(A)$.

As in the proof of (7.6.18), if $\psi \geq 0$,

$$\langle \overline{\operatorname{sgn}(\varphi)}\,\psi, e^{-tA}\varphi\rangle \leq \langle \psi, e^{-tA}|\varphi|\rangle \tag{7.6.20}$$

We have equality at $t=0$, so we get $\geq$ if both e^{-tA} are replaced by $(\mathbb{1}-e^{-tA})$.

If $\varphi \in D(A)$,

$$t^{-1}(\mathbb{1}-e^{-tA})\varphi \to A\varphi \tag{7.6.21}$$

by dominated convergence in a spectral representation. Since $\psi, |\varphi| \in Q(A)$, polarization, and (7.6.17) imply that

$$t^{-1}\langle \psi, (\mathbb{1}-e^{-tA})|\varphi|\rangle \to Q_A(\psi, |\varphi|) \tag{7.6.22}$$

(7.6.20)–(7.6.22) imply (7.6.12) by taking limits.

(4) $\Rightarrow$ (3). Suppose first $\varphi \in D(A)$. Then taking $\psi = |\varphi|$, (7.6.12) becomes (7.6.11) for $\varphi \in D(A)$. (7.6.11) for $\varphi \notin Q(A)$ is trivial since $q_A(\varphi) = \infty$. If $\varphi \in Q(A)$, find $\varphi_n \in D(A)$ converging to φ in $\|\cdot\|_{+1,A}$. Then $q_A(\varphi_n) \to q(\varphi)$, while $q_A(|\varphi|) \leq \liminf q_A(|\varphi_n|)$ by lower semicontinuity and $|\varphi_n| \to |\varphi|$ in L^2. Thus, (7.6.11) for φ_n implies it for φ.

(3) $\Rightarrow$ (2). We begin by noting that if $\varphi, \psi \in D(A)$, we have

$$q_A(\varphi+\psi) = q_A(\varphi) + q_A(\psi) + 2\operatorname{Re}\langle \psi, A\varphi\rangle \tag{7.6.23}$$

By taking limits using the density of $D(A)$ in $\mathcal{H}_{+1,A}$, we see this holds for all $\varphi \in D(A)$ and $\psi \in Q(A)$.

Suppose $\ell_A \equiv \inf \sigma(A) > 0$ so A is invertible. Let $\varphi = A^{-1}\eta$ with $\eta \geq 0$ and pick

$$\psi = |\varphi| - \varphi \tag{7.6.24}$$

which lies in $q(A)$, since (3) implies that $\varphi \in Q(A) \Rightarrow |\varphi| \in Q(A)$. Then

$$\operatorname{Re}\langle \psi, A\varphi \rangle = \operatorname{Re}\langle \psi, \eta \rangle = \langle \operatorname{Re} \psi, \eta \rangle \geq 0 \tag{7.6.25}$$

since $\operatorname{Re} \psi = |\varphi| - \operatorname{Re} \varphi \geq 0$. Thus, by (7.6.23),

$$q_A(|\varphi|) \geq q_A(\varphi) + q_A(|\varphi| - \varphi) \tag{7.6.26}$$

But then, by (7.6.11), we have

$$q_A(|\varphi| - \varphi) = 0 \tag{7.6.27}$$

which implies that $\varphi = |\varphi| \geq 0$ (since $A \geq \ell_A > 0$ means $q_A(\zeta) \geq \ell_A \|\zeta\|^2$), that is, A^{-1} is positivity-preserving.

Returning to general $A > 0$, let $\gamma > 0$. Then $\ell_{A+\gamma} \geq \gamma > 0$ and (7.6.11) holds for $q_{A+\gamma}$, so by the above, $(A+\gamma)^{-1}$ is positivity-preserving. $\square$

Beurling–Deny also have a criteria for e^{-tA} to be a contraction on L^∞. Let $f \geq 0$. Define

$$(f \wedge 1)(x) = \min\bigl(f(x), 1\bigr) \tag{7.6.28}$$

Theorem 7.6.5 (Second Beurling–Deny criterion). *Let $A \geq 0$ be a positive self-adjoint operator on $L^2(M, d\mu)$ Then the following are equivalent:*

(1) *e^{-tA} is a contraction on $L^\infty(M, d\mu)$, i.e., for all $f \in L^\infty \cap L^2$ and all $t > 0$*

$$\|e^{-tA} f\|_\infty \leq \|f\|_\infty \tag{7.6.29}$$

(2) *For all $f \geq 0$ in $L^2(M, d\mu)$*

$$q_A(f \wedge 1) \leq q_A(f) \tag{7.6.30}$$

Remarks. 1. By duality and e^{-tA} self-adjoint, (7.6.29) implies

$$\|e^{-tA} f\|_1 \leq \|f\|_1 \tag{7.6.31}$$

which then (using Riesz–Thorin) implies $\|e^{-tA} f\|_p \leq \|f\|_p$. We said a lot about such semigroups in Section 6.6 of Part 3.

2. (7.6.30) is shorthand for $f \in Q(A) \Rightarrow f \wedge 1 \in Q(A)$ and $\langle (f \wedge 1), A(f \wedge 1)\rangle \leq \langle f, Af \rangle$.

3. If f is C^1,

$$\nabla(f \wedge 1)(x) = \begin{cases} (\nabla f)(x) & \text{if } |f(x)| < 1 \\ 0 & \text{if } |f(x)| > 1 \end{cases} \tag{7.6.32}$$

so formally $\|\nabla(f \wedge 1)\|_2^2 \leq \|\nabla f\|_2^2$ showing, modulo technicalities, when $f(x) = 1$, that $A = -\Delta$ obeys (7.6.30) "explaining" why it obeys (7.6.29).

Proof. (1) $\Rightarrow$ (2). Suppose first that M has only finitely many points $x_1, \dots, x_n$ with $\mu(\{x_j\}) = \alpha_j$ and C is a self-adjoint matrix on $L^2(M, d\mu)$ with nonnegative matrix elements and $0 \le C \le \mathbb{1}$ and that

$$\|Cf\|_1 \le \|f\|_1 \tag{7.6.33}$$

Define δ_j to be the function which is 1 at x_j and zero elsewhere and

$$\widetilde{C}_{ij} = \langle \delta_i, C\delta_j \rangle \tag{7.6.34}$$

Since

$$f = \sum_j f(x_j)\delta_j \tag{7.6.35}$$

and

$$\langle \delta_i, \delta_j \rangle = \alpha_i \delta_{ij} \tag{7.6.36}$$

we see that

$$\langle f, (\mathbb{1} - C)f \rangle = \sum_{i,j} \overline{f(x_i)}(\alpha_i\delta_{ij} - \widetilde{C}_{ij})f(x_j) \tag{7.6.37}$$

(7.6.33) implies $\|C\delta_j\|_1 \le \|\delta_j\|_1 = \alpha_j$ and (by $\langle \delta_i, f\rangle = \alpha_i f(x_i)$)

$$\begin{aligned} \|C\delta_j\|_1 &= \sum_i \alpha_i (C\delta_j)(x_i) = \sum_i \langle \delta_i, C\delta_j \rangle \\ &= \sum_i \widetilde{C}_{ij} \end{aligned} \tag{7.6.38}$$

Thus, $\sum_i \widetilde{C}_{ij} \le \alpha_j$, i.e.,

$$m_j \equiv \alpha_j - \sum_i \widetilde{C}_{ij} = \alpha_j - \sum_i \widetilde{C}_{ji} \ge 0 \tag{7.6.39}$$

We note next that (using $\widetilde{C}_{ij} = \widetilde{C}_{ji}$)

$$\begin{aligned} \frac{1}{2}\sum_{i,j} \widetilde{C}_{ij}|z_i - z_j|^2 &= \sum_j \Big(\sum_i \widetilde{C}_{ij}\Big)|z_j|^2 - \sum_{i,j} \widetilde{C}_{ij}\bar{z}_i z_j \\ &= \sum_j (\alpha_j - m_j)|z_j|^2 - \sum_{i,j} \widetilde{C}_{ij}\bar{z}_i z_j \end{aligned} \tag{7.6.40}$$

Using (7.6.37), we conclude that

$$\langle f, (\mathbb{1} - C)f \rangle = \frac{1}{2}\sum_{i,j} \widetilde{C}_{ij}|f(x_i) - f(x_j)|^2 + \sum_j m_j |f(x_j)|^2$$

In particular, since $a, b \ge 0$ implies $|a \wedge 1|^2 \le a^2$ and $|a\wedge 1 - b \wedge 1|^2 \le |a-b|^2$, we have

$$\langle f \wedge 1, (\mathbb{1} - C) f \wedge 1 \rangle \le \langle f, (\mathbb{1} - C)f \rangle \tag{7.6.41}$$

We return to a semigroup, e^{-tA} on a general $L^2(M, d\mu)$. Pick $S_1, \dots, S_\ell$ disjoint measurable sets in M whose union is M and let P_S be the projection in L^2 onto functions constant on each S_j. P_S is a contraction from $L^p(M, d\mu)$ to $L^p(M_S, d\mu_S)$ where $M_S = \{1, \dots, \ell\}$ with $\mu_S(S_j) = \mu(S_j)$. If $C = P_S e^{-tA} P_S$, it follows that C is a contraction on L^1, positivity-preserving and self-adjoint on $L^2(M_S, d\mu_S)$. Thus, by (7.6.41),

$$\langle (P_S f) \wedge 1, (\mathbb{1} - e^{-tA})(P_S(f) \wedge 1\rangle \leq \langle P_S f, (\mathbb{1} - e^{-tA}) P_S f\rangle \tag{7.6.42}$$

Order $\{S_1, \dots, S_\ell\}$ by $S \lhd T$ if T is a refinement of S, i.e., each S_j is a union of sets in T. Then $\{P_S\}$ is a set and s-lim $P_S = \mathbb{1}$. Taking the limit in (7.6.42)

$$\langle f \wedge 1, (\mathbb{1} - e^{-tA}) f \wedge 1\rangle \leq \langle f, (\mathbb{1} - e^{-tA}) f\rangle \tag{7.6.43}$$

Since (allowing there also $q_A(f) = \infty$ if $f \notin Q(A)$)

$$q_A(f) = \lim_{t \downarrow 0} \langle f, (\mathbb{1} - e^{-tA}) f\rangle t^{-1} \tag{7.6.44}$$

we get (7.6.30).

<u>(2) ⇒ (1).</u> The argument uses a variational principle like the argument behind (3) ⇒ (2) in Theorem 7.6.5. Fix $u \in L^2$ with $0 \leq u \leq 1$. For any v, define

$$\begin{aligned} \psi(v) &= q_A(v) + \|u - v\|^2 \\ &= q_{A+1}(v) + \|u\|^2 - 2\operatorname{Re}\langle u, v\rangle \end{aligned} \tag{7.6.45}$$

We set $w = \left(A + 1\right)^{-1} u$. Since

$$q_{A+1}(w) = \langle w, u\rangle, \quad \psi(w) = \|u\|^2 - \langle w, u\rangle \tag{7.6.46}$$

$$q_{A+1}(w - v) = q_{A+1}(w) + q_{A+1}(v) - 2\operatorname{Re}\langle u, v\rangle \tag{7.6.47}$$

we see that

$$\psi(v) = \psi(w) + q_{A+1}(w - v) \tag{7.6.48}$$

which says $\psi(v)$ is minimized exactly when $v = w$.

Since $u \leq 1$, $|u - (w \wedge 1)| \leq |u - w|$. By (2), $q_A(w \wedge 1) \leq q_A(w)$. It follows that

$$\psi(w \wedge 1) \leq \psi(w) \tag{7.6.49}$$

Thus, by the minimum property of ψ, $w \wedge 1 = w$, i.e.,

$$0 \leq u \leq 1 \Rightarrow 0 \leq (A + 1)^{-1} u \leq 1 \tag{7.6.50}$$

This implies if $\|v\|_\infty \leq 1$

$$|(A + 1)^{-1} v| \leq (A + 1)^{-1} |v| \leq 1 = \|v\|_\infty \tag{7.6.51}$$

So $(A + 1)^{-1}$ is a contraction on L^∞.

By the same argument, $(1 + \frac{A}{n})^{-1}$ is a contraction on L^∞, so $e^{-tA} =$ s-$\lim \left(1 + \frac{A}{n}\right)^{-n}$ is a contraction on L^∞. □

Example 7.6.6 ($-\Delta+V$)**.** We know (see (6.9.28) in Part 1) that on $L^2(\mathbb{R}^\nu)$,

$$(e^{-t\Delta}\varphi)(x) = (4\pi t)^{-\nu/2} \int \exp\biggl(-\frac{(x-y)^2}{4t}\biggr)\varphi(y)\, d^\nu y \tag{7.6.52}$$

This is obviously a positivity-preserving map, proving the original Kato inequality, that as distributions (since $C_0^\infty(\mathbb{R}^\nu) \subset D(-\Delta)$), for $\varphi \in D(-\Delta)$, we have that

$$\Delta|\varphi| \geq \operatorname{Re}(\operatorname{sgn}(\varphi)\Delta\varphi) \tag{7.6.53}$$

(The sign is reversed from (7.6.12) since A is $-\Delta$, not Δ!) By an easy limit argument (Problem 6), this extends, if viewed as a distributional inequality, to all $\varphi \in L^1_{\text{loc}}(\mathbb{R}^\nu)$ with $\Delta\varphi \in L^1_{\text{loc}}(\mathbb{R})$.

The Beurling–Deny criterion is

$$\|\nabla|\varphi|\|_{L^2}^2 \leq \|\nabla\varphi\|_{L^2}^2 \tag{7.6.54}$$

Formally at least, this is obvious, for if we write

$$\varphi = |\varphi|e^{i\theta} \tag{7.6.55}$$

then

$$\nabla\varphi = (\nabla|\varphi| + i|\varphi|\nabla\theta)e^{i\theta} \tag{7.6.56}$$

so

$$|\nabla\varphi|^2 = (\nabla|\varphi|)^2 + (|\varphi|\nabla\theta)^2 \tag{7.6.57}$$

from which (7.6.54) is immediate by integration! This argument is only formal.

If $V \geq 0$, e^{-tV} is multiplication by a positive function, so positivity-preserving. Clearly, if q_1 and q_2 are two closed quadratic forms obeying (7.6.11), so does $q_1 + q_2$, that is, the quadratic form sum of two operators whose semigroups are positivity-preserving, so is the form sum. In particular, if $V \geq 0$, $V \in L^1_{\text{loc}}(\mathbb{R}^\nu)$, then the form sum (on $Q(-\Delta) \cap Q(V)$), $H = -\Delta + V$ has e^{-tH} positivity-preserving. $\square$

Next, we turn to looking at one semigroup dominating another:

Theorem 7.6.7. *Let A and B be positive self-adjoint operators on $L^2(M, d\mu)$ so that e^{-tA} is positivity-preserving. Then*

$$|e^{-tB}\varphi(x)| \leq (e^{-tA}|\varphi|)(x) \tag{7.6.58}$$

for all $\varphi \in L^2$ if and only if

(i) $\varphi \in D(B) \Rightarrow |\varphi| \in Q(A)$

(ii) *For all $\varphi \in D(B)$ and $\psi \geq 0$ in $Q(A)$, we have that*

$$\langle\psi, A|\varphi|\rangle \leq \operatorname{Re}\langle\overline{\operatorname{sgn}(\varphi)}\,\psi, B\varphi\rangle \tag{7.6.59}$$

Remarks. 1. Since e^{-tA} preserves real functions, $A|\varphi|$ is real-valued, so the left side of (7.6.58) is real.

2. This is *Kato's inequality.* Kato did not have a connection to semigroups nor did he consider general pairs, but he had the special case, discussed below in Theorem 7.6.11, where $B = -(\nabla - ia)^2$, $A = -\Delta$.

Proof. (7.6.58) $\Rightarrow$ (i). (7.6.58) implies

$$\langle \varphi, e^{-tB}\varphi\rangle = \langle |\varphi|, |e^{-tB}\varphi|\rangle \leq \langle |\varphi|, e^{-tA}|\varphi|\rangle \tag{7.6.60}$$

By (7.6.17),

$$q_A(|\varphi|) \leq q_B(\varphi) \tag{7.6.61}$$

so $\varphi \in Q(B) \Rightarrow |\varphi| \in Q(|A|)$, which is stronger than (i).

(7.6.58) $\Rightarrow$ (ii). (7.6.58) and $\psi \geq 0$ implies

$$\operatorname{Re}\langle \psi, \operatorname{sgn}(\varphi)e^{-tB}\varphi\rangle \leq \langle \psi, |e^{-tB}\varphi|\rangle \leq \langle \psi, e^{-tA}|\varphi|\rangle \tag{7.6.62}$$

with equality at $t = 0$. Thus, subtracting from the common $t = 0$ value and using (7.6.22) for ψ, $|\varphi| \in Q(A)$ and $t^{-1}(\mathbb{1} - e^{-tB})\varphi \to B\varphi$ for $\varphi \in D(B)$ yields (7.6.59).

(i), (ii) $\Rightarrow$ (7.6.58). Fix $\gamma > 0$. Adding $\gamma\langle\psi, |\varphi|\rangle$ to both sides of (7.6.59) implies

$$\langle \psi, (A+\gamma)|\varphi|\rangle \leq \operatorname{Re}\langle \psi, \operatorname{sgn}(\varphi)(B+\gamma)\varphi\rangle \leq \langle \psi, |(B+\gamma)\varphi|\rangle \tag{7.6.63}$$

for any $\varphi \in D(B)$, $\psi \in Q(A)$, $\psi \geq 0$. Let η, ξ be in $L^2(M, d\mu)$ with $\eta \geq 0$. Let

$$\psi = (A+\gamma)^{-1}\eta, \qquad \varphi = (B+\gamma)^{-1}\xi \tag{7.6.64}$$

Then $\psi \in Q(A)$, $\varphi \in D(B)$, and $\psi \geq 0$ since e^{-tA} is positivity-preserving (and (1) $\Leftrightarrow$ (2) in Theorem 7.6.4), so (7.6.63) becomes

$$\langle \eta, |(B+\gamma)^{-1}\xi|\rangle \leq \langle \eta, (A+\gamma)^{-1}|\xi|\rangle \tag{7.6.65}$$

Since $\eta \geq 0$ is arbitrary, we have for a.e. x,

$$|(B+\gamma)^{-1}\xi(x)| \leq ((A+\gamma)^{-1}|\xi|)(x) \tag{7.6.66}$$

By an easy induction taking $\gamma = n/t$,

$$\left|\left(\mathbb{1} + \frac{tB}{n}\right)^{-n}\xi\right| \leq \left(\mathbb{1} + \frac{tA}{n}\right)^{-n}|\xi| \tag{7.6.67}$$

Taking $n \to \infty$ and using (7.6.18) implies (7.6.58). □

Example 7.6.8 (Comparing $-\Delta$ and $-\Delta+V$). Here we treat $A = -\Delta$ and $B = -\Delta + V$ where, for a.e. x,

$$V(x) \geq 0 \tag{7.6.68}$$

With Theorems 7.6.1 and 7.6.2 in mind, we want to allow $V \in L^1_{\text{loc}}$, but then $D(B)$ is complicated and one needs to prove (7.6.59) for $\varphi \in D(B)$. We'll overcome this by a two-step process. First, $V \in L^\infty$, where $D(B) = D(A)$ by the Kato–Rellich theorem, and then use a limiting argument rather than (7.6.59) for general V.

If $V \in L^\infty$, $V \geq 0$, $D(-\Delta + V) = D(-\Delta)$. By Theorem 7.6.4 and Example 7.6.6, for $\psi \geq 0$, $\psi \in Q(-\Delta)$, $\varphi \in D(-\Delta)$, we have that

$$\langle \psi, (-\Delta)|\varphi| \rangle \leq \text{Re}\langle \overline{\text{sgn}(\varphi)}\, \psi, (-\Delta)\varphi \rangle \tag{7.6.69}$$

Since $V \geq 0$,

$$\langle \overline{\text{sgn}(\varphi)}\, \psi, V\varphi \rangle = \langle |\psi|, V\varphi \rangle \geq 0 \tag{7.6.70}$$

Thus, for $\varphi \in D(-\Delta + V) = D(-\Delta)$ and $\psi \in Q(-\Delta)$, $\psi \geq 0$,

$$\langle \psi, (-\Delta)|\varphi| \rangle \leq \text{Re}\langle \overline{\text{sgn}(\varphi)}\, \psi, (-\Delta + V)\varphi \rangle \tag{7.6.71}$$

By Theorem 7.6.7, for such V and all φ,

$$|e^{-t(-\Delta+V)}\varphi| \leq e^{t\Delta}|\varphi| \tag{7.6.72}$$

For any positive V, even one not in L^1_{loc}, if $V_n = \min(V, n)$, then $|V_n| \leq n$ and (7.6.72) holds if V is replaced by V_n. $-\Delta + V_n$ is monotone in n (as quadratic forms), so by the monotone convergence theorem for forms (Theorem 7.5.18),

$$\operatorname*{s-lim}_{n\to\infty}(-\Delta + V_n + 1)^{-1} \to (-\Delta + V + 1)^{-1} \tag{7.6.73}$$

where $-\Delta + V$ is the form sum. By the continuity of the functional calculus (Theorem 7.2.10),

$$\operatorname*{s-lim}_{n\to\infty} e^{-t(-\Delta+V_n+1)} = e^{-t(-\Delta+V+1)} \tag{7.6.74}$$

so (7.6.72) holds for all positive V. We state the result below for $V \in L^1_{\text{loc}}$. □

Theorem 7.6.9. *If $V \in L^1_{\text{loc}}(\mathbb{R}^\nu)$ with $V \geq 0$ and $H = -\Delta + V$ is the form sum, then with $H_0 = -\Delta$,*

$$|e^{-tH}\varphi(x)| \leq (e^{-tH_0}|\varphi|)(x) \tag{7.6.75}$$

Proof of Theorem 7.6.2. For any positive self-adjoint A, it is easy to see that $\text{Ran}[e^{-A}]$ is both an operator and a form core (Problem 7).

Let H be the form sum $-\Delta + V$ and $H_0 = -\Delta$. e^{-H_0} has a bounded integral kernel, $e^{-H_0}(x, y)$, which is L^1 in y, uniformly in x. Thus, for $\varphi \in L^2$, $e^{-H_0}\varphi \in L^2 \cap L^\infty$. This implies that

$$\mathcal{Q}_\infty \equiv Q(H_0) \cap Q(V) \cap L^\infty(\mathbb{R}^\nu) \tag{7.6.76}$$

is a form core for H.

So let φ lie in $\mathcal{Q}_\infty$ and let $f \in C_0^\infty$. Then

$$\nabla(f\varphi) = (\nabla f)\varphi + f\nabla\varphi \in L^2 \tag{7.6.77}$$

since $\varphi, \nabla\varphi \in L^2$ and $V^{1/2}(f\varphi) \in L^2$ since $|V^{1/2}f\varphi| \le \|f\|_\infty V^{1/2}|\varphi|$, so

$$f \in C_0^\infty \text{ and } \varphi \in \mathcal{Q}_\infty \Rightarrow f\varphi \in \mathcal{Q}_\infty \tag{7.6.78}$$

If $f_0 \in C_0^\infty$ has $f_0(x) \equiv 1$, if $|x| \le 1$, and $\varphi_n = f_0(1/n)\varphi$, then $\nabla\varphi_n \to \nabla\varphi$ and $V^{1/2}\varphi_n \to V^{1/2}\varphi$ in L^2 by the dominated convergence theorem (and $|(\nabla f_0(1/n))\varphi| \le 1/n\|\nabla f_0\|_\infty|\varphi(x)|$). Thus,

$$\mathcal{Q}_{\infty,0} = \{\varphi \in \mathcal{Q}_\infty \mid \text{supp}(\varphi) \text{ is bounded}\} \tag{7.6.79}$$

is a form core also.

If $\varphi \in \mathcal{Q}_{\infty,0}$, pick $j \in C_0^\infty(\mathbb{R}^\nu)$, $j \ge 0$, $\int j(x)\,d^\nu x = 1$ and let $\varphi_n = n^\nu j(\cdot\, n) * \varphi$. Then $\varphi_n \in C_0^\infty(\mathbb{R}^\nu)$, $\nabla\varphi_n \to \nabla\varphi$ in L^2, and since $\varphi \in L^\infty$, the dominated convergence theorem and pointwise convergence a.e. of φ_n to φ implies that $\varphi_n V^{1/2} \to \varphi V^{1/2}$ in L^2. Thus, $C_0^\infty(\mathbb{R}^\nu)$ is a form core for H. □

A key step was (7.6.77). It was easy because $Q(H)$ is so simple. As a preliminary to the proof of Theorem 7.6.1, we need to analyze $D(H)$.

Proposition 7.6.10. *Let $V \in L^2_{\text{loc}}(\mathbb{R}^\nu)$ with $V \ge 0$, and let $H = -\Delta + V$ be the form sum. Then*

(a) $D(H) = \{\varphi \in Q(-\Delta) \cap Q(V) \mid -\Delta\varphi + V\varphi \in L^2\}$ (7.6.80)

(b) $\varphi \in D(H)$ *and* $f \in C_0^\infty(\mathbb{R}^\nu) \Rightarrow f\varphi \in D(H)$

Remark. In (7.6.80), $-\Delta\varphi + V\varphi$ is intended in the distributional sense. Since $V \in L^2_{\text{loc}}$, $\varphi \in L^2$, $V\varphi \in L^1_{\text{loc}}$ defines a distribution.

Proof. (a) By Proposition 7.5.6 and the fact proven in Theorem 7.6.2 that $C_0^\infty(\mathbb{R}^\nu)$ is a form core, $D(H)$ is the set of $\varphi \in Q(H)$ so that there is an η with

$$\langle f, H\varphi\rangle = \langle f, \eta\rangle \tag{7.6.81}$$

for all $f \in C_0^\infty$. This says η is the distributional object $-\Delta\varphi + V\varphi$, proving (7.6.80).

(b) We already proved, while proving Theorem 7.6.2, that $\varphi \in Q(H) \Rightarrow f\varphi \in Q(H)$. As distributions,

$$-\Delta(f\varphi) + Vf\varphi = f(-\Delta\varphi + V\varphi) - (\Delta f)\varphi - 2\nabla f \cdot \nabla\varphi \tag{7.6.82}$$

Since $\varphi \in Q(H)$, $\nabla\varphi \in L^2$, so the right side of (7.6.82) lies in $D(H)$ by (a). □

Proof of Theorem 7.6.1. Given the last proposition, the proof is analogous to that of Theorem 7.6.2. Since $\text{Ran}(e^{-H})$ is an operator core for H,

$$\mathcal{D}_\infty = D(H) \cap L^\infty(\mathbb{R}^\nu) \tag{7.6.83}$$

is an operator core. By the proposition, $f \in C_0^\infty$ and $\varphi \in \mathcal{D}_\infty$ implies $f\varphi \in \mathcal{D}_\infty$, so

$$\mathcal{D}_{\infty,0} = \{\varphi \in \mathcal{D}_\infty \mid \text{supp}(\varphi) \text{ is bounded}\} \tag{7.6.84}$$

is an operator core.

If $\varphi \in \mathcal{D}_{\infty,0}$, $V\varphi \in L^2$, so $-\Delta\varphi \in L^2$. Thus, for any j in $C_0^\infty(\mathbb{R}^\nu)$, with $j \geq 0$ and $\int j(x)\, d^\nu x = 1$, we have that $j_n = n^\nu j(n\,\cdot)$ has $H(j_n * \varphi) \to H\varphi$, so C_0^∞ is an operator core. □

Next, we want to briefly discuss the issue of diamagnetic inequalities. Given $\vec{a} \in C^1(\mathbb{R}^\nu)$, an $\mathbb{R}^\nu$-valued function, called the *gauge potential*, the magnetic field associated to $\vec{a}$ is $\vec{B} = d\vec{a}$, the two-form. In $\mathbb{R}^3$, we are used to using the fact that two-forms are associated to one-forms given a volume form and orientation, so on $\mathbb{R}^3$, $\vec{B} = \vec{\nabla} \times \vec{a}$. In quantum mechanics, the Schrödinger operator in magnetic field with gauge potential, $\vec{a}$, and electric potential, V, is

$$H(\vec{a}, V) = H_0(\vec{a}) + V, \qquad H_0(a) = -(\vec{\nabla} - i\vec{a})^2 \tag{7.6.85}$$

As a formal quadratic form, $H_0(\vec{a})$ makes sense if $\vec{a} \in L^2_{\text{loc}}$. The inequalities below are known to hold with $V \geq 0$, $V \in L^1_{\text{loc}}$, and with this most general $\vec{a}$—see the Notes—but we'll restrict to $\vec{a} \in C_0^1(\mathbb{R}^\nu)$, the C^1-functions of compact support, to avoid technicalities. This is a stronger condition than $\vec{B} \in C_0(\mathbb{R}^\nu)$ since $\vec{B}$'s coming from compact support $\vec{a}$ have zero fluxes.

Theorem 7.6.11 (Diamagnetic Inequalities). *Let $\vec{a} \in C_0^1(\mathbb{R}^\nu)$, $V \in L^1_{\text{loc}}(\mathbb{R}^\nu)$ with $V \geq 0$. Let $H(\vec{a}, V)$ be defined via quadratic forms with*

$$Q(H(\vec{a}, V)) = \{\varphi \in L^2 \mid V^{1/2}\varphi \in L^2,\ (\nabla - ia)\varphi \in L^2\}$$

Then for all $t > 0$ and $\varphi \in L^2(\mathbb{R}^\nu)$,

$$|\exp(-tH(\vec{a}, V))\varphi(x)| \leq (\exp(-tH(\vec{a} = 0, V))|\varphi|)(x) \tag{7.6.86}$$

Remarks. 1. As with the $\vec{a} = 0$ case, this together with the $\exp((-tH)(\vec{a} = 0, V))|\varphi| \leq \exp(-tH_0)|\varphi|$ bound allows one to prove that C_0^∞ is a form core for $H(\vec{a}, V)$. There is also an analog of Theorem 7.6.1; see the Notes.

2. If $\vec{a}_1 = \vec{a} + \nabla\lambda$, the $\vec{B}$'s are the same; this is called a *gauge transformation.* In one dimension, $\vec{B} = 0$ since all two-forms are 0. This is realized by the fact that all $\vec{a}$'s are related to $\vec{a} = 0$ by a gauge transformation. The argument in the proof about the one-dimensional case is essentially a realization of this.

Proof. Since, as forms, $H(\vec{a}, V)$ is monotone in V, the same argument we used in Example 7.6.8 lets us only prove (7.6.86) for $V \in L^\infty$. In that case, $H(\vec{a}, V) - H_0 = -2i\vec{a}\cdot\vec{\nabla} - i\vec{\nabla}\cdot\vec{a} + a^2 + V$ is H_0-bounded with relative bound zero, so

$$D(H(\vec{a}, V)) = D(H_0), \quad Q(H(\vec{a}, V)) = Q(H_0) \tag{7.6.87}$$

Since $\vec{a} \in C^1$, we can find C^2-functions, λ_j, on $\mathbb{R}^\nu$ so that

$$a_j(x) = (\partial_j \lambda_j)(x) \tag{7.6.88}$$

It follows that on C_0^∞-functions, φ,

$$(\partial_j - ia_j)\varphi = (e^{i\lambda_j}\partial_j e^{-i\lambda_j})\varphi \tag{7.6.89}$$

so

$$-(\partial_j - ia_j)^2\varphi = e^{i\lambda_j}(-\partial_j^2)e^{-i\lambda_j}\varphi \tag{7.6.90}$$

Since C_0^∞ is an operator core for $-\partial_j^2$ and for $-(\partial_j - ia_j)^2$, we conclude, for each j and all $\varphi \in L^2$,

$$e^{t(\partial_j - ia_j)^2}\varphi = e^{i\lambda_j}e^{t\partial_j^2}e^{-i\lambda_j}\varphi \tag{7.6.91}$$

Thus, since $e^{t\partial_j^2}$ has an explicit positive integral kernel,

$$|e^{t(\partial_j - ia_j)^2}\varphi| \le e^{t\partial_j^2}|\varphi| \tag{7.6.92}$$

By Theorem 7.6.7, for all $\varphi \in D(H_0) \subset D(\partial_j^2)$ and $\psi \in Q(H_0)$, $\psi \ge 0$, we have

$$\langle \psi, (-\partial_j^2)|\varphi|\rangle \le \operatorname{Re}\langle \overline{\operatorname{sgn}(\varphi)}\,\psi, -(\partial_j - ia_j)^2\varphi\rangle \tag{7.6.93}$$

Summing over j and using $V \ge 0$,

$$\langle \psi, H_0|\varphi|\rangle \le \operatorname{Re}\langle \overline{\operatorname{sgn}(\varphi)}\,\psi, H(\vec{a}, V)\varphi\rangle \tag{7.6.94}$$

By Theorem 7.6.7, we have (7.6.86). □

Our final topic is the Trotter product formula as an aside. We'll use it to recover some results we found above via Kato's inequality methods. We'll also prove the Feynman–Kac formula. For finite matrices, A and B, it is not hard to see (Problem 8) that

$$e^{-t(A+B)} = \lim_{n\to\infty}(e^{-tA/n}e^{-tB/n})^n \tag{7.6.95}$$

We want to obtain a result of this genre for unbounded self-adjoint operators. The version we prove can be improved both with regard to the conditions on A and B and to the kind of limit; see the Notes. We begin with

Proposition 7.6.12 (Chernoff's Theorem). *Let A be a positive self-adjoint operator and $\{G(s)\}_{0<s<\infty}$ be a family of self-adjoint operators with*

$$0 \le G(s) \le 1 \tag{7.6.96}$$

Suppose that as $t \to 0$,

$$S(t) \equiv t^{-1}(\mathbb{1} - G(t)) \longrightarrow A$$

in strong resolvent sense. Then

$$\operatorname*{s-lim}_{n\to\infty} G\left(\frac{t}{n}\right)^n = e^{-tA} \tag{7.6.97}$$

Proof. By Theorem 7.2.10, with the $f(x) = e^{-tx}$,

$$\text{s-lim} \exp\left(-tS\left(\frac{t}{n}\right)\right) = e^{-tA} \tag{7.6.98}$$

Let

$$H_n(x) = \begin{cases} e^{-x} - (1 - \frac{x}{n})^n, & 0 \le x \le n \\ e^{-x}, & x \ge n \end{cases} \tag{7.6.99}$$

Then by a simple argument (Problem 9),

$$\|H_n\|_\infty \to 0 \tag{7.6.100}$$

so $\|H_n(tS(\frac{t}{n}))\| \to 0$. Since $(\mathbb{1} - \frac{tS(t/n)}{n}) = G(\frac{t}{n})$, this says that

$$\left\| \exp\left(-tS\left(\frac{t}{n}\right)\right) - G\left(\frac{t}{n}\right)^n \right\| \to 0 \tag{7.6.101}$$

This plus (7.6.98) implies (7.6.97). □

Theorem 7.6.13 (Trotter Product Formula, Self-adjoint Case). *Let A and B be positive self-adjoint operators and suppose $A+B$ defined on $D(A) \cap D(B)$ is densely defined and essentially self-adjoint. Let C be the closure of $A+B$. Then*

$$\operatorname*{s-lim}_{n\to\infty} (e^{-tA/n} e^{-tB/n}) = e^{-tC} \tag{7.6.102}$$

Remark. For any A, B, this is true if C is the form sum, but the proof is more involved; see the Notes.

Proof. Let

$$G(t) = e^{-tB/2}e^{-tA}e^{-tB/2} \tag{7.6.103}$$

Then, for $\varphi \in \mathcal{D} \equiv D(A) \cap D(B)$, it is easy to see (Problem 10) that

$$S(t)\varphi \equiv t^{-1}(\mathbb{1} - G(t))\varphi \to A\varphi + B\varphi \tag{7.6.104}$$

By Theorem 7.2.11, s-lim $S(t) = C$. Thus,

$$\text{s-lim}[e^{-tB/2n}(e^{-tA/n}e^{-tB/n})^{n-1}e^{-tA/n}e^{-tB/2n}] = e^{-tC} \tag{7.6.105}$$

Since $(e^{-sA} - \mathbb{1})$ and $(e^{-sB} - \mathbb{1})$ go to 0 strongly as $s \downarrow 0$, this implies (7.6.102) (Problem 11). $\square$

Since products and strong limits of positivity-preserving operators are positivity-preserving, if A, B, C are related by (7.6.102) and e^{-tA}, e^{-tB} are positivity-preserving, so is e^{-tC}. Similarly, if A, B, C and A_1, B_1, C_1 are related by (7.6.1) and $|e^{-tA_1}\varphi| \le e^{-tA}|\varphi|$, and similarly for B_1, B, we have the same inequality for C_1, C. This provides alternate ways to obtain the results earlier in this chapter (Theorem 7.6.11 isn't always enough, but the form sum version is! See Problems 12 and 13.).

Finally, we turn to the Feynman–Kac formula. It requires the use of Brownian motion, as discussed in Section 4.16 of Part 1. For this discussion, the key issue is that there is a probability measure, μ, on $C([0,\infty), \mathbb{R}^\nu)$, the $\mathbb{R}^\nu$-valued continuous function on $[0,\infty)$ (with Σ-algebra the Borel subsets of the Polish space $C([0,\infty), \mathbb{R}^\nu)$; see Section 4.14 of Part 1), so that with

$$\mathbb{E}(G(\vec{b})) - \int_{\vec{b}\in C([0,\infty),\mathbb{R}^\nu)} G(\vec{b})\, d\mu(\vec{b}) \tag{7.6.106}$$

we have for any $s_1, \dots, s_n \in [0,\infty)$ that $\{b_j(s_\alpha)\}_{j=1,\dots,\nu;\, \alpha=1,\dots,n}$ have a jointly Gaussian probability distribution with

$$\mathbb{E}(b_j(s_\alpha))b_k(s_\beta)) = \delta_{jk} \min(s_\alpha, s_\beta) \tag{7.6.107}$$

The critical input we need is the following generalization of (4.16.37) of Part 1:

Lemma 7.6.14. *Let $H_0 = -\frac{1}{2}\Delta$ on $L^2(\mathbb{R}^\nu, d^\nu x)$. Let $0 = s_0 \le s_1 \le s_2 \le \dots \le s_n$ and $t_1 = s_1$, $t_j = s_j - s_{j-1}$ for $j = 2, \dots, n$. Let $f_1, \dots, f_n$ be bounded continuous functions on $\mathbb{R}^\nu$ and $\varphi \in L^2(\mathbb{R}^\nu, d^\nu x)$. Then*

$$(f_1 e^{-t_1 H_0} f_2 \dots f_n e^{-t_n H_0}\varphi)(\vec{x}) = \mathbb{E}\biggl(\biggl[\prod_{\alpha=1}^n f_\alpha(\vec{x} + \vec{b}(s_{\alpha-1}))\biggr]\varphi(\vec{x} + \vec{b}(s_n))\biggr) \tag{7.6.108}$$

Proof. Both sides of (7.6.108) are Gaussian integrals, that is, $[\prod_{\alpha=1}^n f_\alpha(x_\alpha)]\varphi(x_{n+1})$ integrated on $d^{(n+1)\nu}(x)$ by inserting a suitable Gaussian factor in $x_1, \dots, x_{n+1}$. On the left, this is because e^{-tH_0} is an operator with Gaussian integral kernel, and on the right, by the construction of Brownian motion. By translation covariance, one can suppose $\vec{x} = 0$ and since these are Gaussians, it suffices to consider the cases where each f_α and φ is 1 or some x_ℓ and exactly one or two are an x_ℓ. (These f's are not bounded and this φ is not in L^2, but the left side can be defined via limits.) Checking this is a straightforward calculation (Problem 14). □

Theorem 7.6.15 (Feynman–Kac Formula). *Let $H_0 = -\frac{1}{2}\Delta$ on $L^2(\mathbb{R}^\nu, d^\nu x)$. Let $V \in L^1_{\text{loc}}(\mathbb{R}^\nu)$ with $V \geq 0$. Let $H = H_0 + V$ be the form sum. Then for all φ, $\psi \in L^2(\mathbb{R}^\nu, d^\nu x)$, we have that*

$$\langle \psi, e^{-tH}\varphi\rangle = \int \mathbb{E}(e^{-\int_0^t V(\vec{x}+\vec{b}(s))\, ds}\varphi(\vec{x} + \vec{b}(t)))\psi(x)\, d^\nu x \qquad (7.6.109)$$

Remarks. 1. (7.6.109) is called the *Feynman–Kac formula.*

2. One can prove that the expectation in (7.6.109) is continuous in x and then obtain a formula for $(e^{-tH}\varphi)(x)$ as the expectation without $\int \dots d^\nu x$. For this and other issues mentioned in the remarks, see Simon [**647**].

3. It is not hard to allow negative parts of V, for example, $V_- \in L^p(\mathbb{R}^\nu)$ with $p > \nu/2$.

Proof. By a use of monotone convergence for forms and dominated convergence in the integral, it is easy to go from $V \in L^\infty$ to $V \in L^1_{\text{loc}}$. By a simple limiting argument (Problem 15), one need only prove (7.6.109) for V a continuous function of compact support and for $\psi, \varphi \in L^1 \cap L^\infty(\mathbb{R}^\nu)$.

By the Trotter product formula,

$$\text{LHS of (7.6.109)} = \lim_{n\to\infty} \langle \psi, (e^{-tH_0/n}e^{-tV/n})^n\varphi\rangle \qquad (7.6.110)$$

By Lemma 7.6.14, the inner product on the right of (7.6.110) is

$$\int \mathbb{E}\biggl(\exp\biggl(-\sum_{j=1}^n \frac{t}{n} V\biggl(\vec{x} + \vec{b}\Bigl(\frac{jt}{n}\Bigr)\biggr)\biggr)\varphi(\vec{x} + \vec{b}(t))\biggr)\psi(x)\, d^\nu x \qquad (7.6.111)$$

Since V and b are continuous, for every b, the sum converges to $\int_0^t V(\vec{x}+\vec{b}(s))\, ds$, so by the dominated convergence theorem, (7.6.110) converges to the right side of (7.6.108). □

Notes and Historical Remarks. Positivity-preserving operators and semigroups have their root in probability theory and in the fact that it is natural to consider matrices with all positive elements (if we think of $\mathbb{C}^n$ as $L^2(\{1, \dots, n\}, d\mu)$ with $\mu(\{j\}) = \frac{1}{n}$, then a matrix, C, is positivity-preserving $\Leftrightarrow C_{ij} \geq 0$ for all i, j). We saw some aspects of this in Section 7.5 of Part 1, especially the Perron–Frobenius theorem, Theorem 7.5.3 of Part 1. We also saw a major role for positivity-preserving operators in the discussion of martingales (Section 2.10) and especially in the theory of hypercontractive semigroups (Section 6.6), both in Part 3.

Historically, the theorems of this section arose from hypercontractivity (although as discussed in the Notes to Section 6.6 of Part 3, there were earlier applications of positivity in quantum theory to get uniqueness of ground states). In [**629**], Simon obtained a precursor of Theorem 7.6.1: he proved that if $V \in L^2(\mathbb{R}^\nu, e^{-x^2} d^\nu x)$, and $V \geq 0$, then $-\Delta + V$ is essentially self-adjoint on $C_0^\infty(\mathbb{R}^\nu)$. His method of proof was to use Theorem 6.6.30 of Part 3 and the hypercontractivity of the intrinsic semigroup associated to $-\Delta + x^2$ to prove $-\Delta + V + x^2$ and used a trick to get rid of the x^2.

The point of this result is that with regard to allowed local singularities, all the previous results didn't distinguish between V_+ and V_-. Because of Example 7.4.26, one can't have an L^p_{loc} result for V_- unless $p \geq \frac{\nu}{2}$ (i.e., not L^2 if $\nu \geq 5$). What Simon showed is that the L^p-condition for V_+ is different from V_-. He conjectured that for $V \geq 0$, $L^2_{\text{loc}}(\mathbb{R}^\nu)$ should be enough.

Motivated by this, Kato [**379**] proved Theorem 7.6.1 by the approach used in Problems 1 and 2 that relied on the distributional inequality (7.6.14). Theorem 7.6.2 is a result of Simon [**640**].

The Beurling–Deny criteria are from their paper [**55**] (see also [**56, 157**]). This paper only looked at spaces with finitely many points but the method they used extends to the infinite-dimensional case. An equivalent form to the first Beurling–Deny criterion is due to Aronszajn–Smith [**29**]. The abstract form of Kato's inequality, (7.6.12) and its equivalence to e^{-tA} positivity-preserving is due to Simon [**636**].

In his paper, Kato [**379**] proved a distributional inequality

$$\Delta\, |u| \geq \operatorname{Re}\big[[\operatorname{sgn}(u)]\,\big(\vec{\nabla} - i\vec{a}\big)^2 u\big] \tag{7.6.112}$$

with $\operatorname{sgn}(u)$ given by (7.6.13). He used this to get self-adjointness results in a magnetic field which he improved in [**381**]. Motivated by this, by the equivalence of (a) and (d) in Theorem 7.6.4 and by Simon's work on diamagnetic inequalities (see below), Hess et al. [**315**] and Simon [**641**] independently proved Theorem 7.6.7 (which had been conjectured in Simon [**636**]).

Theorem 7.6.1 has been extended to allow $V(x) \geq -\nu(\nu-4)/4|x|^2$ in Simon [**630**]. Earlier results related to this (with stronger conditions than L^2_{loc}) are due to Kalf–Walter [**366**] and Schmincke [**603**]. Extensions of Theorem 7.6.1 to magnetic fields with optimal conditions, i.e., $(-i\nabla - a)^2 + V$ with $V \geq 0$, $V \in L^2_{\text{loc}}$, $a \in L^4_{\text{loc}}$ and $\nabla \cdot a \in L^2_{\text{loc}}$ are due to Leinfelder–Simader [**437**]. Theorem 7.6.2 was extended to the optimum condition with magnetic field ($V \geq 0$, $V \in L^1_{\text{loc}}$, $a \in L^2_{\text{loc}}$) by Simon [**640**].

Diamagnetic inequalities (Theorem 7.6.11) go back to Simon [**633**] where he proved $\inf \sigma\big(H(a,V)\big) \geq \inf \sigma\big(H(a=0,V)\big)$ which can be viewed as the

$t \to \infty$ of (7.6.86). The full result (following a suggestion of Nelson) is in Simon [**636**]. For nice a's, it is proven also as a consequence of the results of Hess et al. [**315**] and Simon [**641**]. The idea of using gauge transformation in each variable that we exploit in our proof here is from Simon [**640**]—he uses the Trotter product formula in place of our argument here. This idea follows the result for $a \in L^2_{\text{loc}}$. Hundertmark–Simon [**327**] has extensions to allow Dirichlet and Neumann boundary conditions.

The Lie product formula (7.6.95) is named after Lie's work on the foundation of Lie groups and Lie algebras. More important to him is

$$\exp([A, B]) = \lim_{n\to\infty} \left[\exp\left(-\frac{B}{\sqrt{n}}\right) \exp\left(-\frac{A}{\sqrt{n}}\right) \exp\left(\frac{B}{\sqrt{n}}\right) \exp\left(\frac{A}{\sqrt{n}}\right)\right] \tag{7.6.113}$$

since it justifies the Lie bracket. While Lie did consider differential relations on flows (see, for example, his book with Engel [**446**]) that are related to (7.6.95) and (7.6.113), it seems that he never really wrote them down explicitly.

There is a close connection between these formulae and the Baker–Campbell–Hausdorff fomula

$$\exp(sX)\exp(sY) = \exp\Big(\sum_{j=1}^{\infty} s^j f_j(X, Y)\Big) \tag{7.6.114}$$

whose leading terms are

$$\begin{aligned}\sum_{j=1}^{\infty} s^j f_j(X, Y) &= s(X + Y) + \frac{s^2}{2}[X, Y] \\ &\quad + \frac{s^3}{12}([X, [X, Y]] - [Y, [X, Y]] - \frac{s^4}{24}([Y, [X, [X, Y]]]) \\ &\quad + O(s^5)\end{aligned} \tag{7.6.115}$$

It is easy to define (7.6.95) and (7.6.113) for finite matrices from this (Problem 16) using the fact that for small s the series converges.

Uniformly in dimension it is easy to see for $\|X\| \le 1$, $\|Y\| \le 1$, $\log\big(\exp(sX)\exp(sY)\big)$ is analytic in an X, Y-independent circle $\{s \mid |s| \le R\}$ and to compute the first two terms by hand. The BCH formula is named after 1897–1906 work of Campbell [**97**], Baker [**42**], and Hausdorff [**299**] (Hausdorff had the important result that after the leading term only multiple commutators enter); Poincaré [**531**] also had an early version. For modern presentations of the form of the terms see [**171, 281, 389, 396**].

In 1959, Trotter [**705**] looked at semigroups, T_t, on a Banach space, X, $0 \le t < \infty$, which were strongly continuous at 0 with $T_{t=0} = \mathbb{1}$ and which obeyed $\|T_t\| \le Ce^{\omega t}$. Any such group has a generator, A, with dense domain

$D(A)$ and $A\varphi = \lim_{t\to 0}\big(\frac{T_t - \mathbb{1}}{t}\big)\,\varphi$ for $\varphi \in D(A)$. One writes $T_t = e^{tA}$. He proved if the closure, C, of $A + B$ defined on $D(A) \cap D(B)$ also generates such a semigroup, then $\text{s-lim}_{n\to\infty}\big(e^{tA/n}\, e^{tB/n}\big)^n = e^{tC}$. We've focused on the Hilbert space case e^{-tA}, $A \geq 0$ and self-adjoint, but his result also applies to e^{itA}.

Extensions of this result in the Hilbert space self-adjoint case have focused in four directions:

(1) using more general functions than $e^{tA/n}$;
(2) quadratic form sums;
(3) convergence of the product in a stronger topology;
(4) error estimates.

There is a huge literature on these questions of which we'll hit some high points.

Chernoff's theorem [**122**] (Proposition 7.6.12) provides a tool not only for the traditional form but also for results like

$$\left[\left(1 + \frac{tA}{n}\right)^{-1}\left(1 + \frac{tB}{n}\right)^{-1}\right]^n \to e^{-t(A+B)} \tag{7.6.116}$$

It is also a part of the proof of the following definitive result that shows the limit always exists!

Theorem 7.6.16 (Kato–Trotter Product Formula). *Let A and B be any positive self-adjoint operators on a Hilbert space, $\mathcal{H}$. Let $\mathcal{K} \subset \mathcal{H}$ be the closure of $Q(A) \cap Q(B)$ and P the projection onto $\mathcal{K}$. Let C be the positive self-adjoint operator on $\mathcal{K}$ whose quadratic form is the sum of q_A and q_B. Then*

$$\underset{n\to\infty}{\text{s-lim}}\big(e^{-tA/n}\, e^{-tB/n}\big)^n = e^{-tC}\, P \tag{7.6.117}$$

Remarks. 1. Recall Theorem 7.5.11 that the sum of closed quadratic forms are always quadratic.

2. A and B need not be densely defined—they need only have associated closed quadratic forms. If $\mathcal{K}_1$ is a subspace of $\mathcal{H}$ and A has $q_A(\varphi) = 0$ (respectively, ∞) for $\varphi \in \mathcal{K}_1$ (respectively, $\varphi \notin \mathcal{K}_1$), then e^{-tA} is for all t, just Q_1, the projection on $\mathcal{K}_1$. In this case (for both A and B), (7.6.12) says that if Q_1, Q_2 are two orthogonal projections, then $\text{s-lim}\,(Q_1\, Q_2)^n = Q$ where Q is the projection onto $\operatorname{Ran} Q_1 \cap \operatorname{Ran} Q_2$.

3. There is an appendix in Kato's paper which uses a remark of Simon to extend the result to cases where A and B are not self-adjoint but e^{-tA} and e^{-tB} are contraction semigroups for t in some sector $|\arg z| \leq \alpha \in (0, \frac{\pi}{2})$.

This result is due to Kato [**382**]. For a textbook version, see Reed–Simon [**548**], Supplement to VIII.8.

The first norm convergence results are in Neidhardt–Zagrebnov [**496**] who considered the case where e^{-tA}, e^{-tB} are trace class with $A+B$ self-adjoint on $D(A)\cap D(B)$ and proved the Trotter product converges in trace norm. Operator norm estimates and the earliest explicit $O(n^{-\beta})$ estimates are in Rogava [**576**]. Ichinose et al. [**331**] obtained $O(n^{-1})$ estimates on the norm convergence if $A+B$ is self-adjoint on $D(A)\cap D(B)$ and found examples the convergence was no better than $O(n^{-1})$. For some results on the unitary case, see [**330, 426, 190**].

The Feynman–Kac formula appeared first in Kac [**357**] based on formal considerations for $\exp\bigl(-it(H_0+V)\bigr)$ by Feynman [**201**]. There is a huge literature on trying, with at most partial success, to make sense of Feynman's idea for e^{-itH}. We mention Nelson [**498**] who had the idea of using the Trotter product formula as a key tool (that also works in the e^{-tH} case).

We note that the standard approach to almost all the results in this section relies heavily on the Trotter product formula. The two-step approach to cut off V and then rely only on Kato's inequality in semigroup form seems to be new here.

Problems

1. This problem will prove Kato's inequality for $-\Delta$, (7.6.14), directly without semigroups.

   (a) Let u be C^∞, $u_\varepsilon=\sqrt{|u|^2+\varepsilon^2}$ for some $\varepsilon>0$. prove that
$$2u_\varepsilon\vec{\nabla}u_\varepsilon = 2\operatorname{Re}\bigl(\bar{u}\,\vec{\nabla}u\bigr) \tag{7.6.118}$$
$$u_\varepsilon\Delta u_\varepsilon + |\nabla u_\varepsilon|^2 = \operatorname{Re}(\bar{u}\Delta u)+|\nabla u|^2 \tag{7.6.119}$$

   (b) Prove that $|\nabla u_\varepsilon|\le|\nabla u|$ and conclude that
$$u_\varepsilon\,\Delta u_\varepsilon > \operatorname{Re}(\bar{u}\,\Delta u) \tag{7.6.120}$$

   (c) If $\operatorname{sgn}_\varepsilon(u)=\bar{u}/u_\varepsilon$, prove that
$$\Delta u_\varepsilon \ge \operatorname{Re}\bigl((\operatorname{sgn}_\varepsilon u)\,\Delta u\bigr) \tag{7.6.121}$$

   (d) By convoluting a $u\in L^1_{\mathrm{loc}}$ with an approximate identity, prove (7.6.121) for any $u\in L^1_{\mathrm{loc}}$ with $\Delta u\in L^1_{\mathrm{loc}}$ as a distributional inequality.

   (e) By taking $\varepsilon\downarrow 0$, conclude (7.6.14) for any $u\in L^1_{\mathrm{loc}}$ with $\Delta u\in L^1_{\mathrm{loc}}$.

2. This problem will prove the self-adjointness part of Theorem 7.6.1 directly from (7.6.14). So let $V\ge 0$ be in L^2_{loc} and suppose for $u\in L^2$, we have for all $f\in C_0^\infty(\mathbb{R}^\nu)$ that
$$\langle u,(-\Delta+V+1)f\rangle = 0 \tag{7.6.122}$$

(a) Prove that $\Delta u \in L^1_{\text{loc}}$.

(b) Prove that

$$\Delta\,|u| \geq 0 \tag{7.6.123}$$

as a distribution inequality. (*Hint*: Use Problem 1.)

(c) Let $w_n = j_n * |u|$ where j_n is an approximate identity. Prove that

$$w_n \in D(\Delta) \text{ and } \langle w_n, \Delta w_n\rangle \leq 0 \tag{7.6.124}$$

(d) Prove that $\Delta w_n \geq 0$ so that $\langle w_n, \Delta w_n\rangle \geq 0$.

(e) Conclude that $w_n = 0$ and conclude that $u = 0$.

(f) Prove that $-\Delta + V \restriction C_0^\infty$ is essentially self-adjoint.

3. Let $\vec{a} \in C^1(\mathbb{R}^\nu)$. Prove that as distribution inequalities if $(\nabla - ia)^2 u \in L^1_{\text{loc}}$ and $u \in L^1_{\text{loc}}$, then

$$\Delta\,|u| \geq \operatorname{Re}\bigl(\operatorname{sgn}(u)\,(\nabla - ia)^2\, u\bigr) \tag{7.6.125}$$

by following the proof in Problem 1.

4. Let $\vec{a} \in C^1(\mathbb{R}^\nu)$, $V \geq 0$, $V \in L^2_{\text{loc}}(\mathbb{R}^\nu)$. By following the proof in Problem 2, prove that $(i\nabla - a)^2 + V$ is essentially self-adjoint on $C_0^\infty(\mathbb{R}^\nu)$.

5. (a) Verify (7.6.15) where the integral is a norm convergence Riemann improper integral (Riemann for $\int_0^R$ and the $\lim_{R\to\infty}$).

(b) Prove (7.6.16) holds even with a $\|\cdot\|$-limit. (*Hint*: Spectral theorem.)

6. Knowing (7.6.53) for all $\varphi \in C_0^\infty(\mathbb{R}^\nu)$, extend it to all φ with $\varphi \in L^1_{\text{loc}}$ and $\Delta\varphi \in L^1_{\text{loc}}$.

7. Using the spectral theorem, prove for any positive self-adjoint operator A and any dense subspace $\mathcal{D} \subset \mathcal{H}$, one has that $e^{-tA}\,[\mathcal{D}]$ is both a form and operator core for A.

8. (a) For any finite matrix C, prove that

$$\|e^{-tC} - 1 - tC\| = O(t^2) \tag{7.6.126}$$

(b) For any finite matrices A and B on the same space, prove that

$$\|e^{-(A+B)/n} - e^{-A/n}\, e^{-B/n}\| = O(n^{-2}) \tag{7.6.127}$$

(c) For any finite matrices C, D, prove that

$$\|C^n - D^n\| \leq n\|C - D\|\bigl[\bigl(1 + \|C\|^n\bigr)\,\bigl(1 + \|D\|^n\bigr)\bigr]$$

(d) Prove that $\lim_{n\to\infty}\|e^{-(A+B)} - \bigl(e^{-A/n}\, e^{-B/n}\bigr)^n\| = O(n^{-1})$ for finite matrices.

9. This problem will prove (7.6.100). Let H_n be given by (7.6.99).

   (a) Prove that $|H_n'(x)| \le 1$.

   (b) Prove that $H_n(x) \to 0$ pointwise.

   (c) Conclude for each R

$$\limsup_{n\to\infty} \sup_{0\le x\le R} |H_n(x)| = 0$$

   (*Hint*: Equicontinuity.)

   (d) Prove (7.6.100).

10. Prove (7.6.104).

11. Complete the proof of (7.6.102).

12. For $V \ge 0$ with $V \in L^\infty$ and e^{-tA} positivity-preserving, use the Trotter product formula to prove that for $f \ge 0$, we have

$$0 \le e^{-t(A+V)} f \le e^{-tA} f$$

13. Use the Kato–Trotter product formula (Theorem 7.6.16) to go from (7.6.92) to (7.6.112).

14. Verify the equality (7.6.108) when each f_j is 1 or an $x_\ell (\ell = 1, 2, \dots, \nu)$ and all but one or two are 1's.

15. Show how to go from (7.6.109) for V a continuous function of compact support to $V \ge 0$ in $L^1_{\text{loc}}(\mathbb{R}^\nu)$.

16. Given (7.6.114) and (7.6.115), prove (7.6.95) and (7.6.113) for finite matrices.

17. (a) Let A, B be positive, self-adjoint generators of positivity-preserving semigroups with

$$f \ge 0 \Rightarrow e^{-tA} f \le e^{-tB} f \tag{7.6.128}$$

   Let V be a multiplication operator. Prove that for any function f in L^2 and $\alpha > 0$

$$\|V(A+\alpha)^{-1} f\| \le \|V(B+\alpha)^{-1} f\| \tag{7.6.129}$$

   (b) Prove that if V is relatively B bounded, it is relatively A-bounded with a no larger A-bound.

   (c) Prove if $V_\pm \ge 0$, $V_+ \in L^2_{\text{loc}}(\mathbb{R}^\nu)$, $V_- \in L^q_u(\mathbb{R}^\nu)$ with $q \ge 2$ if $\nu = 3$ and $q > \frac{\nu}{2}$ if $\nu \ge 4$ (see (7.6.94)), then $-\Delta + V_+ - V_-$ is essentially self-adjoint on $C_0^\infty(\mathbb{R}^\nu)$.

   Remark. (b) is called the Davies–Faris theorem after Davies [**146**] and Faris [**193**].

7.7. Self-adjointness and the Moment Problem

In Sections 4.17 and 5.6 of Part 1, we looked at the following question: Given real numbers, $\{c_n\}_{n=0}^\infty$, with $c_0 = 1$, when is there a probability measure, $d\mu$, with $\text{supp}(d\mu)$ not a finite set (aka μ is non-trivial) and

$$\int |x|^n \, d\mu(x) < \infty \tag{7.7.1}$$

for all n so that

$$c_n = \int x^n \, d\mu(x) \tag{7.7.2}$$

For obvious reasons, this is called the moment problem. The *Hausdorff moment problem* is the case $\text{supp}(d\mu) \subset [0,1]$, the *Stieltjes moment problem* is the case $\text{supp}(d\mu) \subset [0,\infty)$, and the *Hamburger moment problem* is $\text{supp}(d\mu) \subset \mathbb{R}$. The Hausdorff problem was studied in Section 4.17 of Part 1, while the other two in Section 5.6 of Part 1. We also suppose $d\mu$ is nondegenerate, that is, not supported on a finite set of points.

Uniqueness of solutions is easy in the Hausdorff case because polynomials are $\|\cdot\|_\infty$-dense in $C([0,1])$, but it is not necessarily true for either the Hamburger or Stieltjes moment problems. Example 5.6.5 of Part 1 noted that $c_n = \exp(\frac{1}{4}n^2 + \frac{1}{2}n)$ had multiple solutions. While we had complete results on existence for these problems, we barely understood the uniqueness (aka determinate) and nonuniqueness (aka indeterminate) problems. In this section, we'll see the Hamburger and Stieltjes moment problems can be rephrased as self-adjointness problems, which will not only allow us to recover the existence results but also to partially understand uniqueness in terms of the von Neumann (for the Hamburger problem) and Birman–Krein–Vishik (for the Stieltjes problem) theories of self-adjoint extensions of real and positive self-adjoint operators.

Given a set of moments, $\{c_n\}_{n=0}^\infty$, we say it obeys the *Hamburger moment condition* if and only if for all n and all $\zeta \in \mathbb{C}^n \setminus \{0\}$, we have

$$\sum_{j,k=1}^n c_{j+k-2}\bar{\zeta}_j \zeta_k > 0 \tag{7.7.3}$$

If it obeys this condition and also

$$\sum_{j,k=1}^n c_{j+k-1}\bar{\zeta}_j \zeta_k > 0 \tag{7.7.4}$$

we say $\{c_n\}_{n=0}^\infty$ obeys the *Stieltjes moment condition.* In Section 4.17 of Part 1 (see Lemma 4.17.2 of Part 1), it is proven these conditions are equivalent to strict positivity of certain determinants.

It is immediate that these conditions are necessary for the Hamburger (respectively, Stieltjes) problems to have solutions, for if c_n obeys (7.7.2), then

$$\sum_{j,k=1}^{n} c_{j+k-2}\bar{\zeta}_j\zeta_j = \int \biggl|\sum_{j=1}^{n} \zeta_j x^{j-1}\biggr|^2 d\mu(x) \tag{7.7.5}$$

$$\sum_{j,k=1}^{n} c_{j+k-1}\bar{\zeta}_j\zeta_j = \int x\biggl|\sum_{j=1}^{n} \zeta_j x^{j-1}\biggr|^2 d\mu(x) \tag{7.7.6}$$

Here is the key that relates these problems to operators on a Hilbert space:

Theorem 7.7.1. *Let $\{c_n\}_{n=0}^\infty$ be a set of moments obeying the Hamburger moment condition. Let $\mathcal{P}$ be the family of complex polynomials $\{\sum_{j=1}^n \zeta_j X^{j-1} \mid \zeta \in \mathbb{C}^n; n = 0,1,2,\dots\}$. Define a sesquilinear form on $\mathcal{P}$ by*

$$\biggl\langle \sum_{j=1}^{n} \zeta_j X^{j-1}, \sum_{k=1}^{m} \omega_k X^{k-1} \biggr\rangle = \sum_{\substack{j=1,\dots,n\\ k=1,\dots,m}} \bar{\zeta}_j \omega_k c_{j+k-2} \tag{7.7.7}$$

Then $\langle\cdot\,,\cdot\rangle$ is an inner product and the completion, $\mathcal{H}$, of $\mathcal{P}$ is an infinite-dimensional Hilbert space. If A is the operator with $D(A) = \mathcal{P}$ and

$$A[P(X)] = XP(X) \tag{7.7.8}$$

then A is a Hermitian operator. If C is the map on $\mathcal{P}$,

$$C\biggl(\sum_{j=1}^{n} \zeta_j X^{j-1}\biggr) = \sum_{j=1}^{n} \bar{\zeta}_j X^{j-1} \tag{7.7.9}$$

then C is a complex conjugation, $C\colon D(A) \to D(A)$, and

$$C(A\varphi) = A(C\varphi) \tag{7.7.10}$$

If $\{c_n\}_{n=0}^\infty$ also obeys the Stieltjes moment condition, then for all $\varphi \in D(A)$,

$$\langle \varphi, A\varphi\rangle \ge 0 \tag{7.7.11}$$

In both cases, if φ_0 is the polynomial, 1, (i.e. X^0), then $\varphi_0 \in D(A^n)$ for all n and

$$\langle \varphi_0, A^n\varphi_0\rangle = c_n \tag{7.7.12}$$

Proof. Positivity of the putative inner product is trivially implied by (7.7.3). This also implies $\zeta \in \mathbb{C}^n\backslash\{0\}$ means $\|\sum_{j=0}^n \zeta_j X^{j-1}\| \neq 0$ so $\{X^j\}_{j=0}^\infty$ are independent, implying that $\mathcal{H}$ has infinite dimension. That A is Hermitian and C a conjugation obeying (7.7.10) is a direct calculation, as are (7.7.11) and (7.7.12). $\square$

From this follows the existence result:

Theorem 7.7.2. *Let $\{c_n\}_{n=0}^{\infty}$ be a sequence of reals that has $c_0 = 1$ and which obeys the Hamburger moment condition. Then there is a nontrivial measure, $d\mu$, on $\mathbb{R}$ obeying (7.7.2), that is, the Hamburger moment problem is solvable. If $\{c_n\}_{n=0}^{\infty}$ obeys the Stieltjes moment condition, there is a nontrivial measure, $d\mu$, on $[0,\infty)$ obeying (7.7.2).*

Remark. If instead of a set of moments, we are given a tridiagonal Jacobi matrix, we can define A on finite sequences and so have (7.7.10) and so an extension of Favard's theorem (Theorem 4.1.3).

Proof. In the Hamburger case, the operator, A, of the last theorem is Hermitian and obeys (7.7.10). Thus, by Corollary 7.4.3, A has a self-adjoint extension, B. Let $d\mu$ be the spectral measure for B with vector φ_0. Then, since $\varphi_0 \in D(A^n)$ for all n, it is in $D(B^n)$ for all n and $A^n\varphi_0 = X^n = B^n\varphi_0$. Thus,

$$c_n = \langle \varphi_0, A^n\varphi_0\rangle = \langle \varphi_0, B^n\varphi_0\rangle = \int x^n\, d\mu(x) \tag{7.7.13}$$

If μ were trivial, say $\operatorname{supp}(d\mu) = \{\lambda_j\}_{j=1}^{J}$, then $P(X) \equiv \prod_{j=1}^{J}(X-\lambda_j)$ has zero $L^2(\mathbb{R}, d\mu)$-norm so $\|P\|^2 = 0$, contrary to strict positivity in the Hamburger condition.

For the Stieltjes case, A is positive, so it has a positive self-adjoint extension (by Theorem 7.5.19). The above argument showing (7.7.13) still holds, but now $d\mu$ is supported on $\sigma(B) \subset [0,\infty)$ since $B \geq 0$. $\square$

Our next goal will be to prove

Theorem 7.7.3. *Let $\{c_n\}_{n=0}^{\infty}$ with $c_0 = 1$ obey the Hamburger moment condition and let A be the operator (7.7.8). Then the solution of (7.7.2) is unique if and only if A has a unique self-adjoint extension (equivalently, A is essentially self-adjoint).*

While at first sight this seems obvious, both directions have subtleties. If A has two different self-adjoint extensions, B and C, then there are spectral measures, $d\mu_{\varphi_0}^{B}$ and $d\mu_{\varphi_0}^{C}$, solving (7.7.2). But just because $B \neq C$, do we know that the measures are distinct? On the other hand, if A has a unique self-adjoint extension, that says there is a unique measure solving (7.7.2) constructed by this procedure. But how do we know there aren't others? Indeed, as we'll see, in case there are multiple self-adjoint extensions, there are many solutions besides those that come from self-adjoint extensions. So we develop some machinery.

Definition. A solution, $d\mu$, of (7.7.2) which comes as the spectral measure of a self-adjoint extension of A is called a *von Neumann solution*.

For now, we note that since $\mathcal{H} = L^2(\mathbb{R}, d\mu)$ and $D(A)$ is the set of polynomials in x and is dense in $\mathcal{H}$, for any von Neumann solution, $d\mu$, the polynomials in x are dense in $L^2(\mathbb{R}, d\mu)$. We'll eventually prove the converse is true and, in particular, if there are multiple solutions of the moment problem, there are solutions where the polynomials are *not* dense in $L^2(\mathbb{R}, d\mu)$.

We begin by writing a matrix form for A. The arguments in Section 4.1 that led to (4.1.7) and (4.1.23) only depended on $p_n(x)$ being the orthonormal set obtained by applying Gram–Schmidt to $1, x, x^2, \dots$. If we look at $\mathcal{H}$ and elements $1, X, X^2, \dots$, we get by the same method an orthonormal basis (since polynomials are dense in $\mathcal{H}$), $\{p_n(X)\}_{n=0}^\infty$, and Jacobi parameters, $\{a_n, b_n\}_{n=1}^\infty$, so that

$$A[p_n(X)] = a_{n+1}p_{n+1}(X) + b_{n+1}p_n(X) + a_n p_{n-1}(X) \tag{7.7.14}$$

where we set

$$a_0 = 1, \qquad p_{-1}(X) = 0 \tag{7.7.15}$$

Thus, A is given by a Jacobi matrix (of the form (4.1.13)), albeit one that might be unbounded.

Thus, we can view $\mathcal{H}$ as $\ell^2(\mathbb{Z}_+)$ so that

$$\{s_n\}_{n=0}^\infty \longleftrightarrow \sum_{n=0}^\infty s_n p_n(X) \tag{7.7.16}$$

and

$$D(A) = \{s \mid \text{for some } N, s_k = 0 \text{ for } k \geq N\} \tag{7.7.17}$$

Notice that for *any* sequence $\{s_n\}_{n=0}^\infty$, we can define a new sequence (where $s_{-1} = 0$)

$$\mathcal{F}(s)_n = a_{n+1}s_{n+1} + b_{n+1}s_n + a_n s_{n-1} \tag{7.7.18}$$

We also want to discuss Wronskians: Given any two sequences, $\{s_n\}_{n=0}^\infty, \{t_n\}_{n=0}^\infty$, we define

$$W(s,t)_n = a_{n+1}(s_{n+1}t_n - s_n t_{n+1}) \tag{7.7.19}$$

We have, as usual,

$$a_{n+1}s_{n+1} + c_n s_n + a_n s_{n-1} = 0, \quad a_{n+1}t_{n+1} + d_n t_n + a_n t_{n-1} = 0 \tag{7.7.20}$$

$$\Rightarrow W(s,t)_n - W(s,t)_{n-1} = (d_n - c_n)s_n t_n \tag{7.7.21}$$

(by multiplying the first equation in (7.7.20) by t_n and the second by s_n and subtracting). In particular, if for all n, $d_n = c_n$, then W is constant.

This lets us describe $D(A^*)$.

Proposition 7.7.4. *We have that*

$$D(A^*) = \{s \in \ell^2 \mid \mathcal{F}(s) \in \ell^2\} \tag{7.7.22}$$

with

$$A^*s = \mathcal{F}(s) \tag{7.7.23}$$

for all $s \in D(A^*)$. *Moreover, if* $s, t \in D(A^*)$, *we have that*

$$\lim_{n\to\infty} W(\bar{s}, t)_n = \langle A^*s, t\rangle - \langle s, A^*t\rangle \tag{7.7.24}$$

Remark. (7.7.24) includes that the limit exists if $s, t \in D(A^*)$.

Proof. For any finite sequence, $\{t_n\}_{n=0}^\infty$ (i.e., for some N, $t_n = 0$ if $n \geq N$), and any sequence $\{s_n\}_{n=0}^\infty$, define

$$\langle s, t\rangle = \sum_{n=0}^\infty \bar{s}_n t_n \tag{7.7.25}$$

Then for any finite sequence t (which lies in $D(A)$), we have

$$\langle s, At\rangle = \langle \mathcal{F}(s), t\rangle \tag{7.7.26}$$

for every sequence s! This is just a summation by parts.

If $s, \mathcal{F}(s) \in \ell^2$, (7.7.26) says $s \in D(A^*)$ and $A^*s = \mathcal{F}(s)$. Conversely, if $s \in D(A^*)$ and $A^*s = \eta$, (7.7.26) implies

$$\langle \mathcal{F}(s) - \eta, t\rangle = 0 \tag{7.7.27}$$

for all finite sequences, so taking $t_n = \delta_{jn}$, we see $\eta = \mathcal{F}(s)$, so $\mathcal{F}(s) \in \ell^2$.

By the calculation that led to (7.7.21), we have for any sequences, s, t, that

$$W(\bar{s}, t)_n - W(\bar{s}, t)_{n-1} = \overline{\mathcal{F}(s)}_n t_n - \bar{s}_n \mathcal{F}(t)_n \tag{7.7.28}$$

(with $W(\bar{s}, t)_{-1}$ interpreted as 0). Thus, for any s, t and any N,

$$\sum_{n=0}^N \overline{(\mathcal{F}s)}_n t_n - \bar{s}_n(\mathcal{F}t)_n = W(\bar{s}, t)_0 + \sum_{n=1}^N W(\bar{s}, t)_n - W(\bar{s}, t)_{n-1} = W(\bar{s}, t)_N \tag{7.7.29}$$

If $s, t \in \mathcal{D}(A)$, the left side of (7.7.29) converges to $\langle A^*s, t\rangle - \langle s, A^*t\rangle$, which yields (7.7.24). □

This looks, of course, like a discrete analog of (7.4.34). Much of the theory below is essentially a discrete version of the Weyl limit point/limit circle analysis of Theorem 7.4.12.

While X in $p_n(X)$ is a dummy variable, we see that putting $z = X$ (with $z \in \mathbb{C}$) that (7.7.14) says that

$$u_n = \begin{cases} p_n(z), & n \geq 0 \\ 0, & n = -1 \end{cases} \tag{7.7.30}$$

obeys ($a_0 = 1$)

$$a_{n+1}u_{n+1} + b_{n+1}u_n + a_n u_{n-1} = zu_n, \qquad n = 0, 1, 2, \dots \tag{7.7.31}$$

and is (by induction) the unique solution with

$$u_{-1} = 0, \qquad u_0 = 1 \tag{7.7.32}$$

We define functions $\{q_n(z)\}_{n=-1}^{\infty}$ so that for each z, $u_n = q_n(z)$ solves (7.7.31) with initial conditions

$$u_{-1} = -1, \qquad u_0 = 0 \tag{7.7.33}$$

Notice that

$$q_1(z) = a_1^{-1} \tag{7.7.34}$$

and then, by induction, $q_n(z)$ is a polynomial of degree $n-1$. They are called *polynomials of the second kind.* We have

Proposition 7.7.5. *$q_n(z)$ is a polynomial of degree $n-1$ for $n \geq 1$. We have, for all $n \geq 0$, that*

$$a_n[q_n(z)p_{n-1}(z) - q_{n-1}(z)p_n(z)] = 1 \tag{7.7.35}$$

Proof. By (7.7.20)/(7.7.21), the left side of (7.7.35) is n-independent. When $n = 0$, $a_0 = 1$, $q_{-1} = -1$, $p_0 = 1$, $q_0 = 0$, so the combination is 1. □

Given a polynomial, $P(X)$, we define $\mathbb{E}_X(P(X))$ to be the result of replacing X^k by c_k (so $\mathbb{E}_X$ is a linear functional on the polynomials).

Proposition 7.7.6. *For $n \geq 0$,*

$$\mathbb{E}_X\left(\frac{p_n(X) - p_n(Y)}{X - Y}\right) = q_n(Y) \tag{7.7.36}$$

Remark. Since

$$\frac{X^k - Y^k}{X - Y} = \sum_{j=0}^{k-1} X^{k-1-j}Y^j \tag{7.7.37}$$

$(p_n(X) - p_n(Y))/(X - Y)$ is a polynomial in X.

Proof. Use $r_n(Y)$ for the left side of (7.7.36) and set $r_{-1}(X) = 0$. Using the recursion relations for $p_n(X)$ and $p_n(Y)$, we see

$$\begin{aligned} &a_{n+1}r_{n+1}(Y) + b_{n+1}r_n(Y) + a_n r_{n-1}(Y) \\ &\quad = \mathbb{E}_X\left(\frac{Xp_n(X) - Yp_n(Y)}{X - Y}\right) = \mathbb{E}_X\left(\frac{Yp_n(X) - Yp_n(Y)}{X - Y} + p_n(X)\right) \\ &\quad = Yr_n(Y) + \delta_{n0} \end{aligned} \tag{7.7.38}$$

Define $\tilde{r}_n$ to be r_n for $n \geq 0$ with $\tilde{r}_n = -1$. Since $a_0 \tilde{r}_{n+1} = 1$, for $n = 0$, we bring δ_{n0} to the left side of (7.7.27) and so that $\tilde{r}_n(z)$ solves (7.7.31) with

$$\tilde{r}_{-1}(z) = -1, \qquad \tilde{r}_0(z) = 0 \tag{7.7.39}$$

(since $p_0(X) - p_0(Y) = 0$). Thus, $\tilde{r}_n = q_n$, so $r_n = q_n$ for $n \geq 0$. □

q enters because of the following result:

Theorem 7.7.7. *Let μ solve (7.7.2). For $z \in \mathbb{C} \setminus \mathbb{R}$, define*

$$m_\mu(z) = \int \frac{d\mu(x)}{x - z} \tag{7.7.40}$$

Then

$$\langle p_n, (\cdot - z)^{-1} \rangle_{L^2(d\mu)} = q_n(z) + m_\mu(z) p_n(z) \tag{7.7.41}$$

In particular,

$$\sum_{n=0}^{\infty} |q_n(z) + m_\mu(z) p_n(z)|^2 \leq \frac{\operatorname{Im} m_\mu(z)}{\operatorname{Im} z} \tag{7.7.42}$$

with equality if μ is a von Neumann solution (indeed, if $\{p_n(x)\}_{n=0}^{\infty}$ is a basis for $L^2(d\mu)$).

Remarks. 1. We'll see later that if μ is not a von Neumann solution, then one has $<$ in (7.7.42) for *every* $z \in \mathbb{C} \setminus \mathbb{R}$ so, in particular, $\{p_n(x)\}_{n=0}^{\infty}$ is a basis if and only if μ is a von Neumann solution.

2. Fix z. Writing $\zeta = \kappa + i\lambda$, we see that any inequality like (7.7.42) if $p_\cdot(z), q_\cdot(z)$ are both ℓ^2 is a quadratic expression of the form (A, B, C, D real, $A > 0$)

$$A(\kappa^2 + \lambda^2) + B\kappa + C\lambda \leq D \tag{7.7.43}$$

Such a quadratic expression either has no solutions, one solution, or a disk of solutions (with equality on the boundary of the disk). We'll see later in the indeterminate case that the solutions are a positive radius disk. The von Neumann solutions will fill the entire circle and all solutions the entire disk.

Proof. Since $p_n(x)$ is a real polynomial, we have that

$$\begin{aligned} \langle p_n, (\cdot - z)^{-1} \rangle &= \int \frac{p_n(x)}{x - z}\, d\mu(x) && (7.7.44) \\ &= \int \frac{p_n(x) - p_n(z)}{x - z}\, d\mu(x) + m_\mu(z) p_n(z) && (7.7.45) \\ &= q_n(z) + m_\mu(z) p_n(z) && (7.7.46) \end{aligned}$$

We claim next that

$$\left\| \frac{1}{x - z} \right\|^2 = \frac{\operatorname{Im} m_\mu(z)}{\operatorname{Im} z} \tag{7.7.47}$$

for $\operatorname{Im}(1/(x-z)) = \operatorname{Im} z/|x-z|^2$. Thus, (7.7.42) is just Bessel's inequality and the equality is just Parseval's equality. □

For any $z \in \mathbb{C}$, we define sequences, $\pi(z), \xi(z)$, by

$$\pi(z)_n = p_n(z), \qquad \xi(z)_n = q_n(z) \tag{7.7.48}$$

The following is a discrete analog of part of the Weyl limit point/limit circle result:

Theorem 7.7.8. *Fix $z_0 \in \mathbb{C} \setminus \mathbb{R}$. Then the following are equivalent:*

(1) *The Jacobi matrix, A, is not essentially self-adjoint.*
(2) $\pi(z_0) \in \ell^2$.
(3) $\xi(z_0) \in \ell^2$.
(4) *Both $\pi(z_0)$ and $\xi(z_0)$ are in ℓ^2.*

If these conditions hold, A has deficiency indices $(1,1)$ and for all $z_0 \in \mathbb{C}\setminus\mathbb{R}$, $\pi(z_0), \xi(z_0)$ lie in $D(A^*) \setminus D(A)$ and

$$D(A^*) = D(A) \dotplus [\pi(z_0)] \dotplus [\xi(z_0)] \tag{7.7.49}$$

Proof. (1)⇔(2). By Proposition 7.7.4, $(A^* - z_0)u = 0$ is equivalent to (7.7.31) for $z = z_0$ and $u_{-1} = 0$. Thus, $u_n = cp_n(z_0)$, that is,

$$\operatorname{Ker}(A^* - z_0) \neq \{0\} \Leftrightarrow \pi(z_0) \in \ell^2 \tag{7.7.50}$$

Since

$$\operatorname{Ker}(A^* - z_0) = \{0\} \Leftrightarrow \operatorname{Ker}(A^* - \bar{z}_0) = \{0\} \tag{7.7.51}$$

(by the existence of a complex conjugation), (7.7.50) shows (1) ⇔ (2).

(2)⇔(3). Since (7.7.2) has (von Neumann) solutions, μ, with $m_\mu(z_0) \neq 0$ (since $\operatorname{Im} m_\mu(z_0) > 0$), we know for some $w_0 \neq 0$,

$$\xi(z_0) + w_0 \pi(z_0) \in \ell^2 \qquad \text{(by (7.7.42))}$$

Thus, $\xi(z_0) \in \ell_2 \Leftrightarrow \pi(z_0) \in \ell_2$.

(2)⇔(4). Immediate from (4)≡(2)+(3) plus (2)⇔(3).

Since $\operatorname{Ker}(A^* - z_0)$ is given by (7.7.50), its dimension is 1 if $\pi(z_0) \in \ell^2$, that is, if A is not essentially self-adjoint, this dimension is always 1. Fix $z \in \mathbb{C}\setminus\mathbb{R}$. Then $\pi(z_0), \xi(x_0) \in \ell_2$. Since $\mathcal{F}(\pi(z_0)) = z_0\pi(z_0)$ and $\mathcal{F}(\xi(z_0)) = z_0\xi(z_0) + \delta_{n0}$, by Proposition 7.7.4, $\pi(z_0), \xi(z_0) \in D(A^*)$. By (7.7.24) and (7.7.35),

$$\langle A^*\xi(\bar{z}_0), \pi(z_0)\rangle - \langle \xi(\bar{z}_0), A^*\pi(z_0)\rangle = 1 \neq 0 \tag{7.7.52}$$

This implies $\pi(z_0), \xi(z_0)$ are not in $D(A)$. Indeed, they are linearly independent in $D(A^*)/D(A)$ (Problem 1). This proves (7.7.49) since $\dim(D(A^*)/D(A))$ is 2. □

Proof of Theorem 7.7.3. If A is essentially self-adjoint, by the above, for any $z_0 \in \mathbb{C} \setminus \mathbb{R}$, $\xi(z_0) + w\pi(z_0) \in \ell^2$ for a unique w (or else $\pi \in \ell^2$). Thus, for any two solutions of the moment problem, μ, ν,

$$\int \frac{d\mu(x)}{x-z} = \int \frac{d\nu(x)}{x-z} \tag{7.7.53}$$

for all $z \in \mathbb{C} \setminus \mathbb{R}$, which implies that $d\mu = d\nu$ by Theorem 2.5.4 of Part 3.

If A is not essentially self-adjoint, and if the spectral measures, μ, ν, for two self-adjoint extensions, B and C, are the same, we have

$$\langle \varphi_0, (B-z)^{-1}\varphi_0 \rangle = m_\mu(z) = \langle \varphi_0, (C-z)^{-1}\varphi_0 \rangle \tag{7.7.54}$$

By (7.7.41) and the completeness of $\{p_n\}_{n=0}^\infty$, we see that

$$(B-z)^{-1}\varphi_0 = (C-z)^{-1}\varphi_0 \tag{7.7.55}$$

Since this is true for all z in $\mathbb{C}\setminus\mathbb{R}$ and polynomials in $(x \pm i)^{-1}$ are $\|\cdot\|_\infty$-dense in $C_\infty(\mathbb{R})$, we see that

$$f(B)\varphi_0 = f(C)\varphi_0 \tag{7.7.56}$$

for all such f. In particular,

$$(B-z)^{-1}p_n(A)\varphi_0 = (C-z)^{-1}p_n(A)\varphi_0 \tag{7.7.57}$$

for all p_n (using $p_n(A)\varphi_0 = p_n(B)\varphi_0 = p_n(C)\varphi_0$). Thus,

$$(B-z)^{-1} = (C-z)^{-1} \tag{7.7.58}$$

implying $B = C$. Thus, distinct self-adjoint extensions must have distinct spectral measures, and so (7.7.2) has multiple solutions. $\square$

We next want to focus on the structure of the solutions in the indeterminate case. We'll show in that case the von Neumann solutions are all pure point measures, that the set of all solutions is infinite-dimensional, and that for non-von Neumann solutions (for which we'll find many), the polynomials are not a basis and one has strict inequality in (7.7.42) at every $z_0 \in \mathbb{C} \setminus \mathbb{R}$.

We start by showing that, in the indeterminate case, $\pi(z_0), \xi(z_0) \in \ell^2$ holds also for $z_0 \in \mathbb{R}$ and this will allow us to parametrize the von Neumann solutions in terms of a real number $t \in \mathbb{R} \cup \{0\}$ and then "explicitly" compute $m_\mu(z)$. Given the method for proving the analog of this in the Weyl limit point/limit circle theory, it is not surprising it is a variation of parameters.

Fix $z_0 \in \mathbb{C}$. Given an arbitrary sequence, $\{x_n\}_{n=-1}^\infty$, we can define $\alpha_n(x), \beta_n(x)$, $n = -1, 0, 1, \dots$, by

$$\begin{pmatrix} x_{n+1} \\ x_n \end{pmatrix} = \alpha_n \begin{pmatrix} p_{n+1}(z_0) \\ p_n(z_0) \end{pmatrix} + \beta_n \begin{pmatrix} q_{n+1}(z_0) \\ q_n(z_0) \end{pmatrix} \tag{7.7.59}$$

Since

$$W(q(z_0), p(z_0))_n = 1 \tag{7.7.60}$$

$\binom{p_{n+1}}{p_n}$ and $\binom{q_{n+1}}{q_n}$ are independent in $\mathbb{C}^2$ and so, such an expansion is possible. Indeed, by (7.7.60)

$$\alpha_n = W(q.(z_0), x.)_n, \qquad \beta_n = -W(p.(z_0), x.)_n \tag{7.7.61}$$

Proposition 7.7.9. *For any moment problem (i.e., not just indeterminate), fix $z, z_0 \in \mathbb{C}$. If $p_n(z)$ or $q_n(z)$ is expanded in α_n, β_n associated to (7.7.59) (i.e., with z_0), then*

$$\begin{pmatrix} \alpha_{n+1} \\ \beta_{n+1} \end{pmatrix} = [\mathbb{1} + (z - z_0) S_n(z_0)] \begin{pmatrix} \alpha_n \\ \beta_n \end{pmatrix} \tag{7.7.62}$$

where

$$S_n(z_0) = \begin{pmatrix} q_n(z_0)p_n(z_0) & q_n(z_0)q_n(z_0) \\ -p_n(z_0)p_n(z_0) & -p_n(z_0)q_n(z_0) \end{pmatrix} \tag{7.7.63}$$

and

$$(\alpha_{-1}, \beta_{-1}) = (1, 0) \text{ for } p.(z); \quad = (0, 1) \text{ for } q.(z) \tag{7.7.64}$$

One has

$$S_n^2 = \det(S_n) = \operatorname{Tr}(S_n) = 0 \tag{7.7.65}$$

Proof. (7.7.62)/(7.7.63) is a simple calculation (Problem 2) using (7.7.19) and the difference equation (7.7.31) for z, z_0. (7.7.64) is immediate from (7.7.59) and that the boundary conditions for p, q are z-independent. (7.7.65) is an immediate calculation. □

Theorem 7.7.10. *Suppose at some z_0, $\pi(z_0)$ and $\xi(z_0)$ are both in ℓ^2. Then*

(a) *With S given by (7.7.63), we have that*

$$\sum_{n=-1}^{\infty} \|S_n(z_0)\| < \infty \tag{7.7.66}$$

(b) *For all $z \in \mathbb{C}$, $\pi(z), \xi(z)$ are in ℓ^2.*

(c) $$T_\infty(z) = \lim_{n\to\infty} (1 + (z - z_0)S_n(z_0)) \dots (1 + (z - z_0)S_{-1}(z_0)) \tag{7.7.67}$$

exists and defines an entire analytic 2×2 matrix-valued function.

(d) $\det(T_\infty(z)) = 1$.

Remark. Thus, if A is not essentially self-adjoint, we have $\pi(z), \xi(z)$ in ℓ^2 for all $z \in \mathbb{C}$, not just $z \in \mathbb{C} \setminus \mathbb{R}$.

Proof. (a) As products of ℓ^2 sequences, the matrix elements in (7.7.63) are ℓ^1.

(b) We note that

$$\|1 + (z - z_0)S_n(z_0)\| \le e^{\|S_n(z_0)\||z - z_0|} \tag{7.7.68}$$

so by (7.7.62),

$$\left\| \begin{pmatrix} \alpha_{n+1} \\ \beta_{n+1} \end{pmatrix} \right\| \le e^{\|S_n(z_0)\| |z-z_0|} \left\| \begin{pmatrix} \alpha_n \\ \beta_n \end{pmatrix} \right\| \tag{7.7.69}$$

Thus, by (7.7.66) and (7.7.67), $C = \sup_n(|\alpha_n| + |\beta_n|) < \infty$, so

$$|p_n(z)| \le C(|p_n(z_0)| + |q_n(z_0)|) \tag{7.7.70}$$

and $p_{\cdot}(z_0), q_{\cdot}(z_0) \in \ell^2$ implies $p_{\cdot}(z), q_{\cdot}(z) \in \ell_2$.

(c) Let $T_n(z)$ be the product in (7.7.13). Then, using (7.7.68),

$$\|T_{n+1}(z) - T_n(z)\| \le |z - z_0| \|S_{n+1}(z_0)\| \exp\left[|z - z_0| \sum_{j=-1}^{n} \|S_j(z_0)\|\right] \tag{7.7.71}$$

proving that T_n is Cauchy (on account of (7.7.66)). The convergence is uniform in z for z in any compact subset of $\mathbb{C}$. Since the T_n are polynomials in z, the limit is entire analytic.

(d) If $\det(C) = \operatorname{Tr}(C) = 0$ for 2×2 matrices, then

$$\det(1 + C) = 1 + \operatorname{Tr}(C) + \det(C) - 1 \tag{7.7.72}$$

so $\det(1 + (z - z_0)S_n(z_0)) = 1$. Thus, $\det(T_n(z)) = 1$ so $\det(T_\infty(z)) = 1$. $\square$

We now specialize to $z_0 = 0$. Thus, $A^*\pi(0) = 0$, $A^*\xi(0) = \delta_{\cdot 0}$, so we can apply Theorem 7.4.7 with $\psi = \pi(0)$, $\varphi = \xi(0)$. The self adjoint extensions, B_t, of A are indexed by $t \in \mathbb{R} \cup \{\infty\}$ with

$$D(B_t) = D(A) + [\xi(0) + t\pi(0)] \tag{7.7.73}$$

(with $[\xi(0) + t\pi(0)]$ interpreted as $[\pi(0)]$ if $t = \infty$). We'll let μ_t be the corresponding von Neumann solution and m_t its Stieltjes transform.

Proposition 7.7.11. (a) *$\varphi \in D(A^*) \setminus D(A)$ lies in $D(B_t)$ if and only if*

$$\lim_{n\to\infty} \frac{W(\xi(0), \varphi)_n}{W(\pi(0), \varphi)_n} = -t \tag{7.7.74}$$

(b) *If $\varphi \in D(A^*) \setminus D(A)$ and α, β are defined by (7.7.59) (with $z_0 = 0$ and $x = \varphi$), then $\lim \alpha_n = \alpha_\infty$ and $\lim \beta_n = \beta_\infty$ exist, both are not zero and $\varphi \in D(B_t)$ if and only if*

$$\alpha_\infty = t\beta_\infty \tag{7.7.75}$$

Proof. (a) Since $D(A^*) = D(A) \dotplus [\xi(0) \dotplus [\pi(0)]$, we can write

$$\varphi = \eta + a\xi(0) + b\pi(0) \tag{7.7.76}$$

for some $\eta \in D(A)$ and $(a, b) \in \mathbb{C}^2 \setminus (0, 0)$. By (7.7.24) and the reality of $\pi(0)$ and $\xi(0)$ and (7.7.35), we get

$$b = \lim_{n\to\infty} W(\xi(0), \varphi)_n, \quad a = \lim_{n\to\infty} -W(\pi(0), \varphi)_n \tag{7.7.77}$$

Thus, $a\xi(0) + b\pi(0) = a(\xi(0) + t\pi(0))$ if and only if (7.7.74) holds.

(b) This follows from (a) and (7.7.61). □

Now define $A(z), B(z), C(z), D(z)$ by

$$T_\infty(z, z_0 = 0)^{-1} = \begin{pmatrix} C(z) & A(z) \\ -D(z) & -B(z) \end{pmatrix} \tag{7.7.78}$$

Since $\det(T_\infty(z)) = 1$, it is always invertible, and since $T_\infty(z)$ are entire, so are A, B, C, D.

Theorem 7.7.12. *We have that*

$$A(z)D(z) - B(z)C(z) = 1 \tag{7.7.79}$$

For each $t \in \mathbb{R} \cup \{\infty\}$ *and* $z \in \mathbb{C} \setminus \mathbb{R}$,

$$m_t(z) = -\frac{C(z)t + A(z)}{D(z)t + B(z)} \tag{7.7.80}$$

Proof. Let

$$(B_t - z)^{-1}\delta_0 = \eta_t(z) \tag{7.7.81}$$

Then

$$(A^* - z)\eta_t(z) = \delta_0 \tag{7.7.82}$$

so

$$\eta_t(z) - \xi(z) \in \operatorname{Ker}(A^* - z) = [\pi(z)] \tag{7.7.83}$$

By (7.7.81)

$$\langle \delta_0, \eta_t(z) \rangle = m_t(z) \tag{7.7.84}$$

Since $\langle \delta_0, \xi(z) \rangle = 0$, $\langle \delta_0, \pi(z) \rangle = 1$, we see that

$$\eta_t(z) = \xi(z) + m_t(z)\pi(z) \tag{7.7.85}$$

It follows that

$$\alpha_{-1} = m_t(z), \qquad \beta_{-1} = 1 \tag{7.7.86}$$

so, by (7.7.75),

$$T_\infty(z, z_0 = 0)\begin{pmatrix} m_t(z) \\ 1 \end{pmatrix} = c_t(z)\begin{pmatrix} t \\ 1 \end{pmatrix} \tag{7.7.87}$$

or, by (7.7.78),

$$\begin{pmatrix} m_t(z) \\ 1 \end{pmatrix} = c_t(z)\begin{pmatrix} C(z) & A(z) \\ -D(z) & -B(z) \end{pmatrix}\begin{pmatrix} t \\ 1 \end{pmatrix} \tag{7.7.88}$$

which immediately yields (7.7.80). □

Theorem 7.7.13. *Suppose that* $\{c_n\}_{n=0}^\infty$ *is a family of indeterminate Hamburger moments. Then*

(a) *For each* t, $d\mu_t$ *is a pure point measure. If* x_0 *is a pure point of* $d\mu_t$, *we have that*

$$\mu_t(\{x_0\}) = \frac{1}{\sum_{n=0}^\infty |p_n(x_0)|^2} \tag{7.7.89}$$

(b) *$\sigma(B_t)$ is the set of pure points and each $\lambda \in \sigma(B_t)$ is an eigenvalue of multiplicity* 1. *Moreover,*

$$\bigcup_{t \in \mathbb{R} \cup \{\infty\}} \sigma(B_t) = \mathbb{R} \tag{7.7.90}$$

(c) *For $t_1 \neq t_2$, $\sigma(B_{t_1}) \cap \sigma(B_2) = \emptyset$.*

(d) *If $x \in \mathbb{R} \setminus [\sigma(B_{t_1}) \cup \sigma(B_{t_2})]$, then $(B_{t_1} - x)^{-1} - (B_{t_2} - x)^{-1}$ is rank-one.*

(e) *The eigenvalues of B_{t_1} and B_{t_2} interlace in the sense that between any two eigenvalues of B_{t_1} is exactly one eigenvalue of B_{t_2}.*

(f) *For all $t \neq \infty$, $0 \notin \sigma(B_t)$ and*

$$m_t(0) = t \tag{7.7.91}$$

(g) *For $t_1, t_2 \neq \infty$,*

$$B_{t_1}^{-1} - B_{t_2}^{-1} = (t_1 - t_2)\langle \pi(0), \cdot \rangle \pi(0) \tag{7.7.92}$$

(h) *The eigenvalues $x_n(t)$ of B_t are monotone decreasing and continuous in t. Indeed,*

$$\frac{dx_n(t)}{dt} = -x_n(t)^2 \left[\sum_{n=0}^{\infty} p_n(0) p_n(x(t)) \right]^2 \tag{7.7.93}$$

Remarks. 1. That (7.7.92) holds, that $\pi(x(t_1))$ is an eigenvector of $B_{t_1}^{-1}$, and no other $B_{t_2}^{-1}$ implies that $\langle \pi(0), \pi(x(t_1)) \rangle \neq 0$. Thus, (7.7.93) shows $\frac{dx_n(t)}{dt} \neq 0$ for all n and all $t \in \mathbb{R}$. A similar analysis looks at $\pi(\varepsilon)$ for ε small, so the parametrization for $t = \infty$ is a finite value, allows one to show at $t^{-1} = 0$, that $\frac{dx_n(t^{-1})}{dt^{-1}} \neq 0$.

2. One can understand the results here in part by looking at the theory of rank-one perturbations of finite matrices.

Proof. (a)–(c) A, B, C, D are all analytic and it is impossible that $D(z)t + B(z)$ and $C(z)t + A(z)$ have a common zero (since if they do, $B(z)/D(z) = -t = A(z)/C(z)$ so $AD - BC = 0$, contrary to $\det(T_\infty^{-1}) = 1 = AD - BC$). Thus, $m_t(z)$ has a continuation from $\mathbb{C}_+$ to an entire meromorphic function. Since also

$$m_t(z) = \int \frac{d\mu_t(x)}{x - z} \tag{7.7.94}$$

we see that $m_t(z)$ has simple poles only on $\mathbb{R}$ with negative residues. It follows that $d\mu_t$ is a pure point measure whose pure points are solutions of

$$\frac{B(z)}{D(z)} = -t \tag{7.7.95}$$

Obviously, distinct t's have distinct solutions of (7.7.95), proving (c) and any $x \in \mathbb{R}$ is a pure point of B_t, where t is given by (7.7.95) (with $z = x$). This proves (7.7.90).

If x_0 is an eigenvalue of B_t, the normalized eigenvector, ψ, must solve $(A^* - x_0)\psi = 0$, that is,

$$\psi = \frac{\pi(x_0)}{\|\pi(x_0)\|^2} \tag{7.7.96}$$

proving that the eigenvalue has multiplicity 1 (since A has deficiency indices $(1,1)$) and the weight $\mu_t(\{x\})$ is given by $|\psi_{n=0}|^2$. Since $p_n(x_0) = 1$, (7.7.96) says

$$|\psi_{n=0}|^2 = \frac{1}{\sum_{n=0}^\infty p_n(x_0)} \tag{7.7.97}$$

proving (7.7.89).

(d)–(e) If $\eta = (A - x)\varphi$, then

$$(B_t - x)^{-1}\eta = \varphi \tag{7.7.98}$$

for all t, proving $(B_{t_1} - x)^{-1} - (B_{t_2} - x)^{-1}$ vanishes on $\text{Ran}(A - x)$ and so on $\overline{\text{Ran}(A - x)}$. Since it is self-adjoint, it is nonvanishing only on $\text{Ran}(A - x)^\perp = \text{Ker}(A^* - x)$, which is one-dimensional. This proves (d). (e) then follows from Theorem 5.8.9.

(f)–(h) $B_t^{-1}\delta_0 = \xi(0) + m_t(0)\pi(0)$ by (7.7.85). But $B_t^{-1}\delta_0$ must obey (7.7.75), so we have (7.7.91).

By the fact that $B_{t_1}^{-1} - B_{t_2}^{-1}$ vanishes on $\text{Ker}(A^*)^\perp$, we see that

$$B_{t_1}^{-1} - B_{t_2}^{-1} = c(t_1, t_2)\langle \pi(0), \cdot \rangle \pi(0)$$

Taking $\langle \delta_0, \cdot \delta_0 \rangle$ matrix elements using $\pi(0) = p_0(0) = 1$ and (7.7.91), we see $c(t_1, t_2) = t_1 - t_2$, proving (7.7.92).

First-order eigenvalue perturbation theory, aka the Feynman–Hellmann theorem (see Theorem 1.4.7), shows that

$$\frac{d}{dt} x_n(t)^{-1} = |\langle \pi(0), \pi(x_n(t)) \rangle|^2 \tag{7.7.99}$$

proving (7.7.93). □

Having understood a lot about the von Neumann solutions, we turn to non-von Neumann solutions.

Theorem 7.7.14. *If $\{c_n\}_{n=0}^\infty$ is a family of indeterminate Hamburger moments, then there are solutions, $d\mu$, of the moment problem in each of the following classes:*

(a) *μ is purely a.c.*
(b) *μ is pure point.*
(c) *μ is purely singular continuous.*
(d) *There is some μ which has nonvanishing components of any pair of type (a)–(c) and μ with all three types.*

Proof. Knowing $\frac{dx_n}{dt} \neq 0$, we conclude if $d\eta(t)$ is a probability measure, then $\int d\eta(t)\, d\mu_t(x) = d\mu^{(\eta)}(x)$ is purely a.c. if η is and purely s.c. if η is. The von Neumann solutions are pure point. To get mixed type, use the fact that the solutions of the moment problem are a convex set and just take a convex combination of the above. □

We are heading towards a wonderful parametrization of all solutions in the indeterminate case. We need the following:

Theorem 7.7.15. *For each $z \in \mathbb{C}_+$, the map*

$$\mathcal{F}_z(w) = -\frac{C(z)w + A(z)}{D(z)w + B(z)} \tag{7.7.100}$$

maps $w \in \mathbb{C}_+$ one–one into the interior of a disk in $\mathbb{C}_+$ and maps $t \in \mathbb{R} \cup \{\infty\}$ one-one onto the boundary of this disk.

Proof. $\mathcal{F}_z(w)$ maps $\mathbb{R} \cup \{\infty\}$ one-one onto a circle in the upper half-plane, since $\mathcal{F}_z(t) = m_t(z)$ and $\mathcal{F}_z(\cdot)$ is a nondegenerate FLT since $-CB+AD = 1$. For $z \in \mathbb{R}$, $\mathcal{F}_z$ is an FLT associated to an element of $\mathbb{SL}(2,\mathbb{R})$ and so maps $\mathbb{C}_+$ to $\mathbb{C}_+$ (see Theorem 7.3.9 of Part 2A).

For each $z \in \mathbb{C}_+$, the image of $\mathbb{R} \cup \{\infty\}$ is a circle and not a straight line since the image is finite for all $t \in \mathbb{R} \cup \{\infty\}$. Thus, for each z, $\mathbb{C}_+$ is mapped onto either the interior or exterior of the circle. But joint continuity in z and w (as a $\mathbb{C} \cup \{\infty\}$-valued map), it must be the same for all $z \in \mathbb{C}_+$ (i.e., all interior or all exterior). If it were always the exterior, in the limit as z approaches a point in $\mathbb{R}$ of some $w \in \mathbb{C}$ would have to lie in $\mathbb{C}_-$ since the interior of the disks expands to the whole of $\mathbb{C}_+$. But we saw above for z real, $\mathcal{F}_z$ maps $\mathbb{C}_+$ to $\mathbb{C}_+$. Thus, the upper half-plane is always mapped to the interior. □

Proposition 7.7.16. *Let $\{c_n\}_{n=0}^\infty$ be a set of Hamburger moments and let $\mathcal{M}(\{c_n\}_{n=0}^\infty)$ be the set of solutions of (7.7.2). Then*

(a) *For all $\mu \in \mathcal{M}(\{c_n\}_{n=0}^\infty)$, we have for all N and $y \in (0,\infty)$,*

$$\int \frac{d\mu(x)}{x - iy} = \sum_{n=0}^{N} \frac{i^{n+1} c_n}{y^{n+1}} + O\biggl(\frac{1}{y^{N+2}}\biggr) \tag{7.7.101}$$

where the $O(y^{-N-2})$ is bounded uniformly in $\mu \in \mathcal{M}(\{c_n\}_{n=0}^\infty)$.

(b) *If $m(z)$ is an arbitrary analytic function on $\mathbb{C}_+$ with $\operatorname{Im} m(x+iy) > 0$ for all $y > 0$, $x \in \mathbb{R}$, and so that, with the left side of (7.7.101) replaced by $m(iy)$, (7.7.101) holds for each N, then $m(z) = \int \frac{d\mu(x)}{x-z}$ where μ has finite moments for all n and lies in $\mathcal{M}(\{c_n\}_{n=0}^\infty)$.*

Proof. (a) By geometric series with remainder,

$$\frac{1}{x-iy}=\sum_{n=0}^{N}\frac{i^{-n+1}x^n}{y^{n+1}}+\frac{i^{N-1}x^{N+1}}{y^{N+1}}\,\frac{1}{x-iy} \tag{7.7.102}$$

Thus, since $|(x-iy)^{-1}|\le y^{-1}$,

$$\left|\int\frac{d\mu(x)}{x-iy}-\sum_{n=0}^{N}\frac{i^{n+1}c_n}{y^{n+1}}\right|\le\frac{c_{N+1}}{y^{N+2}} \tag{7.7.103}$$

which implies (7.7.101) uniformly in μ.

(b) By the Herglotz representation theorem (Theorem 5.9.1 in Part 3),

$$m(z)=cz+d+\int d\mu(x)\left[\frac{1}{x-z}-\frac{x}{1+x^2}\right] \tag{7.7.104}$$

where $c>0$, d is real, and

$$\int\frac{d\mu(x)}{1+x^2}<\infty \tag{7.7.105}$$

Since $y^{-1}m(iy)\to c$ and $m(iy)=O(1/y)$, we have $c=0$. (7.7.105) then implies that

$$y\ \mathrm{Im}\, m(iy)=\int\left(\frac{y^2}{x^2+y^2}\right)d\mu(x) \tag{7.7.106}$$

By the monotone convergence theorem, the right side converges to $\int d\mu(x)$ (which, a priori, might be infinite). Since $y\ \mathrm{Im}\, m(iy)\to c_0$, we see μ is a finite measure.

This in turn implies that

$$\mathrm{Re}\, m(iy)\to d-\int d\mu(x)\left[\frac{x}{1+x^2}\right] \tag{7.7.107}$$

Since $m(iy)=O(1/y)$, we see that

$$m(iy)=\int\frac{d\mu(x)}{x-iy} \tag{7.7.108}$$

Suppose we know $\int|x|^n\,d\mu<\infty$ for $n=0,1,2,\dots,2M-2$ and (7.7.2) holds for $n=0,1,2,\dots,2M-2$. Then (7.7.102) and (7.7.101) for $N=2M$ implies that

$$\int\frac{(iy)^2x^{2M-1}}{x-iy}\,d\mu(x)+iyc_{2M-1}\to -c_{2M} \tag{7.7.109}$$

Taking real and imaginary parts implies

$$c_{2M-1}=\lim_{y\to\infty}\int\frac{y^2x^{2M-1}}{x^2+y^2}\,d\mu(x) \tag{7.7.110}$$

$$c_{2M} = \lim_{y \to \infty} \int \frac{y^2 + x^{2M}}{x^2 + y^2} \, d\mu(x) \tag{7.7.111}$$

The monotone convergence theorem and (7.7.111) implies $\int x^{2M} \, d\mu(x) = c_{2M} < \infty$, and then the dominated convergence theorem and (7.7.110) implies that $\int x^{2M-1} \, d\mu(x) = c_{2M-1}$. Thus, by induction, $\mu \in \mathcal{M}(\{c_n\}_{n=0}^\infty)$. $\square$

Let $\mathcal{H}$ be the set of analytic functions $f \colon \mathbb{C}_+ \to \mathbb{C}$ so

$$\operatorname{Im} f(z) \geq 0 \qquad \text{for all } z \in \mathbb{C}_+ \tag{7.7.112}$$

together with the function $f(z) = \{\infty\}$. Notice $\mathcal{H}$ includes not only Herglotz functions (for which (7.7.112) has $>$, not just ≥ 0) but also the real constants (by the open mapping theorem, if $\operatorname{Im} f(z_0) = 0$, then $f(z)$ is constant).

Theorem 7.7.17 (Nevanlinna Parametrization Theorem). *For any indeterminate Hamburger moment problem, there is a one-one correspondence between $\mathcal{H}$ and solutions of the moment problem $\Phi \in \mathcal{H} \leftrightarrow d\mu_\Phi$ given by*

$$\int \frac{d\mu_\Phi(x)}{x - z} = -\frac{C(z)\Phi(z) + A(z)}{D(z)\Phi(z) + B(z)} \tag{7.7.113}$$

Remarks. 1. We emphasize that (7.7.113) is a nonlinear parametrization. In particular, extreme points in $\mathcal{H}$ are not directly related to extreme points in $\mathcal{M}(\{c_n\}_{n=0}^\infty)$.

2. Computing the inverse of $\left(\begin{smallmatrix} -C & -A \\ B & D \end{smallmatrix}\right)$, we see if $m_\Phi(z)$ is the left side of (7.7.113), then

$$\Phi(z) = -\frac{B(z)m_\Phi(z) + A(z)}{D(z)m_\Phi(z) + C(z)} \tag{7.7.114}$$

Proof. Suppose first $\mu \in \mathcal{M}(\{c_n\}_{n=0}^\infty)$. If m_μ is given by (7.7.40), then by Theorems 7.7.7 and 7.7.15, $\mathcal{F}_z^{-1}(m_\mu(z)) \equiv \Phi(z)$, which is given by (7.7.114), is analytic in z and has $\operatorname{Im} \Phi(z) \geq 0$.

Conversely, if $\Phi \in \mathcal{H}$, the right side of (7.7.113), $m_\Phi(z)$, lies inside the circle $\{m_t(z) \mid t \in \mathbb{R} \cup \{\infty\}\}$. Since this circle lies in $\mathbb{C}_+$, $\operatorname{Im} m_\Phi(z) > 0$. Since $m_t(z)$ obeys (7.7.101) uniformly in t, any function with values in the circle or a convex combination of points on the circle obeys (7.7.101). Thus, by Proposition 7.7.16, m_Φ has the form on the left side of (7.7.113) for $\mu_\Phi \in \mathcal{M}(\{c_n\}_{n=0}^\infty)$. $\square$

Corollary 7.7.18. (a) *If $\mu \in \mathcal{M}(\{c_n\}_{n=0}^\infty)$ is not a von Neumann solution, then for each $z \in \mathbb{C} \setminus \mathbb{R}$, $m_\mu(z)$ lies strictly inside the circle $\{m_t(z) \mid t \in \mathbb{R} \cup \{\infty\}\}$ and, in particular,* (7.7.42) *has strict inequality for all $z \in \mathbb{C} \setminus \mathbb{R}$.*

(b) *For any* $\mu \in \mathcal{M}(\{c_n\}_{n=0}^\infty)$ *which is not a von Neumann solution, the polynomials are not dense in* $L^2(\mathbb{R}, d\mu)$.

(c) *Each von Neumann solution is an extreme point of* $\mathcal{M}(\{c_n\}_{n=0}^\infty)$.

Remarks. 1. There are many more extreme points. For example, further analysis (see the Notes and Problem 3) shows that if $\Phi(z)$ is a rational function, then $d\mu_\Phi$ is an extreme point and the polynomials have finite codimension in $L^2(\mathbb{R}, d\mu)$. The set of measures (all of them the extreme points!) with this codimension exactly N is a $2N+1$-dimensional manifold. These extreme points as N varies are weakly dense in $\mathcal{M}(\{c_n\}_{n=0}^\infty)$!

2. Further analysis (see Problem 4) shows if $\mu \in \mathcal{M}(\{c_n\}_{n=0}^\infty)$ with $\mu(\{x_0\}) \neq 0$ and if μ_t is the von Neumann solution with $\mu_t(\{x_0\}) \neq 0$, then

$$\mu(\{x_0\}) \leq \mu_t(\{x_0\}) \tag{7.7.115}$$

with equality only if $\mu = \mu_t$.

3. Further analysis (see Problem 5) shows if $w \in D(z_0)$, the closed disk whose boundary is $\{m_t(z) \mid t \in \mathbb{R}\cup\{\infty\}\}$, then if $w \in \partial D(z_0)$, then $m_\mu(z) = w \Rightarrow \mu$ is a μ_t. But if $w \in D(z_0)^{\text{int}}$, there are multiple Φ's with $m_\Phi(z) = w$.

Proof. (a) If m_Φ takes a value in the circle $\{m_t(z_0) \mid t \in \mathbb{R} \cup \{\infty\}\}$, then $\Phi(z_0) = t$ for $t \in \mathbb{R} \cup \{\infty\}$. By the open mapping theorem, it follows that $\Phi(z) = t$ for all z. Thus, $\mu = \mu_t$.

(b) The proof of Theorem 7.7.7 implies that one has equality in (7.7.42) if the polynomials are dense.

(c) Given $t_0 \in \mathbb{R}\cup\{\infty\}$, pick $z_0 \in \mathbb{C}_+$. Since the circle $\{m_t(z_0) \mid t \in \mathbb{R}\cup\{\infty\}\}$ bounds a strictly convex set, there is $\omega \in e^{i\theta}$ so that for all w in the disk,

$$\operatorname{Re} \bar{\omega}[w - m_{t_0}(z_0)] \geq 0 \tag{7.7.116}$$

with equality if and only if $w = m_{t_0}(z_0)$.

If μ_t is a convex combination of $\nu_1, \nu_2 \in \mathcal{M}(\{c_n\}_{n=0}^\infty)$, then (7.7.116) implies $m_{\nu_j}(z_0) = m_{t_0}(z_0)$. By the proof of (a), $\nu_1 = \nu_2 = \mu_{t_0}$. $\square$

We now turn to the uniqueness question for the Stieltjes moment problem. If $\{c_n\}_{n=0}^\infty$ is a set of Stieltjes moments, the operator, A, is positive. We thus define

$$\alpha = \inf_{\varphi \in \mathcal{H}, \|\varphi\|=1} \langle \varphi, A\varphi \rangle = \inf_{n, \zeta \in \mathbb{C}^n} \left(\frac{\sum_{j,k=1}^n c_{j+k-1} \bar{\zeta}_j \zeta_k}{\sum_{j,k=1}^n c_{j+k-2} \bar{\zeta} \zeta_k} \right)$$

We are heading towards a proof of

Theorem 7.7.19. *Let* $\{c_n\}_{n=0}^\infty$ *be a set of Stieltjes moments. Then the Stieltjes moment problem is determinate if and only if either*

(i) *The corresponding Hamburger moment problem is determinate.*
(ii) $\alpha = 0$.

This happens if and only if A has a unique positive self-adjoint extension.

As a preliminary, consider a Stieltjes moment problem for which the corresponding Hamburger problem is indeterminate. Let B_t be the associated von Neumann self-adjoint extensions as parametrized above. Since A is positive, it has a Friedrichs extension (see Theorem 7.5.19), that is, for some t_0,

$$A_F = B_{t_0} \tag{7.7.117}$$

Since $\alpha = \inf \sigma(A_F)$ and A_F has discrete spectrum, $\alpha = x_1(t_0)$, the minimum eigenvalue of A_F.

We first claim

$$x_1(t) < \alpha \quad \text{for all } t \neq t_0 \tag{7.7.118}$$

For the B_t have distinct eigenvalues as t varies, so if (7.7.118) fails for $t = t_1 \neq t_0$, then $x_1(t_1) > \alpha$ and thus, $B_{t_0} \geq \alpha$, $B_{t_1} \geq \alpha$. Since $0 \leq B_{t_1} \leq B_{t_0}$, we conclude (see Theorem 7.5.14) that

$$(B_{t_0} + \mathbb{1})^{-1} \leq (B_{t_1} + \mathbb{1})^{-1} \tag{7.7.119}$$

But if $B_t \geq 0$,

$$\|(B_t + \mathbb{1})^{-1}\| = (x_1(t) + 1)^{-1} \tag{7.7.120}$$

so

$$(x_1(t_0) + 1)^{-1} \leq (x_1(t_1) + 1)^{-1} \Rightarrow x_1(t_1) \leq x_1(t_0) \tag{7.7.121}$$

that is, t_1 obeys (7.7.118). This proves (7.7.118).

Define the Krein extension to be $A_K \equiv B_{t=\infty}$. This, by definition, is the extension with $\pi(0) \in D(B_t)$. Since $\text{Ker}(A^*) = [\pi(0)]$, if $\alpha > 0$, this is the Krein extension as defined for strictly positive operators after Theorem 7.5.20. If $\alpha = 0$, it is an exact analog.

If $\alpha = 0$, $0 \in \sigma(A_F)$ so $\text{Ker}(A^*) \subset D(A_F)$, so $A_F = B_{t=\infty}$, and $t_0 = \infty$. If $\alpha > 0$, $t_0 \neq \infty$, and by the fact that $\frac{dx(t)}{dt} < 0$, we see that

$$B_t \geq 0 \Leftrightarrow t = \infty \quad \text{or} \quad t \leq t_0 \tag{7.7.122}$$

As t increases from t_0, $x_1(t)$ moves down. As $t \uparrow \infty$, $x_1(t) \downarrow 0$, then as t moves up from $-\infty$ to t_0, B_t must converge in norm-resolvent sense to B_{t_0} (see Theorem 7.4.1). What happens is that $x_1(t) \downarrow -\infty$ with $x_{n+1}(t) \to x_n(t_0)$ for $n = 1, 2, \dots$.

We summarize in

Proposition 7.7.20. *If A is associated to a Stieltjes moment problem for which the Hamburger moment problem is indeterminate, there is a t_0 with* (7.7.117) *where A_F is the Friedrichs extension. If $t \neq t_0$, $x_1(t) = \inf \sigma(B_t) < x(t_0)$ so that A has a unique positive self-adjoint extension if and only if*

$\alpha = 0$, and in that case, $t_0 = \infty$. If $\alpha > 0$ (equivalently, $t_0 \neq \infty$), B_∞ is the Krein extension. If $d\mu_K, d\mu_F$ are the spectral measures for the Krein and Friedrichs extensions (with vector δ_0), then for all $y > 0$,

$$\int \frac{d\mu_F(x)}{x+y} \leq \int \frac{d\mu_K(x)}{x+y} \tag{7.7.123}$$

Proof. All that remains is (7.7.123). Since $A_F \geq A_K$, for all $y > 0$, $(A_K + y)^{-1} \geq (A_F + y)^{-1}$ (by Theorem 7.5.14), so

$$\langle \delta_0, (A_F + y)^{-1}\delta_0 \rangle \leq \langle \delta_0, (A_K + y)^{-1}\delta_0 \rangle \tag{7.7.124}$$

but this is (7.7.123). □

Here is the theorem we'll prove below that characterizes $\mathcal{S}(\{c_n\}_{n=0}^\infty)$, the set of solutions of a Stieltjes moment problem:

Theorem 7.7.21. *Let $\{c_n\}_{n=0}^\infty$ be a set of Stieltjes moments for which the Hamburger moment problem is indeterminate. Then*

$$\mu \in \mathcal{S}(\{c_n\}_{n=0}^\infty) \Leftrightarrow \forall y > 0, \text{ we have that}$$
$$\int \frac{d\mu_F}{x+y} \leq \int \frac{d\mu(x)}{x+y} \leq \int \frac{d\mu_K(x)}{x+y} \tag{7.7.125}$$

Proof of Theorem 7.7.19 given Theorem 7.7.21. If the Hamburger moment problem is determinate, then, a fortiori, the Stieltjes problem is. Since then, by Theorem 7.7.3, A has a unique self-adjoint extension, it has a unique positive self-adjoint extension.

If $\alpha > 0$, then $d\mu_F \neq d\mu_K$ and so, the Stieltjes problem is indeterminate, and since A_F and A_K are distinct, A has multiple self-adjoint extensions. If $\alpha = 0$, $B_{t=\infty}$ is the only positive self-adjoint extension and $d\mu_F = d\mu_K$, so by (7.7.125), every μ in $\mathcal{S}(\{c_n\}_{n=0}^\infty)$ has

$$\int \frac{d\mu(x)}{x-z} = \int \frac{d\mu_F(x)}{x-z} \tag{7.7.126}$$

for $z \in (-\infty, 0)$. By analyticity, this holds for all $z \in \mathbb{C}_+$ and so $\mu = \mu_F$ by Theorem 2.5.4 of Part 3. □

The proof of Theorem 7.7.21 will rely on finite matrix approximations. We begin with the μ_F lower bound.

Theorem 7.7.22. *Let $A_F^{[n]}$ be the $n \times n$ Jacobi matrix with $b_1, \dots, b_n$ along the diagonal, $a_1, \dots, a_{n-1}$ above and below the diagonal, and 0 otherwise. Then*

(a) $$\langle \delta_0, (A_F^{[n]} - z)^{-1}\delta_0 \rangle \to \langle \delta_0, (A_F - z)^{-1}\delta_0 \rangle \tag{7.7.127}$$

as $n \to \infty$, for all $z \in \mathbb{C} \setminus [0, \infty)$. For $z \in (-\infty, 0)$, the convergence is monotone increasing in n.

(b) $\langle \delta_0, (A_F^{[n]} - z)^{-1}\delta_0 \rangle = -\dfrac{q_n(z)}{p_n(z)}$ (7.7.128)

(c) *For any* $\mu \in \mathcal{S}(\{c_n\}_{n=0}^\infty)$ *and* $y > 0$, *we have for all* n *that*

$$-\frac{q_n(-y)}{p_n(-y)} \le \int \frac{d\mu(x)}{x+y} \tag{7.7.129}$$

(d) *For any* $\mu \in \mathcal{S}(\{c_n\}_{n=0}^\infty)$ *and* $y > 0$, *we have that*

$$\int \frac{d\mu_F(x)}{x+y} \le \int \frac{d\mu(x)}{x+y} \tag{7.7.130}$$

Proof. (a) Let $\mathcal{H}_n$ be the span of $\delta_0, \dots, \delta_{n-1}$. Define

$$a_F^{[n]}(\varphi) = \begin{cases} \langle \varphi, A\varphi \rangle, & \varphi \in \mathcal{H}_n \\ \infty, & \varphi \notin \mathcal{H}_n \end{cases} \tag{7.7.131}$$

Then $a_F^{[n]}$ is a positive quadratic form which is monotone decreasing in n. The limit is the form for A whose closure is the form for A_F. Thus, by the monotone convergence theorem for forms (Theorem 7.5.18), $(A_F^{[n]} - z)^{-1} \oplus 0$ (i.e., the 0 operator on $\mathcal{H}_n^\perp$) converges strongly to $(A_F - z)^{-1}$. In particular, (7.7.127) holds. $a_F^{[n+1]} \le a_K^{[n]}$ and Theorem 7.5.14 shows that for $y \ge 0$, $(A_F^{[n]} + y)^{-1} \le (A_F^{[n+1]} + y)^{-1}$.

(b) We know that $p_n(z) = (a_1 \dots a_n)^{-1} z^n +$ lower order (see (4.1.11)). By an induction (Problem 6) starting with (7.7.34), we have

$$q_n(z) = (a_1 \dots a_n)^{-1} z^{n-1} + \text{lower order} \tag{7.7.132}$$

Thus, if P_n, Q_n are the monic polynomials,

$$\frac{q_n(z)}{p_n(z)} = \frac{Q_n(z)}{P_n(z)} \tag{7.7.133}$$

By (4.1.31), $\det(z - A_F^{[n]}) = P_n(z)$ so

$$P_n(z) = (-1)^n \det(A_F^{[n]} - z), \quad Q_n(z) = (-1)^{n-1} \det(\widehat{A}_F^{[n]} - z) \tag{7.7.134}$$

where $\widehat{A}_F^{[n]}$ is $A_F^{[n]}$ with the top row and leftmost column removed. Cramer's rule plus (7.7.133)/(7.7.134) imply (7.7.128).

(c) For any z, $(p_n(x) - p_n(z))/(x - z)$ is a polynomial in x of degree $n - 1$. Thus,

$$\int d\mu(x) \left(\frac{p_n(x) - p_n(z)}{x - z} \right) p_n(x) = 0 \tag{7.7.135}$$

Taking $z = -y$, we find

$$p_n(-y) \int \frac{d\mu(x)\, p_n(x)}{x+y} = \int \frac{p_n(x)^2}{x+y}\, d\mu(x) \ge 0 \tag{7.7.136}$$

Now use (7.7.41) to see that

$$p_n(-y) \int \frac{d\mu(x)\, p_n(x)}{x+y} = p_n(-y)^2 \int \frac{d\mu(x)}{x+y} + p_n(-y) q_n(-y) \tag{7.7.137}$$

Using (7.7.136) and dividing by $p_n(-y)^2 > 0$ yields (7.7.129) (we have that $p_n(-y) \neq 0$ since the zeros of p_n lie in $[0,\infty)$).

(d) (a) & (b) implies that

$$\lim_{n\to\infty} -\frac{q_n(z)}{p_n(z)} = \int \frac{d\mu_F(x)}{x-z} \tag{7.7.138}$$

so (7.7.129) implies (7.7.130). □

For the finite matrix approximation to A_K, we'll use

$$A_K^{[n]} = \begin{pmatrix} b_1 & a_1 & & & \\ a_1 & b_2 & & & \\ & & \ddots & & \\ & & & b_{n-1} & a_{n-1} \\ & & & a_{n-1} & b_n - \alpha_n \end{pmatrix} \tag{7.7.139}$$

where α_n is chosen via:

Proposition 7.7.23. *Let $A \geq 0$. There is a unique α_n so that the matrix (7.7.139) has a zero eigenvalue. The corresponding eigenvector is $\{p_j(0)\}_{j=0}^{n-1}$. α_n obeys the following formulae:*

$$(b_n - \alpha_n) p_{n-1}(0) + a_{n-1} p_{n-2}(0) = 0 \tag{7.7.140}$$

$$\alpha_n = -\frac{[a_n p_n(0)]}{p_{n-1}(0)} \tag{7.7.141}$$

$$(b_n - \alpha_n)\alpha_{n-1} - a_{n-1}^2 = 0 \tag{7.7.142}$$

$$0 < \alpha_n < b_n \tag{7.7.143}$$

Proof. Let $A^{[n]}(\alpha)$ be the matrix on the right side of (7.7.139) with α_n replaced by α. This is the Jacobi matrix for a measure on n points (see Theorem 4.1.5) which has orthogonal polynomials $p_0(x;\alpha), \dots, p_{n-1}(x;\alpha)$ and monic polynomial $P_n(x;\alpha) = \det(x - A^{[n]}(\alpha))$. In fact, by the recursion relation, $p_j(x;\alpha)$ is α-independent, $j = 0, \dots, n-1$, but $P_n(x;\alpha)$ is linear in α.

The eigenvalues $x_1^{[n]}(\alpha) < x_2^{[n]}(\alpha) < \cdots < x_n^{[n]}(\alpha)$ are all real and simple (again by Theorem 4.1.5) and solve

$$P_n(x_j^{[n]}(\alpha); \alpha) = 0 \tag{7.7.144}$$

and the normalized eigenvectors are

$$u_\ell = p_\ell(x_j(\alpha)) / \left[\sum_{k=0}^{n-1} p_k(x_j(\alpha))^2\right]^{1/2}$$

(by (4.1.31) and (4.1.37)).

By eigenvalue perturbation theory (see Theorem 1.4.7),

$$\frac{d}{d\alpha}\, x_j^{[n]}(\alpha) = -\frac{p_{n-1}(x_j(\alpha))^2}{\sum_{\ell=1}^{n-1} p_\ell(x_j(\alpha))^2} \tag{7.7.145}$$

Since P_n and P_{n-1} have distinct zeros, (7.7.145) implies

$$\frac{d}{d\alpha}\, x_j^{[n]}(\alpha) < 0 \tag{7.7.146}$$

Since $A > 0$ and eigenvalues of $A^{[n]}(0), A^{[n+1]}(0)$ strictly interlace, $A^{[n]}(0)$ has all positive eigenvalues, that is,

$$x_1(0) < x_2(0) < \cdots x_n(0) \tag{7.7.147}$$

Since eigenvalues of $A^{[n]}(\alpha)$ and $A^{[n]}(0)$ interlace ($(A^{[n]}(\alpha) - A^{[n]}(0))$ is rank-one; see Theorem 5.8.9), if $A^{[n]}(\alpha)$ is to have a zero eigenvalue, it must be $x_1(\alpha)$. Since $x_1(\alpha)$ is strictly monotone, there is a unique α with a zero eigenvalue and it has $\alpha > 0$.

$$A^{[n]}(\alpha_n) \geq 0 \text{ and } \langle \varphi, A^{[n]}(\alpha_n)\varphi\rangle = 0 \Rightarrow A^{[n]}(\alpha_n)\varphi = 0 \tag{7.7.148}$$

Since δ_{n-1} is not an eigenvector, we must have $b_n - \alpha_n = \langle \delta_{n-1}, A^{[n]}(\alpha_n)\delta_{n-1}\rangle > 0$, that is, we have proven (7.7.143).

Since $\{p_j(0)\}_{j=0}^{n-1}$ is an eigenvector for $A^{[n]}(\alpha_n)$, the bottom row of $A^{[n]}(\alpha_n)[\pi(0)] = 0$ is (7.7.140). The recursion relation for defining $p_n(0)$ then implies

$$-\alpha_n p_{n-1}(0) = a_n p_n(0) \tag{7.7.149}$$

which is (7.7.141). This equation with n replaced by $n-1$ is

$$p_{n-1}(0) = -\alpha_{n-1} p_{n-2}(0) a_{n-1}^{-1} \tag{7.7.150}$$

Plugging this into (7.7.140) and dividing by $-p_{n-2}(0)a_{n-1}^{-1}$ (since $A^{[n]}(0)$ has no zero eigenvalues, $p_m(0) \neq 0$ for all m) yields (7.7.142). □

We also define as an operator on ℓ^2,

$$\widetilde{A}_K^{[n]} = A_K^{[n]} \oplus 0 \tag{7.7.151}$$

$$\begin{aligned} m_n(z) &= p_n(z) - p_n(0)p_{n-1}(z)p_{n-1}(0)^{-1} \\ r_n(z) &= q_n(z) - p_n(0)q_{n-1}(z)p_{n-1}(0)^{-1} \end{aligned} \tag{7.7.152}$$

Here is the result that completes the proof of Theorem 7.7.21:

Theorem 7.7.24. *For a set of Stieltjes moments that yield an indeterminate Hamburger moment problem, define* $A_K^{[n]}, \widetilde{A}_K^{[n]}, m_n, r_n$ *as above. Then*

(a) $$\langle \delta_0, (A_k^{[n]} - z)^{-1}\delta_0\rangle \to \langle \delta_0, (A_k - z)^{-1}\delta_0\rangle \tag{7.7.153}$$

as $n \to \infty$ *for all* $z \in \mathbb{C} \setminus [0,\infty)$. *For* $z \in (-\infty, 0)$, *the convergence is monotone decreasing in* n.

(b) $$\langle \delta_0, (A_K^{[n]} - z)^{-1}\delta_0\rangle = -\frac{r_n(z)}{m_n(z)} \tag{7.7.154}$$

(c) *For any* $\mu \in \mathcal{S}(\{c_n\}_{n=0}^\infty)$ *and* $y > 0$, *we have for all* n *that*

$$-\frac{r_n(-y)}{m_n(-y)} \geq \int \frac{d\mu(z)}{x+y} \tag{7.7.155}$$

(d) *For any* $\mu \in \mathcal{S}(\{c_n\})$ *and* $y > 0$, *we have that*

$$\int \frac{d\mu_K(x)}{x+y} \geq \int \frac{d\mu(x)}{x+y} \tag{7.7.156}$$

Proof. (a) $\widetilde{A}_K^{[n]} - \widetilde{A}_K^{[n-1]}$ is a 2×2 block in the $(n-1)$-st and n-th column and row of the form

$$C = \begin{pmatrix} \alpha_{n-1} & a_{n-1} \\ a_{n-1} & b_n - \alpha_n \end{pmatrix} \tag{7.7.157}$$

By (7.7.143), $\operatorname{Tr}(C) > 0$ and by (7.7.142), $\det(C) = 0$. Thus, C is a positive rank-one matrix, that is,

$$\widetilde{A}_K^{[n]} \geq \widetilde{A}_K^{[n-1]} \tag{7.7.158}$$

so by Theorem 7.5.14, $\langle \delta_0, (\widetilde{A}_K^{[n]} + y)^{-1}\delta_0\rangle \leq \langle \delta_0, (\widetilde{A}_K^{[n-1]} + y)^{-1}\delta_0\rangle$.

Moreover, we claim that for any $z \in \mathbb{C} \setminus [0,\infty)$ and $\varphi \in \mathcal{H}$,

$$(\widetilde{A}_K^{[n]} - z)^{-1}\varphi \to (\widetilde{A}_K - z)^{-1}\varphi \tag{7.7.159}$$

If $\varphi = \pi(0)$, this is immediate since $\widetilde{A}_K^{[n]}\pi(0) = 0$ and $A_K\pi(0) = 0$. So

$$(\widetilde{A}_K^{[n]} - z)^{-1}\pi(0) = (-z)^{-1}\pi(0) = (A_K - z)^{-1}\pi(0) \tag{7.7.160}$$

On the other hand, if η is a vector in ℓ^2 with finitely many components and $\varphi = (A - z)\eta$, then for n large, $\varphi = (\widetilde{A}_K^{[n]} - z)\eta$, and so for n large,

$$(\widetilde{A}_K^{[n]} - z)^{-1}\varphi = \eta = (A_K - z)^{-1}\varphi \tag{7.7.161}$$

We therefore have (7.7.159) for $\varphi \in \overline{\operatorname{Ran}(A - z)} + [\pi(0)]$. We first note that, by (7.7.52),

$$\pi(0) \in D(A^*) \setminus D(\bar{A}) \tag{7.7.162}$$

Since $(A_K - z)^{-1}(\bar{A} - z)\varphi = \varphi$ for all $\varphi \in D(\bar{A})$ and $(A_K - z)^{-1}\pi(0) = -z^{-1}\pi(0)$, we have

$$(A_K - z)^{-1}[\mathrm{Ran}(\bar{A} - z) + [\pi(0)]] = D(\bar{A}) + [\pi(0)] = D(A_K) \tag{7.7.163}$$

for $D(A_K)/D(A)$ is dimension 1 and, by (7.7.162), $\pi(0) \notin D(A)$. Since $A_K - z$ is a bijection of $D(A_K)$ and ℓ^2, we see

$$\overline{\mathrm{Ran}(A - z)} + [\pi(0)] = \ell^2 \tag{7.7.164}$$

proving (7.7.159).

Because $\widetilde{A}_K^{[n]}$ is a direct sum,

$$\langle \delta_0, (\widetilde{A}_k^{[n]} - z)^{-1}\delta_0 \rangle = \langle \delta_0, (A_K^{[n]} - z)^{-1}\delta_0 \rangle \tag{7.7.165}$$

so we have proven (a).

(b) Expanding $\det(A_K^{[n]} - z)$ in minors proves that

$$\det(A_K^{[n]} - z) = \det(A_F^{[n]} - z) - \alpha_n \det(A_F^{[n-1]} - z) \tag{7.7.166}$$

Thus,

$$(-1)^n (a_1 \dots a_n)^{-1} \det(A_K^{[n]} - z) = p_n(z) - \beta_n p_{n-1}(z) \tag{7.7.167}$$

for $\beta_n = \alpha_n a_n^{-1}$. Since 0 is an eigenvalue of $A_K^{[n]}$, $p_n(0) - \beta_n p_{n-1}(0) = 0$, so

$$\beta_n = \frac{p_n(0)}{p_{n-1}(0)} \tag{7.7.168}$$

and (7.7.167) says

$$\begin{aligned}(-1)^n (a_1 \dots a_n)^{-1} \det(A_K^{[n]} - z) &= p_n(z) - p_n(0)p_{n-1}(z)p_{n-1}(0)^{-1} \\ &= m_n(z)\end{aligned} \tag{7.7.169}$$

If $\widehat{A}_K^{[n]}$ is $A_K^{[n]}$ with the top row and leftmost column removed, the same argument (7.7.134) implies

$$\begin{aligned}(-1)^{n-1} (a_1 \dots a_n)^{-1} \det(\widehat{A}_K^{[n]} - z) &= q_n(z) - \beta_n q_{n-1}(z) \\ &= r_n(z)\end{aligned} \tag{7.7.170}$$

given (7.7.170).

(7.7.169) and (7.7.170) imply (7.7.154).

(c) We first note that (7.7.36) for p_n and p_{n-1} imply that for any $\mu \in \mathcal{S}(\{c_n\}_{n=0}^\infty)$,

$$r_n(z) = \int d\mu(x) \left[\frac{m_n(x) - m_n(z)}{x - z} \right] \tag{7.7.171}$$

Next, note that since $m_n(0) = 0$, $z^{-1}m_n$ is a polynomial of degree $n-1$ and thus, $[x^{-1}m_n(x) - z^{-1}m_n(z)]/[x-z]$ is a polynomial in x of degree $n-2$.

Since m_n is a linear combination of $p_n(x)$ and $p_{n-1}(x)$, it is orthogonal to this polynomial of degree $n-2$. Thus,

$$\int d\mu(x)\,\frac{m_n(x)}{x-z}\left[\frac{m_n(x)}{x}-\frac{m_n(z)}{z}\right]=0 \tag{7.7.172}$$

Therefore, taking $y>0$ and $z=-y$, we see

$$0\le\int\frac{m_n(x)^2}{(x+y)(x)}\,d\mu(x)=\frac{m_n(-y)}{(-y)}\int d\mu(x)\frac{m_n(x)}{x-z} \tag{7.7.173}$$

$$=\frac{m_n(-y)^2}{(-y)}\int\frac{d\mu(x)}{x+y}+\frac{m_n(-y)r_n(-y)}{(-y)} \tag{7.7.174}$$

Dividing by the negative number $m_n(-y)^2/(-y)$, we see that

$$0\ge\int\frac{d\mu(x)}{x+y}+\frac{r_n(-y)}{m_n(-y)} \tag{7.7.175}$$

which is (7.7.155).

(d) Immediate, taking $n\to\infty$ in (7.7.155) and using (7.7.15) and (7.7.154). □

Notes and Historical Remarks. The Notes of Section 5.6 of Part 1 have a comprehensive history of the moment problem. We note that the Hamburger and Stieltjes moment problems are named after their 1920–21 [**286**] and 1894 [**667**] papers, which were the first to prove the conditions of Theorem 7.7.2 implied existence of solutions. The Nevanlinna parametrization is from his 1922 paper [**501**]. For a comprehensive overview from the classical point of view, see the book of Akhiezer [**11**].

The Nevanlinna parametrization has an analog in other moment-like problems. For example, the Nehari theorem discussed in the Notes to and Problem 6 of Section 5.11 of Part 3 can be rephrased as a moment-type problem. Namely, given a Hankel matrix of norm at most 1, find those $\varphi\in L^\infty(\partial\mathbb{D})$ with $\varphi^\sharp_{-n}=C_n$, $n=1,2,\dots$. There is a way of associating the Hankel matrix to a partial isometry with $(d_+,d_-)\equiv\dim(\mathcal{H}_{\mathrm{I}}^\perp,\mathcal{H}_{\mathrm{F}}^\perp)$ (the initial and final subspaces) equal to either $(0,0)$ or $(1,1)$. The solution (i.e., the φ) is unique if and only if these indices are $(0,0)$. If they are $(1,1)$, there are infinitely many solutions where the projection of φ to H^2 is a Schur function given by the right side of (7.7.113) for a suitable matrix like (7.7.78) where now the matrix is in $\mathbb{SU}(1,1)$ (instead of $\mathbb{SL}(2,\mathbb{R})$), z lies in $\mathbb{D}$ and Φ is a Schur function. This theory is due to Adamjan, Arov, and Krein [**3**] and discussed further in Sarason [**589**] and references therein.

The approach via the spectral theorem is due to Stone in his 1932 book [**670**]. In this section, we follow a long article by Simon [**646**] on treating the moment problem as a difference operator. In particular, the names "von

Neumann solution" (Akhiezer calls them N-extremal), Friedrichs solution, and Krein solution are from Simon's paper.

There are explicit formulae for A, B, C, D in terms of p and q (Problem 7):

$$A(z) = z \sum_{n=0}^{\infty} q_n(0) q_n(z) \tag{7.7.176}$$

$$B(z) = -1 + z \sum_{n=0}^{\infty} q_n(0) p_n(z) \tag{7.7.177}$$

$$C(z) = 1 + z \sum_{n=0}^{\infty} p_n(0) q_n(z) \tag{7.7.178}$$

$$D(z) = z \sum_{n=0}^{\infty} p_n(0) p_n(z) \tag{7.7.179}$$

This last formula sheds light on the fact that we saw (see the remark after Theorem 7.7.13) that if $t \neq \infty$, then $\langle \pi(0), \pi(x(t)) \rangle \neq 0$. For if the inner product were 0, then $D(x(t)) = 0$ by (7.7.179). But then $B(x(t)) \neq 0$ (since $AD - BC = 1$), so since it is finite, $D(x(t)) + B(x(t)) \neq 0$, contrary to the fact (given (7.7.80)) that $m_t(z)$ has a pole at $z = x(t)$.

It is easy to see (Problem 8) that for all $\varepsilon > 0$, there is C_ε so that

$$|A(z)| + |B(z)| + |C(z)| + |D(z)| \leq C_\varepsilon c^{\varepsilon |z|} \tag{7.7.180}$$

This implies information on the $x_n(t)$'s.

As we remarked and is proven in Akhiezer [**11**] and Simon [**646**], if Φ is rational, $d\mu_\Phi$ is an extreme point of $\mathcal{M}(\{c_n\}_{n=0}^\infty)$, and this implies the extreme points are dense.

Problems

1. Let A be a symmetric operator. Let $f, g \in D(A^*)$ with $\langle A^* f, g \rangle - \langle f, A^* g \rangle \neq 0$. Prove that f and g are independent in $D(A^*)/D(A)$.
2. Verify (7.7.62)/(7.7.63)
3. We have seen the von Neumann solutions are exactly the solutions, $d\mu$, with the polynomials dense in $\|\cdot\|_2$–norm in $L^2(\mathbb{R}, d\mu)$. In this problem, the reader will prove the extreme points in the set of all solutions of a moment problem are precisely those with the polynomials dense in $\|\cdot\|_1$-norm in $L^1(\mathbb{R}, d\mu)$. This is a result of Naimark [**489**]. It is part of the results that $d\mu_\Phi$ is an extreme point if Φ is a rational Herglotz function and of the density of the extreme points in the weak topology (see the Notes).

(a) If the polynomials are not dense in $L^1(\mathbb{R}, d\mu)$ show that there is a real-valued $F \in L^\infty(\mathbb{R}, d\mu)$, $\|F\|_\infty = 1$ so that $\int x^n F(x) d\mu(x) = 0$ for $n = 0, 1, 2, \dots$.

(b) In that case, prove that $\frac{1}{2}(1 \pm F)d\mu$ also solve the moment problem and conclude that $d\mu$ is not an extreme point.

(c) Let $\mu_1 \neq \mu_2$ solve the moment problem and let $\mu = \frac{1}{2}\mu_1 + \frac{1}{2}\mu_2$. Prove that μ_1 is μ a.c. and, by using Radon–Nikodym derivatives, prove there is F in $L^\infty(\mathbb{R}, d\mu)$ with $\mu_1 = \frac{1}{2}(1+F)d\mu$, $\|F\|_\infty \neq 0$. Conclude that the polynomials are not dense in $L^1(\mathbb{R}, d\mu)$.

4. Let $\mu_t(\{x_0\}) > 0$ and suppose $D(x_0) \neq 0$ and that $\nu = \mu_\Phi$ has $\nu(\{x_0\}) > 0$.
(a) Prove that $\lim_{\varepsilon \downarrow 0} \Phi(x_0 + i\varepsilon) = t$ and that
$$Q_\nu \equiv \lim_{\varepsilon \downarrow 0} (\Phi(x_0 + i\varepsilon) - t)/(i\epsilon) > 0$$
(b) Prove that $\mu_t(\{x_0\}) = [D(x_0)(D'(x_0)t + B(x_0))]^{-1}$ and that
$$\nu(\{x_0\}) = \frac{1}{Q_\nu + D(x_0)[D'(x_0)t + B(x_0)]}$$
and so conclude that $\nu(\{x_0\}) < \mu_t(\{x_0\})$.
(c) If $D(x_0) = 0$, modify the above argument.

5. (a) If $m_\mu(z_0) \in \partial D(z_0)$ for a single z_0, prove that Φ is constant and so that μ is a von-Neumann solution. Thus, if $w \in \partial D(z_0)$, there is a unique solution of the moment problem with $m_\mu(z_0) = w$.

(b) If $w \in D(z_0)^{\text{int}}$, prove there are multiple μ's solving the moment problem with $m_\mu(z_0) = w$. (Hint: Prove first that two distinct nontrivial convex combinations of μ_t's are distinct measures because which μ_t's are involved can be determined from the pure points of the convex combination.)

6. Verify (7.7.132) by induction.

7. Verify (7.7.176)–(7.7.179). (Hint: Extend the product (7.7.67).)

8. Prove (7.7.180). (Hint: Use (7.7.70) only from $n = N$ to ∞ to show $\|T_\infty(z)\|$ is bounded by a polynomial in $|z|$ times $\exp(|z - z_0| \sum_{j=N}^\infty \|S_j(z_0)\|)$.)

7.8. Compact, Rank-One and Trace Class Perturbations

Our goal in this section is to consider a positive (normally unbounded) operator, A, and consider
$$C = A + B$$

where B is one of the following:

(a) relatively compact, in which case we'll show $\sigma_{\text{ess}}(C) = \sigma_{\text{ess}}(A)$;
(b) rank-one, in which case we'll extend the results of Section 5.8;
(c) relatively trace class, in which case we'll prove that $\Sigma_{\text{ac}}(C) = \Sigma_{\text{ac}}(A)$

Of course, these are just analogs of results we have in the bounded case, but as the word "relatively" indicates in (a), (c) (and the formulation we'll give in case (b)), there is a more general aspect. For example, if $V \in C_\infty(\mathbb{R}^\nu)$ is a continuous function vanishing at ∞, multiplication by V is not compact but we've seen that $V(-\Delta+1)^{-1}$ is compact (see Theorem 3.8.8) and that will imply that $\sigma_{\text{ess}}(-\Delta+V) = \sigma_{\text{ess}}(-\Delta) = [0,\infty)$. Similarly, we'll see that for $A = -\frac{d^2}{dx^2}$ on $[0,\infty)$, change of boundary condition can be viewed as a rank-one perturbation. Some proofs will be by analogy, but others will directly use the bounded operator result with a "mapping theorem." We begin by studying these mapping theorems.

Theorem 7.8.1 (Spectral Mapping Theorem). (a) *Let A be a self-adjoint operator and F a bounded continuous function on $\sigma(A)$ if A is bounded and on $\sigma(A) \cup \{\infty\}$ if A is unbounded. Then,*

$$\sigma\big(F(A)\big) = F\big[\sigma(A)\big] \quad \big(\textit{or } F\big[\big(\sigma(A) \cup \{\infty\}\big)\big]\big) \tag{7.8.1}$$

(b) *Suppose there is a $k \in \mathbb{Z}_+$ so for all $\lambda \in \mathbb{R}$, $\sharp\big(F^{-1}\big[|\lambda|\big]\big) \le k$. Then,*

$$\sigma_{\text{ess}}\big(F(A)\big) = F\big[\sigma_{\text{ess}}(A)\big] \quad \big(\textit{ or } F\big(\sigma_{\text{ess}}(A) \cup \{\infty\}\big)\big) \tag{7.8.2}$$

Proof. (a) The proof is identical to the proof of Theorem 5.1.11 (given that if A is unbounded $\sigma(A) \cup \{\infty\}$ is compact).

(b) If $\lambda \in \sigma\big(F(A)\big)$, then, by hypothesis, $F^{-1}\big(\{\lambda\}\big)$ is a finite set. If $\sigma(A) \cap F^{-1}\big(\{\lambda\}\big)$ is in $\sigma_d(A)$, then clearly λ is an isolated eigenvalue of finite multiplicity so $\lambda \in \sigma_d\big(F(A)\big)$. If $\sigma_{\text{ess}}(A) \cap F^{-1}\big(\{\lambda\}\big) \neq \emptyset$, either λ is not isolated in $\sigma\big(F(A)\big)$ (since F is continuous and by the finiteness result μ not isolated in $\sigma(A)$, implies $F(\mu)$ is not isolated in $F\big[\sigma(A)\big]\big)$ or else λ is an eigenvalue of infinite multiplicity. Either way, $\lambda \in \sigma_{\text{ess}}\big(F(A)\big)$. $\square$

Theorem 7.8.2. *Let A be a self-adjoint operator and F a C^1-function in a neighborhood of $\sigma(A)$ (or if A is unbounded of $\sigma(A) \cup \{\infty\}$). Suppose $\{x \in \sigma(A) \mid F'(x) = 0\}$ is finite. Then,*

$$\Sigma_{\text{ac}}\big(F(A)\big) = F\big[\Sigma_{\text{ac}}(A)\big] \tag{7.8.3}$$

Remarks. 1. To say F defined in a neighborhood of ∞ is C^1 at ∞ means $g(y) = F(x^{-1})$ is C^1 at $y = 0$.

2. Because of the hypothesis on F, it is immediate that $S \subset \sigma(A)$ with $|S| = 0 \Rightarrow |\,F[S]\,| = 0$, so if Σ is a family of sets mod sets of Lebesgue

measure zero, $\{F[S] \mid S \in \Sigma\}$ is a family of sets modulo sets of Lebesgue measure zero.

Proof. Let us start with the case where F is C^1 in a neighborhood of $[a,b]$, $F'(x) > 0$ on (a,b) and

$$(A\varphi)(x) = x\varphi(x)$$

on $L^2\big([a,b], g(x)dx\big)$ where $g(x) \in L^1\big((a,b)\big)$.

We claim that $F(A)$ is unitarily equivalent to multiplication by y on $L^2\big(\big[F(a),F(b)\big], h(y)dy\big)$ where $h \in L^1$ and $\{y \mid h(y) \neq 0\} = F\big[\{x \mid g(x) \neq 0\}\big]$. For F is a bijection from $[a,b]$ to $\big[F(a),F(b)\big]$ and if G is its two-sided inverse, by the chain rule,

$$F'\big(G(y)\big)G'(y) = 1 \tag{7.8.4}$$

It follows that with $h(y) \equiv g\big(G(y)\big)G'(y)$, $d\mu(x) = g(x)\,dx$, $d\nu(y) = h(y)dy$ that $\nu\big(F[S]\big) = \mu(S)$ for all Baire S and if

$$(U\varphi)(y) = \big[G'(y)\big]^{-1/2}\varphi\big(G(y)\big) \tag{7.8.5}$$

then U is unitary from $L^2\big([a,b], d\mu\big)$ to $L^2\big(\big[F(a),F(b)\big], d\nu\big)$ and $U\,F(A)\,U^{-1}$ is multiplication by y on $L^2\big(\big[F(a),F(b)\big], d\nu\big)$.

This proves the required result in this case. In general, since F' has only finitely many zeros, we have a direct sum of operators of the above form and a singular piece and a simple argument proves the full result (Problem 1). □

Definition. Let A and B be two self-adjoint operators each bounded from below. We say B is a *relatively compact perturbation* of A if for some $\lambda \in \mathbb{R}$ with $A \geq \lambda, B \geq \lambda$, we have that $(A+\lambda+1)^{-1} - (B+\lambda+1)^{-1}$ is compact.

Remark. It is easy to see (Problem 2) that $(A-z)^{-1} - (B-z)^{-1}$ is compact for one $z \in \mathbb{C} \setminus \big[\sigma(A) \cup \sigma(B)\big]$ if and only if it is compact for all such z.

Theorem 7.8.3. *If B is a relatively compact perturbation of A, then $\sigma_{\text{ess}}(A) = \sigma_{\text{ess}}(B)$.*

Proof. This is an immediate consequence of Weyl's invariance theorem (Theorem 3.14.1) and the spectral mapping theorem (Theorem 7.8.1) applied to $(A+\lambda+1)^{-1}$ and $(B+\lambda+1)^{-1}$. □

Definition. Let A be positive self-adjoint operator. A symmetric operator C is called *relatively A-operator compact* if $D(A) \subset D(C)$ and $C(A+1)^{-1}$ is a compact operator. A symmetric quadratic form, C, is called *relatively A-form compact* if $Q(A) \subset D(C)$ and $(A+1)^{-1/2}C(A+1)^{-1/2}$ is a compact operator.

Remarks. 1. To be more precise in the form compact situation, $Q(A) \subset D(C)$ implies $C(\eta, \kappa)$ is well-defined if $\eta, \kappa \in Q(A)$. Thus $C\big((A+1)^{-1/2}\varphi, (A+1)^{-1/2}\psi\big)$ is defined for all $\varphi, \psi \in \mathcal{H}$. The condition means there is a compact operator, D, so that

$$C\big((A+1)^{-1/2}\varphi, (A+1)^{-1/2}\psi\big) = \langle \varphi, D\psi \rangle \tag{7.8.6}$$

for all $\varphi, \psi \in \mathcal{H}$. We then write $D = (A+1)^{-1/2} C (A+1)^{-1/2}$.

2. It is easy to see (Problem 3) that $(A+1)^{-1/2}$ can be replaced by $(A-z)^{-1/2}$ for any $z \in (-\infty, \alpha)$ with $\alpha = \inf \operatorname{spec}(A)$.

3. It is easy to see (Problem 4) operator (respectively, form) compact perturbations are operator (respectively, form) bounded perturbations of relative bound zero.

4. It is easy to see (Problem 5) that if $C = X^* U X$ where $X \in \mathcal{L}(\mathcal{H}_+, \mathcal{K})$, U is bounded from $\mathcal{K}$ to $\mathcal{K}$, and if X or U is compact (i.e., a norm limit of finite rank operators), then C is relatively form compact and conversely, every such perturbation has this form.

Theorem 7.8.4. *If C is relative A-form (respectively, operator) compact, then $B = A + C$ is a relatively compact perturbation of A. In particular, $\sigma_{\text{ess}}(B) = \sigma_{\text{ess}}(A)$.*

Proof. We discuss the form case. The operator case can be accomplished by a similar argument (Problem 6) or by proving any operator compact perturbation is a form compact perturbation (Problem 7).

As noted above, C is a form bounded perturbation with relative bound zero which implies $\lim_{x\to\infty} \|(A+x)^{-1/2} C (A+x)^{-1/2}\| = 0$. Pick x_0 so the norm is less than 1. Let

$$D = (A+x_0)^{-1/2} C (A+x_0)^{-1/2}, \quad \|D\| < 1 \tag{7.8.7}$$

Write

$$A + C + x_0 = (A+x_0)^{1/2}(1+D)(A+x_0)^{1/2} \tag{7.8.8}$$

Since $\|D\| < 1$, $(1+D)$ is invertible so

$$(B+x_0)^{-1} - (A+x_0)^{-1} = (A+x_0)^{-1/2}\big[(1+D)^{-1} - 1\big](A+x_0)^{-1/2} \tag{7.8.9}$$

$$= -(A+x_0)^{-1/2}\big[D(1+D)^{-1}\big](A+x_0)^{-1/2} \tag{7.8.10}$$

is a product of bounded and one compact operator, so compact. $\square$

Example 7.8.5. $L^p + (L^\infty)_\varepsilon$ is the set of functions f, so that for any ε, we can write $f = f_{1,\varepsilon} + f_{2,\varepsilon}$ where $f_{1,\varepsilon} \in L^p$ and $\|f_{2,\varepsilon}\|_\infty \le \varepsilon$. Fix $\nu \in \mathbb{Z}_+$ and $p \ge 1$, $p > \nu/2$. We claim any $V \in L^p + (L^\infty)_\varepsilon$, as a function on $\mathbb{R}^\nu$, is a form compact perturbation of $-\Delta$ on $L^2(\mathbb{R}^\nu)$. For, by Problem 6 of Section 3.8, and $\int \bigl[(k^2+1)^{-1/2}\bigr]^{2p} d^\nu k < \infty$ (if $p > \nu/2$), we see that $f_1(x) \in L^p(\mathbb{R}^\nu) \Rightarrow |f_1(x)|^{1/2}(-\Delta+1)^{-1/2} \in \mathcal{I}_{2p}$ is compact (since $2p \in [2,\infty)$). The claimed form compactness follows and we get the theorem below. $\square$

Theorem 7.8.6. *Let $p \geq 1$, $p > \nu/2$. Let $V \in L^p(\mathbb{R}^\nu) + L^\infty_\varepsilon(\mathbb{R}^\nu)$. Then, $\sigma_{\text{ess}}(-\Delta+V) = [0,\infty)$.*

Next we turn to rank-one perturbations. $B \geq 0$ is a possibly unbounded perturbation. As in the discussions in Section 5.8, we are interested in the one parameter family, $\alpha \in \mathbb{R}$

$$A_\alpha = B + \alpha\langle\varphi,\,\cdot\,\rangle\,\varphi \tag{7.8.11}$$

The new element is that instead of $\varphi \in \mathcal{H}$, we only demand $\varphi \in \mathcal{H}_{-1}$, the dual of the form domain of B with norm $\|\psi\|_{+1} = \langle\psi,(B+1)\psi\rangle^{1/2}$, equivalently, the completion of $\mathcal{H}$ with norm $\|\psi\|_{-1} = \langle\psi,(B+1)^{-1}\psi\rangle^{1/2}$.

$\langle\varphi,\,\cdot\,\rangle\,\varphi$ is B-form bounded for

$$|\langle\varphi,\psi\rangle|^2 \leq \|\varphi\|_{-1}^2\,\|\psi\|_{+1}^2$$

Writing $\varphi = \varphi_{1,\varepsilon} + \varphi_{2,\varepsilon}$ with $\varphi_{1,\varepsilon} \in \mathcal{H}$ and $\|\varphi_{2,\varepsilon}\|_{-1} \leq \varepsilon$, we see the relative bound is zero, so A_α can be defined for all $\alpha \in \mathbb{R}$ as a closed, bounded from below form with form domain $Q(B) = \mathcal{H}_{+1}$. In fact, $\langle\varphi,\,\cdot\,\rangle\,\varphi$ is clearly B-form compact.

As in Section 5.8, we may as well suppose φ is cyclic for B (i.e., $\{(B-z)^{-1}\varphi \mid z \in \mathbb{C}\setminus[0,\infty)\}$ is total in $\mathcal{H}$) in which case it is not hard to prove (Problem 8) that φ is cyclic for each A_α. We can make this cyclicity assumption since if $\mathcal{H}_\varphi$ is the cyclic subspace generated by B and φ, then $A_\alpha = B$ on $\mathcal{H}_\varphi^\perp$.

By the spectral theorem slightly extended (Problem 9), there is a measure $d\mu_\alpha$ on $\mathbb{R}$ with $\int_0^\infty \frac{d\mu_\alpha(x)}{x+1} < \infty$, so that for $z \in \mathbb{C}\setminus\mathbb{R}$

$$F_\alpha(z) \equiv \langle\varphi,(A_\alpha - z)^{-1}\varphi\rangle = \int \frac{d\mu_\alpha(x)}{x-z} \tag{7.8.12}$$

We can define $\langle\varphi,(A_\alpha - z)^{-1}\varphi\rangle$ since $(A_\alpha - z)^{-1}$ maps $\mathcal{H}_{-1}$ to $\mathcal{H}_{+1}$.

As in Section 5.8, the Aronszajn–Krein formula

$$F_\alpha(z) = \frac{F_0(z)}{1+\alpha F_0(z)} \tag{7.8.13}$$

holds without any change of proof (if one notes that the operator P_φ of (5.8.26) is now a map of $\mathcal{H}_{+1}$ to $\mathcal{H}_{-1}$. Once one has this, all the results of Section 5.8 extend with little change.

Example 7.8.7. This is a rephrasing of Example 7.5.13. Let B have $Q(B) = \{\psi \in L^2\big((0,\infty), dx\big) \mid \psi \text{ is absolutely continuous and } \psi' \in L^2\}$ with $\langle \psi, B\psi\rangle = \int_0^\infty |\psi'(x)|^2\, dx$. As we saw, B is $-\frac{d^2}{dx^2}$ with $\psi'(0) = 0$ boundary conditions. We also saw that there is $\varphi \in \mathcal{H}_{-1}$, so $\langle \varphi, \psi\rangle = \psi(0)$ (formally $\varphi(x) = \delta(x)$). (7.5.63) says that $A_\alpha = B + \alpha\langle\varphi, \cdot\,\rangle\varphi$ with $\alpha = -\cot(\theta)$ is just $-\frac{d^2}{dx^2}$ on $(0,\infty)$ with

$$\cos(\theta)\psi(0) + \sin(\theta)\psi'(0) = 0$$

boundary conditions. Thus, variation of boundary conditions can be rephrased in terms of rank-one perturbations. In a sense, one can make precise (see the Notes), Dirichlet boundary conditions ($\psi(0) = 0$) corresponds to $\alpha = \infty$. □

Next, we discuss trace class perturbations.

Theorem 7.8.8. *Let A, B be self-adjoint operators both bounded from below. Suppose for some $\lambda \in \mathbb{R}$, $\lambda \notin \sigma(A) \cup \sigma(B)$ and $(A+\lambda)^{-1} - (B+\lambda)^{-1}$ is trace class. Then*

$$\Sigma_{\text{ac}}(A) = \Sigma_{\text{ac}}(B) \tag{7.8.14}$$

Proof. By Theorem 5.9.1, $\Sigma_{\text{ac}}\big((A+\lambda)^{-1}\big) = \Sigma_{\text{ac}}\big((B+\lambda)^{-1}\big)$. By Theorem 7.8.2, $\Sigma_{\text{ac}}(A) = \{\mu^{-1} - \lambda \mid \mu \in \Sigma_{\text{ac}}\big((A+\lambda)^{-1}\big)\}$ and similarly for B. (7.8.14) follows. □

Example 7.8.9. Let $V \in L^1(\mathbb{R}^3) \cap \big[L^2(\mathbb{R}^3) + L^\infty(\mathbb{R}^3)\big]$. Then, V is $-\Delta$ bounded with relative bound zero so $-\Delta + V$ is self-adjoint on $D(-\Delta)$ and bounded from below and for x large

$$\begin{aligned}(-\Delta+V+\lambda)^{-1} - (-\Delta+\lambda)^{-1} &= (-\Delta+\lambda)^{-1}(-V)(-\Delta+V+\lambda)^{-1} \\ &= \big[(-\Delta+\lambda)^{-1}|V|^{1/2}\big]\,(\text{sgn}\, V)\big[|V|^{1/2}(-\Delta+\lambda)^{-1}\big] \\ &\qquad \big[(-\Delta+\lambda)(-\Delta+V+\lambda)^{-1}\big]\end{aligned}$$

Since $|V|^{1/2}$ and $f(k) = (k^2+1)^{-1}$ are in L^2, the first and third factors are Hilbert Schmidt (by Theorem 3.8.6) and the other two factors are bounded. We conclude that

$$\Sigma_{\text{ac}}(-\Delta+V) = \Sigma_{\text{ac}}(-\Delta) = [0,\infty) \tag{7.8.15}$$

This is unsatisfying for two reasons. In terms of estimates of the form $|V(x)| \le C(1+|x|)^{-\alpha}$ on $L^2(\mathbb{R}^\nu)$, $V \in L^1$ if and only if $\alpha > \nu$. But it is known that (7.8.15) holds if $\alpha > 1$. This requires non-trace class methods as we'll discuss in the Notes.

But even within the context of trace class methods, this argument is limited for it only works if $\nu \le 3$. If $\nu \ge 4$, $(k^2+1)^{-1} \notin L^2$ so one cannot prove (and it is not true!) that $(-\Delta+V+\lambda)^{-1} - (-\Delta+\lambda)^{-1}$ is trace class.

Limiting to resolvents is not a good approach. We'll say more about this in the Notes. □

Finally, we note a min-max principle which holds for general semibounded self-adjoint operators, but is most useful when one can use the theory of relatively compact perturbations to locate $\Sigma^-(A)$, which is defined by (3.14.26). As in the discussion following that definition, we define $\lambda_n^-(A)$ so that $\lambda_1^-(A) \le \lambda_2^-(A) \le \cdots \le \lambda_n^-(A) \le \Sigma^-(A)$.

Theorem 7.8.10 (Min-Max Principle for Semibounded Operators). *Let A be a self-adjoint operator which is bounded from below. Then*

$$\lambda_n^-(A) = \sup_{\psi_1,\dots,\psi_{n-1}} \Big(\inf_{\substack{\varphi\perp\psi_1,\dots,\psi_{n-1} \\ \|\varphi\|=1, \varphi\in Q(A)}} \langle \varphi, A\varphi\rangle \Big) \tag{7.8.16}$$

Remarks. 1. The proof is the same as Theorem 3.14.5.

2. $Q(A)$ can be replaced by any form core and, in particular, by $D(A)$.

Notes and Historical Remarks. Example 7.8.5 is just the first step in a long discussion of the essential spectrum of Schrödinger operators, Jacobi matrices and CMV matrices in more complicated situations than $\sigma_{\text{ess}}(A) = \sigma_{\text{ess}}(B)$. The earliest such result is the HVZ theorem which identifies the essential spectrum of N-body Schrödinger operators named after work of Hunziker [**328**], van Winter [**713**], and Zhislin [**774**]. Very general results on such spectrum coming from what happens near infinity are due to Georgescu–Iftimovici [**237**], Last–Simon [**429, 430**], Mantoiu [**464**], and Rabinovich [**541**].

One can sometimes make sense of $B + \alpha\langle\varphi, \cdot\rangle\,\varphi$ for φ's more singular than in $\mathcal{H}_{-1}$; see [**12, 297, 391, 406, 416, 733**]. Gesztesy–Simon [**243**] discuss the limit of A_α as $\alpha \to \infty$ when $\varphi \in \mathcal{H}_{-1}$ but $\varphi \notin \mathcal{H}$. Then $Q(A_\infty) = \{\psi \in Q(B) \mid \langle\varphi,\psi\rangle = 0\}$ which is still dense in $\mathcal{H}$ and for $\psi \in Q(A_\infty)$, $\langle\psi, A_\infty\psi\rangle = \langle\psi, B\psi\rangle$. In the case of Example 7.8.7, A_∞ is exactly the Dirichlet boundary condition operator.

Theorem 7.8.8 is due to Kuroda [**418**], Birman [**60, 61**], and deBranges [**151**] using wave operators. The first extensions to unbounded A, B are due to Putnam [**538**] and Kuroda [**418**], with some important extensions by Birman [**60, 61, 62**]. An especially elegant result and proof that includes these earlier results is due to Pearson [**515**]. To get results that cover $|V(x)| \le C\big(1+|x|\big)^{-\alpha}$, $\alpha > \nu$ in $\mathbb{R}^\nu$, one can use a result of Birman [**62**]: We say A is subordinate to B if there is a continuous $f(x) \to \infty$ as $|x| \to \infty$ and a function $g(x)$ so $f(A) \le g(B)$. If A is subordinate to B and vice-versa and if $P_I(A)(A-B)P_I(B)$ is trace class for all bounded intervals, I (with $P_I(\cdot)$ spectral projections), then $\Sigma_{\text{ac}}(A) = \Sigma_{\text{ac}}(B)$.

Results for $|V(x)| \le C\big(1+|x|\big)^{-\alpha}$, $1 < \alpha \le \nu$ are due to Agmon [**6**]; see the exposition in the books of Hörmander [**324**] and Reed–Simon [**551**].

Problems

1. Given the special case of Theorem 7.8.2 where A has a cyclic vector and F is defined and C^1 in a neighborhood of $[a,b]$ with $F'(x) > 0$ on (a,b), prove the general result.

2. Prove that if A and B are arbitrary self-adjoint operators and $(A-z)^{-1} - (B-z)^{-1}$ is compact for one z in $\mathbb{C} \setminus \big[\sigma(A) \cup \sigma(B)\big]$, it is compact for all such z. (Hint: $(A-w)^{-1} - (B-w)^{-1} = \big[1 + (w-z)(A-w)^{-1}\big]\big[(A-z)^{-1} - (B-z)^{-1}\big]\big[1+(w-z)(B-w)^{-1}\big]$.)

3. Let C be a quadratic form with $(A-z)^{-1/2} C (A-z)^{-1/2}$ compact for one $z \in (-\infty, \inf \sigma(A))$. Prove it is compact for all such z. (Hint: $(A-z)^{-1/2}(A-w)^{1/2}$ is bounded on $\mathcal{H}$.)

4. Let B be relatively A-operator compact. Prove that B is A-operator bounded with relative bound zero. Prove the form analog of this statement. (Hint: $\mathcal{H}$ is dense in $\mathcal{H}_{-1}$.)

5. Let $A \ge 0$ be self-adjoint. Prove that C is relatively form compact if and only if $C = X^* U X$ where $X \in \mathcal{L}(\mathcal{H}_{+1}, \mathcal{K})$ is compact and $U \in \mathcal{L}(\mathcal{K}, \mathcal{K})$ is bounded.

6. Prove the operator version of Theorem 7.8.4 by mimicking the form version proof.

7. Prove that any relatively A-operator compact perturbation defines a form which is relatively A-form compact.

8. Let $B \ge 0$ be self-adjoint and $\varphi \in \mathcal{H}_{-1}$. Suppose φ is cyclic for B. Prove that φ is also cyclic for each $A_\alpha = B + \alpha\langle \varphi, \cdot \rangle \varphi$. (Hint: Look at the orthogonal complement of $\{(A_\alpha - z)^{-1}\varphi\}$.)

9. Prove the following extension of the spectral theorem: if $B \ge 0$ is self-adjoint, $\varphi \in \mathcal{H}_{-1}$ and $\{(B-z)^{-1}\varphi \mid z \in \mathbb{C} \setminus [0,\infty)\}$ is total in $\mathcal{H}$, prove there exists a measure $d\mu$ on $[0,\infty)$ with $\int_0^\infty \frac{d\mu(x)}{x+1} < \infty$ and $V : \mathcal{H} \to L^2(\mathbb{R}, d\mu)$ unitary so that
   (i) $Q(B) = \{\eta \in \mathcal{H} \mid \int_0^\infty x|(V\eta)(x)|^2 d\mu(x) < \infty\}$;
   (ii) if $\eta \in Q(B)$, then $\langle \eta, B\eta \rangle = \int x|(V\eta)(x)|^2\, d\mu(x)$;
   (iii) $V\big((B+1)^{-1}\varphi\big) = (x+1)^{-1}$.

7.9. The Birman–Schwinger Principle

Let $A \geq 0$ be a positive self-adjoint operator. Let B be A-form compact and symmetric. Then, as we've seen (see Theorem 7.8.4) $\sigma(A-B)\cap(-\infty,0)$ is only discrete spectrum. In this section, we describe and apply a method for determining if this discrete set (which can be infinite only if 0 is a limit point) is finite or infinite and even for estimating $\dim\bigl[\mathrm{Ran}(E_{(-\infty,0)}(A-B))\bigr]$. This section is a complement to Section 6.7 of Part 3 which also uses this method. Indeed, since the reader should read this section before that one, we use the phrase "as we'll see" when referring to that other section.

Example 7.9.1. Let $N(V)$ denote $\dim\bigl[\mathrm{Ran}(E_{(-\infty,0)}(H))\bigr]$ for $H=-\Delta+V$, a Schrödinger operator on $L^2(\mathbb{R}^3, d^3x)$ where "V" is a shorthand for the operator of multiplication by a measurable function on $\mathbb{R}^3$, also denoted by V. To avoid technical issues, we suppose that $V \in L^\infty$ so that H is self-adjoint on $D(-\Delta)$. Once one has bounds on the quantities $N_e(V)$ and $N(V)$ below, it is easy to extend the bounds to more general V as we'll occasionally point out.

The *Birman–Schwinger bound* which we'll prove below says that

$$N(V) \leq \frac{1}{(4\pi)^2}\int \frac{|V(x)|\,|V(y)|}{|x-y|^2}\,d^3x\,d^3y \tag{7.9.1}$$

Recalling (see Theorem 6.8.2 of Part 1) that $(-\Delta)^{-1}$, at least on $\mathcal{S}(\mathbb{R}^3)$, acts as convolution with $(4\pi)^{-1}|x-y|^{-1}$, we note that the right side of (7.9.1) is the Hilbert–Schmidt norm of the operator with integral kernel

$$K(x,y) = (4\pi)^{-1}\,|V(x)|^{1/2}|x-y|^{-1}|V(y)|^{1/2} \tag{7.9.2}$$

called the *Birman–Schwinger kernel.* This is just the integral kernel of the operator $|V|^{1/2}(-\Delta)^{-1}|V|^{1/2}$. □

This example partly explains why we write $A-B$. If $V=V_+-V_-$ (with $V_\pm = \max(\pm V,0)$), $N(V)\leq N(V_-)$ by the min-max principle and $-\Delta+V\geq -\Delta-V_-$. Motivated by (7.9.2), in the abstract setting where

$$B=C^*D, \quad C,D:\mathcal{H}_{+1}(A)\to\mathcal{K} \tag{7.9.3}$$

a factorizable interaction (see Theorem 7.5.9 which says every B is factorizable), we define the *Birman–Schwinger kernel* for $e\in(-\infty,0)$ by

$$K_e = D(A-e)^{-1}C^* \tag{7.9.4}$$

(*Birman–Schwinger operator* might be better but, given the history, "kernel" is almost universally used even for the abstract theory). Note that $C^*:\mathcal{K}\to\mathcal{H}_{-1}$, $(A-e)^{-1}:\mathcal{H}_{-1}\to\mathcal{H}_{+1}$ and $D:\mathcal{H}_{+1}\to\mathcal{K}$ so K_e is a bounded, indeed (when B is A-form compact) compact, operator from $\mathcal{K}$ to $\mathcal{K}$.

We have the following

Proposition 7.9.2. (a) *$\sigma(K_e)$ is real.*

(b) *e is an eigenvalue of $A - B$ with multiplicity ℓ if and only if 1 is an eigenvalue of K_e of multiplicity ℓ.*

Remark. In many applications, $B \geq 0$ and we can take $C = D$ so K_e is self-adjoint (and positive) and the spectrum is obviously real. We then call the self-adjoint, positive operator $C(A-e)^{-1}C^*$ a *canonical Birman–Schwinger kernel.*

Proof. (a) Let $X = D(A-e)^{-1/2}, Y = C(A-e)^{-1/2}$. Since $(A-e)^{-1/2}$ is bounded from $\mathcal{H}$ to $\mathcal{H}_{+1}$, X, Y are bounded from $\mathcal{H}$ to $\mathcal{K}$, so

$$K_e = XY^* \tag{7.9.5}$$

is a bounded map of $\mathcal{K}$ to $\mathcal{K}$. By Problem 5 of Section 2.2 (extended to maps between Banach spaces—(2.2.66) is still valid)

$$\sigma(K_e) \setminus \{0\} = \sigma(Y^*X) \setminus \{0\} \tag{7.9.6}$$

Since B is symmetric as a form, $Y^*X = (A-e)^{-1/2}\,B\,(A-e)^{-1/2}$ is self-adjoint and so has real spectrum.

(b) For now, think of $A, B : \mathcal{H}_{+1} \to \mathcal{H}_{-}$. $(A - B)\varphi = e\varphi$ (for $\varphi \in \mathcal{H}_{+1}$)

$$\Leftrightarrow (A-e)\varphi = B\varphi \Leftrightarrow (A-e)^{-1}B\varphi = \varphi \tag{7.9.7}$$

Since $\varphi \in \mathcal{H}_{+1}$, we can apply D to both sides and see that

$$D(A-e)^{-1}\,(C^*D)\,\varphi = D\varphi, \quad \text{i.e., } K_e(D\varphi) = D\varphi \tag{7.9.8}$$

Conversely, if $K_e\psi = \psi$ for $\psi \in \mathcal{K}$, $\psi \neq 0 \Rightarrow C^*\psi \neq 0$ (since $\psi = D(A-e)^{-1}(C^*\psi)$). Thus, $\varphi \equiv (A-e)^{-1}C^*\psi \neq 0$. Clearly, $D\varphi = K_e\psi = \psi$ so $\varphi = (A-e)^{-1}C^*D\varphi = (A-e)^{-1}B\varphi$ implies $(A-B)\varphi = e\varphi$ by (7.9.7).

Thus, D is a map of $\{\varphi \mid (A-B)\varphi = C\varphi\}$ onto $\{\psi \mid K_e\psi = \psi\}$ and is a bijection since $(A-e)^{-1}C^*$ is the two-sided inverse of D on this space. □

Now we'll vary the coupling constant. For $\lambda > 0$, let $e_n(\lambda)$ be defined by

$$e_n(\lambda) = \begin{cases} n\text{-th negative eigenvalue of } A - \lambda B \text{ counting multiplicity} \\ 0 \text{ if there are fewer than } n \text{ negative eigenvalues} \end{cases}$$

Proposition 7.9.3. (a) *If $e_n(\lambda_0) < 0$, then for some $\varepsilon > 0$, $e_n(\lambda) < 0$ for $\lambda \in (\lambda_0 - \varepsilon, \lambda_0 + \varepsilon)$ and*

$$\frac{de_n}{d\lambda} < 0 \quad \text{if } e_n(\lambda_0) < 0 \tag{7.9.9}$$

(b) For each n, there is $\lambda^{(n)} \in [0,\infty]$ so that $e_n(\lambda) < 0$ if and only if $\lambda > \lambda^{(n)}$. Moreover, if $\lambda^{(n)} < \infty$, $\lim_{\lambda\to\infty} e_n(\lambda) = -\infty$.

(c) If $\lim_{\lambda\to 1} e_n(\lambda) < e_0 \in (-\infty, 0)$, there is a unique $\lambda \in (0,1)$ so that $e_n(\lambda) = e_0$.

Remarks. 1. It is a basic fact (see Proposition 6.7.10 of Part 3) that $\sum_{j=1}^{n} e_j(\lambda) \equiv T_n(\lambda)$ is a concave function of λ so $T_n(\lambda)$ and thus $e_n = T_n - T_{n-1}$ have derivatives at all but countable many λ's and at all λ, left (respectively, right) derivatives exist and are continuous from the left (respectively, right). (See Theorem 5.3.12 of Part 1). Then (7.9.9) refers to either the right or left derivative.

2. By continuity (which we'll prove), $\lim_{\lambda\uparrow\lambda^{(n)}} e_n(\lambda) = 0$, so if $\lambda^{(n)} < \infty$, $e(\lambda)$ is a bijection of $(\lambda^{(n)}, \infty)$ and $(-\infty, 0)$.

3. $\lambda^{(n)} = \infty$ is shorthand for the statement that $N(A + \lambda B) < n$ for all $\lambda > 0$.

Proof. (a) If $e_n(\lambda_0) < 0$, it is an isolated point of finite multiplicity, so we can find ε and δ both positive so that for $\lambda \in (\lambda_0 - \varepsilon, \lambda_0 + \varepsilon)$, $A - \lambda B$ has exactly ℓ eigenvalues in $\big(e_n(\lambda_0) - \delta, e_n(\lambda_0) + \delta\big) \subset (-\infty, 0)$ and they are given by ℓ real analytic functions, $f_1, \dots f_\ell$, all with $f(\lambda_0) = e_n(\lambda_0)$ on $(\lambda_0 - \varepsilon, \lambda_0 + \varepsilon)$. This follows by Theorems 1.4.4 and 2.3.11. Moreover, there are ℓ analytic $\mathcal{H}$-valued functions, $\varphi_1, \dots, \varphi_\ell$ in $Q(A)$ so that

$$(A - \lambda B)\varphi_j(\lambda) = f_j(\lambda)\varphi_j(\lambda), \quad \|\varphi_j(\lambda)\| = 1 \text{ if } \lambda \text{ is real} \tag{7.9.10}$$

By a Feynman–Hellman type argument (see Theorem 1.4.7),

$$\frac{df_j}{d\lambda} = \langle \varphi_j(\lambda), B\varphi_j(\lambda)\rangle \tag{7.9.11}$$

$$= \lambda^{-1}\big[f_j(\lambda) - \langle \varphi_j(\lambda), A\varphi_j(\lambda)\rangle\big] \tag{7.9.12}$$

$$< 0 \tag{7.9.13}$$

since $f_j < e_n(\lambda_0) + \varepsilon < 0$ and $A \geq 0$. f_j's can cross at λ_0 but there are $j_\pm$ so $e_n(\lambda) = f_{j\pm}(x)$ for $\pm(\lambda - \lambda_0) > 0$. This proves (7.9.9).

(b) By (a), $L = \{\lambda \mid e_n(\lambda) < 0\}$ is open. If it's empty, set $\lambda^{(n)} = \infty$. Otherwise, let (λ_1, λ_2) be a connected component of L. If $\lambda_2 < \infty$, since B is A-form bounded with relative bound zero,

$$\inf_{0\leq\lambda\leq\lambda_2} \big[\inf \sigma(A - \lambda B)\big] > -\infty$$

By (a), $e_n(\lambda)$ is monotone so $\lim_{\lambda\downarrow\lambda_2}$ exists. It is easy to prove (Problem 1(a)) that this limit is $e_n(\lambda_2)$ so $\lambda_2 \in L$ contrary to the assumption that (λ_1, λ_2) was a component. It follows that $\lambda_2 = \infty$. Similarly, one proves (Problem 1(b)) that $\lim_{\lambda\uparrow\lambda_1} e_n(\lambda)$ must be 0.

(c) This is an immediate consequence of continuity, strict monotonicity and $\lim_{\lambda\uparrow\lambda^{(n)}} e_n(\lambda) = 0$. $\square$

These preliminaries immediately yield the main tool of the section.

Theorem 7.9.4 (Birman–Schwinger Principle). *Let $B = C^*D$ be a factorizable form compact perturbation of a nonnegative self-adjoint operator, A. Let $N_e(A+B)$ be $\dim\operatorname{Ran}\bigl(E_{(-\infty,e]}(A+B)\bigr)$ for $e \in (-\infty,0)$, i.e., the number of eigenvalues below e, counting multiplicity. Then*

$$N_e(A+B) = \sharp\{\eta \geq 1 \mid \eta \text{ is an eigenvalue of } K_e \text{ counting multiplicity}\} \tag{7.9.14}$$

If K_e is in some trace ideal, $\mathcal{I}_p$, then

$$N_e(A+B) \leq \|K_e\|_p^p \tag{7.9.15}$$

Remarks. 1. We often abuse notation in situations where A is fixed and write $N_e(B)$.

2. N_e is obviously nondecreasing as e increases, so

$$N(A+B) \equiv \lim_{e\uparrow 0} N_e(A+B)$$

exists. It is clearly $\dim\operatorname{Ran}\bigl((E_{[-\infty,0)}(A+B)\bigr)$, "the total number of bound states" in physicist's parlance.

3. In the canonical case, $C \equiv D$, we don't need the Schur–Lalesco–Weyl inequality in the proof.

Proof. By continuity and the strict monotonicity of $e_j(\lambda)$ (by Proposition 7.9.3), $e_j(1) \leq e_0$ if and only if there is a unique $\lambda_j \in (0,1]$ so that $e_j(\lambda_j) = e_0$. By Proposition 7.9.2, this is true if and only if 1 is an eigenvalue $K_e(A+\lambda_j B) = \lambda_j K_e(A+B)$, if and only if λ_j^{-1} is an eigenvalue of $K_e(A+B)$. Multiplicities are included.

By the Schur–Lalesco–Weyl inequality (Theorem 3.9.1)

$$\sum_\ell |\lambda_\ell(K_e)|^p \leq \|K_e\|_p^p \tag{7.9.16}$$

If $\lambda_\ell(K_e) \geq 1$, it contributes at least 1 to the sum, so

$$\text{RHS of (7.9.15)} \leq \text{RHS of (7.9.16)}$$

which implies (7.9.15) $\square$

The example that both Birman and Schwinger used in discovering their principle was Example 7.9.1 (we do $\mathbb{R}^3$ and will leave $\mathbb{R}^\nu$ to the reader—see Problems 2 and 3).

Theorem 7.9.5 (Birman–Schwinger bound). *For any measurable V on $\mathbb{R}^3$ which is $-\Delta$-form compact, we have*

$$N(V) \leq \frac{1}{(4\pi)^2} \int \frac{|V(x)|\,|V(y)|}{|x-y|^2}\, d^3x\, d^3y \tag{7.9.17}$$

and if $e = -\alpha^2$, $\alpha > 0$, then

$$N_e(V) \leq \frac{1}{(4\pi)^2} \int \frac{|V(x)|\, e^{-2\alpha|x-y|}\, |V(y)|}{|x-y|^2}\, d^3x\, d^3y \tag{7.9.18}$$

Remarks. 1. In fact, if the integral on the right side of (7.9.17) is finite, one can show that V is $-\Delta$-form compact (Problem 4).

2. As usual, we suppose V is nice, e.g., in $L^1 \cap L^\infty$. It is easy to then get the general result by a limiting argument.

3. If $V_- = \max(V, 0)$, then $N(V) \leq N(V_-)$, so in (7.9.17) and similar bounds, one can replace V by V_-.

Proof. As we saw in Section 6.9 of Part 1 (see (6.9.48) of that section), the integral kernel of $(-\Delta + \alpha^2)^{-1}$ is

$$(-\Delta + \alpha^2)(x, y) = (4\pi)^{-1}|x-y|^{-1}e^{-\alpha|x-y|} \tag{7.9.19}$$

so K_e (with $e = -\alpha^2$) has the integral kernel

$$K_e(x, y) = \frac{|V(x)|^{1/2}\, e^{-\alpha|x-y|}\, |V(y)|^{1/2}}{|x-y|} \tag{7.9.20}$$

We recognize the right side of (7.9.18) as the square of the Hilbert–Schmidt norm of K_e, so (7.9.18) is a special case of (7.9.15) with $p = 2$.

By the monotone convergence theorem, as $e \uparrow 1$, RHS of (7.9.18) converges to RHS of (7.9.17) proving (7.9.17). $\square$

By an HLS bound (see Theorem 6.2.1 of Part 3), we have

Corollary 7.9.6. *For a universal constant C, for all $V \in L^{3/2}(\mathbb{R}^3)$, we have*

$$N(V) \leq C\|V\|_{3/2}^2 \tag{7.9.21}$$

Remark. Quasiclassical estimates (see the end of Section 7.5) suggest that $N(\lambda V)$ grows as $\lambda^{3/2}$ as $\lambda \to \infty$, so (7.9.21) has the wrong large λ behavior; this is an issue to discuss further below and in Section 6.7 of Part 3.

Corollary 7.9.7. *Let $e \in (-\infty, 0)$ and V so that $V \in L^2 \cap L^\infty$. Then,*

$$N_e(V) \leq (4\pi)^{-1}(2|e|)^{-1/2} \int |V_{\leq \frac{e}{2}}(x)|^2\, d^3x \tag{7.9.22}$$

where

$$V_{\le e}(x) = \begin{cases} 0 & \text{if } V(x) > e \\ [V(x) - e] & \text{if } V(x) \le e \end{cases} \tag{7.9.23}$$

Proof. We claim that

$$V(x) - e \ge V_{\le \frac{e}{2}}(x) - \frac{e}{2} \tag{7.9.24}$$

For if $V(x) > \frac{e}{2}$, this says $-e \ge -\frac{e}{2}$ obvious since $e < 0$. If $V(x) \le \frac{e}{2}$, this says $V(x) - e \ge V(x) - \frac{e}{2} - \frac{e}{2}$.

(7.9.24) in turn implies, by the variational principle that

$$N_e(V) \le N_{\frac{e}{2}}(V_{\le \frac{e}{2}}) \tag{7.9.25}$$

$$\le \frac{1}{(4\pi)^2} \int |V_{\le \frac{e}{2}}(x)| \, \frac{e^{-\alpha\sqrt{2}|x-y|}}{|x-y|^2} \, |V_{\le \frac{e}{2}}(y)| \tag{7.9.26}$$

by a Birman–Schwinger bound and $\frac{e}{2} = -\big(\frac{\alpha}{\sqrt{2}}\big)^2$.

Young's inequality (see (6.6.10) of Part 1) says that

$$\int |f(x) g(x-y) h(y)| \, d^3x \, d^3y \le \|g\|_1 \, \|f\|_2 \, \|h\|_2 \tag{7.9.27}$$

Noting that (by passing to spherical coordinates and using that for functions of $|x|$, $\int g(|x|) d^3x = 4\pi \int_0^\infty r^2 g(r) dr$) we see

$$\int \frac{e^{-\alpha\sqrt{2}|x|}}{|x|^2} \, d^3x = 4\pi (2|e|)^{-1/2} \tag{7.9.28}$$

and we obtain (7.9.22) from (7.9.26). $\square$

Theorem 7.9.8 (Lieb–Thirring Inequality). *Let V be a bounded function of compact support on $\mathbb{R}^3$. Let $\{\lambda_j(V)\}_{j=1}^J$ be the negative eigenvalues of $-\Delta + V$ counting multiplicity (J is finite or infinite). Let $q > \frac{1}{2}$. Then, there is a constant $C_{q,3}$ so that*

$$S_q(V) \equiv \sum_{j=1}^{J} |\lambda_j(V)|^q \le C_{q,3} \int |V(x)|^{q+3/2} \tag{7.9.29}$$

Remarks. 1. These are the celebrated Lieb–Thirring inequalities. The condition $q > \frac{1}{2}$ is an artifact of the method of proof. They are known to hold (with 3/2 replaced by $\nu/2$) on $\mathbb{R}^\nu$ for $q \ge 0$ if $\nu \ge 3$, $q > 0$ if $\nu = 2$ and for $q \ge \frac{1}{2}$ if $\nu = 1$. It can be seen that the inequalities are false if $\nu = 1$, $q < \frac{1}{2}$ or $\nu = 2$, $q = 0$ (Problem 5). In Problem 3, the reader will use the ideas of this section to get all but the borderline cases ($q = 0$ if $\nu \ge 3$, $q = \frac{1}{2}$ if $\nu = \frac{1}{2}$). Using different ideas, Section 6.7 of Part 3 will prove the borderline cases.

2. $q = 0$, $\nu \geq 3$ are known as the CLR inequalities after Cwickel, Lieb, and Rozenbljum.

3. Since $\lambda_j(-V_-) \leq \lambda_j(V)$, we can replace $|V|^{q+3/2}$ by $|V_-|^{q+3/2}$.

4. Of course, once one has this, it is easy to replace V bounded of compact support by any $V \in L^{q+3/2}$ defining $-\Delta + V$ as a form sum (Problem 6).

Proof. $-dN_e(V)$ is a Stieltjes measure which has only pure points exactly at the eigenvalues of weight the multiplicity of the eigenvalues. Thus

$$S_q(V) = \int |e|^q[-dN_e(V)]$$

$$= q \int_0^\infty y^{q-1} N_{-y}(V)\, dy \tag{7.9.30}$$

$$\leq D_q \int_0^\infty [y^{q-3/2}] \int \big[V(x) + \tfrac{y}{2}\big]_-^2\, d^3x \tag{7.9.31}$$

$$= D_q \int_{\mathbb{R}^3} d^3x \int_0^{2|V_-(x)|} (V(x) + \frac{y}{2})^2 y^{q-\frac{3}{2}}\, dy \tag{7.9.32}$$

$$= D_q \int_{\mathbb{R}^3} [|V_-(x)|^{q+3/2}] \Big[\int_0^2 (1 + \frac{w}{2})^2 w^{q-\frac{3}{2}} dw\Big] \tag{7.9.33}$$

$$= C_{q,3} \int |V_-(x)|^{q+\frac{3}{2}} \tag{7.9.34}$$

In the above, (7.9.30) uses an integration by parts in the Stieltjes integral and (7.9.31) is just (7.9.22). We get (7.9.32) by noting that $V(x) + \frac{y}{2} < 0 \Leftrightarrow y \leq 2|V_-(x)|$ and changing the order of integration. (7.9.33) uses the change of variable $y = w|V(x)|$ and (7.9.34) uses the fact that $q > \frac{1}{2}$, the w integral converges. □

We will use this to deduce a kind of dual relation that generalizes a suitable Gagliardo–Nirenberg inequality. Later, in Section 6.7 of Part 3, we will see that there is a close connection between homogeneous Sobolov inequalities and CLR inequalities and between Lieb–Thirring and Gagliardo–Nirenberg inequalities. We want to consider antisymmetric functions, ψ, in $L^2(\mathbb{R}^{3N}, d^3x)$ (this is suitable for fermions in quantum physics). For now, we consider ψ's, which also obey

$$\int |\psi(x_1, \dots, x_N)|^2\, d^{3N}x = 1 \tag{7.9.35}$$

One defines the *particle density* by

$$\rho(x) = N \int |\psi(x_1, x_2, \dots, x_N)|^2\, d^3x_2 \cdots d^3x_{N-1} \tag{7.9.36}$$

(If we think of ψ as an N-particle wave function, ρ is the physical particle density).

Given $W \geq 0$, let

$$H_N(W) = -\sum_{j=1}^{N} \Delta_j - \sum_{j=1}^{N} W(x_j) \tag{7.9.37}$$

on all of $L^2(\mathbb{R}^{3N}, d^3x)$. It is not hard to see (Problem 7) that eigenvalues of $H_N(W)$ are products $\psi_{j_1}(x_1)\cdots\psi_{j_N}(x_N)$ where $\psi_j(x)$ are eigenvectors of $-\Delta - W$. If we want antisymmetry, the ψ_j's must be distinct (and can be then orthogonal) for which the *Slater determinants*

$$\psi_{j_1\ldots j_N}(x_1,\ldots x_N) = \frac{1}{\sqrt{N!}} \sum_{\pi\in\Sigma_N} \sigma_\pi \psi_{j_{\pi(1)}}(x_1)\cdots\psi_{j_{\pi(N)}}(x_N) \tag{7.9.38}$$

are antisymmetric eigenfunctions with eigenvalues $\sum_{k=1}^N \lambda_{j_k}(-W)$. It follows that the bottom of the spectrum of H_N on the antisymmetric functions is $\sum_{k=1}^N \lambda_k(W)$ (or if $-\Delta - W$ has only $m < N$ eigenvalues $\sum_{k=1}^m \lambda_k(W)$) and so, for all N

$$H_N \geq -S_1(-W) \geq -C_{1,3} \int |W(x)|^{5/2} d^3x \tag{7.9.39}$$

On the other hand, for each j, because of the antisymmetry of ψ and so symmetry of $|\psi|^2$,

$$\begin{aligned}\int W(x_j)|\psi(x_1,\ldots,x_N)|^2 &= \int W(x_1)|\psi(x_1,x_2,\ldots,x_N)|d^{3N}x \\ &= N^{-1}\int \rho(x)W(x)\,d^3x\end{aligned} \tag{7.9.40}$$

It follows that

$$\langle\psi,(-\Delta_N)\psi\rangle \geq \int W(x)\rho(x)d^3x - C_{1,3}\int |W(x)|^{5/2}\,d^3x \tag{7.9.41}$$

Given ψ, pick $W(x) = \alpha\rho^{2/3}(x)$ for some positive α. Then

$$\begin{aligned}\langle\psi,(-\Delta_N)\psi\rangle &\geq (\alpha - C_{1,3}\,\alpha^{5/3})\int \rho^{5/3}(x)d^3x \\ &\geq D_{1,3}\int \rho^{5/3}(x)d^3x\end{aligned} \tag{7.9.42}$$

where $D_{1,3} = \min_\alpha(\alpha - C_{1,3}\,\alpha^{5/3}) < 0$ since $\alpha - C_{1,3}\alpha^{5/2}$ is negative for small α. We have thus proven:

Theorem 7.9.9 (Lieb-Thirring Kinetic Energy Bound)**.** *For any antisymmetric ψ on $L^2(\mathbb{R}^{3N})$ obeying (7.9.35) with ρ given by (7.9.36), we have (7.9.42) for a constant independent of N and ψ.*

Remarks. 1. This is important historically because it was used for an elegant proof of quantum stability of matter depending on the known stability of matter in the Thomas–Fermi theory of atoms; see the Notes.

2. In a quasiclassical limit $\int \rho^{5/3}(x)d^3x$ is the kinetic energy of the ground state, see the end of Section 7.5.

3. By replacing ψ by $\psi/\|\psi\|_2$, we see that for general ψ, (7.9.42) says that

$$\langle \psi, (-\Delta_N)\psi\rangle \, \langle \psi, \psi\rangle^{2/3} \geq D_{1,3} \int |\rho(x)|^{5/3} d^3x \tag{7.9.43}$$

with ρ still given by (7.9.36). Specializing to $N = 1$ where $\rho(x) = |\psi(x)|$, raising both sides to 3/10th power, we get

$$D_{1,3}^{3/10} \, \|\psi\|_{10/3} \leq \|\nabla\psi\|_2^{3/5} \, \|\psi\|_2^{2/5} \tag{7.9.44}$$

which is a special case of the Gagliardo–Nirenberg inequality, (6.3.30) of Part 3 (with $\theta = 3/5$, $s = 1$, $\nu = 3$, $q = r = 2$ so $p = 10/3$). We discuss this further in the Notes.

Having discussed this application of the Birman–Schwinger principle, we turn to when instead of looking at $N_e(V)$ for $e < 0$ and taking $e \uparrow 0$, we can directly look at $\lim_{e\uparrow 0} K_e(V)$. One key is a monotonicity in e. For eigenvalues, this is true for any C^*D factorization.

Proposition 7.9.10. *Let $B = C^*D$ be relatively compact, $\lambda_j^\pm(K_e(B))$ be all the eigenvalues of $K_e(B))$ with $\pm\lambda_j^\pm > 0$ and $\pm\lambda_1^\pm \geq \pm\lambda_2^\pm \geq \dots$. Then as e increases, $\pm\lambda_j^\pm(K_e(B))$ increases.*

Proof. We proved $e_n(\lambda)$ is monotone decreasing in λ (see Proposition 7.9.3). $K_e(B)$ has eigenvalue $\mu = \lambda_j^+(K_e(B))$ if and only if $e_n(\mu^{-1}) = e$. Thus, $(\lambda_j^+)^{-1}$ and e_n are inverse functions and the monotonicity of e_n implies one of λ_j^+. Since $\lambda_j^-(K_e(B)) = -\lambda_j(K_e(B))$, we get the λ_j^- monotonicity. □

Proposition 7.9.11. *In the canonical case $B = C^*C$, $K_e(B)$ is strictly monotone in e as an operator, i.e., $K_e(B) \geq 0$ and for any φ, and $e < \tilde{e} < 0$*

$$\langle \varphi, K_e(B)\varphi\rangle < \langle \varphi, K_{\tilde{e}}(B)\varphi\rangle \tag{7.9.45}$$

Proof. For any finite measure μ on $[0,\infty)$

$$\int \frac{d\mu(x)}{x-e} < \int \frac{d\mu(x)}{x-\tilde{e}} \tag{7.9.46}$$

This fact plus the spectral theorem implies (7.9.45). □

Thus, for any φ, $\lim_{e\uparrow 0}\langle \varphi, K_e(B)\varphi\rangle$ exists although it may be infinite for some φ. If it is always finite, then by polarization, $\lim_{e\uparrow 0}\langle \varphi, K_e(B)\psi\rangle$ always exists and defines an operator which is Hermitian and everywhere defined. Thus, there is a bounded operator, $K_0(B)$, the *zero energy Birman–Schwinger kernel.*

Theorem 7.9.12. *In the canonical case, suppose $K_0(B)$ exists and the spectral projection $P_{(1,\infty)}(K_0)$ has finite dimension. Then $N(B)$ is finite and*

$$N(B) = \dim P_{(1,\infty)}(K_0(B)) \tag{7.9.47}$$

Remarks. 1. It can happen that $K_0(B)$ exists but is not compact (Problem 8).

2. (7.9.47) is sometimes called the *Birman–Schwinger principle.*

3. It can be proven (Problem 9) if $\dim P_{(1,\infty)}(K(B)) = \infty$, then $N(B)$ is infinite.

Proof. Let $d = \dim P_{(1,\infty)}(K_0(B))$. Since $K_e(B) \leq K_0(B)$

$$\dim\bigl(P_{(1,\infty)}(K_e(B))\bigr) \leq d \tag{7.9.48}$$

The Birman–Schwinger principle in the form of Theorem 7.9.4 says

$$N_{<e}(B) = \dim P_{(1,\infty)}(K_e(B)) \tag{7.9.49}$$

Note. We state eigenvalues of $A + B \leq e =$ eigenvalues of $K_e \geq 1$ but the same argument gives the result with $< e$ and > 1.

Thus, since $N(B) = \lim_{e \uparrow 0} N_{<e}(B)$

$$N(B) \leq d \tag{7.9.50}$$

Let $\mathcal{K}$ be the d-dimensional space of eigenvectors of $K_0(B)$ larger than 1. Then, $\inf_{\varphi \in \mathcal{K}, \|\varphi\|=1} \langle \varphi, K_0(B)\varphi \rangle > 1$. It follows for some $e_0 < 0$ that $\inf_{\varphi \in \mathcal{K}, \|\varphi\|=1} \langle \varphi, K_{e_0}(B)\varphi \rangle > 1$ also, so by (7.9.49) and the variational principle, $N_e(B) \geq d$, i.e., $N(B) \geq d$. $\square$

In the case $\mathcal{H} = L^2(\mathbb{R}^3, d^3x)$, $A = -\Delta$ and $B = V \geq 0$, K_0 is exactly (7.9.2).

* * * * *

We now turn to bounds on the number of bound states on $L^2(\mathbb{R}^3)$ when V is spherically symmetric (and on $L^2([0,\infty), dx)$). In that case, as we saw in Theorem 3.5.8 of Part 3 (see the remark after the theorem), $-\Delta$ is unitary equivalent under a natural unitary to

$$\bigoplus_{\ell=0}^{\infty} \bigoplus_{m=-\ell}^{\ell} h_{0,l} \text{ on } \bigoplus_{\ell=0}^{\infty} \bigoplus_{m=-\ell}^{\ell} L^2([0,\infty), dr)$$

where

$$h_{0,\ell} = -\frac{d^2}{dr^2} + \frac{\ell(\ell+1)}{r^2} \tag{7.9.51}$$

If $\ell = 1, 2, \ldots$, the operator is limit point at 0 and infinity as we saw in Example 7.4.24. If $\ell = 0$, one needs $u(0) = 0$ boundary conditions.

So we want to study $\dim P_{(-\infty,0)}(h_{0,\ell}+V)$ for a function, $V(r)$ on $[0,\infty)$. We'll assume initially that V is nonpositive and continuous of compact support. $\ell = 0$ is particularly simple. Using the formalism of Theorem 7.4.17, for $e = -\kappa^2 < 0$

$$u_-(r) = \sinh \kappa r, \quad u_+(r) = e^{-\kappa r}, \quad W(u_+, u_-) = \kappa \tag{7.9.52}$$

$$g_{0,\ell=0,e} = \kappa^{-1} \sinh \kappa r_< \, e^{-\kappa r_>} \tag{7.9.53}$$

Pointwise as $e \uparrow 0$, g_0 has the limit

$$g_{0,\ell=0,e=0} = r_< \tag{7.9.54}$$

(which fits (7.4.44) if we take $u_+(r) \equiv 1$, $u_-(r) \equiv r$). By Mercer's theorem (Theorem 3.11.9) since $K_e \equiv |V|^{1/2}(h_{0,\ell=0}-e)^{-1}\,|V|^{1/2} \geq 0$, K_e is trace class and by Mercer's theorem again (since $K_e - K_{e'} \geq 0$ if $e > e'$), K_e converges as $e \uparrow 0$ in trace class to the trace class operator K_0 where K_0 has kernel

$$K_0 = |V(r)|^{1/2} \min(r, r')|V(r')|^{1/2} \tag{7.9.55}$$

Thus, letting $n_\ell(V)$ be $N_{e=0}(V)$ for $h_{0,\ell} + V$

$$n_{\ell=0}(V) \leq \text{Tr}(K_0)) = \int_0^\infty r|V(r)|dr \tag{7.9.56}$$

The general ℓ case is similar but computationally more complicated since $u_\pm$ for $\ell \neq 0$, $e < 0$ involves Bessel functions (of imaginary argument).

The solutions of

$$-u'' + \frac{\ell(\ell+1)}{r^2} u = 0 \tag{7.9.57}$$

are $r^{\ell+1}$ and $r^{-\ell}$ (as one checks or finds by trying r^α and solving for α). Thus, formally at $e = 0$

$$u_+ = r^{-\ell}, \quad u_- = r^{\ell+1}, \quad W(u_+, u_-) = 2\ell + 1 \tag{7.9.58}$$

so defining $g_{0,\ell,e>0}$ by (7.4.44), we have that

$$g_{0,\ell,e=0}(r, r') = (2\ell+1)^{-1} r_<^{\ell+1} r_>^{-\ell} \tag{7.9.59}$$

and K_0 has integral kernel

$$(2\ell+1)^{-1}|V(r)|^{1/2} r_<^{\ell+1} r_>^{-\ell} |V(r')|^{1/2} \tag{7.9.60}$$

which is trace class by Mercer's theorem.

One can justify that this is indeed $\lim_{e\uparrow 0} K_e$ either by using explicit Bessel functions (Problem 13), or by ODE methods (Problem 14). Either way

$$n_\ell(V) \leq \text{Tr}(K_0) = (2\ell+1)^{-1} \int_0^\infty r|V(r)|dr \tag{7.9.61}$$

By a limiting argument, for $V \leq 0$ and variational principle for V of both signs, one obtains the celebrated

Theorem 7.9.13 (Bargmann's Bound). *If $\int_0^\infty r\,|V(r)|\,dr < 0$, we have*

$$n_\ell(V) \leq (2\ell+1)^{-1} \int_0^\infty r\,|V(r)|\,dr \tag{7.9.62}$$

Problem 10 has a version of this for Jacobi matrices and Problem 11 an ODE proof (the original one of Bargmann).

* * * * *

We discussed the CLR bounds which imply if $V \in C_0^\infty$ (much more generally), $\nu \geq 3$ and λ is small, then $N(-\Delta + \lambda V) = 0$. The analog is false for $\nu = 1, 2$ and one can use the Birman–Schwinger principle to prove and analyze this. We'll discuss the $\nu = 1$ case and leave the $\nu = 2$ case to Problem 12.

As we've seen, $\sqrt{-e}$ enters in K_e so we'll shift to $-\frac{d}{dx^2} + \lambda V$ (not $-\lambda V$), $\alpha > 0$ with $e = -\alpha^2$ and abuse notation and write K_α for $K_{-\alpha^2}$, i.e., K_α has integral kernel

$$K_\alpha(x,y) = -|V(x)|^{1/2}\, \frac{e^{-\alpha|x-y|}}{2\alpha}\, V^{1/2}(y) \tag{7.9.63}$$

with $V^{1/2}(x) \equiv V(x)/|V(x)|^{1/2}$. $e^{-\alpha|x-y|}/2\alpha$ enters since $u_\pm(x) = e^{\mp\alpha x}$ with $W(u_+, u_-) = 2\alpha$. For reasons that will become clear in a moment, we'll assume V obeys

$$\int (1 + |x|^2)\, |V(x)|\, dx < \infty \tag{7.9.64}$$

Notice that as $\alpha \downarrow 0$, $K_\alpha(x,y) \to \infty$, so there is no K_0, but the infinity is only in a rank-one operator. To make this clear, define kernels L_α, $\widetilde{L}$, M_α by

$$\begin{aligned} L_\alpha(x,y) &= (2\alpha)^{-1}\, \widetilde{L}(x,y) \\ \widetilde{L}(x,y) &= -|V(x)|^{1/2}\, V(y)^{1/2}; \\ M_\alpha(x,y) &= -|V(x)|^{1/2} \frac{(-e^{-\alpha|x-y|} - 1)}{2\alpha} V(y)^{1/2} \end{aligned} \tag{7.9.65}$$

We'll see in a moment that L_α, $\widetilde{L}$, M_α define operators and

$$K_\alpha = L_\alpha + M_\alpha \tag{7.9.66}$$

either as kernels or operators.

Proposition 7.9.14. *K_α, L_α, M_α, $\widetilde{L}$ are kernels of trace class operators.*

Proof. K_α has the form $f(x)g(p)h(x)$ where $g(p) = (p^2+\alpha^2)^{-1}$ and so is trace class by Theorem 3.8.6 and $V \in L^1$. Thus, K_α is trace class. L_α and $\widetilde{L}$ are rank-one, so also trace class and $M_\alpha = K_\alpha - L_\alpha$. $\square$

Let M_0 be the integral kernel which is the pointwise limit of $M_\alpha(x,y)$:

$$M_0(x,y) = \tfrac{1}{2}|V(x)|^{1/2}\,|x-y|\,V(y)^{1/2} \tag{7.9.67}$$

Proposition 7.9.15. *M_0 is the kernel of a Hilbert–Schmidt operator and $\|M_\alpha - M_0\|_2 \to 0$ as $\alpha \downarrow 0$ where $\|\cdot\|_2$ is the Hilbert-Schmidt norm.*

Remark. This is presumably also true if "Hilbert–Schmidt" is replaced by "trace class."

Proof.

$$\begin{aligned}\int |M_0(x,y)|^2\,dx\,dy &= \frac{1}{4}\int |V(x)|\,|x-y|^2\,|V(y)|^2\,dx\,dy \\ &\le \frac{1}{2}\int |V(x)|\,(|x|^2+|y|^2)\,|V(y)|^2\,dx\,dy \end{aligned} \tag{7.9.68}$$

is finite by (7.9.64). That

$$\int |M_\alpha(x,y) - M_0(x,y)|^2\,dx\,dy \to 0 \tag{7.9.69}$$

follows from dominated convergence and $|e^{-u}-1| = |\int_0^u e^{-y}\,dy| \le u$ if $u > 0$. $\square$

Theorem 7.9.16. *Let V be a measurable function on $\mathbb{R}$, obeying (7.9.64). Then, for small δ, $H(\delta) = -\frac{d^2}{dx^2} + \delta V(x)$ has at most one bound state.*

Proof. Since $N(\lambda V) \le N(-\lambda|V|)$, we can suppose that $V \le 0$. In that case, K_α, L_α and M_α are all self-adjoint. L_α is rank-one, explicitly, $\langle \varphi, L_\alpha\varphi\rangle = 0$ if $\langle |V|^{1/2}, \varphi\rangle = 0$. Thus, if $0 < \alpha < 1$

$$\varphi \perp |V|^{1/2} \Rightarrow \langle \varphi, K_\alpha\varphi\rangle = \langle \varphi, M_\alpha\varphi\rangle \le \|\varphi\|^2 \sup_{0\le\alpha\le 1}\|M_\alpha\|$$

Then, if $c = \sup_{0\le\alpha\le 1}\|M_\alpha\|$, the set $\sigma(K_\alpha)\setminus[0,c]$ is at most one point by the min-max principle. If $\lambda_0 c < 1$, $\lambda_0 K_\alpha$ has at most one eigenvalue in $[1,\infty)$, i.e., $N_\alpha(\lambda_0 V) \le 1$. Since α is arbitrary in $[0,1]$, $N(\lambda_0 V) \le 1$, i.e., $\lambda < c^{-1} \Rightarrow N(\lambda V) \le 1$. $\square$

Proposition 7.9.17. *Let V be a measurable function on $\mathbb{R}$ obeying (7.9.64). Let $c = \sup_{0\le\alpha<\infty}\|M_\alpha\|$ (which is finite). Let*

$$H(\lambda) = -\frac{d^2}{dx^2} + \lambda V(x) \tag{7.9.70}$$

If $0 < \lambda < c^{-1}$, then $-\alpha^2$ is a negative eigenvalue for $\alpha \in (0, \infty)$ of $H(\lambda)$ if and only if

$$\alpha = -\tfrac{1}{2}\lambda\langle V^{1/2}, (1 - \lambda M_\alpha)^{-1}\, |V|^{1/2}\rangle \tag{7.9.71}$$

Proof. By the Birman–Schwinger principle, we know that $-\alpha^2$ is an eigenvalue if and only if there is $\varphi \in L^2$, $\varphi \neq 0$ so that

$$(\lambda M_\alpha + \lambda L_\alpha)\varphi = \varphi \tag{7.9.72}$$

Since $\|\lambda M_\alpha\| < 1$, this is true if and only if

$$\lambda(1 - \lambda M_\alpha)^{-1} L_\alpha\, \varphi = \varphi \tag{7.9.73}$$

If $\langle V^{1/2}, \varphi\rangle = 0$, $L_\alpha \varphi = 0$, so by (7.9.73), $\varphi = 0$. Thus $\varphi \neq 0 \Rightarrow \langle V^{1/2}, \varphi\rangle \neq 0$. If (7.9.73) holds, taking inner products with $V^{1/2}$, we see that

$$-\lambda(2\alpha)^{-1}\langle V^{1/2}, (1 - \lambda M_\alpha)^{-1}\, |V|^{1/2}\rangle\, \langle V^{1/2}, \varphi\rangle = \langle V^{1/2}, \varphi\rangle$$

so (7.9.71) holds.

For the converse, if (7.9.71) holds for $\alpha > 0$, we set

$$\varphi = -\lambda(2\alpha)^{-1}(1 - \lambda M_\alpha)^{-1}\, |V|^{1/2} \tag{7.9.74}$$

and note that by (7.9.71), $\langle V^{1/2}, \varphi\rangle = 1$ and therefore φ obeys (7.9.73). $\square$

Theorem 7.9.18. *Let V be a measurable function on $\mathbb{R}$ obeying* (7.9.64). *Let*

$$H(\lambda) = -\frac{d^2}{dx^2} + \lambda\, V(x), \qquad a = \int V(x)\, dx \tag{7.9.75}$$

If $a > 0$, for small λ, $H(\lambda)$ has no negative eigenvalues. If $a < 0$, for small λ, $H(\lambda)$ has exactly one negative eigenvalue $e(\lambda)$ with

$$e(\lambda) = -\left(\frac{\lambda a}{2}\right)^2 + o(\lambda^2) \tag{7.9.76}$$

Remarks. 1. We'll discuss $a = 0$ soon; it is more subtle.

2. If $\int e^{A|x|}\, |V(x)|\, dx < \infty$ for some $A > 0$, $e(\lambda)$ in the case $a < 0$ is given by an analytic function of λ (for $\lambda > 0$ and small).

Proof. If (7.9.71) has a solution $\alpha(\lambda)$, then since $M_{\alpha(\lambda)}$ is uniformly bounded as $\alpha \downarrow 0$, $\lim \alpha(\lambda)/\lambda$ exists and equals $-\frac{1}{2}a$. If $a > 0$, this is impossible since $\alpha(\lambda) > 0$, i.e., there is no positive solution, $\alpha(\lambda)$, for all small λ, i.e., $N(\lambda V) = 0$.

Let $\alpha = \beta\lambda$. We are looking for a solution of $F(\beta, \lambda) = 0$ where

$$F(\beta, \lambda) = \beta + \tfrac{1}{2}\langle |V|^{1/2}, (1 + \lambda M_{\beta\lambda})^{-1}\, V^{1/2}\rangle \tag{7.9.77}$$

defined for $\beta > 0$, $\lambda \geq 0$. $F(\beta, \lambda = 0) = \beta + \frac{1}{2}a$ has a positive solution if $a < 0$. Moreover, $\frac{\partial F}{\partial \beta}(\beta = -\frac{1}{2}a, \lambda = 0) = 1$, so by a variant of the implicit function theorem, (7.9.71) has a solution positive for small λ. Of necessity, it obeys $\lim \alpha(\lambda)/\lambda = -\frac{1}{2}a$, i.e., (7.9.76) holds. $\square$

As a preliminary to the case $a = 0$, we need

Proposition 7.9.19. *If f obeys (7.9.64) and $\int f(x)\,dx = 0$, $f \not\equiv 0$, then $\int f(x)|x-y|f(y)\,dx\,dy < 0$.*

Remark. We say $|x-y|$ is a *conditionally strictly negative definite kernel.*

Proof.

$$\begin{aligned}
-\int f(x)|x-y|f(y)\,dx\,dy &= \lim_{\alpha\downarrow 0} 2\int f(x)\frac{(e^{-\alpha|x-y|}-1)}{2\alpha}f(y)\,dx\,dy \\
&= \lim_{\alpha\downarrow 0} 2\int f(x)\frac{e^{-\alpha|x-y|}}{2\alpha}f(y)\,dx\,dy \\
&= \lim_{\alpha\downarrow 0} 2\int |\hat{f}(k)|^2\frac{dk}{k^2+\alpha^2} = 2\int \frac{\hat{f}(k)^2}{k^2}dk > 0
\end{aligned}$$

$\square$

Theorem 7.9.20. *If the hypothesis of Theorem 7.9.18 holds with $a = 0$, $V \not\equiv 0$, then $H(\lambda)$ has exactly one negative eigenvalue for λ small and*

$$e(\lambda) = -\biggl(\frac{\lambda^2}{4}\int V(x)\,|x-y|\,V(y)\,dx\,dy\biggr)^2 + o(\lambda^4)$$

Proof. Since $a = 0$,

$$\begin{aligned}
\langle V^{1/2}, (1-\lambda M_\alpha)^{-1}\,|V|^{1/2}\rangle &= \langle V^{1/2}, \bigl[(1-\lambda M_\alpha)^{-1}-1\bigr]\,|V|^{1/2}\rangle \\
&= -\lambda\langle V^{1/2}, M_\alpha(1-\lambda M_\alpha)^{-1}\,|V|^{1/2}\rangle
\end{aligned}$$

At $\alpha = 0$, $\lambda = 0$, $\langle V^{1/2}, M_\alpha(1-\lambda M_\alpha)^{-1}\,V^{1/2}\rangle\,|_{\alpha=\lambda=0} = \langle V, |x-y|\,V\rangle$. An analysis using $\alpha = \gamma\lambda^2$ similar to the one in the proof of Theorem 7.9.18 proves this result.

$\square$

Notes and Historical Remarks.

The idea of quantitative bounds on $N(V)$ for Schrödinger operators goes back to Bargmann [**45**] in 1952. In 1961, Birman [**59**] and Schwinger [**612**] independently found their method and bound and used the method to recover Bargmann's bound. Birman's paper was largely unknown in the West and the bound was widely referred to as the Schwinger bound, especially in the physics literature. This was changed by Simon's review article [**631**] for a Festschrift in honor of Bargmann's 65th birthday. He introduced the terms Birman–Schwinger principle and Birman–Schwinger bound which by now have wide acceptance and literally hundreds of papers.

We'll say much more about Lieb–Thirring bounds in the Notes to Section 6.7 of Part 3. The method we used in the proof of Theorem 7.9.8 is from the original paper of Lieb–Thirring [**447**] who only considered $q = 1$, $\nu = 3$. The more general result (and method of Problem 3) is from their follow-up paper [**448**] from the same Bargmann Festschrift mentioned above.

The square root of the integral in (7.9.17) is sometimes called the Rollnik norm after its first appearance (even before Birman and Schwinger) in Rollnik [**577**]. Simon [**627, 628**] has systematically developed quantum mechanics for such potentials.

Bargmann (and Lieb–Thirring) estimates for Jacobi matrices on the half-line are due to Hundertmark–Simon [**326**].

The small coupling analysis for $-\Delta + V$ if $\nu = 1, 2$ (Theorem 7.9.16) is due to Simon [**632**]; for extensions to the long range case, see Blankenbecler et al. [**72**] and Klaus [**392**]. In particular, Simon [**632**] proves the analyticity result mentioned after Theorem 7.9.18.

Problems.

1. (a) Let $e_j(\lambda)$ be the negative eigenvalues of $A-\lambda B$ (B is A-form compact, $\sigma(A) \subset [0,\infty)$) counting multiplicity. Let $\lambda \downarrow \lambda_2$ and suppose $A - \lambda B$ has at least n strictly negative eigenvalues for $\lambda \in (\lambda_1, \lambda_2)$. Prove that $A - \lambda_2 B$ has at least n negative eigenvalues and $\lim_{\lambda \downarrow \lambda_2} e_n(\lambda) = e_n(\lambda_2)$. (*Hint*: Prove that $\langle \varphi, B\varphi \rangle > 0$ if φ is an eigenfunction with a negative eigenvalue of any $A - \lambda B$ ($\lambda > 0$) and conclude the number of negative eigenvalues is increasing as λ increases.)

   (b) If $e_n(\lambda) < 0$ for $\lambda < \lambda_1$ but $A - \lambda_1 B$ has fewer than n negative eigenvalues, prove that $\lim_{\lambda \uparrow \lambda_1} e_n(\lambda) = 0$. (*Hint*: If not, use a variational argument to show that $A - \lambda_1 B$ has at least n eigenvalues.)

2. Let $V \in L^p(\mathbb{R}^\nu)$, $\nu \geq 3$, $p > \nu/2$. Prove that $|V|^{1/2}(-\Delta + e)^{-1/2} \in \mathcal{I}_{2p}$ with $\| |V|^{1/2} (-\Delta + e)^{-1/2} \|_{2p}$ bounded uniformly in $e > 0$ and conclude that $N(-\Delta - V) < \infty$. (*Hint*: See Problem 6 in Section 3.8.)

3. (a) Use the ideas of Problem 2 and of our proof of Theorem 7.9.8 to prove (7.9.29) for $\nu \geq 3$, $K > 0$.

   (b) Using the ideas and the analysis in Theorem 7.9.16 prove (7.9.29) for $\nu = 1$, $K > 1/2$ and $\nu = 2$, $K > 0$.

4. If $\int \frac{|V(x)|\,|V(y)|}{|x-y|^2}\, d^3x\, d^3y < \infty$, prove that V is $-\Delta$-form compact.

5. (a) Let $V = \chi_{[-1,1]}(x)$. Prove that for any $\lambda > 0$, $-\frac{d^2}{dx^2} - \lambda V$ has a negative eigenvalue, $e_1(\lambda)$, with $e_1(\lambda) \le c\lambda^2$ for some $c > 0$. Conclude that (7.9.29) fails if $0 < K < \frac{1}{2}$, $\nu = 1$. (*Hint*: Use a trial wave function $\varphi(x) = e^{-\beta|x|}$ for suitable $\beta(\lambda)$.)

   (b) Prove for V the characteristic function of a ball in $\mathbb{R}^2$, that $-\Delta - \lambda V$ has a negative eigenvalue for all $\lambda > 0$.

   Remark. This is, of course, related to Theorem 7.9.16 and Problem 12 below.

6. Knowing Theorem 7.9.8 for V on $\mathbb{R}^3$ which is bounded with compact support, extend the result to $V \in L^{q+3/2}$, $q > 1/2$.

7. Prove that eigenvectors of the $H_N(W)$ of (7.9.37) are sums of products of the form $\psi_{j_1}(x_1) \cdots \psi_{j_N}(x_N)$ where ψ_j is an eigenfunction of $-\Delta + W$. (*Hint*: See Section 3.10.)

8. Let $V(x)$ on $\mathbb{R}^3$ be defined by

$$V(x) = \begin{cases} 1 & \text{if } |x| \le 1 \\ |x|^{-2} & \text{if } |x| \ge 1 \end{cases}$$

   Prove that V is a relatively compact perturbation of $-\Delta$, then that $\lim_{e\uparrow 0} |V|^{1/2}(-\Delta - e)^{-1}|V|^{1/2}$ exists in strong operator sense (and defines a bounded operator), but that the limit is not a compact operator. (*Hint*: Let φ_0 be the characteristic function of $1 < |x| < 2$ and $\varphi_n(x) = 2^{-3\alpha/2}\varphi_n(2^{-n}x)$. Prove that $\varphi_n \to 0$ weakly but $\langle \varphi_n, K_0 \varphi_n \rangle$ does not go to 0.)

9. (a) If $\dim P_{[1,\infty)}(K_0(B)) = \infty$, prove that $N(B) = \infty$.

   (b) Using the example of Problem 8, find an explicit example where $\dim P_{[1,\infty)}(K_0(B)) = N(B) = \infty$.

10. (a) Let $J(a,b)$ be the half-line Jacobi operator with parameter, $\{a_n, b_n\}_{n=1}^\infty$ and $N(a,b)$ the number of eigenvalues in $(-\infty, -2)$. By mimicking the proof of Theorem 7.9.13, prove that

$$N(a \equiv 1, b) \le \sum_{n=1}^\infty n|b_n| \tag{7.9.78}$$

   (b) Prove that

$$\begin{pmatrix} -|a_n - 1| & 1 \\ 1 & -|a_n - 1| \end{pmatrix} \le \begin{pmatrix} 0 & a_n \\ a_n & 0 \end{pmatrix} \tag{7.9.79}$$

and use this to prove that

$$N(a,b) \le \sum_{n=1}^{\infty} n \Big[|b_n| + |a_{n-1} - 1| + |a_n - 1| \Big]$$

Remark. The idea in (b) is due to Hundertmark–Simon [**326**] where details can be found.

11. This problem will sketch an ODE proof of (7.9.62) when $\ell = 0$. We suppose V has compact support and $V \le 0$.

(a) Let u solve $-u'' + Vu = 0$, $u(0) = 0$. Prove that $N(V)$ is the number of zeros of u on $(0, \infty)$.

(b) Let $a(r)$ be defined by $\big(a(r) + r\big)u'(r) = u(r)$. Prove that a obeys

$$a'(r) = -V(r)[r + a(r)]^2$$

and conclude that a is monotone increasing (except at poles).

(c) Prove that the number of poles of a is the number of zeros of u (and so $N(V)$).

(d) Let $b(r) = r^{-1}a(r)$. Let $z_1 = 0 < p_1 < z_2 < p_2 < \dots$ be the zeros and poles of b. On (z_j, p_j) prove that

$$b' \le r|V(r)|\,(b(r) + 1)^2$$

(e) By integrating $-\frac{d}{dr}\big(b(r) + 1\big)^{-1}$, prove that

$$1 \le \int_{z_j}^{p_j} r\,|V(r)|\,dr$$

and conclude that $N(r) \le \int_0^\infty r\,|V(r)|\,dr$.

12. (a) Using the explicit form of $(-\Delta - \alpha^2)^{-1}$ in two dimensions, prove that in that case, $|V|^{1/2}(-\Delta - \alpha^2)^{-1}V^{1/2} = M_\alpha + L_\alpha$ where $L_\alpha(x,y) = -|V(x)|^{1/2}\,V(y)^{1/2}\big[-\frac{1}{2\pi}\log(\alpha)\big]$ and M_α has a limit in Hilbert–Schmidt norm as $\alpha \downarrow 0$.

(b) Prove a two-dimensional variant of Theorem 7.9.16 where

$$e(\lambda) = \Big(-\frac{\lambda}{2}\int_{-\infty}^{\infty} V(x)\,dx\Big)^2 + o(\lambda^2)$$

is replaced by

$$e(\lambda) = -\exp\Big(\Big[\frac{\lambda}{4\pi}\int V(x)d^2x\Big]^{-1}\Big)\,\big[1 + o(1)\big]$$

13. (a) Express solutions of

$$\left(-\frac{d^2}{dr^2} + \frac{\ell(\ell+1)}{r^2} + \kappa^2\right)u = 0 \tag{7.9.80}$$

regular at $r = 0$ in terms of spherical Bessel functions of imaginary argument (see Example 3.5.20 of Part 3).

(b) Find an analogous formula for the solution decaying at infinity in terms of the Hankel function analog.

(c) With these formulae, mimic the proof we gave for $\ell = 0$ to provide a proof of $\ell \neq 0$ Bargmann bounds.

14. Without knowing the explicit solutions of (7.9.80) when $K \neq 0$, use variation of parameters about $K = 0$ to prove continuity, as $K \downarrow 0$, of $g_{0,\ell,-K^2}(r,r')$ in r, r' uniformly for $0 < a < r$, $r' < b < \infty$. Then, use this to prove the Bargmann bound for $\ell \neq 0$ for V bounded and supported in (a,b).

> The White Rabbit put on his spectacles. "Where shall I begin, please your Majesty?" he asked. "Begin at the beginning," the King said very gravely, "and go on till you come to the end: then stop."
>
> —*Lewis Carroll* from *Alice in Wonderland* [**106**][1]

[1]Lewis Carroll is, of course, the pseudonym of Charles Dodgson (1832–1898), a British mathematician.

Bibliography

[1] N. H. Abel, *Untersuchungen über die Reihe:* $1 + \frac{m}{1}x + \frac{m(m-1)}{1\cdot 2}x^2 + \frac{m(m-1)(m-2)}{1\cdot 2\cdot 3}x^3 + \dots$ *u.s.w.*, J. Reine Angew. Math. **1** (1826), 311–339. Cited on 505.)

[2] J. Aczél, *Sur l'équation différentielle des polynômes orthogonaux classiques*, Ann. Univ. Sci. Budapest **2** (1959), 27–29. Cited on 255.)

[3] V. M. Adamjan, D. Z. Arov, and M. G. Krein, *Infinite Hankel matrices and generalized Carathéodory–Fejér and I. Schur problems*, Funk. Anal. i Prilozen **2** (1968), 1–17; English translation: Functional Anal. Appl. 2 (1968), 269–281. Cited on 658.)

[4] J. F. Adams, *Lectures on Lie Groups*, W. A. Benjamin, Inc., New York–Amsterdam 1969. Cited on 443.)

[5] S. Agmon, *Sur un problème de translations*, C. R. Acad. Sci. Paris **229** (1949), 540–542. Cited on 128.)

[6] S. Agmon, *Spectral properties of Schrödinger operators and scattering theory*, Ann. Scuola Norm. Sup. Pisa **2** (1975), 151–218. Cited on 600, 667.)

[7] Y. Aharonov and A. Casher, *Ground state of a spin-12 charged particle in a two-dimensional magnetic field*, Phys. Rev. A **19** (1979), 2461–2462. Cited on 218.)

[8] P. R. Ahern and D. Sarason, *The H^p spaces of a class of function of algebras*, Acta Math. **117** (1967), 123–163. Cited on 490.)

[9] M. Aizenman and B. Simon, *Brownian motion and Harnack's inequality for Schrödinger operators*, Comm. Pure Appl. Math. **35** (1982), 209–273. Cited on 538.)

[10] N. I. Akhiezer, *Theory of Approximation*, Translated from the Russian and with a preface by Charles J. Hyman, reprint of the 1956 English translation, Dover Publications, Inc., New York, 1992 [Russian original: 1947]. Cited on 267.)

[11] N. I. Akhiezer, *The Classical Moment Problem and Some Related Questions in Analysis*, Hafner Publishing, New York, 1965 [Russian original: 1961]. Cited on 658, 659.)

[12] S. Albeverio and P. Kurasov, *Finite rank perturbations and distribution theory*, Proc. Amer. Math. Soc. **127** (1999), 1151–1161. Cited on 666.)

[13] Dž. É. Allahverdiev *On the rate of approximation of completely continuous operators by finite-dimensional operators*, Azerbaidzan. Gos. Univ. Ucen. Zap. Ser. Fiz.-Mat. i Him. Nauk **2** (1957), 27–37 [Russian]. Cited on 136.)

[14] A. Alonso and B. Simon, *The Birman–Krein–Vishik theory of selfadjoint extensions of semibounded operators*, J. Operator Theory **4** (1980), , 251–270. Cited on 601.)

[15] W. Ambrose, *Structure theorems for a special class of Banach algebras*, Trans. Amer. Math. Soc. **57** (1945), 364–386. Cited on 57.)

[16] W. O. Amrein and V. Georgescu, *On the characterization of bound states and scattering states in quantum mechanics*, Helv. Phys. Acta **46** (1973/74), 635–658. Cited on 321.)

[17] W. O. Amrein and K. B. Sinha, *On pairs of projections in a Hilbert space*, Linear Algebra Appl. **208/209** (1994), 425–435. Cited on 218, 354.)

[18] T. Ando, *Lebesgue-type decomposition of positive operators*, Acta Sci. Math. Szeged **38** (1976), 253–260. Cited on 601.)

[19] G. E. Andrews, R. Askey, and R. Roy, *Special Functions*, Encyclopedia of Mathematics and Its Applications, Cambridge University Press, Cambridge, 1999. Cited on 231, 254.)

[20] W. Arendt, R. Nittka, W. Peter, and F. Steiner, *Weyl's Law: Spectral Properties of the Laplacian in Mathematics and Physics*, In Mathematical Analysis of Evolution, Information, and Complexity, Edited by W. Arendt and W. P. Schleich, John Wiley & Sons, New York, 2009. Cited on 604.)

[21] R. Arens, *On a theorem of Gelfand and Neumark*, Proc. Nat. Acad. Sci. USA **32** (1946), 237–239. Cited on 405.)

[22] R. Arens, *The maximal ideals of certain function algebras*, Pacific J. Math. **8** (1958), 641–648. Cited on 389.)

[23] R. Arens and A. P. Calderón, *Analytic functions of several Banach algebra elements*, Ann. Math. **62** (1955), 204–216. Cited on 69.)

[24] N. Aronszajn, *Rayleigh–Ritz and A. Weinstein methods for approximation of eigenvalues, I. Operators in a Hilbert space*, Proc. Nat. Acad. Sci. USA **34**, (1948), 474–480. Cited on 343.)

[25] N. Aronszajn, *The Rayleigh–Ritz and the Weinstein methods for approximation of eigenvalues. II. Differential operators*, Proc. Nat. Acad. Sci. USA **34** (1948), 594–601. Cited on 343.)

[26] N. Aronszajn, *On a problem of Weyl in the theory of singular Sturm–Liouville equations*, Amer. J. Math. **79** (1957), 597–610. Cited on 343.)

[27] N. Aronszajn and W. F. Donoghue, *On exponential representations of analytic functions in the upper half-plane wih positive imaginary part*, J. Anal. Math. **5** (1956–57), 321–388. Cited on 353.)

[28] N. Aronszajn and W. F. Donoghue, *A supplement to the paper on exponential representations of analytic functions in the upper half-plane with positive imaginary part*, J. Anal. Math. **12** (1964), 113–127. Cited on 353.)

[29] N. Aronszajn and K. T. Smith, *Characterization of positive reproducing kernels, Applications to Green's functions*, Amer. J. Math. **79** (1957), 611–622. Cited on 627.)

[30] W. Arveson, *An invitation to C^*-Algebras*, Graduate Texts in Mathematics, Springer-Verlag, New York–Heidelberg, 1976. Cited on 314.)

[31] W. Arveson, *A Short Course on Spectral Theory*, Graduate Texts in Mathematics, Springer-Verlag, New York, 2002. Cited on 217.)

[32] M. S. Ashbaugh, F. Gesztesy, M. Mitrea, R. Shterenberg, and G. Teschl, *A survey on the Krein–von Neumann extension, the corresponding abstract buckling problem, and Weyl-type spectral asymptotics for perturbed Krein Laplacians in nonsmooth domains*, In Mathematical Physics, Spectral Theory and Stochastic Analysis, edited by M. Demuth and W. Kirsch, pp. 1–106, Operator Theory Advances and Applications, Birkhäuser Basel, Basel, 2013. Cited on 602.)

[33] R. Askey and P. Nevai, *Gabor Szegő: 1895–1985*, Math. Intelligencer **18** (1996), 10–22. Cited on 282.)

[34] M. F. Atiyah, R. Bott, and V. K. Patodi, *On the heat equation and the index theorem*, Invent. Math. **19** (1973), 279–330. Cited on 217.)

[35] M. F. Atiyah and I. M. Singer, *The index of elliptic operators on compact manifolds*, Bull. Amer. Math. Soc. **69** (1963), 422–433. Cited on 217.)

[36] M. F. Atiyah and I. M. Singer, *The index of elliptic operators. I*, Ann. Math. (2) **87** (1968), 484–530. Cited on 217.)

[37] F. V. Atkinson, *The normal solubility of linear equations in normed spaces*, Mat. Sb. N.S. **28** (1951), 3–14 [Russian]. Cited on 217.)

[38] L. Autonne, *Sur les groupes linéaires, réels et orthogonaux*, Bull. Soc. Math. France **30** (1902), 121–134. Cited on 82.)

[39] J. Avron and Y. Tomares, unpublished. Cited on 218.)

[40] J. Avron, R. Seiler, and B. Simon, *Charge deficiency, charge transport and comparison of dimensions*, Comm. Math. Phys. **159** (1994), 399–422. Cited on 217.)

[41] J. Avron, R. Seiler, and B. Simon, *The index of a pair of projections*, J. Funct. Anal. **120** (1994), 220–237. Cited on 217, 218.)

[42] H. F. Baker, *Alternants and continuous groups*, Proc. London Math. Soc. (2) **3** (1905), 24–47. Cited on 628.)

[43] S. Banach, *Sur les fonctionelles linéaires. II*, Studia Math. **1** (1929), 223–239. Cited on 43.)

[44] S. Banach, *Théorie des opérations linéaires*, Monograf. Mat, Warsaw, 1932. English translation: *Theory of Linear Operations*, North-Holland Mathematical Library, North-Holland, Amsterdam, 1987. Cited on 43, 100.)

[45] V. Bargmann, *On the number of bound states in a central field of force*, Proc. Nat. Acad. Sci. USA **38** (1952), 961–966. Cited on 682.)

[46] C. W. Barnes, *Remarks on the Bessel polynomials*, Amer. Math. Monthly **80** (1973), 1034–1040. Cited on 255.)

[47] H. Baumgärtel, *Analytic Perturbation Theory for Matrices and Operators*, Revised edition of the 1972 German original, Mathematische Lehrbücher und Monographien, II, Abteilung: Mathematische Monographien, Akademie-Verlag, Berlin, 1984. Cited on 70.)

[48] R. Beals and R. Wong, *Special Functions: A Graduate Text*, Cambridge Studies in Advanced Mathematics, Cambridge University Press, Cambridge, 2010. Cited on 231, 255.)

[49] E. Beltrami, *Sulle funzioni bilineari*, Giornale di Mathematiche **11** (1873), 98–106. Cited on 135.)

[50] J. J. Benedetto, *Spectral Synthesis*, Pure and Applied Mathematics, Academic Press, New York–London, 1975. Cited on 505.)

[51] R. D. Benguria, *Isoperimetric inequalities for eigenvalues of the Laplacian*, In Entropy and the Quantum II, edited by R. Sims and D. Ueltschi, pp. 21–60, Contemporary Mathematics, American Mathematical Society, Providence, RI, 2011. Cited on 605.)

[52] S. Bergman, *Über ausgezeichnete Randflächen in der Theorie der Functionen von Zwei komplexen Veränderlichen*, Math. Ann. **104** (1931), 611–636. Cited on 489.)

[53] A. S. Besicovitch, *Almost Periodic Functions*, Cambridge University Press, Cambridge, 1932. Cited on 419.)

[54] A. Beurling, *Sur les intégrales de Fourier absolument convergentes et leur application à une transformation fonctionnelle*, In Nionde Skandinaviska Matematikerkongr. (Helsingfors, 1938), pp. 345–366. Mercator, Helsingfors, 1939. Cited on 367, 385.)

[55] A. Beurling and J. Deny, *Espaces de Dirichlet. I, Le cas élémentaire*, Acta Math. **99** (1958), 203–224. Cited on 627.)

[56] A. Beurling and J. Deny, *Dirichlet spaces*, Proc. Nat. Acad. Sci. USA **45** (1959), 208–215. Cited on 627.)

[57] E. Bezout, *Recherches sur le degré des équations résultantes de l'évanouissement des inconnues, et sur les moyens qu'il convient d'employer pour trouver ces équations*, Hist. de l'Acad. Roy. des Sciences (1764), 288–338. Cited on 17.)

[58] M. Sh. Birman, *On the theory of self-adjoint extensions of positive definite operators*, Mat. Sb. N.S. **38** (1956), 431–450. Cited on 601.)

[59] M. Sh. Birman, *On the spectrum of singular boundary-value problems*, Mat. Sb. N.S. **55** (1961), 125–174; English translation Trans. Amer. Math. Soc. **53** (1966), 23–80. Cited on 682.)

[60] M. Sh. Birman, *On existence conditions for wave operators*, Dokl. Akad. Nauk. SSSR **143** (1962), 506–509. Cited on 666.)

[61] M. Sh. Birman, *Existence conditions for wave operators*, Izv. Akad. Nauk SSSR Ser. Mat. **27** (1963), 883–906. Cited on 666.)

[62] M. Sh. Birman, *A local test for the existence of wave operators*, Izv. Akad. Nauk SSSR Ser. Mat. **32** (1968), 914–942. Cited on 666.)

[63] M. Sh. Birman,and V. V. Borzov, *The asymptotic behavior of the discrete spectrum of certain singular differential operators*, In Problems of Mathematical Physics, No. 5: Spectral theory, edited by M. Sh. Birman, pp. 24–38, Izdat. Leningrad. Univ., Leningrad, 1971. Cited on 605.)

[64] M. Birman and M. Solomyak, *Estimates of singular numbers of integral operators, III, Operators in unbounded domains*, Vestnik Leningrad Univ. Math. **2** (1975), 9–27; Russian original in Vestnik Leningrad Univ. **24** (1969), 35–48. Cited on 160.)

[65] M. Birman and M. Solomyak, *Remarks on the spectral shift function*, J. Soviet Math. **3** (1975), 408–419; Russian original: —emphBoundary value problems of mathematical physics and related questions in the theory of functions, Zap. Naučn. Sem. Leningrad. Otdel. Mat. Inst. Steklov. (LOMI) **27** (1972), 33–46. Cited on 353.)

[66] M. Sh. Birman and D. R. Yafaev, *The spectral shift function. The work of M. G. Krein and its further development*, St. Petersburg Math. J. **4** (1993), 833–870. Zap. Naučn. Sem. Leningrad. Otdel. Mat. Inst. Steklov. (LOMI) **27** (1972), 33–46. Cited on 353.)

[67] E. Bishop, *The structure of certain measures*, Duke Math. J. **25** (1958), 283–289. Cited on 490.)

[68] E. Bishop, *A minimal boundary for function algebras*, Pacific J. Math. **9** (1959), 629–642. Cited on 490, 492.)

[69] E. Bishop, *Boundary measures of analytic differentials*, Duke Math. J. **27** (1960), 331–340. Cited on 490.)

[70] E. Bishop and K. de Leeuw, *The representation of linear functionals by measures on sets of extreme points*, Ann. Inst. Fourier Grenoble **9** (1959), 305–331. Cited on 490.)

[71] B. Blackadar, *Operator Algebras, Theory of C^*-algebras and von Neumann algebras*, Encyclopaedia of Mathematical Sciences, Operator Algebras and Non-commutative Geometry, III, Springer-Verlag, Berlin, 2006. Cited on 314.)

[72] R. Blankenbecler, M. L. Goldberger, and B. Simon, *The bound states of weakly coupled, long-range, one-dimensional quantum Hamiltonians*, Ann. Phys. **108** (1977), 69–78. Cited on 683.)

[73] S. Bochner, *Beiträge zur Theorie der Fastperiodischen Funktionen, I*, Math. Ann. **96** (1927), 119–147. Cited on 419.)

[74] S. Bochner, *Über Sturm–Liouvillesche Polynomsysteme*, Math. Z. **29** (1929), 730–736. Cited on 254.)

[75] S. Bochner, *Abstrakte Fastperiodische Funktionen*, Acta Math. **61** (1933), 149–184. Cited on 419.)

[76] V. M. Bogdan, *On Frobenius, Mazur, and Gelfand–Mazur theorems on division algebras*, Quaest. Math. **29** (2006), 171–209. Cited on 387.)

[77] H. Bohr, *Zur Theorie der Fastperiodischen Funktionen, II. Zusammenhang der Fastperiodischen Funktionen mit Funktionen von unendlich vielen Variabeln; gleichmassige Approximation durch trigonometrische Summen*, Acta Math. **46** (1925), 101–214. Cited on 419.)

[78] H. Bohr, *Fastperiodischen Funktionen*, Springer-Verlag, Berlin, 1934. Cited on 419.)

[79] H. Bohr, *Almost Periodic Functions*, Chelsea, New York, 1951. Cited on 419.)

[80] F. Bonsall and J. Duncan, *Complete Normed Algebras*, Ergebnisse der Mathematik und ihrer Grenzgebiete, Springer-Verlag, New York–Heidelberg, 1973. Cited on 357.)

[81] A. Borel, *Essays in the History of Lie Groups and Algebraic Groups*, History of Mathematics, American Mathematical Society, Providence, RI; London Mathematical Society, Cambridge, 2001. Cited on 443.)

[82] É. Borel, *Leçons sur les fonctions de variables réelles et les développements en séries de polynômes*, Gauthier-Villars, Paris, 1905. Cited on 266.)

[83] D. Borwein, Book review of Korevaar [**404**], Bull. Amer. Math. Soc. **42** (2005), 401–406. Cited on 506.)

[84] A. Böttcher and B. Silbermann, *Analysis of Toeplitz Operators*, 2nd edition, Springer Monographs in Mathematics, Springer-Verlag, Berlin, 2006. Cited on 218.)

[85] O. Bratteli and D. W. Robinson, *Operator Algebras and Quantum Statistical Mechanics, Vol. 1. C^*- and W^*-algebras, Algebras, Symmetry Groups, Decomposition of States*, Texts and Monographs in Physics, Springer-Verlag, New York-Heidelberg, 1979. Cited on 314.)

[86] O. Bratteli and D. W. Robinson, *Operator Algebras and Quantum-Statistical Mechanics. II. Equilibrium States. Models in Quantum-atatistical Mechanics*, Texts and Monographs in Physics, Springer-Verlag, New York-Berlin, 1981. Cited on 314.)

[87] W. C. Brenke, *On polynomial solutions of a class of linear differential equations of the second order*, Bull. Amer. Math. Soc. **36** (1930), 77–84. Cited on 254.)

[88] J. Brillhart and P. Morton, *Über Summen von Rudin–Shapiroschen Koeffizienten*, Illinois J. Math. **22** (1978), 126–148. Cited on 370.)

[89] J. Brillhart and P. Morton, *A case study in mathematical research: The Golay–Rudin–Shapiro sequence*, Amer. Math. Monthly **103** (1996), 854–869. Cited on 370.)

[90] M. S. Brodskiĭ, *On a problem of I. M. Gel'fand*, Uspehi Matem. Nauk (N.S.) **12** (1957) 129–132. Cited on 128.)

[91] A. Browder, *Introduction to Function Algebras*, W. A. Benjamin, New York–Amsterdam, 1969. Cited on 489.)

[92] L. G. Brown, R. G. Douglas, and P. A. Fillmore, *Unitary equivalence modulo the compact operators and extensions of C^*-algebras*, In Proc. Conf. Operator Theory (Dalhousie Univ., Halifax, N.S., 1973), pp. 58–128, Lecture Notes in Mathematics, Springer, Berlin, 1973. Cited on 199.)

[93] L. G. Brown, R. G. Douglas, and P. A. Fillmore, *Extensions of C^*-algebras and K-homology*, Ann. Math. (2) **105** (1977), 265–324. Cited on 199.)

[94] W. Burnside, *Theory of Groups of Finite Order*, Cambridge, UK, 1897; reprinted, Dover Publications, New York, 1955. Cited on 443.)

[95] J. W. Calkin, *Abstract symmetric boundary conditions*, Trans. Amer. Math. Soc. **45** (1939), 369–442. Cited on 568.)

[96] J. W. Calkin, *Two-sided ideals and congruences in the ring of bounded operators in Hilbert space*, Ann. Math. (2) **42** (1941), 839–873. Cited on 152, 198.)

[97] J. E. Campbell, *On a law of combination of operators bearing on the theory of continuous transformation groups, I, II*, Proc. London Math. Soc. **28**, (1897), 381–390; **29**, (1898), 14–32 Cited on 628.)

[98] M. J. Cantero, L. Moral, and L. Velázquez, *Five-diagonal matrices and zeros of orthogonal polynomials on the unit circle*, Linear Algebra Appl. **362** (2003), 29–56. Cited on 284.)

[99] C. Carathéodory, *Über den Variabilitätsbereich der Koeffizienten von Potenzreihen die gegebene Werte nicht annehmen*, Math. Ann. **64** (1907), 95–115. Cited on 321.)

[100] C. Carathéodory, *Über den Variabilitäatsbereich der Fourier'schen Konstanten von positiven harmonischen Funktionen*, Rend. Circ. Mat. Palermo **32** (1911), 193–217. Cited on 321.)

[101] T. Carleman, *Zur Theorie der linearen Integralgleichungen*, Math. Z. **9** (1921), 196–217. Cited on 160, 186, 192.)

[102] T. Carleman, *Sur les équations Intégrales Singulières a Noyau Réel et Symmétrique*, Uppsala Univ. arsskrift, Uppsala, 1923. Cited on 534.)

[103] L. Carleson, *Interpolations by bounded analytic functions and the corona problem*, Ann. Math. (2) **76** (1962), 547–559. Cited on 389.)

[104] L. Carleson, *Mergelyan's theorem on uniform polynomial approximation*, Math. Scand. **15** (1964), 167–175. Cited on 490.)

[105] L. Carleson, *On H^1 in multiply connected domains*, in Conference on Harmonic Analysis in Honor of Antoni Zygmund, vols. I, II, Chicago, IL, 1981, Wadsworth, Belmont, CA, 1983. Cited on 389.)

[106] L. Carroll, *Alice's Adventures in Wonderland*, MacMillan, London, 1865; available online at `http://www.authorama.com/alice-in-wonderland-1.html`. Cited on 686.)

[107] L. Carroll (C.L. Dodgson), *The Hunting of the Snark*, MacMillan, London 1876; available online at `www.snrk.de/snarkhunt/`. Cited on 430.)

[108] É. Cartan, *Sur la structure des groupes de transformations finis et continus*, These, Paris, Nony, 1894. Cited on 446.)

[109] A. Cauchy, *Mémoire sur les fonctions qui ne peuvent obtenir que deux valeurs égales et de signes contraires par suite des transpositions opérées entre les variables qu'elles renferment*, J. l'Ecole Polytechnique, **17** (1812), 29–112. Cited on 17.)

[110] A. Cayley, *On the Theory of Groups, as Depending on the Symbolic Equation* $\theta_n = 1$, Phil. Mag. **7** (1854) 40–47. Cited on 443.)

[111] A. Cayley, *A memoir on the theory of matrices*, Philos. Trans. Roy. Soc. London **148** (1858), 17–37. Cited on 17.)

[112] E. Čech, *On bicompact spaces*, Ann. Math. **38** (1937), 823–844. Cited on 412.)

[113] R. E. Chandler, *Hausdorff Compactifications*, Lecture Notes in Pure and Applied Mathematics, Marcel Dekker, New York–Basel, 1976. Cited on 412.)

[114] A. Chattopadhyay and K. B. Sinha, *Koplienko trace formula*, Integral Equations Operator Theory **73** (2012), 573–587. Cited on 353.)

[115] I. Chavel, *Eigenvalues in Riemannian geometry*, including a chapter by Burton Randol, with an appendix by Jozef Dodziuk, Pure and Applied Mathematics, Academic Press, Inc., Orlando, FL, 1984. Cited on 605.)

[116] P. L. Chebyshev, *Théorie des mécanismes connus sous le nom de parallélogrammes,* In Mémoires des Savants étrangers présentés à l'Académie de Saint-Pétersbourg, pp. 539–586, St.-Pétersbourg, Impr. de l'Académie, 1854; available online at `http://www.math.umanitoba.ca/HAT/fpapers/cheb11.pdf`. Cited on 266.)

[117] P. L. Chebyshev, *Sur les questions de minima qui se rattachent à la représentation approximative des fonctions*, In Mémoires de l'Académie Impériale des Sciences de St-Pétersbourg, Sixiéme série. Sciences mathématiques et physiques, pp, 199–221, St.-Pétersbourg, Impr. de l'Académie, 1859; available online at `http://www.math.umanitoba.ca/HAT/fpapers/cheb22.pdf`. Cited on 266.)

[118] P. L. Chebyshev, *Sur les fonctions qui différent le moins possible de zéro*, J. Math. Pures et Appl. **19** (1874), 319–346. Cited on 266.)

[119] P. L. Chebyshev, *Sur les valeurs limites des integrales*, J. Math. Pures Appl. (2) **19** (1874), 157–160. Cited on 241.)

[120] E. W. Cheney, *Introduction to Approximation Theory*, Reprint of the second (1982) edition, AMS Chelsea Publishing, Providence, RI, 1998. Cited on 267.)

[121] E. W. Cheney and W. Light, *A Course in Approximation Theory*, Reprint of the 2000 original, Graduate Studies in Mathematics, American Mathematical Society, Providence, RI, 2009. Cited on 267.)

[122] P. R. Chernoff, *Note on product formulas for operator semigroups*, J. Funct. Anal. **2** (1968), 238–242. Cited on 629.)

[123] T. S. Chihara, *An Introduction to Orthogonal Polynomials*, Mathematics and Its Applications, Gordon and Breach, New York–London–Paris, 1978. Cited on 231.)

[124] J. Christiansen, B. Simon, and M. Zinchenko, *Asymptotics of Chebyshev Polynomials, I. Subsets of* $\mathbb{R}$, preprint, 2105. Cited on 267.)

[125] E. A. Coddington and N. Levinson, *Theory of Ordinary Differential Equations*, McGraw-Hill Book Company, Inc., New York–Toronto–London, 1955; Reprint, Krieger Publishing Company, Malabar, FL, 1985. Cited on 569.)

[126] C. Constantinescu, *C*-algebras, Vol. 1. Banach spaces*, North-Holland Mathematical Library, North-Holland Publishing Co., Amsterdam, 2001. Cited on 44, 314.)

[127] C. Constantinescu, *C^*-algebras, Vol. 2. Banach algebras and Compact Operators*, North-Holland Mathematical Library, North-Holland Publishing Co., Amsterdam, 2001. Cited on 314.)

[128] C. Constantinescu, *C^*-algebras, Vol. 3. General Theory of C^*-algebras*, North-Holland Mathematical Library, North-Holland Publishing Co., Amsterdam, 2001. Cited on 314.)

[129] C. Constantinescu, *C^*-algebras, Vol. 4. Hilbert Spaces*, North-Holland Mathematical Library, North-Holland Publishing Co., Amsterdam, 2001. Cited on 314.)

[130] C. Constantinescu, *C^*-algebras, Vol. 5. Selected Topics*, North-Holland Mathematical Library, North-Holland Publishing Co., Amsterdam, 2001. Cited on 314.)

[131] H. Cordes, A. Jensen, S. T. Kuroda, G. Ponce, B. Simon, and M. Taylor, *Tosio Kato (1917–1999)*, Notices Amer. Math. Soc. **47** (2000), 650–657. Cited on 534.)

[132] C. Corduneanu, *Almost Periodic Functions*, Interscience Publishers, New York, 1968. Cited on 419.)

[133] R. Courant, *Über die Eigenwerte bei den Differentialgleichungen der mathematischen Physik*, Math. Z. **7** (1920), 1–57. Cited on 109.)

[134] R. Courant and D. Hilbert, *Methods of Mathematical Physics, Vol. I*, Interscience Publishers, New York, 1953. Cited on 118.)

[135] R. Courant and D. Hilbert, *Methods of Mathematical Physics, Vol. II: Partial Differential Equations*, (Vol. II by R. Courant) Interscience Publishers, New York-London, 1962. Cited on 118, 603.)

[136] R. Courant and F. John, *Introduction to Calculus and Analysis*, Vol. I, Reprint of the 1965 edition, Springer-Verlag, New York, 1989; *Introduction to Calculus and Analysis*, Vol. II, with the assistance of Albert A. Blank and Alan Solomon, Reprint of the 1974 edition, Springer-Verlag, New York, 1989. Cited on 603.)

[137] G. Cramer, *Introduction à l'Analyse des Lignes Courbes algébriques*, Geneve, 1750. Cited on 17.)

[138] M. M. Crum, *On the resultant of two functions*, Quart. J. Math. Oxford **12** (1941), 108–111. Cited on 129.)

[139] A. Császár, *Sur les polynomes orthogonaux classiques*, Ann. Univ. Sci. Budapest **1** (1958), 33–39. Cited on 255.)

[140] C. W. Curtis, *Representation theory of finite groups: from Frobenius to Brauer*, Math. Intelligencer **14** (1992), 48–57. Cited on 446.)

[141] C. W. Curtis, *Pioneers of Representation Theory: Frobenius, Burnside, Schur, and Brauer*, History of Mathematics, American Mathematical Society, Providence, RI; London Mathematical Society, London, 1999. Cited on 443, 446.)

[142] C. W. Curtis and I. Reiner, *Representation Theory of Finite Groups and Associative Algebras*, Pure and Applied Mathematics, Interscience Publishers, a division of John Wiley & Sons, New York–London 1962. Cited on 443.)

[143] M. Cwikel, *Weak type estimates for singular values and the number of bound states of Schrödinger operators*, Ann. Math. (2) **106** (1977), 93–100. Cited on 161.)

[144] H. L. Cycon, R. G. Froese, W. Kirsch, and B. Simon, *Schrödinger Operators with Application to Quantum Mechanics and Global Geometry*, corrected and extended 2nd printing, Springer-Verlag, Berlin, 2008. Cited on 217, 218, 538.)

[145] K. Davidson, *C^*-Algebras by Example*, American Mathematical Society, Providence, Rhode Island, 1996. Cited on 314.)

[146] E. B. Davies, *Properties of the Green's functions of some Schroödinger operators*, J. London Math. Soc. **7** (1973), 473–491. Cited on 632.)

[147] E. B. Davies *A model for absorption or decay*, Helv. Phys. Acta **48** (1975), 365–382. Cited on 601.)

[148] E. B. Davies and B. Simon, *Eigenvalue estimates for non-normal matrices and the zeros of random orthogonal polynomials on the unit circle*, J. Approx. Theory **141** (2006), 189–213. Cited on 323.)

[149] C. Davis, *Separation of two linear subspaces*, Acta Sci. Math. Szeged **19** (1958), 172–187. Cited on 218.)

[150] C. de Boor and E. B. Saff, *Finite sequences of orthogonal polynomials connected by a Jacobi matrix*, Linear Algebra Appl. **75** (1986), 43–55. Cited on 242.)

[151] L. de Branges, *Perturbations of self-adjoint transformations*, Amer. J. Math. **84** (1962), 543–560. Cited on 353, 666.)

[152] P. A. Deift, *Applications of a commutation formula*, Duke Math. J. **45** (1978), 267–310. Cited on 57.)

[153] P. Deift, A. Its, and I. Krasovsky, *Toeplitz matrices and Toeplitz determinants under the impetus of the Ising model, Some history and some recent results*, Comm. Pure Appl. Math. **66** (2013), 1360–1438. Cited on 218.)

[154] Ch. J. de la Vallée Poussin, *Sur les polynômes d'approximation et la répresentation approchée d'un angle*, Bull. Acad. Royale Belgique **12** (1910), 808–845. Cited on 267, 268.)

[155] R. del Rio, S. Jitomirskaya, N. Makarov, and B. Simon, *Singular continuous spectrum is generic*, Bull. Amer. Math. Soc. **31** (1994), 208–212. Cited on 344.)

[156] R. del Rio, N. Makarov, and B. Simon, *Operators with singular continuous spectrum, II. Rank one operators*, Comm. Math. Phys. **165** (1994), 59–67. Cited on 344.)

[157] J. Deny, *Sur les espaces de Dirichlet* in Séminaire Brelot-Choquet-Deny, Théorie du potentiel, pp. 1–14, Institut Henri Pincar'e, Paris, 1957. Cited on 627.)

[158] E. F. Deprettere, *SVD and Signal Processing: Algorithms, Applications, and Architectures*, Elsevier Science, Amsterdam, 1991. Cited on 135.)

[159] P. Diaconis, *Group Representations in Probability and Statistics*, Institute of Mathematical Statistics Lecture Notes–Monograph Series, Institute of Mathematical Statistics, Hayward, CA, 1988. Cited on 443.)

[160] J. Dieudonné, *Sur les homomorphismes d'espaces normés*, Bull. Sci. Math. (2) **67** (1943), 72–84. Cited on 217.)

[161] V. A. Ditkin, *On the structure of ideals in certain normed rings*, Usp. Zapiski Moskov. gosud. Univ. **30** (1939), 83–130. Cited on 504.)

[162] J. Dixmier, *C^*-algebras*, Translated from the French by Francis Jellett, North-Holland Mathematical Library, North-Holland Publishing Co., Amsterdam–New York–Oxford, 1977. Cited on 314.)

[163] J. Dixmier, *von Neumann algebras*, with a preface by E. C. Lance, translated from the second French edition by F. Jellett, North-Holland Mathematical Library, North-Holland Publishing Co., Amsterdam–New York, 1981. Cited on 314.)

[164] A. C. Dixon, *On a class of matrices of infinite order, and on the existence of "metrical" functions of a Riemann surface*, Trans. Cambridge Philos. Soc. **19** (1902), 190–233. Cited on 41, 56.)

[165] W. F. Donoghue, Jr., *The lattice of invariant subspaces of a completely continuous quasi-nilpotent transformation*, Pacific J. Math. **7** (1957), 1031–1035. Cited on 128.)

[166] W. F. Donoghue, Jr., *On the perturbation of spectra*, Comm. Pure Appl. Math. **18** (1965), 559–579. Cited on 343.)

[167] W. F. Donoghue, Jr., *Monotone Matrix Functions and Analytic Continuation*, Die Grundlehren der mathematischen Wissenschaften, Springer, New York–Heidelberg–Berlin, 1974. Cited on 606.)

[168] R. S. Doran (editor), *C^*-algebras: 1943–1993. A Fifty Year Celebration*, Proc. AMS Special Session (San Antonio, TX), 1993, Contemporary Mathematics, American Mathematical Society, Providence, RI, 1994. Cited on 428, 699, 705.)

[169] R. G. Douglas, *Banach Algebra Techniques in the Theory of Toeplitz Operators*, Conference Board of the Mathematical Sciences Regional Conference Series in Mathematics, American Mathematical Society, Providence, RI, 1973. Cited on 218.)

[170] J. Dufresnoy, *Sur le produit de composition de deux fonctions*, C. R. Acad. Sei. Paris **335** (1947), 857–859. Cited on 129.)

[171] J. J. Duistermaat and J. A. C. Kolk, *Lie Groups*, Universitext, Springer-Verlag, Berlin, 2000. Cited on 628.)

[172] J. J. Duistermaat and F. A. Grünbaum, *Differential equations in the spectral parameter*, Comm. Math. Phys. **103** (1986), 177–240. Cited on 255.)

[173] N. Dunford, *Spectral theory*, Bull. Amer. Math. Soc. **49** (1943), 637–651. Cited on 69.)

[174] N. Dunford, *Spectral theory. I. Convergence to projections*, Trans. Amer. Math. Soc. **54** (1943), 185–217. Cited on 69.)

[175] N. Dunford and J. T. Schwartz, *Linear Operators. Part II. Spectral Theory. Selfadjoint Operators in Hilbert Space*, reprint of the 1963 original, Wiley Classics Library, John Wiley & Sons, New York, 1988. Cited on 186, 192, 568, 569.)

[176] A. Duran, *Exceptional Charlier and Hermite orthogonal polynomials*, J. Approx. Theory **182** (2014) 29–58. Cited on 255.)

[177] A. Duran, *Exceptional Meixner and Laguerre orthogonal polynomials*, J. Approx. Theory **184** (2014) 176–208. Cited on 255.)

[178] A. J. Durán and F. A. Grünbaum, *A survey on orthogonal matrix polynomials satisfying second order differential equations*, J. Appl. Math. Comput. **178** (2005), 169–190. Cited on 255.)

[179] M. S. P. Eastham and H. Kalf, *Schrödinger-type Operators with Continuous Spectra*, Research Notes in Mathematics, Pitman (Advanced Publishing Program), Boston—London, 1982. Cited on 569.)

[180] C. Eckart and G. Young, *The approximation of one matrix by another of lower rank*, Psychometrika **1** (1936), 211–218. Cited on 135.)

[181] C. Eckart and G. Young, *A principal axis transformation for non-hermitian matrices*, Bull. Amer. Math. Soc. **45** (1939), 118–121. Cited on 135.)

[182] D. E. Edmunds and W. D. Evans, *Spectral Theory and Differential Operators*, Oxford Mathematical Monographs, Oxford Science Publications, The Clarendon Press, Oxford University Press, New York, 1987. Cited on 198, 602.)

[183] E. G. Effros, *Why the circle is connected: An introduction to quantized topology*, Math. Intelligencer **11** (1989), 27–34. Cited on 218.)

[184] R. W. Emerson, "Self-Reliance" an essay included in *Essays: First Series*, Houghton, Mifflin and Company, Boston, MA, 1883; available online at `http://en.`

wikisource.org/wiki/Essays:_First_Series/Self-Reliance (self-reliance only) and https://archive.org/details/essaysfirstseconx00emer (1883 collection of First and Second Series). Cited on 1.)

[185] P. Enflo, *A counterexample to the approximation property in Banach spaces*, Acta Math. **130** (1973), 309–317. Cited on 100.)

[186] V. Enss, *Asymptotic completeness for quantum mechanical potential scattering, I. Short range potentials*, Comm. Math. Phys. **61** (1978), 285–291. Cited on 321.)

[187] J. A. Erdos, *Operators of finite rank in nest algebras*, J. London Math. Soc. **43** (1968), 391–397. Cited on 128, 187.)

[188] J. A. Erdos, *On the trace of a trace class operator*, Bull. London Math. Soc. **6** (1974), 47–50. Cited on 128, 187.)

[189] L. Erdős and V. Vougalter, *Pauli operator and Aharonov–Casher theorem for measure valued magnetic fields*, Comm. Math. Phys. **225** (2002), 399–421. Cited on 218.)

[190] P. Exner, H. Neidhardt, and V. A. Zagrebnov, *Remarks on the Trotter-Kato product formula for unitary groups*, Integral Equations Operator Theory **69** (2011), 451–478. Cited on 630.)

[191] G. Faber, *Über Tschebyscheffsche Polynome*, J. Reine Angew. Math. **150** (1919), 79–106. Cited on 268.)

[192] K. Fan, *Maximum properties and inequalities for the eigenvalues of completely continuous operators*, Proc. Nat. Acad. Sci. USA **37** (1951), 760–766. Cited on 135.)

[193] W. G. Faris, *Essential self-adjointness of operators in ordered Hilbert space*, Comm. Math. Phys. **30** (1973), 23–34. Cited on 632.)

[194] W. G. Faris, *Self-adjoint Operators*, Lecture Notes in Mathematics, Springer-Verlag, Berlin–New York, 1975. Cited on 601.)

[195] J. Favard, *Sur les polynômes de Tchebycheff*, C. R. Acad. Sci. Paris **200** (1935), 2052–2055. Cited on 241.)

[196] C. Fefferman and D. H. Phong, *The uncertainty principle and sharp Gårding inequalities*, Comm. Pure Appl. Math. **34** (1981), 285–331. Cited on 228.)

[197] L. Fejér, *Sur les fonctions bornées et intègrables*, C. R. Acad. Sci. Paris **131** (1900), 984–987. Cited on 321.)

[198] L. Fejér, *Über die Lage der Nullstellen von Polynomen, die aus Minimumforderungen gewisser Art entspringen*, Math. Ann. **85** (1922), 41–48. Cited on 285.)

[199] M. Fekete, *Über die Verteilung der Wurzeln bei gewissen algebraischen Gleichungen mit ganzzahligen Koeffizienten*, Math. Z. **17** (1923), 228–249. Cited on 268.)

[200] R. P. Feynman, *Forces in molecules*, Phys. Rev. **56** (1939), 340. Cited on 27.)

[201] R. P. Feynman, *Space-time approach to non-relativistic quantum mechanics*, Rev. Modern Phys. **20** (1948), 367–387. Cited on 630.)

[202] E. Fischer, *Über quadratische Formen mit reellen Koeffizienten*, Monatsh. Math. **16** (1905), 234–249. Cited on 109, 604.)

[203] J. W. M. Ford, *A square root lemma for Banach* $(^*)$*-algebras*, J. London Math. Soc. **42** (1967), 521–522. Cited on 83.)

[204] L. Fox and I. B. Parker, *Chebyshev Polynomials in Numerical Analysis*, Oxford University Press, London–New York–Toronto, 1968. Cited on 266.)

[205] J. Fredholm, *Sur une nouvelle méthode pour la résolution du problème de Dirichlet*, Stockh. Öfv. **57** (1900), 39–46. Cited on 41.)

[206] J. Fredholm, *Sur une classe d'équations fonctionnelles*, Acta Math. **27** (1903), 365–390. Cited on 41, 99, 182.)

[207] H. Freudenthal, *Über die Friedrichssche Fortsetzung halbbeschränkter Hermitescher Operatoren*, Proc. Akad. Wet. Amsterdam **39** (1936), 832–833. Cited on 600.)

[208] K. O. Friedrichs, *Spektraltheorie halbbeschränkter Operatoren und Anwendung auf die Spektralzerlegung von Differentialoperatoren. I, II*, Math. Ann. **109** (1934), 465–487, 685–713. Cited on 600.)

[209] K. O. Friedrichs, *Perturbation of Spectra in Hilbert Space*, American Mathematical Society, Providence, RI, 1965. Cited on 27.)

[210] O. Frink, Jr., *Jordan measure and Riemann integration*, Ann. Math. (2) **34** (1933), 518–526. Cited on 603.)

[211] F. G. Frobenius, *Über lineare Substitutionen und bilineare Formen*, J. Reine Angew. Math. **84** (1878), 1–63. Cited on 18, 387.)

[212] G. Frobenius, *Über die Primfactoren der Gruppendeterminante*, Sitzungsber. Preuss. Akad. Wiss. (1896), 343–382. Cited on 444.)

[213] G. Frobenius, *Über vertauschbare Matrizen*, Sitzungsber. Preuss. Akad. Wiss. (1896), 601–614. Cited on 444.)

[214] G. Frobenius, *Über Gruppencharaktere*, Sitzungsber. Preuss. Akad. Wiss. (1896), 985–1021. Cited on 444.)

[215] G. Frobenius, *Über die Darstellung der endlichen Gruppen durch lineare Substitutionen*, Sitzungsber. Preuss. Akad. Wiss. (1897), 994–1015. Cited on 444, 445.)

[216] M. Fukamiya, *On a theorem of Gelfand and Neumark and the B^*-algebra*, Kumamoto J. Sci. Ser. A **1** (1952), 17–22. Cited on 428, 717.)

[217] W. Fulton and J. Harris, *Representation Theory, A First Course*, Graduate Texts in Mathematics, Springer-Verlag, New York, 1991. Cited on 443.)

[218] T. W. Gamelin, *Uniform Algebras*, Prentice-Hall, Englewood Cliffs, NJ, 1969. Cited on 489, 490.)

[219] T. W. Gamelin, *Uniform Algebras and Jensen Measures*, London Mathematical Society Lecture Note Series, Cambridge University Press, Cambridge–New York, 1978. Cited on 489.)

[220] T. W. Gamelin, *Wolff's proof of the corona theorem*, Israel J. Math. **37** (1980), 113–119. Cited on 389.)

[221] D. J. H. Garling, *Inequalities: A Journey Into Linear Analysis*, Cambridge University Press, Cambridge, 2007. Cited on 187.)

[222] J. Garnett, *On a theorem of Mergelyan*, Pacific J. Math. **26** (1968), 461–467. Cited on 490.)

[223] J. B. Garnett, *Bounded analytic functions*, Pure and Applied Mathematics, Academic Press, Inc., New York–London, 1981. Cited on 389.)

[224] J. F. C. Gauss, *Disquisitiones Arithmeticae*, Gerh. Fleishcher, Leipzig, 1801. Cited on 17.)

[225] I. M. Gel'fand, *A problem*, Uspehi Matem. Nauk **5** (1938), 233. Cited on 128.)

[226] I. M. Gel'fand, *On normed rings*, C. R. Acad. Sci. URSS (2) **23** (1939), 430–432. Cited on 387.)

[227] I. M. Gel'fand, *To the theory of normed rings. II. On absolutely convergent trigonometrical series and integrals*, C. R. Acad. Sci. URSS (2) **25** (1939), 570–572. Cited on 388.)

[228] I. M. Gel'fand, *Normierte Ringe*, Rec. Math. [Mat. Sbornik] N.S. **9** (1941), 3–24. Cited on 56, 57, 69, 387.)

[229] I. M. Gel'fand, *On elliptic equations*, Russian Math. Surveys **15** (1960), 113–123. Cited on 217.)

[230] I. M. Gelfand and A. Kolmogorov, *On rings of continuous functions on topological spaces*, C. R. Acad. Sci. URSS (2) **22** (1939), 11–15. Cited on 388.)

[231] I. M. Gel'fand and M. Naimark, *On the imbedding of normed rings into the ring of operators in Hilbert space*, Rec. Math. [Mat. Sbornik] N.S. **12**, (1943), 197-213; English translation in [**168**], 2–19. Cited on 399, 405, 406, 428.)

[232] I. M. Gel'fand and M. Naimark, *Unitary representations of the Lorentz group*, Izv. Akad. Nauk SSSR Ser. Mat. **11** (1947), 411–504. Cited on 447.)

[233] I. M. Gel'fand and D. A. Raikov, *On the theory of characters of commutative topological groups*, C. R. Acad. Sci. URSS (2) **28** (1940), 195–198. Cited on 467.)

[234] I. M. Gel'fand and D. A. Raikov, *Continuous unitary representations of locally bicompact groups*, Mat. Sbornik **55** (1943), 301–316. Cited on 447.)

[235] I. M. Gel'fand, D. A. Raikov, and G. E. Shilov, *Commutative normed rings*, Uspekhi Mat. Nauk **1** (1946), 48–146; Trans. Amer. Math. Soc. **5** (1957), 115–220. Cited on 387, 406, 489.)

[236] I. M. Gel'fand, D. A. Raikov, and G. E. Shilov, *Commutative Normed Rings*, Chelsea, New York, 1964. Cited on 357, 387, 468, 504.)

[237] V. Georgescu, and A. Iftimovici, *Crossed products of C^*-algebras and spectral analysis of quantum Hamiltonians*, Comm. Math. Phys. **228** (2002), 519–560. Cited on 666.)

[238] Ya. L. Geronimus, *On polynomials orthogonal on the circle, on trigonometric moment problem, and on allied Carathéodory and Schur functions*, Mat. Sb. **15** (1944), 99–130. Cited on 283.)

[239] Ya. L. Geronimus, *On the trigonometric moment problem*, Ann. Math. (2) **47** (1946), 742–761. Cited on 283, 284.)

[240] Ya. L. Geronimus, *Orthogonal polynomials*, English translation of the appendix to the Russian translation of Szegő's book [**686**], in Two Papers on Special Functions, American Mathematical Society Translation Series. 2, pp. 37–130, American Mathematical Society, Providence, RI, 1977. Cited on 231, 282.)

[241] F. Gesztesy, M. Malamud, M. Mitrea, and S. Naboko, *Generalized polar decompositions for closed operators in Hilbert spaces and some applications*, Integral Equations Operator Theory **64** (2009), 83–113. Cited on 87.)

[242] F. Gesztesy, A. Pushnitski, and B. Simon, *On the Koplienko spectral shift function, I. Basics*, ZhMat. Fiz. Anal. Geom. **4** (2008), 63–107. Cited on 344, 352, 353.)

[243] F. Gesztesy and B. Simon, *Rank-one perturbations at infinite coupling*, J. Funct. Anal. **128** (1995), 245–252. Cited on 666.)

[244] E. Getzler, *Pseudodifferential operators on supermanifolds and the Atiyah–Singer index theorem*, Comm. Math. Phys. **92** (1983), 163–178. Cited on 217.)

[245] D. J. Gilbert, *On subordinacy and analysis of the spectrum of Schrödinger operators with two singular endpoints*, Proc. Roy. Soc. Edinburgh Sect. A **112** (1989), 213–229. Cited on 570.)

[246] P. B. Gilkey, *Invariance Theory, the Heat Equation, and the Atiyah–Singer Theorem*, 2nd edition, CRC Press, Boca Raton, FL, 1995. Cited on 217.)

[247] L. Gillman and M. Jerison, *Rings of Continuous Functions*, The University Series in Higher Mathematics, Van Nostrand, Princeton, NJ–Toronto–London–New York, 1960. Cited on 412.)

[248] A. M. Gleason, *Function Algebras*, Seminars on Analytic Functions, Institute for Advanced Study, Princeton, NJ, 1957. Cited on 490.)

[249] A. M. Gleason, *A characterization of maximal ideals*, J. Anal. Math. **19** (1967), 171–172. Institute for Advanced Study, Princeton, NJ, 1957. Cited on 392.)

[250] I. Glicksberg, *Dominant representing measures and rational approximation*, Trans. Amer. Math. Soc. **130** (1968), 425–462. Cited on 490.)

[251] I. Glicksberg and J. Wermer, *Measures orthogonal to a Dirichlet algebra*, Duke Math. J. **30** (1963), 661–666. Cited on 490.)

[252] J. Glimm and R. Kadison, *Unitary operators in* C^**-algebras*, Pacific J. Math. **10** (1960), 547–556. Cited on 428.)

[253] R. Godement, *Extension à un groupe abélien quelconque des théorèmes taubériens de N. Wiener et d'un théorème de A. Beurling*, C. R. Acad. Sci. Paris **223** (1946), 16–18. Cited on 504.)

[254] R. Godement, *Les fonctions de type positif et la théorie des groupes*, Trans. Amer. Math. Soc. **63** (1948), 1–84. Cited on 468.)

[255] I. C. Goh'berg *On linear equations in normed spaces*, Dokl. Akad. Nauk SSSR (N.S.) **76** (1951), 477–480. Cited on 217.)

[256] I. C. Goh'berg *On an application of the theory of normed rings to singular integral equations*, Uspehi Matem. Nauk (N.S.) **7** (1952), 149–156. Cited on 218.)

[257] I. C. Goh'berg and M. G. Krein, *Introduction to the theory of linear nonselfadjoint operators*, Translations of Mathematical Monographs, American Mathematical Society, Providence, RI, 1969. Cited on 134, 152, 153, 192, 601.)

[258] I. Goh'berg, S. Goldberg, and N. Krupnik, *Traces and Determinants of Linear Operators*, Operator Theory: Advances and Applications, Birkhäuser Verlag, Basel, 2000. Cited on 187.)

[259] G. H. Golub and W. Kahan, *Calculating the singular values and pseudo-inverse of a matrix*, SIAM J. Numer. Anal. **2** (1965), 205–224. Cited on 135.)

[260] G. H. Golub and C. Reinsch, *Singular value decomposition and least squares solution*, Numer. Math. **14** (1970), 403–420. Cited on 135.)

[261] D. Gomez-Ullate, N. Kamran, and R. Milson, *An extended class of orthogonal polynomials defned by a Sturm–Liouville problem*, J. Math. Anal. Appl. **359** (2009), 352–367. Cited on 255.)

[262] D. Gomez-Ullate, N. Kamran, and R. Milson, *An extension of Bochner's problem: Exceptional invariant subspaces*, J. Approx. Theory, **162** (2010), 987–1006. Cited on 255.)

[263] A. Ya. Gordon, *Exceptional values of the boundary phase for the Schrödinger equation on the semi-axis*, Russian Math. Surveys **47** (1992), 260–261. Cited on 344.)

[264] A. Ya. Gordon, *Pure point spectrum under* 1*-parameter perturbations and instability of Anderson localization*, Comm. Math. Phys. **164** (1994), 489–505. Cited on 344.)

[265] C. Gordon, D. Webb, and S. Wolpert, *Isospectral plane domains and surfaces via Riemannian orbifolds*, Invent. Math. **110** (1992), 1–22. Cited on 605.)

[266] W. T. Gowers and B. Maurey, *The unconditional basic sequence problem*, J. Amer. Math. Soc. **6** (1993), 851–874. Cited on 100.)

[267] W. T. Gowers and B. Maurey, *Banach spaces with small spaces of operators*, Math. Ann. **307** (1997), 543–568. Cited on 44, 100.)

[268] A. Grothendieck, *Produits Tensoriels Topologiques et Espaces Nucléaires*, Mem. Amer. Math. Soc., American Mathematical Society, Providence, RI, 1955. Cited on 144, 186.)

[269] A. Grothendieck, *La théorie de Fredholm*, Bull. Soc. Math. France **84** (1956), 319–384. Cited on 186.)

[270] G. Grubb, *Les problèmes aux limites généraux d'un opérateur elliptique, provenant de le théorie variationnelle*, Bull. Sci. Math. France **94** (1970), 113–157. Cited on 601.)

[271] G. Grubb, *On coerciveness and semiboundedness of general boundary problems*, Israel J. Math. **10** (1971), 32–95. Cited on 601.)

[272] G. Grubb, *Properties of normal boundary problems for elliptic even-order systems*, Ann. Scuola Norm. Sup. Pisa **1** (1975), 1–61. Cited on 601.)

[273] G. Grubb, *Distributions and Operators*, Graduate Texts in Mathematics, Springer, New York, 2009. Cited on 602.)

[274] H. R. Grümm, *Two theorems about $\mathcal{C}_p$*, Rep. Math. Phys. **4** (1973), 211–215. Cited on 153.)

[275] F. A. Grünbaum and L. Haine, *A theorem of Bochner, revisited. Algebraic aspects of integrable systems*, in Algebraic Aspects of Integrable Systems, pp. 143–172, Progress in Nonlinear Differential Equations and their Applications, Birkhäuser Boston, Boston, 1997. Cited on 255.)

[276] K. Gustafson and J. Weidmann, *On the essential spectrum*, J. Math. Anal. Appl. **25** (1969), 121–127. Cited on 198.)

[277] P. Güttinger, *Das Verhalten von Atomen im magnetischen Drehfeld*, Z. Phys. **73** (1932), 169. Cited on 27.)

[278] A. Haar, *Die Minkowskische Geometrie and die Annäherung an stetige Funktionen*, Math. Ann. **78** (1918), 249–311. Cited on 267.)

[279] H. Hahn, *Über die Integrale des Herrn Hellinger und die Orthogonalinvarianten der quadratischen Formen von unendlich vielen Veränderlichen*, Monatsch. für Math. u. Phys. **23** (1912), 161–224. Cited on 313.)

[280] W. Hahn, *Über die Jacobischen Polynome und zwei verwandte Polynomklassen*, Math. Z. **39** (1935), 634–638. Cited on 254, 255.)

[281] B. C. Hall, *Lie Groups, Lie Algebras, and Representations. An Elementary Introduction*, Graduate Texts in Mathematics, Springer-Verlag, New York, 2003. Cited on 628.)

[282] P. R. Halmos, *Introduction to Hilbert Space and the Theory of Spectral Multiplicity*, Chelsea Publishing Co., New York, 1951. Cited on 313, 314.)

[283] P. R. Halmos, *Two subspaces*, Trans. Amer. Math. Soc. **144** (1969), 381–389. Cited on 218.)

[284] P. R. Halmos, *What does the spectral theorem say?*, Amer. Math. Monthly **70** (1963), 241–247. Cited on 299.)

[285] P. R. Halmos, *The legend of John von Neumann*, Amer. Math. Monthly **80** (1973), 382–394. Cited on 536.)

[286] H. Hamburger, *Über eine Erweiterung des Stieltjesschen Momentenproblems*, Math. Ann. **81** (1920), 235–319; **82** (1921), 120–164, 168–187. Cited on 658.)

[287] W. R. Hamilton, *Lectures on Quaternions*, Hodges and Smith, Dublin, 1853; available online at `http://archive.org/details/lecturesonquater00hami`. Cited on 17.)

[288] G. H. Hardy, *Theorems relating to the summability and convergence of slowly oscillating series*, Proc. London Math. Soc. (2) **8** (1910), 301–320. Cited on 506.)

[289] G .H. Hardy, *On the expression of a number as a sum of two squares*, Quart. J. Math. Oxford **46** (1915), 263–283. Cited on 605.)

[290] G. H. Hardy, *Divergent Series*, Clarendon Press, Oxford, 1949. Cited on 505.)

[291] G. H. Hardy and J. E. Littlewood, *Tauberian theorems concerning series of positive terms*, Messenger of Math. **42** (1913), 191–192. Cited on 506.)

[292] G. H. Hardy and J. E. Littlewood, *Tauberian theorems concerning power series and Dirichlet's series whose coefficients are positive*, Proc. London Math. Soc. (2) **13** (1914), 174–191. Cited on 506.)

[293] G. H. Hardy and J. E. Littlewood, *Some problems of Diophantine approximation: A remarkable trigonometrical series*, Proc. Nat. Acad. Sci. USA **2** (1916), 583–586. Cited on 367.)

[294] G. H. Hardy and J. E. Littlewood, *Theorems concerning the summability of series by Borel's exponential method*, Rend. Circ. Mat. Palermo **41** (1916), 36–53. Cited on 506.)

[295] G. H. Hardy and J. E. Littlewood, *On a Tauberian theorem for Lambert's series, and some fundamental theorems in the analytic theory of numbers*, Proc. London Math. Soc. (2) **19** (1921), 21–29. Cited on 506.)

[296] F. Hartogs and A. Rosenthal, *Über Folgen analytischer Funktionen*, Math. Ann. **104** (1931), 101–106. Cited on 489.)

[297] S. Hassi, H. S. V. de Snoo, and A. D. I. Willemsma, *Smooth rank one perturbations of selfadjoint operators*, Proc. Amer. Math. Soc. **126** (1998), 2663–2675. Cited on 666.)

[298] S. Hassi, Z. Sebestyén, and H. de Snoo, *Lebesgue type decompositions for nonnegative forms*, J. Funct. Anal. **257** (2009), 3858–3894. Cited on 601.)

[299] F. Hausdorff, *Die symbolische Exponentialformel in der Gruppentheorie*, Leipz. Ber. **58** (1906), 19–48. Cited on 628.)

[300] T. Hawkins, *The Origins of the Theory of Group Characters*, Arch. Hist. Exact Sci. **7** (1971), 142–170. Cited on 446.)

[301] T. Hawkins, *Hypercomplex Numbers, Lie Groups, and the Creation of Group Representation Theory*, Arch. Hist. Exact Sci. **8** (1972), 243–287. Cited on 446.)

[302] T. Hawkins, *New light on Frobenius' creation of the theory of group characters*, Arch. Hist. Exact Sci. **12** (1974), 217–243. Cited on 446.)

[303] T. Hawkins, *Emergence of the Theory of Lie Groups, An essay in the history of mathematics 1869–1926*, Sources and Studies in the History of Mathematics and Physical Sciences, Springer-Verlag, New York, 2000. Cited on 443.)

[304] T. Hawkins, *The Mathematics of Frobenius in Context, A journey through 18th nto 20th century mathematics*, Sources and Studies in the History of Mathematics and Physical Sciences, Springer, New York, 2013. Cited on 446.)

[305] E. V. Haynsworth, *On the Schur Complement*, Basel Mathematical Notes, Univeristy of Basel, 1968. Cited on 208.)

[306] C. Heil, *A Basis Theory Primer*, expanded edition, Applied and Numerical Harmonic Analysis, Birkhäuser/Springer, New York, 2011. Cited on 100.)

[307] E. Heinz, *Beiträge zur Störunstheorie der Spektralzerlegung*, Math. Ann. **123** (1951), 415–438. Cited on 606.)

[308] A. Ya. Helemskii, *Banach and locally convex algebras*, Translated from the Russian by A. West, Oxford Science Publications, The Clarendon Press, Oxford University Press, New York, 1993. Cited on 44.)

[309] E. Hellinger, *Die Orthogonalinvarianten quadratischer Formen von unendlich vielen Variabeln*, Dissertation, Göttingen, 1907. Cited on 313.)

[310] E. Hellinger, *Neue Begründung der Theorie quadratischer Formen von unendlichvielen Veränderlichen*, J. Reine Angew. Math. **136** (1909), 210–271. Cited on 299, 313.)

[311] E. Hellinger and O. Toeplitz, *Grundlagen für eine Theorie der unendlichen Matrizen*, Math. Ann. **69** (1910), 289–330. Cited on 299.)

[312] H. Hellmann, *Einführung in die Quantenchemie*, Franz Deuticke, Leipzig, 1937. Cited on 27.)

[313] R. Hempel, L. A. Seco, and B. Simon, *The essential spectrum of Neumann Laplacians on some bounded singular domains*, J. Funct. Anal. **102** (1991), 448–483. Cited on 603.)

[314] I. W. Herbst, *Spectral theory of the operator* $(p^2 + m^2)^{1/2} - Ze^2/r$, Comm. Math. Phys. **53** (1977), 285–294. Cited on 572.)

[315] H. Hess, R. Schrader, and D. A. Uhlenbrock, *Domination of semigroups and generalization of Kato's inequality*, Duke Math. J. **44** (1977), 893–904. Cited on 627, 628.)

[316] E. Hewitt and K. A. Ross, *Abstract Harmonic Analysis. Vol. I: Structure of Topological Groups. Integration Theory, Group Representations*, Die Grundlehren der mathematischen Wissenschaften, Springer-Verlag, Berlin–Göttingen–Heidelberg, 1963; *Vol. II: Structure and Analysis for Compact Groups. Analysis on Locally Compact Abelian Groups*, Die Grundlehren der mathematischen Wissenschaften, Springer-Verlag, New York–Berlin, 1970. Cited on 443, 468, 504, 505.)

[317] D. Hilbert, *Grundzüge einer allgemeinen Theorie der linearen Integralgleichungen, I–VI*, Gött. Nachr. (1904), 49–91; (1904), 213–259; (1905), 307–338; (1906), 157–227; (1906), 439–480; (1910), 355–419. Cited on 42, 56, 99, 108, 192, 299.)

[318] T. H. Hildebrandt, *Linear functional transformations in general spaces*, Bull. Amer. Math. Soc. **37** (1931), 185–212. Cited on 43.)

[319] G. W. Hill, *On the part of the motion of the lunar perigee which is a function of the mean motions of the sun and moon*, Cambridge (1877); republished in Acta Math. **8** (1886), 1–36. Cited on 41.)

[320] E. Hille, *Functional Analysis and Semi-Groups*, American Mathematical Society Colloquium Publications, American Mathematical Society, New York, 1948. Cited on 99.)

[321] E. Hille and J. D. Tamarkin, *On the characteristic values of linear integral equations*, Acta Math. **57** (1931), 1–76. Cited on 192.)

[322] K. Hoffman, *Fundamentals of Banach Algebras*, Instituto de Matemática da Universidade do Paraná, Curitiba, 1962. Cited on 357.)

[323] K. Hoffman, *Banach Spaces of Analytic Functions*, reprint of the 1962 original, Dover Publications, New York, 1988. Cited on 489.)

[324] L. Hörmander, *The Analysis of Linear Partial Differential Operators, II. Differential Operators with Constant Coefficients*, Grundlehren der Mathematischen Wissenschaften, Springer-Verlag, Berlin, 1983. Cited on 667.)

[325] J. E. Humphreys, *Introduction to Lie Algebras and Representation Theory*, Second printing, revised, Graduate Texts in Mathematics, . Springer-Verlag, New York–Berlin, 1978. Cited on 443.)

[326] D. Hundertmark,and B. Simon, *Lieb-Thirring inequalities for Jacobi matrices*, J. Approx. Theory **118** (2002), 106–130. Cited on 683, 685.)

[327] D. Hundertmark,and B. Simon, *A diamagnetic inequality for semigroup differences*, J. Reine Angew. Math. **571** (2004), 107–130. Cited on 603, 628.)

[328] W. Hunziker, *On the spectra of Schrödinger multiparticle Hamiltonians*, Helv. Phys. Acta **39** (1966), 451–462. Cited on 666.)

[329] M. N. Huxley, *Exponential Sums and Lattice Points*, Proc. London Math. Soc. **60** (1990), 471–502. Cited on 605.)

[330] T. Ichinose, *A product formula and its application to the Schrödinger equation*, Publ. Res. Inst. Math. Sci. **16** (1980), 585–600. Cited on 630.)

[331] T. Ichinose, Hideo Tamura, Hiroshi Tamura, and V. A. Zagrebnov, *Note on the paper: "The norm convergence of the Trotter-Kato product formula with error bound" by T. Ichinose and H. Tamura*, Comm. Math. Phys. **221** (2001), 499–510. Cited on 630.)

[332] S. Ikehara, *An extension of Landau's theorem in the analytic theory of numbers*, J. Math. Phys. **10** (1931), 1–12. Cited on 506.)

[333] A. E. Ingham, *Some Tauberian theorems connected with the prime number theorem*, J. London Math. Soc. **20** (1945), 171–180. Cited on 504, 513.)

[334] M. E. H. Ismail, *Classical and quantum orthogonal polynomials in one variable*, reprint of the 2005 original, Encyclopedia of Mathematics and its Applications, Cambridge University Press, Cambridge, 2009. Cited on 231, 255.)

[335] M. I. Isaacs, *Character Theory of Finite Groups*, Pure and Applied Mathematics, Academic Press, New York–London, 1976. Cited on 443.)

[336] V. Ja. Ivriĭ, *The second term of the spectral asymptotics for a Laplace–Beltrami operator on manifolds with boundary*, Funktsional. Anal. i Prilozhen **14** (1980), 25–34; English translation: Functional Anal. Appl. **14** (1980), 98–106. Cited on 604.)

[337] V. Ja. Ivriĭ, *Precise spectral asymptotics for elliptic operators acting in fiberings over manifolds with boundary*, Lecture Notes in Mathematics, Springer-Verlag, Berlin, 1984. Cited on 604.)

[338] D. Jackson, *Über die Genauigkeit der Annäherung stetiger Funktionen durch ganze rationale Funktionen gegebenen Grades und trigonometrische Summen gegebener Ordnung*, (On the precision of the approximation of continuous functions by polynomials of given degree and by trigonometric sums of given order). Preisschrift und Dissertation. Univ. Göttingen, June 14, 1911; available online at `http://gdz.sub.uni-goettingen.de/dms/load/img/?IDDOC=40656`. Cited on 267.)

[339] C. Jacobi, *Über die Hauptaxen der Flachen der zweiten Ordnung*, J. Reine Angew. Math. **2** (1827), 227–233. Cited on 17.)

[340] C. Jacobi, *De singulari quadam duplicis Integralis transformatione*, J. Reine Angew. Math. **2** (1827), 234–242. Cited on 17.)

[341] C. Jacobi, *Exercitatio algebraica circa discerptionem singularem fractionum quae plures variabiles inyolvunt*, J. Reine Angew. Math. **5** (1829), 344–364. Cited on 17.)

[342] C. Jacobi, *De transformatione et determinatione integralium duplicium commentatio tertia*, J. Reine Angew. Math. **10** (1832), 101–128. Cited on 17.)

[343] C. Jacobi, *De eliminatione variabilis e duabus a.equationibus algebraicis*, J. Reine Angew. Math. **15** (1835), 101–124. Cited on 17.)

[344] C. Jacobi, *De formatione et proprietatibus Determinantium*, J. Reine Angew. Math. **22** (1841), 285–318. Cited on 17.)

[345] C. Jacobi, *Über die Reduction der quadratischen Formen auf die kleinste Anzahl Glieder*, J. Reine Angew. Math. **39** (1848), 290–292. Cited on 241.)

[346] G. James and A. Kerber, *The Representation Theory of the Symmetric Group*, with a foreword by P. M. Cohn, with an introduction by G. de Beauregard Robinson, Encyclopedia of Mathematics and its Applications, Addison-Wesley Publishing Co., Reading, Mass., 1981. Cited on 443.)

[347] V. A. Javrjan, *A certain inverse problem for Sturm–Liouville operators*, Izv. Akad. Nauk SSSR Ser. Mat. **6** (1971), 246–251. Cited on 344.)

[348] B. E. Johnson, *The uniqueness of the (complete) norm topology*, Bull. Amer. Math. Soc. **73** (1967), 537–539. Cited on 388.)

[349] P. Jonas, *On the trace formula for perturbation theory. I*, Math. Nachr. **137** (1988), 257–281. Cited on 353.)

[350] P. Jonas, *On the trace formula for perturbation theory. II*, Math. Nachr. **197** (1999), 29–49. Cited on 353.)

[351] P. W. Jones and D. Marshall, *Critical points of Green's function, harmonic measure, and the corona problem*, Ark. Mat. **23** (1985), 281–314. Cited on 389.)

[352] C. Jordan, *Mémoire sur les formes bilinéaires*, J. Math. Pures Appl. **19** (1874), 35–54. Cited on 135.)

[353] C. Jordan, *Sur la reduction des formes bilinéaires*, C. R. Acad. Sci. Paris **78** (1874), 614–617. Cited on 135.)

[354] C. Jordan, *Remarques sur les intégrales définies*, J. Math. Pures Appl. **8** (1892), 69–99. Cited on 603.)

[355] I. Kac, *On the spectral multiplicity of a second-order differential operator*, Dokl. Akad. Nauk SSSR **145** (1962), 510–513. Cited on 570.)

[356] I. Kac, *Spectral multiplicity of a second-order differential operator and expansion in eigenfunction*, Izv. Akad. Nauk SSSR Ser. Mat. **27** (1963) 1081–1112. Cited on 570.)

[357] M. Kac, *On some connections between probability theory and differential and integral equations*, In Proceedings of the Second Berkeley Symposium on Mathematical Statistics and Probability, 1950, pp. 189–215, University of California Press, Berkeley and Los Angeles, 1951. Cited on 630.)

[358] M. Kac, *A remark on Wiener's Tauberian theorem*, Proc. Amer. Math. Soc. **16** (1965), 1155–1157. Cited on 504.)

[359] M. Kac, *Can one hear the shape of a drum?* Amer. Math. Monthly **73** (1966), 1–23. Cited on 605.)

[360] R. V. Kadison, *Notes on the Gel'fand–Neumark theorem*, in [**168**], pp. 20–53, Contemporary Mathematics, American Mathematical Society, Providence, RI, 1994. Cited on 428.)

[361] R. V. Kadison and J. R. Ringrose, *Fundamentals of the Theory of Operator Algebras. Vol. I, Elementary Theory*, Pure and Applied Mathematics, Academic Press, Inc., New York, 1983. Cited on 314.)

[362] R. V. Kadison and J.R. Ringrose, *Fundamentals of the Theory of Operator Algebras. Vol. II, Advanced theory*, Pure and Applied Mathematics, Academic Press, Inc., Orlando, FL, 1986. Cited on 314.)

[363] R. V. Kadison and I. M. Singer, *Triangular operator algebras: Fundamentals and hyperreducible theory*, Amer. J. Math. **82** (1960), 227–259. Cited on 128.)

[364] J.-P. Kahane and W. Żelazko, *A characterization of maximal ideals in commutative Banach algebras*, Studia Math. **29** (1968), 339–343. Cited on 392.)

[365] S. Kakutani, *Some characterizations of Euclidean space*, Japan. J. Math. **16**, (1939), 93–97. Cited on 43.)

[366] H. Kalf, and J. Walter, *Strongly singular potentials and essential self-adjointness of singular elliptic operators in $C_0^\infty(R^n \setminus \{0\})$*, J. Functional Analysis **10** (1972), 114–130. Cited on 627.)

[367] G. K. Kalisch, *A functional anaysis proof of Titchmarsh's theorem on convolution*, J. Math. Anal. Appl. **5** (1962), 176–183. Cited on 129.)

[368] E. Kaniuth, *A Course in Commutative Banach Algebras*, Graduate Texts in Mathematics, Springer, New York, 2009. Cited on 357, 389.)

[369] I. Kaplansky, *A theorem on rings of operators*, Pacific J. Math. **1** (1951), 227–232. Cited on 314.)

[370] I. Kaplansky, *Rings of Operators*, W. A. Benjamin, Inc., New York–Amsterdam 1968. Cited on 314.)

[371] J. Karamata, *Über die Hardy–Littlewoodschen Umkehrungen des Abelschen Stetigkeitssatzes*, Math. Z. **32** (1930), 319–320. Cited on 506.)

[372] T. Kato, *On the convergence of the perturbation method, I, II*, Progr. Theor. Phys. **4** (1949), 514–523; **5** (1950), 95–101, 207-212. Cited on 27.)

[373] T. Kato, *Fundamental properties of Hamiltonian operators of Schrödinger type*, Trans. Amer. Math. Soc. **70** (1951), 195–211. Cited on 537.)

[374] T. Kato, *Notes on Projections and Perturbation Theory*, Technical Report No. 9, University of California, 1955 [unpublished]. Cited on 70, 217.)

[375] T. Kato, *Quadratic Forms in Hilbert Spaces and Asymptotic Perturbation Series*, Technical Report, Department of Mathematics, University of California, Berkeley, 1955. Cited on 600.)

[376] T. Kato, *On finite-dimensional perturbations of self-adjoint operators*, J. Math. Soc. Japan **9** (1957), 239–249. Cited on 343, 353, 536.)

[377] T. Kato, *Perturbation of continuous spectra by trace class operators*, Proc. Japan Acad. Tokyo **33** (1957), 260–264. Cited on 352, 353.)

[378] T. Kato, *Wave operators and similarity for some non-selfadjoint operators*, Math. Ann. **162** (1965/1966), 258–279. Cited on 600.)

[379] T. Kato, *Schrödinger operators with singular potentials*, In Proceedings of the International Symposium on Partial Differential Equations and the Geometry of Normed Linear Spaces (Jerusalem, 1972), Israel J. Math. **13** (1972), 135–148. Cited on 627.)

[380] T. Kato, *Perturbation Theory for Linear Operators*, 2nd edition, Grundlehren der Mathematischen Wissenschaften, Springer-Verlag, Berlin–New York, 1976. Cited on 27, 70, 536, 537, 548, 600.)

[381] T. Kato, *Remarks on Schrödinger operators with vector potentials*, Integral Equations Operator Theory **1** (1978), 103–113. Cited on 627.)

[382] T. Kato, *Trotter's product formula for an arbitrary pair of self-adjoint contraction semigroups*, In Topics in Functional Analysis (Essays dedicated to M. G. Krein on the occasion of his 70th birthday), pp. 185–195, Advances in Mathematics, Supplementary Series, Academic Press, New York–London, 1978. Cited on 629.)

[383] T. Kato and S. T. Kuroda, *The abstract theory of scattering*, Rocky Mountain J. Math. **1** (1971), 127–171. Cited on 600.)

[384] Y. Katznelson, *An Introduction to Harmonic Analysis*, 3rd edition, Cambridge Mathematical Library, Cambridge University Press, Cambridge, 2004. Cited on 357, 367, 369.)

[385] J. Kelley and R. Vaught, *The positive cone in Banach algebras*, Trans. Amer. Math. Soc. **74** (1953), 44–55. Cited on 428.)

[386] N. J. Kelton, *A note on pairs of projections*, Boll. Soc. Math. Mexicana **3** (1997), 309–311. Cited on 218.)

[387] R. Killip and B. Simon, *Sum rules and spectral measures of Schrödinger operators with L^2 potentials*, Ann. Math. **170** (2009), 739–782. Cited on 284.)

[388] P. Kirchberger, *Über Tchebychefsche Annäherungsmethoden*, Dissertation. Univ. Göttingen, 1902; available online at `http://gdz.sub.uni-goettingen.de/dms/load/img/?IDDOC=93028`. Cited on 266.)

[389] A. Kirillov Jr., *An Introduction to Lie Groups and Lie Algebras*, Cambridge Studies in Advanced Mathematics, Cambridge University Press, Cambridge, 2008. available online at `www.math.sunysb.edu/~kirillov/mat552/liegroups.pdf`. Cited on 628.)

[390] A. A. Kirillov and A. D. Gvishiani, *Theorems and Problems in Functional Analysis*, translated from the Russian by Harold H. McFaden, Problem Books in Mathematics, Springer-Verlag, New York–Berlin, 1982. (Russian original: Teoremy i zadachi funktsional'nogo analiza, Nauka, Moscow, 1979.) Cited on 217.)

[391] A. Kiselev and B. Simon, *Rank one perturbations with infinitesimal coupling*, J. Funct. Anal. **130** (1995), 345–356. Cited on 666.)

[392] M. Klaus, *On the bound state of Schrödinger operators in one dimension*, Ann. Phys. **108** (1977), 288–300. Cited on 683.)

[393] A. W. Knapp, *Group representations and harmonic analysis from Euler to Langlands, I*, Notices Amer. Math. Soc. **43** (1996), 410–415. Cited on 443.)

[394] A. W. Knapp, *Group representations and harmonic analysis from Euler to Langlands, II*, Notices Amer. Math. Soc. **43** (1996), 537–549. Cited on 443.)

[395] A. W. Knapp, *Representation Theory of Semisimple Groups, An overview based on examples*, Reprint of the 1986 original, Princeton Landmarks in Mathematics. Princeton University Press, Princeton, NJ, 2001. Cited on 443.)

[396] A. W. Knapp, *Lie Groups Beyond an Introduction*, Second edition, Progress in Mathematics, Birkhäuser Boston, Inc., Boston, MA, 2002. Cited on 628.)

[397] L. K. Kodama, *Boundary measures of analytic differentials and uniform approximation on a Riemann surface*, Pacific J. Math. **15** (1965), 1261–1277. Cited on 490.)

[398] R. Koksharov, *Mechanisms by Tchebyshev*, online museum of Chebyshev's devices at `http://tcheb.ru/`, 2014. Cited on 266.)

[399] A. Kolmogorov, *Über Kompaktheit der Funktionenmengen bei der Konvergenz im Mittel*, Gött. Nachr. (1931), 60–63. Cited on 228.)

[400] A. N. Kolmogorov and S. V. Fomin, *Elements of the Theory of Functions and Functional Analysis, Vol. 1. Metric and normed spaces*, Translated from the first Russian edition by L. F. Boron, Graylock Press, Rochester, NY, 1957; *Elements of the Theory of Functions and Functional Analysis. Vol. 2. Measure. The Lebesgue integral. Hilbert space*, Translated from the first (1960) Russian edition by H. Kamel and H. Komm, Graylock Press, Albany, NY, 1961. Cited on 603.)

[401] H. König, *Eigenvalue Distribution of Compact Operators*, Operator Theory: Advances and Applications, Birkhäuser Verlag, Basel, 1986. Cited on 187.)

[402] P. Koosis, *Introduction to H_p Spaces*, With an appendix on Wolff's proof of the corona theorem, London Mathematical Society Lecture Note Series, Cambridge University Press, Cambridge–New York, 1980. Cited on 389.)

[403] L. S. Koplienko, *Trace formula for nontrace-class perturbations*, Siberian Math. J. **25** (1984), 735–743; Russian original: Sibirsk. Mat. Zh. **25** (1984), 62–71. Cited on 353.)

[404] J. Korevaar, *Tauberian Theory. A Century of Developments*, Grundlehren der Mathematischen Wissenschaften, Springer-Verlag, Berlin, 2004. Cited on 506, 691.)

[405] P. P. Korovkin, *Linear Operators and Approximation Theory*, Translated from the (1959) Russian edition, Russian Monographs and Texts on Advanced Mathematics and Physics, Gordon and Breach Publishers, Inc., New York; Hindustan Publishing Corp., Delhi, 1960. Cited on 267.)

[406] Y. D. Koshmanenko, *A contribution to the theory of rank-one singular perturbations of selfadjoint operators*, Ukrainian Math. J. **43** (1991), 1450–1457; Russian original in Ukrain. Mat. Zh. **43** (1991), 1559–1566. Cited on 666.)

[407] S. T. Kowa, *Method of Solving the Dissimulated Problems*, 1683. Cited on 17.)

[408] H. L. Krall and O. Frink, *A new class of orthogonal polynomials: The Bessel polynomials*, Trans. Amer. Math. Soc. **65** (1948), 100–115. Cited on 254.)

[409] M. G. Krein, *Sur une géneralisation du théorème de Plancherel au cas des intégrales de Fourier sur les groupes topologiques commutatifs*, C. R. (Dokl.) Acad. Sci. URSS (N.S.) **30** (1941), 484–488. Cited on 467.)

[410] M. G. Krein, *The theory of self-adjoint extensions of semi-bounded Hermitian transformations and its applications. I*, Rec. Math. [Mat. Sbornik] N.S. **20** (1947), 62, 431–495. Cited on 601.)

[411] M. G. Krein, *A principle of duality for bicompact groups and quadratic block algebras*, Dokl. Akad. Nauk SSSR (N.S.) **69** (1949), 725–728. Cited on 468.)

[412] M. G. Krein, *On the trace formula in perturbation theory*, Mat. Sb. N.S. **33** (1953), 597–626. Cited on 343, 353.)

[413] M. G. Krein, *Integral equations on the half-line with a kernel depending on the difference of the arguments*, Uspehi Matem. Nauk (N.S.) **13** (1958), 3–120. Cited on 218.)

[414] M. G. Krein, *On perturbation determinants and a trace formula for certain classes of pairs of operators*, Trans. Amer. Math. Soc. **145** (1989), 39–84. Cited on 353.)

[415] M. G. Krein and V. A. Yavryan, *Spectral shift functions that arise in perturbations of a positive operator*, J. Operator Theory **6** (1981), 155–191. Cited on 353.)

[416] P. Kurasov, *$\mathcal{H}_{-n}$-perturbations of self-adjoint operators and Krein's resolvent formula*, Integral Equations Operator Theory **45** (2003), 437–460. Cited on 666.)

[417] S. T. Kuroda, *On a theorem of Weyl–von Neumann*, Proc. Japan Acad. **34** (1958), 11–15. Cited on 353.)

[418] S. T. Kuroda, *Perturbation of continuous spectra by unbounded operators. I*, J. Math. Soc. Japan **11** (1959), 246–262. Cited on 666.)

[419] J.-L. Lagrange, *Nouvelle solution du probleme du mouvement de rotation d'un corps de figure quelconque qui n'est anime par aucune force acceleratrice*, Nouv. Mim. de l'Acad. Roy. ... (de Berlin) (1773), 85–128. Cited on 17.)

[420] T. Lalesco, *Une théorème sur les noyaux composés*, Bull. Acad. Roumaine **3** (1915), 271–272. Cited on 163.)

[421] T. Y. Lam, *Representations of finite groups: a hundred years. I*, Notices Amer. Math. Soc. **45** (1998), 361–372. Cited on 446.)

[422] E. Landau, *Zwei neue Herleitungen für die asymptotische Anzahl der Primzahlen unter einer gegebenen Grenze*, Berl. Ber. (1908), 746–764. Cited on 513.)

[423] E. Landau, *Handbuch der Lehre von der Verteilung der Primzahlen*, 2nd edition, Chelsea Publishing, New York, 1953; reprint of the original 1909 edition. Cited on 513.)

[424] E. Landau, *Über die Gitterpunkte in einem Kreise (II)*, Gött. Nachr. (1915), 161–171. Cited on 605.)

[425] E. Landau, *Zur analytischen Zahlentheorie der definiten quadratischen Formen (Über die Gitterpunkte in einem mehrdimensionalen Ellipsoid)*, Sitzungsber. Preuss. Akad. Wiss. **31** (1915), 458–476. Cited on 605.)

[426] M. L. Lapidus, *Product formula for imaginary resolvents with application to a modified Feynman integral*, J. Funct. Anal. **63** (1985),261–275. Cited on 630.)

[427] P. S. Laplace, *Recherches sur le calcul integral et sur le systeme du monde*, Hist. de l'Acad. Roy. des Sciences (1772), 2nd partie, 267–376. Cited on 17.)

[428] R. Larsen, *Banach Algebras. An Introduction*, Pure and Applied Mathematics, Marcel Dekker, New York, 1973. Cited on 357.)

[429] Y. Last and B. Simon, *Eigenfunctions, transfer matrices, and absolutely continuous spectrum of one-dimensional Schrödinger operators*, Invent. Math. **135** (1999), 329–367. Cited on 666.)

[430] Y. Last and B. Simon, *The essential spectrum of Schrödinger, Jacobi, and CMV operators*, J. Anal. Math. **98** (2006), 183–220. Cited on 666.)

[431] M. A. Lavrentiev, *Sur les fonctions d'une variable complexe représentables par des séries de polynômes*, Hermann & Cie, Paris, 1936. Cited on 489.)

[432] P. D. Lax, *Functional Analysis*, Pure and Applied Mathematics, Wiley–Interscience, New York, 2002. Cited on 217.)

[433] P. D. Lax and A. N. Milgram, *Parabolic equations*, In Contributions to the theory of partial differential equations, pp. 167–190, Annals of Mathematics Studies, Princeton University Press, Princeton, NJ, 1954. Cited on 600.)

[434] H. Lebesgue, *Sur le problem de Dirichlet*, Rend. Circ. Mat. Palermo **29** (1907), 371–402. Cited on 489.)

[435] G. W. Leibniz, *Leibnizens Mathematische Schriften*, edited by C. I. Gerhardt, pp. 229, 238–240, 245, H. W. Schmidt, Halle, 1850. Cited on 17.)

[436] G. M. Leibowitz, *Lectures on Complex Function Algebras*, Scott, Foresman and Co., Glenview, IL, 1970. Cited on 489.)

[437] H. Leinfelder and C. G. Simader, *Schrödinger operators with singular magnetic vector potentials*, Math. Z. **176** (1981), 1–19. Cited on 627.)

[438] P. Lesky, *Eine Charakterisierung der klassischen kontinuierlichen, diskreten und q-Orthogonalpolynome*, Shaker, Aachen, 2005. Cited on 255.)

[439] N. Levinson, *Criteria for the limit-point case for second order linear differential operators*, Časopis Pěst. Mat. Fys. **74** (1949), 17–20. Cited on 569.)

[440] N. Levinson, *The prime number theorem from* $\log n!$, Proc. Amer. Math. Soc. **15** (1964), 480–485. Cited on 513.)

[441] B. M. Levitan and I. S. Sargsjan, *Introduction to Spectral Theory: Selfadjoint Ordinary Differential Operators*, Translated from the Russian by Amiel Feinstein, Translations of Mathematical Monographs, American Mathematical Society, Providence, RI, 1975. Cited on 569.)

[442] B. M. Levitan and V. V. Zhikov, *Almost Periodic Functions and Differential Equations*, Cambridge University Press, Cambridge, 1982. Cited on 419.)

[443] P. Lévy, *Sur la convergence absolue des séries de Fourier*, Compos. Math. **1** (1934), 1–14. Cited on 388.)

[444] B. R. Li, *Introduction to Operator Algebras*, World Scientific Publishing Co., Inc., River Edge, NJ, 1992. Cited on 314.)

[445] V. B. Lidskii, *Non-selfadjoint operators with a trace*, Dokl. Akad. Nauk SSSR **125** (1959), 485–487. Cited on 186.)

[446] S. Lie and F. Engel, *Theorie der Transformationsgruppen*, 3 Vols., Teubner, Leipzig, 1888. Cited on 628.)

[447] E. H. Lieb and W. Thirring, *Bound for the kinetic energy of fermions which proves the stability of matter*, Phys. Rev. Lett. **35** (1975), 687–689; Errata **35** (1975), 1116. Cited on 683.)

[448] E. H. Lieb and W. Thirring, *Inequalities for the moments of the eigenvalues of the Schrödinger Hamiltonian and their relation to Sobolev inequalities*, in Studies in Mathematical Physics, Essays in Honor of Valentine Bargmann, edited by E. Lieb, B. Simon, and A. Wightman, pp. 269–303, Princeton University Press, Princeton, NJ, 1976. Cited on 683.)

[449] I. M. Lifshitz, *On a problem of the theory of perturbations connected with quantum statistics*, Usp. Mat. Nauk **7** (1952), 171–180. Cited on 353.)

[450] J. Lindenstrauss and L. Tzafriri, *On the complemented subspaces problem*, Israel J. Math. **9** (1971), 263–269. Cited on 43.)

[451] J.-L. Lions, *Supports de produits de composition, I*, C. R. Acad. Sci. Paris **232** (1951), 1530–1532. Cited on 129.)

[452] J.-L. Lions, *Équations différentielles opérationnelles et problèmes aux limites*, Die Grundlehren der mathematischen Wissenschaften, Springer-Verlag, Berlin–Göttingen–Heidelberg, 1961. Cited on 600.)

[453] J. E. Littlewood, *The converse of Abel's theorem on power series*, Proc. London Math. Soc. (2) **9** (1911), 434–448. Cited on 506.)

[454] K. Loewner, *Über monotone Matrixfunktionen*, Math. Z. **38** (1934), 177–216. Cited on 606.)

[455] V. I. Lomonosov, *Invariant subspaces of the family of operators that commute with a completely continuous operator*, Funkcional. Anal. i Priložen. **7** (1973), 55–56. Cited on 118.)

[456] L. H. Loomis, *An Introduction to Abstract Harmonic Analysis*, Van Nostrand, Toronto–New York–London, 1953. Cited on 468, 504.)

[457] E. R. Lorch, *The spectrum of linear transformations*, Trans. Amer. Math. Soc. **52** (1942), 238–248. Cited on 69.)

[458] E. R. Lorch, *Review of P. R. Halmos, Introduction to Hilbert Space and the Theory of Spectral Multiplicity*, Bull. Amer. Math. Soc. **58** (1952), 412–415. Cited on 313.)

[459] H. A. Lorentz, *Alte und neue Fragen der Physik*, Physikal. Z. **11** (1910), 1234–1257. Cited on 603.)

[460] G. W. Mackey, *Borel structure in groups and their duals*, Trans. Amer. Math. Soc. **85** (1957), 134–165. Cited on 314.)

[461] G. W. Mackey, *Unitary Group Representations in Physics, Probability, and Number Theory*, Second edition, Advanced Book Classics, Addison-Wesley Publishing Company, Advanced Book Program, Redwood City, CA, 1989. Cited on 443.)

[462] G. W. Mackey, *The Scope and History of Commutative and Noncommutative Harmonic Analysis*, History of Mathematics,American Mathematical Society, Providence, RI; London Mathematical Society, London, 1992. Cited on 443.)

[463] C. MacLaurin, *A Treatise of Algebra in Three Parts*, Millar and Nourse, London, 1748; available online at `https://archive.org/details/atreatisealgebr03maclgoog`. Cited on 17.)

[464] M. Mantoiu, *C^*-algebras, dynamical systems at infinity and the essential spectrum of generalized Schrödinger operators*, J. Reine Angew. Math. **550** (2002), 211–229. Cited on 666.)

[465] F. Marcellán, A. Branquinho, and J. Petronilho, *Classical orthogonal polynomials: A functional approach*, Acta Appl. Math. **34** (1994), 283–303. Cited on 255.)

[466] A. A. Markov, *Selected Papers on Continued Fractions and the Theory of Functions Deviating Least from Zero*, OGIZ, Moscow–Leningrad, 1948. Cited on 267.)

[467] A. Markov, *Démonstration de certaines inégalités de Tchebycheff*, Math. Ann. **24** (1884), 172–180. Cited on 241.)

[468] A. Martin, *Bound states in the strong coupling limit*, Helv. Phys. Acta **45** (1972) 140–148. Cited on 605.)

[469] J. C. Mason and D. C. Handscomb, *Chebyshev Polynomials*, Chapman & Hall/CRC, Boca Raton, FL, 2003. Cited on 266.)

[470] S. Mazur, *Sur les anneaux linéaires*, C. R. Acad. Sci. Paris **207** (1938), 1025–1027. Cited on 357, 387.)

[471] V. Maz'ya, *Analytic criteria in the qualitative spectral analysis of the Schrödinger operator*, in Spectral Theory and Mathematical Physics: A Festschrift in Honor of Barry Simon's 60th birthday, pp. 257–288, Proceedings of Symposia in Pure Mathematics, American Mathematical Society, Providence, RI, 2007. Cited on 228, 603.)

[472] R. E. Megginson, *An Introduction to Banach Space Theory*, Graduate Texts in Mathematics, Springer-Verlag, New York, 1998. Cited on 100.)

[473] J. Mercer, *Functions of positive and negative type, and their connection with the theory of integral equations*, Philos. Trans. Roy. Soc. London A **209** (1909), 415–446. Cited on 182.)

[474] S. N. Mergelyan, *Uniform approximations of functions of a complex variable*, Uspehi Matem. Nauk (N.S.) **7** (1952), 31–122; English translation in American Mathematical Society Translations , American Mathematical Society, Providence, RI, 1954. Cited on 489.)

[475] H. N. Mhaskar and D. V. Pai, *Fundamentals of Approximation Theory*, CRC Press, Boca Raton, FL; Narosa Publishing House, New Delhi, 2000. Cited on 267.)

[476] A. J. Michaels, *Hilden's simple proof of Lomonosov's invariant subspace theorem*, Adv. Math. **25** (1977), 56–58. Cited on 119.)

[477] J. G. Mikusinski, *A new proof of Titchmarsh's convolution theorem*, Studia Math. **13** (1952), 56–58. Cited on 129.)

[478] W. J. Miller, *Symmetry Groups and Their Applications*, Pure and Applied Mathematics, Academic Press, New York–London, 1972. Cited on 443.)

[479] J. Milnor, *Eigenvalues of the Laplace operator on certain manifolds*, Proc. Nat. Acad. Sci. USA **51** (1964), 542. Cited on 605.)

[480] T. Molien, *Über Systeme höherer complexer Zahlen*, Math. Ann. **41** (1893), 83–156. Cited on 446.)

[481] M. S. Moonen and B. de Moor, *SVD and Signal Processing, III: Algorithms, Architectures, and Applications*, Elsevier Science, Amsterdam, 1995. Cited on 135.)

[482] R. D. Mosak, *Banach Algebras*, Chicago Lectures in Mathematics, The University of Chicago Press, Chicago-London, 1975. Cited on 357.)

[483] M. S. Moslehian, *A survey on the complemented subspace problem*, Trends Math. **9** (2006) 91–98. Cited on 44.)

[484] T. Muir, *The Theory of Determinants in the Historical Order of Development*, Macmillan, London, 1906. Cited on 17.)

[485] G. Murphy, *C^*-Algebras and Operator Theory*, Academic Press, Inc., San Diego, 1990. Cited on 314.)

[486] F. J. Murray, *On complementary manifolds and projections in spaces L_p and l_p*, Trans. Amer. Math. Soc. **41** (1937) 138–152. Cited on 43.)

[487] F. J. Murray and J. von Neumann, *On rings of operators*, Ann. Math. (2) **37** (1936), 116–229. Cited on 83.)

[488] M. Nagumo, *Einige analytische Untersuchungen in linearen, metrischen Ringen*, Japan. J. Math. **13** (1936), 61–80. Cited on 56, 69, 357.)

[489] M. A. Naimark, *On extremal spectral functions of a symmetric operator*, Dokl. Akad. Nauk SSSR **54** (1946) 7–9. Cited on 659.)

[490] M. A. Naimark, *Normed Rings*, Noordhoff N. V., Groningen, 1959. Cited on 504.)

[491] M. A. Naimark, *Linear Differential Operators, Part II: Linear Differential Operators in Hilbert Space*, With additional material by the author, and a supplement by V. È. Ljance, Translated from the Russian by E. R. Dawson, English translation edited by W. N. Everitt, Frederick Ungar Publishing Co., New York 1968. Cited on 569.)

[492] M. A. Naimark, *Normed Algebras*, 3rd edition, Wolters–Noordhoff Series of Monographs and Textbooks on Pure and Applied Mathematics, Wolters–Noordhoff Publishing, Groningen, 1972. Cited on 357.)

[493] H. Nakano, *Unitärinvariante hypermaximale normale Operatoren*, Ann. Math. (2) **42** (1941), 657–664. Cited on 313.)

[494] R. Narasimhan and Y. Nievergelt, *Complex Analysis in One Variable*, Second edition, Birkhäuser Boston, Inc., Boston, MA, 2001. Cited on 389.)

[495] H. Neidhardt, *Spectral shift function and Hilbert–Schmidt perturbation: Extensions of some work of Koplienko*, Math. Nachr. **138** (1988), 7–25. Cited on 353.)

[496] H. Neidhardt and V. A. Zagrebnov, *On the Trotter product formula for Gibbs semigroups*, Ann. Phys. **502** (1990), 183–191. Cited on 630.)

[497] E. Nelson, *The distinguished boundary of the unit operator ball*, Proc. Amer. Math. Soc. **12** (1961), 994-995. Cited on 323.)

[498] E. Nelson, *Feynman integrals and the Schrödinger equation*, J. Math. Phys. **5** (1964), 332–343. Cited on 630.)

[499] E. Nelson, *Interaction of nonrelativistic particles with a quantized scalar field*, J. Math. Phys. **5** (1964), 1190–1197. Cited on 600.)

[500] C. G. Neumann, *Untersuchungen über das logarithmische und Newtonsche Potential*, Teubner, Leipzig, 1877. Cited on 56, 118.)

[501] R. Nevanlinna, *Asymptotische Entwickelungen beschränkter Funktionen und das Stieltjessche Momentenproblem*, Ann. Acad. Sci. Fenn. A, Suomalaisen tiedeakatemian kustantama, 1922. Cited on 658.)

[502] D. J. Newman, *The nonexistence of projections from L^1 to H^1*. Proc. Amer. Math. Soc. **12** (1961) 98–99. Cited on 43.)

[503] D. J. Newman, *A simple proof of Wiener's $1/f$ theorem*, Proc. Amer. Math. Soc. **48** (1975), 264–265. Cited on 390.)

[504] E. M. Nikishin and V. N. Sorokin, *Rational Aapproximations and Orthogonality*, Translated from the Russian by R. P. Boas, Translations of Mathematical Monographs, American Mathematical Society, Providence, RI, 1991; Russian original, 1988. Cited on 231.)

[505] S. M. Nikolsky, *A Course of Mathematical Analysis. Vol. 1*, Reprint of the 1977 translation by V. M. Volosov, Mir, Moscow, 1985; *A Course of Mathematical Analysis. Vol. 2*, Reprint of the 1977 translation by V. M. Volosov, Mir, Moscow, 1985. Cited on 603.)

[506] F. Noether, *Über eine Klasse singulärer Integralgleichungen*, Math. Ann. **82** (1920), 42–63. Cited on 216.)

[507] S. Nonnenmacher, J. Sjöstrand, and M. Zworski, *Fractal Weyl law for open quantum chaotic maps*, Ann. Math. (2) **179** (2014), 179–251. Cited on 605.)

[508] S. Odake and R. Sasaki, *Infinitely many shape invariant potentials and new orthogonal polynomials*, Phys. Lett. B. **679** (2009), 414–417. Cited on 255.)

[509] R. E. A. C. Paley and N. Wiener, *Characters of Abelian groups*, Proc. Nat. Acad. Sci. USA **19** (1933), 253–257. Cited on 467.)

[510] R. E. A. C. Paley and A. Zygmund, *On some series of functions*, Math. Proc. Cambridge Philos. Soc. **26** (1930), 337–357. Cited on 367.)

[511] T. W. Palmer, *Banach Algebras and the General Theory of $*$-algebras. Vol. I. Algebras and Banach Algebras*, Encyclopedia of Mathematics and Its Applications, Cambridge University Press, Cambridge, 1994. Cited on 357.)

[512] W. Pauli, *Principles of Wave Mechanics* In Handbuch der Physik, Volume 24, p. 162, Julius Springer Verlag, Berlin, 1933. Cited on 27.)

[513] G. Peano, *Applicazioni geometriche del calcolo infinitesimale*, Bocca (1887). Cited on 603.)

[514] C. Pearcy and A. L. Shields, *A survey of the Lomonosov technique in the theory of invariant subspaces* In Topics in Operator Theory, pp. 219–229. Mathematical Surveys, , American Mathematical Society, Providence, RI, 1974. Cited on 119.)

[515] D. B. Pearson, *A generalization of the Birman trace theorem*, J. Funct. Anal. **28** (1978), 182–186. Cited on 666.)

[516] G. K. Pedersen, *C^*-algebras and Their Automorphism Groups*, London Mathematical Society Monographs, Academic Press, Inc., London–New York, 1979. Cited on 314.)

[517] V. V. Peller, *Hankel Operators and Their Applications*, Springer Monographs in Mathematics, Springer-Verlag, New York, 2003. Cited on 218.)

[518] V. V. Peller, *An extension of the Koplienko–Neidhardt trace formulae*, J. Funct. Anal. **221** (2005), 456–481. Cited on 353.)

[519] F. Peter and H. Weyl, *Die Vollständigkeit der primitiven Darstellungen einer geschlossenen kontinuierlichen Gruppe*, Math. Ann. **97** (1927), 737–755. Cited on 447.)

[520] R. R. Phelps, *Lectures on Choquet's Theorem*, 2nd edition, Lecture Notes in Mathematics, Springer-Verlag, Berlin, 2001. Cited on 489.)

[521] R. S. Phillips, *On linear transformations*, Trans. Amer. Math. Soc. **48**, (1940) 516–541. Cited on 43, 45.)

[522] E. Picard, *Sur un thèoréme général relatif aug équations intégrales de prémière espèce et sur quelques problèmes de physique mathématique*, Rend. Circ. Mat. Palermo **29** (1910), 79–97. Cited on 83, 134.)

[523] G. Pisier, *De nouveaux espaces de Banach sans la propriété d'approximation [d'après A. Szankowski]*, Séminaire Bourbaki (1978/79), Exp. No. 542, pp. 312–327, Lecture Notes in Mathematics, Springer, Berlin, 1980. Cited on 100.)

[524] G. Pisier, *Counterexamples to a conjecture of Grothendieck*, Acta Math. **151** (1983), 181–208. Cited on 100.)

[525] G. Pisier, *The Volume of Convex Dodies and Banach Space Geometry*, Cambridge Tracts in Mathematics, Cambridge University Press, Cambridge, 1989. Cited on 187.)

[526] G. Pisier, *The Operator Hilbert Space* OH, *Complex Interpolation and Tensor Norms*, Mem. Amer. Math. Soc. American Mathematical Society, Providence, RI, 1996. Cited on 100.)

[527] H. R. Pitt, *General Tauberian theorems*, Proc. London Math. Soc. (2) **44** (1938), 243–288. Cited on 504.)

[528] J. Plemelj, *Zur theorie Fredholmschen funktionalgleichung*, Monatsh. Math. Phys. **15** (1904), 93–128. Cited on 172.)

[529] A. I. Plesner, and V. A. Rohlin, *Spectral theory of linear operators. II*, Uspehi Matem. Nauk (N.S.) **11** (1946), 71–191. Cited on 313.)

[530] H. Poincaré, *Sur les déterminants d'ordre infini*, Bull. Soc. Math. France **14** (1886), 77–90 Cited on 41.)

[531] H. Poincaré, *Sur les groupes continus*, Trans. Cambridge Philos. Soc. **18** (1900), 220–255. Cited on 628.)

[532] H. Poincaré, *The future of mathematics, an essay contained in Science et méthode*, Paris, E. Flammarion (1908); French original online at `https://archive.org/details/scienceetmthod00poin`; English translation by Frances Maitland as Science and Method, London, T. Nelson (1914), available online at `https://archive.org/details/sciencemethod00poinuoft`. Cited on 355.)

[533] H. Poincaré, *Remarques diverses sur l'équation de Fredholm*, Acta Math. **33** (1909), 57–86. Cited on 192.)

[534] G. Pólya and G. Szegő, *Problems and Theorems in Analysis, I. Series, integral calculus, theory of functions*, Translated from the German by Dorothee Aeppli, Reprint of the 1978 English translation, Classics in Mathematics, Springer-Verlag, Berlin, 1998. Cited on 282.)

[535] G. Pólya and G. Szegő, *Problems and Theorems in Analysis, II. Theory of functions, zeros, polynomials, determinants, number theory, geometry*, Translated from the German by C. E. Billigheimer, Reprint of the 1976 English translation, Classics in Mathematics, Springer-Verlag, Berlin, 1998. Cited on 282.)

[536] L. S. Pontryagin, *The theory of topological commutative groups*, Ann. Math. (2) **35** (1934), 361–388. Cited on 467.)

[537] A. Ya. Povzner, *On positive definite functions on an abelian group*, Dokl. Akad. Nauk SSSR **28** (1940), 294–295. Cited on 467.)

[538] C. R. Putnam, *Continuous spectra and unitary equivalence*, Pacific J. Math. **7** (1957), 993–995. Cited on 666.)

[539] C. Quesne, *Exceptional orthogonal polynomials, exactly solvable potentials and supersymmetry*, J. Phys. A **41**, (2008). Cited on 255.)

[540] C. Quesne, *Solvable rational potentials and exceptional orthogonal polynomials in supersymmetric quantum mechanics*, SIGMA, **5**, 2009. Cited on 255.)

[541] V. S. Rabinovich, *Essential spectrum of perturbed pseudodifferential operators. Applications to the Schrödinger, Klein–Gordon, and Dirac operators*, Russ. J. Math. Phys. **12** (2005), 62–80. Cited on 666.)

[542] H. Radjavi and P. Rosenthal, *Invariant Subspaces*, Ergebnisse der Mathematik und ihrer Grenzgebiete, Springer-Verlag, New York–Heidelberg, 1973. Cited on 119.)

[543] D. A. Raikov, *Positive definite functions on commutative groups with an invariant measure*, C. R. (Doklady) Acad. Sci. URSS (N.S.) **28** (1940), 296–300. Cited on 399, 467, 468.)

[544] D. A. Raikov, *Generalized duality theorem for commutative groups with an invariant measure*, C. R. (Doklady) Acad. Sci. URSS (N.S.) **30** (1941), 589–591. Cited on 468, 469.)

[545] D. A. Raikov, *On the theory of normed rings with involution*, C. R. (Doklady) Acad. Sci. URSS (N.S.) **54** (1946), 391–3940. Cited on 399.)

[546] Lord Rayleigh, *In finding the correction for the open end of an organ-pipe*, Phil. Trans. **161** (1870), 77. Cited on 109.)

[547] Lord Rayleigh, *The Theory of Sound, Vol. I*, MacMillan, London, 1877; 2nd edition, Dover Publications, New York, NY, 1945. Cited on 27, 603.)

[548] M. Reed and B. Simon, *Methods of Modern Mathematical Physics, I: Functional Analysis*, Academic Press, New York, 1972. Cited on 629.)

[549] M. Reed and B. Simon, *Methods of Modern Mathematical Physics, II. Fourier Analysis, Self-Adjointness*, Academic Press, New York, 1975. Cited on 568, 569.)

[550] M. Reed and B. Simon, *Methods of Modern Mathematical Physics, III: Scattering Theory*, Academic Press, New York, 1979. Cited on 353.)

[551] M. Reed and B. Simon, *Methods of Modern Mathematical Physics, IV: Analysis of Operators*, Academic Press, New York, 1977. Cited on 27, 600, 667.)

[552] C. Reid, *Hilbert, With an appreciation of Hilbert's mathematical work by Hermann Weyl*, Springer-Verlag, New York–Berlin, 1970. Cited on 41, 43.)

[553] O. Reinov, *Some more remarks on Grothendieck–Lidskii trace formulas*, J. Prime Res. Math. **8** (2012), 5–11. Cited on 187.)

[554] H. Reiter and J. D. Stegeman, *Classical Harmonic Analysis and Locally Compact Groups*, 2nd edition, London Mathematical Society Monographs (New Series), The Clarendon Press, Oxford University Press, New York, 2000. Cited on 468.)

[555] F. Rellich, *Ein Satz über mittlere Konvergenz*, Gött. Nachr. **1930** (1930), 30–35. Cited on 228.)

[556] F. Rellich, *Störungstheorie der Spektralzerlegung. I–V*, Math. Ann. **113** (1937), 600–619, 677–685; **116** (1939), 555–570; **117** (1940) 356–382; **118** (1942), 462–484. Cited on 27, 536, 548.)

[557] F. Rellich, *Das Eigenwertproblem von* $\Delta u + \lambda u = 0$ *in Halbröhren*, In Studies and Essays Presented to R. Courant on his 60th Birthday, January 8, 1948, pp. 329–344, Interscience Publishers, New York, 1948. Cited on 228, 603.)

[558] F. Rellich, *Perturbation Theory of Eigenvalue Problems*, Gordon and Breach, New York, 1969. Cited on 27.)

[559] J. R. Retherford, *Hilbert Space: Compact Operators and the Trace Theorem*, London Mathematical Society Student Texts, Cambridge University Press, Cambridge, 1993. Cited on 134.)

[560] M. Ribarič and I. Vidav, *Analytic properties of the inverse $A(z) - 1$ of an analytic linear operator valued function $A(z)$*, Arch. Ration. Mech. Anal. **32** (1969), 298–310. Cited on 200.)

[561] C. E. Rickart, Banach algebras with an adjoint operation. Ann. Math. (2) **47** (1946), 528–550. Cited on 405.)

[562] C. E.Rickart, *General Theory of Banach Algebras*, The University Series in Higher Mathematics, Van Nostrand, Princeton–Toronto–London–New York, 1960. Cited on 357.)

[563] F. Riesz, *Untersuchungen über Systeme integrierbarer Funktionen*, Math. Ann. **69** (1910), 449–497. Cited on 43.)

[564] F. Riesz, *Les systèmes d'équations à une infinité d'inconnues*, Gauthier–Villars, Paris, 1913. Cited on 43, 69, 100, 299.)

[565] F. Riesz, *Über ein Problem des Herrn Carathéodory*, J. Reine Angew. Math. 146 (1916), 83–87. Cited on 321.)

[566] F. Riesz, *Über lineare Funktionalgleichungen*, Acta Math. **41** (1916), 71–98. Cited on 100, 118.)

[567] F. Riesz, *Über die linearen Transformationen des komplexen Hilbertschen Raumes*, Acta Sci. Math. Szeged **5** (1930), 23–54. Cited on 69, 83.)

[568] M. Riesz, *Sur les ensembles compacts de fonctions sommables*, Acta Sci. Math. Szeged **6** (1933), 136–142. Cited on 228.)

[569] J. R. Ringrose, *Super-diagonal forms for compact linear operators*, Proc. London Math. Soc. **3–12** (1962), 367–384. Cited on 127, 128.)

[570] J. R. Ringrose, *Compact Non-Self-Adjoint Operators*, Van Nostrand Reinhold Co., 1971. Cited on 128.)

[571] W. Ritz, *Über eine neue Methode zur Lösung gewisser Variationsprobleme der mathematischen Physik*, J. Reine Angew. Math. **135** (1908), 1–61. Cited on 109.)

[572] T. J. Rivlin, *An Introduction to the Approximation of Functions*, Corrected reprint of the 1969 original, Dover Books on Advanced Mathematics, Dover Publications, Inc., New York, 1981. Cited on 267.)

[573] T. J. Rivlin, *Chebyshev Polynomials. From approximation theory to algebra and number theory*, Second edition, Pure and Applied Mathematics, John Wiley & Sons, Inc., New York, 1990. Cited on 266.)

[574] D. Robert, *Comportement asymptotique des valeurs propres d'opérateurs du type Schrödinger à potentiel "dégénéré"*, J. Math. Pures Appl. **61** (1982), 275–300. Cited on 228, 604.)

[575] D. W. Robinson, *The Thermodynamic Pressure in Quantum Statistical Mechanics*, Lecture Notes in Physics, Springer-Verlag, Berlin-New York, 1971. Cited on 601, 605.)

[576] D. L. Rogava, *On the error of Trotter-type formulas in the case of selfadjoint operators*, Funktsional. Anal. i Prilozhen. **27** (1993), 84–86; translation in Functional Anal. Appl. 27 (1993), 217–219. Cited on 630.)

[577] H. Rollnik, *Streumaxima and gebundene Zustände*, Z. Phys. **145** (1956), 639–653. Cited on 683.)

[578] M. Rosenblum, *Perturbation of the continuous spectrum and unitary equivalence*, Pacific J. Math. **7** (1957), 997–1010. Cited on 353.)

[579] A. Roth, *Approximationseigenschaften und Strahlengrenzwerte meromorpher und ganzer Funktionen*, Comment. Math. Helv. **11** (1938), 77–125. Cited on 490.)

[580] E. Routh, *On some properties of certain solutions of a differential equation of the second order*, Proc. London Math. Soc. **16** (1885), 245–261. Cited on 254.)

[581] W. Rudin, *Some theorems on Fourier coefficients*, Proc. Amer. Math. Soc. **10** (1959), 855–859. Cited on 369.)

[582] W. Rudin, *Projections on invariant subspaces*, Proc. Amer. Math. Soc. **13** (1962) 429–432. Cited on 43.)

[583] W. Rudin, *Fourier Analysis on Groups*, reprint of the 1962 original, Wiley Classics Library, John Wiley & Sons, New York, 1990. Cited on 468, 504, 505.)

[584] W. Rudin, *Functional Analysis*, 2nd edition, International Series in Pure and Applied Mathematics, McGraw–Hill, New York, 1991. Cited on 357.)

[585] D. Ruelle, *A remark on bound states in potential-scattering theory*, Il Nuovo Cimento A **61** (1969), 655–662. Cited on 321.)

[586] A. F. Ruston, *Fredholm Theory in Banach Spaces*, Cambridge Tracts in Mathematics, Cambridge University Press, Cambridge, 1986. Cited on 183.)

[587] S. Sakai, *C^*-algebras and W^*-algebras*, Ergebnisse der Mathematik und ihrer Grenzgebiete, Springer-Verlag, New York-Heidelberg, 1971. Cited on 314, 429.)

[588] D. Sarason, *A remark on the Volterra operator*, J. Math. Anal. Appl. **12** (1965), 244–246. Cited on 128.)

[589] D. Sarason, *Moment problems and operators in Hilbert space*, In Moments in Mathematics (San Antonio, Tex., 1987), pp. 54–70, Proc. Sympos. Appl. Math., American Mathematical Society Providence, RI, 1987. Cited on 658.)

[590] R. Schatten, *The cross-space of linear transformations*, Ann. Math. (2) **47** (1946), 73–84. Cited on 152.)

[591] R. Schatten, *A Theory of Cross-Spaces*, Annals of Mathematics Studies, Princeton University Press, Princeton, NJ, 1950. Cited on 152.)

[592] R. Schatten, *Norm Ideals of Completely Continuous Operators*, Ergebnisse der Mathematik und ihrer Grenzgebiete, Springer-Verlag, Berlin, 1960. Cited on 152.)

[593] R. Schatten and J. von Neumann, *The cross-space of linear transformations. II*, Ann. Math. (2) **47** (1946), 608–630. Cited on 143, 152.)

[594] R. Schatten and J. von Neumann, *The cross-space of linear transformations. III*, Ann. Math. (2) **49** (1948), 557–582. Cited on 152.)

[595] J. Schatz, *Review of* [**216**], Math. Rev. **14** (1953), 884. Cited on 428.)

[596] J. Schauder, *Zur Theorie stetiger Abbildungen in Funktionalräumen*, Math. Z. **26** (1927), 47–65. Cited on 100.)

[597] J. Schauder, *Eine Eigenschaft des Haarschen Orthogonalsystems*, Math. Z. **28** (1928), 317–320. Cited on 100.)

[598] J. Schauder, *Über lineare, vollstetige Funktionaloperationen*, Studia Math. **2** (1930), 183–196. Cited on 43, 100, 118.)

[599] M. Schechter, *Principles of Functional Analysis*, Second edition, Graduate Studies in Mathematics, American Mathematical Society, Providence, RI, 2002. Cited on 217.)

[600] K. Schiefermayr, *A lower bound for the minimum deviation of the Chebyshev polynomial on a compact real set*, East J. Approx. **14** (2008), 223–233. Cited on 267.)

[601] E. Schmidt, *Zur Theorie der linearen und nichtlinearen Integralgleichungen. I. Teil: Entwicklung willkürlicher Funktionen nach Systemen vorgeschriebener; II. Teil: Auflösung der allgemeinen linearen Integralgleichung*, Math. Ann. **63** (1907), 433–476; **64** (1907), 161–174. Cited on 83, 99, 108, 134.)

[602] E. Schmidt, *Über die Auflosung lineare Gleichungen mit unendlich vielen Unbekannten*, Rend. Circ. Mat. Palermo, **25** (1908), 53–77. Cited on 299.)

[603] U.-W. Schmincke, *Essential selfadjointness of a Schrödinger operator with strongly singular potential*, Math. Z. **124** (1972), 47–50. Cited on 627.)

[604] K. Schmüdgen, *Unbounded Self-adjoint Operators on Hilbert Space*, Graduate Texts in Mathematics, Springer, Dordrecht, 2012. Cited on 602.)

[605] E. Schrödinger. *Quantisierung als Eigenwertproblem, IV, Storungstheorie mit Anwendung auf den Starkeffekt der Balmerlinien*, Ann. Phys. **80** (1926), 437–490. Cited on 27.)

[606] I. Schur, *Über eine Klasse von Matrizen, die sich einer gegebenen Matrix zuordnen lassen*, Dissertation, Berlin, 1901. Cited on 446.)

[607] I. Schur, *Neue Begründung der Theorie der Gruppencharaktere*, Sitzungsber. Preuss. Akad. Wiss. (1905), 406–432. Cited on 446.)

[608] I. Schur, *Über die charakteristischen Wurzeln einer linearen Substitution mit einer Anwendung auf die Theorie der Integralgleichungen*, Math. Ann. **66** (1909), 488–510. Cited on 163.)

[609] I. Schur, *Über Potenzreihen, die im Innern des Einheitskreises beschränkt sind, I, II*, J. Reine Angew. Math. **147** (1917), 205–232; **148** (1918) 122–145; English translation in I. Schur, *Methods in Operator Theory and Signal Processing*, edited by I. Gohberg, pp. 31–59; 66–88, Operator Theory: Advances and Applications, Birkhäuser, Basel, 1986. Cited on 208.)

[610] I. Schur, *Neue Anwendungen der Integralrechnung auf Probleme der Invariantentheorie. II. Über die Darstellung der Drehungsgruppe durch lineare homogene Substitutionen*, Berl. Ber. (1924), 297–321. Cited on 446.)

[611] J. Schwartz, *Some non-selfadjoint operators*, Comm. Pure Appl. Math. **13** (1960), 609–639. Cited on 600.)

[612] J. Schwinger, *On the bound states of a given potential*, Proc. Nat. Acad. Sci. USA **47** (1961), 122–129. Cited on 682.)

[613] I. E. Segal, *The group ring of a locally compact group. I*, Proc. Nat. Acad. Sci. USA **27** (1941), 348–352. Cited on 447.)

[614] I. E. Segal, *The group algebra of a locally compact group*, Trans. Amer. Math. Soc. **61** (1947), 69–105. Cited on 504.)

[615] I. E. Segal, *Irreducible representations of operator algebras*, Bull. Amer. Math. Soc. **53** (1947), 73–88. Cited on 428, 429, 447.)

[616] I. E. Segal, *Decompositions of Operator Algebras, II. Multiplicity Theory*, Mem. Amer. Math. Soc., American MAthematical Society, Providence, RI 1951. Cited on 299.)

[617] E. Seiler, *Schwinger functions for the Yukawa model in two dimensions with space-time cutoff*, Comm. Math. Phys. **42** (1975), 163–182. Cited on 192.)

[618] E. Seiler and B. Simon, *On finite mass renormalizations in the two dimensional Yukawa model*, J. Math. Phys. **16** (1975), 2289–2293. Cited on 172, 192.)

[619] E. Seiler and B. Simon, *Bounds in the Yukawa$_2$ quantum field theory: upper bound on the pressure, Hamiltonian bound and linear lower bound*, Comm. Math. Phy. **45**, (1975), 99–114. Cited on 161.)

[620] J-P. Serre, *Linear Representations of Finite Groups*, Translated from the second French edition by Leonard L. Scott, Graduate Texts in Mathematics, Springer-Verlag, New York–Heidelberg, 1977. Cited on 443.)

[621] H. S. Shapiro, *Extremal Problems for Polynomials and Power Series*, Master's thesis, Massachusetts Insitute of Technology, Cambridge, MA, 1951. Cited on 369.)

[622] A. Shenitzer and J. Steprans, *The evolution of integration*, Amer. Math. Monthly 101 (1994), 66–72. Cited on 603.)

[623] G. Shilov, *On the theory of ideals in normed rings of functions*, Dokl. Akad. Nauk SSSR **27**, (1940), 900–903. Cited on 389.)

[624] G. Shilov, *On regular normed rings*, Trav. Inst. Math. Stekloff **21** (1947), 118 pp. Cited on 389, 504.)

[625] G. Shilov, *On decomposition of a commutative normed ring in a direct sums of ideals*, Mat. Sb. N.S. **74** (1953), 353–364. Cited on 69.)

[626] W. Sierpinski, *O pewnem zagadnieniu z rachunku funkcyj asymptotycznych*, Prace mat. fyz. **17** (1906), 77–118; Summary in French: Sur un probleme du calcul des fonctions asymptotiques, 115–118. Cited on 605.)

[627] B. Simon, *Quantum Mechanics for Hamiltonians Defined as Quadratic Forms*, Princeton Series in Physics, Princeton University Press, Princeton, NJ, 1971. Cited on 683.)

[628] B. Simon, *Hamiltonians defined as quadratic forms*, Comm. Math. Phys. **21** (1971), 192–210. Cited on 683.)

[629] B. Simon, *Essential self-adjointness of Schrödinger operators with positive potentials*, Math. Ann. **201** (1973), 211–220. Cited on 627.)

[630] B. Simon, *Essential self-adjointness of Schrödinger operators with singular potentials*, Arch. Ration. Mech. Anal. **52** (1973), 44–48. Cited on 627.)

[631] B. Simon, *On the number of bound states of two-body Schrödinger operators: A review*, In Studies in Mathematical Physics, Essays in Honor of Valentine Bargmann, edited by E. H. Lieb, B. Simon and A. S. Wightman, pp. 305–326, Princeton University Press, Princeton, 1976. Cited on 682.)

[632] B. Simon, *The bound state of weakly coupled Schrödinger operators in one and two dimensions*, Ann. Phys. **97** (1976), 279–288. Cited on 683.)

[633] B. Simon, *Universal diamagnetism of spinless Bose systems*, Phys. Rev. Lett. **36** (1976), 1083–1084. Cited on 627.)

[634] B. Simon, *Analysis with weak trace ideals and the number of bound states of Schrödinger operators*, Trans. Amer. Math. Soc. **224** (1976), 367–380. Cited on 161, 609.)

[635] B. Simon, *Notes on infinite determinants of Hilbert space operators*, Adv. in Math. **24** (1977), 244–273. Cited on 187, 192.)

[636] B. Simon, *An abstract Kato's inequality for generators of positivity preserving semigroups*, Indiana Univ. Math. J. **26** (1977), 1067–1073. Cited on 627, 628.)

[637] B. Simon, *Lower semicontinuity of positive quadratic forms*, Proc. Roy. Soc. Edinburgh Sect. A **79** (1977/78), 267–273. Cited on 601.)

[638] B. Simon, *Classical boundary conditions as a technical tool in modern mathematical physics*, Adv. in Math. **30** (1978), 268–281. Cited on 602, 608.)

[639] B. Simon, *A canonical decomposition for quadratic forms with applications to monotone convergence theorems*, J. Funct. Anal. **28** (1978), 377–385. Cited on 601.)

[640] B. Simon, *Maximal and minimal Schrödinger forms*, J. Operator Theory **1** (1979), 37–47. Cited on 627, 628.)

[641] B. Simon, *Kato's inequality and the comparison of semigroups*, J. Funct. Anal. **32** (1979), 97–101. Cited on 627, 628.)

[642] B. Simon, *Nonclassical eigenvalue asymptotics*, J. Funct. Anal. **53** (1983), 84–98. Cited on 228, 604.)

[643] B. Simon, *Some quantum operators with discrete spectrum but classically continuous spectrum*, Ann. Phys. **146** (1983), 209–220. Cited on 228.)

[644] B Simon, *The Neumann Laplacian of a jelly roll*, Proc. Amer. Math. Soc. **114** (1992), 783–785. Cited on 603.)

[645] B. Simon, *Representations of Finite and Compact Groups*, Graduate Studies in Mathematics, American Mathematical Society, Providence, RI, 1996. Cited on 183, 443, 446.)

[646] B. Simon, *The classical moment problem as a self-adjoint finite difference operator*, Adv. in Math. **137** (1998), 82–203. Cited on 658, 659.)

[647] B. Simon, *Functional Integration and Quantum Physics*, 2nd edition, AMS Chelsea Publishing, Providence, RI, 2005. Cited on 509, 604, 608, 626.)

[648] B. Simon, *Orthogonal Polynomials on the Unit Circle, Part 1: Classical Theory*, AMS Colloquium Series, American Mathematical Society, Providence, RI, 2005. Cited on 218, 231, 282, 284, 285, 321, 344.)

[649] B. Simon, *Orthogonal Polynomials on the Unit Circle, Part 2: Spectral Theory*, AMS Colloquium Series, American Mathematical Society, Providence, RI, 2005. Cited on 231, 282, 284.)

[650] B. Simon, *Trace Ideals and Their Applications*, 2nd second, Mathematical Surveys and Monographs, American Mathematical Society, Providence, RI, 2005; 1st edition, Cambridge University Press, 1979. Cited on 134, 152, 153, 161, 192, 344.)

[651] B. Simon, *On a theorem of Kac and Gilbert*, J. Funct. Anal. **223** (2005), 109–115. Cited on 570.)

[652] B. Simon, *Schrödinger operators with purely discrete spectrum*, Methods Funct. Anal. Topology **15** (2009), 61–66. Cited on 228, 603.)

[653] B. Simon, *Szegő's Theorem and Its Descendants: Spectral Theory for L^2 Perturbations of Orthogonal Polynomials*, M. B. Porter Lectures, Princeton University Press, Princeton, NJ, 2011. Cited on 231, 267, 282.)

[654] B. Simon, *Convexity: An Analytic Viewpoint*, Cambridge Tracts in Mathematics, Cambridge University Press, Cambridge, 2011. Cited on 606.)

[655] B. Simon and T. Wolff, *Singular continuous spectrum under rank one perturbations and localization for random Hamiltonians*, Comm. Pure Appl. Math. **39** (1986), 75–90. Cited on 344.)

[656] I. Singer, *Bases in Banach spaces. I*, Die Grundlehren der mathematischen Wissenschaften, Springer-Verlag, New York–Berlin, 1970; *Bases in Banach spaces. II*, Editura Academiei Republicii Socialiste România, Bucharest; Springer-Verlag, Berlin–New York, 1981. Cited on 100.)

[657] K. B. Sinha and A. N. Mohapatra, *Spectral shift function and trace formula*, Proc. Indian Acad. Sci. (Math. Sci.) **104** (1994), 819–853. Cited on 353.)

[658] F. Smithies, *The Fredholm theory of integral equations*, Duke Math. J. **8** (1941), 107–130. Cited on 172, 192.)

[659] M. L. Sodin, and P. M. Yuditskii, *Functions deviating least from zero on closed subsets of the real axis*, St. Petersburg Math. J. **4** (1993), 201–249. Cited on 256.)

[660] M. Z. Solomyak *Asymptotic behavior of the spectrum of a Schrödinger operator with nonregular homogeneous potential*, Soviet Math. Dokl. **30** (1984), 379–383; Russian original in Dokl. Akad. Nauk SSSR **278** (1984), 291–295. Cited on 228.)

[661] A. Sommerfeld, *Die Greensche Funktion der Schwingungsgleichung für ein beliebiges Gebiet*, Physikal. Z. **11** (1910), 1057–1066. Cited on 603.)

[662] H. Stahl and V. Totik, *General Orthogonal Polynomials*, Cambridge University Press, Cambridge, 1992. Cited on 231.)

[663] K-G. Steffens, *The History of Approximation Theory, From Euler to Bernstein*, Birkhäuser Boston, Inc., Boston, MA, 2006. Cited on 267.)

[664] B. Steinberg, *Representation Theory of Finite Groups, An introductory approach*, Universitext, Springer, New York, 2012. Cited on 443.)

[665] G. W. Stewart, *On the early history of the singular value decomposition*, SIAM Rev. **35** (1993), 551–566. Cited on 135.)

[666] T. Stieltjes, *Quelques recherches sur la théorie des quadratures dites mecaniques*, Ann. Sci. Ecole Norm. Sup. (3) **1** (1884), 409–426. Cited on 241.)

[667] T. Stieltjes, *Recherches sur les fractions continues*, Ann. Fac. Sci. Toulouse **8** (1894–1895), J76–J122; **9**, A5–A47. Cited on 241, 658.)

[668] M. H. Stone, *Linear Transformations in Hilbert Space, III*, Proc. Nat. Acad. Sci. USA **15** (1929), 198–200. Cited on 554.)

[669] M. H. Stone, *On One-Parameter Unitary Groups in Hilbert Space*, Ann. Math. (2) **33** (1932), 643–648. Cited on 554.)

[670] M. H. Stone, *Linear Transformations in Hilbert Space*, reprint of the 1932 original, American Mathematical Society Colloquium Publications, American Mathematical Society, Providence, RI, 1990. Cited on 56, 160, 241, 299, 388, 534, 600, 658.)

[671] M. H. Stone, *Applications of the theory of Boolean rings to general topology*, Trans. Amer. Math. Soc. **41** (1937), 375–481. Cited on 412.)

[672] M. H. Stone, *On the theorem of Gelfand–Mazur*, Ann. Soc. Polon. Math. **25** (1952), 238–240. Cited on 387.)

[673] E. L. Stout, *The Theory of Uniform Algebras*, Bogden & Quigley, Tarrytown-on-Hudson, NY, 1971. Cited on 489.)

[674] F. Stummel, *Singuläre elliptische Differential-operatoren in Hilbertschen Räumen*, Math. Ann. **132** (1956), 150–176. Cited on 538.)

[675] C. Sturm and J. Liouville, *Extrait d'un Mémoire sur le développement des fonctions en séries dont les différents termes sont assujettis à satisfaire à une même équation differéntielle linéaire, contenant un paramétre variable*, J. Math. Pures Appl. **2** (1837), 220–223. Cited on 109.)

[676] T. Sunada, *Riemannian coverings and isospectral manifolds*, Ann. of Math. (2) **121** (1985), 169–186. Cited on 605.)

[677] V. S. Sunder, *An Invitation to von Neumann Algebras*, Universitext, Springer-Verlag, New York, 1987. Cited on 314.)

[678] J. J. Sylvester, *A new proof that a general quadric may be reduced to its canonical form (that is, a linear function of squares) by means of a real orthogonal substitution*, Messenger of Math. **19** (1889), 1–5. Cited on 135.)

[679] J. J. Sylvester, *On the reduction of a bilinear quantic of the nth order to the form of a sum of n products by a double orthogonal substitution*, Messenger of Math. **19** (1889), 42–46. Cited on 135.)

[680] J. J. Sylvester, *Sur la reduction biorthogonale d'uneforme lineto-linéaire a saforme cannonique*, C. R. Acad. Sci. Paris **108** (1889), 651–653. Cited on 135.)

[681] A. Szankowski, *$B(\mathcal{H})$ does not have the approximation property*, Acta Math. **147** (1981), 89–108. Cited on 100.)

[682] G. Szegő, *Ein Grenzwertsatz über die Toeplitzschen Determinanten einer reellen positiven Funktion*, Math. Ann. **76** (1915), 490–503. Cited on 285.)

[683] G. Szegő, *Beiträge zur Theorie der Toeplitzschen Formen, I, II*, Math. Z. **6** (1920), 167–202; **9** (1921), 167–190. Cited on 231, 282, 285.)

[684] G. Szegő, *Über den asymptotischen Ausdruck von Polynomen, die durch eine Orthogonalitätseigenschaft definiert sind*, Math. Ann. **86** (1922), 114–139. Cited on 284.)

[685] G. Szegő, *Bemerkungen zu einer Arbeit von Herrn M. Fekete: Über die Verteilung der Wurzeln bei gewissen algebraischen Gleichungen mit ganzzahligen Koeffizienten*, Math. Z. **21** (1924), 203–208. Cited on 268.)

[686] G. Szegő, *Orthogonal Polynomials*, American Mathematical Society Colloquium Publications, American Mathematical Society, Providence, RI, 1939; 3rd edition, 1967. Cited on 231, 282, 699.)

[687] G. Szegő, *On certain Hermitian forms associated with the Fourier series of a positive function*, Comm. Sém. Math. Univ. Lund **1952** (1952), Tome Supplementaire, 228–238. Cited on 285.)

[688] G. Szegő, *Collected Papers*, Vol. 1–3, edited by R. Askey, Birkhäuser, Basel, Switzerland, 1982. Cited on 240.)

[689] B. Sz.-Nagy, *Perturbations des transformations autoadjointes dans l'espace de Hilbert*, Comment. Math. Helv. **19** (1946/47), 347–366. Cited on 27, 70.)

[690] B. Sz.-Nagy, *Sur les contractions de l'espace de Hilbert*, Acta Sci. Math. Szeged **15** (1953), 87–92. Cited on 322.)

[691] B. Sz.-Nagy, C. Foias, H. Bercovici, and L. Kérchy, *Harmonic Analysis of Operators on Hilbert Space*, Second edition, Revised and enlarged edition, Universitext, Springer, New York, 2010. Cited on 322.)

[692] M. Takesaki, *Theory of Operator Algebras. I.*, Reprint of the first (1979) edition, Encyclopaedia of Mathematical Sciences, Operator Algebras and Non-commutative Geometry, Springer-Verlag, Berlin, 2002; *Theory of Operator Algebras. II.*, Encyclopaedia of Mathematical Sciences, Operator Algebras and Non-commutative Geometry, Springer-Verlag, Berlin, 2003; *Theory of Operator algebras. III.*, Encyclopaedia of Mathematical Sciences, Operator Algebras and Non-commutative Geometry, Springer-Verlag, Berlin, 2003. Cited on 314.)

[693] J. Tamarkin, *Review of Aurel Wintner, Spektraltheorie der unendlichen Matrizen, Einführung in den analytischen Apparat der Quantenmechanik*, Bull. Amer. Math. Soc. **37** (1931), 651–652. Cited on 535.)

[694] H. Tamura, *The asymptotic eigenvalue distribution for non-smooth elliptic operators*, Proc. Japan Acad. **50** (1974), 19–22. Cited on 605.)

[695] H. Tamura, *The asymptotic distribution of eigenvalues of the Laplace operator in an unbounded domain*, Nagoya Math. J. **60** (1976), 7–33. Cited on 228, 604.)

[696] T. Tannaka, *Über den Dualitätssatz der nichtkommutativen topologischen Gruppen*, Tohoku Math. J. **45** (1938), 1–12. Cited on 468.)

[697] A. Tauber, *Ein Satz aus der Theorie der unendlichen Reihen*, Monatsh. Math. **8** (1897), 273–277. Cited on 505.)

[698] A. E. Taylor, *Analysis in complex Banach spaces*, Bull. Amer. Math. Soc. **49** (1943), 652–669. Cited on 69.)

[699] J. L. Taylor, *The analytic-functional calculus for several commuting operators*, Acta Math. **125** (1970), 1–38. Cited on 69.)

[700] E. C. Titchmarsh, *The zeros of certain integral functions*, Proc. London Math. Soc. **s2–25**, (1926), 283–302. Cited on 129.)

[701] E. C. Titchmarsh, *Eigenfunction Expansions Associated with Second-order Differential Equations, Part I*, Second Edition, Clarendon Press, Oxford 1962. Cited on 569.)

[702] O. Toeplitz, *Über die Fouriersche Entwickelung positiver Funktionen*, Rend. Circ. Mat. Palermo **32** (1911), 191–192. Cited on 321.)

[703] L. Tonelli, *I polinomi d'approssimazione di Tchébychev*, Jour. Annali di Mat. **3** (1908), 15, 47–119. Cited on 267.)

[704] L. N. Trefethen, *Approximation Theory and Approximation Practice*, Society for Industrial and Applied Mathematics (SIAM), Philadelphia, PA, 2013. Cited on 267.)

[705] H. F. Trotter, *On the Product of Semi-Groups of Operators*, Proc. Amer. Math. Soc. **10**, 545–551. Cited on 628.)

[706] B. S. Tsirelson, *It is impossible to imbed ℓ_p of c_p into an arbitrary Banach space*, Funkcional. Anal. i Priložen. **8** (1974), 57–60. Cited on 100.)

[707] A. Tychonoff, *Über die topologische Erweiterung von Räumen*, Math. Ann. **102** (1930), 544 561. Cited on 412.)

[708] S. M. Ulam, *John von Neumann, 1903–1957*, Bull. Amer. Math. Soc. **64** (1958), 1–49. Cited on 536.)

[709] S. M. Ulam, *Adventures of a Mathematician*, Charles Scribner's Sons, New York, 1976. Cited on 33.)

[710] R. J. Vaccaro, *SVD and Signal processing, II: Algorithms, Analysis, and Applications*, Elsevier Science, Amsterdam, 1991. Cited on 135.)

[711] A. Vandermonde, *Memoire sur l'elimination*, Hist. de l'Acad. Roy. des Sciences (1772), 2nd partie, 516–532. Cited on 17.)

[712] E. R. van Kampen, *Locally bicompact Abelian groups and their character groups*, Ann. Math. (2) **36** (1935), 448–463. Cited on 467.)

[713] C. van Winter, *Theory of finite systems of particles, I. The Green function*, Mat.-Fys. Skr. Danske Vid. Selsk. **2** (1964) 1–60. Cited on 666.)

[714] S. Verblunsky, *On positive harmonic functions: A contribution to the algebra of Fourier series*, Proc. London Math. Soc. **38** (1935), 125–157. Cited on 282, 284.)

[715] S. Verblunsky, *On positive harmonic functions (second paper)*, Proc. London Math. Soc. (2) **40** (1936), 290–320. Cited on 282.)

[716] M. I. Vishik, *On general boundary value problems for elliptic differential operators*, Trudy Mosc. Mat. Obsv. **1** (1952), 187–246; English translation in Amer. Math. Soc. Transl. **24** (1963), 107–172. Cited on 601.)

[717] C. Visser, *Note on linear operators*, Proc. Akad. Wet. Amsterdam **40** (1937), 270–272. Cited on 83.)

[718] H. von Koch, *Sur les déterminants infinis et les équations différentielles linéaires*, Acta Math. **16** (1892), 217–295. Cited on 41.)

[719] H. von Koch, *Sur quelques points de la théorie des déterminants infinis*, Acta Math. **24** (1900), 89–122. Cited on 41.)

[720] J. von Neumann, *Allgemeine Eigenwerttheorie Hermitescher Funktionaloperation*, Math. Ann. **102** (1929), 49–131. Cited on 299, 534, 548, 568, 600.)

[721] J. von Neumann, *Zur Theorie der unbeschränkten Matrizen*, J. Reine Angew. Math. **161** (1929), 208–236. Cited on 535.)

[722] J. von Neumann, *Zur Algebra der Funktionaloperationen und Theorie der Normalen Operatoren*, Math. Ann. **102** (1930), 370–427. Cited on 43, 314.)

[723] J. von Neumann, *Mathematical Foundations of Quantum Mechanics*, 12th printing, Princeton Landmarks in Mathematics, Princeton University Press, Princeton, NJ, 1996; German original, Springer, 1932. Cited on 143, 534.)

[724] J. von Neumann, *Über adjungierte Funktionaloperatoren*, Ann. Math. (2) **33** (1932), 294–310. Cited on 82, 535, 606.)

[725] J. von Neumann, *Charakterisierung des Spektrums eines Integraloperators*, Actualités Sci. Industr. **229** (1935), 3–20. Cited on 353.)

[726] J. von Neumann, *On a certain topology for rings of operators*, Ann. Math. **37** (1936), 111–115. Cited on 43.)

[727] J. von Neumann, *Some matrix inequalities and metrization of matrix spaces*, Tomsk. Univ. Rev. **1** (1937), 286–300. Cited on 152.)

[728] J. von Neumann, *Eine Spektraltheorie für allgemeine Operatoren eines unitären Raumes*, Math. Nachr. **4** (1951) 258–281. Cited on 323.)

[729] L. Waelbroeck, *Le calcul symbolique dans les algébres commutatives*, J. Math. Pures Appl. **33** (1954), 147–186. Cited on 69.)

[730] R. C. Walker, *The Stone–Čech Compactification*, Ergebnisse der Mathematik und ihrer Grenzgebiete, Springer-Verlag, New York–Berlin, 1974. Cited on 412.)

[731] J. L. Walsh, *The approximation of harmonic functions by harmonic polynomials and harmonic rational functions*, Bull. Amer. Math. Soc. **35** (1929), 499–544. Cited on 489.)

[732] F-Y. Wang and J-L. Wu, *Compactness of Schrödinger semigroups with unbounded below potentials*, Bull. Sci. Math. **132** (2008), 679–689. Cited on 603.)

[733] K. Watanabe, *Smooth perturbations the self-adjoint operators defined by the $\mathcal{H}_{-2}$-construction*, Math. Nachr. **250** (2003), 104–114. Cited on 666.)

[734] F. Wecken, *Unitarinvarianten Selbstadjungierter Operatoren*, Math. Ann. **116** (1939), 422–455. Cited on 313.)

[735] A. Weil, *L'intégration dans les groupes topologiques et ses applications*, Actualités Sci. Ind. **869**, Hermann, Paris, 1940. [This book has been republished by the author at Princeton, NJ, 1941.] Cited on 467.)

[736] A. Weinstein, *Étude des spectres des équations aux dérivées partielles de la théorie des plaques élastiques*, Mémorial des Sciences Mathématiques, Gauthiers-Villars, Paris, 1937. Cited on 343.)

[737] S. H. Weintraub, *Representation Theory of Finite Groups: Algebra and Arithmetic*, Graduate Studies in Mathematics, American Mathematical Society, Providence, RI, 2003. Cited on 443.)

[738] B. Wendroff, *On orthogonal polynomials*, Proc. Amer. Math. Soc. **12** (1961), 554–555. Cited on 283.)

[739] J. Wermer, *Banach Algebras and Several Complex Variables*, 2nd edition, Graduate Texts in Mathematics, Springer-Verlag, New York–Heidelberg, 1976. Cited on 489.)

[740] T. T. West, *The decomposition of Riesz operators*, Proc. London Math. Soc. **s3–16**, (1966), 737-752. Cited on 128.)

[741] H. Weyl, *Über beschränkte quadratische Formen, deren Differenz vollstetig ist*, Palermo Rend. **27** (1909), 373–392, 402. Cited on 197, 299, 353.)

[742] H. Weyl, *Über gewöhnliche lineare Differentialgleichungen mit singulären Stellen und ihre Eigenfunktionen I, II*, Gött. Nachr. (1909), 37–63; (1910), 442–467. Cited on 568.)

[743] H. Weyl, *Über gewöhnliche Differentialgleichungen mit Singularitäten und die zugehörigen Entwicklungen willkürlicher Funktionen*, Math. Ann. **68** (1910), 220–269. Cited on 568.)

[744] H. Weyl, *Über die asymptotische Verteilung der Eigenwerte*, Nachr. Kön. Ges. Wiss. Göttingen (1911), 110–117. Cited on 604.)

[745] H. Weyl, *Das asymptotische Verteilungsgesetz der Eigenwerte linearer partieller Differentialgleichungen (mit einer Anwendung auf die Theorie der Hohlraumstrahlung)*, Math. Ann. **71** (1912), 441–479. Cited on 604.)

[746] H. Weyl, *Über die Abhängigkeit der Eigenschwingungen einer Membran von deren Begrenzung*, J. Reine Angew. Math. **141** (1912), 1–11. Cited on 604.)

[747] H. Weyl, *Über die Randwertaufgabe der Strahlungstheorie und asymptotische pektralgeometrie*, J. Reine Angew. Math. **143** (1913), 177–202. Cited on 604.)

[748] H. Weyl, *Theorie der Darstellung kontinuierlicher halbeinfacher Gruppen durch lineare Transformationen. I. II. III.*, Math. Z. **23** (1925), 271–309; **24** (1926), 328–376, 377–395. Cited on 446.)

[749] H. Weyl, *Gruppentheorie und Quantenmechanik*, Reprint of the second edition, Wissenschaftliche Buchgesellschaft, Darmstadt, 1977, German original 1931; English Translation: *The Theory of Groups and Quantum Mechanics*, Dover, New York, 1950. Cited on 443.)

[750] H. Weyl, *The Classical Groups, Their Invariants and Representations*, Fifteenth printing, Princeton Landmarks in Mathematics, Princeton Paperbacks, Princeton University Press, Princeton, NJ, 1997; Original edition 1937. Cited on 443.)

[751] H. Weyl, *Inequalities between the two kinds of eigenvalues of a linear transformation*, Proc. Nat. Acad. Sci. USA. **35** (1949), 408–411. Cited on 134, 164.)

[752] H. Weyl, *Ramifications, old and new, of the eigenvalue problem*, Bull. Amer. Math. Soc. **56** (1950), 115–139. Cited on 604.)

[753] A. N. Whitehead, *The Concept of Nature*, Cambridge University Press, Cambridge, 1920; available online at `http://www.gutenberg.org/ebooks/18835`. Cited on 515.)

[754] R. Whitley, *Projecting m onto c_0*. Amer. Math. Monthly **73** (1966) 285–286. Cited on 45.)

[755] H. Wielandt, *Zur Umkehrung des Abelschen Stetigkeitssatzes*, J. Reine Angew. Math. **56** (1952), 27–39. Cited on 506.)

[756] N. Wiener, *A new method in Tauberian theorems*, J. Math. and Phys. **7** (1928), 161-184. Cited on 506.)

[757] N. Wiener, *Tauberian theorems*, Ann. Math. (2) **33** (1932), 1–100. Cited on 388, 504, 506.)

[758] E. P. Wigner, *Group Theory: And its Application to the Quantum Mechanics of Atomic Spectra*, Expanded and improved ed., translated from the German by J. J. Griffin. Pure and Applied Physics. Vol. 5 Academic Press, New York-London 1959. Cited on 443.)

[759] R. J. Williams, *Reflected Brownian motion: Hunt processes and semimartingale representation*, In Barcelona Seminar on Stochastic Analysis (St. Feliu de Guíxols, 1991), pp. 216–221, Progress in Probability, Birkhäuser, Basel, 1993. Cited on 603.)

[760] J. Williamson, *A polar representation of singular matrices*, Bull. Amer. Math. Soc. **41** (1935), 118–123. Cited on 135.)

[761] A. Wintner, *Spektraltheorie der unendlichen Matrisen, Einfuhrung in den analytischen Apparat der Quantenmechanik*, Leipzig, Verlag S. Hirzel, 1929; available online at `https://archive.org/details/SpektraltheorieDerUnendlichenMatrizen`. Cited on 534.)

[762] A. Wintner, *On the normalization of characteristic differentials in continuous spectra*, Phys. Rev. (2) **72** (1947), 516–517. Cited on 569.)

[763] A. Wintner and F. D. Murnaghan, *On a polar representation of nonsingular square matrices*, Proc. Nat. Acad. Sci. USA **17** (1931), 676–678. Cited on 82.)

[764] W. Wirtinger, *Beiträge zu Riemann's Integrationsmethode für hyperbolische Differentialgleichungen, und deren Anwendungen auf Schwingungsprobleme*, Math. Ann. **48** (1897), 365–389. Cited on 56.)

[765] R. Wüst, *Generalisations of Rellich's theorem on perturbation of (essentially) self-adjoint operators*, Math. Z. **119** (1971), 276–280. Cited on 536.)

[766] D. R. Yafaev, *Perturbation determinants, the spectral shift function, trace identities, and all that*, Functional Anal. Appl. **41** (2007), 217–236; Russian original: Funktsional. Anal. i Prilozhen. **41** (2007), 60–83. Cited on 353.)

[767] B. Yood, *Properties of linear transformations preserved under addition of a completely continuous transformation*, Duke Math. J. **18** (1951), 599–612. Cited on 217.)

[768] K. Yosida, *Functional Analysis*, Sixth edition, Grundlehren der Mathematischen Wissenschaften, Springer-Verlag, Berlin–New York, 1980. Cited on 129.)

[769] K. Yosida and T. Iwamura, *Equivalence on two topologies of Abelian groups*, Proc. Imp. Acad. Tokyo **20** (1944), 451–453. Cited on 467.)

[770] R. M. Young, *An Introduction to Nonharmonic Fourier Series*, Revised first edition, Academic Press, Inc., San Diego, CA, 2001. Cited on 100.)

[771] W. R. Zame, *Existence, uniqueness and continuity of functional calculus homomorphisms*, Proc. London Math. Soc. **39** (1979), 73–92. Cited on 69.)

[772] W. Żelazko, *Banach Algebras*, Elsevier, Amsterdam-London-New York; PWN–Polish Scientific Publishers, Warsaw, 1973. Cited on 357, 387.)

[773] F. Zhang, ed., *The Schur Complement and its Applications*, Numerical Methods and Algorithms, Springer-Verlag, New York, 2005. Cited on 208.)

[774] G. M. Zhislin, *Spectral properties of Hamiltonians with a magnetic field under fixation of pseudomomentum, II*, Teoret. Mat. Fiz. **118** (1999), 15–39; translation in Theoret. and Math. Phys. **118** (1999), 12–31. Cited on 666.)

[775] K. H. Zhu, *An Introduction to Operator Algebras*, Studies in Advanced Mathematics, CRC Press, Boca Raton, FL, 1993. Cited on 314.)

[776] M. Zworski, *Distribution of poles for scattering on the real line*, J. Funct. Anal. **73** (1987), 277–296. Cited on 605.)

[777] A. Zygmund, *Trigonometric Series. Vol. I, II*, 3rd edition, Cambridge Mathematical Library, Cambridge University Press, Cambridge, 2002. Cited on 367, 390.)

Symbol Index

Subject Index

Author Index

Index of Capsule Biographies